Introduction to Mathematical Programming

OPERATIONS RESEARCH:
VOLUME ONE

DUXBURY TITLES OF RELATED INTEREST

Albright, Winston & Zappe, *Data Analysis & Decision Making*
Albright, *VBA for Modelers: Developing Decision Support Systems with Microsoft Excel*
Berger & Maurer, *Experimental Design*
Berk & Carey, *Data Analysis with Microsoft Excel*
Clemen & Reilly, *Making Hard Decisions with DecisionTools*
Derr, *Statistical Consulting: A Guide to Effective Communication*
Devore, *Probability & Statistics for Engineering and the Sciences*
Farnum, *Modern Statistical Quality Control and Improvement*
Fourer, Gay & Kernighan, *AMPL: A Modeling Language for Mathematical Programming*
Higgins & Keller-McNulty, *Concepts in Probability and Stochastic Modeling*
Kao, *Introduction to Stochastic Processes*
Hayter, *Probability & Statistics for Engineers and Scientists*
Hoerl & Snee, *Statistical Thinking: Improving Business Performance*
Kao, *Introduction to Stochastic Processes*
Kenett & Zacks, *Modern Industrial Statistics: Design of Quality and Reliability*
Kirkwood, *Strategic Decision Making: Multiobjective Decision Analysis with Spreadsheets*
Lapin & Whisler, *Quantitative Decision Making with Spreadsheet Applications*
Lattin, Carroll & Green, *Analyzing Multivariate Data*
Lawson & Erjavec, *Engineering and Industrial Statistics*
Lohr, *Sampling: Design and Analysis*
Lunneborg, *Data Analysis by Resampling*
Middleton, *Data Analysis Using Microsoft Excel*
Minh, *Applied Probability Models*
Ramsey, *The Elements of Statistics with Applications to Economics*
SAS Institute Inc., *JMP-IN: Statistical Discovery Software*
Savage, *Decision Making with Insight*
Schrage, *Optimization Modeling Using Lindo*
Seila, Ceric & Tadikamalla, *Applied Simulation Modeling*
Shapiro, *Modeling the Supply Chain*
Vardeman & Jobe, *Basic Engineering Data Collection and Analysis*
Vining, *Statistical Methods for Engineers*
Winston, *Introduction to Probability Models*
Winston, *Operations Research: Applications & Algorithms*
Winston, *Simulation Modeling Using @Risk*
Winston & Albright, *Practical Management Science*

To order copies contact your local bookstore or call 1-800-354-9706.
For more information go to: www.duxbury.com

DUXBURY

Introduction to Mathematical Programming

OPERATIONS RESEARCH:
VOLUME ONE

FOURTH EDITION

Wayne L. Winston

INDIANA UNIVERSITY

Munirpallam Venkataramanan

INDIANA UNIVERSITY

WITH CASES BY
Jeffrey B. Goldberg
UNIVERSITY OF ARIZONA

THOMSON

BROOKS/COLE

Australia ■ Canada ■ Mexico ■ Singapore
Spain ■ United Kingdom ■ United States

THOMSON

BROOKS/COLE

Publisher: Curt Hinrichs

Assistant Editor: Ann Day

Editorial Assistant: Katherine Brayton

Technology Project Manager: Burke Taft

Marketing Manager: Joseph Rogove

Advertising Project Manager: Tami Strang

Print/Media Buyer: Jessica Reed

Permissions Editor: Bob Kauser

Production Service: Penmarin Books

Text Designer: Kaelin Chappell

Copy Editor: Steven Summerlight

Illustrator: Electronic Illustrators Group

Cover Designer: Lisa Langhoff

Cover Image: PhotoDisc

Cover Printer: Lehigh Press

Compositor: ATLIS Graphics

Printer: Quebecor World, Taunton

For more information about our products, contact us at:
Thomson Learning Academic Resource Center
1-800-423-0563
For permission to use material from this text, contact us by:
Phone: 1-800-730-2214   **Fax:** 1-800-730-2215
Web: http://www.thomsonrights.com

Library of Congress Control Number 2002111346
ISBN 0-534-35964-7

Brooks/Cole—Thomson Learning
511 Forest Lodge Road
Pacific Grove, CA 93950
USA

Asia
Thomson Learning
5 Shenton Way #01-01
UIC Building
Singapore 068808

Australia
Nelson Thomson Learning
102 Dodds Street
South Melbourne, Victoria 3205
Australia

Canada
Nelson Thomson Learning
1120 Birchmount Road
Toronto, Ontario M1K 5G4
Canada

Europe/Middle East/Africa
Thomson Learning
High Holborn House
50/51 Bedford Row
London WC1R 4LR
United Kingdom

Latin America
Thomson Learning
Seneca, 53
Colonia Polanco
11560 Mexico D.F.
Mexico

Spain
Paraninfo Thomson Learning
Calle Magallanes, 25
28015 Madrid
Spain

Contents

Preface **ix**

1 An Introduction to Model-Building 1

1.1 An Introduction to Modeling 1

1.2 The Seven-Step Model-Building Process 5

1.3 CITGO Petroleum 6

1.4 San Francisco Police Department Scheduling 7

1.5 GE Capital 9

2 Basic Linear Algebra 11

2.1 Matrices and Vectors 11

2.2 Matrices and Systems of Linear Equations 20

2.3 The Gauss-Jordan Method for Solving Systems of Linear Equations 22

2.4 Linear Independence and Linear Dependence 32

2.5 The Inverse of a Matrix 36

2.6 Determinants 42

3 Introduction to Linear Programming 49

3.1 What Is a Linear Programming Problem? 49

3.2 The Graphical Solution of Two-Variable Linear Programming Problems 56

3.3 Special Cases 63

3.4 A Diet Problem 68

3.5 A Work-Scheduling Problem 72

3.6 A Capital Budgeting Problem 76

3.7 Short-Term Financial Planning 82

3.8 Blending Problems 85

3.9 Production Process Models 95

3.10 Using Linear Programming to Solve Multiperiod Decision Problems: An Inventory Model 100

3.11 Multiperiod Financial Models 105

3.12 Multiperiod Work Scheduling 109

4 The Simplex Algorithm and Goal Programming 127

4.1 How to Convert an LP to Standard Form 127

4.2 Preview of the Simplex Algorithm 130

4.3 Direction of Unboundedness 134

4.4 Why Does an LP Have an Optimal bfs 136

4.5 The Simplex Algorithm 140

4.6 Using the Simplex Algorithm to Solve Minimization Problems 149

4.7 Alternative Optimal Solutions 152

4.8 Unbounded LPs 154

4.9 The LINDO Computer Package 158

4.10 Matrix Generators, LINGO, and Scaling of LPs 163

4.11 Degeneracy and the Convergence of the Simplex Algorithm 168

4.12 The Big M Method 172

4.13 The Two-Phase Simplex Method 178

4.14 Unrestricted-in-Sign Variables 184

4.15 Karmarkar's Method for Solving LPs 190

4.16 Multiattribute Decision Making in the Absence of Uncertainty: Goal Programming 191

4.17 Using the Excel Solver to Solve LPs 202

5 Sensitivity Analysis: An Applied Approach 227

5.1 A Graphical Introduction to Sensitivity Analysis 227

5.2 The Computer and Sensitivity Analysis 232

5.3 Managerial Use of Shadow Prices 246

5.4 What Happens to the Optimal z-Value If the Current Basis Is No Longer Optimal? 248

6 Sensitivity Analysis and Duality 262

6.1 A Graphical Introduction to Sensitivity Analysis 262

6.2 Some Important Formulas 267

6.3 Sensitivity Analysis 275

6.4 Sensitivity Analysis When More Than One Parameter Is Changed: The 100% Rule 289

6.5 Finding the Dual of an LP 295

6.6 Economic Interpretation of the Dual Problem 302

6.7 The Dual Theorem and Its Consequences 304

6.8 Shadow Prices 313

6.9 Duality and Sensitivity Analysis 323

6.10 Complementary Slackness 325

6.11 The Dual Simplex Method 329

6.12 Data Envelopment Analysis 335

7 Transportation, Assignment, and Transshipment Problems 360

7.1 Formulating Transportation Problems 360

7.2 Finding Basic Feasible Solutions for Transportation Problems 373

7.3 The Transportation Simplex Method 382

7.4 Sensitivity Analysis for Transportation Problems 390

7.5 Assignment Problems 393

7.6 Transshipment Problems 400

8 Network Models 413

8.1 Basic Definitions 413

8.2 Shortest Path Problems 414

8.3 Maximum Flow Problems 419

8.4 CPM and PERT 431

8.5 Minimum Cost Network Flow Problems 450

8.6 Minimum Spanning Tree Problems 456

8.7 The Network Simplex Method 459

9 Integer Programming 475

9.1 Introduction to Integer Programming 475

9.2 Formulating Integer Programming Problems 477

9.3 The Branch-and-Bound Method for Solving Pure Integer Programming Problems 512

9.4 The Branch-and-Bound Method for Solving Mixed Integer Programming Problems 523

9.5 Solving Knapsack Problems by the Branch-and-Bound Method 524

9.6 Solving Combinatorial Optimization Problems by the Branch-and-Bound Method 527

9.7 Implicit Enumeration 540

9.8 The Cutting Plane Algorithm 545

10 Advanced Topics in Linear Programming 562

10.1 The Revised Simplex Algorithm 562

10.2 The Product Form of the Inverse 567

10.3 Using Column Generation to Solve Large-Scale LPs 570

10.4 The Dantzig-Wolfe Decomposition Algorithm 576

10.5 The Simplex Method for Upper-Bounded Variables 593

10.6 Karmarkar's Method for Solving LPs 597

11 Game Theory 610

11.1 Two-Person Zero-Sum and Constant-Sum Games: Saddle Points 610

11.2 Two-Person Zero-Sum Games: Randomized Strategies, Domination, and Graphical Solution 614

11.3 Linear Programming and Zero-Sum Games 623

11.4 Two-Person Nonconstant-Sum Games 634

11.5 Introduction to n-Person Game Theory 639

11.6 The Core of an n-Person Game 641

11.7 The Shapley Value 644

12 Nonlinear Programming 653

12.1 Review of Differential Calculus 653

12.2 Introductory Concepts 659

12.3 Convex and Concave Functions 673

12.4 Solving NLPs with One Variable 680

12.5 Golden Section Search 692

12.6 Unconstrained Maximization and Minimization with Several Variables 698

12.7 The Method of Steepest Ascent 703

12.8 Lagrange Multipliers 706

12.9 The Kuhn-Tucker Conditions 713

12.10 Quadratic Programming 723

12.11 Separable Programming 731

12.12 The Method of Feasible Directions 736

12.13 Pareto Optimality and Tradeoff Curves 738

13 Deterministic Dynamic Programming 750

13.1 Two Puzzles 750

13.2 A Network Problem 751

13.3 An Inventory Problem 758

13.4 Resource Allocation Problems 763

13.5 Equipment Replacement Problems 774

13.6 Formulating Dynamic Programming Recursions 778

13.7 Using EXCEL to Solve Dynamic Programming Problems 790

14 Heuristic Techniques 800

14.1 Complexity Theory 800

14.2 Introduction to Heuristic Procedures 804

14.3 Simulated Annealing 805

14.4 Genetic Search 808

14.5 Tabu Search 815

14.6 Comparison of Heuristics 821

15 Solving Optimization Problems with the Evolutionary Solver 823

15.1 Price Bundling, Index Function, Match Function, and Evolutionary Solver 823

15.2 More Nonlinear Pricing Models 830

15.3 Locating Warehouses 836

15.4 Solving Other Combinatorial Problems 839

15.5 Production Scheduling at John Deere 841

15.6 Assigning Workers to Jobs with the Evolutionary Solver 846

15.7 Cluster Analysis 851

15.8 Fitting Curves 857

15.9 Discriminant Analysis 860

16 Neural Networks 866

16.1 Introduction to Neural Networks 866

16.2 Examples of the Use of Neural Networks 870

16.3 Why Neural Nets Can Beat Regression: The XOR Example 871

16.4 Estimating Neural Nets with PREDICT 874

16.5 Using Genetic Algorithms to Optimize a Neural Network 882

16.6 Using Genetic Algorithms to Determine Weights for a Back Propagation Network 884

Appendix: Cases 891

Case 1   Help, I'm Not Getting Any Younger 892

Case 2   Solar Energy for Your Home 892

Case 3   Golf-Sport: Managing Operations 893

Case 4   Vision Corporation: Production Planning and Shipping 896

Case 5   Material Handling in a General Mail-Handling Facility 897

Case 6   Selecting Corporate Training Programs 900

Case 7   Best Chip: Expansion Strategy 903

Case 8   Emergency Vehicle Location in Springfield 905

Case 9   System Design: Project Management 906

Case 10   Modular Design for the Help-You Company 907

Case 11   Brite Power: Capacity Expansion 909

Index 913

Preface

Addressing New Needs

In recent years, mathematical programming software for mainframes and microcomputers has become widely available. Like most tools, however, it is useless unless the user understands its application and purpose. The user must ensure that mathematical input accurately reflects the real-life problems to be solved and that the numerical results are correctly applied to solve them. With this in mind, this book emphasizes model-formulation and model-building skills as well as the interpretation of software output.

Intended Audience

This book is intended to be used as an advanced beginning or intermediate text in linear or mathematical programming. The following groups of students can benefit from using it:

- Undergraduate majors in quantitative methods in business, operations research, management science, or industrial engineering.
- Mathematics, MBA, or Masters of Public Administration students enrolled in applications-oriented mathematical programming courses.
- Graduate students who need an overview of the major topics in mathematical programming.

The book contains enough material for a two-semester course, allowing instructors ample flexibility in adapting the text to their individual course plans.

Making Teaching and Learning Easier

The following features help to make this book reader friendly:

- To provide immediate feedback to students, problems have been placed at the end of each section, and most chapters conclude with review problems. There

are more than 1000 problems, grouped according to difficulty: Group A for practice of basic techniques, Group B for underlying concepts, and Group C for mastering the theory independently.

- The book avoids excessive theoretical formulas in favor of word problems and interesting problem applications. Many problems are based on published applications of mathematical programming. Each chapter includes several examples to guide the student step by step through even the most complex topics. Math programming algorithms still receive comprehensive treatment; for instance, Karmarkar's method for solving linear programming problems is explained in detail.

- To help students review for tests, each chapter has a summary of concepts and formulas. An Instructor's CD-ROM with complete solutions to all problems, Microsoft PowerPoint slides, and advice for instructors is available.

- Each section is as self-contained as possible, which allows the instructor to be extremely flexible in designing a course. The Instructor's Manual identifies which portions of the book must be covered as prerequisites to each section.

- This book contains instruction in using the popular LINDO and LINDO packages and interpreting their output. This edition also includes an EXCEL neural network add-in NeuralWorks® PREDICT and an enhanced version of the EXCEL Solver (Premium Solver for Education) which implements Evolutionary algorithms within EXCEL. The book's CD-ROM includes student versions of LINDO, LINGO, PREDICT, and the Premium Solver for Education as well as all computer files needed for the book's problems and examples.

Organization

The book is completely self-contained, with all necessary mathematical background given in Chapter 2 and

Section 12.1. Students who are familiar with matrix multiplication should have no problems with the book. Chapter 12 is the only chapter that requires differential calculus. All calculus topics needed in Chapter 12 are reviewed in Section 12.1.

Since not all students need a full-blown theoretical treatment of sensitivity analysis, the topic is covered in two chapters. Chapter 5 is an applied approach to sensitivity analysis, emphasizing the interpretation of computer output. Chapter 6 contains a full discussion of sensitivity analysis, duality, and the dual simplex method. The instructor should cover Chapter 5 *or* Chapter 6 but not both. Classes emphasizing model-building and model-formulation skills should cover Chapter 5. Those paying close attention to the algorithms of mathematical programming (particularly classes in which many students will go on to further operations research courses) should study Chapter 6. If Chapter 5 is covered rather than Chapter 6, then Chapter 2 may be omitted.

Changes in the Fourth Edition

The favorable response to the previous editions of this book has been truly gratifying. Many suggestions from users of the text have been incorporated into this edition. There are more than 100 new problems and minor changes in many sections of the book. In addition, there are many substantive changes in the fourth edition. The most significant is a comprehensive discussion of artificial intelligence (AI) methods, including simulated annealing, genetic algorithms, Tabu search, and neural networks. Chapter 14, written by Professor Munirpallam Appadorai Venkataramanan, explains the intuition and methodology underlying simulated annealing, genetic algorithms, and Tabu search. Chapter 15 shows how to use the Evolutionary algorithms embedded in the Premium Solver for Education to solve "hard" optimization problems within the confines of a spreadsheet. Chapter 16 explains the intuition and methodology behind neural networks and explains how the EXCEL neural network add-in PREDICT can be used to fit a neural network to data.

A brief discussion of other major changes in the fourth edition follows:

- All Lotus spreadsheets have been converted to EXCEL.

- More discussion of spreadsheet solution of optimization problems has been added. We have also switched our method of spreadsheet solution from What's Best to the EXCEL Solver.

- Discussion of important EXCEL functions such as MMULT, MINVERSE, MATCH, INDEX, NPV, and XNPV has been added.

- Chapter 4 includes more extensive instruction in the use of LINDO and LINGO.

- Chapter 4 includes more discussion of the geometry of LPs.

- Chapter 12 contains new applications of nonlinear programming to pricing problems.

- Last (but not least!) is the new appendix of cases written by Professor Jeff Goldberg of the University of Arizona.

Acknowledgments

Many people have played significant roles in the development of the fourth edition. My views on teaching math programming were greatly influenced by the many excellent teachers I have had, including Gordon Bradley, Eric Denardo, John Little, Robert Mifflin, Martin Shubik, and Harvey Wagner. In particular, I would like to acknowledge Professor Wagner's *Principles of Operations Research*, which taught me more about mathematical programming than any other single book.

The LINDO and LINGO printouts appear courtesy of Professor Linus Schrage. I would like to thank all the people at LINDO Systems (particularly Mark Wiley, Linus Schrage, and Kevin Cunningham) for their cooperation with the project.

Thanks go to Frontline Systems (particularly Dan Fylstra and Edwin Straver) for letting us include the Premium Solver for Education.

Thanks to Jack Copper of Neuralware, Inc. for allowing us to include PREDICT.

Thanks go to all the people at Duxbury Press and Penmarin Books who worked on the book, especially our outstanding editor, Curt Hinrichs.

The discussions of Lagrange multipliers and the Kuhn-Tucker conditions owe much to the cogent comments of Professor John Hooker of Carnegie-Mellon University.

Finally, I owe a great debt to the reviewers of the second edition, whose comments greatly improved the quality of the manuscript: Esther Arkin, State University of New York at Stony Brook; James W. Chrissis, Air Force Institute of Technology; Rebecca E. Hill, Rochester Institute of Technology; and James G. Morris, University of Wisconsin, Madison.

Thanks to the 146 survey respondents who provided valuable feedback on the course and its emerging needs. They include Nikolaos Adamou, University of Athens & Sage Graduate School; Jeffrey Adler, Rensselaer Polytechnic Institute; Victor K. Akatsa, Chicago State University; Steven Andelin, Kutztown University of Pennsylvania; Badiollah R. Asrabadi, Nicholls State University; Rhonda Aull-Hyde, University of Delaware; Jonathan Bard, Mechanical Engineering; John Barnes, Virginia Commonwealth University; Harold P. Benson, University of Florida; Elinor Berger, Columbus College; Richard H. Bernhard, North Carolina State University; R. L. Bulfin, Auburn University; Laura Burke, Lehigh University; Jonathan Caulkins, Carnegie Mellon University; Beth Chance, University of the Pacific; Alan Chesen, Wright State University; Young Chun, Louisiana State University; Chia-Shin Chung, Cleveland State University; Ken Currie, Tennessee Technological University; Ani Dasgupta, Pennsylvania State University; Nirmil Devi, Embry-Riddle Aeronautical University; James Falk, George Washington University; Kambiz Farahmand, Texas A & M University–Kingsville; Yahya Fathi, North Carolina State University; Steve Fisk, Bowdoin College; William P. Fox, United States Military Academy; Michael C. Fu, University of Maryland; Saul I. Gass, University of Maryland; Ronald Gathro, Western New England College; Perakis Georgia, Massachusetts Institute of Technology; Alan Goldberg, California State University–Hayward; Jerold Griggs, University of South Carolina; David Grimmett, Austin Peay State University; Melike Baykal Gursoy, Rutgers University; Jorge Haddock, Rensselaer Polytechnic Institute; Jane Hagstrom, University of Illinois; Carl Harris, George Mason University; Miriam Heller, University of Houston; Sundresh S. Heragu, Rensselaer Polytechnic Insitute; Rebecca E. Hill, Rochester Institute of Technology; David Holdsworth, Alaska Pacific University; Elained Hubbard, Kennesaw State College; Robert Hull, Western Illinois University; Jeffrey Jarrett, University of Rhode Island; David Kaufman, University of Massachusetts; Davook Khalili, San Jose University; Morton Klein, Columbia University; S. Kumar, Rochester Institute of Technology; David Larsen, University of New Orleans; Mark Lawley, University of Alabama; Kenneth D. Lawrence, New Jersey Institute of Technology; Andreas Lazari, Valdosta State University; Jon Lee, University of Kentucky; Luanne Lohr, University of Georgia; Joseph Malkovich, York College; Masud Mansuri, California State University–Fresno; Steven C. McKelvey, Saint Olaf College; Ojjat Mehri, Youngstown State University; Robert Mifflin, Washington State University; Katya Mints, Columbia University; Rafael Moras, St. Mary's University; James G. Morris, University of Wisconsin, Madison; Frederic Murphy, Temple University; Arthur Neal, Willoughby, Morgan State; David Olson, Texas A & M University; Mufit Ozden, Miami University; R. Gary Parker, Georgia Institute of Technology; Barry Pasternack, California State University, Fullerton; Walter M. Patterson, Lander University; James E Pratt, Cornell University; B. Madhu Rao, Bowling Green State University; T. E. S. Raghavan, University of Illinois–Chicago; Gary Reeves, University of South Carolina; Gaspard Rizzuto, University of Southwestern; David Ronen, University of Missouri–St. Louis; Paul Savory, University of Nebraska, Lincoln; Jon Schlosser, New Mexico Highlands University; Delray Schultz, Millersville University; Richard Serfozo, Georgia Technological Institute; Morteza Shafi-Mousavi, Indiana University–South Bend; Dooyoung Shin, Mankato State University; Ronald L. Shubert, Elizabethtown College; Joel Sobel, University of California, San Diego; Manbir S. Sodhi, University of Rhode Island; Ariela Sofer, George Mason University; Toni M. Somers, Wayne State University; Robert Stark, University of Delaware; Joseph A. Svestka, Cleveland State University; Alexander Sze, Concordia College; Roman Sznajder, University of Maryland–Baltimore County; Bijan Vasigh, Embry-Riddle Aeronautical University; John H. Vande Vate, Georgia Technological University; Richard G. Vinson, University of South Alabama; Jin Wang, Valdodsta State University; Zhongxian Wang, Montclair State University; Robert C. Williams, Alfred University; Shmuel Yahalom, SUNY–Maritime College; James Yates, University of Central Oklahoma; Bill Yurcik, University of Pittsburgh.

Thanks also go to reviewers of the previous editions: Sant Arora, Harold Benson, Warren J. Boe, Bruce Bowerman, Jerald Dauer, S. Selcuk Erenguc, Yahya Fathi, Robert Freund, Irwin Greenberg, John Hooker, Sidney Lawrence, Patrick Lee, Edward Minieka, Joel A. Nachlas, David L. Olson, Sudhakar Pandit, David W. Pentico, Bruce Pollack-Johnson, Michael Richey, Gary D. Schudder, Lawrence Seiford, Michael Sinchcomb, and Paul Stiebitz.

We retain responsibility for all errors and would love to hear from users of the book. We may both be reached at

Indiana University
Department of Operations and Decision Technology
Kelley School of Business
Room 570
Bloomington, IN 47405

Wayne Winston (Winston@indiana.edu)
M. A. Venkataramanan (Venkatar@indiana.edu)

1

An Introduction to Model Building

1.1 An Introduction to Modeling

Operations research (often referred to as **management science**) is simply a scientific approach to decision making that seeks to best design and operate a system, usually under conditions requiring the allocation of scarce resources.

By a **system,** we mean an organization of interdependent components that work together to accomplish the goal of the system. For example, Ford Motor Company is a system whose goal consists of maximizing the profit that can be earned by producing quality vehicles.

The term *operations research* was coined during World War II when British military leaders asked scientists and engineers to analyze several military problems such as the deployment of radar and the management of convoy, bombing, antisubmarine, and mining operations.

The scientific approach to decision making usually involves the use of one or more **mathematical models.** A mathematical model is a mathematical representation of an actual situation that may be used to make better decisions or simply to understand the actual situation better. The following example should clarify many of the key terms used to describe mathematical models.

EXAMPLE 1   **Maximizing Wozac Yield**

Eli Daisy produces Wozac in huge batches by heating a chemical mixture in a pressurized container. Each time a batch is processed, a different amount of Wozac is produced. The amount produced is the *process yield* (measured in pounds). Daisy is interested in understanding the factors that influence the yield of the Wozac production process. Describe a model-building process for this situation.

Solution Daisy is first interested in determining the factors that influence the yield of the process. This would be referred to as a *descriptive model,* because it describes the behavior of the actual yield as a function of various factors. Daisy might determine (using regression methods discussed in *Stochastic Models in Operations Research: Applications and Algorithms*) that the following factors influence yield:

- container volume in liters (V)
- container pressure in milliliters (P)
- container temperature in degrees centigrade (T)
- chemical composition of the processed mixture

If we let A, B, and C be percentage of mixture made up of chemicals A, B, and C, then Daisy might find, for example, that

$$(1) \quad \text{yield} = 300 + .8V + .01P + .06T + .001T*P - .01T^2 - .001P^2$$
$$+ 11.7A + 9.4B + 16.4C + 19A*B + 11.4A*C - 9.6B*C.$$

To determine this relationship, the yield of the process would have to be measured for many different combinations of the previously listed factors. Knowledge of this equation would enable Daisy to describe the yield of the production process once volume, pressure, temperature, and chemical composition were known.

Prescriptive or Optimization Models

Most of the models discussed in this book will be **prescriptive** or **optimization** models. A prescriptive model "prescribes" behavior for an organization that will enable it to best meet its goal(s). The components of a prescriptive model include

- objective function(s)
- decision variables
- constraints

In short, an optimization model seeks to find values of the decision variables that optimize (maximize or minimize) an objective function among the set of all values for the decision variables that satisfy the given constraints.

The Objective Function

Naturally, Daisy would like to maximize the yield of the process. In most models, there will be a function we wish to maximize or minimize. This function is called the model's *objective function*. Of course, to maximize the process yield we need to find the values of V, P, T, A, B, and C that make (1) as large as possible.

In many situations, an organization may have more than one objective. For example, in assigning students to the two high schools in Bloomington, Indiana, the Monroe County School Board stated that the assignment of students involved the following objectives:

- Equalize the number of students at the two high schools.
- Minimize the average distance students travel to school.
- Have a diverse student body at both high schools.

Multiple objective decision-making problems are discussed in Sections 4.14 and 12.13.

The Decision Variables

The variables whose values are under our control and influence the performance of the system are called *decision variables*. In our example, V, P, T, A, B, and C are decision variables. Most of this book will be devoted to a discussion of how to determine the value of decision variables that maximize (sometimes minimize) an objective function.

Constraints

In most situations, only certain values of decision variables are possible. For example, certain volume, pressure, and temperature combinations might be unsafe. Also, A B, and C must be non-negative numbers that add to one. Restrictions on the values of decision variables are called *constraints*. Suppose the following:

- Volume must be between 1 and 5 liters.
- Pressure must be between 200 and 400 milliliters.
- Temperature must be between 100 and 200 degrees centigrade.
- Mixture must be made up entirely of A, B, and C.
- For the drug to properly perform, only half the mixture at most can be product A.

These constraints can be expressed mathematically by the following constraints:

$$V \leq 5$$
$$V \geq 1$$
$$P \leq 400$$
$$P \geq 200$$
$$T \leq 200$$
$$T \geq 100$$
$$A \geq 0$$
$$B \geq 0$$
$$A + B + C = 1$$
$$A \leq 5$$

The Complete Optimization Model

After letting z represent the value of the objective function, our entire optimization model may be written as follows:

Maximize $z = 300 + .8V + .01P + .06T + .001T*P - .01T^2 - .001P^2$
$$+ 11.7A + 9.4B + 16.4C + 19A*B + 11.4A*C - 9.6B*C$$

Subject to (s.t.)

$$V \leq 5$$
$$V \geq 1$$
$$P \leq 400$$
$$P \geq 200$$
$$T \leq 200$$
$$T \geq 100$$
$$A \geq 0$$
$$B \geq 0$$
$$C \geq 0$$
$$A + B + C = 1$$
$$A \leq 5$$

Any specification of the decision variables that satisfies all of the model's constraints is said to be in the **feasible region.** For example, $V = 2$, $P = 300$, $T = 150$, $A = .4$, $B = .3$, and $C = .1$ is in the feasible region. An **optimal solution** to an optimization model is any point in the feasible region that optimizes (in this case, *maximizes*) the objective function. Using the LINGO package that comes with this book, it can be determined that the optimal solution to this model is $V = 5$, $P = 200$, $T = 100$, $A = .294$, $B = 0$, $C = .706$, and $z = 183.38$. Thus, a maximum yield of 183.38 pounds can be obtained with a 5-liter container, pressure of 200 milliliters, temperature of 100 degrees centigrade, and 29% A and 71% C. This means no other feasible combination of decision variables can obtain a yield exceeding 183.38 pounds.

Static and Dynamic Models

A **static model** is one in which the decision variables do not involve sequences of decisions over multiple periods. A **dynamic model** is a model in which the decision variables *do* involve sequences of decisions over multiple periods. Basically, in a static model we solve a "one-shot" problem whose solutions prescribe optimal values of decision variables at all points in time. Example 1 is an example of a static model; the optimal solution will tell Daisy how to maximize yield at all points in time.

For an example of a dynamic model, consider a company (call it Sailco) that must determine how to minimize the cost of meeting (on-time) the demand for sailboats during the next year. Clearly Sailco's must determine how many sailboats it will produce during each of the next four quarters. Sailco's decisions involve decisions made over multiple periods, hence a model of Sailco's problem (see Section 3.10) would be a dynamic model.

Linear and Nonlinear Models

Suppose that whenever decision variables appear in the objective function and in the constraints of an optimization model, the decision variables are always multiplied by constants and added together. Such a model is a **linear model.** If an optimization model is not linear, then it is a **nonlinear model.** In the constraints of Example 1, the decision variables are always multiplied by constants and added together. Thus, Example 1's constraints pass the test for a linear model. However, in the objective function for Example 1, the terms .001T*P, $-.01T^2$, 19A*B, 11.4A*C, and $-9.6B*C$ make the model nonlinear. In general, nonlinear models are much harder to solve than linear models. We will discuss linear models in Chapters 2 through 10. Nonlinear models will be discussed in Chapter 12.

Integer and Noninteger Models

If one or more decision variables must be integer, then we say that an optimization model is an **integer model.** If all the decision variables are free to assume fractional values, then the optimization model is a **noninteger** model. Clearly, volume, temperature, pressure, and percentage composition of our inputs may all assume fractional values. Thus, Example 1 is a noninteger model. If the decision variables in a model represent the number of workers starting work during each shift at a fast-food restaurant, then clearly we have an integer model. Integer models are much harder to solve than nonlinear models. They will be discussed in detail in Chapter 9.

Deterministic and Stochastic Models

Suppose that for any value of the decision variables the value of the objective function and whether or not the constraints are satisfied is known with certainty. We then have a **deterministic model.** If this is not the case, then we have a **stochastic model.** All models in this volume will be deterministic models. Stochastic models will be covered in *Stochastic Models in Operations Research: Applications and Algorithms.*

If we view Example 1 as a deterministic model, then we are making the (unrealistic) assumption that for given values of V, P, T, A, B, and C the process yield will always be the same. This is highly unlikely. We can view (1) as a representation of the **average** yield of the process for given values of the decision variables. Then our objective is to find values of the decision variables that maximize the average yield of the process.

We can often gain useful insights into optimal decisions by using a deterministic model in a situation where a stochastic model is more appropriate. Consider Sailco's problem of minimizing the cost of meeting the demand (on-time) for sailboats. The uncertainty about future demand for sailboats implies that for a given production schedule, we do not know whether demand is met on-time. This leads us to believe that a stochastic model is needed to model Sailco's situation. We will see in Section 3.10, however, that we can develop a deterministic model for this situation that yields good decisions for Sailco.

1.2 The Seven-Step Model-Building Process

When operations research is used to solve an organization's problem, the following seven-step model-building procedure should be followed:

Step 1: Formulate the Problem The operations researcher first defines the organization's problem. Defining the problem includes specifying the organization's objectives and the parts of the organization that must be studied before the problem can be solved. In Example 1, the problem was to determine how to maximize the yield from a batch of Wozac.

Step 2: Observe the System Next the operations researcher collects data to estimate the value of parameters that affect the organization's problem. These estimates are used to develop (in step 3) and evaluate (in step 4) a mathematical model of the organization's problem. For example, in Example 1 data would be collected in an attempt to determine how the values of T, P, V, A, B, and C influence process yield.

Step 3: Formulate a Mathematical Model of the Problem In this step, the operations researcher develops a mathematical model of the problem. In this book, we will describe many mathematical techniques that can be used to model systems. For Example 1, our optimization model would be the result of step 3.

Step 4: Verify the Model and Use the Model for Prediction The operations researcher now tries to determine if the mathematical model developed in step 3 is an accurate representation of reality. For example, to validate our model, we might check and see if (1) accurately represents yield for values of the decision variables that were not used to estimate (1). Even if a model is valid for the current situation, we must be aware of blindly applying it. For example, if the government placed new restrictions on Wozac, then we might have to add new constraints to our model, and the yield of the process [and equation (1)] might change.

Step 5: Select a Suitable Alternative Given a model and a set of alternatives, the operations researcher now chooses the alternative that best meets the organization's objectives. (There may be more than one!) For instance, our model enabled us to determine that yield was maximized with V = 5, P = 200, T = 100, A = .294, B = 0, C = .706, and z = 183.38.

Step 6: Present the Results and Conclusion of the Study to the Organization In this step, the operations researcher presents the model and recommendation from step 5 to the decision-making individual or group. In some situations, one might present several alternatives and let the organization choose the one that best meets its needs. After presenting the results of the operations research study, the analyst may find that the organization does not approve of the recommendation. This may result from incorrect definition of the organization's problems or from failure to involve the decision maker from the start of the project. In this case, the operations researcher should return to step 1, 2, or 3.

Step 7: Implement and Evaluate Recommendations   If the organization has accepted the study, then the analyst aids in implementing the recommendations. The system must be constantly monitored (and updated dynamically as the environment changes) to ensure that the recommendations enable the organization to meet its objectives.

In what follows, we discuss three successful management science applications. We will give a detailed (but nonquantitative) description of each application. We will tie our discussion of each application to the seven-step model-building process described in Section 1.2.

1.3   CITGO Petroleum

Klingman et al. (1987) applied a variety of management-science techniques to CITGO Petroleum. Their work saved the company an estimated $70 million per year. CITGO is an oil-refining and -marketing company that was purchased by Southland Corporation (the owners of the 7-Eleven stores). We will focus on two aspects of the CITGO team's work:

1   a mathematical model to optimize operation of CITGO's refineries, and

2   a mathematical model—supply distribution marketing (SDM) system—that was used to develop an 11-week supply, distribution, and marketing plan for the entire business.

Optimizing Refinery Operations

Step 1   Klingman et al. wanted to minimize the cost of operating CITGO's refineries.

Step 2   The Lake Charles, Louisiana, refinery was closely observed in an attempt to estimate key relationships such as:

1   How the cost of producing each of CITGO's products (motor fuel, no. 2 fuel oil, turbine fuel, naptha, and several blended motor fuels) depends on the inputs used to produce each product.

2   The amount of energy needed to produce each product. This required the installation of a new metering system.

3   The yield associated with each input–output combination. For example, if 1 gallon of crude oil would yield .52 gallons of motor fuel, then the yield would equal 52%.

4   To reduce maintenance costs, data were collected on parts inventories and equipment breakdowns. Obtaining accurate data required the installation of a new database-management system and integrated maintenance-information system. A process control system was also installed to accurately monitor the inputs and resources used to manufacture each product.

Step 3   Using linear programming (LP), a model was developed to optimize refinery operations. The model determines the cost-minimizing method for mixing or blending together inputs to produce desired outputs. The model contains **constraints** that ensure that inputs are blended so that each output is of the desired quality. Blending constraints are discussed in Section 3.8. The model ensures that plant capacities are not exceeded and allows for the fact that each refinery may carry an inventory of each end product. Sections 3.10 and 4.12 discuss inventory constraints.

Step 4   To validate the model, inputs and outputs from the Lake Charles refinery were collected for one month. Given the actual inputs used at the refinery during that month, the actual outputs were compared to those predicted by the model. After extensive changes, the model's predicted outputs were close to the actual outputs.

Step 5   Running the LP yielded a daily strategy for running the refinery. For instance, the model might, say, produce 400,000 gallons of turbine fuel using 300,000 gallons of crude 1 and 200,000 gallons of crude 2.

Steps 6 and 7   Once the data base and process control were in place, the model was used to guide day-to-day refinery operations. CITGO estimated that the overall benefits of the refinery system exceeded $50 million dollars annually.

The Supply Distribution Marketing (SDM) System

Step 1   CITGO wanted a mathematical model that could be used to make supply, distribution, and marketing decisions such as:

1   Where should crude oil be purchased?

2   Where should products be sold?

3   What price should be charged for products?

4   How much of each product should be held in inventory?

The goal, of course, was to maximize the profitability associated with these decisions.

Step 2   A database that kept track of sales, inventory, trades, and exchanges of all refined products was installed. Also, regression analysis (see *Stochastic Models in Operations Research: Applications and Algorithms*) was used to develop forecasts for wholesale prices and wholesale demand for each CITGO product.

Steps 3 and 5   A minimum-cost network flow model (MCNFM) (see Section 7.4) is used to determine an 11-week supply, marketing, and distribution strategy. The model makes all decisions discussed in our step 1 discussion. A typical model run that involved 3,000 equations and 15,000 decision variables required only 30 seconds on an IBM 4381.

Step 4   The forecasting modules are continuously evaluated to ensure that they continue to give accurate forecasts.

Steps 6 and 7   Implementing the SDM required several organizational changes. A new vice-president was appointed to coordinate the operation of the SDM and LP refinery model. The product supply and product scheduling departments were combined to improve communication and information flow.

1.4   San Francisco Police Department Scheduling

Taylor and Huxley (1989) developed a police patrol scheduling system (PPSS). All San Francisco (SF) police precincts use PPSS to schedule their officers. It is estimated that PPSS saves the SF police more than $5 million annually. Other cities such as Virginia

Beach, Virginia, and Richmond, California, have also adopted PPSS. Following our seven-step model-building procedure, here is a description of PPSS.

Step 1 The SFPD police department wanted a method to schedule patrol officers in each precinct that would quickly produce (in less than one hour) a schedule and graphically display it. The program should first determine the personnel requirements for each hour of the week. For example, 38 officers might be needed between 1 A.M. and 2 A.M. Sunday but only 14 officers might be needed from 4 A.M. to 5 A.M. Sunday. Officers should then be scheduled to minimize the sum over each hour of the week of the shortages and surpluses relative to the needed number of officers. For example, if 20 officers were assigned to the midnight to 8 A.M. Sunday shift, we would have a shortage of $38 - 20 = 18$ officers from 1 to 2 A.M. and a surplus of $20 - 14 = 6$ officers from 4 to 5 A.M. A secondary criterion was to minimize the maximum shortage because a shortage of 10 officers during a single hour is far more serious than a shortage of one officer during 10 different hours. The SFPD also wanted a scheduler that precinct captains could easily fine-tune to produce the optimal schedule.

Step 2 The SFPD had a sophisticated computer-aided dispatch (CAD) system to keep track of all calls for police help, police travel time, police response time, and so on. SFPD had a standard percentage of time that administrators felt each officer should be busy. Using CAD, it is easy to determine the number of workers needed each hour. Suppose, for example, an officer should be busy 80% of the time and CAD indicates that 30.4 hours of work come in from 4 to 5 A.M. Sunday. Then we need 38 officers from 4 to 5 A.M. on Sunday $[.8*(38) = 30.4$ hours$]$.

Step 3 An LP model was formulated (see Section 3.5 for a discussion of scheduling models). As discussed in step 1, the primary objective was to minimize the sum of hourly shortages and surpluses. At first, schedulers assumed that officers worked five consecutive days for eight hours a day (this was the policy prior to PPSS) and that there were three shift starting times (say, 6 A.M., 2 P.M., and 10 A.M.). The constraints in the PPSS model reflected the limited number of officers available and the relationship of the number of officers working each hour to the shortages and surpluses for that hour. Then PPSS would produce a schedule that would tell the precinct captain how many officers should start work at each possible shift time. For example, PPSS might say that 20 officers should start work at 6 A.M. Monday (working 6 A.M.–2 P.M. Monday–Friday) and 30 officers should start work at 2 P.M. Saturday (working 2 P.M.–10 P.M. Saturday–Wednesday). The fact that the number of officers assigned to a start time must be an integer made it far more difficult to find an optimal schedule. (Problems in which decision variables must be integers are discussed in Chapter 9.)

Step 4 Before implementing PPSS, the SFPD tested the PPSS schedules against manually created schedules. PPSS produced an approximately 50% reduction in both surpluses and shortages. This convinced the department to implement PPSS.

Step 5 Given the starting times for shifts and the type of work schedule [four consecutive days for 10 hours per day (the 4/10 schedule) or five consecutive days for eight hours per day (the 5/8 schedule)], PPSS can produce a schedule that minimizes the sum of shortages and surpluses. More important, PPSS can be used to experiment with shift times and work rules. Using PPSS, it was found that if only three shift times are allowed, then a 5/8 schedule was superior to a 4/10 schedule. If, however, five shift times were allowed, then a 4/10 schedule was found to be superior. This finding was of critical importance because police officers had wanted to switch to a 4/10 schedule for years. The city had resisted 4/10 schedules because they appeared to reduce productivity. PPSS showed that 4/10 schedules need not reduce productivity. After the introduction of PPSS, the SFPD went to 4/10 schedules and *improved productivity!* PPSS also enables the department to experiment with a mix of one officer and two officer patrol cars.

Steps 6 and 7   It is estimated that PPSS created an extra 170,000 productive hours per year, thereby saving the city of San Francisco $5.2 million dollars per year. Ninety-six percent of all workers preferred PPSS generated schedules to manually generated schedules. PPSS enabled SFPD to make strategic changes (such as adopting the 4/10 schedule), which made officers happier and increased productivity. Response times to calls improved by 20% after PPSS was adopted.

A major reason for the success of PPSS was that the system allowed precinct captains to fine-tune the computer-generated schedule and obtain a new schedule in less than one minute. For example, precinct captains could easily add or delete officers and add or delete shifts and quickly see how these changes modified the master schedule.

1.5   GE Capital

GE Capital provides credit card service to 50 million accounts. The average total outstanding balance exceeds $12 billion. GE Capital led by Makuch et al. (1989) developed the PAYMENT system to reduce delinquent accounts and the cost of collecting from delinquent accounts.

Step 1   At any one time, GE Capital has more than $1 billion dollars in delinquent accounts. The company spends $100 million per year processing these accounts. Each day, workers contact more than 200,000 delinquent credit card holders with letters, messages, or live calls. The company's goal was to reduce delinquent accounts and the cost of processing them. To do this, GE Capital needed to come up with a method of assigning scarce labor resources to delinquent accounts. For example, PAYMENT determines which delinquent accounts receive live phone calls and which delinquent accounts receive no contact.

Step 2   The key to modeling delinquent accounts is the concept of a **delinquency movement matrix (DMM).** The DMM determines how the probability of the payment on a delinquent account during the current month depends on the following factors: size of unpaid balance (either $<$300 or $\geq$300), action taken (no action, live phone call, taped message, letters), and a performance score (high, medium, or low). The higher the performance score associated with a delinquent account, the more likely the account is to be collected. For example, if a $250 account is two months delinquent, has a high performance score, and is contacted with a phone message, then the following may occur:

TABLE 1

Sample Entries in DMM

Event	Probability
Account completely paid	.30
One month is paid	.40
Nothing is paid	.30

Because GE Capital has millions of delinquent accounts, there is ample data to accurately estimate the DMM. For example, suppose there were 10,000 two-month delinquent accounts with balances under $300 that have a high performance score and are contacted with phone messages. If 3,000 of those accounts were completely paid off during the current month, then we would estimate the probability of an account being completely paid off during the current month as 3,000/10,000 = .30.

Step 3 GE Capital developed a linear optimization model. The objective function for the PAYMENT model was to maximize the expected delinquent accounts collected during the next six months. The decision variables represented the fraction of each type of delinquent account (accounts are classified by payment balance, performance score, and months delinquent) that experienced each type of contact (no action, live phone call, taped message, or letter). The constraints in the PAYMENT model ensure that available resources are not overused. Constraints also relate the number of each type of delinquent account present in, say, January to the number of delinquent accounts of each type present during the next month (February). This **dynamic** aspect of the PAYMENT model is crucial to its success. Without this aspect, the model would simply "skim" the accounts that are easiest to collect each month. This would result in few collections during later months.

Step 4 PAYMENT was piloted on a $62 million portfolio for a single department store. GE Capital managers came up with their own strategies for allocating resources (collectively called CHAMPION). The store's delinquent accounts were randomly assigned to the CHAMPIONS and PAYMENT strategies. PAYMENT used more live phone calls and more "no action" than the CHAMPION strategies. PAYMENT also collected $180,000 per month more than any of the CHAMPION strategies, a 5% to 7% improvement. Note that using more of the no-action strategy certainly leads to a long-run increase in customer goodwill!

Step 5 As described in step 3, for each type of account PAYMENT tells the credit managers the fraction that should receive each type of contact. For example, for three-month delinquent accounts with a small (<$300) unpaid balance and high performance score, PAYMENT might prescribe 30% no action, 20% letters, 30% phone messages, and 20% live phone calls.

Steps 6 and 7 PAYMENT was next applied to the 18 million accounts of the $4.6 billion Montgomery-Ward department store portfolio. Comparing the collection results to the same time period a year earlier, it was found that PAYMENT increased collections by $1.6 million per month (more than $19 million per year). This is actually a conservative estimate of the benefit obtained from PAYMENT, because PAYMENT was first applied to the Montgomery-Ward portfolio during the depths of a recession—and a recession makes it much more difficult to collect delinquent accounts.

Overall, GE Capital estimates that PAYMENT increased collections by $37 million per year and used fewer resources than previous strategies.

REFERENCES

Klingman, D., N. Phillips, D. Steiger, and W. Young, "The Successful Deployment of Management Science Throughout Citgo Corporation," *Interfaces* 17 (1987, no. 1):4–25.

Makuch, W., J. Dodge, J. Ecker, D. Granfors, and G. Hahn, "Managing Consumer Credit Delinquency in the US Economy: A Multi-Billion Dollar Management Science Application," *Interfaces* 22 (1992, no. 1):90–109.

Taylor, P., and S. Huxley, "A Break from Tradition for the San Francisco Police: Patrol Officer Scheduling Using an Optimization-Based Decision Support Tool," *Interfaces* 19 (1989, no. 1):4–24.

2

Basic Linear Algebra

In this chapter, we study the topics in linear algebra that will be needed in the rest of the book. We begin by discussing the building blocks of linear algebra: matrices and vectors. Then we use our knowledge of matrices and vectors to develop a systematic procedure (the Gauss–Jordan method) for solving linear equations, which we then use to invert matrices. We close the chapter with an introduction to determinants.

The material covered in this chapter will be used in our study of linear and nonlinear programming.

2.1 Matrices and Vectors

Matrices

DEFINITION ■ A **matrix** is any rectangular array of numbers. ■

For example,

$$\begin{bmatrix} 1 & 2 \\ 3 & 4 \end{bmatrix}, \quad \begin{bmatrix} 1 & 2 & 3 \\ 4 & 5 & 6 \end{bmatrix}, \quad \begin{bmatrix} 1 \\ -2 \end{bmatrix}, \quad \begin{bmatrix} 2 & 1 \end{bmatrix}$$

are all matrices.

If a matrix A has m rows and n columns, we call A an $m \times n$ matrix. We refer to $m \times n$ as the **order** of the matrix. A typical $m \times n$ matrix A may be written as

$$A = \begin{bmatrix} a_{11} & a_{12} & \cdots & a_{1n} \\ a_{21} & a_{22} & \cdots & a_{2n} \\ \vdots & \vdots & & \vdots \\ a_{m1} & a_{m2} & \cdots & a_{mn} \end{bmatrix}$$

DEFINITION ■ The number in the ith row and jth column of A is called the **ijth element** of A and is written a_{ij}. ■

For example, if

$$A = \begin{bmatrix} 1 & 2 & 3 \\ 4 & 5 & 6 \\ 7 & 8 & 9 \end{bmatrix}$$

then $a_{11} = 1$, $a_{23} = 6$, and $a_{31} = 7$.

Sometimes we will use the notation $A = [a_{ij}]$ to indicate that A is the matrix whose ijth element is a_{ij}.

Two matrices $A = [a_{ij}]$ and $B = [b_{ij}]$ are **equal** if and only if A and B are of the same order and for all i and j, $a_{ij} = b_{ij}$. ■

For example, if

$$A = \begin{bmatrix} 1 & 2 \\ 3 & 4 \end{bmatrix} \quad \text{and} \quad B = \begin{bmatrix} x & y \\ w & z \end{bmatrix}$$

then $A = B$ if and only if $x = 1$, $y = 2$, $w = 3$, and $z = 4$.

Vectors

Any matrix with only one column (that is, any $m \times 1$ matrix) may be thought of as a **column vector.** The number of rows in a column vector is the **dimension** of the column vector. Thus,

$$\begin{bmatrix} 1 \\ 2 \end{bmatrix}$$

may be thought of as a 2×1 matrix or a two-dimensional column vector. R^m will denote the set of all m-dimensional column vectors.

In analogous fashion, we can think of any vector with only one row (a $1 \times n$ matrix as a **row vector.** The dimension of a row vector is the number of columns in the vector. Thus, [9 2 3] may be viewed as a 1×3 matrix or a three-dimensional row vector. In this book, vectors appear in boldface type: for instance, vector **v.** An m-dimensional vector (either row or column) in which all elements equal zero is called a **zero vector** (written **0**). Thus,

$$[0 \quad 0] \quad \text{and} \quad \begin{bmatrix} 0 \\ 0 \end{bmatrix}$$

are two-dimensional zero vectors.

Any m-dimensional vector corresponds to a directed line segment in the m-dimensional plane. For example, in the two-dimensional plane, the vector

$$\mathbf{u} = \begin{bmatrix} 1 \\ 2 \end{bmatrix}$$

corresponds to the line segment joining the point

$$\begin{bmatrix} 0 \\ 0 \end{bmatrix}$$

to the point

$$\begin{bmatrix} 1 \\ 2 \end{bmatrix}$$

The directed line segments corresponding to

$$\mathbf{u} = \begin{bmatrix} 1 \\ 2 \end{bmatrix}, \quad \mathbf{v} = \begin{bmatrix} 1 \\ -3 \end{bmatrix}, \quad \mathbf{w} = \begin{bmatrix} -1 \\ -2 \end{bmatrix}$$

are drawn in Figure 1.

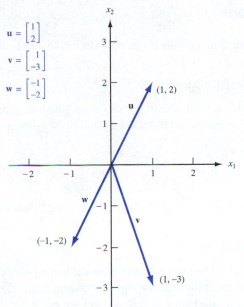

$\mathbf{u} = \begin{bmatrix} 1 \\ 2 \end{bmatrix}$

$\mathbf{v} = \begin{bmatrix} 1 \\ -3 \end{bmatrix}$

$\mathbf{w} = \begin{bmatrix} -1 \\ -2 \end{bmatrix}$

FIGURE 1
Vectors Are Directed
Line Segments

The Scalar Product of Two Vectors

An important result of multiplying two vectors is the *scalar product*. To define the scalar product of two vectors, suppose we have a row vector $\mathbf{u} = [u_1 \quad u_2 \quad \cdots \quad u_n]$ and a column vector

$$\mathbf{v} = \begin{bmatrix} v_1 \\ v_2 \\ \vdots \\ v_n \end{bmatrix}$$

of the same dimension. The **scalar product** of $\mathbf{u}$ and $\mathbf{v}$ (written $\mathbf{u} \cdot \mathbf{v}$) is the number $u_1v_1 + u_2v_2 + \cdots + u_nv_n$.

For the scalar product of two vectors to be defined, the first vector must be a row vector and the second vector must be a column vector. For example, if

$$\mathbf{u} = [1 \quad 2 \quad 3] \qquad \text{and} \qquad \mathbf{v} = \begin{bmatrix} 2 \\ 1 \\ 2 \end{bmatrix}$$

then $\mathbf{u} \cdot \mathbf{v} = 1(2) + 2(1) + 3(2) = 10$. By these rules for computing a scalar product, if

$$\mathbf{u} = \begin{bmatrix} 1 \\ 2 \end{bmatrix} \qquad \text{and} \qquad \mathbf{v} = [2 \quad 3]$$

then $\mathbf{u} \cdot \mathbf{v}$ is not defined. Also, if

$$\mathbf{u} = [1 \quad 2 \quad 3] \qquad \text{and} \qquad \mathbf{v} = \begin{bmatrix} 3 \\ 4 \end{bmatrix}$$

then $\mathbf{u} \cdot \mathbf{v}$ is not defined because the vectors are of two different dimensions.

Note that two vectors are perpendicular if and only if their scalar product equals 0. Thus, the vectors $[1 \quad -1]$ and $[1 \quad 1]$ are perpendicular.

We note that $\mathbf{u} \cdot \mathbf{v} = \|\mathbf{u}\| \, \|\mathbf{v}\| \cos \theta$, where $\|\mathbf{u}\|$ is the length of the vector $\mathbf{u}$ and θ is the angle between the vectors $\mathbf{u}$ and $\mathbf{v}$.

Matrix Operations

We now describe the arithmetic operations on matrices that are used later in this book.

The Scalar Multiple of a Matrix

Given any matrix A and any number c (a *number* is sometimes referred to as a *scalar*), the matrix cA is obtained from the matrix A by multiplying each element of A by c. For example,

$$\text{if} \quad A = \begin{bmatrix} 1 & 2 \\ -1 & 0 \end{bmatrix}, \quad \text{then} \quad 3A = \begin{bmatrix} 3 & 6 \\ -3 & 0 \end{bmatrix}$$

For $c = -1$, scalar multiplication of the matrix A is sometimes written as $-A$.

Addition of Two Matrices

Let $A = [a_{ij}]$ and $B = [b_{ij}]$ be two matrices with the same order (say, $m \times n$). Then the matrix $C = A + B$ is defined to be the $m \times n$ matrix whose ijth element is $a_{ij} + b_{ij}$. Thus, to obtain the sum of two matrices A and B, we add the corresponding elements of A and B. For example, if

$$A = \begin{bmatrix} 1 & 2 & 3 \\ 0 & -1 & 1 \end{bmatrix} \quad \text{and} \quad B = \begin{bmatrix} -1 & -2 & -3 \\ 2 & 1 & -1 \end{bmatrix}$$

then

$$A + B = \begin{bmatrix} 1-1 & 2-2 & 3-3 \\ 0+2 & -1+1 & 1-1 \end{bmatrix} = \begin{bmatrix} 0 & 0 & 0 \\ 2 & 0 & 0 \end{bmatrix}.$$

This rule for matrix addition may be used to add vectors of the same dimension. For example, if $\mathbf{u} = [1 \ \ 2]$ and $\mathbf{v} = [2 \ \ 1]$, then $\mathbf{u} + \mathbf{v} = [1 + 2 \ \ 2 + 1] = [3 \ \ 3]$. Vectors may be added geometrically by the parallelogram law (see Figure 2).

We can use scalar multiplication and the addition of matrices to define the concept of a line segment. A glance at Figure 1 should convince you that any point u in the m-dimensional plane corresponds to the m-dimensional vector $\mathbf{u}$ formed by joining the origin to the point u. For any two points u and v in the m-dimensional plane, the **line segment** joining u and v (called the line segment uv) is the set of all points in the m-dimensional plane that correspond to the vectors $c\mathbf{u} + (1 - c)\mathbf{v}$, where $0 \le c \le 1$ (Figure 3). For example, if $u = (1, 2)$ and $v = (2, 1)$, then the line segment uv consists

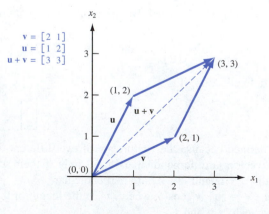

$\mathbf{v} = [2 \ \ 1]$
$\mathbf{u} = [1 \ \ 2]$
$\mathbf{u} + \mathbf{v} = [3 \ \ 3]$

FIGURE 2
Addition of Vectors

14   CHAPTER **2** Basic Linear Algebra

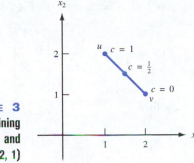

FIGURE 3
Line Segment Joining
$u = (1, 2)$ **and**
$v = (2, 1)$

of the points corresponding to the vectors $c[1 \quad 2] + (1 - c)[2 \quad 1] = [2 - c \quad 1 + c]$, where $0 \le c \le 1$. For $c = 0$ and $c = 1$, we obtain the endpoints of the line segment uv; for $c = \frac{1}{2}$, we obtain the midpoint $(0.5\mathbf{u} + 0.5\mathbf{v})$ of the line segment uv.

Using the parallelogram law, the line segment uv may also be viewed as the points corresponding to the vectors $\mathbf{u} + c(\mathbf{v} - \mathbf{u})$, where $0 \le c \le 1$ (Figure 4). Observe that for $c = 0$, we obtain the vector $\mathbf{u}$ (corresponding to point u), and for $c = 1$, we obtain the vector $\mathbf{v}$ (corresponding to point v).

The Transpose of a Matrix

Given any $m \times n$ matrix

$$A = \begin{bmatrix} a_{11} & a_{12} & \cdots & a_{1n} \\ a_{21} & a_{22} & \cdots & a_{2n} \\ \vdots & \vdots & & \vdots \\ a_{m1} & a_{m2} & \cdots & a_{mn} \end{bmatrix}$$

the **transpose** of A (written A^T) is the $n \times m$ matrix

$$A^T = \begin{bmatrix} a_{11} & a_{21} & \cdots & a_{m1} \\ a_{12} & a_{22} & \cdots & a_{m2} \\ \vdots & \vdots & & \vdots \\ a_{1n} & a_{2n} & \cdots & a_{mn} \end{bmatrix}$$

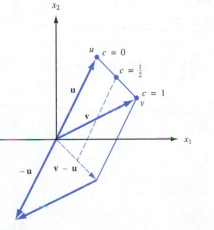

FIGURE 4
Representation of Line
Segment uv

Thus, A^T is obtained from A by letting row 1 of A be column 1 of A^T, letting row 2 of A be column 2 of A^T, and so on. For example,

$$\text{if} \quad A = \begin{bmatrix} 1 & 2 & 3 \\ 4 & 5 & 6 \end{bmatrix}, \quad \text{then} \quad A^T = \begin{bmatrix} 1 & 4 \\ 2 & 5 \\ 3 & 6 \end{bmatrix}$$

Observe that $(A^T)^T = A$. Let $B = [1 \quad 2]$; then

$$B^T = \begin{bmatrix} 1 \\ 2 \end{bmatrix} \quad \text{and} \quad (B^T)^T = [1 \quad 2] = B$$

As indicated by these two examples, for any matrix A, $(A^T)^T = A$.

Matrix Multiplication

Given two matrices A and B, the matrix product of A and B (written AB) is defined if and only if

$$\text{Number of columns in } A = \text{number of rows in } B \tag{1}$$

For the moment, assume that for some positive integer r, A has r columns and B has r rows. Then for some m and n, A is an $m \times r$ matrix and B is an $r \times n$ matrix.

DEFINITION ■ The **matrix product** $C = AB$ of A and B is the $m \times n$ matrix C whose ijth element is determined as follows:

ijth element of C = scalar product of row i of $A \times$ column j of B ■ (2)

If Equation (1) is satisfied, then each row of A and each column of B will have the same number of elements. Also, if (1) is satisfied, then the scalar product in Equation (2) will be defined. The product matrix $C = AB$ will have the same number of rows as A and the same number of columns as B.

EXAMPLE 1   **Matrix Multiplication**

Compute $C = AB$ for

$$A = \begin{bmatrix} 1 & 1 & 2 \\ 2 & 1 & 3 \end{bmatrix} \quad \text{and} \quad B = \begin{bmatrix} 1 & 1 \\ 2 & 3 \\ 1 & 2 \end{bmatrix}$$

Solution   Because A is a 2×3 matrix and B is a 3×2 matrix, AB is defined, and C will be a 2×2 matrix. From Equation (2),

$$c_{11} = [1 \quad 1 \quad 2] \begin{bmatrix} 1 \\ 2 \\ 1 \end{bmatrix} = 1(1) + 1(2) + 2(1) = 5$$

$$c_{12} = [1 \quad 1 \quad 2] \begin{bmatrix} 1 \\ 3 \\ 2 \end{bmatrix} = 1(1) + 1(3) + 2(2) = 8$$

$$c_{21} = [2 \quad 1 \quad 3] \begin{bmatrix} 1 \\ 2 \\ 1 \end{bmatrix} = 2(1) + 1(2) + 3(1) = 7$$

$$c_{22} = \begin{bmatrix} 2 & 1 & 3 \end{bmatrix} \begin{bmatrix} 1 \\ 3 \\ 2 \end{bmatrix} = 2(1) + 1(3) + 3(2) = 11$$

$$C = AB = \begin{bmatrix} 5 & 8 \\ 7 & 11 \end{bmatrix}$$

EXAMPLE 2   Column Vector Times Row Vector

Find AB for

$$A = \begin{bmatrix} 3 \\ 4 \end{bmatrix} \quad \text{and} \quad B = \begin{bmatrix} 1 & 2 \end{bmatrix}$$

Solution   Because A has one column and B has one row, $C = AB$ will exist. From Equation (2), we know that C is a 2×2 matrix with

$$c_{11} = 3(1) = 3 \qquad c_{21} = 4(1) = 4$$
$$c_{12} = 3(2) = 6 \qquad c_{22} = 4(2) = 8$$

Thus,

$$C = \begin{bmatrix} 3 & 6 \\ 4 & 8 \end{bmatrix}$$

EXAMPLE 3   Row Vector Times Column Vector

Compute $D = BA$ for the A and B of Example 2.

Solution   In this case, D will be a 1×1 matrix (or a scalar). From Equation (2),

$$d_{11} = \begin{bmatrix} 1 & 2 \end{bmatrix} \begin{bmatrix} 3 \\ 4 \end{bmatrix} = 1(3) + 2(4) = 11$$

Thus, $D = [11]$. In this example, matrix multiplication is equivalent to scalar multiplication of a row and column vector.

Recall that if you multiply two real numbers a and b, then $ab = ba$. This is called the *commutative property of multiplication.* Examples 2 and 3 show that for matrix multiplication it may be that $AB \neq BA$. Matrix multiplication is not necessarily commutative. (In some cases, however, $AB = BA$ will hold.)

EXAMPLE 4   Undefined Matrix Product

Show that AB is undefined if

$$A = \begin{bmatrix} 1 & 2 \\ 3 & 4 \end{bmatrix} \quad \text{and} \quad B = \begin{bmatrix} 1 & 1 \\ 0 & 1 \\ 1 & 2 \end{bmatrix}$$

Solution   This follows because A has two columns and B has three rows. Thus, Equation (1) is not satisfied.

TABLE **1**

Gallons of Crude Oil Required to Produce 1 Gallon
of Gasoline

Crude Oil	Premium Unleaded	Regular Unleaded	Regular Leaded
1	$\frac{3}{4}$	$\frac{2}{3}$	$\frac{1}{4}$
2	$\frac{1}{4}$	$\frac{1}{3}$	$\frac{3}{4}$

Many computations that commonly occur in operations research (and other branches of mathematics) can be concisely expressed by using matrix multiplication. To illustrate this, suppose an oil company manufactures three types of gasoline: premium unleaded, regular unleaded, and regular leaded. These gasolines are produced by mixing two types of crude oil: crude oil 1 and crude oil 2. The number of gallons of crude oil required to manufacture 1 gallon of gasoline is given in Table 1.

From this information, we can find the amount of each type of crude oil needed to manufacture a given amount of gasoline. For example, if the company wants to produce 10 gallons of premium unleaded, 6 gallons of regular unleaded, and 5 gallons of regular leaded, then the company's crude oil requirements would be

$$\text{Crude 1 required} = (\tfrac{3}{4})(10) + (\tfrac{2}{3})(6) + (\tfrac{1}{4})5 = 12.75 \text{ gallons}$$
$$\text{Crude 2 required} = (\tfrac{1}{4})(10) + (\tfrac{1}{3})(6) + (\tfrac{3}{4})5 = 8.25 \text{ gallons}$$

More generally, we define

$$p_U = \text{gallons of premium unleaded produced}$$
$$r_U = \text{gallons of regular unleaded produced}$$
$$r_L = \text{gallons of regular leaded produced}$$
$$c_1 = \text{gallons of crude 1 required}$$
$$c_2 = \text{gallons of crude 2 required}$$

Then the relationship between these variables may be expressed by

$$c_1 = (\tfrac{3}{4}) p_U + (\tfrac{2}{3}) r_U + (\tfrac{1}{4}) r_L$$
$$c_2 = (\tfrac{1}{4}) p_U + (\tfrac{1}{3}) r_U + (\tfrac{3}{4}) r_L$$

Using matrix multiplication, these relationships may be expressed by

$$\begin{bmatrix} c_1 \\ c_2 \end{bmatrix} = \begin{bmatrix} \frac{3}{4} & \frac{2}{3} & \frac{1}{4} \\ \frac{1}{4} & \frac{1}{3} & \frac{3}{4} \end{bmatrix} \begin{bmatrix} p_U \\ r_U \\ r_L \end{bmatrix}$$

Properties of Matrix Multiplication

To close this section, we discuss some important properties of matrix multiplication. In what follows, we assume that all matrix products are defined.

1 Row i of AB = (row i of A)B. To illustrate this property, let

$$A = \begin{bmatrix} 1 & 1 & 2 \\ 2 & 1 & 3 \end{bmatrix} \quad \text{and} \quad B = \begin{bmatrix} 1 & 1 \\ 2 & 3 \\ 1 & 2 \end{bmatrix}$$

Then row 2 of the 2×2 matrix AB is equal to

$$[2 \quad 1 \quad 3] \begin{bmatrix} 1 & 1 \\ 2 & 3 \\ 1 & 2 \end{bmatrix} = [7 \quad 11]$$

This answer agrees with Example 1.

2 Column j of AB = A(column j of B). Thus, for A and B as given, the first column of AB is

$$\begin{bmatrix} 1 & 1 & 2 \\ 2 & 1 & 3 \end{bmatrix} \begin{bmatrix} 1 \\ 2 \\ 1 \end{bmatrix} = \begin{bmatrix} 5 \\ 7 \end{bmatrix}$$

Properties 1 and 2 are helpful when you need to compute only *part* of the matrix AB.

3 Matrix multiplication is associative. That is, $A(BC) = (AB)C$. To illustrate, let

$$A = [1 \quad 2], \qquad B = \begin{bmatrix} 2 & 3 \\ 4 & 5 \end{bmatrix}, \qquad C = \begin{bmatrix} 2 \\ 1 \end{bmatrix}$$

Then $AB = [10 \quad 13]$ and $(AB)C = 10(2) + 13(1) = [33]$.
   On the other hand,

$$BC = \begin{bmatrix} 7 \\ 13 \end{bmatrix}$$

so $A(BC) = 1(7) + 2(13) = [33]$. In this case, $A(BC) = (AB)C$ does hold.

4 Matrix multiplication is distributive. That is, $A(B + C) = AB + AC$ and $(B + C)D = BD + CD$.

Matrix Multiplication with EXCEL

Using the EXCEL MMULT function, it is easy to multiply matrices. To illustrate, let's use EXCEL to find the matrix product AB that we found in Example 1 (see Figure 5 and file Mmult.xls). We proceed as follows:

Step 1 Enter A and B in D2:F3 and D5:E7, respectively.

Step 2 Select the range (D9:E10) in which the product AB will be computed.

Step 3 In the upper left-hand corner (D9) of the selected range, type the formula

$$= \text{MMULT(D2:F3,D5:E7)}.$$

Then hit **Control Shift Enter** (not just enter), and the desired matrix product will be computed. Note that MMULT is an **array** function and not an ordinary spreadsheet function. This explains why we must preselect the range for AB and use Control Shift Enter.

	A	B	C	D	E	F
1	MatrixMultiplication					
2				1	1	2
3			A	2	1	3
4						
5			B	1	1	
6				2	3	
7				1	2	
8						
9				5	8	
10			C	7	11	
11						

FIGURE 5

PROBLEMS

Group A $3 \cdot 3$

1 For $A = \begin{bmatrix} 1 & 2 & 3 \\ 4 & 5 & 6 \\ 7 & 8 & 9 \end{bmatrix}$ and $\quad 3 \cdot 2 \quad B = \begin{bmatrix} 1 & 2 \\ 0 & -1 \\ 1 & 2 \end{bmatrix}$, find:

    a $-A$ **b** $3A$ **c** $A + 2B$

    d A^T **e** B^T **f** AB

    g BA

2 Only three brands of beer (beer 1, beer 2, and beer 3) are available for sale in Metropolis. From time to time, people try one or another of these brands. Suppose that at the beginning of each month, people change the beer they are drinking according to the following rules:

30% of the people who prefer beer 1 switch to beer 2.

20% of the people who prefer beer 1 switch to beer 3.

30% of the people who prefer beer 2 switch to beer 3.

30% of the people who prefer beer 3 switch to beer 2.

10% of the people who prefer beer 3 switch to beer 1.

For $i = 1, 2, 3$, let x_i be the number who prefer beer i at the beginning of this month and y_i be the number who prefer beer i at the beginning of next month. Use matrix multiplication to relate the following:

$$\begin{bmatrix} y_1 \\ y_2 \\ y_3 \end{bmatrix} \qquad \begin{bmatrix} x_1 \\ x_2 \\ x_3 \end{bmatrix}$$

Group B

3 Prove that matrix multiplication is associative.

4 Show that for any two matrices A and B, $(AB)^T = B^T A^T$.

5 An $n \times n$ matrix A is symmetric if $A = A^T$.

    a Show that for any $n \times n$ matrix, AA^T is a symmetric matrix.

    b Show that for any $n \times n$ matrix A, $(A + A^T)$ is a symmetric matrix.

6 Suppose that A and B are both $n \times n$ matrices. Show that computing the matrix product AB requires n^3 multiplications and $n^3 - n^2$ additions.

7 The **trace of a matrix** is the sum of its diagonal elements.

    a For any two matrices A and B, show that trace $(A + B) = $ trace $A + $ trace B.

    b For any two matrices A and B for which the products AB and BA are defined, show that trace $AB = $ trace BA.

2.2 Matrices and Systems of Linear Equations

Consider a system of linear equations given by

$$\begin{aligned}
a_{11}x_1 + a_{12}x_2 + \cdots + a_{1n}x_n &= b_1 \\
a_{21}x_1 + a_{22}x_2 + \cdots + a_{2n}x_n &= b_2 \\
&\ \ \vdots \\
a_{m1}x_1 + a_{m2}x_2 + \cdots + a_{mn}x_n &= b_m
\end{aligned} \qquad (3)$$

In Equation (3), $x_1, x_2, \ldots, x_n$ are referred to as **variables,** or unknowns, and the a_{ij}'s and b_i's are **constants.** A set of equations such as (3) is called a linear system of m equations in n variables.

DEFINITION ■   A **solution** to a linear system of m equations in n unknowns is a set of values for the unknowns that satisfies each of the system's m equations. ■

To understand linear programming, we need to know a great deal about the properties of solutions to linear equation systems. With this in mind, we will devote much effort to studying such systems.

We denote a possible solution to Equation (3) by an n-dimensional column vector **x,** in which the ith element of **x** is the value of x_i. The following example illustrates the concept of a solution to a linear system.

EXAMPLE 5   **Solution to Linear System**

Show that

$$\mathbf{x} = \begin{bmatrix} 1 \\ 2 \end{bmatrix}$$

is a solution to the linear system

$$\begin{aligned} x_1 + 2x_2 &= 5 \\ 2x_1 - x_2 &= 0 \end{aligned}$$
(4)

and that

$$\mathbf{x} = \begin{bmatrix} 3 \\ 1 \end{bmatrix}$$

is not a solution to linear system (4).

Solution   To show that

$$\mathbf{x} = \begin{bmatrix} 1 \\ 2 \end{bmatrix}$$

is a solution to Equation (4), we substitute $x_1 = 1$ and $x_2 = 2$ in both equations and check that they are satisfied: $1 + 2(2) = 5$ and $2(1) - 2 = 0$.

The vector

$$\mathbf{x} = \begin{bmatrix} 3 \\ 1 \end{bmatrix}$$

is not a solution to (4), because $x_1 = 3$ and $x_2 = 1$ fail to satisfy $2x_1 - x_2 = 0$.

Using matrices can greatly simplify the statement and solution of a system of linear equations. To show how matrices can be used to compactly represent Equation (3), let

$$A = \begin{bmatrix} a_{11} & a_{12} & \cdots & a_{1n} \\ a_{21} & a_{22} & \cdots & a_{2n} \\ \vdots & \vdots & & \vdots \\ a_{m1} & a_{m2} & \cdots & a_{mn} \end{bmatrix}, \qquad \mathbf{x} = \begin{bmatrix} x_1 \\ x_2 \\ \vdots \\ x_n \end{bmatrix}, \qquad \mathbf{b} = \begin{bmatrix} b_1 \\ b_2 \\ \vdots \\ b_m \end{bmatrix}$$

Then (3) may be written as

$$A\mathbf{x} = \mathbf{b}$$
(5)

Observe that both sides of Equation (5) will be $m \times 1$ matrices (or $m \times 1$ column vectors). For the matrix $A\mathbf{x}$ to equal the matrix $\mathbf{b}$ (or for the vector $A\mathbf{x}$ to equal the vector $\mathbf{b}$), their corresponding elements must be equal. The first element of $A\mathbf{x}$ is the scalar product of row 1 of A with $\mathbf{x}$. This may be written as

$$\begin{bmatrix} a_{11} & a_{12} & \cdots & a_{1n} \end{bmatrix} \begin{bmatrix} x_1 \\ x_2 \\ \vdots \\ x_n \end{bmatrix} = a_{11}x_1 + a_{12}x_2 + \cdots + a_{1n}x_n$$

This must equal the first element of $\mathbf{b}$ (which is b_1). Thus, (5) implies that $a_{11}x_1 + a_{12}x_2 + \cdots + a_{1n}x_n = b_1$. This is the first equation of (3). Similarly, (5) implies that the scalar product of row i of A with $\mathbf{x}$ must equal b_i, and this is just the ith equation of (3). Our discussion shows that (3) and (5) are two different ways of writing the same linear system. We call (5) the **matrix representation** of (3). For example, the matrix representation of (4) is

$$\begin{bmatrix} 1 & 2 \\ 2 & -1 \end{bmatrix} \begin{bmatrix} x_1 \\ x_2 \end{bmatrix} = \begin{bmatrix} 5 \\ 0 \end{bmatrix}$$

Sometimes we abbreviate (5) by writing

$$A|\mathbf{b} \qquad\qquad (6)$$

If A is an $m \times n$ matrix, it is assumed that the variables in (6) are $x_1, x_2, \ldots, x_n$. Then (6) is still another representation of (3). For instance, the matrix

$$\begin{bmatrix} 1 & 2 & 3 & | & 2 \\ 0 & 1 & 2 & | & 3 \\ 1 & 1 & 1 & | & 1 \end{bmatrix}$$

represents the system of equations

$$x_1 + 2x_2 + 3x_3 = 2$$
$$x_2 + 2x_3 = 3$$
$$x_1 + x_2 + x_3 = 1$$

PROBLEM

Group A

1 Use matrices to represent the following system of equations in two different ways:

$$x_1 - x_2 = 4$$
$$2x_1 + x_2 = 6$$
$$x_1 + 3x_2 = 8$$

2.3 The Gauss–Jordan Method for Solving Systems of Linear Equations

We develop in this section an efficient method (the Gauss–Jordan method) for solving a system of linear equations. Using the Gauss–Jordan method, we show that any system of linear equations must satisfy one of the following three cases:

Case 1 The system has no solution.

Case 2 The system has a unique solution.

Case 3 The system has an infinite number of solutions.

The Gauss–Jordan method is also important because many of the manipulations used in this method are used when solving linear programming problems by the simplex algorithm (see Chapter 4).

Elementary Row Operations

Before studying the Gauss–Jordan method, we need to define the concept of an **elementary row operation** (ERO). An ERO transforms a given matrix A into a new matrix A' via one of the following operations.

Type 1 ERO

A' is obtained by multiplying any row of A by a nonzero scalar. For example, if

$$A = \begin{bmatrix} 1 & 2 & 3 & 4 \\ 1 & 3 & 5 & 6 \\ 0 & 1 & 2 & 3 \end{bmatrix}$$

then a Type 1 ERO that multiplies row 2 of A by 3 would yield

$$A' = \begin{bmatrix} 1 & 2 & 3 & 4 \\ 3 & 9 & 15 & 18 \\ 0 & 1 & 2 & 3 \end{bmatrix}$$

Type 2 ERO

Begin by multiplying any row of A (say, row i) by a nonzero scalar c. For some $j \neq i$, let row j of $A' = c(\text{row } i \text{ of } A) + \text{row } j \text{ of } A$, and let the other rows of A' be the same as the rows of A.

For example, we might multiply row 2 of A by 4 and replace row 3 of A by $4(\text{row } 2$ of $A) + \text{row } 3$ of A. Then row 3 of A' becomes

$$4 \begin{bmatrix} 1 & 3 & 5 & 6 \end{bmatrix} + \begin{bmatrix} 0 & 1 & 2 & 3 \end{bmatrix} = \begin{bmatrix} 4 & 13 & 22 & 27 \end{bmatrix}$$

and

$$A' = \begin{bmatrix} 1 & 2 & 3 & 4 \\ 1 & 3 & 5 & 6 \\ 4 & 13 & 22 & 27 \end{bmatrix}$$

Type 3 ERO

Interchange any two rows of A. For instance, if we interchange rows 1 and 3 of A, we obtain

$$A' = \begin{bmatrix} 0 & 1 & 2 & 3 \\ 1 & 3 & 5 & 6 \\ 1 & 2 & 3 & 4 \end{bmatrix}$$

Type 1 and Type 2 EROs formalize the operations used to solve a linear equation system. To solve the system of equations

$$\begin{aligned} x_1 + x_2 &= 2 \\ 2x_1 + 4x_2 &= 7 \end{aligned} \tag{7}$$

we might proceed as follows. First replace the second equation in (7) by $-2(\text{first equation in (7)}) + \text{second equation in (7)}$. This yields the following linear system:

$$\begin{aligned} x_1 + x_2 &= 2 \\ 2x_2 &= 3 \end{aligned} \tag{7.1}$$

Then multiply the second equation in (7.1) by $\frac{1}{2}$, yielding the system

$$\begin{aligned} x_1 + x_2 &= 2 \\ x_2 &= \tfrac{3}{2} \end{aligned} \tag{7.2}$$

Finally, replace the first equation in (7.2) by $-1[\text{second equation in (7.2)}] + \text{first equation in (7.2)}$. This yields the system

$$x_1 \quad = \frac{1}{2}$$
$$x_2 = \frac{3}{2}$$

(7.3)

System (7.3) has the unique solution $x_1 = \frac{1}{2}$ and $x_2 = \frac{3}{2}$. The systems (7), (7.1), (7.2), and (7.3) are *equivalent* in that they have the same set of solutions. This means that $x_1 = \frac{1}{2}$ and $x_2 = \frac{3}{2}$ is also the unique solution to the original system, (7).

If we view (7) in the augmented matrix form $(A|\mathbf{b})$, we see that the steps used to solve (7) may be seen as Type 1 and Type 2 EROs applied to $A|\mathbf{b}$. Begin with the augmented matrix version of (7):

$$\begin{bmatrix} 1 & 1 & | & 2 \\ 2 & 4 & | & 7 \end{bmatrix}$$

(7′)

Now perform a Type 2 ERO by replacing row 2 of (7′) by -2(row 1 of (7′)) + row 2 of (7′). The result is

$$\begin{bmatrix} 1 & 1 & | & 2 \\ 0 & 2 & | & 3 \end{bmatrix}$$

(7.1′)

which corresponds to (7.1). Next, we multiply row 2 of (7.1′) by $\frac{1}{2}$ (a Type 1 ERO), resulting in

$$\begin{bmatrix} 1 & 1 & | & 2 \\ 0 & 1 & | & \frac{3}{2} \end{bmatrix}$$

(7.2′)

which corresponds to (7.2). Finally, perform a Type 2 ERO by replacing row 1 of (7.2′) by -1(row 2 of (7.2′)) + row 1 of (7.2′). The result is

$$\begin{bmatrix} 1 & 0 & | & \frac{1}{2} \\ 0 & 1 & | & \frac{3}{2} \end{bmatrix}$$

(7.3′)

which corresponds to (7.3). Translating (7.3′) back into a linear system, we obtain the system $x_1 = \frac{1}{2}$ and $x_2 = \frac{3}{2}$, which is identical to (7.3).

Finding a Solution by the Gauss–Jordan Method

The discussion in the previous section indicates that if the matrix $A'|\mathbf{b}'$ is obtained from $A|\mathbf{b}$ via an ERO, the systems $A\mathbf{x} = \mathbf{b}$ and $A'\mathbf{x} = \mathbf{b}'$ are equivalent. Thus, any sequence of EROs performed on the augmented matrix $A|\mathbf{b}$ corresponding to the system $A\mathbf{x} = \mathbf{b}$ will yield an equivalent linear system.

The Gauss–Jordan method solves a linear equation system by utilizing EROs in a systematic fashion. We illustrate the method by finding the solution to the following linear system:

$$2x_1 + 2x_2 + x_3 = 9$$
$$2x_1 - x_2 + 2x_3 = 6$$
$$x_1 - x_2 + 2x_3 = 5$$

(8)

The augmented matrix representation is

$$A|\mathbf{b} = \begin{bmatrix} 2 & 2 & 1 & | & 9 \\ 2 & -1 & 2 & | & 6 \\ 1 & -1 & 2 & | & 5 \end{bmatrix}$$

(8′)

Suppose that by performing a sequence of EROs on (8′) we could transform (8′) into

$$\begin{bmatrix} 1 & 0 & 0 & | & 1 \\ 0 & 1 & 0 & | & 2 \\ 0 & 0 & 1 & | & 3 \end{bmatrix} \qquad (9')$$

We note that the result obtained by performing an ERO on a system of equations can also be obtained by multiplying both sides of the matrix representation of the system of equations by a particular matrix. This explains why EROs do not change the set of solutions to a system of equations.

Matrix $(9')$ corresponds to the following linear system:

$$\begin{aligned} x_1 \quad &= 1 \\ x_2 \quad &= 2 \\ x_3 &= 3 \end{aligned} \qquad (9)$$

System (9) has the unique solution $x_1 = 1$, $x_2 = 2$, $x_3 = 3$. Because $(9')$ was obtained from $(8')$ by a sequence of EROs, we know that (8) and (9) are equivalent linear systems. Thus, $x_1 = 1$, $x_2 = 2$, $x_3 = 3$ must also be the unique solution to (8). We now show how we can use EROs to transform a relatively complicated system such as (8) into a relatively simple system like (9). This is the essence of the Gauss–Jordan method.

We begin by using EROs to transform the first column of $(8')$ into

$$\begin{bmatrix} 1 \\ 0 \\ 0 \end{bmatrix}$$

Then we use EROs to transform the second column of the resulting matrix into

$$\begin{bmatrix} 0 \\ 1 \\ 0 \end{bmatrix}$$

Finally, we use EROs to transform the third column of the resulting matrix into

As a final result, we will have obtained $(9')$. We now use the Gauss–Jordan method to solve (8). We begin by using a Type 1 ERO to change the element of $(8')$ in the first row and first column into a 1. Then we add multiples of row 1 to row 2 and then to row 3 (these are Type 2 EROs). The purpose of these Type 2 EROs is to put zeros in the rest of the first column. The following sequence of EROs will accomplish these goals.

Step 1 Multiply row 1 of $(8')$ by $\frac{1}{2}$. This Type 1 ERO yields

$$A_1|\mathbf{b}_1 = \begin{bmatrix} 1 & 1 & \frac{1}{2} & | & \frac{9}{2} \\ 2 & -1 & 2 & | & 6 \\ 1 & -1 & 2 & | & 5 \end{bmatrix}$$

Step 2 Replace row 2 of $A_1|\mathbf{b}_1$ by $-2(\text{row 1 of } A_1|\mathbf{b}_1) + \text{row 2 of } A_1|\mathbf{b}_1$. The result of this Type 2 ERO is

$$A_2|\mathbf{b}_2 = \begin{bmatrix} 1 & 1 & \frac{1}{2} & | & \frac{9}{2} \\ 0 & -3 & 1 & | & -3 \\ 1 & -1 & 2 & | & 5 \end{bmatrix}$$

Step 3 Replace row 3 of $A_2|\mathbf{b}_2$ by -1(row 1 of $A_2|\mathbf{b}_2$ + row 3 of $A_2|\mathbf{b}_2$. The result of this Type 2 ERO is

$$A_3|\mathbf{b}_3 = \begin{bmatrix} 1 & 1 & \frac{1}{2} & \Big| & \frac{9}{2} \\ 0 & -3 & 1 & \Big| & -3 \\ 0 & -2 & \frac{3}{2} & \Big| & \frac{1}{2} \end{bmatrix}$$

The first column of (8′) has now been transformed into

$$\begin{bmatrix} 1 \\ 0 \\ 0 \end{bmatrix}$$

By our procedure, we have made sure that the variable x_1 occurs in only a single equation and in that equation has a coefficient of 1. We now transform the second column of $A_3|\mathbf{b}_3$ into

$$\begin{bmatrix} 0 \\ 1 \\ 0 \end{bmatrix}$$

We begin by using a Type 1 ERO to create a 1 in row 2 and column 2 of $A_3|\mathbf{b}_3$. Then we use the resulting row 2 to perform the Type 2 EROs that are needed to put zeros in the rest of column 2. Steps 4–6 accomplish these goals.

Step 4 Multiply row 2 of $A_3|\mathbf{b}_3$ by $-\frac{1}{3}$.The result of this Type 1 ERO is

$$A_4|\mathbf{b}_4 = \begin{bmatrix} 1 & 1 & \frac{1}{2} & \Big| & \frac{9}{2} \\ 0 & 1 & -\frac{1}{3} & \Big| & 1 \\ 0 & -2 & \frac{3}{2} & \Big| & \frac{1}{2} \end{bmatrix}$$

Step 5 Replace row 1 of $A_4|\mathbf{b}_4$ by -1(row 2 of $A_4|\mathbf{b}_4$) + row 1 of $A_4|\mathbf{b}_4$. The result of this Type 2 ERO is

$$A_5|\mathbf{b}_5 = \begin{bmatrix} 1 & 0 & \frac{5}{6} & \Big| & \frac{7}{2} \\ 0 & 1 & -\frac{1}{3} & \Big| & 1 \\ 0 & -2 & \frac{3}{2} & \Big| & \frac{1}{2} \end{bmatrix}$$

Step 6 Replace row 3 of $A_5|\mathbf{b}_5$ by 2(row 2 of $A_5|\mathbf{b}_5$) + row 3 of $A_5|\mathbf{b}_5$. The result of this Type 2 ERO is

$$A_6|\mathbf{b}_6 = \begin{bmatrix} 1 & 0 & \frac{5}{6} & \Big| & \frac{7}{2} \\ 0 & 1 & -\frac{1}{3} & \Big| & 1 \\ 0 & 0 & \frac{5}{6} & \Big| & \frac{5}{2} \end{bmatrix}$$

Column 2 has now been transformed into

$$\begin{bmatrix} 0 \\ 1 \\ 0 \end{bmatrix}$$

Observe that our transformation of column 2 did not change column 1.

To complete the Gauss–Jordan procedure, we must transform the third column of $A_6|\mathbf{b}_6$ into

$$\begin{bmatrix} 0 \\ 0 \\ 1 \end{bmatrix}$$

We first use a Type 1 ERO to create a 1 in the third row and third column of $A_6|\mathbf{b}_6$. Then we use Type 2 EROs to put zeros in the rest of column 3. Steps 7–9 accomplish these goals.

Step 7   Multiply row 3 of $A_6|\mathbf{b}_6$ by $\frac{6}{5}$. The result of this Type 1 ERO is

$$A_7|\mathbf{b}_7 = \begin{bmatrix} 1 & 0 & \frac{5}{6} & \Big| & \frac{7}{2} \\ 0 & 1 & -\frac{1}{3} & \Big| & 1 \\ 0 & 0 & 3 & \Big| & 3 \end{bmatrix}$$

Step 8   Replace row 1 of $A_7|\mathbf{b}_7$ by $-\frac{5}{6}$(row 3 of $A_7|\mathbf{b}_7$) + row 1 of $A_7|\mathbf{b}_7$. The result of this Type 2 ERO is

$$A_8|\mathbf{b}_8 = \begin{bmatrix} 1 & 0 & 0 & \Big| & 1 \\ 0 & 1 & -\frac{1}{3} & \Big| & 1 \\ 0 & 0 & 1 & \Big| & 3 \end{bmatrix}$$

Step 9   Replace row 2 of $A_8|\mathbf{b}_8$ by $\frac{1}{3}$(row 3 of $A_8|\mathbf{b}_8$) + row 2 of $A_8|\mathbf{b}_8$. The result of this Type 2 ERO is

$$A_9|\mathbf{b}_9 = \begin{bmatrix} 1 & 0 & 0 & \Big| & 1 \\ 0 & 1 & 0 & \Big| & 2 \\ 0 & 0 & 1 & \Big| & 3 \end{bmatrix}$$

$A_9|\mathbf{b}_9$ represents the system of equations

$$\begin{aligned} x_1 & & & = 1 \\ & x_2 & & = 2 \\ & & x_3 & = 3 \end{aligned} \tag{9}$$

Thus, (9) has the unique solution $x_1 = 1$, $x_2 = 2$, $x_3 = 3$. Because (9) was obtained from (8) via EROs, the unique solution to (8) must also be $x_1 = 1$, $x_2 = 2$, $x_3 = 3$.

The reader might be wondering why we defined Type 3 EROs (interchanging of rows). To see why a Type 3 ERO might be useful, suppose you want to solve

$$\begin{aligned} 2x_2 + x_3 &= 6 \\ x_1 + x_2 - x_3 &= 2 \\ 2x_1 + x_2 + x_3 &= 4 \end{aligned} \tag{10}$$

To solve (10) by the Gauss–Jordan method, first form the augmented matrix

$$A|\mathbf{b} = \begin{bmatrix} 0 & 2 & 1 & \Big| & 6 \\ 1 & 1 & -1 & \Big| & 2 \\ 2 & 1 & 1 & \Big| & 4 \end{bmatrix}$$

The 0 in row 1 and column 1 means that a Type 1 ERO cannot be used to create a 1 in row 1 and column 1. If, however, we interchange rows 1 and 2 (a Type 3 ERO), we obtain

$$\begin{bmatrix} 1 & 1 & -1 & \Big| & 2 \\ 0 & 2 & 1 & \Big| & 6 \\ 2 & 1 & 1 & \Big| & 4 \end{bmatrix} \tag{10'}$$

Now we may proceed as usual with the Gauss–Jordan method.

Special Cases: No Solution or an Infinite Number of Solutions

Some linear systems have no solution, and some have an infinite number of solutions. The following two examples illustrate how the Gauss–Jordan method can be used to recognize these cases.

EXAMPLE 6   **Linear System with No Solution**

Find all solutions to the following linear system:

$$x_1 + 2x_2 = 3$$
$$2x_1 + 4x_2 = 4 \qquad\qquad (11)$$

Solution We apply the Gauss–Jordan method to the matrix

$$A|\mathbf{b} = \begin{bmatrix} 1 & 2 & | & 3 \\ 2 & 4 & | & 4 \end{bmatrix}$$

We begin by replacing row 2 of $A|\mathbf{b}$ by -2(row 1 of $A|\mathbf{b}$) + row 2 of $A|\mathbf{b}$. The result of this Type 2 ERO is

$$\begin{bmatrix} 1 & 2 & | & 3 \\ 0 & 0 & | & -2 \end{bmatrix} \qquad\qquad (12)$$

We would now like to transform the second column of (12) into

$$\begin{bmatrix} 0 \\ 1 \end{bmatrix}$$

but this is not possible. System (12) is equivalent to the following system of equations:

$$x_1 + 2x_2 = 3$$
$$0x_1 + 0x_2 = -2 \qquad\qquad (12')$$

Whatever values we give to x_1 and x_2, the second equation in (12') can never be satisfied. Thus, (12') has no solution. Because (12') was obtained from (11) by use of EROs, (11) also has no solution.

Example 6 illustrates the following idea: *If you apply the Gauss–Jordan method to a linear system and obtain a row of the form* $[\,0 \quad 0 \quad \cdots \quad 0\,|\,c\,]$ $(c \neq 0)$, *then the original linear system has no solution.*

EXAMPLE 7   **Linear System with Infinite Number of Solutions**

Apply the Gauss–Jordan method to the following linear system:

$$x_1 + x_2 \qquad\quad = 1$$
$$x_2 + x_3 = 3 \qquad\qquad (13)$$
$$x_1 + 2x_2 + x_3 = 4$$

Solution The augmented matrix form of (13) is

$$A|\mathbf{b} = \begin{bmatrix} 1 & 1 & 0 & | & 1 \\ 0 & 1 & 1 & | & 3 \\ 1 & 2 & 1 & | & 4 \end{bmatrix}$$

We begin by replacing row 3 (because the row 2, column 1 value is already 0) of $A|\mathbf{b}$ by -1(row 1 of $A|\mathbf{b}$) + row 3 of $A|\mathbf{b}$. The result of this Type 2 ERO is

$$A_1|\mathbf{b}_1 = \begin{bmatrix} 1 & 1 & 0 & | & 1 \\ 0 & 1 & 1 & | & 3 \\ 0 & 1 & 1 & | & 3 \end{bmatrix} \tag{14}$$

Next we replace row 1 of $A_1|\mathbf{b}_1$ by -1(row 2 of $A_1|\mathbf{b}_1$) + row 1 of $A_1|\mathbf{b}_1$. The result of this Type 2 ERO is

$$A_2|\mathbf{b}_2 = \begin{bmatrix} 1 & 0 & -1 & | & -2 \\ 0 & 1 & 1 & | & 3 \\ 0 & 1 & 1 & | & 3 \end{bmatrix}$$

Now we replace row 3 of $A_2|\mathbf{b}_2$ by -1(row 2 of $A_2|\mathbf{b}_2$) + row 3 of $A_2|\mathbf{b}_2$. The result of this Type 2 ERO is

$$A_3|\mathbf{b}_3 = \begin{bmatrix} 1 & 0 & -1 & | & -2 \\ 0 & 1 & 1 & | & 3 \\ 0 & 0 & 0 & | & 0 \end{bmatrix}$$

We would now like to transform the third column of $A_3|\mathbf{b}_3$ into

$$\begin{bmatrix} 0 \\ 0 \\ 1 \end{bmatrix}$$

but this is not possible. The linear system corresponding to $A_3|\mathbf{b}_3$ is

$$x_1 \qquad - \quad x_3 = -2 \tag{14.1}$$
$$x_2 + \quad x_3 = 3 \tag{14.2}$$
$$0x_1 + 0x_2 + 0x_3 = 0 \tag{14.3}$$

Suppose we assign an arbitrary value k to x_3. Then (14.1) will be satisfied if $x_1 - k = -2$, or $x_1 = k - 2$. Similarly, (14.2) will be satisfied if $x_2 + k = 3$, or $x_2 = 3 - k$. Of course, (14.3) will be satisfied for any values of x_1, x_2, and x_3. Thus, for any number k, $x_1 = k - 2$, $x_2 = 3 - k$, $x_3 = k$ is a solution to (14). Thus, (14) has an infinite number of solutions (one for each number k). Because (14) was obtained from (13) via EROs, (13) also has an infinite number of solutions. A more formal characterization of linear systems that have an infinite number of solutions will be given after the following summary of the Gauss–Jordan method.

Summary of the Gauss–Jordan Method

Step 1   To solve $A\mathbf{x} = \mathbf{b}$, write down the augmented matrix $A|\mathbf{b}$.

Step 2   At any stage, define a current row, current column, and current entry (the entry in the current row and column). Begin with row 1 as the current row, column 1 as the current column, and a_{11} as the current entry. (**a**) If a_{11} (the current entry) is nonzero, then use EROs to transform column 1 (the current column) to

$$\begin{bmatrix} 1 \\ 0 \\ \vdots \\ 0 \end{bmatrix}$$

Then obtain the new current row, column, and entry by moving down one row and one column to the right, and go to step 3. (**b**) If a_{11} (the current entry) equals 0, then do a Type 3 ERO involving the current row and any row that contains a nonzero number in the current column. Use EROs to transform column 1 to

Then obtain the new current row, column, and entry by moving down one row and one column to the right. Go to Step 3. (**c**) If there are no nonzero numbers in the first column, then obtain a new current column and entry by moving one column to the right. Then go to step 3.

Step 3   (**a**) If the new current entry is nonzero, then use EROs to transform it to 1 and the rest of the current column's entries to 0. When finished, obtain the new current row, column, and entry. If this is impossible, then stop. Otherwise, repeat step 3. (**b**) If the current entry is 0, then do a Type 3 ERO with the current row and any row that contains a nonzero number in the current column. Then use EROs to transform that current entry to 1 and the rest of the current column's entries to 0. When finished, obtain the new current row, column, and entry. If this is impossible, then stop. Otherwise, repeat step 3. (**c**) If the current column has no nonzero numbers below the current row, then obtain the new current column and entry, and repeat step 3. If it is impossible, then stop.

This procedure may require "passing over" one or more columns without transforming them (see Problem 8).

Step 4   Write down the system of equations $A'\mathbf{x} = \mathbf{b}'$ that corresponds to the matrix $A'|\mathbf{b}'$ obtained when step 3 is completed. Then $A'\mathbf{x} = \mathbf{b}'$ will have the same set of solutions as $A\mathbf{x} = \mathbf{b}$.

Basic Variables and Solutions to Linear Equation Systems

To describe the set of solutions to $A'\mathbf{x} = \mathbf{b}'$ (and $A\mathbf{x} = \mathbf{b}$), we need to define the concepts of basic and nonbasic variables.

DEFINITION ■   After the Gauss–Jordan method has been applied to any linear system, a variable that appears with a coefficient of 1 in a single equation and a coefficient of 0 in all other equations is called a **basic variable** (BV). ■

Any variable that is not a basic variable is called a **nonbasic variable** (NBV). ■

Let BV be the set of basic variables for $A'\mathbf{x} = \mathbf{b}'$ and NBV be the set of nonbasic variables for $A'\mathbf{x} = \mathbf{b}'$. The character of the solutions to $A'\mathbf{x} = \mathbf{b}'$ depends on which of the following cases occurs.

Case 1 $A'\mathbf{x} = \mathbf{b}'$ has at least one row of form $[0 \quad 0 \quad \cdots \quad 0|c]$ $(c \neq 0)$. Then $A\mathbf{x} = \mathbf{b}$ has no solution (recall Example 6). As an example of Case 1, suppose that when the Gauss–Jordan method is applied to the system $A\mathbf{x} = \mathbf{b}$, the following matrix is obtained:

$$A'|\mathbf{b}' = \begin{bmatrix} 1 & 0 & 0 & 1 & | & 1 \\ 0 & 1 & 0 & 2 & | & 1 \\ 0 & 0 & 1 & 3 & | & -1 \\ 0 & 0 & 0 & 0 & | & 0 \\ 0 & 0 & 0 & 0 & | & 2 \end{bmatrix}$$

In this case, $A'\mathbf{x} = \mathbf{b}'$ (and $A\mathbf{x} = \mathbf{b}$) has no solution.

Case 2 Suppose that Case 1 does not apply and NBV, the set of nonbasic variables, is empty. Then $A'\mathbf{x} = \mathbf{b}'$ (and $A\mathbf{x} = \mathbf{b}$) will have a unique solution. To illustrate this, we recall that in solving

$$2x_1 + 2x_2 + x_3 = 9$$
$$2x_1 - x_2 + 2x_3 = 6$$
$$x_1 - x_2 + 2x_3 = 5$$

the Gauss–Jordan method yielded

$$A'|\mathbf{b}' = \begin{bmatrix} 1 & 0 & 0 & | & 1 \\ 0 & 1 & 0 & | & 2 \\ 0 & 0 & 1 & | & 3 \end{bmatrix}$$

In this case, BV $= \{x_1, x_2, x_3\}$ and NBV is empty. Then the unique solution to $A'\mathbf{x} = \mathbf{b}'$ (and $A\mathbf{x} = \mathbf{b}$) is $x_1 = 1$, $x_2 = 2$, $x_3 = 3$.

Case 3 Suppose that Case 1 does not apply and NBV is nonempty. Then $A'\mathbf{x} = \mathbf{b}'$ (and $A\mathbf{x} = \mathbf{b}$) will have an infinite number of solutions. To obtain these, first assign each nonbasic variable an arbitrary value. Then solve for the value of each basic variable in terms of the nonbasic variables. For example, suppose

$$A'|\mathbf{b}' = \begin{bmatrix} 1 & 0 & 0 & 1 & 1 & | & 3 \\ 0 & 1 & 0 & 2 & 0 & | & 2 \\ 0 & 0 & 1 & 0 & 1 & | & 1 \\ 0 & 0 & 0 & 0 & 0 & | & 0 \end{bmatrix} \qquad (15)$$

Because Case 1 does not apply, and BV $= \{x_1, x_2, x_3\}$ and NBV $= \{x_4, x_5\}$, we have an example of Case 3: $A'\mathbf{x} = \mathbf{b}'$ (and $A\mathbf{x} = \mathbf{b}$) will have an infinite number of solutions. To see what these solutions look like, write down $A'\mathbf{x} = \mathbf{b}'$:

$$x_1 \qquad\qquad + x_4 + x_5 = 3 \qquad\qquad (15.1)$$
$$x_2 \qquad 2x_4 \qquad = 2 \qquad\qquad (15.2)$$
$$x_3 \qquad + x_5 = 1 \qquad\qquad (15.3)$$
$$0x_1 + 0x_2 + 0x_3 + 0x_4 + 0x_5 = 0 \qquad\qquad (15.4)$$

Now assign the nonbasic variables (x_4 and x_5) arbitrary values c and k, with $x_4 = c$ and $x_5 = k$. From (15.1), we find that $x_1 = 3 - c - k$. From (15.2), we find that $x_2 = 2 - 2c$. From (15.3), we find that $x_3 = 1 - k$. Because (15.4) holds for all values of the variables, $x_1 = 3 - c - k$, $x_2 = 2 - 2c$, $x_3 = 1 - k$, $x_4 = c$, and $x_5 = k$ will, for any values of c and k, be a solution to $A'\mathbf{x} = \mathbf{b}'$ (and $A\mathbf{x} = \mathbf{b}$).

Our discussion of the Gauss–Jordan method is summarized in Figure 6. We have devoted so much time to the Gauss–Jordan method because, in our study of linear programming, examples of Case 3 (linear systems with an infinite number of solutions) will occur repeatedly. Because the end result of the Gauss–Jordan method must always be one of Cases 1–3, we have shown that any linear system will have no solution, a unique solution, or an infinite number of solutions.

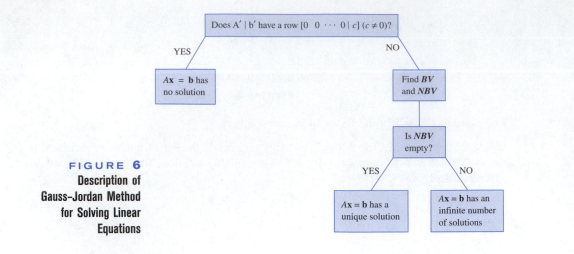

PROBLEMS

Group A

Use the Gauss–Jordan method to determine whether each of the following linear systems has no solution, a unique solution, or an infinite number of solutions. Indicate the solutions (if any exist).

1
$$x_1 + x_2 \qquad + x_4 = 3$$
$$x_2 + x_3 \qquad = 4$$
$$x_1 + 2x_2 + x_3 + x_4 = 8$$

2
$$x_1 + x_2 + x_3 = 4$$
$$x_1 + 2x_2 \qquad = 6$$

3
$$x_1 + x_2 = 1$$
$$2x_1 + x_2 = 3$$
$$3x_1 + 2x_2 = 4$$

4
$$2x_1 - x_2 + x_3 + x_4 = 6$$
$$x_1 + x_2 + x_3 \qquad = 4$$

5
$$x_1 \qquad + \qquad x_4 = 5$$
$$x_2 \qquad + 2x_4 = 5$$
$$x_3 + 0.5x_4 = 1$$
$$2x_3 + \qquad x_4 = 3$$

6
$$2x_2 + 2x_3 = 4$$
$$x_1 + 2x_2 + x_3 = 4$$
$$x_2 - x_3 = 0$$

7
$$x_1 + x_2 \qquad = 2$$
$$-x_2 + 2x_3 = 3$$
$$x_2 + x_3 = 3$$

8
$$x_1 + x_2 + x_3 \qquad = 1$$
$$x_2 + 2x_3 + x_4 = 2$$
$$x_4 = 3$$

Group B

9 Suppose that a linear system $A\mathbf{x} = \mathbf{b}$ has more variables than equations. Show that $A\mathbf{x} = \mathbf{b}$ cannot have a unique solution.

2.4 Linear Independence and Linear Dependence[†]

In this section, we discuss the concepts of a linearly independent set of vectors, a linearly dependent set of vectors, and the rank of a matrix. These concepts will be useful in our study of matrix inverses.

Before defining a linearly independent set of vectors, we need to define a linear combination of a set of vectors. Let $V = \{\mathbf{v}_1, \mathbf{v}_2, \ldots, \mathbf{v}_k\}$ be a set of row vectors all of which have the same dimension.

[†]This section covers topics that may be omitted with no loss of continuity.

DEFINITION ■ A **linear combination** of the vectors in V is any vector of the form $c_1\mathbf{v}_1 + c_2\mathbf{v}_2 \cdots + c_k\mathbf{v}_k$, where $c_1, c_2, \ldots, c_k$ are arbitrary scalars. ■

For example, if $V = \{[1\ \ 2], [2\ \ 1]\}$, then

$$2\mathbf{v}_1 - \mathbf{v}_2 = 2([1\ \ 2]) - [2\ \ 1] = [0\ \ 3]$$
$$\mathbf{v}_1 + 3\mathbf{v}_2 = [1\ \ 2] + 3([2\ \ 1]) = [7\ \ 5]$$
$$0\mathbf{v}_1 + 3\mathbf{v}_2 = [0\ \ 0] + 3([2\ \ 1]) = [6\ \ 3]$$

are linear combinations of vectors in V. The foregoing definition may also be applied to a set of column vectors.

Suppose we are given a set $V = \{\mathbf{v}_1, \mathbf{v}_2, \ldots, \mathbf{v}_k\}$ of m-dimensional row vectors. Let $\mathbf{0} = [0\ \ 0\ \cdots\ 0]$ be the m-dimensional $\mathbf{0}$ vector. To determine whether V is a linearly independent set of vectors, we try to find a linear combination of the vectors in V that adds up to $\mathbf{0}$. Clearly, $0\mathbf{v}_1 + 0\mathbf{v}_2 + \cdots + 0\mathbf{v}_k$ is a linear combination of vectors in V that adds up to $\mathbf{0}$. We call the linear combination of vectors in V for which $c_1 = c_2 = \cdots = c_k = 0$ the *trivial* linear combination of vectors in V. We may now define linearly independent and linearly dependent sets of vectors.

DEFINITION ■ A set V of m-dimensional vectors is **linearly independent** if the only linear combination of vectors in V that equals $\mathbf{0}$ is the trivial linear combination. ■

A set V of m-dimensional vectors is **linearly dependent** if there is a nontrivial linear combination of the vectors in V that adds up to $\mathbf{0}$. ■

The following examples should clarify these definitions.

EXAMPLE 8   0 Vector Makes Set LD

Show that any set of vectors containing the $\mathbf{0}$ vector is a linearly dependent set.

Solution To illustrate, we show that if $V = \{[0\ \ 0], [1\ \ 0], [0\ \ 1]\}$, then V is linearly dependent, because if, say, $c_1 \neq 0$, then $c_1([0\ \ 0]) + 0([1\ \ 0]) + 0([0\ \ 1]) = [0\ \ 0]$. Thus, there is a nontrivial linear combination of vectors in V that adds up to $\mathbf{0}$.

EXAMPLE 9   LI Set of Vectors

Show that the set of vectors $V = \{[1\ \ 0], [0\ \ 1]\}$ is a linearly independent set of vectors.

Solution We try to find a nontrivial linear combination of the vectors in V that yields $\mathbf{0}$. This requires that we find scalars c_1 and c_2 (at least one of which is nonzero) satisfying $c_1([1\ \ 0]) + c_2([0\ \ 1]) = [0\ \ 0]$. Thus, c_1 and c_2 must satisfy $[c_1\ \ c_2] = [0\ \ 0]$. This implies $c_1 = c_2 = 0$. The only linear combination of vectors in V that yields $\mathbf{0}$ is the trivial linear combination. Therefore, V is a linearly independent set of vectors.

EXAMPLE 10   LD Set of Vectors

Show that $V = \{[1\ \ 2], [2\ \ 4]\}$ is a linearly dependent set of vectors.

Solution Because $2([1\ \ 2]) - 1([2\ \ 4]) = [0\ \ 0]$, there is a nontrivial linear combination with $c_1 = 2$ and $c_2 = -1$ that yields $\mathbf{0}$. Thus, V is a linearly dependent set of vectors.

Intuitively, what does it mean for a set of vectors to be linearly dependent? To understand the concept of linear dependence, observe that a set of vectors V is linearly dependent (as

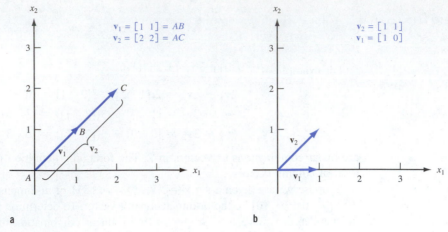

FIGURE 7
(a) Two Linearly Dependent Vectors
(b) Two Linearly Independent Vectors

a   b

long as $\mathbf{0}$ is not in V) if and only if some vector in V can be written as a nontrivial linear combination of other vectors in V (see Problem 9 at the end of this section). For instance, in Example 10, $[2 \quad 4] = 2([1 \quad 2])$. Thus, if a set of vectors V is linearly dependent, the vectors in V are, in some way, not all "different" vectors. By "different" we mean that the direction specified by any vector in V cannot be expressed by adding together multiples of other vectors in V. For example, in two dimensions it can be shown that two vectors are linearly dependent if and only if they lie on the same line (see Figure 7).

The Rank of a Matrix

The Gauss–Jordan method can be used to determine whether a set of vectors is linearly independent or linearly dependent. Before describing how this is done, we define the concept of the rank of a matrix.

Let A be any $m \times n$ matrix, and denote the rows of A by $\mathbf{r}_1, \mathbf{r}_2, \ldots, \mathbf{r}_m$. Also define $R = \{\mathbf{r}_1, \mathbf{r}_2, \ldots, \mathbf{r}_m\}$.

DEFINITION ■ The **rank** of A is the number of vectors in the largest linearly independent subset of R. ■

The following three examples illustrate the concept of rank.

EXAMPLE 11   **Matrix with 0 Rank**

Show that rank $A = 0$ for the following matrix:

$$A = \begin{bmatrix} 0 & 0 \\ 0 & 0 \end{bmatrix}$$

Solution   For the set of vectors $R = \{[0 \quad 0], [0, \quad 0]\}$, it is impossible to choose a subset of R that is linearly independent (recall Example 8).

EXAMPLE 12   **Matrix with Rank of 1**

Show that rank $A = 1$ for the following matrix:

$$A = \begin{bmatrix} 1 & 1 \\ 2 & 2 \end{bmatrix}$$

Solution Here $R = \{[1 \quad 1], [2 \quad 2]\}$. The set $\{[1 \quad 1]\}$ is a linearly independent subset of R, so rank A must be at least 1. If we try to find two linearly independent vectors in R, we fail because $2([1 \quad 1]) - [2 \quad 2] = [0 \quad 0]$. This means that rank A cannot be 2. Thus, rank A must equal 1.

EXAMPLE 13 **Matrix with Rank of 2**

Show that rank $A = 2$ for the following matrix:

$$A = \begin{bmatrix} 1 & 0 \\ 0 & 1 \end{bmatrix}$$

Solution Here $R = \{[1 \quad 0], [0 \quad 1]\}$. From Example 9, we know that R is a linearly independent set of vectors. Thus, rank $A = 2$.

To find the rank of a given matrix A, simply apply the Gauss–Jordan method to the matrix A. Let the final result be the matrix $\bar{A}$. It can be shown that performing a sequence of EROs on a matrix does not change the rank of the matrix. This implies that rank $A = $ rank $\bar{A}$. It is also apparent that the rank of $\bar{A}$ will be the number of nonzero rows in $\bar{A}$. Combining these facts, we find that rank $A = $ rank $\bar{A} = $ number of nonzero rows in $\bar{A}$.

EXAMPLE 14 **Using Gauss–Jordan Method to Find Rank of Matrix**

Find

$$\text{rank } A = \begin{bmatrix} 1 & 0 & 0 \\ 0 & 2 & 1 \\ 0 & 2 & 3 \end{bmatrix}$$

Solution The Gauss–Jordan method yields the following sequence of matrices:

$$A = \begin{bmatrix} 1 & 0 & 0 \\ 0 & 2 & 1 \\ 0 & 2 & 3 \end{bmatrix} \rightarrow \begin{bmatrix} 1 & 0 & 0 \\ 0 & 1 & \frac{1}{2} \\ 0 & 2 & 3 \end{bmatrix} \rightarrow \begin{bmatrix} 1 & 0 & 0 \\ 0 & 1 & \frac{1}{2} \\ 0 & 0 & 2 \end{bmatrix} \rightarrow \begin{bmatrix} 1 & 0 & 0 \\ 0 & 1 & \frac{1}{2} \\ 0 & 0 & 1 \end{bmatrix} \rightarrow \begin{bmatrix} 1 & 0 & 0 \\ 0 & 1 & 0 \\ 0 & 0 & 1 \end{bmatrix}$$

$$= \bar{A}$$

Thus, rank $A = $ rank $\bar{A} = 3$.

How to Tell Whether a Set of Vectors Is Linearly Independent

We now describe a method for determining whether a set of vectors $V = \{\mathbf{v}_1, \mathbf{v}_2, \ldots, \mathbf{v}_m\}$ is linearly independent.

Form the matrix A whose ith row is $\mathbf{v}_i$. A will have m rows. If rank $A = m$, then V is a linearly independent set of vectors, whereas if rank $A < m$, then V is a linearly dependent set of vectors.

EXAMPLE 15 **A Linearly Dependent Set of Vectors**

Determine whether $V = \{[1 \quad 0 \quad 0], [0 \quad 1 \quad 0], [1 \quad 1 \quad 0]\}$ is a linearly independent set of vectors.

Solution The Gauss–Jordan method yields the following sequence of matrices:

$$A = \begin{bmatrix} 1 & 0 & 0 \\ 0 & 1 & 0 \\ 0 & 1 & 0 \end{bmatrix} \rightarrow \begin{bmatrix} 1 & 0 & 0 \\ 0 & 1 & 0 \\ 0 & 1 & 0 \end{bmatrix} \rightarrow \begin{bmatrix} 1 & 0 & 0 \\ 0 & 1 & 0 \\ 0 & 0 & 0 \end{bmatrix} = \bar{A}$$

Thus, rank A = rank $\bar{A}$ = 2 < 3. This shows that V is a linearly dependent set of vectors. In fact, the EROs used to transform A to $\bar{A}$ can be used to show that [1 1 0] = [1 0 0] + [0 1 0]. This equation also shows that V is a linearly dependent set of vectors.

PROBLEMS

Group A

Determine if each of the following sets of vectors is linearly independent or linearly dependent.

1 $V = \{[1 \ \ 0 \ \ 1], [1 \ \ 2 \ \ 1], [2 \ \ 2 \ \ 2]\}$

2 $V = \{[2 \ \ 1 \ \ 0], [1 \ \ 2 \ \ 0], [3 \ \ 3 \ \ 1]\}$

3 $V = \{[2 \ \ 1], [1 \ \ 2]\}$

4 $V = \{[2 \ \ 0], [3 \ \ 0]\}$

5 $V = \left\{ \begin{bmatrix} 1 \\ 2 \\ 3 \end{bmatrix}, \begin{bmatrix} 4 \\ 5 \\ 6 \end{bmatrix}, \begin{bmatrix} 5 \\ 7 \\ 9 \end{bmatrix} \right\}$

6 $V = \left\{ \begin{bmatrix} 1 \\ 0 \\ 0 \end{bmatrix}, \begin{bmatrix} 0 \\ 2 \\ 1 \end{bmatrix}, \begin{bmatrix} 1 \\ 0 \\ 1 \end{bmatrix} \right\}$

Group B

7 Show that the linear system $A\mathbf{x} = \mathbf{b}$ has a solution if and only if $\mathbf{b}$ can be written as a linear combination of the columns of A.

8 Suppose there is a collection of three or more two-dimensional vectors. Provide an argument showing that the collection must be linearly dependent.

9 Show that a set of vectors V (not containing the $\mathbf{0}$ vector) is linearly dependent if and only if there exists some vector in V that can be written as a nontrivial linear combination of other vectors in V.

2.5 The Inverse of a Matrix

To solve a single linear equation such as $4x = 3$, we simply multiply both sides of the equation by the multiplicative inverse of 4, which is 4^{-1}, or $\frac{1}{4}$. This yields $4^{-1}(4x) = (4^{-1})3$, or $x = \frac{3}{4}$. (Of course, this method fails to work for the equation $0x = 3$, because zero has no multiplicative inverse.) In this section, we develop a generalization of this technique that can be used to solve "square" (number of equations = number of unknowns) linear systems. We begin with some preliminary definitions.

DEFINITION ■ A **square matrix** is any matrix that has an equal number of rows and columns. ■

The **diagonal elements** of a square matrix are those elements a_{ij} such that $i = j$. ■

A square matrix for which all diagonal elements are equal to 1 and all nondiagonal elements are equal to 0 is called an **identity matrix.** ■

The $m \times m$ identity matrix will be written as I_m. Thus,

$$I_2 = \begin{bmatrix} 1 & 0 \\ 0 & 1 \end{bmatrix}, \qquad I_3 = \begin{bmatrix} 1 & 0 & 0 \\ 0 & 1 & 0 \\ 0 & 0 & 1 \end{bmatrix}, \qquad \cdots$$

If the multiplications $I_m A$ and AI_m are defined, it is easy to show that $I_m A = AI_m = A$. Thus, just as the number 1 serves as the unit element for multiplication of real numbers, I_m serves as the unit element for multiplication of matrices.

Recall that $\frac{1}{4}$ is the multiplicative inverse of 4. This is because $4(\frac{1}{4}) = (\frac{1}{4})4 = 1$. This motivates the following definition of the inverse of a matrix.

DEFINITION ■ For a given $m \times m$ matrix A, the $m \times m$ matrix B is the **inverse** of A if

$$BA = AB = I_m \tag{16}$$

(It can be shown that if $BA = I_m$ or $AB = I_m$, then the other quantity will also equal I_m.) ■

Some square matrices do not have inverses. If there does exist an $m \times m$ matrix B that satisfies Equation (16), then we write $B = A^{-1}$. For example, if

$$A = \begin{bmatrix} 2 & 0 & -1 \\ 3 & 1 & 2 \\ -1 & 0 & 1 \end{bmatrix}$$

the reader can verify that

$$\begin{bmatrix} 2 & 0 & -1 \\ 3 & 1 & 2 \\ -1 & 0 & 1 \end{bmatrix}\begin{bmatrix} 1 & 0 & 1 \\ -5 & 1 & -7 \\ 1 & 0 & 2 \end{bmatrix} = \begin{bmatrix} 1 & 0 & 0 \\ 0 & 1 & 0 \\ 0 & 0 & 1 \end{bmatrix}$$

and

$$\begin{bmatrix} 1 & 0 & 1 \\ -5 & 1 & -7 \\ 1 & 0 & 2 \end{bmatrix}\begin{bmatrix} 2 & 0 & -1 \\ 3 & 1 & 2 \\ -1 & 0 & 1 \end{bmatrix} = \begin{bmatrix} 1 & 0 & 0 \\ 0 & 1 & 0 \\ 0 & 0 & 1 \end{bmatrix}$$

Thus,

$$A^{-1} = \begin{bmatrix} 1 & 0 & 1 \\ -5 & 1 & -7 \\ 1 & 0 & 2 \end{bmatrix}$$

To see why we are interested in the concept of a matrix inverse, suppose we want to solve a linear system $A\mathbf{x} = \mathbf{b}$ that has m equations and m unknowns. Suppose that A^{-1} exists. Multiplying both sides of $A\mathbf{x} = \mathbf{b}$ by A^{-1}, we see that any solution of $A\mathbf{x} = \mathbf{b}$ must also satisfy $A^{-1}(A\mathbf{x}) = A^{-1}\mathbf{b}$. Using the associative law and the definition of a matrix inverse, we obtain

$$(A^{-1}A)\mathbf{x} = A^{-1}\mathbf{b}$$
or
$$I_m\mathbf{x} = A^{-1}\mathbf{b}$$
or
$$\mathbf{x} = A^{-1}\mathbf{b}$$

This shows that knowing A^{-1} enables us to find the unique solution to a square linear system. This is the analog of solving $4x = 3$ by multiplying both sides of the equation by 4^{-1}.

The Gauss–Jordan method may be used to find A^{-1} (or to show that A^{-1} does not exist). To illustrate how we can use the Gauss–Jordan method to invert a matrix, suppose we want to find A^{-1} for

$$A = \begin{bmatrix} 2 & 5 \\ 1 & 3 \end{bmatrix}$$

This requires that we find a matrix

$$\begin{bmatrix} a & b \\ c & d \end{bmatrix} = A^{-1}$$

that satisfies

$$\begin{bmatrix} 2 & 5 \\ 1 & 3 \end{bmatrix} \begin{bmatrix} a & b \\ c & d \end{bmatrix} = \begin{bmatrix} 1 & 0 \\ 0 & 1 \end{bmatrix} \tag{17}$$

From Equation (17), we obtain the following pair of simultaneous equations that must be satisfied by a, b, c, and d:

$$\begin{bmatrix} 2 & 5 \\ 1 & 3 \end{bmatrix} \begin{bmatrix} a \\ c \end{bmatrix} = \begin{bmatrix} 1 \\ 0 \end{bmatrix}; \qquad \begin{bmatrix} 2 & 5 \\ 1 & 3 \end{bmatrix} \begin{bmatrix} b \\ d \end{bmatrix} = \begin{bmatrix} 0 \\ 1 \end{bmatrix}$$

Thus, to find

$$\begin{bmatrix} a \\ c \end{bmatrix}$$

(the first column of A^{-1}), we can apply the Gauss–Jordan method to the augmented matrix

$$\begin{bmatrix} 2 & 5 & | & 1 \\ 1 & 3 & | & 0 \end{bmatrix}$$

Once EROs have transformed

$$\begin{bmatrix} 2 & 5 \\ 1 & 3 \end{bmatrix}$$

to I_2,

$$\begin{bmatrix} 1 \\ 0 \end{bmatrix}$$

will have been transformed into the first column of A^{-1}. To determine

$$\begin{bmatrix} b \\ d \end{bmatrix}$$

(the second column of A^{-1}), we apply EROs to the augmented matrix

$$\begin{bmatrix} 2 & 5 & | & 0 \\ 1 & 3 & | & 1 \end{bmatrix}$$

When

$$\begin{bmatrix} 2 & 5 \\ 1 & 3 \end{bmatrix}$$

has been transformed into I_2,

$$\begin{bmatrix} 0 \\ 1 \end{bmatrix}$$

will have been transformed into the second column of A^{-1}. Thus, to find each column of A^{-1}, we must perform a sequence of EROs that transform

$$\begin{bmatrix} 2 & 5 \\ 1 & 3 \end{bmatrix}$$

into I_2. This suggests that we can find A^{-1} by applying EROs to the 2×4 matrix

$$A|I_2 = \begin{bmatrix} 2 & 5 & | & 1 & 0 \\ 1 & 3 & | & 0 & 1 \end{bmatrix}$$

When

$$\begin{bmatrix} 2 & 5 \\ 1 & 3 \end{bmatrix}$$

has been transformed to I_2,

$$\begin{bmatrix} 1 \\ 0 \end{bmatrix}$$

will have been transformed into the first column of A^{-1}, and

$$\begin{bmatrix} 0 \\ 1 \end{bmatrix}$$

will have been transformed into the second column of A^{-1}. Thus, *as A is transformed into I_2, I_2 is transformed into A^{-1}.* The computations to determine A^{-1} follow.

Step 1   Multiply row 1 of $A|I_2$ by $\frac{1}{2}$. This yields

$$A'|I_2' = \begin{bmatrix} 1 & \frac{5}{2} & | & \frac{1}{2} & 0 \\ 1 & 3 & | & 0 & 1 \end{bmatrix}$$

Step 2   Replace row 2 of $A'|I_2'$ by -1(row 1 of $A'|I_2'$) + row 2 of $A'|I_2'$. This yields

$$A''|I_2'' = \begin{bmatrix} 1 & \frac{5}{2} & | & \frac{1}{2} & 0 \\ 0 & \frac{1}{2} & | & -\frac{1}{2} & 1 \end{bmatrix}$$

Step 3   Multiply row 2 of $A''|I_2''$ by 2. This yields

$$A'''|I_2''' = \begin{bmatrix} 1 & \frac{5}{2} & | & \frac{1}{2} & 0 \\ 0 & 1 & | & -1 & 2 \end{bmatrix}$$

Step 4   Replace row 1 of $A'''|I_2'''$ by $-\frac{5}{2}$(row 2 of $A'''|I_2'''$) + row 1 of $A'''|I_2'''$. This yields

$$\begin{bmatrix} 1 & 0 & | & 3 & -5 \\ 0 & 1 & | & -1 & 2 \end{bmatrix}$$

Because A has been transformed into I_2, I_2 will have been transformed into A^{-1}. Hence,

$$A^{-1} = \begin{bmatrix} 3 & -5 \\ -1 & 2 \end{bmatrix}$$

The reader should verify that $AA^{-1} = A^{-1}A = I_2$.

A Matrix May Not Have an Inverse

Some matrices do not have inverses. To illustrate, let

$$A = \begin{bmatrix} 1 & 2 \\ 2 & 4 \end{bmatrix} \quad \text{and} \quad A^{-1} = \begin{bmatrix} e & f \\ g & h \end{bmatrix} \tag{18}$$

To find A^{-1} we must solve the following pair of simultaneous equations:

$$\begin{bmatrix} 1 & 2 \\ 2 & 4 \end{bmatrix} \begin{bmatrix} e \\ g \end{bmatrix} = \begin{bmatrix} 1 \\ 0 \end{bmatrix} \tag{18.1}$$

$$\begin{bmatrix} 1 & 2 \\ 2 & 4 \end{bmatrix} \begin{bmatrix} f \\ h \end{bmatrix} = \begin{bmatrix} 0 \\ 1 \end{bmatrix} \tag{18.2}$$

When we try to solve (18.1) by the Gauss–Jordan method, we find that

$$\begin{bmatrix} 1 & 2 & | & 1 \\ 2 & 4 & | & 0 \end{bmatrix}$$

is transformed into

$$\begin{bmatrix} 1 & 2 & | & 1 \\ 0 & 0 & | & -2 \end{bmatrix}$$

This indicates that (18.1) has no solution, and A^{-1} cannot exist.

Observe that (18.1) fails to have a solution, because the Gauss–Jordan method transforms A into a matrix with a row of zeros on the bottom. This can only happen if rank $A < 2$. If $m \times m$ matrix A has rank $A < m$, then A^{-1} will not exist.

The Gauss–Jordan Method for Inverting an $m \times m$ Matrix A

Step 1   Write down the $m \times 2m$ matrix $A|I_m$.

Step 1   Use EROs to transform $A|I_m$ into $I_m|B$. This will be possible only if rank $A = m$. In this case, $B = A^{-1}$. If rank $A < m$, then A has no inverse.

Using Matrix Inverses to Solve Linear Systems

As previously stated, matrix inverses can be used to solve a linear system $A\mathbf{x} = \mathbf{b}$ in which the number of variables and equations are equal. Simply multiply both sides of $A\mathbf{x} = \mathbf{b}$ by A^{-1} to obtain the solution $\mathbf{x} = A^{-1}\mathbf{b}$. For example, to solve

$$\begin{aligned} 2x_1 + 5x_2 &= 7 \\ x_1 + 3x_2 &= 4 \end{aligned} \tag{19}$$

write the matrix representation of (19):

$$\begin{bmatrix} 2 & 5 \\ 1 & 3 \end{bmatrix} \begin{bmatrix} x_1 \\ x_2 \end{bmatrix} = \begin{bmatrix} 7 \\ 4 \end{bmatrix} \tag{20}$$

Let

$$A = \begin{bmatrix} 2 & 5 \\ 1 & 3 \end{bmatrix}$$

We found in the previous illustration that

$$A^{-1} = \begin{bmatrix} 3 & -5 \\ -1 & 2 \end{bmatrix}$$

FIGURE 8

	A	B	C	D	E	F	G	H
1		Inverting						
2		a						
3		Matrix			2	0	-1	
4				A	3	1	2	
5					-1	0	1	
6								
7					1	0	1	
8				A^{-1}	-5	1	-7	
9					1	0	2	

Multiplying both sides of (20) by A^{-1}, we obtain

$$\begin{bmatrix} 3 & -5 \\ -1 & 2 \end{bmatrix}\begin{bmatrix} 2 & 5 \\ 1 & 3 \end{bmatrix}\begin{bmatrix} x_1 \\ x_2 \end{bmatrix} = \begin{bmatrix} 3 & -5 \\ -1 & 2 \end{bmatrix}\begin{bmatrix} 7 \\ 4 \end{bmatrix}$$

$$\begin{bmatrix} x_1 \\ x_2 \end{bmatrix} = \begin{bmatrix} 1 \\ 1 \end{bmatrix}$$

Thus, $x_1 = 1$, $x_2 = 1$ is the unique solution to system (19).

Inverting Matrices with EXCEL

The EXCEL = MINVERSE command makes it easy to invert a matrix. See Figure 8 and file Minverse.xls. Suppose we want to invert the matrix

$$A = \begin{bmatrix} 2 & 0 & -1 \\ 3 & 1 & 2 \\ -1 & 0 & 1 \end{bmatrix}.$$

Simply enter the matrix in E3:G5 and select the range (we chose E7:G9) where you want A^{-1} to be computed. In the upper left-hand corner of the range E7:G9 (cell E7) we enter the formula

$$= \text{MINVERSE(E3:G5)}$$

and select **Control Shift Enter.** This enters an "array function" that computes A^{-1} in the range E7:G9. You cannot edit part of an array function, so if you want to delete A^{-1}, you must delete the entire range where A^{-1} is present.

PROBLEMS

Group A

Find A^{-1} (if it exists) for the following matrices:

1 $\begin{bmatrix} 1 & 3 \\ 2 & 5 \end{bmatrix}$ **2** $\begin{bmatrix} 1 & 0 & 1 \\ 4 & 1 & -2 \\ 3 & 1 & -1 \end{bmatrix}$

3 $\begin{bmatrix} 1 & 0 & 1 \\ 1 & 1 & 1 \\ 2 & 1 & 2 \end{bmatrix}$ **4** $\begin{bmatrix} 1 & 2 & 1 \\ 1 & 2 & 0 \\ 2 & 4 & 1 \end{bmatrix}$

5 Use the answer to Problem 1 to solve the following linear system:

$$x_1 + 3x_2 = 4$$
$$2x_1 + 5x_2 = 7$$

6 Use the answer to Problem 2 to solve the following linear system:

$$x_1 + \quad x_3 = 4$$
$$4x_1 + x_2 - 2x_3 = 0$$
$$3x_1 + x_2 - \quad x_3 = 2$$

Group B

7 Show that a square matrix has an inverse if and only if its rows form a linearly independent set of vectors.

8 Consider a square matrix B whose inverse is given by B^{-1}.

a In terms of B^{-1}, what is the inverse of the matrix $100B$?

b Let B' be the matrix obtained from B by doubling every entry in row 1 of B. Explain how we could obtain the inverse of B' from B^{-1}.

c Let B' be the matrix obtained from B by doubling every entry in column 1 of B. Explain how we could obtain the inverse of B' from B^{-1}.

9 Suppose that A and B both have inverses. Find the inverse of the matrix AB.

10 Suppose A has an inverse. Show that $(A^T)^{-1} = (A^{-1})^T$. (*Hint:* Use the fact that $AA^{-1} = I$, and take the transpose of both sides.)

11 A square matrix A is *orthogonal* if $AA^T = I$. What properties must be possessed by the columns of an orthogonal matrix?

2.6 Determinants

Associated with any square matrix A is a number called the *determinant* of A (often abbreviated as det A or $|A|$). Knowing how to compute the determinant of a square matrix will be useful in our study of nonlinear programming.

For a 1×1 matrix $A = [a_{11}]$,

$$\det A = a_{11} \tag{21}$$

For a 2×2 matrix

$$A = \begin{bmatrix} a_{11} & a_{12} \\ a_{21} & a_{22} \end{bmatrix} \tag{22}$$

$$\det A = a_{11}a_{22} - a_{21}a_{12}$$

For example,

$$\det \begin{bmatrix} 2 & 4 \\ 3 & 5 \end{bmatrix} = 2(5) - 3(4) = -2$$

Before we learn how to compute det A for larger square matrices, we need to define the concept of the *minor* of a matrix.

DEFINITION ■ If A is an $m \times m$ matrix, then for any values of i and j, the *ij*th **minor** of A (written A_{ij}) is the $(m - 1) \times (m - 1)$ submatrix of A obtained by deleting row i and column j of A. ■

For example,

$$\text{if} \quad A = \begin{bmatrix} 1 & 2 & 3 \\ 4 & 5 & 6 \\ 7 & 8 & 9 \end{bmatrix}, \quad \text{then} \quad A_{12} = \begin{bmatrix} 4 & 6 \\ 7 & 9 \end{bmatrix} \quad \text{and} \quad A_{32} = \begin{bmatrix} 1 & 3 \\ 4 & 6 \end{bmatrix}$$

Let A be any $m \times m$ matrix. We may write A as

$$A = \begin{bmatrix} a_{11} & a_{12} & \cdots & a_{1n} \\ a_{21} & a_{22} & \cdots & a_{2n} \\ \vdots & \vdots & & \vdots \\ a_{m1} & a_{m2} & \cdots & a_{mn} \end{bmatrix}$$

To compute det A, pick any value of i ($i = 1, 2, \ldots, m$) and compute det A:

$$\det A = (-1)^{i+1}a_{i1}(\det A_{i1}) + (-1)^{i+2}a_{i2}(\det A_{i2}) + \cdots + (-1)^{i+m}a_{im}(\det A_{im}) \tag{23}$$

$i = $ row

Formula (23) is called the expansion of det A by the cofactors of row i. The virtue of (23) is that it reduces the computation of det A for an $m \times m$ matrix to computations involving only $(m - 1) \times (m - 1)$ matrices. Apply (23) until det A can be expressed in terms of 2×2 matrices. Then use Equation (22) to find the determinants of the relevant 2×2 matrices.

To illustrate the use of (23), we find det A for

$$A = \begin{bmatrix} 1 & 2 & 3 \\ 4 & 5 & 6 \\ 7 & 8 & 9 \end{bmatrix}$$

We expand det A by using row 1 cofactors. Notice that $a_{11} = 1$, $a_{12} = 2$, and $a_{13} = 3$. Also

$$A_{11} = \begin{bmatrix} 5 & 6 \\ 8 & 9 \end{bmatrix}$$

so by (22), det $A_{11} = 5(9) - 8(6) = -3$;

$$A_{12} = \begin{bmatrix} 4 & 6 \\ 7 & 9 \end{bmatrix}$$

so by (22), det $A_{12} = 4(9) - 7(6) = -6$; and

$$A_{13} = \begin{bmatrix} 4 & 5 \\ 7 & 8 \end{bmatrix}$$

so by (22), det $A_{13} = 4(8) - 7(5) = -3$. Then by (23),

$$\det A = (-1)^{1+1}a_{11}(\det A_{11}) + (-1)^{1+2}a_{12}(\det A_{12}) + (-1)^{1+3}a_{13}(\det A_{13})$$
$$= (1)(1)(-3) + (-1)(2)(-6) + (1)(3)(-3) = -3 + 12 - 9 = 0$$

The interested reader may verify that expansion of det A by either row 2 or row 3 cofactors also yields det $A = 0$.

We close our discussion of determinants by noting that they can be used to invert square matrices and to solve linear equation systems. Because we already have learned to use the Gauss–Jordan method to invert matrices and to solve linear equation systems, we will not discuss these uses of determinants.

PROBLEMS

Group A

1 Verify that det $\begin{bmatrix} 1 & 2 & 3 \\ 4 & 5 & 6 \\ 7 & 8 & 9 \end{bmatrix} = 0$ by using expansions by row 2 and row 3 cofactors.

2 Find det $\begin{bmatrix} 1 & 0 & 0 & 0 \\ 0 & 2 & 0 & 0 \\ 0 & 0 & 3 & 0 \\ 0 & 0 & 0 & 5 \end{bmatrix}$

3 A matrix is said to be upper triangular if for $i > j$, $a_{ij} = 0$. Show that the determinant of any upper triangular 3×3 matrix is equal to the product of the matrix's diagonal elements. (This result is true for any upper triangular matrix.)

Group B

4 a Show that for any 1×1 and 3×3 matrix, det $-A = -\det A$.

b Show that for any 2×2 and 4×4 matrix, det $-A = \det A$.

c Generalize the results of parts (a) and (b).

Matrices

A **matrix** is any rectangular array of numbers. For the matrix A, we let a_{ij} represent the element of A in row i and column j.

A matrix with only one row or one column may be thought of as a **vector**. Vectors appear in boldface type ($\mathbf{v}$). Given a row vector $\mathbf{u} = [u_1 \quad u_2 \quad \cdots \quad u_n]$ and a column

$$\mathbf{v} = \begin{bmatrix} v_1 \\ v_2 \\ \vdots \\ v_n \end{bmatrix}$$

of the same dimension, the **scalar product** of $\mathbf{u}$ and $\mathbf{v}$ (written $\mathbf{u} \cdot \mathbf{v}$) is the number $u_1 v_1 + u_2 v_2 + \cdots + u_n v_n$.

Given two matrices A and B, the **matrix product** of A and B (written AB) is defined if and only if the number of columns in A = the number of rows in B. Suppose this is the case and A has m rows and B has n columns. Then the matrix product $C = AB$ of A and B is the $m \times n$ matrix C whose ijth element is determined as follows: The ijth element of C = the scalar product of row i of A with column j of B.

Matrices and Linear Equations

The **linear equation system**

$$
\begin{aligned}
a_{11}x_1 + a_{12}x_2 + \cdots + a_{1n}x_n &= b_1 \\
a_{21}x_1 + a_{22}x_2 + \cdots + a_{2n}x_n &= b_2 \\
\vdots \qquad \vdots \qquad\quad \vdots \ &= \ \vdots \\
a_{m1}x_1 + a_{m2}x_2 + \cdots + a_{mn}x_n &= b_m
\end{aligned}
$$

may be written as $A\mathbf{x} = \mathbf{b}$ or $A|\mathbf{b}$, where

$$
A = \begin{bmatrix} a_{11} & a_{12} & \cdots & a_{1n} \\ a_{21} & a_{22} & \cdots & a_{2n} \\ \vdots & \vdots & & \vdots \\ a_{m1} & a_{m2} & \ldots & a_{mn} \end{bmatrix}, \qquad
\mathbf{x} = \begin{bmatrix} x_1 \\ x_2 \\ \vdots \\ x_n \end{bmatrix}, \qquad
\mathbf{b} = \begin{bmatrix} b_1 \\ b_2 \\ \vdots \\ b_m \end{bmatrix}
$$

The Gauss–Jordan Method

Using **elementary row operations** (EROs), we may solve any linear equation system. From a matrix A, an ERO yields a new matrix A' via one of three procedures.

Type 1 ERO

Obtain A' by multiplying any row of A by a nonzero scalar.

Type 2 ERO

Multiply any row of A (say, row i) by a nonzero scalar c. For some $j \neq i$, let row j of $A' = c(\text{row } i \text{ of } A) + \text{row } j \text{ of } A$, and let the other rows of A' be the same as the rows of A.

Type 3 ERO

Interchange any two rows of A.

The Gauss–Jordan method uses EROs to solve linear equation systems, as shown in the following steps.

Step 1 To solve $A\mathbf{x} = \mathbf{b}$, write down the augmented matrix $A|\mathbf{b}$.

Step 2 Begin with row 1 as the current row, column 1 as the current column, and a_{11} as the current entry. (**a**) If a_{11} (the current entry) is nonzero, then use EROs to transform column 1 (the current column) to

Then obtain the new current row, column, and entry by moving down one row and one column to the right, and go to step 3. (**b**) If a_{11} (the current entry) equals 0, then do a Type 3 ERO switch with any row with a nonzero value in the same column. Use EROs to transform column 1 to

and proceed to step 3 after moving into a new current row, column, and entry. (**c**) If there are no nonzero numbers in the first column, then proceed to a new current column and entry. Then go to step 3.

Step 3 (**a**) If the current entry is nonzero, use EROs to transform it to 1 and the rest of the current column's entries to 0. Obtain the new current row, column, and entry. If this is impossible, then stop. Otherwise, repeat step 3. (**b**) If the current entry is 0, then do a Type 3 ERO switch with any row with a nonzero value in the same column. Transform the column using EROs and move to the next current entry. If this is impossible, then stop. Otherwise, repeat step 3. (**c**) If the current column has no nonzero numbers below the current row, then obtain the new current column and entry, and repeat step 3. If it is impossible, then stop.

This procedure may require "passing over" one or more columns without transforming them.

Step 4 Write down the system of equations $A'\mathbf{x} = \mathbf{b}'$ that corresponds to the matrix $A'|\mathbf{b}'$ obtained when step 3 is completed. Then $A'\mathbf{x} = \mathbf{b}'$ will have the same set of solutions as $A\mathbf{x} = \mathbf{b}$.

To describe the set of solutions to $A'\mathbf{x} = \mathbf{b}'$ (and $A\mathbf{x} = \mathbf{b}$), we define the concepts of basic and nonbasic variables. After the Gauss–Jordan method has been applied to any linear system, a variable that appears with a coefficient of 1 in a single equation and a coefficient of 0 in all other equations is called a **basic variable.** Any variable that is not a basic variable is called a **nonbasic variable.**

Let BV be the set of basic variables for $A'\mathbf{x} = \mathbf{b}'$ and NBV be the set of nonbasic variables for $A'\mathbf{x} = \mathbf{b}'$.

Case 1 $A'\mathbf{x} = \mathbf{b}'$ contains at least one row of the form $[0 \quad 0 \quad \cdots \quad 0|c](c \neq 0)$. In this case, $A\mathbf{x} = \mathbf{b}$ has no solution.

Case 2 If Case 1 does not apply and NBV, the set of nonbasic variables, is empty, then $A\mathbf{x} = \mathbf{b}$ will have a unique solution.

Case 3 If Case 1 does not hold and NBV is nonempty, then $A\mathbf{x} = \mathbf{b}$ will have an infinite number of solutions.

Linear Independence, Linear Dependence, and the Rank of a Matrix

A set V of m-dimensional vectors is **linearly independent** if the only linear combination of vectors in V that equals $\mathbf{0}$ is the trivial linear combination. A set V of m-dimensional vectors is **linearly dependent** if there is a nontrivial linear combination of the vectors in V that adds to $\mathbf{0}$.

Let A be any $m \times n$ matrix, and denote the rows of A by $\mathbf{r}_1, \mathbf{r}_2, \ldots, \mathbf{r}_m$. Also define $R = \{\mathbf{r}_1, \mathbf{r}_2, \ldots, \mathbf{r}_m\}$. The **rank** of A is the number of vectors in the largest linearly independent subset of R. To find the rank of a given matrix A, apply the Gauss–Jordan method to the matrix A. Let the final result be the matrix $\bar{A}$. Then rank A = rank $\bar{A}$ = number of nonzero rows in $\bar{A}$.

To determine if a set of vectors $V = \{\mathbf{v}_1, \mathbf{v}_2, \ldots, \mathbf{v}_m\}$ is linearly dependent, form the matrix A whose ith row is $\mathbf{v}_i$. A will have m rows. If rank $A = m$, then V is a linearly independent set of vectors; if rank $A < m$, then V is a linearly dependent set of vectors.

Inverse of a Matrix

For a given square $(m \times m)$ matrix A, if $AB = BA = I_m$, then B is the **inverse** of A (written $B = A^{-1}$). The Gauss–Jordan method for inverting an $m \times m$ matrix A to get A^{-1} is as follows:

Step 1 Write down the $m \times 2m$ matrix $A|I_m$.

Step 2 Use EROs to transform $A|I_m$ into $I_m|B$. This will only be possible if rank $A = m$. In this case, $B = A^{-1}$. If rank $A < m$, then A has no inverse.

Determinants

Associated with any square $(m \times m)$ matrix A is a number called the **determinant** of A (written det A or $|A|$). For a 1×1 matrix, det $A = a_{11}$. For a 2×2 matrix, det $A = a_{11}a_{22} - a_{21}a_{12}$. For a general $m \times m$ matrix, we can find det A by repeated application of the following formula (valid for $i = 1, 2, \ldots, m$):

$$\det A = (-1)^{i+1}a_{i1}(\det A_{i1}) + (-1)^{i+2}a_{i2}(\det A_{i2}) + \cdots + (-1)^{i+m}a_{im}(\det A_{im})$$

Here A_{ij} is the ijth **minor** of A, which is the $(m - 1) \times (m - 1)$ matrix obtained from A after deleting the ith row and jth column of A.

REVIEW PROBLEMS

Group A

1 Find all solutions to the following linear system:

$$
\begin{aligned}
x_1 + x_2 \quad\quad &= 2 \\
x_2 + x_3 &= 3 \\
x_1 + 2x_2 + x_3 &= 5
\end{aligned}
$$

2 Find the inverse of the matrix $\begin{bmatrix} 0 & 3 \\ 2 & 1 \end{bmatrix}$.

3 Each year, 20% of all untenured State University faculty become tenured, 5% quit, and 75% remain untenured. Each year, 90% of all tenured S.U. faculty remain tenured and 10% quit. Let U_t be the number of untenured S.U. faculty at the beginning of year t, and T_t the tenured number.

Use matrix multiplication to relate the vector $\begin{bmatrix} U_{t+1} \\ T_{t+1} \end{bmatrix}$ to the vector $\begin{bmatrix} U_t \\ T_t \end{bmatrix}$.

4 Use the Gauss–Jordan method to determine all solutions to the following linear system:

$$
\begin{aligned}
2x_1 + 3x_2 &= 3 \\
x_1 + x_2 &= 1 \\
x_1 + 2x_2 &= 2
\end{aligned}
$$

5 Find the inverse of the matrix $\begin{bmatrix} 0 & 2 \\ 1 & 3 \end{bmatrix}$.

6 The grades of two students during their last semester at S.U. are shown in Table 2.

Courses 1 and 2 are four-credit courses, and courses 3 and 4 are three-credit courses. Let GPA_i be the semester grade point average for student i. Use matrix multiplication to express the vector $\begin{bmatrix} GPA_1 \\ GPA_2 \end{bmatrix}$ in terms of the information given in the problem.

7 Use the Gauss–Jordan method to find all solutions to the following linear system:

$$
\begin{aligned}
2x_1 + x_2 &= 3 \\
3x_1 + x_2 &= 4 \\
x_1 - x_2 &= 0
\end{aligned}
$$

8 Find the inverse of the matrix $\begin{bmatrix} 2 & 3 \\ 3 & 5 \end{bmatrix}$.

9 Let C_t = number of children in Indiana at the beginning of year t, and A_t = number of adults in Indiana at the beginning of year t. During any given year, 5% of all children

become adults, and 1% of all children die. Also, during any given year, 3% of all adults die. Use matrix multiplication to express the vector $\begin{bmatrix} C_{t+1} \\ A_{t+1} \end{bmatrix}$ in terms of $\begin{bmatrix} C_t \\ A_t \end{bmatrix}$.

10 Use the Gauss–Jordan method to find all solutions to the following linear equation system:

$$
\begin{aligned}
x_1 \quad\quad - x_3 &= 4 \\
x_2 + x_3 &= 2 \\
x_1 + x_2 \quad\quad &= 5
\end{aligned}
$$

11 Use the Gauss–Jordan method to find the inverse of the matrix $\begin{bmatrix} 1 & 0 & 2 \\ 0 & 1 & 0 \\ 0 & 1 & 1 \end{bmatrix}$.

12 During any given year, 10% of all rural residents move to the city, and 20% of all city residents move to a rural area (all other people stay put!). Let R_t be the number of rural residents at the beginning of year t, and C_t be the number of city residents at the beginning of year t. Use matrix multiplication to relate the vector $\begin{bmatrix} R_{t+1} \\ C_{t+1} \end{bmatrix}$ to the vector $\begin{bmatrix} R_t \\ C_t \end{bmatrix}$.

13 Determine whether the set $V = \{[1 \quad 2 \quad 1], [2 \quad 0 \quad 0]\}$ is a linearly independent set of vectors.

14 Determine whether the set $V = \{[1 \quad 0 \quad 0], [0 \quad 1 \quad 0], [-1 \quad -1 \quad 0]\}$ is a linearly independent set of vectors.

15 Let $A = \begin{bmatrix} a & 0 & 0 & 0 \\ 0 & b & 0 & 0 \\ 0 & 0 & c & 0 \\ 0 & 0 & 0 & d \end{bmatrix}$.

a For what values of a, b, c, and d will A^{-1} exist?

b If A^{-1} exists, then find it.

16 Show that the following linear system has an infinite number of solutions:

$$
\begin{bmatrix} 1 & 1 & 0 & 0 \\ 0 & 0 & 1 & 1 \\ 1 & 0 & 1 & 0 \\ 0 & 1 & 0 & 1 \end{bmatrix}
\begin{bmatrix} x_1 \\ x_2 \\ x_3 \\ x_4 \end{bmatrix} =
\begin{bmatrix} 2 \\ 3 \\ 4 \\ 1 \end{bmatrix}
$$

17 Before paying employee bonuses and state and federal taxes, a company earns profits of $60,000. The company pays employees a bonus equal to 5% of after-tax profits. State tax is 5% of profits (after bonuses are paid). Finally, federal tax is 40% of profits (after bonuses and state tax are paid). Determine a linear equation system to find the amounts paid in bonuses, state tax, and federal tax.

18 Find the determinant of the matrix $A = \begin{bmatrix} 2 & 4 & 6 \\ 1 & 0 & 0 \\ 0 & 0 & 1 \end{bmatrix}$.

19 Show that any 2×2 matrix A that does not have an inverse will have det $A = 0$.

TABLE 2

Student	Course			
	1	**2**	**3**	**4**
1	3.6	3.8	2.6	3.4
2	2.7	3.1	2.9	3.6

Group B

20 Let A be an $m \times m$ matrix.

a Show that if rank $A = m$, then $A\mathbf{x} = \mathbf{0}$ has a unique solution. What is the unique solution?

b Show that if rank $A < m$, then $A\mathbf{x} = \mathbf{0}$ has an infinite number of solutions.

21 Consider the following linear system:

$$[x_1 \quad x_2 \quad \cdots \quad x_n] = [x_1 \quad x_2 \quad \cdots \quad x_n]P$$

where

$$P = \begin{bmatrix} p_{11} & p_{12} & \cdots & p_{1n} \\ p_{21} & p_{22} & \cdots & p_{2n} \\ \vdots & \vdots & & \vdots \\ p_{n1} & p_{n2} & \cdots & p_{nn} \end{bmatrix}$$

If the sum of each row of the P matrix equals 1, then use Problem 20 to show that this linear system has an infinite number of solutions.

22[†] The national economy of Seriland manufactures three products: steel, cars, and machines. (1) To produce \$1 of steel requires 30¢ of steel, 15¢ of cars, and 40¢ of machines. (2) To produce \$1 of cars requires 45¢ of steel, 20¢ of cars, and 10¢ of machines. (3) To produce \$1 of machines requires 40¢ of steel, 10¢ of cars, and 45¢ of machines. During the coming year, Seriland wants to consume d_s dollars of steel, d_c dollars of cars, and d_m dollars of machinery.

For the coming year, let

s = dollar value of steel produced

c = dollar value of cars produced

m = dollar value of machines produced

Define A to be the 3×3 matrix whose ijth element is the dollar value of product i required to produce \$1 of product j (steel = product 1, cars = product 2, machinery = product 3).

a Determine A.

b Show that

$$\begin{bmatrix} s \\ c \\ m \end{bmatrix} = A \begin{bmatrix} s \\ c \\ m \end{bmatrix} + \begin{bmatrix} d_s \\ d_c \\ d_m \end{bmatrix} \tag{24}$$

(*Hint:* Observe that the value of next year's steel production = (next year's consumer steel demand) + (steel needed to make next year's steel) + (steel needed to make next year's cars) + (steel needed to make next year's machines). This should give you the general idea.)

c Show that Equation (24) may be rewritten as

$$(I - A) \begin{bmatrix} s \\ c \\ m \end{bmatrix} = \begin{bmatrix} d_s \\ d_c \\ d_m \end{bmatrix}$$

d Given values for d_s, d_c, and d_m, describe how you can use $(I - A)^{-1}$ to determine if Seriland can meet next year's consumer demand.

e Suppose next year's demand for steel increases by \$1. This will increase the value of the steel, cars, and machines that must be produced next year. In terms of $(I - A)^{-1}$, determine the change in next year's production requirements.

REFERENCES

The following references contain more advanced discussions of linear algebra. To understand the theory of linear and nonlinear programming, master at least one of these books:

Dantzig, G. *Linear Programming and Extensions.* Princeton, N.J.: Princeton University Press, 1963.

Hadley, G. *Linear Algebra.* Reading, Mass.: Addison-Wesley, 1961.

Strang, G. *Linear Algebra and Its Applications,* 3d ed. Orlando, Fla.: Academic Press, 1988.

Leontief, W. *Input–Output Economics.* New York: Oxford University Press, 1966.

Teichroew, D. *An Introduction to Management Science: Deterministic Models.* New York: Wiley, 1964. A more extensive discussion of linear algebra than this chapter gives (at a comparable level of difficulty).

[†]Based on Leontief (1966). See references at end of chapter.

3

Introduction to Linear Programming

Linear programming (LP) is a tool for solving optimization problems. In 1947, George Dantzig developed an efficient method, the simplex algorithm, for solving linear programming problems (also called LP). Since the development of the simplex algorithm, LP has been used to solve optimization problems in industries as diverse as banking, education, forestry, petroleum, and trucking. In a survey of Fortune 500 firms, 85% of the respondents said they had used linear programming. As a measure of the importance of linear programming in operations research, approximately 70% of this book will be devoted to linear programming and related optimization techniques.

In Section 3.1, we begin our study of linear programming by describing the general characteristics shared by all linear programming problems. In Sections 3.2 and 3.3, we learn how to solve graphically those linear programming problems that involve only two variables. Solving these simple LPs will give us useful insights for solving more complex LPs. The remainder of the chapter explains how to formulate linear programming models of real-life situations.

3.1 What Is a Linear Programming Problem?

In this section, we introduce linear programming and define important terms that are used to describe linear programming problems.

EXAMPLE 1   Giapetto's Woodcarving

Giapetto's Woodcarving, Inc., manufactures two types of wooden toys: soldiers and trains. A soldier sells for $27 and uses $10 worth of raw materials. Each soldier that is manufactured increases Giapetto's variable labor and overhead costs by $14. A train sells for $21 and uses $9 worth of raw materials. Each train built increases Giapetto's variable labor and overhead costs by $10. The manufacture of wooden soldiers and trains requires two types of skilled labor: carpentry and finishing. A soldier requires 2 hours of finishing labor and 1 hour of carpentry labor. A train requires 1 hour of finishing and 1 hour of carpentry labor. Each week, Giapetto can obtain all the needed raw material but only 100 finishing hours and 80 carpentry hours. Demand for trains is unlimited, but at most 40 soldiers are bought each week. Giapetto wants to maximize weekly profit (revenues − costs). Formulate a mathematical model of Giapetto's situation that can be used to maximize Giapetto's weekly profit.

Solution   In developing the Giapetto model, we explore characteristics shared by all linear programming problems.

Decision Variables   We begin by defining the relevant **decision variables.** In any linear programming model, the decision variables should completely describe the decisions to be made (in this case, by Giapetto). Clearly, Giapetto must decide how many soldiers and trains should be manufactured each week. With this in mind, we define

$$x_1 = \text{number of soldiers produced each week}$$
$$x_2 = \text{number of trains produced each week}$$

Objective Function In any linear programming problem, the decision maker wants to maximize (usually revenue or profit) or minimize (usually costs) some function of the decision variables. The function to be maximized or minimized is called the **objective function.** For the Giapetto problem, we note that *fixed costs* (such as rent and insurance) do not depend on the values of x_1 and x_2. Thus, Giapetto can concentrate on maximizing (weekly revenues) − (raw material purchase costs) − (other variable costs).

Giapetto's weekly revenues and costs can be expressed in terms of the decision variables x_1 and x_2. It would be foolish for Giapetto to manufacture more soldiers than can be sold, so we assume that all toys produced will be sold. Then

$$\text{Weekly revenues} = \text{weekly revenues from soldiers}$$
$$+ \text{ weekly revenues from trains}$$
$$= \left(\frac{\text{dollars}}{\text{soldier}}\right)\left(\frac{\text{soldiers}}{\text{week}}\right) + \left(\frac{\text{dollars}}{\text{train}}\right)\left(\frac{\text{trains}}{\text{week}}\right)$$
$$= 27x_1 + 21x_2$$

Also,

$$\text{Weekly raw material costs} = 10x_1 + 9x_2$$
$$\text{Other weekly variable costs} = 14x_1 + 10x_2$$

Then Giapetto wants to maximize

$$(27x_1 + 21x_2) - (10x_1 + 9x_2) - (14x_1 + 10x_2) = 3x_1 + 2x_2$$

Another way to see that Giapetto wants to maximize $3x_1 + 2x_2$ is to note that

$$\text{Weekly revenues} = \text{weekly contribution to profit from soldiers}$$
$$- \text{ weekly nonfixed costs} + \text{weekly contribution to profit from trains}$$
$$= \left(\frac{\text{contribution to profit}}{\text{soldier}}\right)\left(\frac{\text{soldiers}}{\text{week}}\right)$$
$$+ \left(\frac{\text{contribution to profit}}{\text{train}}\right)\left(\frac{\text{trains}}{\text{week}}\right)$$

Also,

$$\frac{\text{Contribution to profit}}{\text{Soldier}} = 27 - 10 - 14 = 3$$
$$\frac{\text{Contribution to profit}}{\text{Train}} = 21 - 9 - 10 = 2$$

Then, as before, we obtain

$$\text{Weekly revenues} - \text{weekly nonfixed costs} = 3x_1 + 2x_2$$

Thus, Giapetto's objective is to choose x_1 and x_2 to maximize $3x_1 + 2x_2$. We use the variable z to denote the objective function value of any LP. Giapetto's objective function is

$$\text{Maximize } z = 3x_1 + 2x_2 \tag{1}$$

(In the future, we will abbreviate "maximize" by *max* and "minimize" by *min*.) The coefficient of a variable in the objective function is called the **objective function coefficient** of the variable. For example, the objective function coefficient for x_1 is 3, and the objective function coefficient for x_2 is 2. In this example (and in many other problems), the objective function coefficient for each variable is simply the contribution of the variable to the company's profit.

Constraints   As x_1 and x_2 increase, Giapetto's objective function grows larger. This means that if Giapetto were free to choose any values for x_1 and x_2, the company could make an arbitrarily large profit by choosing x_1 and x_2 to be very large. Unfortunately, the values of x_1 and x_2 are limited by the following three restrictions (often called **constraints**):

Constraint 1   Each week, no more than 100 hours of finishing time may be used.

Constraint 2   Each week, no more than 80 hours of carpentry time may be used.

Constraint 3   Because of limited demand, at most 40 soldiers should be produced each week.

The amount of raw material available is assumed to be unlimited, so no restrictions have been placed on this.

The next step in formulating a mathematical model of the Giapetto problem is to express Constraints 1–3 in terms of the decision variables x_1 and x_2. To express Constraint 1 in terms of x_1 and x_2, note that

$$\frac{\text{Total finishing hrs.}}{\text{Week}} = \left(\frac{\text{finishing hrs.}}{\text{soldier}}\right)\left(\frac{\text{soldiers made}}{\text{week}}\right)$$

$$+ \left(\frac{\text{finishing hrs.}}{\text{train}}\right)\left(\frac{\text{trains made}}{\text{week}}\right)$$

$$= 2(x_1) + 1(x_2) = 2x_1 + x_2.$$ To express Constraint

Now Constraint 1 may be expressed by

$$2x_1 + x_2 \leq 100 \tag{2}$$

Note that the units of each term in (2) are finishing hours per week. *For a constraint to be reasonable, all terms in the constraint must have the same units.* Otherwise one is adding apples and oranges, and the constraint won't have any meaning.

To express Constraint 2 in terms of x_1 and x_2, note that

$$\frac{\text{Total carpentry hrs.}}{\text{Week}} = \left(\frac{\text{carpentry hrs.}}{\text{solider}}\right)\left(\frac{\text{soldiers}}{\text{week}}\right)$$

$$+ \left(\frac{\text{carpentry hrs.}}{\text{train}}\right)\left(\frac{\text{trains}}{\text{week}}\right)$$

$$= 1(x_1) + 1(x_2) = x_1 + x_2$$

Then Constraint 2 may be written as

$$x_1 + x_2 \leq 80 \tag{3}$$

Again, note that the units of each term in (3) are the same (in this case, carpentry hours per week).

Finally, we express the fact that at most 40 soldiers per week can be sold by limiting the weekly production of soldiers to at most 40 soldiers. This yields the following constraint:

$$x_1 \leq 40 \tag{4}$$

Thus (2)–(4) express Constraints 1–3 in terms of the decision variables; they are called the *constraints* for the Giapetto linear programming problem. The coefficients of the decision variables in the constraints are called **technological coefficients.** This is because the technological coefficients often reflect the technology used to produce different products. For example, the technological coefficient of x_2 in (3) is 1, indicating that a soldier requires 1 carpentry hour. The number on the right-hand side of each constraint is called the constraint's **right-hand side** (or **rhs**). Often the rhs of a constraint represents the quantity of a resource that is available.

Sign Restrictions   To complete the formulation of a linear programming problem, the following question must be answered for each decision variable: Can the decision variable only assume nonnegative values, or is the decision variable allowed to assume both positive and negative values?

If a decision variable x_i can only assume nonnegative values, then we add the **sign restriction** $x_i \geq 0$. If a variable x_i can assume both positive and negative (or zero) values, then we say that x_i is **unrestricted in sign** (often abbreviated **urs**). For the Giapetto problem, it is clear that $x_1 \geq 0$ and $x_2 \geq 0$. In other problems, however, some variables may be urs. For example, if x_i represented a firm's cash balance, then x_i could be considered negative if the firm owed more money than it had on hand. In this case, it would be appropriate to classify x_i as urs. Other uses of urs variables are discussed in Section 4.12.

Combining the sign restrictions $x_1 \geq 0$ and $x_2 \geq 0$ with the objective function (1) and Constraints (2)–(4) yields the following optimization model:

$$\max z = 3x_1 + 2x_2 \qquad \text{(Objective function)} \tag{1}$$

subject to (s.t.)

$$2x_1 + x_2 \leq 100 \qquad \text{(Finishing constraint)} \tag{2}$$

$$x_1 + x_2 \leq 80 \qquad \text{(Carpentry constraint)} \tag{3}$$

$$x_1 \qquad \leq 40 \qquad \text{(Constraint on demand for soldiers)} \tag{4}$$

$$x_1 \qquad \geq 0 \qquad \text{(Sign restriction)}^\dagger \tag{5}$$

$$x_2 \geq 0 \qquad \text{(Sign restriction)} \tag{6}$$

"Subject to" (s.t.) means that the values of the decision variables x_1 and x_2 must satisfy all constraints and all sign restrictions.

Before formally defining a linear programming problem, we define the concepts of linear function and linear inequality.

DEFINITION ■   A function $f(x_1, x_2, \ldots, x_n)$ of $x_1, x_2, \ldots, x_n$ is a **linear function** if and only if for some set of constants $c_1, c_2, \ldots, c_n$, $f(x_1, x_2, \ldots, x_n) = c_1 x_1 + c_2 x_2 + \cdots + c_n x_n$. ■

For example, $f(x_1, x_2) = 2x_1 + x_2$ is a linear function of x_1 and x_2, but $f(x_1, x_2) = x_1^2 x_2$ is not a linear function of x_1 and x_2.

DEFINITION ■   For any linear function $f(x_1, x_2, \ldots, x_n)$ and any number b, the inequalities $f(x_1, x_2, \ldots, x_n) \leq b$ and $f(x_1, x_2, \ldots, x_n) \geq b$ are **linear inequalities.** ■

Thus, $2x_1 + 3x_2 \leq 3$ and $2x_1 + x_2 \geq 3$ are linear inequalities, but $x_1^2 x_2 \geq 3$ is not a linear inequality.

---

†The sign restrictions do constrain the values of the decision variables, but we choose to consider the sign restrictions as being separate from the constraints. The reason for this will become apparent when we study the simplex algorithm in Chapter 4.

DEFINITION ■ A **linear programming problem** (LP) is an optimization problem for which we do the following:

1 We attempt to maximize (or minimize) a *linear* function of the decision variables. The function that is to be maximized or minimized is called the *objective function*.

2 The values of the decision variables must satisfy a set of *constraints*. Each constraint must be a linear equation or linear inequality.

3 A *sign restriction* is associated with each variable. For any variable x_i, the sign restriction specifies that x_i must be either nonnegative ($x_i \geq 0$) or unrestricted in sign (urs). ■

Because Giapetto's objective function is a linear function of x_1 and x_2, and all of Giapetto's constraints are linear inequalities, the Giapetto problem is a linear programming problem. Note that the Giapetto problem is typical of a wide class of linear programming problems in which a decision maker's goal is to maximize profit subject to limited resources.

The Proportionality and Additivity Assumptions

The fact that the objective function for an LP must be a linear function of the decision variables has two implications.

1 The contribution of the objective function from each decision variable is proportional to the value of the decision variable. For example, the contribution to the objective function from making four soldiers ($4 \times 3 = \$12$) is exactly four times the contribution to the objective function from making one soldier ($\$3$).

2 The contribution to the objective function for any variable is independent of the values of the other decision variables. For example, no matter what the value of x_2, the manufacture of x_1 soldiers will always contribute $3x_1$ dollars to the objective function.

Analogously, the fact that each LP constraint must be a linear inequality or linear equation has two implications.

1 The contribution of each variable to the left-hand side of each constraint is proportional to the value of the variable. For example, it takes exactly three times as many finishing hours ($2 \times 3 = 6$ finishing hours) to manufacture three soldiers as it takes to manufacture one soldier (2 finishing hours).

2 The contribution of a variable to the left-hand side of each constraint is independent of the values of the variable. For example, no matter what the value of x_1, the manufacture of x_2 trains uses x_2 finishing hours and x_2 carpentry hours.

The first implication given in each list is called the **Proportionality Assumption of Linear Programming.** Implication 2 of the first list implies that the value of the objective function is the sum of the contributions from individual variables, and implication 2 of the second list implies that the left-hand side of each constraint is the sum of the contributions from each variable. For this reason, the second implication in each list is called the **Additivity Assumption of Linear Programming.**

For an LP to be an appropriate representation of a real-life situation, the decision variables must satisfy both the Proportionality and Additivity Assumptions. Two other assumptions must also be satisfied before an LP can appropriately represent a real situation: the Divisibility and Certainty Assumptions.

The Divisibility Assumption

The **Divisibility Assumption** requires that each decision variable be allowed to assume fractional values. For example, in the Giapetto problem, the Divisibility Assumption implies that it is acceptable to produce 1.5 soldiers or 1.63 trains. Because Giapetto cannot actually produce a fractional number of trains or soldiers, the Divisibility Assumption is not satisfied in the Giapetto problem. A linear programming problem in which some or all of the variables must be nonnegative integers is called an **integer programming problem.** The solution of integer programming problems is discussed in Chapter 9.

In many situations where divisibility is not present, rounding off each variable in the optimal LP solution to an integer may yield a reasonable solution. Suppose the optimal solution to an LP stated that an auto company should manufacture 150,000.4 compact cars during the current year. In this case, you could tell the auto company to manufacture 150,000 or 150,001 compact cars and be fairly confident that this would reasonably approximate an optimal production plan. On the other hand, if the number of missile sites that the United States should use were a variable in an LP and the optimal LP solution said that 0.4 missile sites should be built, it would make a big difference whether we rounded the number of missile sites down to 0 or up to 1. In this situation, the integer programming methods of Chapter 9 would have to be used, because the number of missile sites is definitely not divisible.

The Certainty Assumption

The **Certainty Assumption** is that each parameter (objective function coefficient, right-hand side, and technological coefficient) is known with certainty. If we were unsure of the exact amount of carpentry and finishing hours required to build a train, the Certainty Assumption would be violated.

Feasible Region and Optimal Solution

Two of the most basic concepts associated with a linear programming problem are feasible region and optimal solution. For defining these concepts, we use the term *point* to mean a specification of the value for each decision variable.

DEFINITION ■ The **feasible region** for an LP is the set of all points that satisfies all the LP's constraints and sign restrictions. ■

For example, in the Giapetto problem, the point $(x_1 = 40, x_2 = 20)$ is in the feasible region. Note that $x_1 = 40$ and $x_2 = 20$ satisfy the constraints (2)–(4) and the sign restrictions (5)–(6):

Constraint (2), $2x_1 + x_2 \leq 100$, is satisfied, because $2(40) + 20 \leq 100$.

Constraint (3), $x_1 + x_2 \leq 80$, is satisfied, because $40 + 20 \leq 80$.

Constraint (4), $x_1 \leq 40$, is satisfied, because $40 \leq 40$.

Restriction (5), $x_1 \geq 0$, is satisfied, because $40 \geq 0$.

Restriction (6), $x_2 \geq 0$, is satisfied, because $20 \geq 0$.

On the other hand, the point ($x_1 = 15$, $x_2 = 70$) is not in the feasible region, because even though $x_1 = 15$ and $x_2 = 70$ satisfy (2), (4), (5), and (6), they fail to satisfy (3): $15 + 70$ is not less than or equal to 80. Any point that is not in an LP's feasible region is said to be an **infeasible point.** As another example of an infeasible point, consider ($x_1 = 40$, $x_2 = -20$). Although this point satisfies all the constraints and the sign restriction (5), it is infeasible because it fails to satisfy the sign restriction (6), $x_2 \geq 0$. The feasible region for the Giapetto problem is the set of possible production plans that Giapetto must consider in searching for the optimal production plan.

DEFINITION ■ For a maximization problem, an **optimal solution** to an LP is a point in the feasible region with the largest objective function value. Similarly, for a minimization problem, an optimal solution is a point in the feasible region with the smallest objective function value. ■

Most LPs have only one optimal solution. However, some LPs have no optimal solution, and some LPs have an infinite number of solutions (these situations are discussed in Section 3.3). In Section 3.2, we show that the unique optimal solution to the Giapetto problem is ($x_1 = 20$, $x_2 = 60$). This solution yields an objective function value of

$$z = 3x_1 + 2x_2 = 3(20) + 2(60) = \$180$$

When we say that ($x_1 = 20$, $x_2 = 60$) is the optimal solution to the Giapetto problem, we are saying that no point in the feasible region has an objective function value that exceeds 180. Giapetto can maximize profit by building 20 soldiers and 60 trains each week. If Giapetto were to produce 20 soldiers and 60 trains each week, the weekly profit would be $180 less weekly fixed costs. For example, if Giapetto's only fixed cost were rent of $100 per week, then weekly profit would be $180 - 100 = \$80$ per week.

PROBLEMS

Group A

1 Farmer Jones must determine how many acres of corn and wheat to plant this year. An acre of wheat yields 25 bushels of wheat and requires 10 hours of labor per week. An acre of corn yields 10 bushels of corn and requires 4 hours of labor per week. All wheat can be sold at $4 a bushel, and all corn can be sold at $3 a bushel. Seven acres of land and 40 hours per week of labor are available. Government regulations require that at least 30 bushels of corn be produced during the current year. Let x_1 = number of acres of corn planted, and x_2 = number of acres of wheat planted. Using these decision variables, formulate an LP whose solution will tell Farmer Jones how to maximize the total revenue from wheat and corn.

2 Answer these questions about Problem 1.

  a Is ($x_1 = 2$, $x_2 = 3$) in the feasible region? nope

  b Is ($x_1 = 4$, $x_2 = 3$) in the feasible region? nope

  c Is ($x_1 = 2$, $x_2 = -1$) in the feasible region? nope

  d Is ($x_1 = 3$, $x_2 = 2$) in the feasible region? ✓

3 Using the variables x_1 = number of bushels of corn

produced and x_2 = number of bushels of wheat produced, reformulate Farmer Jones's LP.

4 Truckco manufactures two types of trucks: 1 and 2. Each truck must go through the painting shop and assembly shop. If the painting shop were completely devoted to painting Type 1 trucks, then 800 per day could be painted; if the painting shop were completely devoted to painting Type 2 trucks, then 700 per day could be painted. If the assembly shop were completely devoted to assembling truck 1 engines, then 1,500 per day could be assembled; if the assembly shop were completely devoted to assembling truck 2 engines, then 1,200 per day could be assembled. Each Type 1 truck contributes $300 to profit; each Type 2 truck contributes $500. Formulate an LP that will maximize Truckco's profit.

Group B

5 Why don't we allow an LP to have $<$ or $>$ constraints?

3.2 The Graphical Solution of Two-Variable Linear Programming Problems

Any LP with only two variables can be solved graphically. We always label the variables x_1 and x_2 and the coordinate axes the x_1 and x_2 axes. Suppose we want to graph the set of points that satisfies

$$2x_1 + 3x_2 \leq 6 \tag{7}$$

The same set of points (x_1, x_2) satisfies

$$3x_2 \leq 6 - 2x_1$$

This last inequality may be rewritten as

$$x_2 \leq \tfrac{1}{3}(6 - 2x_1) = 2 - \tfrac{2}{3}x_1 \tag{8}$$

Because moving downward on the graph decreases x_2 (see Figure 1), the set of points that satisfies (8) and (7) lies on or below the line $x_2 = 2 - \tfrac{2}{3}x_1$. This set of points is indicated by darker shading in Figure 1. Note, however, that $x_2 = 2 - \tfrac{2}{3}x_1$, $3x_2 = 6 - 2x_1$, and $2x_1 + 3x_2 = 6$ are all the same line. This means that the set of points satisfying (7) lies on or below the line $2x_1 + 3x_2 = 6$. Similarly, the set of points satisfying $2x_1 + 3x_2 \geq 6$ lies on or above the line $2x_1 + 3x_2 = 6$. (These points are shown by lighter shading in Figure 1.)

Consider a linear inequality constraint of the form $f(x_1, x_2) \geq b$ or $f(x_1, x_2) \leq b$. In general, it can be shown that in two dimensions, the set of points that satisfies a linear inequality includes the points on the line $f(x_1, x_2) = b$, defining the inequality plus all points on one side of the line.

There is an easy way to determine the side of the line for which an inequality such as $f(x_1, x_2) \leq b$ or $f(x_1, x_2) \geq b$ is satisfied. Just choose any point P that does not satisfy the line $f(x_1, x_2) = b$. Determine whether P satisfies the inequality. If it does, then all points on the *same* side as P of $f(x_1, x_2) = b$ will satisfy the inequality. If P does not satisfy the inequality, then all points on the *other* side of $f(x_1, x_2) = b$, which does not contain P, will satisfy the inequality. For example, to determine whether $2x_1 + 3x_2 \geq 6$ is satisfied by points above or below the line $2x_1 + 3x_2 = 6$, we note that $(0, 0)$ does not satisfy $2x_1 + 3x_2 \geq 6$. Because $(0, 0)$ is *below* the line $2x_1 + 3x_2 = 6$, the set of points satisfying $2x_1 + 3x_2 \geq 6$ includes the line $2x_1 + 3x_2 = 6$ and the points *above* the line $2x_1 + 3x_2 = 6$. This agrees with Figure 1.

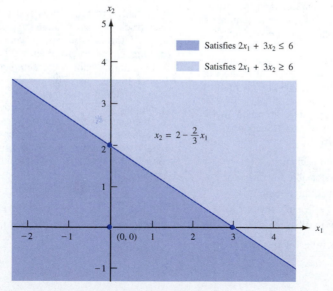

FIGURE 1
Graphing a Linear Inequality

Finding the Feasible Solution

We illustrate how to solve two-variable LPs graphically by solving the Giapetto problem. To begin, we graphically determine the feasible region for Giapetto's problem. The feasible region for the Giapetto problem is the set of all points (x_1, x_2) satisfying

$$2x_1 + x_2 \leq 100 \qquad \text{(Constraints)} \qquad \text{(2)}$$
$$x_1 + x_2 \leq 80 \qquad \text{(3)}$$
$$x_1 \qquad \leq 40 \qquad \text{(4)}$$
$$x_1 \qquad \geq 0 \qquad \text{(Sign restrictions)} \qquad \text{(5)}$$
$$x_2 \geq 0 \qquad \text{(6)}$$

For a point (x_1, x_2) to be in the feasible region, (x_1, x_2) must satisfy *all* the inequalities (2)–(6). Note that the only points satisfying (5) and (6) lie in the first quadrant of the x_1–x_2 plane. This is indicated in Figure 2 by the arrows pointing to the right from the x_2 axis and upward from the x_1 axis. Thus, any point that is outside the first quadrant cannot be in the feasible region. This means that the feasible region will be the set of points in the first quadrant that satisfies (2)–(4).

Our method for determining the set of points that satisfies a linear inequality will also identify those that meet (2)–(4). From Figure 2, we see that (2) is satisfied by all points below or on the line AB (AB is the line $2x_1 + x_2 = 100$). Inequality (3) is satisfied by all points on or below the line CD (CD is the line $x_1 + x_2 = 80$). Finally, (4) is satisfied by all points on or to the left of line EF (EF is the line $x_1 = 40$). The side of a line that satisfies an inequality is indicated by the direction of the arrows in Figure 2.

From Figure 2, we see that the set of points in the first quadrant that satisfies (2), (3), and (4) is bounded by the five-sided polygon $DGFEH$. Any point on this polygon or in its interior is in the feasible region. Any other point fails to satisfy at least one of the inequalities (2)–(6). For example, the point (40, 30) lies outside $DGFEH$ because it is above the line segment AB. Thus (40, 30) is infeasible, because it fails to satisfy (2).

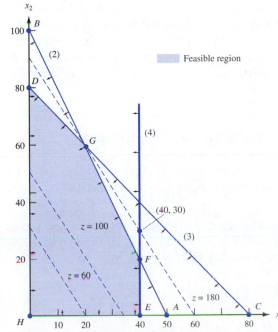

FIGURE 2
Graphical Solution of Giapetto Problem

An easy way to find the feasible region is to determine the set of infeasible points. Note that all points above line AB in Figure 2 are infeasible, because they fail to satisfy (2). Similarly, all points above CD are infeasible, because they fail to satisfy (3). Also, all points to the right of the vertical line EF are infeasible, because they fail to satisfy (4). After these points are eliminated from consideration, we are left with the feasible region (*DGFEH*).

Finding the Optimal Solution

Having identified the feasible region for the Giapetto problem, we now search for the optimal solution, which will be the point in the feasible region with the largest value of $z = 3x_1 + 2x_2$. To find the optimal solution, we need to graph a line on which all points have the same z-value. In a max problem, such a line is called an **isoprofit line** (in a min problem, an **isocost line**). To draw an isoprofit line, choose any point in the feasible region and calculate its z-value. Let us choose (20, 0). For (20, 0), $z = 3(20) + 2(0) = 60$. Thus, (20, 0) lies on the isoprofit line $z = 3x_1 + 2x_2 = 60$. Rewriting $3x_1 + 2x_2 = 60$ as $x_2 = 30 - \frac{3}{2}x_1$, we see that the isoprofit line $3x_1 + 2x_2 = 60$ has a slope of $-\frac{3}{2}$. Because all isoprofit lines are of the form $3x_1 + 2x_2 = $ constant, all isoprofit lines have the same slope. *This means that once we have drawn one isoprofit line, we can find all other isoprofit lines by moving parallel to the isoprofit line we have drawn.*

It is now clear how to find the optimal solution to a two-variable LP. After you have drawn a single isoprofit line, generate other isoprofit lines by moving parallel to the drawn isoprofit line in a direction that increases z (for a max problem). After a point, the isoprofit lines will no longer intersect the feasible region. The last isoprofit line intersecting (touching) the feasible region defines the largest z-value of any point in the feasible region and indicates the optimal solution to the LP. In our problem, the objective function $z = 3x_1 + 2x_2$ will increase if we move in a direction for which both x_1 and x_2 increase. Thus, we construct additional isoprofit lines by moving parallel to $3x_1 + 2x_2 = 60$ in a northeast direction (upward and to the right). From Figure 2, we see that the isoprofit line passing through point G is the last isoprofit line to intersect the feasible region. Thus, G is the point in the feasible region with the largest z-value and is therefore the optimal solution to the Giapetto problem. Note that point G is where the lines $2x_1 + x_2 = 100$ and $x_1 + x_2 = 80$ intersect. Solving these two equations simultaneously, we find that ($x_1 = 20, x_2 = 60$) is the optimal solution to the Giapetto problem. The optimal value of z may be found by substituting these values of x_1 and x_2 into the objective function. Thus, the optimal value of z is $z = 3(20) + 2(60) = 180$.

Binding and Nonbinding Constraints

Once the optimal solution to an LP has been found, it is useful (see Chapters 5 and 6) to classify each constraint as being a binding constraint or a nonbinding constraint.

DEFINITION ■ A constraint is **binding** if the left-hand side and the right-hand side of the constraint are equal when the optimal values of the decision variables are substituted into the constraint. ■

Thus, (2) and (3) are binding constraints.

A constraint is **nonbinding** if the left-hand side and the right-hand side of the constraint are unequal when the optimal values of the decision variables are substituted into the constraint. ■

Because $x_1 = 20$ is less than 40, (4) is a nonbinding constraint.

Convex Sets, Extreme Points, and LP

The feasible region for the Giapetto problem is an example of a convex set.

A set of points S is a **convex set** if the line segment joining any pair of points in S is wholly contained in S. ■

Figure 3 gives four illustrations of this definition. In Figures 3a and 3b, each line segment joining two points in S contains only points in S. Thus, in both these figures, S is convex. In Figures 3c and 3d, S is not convex. In each figure, points A and B are in S, but there are points on the line segment AB that are not contained in S. In our study of linear programming, a certain type of point in a convex set (called an *extreme point*) is of great interest.

For any convex set S, a point P in S is an **extreme point** if each line segment that lies completely in S and contains the point P has P as an endpoint of the line segment. ■

For example, in Figure 3a, each point on the circumference of the circle is an extreme point of the circle. In Figure 3b, points A, B, C, and D are extreme points of S. Although point E is on the boundary of S in Figure 3b, E is not an extreme point of S. This is because E lies on the line segment AB (AB lies completely in S), and E is not an endpoint of the line segment AB. Extreme points are sometimes called **corner points,** because if the set S is a polygon, the extreme points of S will be the vertices, or corners, of the polygon.

The feasible region for the Giapetto problem is a convex set. This is no accident: It can be shown that the feasible region for any LP will be a convex set. From Figure 2, we see that the extreme points of the feasible region are simply points D, F, E, G, and H. It can be shown that the feasible region for any LP has only a finite number of extreme points. Also note that the optimal solution to the Giapetto problem (point G) is an extreme point of the feasible region. It can be shown that *any LP that has an optimal solution has an extreme point that is optimal*. This result is very important, because it reduces the set of points that yield an optimal solution from the entire feasible region (which generally contains an *infinite* number of points) to the set of extreme points (a *finite* set).

S = shaded area

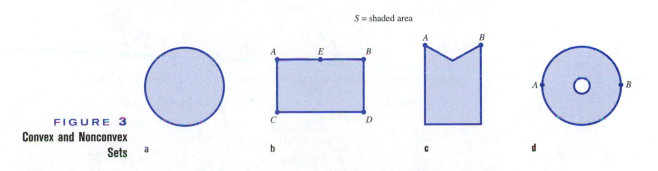

FIGURE 3
Convex and Nonconvex Sets

a   b   c   d

For the Giapetto problem, it is easy to see why the optimal solution must be an extreme point of the feasible region. We note that z increases as we move isoprofit lines in a northeast direction, so the largest z-value in the feasible region must occur at some point P that has no points in the feasible region northeast of P. This means that the optimal solution must lie somewhere on the boundary of the feasible region $DGFEH$. The LP must have an extreme point that is optimal, because for any line segment on the boundary of the feasible region, the largest z-value on that line segment must be assumed at one of the endpoints of the line segment.

To see this, look at the line segment FG in Figure 2. FG is part of the line $2x_1 + x_2 = 100$ and has a slope of -2. If we move along FG and decrease x_1 by 1, then x_2 will increase by 2, and the value of z changes as follows: $3x_1$ goes down by $3(1) = 3$, and $2x_2$ goes up by $2(2) = 4$. Thus, in total, z increases by $4 - 3 = 1$. This means that moving along FG in a direction of decreasing x_1 increases z. Thus, the value of z at point G must exceed the value of z at any other point on the line segment FG.

A similar argument shows that for any objective function, the maximum value of z on a given line segment must occur at an endpoint of the line segment. Therefore, for any LP, the largest z-value in the feasible region must be attained at an endpoint of one of the line segments forming the boundary of the feasible region. In short, one of the extreme points of the feasible region must be optimal. (To test your understanding, show that if Giapetto's objective function were $z = 6x_1 + x_2$, point F would be optimal, whereas if Giapetto's objective function were $z = x_1 + 6x_2$, point D would be optimal.)

Our proof that an LP always has an optimal extreme point depended heavily on the fact that both the objective function and the constraints were linear functions. In Chapter 12, we show that for an optimization problem in which the objective function or some of the constraints are not linear, the optimal solution to the optimization problem may not occur at an extreme point.

The Graphical Solution of Minimization Problems

EXAMPLE 2	Dorian Auto

Dorian Auto manufactures luxury cars and trucks. The company believes that its most likely customers are high-income women and men. To reach these groups, Dorian Auto has embarked on an ambitious TV advertising campaign and has decided to purchase 1-minute commercial spots on two types of programs: comedy shows and football games. Each comedy commercial is seen by 7 million high-income women and 2 million high-income men. Each football commercial is seen by 2 million high-income women and 12 million high-income men. A 1-minute comedy ad costs \$50,000, and a 1-minute football ad costs \$100,000. Dorian would like the commercials to be seen by at least 28 million high-income women and 24 million high-income men. Use linear programming to determine how Dorian Auto can meet its advertising requirements at minimum cost.

Solution Dorian must decide how many comedy and football ads should be purchased, so the decision variables are

$$x_1 = \text{number of 1-minute comedy ads purchased}$$
$$x_2 = \text{number of 1-minute football ads purchased}$$

Then Dorian wants to minimize total advertising cost (in thousands of dollars).

Total advertising cost $=$ cost of comedy ads $+$ cost of football ads

$$= \left(\frac{\text{cost}}{\text{comedy ad}}\right)\left(\begin{matrix}\text{total}\\\text{comedy ads}\end{matrix}\right) + \left(\frac{\text{cost}}{\text{football ad}}\right)\left(\begin{matrix}\text{total}\\\text{football ads}\end{matrix}\right)$$

$$= 50x_1 + 100x_2$$

Thus, Dorian's objective function is

$$\min z = 50x_1 + 100x_2 \qquad (9)$$

Dorian faces the following constraints:

Constraint 1 Commercials must reach at least 28 million high-income women.

Constraint 2 Commercials must reach at least 24 million high-income men.

To express Constraints 1 and 2 in terms of x_1 and x_2, let HIW stand for high-income women viewers and HIM stand for high-income men viewers (in millions).

$$\text{HIW} = \left(\frac{\text{HIW}}{\text{comedy ad}}\right)\left(\frac{\text{total}}{\text{comedy ads}}\right) + \left(\frac{\text{HIW}}{\text{football ad}}\right)\left(\frac{\text{total}}{\text{football ads}}\right)$$
$$= 7x_1 + 2x_2$$

$$\text{HIM} = \left(\frac{\text{HIM}}{\text{comedy ad}}\right)\left(\frac{\text{total}}{\text{comedy ads}}\right) + \left(\frac{\text{HIM}}{\text{football ad}}\right)\left(\frac{\text{total}}{\text{football ads}}\right)$$
$$= 2x_1 + 12x_2$$

Constraint 1 may now be expressed as

$$7x_1 + 2x_2 \geq 28 \qquad (10)$$

and Constraint 2 may be expressed as

$$2x_1 + 12x_2 \geq 24 \qquad (11)$$

The sign restrictions $x_1 \geq 0$ and $x_2 \geq 0$ are necessary, so the Dorian LP is given by:

$$\min z = 50x_1 + 100x_2$$
$$\text{s.t.} \quad 7x_1 + 2x_2 \geq 28 \quad \text{(HIW)}$$
$$2x_1 + 12x_2 \geq 24 \quad \text{(HIM)}$$
$$x_1, x_2 \geq 0$$

This problem is typical of a wide range of LP applications in which a decision maker wants to minimize the cost of meeting a certain set of requirements. To solve this LP graphically, we begin by graphing the feasible region (Figure 4). Note that (10) is satisfied by points on or above the line AB (AB is part of the line $7x_1 + 2x_2 = 28$) and that

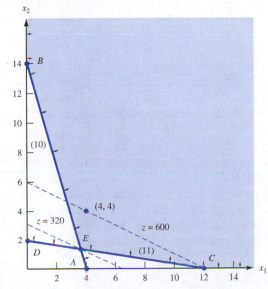

FIGURE 4
Graphical Solution of Dorian Problem

(11) is satisfied by the points on or above the line CD (CD is part of the line $2x_1 + 12x_2 = 24$). From Figure 4, we see that the only first-quadrant points satisfying both (10) and (11) are the points in the shaded region bounded by the x_1 axis, CEB, and the x_2 axis.

Like the Giapetto problem, the Dorian problem has a convex feasible region, but the feasible region for Dorian, unlike Giapetto's, contains points for which the value of at least one variable can assume arbitrarily large values. Such a feasible region is called an **unbounded feasible region.**

Because Dorian wants to minimize total advertising cost, the optimal solution to the problem is the point in the feasible region with the *smallest z*-value. To find the optimal solution, we need to draw an *isocost line* that intersects the feasible region. An isocost line is any line on which all points have the same z-value (or same cost). We arbitrarily choose the isocost line passing through the point ($x_1 = 4$, $x_2 = 4$). For this point, $z = 50(4) + 100(4) = 600$, and we graph the isocost line $z = 50x_1 + 100x_2 = 600$.

We consider lines parallel to the isocost line $50x_1 + 100x_2 = 600$ in the direction of decreasing z (southwest). The last point in the feasible region that intersects an isocost line will be the point in the feasible region having the *smallest z*-value. From Figure 4, we see that point E has the smallest z-value of any point in the feasible region; this is the optimal solution to the Dorian problem. Note that point E is where the lines $7x_1 + 2x_2 = 28$ and $2x_1 + 12x_2 = 24$ intersect. Simultaneously solving these equations yields the optimal solution ($x_1 = 3.6$, $x_2 = 1.4$). The optimal z-value can then be found by substituting these values of x_1 and x_2 into the objective function. Thus, the optimal z-value is $z = 50(3.6) + 100(1.4) = 320 = \$320,000$. Because at point E both the HIW and HIM constraints are satisfied with equality, both constraints are binding.

Does the Dorian model meet the four assumptions of linear programming outlined in Section 3.1?

For the Proportionality Assumption to be valid, each extra comedy commercial must add exactly 7 million HIW and 2 million HIM. This contradicts empirical evidence, which indicates that after a certain point advertising yields diminishing returns. After, say, 500 auto commercials have been aired, most people have probably seen one, so it does little good to air more commercials. Thus, the Proportionality Assumption is violated.

We used the Additivity Assumption to justify writing (total HIW viewers) = (HIW viewers from comedy ads) + (HIW viewers from football ads). In reality, many of the same people will see a Dorian comedy commercial and a Dorian football commercial. We are double-counting such people, and this creates an inaccurate picture of the total number of people seeing Dorian commercials. The fact that the same person may see more than one type of commercial means that the effectiveness of, say, a comedy commercial depends on the number of football commercials. This violates the Additivity Assumption.

If only 1-minute commercials are available, then it is unreasonable to say that Dorian should buy 3.6 comedy commercials and 1.4 football commercials, so the Divisibility Assumption is violated, and the Dorian problem should be considered an integer programming problem. In Section 9.3, we show that if the Dorian problem is solved as an integer programming problem, then the minimum cost is attained by choosing ($x_1 = 6$, $x_2 = 1$) or ($x_1 = 4$, $x_2 = 2$). For either solution, the minimum cost is \$400,000. This is 25% higher than the cost obtained from the optimal LP solution.

Because there is no way to know with certainty how many viewers are added by each type of commercial, the Certainty Assumption is also violated. Thus, all the assumptions of linear programming seem to be violated by the Dorian Auto problem. Despite these drawbacks, analysts have used similar models to help companies determine their optimal media mix.[†]

[†]Lilien and Kotler (1983).

PROBLEMS

Group A

1 Graphically solve Problem 1 of Section 3.1.

2 Graphically solve Problem 4 of Section 3.1.

3 Leary Chemical manufactures three chemicals: A, B, and C. These chemicals are produced via two production processes: 1 and 2. Running process 1 for an hour costs $4 and yields 3 units of A, 1 of B, and 1 of C. Running process 2 for an hour costs $1 and produces 1 unit of A and 1 of B. To meet customer demands, at least 10 units of A, 5 of B, and 3 of C must be produced daily. Graphically determine a daily production plan that minimizes the cost of meeting Leary Chemical's daily demands.

4 For each of the following, determine the direction in which the objective function increases:

    a $z = 4x_1 - x_2$

    b $z = -x_1 + 2x_2$

    c $z = -x_1 - 3x_2$

5 Furnco manufactures desks and chairs. Each desk uses 4 units of wood, and each chair uses 3. A desk contributes $40 to profit, and a chair contributes $25. Marketing restrictions require that the number of chairs produced be at least twice the number of desks produced. If 20 units of wood are available, formulate an LP to maximize Furnco's profit. Then graphically solve the LP.

6 Farmer Jane owns 45 acres of land. She is going to plant each with wheat or corn. Each acre planted with wheat yields $200 profit; each with corn yields $300 profit. The labor and fertilizer used for each acre are given in Table 1. One hundred workers and 120 tons of fertilizer are available. Use linear programming to determine how Jane can maximize profits from her land.

TABLE 1

	Wheat	Corn
Labor	3 workers	2 workers
Fertilizer	2 tons	4 tons

3.3 Special Cases

The Giapetto and Dorian problems each had a unique optimal solution. In this section, we encounter three type of LPs that do not have unique optimal solutions.

1 Some LPs have an infinite number of optimal solutions (*alternative* or *multiple optimal solutions*).

2 Some LPs have no feasible solutions (*infeasible* LPs).

3 Some LPs are *unbounded:* There are points in the feasible region with arbitrarily large (in a max problem) z-values.

Alternative or Multiple Optimal Solutions

EXAMPLE 3   **Alternative Optimal Solutions**

An auto company manufactures cars and trucks. Each vehicle must be processed in the paint shop and body assembly shop. If the paint shop were only painting trucks, then 40 per day could be painted. If the paint shop were only painting cars, then 60 per day could be painted. If the body shop were only producing cars, then it could process 50 per day. If the body shop were only producing trucks, then it could process 50 per day. Each truck contributes $300 to profit, and each car contributes $200 to profit. Use linear programming to determine a daily production schedule that will maximize the company's profits.

Solution   The company must decide how many cars and trucks should be produced daily. This leads us to define the following decision variables:

$$x_1 = \text{number of trucks produced daily}$$

$$x_2 = \text{number of cars produced daily}$$

The company's daily profit (in hundreds of dollars) is $3x_1 + 2x_2$, so the company's objective function may be written as

$$\max z = 3x_1 + 2x_2 \tag{12}$$

The company's two constraints are the following:

Constraint 1   The fraction of the day during which the paint shop is busy is less than or equal to 1.

Constraint 2   The fraction of the day during which the body shop is busy is less than or equal to 1.

We have

$$\text{Fraction of day paint shop works on trucks} = \left(\frac{\text{fraction of day}}{\text{truck}}\right)\left(\frac{\text{trucks}}{\text{day}}\right)$$

$$= \tfrac{1}{40}\, x_1$$

$$\text{Fraction of day paint shop works on cars} = \tfrac{1}{60}\, x_2$$

$$\text{Fraction of day body shop works on trucks} = \tfrac{1}{50}\, x_1$$

$$\text{Fraction of day body shop works on cars} = \tfrac{1}{50}\, x_2$$

Thus, Constraint 1 may be expressed by

$$\frac{1}{40} x_1 + \frac{1}{60} x_2 \le 1 \qquad \text{(Paint shop constraint)} \tag{13}$$

and Constraint 2 may be expressed by

$$\frac{1}{50} x_1 + \frac{1}{50} x_2 \le 1 \qquad \text{(Body shop constraint)} \tag{14}$$

Because $x_1 \ge 0$ and $x_2 \ge 0$ must hold, the relevant LP is

$$\max z = 3x_1 + 2x_2 \tag{12}$$

$$\text{s.t.} \quad \frac{1}{40} x_1 + \frac{1}{60} x_2 \le 1 \tag{13}$$

$$\frac{1}{50} x_1 + \frac{1}{50} x_2 \le 1 \tag{14}$$

$$x_1, x_2 \ge 0$$

The feasible region for this LP is the shaded region in Figure 5 bounded by $AEDF$.[†]

For our isoprofit line, we choose the line passing through the point (20, 0). Because (20, 0) has a z-value of $3(20) + 2(0) = 60$, this yields the isoprofit line $z = 3x_1 + 2x_2 = 60$. Examining lines parallel to this isoprofit line in the direction of increasing z (northeast), we find that the last "point" in the feasible region to intersect an isoprofit line is the *entire* line segment AE. This means that any point on the line segment AE is optimal. We can use any point on AE to determine the optimal z-value. For example, point A, (40, 0), gives $z = 3(40) = 120$.

In summary, the auto company's LP has an infinite number of optimal solutions, or *multiple* or *alternative optimal solutions*. This is indicated by the fact that as an isoprofit

[†]Constraint (13) is satisfied by all points on or below AB (AB is $\frac{1}{40} x_1 + \frac{1}{60} x_2 = 1$), and (14) is satisfied by all points on or below CD (CD is $\frac{1}{50} x_1 + \frac{1}{50} x_2 = 1$).

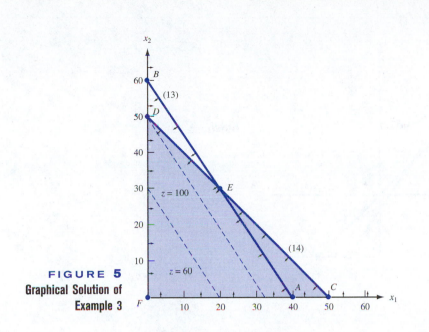

FIGURE 5
Graphical Solution of
Example 3

line leaves the feasible region, it will intersect an entire line segment corresponding to the binding constraint (in this case, AE).

From our current example, it seems reasonable (and can be shown to be true) that if two points (A and E here) are optimal, then *any* point on the line segment joining these two points will also be optimal.

If an alternative optimum occurs, then the decision maker can use a secondary criterion to choose between optimal solutions. The auto company's managers might prefer point A because it would simplify their business (and still allow them to maximize profits) by allowing them to produce only one type of product (trucks).

The technique of **goal programming** (see Section 4.14) is often used to choose among alternative optimal solutions.

Infeasible LP

It is possible for an LP's feasible region to be empty (contain no points), resulting in an *infeasible* LP. Because the optimal solution to an LP is the best point in the feasible region, an infeasible LP has no optimal solution.

EXAMPLE 4   Infeasible LP

Suppose that auto dealers require that the auto company in Example 3 produce at least 30 trucks and 20 cars. Find the optimal solution to the new LP.

Solution   After adding the constraints $x_1 \geq 30$ and $x_2 \geq 20$ to the LP of Example 3, we obtain the following LP:

$$\max z = 3x_1 + 2x_2$$

$$\text{s.t.} \quad \frac{1}{40}x_1 + \frac{1}{60}x_2 \leq 1 \tag{15}$$

$$\frac{1}{50}x_1 + \frac{1}{50}x_2 \leq 1 \tag{16}$$

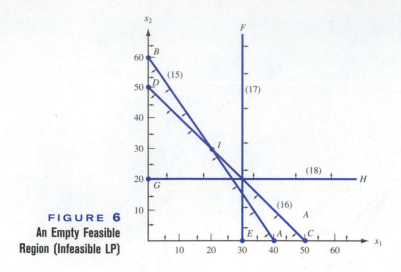

FIGURE 6
An Empty Feasible Region (Infeasible LP)

$$x_1 \qquad\qquad \ge 30 \qquad\qquad\qquad \textbf{(17)}$$
$$x_2 \ge 20 \qquad\qquad\qquad \textbf{(18)}$$
$$x_1, x_2 \ge 0$$

The graph of the feasible region for this LP is Figure 6.

Constraint (15) is satisfied by all points on or below AB (AB is $\frac{1}{40}x_1 + \frac{1}{60}x_2 = 1$).

Constraint (16) is satisfied by all points on or below CD (CD is $\frac{1}{50}x_1 + \frac{1}{50}x_2 = 1$).

Constraint (17) is satisfied by all points on or to the right of EF (EF is $x_1 = 30$).

Constraint (18) is satisfied by all points on or above GH (GH is $x_2 = 20$).

From Figure 6 it is clear that no point satisfies all of (15)–(18). This means that Example 4 has an empty feasible region and is an infeasible LP.

In Example 4, the LP is infeasible because producing 30 trucks and 20 cars requires more paint shop time than is available.

Unbounded LP

Our next special LP is an *unbounded* LP. For a max problem, an unbounded LP occurs if it is possible to find points in the feasible region with arbitrarily large z-values, which corresponds to a decision maker earning arbitrarily large revenues or profits. This would indicate that an unbounded optimal solution should not occur in a correctly formulated LP. Thus, if the reader ever solves an LP on the computer and finds that the LP is unbounded, then an error has probably been made in formulating the LP or in inputting the LP into the computer.

For a minimization problem, an LP is unbounded if there are points in the feasible region with arbitrarily small z-values. When graphically solving an LP, we can spot an unbounded LP as follows: A max problem is unbounded if, when we move parallel to our original isoprofit line in the direction of increasing z, we never entirely leave the feasible region. A minimization problem is unbounded if we never leave the feasible region when moving in the direction of decreasing z.

EXAMPLE 5   Unbounded LP

Graphically solve the following LP:

$$\max z = 2x_1 - x_2$$

$$\text{s.t.} \quad x_1 - x_2 \leq 1 \tag{19}$$

$$2x_1 + x_2 \geq 6 \tag{20}$$

$$x_1, x_2 \geq 0$$

Solution   From Figure 7, we see that (19) is satisfied by all points on or above AB (AB is the line $x_1 - x_2 = 1$). Also, (20) is satisfied by all points on or above CD (CD is $2x_1 + x_2 = 6$). Thus, the feasible region for Example 5 is the (shaded) unbounded region in Figure 7, which is bounded only by the x_2 axis, line segment DE, and the part of line AB beginning at E. To find the optimal solution, we draw the isoprofit line passing through $(2, 0)$. This isoprofit line has $z = 2x_1 - x_2 = 2(2) - 0 = 4$. The direction of increasing z is to the southeast (this makes x_1 larger and x_2 smaller). Moving parallel to $z = 2x_1 - x_2$ in a southeast direction, we see that any isoprofit line we draw will intersect the feasible region. (This is because any isoprofit line is steeper than the line $x_1 - x_2 = 1$.)

Thus, there are points in the feasible region that have arbitrarily large z-values. For example, if we wanted to find a point in the feasible region that had $z \geq 1,000,000$, we could choose any point in the feasible region that is southeast of the isoprofit line $z = 1,000,000$.

From the discussion in the last two sections, we see that every LP with two variables must fall into one of the following four cases:

Case 1   The LP has a unique optimal solution.

Case 2   The LP has alternative or multiple optimal solutions: Two or more extreme points are optimal, and the LP will have an infinite number of optimal solutions.

Case 3   The LP is infeasible: The feasible region contains no points.

Case 4   The LP is unbounded: There are points in the feasible region with arbitrarily large z-values (max problem) or arbitrarily small z-values (min problem).

In Chapter 4, we show that every LP (not just LPs with two variables) must fall into one of Cases 1–4.

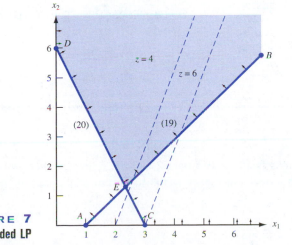

FIGURE 7
An Unbounded LP

In the rest of this chapter, we lead the reader through the formulation of several more complicated linear programming models. The most important step in formulating an LP model is the proper choice of decision variables. If the decision variables have been properly chosen, the objective function and constraints should follow without much difficulty. Trouble in determining an LP's objective function and constraints is usually the result of an incorrect choice of decision variables.

PROBLEMS

Group A

Identify which of Cases 1–4 apply to each of the following LPs:

1
$$\max z = x_1 + x_2$$
$$\text{s.t.} \quad x_1 + x_2 \leq 4$$
$$x_1 - x_2 \geq 5$$
$$x_1, x_2 \geq 0$$

2
$$\max z = 4x_1 + x_2$$
$$\text{s.t.} \quad 8x_1 + 2x_2 \leq 16$$
$$5x_1 + 2x_2 \leq 12$$
$$x_1, x_2 \geq 0$$

3
$$\max z = -x_1 + 3x_2$$
$$\text{s.t.} \quad x_1 - x_2 \leq 4$$
$$x_1 + 2x_2 \geq 4$$
$$x_1, x_2 \geq 0$$

4
$$\max z = 3x_1 + x_2$$
$$\text{s.t.} \quad 2x_1 + x_2 \leq 6$$
$$x_1 + 3x_2 \leq 9$$
$$x_1, x_2 \geq 0$$

5 True or false: For an LP to be unbounded, the LP's feasible region must be unbounded.

6 True or false: Every LP with an unbounded feasible region has an unbounded optimal solution.

7 If an LP's feasible region is not unbounded, we say the LP's feasible region is bounded. Suppose an LP has a bounded feasible region. Explain why you can find the optimal solution to the LP (without an isoprofit or isocost line) by simply checking the z-values at each of the feasible region's extreme points. Why might this method fail if the LP's feasible region is unbounded?

8 Graphically find all optimal solutions to the following LP:
$$\min z = x_1 - x_2$$
$$\text{s.t.} \quad x_1 + x_2 \leq 6$$
$$x_1 - x_2 \geq 0$$
$$x_2 - x_1 \geq 3$$
$$x_1, x_2 \geq 0$$

9 Graphically determine two optimal solutions to the following LP:
$$\min z = 3x_1 + 5x_2$$
$$\text{s.t.} \quad 3x_1 + 2x_2 \geq 36$$
$$3x_1 + 5x_2 \geq 45$$
$$x_1, x_2 \geq 0$$

Group B

10 Money manager Boris Milkem deals with French currency (the franc) and American currency (the dollar). At 12 midnight, he can buy francs by paying .25 dollars per franc and dollars by paying 3 francs per dollar. Let $x_1 =$ number of dollars bought (by paying francs) and $x_2 =$ number of francs bought (by paying dollars). Assume that both types of transactions take place simultaneously, and the only constraint is that at 12:01 A.M. Boris must have a nonnegative number of francs and dollars.

a Formulate an LP that enables Boris to maximize the number of dollars he has after all transactions are completed.

b Graphically solve the LP and comment on the answer.

3.4 A Diet Problem

Many LP formulations (such as Example 2 and the following diet problem) arise from situations in which a decision maker wants to minimize the cost of meeting a set of requirements.

EXAMPLE 6   **Diet Problem**

My diet requires that all the food I eat come from one of the four "basic food groups" (chocolate cake, ice cream, soda, and cheesecake). At present, the following four foods are available for consumption: brownies, chocolate ice cream, cola, and pineapple cheesecake. Each brownie costs 50¢, each scoop of chocolate ice cream costs 20¢, each bottle of cola costs 30¢, and each piece of pineapple cheesecake costs 80¢. Each day, I must ingest at least 500 calories, 6 oz of chocolate, 10 oz of sugar, and 8 oz of fat. The nutritional content per unit of each food is shown in Table 2. Formulate a linear programming model that can be used to satisfy my daily nutritional requirements at minimum cost.

Solution   As always, we begin by determining the decisions that must be made by the decision maker: how much of each type of food should be eaten daily. Thus, we define the decision variables:

$$x_1 = \text{number of brownies eaten daily}$$
$$x_2 = \text{number of scoops of chocolate ice cream eaten daily}$$
$$x_3 = \text{bottles of cola drunk daily}$$
$$x_4 = \text{pieces of pineapple cheesecake eaten daily}$$

My objective is to minimize the cost of my diet. The total cost of any diet may be determined from the following relation: (total cost of diet) = (cost of brownies) + (cost of ice cream) + (cost of cola) + (cost of cheesecake). To evaluate the total cost of a diet, note that, for example,

$$\text{Cost of cola} = \left(\frac{\text{cost}}{\text{bottle of cola}}\right)\left(\begin{array}{c}\text{bottles of}\\\text{cola drunk}\end{array}\right) = 30x_3$$

Applying this to the other three foods, we have (in cents)

$$\text{Total cost of diet} = 50x_1 + 20x_2 + 30x_3 + 80x_4$$

Thus, the objective function is

$$\min z = 50x_1 + 20x_2 + 30x_3 + 80x_4$$

The decision variables must satisfy the following four constraints:

Constraint 1   Daily calorie intake must be at least 500 calories.

Constraint 2   Daily chocolate intake must be at least 6 oz.

Constraint 3   Daily sugar intake must be at least 10 oz.

Constraint 4   Daily fat intake must be at least 8 oz.

TABLE 2
Nutritional Values for Diet

Type of Food	Calories	Chocolate (Ounces)	Sugar (Ounces)	Fat (Ounces)
Brownie	400	3	2	2
Chocolate ice cream (1 scoop)	200	2	2	4
Cola (1 bottle)	150	0	4	1
Pineapple cheesecake (1 piece)	500	0	4	5

To express Constraint 1 in terms of the decision variables, note that (daily calorie intake) = (calories in brownies) + (calories in chocolate ice cream) + (calories in cola) + (calories in pineapple cheesecake).

The calories in the brownies consumed can be determined from

$$\text{Calories in brownies} = \left(\frac{\text{calories}}{\text{brownie}}\right)\left(\frac{\text{brownies}}{\text{eaten}}\right) = 400x_1$$

Applying similar reasoning to the other three foods shows that

$$\text{Daily calorie intake} = 400x_1 + 200x_2 + 150x_3 + 500x_4$$

Constraint 1 may be expressed by

$$400x_1 + 200x_2 + 150x_3 + 500x_4 \geq 500 \qquad \text{(Calorie constraint)} \qquad \text{(21)}$$

Constraint 2 may be expressed by

$$3x_1 + 2x_2 \geq 6 \qquad \text{(Chocolate constraint)} \qquad \text{(22)}$$

Constraint 3 may be expressed by

$$2x_1 + 2x_2 + 4x_3 + 4x_4 \geq 10 \qquad \text{(Sugar constraint)} \qquad \text{(23)}$$

Constraint 4 may be expressed by

$$2x_1 + 4x_2 + x_3 + 5x_4 \geq 8 \qquad \text{(Fat constraint)} \qquad \text{(24)}$$

Finally, the sign restrictions $x_i \geq 0$ ($i = 1, 2, 3, 4$) must hold.

Combining the objective function, constraints (21)–(24), and the sign restrictions yields the following:

$$
\begin{aligned}
\min z = 50x_1 &+ 20x_2 + 30x_3 + 80x_4 \\
\text{s.t.} \quad 400x_1 &+ 200x_2 + 150x_3 + 500x_4 \geq 500 &&\text{(Calorie constraint)} &&\text{(21)} \\
3x_1 &+ 2x_2 \geq 6 &&\text{(Chocolate constraint)} &&\text{(22)} \\
2x_1 &+ 2x_2 + 4x_3 + 4x_4 \geq 10 &&\text{(Sugar constraint)} &&\text{(23)} \\
2x_1 &+ 4x_2 + x_3 + 5x_4 \geq 8 &&\text{(Fat constraint)} &&\text{(24)} \\
x_i &\geq 0 \ (i = 1, 2, 3, 4) &&\text{(Sign restrictions)}
\end{aligned}
$$

The optimal solution to this LP is $x_1 = x_4 = 0$, $x_2 = 3$, $x_3 = 1$, $z = 90$. Thus, the minimum-cost diet incurs a daily cost of 90¢ by eating three scoops of chocolate ice cream and drinking one bottle of cola. The optimal z-value may be obtained by substituting the optimal value of the decision variables into the objective function. This yields a total cost of $z = 3(20) + 1(30) = 90$¢. The optimal diet provides

$$200(3) + 150(1) = 750 \text{ calories}$$

$$2(3) = 6 \text{ oz of chocolate}$$

$$2(3) + 4(1) = 10 \text{ oz of sugar}$$

$$4(3) + 1(1) = 13 \text{ oz of fat}$$

Thus, the chocolate and sugar constraints are binding, but the calories and fat constraints are nonbinding.

A version of the diet problem with a more realistic list of foods and nutritional requirements was one of the first LPs to be solved by computer. Stigler (1945) proposed a diet problem in which 77 types of food were available and 10 nutritional requirements (vitamin A, vitamin C, and so on) had to be satisfied. When solved by computer, the optimal solution yielded a diet consisting of corn meal, wheat flour, evaporated milk, peanut butter, lard, beef, liver, potatoes, spinach, and cabbage. Although such a diet is clearly high in vital nutrients, few people would be satisfied with it because it does not seem to meet a minimum standard of tastiness (and Stigler required that the same diet be eaten each day). The optimal solution to any LP model will reflect only those aspects of reality that are captured by the objective function and constraints. Stigler's (and our) formulation of the diet problem did not reflect people's desire for a tasty and varied diet. Integer programming has been used to plan institutional menus for a weekly or monthly period.[†] Menu-planning models do contain constraints that reflect tastiness and variety requirements.

PROBLEMS

Group A

1 There are three factories on the Momiss River (1, 2, and 3). Each emits two types of pollutants (1 and 2) into the river. If the waste from each factory is processed, the pollution in the river can be reduced. It costs $15 to process a ton of factory 1 waste, and each ton processed reduces the amount of pollutant 1 by 0.10 ton and the amount of pollutant 2 by 0.45 ton. It costs $10 to process a ton of factory 2 waste, and each ton processed will reduce the amount of pollutant 1 by 0.20 ton and the amount of pollutant 2 by 0.25 ton. It costs $20 to process a ton of factory 3 waste, and each ton processed will reduce the amount of pollutant 1 by 0.40 ton and the amount of pollutant 2 by 0.30 ton. The state wants to reduce the amount of pollutant 1 in the river by at least 30 tons and the amount of pollutant 2 in the river by at least 40 tons. Formulate an LP that will minimize the cost of reducing pollution by the desired amounts. Do you think that the LP assumptions (Proportionality, Additivity, Divisibility, and Certainty) are reasonable for this problem?

2[‡] U.S. Labs manufactures mechanical heart valves from the heart valves of pigs. Different heart operations require valves of different sizes. U.S. Labs purchases pig valves from three different suppliers. The cost and size mix of the valves purchased from each supplier are given in Table 3. Each month, U.S. Labs places one order with each supplier. At least 500 large, 300 medium, and 300 small valves must be purchased each month. Because of limited availability of pig valves, at most 700 valves per month can be purchased from each supplier. Formulate an LP that can be used to minimize the cost of acquiring the needed valves.

3 Peg and Al Fundy have a limited food budget, so Peg is trying to feed the family as cheaply as possible. However, she still wants to make sure her family members meet their daily nutritional requirements. Peg can buy two foods. Food

TABLE 3

Supplier	Cost Per Value ($)	Percent Large	Percent Medium	Percent Small
1	5	40	40	20
2	4	30	35	35
3	3	20	20	60

1 sells for $7 per pound, and each pound contains 3 units of vitamin A and 1 unit of vitamin C. Food 2 sells for $1 per pound, and each pound contains 1 unit of each vitamin. Each day, the family needs at least 12 units of vitamin A and 6 units of vitamin C.

a Verify that Peg should purchase 12 units of food 2 each day and thus oversatisfy the vitamin C requirement by 6 units.

b Al has put his foot down and demanded that Peg fulfill the family's daily nutritional requirement exactly by obtaining precisely 12 units of vitamin A and 6 units of vitamin C. The optimal solution to the new problem will involve ingesting less vitamin C, but it will be more expensive. Why?

4 Goldilocks needs to find at least 12 lb of gold and at least 18 lb of silver to pay the monthly rent. There are two mines in which Goldilocks can find gold and silver. Each day that Goldilocks spends in mine 1, she finds 2 lb of gold and 2 lb of silver. Each day that Goldilocks spends in mine 2, she finds 1 lb of gold and 3 lb of silver. Formulate an LP to help Goldilocks meet her requirements while spending as little time as possible in the mines. Graphically solve the LP.

[†]Balintfy (1976).
[‡]Based on Hilal and Erickson (1981).

3.5 A Work-Scheduling Problem

Many applications of linear programming involve determining the minimum-cost method for satisfying workforce requirements. The following example illustrates the basic features common to many of these applications.

EXAMPLE 7   **Post Office Problem**

A post office requires different numbers of full-time employees on different days of the week. The number of full-time employees required on each day is given in Table 4. Union rules state that each full-time employee must work five consecutive days and then receive two days off. For example, an employee who works Monday to Friday must be off on Saturday and Sunday. The post office wants to meet its daily requirements using only full-time employees. Formulate an LP that the post office can use to minimize the number of full-time employees who must be hired.

Solution   Before giving the correct formulation of this problem, let's begin by discussing an *incorrect* solution. Many students begin by defining x_i to be the number of employees working on day i (day 1 = Monday, day 2 = Tuesday, and so on). Then they reason that (number of full-time employees) = (number of employees working on Monday) + (number of employees working on Tuesday) + $\cdots$ + (number of employees working on Sunday). This reasoning leads to the following objective function:

$$\min z = x_1 + x_2 + \cdots + x_6 + x_7$$

To ensure that the post office has enough full-time employees working on each day, they add the constraints $x_i \geq$ (number of employees required on day i). For example, for Monday add the constraint $x_1 \geq 17$. Adding the sign restrictions $x_i \geq 0$ ($i = 1, 2, \ldots, 7$) yields the following LP:

$$\min z = x_1 + x_2 + x_3 + x_4 + x_5 + x_6 + x_7$$
$$
\begin{aligned}
\text{s.t.} \quad x_1 & \geq 17 \\
x_2 & \geq 13 \\
x_3 & \geq 15 \\
x_4 & \geq 19 \\
x_5 & \geq 14 \\
x_6 & \geq 16 \\
x_7 & \geq 11 \\
x_i & \geq 0 \quad (i = 1, 2, \ldots, 7)
\end{aligned}
$$

There are at least two flaws in this formulation. First, the objective function is *not* the number of full-time post office employees. The current objective function counts each employee five times, not once. For example, each employee who starts work on Monday works Monday to Friday and is included in x_1, x_2, x_3, x_4, and x_5. Second, the variables x_1, $x_2, \ldots, x_7$ are interrelated, and the interrelation between the variables is not captured by the current set of constraints. For example, some of the people who are working on Monday (the x_1 people) will be working on Tuesday. This means that x_1 and x_2 are interrelated, but our constraints do not indicate that the value of x_1 has any effect on the value of x_2.

The key to correctly formulating this problem is to realize that the post office's primary decision is not how many people are working each day but rather how many people *begin* work on each day of the week. With this in mind, we define

TABLE 4
Requirements for Post Office

Day	Number of Full-time Employees Required
1 = Monday	17
2 = Tuesday	13
3 = Wednesday	15
4 = Thursday	19
5 = Friday	14
6 = Saturday	16
7 = Sunday	11

$$x_i = \text{number of employees beginning work on day } i$$

For example, x_1 is the number of people beginning work on Monday (these people work Monday to Friday). With the variables properly defined, it is easy to determine the correct objective function and constraints. To determine the objective function, note that (number of full-time employees) = (number of employees who start work on Monday) + (number of employees who start work on Tuesday) $+\cdots+$ (number of employees who start work on Sunday). Because each employee begins work on exactly one day of the week, this expression does not double-count employees. Thus, when we correctly define the variables, the objective function is

$$\min z = x_1 + x_2 + x_3 + x_4 + x_5 + x_6 + x_7$$

The post office must ensure that enough employees are working on each day of the week. For example, at least 17 employees must be working on Monday. Who is working on Monday? Everybody except the employees who begin work on Tuesday or on Wednesday (they get, respectively, Sunday and Monday, and Monday and Tuesday off). This means that the number of employees working on Monday is $x_1 + x_4 + x_5 + x_6 + x_7$. To ensure that at least 17 employees are working on Monday, we require that the constraint

$$x_1 + x_4 + x_5 + x_6 + x_7 \geq 17$$

be satisfied. Adding similar constraints for the other six days of the week and the sign restrictions $x_i \geq 0$ $(i = 1, 2, \ldots, 7)$ yields the following formulation of the post office's problem:

$$\min z = x_1 + x_2 + x_3 + x_4 + x_5 + x_6 + x_7$$

$$
\begin{aligned}
\text{s.t.} \quad & x_1 \quad\quad\quad + x_4 + x_5 + x_6 + x_7 \geq 17 && \text{(Monday constraint)} \\
& x_1 + x_2 \quad\quad\; + x_5 + x_6 + x_7 \geq 13 && \text{(Tuesday constraint)} \\
& x_1 + x_2 + x_3 \quad\quad + x_6 + x_7 \geq 15 && \text{(Wednesday constraint)} \\
& x_1 + x_2 + x_3 + x_4 \quad\quad + x_7 \geq 19 && \text{(Thursday constraint)} \\
& x_1 + x_2 + x_3 + x_4 + x_5 \quad\quad \geq 14 && \text{(Friday constraint)} \\
& \quad\; x_2 + x_3 + x_4 + x_5 + x_6 \quad \geq 16 && \text{(Saturday constraint)} \\
& \quad\quad\quad x_3 + x_4 + x_5 + x_6 + x_7 \geq 11 && \text{(Sunday constraint)} \\
& \quad\quad\quad x_i \geq 0 \quad (i = 1, 2, \ldots, 7) && \text{(Sign restrictions)}
\end{aligned}
$$

The optimal solution to this LP is $z = \frac{67}{3}$, $x_1 = \frac{4}{3}$, $x_2 = \frac{10}{3}$, $x_3 = 2$, $x_4 = \frac{22}{3}$, $x_5 = 0$, $x_6 = \frac{10}{3}$, $x_7 = 5$. Because we are only allowing full-time employees, however, the variables must be integers, and the Divisibility Assumption is not satisfied. To find a reasonable answer in which all variables are integers, we could try to round the fractional variables up, yielding the feasible solution $z = 25$, $x_1 = 2$, $x_2 = 4$, $x_3 = 2$, $x_4 = 8$, $x_5 = 0$, $x_6 = 4$, $x_7 = 5$. It turns out, however, that integer programming can be used to show that an optimal solution to the post office problem is $z = 23$, $x_1 = 4$, $x_2 = 4$, $x_3 = 2$, $x_4 = 6$, $x_5 = 0$, $x_6 = 4$, $x_7 = 3$. Notice that there is no way that the optimal linear programming solution could have been rounded to obtain the optimal all-integer solution.

Baker (1974) has developed an efficient technique (that does not use linear programming) to determine the minimum number of employees required when each worker receives two consecutive days off.

If you solve this problem using LINDO, LINGO, or the Excel Solver, you may get a different workforce schedule that uses 23 employees. This shows that Example 7 has alternative optimal solutions.

Creating a Fair Schedule for Employees

The optimal solution we found requires 4 workers to start on Monday, 4 on Tuesday, 2 on Wednesday, 6 on Thursday, 4 on Saturday, and 3 on Sunday. The workers who start on Saturday will be unhappy because they never receive a weekend day off. By rotating the schedules of the employees over a 23-week period, a fairer schedule can be obtained. To see how this is done, consider the following schedule:

- weeks 1–4: start on Monday
- weeks 5–8: start on Tuesday
- weeks 9–10: start on Wednesday
- weeks 11–16: start on Thursday
- weeks 17–20: start on Saturday
- weeks 21–23: start on Sunday

Employee 1 follows this schedule for a 23-week period. Employee 2 starts with week 2 of this schedule (starting on Monday for 3 weeks, then on Tuesday for 4 weeks, and closing with 3 weeks starting on Sunday and 1 week on Monday). We continue in this fashion to generate a 23-week schedule for each employee. For example, employee 13 will have the following schedule:

- weeks 1–4: start on Thursday
- weeks 5–8: start on Saturday
- weeks 9–11: start on Sunday
- weeks 12–15: start on Monday
- weeks 16–19: start on Tuesday
- weeks 20–21: start on Wednesday
- weeks 22–23 start on Thursday

This method of scheduling treats each employee equally.

Modeling Issues

1 This example is a **static scheduling problem,** because we assume that the post office faces the same schedule each week. In reality, demands change over time, workers take vacations in the summer, and so on, so the post office does not face the same situation each week. A **dynamic scheduling problem** will be discussed in Section 3.12.

2 If you wanted to set up a weekly scheduling model for a supermarket or a fast-food restaurant, the number of variables could be very large and the computer might have difficulty finding an exact solution. In this case, **heuristic methods** can be used to find a good solution to the problem. See Love and Hoey (1990) for an example of scheduling a fast-food restaurant.

3 Our model can easily be expanded to handle part-time employees, the use of overtime, and alternative objective functions such as maximizing the number of weekend days off. (See Problems 1, 3, and 4.)

4 How did we determine the number of workers needed each day? Perhaps the post office wants to have enough employees to ensure that 95% of all letters are sorted within an hour. To determine the number of employees needed to provide adequate service, the post office would use queuing theory, which is discussed in *Stochastic Models in Operations Research: Applications and Algorithms;* and forecasting, which is discussed in Chapter 14 of this book.

Real-World Application

Krajewski, Ritzman, and McKenzie (1980) used LP to schedule clerks who processed checks at the Ohio National Bank. Their model determined the minimum-cost combination of part-time employees, full-time employees, and overtime labor needed to process each day's checks by the end of the workday (10 P.M.). The major input to their model was a forecast of the number of checks arriving at the bank each hour. This forecast was produced using multiple regression (see *Stochastic Models in Operations Research: Applications and Algorithms*). The major output of the LP was a work schedule. For example, the LP might suggest that 2 full-time employees work daily from 11 A.M. to 8 P.M., 33 part-time employees work every day from 6 P.M. to 10 P.M., and 27 part-time employees work from 6 P.M. to 10 P.M. on Monday, Tuesday, and Friday.

P R O B L E M S

Group A

1 In the post office example, suppose that each full-time employee works 8 hours per day. Thus, Monday's requirement of 17 workers may be viewed as a requirement of $8(17) = 136$ hours. The post office may meet its daily labor requirements by using both full-time and part-time employees. During each week, a full-time employee works 8 hours a day for five consecutive days, and a part-time employee works 4 hours a day for five consecutive days. A full-time employee costs the post office $15 per hour, whereas a part-time employee (with reduced fringe benefits) costs the post office only $10 per hour. Union requirements limit part-time labor to 25% of weekly labor requirements. Formulate an LP to minimize the post office's weekly labor costs.

2 During each 4-hour period, the Smalltown police force requires the following number of on-duty police officers: 12 midnight to 4 A.M.—8; 4 to 8 A.M.—7; 8 A.M. to 12 noon—6; 12 noon to 4 P.M.—6; 4 to 8 P.M.—5; 8 P.M. to 12 midnight—4. Each police officer works two consecutive 4-hour shifts. Formulate an LP that can be used to minimize the number of police officers needed to meet Smalltown's daily requirements.

Group B

3 Suppose that the post office can force employees to work one day of overtime each week. For example, an employee whose regular shift is Monday to Friday can also be required to work on Saturday. Each employee is paid $50 a day for each of the first five days worked during a week and $62 for the overtime day (if any). Formulate an LP whose solution will enable the post office to minimize the cost of meeting its weekly work requirements.

4 Suppose the post office had 25 full-time employees and was not allowed to hire or fire any employees. Formulate an LP that could be used to schedule the employees in order to maximize the number of weekend days off received by the employees.

5 Each day, workers at the Gotham City Police Department work two 6-hour shifts chosen from 12 A.M. to 6 A.M., 6 A.M. to 12 P.M., 12 P.M. to 6 P.M., and 6 P.M. to 12 A.M. The following number of workers are needed during each shift: 12 A.M. to 6 A.M.—15 workers; 6 A.M. to 12 P.M.—5 workers; 12 P.M. to 6 P.M.—12 workers; 6 P.M. to 12 A.M.—6 workers. Workers whose two shifts are consecutive are paid $12 per hour; workers whose shifts are not consecutive are paid $18 per hour. Formulate an LP that can be used to minimize the cost of meeting the daily workforce demands of the Gotham City Police Department.

6 During each 6-hour period of the day, the Bloomington Police Department needs at least the number of policemen shown in Table 5. Policemen can be hired to work either 12 consecutive hours or 18 consecutive hours. Policemen are paid $4 per hour for each of the first 12 hours a day they work and are paid $6 per hour for each of the next 6 hours they work in a day. Formulate an LP that can be used to minimize the cost of meeting Bloomington's daily police requirements.

7 Each hour from 10 A.M. to 7 P.M., Bank One receives checks and must process them. Its goal is to process all the checks the same day they are received. The bank has 13 check-processing machines, each of which can process up to 500 checks per hour. It takes one worker to operate each machine. Bank One hires both full-time and part-time workers. Full-time workers work 10 A.M.–6 P.M., 11 A.M.–7 P.M., or Noon–8 P.M. and are paid $160 per day. Part-time workers work either 2 P.M.–7 P.M. or 3 P.M.–8 P.M. and are paid $75 per day. The number of checks received each hour is given in Table 6. In the interest of maintaining continuity, Bank One believes it must have at least three full-time workers under contract. Develop a cost-minimizing work schedule that processes all checks by 8 P.M.

TABLE 6

Time	Checks Received
10 A.M.	5,000
11 A.M.	4,000
Noon	3,000
1 P.M.	4,000
2 P.M.	2,500
3 P.M.	3,000
4 P.M.	4,000
5 P.M.	4,500
6 P.M.	3,500
7 P.M.	3,000

TABLE 5

Time Period	Number of Policemen Required
12 A.M.–6 A.M.	12
6 A.M.–12 P.M.	8
12 P.M.–6 P.M.	6
6 P.M.–12 A.M.	15

3.6 A Capital Budgeting Problem

In this section (and in Sections 3.7 and 3.11), we discuss how linear programming can be used to determine optimal financial decisions. This section considers a simple capital budgeting model.[†]

We first explain briefly the concept of net present value (NPV), which can be used to compare the desirability of different investments. Time 0 is the present.

Suppose investment 1 requires a cash outlay of $10,000 at time 0 and a cash outlay of $14,000 two years from now and yields a cash flow of $24,000 one year from now. Investment 2 requires a $6,000 cash outlay at time 0 and a $1,000 cash outlay two years from now and yields a cash flow of $8,000 one year from now. Which investment would you prefer?

Investment 1 has a net cash flow of

$$-10,000 + 24,000 - 14,000 = \$0$$

and investment 2 has a net cash flow of

$$-6000 + 8000 - 1000 = \$1000$$

On the basis of net cash flow, investment 2 is superior to investment 1. When we compare investments on the basis of net cash flow, we are assuming that a dollar received at

[†]This section is based on Weingartner (1963).

any point in time has the same value. This is not true! Suppose that there exists an investment (such as a money market fund) for which \$1 invested at a given time will yield (with certainty) $\$(1 + r)$ one year later. We call r the *annual interest rate*. Because \$1 now can be transformed into $\$(1 + r)$ one year from now, we may write

$$\$1 \text{ now} = \$(1 + r) \text{ one year from now}$$

Applying this reasoning to the $\$(1 + r)$ obtained one year from now shows that

$$\$1 \text{ now} = \$(1 + r) \text{ one year from now} = \$(1 + r)^2 \text{ two years from now}$$

and

$$\$1 \text{ now} = \$(1 + r)^k \ k \text{ years from now}$$

Dividing both sides of this equality by $(1 + r)^k$ shows that

$$\$1 \text{ received } k \text{ years from now} = \$(1 + r)^{-k} \text{ now}$$

In other words, a dollar received k years from now is equivalent to receiving $\$(1 + r)^{-k}$ now.

We can use this idea to express all cash flows in terms of time 0 dollars (this process is called *discounting cash flows to time* 0). Using discounting, we can determine the total value (in time 0 dollars) of the cash flows for any investment. The total value (in time 0 dollars) of the cash flows for any investment is called the **net present value,** or **NPV,** of the investment. The NPV of an investment is the amount by which the investment will increase the firm's value (as expressed in time 0 dollars).

Assuming that $r = 0.20$, we can compute the NPV for investments 1 and 2.

$$\text{NPV of investment 1} = -10{,}000 + \frac{24{,}000}{1 + 0.20} - \frac{14{,}000}{(1 + 0.20)^2}$$

$$= \$277.78$$

This means that if a firm invested in investment 1, then the value of the firm (in time 0 dollars) would increase by \$277.78. For investment 2,

$$\text{NPV of investment 2} = -6000 + \frac{8000}{1 + 0.20} - \frac{1000}{(1 + 0.20)^2}$$

$$= -\$27.78$$

If a firm invested in investment 2, then the value of the firm (in time 0 dollars) would be reduced by \$27.78.

Thus, the NPV concept says that investment 1 is superior to investment 2. This conclusion is contrary to the one reached by comparing the net cash flows of the two investments. Note that the comparison between investments often depends on the value of r. For example, the reader is asked to show in Problem 1 at the end of this section that for $r = 0.02$, investment 2 has a higher NPV than investment 1. Of course, our analysis assumes that the future cash flows of an investment are known with certainty.

Computing NPV with EXCEL

If we receive a cash flow of c_t in t years from now ($t = 1, 2, \ldots T$) and we discount cash flows at a rate r, then the NPV of our cash flows is given by

$$\sum_{t=1}^{t=T} \frac{c_t}{(1 + r)^t}.$$

The basic idea is that $1 today equals $(1 + r)$ a year from now, so

$$\frac{1}{1 + r} \text{ today} = \$1 \text{ a year from now.}$$

The Excel = NPV function makes this computation easy. Syntax is

$$= \text{NPV }(r, \text{ range of cash flows}).$$

Formula assumes that cash flows occur at end of year.

Projects with NPV > 0 add value to company while projects with negative NPV reduce the company's value.

We illustrate the computation of NPV in the file NPV.xls.

EXAMPLE 8   Computing NPV

For a discount rate of 15%, consider a project with the cash flows shown in Figure 8.

a Compute project NPV if cash flows are at the end of the year.

b Compute project NPV if cash flows are at the beginning of the year.

c Compute project NPV if cash flows are at the middle of the year.

Solution **a** We enter in cell C7 the formula

$$= \text{NPV}(C1,C4:I4)$$

and obtain $375.06.

b Because all cash flows are received a year earlier, we multiply each cash flow's value by $(1 + 1.15)$, so the answer is obtained in C8 with formula

$$= (1 + C1) \cdot C7.$$

NPV is now larger: $431.32.

We checked this in cell D8 with the formula

$$= C4 + \text{NPV}(C1,D4:I4).$$

c Because all cash flows are received six months earlier we multiply each cash flow's value by $\sqrt{1.15}$. NPV is now computed in C9 with the formula

$$= (1.15)\text{\textasciicircum}0.5 \cdot C7.$$

Now NPV is $402.21.

	A	B	C	D	E	F	G	H	I
1		dr	0.15						
2									
3		Time	1	2	3	4	5	6	7
4			-400	200	600	-900	1000	250	230
5									
6									
7	end of year	end of yr.	$375.06						
8	beginning of yr.	beg. of yr.	$431.32	$431.32					
9	middle of year	middle of yr.	$402.21						

FIGURE 8

The XNPV Function

Often cash flows occur at irregular intervals. This makes it difficult to compute the NPV of these cash flows. Fortunately, the Excel XNPV function makes computing NPV's of irregularly timed cash flows a snap. To use the XNPV function, you must first have added the Analysis Toolpak. To do this, select Tools Add-Ins and check the Analysis Toolpak and Analysis Tookpak VBA boxes. Here is an example of XNPV in action.

EXAMPLE 9 | **Finding NPV of Nonperiodic Cash Flows**

Suppose on April 8, 2001, we paid out $900. Then we receive

- $300 on 8/15/01
- $400 on 1/15/02
- $200 on 6/25/02
- $100 on 7/03/03.

If the annual discount rate is 10%, what is the NPV of these cash flows?

Solution We enter the dates (in Excel date format) in D3:D7 and the cash flows in E3:E7 (see Figure 9). Entering the formula

$$= \text{XNPV}(A9,E3:E7,D3:D7)$$

in cell D11 computes the project's NPV in terms of April 8, 2001, dollars because that is the first date chronologically. What Excel did was as follows:

1 Compute the number of years after April 8, 2001, that each date occurred. (We did this in column F). For example, August 15, 2001, is .3534 years after April 8.

2 Then discount cash flows at a rate $\left(\dfrac{1}{1 + \text{rate}}\right)^{\text{years after}}$. For example, the August 15, 2001, cash flow is discounted by $\left(\dfrac{1}{1 + .1}\right)^{.3534} = .967$.

3 We obtained Excel dates in serial number form by changing format to General.

If you want the XNPV function to determine a project's NPV in today's dollars, insert a $0 cash flow on today's date and include this row in the XNPV calculation. Excel will then return the project's NPV as of today's date.

	A	B	C	D	E	F	G
1							
2	XNPV Function		Code	Date	Cash Flow	Time	df
3			36989.00	4/8/01	-900		1
4			37118.00	8/15/01	300	0.353425	0.966876
5			37271.00	1/15/02	400	0.772603	0.929009
6			37432.00	6/25/02	200	1.213699	0.890762
7			37805.00	7/3/03	100	2.235616	0.808094
8	Rate						
9		0.1					
10				XNPV	Direct		
11				20.62822	20.628217		
12							
13				XIRR			
14				12.97%			

FIGURE 9
Example of XNPV Function

With this background information, we are ready to explain how linear programming can be applied to problems in which limited investment funds must be allocated to investment projects. Such problems are called **capital budgeting problems.**

EXAMPLE 10 — Project Selection

Star Oil Company is considering five different investment opportunities. The cash outflows and net present values (in millions of dollars) are given in Table 7. Star Oil has $40 million available for investment now (time 0); it estimates that one year from now (time 1) $20 million will be available for investment. Star Oil may purchase any fraction of each investment. In this case, the cash outflows and NPV are adjusted accordingly. For example, if Star Oil purchases one-fifth of investment 3, then a cash outflow of $\frac{1}{5}(5) = \$1$ million would be required at time 0, and a cash outflow of $\frac{1}{5}(5) = \$1$ million would be required at time 1. The one-fifth share of investment 3 would yield an NPV of $\frac{1}{5}(16) = \$3.2$ million. Star Oil wants to maximize the NPV that can be obtained by investing in investments 1–5. Formulate an LP that will help achieve this goal. Assume that any funds left over at time 0 cannot be used at time 1.

Solution Star Oil must determine what fraction of each investment to purchase. We define

$$x_i = \text{fraction of investment } i \text{ purchased by Star Oil} \qquad (i = 1, 2, 3, 4, 5)$$

Star's goal is to maximize the NPV earned from investments. Now, (total NPV) = (NPV earned from investment 1) + (NPV earned from investment 2) + $\cdots$ + (NPV earned from investment 5). Note that

NPV from investment 1 = (NPV from investment 1)(fraction of investment 1 purchased)
$$= 13x_1$$

Applying analogous reasoning to investments 2–5 shows that Star Oil wants to maximize

$$z = 13x_1 + 16x_2 + 16x_3 + 14x_4 + 39x_5 \tag{25}$$

Star Oil's constraints may be expressed as follows:

Constraint 1 Star cannot invest more than $40 million at time 0.

Constraint 2 Star cannot invest more than $20 million at time 1.

Constraint 3 Star cannot purchase more than 100% of investment i ($i = 1, 2, 3, 4, 5$).

To express Constraint 1 mathematically, note that (dollars invested at time 0) = (dollars invested in investment 1 at time 0) + (dollars invested in investment 2 at time 0) + $\cdots$ + (dollars invested in investment 5 at time 0). Also, in millions of dollars,

$$\begin{pmatrix} \text{Dollars invested in investment 1} \\ \text{at time 0} \end{pmatrix} = \begin{pmatrix} \text{dollars required for} \\ \text{investment 1 at time 0} \end{pmatrix}\begin{pmatrix} \text{fraction of} \\ \text{investment 1 purchased} \end{pmatrix}$$
$$= 11x_1$$

TABLE 7
Cash Flows and Net Present Value for Investments in Capital Budgeting

	Investment ($)				
	1	2	3	4	5
Time 0 cash outflow	11	53	5	5	29
Time 1 cash outflow	3	6	5	1	34
NPV	13	16	16	14	39

Similarly, for investments 2–5,

$$\text{Dollars invested at time } 0 = 11x_1 + 53x_2 + 5x_3 + 5x_4 + 29x_5$$

Then Constraint 1 reduces to

$$11x_1 + 53x_2 + 5x_3 + 5x_4 + 29x_5 \leq 40 \qquad \text{(Time 0 constraint)} \qquad \textbf{(26)}$$

Constraint 2 reduces to

$$3x_1 + 6x_2 + 5x_3 + x_4 + 34x_5 \leq 20 \qquad \text{(Time 1 constraint)} \qquad \textbf{(27)}$$

Constraints 3–7 may be represented by

$$x_i \leq 1 \qquad (i = 1, 2, 3, 4, 5) \qquad \textbf{(28-32)}$$

Combining (26)–(32) with the sign restrictions $x_i \geq 0$ ($i = 1, 2, 3, 4, 5$) yields the following LP:

$$\begin{aligned}
\max z = \ & 13x_1 + 16x_2 + 16x_3 + 14x_4 + 39x_5 \\
\text{s.t.} \quad & 11x_1 + 53x_2 + 5x_3 + 5x_4 + 29x_5 \leq 40 \qquad \text{(Time 0 constraint)} \\
& 3x_1 + 6x_2 + 5x_3 + x_4 + 34x_5 \leq 20 \qquad \text{(Time 1 constraint)} \\
& x_1 \qquad\qquad\qquad\qquad\qquad \leq 1 \\
& \quad x_2 \qquad\qquad\qquad\qquad \leq 1 \\
& \qquad\quad x_3 \qquad\qquad\qquad \leq 1 \\
& \qquad\qquad\quad x_4 \qquad\qquad \leq 1 \\
& \qquad\qquad\qquad\quad x_5 \leq 1 \\
& x_i \geq 0 \qquad (i = 1, 2, 3, 4, 5)
\end{aligned}$$

The optimal solution to this LP is $x_1 = x_3 = x_4 = 1$, $x_2 = 0.201$, $x_5 = 0.288$, $z = 57.449$. Star Oil should purchase 100% of investments 1, 3, and 4; 20.1% of investment 2; and 28.8% of investment 5. A total NPV of $57,449,000 will be obtained from these investments.

It is often impossible to purchase only a fraction of an investment without sacrificing the investment's favorable cash flows. Suppose it costs $12 million to drill an oil well just deep enough to locate a $30-million gusher. If there were a sole investor in this project who invested $6 million to undertake half of the project, then hc or she would lose the entire investment and receive no positive cash flows. Because, in this example, reducing the money invested by 50% reduces the return by more than 50%, this situation would violate the Proportionality Assumption.

In many capital budgeting problems, it is unreasonable to allow the x_i to be fractions: Each x_i should be restricted to 0 (not investing at all in investment i) or 1 (purchasing all of investment i). Thus, many capital budgeting problems violate the Divisibility Assumption.

A capital budgeting model that allows each x_i to be only 0 or 1 is discussed in Section 9.2.

PROBLEMS

Group A

1 Show that if $r = 0.02$, investment 2 has a larger NPV than investment 1.

2 Two investments with varying cash flows (in thousands of dollars) are available, as shown in Table 8. At time 0, $10,000 is available for investment, and at time 1, $7,000 is available. Assuming that $r = 0.10$, set up an LP whose solution maximizes the NPV obtained from these investments. Graphically find the optimal solution to the LP.

TABLE **8**

Investment	Cash Flow (in $ Thousands) at Time			
	0	1	2	3
1	−6	−5	7	9
2	−8	−3	9	7

(Assume that any fraction of an investment may be purchased.)

3 Suppose that r, the annual interest rate, is 0.20, and that all money in the bank earns 20% interest each year (that is, after being in the bank for one year, $1 will increase to $1.20). If we place $100 in the bank for one year, what is the NPV of this transaction?

4 A company has nine projects under consideration. The NPV added by each project and the capital required by each project during the next two years is given in Table 9. All

figures are in millions. For example, Project 1 will add $14 million in NPV and require expenditures of $12 million during year 1 and $3 million during year 2. Fifty million is available for projects during year 1 and $20 million is available during year 2. Assuming we may undertake a fraction of each project, how can we maximize NPV?

Group B

5[†] Finco must determine how much investment and debt to undertake during the next year. Each dollar invested reduces the NPV of the company by 10¢, and each dollar of debt increases the NPV by 50¢ (due to deductibility of interest payments). Finco can invest at most $1 million during the coming year. Debt can be at most 40% of investment. Finco now has $800,000 in cash available. All investment must be paid for from current cash or borrowed money. Set up an LP whose solution will tell Finco how to maximize its NPV. Then graphically solve the LP.

TABLE **9**

	Project								
	1	2	3	4	5	6	7	8	9
Year 1 Outflow	12	54	6	6	30	6	48	36	18
Year 2 Outflow	3	7	6	2	35	6	4	3	3
NPV	14	17	17	15	40	12	14	10	12

3.7 Short-Term Financial Planning[‡]

LP models can often be used to aid a firm in short- or long-term financial planning (also see Section 3.11). Here we consider a simple example that illustrates how linear programming can be used to aid a corporation's short-term financial planning.[§]

EXAMPLE 11 **Short-Term Financial Planning**

Semicond is a small electronics company that manufactures tape recorders and radios. The per-unit labor costs, raw material costs, and selling price of each product are given in Table 10. On December 1, 2002, Semicond has available raw material that is sufficient to manufacture 100 tape recorders and 100 radios. On the same date, the company's balance sheet is as shown in Table 11, and Semicond's asset–liability ratio (called the current ratio) is $20,000/10,000 = 2$.

Semicond must determine how many tape recorders and radios should be produced during December. Demand is large enough to ensure that all goods produced will be sold. All sales are on credit, however, and payment for goods produced in December will not

[†]Based on Myers and Pogue (1974).
[‡]This section covers material that may be omitted with no loss of continuity.
[§]This section is based on an example in Neave and Wiginton (1981).

TABLE 10

Cost Information for Semicond

	Tape Recorder	Radio
Selling price	$100	$90
Labor cost	$ 50	$35
Raw material cost	$ 30	$40

TABLE 11

Balance Sheet for Semicond

	Assets	Liabilities
Cash	$10,000	
Accounts receivable[§]	$ 3,000	
Inventory outstanding[¶]	$ 7,000	
Bank loan		$10,000

[§]Accounts receivable is money owed to Semicond by customers who have previously purchased Semicond products.

[¶]Value of December 1, 2002, inventory = 30(100) + 40(100) = $7,000.

be received until February 1, 2003. During December, Semicond will collect $2,000 in accounts receivable, and Semicond must pay off $1,000 of the outstanding loan and a monthly rent of $1,000. On January 1, 2003, Semicond will receive a shipment of raw material worth $2,000, which will be paid for on February 1, 2003. Semicond's management has decided that the cash balance on January 1, 2003, must be at least $4,000. Also, Semicond's bank requires that the current ratio at the beginning of January be at least 2. To maximize the contribution to profit from December production, (revenues to be received) − (variable production costs), what should Semicond produce during December?

Solution Semicond must determine how many tape recorders and radios should be produced during December. Thus, we define

$$x_1 = \text{number of tape recorders produced during December}$$
$$x_2 = \text{number of radios produced during December}$$

To express Semicond's objective function, note that

$$\frac{\text{Contribution to profit}}{\text{Tape recorder}} = 100 - 50 - 30 = \$20$$

$$\frac{\text{Contribution to profit}}{\text{Radio}} = 90 - 35 - 40 = \$15$$

As in the Giapetto example, this leads to the objective function

$$\max z = 20x_1 + 15x_2 \tag{33}$$

Semicond faces the following constraints:

Constraint 1 Because of limited availability of raw material, at most 100 tape recorders can be produced during December.

Constraint 2 Because of limited availability of raw material, at most 100 radios can be produced during December.

Constraint 3 Cash on hand on January 1, 2002, must be at least $4,000.

Constraint 4 (January 1 assets)/(January 1 liabilities) ≥ 2 must hold.

Constraint 1 is described by

$$x_1 \leq 100 \tag{34}$$

Constraint 2 is described by

$$x_2 \leq 100 \tag{35}$$

To express Constraint 3, note that

$$
\begin{aligned}
\text{January 1 cash on hand} = {} & \text{December 1 cash on hand} \\
& + \text{accounts receivable collected during December} \\
& - \text{portion of loan repaid during December} \\
& - \text{December rent} - \text{December labor costs} \\
= {} & 10{,}000 + 2{,}000 - 1{,}000 - 1{,}000 - 50x_1 - 35x_2 \\
= {} & 10{,}000 - 50x_1 - 35x_2
\end{aligned}
$$

Now Constraint 3 may be written as

$$10{,}000 - 50x_1 - 35x_2 \geq 4{,}000 \tag{36'}$$

Most computer codes require each LP constraint to be expressed in a form in which all variables are on the left-hand side and the constant is on the right-hand side. Thus, for computer solution, we should write (36′) as

$$50x_1 + 35x_2 \leq 6{,}000 \tag{36}$$

To express Constraint 4, we need to determine Semicond's January 1 cash position, accounts receivable, inventory position, and liabilities in terms of x_1 and x_2. We have already shown that

$$\text{January 1 cash position} = 10{,}000 - 50x_1 - 35x_2$$

Then

$$
\begin{aligned}
\text{January 1 accounts receivable} = {} & \text{December 1 accounts receivable} \\
& + \text{accounts receivable from December sales} \\
& - \text{accounts receivable collected during December} \\
= {} & 3{,}000 + 100x_1 + 90x_2 - 2000 \\
= {} & 1{,}000 + 100x_1 + 90x_2
\end{aligned}
$$

It now follows that

$$
\begin{aligned}
\text{Value of January 1 inventory} = {} & \text{value of December 1 inventory} \\
& - \text{value of inventory used in December} \\
& + \text{value of inventory received on January 1} \\
= {} & 7{,}000 - (30x_1 + 40x_2) + 2{,}000 \\
= {} & 9{,}000 - 30x_1 - 40x_2
\end{aligned}
$$

We can now compute the January 1 asset position:

$$
\begin{aligned}
\text{January 1 asset position} = {} & \text{January 1 cash position} + \text{January 1 accounts receivable} \\
& + \text{January 1 inventory position} \\
= {} & (10{,}000 - 50x_1 - 35x_2) + (1{,}000 + 100x_1 + 90x_2) \\
& + (9{,}000 - 30x_1 - 40x_2) \\
= {} & 20{,}000 + 20x_1 + 15x_2
\end{aligned}
$$

Finally,

$$\text{January 1 liabilities} = \text{December 1 liabilities} - \text{December loan payment}$$
$$+ \text{ amount due on January 1 inventory shipment}$$
$$= 10,000 - 1,000 + 2,000$$
$$= \$11,000$$

Constraint 4 may now be written as

$$\frac{20,000 + 20x_1 + 15x_2}{11,000} \geq 2$$

Multiplying both sides of this inequality by 11,000 yields

$$20,000 + 20x_1 + 15x_2 \geq 22,000$$

Putting this in a form appropriate for computer input, we obtain

$$20x_1 + 15x_2 \geq 2,000 \tag{37}$$

Combining (33)–(37) with the sign restrictions $x_1 \geq 0$ and $x_2 \geq 0$ yields the following LP:

$$\max z = 20x_1 + 15x_2$$

$$
\begin{array}{lll}
\text{s.t.} \quad x_1 & \leq 100 & \text{(Tape recorder constraint)} \\
x_2 & \leq 100 & \text{(Radio constraint)} \\
50x_1 + 35x_2 & \leq 6,000 & \text{(Cash position constraint)} \\
20x_1 + 15x_2 & \geq 2,000 & \text{(Current ratio constraint)} \\
x_1, x_2 & \geq 0 & \text{(Sign restrictions)}
\end{array}
$$

When solved graphically (or by computer), the following optimal solution is obtained: $z = 2,500$, $x_1 = 50$, $x_2 = 100$. Thus, Semicond can maximize the contribution of December's production to profits by manufacturing 50 tape recorders and 100 radios. This will contribute $20(50) + 15(100) = \$2,500$ to profits.

PROBLEMS

Group A

1 Graphically solve the Semicond problem.

2 Suppose that the January 1 inventory shipment had been valued at $7,000. Show that Semicond's LP is now infeasible.

3.8 Blending Problems

Situations in which various inputs must be blended in some desired proportion to produce goods for sale are often amenable to linear programming analysis. Such problems are called **blending problems.** The following list gives some situations in which linear programming has been used to solve blending problems.

1 Blending various types of crude oils to produce different types of gasoline and other outputs (such as heating oil)

2 Blending various chemicals to produce other chemicals

3 Blending various types of metal alloys to produce various types of steels

4 Blending various livestock feeds in an attempt to produce a minimum-cost feed mixture for cattle

5 Mixing various ores to obtain ore of a specified quality

6 Mixing various ingredients (meat, filler, water, and so on) to produce a product like bologna

7 Mixing various types of papers to produce recycled paper of varying quality

The following example illustrates the key ideas that are used in formulating LP models of blending problems.

EXAMPLE 12   **Oil Blending**

Sunco Oil manufactures three types of gasoline (gas 1, gas 2, and gas 3). Each type is produced by blending three types of crude oil (crude 1, crude 2, and crude 3). The sales price per barrel of gasoline and the purchase price per barrel of crude oil are given in Table 12. Sunco can purchase up to 5,000 barrels of each type of crude oil daily.

The three types of gasoline differ in their octane rating and sulfur content. The crude oil blended to form gas 1 must have an average octane rating of at least 10 and contain at most 1% sulfur. The crude oil blended to form gas 2 must have an average octane rating of at least 8 and contain at most 2% sulfur. The crude oil blended to form gas 3 must have an octane rating of at least 6 and contain at most 1% sulfur. The octane rating and the sulfur content of the three types of oil are given in Table 13. It costs $4 to transform one barrel of oil into one barrel of gasoline, and Sunco's refinery can produce up to 14,000 barrels of gasoline daily.

Sunco's customers require the following amounts of each gasoline: gas 1—3,000 barrels per day; gas 2—2,000 barrels per day; gas 3—1,000 barrels per day. The company considers it an obligation to meet these demands. Sunco also has the option of advertising to stimulate demand for its products. Each dollar spent daily in advertising a particular type of gas increases the daily demand for that type of gas by 10 barrels. For example, if Sunco decides to spend $20 daily in advertising gas 2, then the daily demand for gas 2 will increase by 20(10) = 200 barrels. Formulate an LP that will enable Sunco to maximize daily profits (profits = revenues − costs).

Solution   Sunco must make two types of decisions: first, how much money should be spent in advertising each type of gas, and second, how to blend each type of gasoline from the three types of crude oil available. For example, Sunco must decide how many barrels of crude 1 should be used to produce gas 1. We define the decision variables

a_i = dollars spent daily on advertising gas i ($i = 1, 2, 3$)

x_{ij} = barrels of crude oil i used daily to produce gas j ($i = 1, 2, 3; j = 1, 2, 3$)

For example, x_{21} is the number of barrels of crude 2 used each day to produce gas 1.

TABLE 12
Gas and Crude Oil Prices for Blending

Gas	Sales Price per Barrel ($)	Crude	Purchase Price per Barrel ($)
1	70	1	45
2	60	2	35
3	50	3	25

TABLE 13
Octane Ratings and Sulfur Requirements
for Blending

Crude	Octane Rating	Sulfur Content (%)
1	12	0.5
2	6	2.0
3	8	3.0

Knowledge of these variables is sufficient to determine Sunco's objective function and constraints, but before we do this, we note that the definition of the decision variables implies that

$$x_{11} + x_{12} + x_{13} = \text{barrels of crude 1 used daily}$$
$$x_{21} + x_{22} + x_{23} = \text{barrels of crude 2 used daily} \tag{38}$$
$$x_{31} + x_{32} + x_{33} = \text{barrels of crude 3 used daily}$$

$$x_{11} + x_{21} + x_{31} = \text{barrels of gas 1 produced daily}$$
$$x_{12} + x_{22} + x_{32} = \text{barrels of gas 2 produced daily} \tag{39}$$
$$x_{13} + x_{23} + x_{33} = \text{barrels of gas 3 produced daily}$$

To simplify matters, let's assume that gasoline cannot be stored, so it must be sold on the day it is produced. This implies that for $i = 1, 2, 3$, the amount of gas i produced daily should equal the daily demand for gas i. Suppose that the amount of gas i produced daily exceeded the daily demand. Then we would have incurred unnecessary purchasing and production costs. On the other hand, if the amount of gas i produced daily is less than the daily demand for gas i, then we are failing to meet mandatory demands or incurring unnecessary advertising costs.

We are now ready to determine Sunco's objective function and constraints. We begin with Sunco's objective function. From (39),

$$\text{Daily revenues from gas sales} = 70(x_{11} + x_{21} + x_{31}) + 60(x_{12} + x_{22} + x_{32})$$
$$+ 50(x_{13} + x_{23} + x_{33})$$

From (38),

$$\text{Daily cost of purchasing crude oil} = 45(x_{11} + x_{12} + x_{13}) + 35(x_{21} + x_{22} + x_{23})$$
$$+ 25(x_{31} + x_{32} + x_{33})$$

Also,

$$\text{Daily advertising costs} = a_1 + a_2 + a_3$$
$$\text{Daily production costs} = 4(x_{11} + x_{12} + x_{13} + x_{21} + x_{22} + x_{23} + x_{31} + x_{32} + x_{33})$$

Then,

$$\begin{aligned}
\text{Daily profit} = {}& \text{daily revenue from gas sales} \\
& - \text{daily cost of purchasing crude oil} \\
& - \text{daily advertising costs} - \text{daily production costs} \\
= {}& (70 - 45 - 4)x_{11} + (60 - 45 - 4)x_{12} + (50 - 45 - 4)x_{13} \\
& + (70 - 35 - 4)x_{21} + (60 - 35 - 4)x_{22} + (50 - 35 - 4)x_{23} \\
& + (70 - 25 - 4)x_{31} + (60 - 25 - 4)x_{32} \\
& + (50 - 25 - 4)x_{33} - a_1 - a_2 - a_3
\end{aligned}$$

Thus, Sunco's goal is to maximize

$$z = 21x_{11} + 11x_{12} + x_{13} + 31x_{21} + 21x_{22} + 11x_{23} + 41x_{31}$$
$$+ 31x_{32} + 21x_{33} - a_1 - a_2 - a_3 \tag{40}$$

Regarding Sunco's constraints, we see that the following 13 constraints must be satisfied:

Constraint 1 Gas 1 produced daily should equal its daily demand.

Constraint 2 Gas 2 produced daily should equal its daily demand.

Constraint 3 Gas 3 produced daily should equal its daily demand.

Constraint 4 At most 5,000 barrels of crude 1 can be purchased daily.

Constraint 5 At most 5,000 barrels of crude 2 can be purchased daily.

Constraint 6 At most 5,000 barrels of crude 3 can be purchased daily.

Constraint 7 Because of limited refinery capacity, at most 14,000 barrels of gasoline can be produced daily.

Constraint 8 Crude oil blended to make gas 1 must have an average octane level of at least 10.

Constraint 9 Crude oil blended to make gas 2 must have an average octane level of at least 8.

Constraint 10 Crude oil blended to make gas 3 must have an average octane level of at least 6.

Constraint 11 Crude oil blended to make gas 1 must contain at most 1% sulfur.

Constraint 12 Crude oil blended to make gas 2 must contain at most 2% sulfur.

Constraint 13 Crude oil blended to make gas 3 must contain at most 1% sulfur.

To express Constraint 1 in terms of decision variables, note that

$$\text{Daily demand for gas 1} = 3{,}000 + \text{gas 1 demand generated by advertising}$$

$$\text{Gas 1 demand generated by advertising} = \left(\frac{\text{gas 1 demand}}{\text{dollar spent}}\right)\left(\frac{\text{dollars}}{\text{spent}}\right)$$
$$= 10a_1^{\dagger}$$

Thus, daily demand for gas $1 = 3{,}000 + 10a_1$. Constraint 1 may now be written as

$$x_{11} + x_{21} + x_{31} = 3{,}000 + 10a_1 \tag{41'}$$

which we rewrite as

$$x_{11} + x_{21} + x_{31} - 10a_1 = 3{,}000 \tag{41}$$

Constraint 2 is expressed by

$$x_{12} + x_{22} + x_{32} - 10a_2 = 2{,}000 \tag{42}$$

---

†Many students believe that gas 1 demand generated by advertising should be written as $\frac{1}{10}a_1$. Analyzing the units of this term will show that this is not correct. $\frac{1}{10}$ has units of dollars spent per barrel of demand, and a_1 has units of dollars spent. Thus, the term $\frac{1}{10}a_1$ would have units of (dollars spent)^2 per barrel of demand. This cannot be correct!

Constraint 3 is expressed by

$$x_{13} + x_{23} + x_{33} - 10a_3 = 1,000 \tag{43}$$

From (38), Constraint 4 reduces to

$$x_{11} + x_{12} + x_{13} \leq 5,000 \tag{44}$$

Constraint 5 reduces to

$$x_{21} + x_{22} + x_{23} \leq 5,000 \tag{45}$$

Constraint 6 reduces to

$$x_{31} + x_{32} + x_{33} \leq 5,000 \tag{46}$$

Note that

Total gas produced = gas 1 produced + gas 2 produced + gas 3 produced

$$= (x_{11} + x_{21} + x_{31}) + (x_{12} + x_{22} + x_{32}) + (x_{13} + x_{23} + x_{33})$$

Then Constraint 7 becomes

$$x_{11} + x_{21} + x_{31} + x_{12} + x_{22} + x_{32} + x_{13} + x_{23} + x_{33} \leq 14,000 \tag{47}$$

To express Constraints 8–10, we must be able to determine the "average" octane level in a mixture of different types of crude oil. We assume that the octane levels of different crudes blend linearly. For example, if we blend two barrels of crude 1, three barrels of crude 2, and one barrel of crude 3, the average octane level in this mixture would be

$$\frac{\text{Total octane value in mixture}}{\text{Number of barrels in mixture}} = \frac{12(2) + 6(3) + 8(1)}{2 + 3 + 1} = \frac{50}{6} = 8\frac{1}{3}$$

Generalizing, we can express Constraint 8 by

$$\frac{\text{Total octane value in gas 1}}{\text{Gas 1 in mixture}} = \frac{12x_{11} + 6x_{21} + 8x_{31}}{x_{11} + x_{21} + x_{31}} \geq 10 \tag{48'}$$

Unfortunately, (48′) is not a linear inequality. To transform (48′) into a linear inequality, all we have to do is multiply both sides by the denominator of the left-hand side. The resulting inequality is

$$12x_{11} + 6x_{21} + 8x_{31} \geq 10(x_{11} + x_{21} + x_{31})$$

which may be rewritten as

$$2x_{11} - 4x_{21} - 2x_{31} \geq 0 \tag{48}$$

Similarly, Constraint 9 yields

$$\frac{12x_{12} + 6x_{22} + 8x_{32}}{x_{12} + x_{22} + x_{32}} \geq 8$$

Multiplying both sides of this inequality by $x_{12} + x_{22} + x_{32}$ and simplifying yields

$$4x_{12} - 2x_{22} \geq 0 \tag{49}$$

Because each type of crude oil has an octane level of 6 or higher, whatever we blend to manufacture gas 3 will have an average octane level of at least 6. This means that any values of the variables will satisfy Constraint 10. To verify this, we may express Constraint 10 by

$$\frac{12x_{13} + 6x_{23} + 8x_{33}}{x_{13} + x_{23} + x_{33}} \geq 6$$

Multiplying both sides of this inequality by $x_{13} + x_{23} + x_{33}$ and simplifying, we obtain

$$6x_{13} + 2x_{33} \geq 0 \tag{50}$$

Because $x_{13} \geq 0$ and $x_{33} \geq 0$ are always satisfied, (50) will automatically be satisfied and thus need not be included in the model. A constraint such as (50) that is implied by other constraints in the model is said to be a **redundant constraint** and need not be included in the formulation.

Constraint 11 may be written as

$$\frac{\text{Total sulfur in gas 1 mixture}}{\text{Number of barrels in gas 1 mixture}} \leq 0.01$$

Then, using the percentages of sulfur in each type of oil, we see that

$$\text{Total sulfur in gas 1 mixture} = \text{Sulfur in oil 1 used for gas 1}$$
$$+ \text{ sulfur in oil 2 used for gas 1}$$
$$+ \text{ sulfur in oil 3 used for gas 1}$$
$$= 0.005x_{11} + 0.02x_{21} + 0.03x_{31}$$

Constraint 11 may now be written as

$$\frac{0.005x_{11} + 0.02x_{21} + 0.03x_{31}}{x_{11} + x_{21} + x_{31}} \leq 0.01$$

Again, this is not a linear inequality, but we can multiply both sides of the inequality by $x_{11} + x_{21} + x_{31}$ and simplify, obtaining

$$-0.005x_{11} + 0.01x_{21} + 0.02x_{31} \leq 0 \tag{51}$$

Similarly, Constraint 12 is equivalent to

$$\frac{0.005x_{12} + 0.02x_{22} + 0.03x_{32}}{x_{12} + x_{22} + x_{32}} \leq 0.02$$

Multiplying both sides of this inequality by $x_{12} + x_{22} + x_{32}$ and simplifying yields

$$-0.015x_{12} + 0.01x_{32} \leq 0 \tag{52}$$

Finally, Constraint 13 is equivalent to

$$\frac{0.005x_{13} + 0.02x_{23} + 0.03x_{33}}{x_{13} + x_{23} + x_{33}} \leq 0.01$$

Multiplying both sides of this inequality by $x_{13} + x_{23} + x_{33}$ and simplifying yields the LP constraint

$$-0.005x_{13} + 0.01x_{23} + 0.02x_{33} \leq 0 \tag{53}$$

Combining (40)–(53), except the redundant constraint (50), with the sign restrictions $x_{ij} \geq 0$ and $a_i \geq 0$ yields an LP that may be expressed in tabular form (see Table 14). In Table 14, the first row (max) represents the objective function, the second row represents the first constraint, and so on. When solved on a computer, an optimal solution to Sunco's LP is found to be

$$z = 287{,}500$$

$$x_{11} = 2222.22 \qquad x_{12} = 2111.11 \qquad x_{13} = 666.67$$
$$x_{21} = 444.44 \qquad x_{22} = 4222.22 \qquad x_{23} = 333.34$$
$$x_{31} = 333.33 \qquad x_{32} = 3166.67 \qquad x_{33} = 0$$
$$a_1 = 0 \qquad a_2 = 750 \qquad a_3 = 0$$

TABLE **14**
Objective Function and Constraints for Blending

x_{11}	x_{12}	x_{13}	x_{21}	x_{22}	x_{23}	x_{31}	x_{32}	x_{33}	a_1	a_2	a_3	
21	11	1	31	21	11	41	31	21	−1	−1	−1	(max)
1	0	0	1	0	0	1	0	0	−10	0	0	= 3,000
0	1	0	0	1	0	0	1	0	0	−10	0	= 2,000
0	0	1	0	0	1	0	0	1	0	0	−10	= 1,000
1	1	1	0	0	0	0	0	0	0	0	0	≤ 5,000
0	0	0	1	1	1	0	0	0	0	0	0	≤ 5,000
0	0	0	0	0	0	1	1	1	0	0	0	≤ 5,000
1	1	1	1	1	1	1	1	1	0	0	0	≤ 14,000
2	0	0	−4	0	0	−2	0	0	0	0	0	≥ 0
0	4	0	0	−2	0	0	0	0	0	0	0	≥ 0
−0.005	0	0	0.01	0	0	0.02	0	0	0	0	0	≤ 0
0	−0.015	0	0	0	0	0	0.01	0	0	0	0	≤ 0
0	0	−0.005	0	0	0.01	0	0	0.02	0	0	0	≤ 0

Thus, Sunco should produce $x_{11} + x_{21} + x_{31} = 3,000$ barrels of gas 1, using 2222.22 barrels of crude 1, 444.44 barrels of crude 2, and 333.33 barrels of crude 3. The firm should produce $x_{12} + x_{22} + x_{32} = 9,500$ barrels of gas 2, using 2,111.11 barrels of crude 1, 4222.22 barrels of crude 2, and 3,166.67 barrels of crude 3. Sunco should also produce $x_{13} + x_{23} + x_{33} = 1,000$ barrels of gas 3, using 666.67 barrels of crude 1 and 333.34 barrels of crude 2. The firm should also spend \$750 on advertising gas 2. Sunco will earn a profit of \$287,500.

Observe that although gas 1 appears to be most profitable, we stimulate demand for gas 2, not gas 1. The reason for this is that given the quality (with respect to octane level and sulfur content) of the available crude, it is difficult to produce gas 1. Therefore, Sunco can make more money by producing more of the lower-quality gas 2 than by producing extra quantities of gas 1.

Modeling Issues

1 We have assumed that the quality level of a mixture is a **linear** function of each input used in the mixture. For example, we have assumed that if gas 3 is made with $\frac{2}{3}$ crude 1 and $\frac{1}{3}$ crude 2, then octane level for gas 3 $= (\frac{2}{3}) \cdot$ (octane level for crude 1) $+ (\frac{1}{3}) \cdot$ (octane level for crude 2). If the octane level of a gas is not a linear function of the fraction of each input used to produce the gas, then we no longer have a linear programming problem; we have a **nonlinear programming** problem. For example, let g_{i3} = fraction of gas 3 made with oil i. Suppose that the octane level for gas 3 is given by gas 3 octane level = $g_{13}^{.5} \cdot$ (oil 1 octane level) $+ g_{23}^{.4} \cdot$ (oil 2 octane level) $+ g_{33}^{.3} \cdot$ (oil 3 octane level). Then we do not have an LP problem. The reason for this is that the octane level of gas 3 is not a linear function of g_{13}, g_{23}, and g_{33}. We discuss nonlinear programming in Chapter 12.

2 In reality, a company using a blending model would run the model periodically (each day, say) and set production on the basis of the current inventory of inputs and current demand forecasts. Then the forecast levels and input levels would be updated, and the model would be run again to determine the next day's production.

Real-World Applications

Blending at Texaco

Texaco (see Dewitt et al., 1980) uses a nonlinear programming model (OMEGA) to plan and schedule its blending applications. The company's model is nonlinear because blend volatilities and octanes are nonlinear functions of the amount of each input used to produce a particular gasoline.

Blending in the Steel Industry

Fabian (1958) describes a complex LP model that can be used to optimize the production of iron and steel. For each product produced there are several blending constraints. For example, basic pig iron must contain at most 1.5% silicon, at most .05% sulphur, between .11% and .90% phosphorus, between .4% and 2% manganese, and between 4.1% and 4.4% carbon. See Problem 6 (in the Review Problems section) for a simple example of blending in the steel industry.

Blending in the Oil Industry

Many oil companies use LP to optimize their refinery operations. Problem 14 contains an example (based on Magoulas and Marinos-Kouris [1988]) of a blending model that can be used to maximize a refinery's profit.

PROBLEMS

Group A

1 You have decided to enter the candy business. You are considering producing two types of candies: Slugger Candy and Easy Out Candy, both of which consist solely of sugar, nuts, and chocolate. At present, you have in stock 100 oz of sugar, 20 oz of nuts, and 30 oz of chocolate. The mixture used to make Easy Out Candy must contain at least 20% nuts. The mixture used to make Slugger Candy must contain at least 10% nuts and 10% chocolate. Each ounce of Easy Out Candy can be sold for 25¢ , and each ounce of Slugger Candy for 20¢. Formulate an LP that will enable you to maximize your revenue from candy sales.

2 O.J. Juice Company sells bags of oranges and cartons of orange juice. O.J. grades oranges on a scale of 1 (poor) to 10 (excellent). O.J. now has on hand 100,000 lb of grade 9 oranges and 120,000 lb of grade 6 oranges. The average quality of oranges sold in bags must be at least 7, and the average quality of the oranges used to produce orange juice must be at least 8. Each pound of oranges that is used for juice yields a revenue of $1.50 and incurs a variable cost (consisting of labor costs, variable overhead costs, inventory costs, and so on) of $1.05. Each pound of oranges sold in bags yields a revenue of 50¢ and incurs a variable cost of 20¢. Formulate an LP to help O.J. maximize profit.

3 A bank is attempting to determine where its assets should be invested during the current year. At present, $500,000 is available for investment in bonds, home loans, auto loans, and personal loans. The annual rate of return on each type of investment is known to be: bonds, 10%; home loans, 16%; auto loans, 13%; personal loans, 20%. To ensure that the bank's portfolio is not too risky, the bank's investment manager has placed the following three restrictions on the bank's portfolio:

  a The amount invested in personal loans cannot exceed the amount invested in bonds.

  b The amount invested in home loans cannot exceed the amount invested in auto loans.

  c No more than 25% of the total amount invested may be in personal loans.

The bank's objective is to maximize the annual return on its investment portfolio. Formulate an LP that will enable the bank to meet this goal.

4 Young MBA Erica Cudahy may invest up to $1,000. She can invest her money in stocks and loans. Each dollar invested in stocks yields 10¢ profit, and each dollar invested in a loan yields 15¢ profit. At least 30% of all money invested must be in stocks, and at least $400 must be in loans. Formulate an LP that can be used to maximize total profit earned from Erica's investment. Then graphically solve the LP.

5 Chandler Oil Company has 5,000 barrels of oil 1 and 10,000 barrels of oil 2. The company sells two products: gasoline and heating oil. Both products are produced by combining oil 1 and oil 2. The quality level of each oil is as follows: oil 1—10; oil 2—5. Gasoline must have an average quality level of at least 8, and heating oil at least 6. Demand for each product must be created by advertising. Each dollar spent advertising gasoline creates 5 barrels of demand and each spent on heating oil creates 10 barrels of demand. Gasoline is sold for $25 per barrel, heating oil for $20. Formulate an LP to help Chandler maximize profit. Assume that no oil of either type can be purchased.

6 Bullco blends silicon and nitrogen to produce two types of fertilizers. Fertilizer 1 must be at least 40% nitrogen and sells for $70/lb. Fertilizer 2 must be at least 70% silicon and sells for $40/lb. Bullco can purchase up to 80 lb of nitrogen at $15/lb and up to 100 lb of silicon at $10/lb. Assuming that all fertilizer produced can be sold, formulate an LP to help Bullco maximize profits.

7 Eli Daisy uses chemicals 1 and 2 to produce two drugs. Drug 1 must be at least 70% chemical 1, and drug 2 must be at least 60% chemical 2. Up to 40 oz of drug 1 can be sold at $6 per oz; up to 30 oz of drug 2 can be sold at $5 per oz. Up to 45 oz of chemical 1 can be purchased at $6 per oz, and up to 40 oz of chemical 2 can be purchased at $4 per oz. Formulate an LP that can be used to maximize Daisy's profits.

8 Highland's TV-Radio Store must determine how many TVs and radios to keep in stock. A TV requires 10 sq ft of floorspace, whereas a radio requires 4 sq ft; 200 sq ft of floorspace is available. A TV will earn Highland $60 in profits, and a radio will earn $20. The store stocks only TVs and radios. Marketing requirements dictate that at least 60% of all appliances in stock be radios. Finally, a TV ties up $200 in capital, and a radio, $50. Highland wants to have at most $3,000 worth of capital tied up at any time. Formulate an LP that can be used to maximize Highland's profit.

9 Linear programming models are used by many Wall Street firms to select a desirable bond portfolio. The following is a simplified version of such a model. Solodrex is considering investing in four bonds; $1,000,000 is available for investment. The expected annual return, the worst-case annual return on each bond, and the "duration" of each bond are given in Table 15. The duration of a bond is a measure of the bond's sensitivity to interest rates. Solodrex wants to maximize the expected return from its bond investments, subject to three constraints.

Constraint 1 The worst-case return of the bond portfolio must be at least 8%.
Constraint 2 The average duration of the portfolio must be at most 6. For example, a portfolio that invested $600,000

in bond 1 and $400,000 in bond 4 would have an average duration of

$$\frac{600,000(3) + 400,000(9)}{1,000,000} = 5.4$$

Constraint 3 Because of diversification requirements, at most 40% of the total amount invested can be invested in a single bond.

Formulate an LP that will enable Solodrex to maximize the expected return on its investment.

10 Coalco produces coal at three mines and ships it to four customers. The cost per ton of producing coal, the ash and sulfur content (per ton) of the coal, and the production capacity (in tons) for each mine are given in Table 16. The number of tons of coal demanded by each customer are given in Table 17.

The cost (in dollars) of shipping a ton of coal from a mine to each customer is given in Table 18. It is required that the total amount of coal shipped contain at most 5% ash and at most 4% sulfur. Formulate an LP that minimizes the cost of meeting customer demands.

11 Eli Daisy produces the drug Rozac from four chemicals. Today they must produce 1,000 lb of the drug. The three active ingredients in Rozac are A, B, and C. By weight, at least 8% of Rozac must consist of A, at least 4% of B, and at least 2% of C. The cost per pound of each chemical and the amount of each active ingredient in 1 lb of each chemical are given in Table 19.

It is necessary that at least 100 lb of chemical 2 be used. Formulate an LP whose solution would determine the cheapest way of producing today's batch of Rozac.

TABLE 16

Mine	Production Cost ($)	Capacity	Ash Content (Tons)	Sulfur Content (Tons)
1	50	120	.08	.05
2	55	100	.06	.04
3	62	140	.04	.03

TABLE 17

Customer 1	Customer 2	Customer 3	Customer 4
80	70	60	40

TABLE 15

Bond	Expected Return (%)	Worst-Case Return (%)	Duration
1	13	6%	3
2	8	8%	4
3	12	10%	7
4	14	9%	9

TABLE 18

| Mine | Customer | | | |
	1	2	3	4
1	4	6	8	12
2	9	6	7	11
3	8	12	3	5

TABLE 19

Chemical	Cost ($ per Lb)	A	B	C
1	8	.03	.02	.01
2	10	.06	.04	.01
3	11	.10	.03	.04
4	14	.12	.09	.04

12 (A spreadsheet might be helpful on this problem.) The *risk index* of an investment can be obtained from return on investment (ROI) by taking the percentage of change in the value of the investment (in absolute terms) for each year, and averaging them.

Suppose you are trying to determine what percentage of your money should be invested in T-bills, gold, and stocks. In Table 20 (or File Inv68.xls) you are given the annual returns (change in value) for these investments for the years 1968–1988. Let the risk index of a portfolio be the weighted (according to the fraction of your money assigned to each investment) average of the risk index of each individual investment. Suppose that the amount of each investment must be between 20% and 50% of the total invested. You would like the risk index of your portfolio to equal .15, and your goal is to maximize the expected return on your portfolio. Formulate an LP whose solution will maximize the expected return on your portfolio, subject to the given constraints. Use the average return earned by each investment during the years 1968–1988 as your estimate of expected return.[†]

Group B

13 The owner of Sunco does not believe that our LP optimal solution will maximize daily profit. He reasons, "We have 14,000 barrels of daily refinery capacity, but your optimal solution produces only 13,500 barrels. Therefore, it cannot be maximizing profit." How would you respond?

14 Oilco produces two products: regular and premium gasoline. Each product contains .15 gram of lead per liter. The two products are produced from six inputs: reformate, fluid catalytic cracker gasoline (FCG), isomerate (ISO), polymer (POL), MTBE (MTB), and butane (BUT). Each input has four attributes:

Attribute 1 Research octane number (RON)
Attribute 2 RVP
Attribute 3 ASTM volatility at 70°C
Attribute 4 ASTM volatility at 130°C

The attributes and daily availability (in liters) of each input are given in Table 21.

The requirements for each output are given in Table 22.

The daily demand (in thousands of liters) for each product must be met, but more can be produced if desired. The RON and ASTM requirements are minimums. Regular gasoline sells for 29.49¢/liter, premium gasoline for 31.43¢. Before being ready for sale, .15 gram/liter of lead must be re-

moved from each product. The cost of removing .1 gram/liter is 8.5¢. At most 38% of each type of gasoline can consist of FCG. Formulate and solve an LP whose solution will tell Oilco how to maximize their daily profit.[‡]

TABLE 20

Year	Stocks	Gold	T-Bills
1968	11	11	5
1969	−9	8	7
1970	4	−14	7
1971	14	14	4
1972	19	44	4
1973	−15	66	7
1974	−27	64	8
1975	37	0	6
1976	24	−22	5
1977	−7	18	5
1978	7	31	7
1979	19	59	10
1980	33	99	11
1981	−5	−25	15
1982	22	4	11
1983	23	−11	9
1984	6	−15	10
1985	32	−12	8
1986	19	16	6
1987	5	22	5
1988	17	−2	6

TABLE 21

	Availability	RON	RVP	ASTM(70)	ASTM(130)
Reformate	15,572	98.9	7.66	−5	46
FCG	15,434	93.2	9.78	57	103
ISO	6,709	86.1	29.52	107	100
POL	1,190	97	14.51	7	73
MTB	748	117	13.45	98	100
BUT	Unlimited	98	166.99	130	100

TABLE 22

	Demand	RON	RVP	ASTM(70)	ASTM(130)
Regular	9.8	90	21.18	10	50
Premium	30	96	21.18	10	50

[†]Based on Chandy (1987).

[‡]Based on Magoulas and Marinos-Kouris (1988).

3.9 Production Process Models

We now explain how to formulate an LP model of a simple production process.[†] The key step is to determine how the outputs from a later stage of the process are related to the outputs from an earlier stage.

EXAMPLE 13   **Brute Production Process**

Rylon Corporation manufactures Brute and Chanelle perfumes. The raw material needed to manufacture each type of perfume can be purchased for $3 per pound. Processing 1 lb of raw material requires 1 hour of laboratory time. Each pound of processed raw material yields 3 oz of Regular Brute Perfume and 4 oz of Regular Chanelle Perfume. Regular Brute can be sold for $7/oz and Regular Chanelle for $6/oz. Rylon also has the option of further processing Regular Brute and Regular Chanelle to produce Luxury Brute, sold at $18/oz, and Luxury Chanelle, sold at $14/oz. Each ounce of Regular Brute processed further requires an additional 3 hours of laboratory time and $4 processing cost and yields 1 oz of Luxury Brute. Each ounce of Regular Chanelle processed further requires an additional 2 hours of laboratory time and $4 processing cost and yields 1 oz of Luxury Chanelle. Each year, Rylon has 6,000 hours of laboratory time available and can purchase up to 4,000 lb of raw material. Formulate an LP that can be used to determine how Rylon can maximize profits. Assume that the cost of the laboratory hours is a fixed cost.

Solution   Rylon must determine how much raw material to purchase and how much of each type of perfume should be produced. We therefore define our decision variables to be

$$x_1 = \text{number of ounces of Regular Brute sold annually}$$
$$x_2 = \text{number of ounces of Luxury Brute sold annually}$$
$$x_3 = \text{number of ounces of Regular Chanelle sold annually}$$
$$x_4 = \text{number of ounces of Luxury Chanelle sold annually}$$
$$x_5 = \text{number of pounds of raw material purchased annually}$$

Rylon wants to maximize

$$\text{Contribution to profit} = \text{revenues from perfume sales} - \text{processing costs}$$
$$- \text{ costs of purchasing raw material}$$
$$= 7x_1 + 18x_2 + 6x_3 + 14x_4 - (4x_2 + 4x_4) - 3x_5$$
$$= 7x_1 + 14x_2 + 6x_3 + 10x_4 - 3x_5$$

Thus, Rylon's objective function may be written as

$$\max z = 7x_1 + 14x_2 + 6x_3 + 10x_4 - 3x_5 \tag{54}$$

Rylon faces the following constraints:

Constraint 1   No more than 4,000 lb of raw material can be purchased annually.

Constraint 2   No more than 6,000 hours of laboratory time can be used each year.

Constraint 1 is expressed by

$$x_5 \leq 4,000 \tag{55}$$

[†]This section is based on Hartley (1971).

To express Constraint 2, note that

Total lab time used annually = time used annually to process raw material

+ time used annually to process Luxury Brute

+ time used annually to process Luxury Chanelle

$= x_5 + 3x_2 + 2x_4$

Then Constraint 2 becomes

$$3x_2 + 2x_4 + x_5 \le 6{,}000 \tag{56}$$

After adding the sign restrictions $x_i \ge 0$ ($i = 1, 2, 3, 4, 5$), many students claim that Rylon should solve the following LP:

$$\max z = 7x_1 + 14x_2 + 6x_3 + 10x_4 - 3x_5$$
$$\text{s.t.} \qquad x_5 \le 4{,}000$$
$$3x_2 + 2x_4 + x_5 \le 6{,}000$$
$$x_i \ge 0 \qquad (i = 1, 2, 3, 4, 5)$$

This formulation is incorrect. Observe that the variables x_1 and x_3 do not appear in any of the constraints. This means that any point with $x_2 = x_4 = x_5 = 0$ and x_1 and x_3 very large is in the feasible region. Points with x_1 and x_3 large can yield arbitrarily large profits. Thus, this LP is unbounded. Our mistake is that the current formulation does not indicate that the amount of raw material purchased determines the amount of Brute and Chanelle that is available for sale or further processing. More specifically, from Figure 10 (and the fact that 1 oz of processed Brute yields exactly 1 oz of Luxury Brute), it follows that

$$\begin{array}{l} \text{Ounces of Regular Brute Sold} \\ \text{+ ounces of Luxury Brute sold} \end{array} = \left(\frac{\text{ounces of Brute produced}}{\text{pound of raw material}} \right) \left(\begin{array}{l} \text{pounds of raw} \\ \text{material purchased} \end{array} \right)$$

$$= 3x_5$$

This relation is reflected in the constraint

$$x_1 + x_2 = 3x_5 \qquad \text{or} \qquad x_1 + x_2 - 3x_5 = 0 \tag{57}$$

Similarly, from Figure 10 it is clear that

Ounces of Regular Chanelle sold + ounces of Luxury Chanelle sold $= 4x_5$

This relation yields the constraint

$$x_3 + x_4 = 4x_5 \qquad \text{or} \qquad x_3 + x_4 - 4x_5 = 0 \tag{58}$$

Constraints (57) and (58) relate several decision variables. Students often omit constraints of this type. As this problem shows, leaving out even one constraint may very well

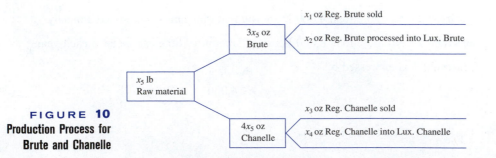

FIGURE **10**
Production Process for
Brute and Chanelle

lead to an unacceptable answer (such as an unbounded LP). If we combine (53)–(58) with the usual sign restrictions, we obtain the *correct* LP formulation.

$$\max z = 7x_1 + 14x_2 + 6x_3 + 10x_4 - 3x_5$$

$$\text{s.t.} \qquad\qquad\qquad\qquad\qquad\qquad x_5 \le 4{,}000$$

$$3x_2 \qquad\quad + 2x_4 + x_5 \le 6{,}000$$

$$x_1 + x_2 \qquad\qquad\quad - 3x_5 = 0$$

$$x_3 + x_4 - 4x_5 = 0$$

$$x_i \ge 0 \qquad (i = 1, 2, 3, 4, 5)$$

The optimal solution is $z = 172{,}666.667$, $x_1 = 11{,}333.333$ oz, $x_2 = 666.667$ oz, $x_3 = 16{,}000$ oz, $x_4 = 0$, and $x_5 = 4{,}000$ lb. Thus, Rylon should purchase all 4,000 lb of available raw material and produce 11,333.333 oz of Regular Brute, 666.667 oz of Luxury Brute, and 16,000 oz of Regular Chanelle. This production plan will contribute $172,666.667 to Rylon's profits. In this problem, a fractional number of ounces seems reasonable, so the Divisibility Assumption holds.

We close our discussion of the Rylon problem by discussing an error that is made by many students. They reason that

$$1 \text{ lb raw material} = 3 \text{ oz Brute} + 4 \text{ oz Chanelle}$$

Because $x_1 + x_2 =$ total ounces of Brute produced, and $x_3 + x_4 =$ total ounces of Chanelle produced, students conclude that

$$x_5 = 3(x_1 + x_2) + 4(x_3 + x_4) \qquad\qquad (59)$$

This equation might make sense as a statement for a computer program; in a sense, the variable x_5 is replaced by the right side of (59). As an LP constraint, however, (59) makes no sense. To see this, note that the left side has the units "pounds of raw material," and the term $3x_1$ on the right side has the units

$$\left(\frac{\text{Ounces of Brute}}{\text{Pounds of raw material}}\right) (\text{ounces of Brute})$$

Because some of the terms do not have the same units, (59) cannot be correct. *If there are doubts about a constraint, then make sure that all terms in the constraint have the same units.* This will avoid many formulation errors. (Of course, even if the units on both sides of a constraint are the same, the constraint may still be wrong.)

PROBLEMS

Group A

1 Sunco Oil has three different processes that can be used to manufacture various types of gasoline. Each process involves blending oils in the company's catalytic cracker. Running process 1 for an hour costs $5 and requires 2 barrels of crude oil 1 and 3 barrels of crude oil 2. The output from running process 1 for an hour is 2 barrels of gas 1 and 1 barrel of gas 2. Running process 2 for an hour costs $4 and requires 1 barrel of crude 1 and 3 barrels of crude 2. The output from running process 2 for an hour is 3 barrels of gas 2. Running process 3 for an hour costs $1 and requires 2 barrels of crude 2 and 3 barrels of gas 2. The output from running process 3 for an hour is 2 barrels of gas 3. Each week, 200 barrels of crude 1, at $2/barrel, and 300 barrels of crude 2, at $3/barrel, may be purchased. All gas produced can be sold at the following per-barrel prices: gas 1, $9; gas 2, $10; gas 3, $24. Formulate an LP whose solution will maximize revenues less costs. Assume that only 100 hours of time on the catalytic cracker are available each week.

2 Furnco manufactures tables and chairs. A table requires 40 board ft of wood, and a chair requires 30 board ft of wood. Wood may be purchased at a cost of $1 per board ft, and 40,000 board ft of wood are available for purchase. It takes 2 hours of skilled labor to manufacture an unfinished table or an unfinished chair. Three more hours of skilled labor will turn an unfinished table into a finished table, and 2 more hours of skilled labor will turn an unfinished chair into a finished chair. A total of 6,000 hours of skilled labor are available (and have already been paid for). All furniture produced can be sold at the following unit prices: unfinished table, $70; finished table, $140; unfinished chair, $60; finished chair, $110. Formulate an LP that will maximize the contribution to profit from manufacturing tables and chairs.

3 Suppose that in Example 11, 1 lb of raw material could be used to produce either 3 oz of Brute *or* 4 oz of Chanelle. How would this change the formulation?

4 Chemco produces three products: 1, 2, and 3. Each pound of raw material costs $25. It undergoes processing and yields 3 oz of product 1 and 1 oz of product 2. It costs $1 and takes 2 hours of labor to process each pound of raw material. Each ounce of product 1 can be used in one of three ways.

It can be sold for $10/oz.

It can be processed into 1 oz of product 2. This requires 2 hours of labor and costs $1.

It can be processed into 1 oz of product 3. This requires 3 hours of labor and costs $2.

Each ounce of product 2 can be used in one of two ways.

It can be sold for $20/oz.

It can be processed into 1 oz of product 3. This requires 1 hour of labor and costs $6.

Product 3 is sold for $30/oz. The maximum number of ounces of each product that can be sold is given in Table 23. A maximum of 25,000 hours of labor are available. Determine how Chemco can maximize profit.

TABLE 23

Product	Oz
1	5,000
2	5,000
3	3,000

Group B

5 A company produces A, B, and C and can sell these products in unlimited quantities at the following unit prices: A, $10; B, $56; C, $100. Producing a unit of A requires 1 hour of labor; a unit of B, 2 hours of labor plus 2 units of A; and a unit of C, 3 hours of labor plus 1 unit of B. Any A that is used to produce B cannot be sold. Similarly, any B that is used to produce C cannot be sold. A total of 40 hours of labor are available. Formulate an LP to maximize the company's revenues.

6 Daisy Drugs manufactures two drugs: 1 and 2. The drugs are produced by blending two chemicals: 1 and 2. By weight, drug 1 must contain at least 65% chemical 1, and drug 2 must contain at least 55% chemical 1. Drug 1 sells for $6/oz, and drug 2 sells for $4/oz. Chemicals 1 and 2 can be produced by one of two production processes. Running process 1 for an hour requires 3 oz of raw material and 2 hours skilled labor and yields 3 oz of each chemical. Running process 2 for an hour requires 2 oz of raw material and 3 hours of skilled labor and yields 3 oz of chemical 1 and 1 oz of chemical 2. A total of 120 hours of skilled labor and 100 oz of raw material are available. Formulate an LP that can be used to maximize Daisy's sales revenues.

7[†] Lizzie's Dairy produces cream cheese and cottage cheese. Milk and cream are blended to produce these two products. Both high-fat and low-fat milk can be used to produce cream cheese and cottage cheese. High-fat milk is 60% fat; low-fat milk is 30% fat. The milk used to produce cream cheese must average at least 50% fat and that for cottage cheese, at least 35% fat. At least 40% (by weight) of the inputs to cream cheese and at least 20% (by weight) of the inputs to cottage cheese must be cream. Both cottage cheese and cream cheese are produced by putting milk and cream through the cheese machine. It costs 40¢ to process 1 lb of inputs into a pound of cream cheese. It costs 40¢ to produce 1 lb of cottage cheese, but every pound of input for cottage cheese yields 0.9 lb of cottage cheese and 0.1 lb of waste. Cream can be produced by evaporating high-fat and low-fat milk. It costs 40¢ to evaporate 1 lb of high-fat milk. Each pound of high-fat milk that is evaporated yields 0.6 lb of cream. It costs 40¢ to evaporate 1 lb of low-fat milk. Each pound of low-fat milk that is evaporated yields 0.3 lb of cream. Each day, up to 3,000 lb of input may be sent through the cheese machine. Each day, at least 1,000 lb of cottage cheese and 1,000 lb of cream cheese must be produced. Up to 1,500 lb of cream cheese and 2,000 lb of cottage cheese can be sold each day. Cottage cheese is sold for $1.20/lb and cream cheese for $1.50/lb. High-fat milk is purchased for 80¢/lb and low-fat milk for 40¢/lb. The evaporator can process at most 2,000 lb of milk daily. Formulate an LP that can be used to maximize Lizzie's daily profit.

8 A company produces six products in the following fashion. Each unit of raw material purchased yields four units of product 1, two units of product 2, and one unit of product 3. Up to 1,200 units of product 1 can be sold, and up to 300 units of product 2 can be sold. Each unit of product 1 can be sold or processed further. Each unit of product 1 that is processed yields a unit of product 4. Demand for products 3 and 4 is unlimited. Each unit of product 2 can be sold or processed further. Each unit of product 2 that is processed further yields 0.8 unit of product 5 and 0.3 unit of product 6. Up to 1,000 units of product 5 can be sold, and up to 800 units of product 6 can be sold. Up to 3,000 units of raw material can be purchased at $6 per unit. Leftover units of products 5 and 6 must be destroyed. It costs $4 to destroy each leftover unit of product 5 and $3

[†]Based on Sullivan and Secrest (1985).

TABLE 24

Product	Sales Price ($)	Production Cost ($)
1	7	4
2	6	4
3	4	2
4	3	1
5	20	5
6	35	5

to destroy each leftover unit of product 6. Ignoring raw material purchase costs, the per-unit sales price and production costs for each product are shown in Table 24. Formulate an LP whose solution will yield a profit-maximizing production schedule.

9 Each week Chemco can purchase unlimited quantities of raw material at $6/lb. Each pound of purchased raw material can be used to produce either input 1 or input 2. Each pound of raw material can yield 2 oz of input 1, requiring 2 hours of processing time and incurring $2 in processing costs. Each pound of raw material can yield 3 oz of input 2, requiring 2 hours of processing time and incurring $4 in processing costs.

Two production processes are available. It takes 2 hours to run process 1, requiring 2 oz of input 1 and 1 oz of input 2. It costs $1 to run process 1. Each time process 1 is run 1 oz of product A and 1 oz of liquid waste are produced. Each time process 2 is run requires 3 hours of processing time, 2 oz of input 2 and 1 oz of input 1. Process 2 yields 1 oz of product B and .8 oz of liquid waste. Process 2 incurs $8 in costs.

Chemco can dispose of liquid waste in the Port Charles River or use the waste to produce product C or product D. Government regulations limit the amount of waste Chemco is allowed to dump into the river to 1,000 oz/week. One ounce of product C costs $4 to produce and sells for $11. One hour of processing time, 2 oz of input 1, and .8 oz of liquid waste are needed to produce an ounce of product C. One unit of product D costs $5 to produce and sells for $7. One hour of processing time, 2 oz of input 2, and 1.2 oz of liquid waste are needed to produce an ounce of product D.

At most 5,000 oz of product A and 5,000 oz of product B can be sold each week, but weekly demand for products C and D is unlimited. Product A sells for $18/oz and product B sells for $24/oz. Each week 6,000 hours of processing time is available. Formulate an LP whose solution will tell Chemco how to maximize weekly profit.

10 LIMECO owns a lime factory and sells six grades of lime (grades 1 through 6). The sales price per pound is given in Table 25. Lime is produced by kilns. If a kiln is run for an 8-hour shift, the amounts (in pounds) of each grade

TABLE 25

Grade	1	2	3	4	5	6
Price($)	12	14	10	18	20	25

TABLE 26

Grade	1	2	3	4	5	6
Amount produced	2	3	1	1.5	2	3

TABLE 27

Grade	1	2	3	4	5	6
Maximum demand	20	30	40	35	25	50

of lime given in Table 26 are produced. It costs $150 to run a kiln for an 8-hour shift. Each day the factory believes it can sell up to the amounts (in pounds) of lime given in Table 27.

Lime that is produced by the kiln may be reprocessed by using any one of the five processes described in Table 28.

For example, at a cost of $1/lb, a pound of grade 4 lime may be transformed into .5 lb of grade 5 lime and .5 lb of grade 6 lime.

Any extra lime leftover at the end of each day must be disposed of, with the disposal costs (per pound) given in Table 29.

Formulate an LP whose solution will tell LIMECO how to maximize their daily profit.

11 Chemco produces three products: A, B, and C. They can sell up to 30 pounds of each product at the following prices (per pound): product A, $10; product B, $12; product C, $20. Chemco purchases raw material at $5/lb. Each pound of raw material can be used to produce *either* 1 lb of A or 1 lb of B. For a cost of $3/lb processed, product A can be converted to .6 lb of product B and .4 lb of product C. For a cost of $2/lb processed, product B can be converted to .8 lb of product C. Formulate an LP whose solution will tell Chemco how to maximize their profit.

12 Chemco produces 3 chemicals: B, C, and D. They begin by purchasing chemical A for a cost of $6/100 liters. For an

TABLE 28

Input (1 Lb)	Output	Cost ($ per Lb of Input)
Grade 1	.3 lb Grade 3	
	.2 lb Grade 4	2
	.3 lb Grade 5	
	.2 lb Grade 6	
Grade 2	1 lb Grade 6	1
Grade 3	.8 lb Grade 4	1
Grade 4	.5 lb Grade 5	1
	.5 lb Grade 6	
Grade 5	.9 lb Grade 6	2

TABLE 29

Grade	1	2	3	4	5	6
Cost of Disposition ($)	3	2	3	2	4	2

additional cost of $3 and the use of 3 hours of skilled labor, 100 liters of A can be transformed into 40 liters of C and 60 liters of B. Chemical C can either be sold or processed further. It costs $1 and takes 1 hour of skilled labor to process 100 liters of C into 60 liters of D and 40 liters of B. For each chemical the sales price per 100 liters and the maximum amount (in 100s of liters) that can be sold are given in Table 30.

A maximum of 200 labor hours are available. Formulate an LP whose solution will tell Chemco how to maximize their profit.

13 Carrington Oil produces two types of gasoline, gas 1 and gas 2, from two types of crude oil, crude 1 and crude 2. Gas 1 is allowed to contain up to 4% impurities, and gas 2 is allowed to contain up to 3% impurities. Gas 1 sells for $8 per barrel, whereas gas 2 sells for $12 per barrel. Up to 4,200 barrels of gas 1 and up to 4,300 barrels of gas 2 can be sold. The cost per barrel of each crude, availability, and the level of impurities in each crude are as shown in Table 31. Before blending the crude oil into gas, any amount of each crude can be "purified" for a cost of $0.50 per barrel. Purification eliminates half the impurities in the crude oil. Determine how to maximize profit.

14 You have been put in charge of the Melrose oil refinery. The refinery produces gas and heating oil from crude oil. Gas sells for $8 per barrel and must have an average "grade level" of at least 9. Heating oil sells for $6 a barrel and must

have an average grade level of at least 7. At most, 2,000 barrels of gas and 600 barrels of heating oil can be sold. Incoming crude can be processed by one of three methods. The per barrel yield and per barrel cost of each processing method are shown in Table 32. For example, if we refine 1 barrel of incoming crude by method 1, it costs us $3.40 and yields .2 barrels of grade 6, .2 barrels of grade 8, and .6 barrels of grade 10.

Before being processed into gas and heating oil, processed grades 6 and 8 may be sent through the catalytic cracker to improve their quality. For $1.30 per barrel, a barrel of grade 6 may be "cracked" into a barrel of grade 8. For $2 per barrel, a barrel of grade 8 may be cracked into a barrel of grade 10. Any leftover processed or cracked oil that cannot be used for heating oil or gas must be disposed of at a cost of $0.20 per barrel. Determine how to maximize the refinery's profit.

TABLE 31

Oil	Cost per Barrel ($)	Impurity Level (%)	Availability (Barrels)
Crude 1	6	10%	5,000
Crude 2	8	2%	4,500

TABLE 30

	B	C	D
Price ($)	12	16	26
Maximum demand	30	60	40

TABLE 32

Method	Grade 6	Grade 8	Grade 10	Cost ($)
1	.2	.2	.6	3.40
2	.3	.3	.4	3.00
3	.4	.4	.2	2.60

3.10 Using Linear Programming to Solve Multiperiod Decision Problems: An Inventory Model

Up to this point, all the LP formulations we have discussed are examples of *static,* or *one-period, models.* In a static model, we assume that all decisions are made at a single point in time. The rest of the examples in this chapter show how linear programming can be used to determine optimal decisions in **multiperiod,** or **dynamic, models.** Dynamic models arise when the decision maker makes decisions at more than one point in time. In a dynamic model, decisions made during the current period influence decisions made during future periods. For example, consider a company that must determine how many units of a product should be produced during each month. If it produced a large number of units during the current month, this would reduce the number of units that should be produced during future months. The examples discussed in Sections 3.10–3.12 illustrate how earlier decisions affect later decisions. We will return to dynamic decision models when we study dynamic programming in Chapter 13.

EXAMPLE 14 | **Sailco Inventory**

Sailco Corporation must determine how many sailboats should be produced during each of the next four quarters (one quarter = three months). The demand during each of the next four quarters is as follows: first quarter, 40 sailboats; second quarter, 60 sailboats; third quarter, 75 sailboats; fourth quarter, 25 sailboats. Sailco must meet demands on time. At the beginning of the first quarter, Sailco has an inventory of 10 sailboats. At the beginning of each quarter, Sailco must decide how many sailboats should be produced during that quarter. For simplicity, we assume that sailboats manufactured during a quarter can be used to meet demand for that quarter. During each quarter, Sailco can produce up to 40 sailboats with regular-time labor at a total cost of $400 per sailboat. By having employees work overtime during a quarter, Sailco can produce additional sailboats with overtime labor at a total cost of $450 per sailboat.

At the end of each quarter (after production has occurred and the current quarter's demand has been satisfied), a carrying or holding cost of $20 per sailboat is incurred. Use linear programming to determine a production schedule to minimize the sum of production and inventory costs during the next four quarters.

Solution   For each quarter, Sailco must determine the number of sailboats that should be produced by regular-time and by overtime labor. Thus, we define the following decision variables:

x_t = number of sailboats produced by regular-time labor (at $400/boat) during quarter t ($t = 1, 2, 3, 4$)

y_t = number of sailboats produced by overtime labor (at $450/boat) during quarter t ($t = 1, 2, 3, 4$)

It is convenient to define decision variables for the inventory (number of sailboats on hand) at the end of each quarter:

i_t = number of sailboats on hand at end of quarter t ($t = 1, 2, 3, 4$)

Sailco's total cost may be determined from

Total cost = cost of producing regular-time boats

   + cost of producing overtime boats + inventory costs

= $400(x_1 + x_2 + x_3 + x_4) + 450(y_1 + y_2 + y_3 + y_4)$

   + $20(i_1 + i_2 + i_3 + i_4)$

Thus, Sailco's objective function is

$$\min z = 400x_1 + 400x_2 + 400x_3 + 400x_4 + 450y_1 + 450y_2$$
$$+ 450y_3 + 450y_4 + 20i_1 + 20i_2 + 20i_3 + 20i_4 \tag{60}$$

Before determining Sailco's constraints, we make two observations that will aid in formulating multiperiod production-scheduling models.

For quarter t,

Inventory at end of quarter t = inventory at end of quarter $(t - 1)$

   + quarter t production − quarter t demand

This relation plays a key role in formulating almost all multiperiod production-scheduling models. If we let d_t be the demand during period t (thus, $d_1 = 40$, $d_2 = 60$, $d_3 = 75$, and $d_4 = 25$), our observation may be expressed in the following compact form:

$$i_t = i_{t-1} + (x_t + y_t) - d_t \qquad (t = 1, 2, 3, 4) \tag{61}$$

In (61), i_0 = inventory at end of quarter 0 = inventory at beginning of quarter 1 = 10. For example, if we had 20 sailboats on hand at the end of quarter 2 (i_2 = 20) and produced 65 sailboats during quarter 3 (this means $x_3 + y_3$ = 65), what would be our ending third-quarter inventory? Simply the number of sailboats on hand at the end of quarter 2 plus the sailboats produced during quarter 3, less quarter 3's demand of 75. In this case, $i_3 = 20 + 65 - 75 = 10$, which agrees with (61). Equation (61) relates decision variables associated with different time periods. In formulating any multiperiod LP model, the hardest step is usually finding the relation (such as (61)) that relates decision variables from different periods.

We also note that quarter t's demand will be met on time if and only if (sometimes written *iff*) $i_t \geq 0$. To see this, observe that $i_{t-1} + (x_t + y_t)$ is available to meet period t's demand, so that period t's demand will be met if and only if

$$i_{t-1} + (x_t + y_t) \geq d_t \quad \text{or} \quad i_t = i_{t-1} + (x_t + y_t) - d_t \geq 0$$

This means that the sign restrictions $i_t \geq 0$ (t = 1, 2, 3, 4) will ensure that each quarter's demand will be met on time.

We can now determine Sailco's constraints. First, we use the following four constraints to ensure that each period's regular-time production will not exceed 40: $x_1, x_2, x_3, x_4 \leq 40$. Then we add constraints of the form (61) for each time period (t = 1, 2, 3, 4). This yields the following four constraints:

$$i_1 = 10 + x_1 + y_1 - 40 \qquad i_2 = i_1 + x_2 + y_2 - 60$$
$$i_3 = i_2 + x_3 + y_3 - 75 \qquad i_4 = i_3 + x_4 + y_4 - 25$$

Adding the sign restrictions $x_t \geq 0$ (to rule out negative production levels) and $i_t \geq 0$ (to ensure that each period's demand is met on time) yields the following formulation:

$$\min z = 400x_1 + 400x_2 + 400x_3 + 400x_4 + 450y_1 + 450y_2 + 450y_3 + 450y_4$$
$$+ \ 20i_1 + 20i_2 + 20i_3 + 20i_4$$

s.t. $\quad x_1 \leq 40, \qquad x_2 \leq 40, \qquad x_3 \leq 40, \qquad x_4 \leq 40$

$$\qquad i_1 = 10 + x_1 + y_1 - 40, \qquad i_2 = i_1 + x_2 + y_2 - 60$$
$$\qquad i_3 = i_2 + x_3 + y_3 - 75, \qquad i_4 = i_3 + x_4 + y_4 - 25$$
$$\qquad i_t \geq 0, \qquad y_t \geq 0, \qquad \text{and} \qquad x_t \geq 0 \qquad (t = 1, 2, 3, 4)$$

The optimal solution to this problem is z = 78,450; $x_1 = x_2 = x_3$ = 40; x_4 = 25; y_1 = 0; y_2 = 10; y_3 = 35; y_4 = 0; i_1 = 10; $i_2 = i_3 = i_4$ = 0. Thus, the minimum total cost that Sailco can incur is $78,450. To incur this cost, Sailco should produce 40 sailboats with regular-time labor during quarters 1–3 and 25 sailboats with regular-time labor during quarter 4. Sailco should also produce 10 sailboats with overtime labor during quarter 2 and 35 sailboats with overtime labor during quarter 3. Inventory costs will be incurred only during quarter 1.

Some readers might worry that our formulation allows Sailco to use overtime production during quarter t even if period t's regular production is less than 40. True, our formulation does not make such a schedule infeasible, but any production plan that had $y_t > 0$ and $x_t < 40$ could not be optimal. For example, consider the following two production schedules:

Production schedule A = $x_1 = x_2 = x_3$ = 40; $\qquad x_4$ = 25;

$\qquad\qquad\qquad\qquad\qquad y_2$ = 10; $\qquad y_3$ = 25; $\qquad y_4$ = 0

Production schedule B = x_1 = 40; $\qquad x_2$ = 30; $\qquad x_3$ = 30; $\qquad x_4$ = 25;

$\qquad\qquad\qquad\qquad\qquad y_2$ = 20; $\qquad y_3$ = 35; $\qquad y_4$ = 0

Schedules A and B both have the same production level during each period. This means that both schedules will have identical inventory costs. Also, both schedules are feasible, but schedule B incurs more overtime costs than schedule A. Thus, in minimizing costs, schedule B (or any schedule having $y_t > 0$ and $x_t < 40$) would never be chosen.

In reality, an LP such as Example 14 would be implemented by using a **rolling horizon,** which works in the following fashion. After solving Example 14, Sailco would implement only the quarter 1 production strategy (produce 40 boats with regular-time labor). Then the company would observe quarter 1's actual demand. Suppose quarter 1's actual demand is 35 boats. Then quarter 2 begins with an inventory of $10 + 40 - 35 = 15$ boats. We now make a forecast for quarter 5 demand (suppose the forecast is 36). Next determine production for quarter 2 by solving an LP in which quarter 2 is the first quarter, quarter 5 is the final quarter, and beginning inventory is 15 boats. Then quarter 2's production would be determined by solving the following LP:

$$\min z = 400(x_2 + x_3 + x_4 + x_5) + 450(y_2 + y_3 + y_4 + y_5) + 20(i_2 + i_3 + i_4 + i_5)$$

$$\text{s.t.} \quad x_2 \le 40, \quad x_3 \le 40, \quad x_4 \le 40, \quad x_5 \le 40$$

$$i_2 = 15 + x_2 + y_2 - 60, \quad i_3 = i_2 + x_3 + y_3 - 75$$

$$i_4 = i_3 + x_4 + y_4 - 25, \quad i_5 = i_4 + x_5 + y_5 - 36$$

$$i_t \ge 0, \quad y_t \ge 0, \quad \text{and} \quad x_t \ge 0 \quad (t = 2, 3, 4, 5)$$

Here, $x_5 =$ quarter 5's regular-time production, $y_5 =$ quarter 5's overtime production, and $i_5 =$ quarter 5's ending inventory. The optimal values of x_2 and y_2 for this LP are then used to determine quarter 2's production. Thus, each quarter, an LP (with a planning horizon of four quarters) is solved to determine the current quarter's production. Then current demand is observed, demand is forecasted for the next four quarters, and the process repeats itself. This technique of "rolling planning horizon" is the method by which most dynamic or multiperiod LP models are implemented in real-world applications.

Our formulation of the Sailco problem has several other limitations.

1 Production cost may not be a linear function of the quantity produced. This would violate the Proportionality Assumption. We discuss how to deal with this problem in Chapters 9 and 13.

2 Future demands may not be known with certainty. In this situation, the Certainty Assumption is violated.

3 We have required Sailco to meet all demands on time. Often companies can meet demands during later periods but are assessed a penalty cost for demands that are not met on time. For example, if demand is not met on time, then customer displeasure may result in a loss of future revenues. If demand can be met during later periods, then we say that demands can be **backlogged.** Our current LP formulation can be modified to incorporate backlogging (see Problem 1 of Section 4.12).

4 We have ignored the fact that quarter-to-quarter variations in the quantity produced may result in extra costs (called **production-smoothing costs.**) For example, if we increase production a great deal from one quarter to the next, this will probably require the costly training of new workers. On the other hand, if production is greatly decreased from one quarter to the next, extra costs resulting from laying off workers may be incurred. In Section 4.12, we modify the present model to account for smoothing costs.

5 If any sailboats are left at the end of the last quarter, we have assigned them a value of zero. This is clearly unrealistic. In any inventory model with a finite horizon, the inventory left at the end of the last period should be assigned a **salvage value** that is indicative of the worth of the final period's inventory. For example, if Sailco feels that each sailboat left at the end of quarter 4 is worth \$400, then a term $-400i_4$ (measuring the worth of quarter 4's inventory) should be added to the objective function.

PROBLEMS

Group A

1 A customer requires during the next four months, respectively, 50, 65, 100, and 70 units of a commodity (no backlogging is allowed). Production costs are $5, $8, $4, and $7 per unit during these months. The storage cost from one month to the next is $2 per unit (assessed on ending inventory). It is estimated that each unit on hand at the end of month 4 could be sold for $6. Formulate an LP that will minimize the net cost incurred in meeting the demands of the next four months.

2 A company faces the following demands during the next three periods: period 1, 20 units; period 2, 10 units; period 3, 15 units. The unit production cost during each period is as follows: period 1—$13; period 2—$14; period 3—$15. A holding cost of $2 per unit is assessed against each period's ending inventory. At the beginning of period 1, the company has 5 units on hand.

In reality, not all goods produced during a month can be used to meet the current month's demand. To model this fact, we assume that only one half of the goods produced during a period can be used to meet the current period's demands. Formulate an LP to minimize the cost of meeting the demand for the next three periods. (*Hint:* Constraints such as $i_1 = x_1 + 5 - 20$ are certainly needed. Unlike our example, however, the constraint $i_1 \geq 0$ will not ensure that period 1's demand is met. For example, if $x_1 = 20$, then $i_1 \geq 0$ will hold, but because only $\frac{1}{2}(20) = 10$ units of period 1 production can be used to meet period 1's demand, $x_1 = 20$ would not be feasible. Try to think of a type of constraint that will ensure that what is available to meet each period's demand is at least as large as that period's demand.)

Group B

3 James Beerd bakes cheesecakes and Black Forest cakes. During any month, he can bake at most 65 cakes. The costs per cake and the demands for cakes, which must be met on time, are listed in Table 33. It costs 50¢ to hold a cheesecake, and 40¢ to hold a Black Forest cake, in inventory for a month. Formulate an LP to minimize the total cost of meeting the next three months' demands.

4 A manufacturing company produces two types of products: A and B. The company has agreed to deliver the products on the schedule shown in Table 34. The company has two assembly lines, 1 and 2, with the available production hours shown in Table 35. The production rates for each assembly line and product combination, in terms of

TABLE 34

Date	A	B
March 31	5,000	2,000
April 30	8,000	4,000

TABLE 35

	Production Hours Available	
Month	Line 1	Line 2
March	800	2,000
April	400	1,200

TABLE 36

	Production Rate	
Product	Line 1	Line 2
A	0.15	0.16
B	0.12	0.14

hours per product, are shown in Table 36. It takes 0.15 hour to manufacture 1 unit of product A on line 1, and so on. It costs $5 per hour of line time to produce any product. The inventory carrying cost per month for each product is 20¢ per unit (charged on each month's ending inventory). Currently, there are 500 units of A and 750 units of B in inventory. Management would like at least 1,000 units of each product in inventory at the end of April. Formulate an LP to determine the production schedule that minimizes the total cost incurred in meeting demands on time.

5 During the next two months, General Cars must meet (on time) the following demands for trucks and cars: month 1—400 trucks, 800 cars; month 2—300 trucks, 300 cars. During each month, at most 1,000 vehicles can be produced. Each truck uses 2 tons of steel, and each car uses 1 ton of steel. During month 1, steel costs $400 per ton; during month 2, steel costs $600 per ton. At most, 1,500 tons of steel may be purchased each month (steel may only be used

TABLE 33

Item	Month 1 Demand	Month 1 Cost/Cake ($)	Month 2 Demand	Month 2 Cost/Cake ($)	Month 3 Demand	Month 3 Cost/Cake ($)
Cheesecake	40	3.00	30	3.40	20	3.80
Black Forest	20	2.50	30	2.80	10	3.40

during the month in which it is purchased). At the beginning of month 1, 100 trucks and 200 cars are in inventory. At the end of each month, a holding cost of $150 per vehicle is assessed. Each car gets 20 mpg, and each truck gets 10 mpg. During each month, the vehicles produced by the company must average at least 16 mpg. Formulate an LP to meet the demand and mileage requirements at minimum cost (include steel costs and holding costs).

6 Gandhi Clothing Company produces shirts and pants. Each shirt requires 2 sq yd of cloth, each pair of pants, 3. During the next two months, the following demands for shirts and pants must be met (on time): month 1—10 shirts, 15 pairs of pants; month 2—12 shirts, 14 pairs of pants. During each month, the following resources are available: month 1—90 sq yd of cloth; month 2—60 sq yd. (Cloth that is available during month 1 may, if unused during month 1, be used during month 2.)

During each month, it costs $4 to make an article of clothing with regular-time labor and $8 with overtime labor. During each month, a total of at most 25 articles of clothing may be produced with regular-time labor, and an unlimited number of articles of clothing may be produced with overtime labor. At the end of each month, a holding cost of $3 per article of clothing is assessed. Formulate an LP that can be used to meet demands for the next two months (on time) at minimum cost. Assume that at the beginning of month 1, 1 shirt and 2 pairs of pants are available.

7 Each year, Paynothing Shoes faces demands (which must be met on time) for pairs of shoes as shown in Table 37. Workers work three consecutive quarters and then receive one quarter off. For example, a worker may work during quarters 3 and 4 of one year and quarter 1 of the next year. During a quarter in which a worker works, he or she can produce up to 50 pairs of shoes. Each worker is paid $500 per quarter. At the end of each quarter, a holding cost of $50 per pair of shoes is assessed. Formulate an LP that can be used to minimize the cost per year (labor + holding) of meeting the demands for shoes. To simplify matters, assume

TABLE 37

Quarter 1	Quarter 2	Quarter 3	Quarter 4
600	300	800	100

that at the end of each year, the ending inventory is zero. (*Hint:* It is allowable to assume that a given worker will get the same quarter off during each year.)

8 A company must meet (on time) the following demands: quarter 1—30 units; quarter 2—20 units; quarter 3—40 units. Each quarter, up to 27 units can be produced with regular-time labor, at a cost of $40 per unit. During each quarter, an unlimited number of units can be produced with overtime labor, at a cost of $60 per unit. Of all units produced, 20% are unsuitable and cannot be used to meet demand. Also, at the end of each quarter, 10% of all units on hand spoil and cannot be used to meet any future demands. After each quarter's demand is satisfied and spoilage is accounted for, a cost of $15 per unit is assessed against the quarter's ending inventory. Formulate an LP that can be used to minimize the total cost of meeting the next three quarters' demands. Assume that 20 usable units are available at the beginning of quarter 1.

9 Donovan Enterprises produces electric mixers. During the next four quarters, the following demands for mixers must be met on time: quarter 1—4,000; quarter 2—2,000; quarter 3—3,000; quarter 4—10,000. Each of Donovan's workers works three quarters of the year and gets one quarter off. Thus, a worker may work during quarters 1, 2, and 4 and get quarter 3 off. Each worker is paid $30,000 per year and (if working) can produce up to 500 mixers during a quarter. At the end of each quarter, Donovan incurs a holding cost of $30 per mixer on each mixer in inventory. Formulate an LP to help Donovan minimize the cost (labor and inventory) of meeting the next year's demand (on time). At the beginning of quarter 1, 600 mixers are available.

3.11 Multiperiod Financial Models

The following example illustrates how linear programming can be used to model multiperiod cash management problems. The key is to determine the relations of cash on hand during different periods.

EXAMPLE 15   **Finco Multiperiod Investment**

Finco Investment Corporation must determine investment strategy for the firm during the next three years. Currently (time 0), $100,000 is available for investment. Investments A, B, C, D, and E are available. The cash flow associated with investing $1 in each investment is given in Table 38.

For example, $1 invested in investment B requires a $1 cash outflow at time 1 and returns 50¢ at time 2 and $1 at time 3. To ensure that the company's portfolio is diversified, Finco requires that at most $75,000 be placed in any single investment. In addition to investments A–E, Finco can earn interest at 8% per year by keeping uninvested cash in

TABLE **38**

	Cash Flow ($) at Time*			
	0	**1**	**2**	**3**
A	−1	+0.50	+1	0
B	0	−1	+0.50	+1
C	−1	+1.2	0	0
D	−1	0	0	+1.9
E	0	0	−1	+1.5

Note: Time 0 = present; time 1 = 1 year from now; time 2 = 2 years from now; time 3 = 3 years from now.

money market funds. Returns from investments may be immediately reinvested. For example, the positive cash flow received from investment C at time 1 may immediately be reinvested in investment B. Finco cannot borrow funds, so the cash available for investment at any time is limited to cash on hand. Formulate an LP that will maximize cash on hand at time 3.

Solution   Finco must decide how much money should be placed in each investment (including money market funds). Thus, we define the following decision variables:

A = dollars invested in investment A

B = dollars invested in investment B

C = dollars invested in investment C

D = dollars invested in investment D

E = dollars invested in investment E

S_t = dollars invested in money market funds at time t $(t = 0, 1, 2)$

Finco wants to maximize cash on hand at time 3. At time 3, Finco's cash on hand will be the sum of all cash inflows at time 3. From the description of investments A–E and the fact that from time 2 to time 3, S_2 will increase to $1.08S_2$,

$$\text{Time 3 cash on hand} = B + 1.9D + 1.5E + 1.08S_2$$

Thus, Finco's objective function is

$$\max z = B + 1.9D + 1.5E + 1.08S_2 \tag{62}$$

In multiperiod financial models, the following type of constraint is usually used to relate decision variables from different periods:

Cash available at time t = cash invested at time t

+ uninvested cash at time t that is carried over to time $t + 1$

If we classify money market funds as investments, we see that

$$\text{Cash available at time } t = \text{cash invested at time } t \tag{63}$$

Because investments A, C, D, and S_0 are available at time 0, and $100,000 is available at time 0, (63) for time 0 becomes

$$100,000 = A + C + D + S_0 \tag{64}$$

At time 1, $0.5A + 1.2C + 1.08S_0$ is available for investment, and investments B and S_1 are available. Then for $t = 1$, (63) becomes

$$0.5A + 1.2C + 1.08S_0 = B + S_1 \tag{65}$$

At time 2, $A + 0.5B + 1.08S_1$ is available for investment, and investments E and S_2 are available. Thus, for $t = 2$, (63) reduces to

$$A + 0.5B + 1.08S_1 = E + S_2 \tag{66}$$

Let's not forget that at most \$75,000 can be placed in any of investments A–E. To take care of this, we add the constraints

$$A \leq 75,000 \tag{67}$$
$$B \leq 75,000 \tag{68}$$
$$C \leq 75,000 \tag{69}$$
$$D \leq 75,000 \tag{70}$$
$$E \leq 75,000 \tag{71}$$

Combining (62) and (64)–(71) with the sign restrictions (all variables ≥ 0) yields the following LP:

$$
\begin{aligned}
\max z = {} & B + 1.9D + 1.5E + 1.08S_2 \\
\text{s.t.} \quad & A + C + D + S_0 = 100,000 \\
& 0.5A + 1.2C + 1.08S_0 = B + S_1 \\
& A + 0.5B + 1.08S_1 = E + S_2 \\
& A \leq 75,000 \\
& B \leq 75,000 \\
& C \leq 75,000 \\
& D \leq 75,000 \\
& E \leq 75,000 \\
& A, B, C, D, E, S_0, S_1, S_2 \geq 0
\end{aligned}
$$

We find the optimal solution to be $z = 218,500$, $A = 60,000$, $B = 30,000$, $D = 40,000$, $E = 75,000$, $C = S_0 = S_1 = S_2 = 0$. Thus, Finco should not invest in money market funds. At time 0, Finco should invest \$60,000 in A and \$40,000 in D. Then, at time 1, the \$30,000 cash inflow from A should be invested in B. Finally, at time 2, the \$60,000 cash inflow from A and the \$15,000 cash inflow from B should be invested in E. At time 3, Finco's \$100,000 will have grown to \$218,500.

You might wonder how our formulation ensures that Finco never invests more money at any time than the firm has available. This is ensured by the fact that each variable S_i must be nonnegative. For example, $S_0 \geq 0$ is equivalent to $100,000 - A - C - D \geq 0$, which ensures that at most \$100,000 will be invested at time 0.

Real-World Application

Using LP to Optimize Bond Portfolios

Many Wall Street firms buy and sell bonds. Rohn (1987) discusses a bond selection model that maximizes profit from bond purchases and sales subject to constraints that minimize the firm's risk exposure. See Problem 4 for a simplified version of this model.

PROBLEMS

Group A

1 A consultant to Finco claims that Finco's cash on hand at time 3 is the sum of the cash inflows from all investments, not just those investments yielding a cash inflow at time 3. Thus, the consultant claims that Finco's objective function should be

$$\max z = 1.5A + 1.5B + 1.2C + 1.9D + 1.5E$$
$$+ 1.08S_0 + 1.08S_1 + 1.08S_2$$

Explain why the consultant is incorrect.

2 Show that Finco's objective function may also be written as

$$\max z = 100,000 + 0.5A + 0.5B + 0.2C + 0.9D + 0.5E$$
$$+ 0.08S_0 + 0.08S_1 + 0.08S_2$$

3 At time 0, we have $10,000. Investments A and B are available; their cash flows are shown in Table 39. Assume that any money not invested in A or B earns *no* interest. Formulate an LP that will maximize cash on hand at time 3. Can you guess the optimal solution to this problem?

Group B

4[†] Broker Steve Johnson is currently trying to maximize his profit in the bond market. Four bonds are available for purchase and sale, with the bid and ask price of each bond as shown in Table 40. Steve can buy up to 1,000 units of each bond at the ask price or sell up to 1,000 units of each bond at the bid price. During each of the next three years, the person who sells a bond will pay the owner of the bond the cash payments shown in Table 41.

Steve's goal is to maximize his revenue from selling bonds less his payment for buying bonds, subject to the constraint that after each year's payments are received, his current cash position (due only to cash payments from bonds and not purchases or sale of bonds) is nonnegative. Assume

TABLE 39

Time	A	B
0	−$1	$0
1	$0.2	−$1
2	$1.5	$0
3	$0	$1.0

TABLE 40

Bond	Bid Price	Ask Price
1	980	990
2	970	985
3	960	972
4	940	954

[†]Based on Rohn (1987).

TABLE 41

Year	Bond 1	Bond 2	Bond 3	Bond 4
1	100	80	70	60
2	110	90	80	50
3	1,100	1,120	1,090	1,110

TABLE 42

Month	Cash Flow	Month	Cash Flow
January	−12	July	−7
February	−10	August	−2
March	−8	September	15
April	−10	October	12
May	−4	November	−7
June	5	December	45

that cash payments are discounted, with a payment of $1 one year from now being equivalent to a payment of 90¢ now. Formulate an LP to maximize net profit from buying and selling bonds, subject to the arbitrage constraints previously described. Why do you think we limit the number of units of each bond that can be bought or sold?

5 A small toy store, Toyco projects the monthly cash flows (in thousands of dollars) in Table 42 during the year 2003. A negative cash flow means that cash outflows exceed cash inflows to the business. To pay its bills, Toyco will need to borrow money early in the year. Money can be borrowed in two ways:

a Taking out a long-term one-year loan in January. Interest of 1% is charged each month, and the loan must be paid back at the end of December.

b Each month money can be borrowed from a short-term bank line of credit. Here, a monthly interest rate of 1.5% is charged. All short-term loans must be paid off at the end of December.

At the end of each month, excess cash earns 0.4% interest. Formulate an LP whose solution will help Toyco maximize its cash position at the beginning of January, 2004.

6 Consider Problem 5 with the following modification: Each month Toyco can delay payments on some or all of the cash owed for the current month. This is called "stretching payments." Payments may be stretched for only one month, and a 1% penalty is charged on the amount stretched. Thus, if it stretches payments on $10,000 cash owed in January, then it must pay $10,000(1.01) = $10,100 in February. With this modification, formulate an LP that would help Toyco maximize its cash on hand at the beginning of January 1, 2004.

7 Suppose we are borrowing $1,000 at 12% annual interest with 60 monthly payments. Assume equal payments are made at the end of month 1, month 2, . . . month 60. We know that entering into Excel the function

$$= \text{PMT}(.01, 60, 1,000)$$

would yield the monthly payment ($22.24).

It is instructive to use LP to determine the montly payment. Let p be the (unknown) monthly payment. Each month we owe $.01 \cdot$ (our current unpaid balance) in interest. The remainder of our monthly payment is used to reduce the unpaid balance. For example, suppose we paid $30 each month. At the beginning of month 1, our unpaid balance is $1,000. Of our month 1 payment, $10 goes to interest and $20 to paying off the unpaid balance. Then we would begin month 2 with an unpaid balance of $980. The trick is to use LP to determine the monthly payment that will pay off the loan at the end of month 60.

8 You are a CFA (chartered financial analyst). Madonna has come to you because she needs help paying off her credit card bills. She owes the amounts on her credit cards shown in Table 43. Madonna is willing to allocate up to $5,000 per month to pay off these credit cards. All cards must be paid off within 36 months. Madonna's goal is to minimize the total of all her payments. To solve this problem, you must understand how interest on a loan works. To illustrate, suppose Madonna pays $5,000 on Saks during month 1. Then her Saks balance at the beginning of month 2 is

$$20,000 - (5000 - .005(20,000))$$

This follows because during month 1 Madonna incurs .005(20,000) in interest charges on her Saks card. Help Madonna solve her problems!

9 Winstonco is considering investing in three projects. If we fully invest in a project, the realized cash flows (in millions of dollars) will be as shown in Table 44. For example, project 1 requires cash outflow of $3 million today

TABLE 43

Card	Balance ($)	Monthly Rate (%)
Saks Fifth Avenue	20,000	.5
Bloomingdale's	50,000	1
Macys	40,000	1.5

TABLE 44

Time (Years)	Cash Flow		
	Project 1	Project 2	Project 3
0	−3	−2	−2
.5	−1	−.5	−2
1	+1.8	1.5	−1.8
1.5	1.4	1.5	1
2	1.8	1.5	1
2.5	1.8	.2	1
3	5.5	−1	6

and returns $5.5 million 3 years from now. Today we have $2 million in cash. At each time point (0, .5, 1, 1.5, 2, and 2.5 years from today) we may, if desired, borrow up to $2 million at 3.5% (per 6 months) interest. Leftover cash earns 3% (per 6 months) interest. For example, if after borrowing and investing at time 0 we have $1 million we would receive $30,000 in interest at time .5 years. Winstonco's goal is to maximize cash on hand after it accounts for time 3 cash flows. What investment and borrowing strategy should be used? Remember that we may invest in a fraction of a project. For example, if we invest in .5 of project 3, then we have cash outflows of −$1 million at time 0 and .5.

3.12 Multiperiod Work Scheduling

In Section 3.5, we saw that linear programming could be used to schedule employees in a static environment where demand did not change over time. The following example (a modified version of a problem from Wagner [1975]) shows how LP can be used to schedule employee training when a firm faces demand that changes over time.

EXAMPLE 16   Multiperiod Work Scheduling

CSL is a chain of computer service stores. The number of hours of skilled repair time that CSL requires during the next five months is as follows:

Month 1 (January): 6,000 hours

Month 2 (February): 7,000 hours

Month 3 (March): 8,000 hours

Month 4 (April): 9,500 hours

Month 5 (May): 11,000 hours

At the beginning of January, 50 skilled technicians work for CSL. Each skilled technician can work up to 160 hours per month. To meet future demands, new technicians must be trained. It takes one month to train a new technician. During the month of training, a trainee must be supervised for 50 hours by an experienced technician. Each experienced technician is paid $2,000 a month (even if he or she does not work the full 160 hours). During the month of training, a trainee is paid $1,000 a month. At the end of each month, 5% of CSL's experienced technicians quit to join Plum Computers. Formulate an LP whose solution will enable CSL to minimize the labor cost incurred in meeting the service requirements for the next five months.

Solution CSL must determine the number of technicians who should be trained during month t ($t = 1, 2, 3, 4, 5$). Thus, we define

$$x_t = \text{number of technicians trained during month } t \qquad (t = 1, 2, 3, 4, 5)$$

CSL wants to minimize total labor cost during the next five months. Note that

Total labor cost = cost of paying trainees + cost of paying experienced technicians

To express the cost of paying experienced technicians, we need to define, for $t = 1, 2, 3, 4, 5$,

$$y_t = \text{number of experienced technicians at the beginning of month } t$$

Then

$$\text{Total labor cost} = (1{,}000x_1 + 1{,}000x_2 + 1{,}000x_3 + 1{,}000x_4 + 1{,}000x_5)$$
$$+ (2{,}000y_1 + 2000y_2 + 2{,}000y_3 + 2{,}000y_4 + 2{,}000y_5)$$

Thus, CSL's objective function is

$$\min z = 1{,}000x_1 + 1{,}000x_2 + 1{,}000x_3 + 1{,}000x_4 + 1{,}000x_5$$
$$+ 2{,}000y_1 + 2{,}000y_2 + 2{,}000y_3 + 2{,}000y_4 + 2{,}000y_5$$

What constraints does CSL face? Note that we are given $y_1 = 50$, and that for $t = 1, 2, 3, 4, 5$, CSL must ensure that

Number of available technician hours during month t

$\geq$ Number of technician hours required during month t **(72)**

Because each trainee requires 50 hours of experienced technician time, and each skilled technician is available for 160 hours per month,

Number of available technician hours during month $t = 160y_t - 50x_t$

Now (72) yields the following five constraints:

$$160y_1 - 50x_1 \geq 6{,}000 \qquad \text{(month 1 constraint)}$$
$$160y_2 - 50x_2 \geq 7{,}000 \qquad \text{(month 2 constraint)}$$
$$160y_3 - 50x_3 \geq 8{,}000 \qquad \text{(month 3 constraint)}$$
$$160y_4 - 50x_4 \geq 9{,}500 \qquad \text{(month 4 constraint)}$$
$$160y_5 - 50x_5 \geq 11{,}000 \qquad \text{(month 5 constraint)}$$

As in the other multiperiod formulations, we need constraints that relate variables from different periods. In the CSL problem, it is important to realize that the number of skilled technicians available at the beginning of any month is determined by the number of skilled technicians available during the previous month and the number of technicians trained during the previous month:

Experienced technicians available at beginning of month t = Experienced technicians available at beginning of month $(t-1)$

+ technicians trained during month $(t-1)$

− experienced technicians who quit during month $(t-1)$

(73)

For example, for February, (73) yields

$$y_2 = y_1 + x_1 - 0.05y_1 \quad \text{or} \quad y_2 = 0.95y_1 + x_1$$

Similarly, for March, (73) yields

$$y_3 = 0.95y_2 + x_2$$

and for April,

$$y_4 = 0.95y_3 + x_3$$

and for May,

$$y_5 = 0.95y_4 + x_4$$

Adding the sign restrictions $x_t \geq 0$ and $y_t \geq 0$ ($t = 1, 2, 3, 4, 5$), we obtain the following LP:

$$\min z = 1{,}000x_1 + 1{,}000x_2 + 1{,}000x_3 + 1{,}000x_4 + 1{,}000x_5$$
$$+ 2{,}000y_1 + 2{,}000y_2 + 2{,}000y_3 + 2{,}000y_4 + 2{,}000y_5$$

s.t.

$$160y_1 - 50x_1 \geq 6{,}000 \qquad y_1 = 50$$
$$160y_2 - 50x_2 \geq 7{,}000 \qquad 0.95y_1 + x_1 = y_2$$
$$160y_3 - 50x_3 \geq 8{,}000 \qquad 0.95y_2 + x_2 = y_3$$
$$160y_4 - 50x_4 \geq 9{,}500 \qquad 0.95y_3 + x_3 = y_4$$
$$160y_5 - 50x_5 \geq 11{,}000 \qquad 0.95y_4 + x_4 = y_5$$
$$x_t, y_t \geq 0 \qquad (t = 1, 2, 3, 4, 5)$$

The optimal solution is $z = 593{,}777$; $x_1 = 0$; $x_2 = 8.45$; $x_3 = 11.45$; $x_4 = 9.52$; $x_5 = 0$; $y_1 = 50$; $y_2 = 47.5$; $y_3 = 53.58$; $y_4 = 62.34$; and $y_5 = 68.75$.

In reality, the y_t's must be integers, so our solution is difficult to interpret. The problem with our formulation is that assuming that exactly 5% of the employees quit each month can cause the number of employees to change from an integer during one month to a fraction during the next month. We might want to assume that the number of employees quitting each month is the integer closest to 5% of the total workforce, but then we do not have a linear programming problem!

PROBLEMS

Group A

1 If $y_1 = 38$, then what would be the optimal solution to CSL's problem?

2 An insurance company believes that it will require the following numbers of personal computers during the next six months: January, 9; February, 5; March, 7; April, 9; May, 10; June, 5. Computers can be rented for a period of one, two, or three months at the following unit rates: one-month rate, $200; two-month rate, $350; three-month rate, $450. Formulate an LP that can be used to minimize the cost of renting the required computers. You may assume that if a machine is rented for a period of time extending beyond June, the cost of the rental should be prorated. For example, if a computer is rented for three months at the beginning of May, then a rental fee of $\frac{2}{3}(450) = \$300$, not $450, should be assessed in the objective function.

3 The IRS has determined that during each of the next 12 months it will need the number of supercomputers given in Table 45. To meet these requirements, the IRS rents

TABLE 45

Month	Computer Requirements
1	800
2	1,000
3	600
4	500
5	1,200
6	400
7	800
8	600
9	400
10	500
11	800
12	600

TABLE 46

Month	Selling Price ($)	Purchase Price ($)
1	3	8
2	6	8
3	7	2
4	1	3
5	4	4
6	5	3
7	5	3
8	1	2
9	3	5
10	2	5

supercomputers for a period of one, two, or three months. It costs $100 to rent a supercomputer for one month, $180 for two months, and $250 for three months. At the beginning of month 1, the IRS has no supercomputers. Determine the rental plan that meets the next 12 months' requirements at minimum cost. *Note:* You may assume that fractional rentals are okay, so if your solution says to rent 140.6 computers for one month we can round this up or down (to 141 or 140) without having much effect on the total cost.

Group B

4 You own a wheat warehouse with a capacity of 20,000 bushels. At the beginning of month 1, you have 6,000 bushels

of wheat. Each month, wheat can be bought and sold at the price per 1000 bushels given in Table 46.

The sequence of events during each month is as follows:

a You observe your initial stock of wheat.

b You can sell any amount of wheat up to your initial stock at the current month's selling price.

c You can buy (at the current month's buying price) as much wheat as you want, subject to the warehouse size limitation.

Your goal is to formulate an LP that can be used to determine how to maximize the profit earned over the next 10 months.

SUMMARY   Linear Programming Definitions

A **linear programming problem (LP)** consists of three parts:

1 A linear function (the **objective function**) of decision variables (say, $x_1, x_2, \ldots, x_n$) that is to be maximized or minimized.

2 A set of **constraints** (each of which must be a linear equality or linear inequality) that restrict the values that may be assumed by the decision variables.

3 The **sign restrictions,** which specify for each decision variable x_j either (1) variable x_j must be nonnegative—$x_j \geq 0$; or (2) variable x_j may be positive, zero, or negative—x_j is **unrestricted in sign (urs).**

The coefficient of a variable in the objective function is the variable's **objective function coefficient.** The coefficient of a variable in a constraint is a **technological coefficient.** The right-hand side of each constraint is called a **right-hand side (rhs).**

A *point* is simply a specification of the values of each decision variable. The **feasible region** of an LP consists of all points satisfying the LP's constraints and sign restrictions. Any point in the feasible region that has the largest z-value of all points in the feasible region (for a max problem) is an **optimal solution** to the LP. An LP may have no optimal solution, one optimal solution, or an infinite number of optimal solutions.

A constraint in an LP is **binding** if the left-hand side and the right-hand side are equal when the values of the variables in the optimal solution are substituted into the constraint.

Graphical Solution of Linear Programming Problems

The feasible region for any LP is a **convex set.** If an LP has an optimal solution, there is an extreme (or corner) point of the feasible region that is an optimal solution to the LP.

We may graphically solve an LP (max problem) with two decision variables as follows:

Step 1 Graph the feasible region.

Step 2 Draw an isoprofit line.

Step 3 Move parallel to the isoprofit line in the direction of increasing z. The last point in the feasible region that contacts an isoprofit line is an optimal solution to the LP.

LP Solutions: Four Cases

When an LP is solved, one of the following four cases will occur:

Case 1 The LP has a unique solution.

Case 2 The LP has more than one (actually an infinite number of) optimal solutions. This is the case of **alternative optimal solutions.** Graphically, we recognize this case when the isoprofit line last hits an entire line segment before leaving the feasible region.

Case 3 The LP is **infeasible** (it has no feasible solution). This means that the feasible region contains no points.

Case 4 The LP is unbounded. This means (in a max problem) that there are points in the feasible region with arbitrarily large z-values. Graphically, we recognize this case by the fact that when we move parallel to an isoprofit line in the direction of increasing z, we never lose contact with the LP's feasible region.

Formulating LPs

The most important step in formulating most LPs is to determine the decision variables correctly.

In any constraint, the terms must have the same units. For example, one term cannot have the units "pounds of raw material" while another term has the units "ounces of raw material."

REVIEW PROBLEMS

Group A

1 Bloomington Breweries produces beer and ale. Beer sells for $5 per barrel, and ale sells for $2 per barrel. Producing a barrel of beer requires 5 lb of corn and 2 lb of hops. Producing a barrel of ale requires 2 lb of corn and 1 lb of hops. Sixty pounds of corn and 25 lb of hops are available. Formulate an LP that can be used to maximize revenue. Solve the LP graphically.

2 Farmer Jones bakes two types of cake (chocolate and vanilla) to supplement his income. Each chocolate cake can be sold for $1, and each vanilla cake can be sold for 50¢. Each chocolate cake requires 20 minutes of baking time and uses 4 eggs. Each vanilla cake requires 40 minutes of baking time and uses 1 egg. Eight hours of baking time and 30 eggs are available. Formulate an LP to maximize Farmer Jones's revenue, then graphically solve the LP. (A fractional number of cakes is okay.)

3 I now have $100. The following investments are available during the next three years:

Investment A Every dollar invested now yields $0.10 a year from now and $1.30 three years from now.
Investment B Every dollar invested now yields $0.20 a year from now and $1.10 two years from now.
Investment C Every dollar invested a year from now yields $1.50 three years from now.

During each year, uninvested cash can be placed in money market funds, which yield 6% interest per year. At most $50 may be placed in each of investments A, B, and C. Formulate an LP to maximize my cash on hand three years from now.

4 Sunco processes oil into aviation fuel and heating oil. It costs $40 to purchase each 1,000 barrels of oil, which is then distilled and yields 500 barrels of aviation fuel and 500 barrels of heating oil. Output from the distillation may be sold directly or processed in the catalytic cracker. If sold after distillation without further processing, aviation fuel sells for $60 per 1,000 barrels, and heating oil sells for $40 per 1,000 barrels. It takes 1 hour to process 1,000 barrels of aviation fuel in the catalytic cracker, and these 1,000 barrels can be sold for $130. It takes 45 minutes to process 1,000 barrels of heating oil in the cracker, and these 1,000 barrels can be sold for $90. Each day, at most 20,000 barrels of oil can be purchased, and 8 hours of cracker time are available. Formulate an LP to maximize Sunco's profits.

5 Finco has the following investments available:

Investment A For each dollar invested at time 0, we receive $0.10 at time 1 and $1.30 at time 2. (Time 0 = now; time 1 = one year from now; and so on.)
Investment B For each dollar invested at time 1, we receive $1.60 at time 2.
Investment C For each dollar invested at time 2, we receive $1.20 at time 3.

At any time, leftover cash may be invested in T-bills, which pay 10% per year. At time 0, we have $100. At most, $50 can be invested in each of investments A, B, and C. Formulate an LP that can be used to maximize Finco's cash on hand at time 3.

6 All steel manufactured by Steelco must meet the following requirements: 3.2–3.5% carbon; 1.8–2.5% silicon; 0.9–1.2% nickel; tensile strength of at least 45,000 pounds per square inch (psi). Steelco manufactures steel by combining two alloys. The cost and properties of each alloy are given in Table 47. Assume that the tensile strength of a

mixture of the two alloys can be determined by averaging that of the alloys that are mixed together. For example, a one-ton mixture that is 40% alloy 1 and 60% alloy 2 has a tensile strength of 0.4(42,000) + 0.6(50,000). Use linear programming to determine how to minimize the cost of producing a ton of steel.

7 Steelco manufactures two types of steel at three different steel mills. During a given month, each steel mill has 200 hours of blast furnace time available. Because of differences in the furnaces at each mill, the time and cost to produce a ton of steel differs for each mill. The time and cost for each mill are shown in Table 48. Each month, Steelco must manufacture at least 500 tons of steel 1 and 600 tons of steel 2. Formulate an LP to minimize the cost of manufacturing the desired steel.

8[†] Walnut Orchard has two farms that grow wheat and corn. Because of differing soil conditions, there are differences in the yields and costs of growing crops on the two farms. The yields and costs are shown in Table 49. Each farm has 100 acres available for cultivation; 11,000 bushels of wheat and 7,000 bushels of corn must be grown. Determine a planting plan that will minimize the cost of meeting these demands. How could an extension of this model be used to allocate crop production efficiently throughout a nation?

9 Candy Kane Cosmetics (CKC) produces Leslie Perfume, which requires chemicals and labor. Two production processes are available: Process 1 transforms 1 unit of labor and 2 units of chemicals into 3 oz of perfume. Process 2 transforms 2 units of labor and 3 units of chemicals into 5 oz of perfume. It costs CKC $3 to purchase a unit of labor and $2 to purchase a unit of chemicals. Each year, up to 20,000 units of labor and 35,000 units of chemicals can be purchased. In the absence of advertising, CKC believes it can sell 1,000 oz of perfume. To stimulate demand for

TABLE 48
Producing a Ton of Steel

Mill	Steel 1		Steel 2	
	Cost	Time (Minutes)	Cost	Time (Minutes)
1	$10	20	$11	22
2	$12	24	$ 9	18
3	$14	28	$10	30

TABLE 47

	Alloy 1	Alloy 2
Cost per ton ($)	$190	$200
Percent silicon	2	2.5
Percent nickel	1	1.5
Percent carbon	3	4
Tensile strength (psi)	42,000	50,000

TABLE 49

	Farm 1	Farm 2
Corn yield/acre (bushels)	500	650
Cost/acre of corn ($)	100	120
Wheat yield/acre (bushels)	400	350
Cost/acre of wheat ($)	90	80

†Based on Heady and Egbert (1964).

Leslie, CKC can hire the lovely model Jenny Nelson. Jenny is paid $100/hour. Each hour Jenny works for the company is estimated to increase the demand for Leslie Perfume by 200 oz. Each ounce of Leslie Perfume sells for $5. Use linear programming to determine how CKC can maximize profits.

10 Carco has a $150,000 advertising budget. To increase automobile sales, the firm is considering advertising in newspapers and on television. The more Carco uses a particular medium, the less effective is each additional ad. Table 50 shows the number of new customers reached by each ad. Each newspaper ad costs $1,000, and each television ad costs $10,000. At most, 30 newspaper ads and 15 television ads can be placed. How can Carco maximize the number of new customers created by advertising?

11 Sunco Oil has refineries in Los Angeles and Chicago. The Los Angeles refinery can refine up to 2 million barrels of oil per year, and the Chicago refinery up to 3 million. Once refined, oil is shipped to two distribution points: Houston and New York City. Sunco estimates that each distribution point can sell up to 5 million barrels per year. Because of differences in shipping and refining costs, the profit earned (in dollars) per million barrels of oil shipped depends on where the oil was refined and on the point of distribution (see Table 51). Sunco is considering expanding the capacity of each refinery. Each million barrels of annual refining capacity that is added will cost $120,000 for the Los Angeles refinery and $150,000 for the Chicago refinery. Use linear programming to determine how Sunco can maximize its profits less expansion costs over a ten-year period.

12 For a telephone survey, a marketing research group needs to contact at least 150 wives, 120 husbands, 100 single adult males, and 110 single adult females. It costs $2 to make a daytime call and (because of higher labor costs) $5 to make an evening call. Table 52 lists the results. Because of limited staff, at most half of all phone calls can be evening calls. Formulate an LP to minimize the cost of completing the survey.

TABLE 50

	Number of Ads	New Customers
Newspaper	1–10	900
	11–20	600
	21–30	300
Television	1–5	10,000
	6–10	5,000
	11–15	2,000

TABLE 51

	Profit per Million Barrels ($)	
From	To Houston	To New York
Los Angeles	20,000	15,000
Chicago	18,000	17,000

TABLE 52

Person Responding	Percent of Daytime Calls	Percent of Evening Calls
Wife	30	30
Husband	10	30
Single male	10	15
Single female	10	20
None	40	5

13 Feedco produces two types of cattle feed, both consisting totally of wheat and alfalfa. Feed 1 must contain at least 80% wheat, and feed 2 must contain at least 60% alfalfa. Feed 1 sells for $1.50/lb, and feed 2 sells for $1.30/lb. Feedco can purchase up to 1,000 lb of wheat at 50¢/lb and up to 800 lb of alfalfa at 40¢/lb. Demand for each type of feed is unlimited. Formulate an LP to maximize Feedco's profit.

14 Feedco (see Problem 13) has decided to give its customer (assume it has only one customer) a quantity discount. If the customer purchases more than 300 lb of feed 1, each pound over the first 300 lb will sell for only $1.25/lb. Similarly, if the customer purchases more than 300 pounds of feed 2, each pound over the first 300 lb will sell for $1.00/lb. Modify the LP of Problem 13 to account for the presence of quantity discounts. (*Hint:* Define variables for the feed sold at each price.)

15 Chemco produces two chemicals: A and B. These chemicals are produced via two manufacturing processes. Process 1 requires 2 hours of labor and 1 lb of raw material to produce 2 oz of A and 1 oz of B. Process 2 requires 3 hours of labor and 2 lb of raw material to produce 3 oz of A and 2 oz of B. Sixty hours of labor and 40 lb of raw material are available. Demand for A is unlimited, but only 20 oz of B can be sold. A sells for $16/oz, and B sells for $14/oz. Any B that is unsold must be disposed of at a cost of $2/oz. Formulate an LP to maximize Chemco's revenue less disposal costs.

16 Suppose that in the CSL computer example of Section 3.12, it takes two months to train a technician and that during the second month of training, each trainee requires 10 hours of experienced technician time. Modify the formulation in the text to account for these changes.

17 Furnco manufactures tables and chairs. Each table and chair must be made entirely out of oak or entirely out of pine. A total of 150 board ft of oak and 210 board ft of pine are available. A table requires either 17 board ft of oak or 30 board ft of pine, and a chair requires either 5 board ft of oak or 13 board ft of pine. Each table can be sold for $40, and each chair for $15. Formulate an LP that can be used to maximize revenue.

18[†] The city of Busville contains three school districts. The number of minority and nonminority students in each district is given in Table 53. Of all students, 25% $\left(\frac{200}{500}\right)$ are minority students.

[†]Based on Franklin and Koenigsberg (1973).

TABLE 53

District	Minority Students	Nonminority Students
1	50	200
2	50	250
3	100	150

TABLE 54

District	Cooley High	Walt Whitman High
1	1	2
2	2	1
3	1	1

The local court has decided that both of the town's two high schools (Cooley High and Walt Whitman High) must have approximately the same percentage of minority students (within ±5%) as the entire town. The distances (in miles) between the school districts and the high schools are given in Table 54. Each high school must have an enrollment of 300–500 students. Use linear programming to determine an assignment of students to schools that minimizes the total distance students must travel to school.

19[†] Brady Corporation produces cabinets. Each week, it requires 90,000 cu ft of processed lumber. The company may obtain lumber in two ways. First, it may purchase lumber from an outside supplier and then dry it in the supplier's kiln. Second, it may chop down logs on its own land, cut them into lumber at its sawmill, and finally dry the lumber in its own kiln. Brady can purchase grade 1 or grade 2 lumber. Grade 1 lumber costs $3 per cu ft and when dried yields 0.7 cu ft of useful lumber. Grade 2 lumber costs $7 per cubic foot and when dried yields 0.9 cu ft of useful lumber. It costs the company $3 to chop down a log. After being cut and dried, a log yields 0.8 cu ft of lumber. Brady incurs costs of $4 per cu ft of lumber dried. It costs $2.50 per cu ft of logs sent through the sawmill. Each week, the sawmill can process up to 35,000 cu ft of lumber. Each week, up to 40,000 cu ft of grade 1 lumber and up to 60,000 cu ft of grade 2 lumber can be purchased. Each week, 40 hours of time are available for drying lumber. The time it takes to dry 1 cu ft of grade 1 lumber, grade 2 lumber, or logs is as follows: grade 1—2 seconds; grade 2—0.8 second; log—1.3 seconds. Formulate an LP to help Brady minimize the weekly cost of meeting the demand for processed lumber.

20[‡] The Canadian Parks Commission controls two tracts of land. Tract 1 consists of 300 acres and tract 2, 100 acres. Each acre of tract 1 can be used for spruce trees or hunting, or both. Each acre of tract 2 can be used for spruce trees or camping, or both. The capital (in hundreds of dollars) and labor (in worker-days) required to maintain one acre of each tract, and the profit (in thousands of dollars) per acre for each possible use of land are given in Table 55. Capital of $150,000 and 200 man-days of labor are available. How should the land be allocated to various uses to maximize profit received from the two tracts?

21[§] Chandler Enterprises produces two competing products: A and B. The company wants to sell these products to two groups of customers: group 1 and group 2. The value each customer places on a unit of A and B is as shown in Table 56. Each customer will buy either product A or product B, but not both. A customer is willing to buy product A if she believes that

Value of product A − price of product A

$\geq$ Value of product B − price of product B

and

Value of product A − price of product A $\geq$ 0

A customer is willing to buy product B if she believes that

Value of product B − price of product B

$\geq$ value of product A − price of product A

and

Value of product B − price of product B $\geq$ 0

Group 1 has 1,000 members, and group 2 has 1,500 members. Chandler wants to set prices for each product that ensure that group 1 members purchase product A and group 2 members purchase product B. Formulate an LP that will help Chandler maximize revenues.

22[¶] Alden Enterprises produces two products. Each product can be produced on one of two machines. The length of time needed to produce each product (in hours) on each machine is as shown in Table 57. Each month, 500 hours of time are available on each machine. Each month, customers are willing to buy up to the quantities of each product at the

TABLE 55

Tract	Capital	Labor	Profit
1 Spruce	3	0.1	0.2
1 Hunting	3	0.2	0.4
1 Both	4	0.2	0.5
2 Spruce	1	0.05	0.06
2 Camping	30	5	0.09
2 Both	10	1.01	1.1

TABLE 56

	Group 1 Customer	Group 2 Customer
Value of A to	$10	$12
Value of B to	$8	$15

[§]Based on Dobson and Kalish (1988).
[¶]Based on Jain, Stott, and Vasold (1978).

[†]Based on Carino and Lenoir (1988).
[‡]Based on Cheung and Auger (1976).

TABLE 57

Product	Machine 1	Machine 2
1	4	3
2	7	4

TABLE 58

	Demands		Prices	
Product	Month 1	Month 2	Month 1	Month 2
1	100	190	$55	$12
2	140	130	$65	$32

prices given in Table 58. The company's goal is to maximize the revenue obtained from selling units during the next two months. Formulate an LP to help meet this goal.

23 Kiriakis Electronics produces three products. Each product must be processed on each of three types of machines. When a machine is in use, it must be operated by a worker. The time (in hours) required to process each product on each machine and the profit associated with each product are shown in Table 59. At present, five type 1 machines, three type 2 machines, and four type 3 machines are available. The company has 10 workers available and must determine how many workers to assign to each machine. The plant is open 40 hours per week, and each worker works 35 hours per week. Formulate an LP that will enable Kiriakis to assign workers to machines in a way that maximizes weekly profits. (*Note:* A worker need not spend the entire work week operating a single machine.)

24 Gotham City Hospital serves cases from four diagnostic-related groups (DRGs). The profit contribution, diagnostic service use (in hours), bed-day use (in days), nursing care use (in hours), and drug use (in dollars) are

TABLE 59

	Product 1	Product 2	Product 3
Machine 1	2	3	4
Machine 2	3	5	6
Machine 3	4	7	9
Profit ($)	6	8	10

TABLE 60

DRG	Profit	Diagnostic Services	Bed-Day	Nursing Use	Drugs
1	2,000	7	5	30	800
2	1,500	4	2	10	500
3	500	2	1	5	150
4	300	1	0	1	50

given in Table 60. The hospital now has available each week 570 hours of diagnostic services, 1,000 bed-days, 50,000 nursing hours, and $50,000 worth of drugs. To meet the community's minimum health care demands at least 10 DRG1, 15 DRG2, 40 DRG3, and 160 DRG4 cases must be handled each week. Use LP to determine the hospital's optimal mix of DRGs.[†]

25 Oliver Winery produces four award-winning wines in Bloomington, Indiana. The profit contribution, labor hours, and tank usage (in hours) per gallon for each type of wine are given in Table 61. By law, at most 100,000 gallons of wine can be produced each year. A maximum of 12,000 labor hours and 32,000 tank hours are available annually. Each gallon of wine 1 spends an average of $\frac{1}{3}$ year in inventory; wine 2, an average of 1 year; wine 3, an average of 2 years; wine 4, an average of 3.333 years. The winery's warehouse can handle an average inventory level of 50,000 gallons. Determine how much of each type of wine should be produced annually to maximize Oliver Winery's profit.

26 Graphically solve the following LP:

$$\min z = 5x_1 + x_2$$
$$\text{s.t.} \quad 2x_1 + x_2 \geq 6$$
$$x_1 + x_2 \geq 4$$
$$2x_1 + 10x_2 \geq 20$$
$$x_1, x_2 \geq 0$$

27 Grummins Engine produces diesel trucks. New government emission standards have dictated that the average pollution emissions of all trucks produced in the next three years cannot exceed 10 grams per truck. Grummins produces two types of trucks. Each type 1 truck sells for $20,000, costs $15,000 to manufacture, and emits 15 grams of pollution. Each type 2 truck sells for $17,000, costs $14,000 to manufacture, and emits 5 grams of pollution. Production capacity limits total truck production during each year to at most 320 trucks. Grummins knows that the maximum number of each truck type that can be sold during each of the next three years is given in Table 62.

Thus, *at most,* 300 type 1 trucks can be sold during year 3. Demand may be met from previous production or the current year's production. It costs $2,000 to hold 1 truck (of any type) in inventory for one year. Formulate an LP to help Grummins maximize its profit during the next three years.

TABLE 61

Wine	Profit ($)	Labor (Hr)	Tank (Hr)
1	6	.2	.5
2	12	.3	.5
3	20	.3	1
4	30	.5	1.5

[†]Based on Robbins and Tuntiwonpiboon (1989).

TABLE **62**
Maximum Demand for Trucks

Year	Type 1	Type 2
1	100	200
2	200	100
3	300	150

TABLE **63**

Shift	Hourly Salary	Defects (per Capacitor)	Price
8 A.M.–4 P.M.	$12	4	$18
4 P.M.–Midnight	$16	3	$22
Midnight–8 A.M.	$20	2	$24

28 Describe all optimal solutions to the following LP:

$$\min z = 4x_1 + x_2$$
$$\text{s.t.} \quad 3x_1 + x_2 \geq 6$$
$$4x_1 + x_2 \geq 12$$
$$x_1 \geq 2$$
$$x_1, x_2 \geq 0$$

29 Juiceco manufactures two products: premium orange juice and regular orange juice. Both products are made by combining two types of oranges: grade 6 and grade 3. The oranges in premium juice must have an average grade of at least 5, those in regular juice, at least 4. During each of the next two months Juiceco can sell up to 1,000 gallons of premium juice and up to 2,000 gallons of regular juice. Premium juice sells for $1.00 per gallon, while regular juice sells for 80¢ per gallon. At the beginning of month 1, Juiceco has 3,000 gallons of grade 6 oranges and 2,000 gallons of grade 3 oranges. At the beginning of month 2, Juiceco may purchase additional grade 3 oranges for 40¢ per gallon and additional grade 6 oranges for 60¢ per gallon. Juice spoils at the end of the month, so it makes no sense to make extra juice during month 1 in the hopes of using it to meet month 2 demand. Oranges left at the end of month 1 may be used to produce juice for month 2. At the end of month 1 a holding cost of 5¢ is assessed against each gallon of leftover grade 3 oranges, and 10¢ against each gallon of leftover grade 6 oranges. In addition to the cost of the oranges, it costs 10¢ to produce each gallon of (regular or premium) juice. Formulate an LP that could be used to maximize the profit (revenues − costs) earned by Juiceco during the next two months.

30 Graphically solve the following linear programming problem:

$$\max z = 5x_1 - x_2$$
$$\text{s.t.} \quad 2x_1 + 3x_2 \geq 12$$
$$x_1 - 3x_2 \geq 0$$
$$x_1 \geq 0, x_2 \geq 0$$

31 Graphically find all solutions to the following LP:

$$\min z = x_1 - 2x_2$$
$$\text{s.t.} \quad x_1 \geq 4$$
$$x_1 + x_2 \geq 8$$
$$x_1 - x_2 \leq 6$$
$$x_1, x_2 \geq 0$$

32 Each day Eastinghouse produces capacitors during three shifts: 8 A.M.–4 P.M., 4 P.M.–midnight, midnight–8 A.M. The hourly salary paid to the employees on each shift, the price charged for each capacitor made during each shift, and the number of defects in each capacitor produced during a given shift are shown in Table 63. Each of the company's 25 workers can be assigned to one of the three shifts. A worker produces 10 capacitors during a shift, but because of machinery limitations, no more than 10 workers can be assigned to any shift. Each day, at most 250 capacitors can be sold, and the average number of defects per capacitor for the day's production cannot exceed three. Formulate an LP to maximize Eastinghouse's daily profit (sales revenue − labor cost).

33 Graphically find all solutions to the following LP:

$$\max z = 4x_1 + x_2$$
$$\text{s.t.} \quad 8x_1 + 2x_2 \leq 16$$
$$x_1 + x_2 \leq 12$$
$$x_1, x_2 \geq 0$$

34 During the next three months Airco must meet (on time) the following demands for air conditioners: month 1, 300; month 2, 400; month 3, 500. Air conditioners can be produced in either New York or Los Angeles. It takes 1.5 hours of skilled labor to produce an air conditioner in Los Angeles, and 2 hours in New York. It costs $400 to produce an air conditioner in Los Angeles, and $350 in New York. During each month, each city has 420 hours of skilled labor available. It costs $100 to hold an air conditioner in inventory for a month. At the beginning of month 1, Airco has 200 air conditioners in stock. Formulate an LP whose solution will tell Airco how to minimize the cost of meeting air conditioner demands for the next three months.

35 Formulate the following as a linear programming problem: A greenhouse operator plans to bid for the job of providing flowers for city parks. He will use tulips, daffodils, and flowering shrubs in three types of layouts. A Type 1 layout uses 30 tulips, 20 daffodils, and 4 flowering shrubs. A Type 2 layout uses 10 tulips, 40 daffodils, and 3 flowering shrubs. A Type 3 layout uses 20 tulips, 50 daffodils, and 2 flowering shrubs. The net profit is $50 for each Type 1 layout, $30 for each Type 2 layout, and $60 for each Type 3 layout. He has 1,000 tulips, 800 daffodils, and 100 flowering shrubs. How many layouts of each type should be used to yield maximum profit?

36 Explain how your formulation in Problem 35 changes if both of the following conditions are added:

a The number of Type 1 layouts cannot exceed the number of Type 2 layouts.

b There must be at least five layouts of each type.

37 Graphically solve the following LP problem:

$$\min z = 6x_1 + 2x_2$$
$$\text{s.t.} \quad 3x_1 + 2x_2 \geq 12$$
$$2x_1 + 4x_2 \geq 12$$
$$x_2 \geq 1$$
$$x_1, x_2 \geq 0$$

38 We produce two products: product 1 and product 2 on two machines (machine 1 and machine 2). The number of hours of machine time and labor depends on the machine and the product as shown in Table 64.

The cost of producing a unit of each product is shown in Table 65.

The number of labor hours and machine time available this month are in Table 66.

This month, at least 200 units of product 1 and at least 240 units of product 2 must be produced. Also, at least half of product 1 must be made on machine 1, and at least half of product 2 must be made on machine 2. Determine how we can minimize the cost of meeting our monthly demands.

39 Carrotco manufactures two products: 1 and 2. Each unit of each product must be processed on machine 1 and machine 2 and uses raw material 1 and raw material 2. The resource usage is as in Table 67.

TABLE 64

	Product 1 Machine 1	Product 2 Machine 1	Product 1 Machine 2	Product 2 Machine 2
Machine time	0.7	0.75	0.8	0.9
Labor	0.75	0.75	1.2	1

TABLE 65

Product 1 Machine 1	Product 2 Machine 1	Product 1 Machine 2	Product 2 Machine 2
$1.50	$0.40	$2.20	$4.00

TABLE 66

Resource	Hours Available
Machine 1	200
Machine 2	200
Labor	400

TABLE 67

	Product 1	Product 2
Machine 1	0.6	0.4
Machine 2	0.4	0.3
Raw material 1	2	1
Raw material 2	1	2

TABLE 68

	Product 1	Product 2
Demand	400	300
Sales Price	30	35

Thus, producing one unit of product 1 uses .6 unit of machine 1 time, .4 unit of machine 2 time, 2 units of raw material 1, and 1 unit of raw material 2. The sales price per unit and demand for each product are in Table 68.

It costs $4 to purchase each unit of raw material 1 and $5 to produce each unit of raw material 2. Unlimited amounts of raw material can be purchased. Two hundred units of machine 1 time and 300 units of machine 2 time are available. Determine how Carrotco can maximize its profit.

40 A company assembles two products: A and B. Product A sells for $11 per unit, and product B sells for $23 per unit. A unit of product A requires 2 hours on assembly line 1 and 1 unit of raw material. A unit of product B requires 2 units of raw material, 1 unit of A, and 2 hours on line 2. For line 1, 1,300 hours of time are available and 500 hours of time are available on line 2. A unit of raw material may be bought (for $5 a unit) or produced (at no cost) by using 2 hours of time on line 1. Determine how to maximize profit.

41 Ann and Ben are getting divorced and want to determine how to divide their joint property: retirement account, home, summer cottage, investments, and miscellaneous assets. To begin, Ann and Ben are told to allocate 100 total points to the assets. Their allocation is as shown in Table 69.

Assuming that all assets are divisible (that is, a fraction of each asset may be given to each person), how should the assets be allocated? Two criteria should govern the asset allocation:

Criteria 1 Each person should end up with the same number of points. This prevents Ann from envying Ben and Ben from envying Ann.

Criteria 2 The total number of points received by Ann and Ben should be maximized.

If assets could not be split between people, what problem arises?

42 Eli Daisy manufactures two drugs in Los Angeles and Indianapolis. The cost of manufacturing a pound of each drug is shown in Table 70.

TABLE 69

	Points	
Item	Ann's	Ben's
Retirement account	50	40
Home	20	30
Summer cottage	15	10
Investments	10	10
Miscellaneous	5	10

TABLE 70

City	Drug 1 Cost ($)	Drug 2 Cost ($)
Indianapolis	4.10	4.50
Los Angeles	4.00	5.20

TABLE 71

City	Drug 1 Time (Hr)	Drug 2 Time (Hr)
Indianapolis	.2	.3
Los Angeles	.24	.33

The machine time (in hours) required to produce a pound of each drug at each city is as in Table 71.

Daisy needs to produce at least 1,000 pounds of drug 1 and 2,000 pounds of drug 2 per week. The company has 500 hours per week of machine time in Indianapolis and 400 hours per week of machine time in Los Angeles. Determine how Lilly can minimize the cost of producing the needed drugs.

43 Daisy also produces Wozac in New York and Chicago. Each month, it can produce up to 30 units in New York and up to 35 units in Chicago. The cost of producing a unit each month at each location is shown in Table 72.

The customer demands shown in Table 73 must be met on time.

The cost of holding a unit in inventory (measured against ending inventory) is shown in Table 74.

TABLE 72

	Cost ($)	
Month	New York	Chicago
1	8.62	8.40
2	8.70	8.75
3	8.90	9.00

TABLE 73

Month	Demand (Units)
1	50
2	60
3	40

TABLE 74

Month	Holding Cost ($)
1	0.26
2	0.12
3	0.12

At the beginning of month 1, we have 10 units of Wozac in inventory. Determine a cost-minimizing schedule for the next three months.

44 You have been put in charge of the Dawson Creek oil refinery. The refinery produces gas and heating oil from crude oil. Gas sells for $11 per barrel and must have an average grade level of at least 9. Heating oil sells for $6 a barrel and must have an average grade level of at least 7. At most, 2,000 barrels of gas and 600 barrels of heating oil can be sold.

Incoming crude can be processed by one of three methods. The per barrel yield and per barrel cost of each processing method are shown in Table 75.

For example, if we refine one barrel of incoming crude by method 1, it costs us $3.40 and yields .2 barrels of grade 6, .2 barrels of grade 8, and .6 barrels of grade 10. These costs include the costs of buying the crude oil.

Before being processed into gas and heating oil, grades 6 and 8 may be sent through the catalytic cracker to improve their quality. For $1 per barrel, one barrel of grade 6 can be "cracked" into a barrel of grade 8. For $1.50 per barrel, a barrel of grade 8 can be cracked into a barrel of grade 10. Determine how to maximize the refinery's profit.

45 Currently we own 100 shares each of stocks 1 through 10. The original price we paid for these stocks, today's price, and the expected price in one year for each stock is shown in Table 76.

We need money today and are going to sell some of our stocks. The tax rate on capital gains is 30%. If we sell 50 shares of stock 1, then we must pay tax of $.3 \cdot 50(30 - 20) = \$150$. We must also pay transaction costs of 1% on each transaction. Thus, our sale of 50 shares of stock 1 would incur transaction costs of $.01 \cdot 50 \cdot 30 = \$15$. After taxes and transaction costs, we must be left with $30,000 from our stock sales. Our goal is to maximize the expected (before-tax) value in one year of our remaining stock. What stocks should we sell? Assume it is all right to sell a fractional share of stock.

Group B

46 Gotham City National Bank is open Monday–Friday from 9 A.M. to 5 P.M. From past experience, the bank knows that it needs the number of tellers shown in Table 77. The bank hires two types of tellers. Full-time tellers work 9–5 five days a week, except for 1 hour off for lunch. (The bank determines when a full-time employee takes lunch hour, but each teller must go between noon and 1 P.M. or between 1 P.M. and 2 P.M.) Full-time employees are paid (including fringe benefits) $8/hour (this includes payment for lunch hour). The bank may also hire part-time tellers. Each part-

TABLE 75

		Grade		
Method	6	8	10	Cost ($ per Barrel)
1	.2	.3	.5	3.40
2	.3	.4	.3	3.00
3	.4	.4	.2	2.60

TABLE **76**

Stock	Shares Owned	Price ($)		
		Purchase	Current	In One Year
1	100	20	30	36
2	100	25	34	39
3	100	30	43	42
4	100	35	47	45
5	100	40	49	51
6	100	45	53	55
7	100	50	60	63
8	100	55	62	64
9	100	60	64	66
10	100	65	66	70
Tax rate (%)	0.3			
Transaction cost (%)	0.01			

TABLE **77**

Time Period	Tellers Required
9–10	4
10–11	3
11–Noon	4
Noon–1	6
1–2	5
2–3	6
3–4	8
4–5	8

time teller must work exactly 3 consecutive hours each day. A part-time teller is paid $5/hour (and receives no fringe benefits). To maintain adequate quality of service, the bank has decided that at most five part-time tellers can be hired. Formulate an LP to meet the teller requirements at minimum cost. Solve the LP on a computer. Experiment with the LP answer to determine an employment policy that comes close to minimizing labor cost.

47[†] The Gotham City Police Department employs 30 police officers. Each officer works 5 days per week. The crime rate fluctuates with the day of the week, so the number of police officers required each day depends on which day of the week it is: Saturday, 28; Sunday, 18; Monday, 18; Tuesday, 24; Wednesday, 25; Thursday, 16; Friday, 21. The police department wants to schedule police officers to minimize the number whose days off are not consecutive. Formulate an LP that will accomplish this goal. (*Hint:* Have a constraint for each day of the week that ensures that the proper number of officers are *not* working on the given day.)

[†]Based on Rothstein (1973).

48[‡]Alexis Cornby makes her living buying and selling corn. On January 1, she has 50 tons of corn and $1,000. On the first day of each month Alexis can buy corn at the following prices per ton: January, $300; February, $350; March, $400; April, $500. On the last day of each month, Alexis can sell corn at the following prices per ton: January, $250; February, $400; March, $350; April, $550. Alexis stores her corn in a warehouse that can hold at most 100 tons of corn. She must be able to pay cash for all corn at the time of purchase. Use linear programming to determine how Alexis can maximize her cash on hand at the end of April.

49[§]At the beginning of month 1, Finco has $400 in cash. At the beginning of months 1, 2, 3, and 4, Finco receives certain revenues, after which it pays bills (see Table 78). Any money left over may be invested for one month at the interest rate of 0.1% per month; for two months at 0.5% per month; for three months at 1% per month; or for four months at 2% per month. Use linear programming to determine an investment strategy that maximizes cash on hand at the beginning of month 5.

50 City 1 produces 500 tons of waste per day, and city 2 produces 400 tons of waste per day. Waste must be incinerated at incinerator 1 or 2, and each incinerator can process up to 500 tons of waste per day. The cost to incinerate waste is $40/ton at incinerator 1 and $30/ton at 2.

TABLE **78**

Month	Revenues ($)	Bills ($)
1	400	600
2	800	500
3	300	500
4	300	250

[‡]Based on Charnes and Cooper (1955).
[§]Based on Robichek, Teichroew, and Jones (1965).

Incineration reduces each ton of waste to 0.2 tons of debris, which must be dumped at one of two landfills. Each landfill can receive at most 200 tons of debris per day. It costs $3 per mile to transport a ton of material (either debris or waste). Distances (in miles) between locations are shown in Table 79. Formulate an LP that can be used to minimize the total cost of disposing of the waste of both cities.

51[†] Silicon Valley Corporation (Silvco) manufactures transistors. An important aspect of the manufacture of transistors is the melting of the element germanium (a major component of a transistor) in a furnace. Unfortunately, the melting process yields germanium of highly variable quality.

Two methods can be used to melt germanium; method 1 costs $50 per transistor, and method 2 costs $70 per transistor. The qualities of germanium obtained by methods 1 and 2 are shown in Table 80. Silvco can refire melted germanium in an attempt to improve its quality. It costs $25 to refire the melted germanium for one transistor. The results of the refiring process are shown in Table 81. Silvco has sufficient furnace capacity to melt or refire germanium for at most 20,000 transistors per month. Silvco's monthly demands are for 1,000 grade 4 transistors, 2,000 grade 3 transistors, 3,000 grade 2 transistors, and 3,000 grade 1 transistors. Use linear programming to minimize the cost of producing the needed transistors.

TABLE 79

City	Incinerator	
	1	2
1	30	5
2	36	42

Incinerator	Landfill	
	1	2
1	5	8
2	9	6

TABLE 80

Grade of[‡] Melted Germanium	Percent Yielded by Melting	
	Method 1	Method 2
Defective	30	20
1	30	20
2	20	25
3	15	20
4	5	15

[‡]*Note:* Grade 1 is poor; grade 4 is excellent. The quality of the germanium dictates the quality of the manufactured transistor.

[†]Based on Smith (1965).

TABLE 81

Refired Grade of Germanium	Percent Yielded by Refiring			
	Defective	Grade 1	Grade 2	Grade 3
Defective	30	0	0	0
1	25	30	0	0
2	15	30	40	0
3	20	20	30	50
4	10	20	30	50

TABLE 82

Input	Cost ($)	Pulp Content (%)
Box board	5	15
Tissue paper	6	20
Newsprint	8	30
Book paper	10	40

52[‡] A paper-recycling plant processes box board, tissue paper, newsprint, and book paper into pulp that can be used to produce three grades of recycled paper (grades 1, 2, and 3). The prices per ton and the pulp contents of the four inputs are shown in Table 82. Two methods, de-inking and asphalt dispersion, can be used to process the four inputs into pulp. It costs $20 to de-ink a ton of any input. The process of de-inking removes 10% of the input's pulp, leaving 90% of the original pulp. It costs $15 to apply asphalt dispersion to a ton of material. The asphalt dispersion process removes 20% of the input's pulp. At most, 3,000 tons of input can be run through the asphalt dispersion process or the de-inking process. Grade 1 paper can only be produced with newsprint or book paper pulp; grade 2 paper, only with book paper, tissue paper, or box board pulp; and grade 3 paper, only with newsprint, tissue paper, or box board pulp. To meet its current demands, the company needs 500 tons of pulp for grade 1 paper, 500 tons of pulp for grade 2 paper, and 600 tons of pulp for grade 3 paper. Formulate an LP to minimize the cost of meeting the demands for pulp.

53 Turkeyco produces two types of turkey cutlets for sale to fast-food restaurants. Each type of cutlet consists of white meat and dark meat. Cutlet 1 sells for $4/lb and must consist of at least 70% white meat. Cutlet 2 sells for $3/lb and must consist of at least 60% white meat. At most, 50 lb of cutlet 1 and 30 lb of cutlet 2 can be sold. The two types of turkey used to manufacture the cutlets are purchased from the GobbleGobble Turkey Farm. Each type 1 turkey costs $10 and yields 5 lb of white meat and 2 lb of dark meat. Each type 2 turkey costs $8 and yields 3 lb of white meat and 3 lb of dark meat. Formulate an LP to maximize Turkeyco's profit.

54 Priceler manufactures sedans and wagons. The number of vehicles that can be sold each of the next three months

[‡]Based on Glassey and Gupta (1975).

TABLE **83**

Month	Sedans	Wagons
1	1,100	600
2	1,500	700
3	1,200	50

are listed in Table 83. Each sedan sells for $8,000, and each wagon sells for $9,000. It costs $6,000 to produce a sedan and $7,500 to produce a wagon. To hold a vehicle in inventory for one month costs $150 per sedan and $200 per wagon. During each month, at most 1,500 vehicles can be produced. Production line restrictions dictate that during month 1 at least two-thirds of all cars produced must be sedans. At the beginning of month 1, 200 sedans and 100 wagons are available. Formulate an LP that can be used to maximize Priceler's profit during the next three months.

55 The production-line employees at Grummins Engine work four days a week, 10 hours a day. Each day of the week, (at least) the following numbers of line employees are needed: Monday–Friday, 7 employees; Saturday and Sunday, 3 employees. Grummins has 11 production-line employees. Formulate an LP that can be used to maximize the number of consecutive days off received by the employees. For example, a worker who gets Sunday, Monday, and Wednesday off receives two consecutive days off.

56 Bank 24 is open 24 hours per day. Tellers work two consecutive 6-hour shifts and are paid $10 per hour. The possible shifts are as follows: midnight–6 A.M., 6 A.M.–noon, noon–6 P.M., 6 P.M.–midnight. During each shift, the following numbers of customers enter the bank: midnight–6 A.M., 100; 6 A.M.–noon, 200; noon–6 P.M., 300; 6 P.M.–midnight, 200. Each teller can serve up to 50 customers per shift. To model a cost for customer impatience, we assume that any customer who is present at the end of a shift "costs" the bank $5. We assume that by midnight of each day, all customers must be served, so each day's midnight–6 A.M. shift begins with 0 customers in the bank. Formulate an LP that can be used to minimize the sum of the bank's labor and customer impatience costs.

57[†] Transeast Airlines flies planes on the following route: L.A.–Houston–N.Y.–Miami–L.A. The length (in miles) of each segment of this trip is as follows: L.A.–Houston, 1,500 miles; Houston–N.Y., 1,700 miles; N.Y.–Miami, 1,300 miles; Miami–L.A., 2,700 miles. At each stop, the plane may purchase up to 10,000 gallons of fuel. The price of fuel at each city is as follows: L.A., 88¢; Houston, 15¢; N.Y., $1.05; Miami, 95¢. The plane's fuel tank can hold at most 12,000 gallons. To allow for the possibility of circling over a landing site, we require that the ending fuel level for each leg of the flight be at least 600 gallons. The number of gallons used per mile on each leg of the flight is

    1 + (average fuel level on leg of flight/2000)

[†]Based on Darnell and Loflin (1977).

To simplify matters, assume that the average fuel level on any leg of the flight is

$$\frac{\text{(Fuel level at start of leg)} + \text{(fuel level at end of leg)}}{2}$$

Formulate an LP that can be used to minimize the fuel cost incurred in completing the schedule.

58[‡] To process income tax forms, the IRS first sends each form through the data preparation (DP) department, where information is coded for computer entry. Then the form is sent to data entry (DE), where it is entered into the computer. During the next three weeks, the following number of forms will arrive: week 1, 40,000; week 2, 30,000; week 3, 60,000. The IRS meets the crunch by hiring employees who work 40 hours per week and are paid $200 per week. Data preparation of a form requires 15 minutes, and data entry of a form requires 10 minutes. Each week, an employee is assigned to either data entry or data preparation. The IRS must complete processing of all forms by the end of week 5 and wants to minimize the cost of accomplishing this goal. Formulate an LP that will determine how many workers should be working each week and how the workers should be assigned over the next five weeks.

59 In the electrical circuit of Figure 11, I_t = current (in amperes) flowing through resistor t, V_t = voltage drop (in volts) across resistor t, and R_t = resistance (in ohms) of resistor t. Kirchoff's Voltage and Current Laws imply that $V_1 = V_2 = V_3$ and $I_1 + I_2 + I_3 = I_4$. The power dissipated by the current flowing through resistor t is $I_t^2 R_t$. Ohm's Law implies that $V_t = I_t R_t$. The two parts of this problem should be solved independently.

  a Suppose you are told that $I_1 = 4$, $I_2 = 6$, $I_3 = 8$, and $I_4 = 18$ are required. Also, the voltage drop across each resistor must be between 2 and 10 volts. Choose the R_t's to minimize the total dissipated power. Formulate an LP whose solution will solve your problem.

  b Suppose you are told that $V_1 = 6$, $V_2 = 6$, $V_3 = 6$, and $V_4 = 4$ are required. Also, the current flowing through each resistor must be between 2 and 6 amperes. Choose the R_t's to minimize the total dissipated power. Formulate an LP whose solution will solve your problem. (*Hint:* Let $\dfrac{1}{R_t}$ ($t = 1, 2, 3, 4$) be your decision variables.)

60 The mayor of Llanview is trying to determine the number of judges needed to handle the judicial caseload.

FIGURE 11

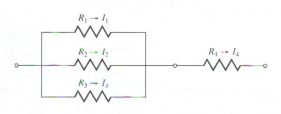

[‡]Based on Lanzenauer et al. (1987).

TABLE 84

Month	Hours
January	400
February	300
March	200
April	600
May	800
June	300
July	200
August	400
September	300
October	200
November	100
December	300

During each month of the year it is estimated that the number of judicial hours needed is as given in Table 84.

a Each judge works all 12 months and can handle as many as 120 hours per month of casework. To avoid creating a backlog, all cases must be handled by the end of December. Formulate an LP whose solution will determine how many judges Llanview needs.

b If each judge received one month of vacation each year, how would your answer change?

Group C

61[†] E.J. Korvair Department Store has $1,000 in available cash. At the beginning of each of the next six months, E.J. will receive revenues and pay bills as shown in Table 85. It is clear that E.J. will have a short-term cash flow problem until the store receives revenues from the Christmas shopping season. To solve this problem, E.J. must borrow money.

At the beginning of July, E.J. may take out a six-month loan. Any money borrowed for a six-month period must be paid back at the end of December along with 9% interest (early payback does not reduce the interest cost of the loan). E.J. may also meet cash needs through month-to-month borrowing. Any money borrowed for a one-month period incurs an interest cost of 4% per month. Use linear programming

to determine how E.J. can minimize the cost of paying its bills on time.

62[‡] Olé Oil produces three products: heating oil, gasoline, and jet fuel. The average octane levels must be at least 4.5 for heating oil, 8.5 for gas, and 7.0 for jet fuel. To produce these products Olé purchases two types of oil: crude 1 (at $12 per barrel) and crude 2 (at $10 per barrel). Each day, at most 10,000 barrels of each type of oil can be purchased.

Before crude can be used to produce products for sale, it must be distilled. Each day, at most 15,000 barrels of oil can be distilled. It costs 10¢ to distill a barrel of oil. The result of distillation is as follows: (1) Each barrel of crude 1 yields 0.6 barrel of naphtha, 0.3 barrel of distilled 1, and 0.1 barrel of distilled 2. (2) Each barrel of crude 2 yields 0.4 barrel of naphtha, 0.2 barrel of distilled 1, and 0.4 barrel of distilled 2. Distilled naphtha can be used only to produce gasoline or jet fuel. Distilled oil can be used to produce heating oil or it can be sent through the catalytic cracker (at a cost of 15¢ per barrel). Each day, at most 5,000 barrels of distilled oil can be sent through the cracker. Each barrel of distilled 1 sent through the cracker yields 0.8 barrel of cracked 1 and 0.2 barrel of cracked 2. Each barrel of distilled 2 sent through the cracker yields 0.7 barrel of cracked 1 and 0.3 barrel of cracked 2. Cracked oil can be used to produce gasoline and jet fuel but not to produce heating oil.

The octane level of each type of oil is as follows: naphtha, 8; distilled 1, 4; distilled 2, 5; cracked 1, 9; cracked 2, 6.

All heating oil produced can be sold at $14 per barrel; all gasoline produced, $18 per barrel; and all jet fuel produced, $16 per barrel. Marketing considerations dictate that at least 3,000 barrels of each product must be produced daily. Formulate an LP to maximize Olé's daily profit.

63 Donald Rump is the international funds manager for Countribank. Each day Donald's job is to determine how the bank's current holdings of dollars, pounds, marks, and yen should be adjusted to meet the day's currency needs. Today the exchange rates between the various currencies are given in Table 86. For example, one dollar can be converted to .58928 pounds, or one pound can be converted to 1.697 dollars.

At the beginning of the day, Countribank has the currency holdings given in Table 87.

At the end of the day, Countribank must have at least the amounts of each currency given in Table 88.

Donald's goal is to each day transfer funds in a way that makes currency holdings satisfy the previously listed mini-

TABLE 85

Month	Revenues ($)	Bills ($)
July	1,000	5,000
August	2,000	5,000
September	2,000	6,000
October	4,000	2,000
November	7,000	2,000
December	9,000	1,000

TABLE 86

| From | To | | | |
	Dollars	Pounds	Marks	Yen
Dollars	1	.58928	1.743	138.3
Pounds	1.697	1	2.9579	234.7
Marks	.57372	.33808	1	79.346
Yen	.007233	.00426	.0126	1

[†]Based on Robichek, Teichroew, and Jones (1965).

[‡]Based on Garvin et al. (1957).

TABLE 87			TABLE 88		
Currency	Units (in Billions)		Currency	Units (in Billions)	
Dollars	8		Dollars	6	
Pounds	1		Pounds	3	
Marks	8		Marks	1	
Yen	0		Yen	10	

mums, and maximizes the dollar value of the currency holdings at the end of the day.

To figure out the dollar value of, say, one pound, average the two conversion rates. Thus, one pound is worth approximately

$$\frac{1.697 + (1/.58928)}{2} = 1.696993 \text{ dollars}$$

REFERENCES

Each of the following seven books is a cornucopia of interesting LP formulations:

Bradley, S., A. Hax, and T. Magnanti. *Applied Mathematical Programming.* Reading, Mass.: Addison-Wesley, 1977.

Lawrence, K., and S. Zanakis. *Production Planning and Scheduling: Mathematical Programming Applications.* Atlanta, Ga: Industrial Engineering and Management Press, 1984.

Murty, K. *Operations Research: Deterministic Optimization Models.* Saddle River, N.J.: Prentice-Hall, 1995.

Schrage, L. *Linear Integer and Quadratic Programming With LINDO.* Palo Alto, Calif.: Scientific Press, 1986.

Shapiro, J. *Optimization Models for Planning and Allocation: Text and Cases in Mathematical Programming.* New York: Wiley, 1984.

Wagner, H. *Principles of Operations Research,* 2d ed. Englewood Cliffs, N.J.: Prentice Hall, 1975.

Williams, H. *Model Building in Mathematical Programming,* 2d ed. New York: Wiley, 1985.

Baker, K. "Scheduling a Full-Time Work Force to Meet Cyclic Staffing Requirements," *Management Science* 20(1974):1561–1568. Presents a method (other than LP) for scheduling personnel to meet cyclic workforce requirements.

Balintfy, J. "A Mathematical Programming System for Food Management Applications," *Interfaces* 6(no. 1, pt 2, 1976):13–31. Discusses menu planning models.

Carino, H., and C. Lenoir. "Optimizing Wood Procurement in Cabinet Manufacturing," *Interfaces* 18(no. 2, 1988):11–19.

Chandy, K. "Pricing in the Government Bond Market," *Interfaces* 16(1986):65–71.

Charnes, A., and W. Cooper. "Generalization of the Warehousing Model," *Operational Research Quarterly* 6(1955):131–172.

Cheung, H., and J. Auger. "Linear Programming and Land Use Allocation," *Socio-Economic Planning Science* 10(1976):43–45.

Darnell, W., and C. Loflin. "National Airlines Fuel Management and Allocation Model," *Interfaces* 7(no. 3, 1977):1–15.

Dobson, G., and S. Kalish. "Positioning and Pricing a Product Line," *Marketing Science* 7(1988):107–126.

Fabian, T. "A Linear Programming Model of Integrated Iron and Steel Production," *Management Science,* 4(1958):415–449.

Forgionne, G. "Corporate MS Activities: An Update," *Interfaces* 13(1983):20–23. Concerns the fraction of large firms using linear programming (and other operations research techniques).

Franklin, A., and E. Koenigsberg. "Computed School Assignments in a Large District," *Operations Research* 21(1973):413–426.

Garvin, W., et al. "Applications of Linear Programming in the Oil Industry," *Management Science* 3(1957): 407–430.

Glassey, R., and V. Gupta. "An LP Analysis of Paper Recycling." In *Studies in Linear Programming,* ed. H. Salkin and J. Saha. New York: North-Holland, 1975.

Hartley, R. "Decision Making When Joint Products Are Involved," *Accounting Review* (1971):746–755.

Heady, E., and A. Egbert. "Regional Planning of Efficient Agricultural Patterns," *Econometrica* 32(1964):374–386.

Hilal, S., and W. Erickson. "Matching Supplies to Save Lives: Linear Programming the Production of Heart Valves," *Interfaces* 11(1981):48–56.

Jain, S., K. Stott, and E. Vasold. "Orderbook Balancing Using a Combination of LP and Heuristic Techniques," *Interfaces* 9(no. 1, 1978):55–67.

Krajewski, L., L. Ritzman, and P. McKenzie. "Shift Scheduling in Banking Operations: A Case Application," *Interfaces,* 10(no. 2, 1980):1–8.

Love, R., and J. Hoey, "Management Science Improves Fast Food Operations," *Interfaces,* 20(no. 2, 1990): 21–29.

Magoulas, K., and D. Marinos-Kouris. "Gasoline Blending LP," *Oil and Gas Journal* (July 18, 1988):44–48.

Moondra, S. "An LP Model for Workforce Scheduling in Banks," *Journal of Bank Research* (1976).

Myers, S., and C. Pogue. "A Programming Approach to Corporate Financial Management," *Journal of Finance* 29(1974):579–599.

Neave, E., and J. Wiginton. *Financial Management: Theory and Strategies.* Englewood Cliffs, N.J.: Prentice Hall, 1981.

Robbins, W., and N. Tuntiwonpiboon. "Linear Programming a Useful Tool in Case-Mix Management," *HealthCare Financial Management* (1989):114–117.

Robichek, A., D. Teichroew, and M. Jones. "Optimal Short-Term Financing Decisions," *Management Science* 12(1965):1–36.

Rohn, E. "A New LP Approach to Bond Portfolio Management," *Journal of Financial and Quantitative Analysis* 22(1987):439–467.

Rothstein, M. "Hospital Manpower Shift Scheduling by Mathematical Programming," *Health Services Research* (1973).

Smith, S. "Planning Transistor Production by Linear Programming," *Operations Research* 13(1965): 132–139.

Stigler, G. "The Cost of Subsistence," *Journal of Farm Economics* 27(1945). Discusses the diet problem.

Sullivan, R., and S. Secrest. "A Simple Optimization DSS for Production Planning at Dairyman's Cooperative Creamery Association," *Interfaces* 15(no. 5, 1985): 46–54.

Weingartner, H. *Mathematical Programming and the Analysis of Capital Budgeting.* Englewood Cliffs, N.J.: Prentice Hall, 1963.

4

The Simplex Algorithm and Goal Programming

In Chapter 3, we saw how to solve two-variable linear programming problems graphically. Unfortunately, most real-life LPs have many variables, so a method is needed to solve LPs with more than two variables. We devote most of this chapter to a discussion of the simplex algorithm, which is used to solve even very large LPs. In many industrial applications, the simplex algorithm is used to solve LPs with thousands of constraints and variables.

In this chapter, we explain how the simplex algorithm can be used to find optimal solutions to LPs. We also detail how two state-of-the-art computer packages (LINDO and LINGO) can be used to solve LPs. Briefly, we also discuss Karmarkar's exciting new approach for solving LPs. We close the chapter with an introduction to goal programming, which enables the decision maker to consider more than one objective function.

4.1 How to Convert an LP to Standard Form

We have seen that an LP can have both equality and inequality constraints. It also can have variables that are required to be non-negative as well as those allowed to be unrestricted in sign (urs). Before the simplex algorithm can be used to solve an LP, the LP must be converted into an equivalent problem in which all constraints are equations and all variables are non-negative. An LP in this form is said to be in **standard form.**[†]

To convert an LP into standard form, each inequality constraint must be replaced by an equality constraint. We illustrate this procedure using the following problem.

EXAMPLE 1   Leather Limited

Leather Limited manufactures two types of belts: the deluxe model and the regular model. Each type requires 1 sq yd of leather. A regular belt requires 1 hour of skilled labor, and a deluxe belt requires 2 hours. Each week, 40 sq yd of leather and 60 hours of skilled labor are available. Each regular belt contributes \$3 to profit and each deluxe belt, \$4. If we define

$$x_1 = \text{number of deluxe belts produced weekly}$$
$$x_2 = \text{number of regular belts produced weekly}$$

[†]Throughout the first part of the chapter we assume that all variables must be non-negative (≥ 0). The conversion of urs variables to non-negative variables is discussed in Section 4.12.

the appropriate LP is

$$\max z = 4x_1 + 3x_2 \qquad \text{(LP 1)}$$
$$\text{s.t.} \qquad x_1 + x_2 \le 40 \qquad \text{(Leather constraint)} \qquad (1)$$
$$2x_1 + x_2 \le 60 \qquad \text{(Labor constraint)} \qquad (2)$$
$$x_1, x_2 \ge 0$$

How can we convert (1) and (2) to equality constraints? We define for each $\le$ constraint a **slack variable** s_i (s_i = slack variable for ith constraint), which is the amount of the resource unused in the ith constraint. Because $x_1 + x_2$ sq yd of leather are being used, and 40 sq yd are available, we define s_1 by

$$s_1 = 40 - x_1 - x_2 \qquad \text{or} \qquad x_1 + x_2 + s_1 = 40$$

Similarly, we define s_2 by

$$s_2 = 60 - 2x_1 - x_2 \qquad \text{or} \qquad 2x_1 + x_2 + s_2 = 60$$

Observe that a point (x_1, x_2) satisfies the ith constraint if and only if $s_i \ge 0$. For example, $x_1 = 15$, $x_2 = 20$ satisfies (1) because $s_1 = 40 - 15 - 20 = 5 \ge 0$.

Intuitively, (1) is satisfied by the point (15, 20), because $s_1 = 5$ sq yd of leather are unused. Similarly, (15, 20) satisfies (2), because $s_2 = 60 - 2(15) - 20 = 10$ labor hours are unused. Finally, note that the point $x_1 = x_2 = 25$ fails to satisfy (2), because $s_2 = 60 - 2(25) - 25 = -15$ indicates that (25, 25) uses more labor than is available.

In summary, to convert (1) to an equality constraint, we replace (1) by $s_1 = 40 - x_1 - x_2$ (or $x_1 + x_2 + s_1 = 40$) and $s_1 \ge 0$. To convert (2) to an equality constraint, we replace (2) by $s_2 = 60 - 2x_1 - x_2$ (or $2x_1 + x_2 + s_2 = 60$) and $s_2 \ge 0$. This converts LP 1 to

$$\max z = 4x_1 + 3x_2$$
$$\text{s.t.} \qquad x_1 + x_2 + s_1 \qquad = 40$$
$$2x_1 + x_2 \qquad + s_2 = 60 \qquad \text{(LP 1$'$)}$$
$$x_1, x_2, s_1, s_2 \ge 0$$

Note that LP 1$'$ is in standard form. In summary, *if constraint i of an LP is a $\le$ constraint, then we convert it to an equality constraint by adding a slack variable s_i to the ith constraint and adding the sign restriction $s_i \ge 0$.*

To illustrate how a $\ge$ constraint can be converted to an equality constraint, let's consider the diet problem of Section 3.4.

$$\min z = 50x_1 + 20x_2 + 30x_3 + 80x_4$$
$$\text{s.t.} \quad 400x_1 + 200x_2 + 150x_3 + 500x_4 \ge 500 \qquad \text{(Calorie constraint)} \qquad (3)$$
$$3x_1 + 2x_2 \qquad\qquad\qquad \ge 6 \qquad \text{(Chocolate constraint)} \qquad (4)$$
$$2x_1 + 2x_2 + 4x_3 + 4x_4 \ge 10 \qquad \text{(Sugar constraint)} \qquad (5)$$
$$2x_1 + 4x_2 + x_3 + 5x_4 \ge 8 \qquad \text{(Fat constraint)} \qquad (6)$$
$$x_1, x_2, x_3, x_4 \ge 0$$

To convert the ith $\ge$ constraint to an equality constraint, we define an **excess variable** (sometimes called a surplus variable) e_i. (e_i will always be the excess variable for the ith constraint.) We define e_i to be the amount by which the ith constraint is oversatisfied. Thus, for the diet problem,

$$e_1 = 400x_1 + 200x_2 + 150x_3 + 500x_4 - 500, \quad \text{or} \tag{3'}$$
$$400x_1 + 200x_2 + 150x_3 + 500x_4 - e_1 = 500$$

$$e_2 = 3x_1 + 2x_2 - 6, \quad \text{or} \quad 3x_1 + 2x_2 - e_2 = 6 \tag{4'}$$
$$e_3 = 2x_1 + 2x_2 + 4x_3 + 4x_4 - 10, \quad \text{or} \quad 2x_1 + 2x_2 + 4x_3 + 4x_4 - e_3 = 10 \tag{5'}$$
$$e_4 = 2x_1 + 4x_2 + x_3 + 5x_4 - 8, \quad \text{or} \quad 2x_1 + 4x_2 + x_3 + 5x_4 - e_4 = 8 \tag{6'}$$

A point (x_1, x_2, x_3, x_4) satisfies the ith $\geq$ constraint if and only if e_i is non-negative. For example, from (4'), $e_2 \geq 0$ if and only if $3x_1 + 2x_2 \geq 6$. For a numerical example, consider the point $x_1 = 2, x_3 = 4, x_2 = x_4 = 0$, which satisfies all four of the diet problem's constraints. For this point,

$$e_1 = 400(2) + 150(4) - 500 = 900 \geq 0$$
$$e_2 = 3(2) - 6 = 0 \geq 0$$
$$e_3 = 2(2) + 4(4) - 10 = 10 \geq 0$$
$$e_4 = 2(2) + 4 - 8 = 0 \geq 0$$

As another example, consider $x_1 = x_2 = 1, x_3 = x_4 = 0$. This point is infeasible; it violates the chocolate, sugar, and fat constraints. The infeasibility of this point is indicated by

$$e_2 = 3(1) + 2(1) - 6 = -1 < 0$$
$$e_3 = 2(1) + 2(1) - 10 = -6 < 0$$
$$e_4 = 2(1) + 4(1) - 8 = -2 < 0$$

Thus, to transform the diet problem into standard form, replace (3) by (3'); (4) by (4'); (5) by (5'); and (6) by (6'). We must also add the sign restrictions $e_i \geq 0$ ($i = 1, 2, 3, 4$). The resulting LP is in standard form and may be written as

$$\min z = 50x_1 + 20x_2 + 30x_3 + 80x_4$$
$$\text{s.t.} \quad 400x_1 + 200x_2 + 150x_3 + 500x_4 - e_1 \qquad\qquad = 500$$
$$3x_1 + 2x_2 \qquad\qquad\qquad - e_2 \qquad\qquad = 6$$
$$2x_1 + 2x_2 + 4x_3 + 4x_4 \qquad - e_3 \qquad = 10$$
$$2x_1 + 4x_2 + x_3 + 5x_4 \qquad\qquad - e_4 = 8$$
$$x_i, e_i \geq 0 \quad (i = 1, 2, 3, 4)$$

In summary, *if the ith constraint of an LP is a $\geq$ constraint, then it can be converted to an equality constraint by subtracting an excess variable e_i from the ith constraint and adding the sign restriction $e_i \geq 0$.*

If an LP has both $\leq$ and $\geq$ constraints, then simply apply the procedures we have described to the individual constraints. As an example, let's convert the short-term financial planning model of Section 3.7 to standard form. Recall that the original LP was

$$\max z = 20x_1 + 15x_2$$
$$\text{s.t.} \quad x_1 \qquad\quad \leq 100$$
$$x_2 \leq 100$$
$$50x_1 + 35x_2 \leq 6{,}000$$
$$20x_1 + 15x_2 \geq 2{,}000$$
$$x_1, x_2 \geq 0$$

Following the procedures described previously, we transform this LP into standard form by adding slack variables s_1, s_2, and s_3, respectively, to the first three constraints and subtracting an excess variable e_4 from the fourth constraint. Then we add the sign restrictions $s_1 \geq 0$, $s_2 \geq 0$, $s_3 \geq 0$, and $e_4 \geq 0$. This yields the following LP in standard form:

$$\max z = 20x_1 + 15x_2$$
$$\text{s.t.} \quad x_1 + \quad\quad + s_1 \quad\quad\quad\quad = 100$$
$$x_2 \quad + s_2 \quad\quad\quad = 100$$
$$50x_1 + 35x_2 \quad\quad + s_3 \quad\quad = 6{,}000$$
$$20x_1 + 15x_2 \quad\quad\quad - e_4 = 2{,}000$$
$$x_i \geq 0 \quad (i = 1, 2); \quad s_i \geq 0 \quad (i = 1, 2, 3); \quad e_4 \geq 0$$

Of course, we could easily have labeled the excess variable for the fourth constraint e_1 (because it is the first excess variable). We chose to call it e_4 rather than e_1 to indicate that e_4 is the excess variable for the fourth constraint.

PROBLEMS

Group A

1 Convert the Giapetto problem (Example 1 in Chapter 3) to standard form.

2 Convert the Dorian problem (Example 2 in Chapter 3) to standard form.

3 Convert the following LP to standard form:

$$\min z = 3x_1 + x_2$$
$$\text{s.t.} \quad x_1 \quad\quad \geq 3$$
$$x_1 + x_2 \leq 4$$
$$2x_1 - x_2 = 3$$
$$x_1, x_2 \geq 0$$

4.2 Preview of the Simplex Algorithm

Suppose we have converted an LP with m constraints into standard form. Assuming that the standard form contains n variables (labeled for convenience $x_1, x_2, \ldots, x_n$), the standard form for such an LP is

$$\max z = c_1x_1 + c_2x_2 + \cdots + c_nx_n$$

(or min)

$$\text{s.t.} \quad a_{11}x_1 + a_{12}x_2 + \cdots + a_{1n}x_n = b_1$$
$$a_{21}x_1 + a_{22}x_2 + \cdots + a_{2n}x_n = b_2 \qquad (7)$$
$$\vdots \quad\quad \vdots \quad\quad\quad \vdots$$
$$a_{m1}x_1 + a_{m2}x_2 + \cdots + a_{mn}x_n = b_m$$
$$x_i \geq 0 \quad (i = 1, 2, \ldots, n)$$

If we define

$$A = \begin{bmatrix} a_{11} & a_{12} & \ldots & a_{1n} \\ a_{12} & a_{22} & \ldots & a_{2n} \\ \vdots & \vdots & & \vdots \\ a_{1n} & a_{m2} & \ldots & a_{mn} \end{bmatrix}$$

and

$$\mathbf{x} = \begin{bmatrix} x_1 \\ x_2 \\ \vdots \\ x_n \end{bmatrix}, \qquad \mathbf{b} = \begin{bmatrix} b_1 \\ b_2 \\ \vdots \\ b_m \end{bmatrix}$$

the constraints for (7) may be written as the system of equations $A\mathbf{x} = \mathbf{b}$. Before proceeding further with our discussion of the simplex algorithm, we must define the concept of a basic solution to a linear system.

Basic and Nonbasic Variables

Consider a system $A\mathbf{x} = \mathbf{b}$ of m linear equations in n variables (assume $n \geq m$).

DEFINITION ■ A basic solution to $A\mathbf{x} = \mathbf{b}$ is obtained by setting $n - m$ variables equal to 0 and solving for the values of the remaining m variables. This assumes that setting the $n - m$ variables equal to 0 yields unique values for the remaining m variables or, equivalently, the columns for the remaining m variables are linearly independent. ■

To find a basic solution to $A\mathbf{x} = \mathbf{b}$, we choose a set of $n - m$ variables (the **nonbasic variables,** or **NBV**) and set each of these variables equal to 0. Then we solve for the values of the remaining $n - (n - m) = m$ variables (the **basic variables,** or **BV**) that satisfy $A\mathbf{x} = \mathbf{b}$.

Of course, the different choices of nonbasic variables will lead to different basic solutions. To illustrate, we find all the basic solutions to the following system of two equations in three variables:

$$\begin{aligned} x_1 + x_2 \phantom{{}+ x_3} &= 3 \\ - x_2 + x_3 &= -1 \end{aligned} \tag{8}$$

We begin by choosing a set of $3 - 2 = 1$ (3 variables, 2 equations) nonbasic variables. For example, if NBV $= \{x_3\}$, then BV $= \{x_1, x_2\}$. We obtain the values of the basic variables by setting $x_3 = 0$ and solving

$$\begin{aligned} x_1 + x_2 &= 3 \\ -x_2 &= -1 \end{aligned}$$

We find that $x_1 = 2$, $x_2 = 1$. Thus, $x_1 = 2$, $x_2 = 1$, $x_3 = 0$ is a basic solution to (8). However, if we choose NBV $= \{x_1\}$ and BV $= \{x_2, x_3\}$, we obtain the basic solution $x_1 = 0$, $x_2 = 3$, $x_3 = 2$. If we choose NBV $= \{x_2\}$, we obtain the basic solution $x_1 = 3$, $x_2 = 0$, $x_3 = -1$. The reader should verify these results.

Some sets of m variables do not yield a basic solution. For example, consider the following linear system:

$$\begin{aligned} x_1 + 2x_2 + x_3 &= 1 \\ 2x_1 + 4x_2 + x_3 &= 3 \end{aligned}$$

If we choose NBV $= \{x_3\}$ and BV $= \{x_1, x_2\}$, the corresponding basic solution would be obtained by solving

$$\begin{aligned} x_1 + 2x_2 &= 1 \\ 2x_1 + 4x_2 &= 3 \end{aligned}$$

Because this system has no solution, there is no basic solution corresponding to BV = $\{x_1, x_2\}$.

Feasible Solutions

A certain subset of the basic solutions to the constraints $A\mathbf{x} = \mathbf{b}$ of an LP plays an important role in the theory of linear programming.

DEFINITION ■ Any basic solution to (7) in which all variables are non-negative is a **basic feasible solution** (or **bfs**). ■

Thus, for an LP with the constraints given by (8), the basic solutions $x_1 = 2$, $x_2 = 1$, $x_3 = 0$, and $x_1 = 0$, $x_2 = 3$, $x_3 = 2$ are basic *feasible* solutions, but the basic solution $x_1 = 3$, $x_2 = 0$, $x_3 = -1$ fails to be a basic solution (because $x_3 < 0$).

In the rest of this section, we assume that all LPs are in standard form. Recall from Section 3.2 that the feasible region for any LP is a convex set. Let S be the feasible region for an LP in standard form. Recall that a point P is an extreme point of S if all line segments that contain P and are completely contained in S have P as an endpoint. It turns out that the extreme points of an LP's feasible region and the LP's basic feasible solutions are actually one and the same. More formally,

THEOREM 1

A point in the feasible region of an LP is an extreme point if and only if it is a basic feasible solution to the LP.
    See Luenburger (1984) for a proof of Theorem 1.

To illustrate the correspondence between extreme points and basic feasible solutions outlined in Theorem 1, let's look at the Leather Limited example of Section 4.1. Recall that the LP was

$$\max z = 4x_1 + 3x_2$$
$$\text{s.t.} \quad x_1 + x_2 \leq 40 \quad \text{(LP 1)}$$
$$2x_1 + x_2 \leq 60 \quad \text{(1)}$$
$$x_1, x_2 \geq 0 \quad \text{(2)}$$

By adding slack variables s_1 and s_2, respectively, to (1) and (2), we obtain LP 1 in standard form:

$$\max z = 4x_1 + 3x_2$$
$$\text{s.t.} \quad x_1 + x_2 + s_1 \quad\quad = 40$$
$$2x_1 + x_2 \quad\quad + s_2 = 60 \quad \text{(LP 1′)}$$
$$x_1, x_2, s_1, s_2 \geq 0$$

The feasible region for the Leather Limited problem is graphed in Figure 1. Both inequalities are satisfied: (1) by all points below or on the line $AB(x_1 + x_2 = 40)$, and (2) by all points on or below the line $CD(2x_1 + x_2 = 60)$. Thus, the feasible region for LP 1 is the shaded region bounded by the quadrilateral $BECF$. The extreme points of the feasible region are $B = (0, 40)$, $C = (30, 0)$, $E = (20, 20)$, and $F = (0, 0)$.

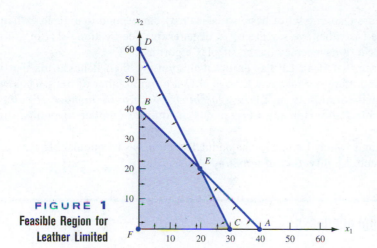

FIGURE 1

Feasible Region for
Leather Limited

Table 1 shows the correspondence between the basic feasible solutions to LP 1′ and the extreme points of the feasible region for LP 1. This example should make it clear that the basic feasible solutions to the standard form of an LP correspond in a natural fashion to the LP's extreme points.

In the context of the Leather Limited example, it is easy to show why any bfs is an extreme point. The converse is harder! We now show that for the LL problem, any bfs is an extreme point. Any point in the feasible region for LL may be specified as a four-dimensional column vector with the four elements of the vector denoting x_1, x_2, s_1, and s_2, respectively. Consider the bfs B with BV = $\{x_2, s_2\}$. If B is not an extreme point, then there exists two distinct feasible points v_1 and v_2 and non-negative numbers σ_1 and σ_2 satisfying $0 < \sigma_i < 1$ and $\sigma_1 + \sigma_2 = 1$ such that

$$\begin{bmatrix} 0 \\ 40 \\ 0 \\ 20 \end{bmatrix} = \sigma_1 v_1 + \sigma_2 v_2$$

Clearly, both v_1 and v_2 must both have $x_1 = s_2 = 0$. But because v_1 and v_2 are both feasible, the values of x_2 and s_2 for both v_1 and v_2 can be determined by solving $x_2 = 40$ and $x_2 + s_2 = 60$. These equations have a unique solution (because columns corresponding to basic variables x_2 and s_2 are linearly independent). This shows that $v_1 = v_2$, so B is indeed an extreme point.

TABLE 1

Correspondence between Basic Feasible Solutions and Corner Points for Leather Limited

Basic Variables	Nonbasic Variables	Basic Feasible Solution	Corresponds to Corner Point
x_1, x_2	s_1, s_2	$s_1 = s_2 = 0, x_1 = x_2 = 20$	E
x_1, s_1	x_2, s_2	$x_2 = s_2 = 0, x_1 = 30, s_1 = 10$	C
x_1, s_2	x_2, s_1	$x_2 = s_1 = 0, x_1 = 40, s_2 = -20$	Not a bfs because $s_2 < 0$
x_2, s_1	x_1, s_2	$x_1 = s_2 = 0, s_1 = -20, x_2 = 60$	Not a bfs because $s_1 < 0$
x_2, s_2	x_1, s_1	$x_1 = s_1 = 0, x_2 = 40, s_2 = 20$	B
s_1, s_2	x_1, x_2	$x_1 = x_2 = 0, s_1 = 40, s_2 = 60$	F

We note that more than one set of basic variables may correspond to a given extreme point. If this is the case, then we say the LP is **degenerate.** See Section 4.11 for a discussion of the impact of degeneracy on the simplex algorithm.

We will soon see that if an LP has an optimal solution, then it has a bfs that is optimal. This is important because any LP has only a finite number of bfs's. Thus we can find the optimal solution to an LP by *searching only a finite number of points.* Because the feasible region for any LP contains an infinite number of points, this helps us a lot!

Before explaining why any LP that has an optimal solution has an optimal bfs, we need to define the concept of a **direction of unboundedness.**

4.3   Direction of Unboundedness

Consider an LP in standard form with feasible region S and constraints $A\mathbf{x} = \mathbf{b}$ and $\mathbf{x} \geq \mathbf{0}$. Assuming that our LP has n variables, $\mathbf{0}$ represents an n-dimensional column vector consisting of all 0's. A nonzero vector $\mathbf{d}$ is a **direction of unboundedness** if for all $\mathbf{x} \in S$ and any $c \geq 0$, $x + c\mathbf{d} \in S$. In short, if we are in the LP's feasible region, then we can move as far as we want in the direction $\mathbf{d}$ and remain in the feasible region. Figure 2 displays the feasible region for the Dorian Auto example (Example 2 of Chapter 3). In standard form, the Dorian example is

$$\min z = 50x_1 + 100x_2$$
$$7x_1 + 2x_2 - e_1 = 28$$
$$2x_1 + 12x_2 - e_2 = 24$$
$$x_1, x_2, e_1, e_2 \geq 0$$

Looking at Figure 2 it is clear that if we start at any feasible point and move up and to the right at a 45-degree angle, we will remain in the feasible region. This means that

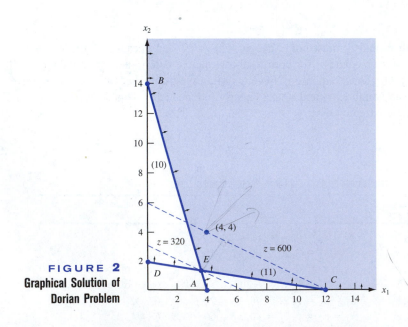

FIGURE 2
Graphical Solution of Dorian Problem

$$d = \begin{bmatrix} 1 \\ 1 \\ 9 \\ 14 \end{bmatrix}$$

is a direction of unboundedness for this LP. It is easy to show (see Problem 6) that d is a direction of unboundedness if and only if $Ad = 0$ and $d \geq 0$.

The following Representation Theorem [for a proof, see Nash and Sofer (1996)] is the key insight needed to show why any LP with an optimal solution has an optimal bfs.

THEOREM 2

Consider an LP in standard form, having bfs $b_1, b_2, \ldots, b_k$. Any point x in the LP's feasible region may be written in the form

$$x = d + \sum_{i=1}^{i=k} \sigma_i b_i$$

where d is 0 or a direction of unboundedness and $\sum_{i=1}^{i=k} \sigma_i = 1$ and $\sigma_i \geq 0$.

If the LP's feasible region is bounded, then $d = 0$, and we may write $x = \sum_{i=1}^{i=k} \sigma_i b_i$, where the σ_i are non-negative weights adding to 1. In this case, we see that any feasible x may be written as a **convex combination** of the LP's bfs. We now give two illustrations of Theorem 2.

Consider the Leather Limited example. The feasible region is bounded. To illustrate Theorem 2, we can write the point $G = (20, 10)$ (G is not a bfs!) in Figure 3 as a convex combination of the LP's bfs. Note from Figure 3 that point G may be written as $\frac{1}{6}F + \frac{5}{6}H$ [here $H = (24, 12)$]. Then note that point H may be written as $.6E + .4C$. Putting these two relationships together, we may write point G as $\frac{1}{6}F + \frac{5}{6}(.6E + .4C) = \frac{1}{6}F + \frac{1}{2}E + \frac{1}{3}C$. This expresses point G as a convex combination of the LP's extreme points.

To illustrate Theorem 2 for an unbounded LP, let's consider Example 2 of Chapter 3 (the Dorian example; see Figure 4) and try to express the point $F = (14, 4)$ in the representation given in Theorem 2. Recall that in standard form the constraints for the Dorian example are given by

$$7x_1 + 2x_2 - e_1 = 28$$
$$2x_1 + 12x_2 - e_2 = 24$$

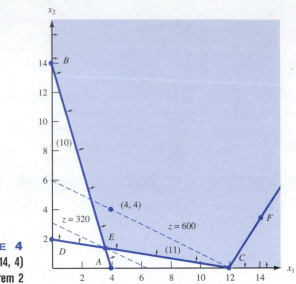

FIGURE 4
Expressing $F = (14, 4)$
Using Theorem 2

From Figure 4, we see that to move from bfs C to point F we need to move up and to the right along a line having slope $\frac{4 - 0}{14 - 12} = 2$. This line corresponds to the **direction of unboundedness**

$$\mathbf{d} = \begin{bmatrix} 2 \\ 4 \\ 22 \\ 52 \end{bmatrix}.$$

Letting

$$\mathbf{b}_1 = \begin{bmatrix} 12 \\ 0 \\ 56 \\ 0 \end{bmatrix} \quad \text{and} \quad \mathbf{x} = \begin{bmatrix} 14 \\ 4 \\ 78 \\ 52 \end{bmatrix}$$

we may write $\mathbf{x} = \mathbf{d} + \mathbf{b}_1$, which is the desired representation.

4.4 Why Does an LP Have an Optimal bfs?

Consider an LP with objective function max $\mathbf{cx}$ and constraints $A\mathbf{x} = \mathbf{b}$. Suppose this LP has an optimal solution. We now sketch a proof of the fact that the LP has an optimal bfs.

> **THEOREM 3**
>
> If an LP has an optimal solution, then it has an optimal bfs.
>
> **Proof**   Let $\mathbf{x}$ be an optimal solution to our LP. Because $\mathbf{x}$ is feasible, Theorem 2 tells us that we may write $\mathbf{x} = \mathbf{d} + \sum_{i=1}^{i=k} \sigma_i \mathbf{b}_i$, where $\mathbf{d}$ is $\mathbf{0}$ or a direction of unboundedness and $\mathbf{b}_1, \mathbf{b}_2, \ldots, \mathbf{b}_k$ are the LP's bfs. Also, $\sum_{i=1}^{i=k} \sigma_i = 1$ and $\sigma_i \geq 0$. If $\mathbf{cd} > 0$, then for any $k > 0$, $k\mathbf{d} + \sum_{i=1}^{i=k} \sigma_i \mathbf{b}_i$ is feasible, and as k grows larger and larger, the objective function value approaches infinity. This contradicts the fact that the LP has an optimal solution. If $\mathbf{cd} < \mathbf{0}$, then the feasible point $\sum_{i=1}^{i=k} \sigma_i \mathbf{b}_i$ has a larger ob-

jective function value than $\mathbf{x}$. This contradicts the optimality of $\mathbf{x}$. In short, we have shown that if $\mathbf{x}$ is optimal, then $\mathbf{cd} = 0$. Now the objective function value for $\mathbf{x}$ is given by

$$\mathbf{cx} = \mathbf{cd} + \textstyle\sum_{i=1}^{i=k} \sigma_i \mathbf{cb}_i = \sum_{i=1}^{i=k} \sigma_i \mathbf{cb}_i$$

Suppose that $\mathbf{b}_1$ is the bfs with the largest objective function value. Because $\sum_{i=1}^{i=k} \sigma_i = 1$ and $\sigma_i \geq 0$,

$$\mathbf{cb}_1 \geq \mathbf{cx}$$

Because $\mathbf{x}$ is optimal, this shows that $\mathbf{b}_1$ is also optimal, and the LP does indeed have an optimal bfs.

Adjacent Basic Feasible Solutions

Before describing the simplex algorithm in general terms, we need to define the concept of an adjacent basic feasible solution.

DEFINITION ■ For any **LP** with m constraints, two basic feasible solutions are said to be **adjacent** if their sets of basic variables have $m - 1$ basic variables in common. ■

For example, in Figure 3, two basic feasible solutions will be adjacent if they have $2 - 1 = 1$ basic variable in common. Thus, the bfs corresponding to point E in Figure 3 is adjacent to the bfs corresponding to point C. Point E is not, however, adjacent to bfs F. Intuitively, two basic feasible solutions are adjacent if they both lie on the same edge of the boundary of the feasible region.

We now give a general description of how the simplex algorithm solves LPs in a max problem.

Step 1 Find a bfs to the LP. We call this bfs the initial basic feasible solution. In general, the most recent bfs will be called the current bfs, so at the beginning of the problem the initial bfs is the current bfs.

Step 2 Determine if the current bfs is an optimal solution to the LP. If it is not, then find an adjacent bfs that has a larger z-value.

Step 3 Return to step 2, using the new bfs as the current bfs.

If an LP in standard form has m constraints and n variables, then there may be a basic solution for each choice of nonbasic variables. From n variables, a set of $n - m$ nonbasic variables (or equivalently, m basic variables) can be chosen in

$$\binom{n}{m} = \frac{n!}{(n - m)!m!}$$

different ways. Thus, an LP can have at most

$$\binom{n}{m}$$

basic solutions. Because some basic solutions may not be feasible, an LP can have at most

$$\binom{n}{m}$$

basic feasible solutions. If we were to proceed from the current bfs to a better bfs (without ever repeating a bfs), then we would surely find the optimal bfs after examining at most

$$\binom{n}{m}$$

basic feasible solutions. This means (assuming that no bfs is repeated) that the simplex algorithm will find the optimal bfs after a finite number of calculations. We return to this discussion in Section 4.11.

In principle, we could enumerate all basic feasible solutions to an LP and find the bfs with the largest z-value. The problem with this approach is that even small LPs have a very large number of basic feasible solutions. For example, an LP in standard form that has 20 variables and 10 constraints might have (if each basic solution were feasible) up to

$$\binom{20}{10} = 184{,}756$$

basic feasible solutions. Fortunately, vast experience with the simplex algorithm indicates that when this algorithm is applied to an n-variable, m-constraint LP in standard form, an optimal solution is usually found after examining fewer than $3m$ basic feasible solutions. Thus, for a 20-variable, 10-constraint LP in standard form, the simplex will usually find the optimal solution after examining fewer than $3(10) = 30$ basic feasible solutions. Compared with the alternative of examining 184,756 basic solutions, the simplex is quite efficient![†]

Geometry of Three-Dimensional LPs

Consider the following LP:

$$\begin{aligned}
\max z = \;& x_1 + 2x_2 + 2x_3 \\
\text{s.t.} \quad & 2x_1 + x_2 \qquad\; \leq 8 \\
& \qquad\qquad\quad x_3 \leq 10 \\
& x_1, x_2, x_3 \geq 0
\end{aligned}$$

The set of points satisfying a linear inequality in three (or any number of) dimensions is a **half-space.** For example, the set of points in three dimensions satisfying $2x_1 + x_2 \leq 8$ is a half-space. Thus, the feasible region for our LP is the intersection of the following five half-spaces: $2x_1 + x_2 \leq 8$, $x_3 \leq 10$, $x_1 \geq 0$, $x_2 \geq 0$, and $x_3 \geq 0$. The intersection of half-spaces is called a **polyhedron.** The feasible region for our LP is the prism pictured in Figure 5.

On each face (or facet) of the feasible region, one constraint (or sign restriction) is binding for all points on that face. For example, the constraint $2x_1 + x_2 \leq 8$ is binding for all points on the face *ABCD*; $x_3 \geq 0$ is binding on face *ABF*; $x_3 \leq 10$ is binding on face *DEC*; $x_2 \geq 0$ is binding on face *ADEF*; $x_1 \geq 0$ is binding on face *CBFE*.

Clearly, the corner (or extreme) points of the LP's feasible region are *A*, *B*, *C*, *D*, *E*, and *F*. In this case, the correspondence between the bfs and corner points is as shown in Table 2.

To illustrate the concept of adjacent basic feasible solutions, note that corner points *A*, *E*, and *B* are adjacent to corner point *F*. Thus, if the simplex algorithm begins at *F*, then we can be sure that the next bfs to be considered will be *A*, *E*, or *B*.

[†]In solving many LPs with 50 variables and $m \leq 50$ constraints, Chvàtal (1983) found that the simplex algorithm examined an average of $2m$ basic feasible solutions before finding an LP's optimal solution.

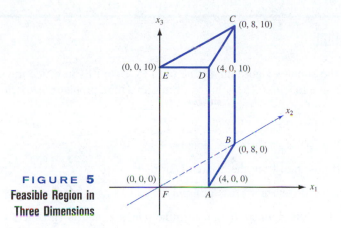

FIGURE 5
Feasible Region in Three Dimensions

TABLE 2
Correspondence between bfs and Corner Points

Basic Variables	Basic Feasible Solution	Corresponds to Corner Point
x_1, x_3	$x_1 = 4, x_3 = 10, x_2 = s_1 = s_2 = 0$	D
s_1, s_2	$s_1 = 8, s_2 = 10, x_1 = x_2 = x_3 = 0$	F
s_1, x_3	$s_1 = 8, x_3 = 10, x_1 = x_2 = s_2 = 0$	E
x_2, x_3	$x_2 = 8, x_3 = 10, x_1 = s_1 = s_2 = 0$	C
x_2, s_2	$x_2 = 8, s_2 = 10, x_1 = x_3 = s_1 = 0$	B
x_1, s_2	$x_1 = 4, s_2 = 10, x_2 = x_3 = s_1 = 0$	A

PROBLEMS

Group A

1 For the Giapetto problem (Example 1 in Chapter 3), show how the basic feasible solutions to the LP in standard form correspond to the extreme points of the feasible region.

2 For the Dorian problem (Example 2 in Chapter 3), show how the basic feasible solutions to the LP in standard form correspond to the extreme points of the feasible region.

3 Widgetco produces two products: 1 and 2. Each requires the amounts of raw material and labor, and sells for the price given in Table 3.

Up to 350 units of raw material can be purchased at $2 per unit, while up to 400 hours of labor can be purchased at $1.50 per hour. To maximize profit, Widgetco must solve the following LP:

$$\max z = 2x_1 + 2.5x_2$$
$$\text{s.t.} \quad x_1 + 2x_2 \le 350 \quad \text{(Raw material)}$$
$$2x_1 + x_2 \le 400 \quad \text{(Labor)}$$
$$x_1, x_2 \ge 0$$

Here, x_i = number of units of product i produced. Demonstrate the correspondence between corner points and basic feasible solutions.

TABLE 3

	Product 1	Product 2
Raw material	1 unit	2 units
Labor	2 hours	1 hour
Sales price	$7	$8

4 For the Leather Limited problem, represent the point $(10, 20)$ in the form $\mathbf{cd} + \sum_{i=1}^{i=k} \sigma_i \mathbf{cb}_i$.

5 For the Dorian problem, represent the point $(10,40)$ in the form $\mathbf{cd} + \sum_{i=1}^{i=k} \sigma_i \mathbf{cb}_i$.

Group B

6 For an LP in standard form with constraints $A\mathbf{x} = \mathbf{b}$ and $\mathbf{x} \ge \mathbf{0}$, show that $\mathbf{d}$ is a direction of unboundedness if and only if $A\mathbf{d} = \mathbf{0}$ and $\mathbf{d} \ge \mathbf{0}$.

7 Recall that Example 5 of Chapter 2 is an unbounded LP. Find a direction of unboundedness along which we can move for which the objective function becomes arbitrarily large.

4.5 The Simplex Algorithm

We now describe how the simplex algorithm can be used to solve LPs in which the goal is to maximize the objective function. The solution of minimization problems is discussed in Section 4.4.

The simplex algorithm proceeds as follows:

Step 1 Convert the LP to standard form (see Section 4.1).

Step 2 Obtain a bfs (if possible) from the standard form.

Step 3 Determine whether the current bfs is optimal.

Step 4 If the current bfs is not optimal, then determine which nonbasic variable should become a basic variable and which basic variable should become a nonbasic variable to find a new bfs with a better objective function value.

Step 5 Use EROs to find the new bfs with the better objective function value. Go back to step 3.

In performing the simplex algorithm, write the objective function

$$z = c_1 x_1 + c_2 x_2 + \cdots + c_n x_n$$

in the form

$$z - c_1 x_1 - c_2 x_2 - \cdots - c_n x_n = 0$$

We call this format the **row 0 version** of the objective function (row 0 for short).

EXAMPLE 2   **Dakota Furniture Company**

The Dakota Furniture Company manufactures desks, tables, and chairs. The manufacture of each type of furniture requires lumber and two types of skilled labor: finishing and carpentry. The amount of each resource needed to make each type of furniture is given in Table 4.

Currently, 48 board feet of lumber, 20 finishing hours, and 8 carpentry hours are available. A desk sells for \$60, a table for \$30, and a chair for \$20. Dakota believes that demand for desks and chairs is unlimited, but at most five tables can be sold. Because the available resources have already been purchased, Dakota wants to maximize total revenue. Defining the decision variables as

$$x_1 = \text{number of desks produced}$$
$$x_2 = \text{number of tables produced}$$
$$x_3 = \text{number of chairs produced}$$

TABLE 4
Resource Requirements for Dakota Furniture

Resource	Desk	Table	Chair
Lumber (board ft)	8	6	1
Finishing hours	4	2	1.5
Carpentry hours	2	1.5	0.5

it is easy to see that Dakota should solve the following LP:

$$\max z = 60x_1 + 30x_2 + 20x_3$$

$$
\begin{aligned}
\text{s.t.} \quad 8x_1 + 6x_2 + x_3 &\le 48 \quad &&\text{(Lumber constraint)} \\
4x_1 + 2x_2 + 1.5x_3 &\le 20 \quad &&\text{(Finishing constraint)} \\
2x_1 + 1.5x_2 + 0.5x_3 &\le 8 \quad &&\text{(Carpentry constraint)} \\
x_2 &\le 5 \quad &&\text{(Limitation on table demand)} \\
x_1, x_2, x_3 &\ge 0
\end{aligned}
$$

Convert the LP to Standard Form

We begin the simplex algorithm by converting the constraints of the LP to the standard form discussed in Section 4.1. Then we convert the LP's objective function to the row 0 format. To put the constraints in standard form, we simply add slack variables s_1, s_2, s_3, and s_4, respectively, to the four constraints. We label the constraints row 1, row 2, row 3, and row 4, and add the sign restrictions $s_i \ge 0$ ($i = 1, 2, 3, 4$). Note that the row 0 format for our objective function is

$$z - 60x_1 - 30x_2 - 20x_3 = 0$$

Putting rows 1–4 together with row 0 and the sign restrictions yields the equations and basic variables shown in Table 5. A system of linear equations (such as canonical form 0, shown in Table 5) in which each equation has a variable with a coefficient of 1 in that equation (and a zero coefficient in all other equations) is said to be in *canonical form*. We will soon see that if the right-hand side of each constraint in a canonical form is non-negative, a basic feasible solution can be obtained by inspection.[†]

From Section 4.2, we know that the simplex algorithm begins with an initial basic feasible solution and attempts to find better ones. After obtaining a canonical form, we therefore search for the initial bfs. By inspection, we see that if we set $x_1 = x_2 = x_3 = 0$, we can solve for the values of s_1, s_2, s_3, and s_4 by setting s_i equal to the right-hand side of row i.

$$\text{BV} = \{s_1, s_2, s_3, s_4\} \quad \text{and} \quad \text{NBV} = \{x_1, x_2, x_3\}$$

TABLE 5
Canonical Form 0

Row			Basic Variable
0	$z - 60x_1 - 30x_2 - 20x_3$	$= 0$	$z = 0$
1	$8x_1 + 6x_2 + x_3 + s_1$	$= 48$	$s_1 = 48$
2	$4x_1 + 2x_2 + 1.5x_3 + s_2$	$= 20$	$s_2 = 20$
3	$2x_1 + 1.5x_2 + 0.5x_3 + s_3$	$= 8$	$s_3 = 8$
4	$x_2 + s_4 = 5$		$s_4 = 5$

[†]If a canonical form with non-negative right-hand sides is not readily available, however, then the techniques described in Sections 4.12 and 4.13 can be used to find a canonical form and a basic feasible solution.

The basic feasible solution for this set of basic variables is $s_1 = 48$, $s_2 = 20$, $s_3 = 8$, $s_4 = 5$, $x_1 = x_2 = x_3 = 0$. Observe that each basic variable may be associated with the row of the canonical form in which the basic variable has a coefficient of 1. Thus, for canonical form 0, s_1 may be thought of as the basic variable for row 1, as may s_2 for row 2, s_3 for row 3, and s_4 for row 4.

To perform the simplex algorithm, we also need a basic (although not necessarily nonnegative) variable for row 0. Because z appears in row 0 with a coefficient of 1, and z does not appear in any other row, we use z as its basic variable. With this convention, the basic feasible solution for our initial canonical form has

$$\text{BV} = \{z, s_1, s_2, s_3, s_4\} \qquad \text{and} \qquad \text{NBV} = \{x_1, x_2, x_3\}$$

For this basic feasible solution, $z = 0$, $s_1 = 48$, $s_2 = 20$, $s_3 = 8$, $s_4 = 5$, $x_1 = x_2 = x_3 = 0$.

As this example indicates, a slack variable can be used as a basic variable for an equation if the right-hand side of the constraint is non-negative.

Is the Current Basic Feasible Solution Optimal?

Once we have obtained a basic feasible solution, we need to determine whether it is optimal; if the bfs is not optimal, then we try to find a bfs adjacent to the initial bfs with a larger z-value. To do this, we try to determine whether there is any way that z can be increased by increasing some nonbasic variable from its current value of zero while holding all other nonbasic variables at their current values of zero. If we solve for z by rearranging row 0, then we obtain

$$z = 60x_1 + 30x_2 + 20x_3 \tag{9}$$

For each nonbasic variable, we can use (9) to determine whether increasing a nonbasic variable (and holding all other nonbasic variables at zero) will increase z. For example, suppose we increase x_1 by 1 (holding the other nonbasic variables x_2 and x_3 at zero). Then (9) tells us that z will increase by 60. Similarly, if we choose to increase x_2 by 1 (holding x_1 and x_3 at zero), then (9) tells us that z will increase by 30. Finally, if we choose to increase x_3 by 1 (holding x_1 and x_2 at zero), then (9) tells us that z will increase by 20. Thus, increasing any of the nonbasic variables will increase z. Because a unit increase in x_1 causes the largest rate of increase in z, we choose to increase x_1 from its current value of zero. If x_1 is to increase from its current value of zero, then it will have to become a basic variable. For this reason, we call x_1 the **entering variable.** Observe that x_1 has the most negative coefficient in row 0.

Determine the Entering Variable

We choose the entering variable (in a max problem) to be the nonbasic variable with the most negative coefficient in row 0 (ties may be broken in an arbitrary fashion). Because each one-unit increase of x_1 increases z by 60, we would like to make x_1 as large as possible. What limits how large we can make x_1? Note that as x_1 increases, the values of the current basic variables (s_1, s_2, s_3, and s_4) will change. This means that increasing x_1 may cause a basic variable to become negative. With this in mind, we look at how increasing x_1 (while holding $x_2 = x_3 = 0$) changes the values of the current set of basic variables. From row 1, we see that $s_1 = 48 - 8x_1$ (remember that $x_2 = x_3 = 0$). Because the sign restriction $s_1 \geq 0$ must be satisfied, we can only increase x_1 as long as $s_1 \geq 0$, or $48 - 8x_1 \geq 0$, or $x_1 \leq \frac{48}{8} = 6$. From row 2, $s_2 = 20 - 4x_1$. We can only increase x_1 as long as

$s_2 \geq 0$, so x_1 must satisfy $20 - 4x_1 \geq 0$ or $x_1 \leq \frac{20}{4} = 5$. From row 3, $s_3 = 8 - 2x_1$ so $x_1 \leq \frac{8}{2} = 4$. Similarly, we see from row 4 that $s_4 = 5$. Thus, whatever the value of x_1, s_4 will be non-negative. Summarizing,

$$s_1 \geq 0 \qquad \text{for} \qquad x_1 \leq \frac{48}{8} = 6$$

$$s_2 \geq 0 \qquad \text{for} \qquad x_1 \leq \frac{20}{4} = 5$$

$$s_3 \geq 0 \qquad \text{for} \qquad x_1 \leq \frac{8}{2} = 4$$

$$s_4 \geq 0 \qquad \text{for all values of } x_1$$

This means that to keep all the basic variables non-negative, the largest that we can make x_1 is min $\{\frac{48}{8}, \frac{20}{4}, \frac{8}{2}\} = 4$. If we make $x_1 > 4$, then s_3 will become negative, and we will no longer have a basic feasible solution. Notice that each row in which the entering variable had a positive coefficient restricted how large the entering variable could become. Also, for any row in which the entering variable had a positive coefficient, the row's basic variable became negative when the entering variable exceeded

$$\frac{\text{Right-hand side of row}}{\text{Coefficient of entering variable in row}} \qquad \text{(10)}$$

If the entering variable has a nonpositive coefficient in a row (such as x_1 in row 4), the row's basic variable will remain positive for all values of the entering variable. Using (10), we can quickly compute how large x_1 can become before a basic variable becomes negative.

Row 1 limit on $x_1 = \dfrac{48}{8} = 6$

Row 2 limit on $x_1 = \dfrac{20}{4} = 5$

Row 3 limit on $x_1 = \dfrac{8}{2} = 4$

Row 4 limit on $x_1 = $ no limit   (Because coefficient of x_1 in row 4 is nonpositive)

We can state the following rule for determining how large we can make an entering variable.

The Ratio Test

When entering a variable into the basis, compute the ratio in (10) for every constraint in which the entering variable has a positive coefficient. The constraint with the smallest ratio is called the **winner of the ratio test.** The smallest ratio is the largest value of the entering variable that will keep all the current basic variables nonnegative. In our example, row 3 was the winner of the ratio test for entering x_1 into the basis.

Find a New Basic Feasible Solution: Pivot in the Entering Variable

Returning to our example, we know that the largest we can make x_1 is 4. For x_1 to equal 4, it must become a basic variable. Looking at rows 1–4, we see that if we make x_1 a basic variable in row 1, then x_1 will equal $\frac{48}{8} = 6$; in row 2, x_1 will equal $\frac{20}{4} = 5$; in row 3, x_1 will equal $\frac{8}{2} = 4$. Also, because x_1 does not appear in row 4, x_1 cannot be made a basic variable in row 4. Thus, if we want to make $x_1 = 4$, then we have to make it a basic variable in row 3. The fact that row 3 was the winner of the ratio test illustrates the following rule.

In Which Row Does the Entering Variable Become Basic?

Always make the entering variable a basic variable in a row that wins the ratio test (ties may be broken arbitrarily).

To make x_1 a basic variable in row 3, we use elementary row operations to make x_1 have a coefficient of 1 in row 3 and a coefficient of 0 in all other rows. This procedure is called **pivoting** on row 3; and row 3 is the **pivot row**. The final result is that x_1 replaces s_3 as the basic variable for row 3. The term in the pivot row that involves the entering basic variable is called the **pivot term.** Proceeding as we did when we studied the Gauss–Jordan method in Chapter 2, we make x_1 a basic variable in row 3 by performing the following EROs.

ERO 1 Create a coefficient of 1 for x_1 in row 3 by multiplying row 3 by $\frac{1}{2}$. The resulting row (marked with a prime to show it is the first iteration) is

$$x_1 + 0.75x_2 + 0.25x_3 + 0.5s_3 = 4 \qquad \text{(row 3')}$$

ERO 2 To create a zero coefficient for x_1 in row 0, replace row 0 with 60(row 3') + row 0.

$$z + 15x_2 - 5x_3 + 30s_3 = 240 \qquad \text{(row 0')}$$

ERO 3 To create a zero coefficient for x_1 in row 1, replace row 1 with -8(row 3') + row 1.

$$-x_3 + s_1 - 4s_3 = 16 \qquad \text{(row 1')}$$

ERO 4 To create a zero coefficient for x_1 in row 2, replace row 2 with -4(row 3') + row 2.

$$-x_2 + 0.5x_3 + s_2 - 2s_3 = 4 \qquad \text{(row 2')}$$

Because x_1 does not appear in row 4, we don't need to perform an ero to eliminate x_1 from row 4. Thus, we may write the "new" row 4 (call it row 4' to be consistent with other notation) as

$$x_2 + s_4 = 5 \qquad \text{(row 4')}$$

Putting rows 0'–4' together, we obtain the canonical form shown in Table 6.

Looking for a basic variable in each row of the current canonical form, we find that

$$BV = \{z, s_1, s_2, \mathbf{x}_1, s_4\} \qquad \text{and} \qquad NBV = \{\mathbf{s}_3, x_2, x_3\}$$

Thus, canonical form 1 yields the basic feasible solution $z = 240$, $s_1 = 16$, $s_2 = 4$, $x_1 = 4$, $s_4 = 5$, $x_2 = x_3 = s_3 = 0$. We could have predicted that the value of z in canonical form 1 would be 240 from the fact that each unit by which x_1 is increased increases z by 60. Because x_1 was increased by 4 units (from $x_1 = 0$ to $x_1 = 4$), we would expect that

$$\text{Canonical form 1 } z\text{-value} = \text{initial } z\text{-value} + 4(60)$$
$$= 0 + 240 = 240$$

TABLE 6
Canonical Form 1

Row								Basic Variable
Row 0'	z	$+ \ 15x_2 -$	$5x_3$		$+ \ 30s_3$	$= 240$		$z = 240$
Row 1'		$-$	$x_3 + s_1$		$- \ 4s_3$	$= 16$		$s_1 = 16$
Row 2'		$- \ x_2 +$	$0.5x_3$	$+ \ s_2 -$	$2s_3$	$= 4$		$s_2 = 4$
Row 3'	x_1	$+ \ 0.75x_2 +$	$0.25x_3$		$+ \ 0.5s_3$	$= 4$		$x_1 = 4$
Row 4'		x_2			$+ \ s_4$	$= 5$		$s_4 = 5$

In obtaining canonical form 1 from the initial canonical form, we have gone from one bfs to a better (larger z-value) bfs. Note that the initial bfs and the improved bfs are adjacent. This follows because the two basic feasible solutions have $4 - 1 = 3$ basic variables (s_1, s_2, and s_4) in common (excluding z, which is a basic variable in every canonical form). Thus, we see that in going from one canonical form to the next, we have proceeded from one bfs to a better adjacent bfs. The procedure used to go from one bfs to a better adjacent bfs is called an **iteration** (or sometimes, a *pivot*) of the simplex algorithm.

We now try to find a bfs that has a still larger z-value. We begin by examining canonical form 1 (Table 6) to see if we can increase z by increasing the value of some nonbasic variable (while holding all other nonbasic variables equal to zero). Rearranging row $0'$ to solve for z yields

$$z = 240 - 15x_2 + 5x_3 - 30s_3 \qquad \text{(11)}$$

From (11), we see that increasing the nonbasic variable x_2 by 1 (while holding $x_3 = s_3 = 0$) will decrease z by 15. We don't want to do that! Increasing the nonbasic variable s_3 by 1 (holding $x_2 = x_3 = 0$) will decrease z by 30. Again, we don't want to do that. On the other hand, increasing x_3 by 1 (holding $x_2 = s_3 = 0$) will increase z by 5. Thus, we choose to enter x_3 into the basis. Recall that our rule for determining the entering variable is to choose the variable with the most negative coefficient in the current row 0. Because x_3 is the only variable with a negative coefficient in row $0'$, it should be entered into the basis.

Increasing x_3 by 1 will increase z by 5, so it is to our advantage to make x_3 as large as possible. We can increase x_3 as long as the current basic variables (s_1, s_2, x_1, and s_4) remain non-negative. To determine how large x_3 can be, we must solve for the values of the current basic variables in terms of x_3 (holding $x_2 = s_3 = 0$). We obtain

From row $1'$: $s_1 = 16 + x_3$

From row $2'$: $s_2 = 4 - 0.5x_3$

From row $3'$: $x_1 = 4 - 0.25x_3$

From row $4'$: $s_4 = 5$

These equations tell us that $s_1 \geq 0$ and $s_4 \geq 0$ will hold for all values of x_3. From row $2'$, we see that $s_2 \geq 0$ will hold if $4 - 0.5x_3 \geq 0$, or $x_3 \leq \frac{4}{0.5} = 8$. From row $3'$, $x_1 \geq 0$ will hold if $4 - 0.25x_3 \geq 0$, or $x_3 \leq \frac{4}{0.25} = 16$. This shows that the largest we can make x_3 is min $\{\frac{4}{0.5}, \frac{4}{0.25}\} = 8$. This fact could also have been discovered by using (10) and the ratio test, as follows:

Row $1'$: no ratio $\qquad$ (x_3 has negative coefficient in row 1)

Row $2'$: $\dfrac{4}{0.5} = 8$

Row $3'$: $\dfrac{4}{0.25} = 16$

Row $4'$: no ratio $\qquad$ (x_3 has a nonpositive coefficient in row 4)

Thus, the smallest ratio occurs in row $2'$, and row $2'$ wins the ratio test. This means that we should use EROs to make x_3 a basic variable in row $2'$.

ERO 1 Create a coefficient of 1 for x_3 in row $2'$ by replacing row $2'$ with 2(row $2'$):

$$-2x_2 + x_3 + 2s_2 - 4s_3 = 8 \qquad \text{(row } 2'')$$

ERO 2 Create a coefficient of 0 for x_3 in row $0'$ by replacing row $0'$ with 5(row 2)$''$ + row $0'$:

$$z + 5x_2 + 10s_2 + 10s_3 = 280 \qquad \text{(row } 0'')$$

TABLE 7
Canonical Form 2

Row						Basic Variable
0″	z	$+ \quad 5x_2$	$+ \; 10s_2 +$	$10s_3$	$= 280$	$z = 280$
1″		$- \quad 2x_2$	$+ s_1 + \quad 2s_2 -$	$8s_3$	$= 24$	$s_1 = 24$
2″		$- \quad 2x_2 + x_3$	$+ \quad 2s_2 -$	$4s_3$	$= 8$	$x_3 = 8$
3″		$x_1 + 1.25x_2$	$- \; 0.5s_2 +$	$1.5s_3$	$= 2$	$x_1 = 2$
4″		x_2		$+ s_4$	$= 5$	$s_4 = 5$

ERO 3 Create a coefficient of 0 for x_3 in row 1′ by replacing row 1′ with row 2″ + row 1′:

$$-2x_2 + s_1 + 2s_2 - 8s_3 = 24 \qquad \text{(row 1″)}$$

ERO 4 Create a coefficient of 0 for x_3 in row 3′, by replacing row 3′ with $-\frac{1}{4}$(row 2″) + 3′:

$$x_1 + 1.25x_2 - 0.5s_2 + 1.5s_3 = 2 \qquad \text{(row 3″)}$$

Because x_3 already has a zero coefficient in row 4′, we may write

$$x_2 + s_4 = 5 \qquad \text{(row 4″)}$$

Combining rows 0″–4″ gives the canonical form shown in Table 7.

Looking for a basic variable in each row of canonical form 2, we find

$$\text{BV} = \{z, s_1, x_3, x_1, s_4\} \qquad \text{and} \qquad \text{NBV} = \{s_2, s_3, x_2\}$$

Canonical form 2 yields the following bfs: $z = 280$, $s_1 = 24$, $x_3 = 8$, $x_1 = 2$, $s_4 = 5$, $s_2 = s_3 = x_2 = 0$. We could have predicted that canonical form 2 would have $z = 280$ from the fact that each unit of the entering variable x_3 increased z by 5, and we have increased x_3 by 8 units. Thus,

$$\text{Canonical form 2 } z\text{-value} = \text{canonical form 1 } z\text{-value} + 8(5)$$
$$= 240 + 40 = 280$$

Because the bfs's for canonical forms 1 and 2 have (excluding z) $4 - 1 = 3$ basic variables in common (s_1, s_4, x_1), they are adjacent basic feasible solutions.

Now that the second iteration (or pivot) of the simplex algorithm has been completed, we examine canonical form 2 to see if we can find a better bfs. If we rearrange row 0″ and solve for z, we obtain

$$z = 280 - 5x_2 - 10s_2 - 10s_3 \qquad \text{(12)}$$

From (12), we see that increasing x_2 by 1 (while holding $s_2 = s_3 = 0$) will decrease z by 5; increasing s_2 by 1 (holding $s_3 = x_2 = 0$) will decrease z by 10; increasing s_3 by 1 (holding $x_2 = s_2 = 0$) will decrease z by 10. Thus, increasing any nonbasic variable will cause z to decrease. This might lead us to believe that our current bfs from canonical form 2 is an optimal solution. This is indeed correct! To see why, look at (12). We know that any feasible solution to the Dakota Furniture problem must have $x_2 \geq 0$, $s_2 \geq 0$, and $s_3 \geq 0$, and $-5x_2 \leq 0$, $-10s_2 \leq 0$, and $-10s_3 \leq 0$. Combining these inequalities with (12), it is clear that any feasible solution must have $z = 280 +$ terms that are ≤ 0, and $z \leq 280$. Our current bfs from canonical form 2 has $z = 280$, so it must be optimal.

The argument that we just used to show that canonical form 2 is optimal revolved around the fact that each of its nonbasic variables had a nonnegative coefficient in row 0″.

This means that we can determine whether a canonical form's bfs is optimal by applying the following simple rule.

Is a Canonical Form Optimal (Max Problem)?

A canonical form is optimal (for a max problem) if each nonbasic variable has a non-negative coefficient in the canonical form's row 0.

REMARKS
1 The coefficient of a decision variable in row 0 is often referred to as the variable's **reduced cost.** Thus, in our optimal canonical form, the reduced costs for x_1 and x_3 are 0, and the reduced cost for x_2 is 5. The reduced cost of a nonbasic variable is the amount by which the value of z will decrease if we increase the value of the nonbasic variable by 1 (while all the other nonbasic variables remain equal to 0). For example, the reduced cost for the variable "tables" (x_2) in canonical form 2 is 5. From (12), we see that increasing x_2 by 1 will reduce z by 5. Note that because all basic variables (except z, of course) must have zero coefficients in row 0, the reduced cost for a basic variable will always be 0. In Chapters 5 and 6, we discuss the concept of reduced costs in much greater detail.

These comments are correct only if the values of all the basic variables remain nonnegative after the nonbasic variable is increased by 1. Increasing x_2 to 1 leaves x_1, x_3, and s_1 all nonnegative, so our comments are valid.

2 From canonical form 2, we see that the optimal solution to the Dakota Furniture problem is to manufacture 2 desks ($x_1 = 2$) and 8 chairs ($x_3 = 8$). Because $x_2 = 0$, no tables should be made. Also, $s_1 = 24$ is reasonable because only $8 + 8(2) = 24$ board feet of lumber are being used. Thus, $48 - 24 = 24$ board feet of lumber are not being used. Similarly, $s_4 = 5$ makes sense because, although up to 5 tables could have been produced, 0 tables are actually being produced. Thus, the slack in constraint 4 is $5 - 0 = 5$. Because $s_2 = s_3 = 0$, all available finishing and carpentry hours are being utilized, so the finishing and carpentry constraints are binding.

3 We have chosen the entering variable to be the one with the most negative coefficient in row 0, but this may not always lead us quickly to the optimal bfs (see Review Problem 11). Actually, even if we choose the variable with the smallest (in absolute value) negative coefficient, the simplex algorithm will eventually find the LP's optimal solution.

4 Although any variable with a negative row 0 coefficient may be chosen to enter the basis, the pivot row *must* be chosen by the ratio test. To show this formally, suppose that we have chosen to enter x_i into the basis, and in the current tableau x_i is a basic variable in row k. Then row k may be written as

$$\bar{a}_{ki}x_i + \cdots = \bar{b}_k$$

Consider any other constraint (say, row j) in the canonical form. Row j in the current canonical form may be written as

$$\bar{a}_{ji}x_i + \cdots = \bar{b}_j$$

If we pivot on row k, row k becomes

$$x_i + \cdots = \frac{\bar{b}_k}{\bar{a}_{ki}}$$

The new row j after the pivot will be obtained by adding $-\bar{a}_{ji}$ times the last equation to row j of the current canonical form. This yields a new row j of

$$0x_i + \cdots = \bar{b}_j - \frac{\bar{b}_k \bar{a}_{ji}}{\bar{a}_{ki}}$$

We know that after the pivot, each constraint must have a non-negative right-hand side. Thus, $\bar{a}_{ki} > 0$ must hold to ensure that row k has a non-negative right-hand side after the pivot. Suppose $\bar{a}_{ji} > 0$. Then, to ensure that row j will have a non-negative right-hand side after the pivot, we must have

$$\frac{\bar{b}_j - \bar{b}_k \bar{a}_{ji}}{\bar{a}_{ki}} \geq 0$$

or (because $\bar{a}_{ji} > 0$)

$$\frac{\bar{b}_j}{\bar{a}_{ji}} \geq \frac{\bar{b}_k}{\bar{a}_{ki}}$$

Thus, row k must be a "winner" of the ratio test to ensure that row j will have a non-negative right-hand side after the pivot is completed.

If $\bar{a}_{ji} \leq 0$, then the right-hand side of row j will surely be non-negative after the pivot. This follows because

$$-\frac{\bar{b}_k \bar{a}_{ji}}{\bar{a}_{ki}} \geq 0$$

will now hold.

As promised earlier, we have outlined an algorithm that proceeds from one bfs to a better bfs. The algorithm stops when an optimal solution has been found. The convergence of the simplex algorithm is discussed further in Section 4.11.

Summary of the Simplex Algorithm for a Max Problem

Step 1 Convert the LP to standard form.

Step 2 Find a basic feasible solution. This is easy if all the constraints are $\leq$ with non-negative right-hand sides. Then the slack variable s_i may be used as the basic variable for row i. If no bfs is readily apparent, then use the techniques discussed in Sections 4.12 and 4.13 to find a bfs.

Step 3 If all nonbasic variables have non-negative coefficients in row 0, then the current bfs is optimal. If any variables in row 0 have negative coefficients, then choose the variable with the most negative coefficient in row 0 to enter the basis. We call this variable the *entering variable.*

Step 4 Use EROs to make the entering variable the basic variable in any row that wins the ratio test (ties may be broken arbitrarily). After the EROs have been used to create a new canonical form, return to step 3, using the current canonical form.

When using the simplex algorithm to solve problems, there should never be a constraint with a negative right-hand side (it is okay for row 0 to have a negative right-hand side; see Section 4.6). A constraint with a negative right-hand side is usually the result of an error in the ratio test or in performing one or more EROs. If one (or more) of the constraints has a negative right-hand side, then there is no longer a bfs, and the rules of the simplex algorithm may not lead to a better bfs.

Representing Simplex Tableaus

Rather than writing each variable in every constraint, we often used a shorthand display called a **simplex tableau.** For example, the canonical form

$$
\begin{aligned}
z + 3x_1 + x_2 \qquad\qquad &= 6 \\
x_1 \qquad + s_1 \quad\; &= 4 \\
2x_1 + x_2 \qquad + s_2 &= 3
\end{aligned}
$$

would be written in abbreviated form as shown in Table 8 (rhs = right-hand side). This format makes it very easy to spot basic variables: Just look for columns having a single entry of 1 and all other entries equal to 0 (s_1 and s_2). In our use of simplex tableaus, we will encircle the pivot term and denote the winner of the ratio test by *.

TABLE 8
A Simplex Tableau

z	x_1	x_2	s_1	s_2	rhs	Basic Variable
1	3	1	0	0	6	$z = 6$
0	1	0	1	0	4	$s_1 = 4$
0	2	1	0	1	3	$s_2 = 3$

PROBLEMS

Group A

1 Use the simplex algorithm to solve the Giapetto problem (Example 1 in Chapter 3).

2 Use the simplex algorithm to solve the following LP:

$$\max z = 2x_1 + 3x_2$$
$$\text{s.t.} \quad x_1 + 2x_2 \leq 6$$
$$2x_1 + x_2 \leq 8$$
$$x_1, x_2 \geq 0$$

3 Use the simplex algorithm to solve the following problem:

$$\max z = 2x_1 - x_2 + x_3$$
$$\text{s.t.} \quad 3x_1 + x_2 + x_3 \leq 60$$
$$x_1 - x_2 + 2x_3 \leq 10$$
$$x_1 + x_2 - x_3 \leq 20$$
$$x_1, x_2, x_3 \geq 0$$

4 Suppose you want to solve the Dorian problem (Example 2 in Chapter 3) by the simplex algorithm. What difficulty would occur?

5 Use the simplex algorithm to solve the following LP:

$$\max z = x_1 + x_2$$
$$\text{s.t.} \quad 4x_1 + x_2 \leq 100$$
$$x_1 + x_2 \leq 80$$
$$x_1 \leq 40$$
$$x_1, x_2 \geq 0$$

6 Use the simplex algorithm to solve the following LP:

$$\max z = x_1 + x_2 + x_3$$
$$\text{s.t.} \quad x_1 + 2x_2 + 2x_3 \leq 20$$
$$2x_1 + x_2 + 2x_3 \leq 20$$
$$2x_1 + 2x_2 + x_3 \leq 20$$
$$x_1, x_2, x_3 \geq 0$$

Group B

7 It has been suggested that at each iteration of the simplex algorithm, the entering variable should be (in a maximization problem) the variable that would bring about the greatest increase in the objective function. Although this usually results in fewer pivots than the rule of entering the most negative row 0 entry, the greatest increase rule is hardly ever used. Why not?

4.6 Using the Simplex Algorithm to Solve Minimization Problems

There are two different ways that the simplex algorithm can be used to solve minimization problems. We illustrate these methods by solving the following LP:

$$\min z = 2x_1 - 3x_2$$
$$\text{s.t.} \quad x_1 + x_2 \leq 4$$
$$x_1 - x_2 \leq 6$$
$$x_1, x_2 \geq 0$$

(LP 2)

Method 1

The optimal solution to LP 2 is the point (x_1, x_2) in the feasible region for LP 2 that makes $z = 2x_1 - 3x_2$ the smallest. Equivalently, we may say that the optimal solution to LP 2 is the point in the feasible region that makes $-z = -2x_1 + 3x_2$ the largest. This means that we can find the optimal solution to LP 2 by solving LP 2′:

$$\max - z = -2x_1 + 3x_2$$
$$\text{s.t.} \quad x_1 + x_2 \le 4$$
$$x_1 - x_2 \le 6 \qquad \text{(LP 2′)}$$
$$x_1, x_2 \ge 0$$

In solving LP 2′, we will use $-z$ as the basic variable for row 0. After adding slack variables s_1 and s_2 to the two constraints, we obtain the initial tableau in Table 9. Because x_2 is the only variable with a negative coefficient in row 0, we enter x_2 into the basis. The ratio test indicates that x_2 should enter the basis in the first constraint, row 1. The resulting tableau is shown in Table 10. Because each variable in row 0 has a non-negative coefficient, this is an optimal tableau. Thus, the optimal solution to LP 2′ is $-z = 12$, $x_2 = 4$, $s_2 = 10$, $x_1 = s_1 = 0$. Then the optimal solution to LP 2 is $z = -12$, $x_2 = 4$, $s_2 = 10$, $x_1 = s_1 = 0$. Substituting the values of x_1 and x_2 into LP 2's objective function, we obtain

$$z = 2x_1 - 3x_2 = 2(0) - 3(4) = -12$$

In summary, multiply the objective function for the min problem by -1 and solve the problem as a maximization problem with objective function $-z$. The optimal solution to the max problem will give you the optimal solution to the min problem. Remember that (optimal z-value for min problem) $= -$(optimal objective function value z for max problem).

Method 2

A simple modification of the simplex algorithm can be used to solve min problems directly. Modify Step 3 of the simplex as follows: If all nonbasic variables in row 0 have nonpositive coefficients, then the current bfs is optimal. If any nonbasic variable in row

TABLE 9
Initial Tableau for LP 2—Method 1

$-z$	x_1	x_2	s_1	s_2	rhs	Basic Variable	Ratio
1	2	-3	0	0	0	$-z = 0$	
0	1	①	1	0	4	$s_1 = 4$	$\frac{4}{1} = 4^*$
0	1	-1	0	1	6	$s_2 = 6$	None

TABLE 10
Optimal Tableau for LP 2—Method 1

$-z$	x_1	x_2	s_1	s_2	rhs	Basic Variable
1	5	0	3	0	12	$-z = 12$
0	1	1	1	0	4	$x_2 = 4$
0	2	0	1	1	10	$s_2 = 10$

TABLE 11
Initial Tableau for LP 2—Method 2

z	x_1	x_2	s_1	s_2	rhs	Basic Variable	Ratio
1	-2	3	0	0	0	$z = 0$	
0	1	①	1	0	4	$s_1 = 4$	$\frac{4}{1} = 4^*$
0	1	-1	0	1	6	$s_2 = 6$	None

TABLE 12
Optimal Tableau for LP 2—Method 2

z	x_1	x_2	s_1	s_2	rhs	Basic Variable
1	-5	0	-3	0	-12	$z = -12$
0	1	1	1	0	4	$x_2 = 4$
0	2	0	1	1	10	$s_2 = 10$

0 has a positive coefficient, choose the variable with the "most positive" coefficient in row 0 to enter the basis.

This modification of the simplex algorithm works because increasing a nonbasic variable with a positive coefficient in row 0 will *decrease z*. If we use this method to solve LP 2, then our initial tableau will be as shown in Table 11. Because x_2 has the most positive coefficient in row 0, we enter x_2 into the basis. The ratio test says that x_2 should enter the basis in row 1, resulting in Table 12. Because each variable in row 0 has a non-positive coefficient, this is an optimal tableau.[†] Thus, the optimal solution to LP 2 is (as we have already seen) $z = -12$, $x_2 = 4$, $s_2 = 10$, $x_1 = s_1 = 0$.

PROBLEMS

Group A

1 Use the simplex algorithm to find the optimal solution to the following LP:

$$\min z = 4x_1 - x_2$$
$$\text{s.t.} \quad 2x_1 + x_2 \le 8$$
$$x_2 \le 5$$
$$x_1 - x_2 \le 4$$
$$x_1, x_2 \ge 0$$

2 Use the simplex algorithm to find the optimal solution to the following LP:

$$\min z = -x_1 - x_2$$
$$\text{s.t.} \quad x_1 - x_2 \le 1$$
$$x_1 + x_2 \le 2$$
$$x_1, x_2 \ge 0$$

3 Use the simplex algorithm to find the optimal solution to the following LP:

$$\min z = 2x_1 - 5x_2$$
$$\text{s.t.} \quad 3x_1 + 8x_2 \le 12$$
$$2x_1 + 3x_2 \le 6$$
$$x_1, x_2 \ge 0$$

4 Use the simplex algorithm to find the optimal solution to the following LP:

$$\min z = -3x_1 + 8x_2$$
$$\text{s.t.} \quad 4x_1 + 2x_2 \le 12$$
$$2x_1 + 3x_2 \le 6$$
$$x_1, x_2 \ge 0$$

[†]To see that this tableau is optimal, note that from row 0, $z = -12 + 5x_1 + 3s_1$. Because $x_1 \ge 0$ and $s_1 \ge 0$, this shows that $z \ge -12$. Thus, the current bfs (which has $z = -12$) must be optimal.

4.7 Alternative Optimal Solutions

Recall from Example 3 of Section 3.3 that for some LPs, more than one extreme point is optimal. If an LP has more than one optimal solution, then we say that it has multiple or **alternative optimal solutions.** We show now how the simplex algorithm can be used to determine whether an LP has alternative optimal solutions.

Reconsider the Dakota Furniture example of Section 4.3, with the modification that tables sell for \$35 instead of \$30 (see Table 13). Because x_1 has the most negative coefficient in row 0, we enter x_1 into the basis. The ratio test indicates that x_1 should be entered in row 3. Now only x_3 has a negative coefficient in row 0, so we enter x_3 into the basis (see Table 14). The ratio test indicates that x_3 should enter the basis in row 2. The resulting, optimal, tableau is given in Table 15. As in Section 4.3, this tableau indicates that the optimal solution to the Dakota Furniture problem is $s_1 = 24$, $x_3 = 8$, $x_1 = 2$, $s_4 = 5$, and $x_2 = s_2 = s_3 = 0$.

TABLE 13
Initial Tableau for Dakota Furniture (\$35/Table)

z	x_1	x_2	x_3	s_1	s_2	s_3	s_4	rhs	Basic Variable	Ratio
1	−60	−35	−20	0	0	0	0	0	$z = 0$	
0	8	6	1	1	0	0	0	48	$s_1 = 48$	$\frac{48}{8} = 6$
0	4	2	1.5	0	1	0	0	20	$s_2 = 20$	$\frac{20}{4} = 5$
0	②	1.5	0.5	0	0	1	0	8	$s_3 = 8$	$\frac{8}{2} = 4*$
0	0	1	0	0	0	0	1	5	$s_4 = 5$	None

TABLE 14
First Tableau for Dakota Furniture (\$35/Table)

z	x_1	x_2	x_3	s_1	s_2	s_3	s_4	rhs	Basic Variable	Ratio
1	0	10	−5	0	0	30	0	240	$z = 240$	
0	0	0	−1	1	0	−4	0	16	$s_1 = 16$	None
0	0	−1	⑤0.5	0	1	−2	0	4	$s_2 = 4$	$\frac{4}{0.5} = 8*$
0	1	0.75	0.25	0	0	0.5	0	4	$x_1 = 4$	$\frac{4}{0.25} = 16$
0	0	1	0	0	0	0	1	5	$s_4 = 5$	None

TABLE 15
Second (and Optimal) Tableau for Dakota Furniture (\$35/Table)

z	x_1	x_2	x_3	s_1	s_2	s_3	s_4	rhs	Basic Variable
1	0	0	0	0	10	10	0	280	$z = 280$
0	0	−2	0	1	2	−8	0	24	$s_1 = 24$
0	0	−2	1	0	2	−4	0	8	$x_3 = 8$
0	1	①1.25	0	0	−0.5	1.5	0	2	$x_1 = 2*$
0	0	1	0	0	0	0	1	5	$s_4 = 5$

TABLE 16
Another Optimal Tableau for Dakota Furniture ($35/Table)

z	x_1	x_2	x_3	s_1	s_2	s_3	s_4	rhs	Basic Variable
1	0	0	0	0	10	10	0	280	$z = 280$
0	1.6	0	0	1	1.2	−5.6	0	27.2	$s_1 = 27.2$
0	1.6	0	1	0	1.2	−1.6	0	11.2	$x_3 = 11.2$
0	0.8	1	0	0	−0.4	1.2	0	1.6	$x_2 = 1.6$
0	−0.8	0	0	0	0.4	−1.2	1	3.4	$s_4 = 3.4$

Recall that all basic variables must have a zero coefficient in row 0 (or else they wouldn't be basic variables). However, in our optimal tableau, there is a nonbasic variable, x_2, which also has a zero coefficient in row 0. Let us see what happens if we enter x_2 into the basis. The ratio test indicates that x_2 should enter the basis in row 3 (check this). The resulting tableau is given in Table 16. The important thing to notice is that *because x_2 has a zero coefficient in the optimal tableau's row 0, the pivot that enters x_2 into the basis does not change row 0*. This means that all variables in our new row 0 will still have non-negative coefficients. Thus, our new tableau is also optimal. Because the pivot has not changed the value of z, an alternative optimal solution for the Dakota example is $z = 280$, $s_1 = 27.2$, $x_3 = 11.2$, $x_2 = 1.6$, $s_4 = 3.4$, and $x_1 = s_3 = s_2 = 0$.

In summary, if tables sell for $35, Dakota can obtain $280 in sales revenue by manufacturing 2 desks and 8 chairs or by manufacturing 1.6 tables and 11.2 chairs. Thus, Dakota has multiple (or alternative) optimal extreme points.

As stated in Chapter 3, it can be shown that any point on the line segment joining two optimal extreme points will also be optimal. To illustrate this idea, let's write our two optimal extreme points:

$$\text{Optimal extreme point 1} = \begin{bmatrix} x_1 \\ x_2 \\ x_3 \end{bmatrix} = \begin{bmatrix} 2 \\ 0 \\ 8 \end{bmatrix}$$

$$\text{Optimal extreme point 2} = \begin{bmatrix} x_1 \\ x_2 \\ x_3 \end{bmatrix} = \begin{bmatrix} 0 \\ 1.6 \\ 11.2 \end{bmatrix}$$

Thus, for $0 \le c \le 1$,

$$\begin{bmatrix} x_1 \\ x_2 \\ x_3 \end{bmatrix} = c \begin{bmatrix} 2 \\ 0 \\ 8 \end{bmatrix} + (1 - c) \begin{bmatrix} 0 \\ 1.6 \\ 11.2 \end{bmatrix} = \begin{bmatrix} 2c \\ 1.6 - 1.6c \\ 11.2 - 3.2c \end{bmatrix}$$

will be optimal. This shows that although the Dakota Furniture example has only two optimal extreme points, there are an infinite number of optimal solutions to the Dakota problem. For example, by choosing $c = 0.5$, we obtain the optimal solution $x_1 = 1$, $x_2 = 0.8$, $x_3 = 9.6$.

If there is no nonbasic variable with a zero coefficient in row 0 of the optimal tableau, then the LP has a unique optimal solution (see Problem 3). Even if there is a nonbasic variable with a zero coefficient in row 0 of the optimal tableau, it is possible that the LP may not have alternative optimal solutions (see Review Problem 25).

PROBLEMS

Group A

1 Show that if a toy soldier sold for $28, then the Giapetto problem would have alternative optimal solutions.

2 Show that the following LP has alternative optimal solutions; find three of them.

$$\max z = -3x_1 + 6x_2$$
$$\text{s.t.} \quad 5x_1 + 7x_2 \leq 35$$
$$-x_1 + 2x_2 \leq 2$$
$$x_1, x_2 \geq 0$$

3 Find alternative optimal solutions to the following LP:

$$\max z = x_1 + x_2$$
$$\text{s.t.} \quad x_1 + x_2 + x_3 \leq 1$$
$$x_1 \quad\quad + 2x_3 \leq 1$$
$$\text{All } x_i \geq 0$$

4 Find all optimal solutions to the following LP:

$$\max z = 3x_1 + 3x_2$$
$$\text{s.t.} \quad x_1 + x_2 \leq 1$$
$$\text{All } x_i \geq 0$$

5 How many optimal basic feasible solutions does the following LP have?

$$\max z = 2x_1 + 2x_2$$
$$\text{s.t.} \quad x_1 + x_2 \leq 6$$
$$2x_1 + x_2 \leq 13$$
$$\text{All } x_i \geq 0$$

Group B

6 Suppose you have found this optimal tableau (Table 17) for a maximization problem. Use the fact that each nonbasic variable has a strictly positive coefficient in row 0 to show that $x_1 = 4$, $x_2 = 3$, $s_1 = s_2 = 0$ is the unique optimal solution to this LP. (*Hint:* Can any extreme point having $s_1 > 0$ or $s_2 > 0$ have $z = 10$?)

7 Explain why the set of optimal solutions to an LP is a convex set.

8 Consider an LP with the optimal tableau shown in Table 18.

   a Does this LP have more than one bfs that is optimal?

   b How many optimal solutions does this LP have? (*Hint:* If the value of x_3 is increased, then how does this change the values of the basic variables and the z-value?)

9 Characterize all optimal solutions to the following LP:

$$\max z = -8x_5$$
$$\text{s.t.} \quad x_1 + \quad x_3 + 3x_4 + 2x_5 = 2$$
$$x_2 + 2x_3 + 4x_4 + 5x_5 = 5$$
$$\text{All } x_i \geq 0$$

TABLE 17

z	x_1	x_2	s_1	s_2	rhs
1	0	0	2	3	10
0	1	0	3	2	4
0	0	1	1	1	3

TABLE 18

z	x_1	x_2	x_3	x_4	rhs
1	0	0	0	2	2
0	1	0	-1	1	2
0	0	1	-2	3	3

4.8 Unbounded LPs

Recall from Section 3.3 that for some LPs, there exist points in the feasible region for which z assumes arbitrarily large (in max problems) or arbitrarily small (in min problems) values. When this situation occurs, we say that LP is unbounded. In this section, we show how the simplex algorithm can be used to determine whether an LP is unbounded.

EXAMPLE 3   **Breadco Bakeries: An Unbounded LP**

Breadco Bakeries bakes two kinds of bread: French and sourdough. Each loaf of French bread can be sold for 36¢, and each loaf of sourdough bread for 30¢. A loaf of French bread requires 1 yeast packet and 6 oz of flour; sourdough requires 1 yeast packet and 5 oz of flour. At present, Breadco has 5 yeast packets and 10 oz of flour. Additional yeast packets can be purchased at 3¢ each, and additional flour at 4¢/oz. Formulate and solve an LP that can be used to maximize Breadco's profits (= revenues − costs).

Solution Define

$$x_1 = \text{number of loaves of French bread baked}$$
$$x_2 = \text{number of loaves of sourdough bread baked}$$
$$x_3 = \text{number of yeast packets purchased}$$
$$x_4 = \text{number of ounces of flour purchased}$$

Then Breadco's objective is to maximize z = revenues − costs, where

$$\text{Revenues} = 36x_1 + 30x_2 \quad \text{and} \quad \text{Costs} = 3x_3 + 4x_4$$

Thus, Breadco's objective function is

$$\max z = 36x_1 + 30x_2 - 3x_3 - 4x_4$$

Breadco faces the following two constraints:

Constraint 1 Number of yeast packages used to bake bread cannot exceed available yeast plus purchased yeast.

Constraint 2 Ounces of flour used to bake breads cannot exceed available flour plus purchased flour.

Because

$$\text{Available yeast} + \text{purchased yeast} = 5 + x_3$$
$$\text{Available flour} + \text{purchased flour} = 10 + x_4$$

Constraint 1 may be written as

$$x_1 + x_2 \leq 5 + x_3 \quad \text{or} \quad x_1 + x_2 - x_3 \leq 5$$

and Constraint 2 may be written as

$$6x_1 + 5x_2 \leq 10 + x_4 \quad \text{or} \quad 6x_1 + 5x_2 - x_4 \leq 10$$

Adding the sign restrictions $x_i \geq 0$ ($i = 1, 2, 3, 4$) yields the following LP:

$$\max z = 36x_1 + 30x_2 - 3x_3 - 4x_4$$
$$\text{s.t.} \quad x_1 + x_2 - x_3 \qquad \leq 5 \qquad \text{(Yeast constraint)}$$
$$6x_1 + 5x_2 \qquad - x_4 \leq 10 \qquad \text{(Flour constraint)}$$
$$x_1, x_2, x_3, x_4 \geq 0$$

Adding slack variables s_1 and s_2 to the two constraints, we obtain the tableau in Table 19.

TABLE 19
Initial Tableau for Breadco

z	x_1	x_2	x_3	x_4	s_1	s_2	rhs	Basic Variable	Ratio
1	−36	−30	3	4	0	0	0	$z = 0$	
0	1	1	−1	0	1	0	5	$s_1 = 5$	$\frac{5}{1} = 5$
0	⑥	5	0	−1	0	1	10	$s_2 = 10$	$\frac{10}{6} = \frac{5}{3}*$

TABLE 20
First Tableau for Breadco

z	x_1	x_2	x_3	x_4	s_1	s_2	rhs	Basic Variable	Ratio
1	0	0	3	-2	0	6	60	$z = 60$	
0	0	$\frac{1}{6}$	-1	$\boxed{\frac{1}{6}}$	1	$-\frac{1}{6}$	$\frac{10}{3}$	$s_1 = \frac{10}{3}$	$(\frac{10}{3})/(\frac{1}{6}) = 20^*$
0	1	$\frac{5}{6}$	0	$-\frac{1}{6}$	0	$\frac{1}{6}$	$\frac{5}{3}$	$s_2 = \frac{5}{3}$	None

TABLE 21
Second Tableau for Breadco

z	x_1	x_2	x_3	x_4	s_1	s_2	rhs	Basic Variable	Ratio
1	0	2	-9	0	12	4	100	$z = 100$	
0	0	1	-6	1	6	-1	20	$x_4 = 20$	None
0	1	1	-1	0	1	0	5	$x_1 = 5$	None

Because $-36 < -30$, we enter x_1 into the basis. The ratio test indicates that x_1 should enter the basis in row 2. Entering x_1 into the basis in row 2 yields the tableau in Table 20. Because x_4 has the only negative coefficient in row 0, we enter x_4 into the basis. The ratio test indicates that x_4 should enter the basis in row 1, with the resulting tableau in Table 21. Because x_3 has the most negative coefficient in row 0, we would like to enter x_3 into the basis. The ratio test, however, fails to indicate the row in which x_3 should enter the basis. What is happening? Going back to the basic ideas that led us to the ratio test, we see that as x_3 is increased (holding the other nonbasic variables at zero), the current basic variables, x_4 and x_1, change as follows:

$$x_4 = 20 + 6x_3 \tag{13}$$
$$x_1 = 5 + x_3 \tag{14}$$

As x_3 is increased, both x_4 and x_1 increase. This means that no matter how large we make x_3, the inequalities $x_4 \geq 0$ and $x_1 \geq 0$ will still be true. Because each unit by which we increase x_3 will increase z by 9, we can find points in the feasible region for which z assumes an arbitrarily large value. For example, can we find a feasible point with $z \geq 1,000$? To do this, we need to increase z by $1,000 - 100 = 900$. Each unit by which x_3 is increased will increase z by 9, so increasing x_3 by $\frac{900}{9} = 100$ should give us $z = 1,000$. If we set $x_3 = 100$ (and hold the other nonbasic variables at zero), then (13) and (14) show that x_4 and x_1 must now equal

$$x_4 = 20 + 6(100) = 620$$
$$x_1 = 5 + (100) = 105$$

Thus, $x_1 = 105$, $x_3 = 100$, $x_4 = 620$, $x_2 = 0$ is a point in the feasible region with $z = 1,000$. In a similar fashion, we can find points in the feasible region having arbitrarily large z-values. This means the Breadco problem is an unbounded LP.

From the Breadco example, we see that an unbounded LP occurs in a max problem if there is a nonbasic variable with a negative coefficient in row 0 and there is no constraint that limits how large we can make the nonbasic variable. This situation will occur if a nonbasic variable (such as x_3) has a negative coefficient in row 0 and nonpositive coefficients in each constraint. To summarize, *an unbounded LP for a max problem occurs when a variable with a negative coefficient in row 0 has a nonpositive coefficient in each constraint.*

If an LP is unbounded, one will eventually come to a tableau where one wants to enter a variable (such as x_3) into the basis, but the ratio test will fail. This is probably the easiest way to spot an unbounded LP.

As we noted in Chapter 3, an unbounded LP is usually caused by an incorrect formulation. In the Breadco example, we obtained an unbounded LP because we allowed Breadco to pay $3 + 6(4) = 27$¢ for the ingredients in a loaf of French bread and then sell the loaf for 36¢. Thus, each loaf of French bread earns a profit of 9¢. Because unlimited purchases of yeast and flour are allowed, it is clear that our model allows Breadco to manufacture as much French bread as it desires, thereby earning arbitrarily large profits. This is the cause of the unbounded LP.

Of course, our formulation of the Breadco example ignored several aspects of reality. First, we assumed that demand for Breadco's products is unlimited. Second, we ignored the fact that certain resources to make bread (such as ovens and labor) are in limited supply. Finally, we made the unrealistic assumption that unlimited quantities of yeast and flour could be purchased.

Unbounded LPs and Directions of Unboundedness

Consider an LP with an objective function $c_1x_1 + c_2x_2 + \cdots + c_nx_n$. Let $\mathbf{c} = [c_1 \quad c_2 \ldots c_n]$. If the LP is a maximization problem, then the LP will be unbounded if and only if it has a direction of unboundedness $\mathbf{d}$ satisfying $\mathbf{cd} > 0$. If the LP is a minimization problem, then the LP will be unbounded if and only if it has a direction of unboundedness $\mathbf{d}$ satisfying $\mathbf{cd} < 0$. I[n Example 3, the last tableau shows us that if we start at the point

$$\begin{bmatrix} 5 \\ 0 \\ 0 \\ 20 \\ 0 \\ 0 \end{bmatrix}$$

(the variables are listed in the same order they are listed in the tableau), we can find a direction of unboundedness as follows. Every unit by which x_3 is increased will maintain feasibility if we increase x_1 by one unit and x_4 by six units and leave x_2, s_1, and s_2 unchanged. Because we can increase x_3 without limit, this indicates that

$$\mathbf{d} = \begin{bmatrix} 1 \\ 0 \\ 1 \\ 6 \\ 0 \\ 0 \end{bmatrix}$$

is a direction of unboundedness. Because

$$\mathbf{cd} = \begin{bmatrix} 36 & 30 & -3 & -4 & 0 & 0 \end{bmatrix} \begin{bmatrix} 1 \\ 0 \\ 1 \\ 6 \\ 0 \\ 0 \end{bmatrix} = 9$$

we know that LP is unbounded. This follows because each time we move in the direction **d** an amount that increases x_3 by one unit, we increase z by 9, and we can move as far as we want in the direction **d**.

PROBLEMS

Group A

1 Show that the following LP is unbounded:

$$\max z = 2x_2$$
$$\text{s.t.} \quad x_1 - x_2 \le 4$$
$$-x_1 + x_2 \le 1$$
$$x_1, x_2 \ge 0$$

Find a point in the feasible region with $z \ge 10,000$.

2 State a rule that can be used to determine if a min problem has an unbounded optimal solution (that is, z can be made arbitrarily small). Use the rule to show that

$$\min z = -2x_1 - 3x_2$$
$$\text{s.t.} \quad x_1 - x_2 \le 1$$
$$x_1 - 2x_2 \le 2$$
$$x_1, x_2 \ge 0$$

is an unbounded LP.

3 Suppose that in solving an LP, we obtain the tableau in Table 22. Although x_1 can enter the basis, this LP is unbounded. Why?

4 Use the simplex method to solve Problem 10 of Section 3.3.

TABLE 22

z	x_1	x_2	x_3	x_4	rhs
1	−3	−2	0	0	0
0	1	−1	1	0	3
0	2	0	0	1	4

5 Show that the following LP is unbounded:

$$\max z = x_1 + 2x_2$$
$$\text{s.t.} \quad -x_1 + x_2 \le 2$$
$$-2x_1 + x_2 \le 1$$
$$x_1, x_2 \ge 0$$

6 Show that the following LP is unbounded:

$$\min z = -x_1 - 3x_2$$
$$\text{s.t.} \quad x_1 - 2x_2 \le 4$$
$$-x_1 + x_2 \le 3$$
$$x_1, x_2 \ge 0$$

4.9 The LINDO Computer Package

LINDO (Linear Interactive and Discrete Optimizer) was developed by Linus Schrage (1986). It is a user-friendly computer package that can be used to solve linear, integer, and quadratic programming problems.[†] Appendix A to this chapter gives a brief explanation of how LINDO can be used to solve LP's. In this section, we explain how the information on a LINDO printout is related to our discussion of the simplex algorithm.

We begin by discussing the LINDO ouput for the Dakota Furniture example (see Figure 6). LINDO allows the user to name the variables, so we define

$$\text{DESKS} = \text{number of desks produced}$$
$$\text{TABLES} = \text{number of tables produced}$$
$$\text{CHAIRS} = \text{number of chairs produced}$$

Then the Dakota formulation in the first block of Figure 6 is

[†]See Chapter 9 for a discussion of integer programming and Chapter 12 for a discussion of quadratic programming.

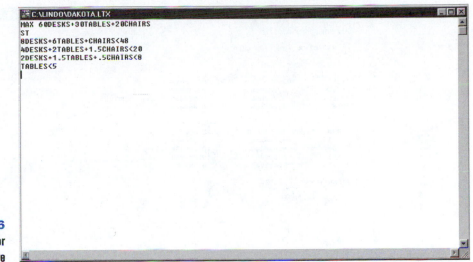

FIGURE 6
LINDO Output for
Dakota Furniture

```
C:\LINDO\DAKOTA.LTX
MAX 60DESKS+30TABLES+20CHAIRS
ST
8DESKS+6TABLES+CHAIRS<48
4DESKS+2TABLES+1.5CHAIRS<20
2DESKS+1.5TABLES+.5CHAIRS<8
TABLES<5
```

$$\max 60 \text{ DESKS} + 30 \text{ TABLES} + 20 \text{ CHAIRS} \quad \text{(Row 1)}$$

$$
\begin{array}{llll}
\text{s.t.} & 8 \text{ DESKS} + 6 \text{ TABLES} + \text{CHAIRS} \le 48 & \text{(Row 2)} & \text{(Lumber constraint)} \\
& 4 \text{ DESKS} + 2 \text{ TABLES} + 1.5 \text{ CHAIRS} \le 20 & \text{(Row 3)} & \text{(Finishing constraint)} \\
& 2 \text{ DESKS} + 1.5 \text{ TABLES} + 0.5 \text{ CHAIRS} \le 8 & \text{(Row 4)} & \text{(Carpentry constraint)} \\
& \phantom{2 \text{ DESKS} + 1.5 \text{ TABLES} + 0.5 \text{ CHAIRS}} \text{TABLES} \le 5 & \text{(Row 5)} &
\end{array}
$$

$$\text{DESKS, TABLES, CHAIRS} \ge 0$$

(LINDO assumes that all variables are non-negative, so the non-negativity constraints need not be input to the computer.) To be consistent with LINDO, we have labeled the objective function row 1 and the constraint rows 2–5.

To enter this problem in LINDO, make sure the screen contains a blank window, or work area, with "Untitled" at the top of the work area. If necessary, a new window can be opened by selecting New from the file menu or by clicking on the New File button.

The first statement in a LINDO model is always the objective. Enter the objective much like you would write it in equation form:

$$\text{MAX } 60 \text{ DESKS} + 30 \text{ TABLES} + 20 \text{ CHAIRS}$$

This tells LINDO to maximize the objective function. Proceed by entering the constraints as follows:

$$
\begin{array}{l}
\text{SUBJECT TO (OR s.t.)} \\
8 \text{ DESKS} + 6 \text{ TABLES} + \text{CHAIRS} < 48 \\
4 \text{ DESKS} + 2 \text{ TABLES} + 1.5 \text{ CHAIRS} < 20 \\
2 \text{ DESKS} + 1.5 \text{ TABLES} + .5 \text{ CHAIRS} < 8 \\
\phantom{2 \text{ DESKS} + 1.5 \text{ TABLES} + .5 \text{ CHAIRS}} \text{TABLES} < 5
\end{array}
$$

Your screen will now look like the one in Figure 6. Note that LINDO automatically assumes that all decision variables are non-negative.

Dakota

To save the file for later use, select Save from the File menu and when asked for a file name replace the * symbol with a name of your choice (we chose *Dakota*). Do not type over the characters .LTX. You may now use the File Open command to retrieve the problem.

To solve the model, proceed as follows:

1 From the Solve menu, select the Solve command or click the button with a bull's-eye.

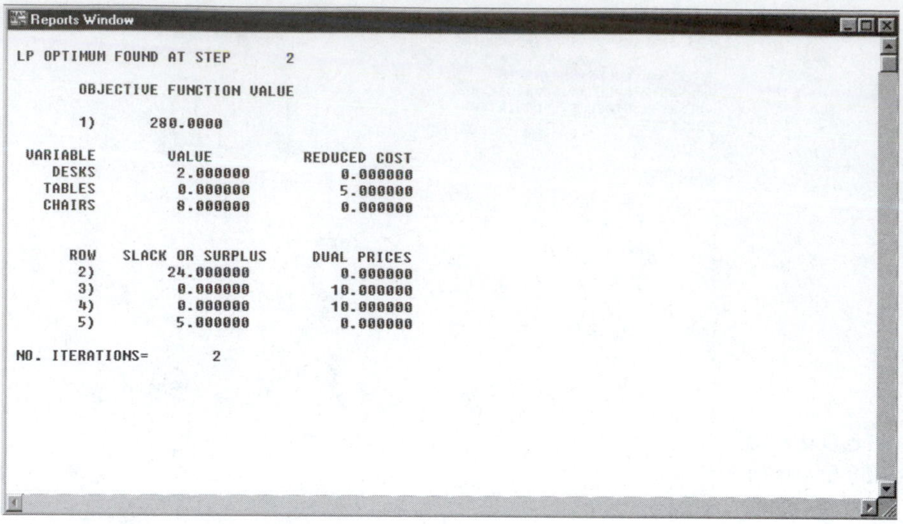

```
Reports Window

LP OPTIMUM FOUND AT STEP      2

      OBJECTIVE FUNCTION VALUE

      1)      280.0000

  VARIABLE         VALUE          REDUCED COST
     DESKS         2.000000           0.000000
    TABLES         0.000000           5.000000
    CHAIRS         8.000000           0.000000

       ROW   SLACK OR SURPLUS     DUAL PRICES
        2)        24.000000           0.000000
        3)         0.000000          10.000000
        4)         0.000000          10.000000
        5)         5.000000           0.000000

NO. ITERATIONS=       2
```

FIGURE 7

2 When asked if you want to do a range (sensitivity analysis) choose No. We will explain how to interpret a range or sensitivity analysis in Chapter 6.

3 When the solution is completed, a display showing the status of the Solve command will be present. After reviewing the displayed information, select Close.

4 You should now see your input data overlaying a display labeled "Reports Window." Click anywhere in the Reports window and your input data will be removed from the foreground. Move to the top of the screen using the single arrow at the right of the screen, and your screen should now look like that in Figure 7.

Looking now at the LINDO output in Figure 7, we see

LP OPTIMUM FOUND AT STEP 2

indicating that LINDO found the optimal solution after two iterations (or pivots) of the simplex algorithm.

OBJECTIVE FUNCTION VALUE 280.000000

indicates that the optimal z-value is 280.

VALUE

gives the value of the variable in the optimal LP solution. Thus, the optimal solution calls for Dakota to produce 2 desks, 0 tables, and 8 chairs.

SLACK OR SURPLUS

gives the value of the slack or excess ("surplus variable" is another name for excess variable) in the optimal solution. Thus,

$$s_1 = \text{slack for row 2 on LINDO output} = 24$$
$$s_2 = \text{slack for row 3 on LINDO output} = 0$$
$$s_3 = \text{slack for row 4 on LINDO output} = 0$$
$$s_4 = \text{slack for row 5 on LINDO output} = 5$$

REDUCED COST

gives the coefficient of the variable in row 0 of the optimal tableau (in a max problem). As discussed in Section 4.3, the reduced cost for each basic variable must be 0. For a non-

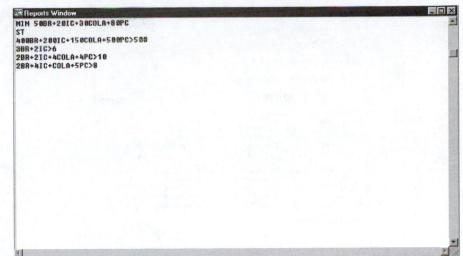

FIGURE 8

```
Reports Window
MIN 50BR+20IC+30COLA+80PC
ST
400BR+200IC+150COLA+500PC>500
3BR+2IC>6
2BR+2IC+4COLA+4PC>10
2BR+4IC+COLA+5PC>8
```

basic variable x_j, the reduced cost is the amount by which the optimal z-value is decreased if x_j is increased by 1 unit (and all other nonbasic variables remain equal to 0). In the LINDO output for the Dakota problem, the reduced cost is 0 for each of the basic variables (DESKS and CHAIRS). Also, the reduced cost for TABLES is 5. This means that if Dakota were forced to produce a table, revenue would decrease by $5.

For a minimization problem, the LP Optimum, Objective Function Value, and Slack and Surplus columns are interpreted as described. But the reduced cost for a variable is $-$(coefficient of variable in optimal row 0). Thus, in a min problem, the reduced cost for a basic variable will again be zero, but the reduced cost for a nonbasic variable x_j will be the amount by which the optimal z-value increases if x_j is increased by 1 unit (and all other nonbasic variables remain equal to 0).

To illustrate the interpretation of the LINDO output for a minimization problem, let's look at the LINDO output for the diet problem of Section 3.4 (see Figure 9). If we let

$$BR = \text{brownies eaten daily}$$
$$IC = \text{scoops of chocolate ice cream eaten daily}$$
$$COLA = \text{number of bottles of soda drunk daily}$$
$$PC = \text{pieces of pineapple cheesecake eaten daily}$$

then the diet problem may be formulated as

$$
\begin{aligned}
\min \quad & 50\,BR + 20\,IC + 30\,COLA + 80\,PC \\
\text{s.t.} \quad & 400\,BR + 200\,IC + 150\,COLA + 500\,PC \geq 500 && \text{(Calorie constraint)} \\
& 3\,BR + 2\,IC \geq 6 && \text{(Chocolate constraint)} \\
& 2\,BR + 2\,IC + 4\,COLA + 4\,PC \geq 10 && \text{(Sugar constraint)} \\
& 2\,BR + 4\,IC + COLA + 5\,PC \geq 8 && \text{(Fat constraint)} \\
& BR,\ IC,\ COLA,\ PC \geq 0
\end{aligned}
$$

The Value column shows that the optimal solution is to eat three scoops of chocolate ice cream daily and drink one bottle of soda daily. The Objective Function Value on the LINDO output indicates that the cost of this diet is 90¢. The Slack or Surplus column shows that the first constraint (calories) has an excess of 250 calories and that the fourth

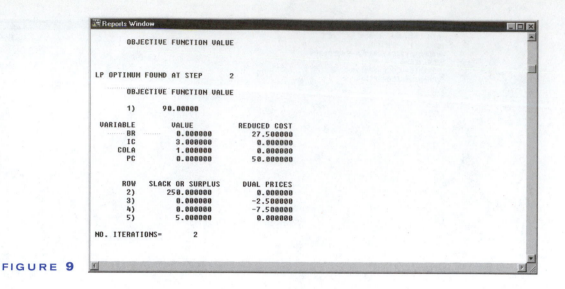

FIGURE 9

constraint (fat) has an excess of 5 oz. Thus, the calorie and fat constraints are nonbinding. The chocolate and sugar constraints have no excess and are therefore binding constraints.

From the Reduced Cost column, we see that if we were forced to eat a brownie (while keeping PC = 0), the minimum cost of the daily diet would increase by 27.5¢, and if we were forced to eat a piece of pineapple cheesecake (while holding BR = 0), the minimum cost of the daily diet would increase by 50¢.

The Tableau Command

If, after obtaining the optimal solution to the Dakota furniture problem, you close the Reports window and select the Tableau command (under the Reports menu), LINDO will display the optimal tableau (see Figure 10). Remembering that the first constraint is row 2 in LINDO, we see that BV = $\{s_1, \text{CHAIRS}, \text{DESKS}, s_4\}$. Thus, for example, SLK5 on the LINDO output corresponds to s_4. The artificial variable (ART) listed as basic in row 1 is z; thus, row 0 of the optimal tableau is $z + 5\text{TABLES} + 10s_2 + 10s_3 = 280$.

When you have installed LINDO on your hard drive, the LINDO formulation for the Dakota and Diet problems will be in the directory C:\WINSTON\LINDO\SAMPLES.

See Appendix A of Chapter 4 for further discussion of LINDO.

```
THE TABLEAU
   ROW   (BASIS)      DESKS      TABLES      CHAIRS     SLK   2     SLK   3
     1   ART           .000       5.000        .000        .000      10.000
     2   SLK   2       .000      -2.000        .000       1.000       2.000
     3     CHAIRS      .000      -2.000       1.000        .000       2.000
     4     DESKS      1.000       1.250        .000        .000       -.500
     5   SLK   5       .000       1.000        .000        .000        .000

   ROW   SLK   4     SLK   5
     1    10.000        .000     280.000
     2    -8.000        .000      24.000
     3    -4.000        .000       8.000
     4     1.500        .000       2.000
     5      .000       1.000       5.000
```

FIGURE 10
Example of TABLEAU
Command

4.10 Matrix Generators, LINGO, and Scaling of LPs

Many LPs solved in practice contain thousands of constraints and decision variables. Few users of linear programming would want to input the constraints and objective function each time such an LP is to be solved. For this reason, most actual applications of LP use a **matrix generator** to simplify the inputting of the LP. A matrix generator allows the user to input the relevant parameters that determine the LP's objective function and constraints; it then generates the LP formulation from that information. For example, let's consider the Sailco example from Section 3.10. If we were dealing with a planning horizon of 200 periods, then this problem would involve 400 constraints and 600 decision variables— clearly too many for convenient input. A matrix generator for this problem would require the user to input only the following information for each period: cost of producing a sailboat with regular-time labor, cost with overtime labor, demand, and holding costs. From this information, the matrix generator would generate the LP's objective function and constraints, call up an LP software package (such as LINDO) and solve the problem. Finally, an output analyzer would be written to display the output in a user-friendly format.

The LINGO Package

The package LINGO is an example of a sophisticated matrix generator (and much more!). LINGO is an optimization modeling language that enables the user to create many (perhaps thousands) of constraints or objective function terms by typing one line. To illustrate how LINGO works, we will solve the Sailco problem (Example 12 of Chapter 3).

Solving the Sailco Problem

Sail.ling

The LINGO model follows (it is the file "Sail.lng" on your disk).

```
MODEL:
 1] SETS:
 2] QUARTERS/Q1,Q2,Q3,Q4/:TIME,DEM,RP,OP,INV;
 3] ENDSETS
 4] MIN=@SUM(QUARTERS:400*RP+450*OP+20*INV);
 5] @FOR(QUARTERS(I):RP(I)<40);
 6] @FOR(QUARTERS(I)|TIME(I)#GT#1:
 7] INV(I)=INV(I-1)+RP(I)+OP(I)-DEM(I););
 8] INV(1)=10+RP(1)+OP(1)-DEM(1);
 9] DATA:
10] DEM=40,60,75,25;
11] TIME=1,2,3,4;
12] ENDDATA
END
```

To begin setting up a model with LINGO, think of the objects or sets that define the problem. For Sailco, the four quarters (Q1, Q2, Q3, and Q4) help define the problem. For each quarter we determine the objects that must be known to find an optimal production schedule—demand (DEM), regular-time production (RP), overtime production (OP), and end-of-quarter inventory (INV). The first three lines of the Sailco program define these objects. **SETS:** begins the definition of the sets needed to model the problem and **END-SETS** ends it. The effect of line 2 is to define four quarters: Q1, Q2, Q3, and Q4. For each quarter, line 2 creates time (indicating if the quarter is the first, second, third, or fourth quarter); the demand for sailboats; the regular-time and overtime production levels; and the ending inventory. Now that these sets and objects have been defined, we can use them to build a model (containing an objective function and constraints). LINGO will solve for the RP, OP, and INV once we input (in the DATA section of the program) the demands and numbers of the quarters.

Line 4 creates the objective function; **MIN** = indicates that we are minimizing. @**SUM**(QUARTERS: followed by 400*RP + 450*OP + 20*INV means sum 400*RP + 450*OP + 20*INV over all quarters. Thus for each quarter we compute 400*(regular-time production) + 450*(overtime production) + 20*(ending inventory). Notice that line 4 creates the proper objective function whether there are 4, 40, 400, or 4,000 quarters!

Line 5 says that for each quarter, RP cannot exceed 40. Again, if there were 400 quarters in the planning horizon, this statement would generate 400 constraints.

Together, lines 6 and 7 create constraints for all quarters (except the first) that ensure that

Ending Inventory for Quarter i = (Ending Inventory for Quarter $i - 1$)

$$+ \text{(Quarter } i \text{ Production)} - \text{(Quarter } i \text{ Demand)}$$

Notice that unlike LINDO, variables are allowed on the right side of a constraint (and numbers are allowed on the left side).

Line 8 creates the constraint ensuring that

(Ending Quarter 1 Inventory) = (Beginning Quarter 1 Inventory)

$$+ \text{(Quarter 1 Production)} - \text{(Quarter 1 Demand)}$$

Lines 9–12 input the needed data (the number of the quarter and the demand for each quarter). The DATA section must begin with a **DATA:** statement and end with an **END-DATA** statement. As with LINDO, a LINGO program ends with an **END** statement.

Notice that once we have created the LINGO model to solve the Sailco example, we can easily edit the model to solve any n-period production-scheduling model. If we were solving a 12-quarter problem, we would simply edit (see Remark 3 later) line 2 to QUARTERS/1..12/:TIME,DEM,RP,OP,INV;. Then enter the 12 quarterly demands in line 10 and change Line 11 to TIME=1,2,3,4,5,6,7,8,9,10,11,12;. To find the optimal solution to the problem either select the Solve command from the LINGO menu or click the button with a bull's-eye.

In this example, we will also look at how to use some of the editing capabilities of LINGO. Type the first four lines of this model just as you normally would. This will define the sets section and the objective function, and should appear as follows:

```
SETS:
  QUARTERS/Q1,Q2,Q3,Q4/:TIME,DEM,RP,OP,INV;
ENDSETS
MIN = @SUM(QUARTERS:400*RP+450*OP+20*INV);
```

The next line required is the @FOR statement that restricts regular-time production (RP) to values less than 40. Instead of typing in this entire statement, use LINGO's Paste Function command as follows:

1 From the Edit menu, select Paste Function. Notice that you do not have to click on this, but only highlight it, and a submenu appears.

2 From the submenu, select Set, and another submenu appears listing various @ functions.

3 Select the @FOR function, and a general form of the @FOR statement will appear in your input window.

4 Replace the general terms of the function with your specific parameters. This statement should then appear as follows:

```
@FOR(QUARTERS(I):RP(I)<40);
```

Because another @FOR statement is needed to further define constraints on all quarters, you could type this in or use the Paste Function command again. Using additional Edit commands, however, will allow you to copy and paste a portion of the previous @FOR statement instead of retyping it. Do this as follows:

1 Place your cursor at the beginning of the @FOR statement previously typed.

2 Hold down the left mouse button and drag the mouse to highlight the portion of the statement that can be reused, as shown below:

```
@FOR(QUARTERS(I):RP(I)<40);
```

3 From the Edit menu, select Copy (or use the shortcut Ctrl+C) to copy the highlighted text.

4 Place the cursor at the beginning of the next blank line and press Ctrl+V to paste the copied text.

You can now type in the remainder of this line, and the following lines as shown below.

```
@FOR(QUARTERS(I)|TIME(I) #GT#1:
INV(I)=INV(I-1)+RP(I)+OP(I)-DEM(I););
INV(1)=10+RP(1)+OP(1)-DEM(1);
DATA:
  DEM=40,60,75,25;
  TIME=1,2,3,4;
ENDDATA
END
```

While the Copy command only saved a few keystrokes in this example, it can save significantly more steps when you have repetition within a model. In a similar manner, the Cut command can remove highlighted portions for placement elsewhere within a model. The input for this example is saved in the file SAIL.LNG.

SAIL.LING

After solving the model, the first portion of the Reports window should indicate an objective value of $78,450, as shown in the output screen in Figure 11.

LINGO and the Post Office Problem

Post.lng

We now show how to use LINGO to solve the Post Office Scheduling example (Example 7) from Chapter 3. The following LINGO model (file "Post.lng") can be used to solve this problem.

```
MODEL:
  1]  SETS:
  2]  DAYS/1..7/:RQMT,START;
  3]  ENDSETS
  4]  MIN=@SUM(DAYS:START);
  5]  @FOR(DAYS(I):@SUM(DAYS(J)|
  6]  (J#GT#I+2)#OR#(J#LE#I#AND#J#GT#I-5):
  7]  START(J))>RQMT(I););
  8]  DATA:
  9]  RQMT=17,13,15,19,14,16,11;
 10]  ENDDATA
END
```

Line 1 defines the sets needed to solve the problem. Line 2 defines the days of the week (Monday, Tuesday, ..., Sunday) and associates each with two quantities: the number of workers needed (RQMT) and the number of workers that will begin work on that day of the week (START). Line 3 ends the definitions of the sets.

In Line 4, we create an objective function by summing the number of workers starting work on each day of the week. Lines 5–7 create for each day of the week the constraint that ensures the number of employees working that day is at least as large as the day's requirement. For DAY(I), lines 5 and 6 sum the number of employees starting work over the values of J satisfying $J > I + 2$ or $J \le I$ and $J > I - 5$. For instance for $I = 1$, this

```
MODEL:
SETS:
QUARTERS/Q1,Q2,Q3,Q4/:TIME,DEM,RP,OP,INV;
ENDSETS
MIN=@SUM(QUARTERS:400*RP+450*OP+20*INV);
@FOR(QUARTERS(I):RP(I)<40);
@FOR(QUARTERS(I)|TIME(I)#GT#1:
INV(I)=INV(I-1)+RP(I)+OP(I)-DEM(I););
INV(1)=10+RP(1)+OP(1)-DEM(1);
DATA:
DEM=40,60,75,25;
TIME=1,2,3,4;
ENDDATA
 END
```

```
MIN     400 RP( Q1) + 450 OP( Q1) + 20 INV( Q1) + 400 RP( Q2)
      + 450 OP( Q2) + 20 INV( Q2) + 400 RP( Q3) + 450 OP( Q3)
      + 20 INV( Q3) + 400 RP( Q4) + 450 OP( Q4) + 20 INV( Q4)
 SUBJECT TO
2]   RP( Q1) <=   40
3]   RP( Q2) <=   40
4]   RP( Q3) <=   40
5]   RP( Q4) <=   40
6]- INV( Q1) - RP( Q2) - OP( Q2) + INV( Q2) =  - 60
7]- INV( Q2) - RP( Q3) - OP( Q3) + INV( Q3) =  - 75
8]- INV( Q3) - RP( Q4) - OP( Q4) + INV( Q4) =  - 25
9]- RP( Q1) - OP( Q1) + INV( Q1) =  - 30
 END
```

```
Global optimal solution found at step:            7
 Objective value:                           78450.00
```

Variable	Value	Reduced Cost
TIME(Q1)	1.000000	0.0000000
TIME(Q2)	2.000000	0.0000000
TIME(Q3)	3.000000	0.0000000
TIME(Q4)	4.000000	0.0000000
DEM(Q1)	40.00000	0.0000000
DEM(Q2)	60.00000	0.0000000
DEM(Q3)	75.00000	0.0000000
DEM(Q4)	25.00000	0.0000000
RP(Q1)	40.00000	0.0000000
RP(Q2)	40.00000	0.0000000
RP(Q3)	40.00000	0.0000000
RP(Q4)	25.00000	0.0000000
OP(Q1)	0.0000000	20.00000
OP(Q2)	10.00000	0.0000000
OP(Q3)	35.00000	0.0000000
OP(Q4)	0.0000000	50.00000
INV(Q1)	10.00000	0.0000000
INV(Q2)	0.0000000	20.00000
INV(Q3)	0.0000000	70.00000
INV(Q4)	0.0000000	420.0000

Row	Slack or Surplus	Dual Price
1	78450.00	1.000000
2	0.0000000	30.00000
3	0.0000000	50.00000
4	0.0000000	50.00000
5	15.00000	0.0000000
6	0.0000000	450.0000
7	0.0000000	450.0000
8	0.0000000	400.0000
9	0.0000000	430.0000

FIGURE 11

generates the sum START(1) + START(4) + START(5) + START(6) + START(7), which is indeed the number of workers working on Day 1 (Monday). Line 7 (in concert with lines 5 and 6) then ensures that the number of employees working on Day I is at least as large as the number needed on Day I [RQMT(I)]. Line 8 begins the DATA section of the program. In Line 9, we input the requirements for each day of the week.

See Appendix B of Chapter 4 for further discussion of LINGO. Chapters 7, 8, 9, 11, and 12 contain many more examples of problems solved with LINGO.

Scaling of LPs

We close our discussion of computer packages by noting that an LP package may have trouble solving LPs in which there are nonzero coefficients that are either very small or very large in absolute value. If such coefficients are present, then LINDO will respond with a message that the LP is poorly scaled. The LINDO manual recommends that the user define the units of the objective function, right-hand sides, and decision variables so that no nonzero coefficients have absolute values of more than 100,000 or less than 0.0001.

PROBLEMS

Group A

1 A company produces three products. The per-unit profit, labor usage, and pollution produced per unit are given in Table 23. At most, 3 million labor hours can be used to produce the three products, and government regulations require that the company produce at most 2 lb of pollution. If we let x_i = units produced of product i, then the appropriate LP is

$$\max z = 6x_1 + 4x_2 + 3x_3$$

$$\text{s.t.} \quad 4x_1 + 3x_2 + 2x_3 \le 3{,}000{,}000$$

$$0.000003x_1 + 0.000002x_2 + 0.000001x_3 \le 2$$

$$x_1, x_2, x_3 \ge 0$$

a Explain why this LP is poorly scaled.

b Eliminate the scaling problem by redefining the units of the objective function, decision variables, and right-hand sides.

2 Use LINGO to solve Problem 1 in Section 3.5.

3 Use LINGO to solve Example 14 of Chapter 3.

4 The **product mix** problem occurs when we manufacture N products. Each unit produced of a given product uses a given amount of M resources. Each unit produced of product j earns a profit p_j. A quantity r_i of resource i is available. Formulate a LINGO model that could be used to maximize profit in this situation. Then use it to solve the product mix problem defined by the data in Tables 24 and 25. Assume that a fractional number of vehicles are allowed.

5 The **media mix** problem occurs when a company has N media in which the company can place an ad. There are K groups of people the company wishes to reach, and the company wishes its ads to be seen at least e_i times by members of group i. An ad on media j costs c_j dollars and reaches a_{ij} members of group i. The goal is to minimize the cost of ensuring that the desired number of people in each group see the ads. Set up a LINGO model that can be used to solve any media mix problem. Then solve the media mix problem defined by the data in Tables 26 and 27. Assume that a fractional number of ads is feasible.

TABLE 25

Resource	Quantity Available
Steel	50 tons
Rubber	10 tons
Labor	150 hours

TABLE 23

Product	Profit ($)	Labor Usage (Hrs)	Pollution (Lb)
1	6	4	0.000003 lb
2	4	3	0.000002 lb
3	3	2	0.000001 lb

TABLE 26

Group	Needed Exposures (in Millions)
Children	15
Men	40
Women	50

TABLE 24

	Cars	Trucks	Trains
Steel used (tons)	2	3	5
Rubber used (tons)	.3	.7	.2
Labor used (hrs)	10	12	20
Unit profit ($)	800	1,500	2,500

TABLE 27

No. Watching (million)	Program Sponge Bob	Program Friends	Program Dawson's Creek
Children	3	1	0
Men	1	15	4
Women	2	20	9
Unit cost ($)	30,000	360,000	80,000

TABLE 28

	District									
	1	2	3	4	5	6	7	8	9	10
Whites	400	200	150	300	400	100	200	300	250	150
Blacks	200	150	100	120	80	90	140	160	100	60

	Distance (Miles)									
	1	2	3	4	5	6	7	8	9	10
High School 1	1	2	3	2	3	4	2	3	1	2
High School 2	2	1	3	3	4	2	1	2	2	3
High School 3	3	3	2	1	2	3	2	2	3	1

6 Consider the following **school redistricting problem.** There are I districts in a city and J high schools in the city. The distance between District i and High School j is d_{ij} miles. District i has w_i white and b_i black residents. Each high school must have between L and U students. In the interests of racial harmony, the percentages of blacks at each high school must be between 80% and 120% of the percentage of black students in the entire city.

a Set up a LINGO model that can be used to minimize the total distance that students will have to travel in order to meet the racial balance requirements.

b Use your model to solve the problem defined by the data in Table 28.

c What might be some alternative objective functions for this situation?

d Do you see any other problems with our model?

4.11 Degeneracy and the Convergence of the Simplex Algorithm

Theoretically, the simplex algorithm (as we have described it) can fail to find the optimal solution to an LP. However, LPs arising from actual applications seldom exhibit this unpleasant behavior. For the sake of completeness, however, we now discuss the type of situation in which the simplex can fail. Our discussion depends crucially on the following relationship (for a max problem) between the z-values for the current bfs and the new bfs (that is, the bfs after the next pivot):

$$z\text{-value for new bfs} = z\text{-value of current bfs}$$
$$- \text{(value of entering variable in new bfs)(coefficient} \quad \text{(15)}$$
$$\text{of entering variable in row 0 of current bfs)}$$

Equation (15) follows, because each unit by which the entering variable is increased will increase z by $-$ (coefficient of entering variable in row 0 of current bfs). Recall that (coefficient of entering variable in row 0) < 0 and (value of entering variable in new bfs) ≥ 0. Combining these facts with (15), we can deduce the following facts:

1 If (value of entering variable in new bfs) > 0, then (z-value for new bfs) $>$ (z-value for current bfs).

2 If (value of entering variable in new bfs) $= 0$, then (z-value for new bfs) $=$ (z-value for current bfs).

For the moment, assume that the LP we are solving has the following property: In each of the LP's basic feasible solutions, all of the basic variables are positive (positive means > 0). An LP with this property is a **nondegenerate LP.**

If we are using the simplex to solve a nondegenerate LP, fact 1 in the foregoing list tells us that each iteration of the simplex will *increase* z. This implies that when the simplex is used to solve a nondegenerate LP, it is impossible to encounter the same bfs twice.

To see this, suppose that we are at a basic feasible solution (call it bfs 1) that has $z = 20$. Fact 1 shows that our next pivot will take us to a bfs (call it bfs 2) and has $z > 20$. Because no future pivot can decrease z, we can never return to a bfs having $z = 20$. Thus, we can never return to bfs 1. Now recall that every LP has only a finite number of basic feasible solutions. Because we can never repeat a bfs, this argument shows that when we use the simplex algorithm to solve a nondegenerate LP, we are guaranteed to find the optimal solution in a finite number of iterations. For example, suppose we are solving a nondegenerate LP with 10 variables and 5 constraints. Such an LP has at most

$$\binom{10}{5} = 252$$

basic feasible solutions. We will never repeat a bfs, so we know that for this problem, the simplex is guaranteed to find an optimal solution after at most 252 pivots.

However, the simplex may fail for a degenerate LP.

DEFINITION ■ An LP is **degenerate** if it has at least one bfs in which a basic variable is equal to zero. ■

The following LP is degenerate:

$$\max z = 5x_1 + 2x_2$$
$$\text{s.t.} \quad x_1 + x_2 \leq 6$$
$$x_1 - x_2 \leq 0 \tag{16}$$
$$x_1, x_2 \geq 0$$

What happens when we use the simplex algorithm to solve (16)? After adding slack variables s_1 and s_2 to the two constraints, we obtain the initial tableau in Table 29. In this bfs, the basic variable $s_2 = 0$. Thus, (16) is a degenerate LP. Any bfs that has at least one basic variable equal to zero (or, equivalently, at least one constraint with a zero right-hand side) is a **degenerate bfs.** Because $-5 < -2$, we enter x_1 into the basis. The winning ratio is 0. This means that after x_1 enters the basis, x_1 will equal zero in the new bfs. After doing the pivot, we obtain the tableau in Table 30. Our new bfs has the same z-value as

TABLE 29
A Degenerate LP

z	x_1	x_2	s_1	s_2	rhs	Basic Variable	Ratio
1	-5	-2	0	0	0	$z = 0$	
0	1	1	1	0	6	$s_1 = 6$	6
0	①	-1	0	1	0	$s_2 = 0$	0*

TABLE 30
First Tableau for (16)

z	x_1	x_2	s_1	s_2	rhs	Basic Variable	Ratio
1	0	-7	0	5	0	$z = 0$	
0	0	②	1	-1	6	$s_1 = 6$	$\frac{6}{2} = 3*$
0	1	-1	0	1	0	$x_1 = 0$	None

TABLE 31
Optimal Tableau for (16)

z	x_1	x_2	s_1	s_2	rhs	Basic Variable
1	0	0	3.5	1.5	21	$z = 21$
0	0	1	0.5	−0.5	3	$x_2 = 3$
0	1	0	0.5	0.5	3	$x_1 = 3$

the old bfs. This is consistent with fact 2. In the new bfs, all variables have exactly the same values as they had before the pivot! Thus, our new bfs is also degenerate. Continuing with the simplex, we enter x_2 in row 1. The resulting tableau is shown in Table 31. This is an optimal tableau, so the optimal solution to (16) is $z = 21$, $x_2 = 3$, $x_1 = 3$, $s_1 = s_2 = 0$.

We can now explain why the simplex may have problems in solving a degenerate LP. Suppose we are solving a degenerate LP for which the optimal z-value is $z = 30$. If we begin with a bfs that has, say, $z = 20$, we know (look at the LP we just solved) that it is possible for a pivot to leave the value of z unchanged. This means that it is possible for a sequence of pivots like the following to occur:

$$\text{Initial bfs (bfs 1): } z = 20$$
$$\text{After first pivot (bfs 2): } z = 20$$
$$\text{After second pivot (bfs 3): } z = 20$$
$$\text{After third pivot (bfs 4): } z = 20$$
$$\text{After fourth pivot (bfs 1): } z = 20$$

In this situation, we encounter the same bfs twice. This occurrence is called **cycling.** If cycling occurs, then we will loop, or cycle, forever among a set of basic feasible solutions and never get to the optimal solution ($z = 30$, in our example). Cycling can indeed occur (see Problem 3 at the end of this section). Fortunately, the simplex algorithm can be modified to ensure that cycling will never occur [see Bland (1977) or Dantzig (1963) for details].[†] For a practical example of cycling, see Kotiah and Slater (1973).

If an LP has many degenerate basic feasible solutions (or a bfs with many basic variables equal to zero), then the simplex algorithm is often very inefficient. To see why, look at the feasible region for (16) in Figure 12, the shaded triangle BCD. The extreme points of the feasible region are B, C, and D. Following the procedure outlined in Section 4.2, let's look at the correspondence between the basic feasible solutions to (16) and the extreme points of its feasible region (see Table 32). Three sets of basic variables correspond to extreme point C. It can be shown that for an LP with n decision variables to be degenerate, $n + 1$ or more of the LP's constraints (including the sign restrictions $x_i \geq 0$ as constraints) must be binding at an extreme point.

In (16), the constraints $x_1 - x_2 \leq 0$, $x_1 \geq 0$, and $x_2 \geq 0$ are all binding at point C. Each extreme point at which three or more constraints are binding will correspond to more than one set of basic variables. For example, at point C, s_1 must be one of the basic variables, but the other basic variable may be x_2, x_1, or s_2.

[†]Bland showed that cycling can be avoided by applying the following rules (assume that slack and excess variables are numbered $x_{n+1}, x_{n+2}, \ldots$):
1 Choose as the entering variable (in a max problem) the variable with a negative coefficient in row 0 that has the smallest subscript.
2 If there is a tie in the ratio test, then break the tie by choosing the winner of the ratio test so that the variable leaving the basis has the smallest subscript.

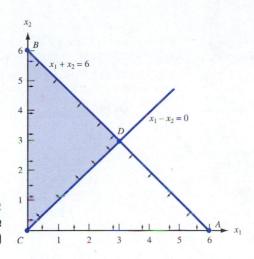

FIGURE 12
Feasible Region for the
LP (16)

TABLE 32
Three Sets of Basic Variables Correspond to Corner Point C

Basic Variables	Basic Feasible Solution	Corresponds to Extreme Point
x_1, x_2	$x_1 = x_2 = 3, s_1 = s_2 = 0$	D
x_1, s_1	$x_1 = 0, s_1 = 6, x_2 = s_2 = 0$	C
x_1, s_2	$x_1 = 6, s_2 = -6, x_2 = s_1 = 0$	Infeasible
x_2, s_1	$x_2 = 0, s_1 = 6, x_1 = s_2 = 0$	C
x_2, s_2	$x_2 = 6, s_2 = 6, s_1 = x_1 = 0$	B
s_1, s_2	$s_1 = 6, s_2 = 0, x_1 = x_2 = 0$	C

We can now discuss why the simplex algorithm often is an inefficient method for solving degenerate LPs. Suppose an LP is degenerate. Then there may be many sets (maybe hundreds) of basic variables that correspond to some nonoptimal extreme point. The simplex algorithm might encounter all these sets of basic variables before it finds that it was at a nonoptimal extreme point. This problem was illustrated (on a small scale) in solving (16): The simplex took two pivots before it found that point C was suboptimal. Fortunately, some degenerate LPs have a special structure that enables us to solve them by methods other than the simplex (see, for example, the discussion of the assignment problem in Chapter 7).

PROBLEMS

Group A

1 Even if an LP's initial tableau is nondegenerate, later tableaus may exhibit degeneracy. Degenerate tableaus often occur in the tableau following a tie in the ratio test. To illustrate this, solve the following LP:

$$\max z = 5x_1 + 3x_2$$
$$\text{s.t.} \quad 4x_1 + 2x_2 \le 12$$
$$4x_1 + x_2 \le 10$$
$$x_1 + x_2 \le 4$$
$$x_1, x_2 \ge 0$$

Also graph the feasible region and show which extreme points correspond to more than one set of basic variables.

2 Find the optimal solution to the following LP:

$$\min z = -x_1 - x_2$$
$$\text{s.t.} \quad x_1 + x_2 \le 1$$
$$-x_1 + x_2 \le 0$$
$$x_1, x_2 \ge 0$$

Group B

3 Show that if ties in the ratio test are broken by favoring row 1 over row 2, then cycling occurs when the following LP is solved by the simplex:

$$\max z = 2x_1 + 3x_2 - x_3 - 12x_4$$
$$\text{s.t} \quad -2x_1 - 9x_2 + x_3 + 9x_4 \leq 0$$
$$\tfrac{x_1}{3} + x_2 - \tfrac{x_3}{3} - 2x_4 \leq 0$$
$$x_i \geq 0 \quad (i = 1, 2, 3, 4)$$

4 Show that if ties are broken in favor of lower numbered rows, then cycling occurs when the simplex method is used to solve the following LP:

$$\max z = -3x_1 + x_2 - 6x_3$$
$$9x_1 + x_2 - 9x_3 - 2x_4 \leq 0$$
$$x_1 + \tfrac{x_2}{3} - 2x_3 - \tfrac{x_4}{3} \leq 0$$
$$-9x_1 - x_2 + 9x_3 + 2x_4 \leq 1$$
$$x_i \geq 0 \quad (i = 1, 2, 3, 4)$$

5 Show that if Bland's Rule to prevent cycling is applied to Problem 4, then cycling does not occur.

6 Consider an LP (maximization problem) in which each basic feasible solution is nondegenerate. Suppose that x_i is the only variable in our current tableau having a negative coefficient in row 0. Show that any optimal solution to the LP must have $x_i > 0$.

4.12 The Big M Method

Recall that the simplex algorithm requires a starting bfs. In all the problems we have solved so far, we found a starting bfs by using the slack variables as our basic variables. If an LP has any $\geq$ or equality constraints, however, a starting bfs may not be readily apparent. Example 4 will illustrate that a bfs may be hard to find. When a bfs is not readily apparent, the Big M method (or the two-phase simplex method of Section 4.13) may be used to solve the problem. In this section, we discuss the **Big M method,** a version of the simplex algorithm that first finds a bfs by adding "artificial" variables to the problem. The objective function of the original LP must, of course, be modified to ensure that the artificial variables are all equal to 0 at the conclusion of the simplex algorithm. The following example illustrates the Big M method.

EXAMPLE 4   Bevco

Bevco manufactures an orange-flavored soft drink called Oranj by combining orange soda and orange juice. Each ounce of orange soda contains 0.5 oz of sugar and 1 mg of vitamin C. Each ounce of orange juice contains 0.25 oz of sugar and 3 mg of vitamin C. It costs Bevco 2¢ to produce an ounce of orange soda and 3¢ to produce an ounce of orange juice. Bevco's marketing department has decided that each 10-oz bottle of Oranj must contain at least 20 mg of vitamin C and at most 4 oz of sugar. Use linear programming to determine how Bevco can meet the marketing department's requirements at minimum cost.

Solution   Let

$$x_1 = \text{number of ounces of orange soda in a bottle of Oranj}$$
$$x_2 = \text{number of ounces of orange juice in a bottle of Oranj}$$

Then the appropriate LP is

$$\min z = 2x_1 + 3x_2$$
$$\text{s.t.} \quad \tfrac{1}{2}x_1 + \tfrac{1}{4}x_2 \leq 4 \qquad \text{(Sugar constraint)} \tag{17}$$
$$x_1 + 3x_2 \geq 20 \qquad \text{(Vitamin C constraint)}$$

$$x_1 + x_2 = 10 \qquad \text{(10 oz in bottle of Oranj)}$$
$$x_1, x_2 \geq 0$$

(The solution will be continued later in this section.)

To put (17) into standard form, we add a slack variable s_1 to the sugar constraint and subtract an excess variable e_2 from the vitamin C constraint. After writing the objective function as $z - 2x_1 - 3x_2 = 0$, we obtain the following standard form:

$$
\begin{array}{llll}
\text{Row 0:} & z - 2x_1 - 3x_2 & = 0 & \\
\text{Row 1:} & \tfrac{1}{2}x_1 + \tfrac{1}{4}x_2 + s_1 & = 4 & \text{(18)} \\
\text{Row 2:} & x_1 + 3x_2 \quad - e_2 & = 20 & \\
\text{Row 3:} & x_1 + x_2 & = 10 &
\end{array}
$$

All variables non-negative

In searching for a bfs, we see that $s_1 = 4$ could be used as a basic (and feasible) variable for row 1. If we multiply row 2 by -1, we see that $e_2 = -20$ could be used as a basic variable for row 2. Unfortunately, $e_2 = -20$ violates the sign restriction $e_2 \geq 0$. Finally, in row 3 there is no readily apparent basic variable. Thus, in order to use the simplex to solve (17), rows 2 and 3 each need a basic (and feasible) variable. To remedy this problem, we simply "invent" a basic feasible variable for each constraint that needs one. Because these variables are created by us and are not real variables, we call them **artificial variables.** If an artificial variable is added to row i, we label it a_i. In the current problem, we need to add an artificial variable a_2 to row 2 and an artificial variable a_3 to row 3. The resulting set of equations is

$$
\begin{array}{ll}
z - 2x_1 - 3x_2 & = 0 \\
\tfrac{1}{2}x_1 + \tfrac{1}{4}x_2 + s_1 & = 4 \\
x_1 + 3x_2 \quad - e_2 + a_2 & = 20 \\
x_1 + x_2 \quad + a_3 & = 10
\end{array}
\qquad \text{(18)}
$$

We now have a bfs: $z = 0$, $s_1 = 4$, $a_2 = 20$, $a_3 = 10$. Unfortunately, there is no guarantee that the optimal solution to (18) will be the same as the optimal solution to (17). In solving (18), we might obtain an optimal solution in which one or more artificial variables are positive. Such a solution may not be feasible in the original problem (17). For example, in solving (18), the optimal solution may easily be shown to be $z = 0$, $s_1 = 4$, $a_2 = 20$, $a_3 = 10$, $x_1 = x_2 = 0$. This "solution" contains no vitamin C and puts 0 ounces of soda in a bottle, so it cannot possibly solve our original problem! If the optimal solution to (18) is to solve (17), then we must make sure that the optimal solution to (18) sets all artificial variables equal to zero. In a min problem, we can ensure that all the artificial variables will be zero by adding a term Ma_i to the objective function for each artificial variable a_i. (In a max problem, add a term $-Ma_i$ to the objective function.) Here M represents a "very large" positive number. Thus, in (18), we would change our objective function to

$$\min z = 2x_1 + 3x_2 + Ma_2 + Ma_3$$

Then row 0 will change to

$$z - 2x_1 - 3x_2 - Ma_2 - Ma_3 = 0$$

Modifying the objective function in this way makes it extremely costly for an artificial variable to be positive. With this modified objective function, it seems reasonable that the optimal solution to (18) will have $a_2 = a_3 = 0$. In this case, the optimal solution to (18) will solve the original problem (17). It sometimes happens, however, that in solving the analog of (18), some of the artificial variables may assume positive values in the optimal solution. If this occurs, the original problem has no feasible solution.

For obvious reasons, the method we have just outlined is often called the Big M method. We now give a formal description of the Big M method.

Description of Big M Method

Step 1   Modify the constraints so that the right-hand side of each constraint is non-negative. This requires that each constraint with a negative right-hand side be multiplied through by -1. Remember that if you multiply an inequality by any negative number, the direction of the inequality is reversed. For example, our method would transform the inequality $x_1 + x_2 \geq -1$ into $-x_1 - x_2 \leq 1$. It would also transform $x_1 - x_2 \leq -2$ into $-x_1 + x_2 \geq 2$.

Step 1'   Identify each constraint that is now (after step 1) an $=$ or $\geq$ constraint. In Step 3, we will add an artificial variable to each of these constraints.

Step 2   Convert each inequality constraint to standard form. This means that if constraint i is a $\leq$ constraint, we add a slack variable s_i, and if constraint i is a $\geq$ constraint, we subtract an excess variable e_i.

Step 3   If (after step 1 has been completed) constraint i is a $\geq$ or $=$ constraint, add an artificial variable a_i. Also add the sign restriction $a_i \geq 0$.

Step 4   Let M denote a very large positive number. If the LP is a min problem, add (for each artificial variable) Ma_i to the objective function. If the LP is a max problem, add (for each artificial variable) $-Ma_i$ to the objective function.

Step 5   Because each artificial variable will be in the starting basis, all artificial variables must be eliminated from row 0 before beginning the simplex. This ensures that we begin with a canonical form. In choosing the entering variable, remember that M is a very large positive number. For example, $4M - 2$ is more positive than $3M + 900$, and $-6M - 5$ is more negative than $-5M - 40$. Now solve the transformed problem by the simplex. If all artificial variables are equal to zero in the optimal solution, then we have found the optimal solution to the original problem. If any artificial variables are positive in the optimal solution, then the original problem is infeasible.[†]

When an artificial variable leaves the basis, its column may be dropped from future tableaus because the purpose of an artificial variable is only to get a starting basic feasible solution. Once an artificial variable leaves the basis, we no longer need it. Despite this fact, we often maintain the artificial variables in all tableaus. The reason for this will become apparent in Section 6.7.

Solution   **Example 4 (Continued)**

Step 1   Because none of the constraints has a negative right-hand side, we don't have to multiply any constraint through by -1.

[†]We have ignored the possibility that when the LP (with the artificial variables) is solved, the final tableau may indicate that the LP is unbounded. If the final tableau indicates the LP is unbounded and all artificial variables in this tableau equal zero, then the original LP is unbounded. If the final tableau indicates that the LP is unbounded and at least one artificial variable is positive, then the original LP is infeasible. See Bazaraa and Jarvis (1990) for details.

Step 1′ Constraints 2 and 3 will require artificial variables.

Step 2 Add a slack variable s_1 to row 1 and subtract an excess variable e_2 from row 2. The result is

$$\min z = 2x_1 + 3x_2$$

$$\text{Row 1:} \quad \tfrac{1}{2}x_1 + \tfrac{1}{4}x_2 + s_1 \qquad\qquad = 4$$

$$\text{Row 2:} \quad x_1 + 3x_2 \qquad\quad - e_2 = 20$$

$$\text{Row 3:} \quad x_1 + x_2 \qquad\qquad\quad = 10$$

Step 3 Add an artificial variable a_2 to row 2 and an artificial variable a_3 to row 3. The result is

$$\min z = 2x_1 + 3x_2$$

$$\text{Row 1:} \quad \tfrac{1}{2}x_1 + \tfrac{1}{4}x_2 + s_1 \qquad\qquad\qquad = 4$$

$$\text{Row 2:} \quad x_1 + 3x_2 \qquad\quad - e_2 + a_2 \qquad = 20$$

$$\text{Row 3:} \quad x_1 + x_2 \qquad\qquad\qquad + a_3 = 10$$

From this tableau, we see that our initial bfs will be $s_1 = 4$, $a_2 = 20$, and $a_3 = 10$.

Step 4 Because we are solving a min problem, we add $Ma_2 + Ma_3$ to the objective function (if we were solving a max problem, we would add $-Ma_2 - Ma_3$). This makes a_2 and a_3 very unattractive, and the act of minimizing z will cause a_2 and a_3 to be zero. The objective function is now

$$\min z = 2x_1 + 3x_2 + Ma_2 + Ma_3$$

Step 5 Row 0 is now

$$z - 2x_1 - 3x_2 - Ma_2 - Ma_3 = 0$$

Because a_2 and a_3 are in our starting bfs (that's why we introduced them), they must be eliminated from row 0. To eliminate a_2 and a_3 from row 0, simply replace row 0 by row $0 + M(\text{row 2}) + M(\text{row 3})$. This yields

$$\text{Row 0:} \quad z - \qquad 2x_1 - \qquad 3x_2 \qquad - Ma_2 - Ma_3 = 0$$

$$M(\text{row 2):} \qquad\quad Mx_1 + \quad 3Mx_2 - Me_2 + Ma_2 \qquad\quad = 20M$$

$$M(\text{row 3):} \qquad\quad Mx_1 + \quad Mx_2 \qquad\qquad\quad + Ma_3 = 10M$$

$$\text{New row 0:} \quad z + (2M - 2)x_1 + (4M - 3)x_2 - Me_2 \qquad\qquad = 30M$$

Combining the new row 0 with rows 1–3 yields the initial tableau shown in Table 33.

We are solving a min problem, so the variable with the most positive coefficient in row 0 should enter the basis. Because $4M - 3 > 2M - 2$, variable x_2 should enter the basis. The ratio test indicates that x_2 should enter the basis in row 2, which means the artificial variable a_2 will leave the basis. The most difficult part of doing the pivot is eliminating

TABLE 33
Initial Tableau for Bevco

z	x_1	x_2	s_1	e_2	a_2	a_3	rhs	Basic Variable	Ratio
1	$2M - 2$	$4M - 3$	0	$-M$	0	0	$30M$	$z = 30M$	
0	$\tfrac{1}{2}$	$\tfrac{1}{4}$	1	0	0	0	4	$s_1 = 4$	16
0	1	③	0	-1	1	0	20	$a_2 = 20$	$\tfrac{20}{3}$*
0	1	1	0	0	0	1	10	$a_3 = 10$	10

TABLE 34

First Tableau for Bevco

z	x_1	x_2	s_1	e_2	a_2	a_3	rhs	Basic Variable	Ratio
1	$\frac{2M-3}{3}$	0	0	$\frac{M-3}{3}$	$\frac{3-4M}{3}$	0	$\frac{60+10M}{3}$	$z = \frac{60+10M}{3}$	
0	$\frac{5}{12}$	0	1	$\frac{1}{12}$	$-\frac{1}{12}$	0	$\frac{7}{3}$	$s_1 = \frac{7}{3}$	$\frac{28}{5}$
0	$\frac{1}{3}$	1	0	$-\frac{1}{3}$	$\frac{1}{3}$	0	$\frac{20}{3}$	$x_2 = \frac{20}{3}$	20
0	$\left(\frac{2}{3}\right)$	0	0	$\frac{1}{3}$	$-\frac{1}{3}$	1	$\frac{10}{3}$	$a_3 = \frac{10}{3}$	5*

x_2 from row 0. First, replace row 2 by $\frac{1}{3}$(row 2). Thus, the new row 2 is

$$\tfrac{1}{3}x_1 + x_2 - \tfrac{1}{3}e_2 + \tfrac{1}{3}a_2 = \tfrac{20}{3}$$

We can now eliminate x_2 from row 0 by adding $-(4M - 3)$(new row 2) to row 0 or $(3 - 4M)$(new row 2) + row 0. Now

$$(3 - 4M)(\text{new row 2}) \quad =$$

$$\frac{(3 - 4M)x_1}{3} + (3 - 4M)x_2 - \frac{(3 - 4M)e_2}{3} + \frac{(3 - 4M)a_2}{3} = \frac{20(3 - 4M)}{3}$$

Row 0: $\qquad z + (2M - 2)x_1 + (4M - 3)x_2 - Me_2 = 30M$

New row 0: $z + \dfrac{(2M - 3)x_1}{3} + \dfrac{(M - 3)e_2}{3} + \dfrac{(3 - 4M)a_2}{3} = \dfrac{60 + 10M}{3}$

After using EROs to eliminate x_2 from row 1 and row 3, we obtain the tableau in Table 34. Because $\frac{2M-3}{3} > \frac{M-3}{3}$, we next enter x_1 into the basis. The ratio test indicates that x_1 should enter the basis in the third row of the current tableau. Then a_3 will leave the basis, and our next tableau will have $a_2 = a_3 = 0$. To enter x_1 into the basis in row 3, we first replace row 3 by $\frac{3}{2}$(row 3). Thus, new row 3 will be

$$x_1 + \frac{e_2}{2} - \frac{a_2}{2} + \frac{3a_3}{2} = 5$$

To eliminate x_1 from row 0, we replace row 0 by row 0 + $(3 - 2M)$(new row 3)/3.

Row 0: $\qquad z + \dfrac{(2M - 3)x_1}{3} + \dfrac{(M - 3)e_2}{3} + \dfrac{(3 - 4M)a_2}{3} = \dfrac{60 + 10M}{3}$

$\dfrac{(3 - 2M)(\text{new row 3})}{3} : \dfrac{(3 - 2M)x_1}{3} + \dfrac{(3 - 2M)e_2}{6} + \dfrac{(2M - 3)a_2}{6}$

$$+ \frac{(3 - 2M)a_3}{2} = \frac{15 - 10M}{3}$$

New row 0: $\qquad z - \dfrac{e_2}{2} + \dfrac{(1 - 2M)a_2}{2} + \dfrac{(3 - 2M)a_3}{2} = 25$

New row 1 and new row 2 are computed as usual, yielding the tableau in Table 35. Because all variables in row 0 have nonpositive coefficients, this is an optimal tableau; all artificial variables are equal to zero in this tableau, so we have found the optimal solution to the Bevco problem: $z = 25$, $x_1 = x_2 = 5$, $s_1 = \frac{1}{4}$, $e_2 = 0$. This means that Bevco can hold the cost of producing a 10-oz bottle of Oranj to 25¢ by mixing 5 oz of orange soda and 5 oz of orange juice. Note that the a_2 column could have been dropped after a_2 left the basis (at the conclusion of the first pivot), and the a_3 column could have been dropped after a_3 left the basis (at the conclusion of the second pivot).

TABLE 35

Optimal Tableau for Bevco

z	x_1	x_2	s_1	e_2	a_2	a_3	rhs	Basic Variable
1	0	0	0	$-\frac{1}{2}$	$\frac{1-2M}{2}$	$\frac{3-2M}{2}$	25	$z = 25$
0	0	0	1	$-\frac{1}{8}$	$\frac{1}{8}$	$-\frac{5}{8}$	$\frac{1}{4}$	$s_1 = \frac{1}{4}$
0	0	1	0	$-\frac{1}{2}$	$\frac{1}{2}$	$-\frac{1}{2}$	5	$x_2 = 5$
0	1	0	0	$\frac{1}{2}$	$-\frac{1}{2}$	$\frac{3}{2}$	5	$x_1 = 5$

How to Spot an Infeasible LP

We now modify the Bevco problem by requiring that a 10-oz bottle of Oranj contain at least 36 mg of vitamin C. Even 10 oz of orange juice contain only $3(10) = 30$ mg of vitamin C, so we know that Bevco cannot possibly meet the new vitamin C requirement. This means that Bevco's LP should now have no feasible solution. Let's see how the Big M method reveals the LP's infeasibility. We have changed Bevco's LP to

$$\min z = 2x_1 + 3x_2$$
$$\text{s.t.} \quad \tfrac{1}{2}x_1 + \tfrac{1}{4}x_2 \leq 4 \qquad \text{(Sugar constraint)}$$
$$x_1 + 3x_2 \geq 36 \qquad \text{(Vitamin C constraint)} \qquad \text{(19)}$$
$$x_1 + x_2 = 10 \qquad \text{(10 oz constraint)}$$
$$x_1, x_2 \geq 0$$

After going through Steps 1–5 of the Big M method, we obtain the initial tableau in Table 36. Because $4M - 3 > 2M - 2$, we enter x_2 into the basis. The ratio test indicates that x_2 should be entered in row 3, causing a_3 to leave the basis. After entering x_2 into the basis, we obtain the tableau in Table 37. Because each variable has a nonpositive coefficient in row 0, this is an optimal tableau. The optimal solution indicated by this tableau is $z = 30 + 6M$, $s_1 = \frac{3}{2}$, $a_2 = 6$, $x_2 = 10$, $a_3 = e_2 = x_1 = 0$. An artificial variable (a_2) is positive in the optimal tableau, so Step 5 shows that the original LP has no feasible solution.[†] In summary, *if any artificial variable is positive in the optimal Big M tableau, then the original LP has no feasible solution.*

TABLE 36

Initial Tableau for Bevco (Infeasible)

z	x_1	x_2	s_1	e_2	a_2	a_3	rhs	Basic Variable	Ratio
1	$2M - 2$	$4M - 3$	0	$-M$	0	0	$46M$	$z = 46M$	
0	$\frac{1}{2}$	$\frac{1}{4}$	1	0	0	0	4	$s_1 = 4$	16
0	1	3	0	-1	1	0	36	$a_2 = 36$	12
0	1	①	0	0	0	1	10	$a_3 = 10$	10*

[†]To explain why (19) can have no feasible solution, suppose that it does $(\bar{x}_1, \bar{x}_2)$. Clearly, if we set $a_3 = a_2 = 0$, $(\bar{x}_1, \bar{x}_2)$ will be feasible for our modified LP (the LP with artificial variables). If we substitute $(\bar{x}_1, \bar{x}_2)$ into the modified objective function ($z = 2\bar{x}_1 + 3\bar{x}_2 + Ma_2 + Ma_3$), we obtain $z = 2\bar{x}_1 + 3\bar{x}_2$ (this follows because $a_3 = a_2 = 0$). Because M is large, this z-value is certainly less than $6M + 30$. This contradicts the fact that the best z-value for our modified objective function is $6M + 30$. This means that our original LP (19) must have no feasible solution.

TABLE **37**
Tableau Indicating Infeasibility for Bevco (Infeasible)

z	x_1	s_2	s_1	e_2	a_2	a_3	rhs	Basic Variable
1	$1 - 2M$	0	0	$-M$	0	$3 - 4M$	$30 + 6M$	$z = 6M + 30$
0	$\frac{1}{4}$	0	1	0	0	$-\frac{1}{4}$	$\frac{3}{2}$	$s_1 = \frac{3}{2}$
0	-2	0	0	-1	1	-3	6	$a_2 = 6$
0	1	1	0	0	0	1	10	$x_2 = 10$

Note that when the Big M method is used, it is difficult to determine how large M should be. Generally, M is chosen to be at least 100 times larger than the largest coefficient in the original objective function. The introduction of such large numbers into the problem can cause roundoff errors and other computational difficulties. For this reason, most computer codes solve LPs by using the two-phase simplex method (described in Section 4.13).

PROBLEMS

Group A

Use the Big M method to solve the following LPs:

1 $\min z = 4x_1 + 4x_2 + x_3$
s.t. $\quad x_1 + x_2 + \ x_3 \leq 2$
$\quad\quad 2x_1 + x_2 \quad\quad \leq 3$
$\quad\quad 2x_1 + x_2 + 3x_3 \geq 3$
$\quad\quad\quad x_1, x_2, x_3 \geq 0$

2 $\min z = 2x_1 + 3x_2$
s.t. $\quad 2x_1 + x_2 \geq 4$
$\quad\quad x_1 - x_2 \geq -1$
$\quad\quad\quad x_1, x_2 \geq 0$

3 $\max z = 3x_1 + x_2$
s.t. $\quad x_1 + x_2 \geq 3$
$\quad\quad 2x_1 + x_2 \leq 4$
$\quad\quad x_1 + x_2 = 3$
$\quad\quad\quad x_1, x_2 \geq 0$

4 $\min z = 3x_1$
s.t. $\quad 2x_1 + \ x_2 \geq 6$
$\quad\quad 3x_1 + 2x_2 = 4$
$\quad\quad\quad x_1, x_2 \geq 0$

5 $\min z = x_1 + x_2$
s.t. $\quad 2x_1 + x_2 + \ x_3 = 4$
$\quad\quad x_1 + x_2 + 2x_3 = 2$
$\quad\quad\quad x_1, x_2, x_3 \geq 0$

6 $\min z = x_1 + x_2$
s.t. $\quad x_1 + \ x_2 = 2$
$\quad\quad 2x_1 + 2x_2 = 4$
$\quad\quad\quad x_1, x_2 \geq 0$

4.13 The Two-Phase Simplex Method[†]

When a basic feasible solution is not readily available, the two-phase simplex method may be used as an alternative to the Big M method. In the two-phase simplex method, we add artificial variables to the same constraints as we did in the Big M method. Then we find a bfs to the original LP by solving the Phase I LP. In the Phase I LP, the objective function is to minimize the sum of all artificial variables. At the completion of Phase I, we reintroduce the original LP's objective function and determine the optimal solution to the original LP.

The following steps describe the two-phase simplex method. Note that steps 1–3 for the two-phase simplex are identical to steps 1–3 for the Big M method.

[†]This section covers topics that may be omitted with no loss of continuity.

Step 1 Modify the constraints so that the right-hand side of constraint is non-negative. This requires that each constraint with a negative right-hand side be multiplied through by -1.

Step 1' Identify each constraint that is now (after step 1) an $=$ or $\geq$ constraint. In step 3, we will add an artificial variable to each constraint.

Step 2 Convert each inequality constraint to the standard form. If constraint i is a $\leq$ constraint, then add a slack variable s_i. If constraint i is a $\geq$ constraint, subtract an excess variable e_i.

Step 3 If (after step 1') constraint i is a $\geq$ or $=$ constraint, add an artificial variable a_i. Also add the sign restriction $a_i \geq 0$.

Step 4 For now, ignore the original LP's objective function. Instead solve an LP whose objective function is min $w' = $ (sum of all the artificial variables). This is called the **Phase I LP.** The act of solving the Phase I LP will force the artificial variables to be zero.

Because each $a_i \geq 0$, solving the Phase I LP will result in one of the following three cases:

Case 1 The optimal value of w' is greater than zero. In this case, the original LP has no feasible solution.

Case 2 The optimal value of w' is equal to zero, and no artificial variables are in the optimal Phase I basis. In this case, we drop all columns in the optimal Phase I tableau that correspond to the artificial variables. We now combine the original objective function with the constraints from the optimal Phase I tableau. This yields the **Phase II LP.** The optimal solution to the Phase II LP is the optimal solution to the original LP.

Case 3 The optimal value of w' is equal to zero and at least one artificial variable is in the optimal Phase I basis. In this case, we can find the optimal solution to the original LP if at the end of Phase I we drop from the optimal Phase I tableau all nonbasic artificial variables and any variable from the original problem that has a negative coefficient in row 0 of the optimal Phase I tableau.

Before solving examples illustrating Cases 1–3, we briefly discuss why $w' > 0$ corresponds to the original LP having no feasible solution and $w' = 0$ corresponds to the original LP having at least one feasible solution.

Phases I and II Feasible Solutions

Suppose the original LP is infeasible. Then the only way to obtain a feasible solution to the Phase I LP is to let at least one artificial variable be positive. In this situation, $w' > 0$ (Case 1) will result. On the other hand, if the original LP has a feasible solution, then this feasible solution (with all $a_i = 0$) is feasible in the Phase I LP and yields $w' = 0$. This means that if the original LP has a feasible solution, the optimal Phase I solution will have $w' = 0$. We now work through examples of Cases 1 and 2 of the two-phase simplex method.

EXAMPLE 5 **Two-Phase Simplex: Case II**

First we use the two-phase simplex to solve the Bevco problem of Section 4.12. Recall that the Bevco problem was

$$\min z = 2x_1 + 3x_2$$
$$\text{s.t.} \quad \tfrac{1}{2}x_1 + \tfrac{1}{4}x_2 \leq 4$$
$$x_1 + 3x_2 \geq 20$$
$$x_1 + x_2 = 10$$
$$x_1, x_2 \geq 0$$

Solution As in the Big M method, Steps 1–3 transform the constraints into

$$\frac{1}{2}x_1 + \frac{1}{4}x_2 + s_1 \qquad\qquad\qquad = 4$$
$$x_1 + 3x_2 \qquad - e_2 + a_2 \qquad\quad = 20$$
$$x_1 + x_2 \qquad\qquad\qquad + a_3 = 10$$

Step 4 yields the following Phase I LP:

$$\min w' = a_2 + a_3$$
$$\text{s.t.} \quad \frac{1}{2}x_1 + \frac{1}{4}x_2 + s_1 \qquad\qquad\qquad = 4$$
$$x_1 + 3x_2 \qquad - e_2 + a_2 \qquad\quad = 20$$
$$x_1 + x_2 \qquad\qquad\qquad + a_3 = 10$$

This set of equations yields a starting bfs for Phase I ($s_1 = 4$, $a_2 = 20$, $a_3 = 10$).

Note, however, that the row 0 for this tableau ($w' - a_2 - a_3 = 0$) contains the basic variables a_2 and a_3. As in the Big M method, a_2 and a_3 must be eliminated from row 0 before we can solve Phase I. To eliminate a_2 and a_3 from row 0, simply add row 2 and row 3 to row 0:

$$\begin{aligned}
\text{Row 0:} &\quad w' \qquad\qquad\qquad\qquad - a_2 - a_3 = 0 \\
+ \text{ Row 2:} &\quad \quad x_1 + 3x_2 - e_2 + a_2 \qquad\quad = 20 \\
+ \text{ Row 3:} &\quad \quad x_1 + x_2 \qquad\qquad + a_3 = 10 \\
= \text{ New row 0:} &\quad w' + 2x_1 + 4x_2 - e_2 \qquad\qquad = 30
\end{aligned}$$

Combining the new row 0 with the Phase I constraints yields the initial Phase I tableau in Table 38. Because the Phase I problem is *always* a min problem (even if the original LP is a max problem), we enter x_2 into the basis. The ratio test indicates that x_2 will enter the basis in row 2, with a_2 exiting the basis. After performing the necessary EROs, we obtain the tableau in Table 39. Because $5 < 20$ and $5 < \frac{28}{5}$, x_1 enters the basis in row 3. Thus, a_3 will leave the basis. Because a_2 and a_3 will be nonbasic after the current pivot is completed, we already know that the next tableau will be optimal for Phase I. A glance at the tableau in Table 40 confirms this fact.

Because $w' = 0$, Phase I has been concluded. The basic feasible solution $s_1 = \frac{1}{4}$, $x_2 = 5$, $x_1 = 5$ has been found. No artificial variables are in the optimal Phase I basis, so the problem is an example of Case 2. We now drop the columns for the artificial variables a_2 and a_3 (we no longer need them) and reintroduce the original objective function.

$$\min z = 2x_1 + 3x_2 \qquad \text{or} \qquad z - 2x_1 - 3x_2 = 0$$

Because x_1 and x_2 are both in the optimal Phase I basis, they must be eliminated from the Phase II row 0. We add 3(row 2) + 2(row 3) of the optimal Phase I tableau to row 0.

$$\begin{aligned}
\text{Phase II row 0:} &\quad z - 2x_1 - 3x_2 \qquad\qquad = 0 \\
+ 3(\text{row 2}): &\quad \qquad\qquad 3x_2 - \tfrac{3}{2}e_2 = 15 \\
+ 2(\text{row 3}): &\quad \qquad 2x_1 \qquad + e_2 = 10 \\
= \text{ New Phase II row 0:} &\quad z \qquad\qquad\qquad - \tfrac{1}{2}e_2 = 25
\end{aligned}$$

We now begin Phase II with the following set of equations:

$$\min z - \tfrac{1}{2}e_2 = 25$$
$$s_1 - \tfrac{1}{8}e_2 = \tfrac{1}{4}$$
$$x_2 - \tfrac{1}{2}e_2 = 5$$
$$x_1 + \tfrac{1}{2}e_2 = 5$$

TABLE 38
Initial Phase I Tableau for Bevco

w'	x_1	x_2	s_1	e_2	a_2	a_3	rhs	Basic Variable	Ratio
1	2	4	0	-1	0	0	30	$w' = 30$	
0	$\frac{1}{2}$	$\frac{1}{4}$	1	0	0	0	4	$s_1 = 4$	16
0	1	③	0	-1	1	0	20	$a_2 = 20$	$\frac{20}{3}$*
0	1	1	0	0	0	1	10	$a_3 = 10$	10

TABLE 39

Phase I Tableau for Bevco after One Iteration

w'	x_1	x_2	s_1	e_2	a_2	a_3	rhs	Basic Variable	Ratio
1	$\frac{2}{3}$	0	0	$\frac{1}{3}$	$-\frac{4}{3}$	0	$\frac{10}{3}$	$w' = \frac{10}{3}$	
0	$\frac{5}{12}$	0	1	$\frac{1}{12}$	$-\frac{1}{12}$	0	$\frac{7}{3}$	$s_1 = \frac{7}{3}$	$\frac{28}{5}$
0	$\frac{1}{3}$	1	0	$-\frac{1}{3}$	$\frac{1}{3}$	0	$\frac{20}{3}$	$x_2 = \frac{20}{3}$	20
0	$\left(\frac{2}{3}\right)$	0	0	$\frac{1}{3}$	$-\frac{1}{3}$	1	$\frac{10}{3}$	$a_3 = \frac{10}{3}$	5*

TABLE 40

Optimal Phase I Tableau for Bevco

w'	x_1	x_2	s_1	e_2	a_2	a_3	rhs	Basic Variable
1	0	0	0	0	-1	-1	0	$w' = 0$
0	0	0	1	$-\frac{1}{8}$	$\frac{1}{8}$	$-\frac{5}{8}$	$\frac{1}{4}$	$s_1 = \frac{1}{4}$
0	0	1	0	$-\frac{1}{2}$	$\frac{1}{2}$	$-\frac{1}{2}$	5	$x_2 = 5$
0	1	0	0	$\frac{1}{2}$	$-\frac{1}{2}$	$\frac{3}{2}$	5	$x_1 = 5$

This is optimal. Thus, in this problem, Phase II requires no pivots to find an optimal solution. If the Phase II row 0 does not indicate an optimal tableau, then simply continue with the simplex until an optimal row 0 is obtained. In summary, our optimal Phase II tableau shows that the optimal solution to the Bevco problem is $z = 25$, $x_1 = 5$, $x_2 = 5$, $s_1 = \frac{1}{4}$, and $e_2 = 0$. This agrees, of course, with the optimal solution found by the Big M method in Section 4.12.

EXAMPLE 6   Two-Phase Simplex: Case I

To illustrate Case 2, we now modify Bevco's problem so that 36 mg of vitamin C are required. From Section 4.12, we know that this problem is infeasible. This means that the optimal Phase I solution should have $w' > 0$ (Case 1). To show that this is true, we begin with the original problem:

$$\min z = 2x_1 + 3x_2$$
$$\text{s.t.} \quad \tfrac{1}{2}x_1 + \tfrac{1}{4}x_2 \le 4$$
$$x_1 + 3x_2 \ge 36$$
$$x_1 + x_2 = 10$$
$$x_1, x_2 \ge 0$$

TABLE 41
Initial Phase I Tableau for Bevco (Infeasible)

w'	x_1	x_2	s_1	e_2	a_2	a_3	rhs	Basic Variable	Ratio
1	2	4	0	-1	0	0	46	$w' = 46$	
0	$\frac{1}{2}$	$\frac{1}{4}$	1	0	0	0	4	$s_1 = 4$	16
0	1	3	0	-1	1	0	36	$a_2 = 36$	12
0	1	①	0	0	0	1	10	$a_3 = 0$	10*

TABLE 42
Tableau Indicating Infeasibility for Bevco (Infeasible)

w'	x_1	x_2	s_1	e_2	a_2	a_3	rhs	Basic Variable
1	-2	0	0	-1	0	-4	6	$w' = 6$
0	$\frac{1}{4}$	0	1	0	0	$-\frac{1}{4}$	$\frac{3}{2}$	$s_1 = \frac{3}{2}$
0	-2	0	0	-1	1	-3	6	$a_2 = 6$
0	1	1	0	0	0	1	10	$x_2 = 10$

Solution After completing steps 1–4 of the two-phase simplex, we obtain the following Phase I problem:

$$\min w' = a_2 + a_3$$
$$\text{s.t.} \quad \tfrac{1}{2}x_1 + \tfrac{1}{4}x_2 + s_1 \qquad\qquad\qquad = 4$$
$$x_1 + 3x_2 \qquad - e_2 + a_2 \qquad = 36$$
$$x_1 + x_2 \qquad\qquad\qquad + a_3 = 10$$

From this set of equations, we see that the initial Phase I bfs is $s_1 = 4$, $a_2 = 36$, and $a_3 = 10$. Because the basic variables a_2 and a_3 occur in the Phase I objective function, they must be eliminated from the Phase I row 0. To do this, we add rows 2 and 3 to row 0:

$$\text{Row 0:} \qquad w' \qquad\qquad\qquad\qquad - a_2 - a_3 = 0$$
$$+ \text{Row 2:} \qquad\qquad x_1 + 3x_2 - e_2 + a_2 \qquad\quad = 36$$
$$+ \text{Row 3:} \qquad\qquad x_1 + x_2 \qquad\qquad + a_3 = 10$$
$$= \text{New row 0:} \quad w' + 2x_1 + 4x_2 - e_2 \qquad\qquad\qquad = 46$$

With the new row 0, the initial Phase I tableau is as shown in Table 41. Because $4 > 2$, we should enter x_2 into the basis. The ratio test indicates that x_2 should enter the basis in row 3, forcing a_3 to leave the basis. The resulting tableau is shown in Table 42. No variable in row 0 has a positive coefficient, so this is an optimal Phase I tableau, and since the optimal value of w' is $6 > 0$, the original LP must have no feasible solution. This is reasonable, because if the original LP had a feasible solution, it would have been feasible in the Phase I LP (after setting $a_2 = a_3 = 0$). This feasible solution would have yielded $w' = 0$. Because the simplex could not find a Phase I solution with $w' = 0$, the original LP must have no feasible solution.

REMARKS **1** As with the Big M method, the column for any artificial variable may be dropped from future tableaus as soon as the artificial variable leaves the basis. Thus, when we solved the Bevco problem, a_2's column could have been dropped after the first Phase I pivot, and a_3's column could have been dropped after the second Phase I pivot.

2 It can be shown that (barring ties for the entering variable and in the ratio test) the Big M method and Phase I of the two-phase method make the same sequence of pivots. Despite this equivalence, most computer codes utilize the two-phase method to find a bfs. This is because M, being a large positive number, may cause roundoff errors and other computational difficulties. The two-phase method does not introduce any large numbers into the objective function, so it avoids this problem.

EXAMPLE 7　**Two-Phase Simplex: Case III**

Use the two-phase simplex method to solve the following LP:

$$\min z = 40x_1 + 10x_2 + 7x_5 + 14x_6$$

$$
\begin{aligned}
\text{s.t.} \quad x_1 - x_2 \quad\quad\quad + 2x_5 \quad\quad &= 0 \\
-2x_1 + x_2 \quad\quad\quad - 2x_5 \quad\quad &= 0 \\
x_1 \quad\quad + x_3 \quad\quad + x_5 - x_6 &= 3 \\
2x_2 + x_3 + x_4 + 2x_5 + x_6 &= 4
\end{aligned}
$$

$$\text{All } x_i \geq 0$$

Solution　We may use x_4 as a basic variable for the fourth constraint and use artificial variables a_1, a_2, and a_3 as basic variables for the first three constraints. Our Phase I objective is to minimize $w = a_1 + a_2 + a_3$. After adding the first three constraints to $w - a_1 - a_2 - a_3 = 0$, we obtain the initial Phase I tableau shown in Table 43.

Even though x_5 has the most positive coefficient in row 0, we choose to enter x_3 into the basis (as a basic variable in Row 3). We see that this will immediately yield $w = 0$. Our final Phase I tableau is shown in Table 44.

Because $w = 0$, we now have an optimal Phase I tableau. Two artificial variables remain in the basis (a_1 and a_2) at a zero level. We may now drop the artificial variable a_3 from our first Phase II tableau. The only original variable with a negative coefficient in the optimal Phase I tableau is x_1, so we may drop x_1 from all future tableaus. This is because from the optimal Phase I tableau we find $w = x_1$. This implies that x_1 can never become positive during Phase II, so we may drop x_1 from all future tableaus. Because $z - 40x_1 - 10x_2 - 7x_5 - 14x_6 = 0$ contains no basic variables our initial tableau for Phase II is as in Table 45.

TABLE 43

w	x_1	x_2	x_3	x_4	x_5	x_6	a_1	a_2	a_3	rhs	Basic Variable
1	0	0	1	0	1	−1	0	0	0	3	$w = 3$
0	1	−1	0	0	2	0	1	0	0	0	$a_1 = 0$
0	−2	1	0	0	−2	0	0	1	0	0	$a_2 = 0$
0	1	0	①	0	1	−1	0	0	1	3	$a_3 = 3$
0	0	2	1	1	2	1	0	0	0	4	$x_4 = 4$

TABLE 44

w	x_1	x_2	x_3	x_4	x_5	x_6	a_1	a_2	a_3	rhs	Basic Variable
1	−1	0	0	0	0	0	0	0	−1	0	$w = 0$
0	1	−1	0	0	2	0	1	0	0	0	$a_1 = 0$
0	−2	1	0	0	−2	0	0	1	0	0	$a_2 = 0$
0	1	0	1	0	1	−1	0	0	1	3	$x_3 = 3$
0	−1	2	0	1	1	2	0	0	−1	1	$x_4 = 1$

TABLE 45

z	x_2	x_3	x_4	x_5	x_6	a_1	a_2	rhs	Basic Variables
1	-10	0	0	-7	-14	0	0	0	$z = 0$
0	-1	0	0	2	0	1	0	0	$a_1 = 0$
0	1	0	0	-2	0	0	1	0	$a_2 = 0$
0	0	1	0	1	-1	0	0	3	$x_3 = 3$
0	2	0	1	1	②	0	0	1	$x_4 = 1$

TABLE 46

z	x_2	x_3	x_4	x_5	x_6	a_1	a_2	rhs	Basic Variables
1	4	0	7	0	0	0	0	7	$z = 7$
0	0	0	0	2	0	1	0	0	$a_1 = 0$
0	1	0	0	0	0	0	1	0	$a_2 = 0$
0	1	1	$\frac{1}{2}$	$\frac{3}{2}$	0	0	0	$\frac{7}{2}$	$x_3 = \frac{7}{2}$
0	0	0	$\frac{1}{2}$	$\frac{1}{2}$	1	0	0	$\frac{1}{2}$	$x_4 = \frac{1}{2}$

We now enter x_6 into the basis in row 4 and obtain the optimal tableau shown in Table 46.

The optimal solution to our original LP is $z = 7$, $x_3 = 7/2$, $x_4 = \frac{1}{2}$, $x_2 = x_5 = x_6 = x_3 = 0$.

PROBLEMS

Group A

1 Use the two-phase simplex method to solve the Section 4.12 problems.

2 Explain why the Phase I LP will usually have alternative optimal solutions.

4.14 Unrestricted-in-Sign Variables

In solving LPs with the simplex algorithm, we used the ratio test to determine the row in which the entering variable became a basic variable. Recall that the ratio test depended on the fact that any feasible point required all variables to be non-negative. Thus, if some variables are allowed to be unrestricted in sign (urs), the ratio test and therefore the simplex algorithm are no longer valid. In this section, we show how an LP with unrestricted-in-sign variables can be transformed into an LP in which all variables are required to be non-negative.

For each urs variable x_i, we begin by defining two new variables x_i' and x_i''. Then, substitute $x_i' - x_i''$ for x_i in each constraint and in the objective function. Also add the sign restrictions $x_i' \geq 0$ and $x_i'' \geq 0$. The effect of this substitution is to express x_i as the difference of the two nonnegative variables x_i' and x_i''. Because all variables are now required to be non-negative, we can proceed with the simplex. As we will soon see, no basic feasible solution can have both $x_i' > 0$ and $x_i'' > 0$. This means that for any basic feasible solution, each urs variable x_i must fall into one of the following three cases:

Case 1 $x_i' > 0$ and $x_i'' = 0$. This case occurs if a bfs has $x_i > 0$. In this case, $x_i = x_i' - x_i'' = x_i'$. Thus, $x_i = x_i'$. For example, if $x_i = 3$ in a bfs, this will be indicated by $x_i' = 3$ and $x_i'' = 0$.

Case 2 $x_i' = 0$ and $x_i'' > 0$. This case occurs if $x_i < 0$. Because $x_i = x_i' - x_i''$, we obtain $x_i = -x_i''$. For example, if $x_i = -5$ in a bfs, we will have $x_i' = 0$ and $x_i'' = 5$. Then $x_i = 0 - 5 = -5$.

Case 3 $x_i' = x_i'' = 0$. In this case, $x_i = 0 - 0 = 0$.

In solving the following example, we will learn why no bfs can ever have both $x_i' > 0$ and $x_i'' > 0$.

EXAMPLE 8   **Using urs Variables**

A baker has 30 oz of flour and 5 packages of yeast. Baking a loaf of bread requires 5 oz of flour and 1 package of yeast. Each loaf of bread can be sold for 30¢. The baker may purchase additional flour at 4¢/oz or sell leftover flour at the same price. Formulate and solve an LP to help the baker maximize profits (revenues − costs).

Solution   Define

x_1 = number of loaves of bread baked

x_2 = number of ounces by which flour supply is increased by cash transactions

Therefore, $x_2 > 0$ means that x_2 oz of flour were purchased, and $x_2 < 0$ means that $-x_2$ ounces of flour were sold ($x_2 = 0$ means no flour was bought or sold). After noting that $x_1 \geq 0$ and x_2 is urs, the appropriate LP is

$$\max z = 30x_1 - 4x_2$$
$$\text{s.t.} \quad 5x_1 \leq 30 + x_2 \quad \text{(Flour constraint)}$$
$$x_1 \leq 5 \quad \text{(Yeast constraint)}$$
$$x_1 \geq 0, x_2 \text{ urs}$$

Because x_2 is urs, we substitute $x_2' - x_2''$ for x_2 in the objective function and constraints. This yields

$$\max z = 30x_1 - 4x_2' + 4x_2''$$
$$\text{s.t.} \quad 5x_1 \leq 30 + x_2' - x_2''$$
$$x_1 \leq 5$$
$$x_1, x_2', x_2'' \geq 0$$

After transforming the objective function to row 0 form and adding slack variables s_1 and s_2 to the two constraints, we obtain the initial tableau in Table 47. Notice that the x_2' column is simply the negative of the x_2'' column. We will see that *no matter how many pivots we make, the x_2' column will always be the negative of the x_2'' column.* (See Problem 6 for a proof of this assertion.)

Because x_1 has the most negative coefficient in row 0, x_1 enters the basis—in row 2. The resulting tableau is shown in Table 48. Again note that the x_2' column is the negative of the x_2'' column.

Because x_2'' now has the most negative coefficient in row 0, we enter x_2'' into the basis in row 1. The resulting tableau is shown in Table 49. Observe that the x_2' column is still the negative of the x_2'' column. This is an optimal tableau, so the optimal solution to the baker's problem is $z = 170$, $x_1 = 5$, $x_2'' = 5$, $x_2' = 0$, $s_1 = s_2 = 0$. Thus, the baker can earn a profit of 170¢ by baking 5 loaves of bread. Because $x_2 = x_2' - x_2'' = 0 - 5 = -5$,

TABLE 47
Initial Tableau for urs LP

z	x_1	x_2'	x_2''	s_1	s_2	rhs	Basic Variable	Ratio
1	−30	4	−4	0	0	0	$z = 0$	
0	5	−1	1	1	0	30	$s_1 = 30$	6
0	①	0	0	0	1	5	$s_2 = 5$	5*

TABLE 48
First Tableau for urs LP

z	x_1	x_2'	x_2''	s_1	s_2	rhs	Basic Variable	Ratio
1	0	4	−4	0	30	150	$z = 150$	
0	0	−1	①	1	−5	5	$s_1 = 5$	5*
0	1	0	0	0	1	5	$x_1 = 5$	None

TABLE 49
Optimal Tableau for urs LP

z	x_1	x_2'	x_2''	s_1	s_2	rhs	Basic Variable
1	0	0	0	4	10	170	$z = 170$
0	0	−1	1	1	−5	5	$x_2'' = 5$
0	1	0	0	0	1	5	$x_1 = 5$

the baker should sell 5 oz of flour. It is optimal for the baker to sell flour, because having 5 packages of yeast limits the baker to manufacturing at most 5 loaves of bread. These 5 loaves of bread use 5(5) = 25 oz of flour, so 30 − 25 = 5 oz of flour are left to sell.

The variables x_2' and x_2'' will never both be basic variables in the same tableau. To see why, suppose that x_2'' is basic (as it is in the optimal tableau). Then the x_2'' column will contain a single 1 and have every other entry equal to 0. The x_2' column is always the negative of the x_2'' column, so the x_2' column will contain a single −1 and have all other entries equal to 0. Such a tableau cannot also have x_2' as a basic feasible variable. The same reasoning shows that if x_i is urs, then x_i' and x_i'' cannot both be basic variables in the same tableau. This means that in any tableau, x_i', x_i'', or both must equal 0 and that one of Cases 1–3 must always occur.

The following example shows how urs variables can be used to model the production-smoothing costs discussed in the Sailco example of Section 3.10.

EXAMPLE 9   Modeling Production-Smoothing Costs

Mondo Motorcycles is determining its production schedule for the next four quarters. Demand for motorcycles will be as follows: quarter 1—40; quarter 2—70; quarter 3—50; quarter 4—20. Mondo incurs four types of costs.

1   It costs Mondo $400 to manufacture each motorcycle.

2 At the end of each quarter, a holding cost of $100 per motorcycle is incurred.

3 Increasing production from one quarter to the next incurs costs for training employees. It is estimated that a cost of $700 per motorcycle is incurred if production is increased from one quarter to the next.

4 Decreasing production from one quarter to the next incurs costs for severance pay, decreasing morale, and so forth. It is estimated that a cost of $600 per motorcycle is incurred if production is decreased from one quarter to the next.

All demands must be met on time, and a quarter's production may be used to meet demand for the current quarter. During the quarter immediately preceding quarter 1, 50 Mondos were produced. Assume that at the beginning of quarter 1, no Mondos are in inventory. Formulate an LP that minimizes Mondo's total cost during the next four quarters.

Solution To express inventory and production costs, we define for $t = 1, 2, 3, 4$,

$$p_t = \text{number of motorcycles produced during quarter } t$$
$$i_t = \text{inventory at end of quarter } t$$

To determine smoothing costs (costs 3 and 4), we define

$$x_t = \text{amount by which quarter } t \text{ production exceeds quarter } t - 1 \text{ production}$$

Because x_t is unrestricted in sign, we may write $x_t = x_t' - x_t''$, where $x_t' \geq 0$ and $x_t'' \geq 0$. We know that if $x_t \geq 0$, then $x_t = x_t'$ and $x_t'' = 0$. Also, if $x_t \leq 0$, then $x_t = -x_t''$ and $x_t' = 0$. This means that

$$x_t' = \text{increase in quarter } t \text{ production over quarter } t - 1 \text{ production}$$
$$(x_t' = 0 \text{ if period } t \text{ production is less than period } t - 1 \text{ production})$$
$$x_t'' = \text{decrease in quarter } t \text{ production from quarter } t - 1 \text{ production}$$
$$(x_t' = 0 \text{ if period } t \text{ production is more than period } t - 1 \text{ production})$$

For example, if $p_1 = 30$ and $p_2 = 50$, we have $x_2 = 50 - 30 = 20$, $x_2' = 20$, $x_2'' = 0$. Similarly, if $p_1 = 30$ and $p_2 = 15$, we have $x_2 = 15 - 30 = -15$, $x_2' = 0$, and $x_2'' = 15$. The variables x_t' and x_t'' can now be used to express the smoothing costs for quarter t.

We may now express Mondo's total cost as

$$\text{Total cost} = \text{production cost} + \text{inventory cost}$$
$$+ \text{ smoothing cost due to increasing production}$$
$$+ \text{ smoothing cost due to decreasing production}$$
$$= 400(p_1 + p_2 + p_3 + p_4) + 100(i_1 + i_2 + i_3 + i_4)$$
$$+ 700(x_1' + x_2' + x_3' + x_4') + 600(x_1'' + x_2'' + x_3'' + x_4'')$$

To complete the formulation, we add two types of constraints. First we need inventory constraints (as in the Sailco problem of Section 3.10) that relate the inventory from the current quarter to the past quarter's inventory and the current quarter's production. For quarter t, the inventory constraint takes the form

$$\text{Quarter } t \text{ inventory} = (\text{quarter } t - 1 \text{ inventory}) + (\text{quarter } t \text{ production})$$
$$- (\text{quarter } t \text{ demand})$$

For $t = 1, 2, 3, 4$, respectively, this yields the following four constraints:

$$i_1 = 0 + p_1 - 40 \qquad i_2 = i_1 + p_2 - 70$$
$$i_3 = i_2 + p_3 - 50 \qquad i_4 = i_3 + p_4 - 20$$

The sign restrictions $i_t \geq 0$ ($t = 1, 2, 3, 4$) ensure that each quarter's demands will be met on time.

The second type of constraint reflects the fact that p_t, p_{t-1}, x_t', and x_t'' are related. This relationship is captured by

$$(\text{quarter } t \text{ production}) - (\text{quarter } t - 1 \text{ production}) = x_t = x_t' - x_t''$$

For $t = 1, 2, 3, 4$, this relation yields the following four constraints:

$$p_1 - 50 = x_1' - x_1'' \qquad p_2 - p_1 = x_2' - x_2''$$
$$p_3 - p_2 = x_3' - x_3'' \qquad p_4 - p_3 = x_4' - x_4''$$

Combining the objective function, the four inventory constraints, the last four constraints, and the sign restrictions ($i_t, p_t, x_t', x_t'' \geq 0$ for $t = 1, 2, 3, 4$), we obtain the following LP:

$$\min z = 400p_1 + 400p_2 + 400p_3 + 400p_4 + 100i_1 + 100i_2 + 100i_3 + 100i_4$$
$$+ 700x_1' + 700x_2' + 700x_3' + 700x_4' + 600x_1'' + 600x_2'' + 600x_3'' + 600x_4''$$

$$\text{s.t.} \quad i_1 = 0 + p_1 - 40$$
$$i_2 = i_1 + p_2 - 70$$
$$i_3 = i_2 + p_3 - 50$$
$$i_4 = i_3 + p_4 - 20$$
$$p_1 - 50 = x_1' - x_1''$$
$$p_2 - p_1 = x_2' - x_2''$$
$$p_3 - p_2 = x_3' - x_3''$$
$$p_4 - p_3 = x_4' - x_4''$$
$$i_t, p_t, x_t', x_t'' \geq 0 \quad (t = 1, 2, 3, 4)$$

As in Example 7, the column for x_t' in the constraints is the negative of the x_t'' column. Thus, as in Example 7, no bfs to Mondo's LP can have both $x_t' > 0$ and $x_t'' > 0$. This means that x_t' actually is the increase in production during quarter t, and x_t'' actually is the amount by which production decreases during quarter t.

There is another way to show that the optimal solution will not have both $x_t' > 0$ and $x_t'' > 0$. Suppose, for example, that $p_2 = 70$ and $p_1 = 60$. Then the constraint

$$p_2 - p_1 = 70 - 60 = x_2' - x_2'' \tag{20}$$

can be satisfied by many combinations of x_2' and x_2''. For example, $x_2' = 10$ and $x_2'' = 0$ will satisfy (20), as will $x_2' = 20$, and $x_2'' = 10$; $x_2' = 40$ and $x_2'' = 30$; and so on. If $p_2 - p_1 = 10$, the optimal LP solution will always choose $x_2' = 10$ and $x_2'' = 0$ over any other possibility. To see why, look at Mondo's objective function. If $x_2' = 10$ and $x_2'' = 0$, then x_2' and x_2'' contribute $10(700) = \$7,000$ in smoothing costs. On the other hand, any other choice of x_2' and x_2'' satisfying (20) will contribute more than $\$7,000$ in smoothing costs. For example, $x_2' = 20$ and $x_2'' = 10$ contributes $20(700) + 10(600) = \$20,000$ in smoothing costs. We are minimizing total cost, so the simplex will never choose a solution where $x_t' > 0$ and $x_t'' > 0$ both hold.

The optimal solution to Mondo's problem is $p_1 = 55$, $p_2 = 55$, $p_3 = 50$, $p_4 = 50$. This solution incurs a total cost of $\$95,000$. The optimal production schedule produces a total of 210 Mondos. Because total demand for the four quarters is only 180 Mondos, there will be an ending inventory of $210 - 180 = 30$ Mondos. Note that this is in contrast to the Sailco inventory model of Section 3.10, in which ending inventory was always 0. The optimal solution to the Mondo problem has a nonzero inventory in quarter 4, because for the quarter 4 inventory to be 0, quarter 4 production must be lower than quarter 3 production. Rather than incur the excessive smoothing costs associated with this strategy, the optimal solution opts for holding 30 Mondos in inventory at the end of quarter 4.

PROBLEMS

Group A

1 Suppose that Mondo no longer must meet demands on time. For each quarter that demand for a motorcycle is unmet, a penalty or shortage cost of $110 per motorcycle short is assessed. Thus, demand can now be backlogged. All demands must be met, however, by the end of quarter 4. Modify the formulation of the Mondo problem to allow for backlogged demand. (*Hint:* Unmet demand corresponds to $i_t \leq 0$. Thus, i_t is now urs, and we must substitute $i_t = i_t' - i_t''$. Now i_t'' will be the amount of demand that is unmet at the end of quarter t.)

2 Use the simplex algorithm to solve the following LP:

$$\max z = 2x_1 + x_2$$
$$\text{s.t.} \quad 3x_1 + x_2 \leq 6$$
$$x_1 + x_2 \leq 4$$
$$x_1 \geq 0, x_2 \text{ urs}$$

Group B

3 During the next three months, Steelco faces the following demands for steel: 100 tons (month 1); 200 tons (month 2); 50 tons (month 3). During any month, a worker can produce up to 15 tons of steel. Each worker is paid $5,000 per month. Workers can be hired or fired at a cost of $3,000 per worker fired and $4,000 per worker hired (it takes 0 time to hire a worker). The cost of holding a ton of steel in inventory for one month is $100. Demand may be backlogged at a cost of $70 per ton month. That is, if 1 ton of month 1 demand is met during month 3, then a backlogging cost of $140 is incurred. At the beginning of month 1, Steelco has 8 workers. During any month, at most 2 workers can be hired. All demand must be met by the end of month 3. The raw material used to produce a ton of steel costs $300. Formulate an LP to minimize Steelco's costs.

4 Show how you could use linear programming to solve the following problem:

$$\max z = |2x_1 - 3x_2|$$
$$\text{s.t.} \quad 4x_1 + x_2 \leq 4$$
$$2x_1 - x_2 \leq 0.5$$
$$x_1, x_2 \geq 0$$

FIGURE 13

Steel manufacturing area
• (700, 600)

Shipping area

(0, 0)   (1000, 0)

5[†] Steelco's main plant currently has a steel manufacturing area and shipping area located as shown in Figure 13 (distances are in feet). The company must determine where to locate a casting facility and an assembly and storage facility to minimize the daily cost of moving material through the plant. The number of trips made each day are as shown in Table 50.

Assuming that all travel is in only an east–west or north–south direction, formulate an LP that can be used to determine where the casting and assembly and storage plants should be located in order to minimize daily transportation costs. (*Hint:* If the casting facility has coordinates (c1, c2), how should the constraint $c1 - 700 = e_1 - w_1$ be interpreted?)

6 Show that after any number of pivots the coefficient of x_i' in each row of the simplex tableau will equal the negative of the coefficient of x_i'' in the same row.

7 Clothco manufactures pants. During each of the next six months they can sell *up to* the numbers of pants given in Table 51.

Demand that is not met during a month is lost. Thus, for example, Clothco can sell up to 500 pants during month 1. A pair of pants sells for $40, requires 2 hours of labor, and uses $10 of raw material. At the beginning of month 1, Clothco has 4 workers. A worker can work at making pants up to 200 hours per month, and is paid $2,000 per month (irrespective of how many hours worked). At the beginning of each month, workers can be hired and fired. It costs

TABLE 50

From	To	Daily Number of Trips	Cost (¢) Per 100 Feet Traveled
Casting	Assembly and storage	40	10
Steel manufacturing	Casting	8	10
Steel manufacturing	Assembly and storage	8	10
Shipping	Assembly and storage	2	20

TABLE 51

Month	Maximum Demand
1	500
2	600
3	300
4	400
5	300
6	800

[†]Based on Love and Yerex (1976).

$1,500 to hire and $1,000 to fire a worker. A holding cost of $5 per pair of pants is assessed against each month's ending inventory.

Determine how Clothco can maximize its profit for the next six months. Ignore the fact that during each month the number of hired and fired workers must be an integer.

4.15 Karmarkar's Method for Solving LPs

We now give a brief description of Karmarkar's method for solving LPs. For a more detailed explanation, see Section 10.6. Karmarkar's method requires that the LP be placed in the following form:

$$\min z = \mathbf{cx}$$
$$\text{s.t.} \quad K\mathbf{x} = 0$$
$$x_1 + x_2 + \cdots x_n = 1$$
$$x_i \geq 0$$

and that

1 The point $\mathbf{x}^0 = [\frac{1}{n} \quad \frac{1}{n} \quad \cdots \quad \frac{1}{n}]$ be feasible for this LP.

2 The optimal z-value for the LP equals 0.

Surprisingly, any LP can be put in this form. Karmarkar's method uses a transformation from projective geometry to create a set of transformed variables $y_1, y_2, \ldots, y_n$. This transformation (call it f) will always transform the current point into the "center" of the feasible region in the space defined by the transformed variables. If the transformation takes the point $\mathbf{x}$ into the point $\mathbf{y}$, we write $f(\mathbf{x}) = \mathbf{y}$. The algorithm begins in the transformed space by moving from $f(\mathbf{x}^0)$ in the transformed space in a "good" direction (a direction that tends to improve z and maintains feasibility). This yields a point $\mathbf{y}^1$ in the transformed space, which is close to the boundary of the feasible region. Our new point is $\mathbf{x}^1$, satisfying $f(\mathbf{x}^1) = \mathbf{y}^1$. The procedure is repeated (this time $\mathbf{x}^1$ replaces $\mathbf{x}^0$) until the z-value for $\mathbf{x}^k$ is sufficiently close to 0.

If our current point is $\mathbf{x}^k$, then the transformation will have the property that $f(\mathbf{x}^k) = [\frac{1}{n} \quad \frac{1}{n} \quad \cdots \quad \frac{1}{n}]$. Thus, in transformed space, we are always moving away from the "center" of the feasible region.

Karmarkar's method has been shown to be a **polynomial time algorithm.** This implies that if an LP of size n is solved by Karmarkar's method, then there exist positive numbers a and b such that for any n, an LP of size n can be solved in a time of at most an^b.[†]

In contrast to Karmarkar's method, the simplex algorithm is an **exponential time algorithm** for solving LPs. If an LP of size n is solved by the simplex, then there exists a positive number c such that for any n, the simplex algorithm will find the optimal solution in a time of at most $c2^n$. For large enough n (for positive a, b, and c), $c2^n > an^b$. This means that, in theory, a polynomial time algorithm is superior to an exponential time algorithm. Preliminary testing of Karmarkar's method (by Karmarkar) has shown that for large LPs arising in actual application, this method may be up to 50 times as fast as the simplex algorithm. Hopefully, Karmarkar's method will enable researchers to solve many large LPs that currently require a prohibitively large amount of computer time when solved by the simplex. If Karmarkar's method lives up to its early promise, the ability to formulate LP models will be even more important in the near future than it is today.

Karmarkar's method has recently been utilized by the Military Airlift Command to determine how often to fly various routes, and which aircraft to use. The resulting LP con-

---

[†]The size of an LP may be defined as the number of symbols needed to represent the LP in binary notation.

tained 150,000 variables and 12,000 constraints and was solved in one hour of computer time using Karmarkar's method. Using the simplex method, an LP with similar structure containing 36,000 variables and 10,000 constraints required four hours of computer time. Delta Airlines has used Karmarkar's method to develop monthly schedules for 7,000 pilots and more than 400 aircraft. When the project is completed, Delta expects to have saved millions of dollars.

4.16 Multiattribute Decision Making in the Absence of Uncertainty: Goal Programming

In some situations, a decision maker may face multiple objectives, and there may be no point in an LP's feasible region satisfying all objectives. In such a case, how can the decision maker choose a satisfactory decision? **Goal programming** is one technique that can be used in such situations. The following example illustrates the main ideas of goal programming.

EXAMPLE 10 | **Burnit Goal Programming**

The Leon Burnit Advertising Agency is trying to determine a TV advertising schedule for Priceler Auto Company. Priceler has three goals:

Goal 1 Its ads should be seen by at least 40 million high-income men (HIM).

Goal 2 Its ads should be seen by at least 60 million low-income people (LIP).

Goal 3 Its ads should be seen by at least 35 million high-income women (HIW).

Leon Burnit can purchase two types of ads: those shown during football games and those shown during soap operas. At most, $600,000 can be spent on ads. The advertising costs and potential audiences of a one-minute ad of each type are shown in Table 52. Leon Burnit must determine how many football ads and soap opera ads to purchase for Priceler.

Solution Let

$$x_1 = \text{number of minutes of ads shown during football games}$$
$$x_2 = \text{number of minutes of ads shown during soap operas}$$

Then any feasible solution to the following LP would meet Priceler's goals:

$$\min (\text{or max}) \ z = 0x_1 + 0x_2 \quad (\text{or any other objective function})$$
$$\text{s.t.} \quad 7x_1 + 3x_2 \geq 40 \quad (\text{HIM constraint})$$
$$10x_1 + 5x_2 \geq 60 \quad (\text{LIP constraint})$$
$$5x_1 + 4x_2 \geq 35 \quad (\text{HIW constraint})$$
$$100x_1 + 60x_2 \leq 600 \quad (\text{Budget constraint})$$
$$x_1, x_2 \geq 0$$

(21)

From Figure 14, we find that no point that satisfies the budget constraint meets all three of Priceler's goals. Thus, (21) has no feasible solution. It is impossible to meet all of Priceler's goals, so Burnit might ask Priceler to identify, for each goal, a cost (per-unit short of meeting each goal) that is incurred for failing to meet the goal. Suppose Priceler determines that

TABLE 52
Cost and Number of Viewers of Ads for Priceler

Ad	Millions of Viewers			Cost ($)
	HIM	LIP	HIW	
Football	7	10	5	100,000
Soap opera	3	5	4	60,000

Each million exposures by which Priceler falls short of the HIM goal costs Priceler a $200,000 penalty because of lost sales.

Each million exposures by which Priceler falls short of the LIP goal costs Priceler a $100,000 penalty because of lost sales.

Each million exposures by which Priceler falls short of the HIW goal costs Priceler a $50,000 penalty because of lost sales.

Burnit can now formulate an LP that minimizes the cost incurred in deviating from Priceler's three goals. The trick is to transform each inequality constraint in (21) that represents one of Priceler's goals into an equality constraint. Because we don't know whether the cost-minimizing solution will undersatisfy or oversatisfy a given goal, we need to define the following variables:

$$s_i^+ = \text{amount by which we numerically exceed the } i\text{th goal}$$
$$s_i^- = \text{amount by which we are numerically under the } i\text{th goal}$$

The s_i^+ and s_i^- are referred to as **deviational variables.** For the Priceler problem, we assume that each s_i^+ and s_i^- is measured in millions of exposures. Using the deviational variables, we can rewrite the first three constraints in (21) as

$$7x_1 + 3x_2 + s_1^- - s_1^+ = 40 \qquad \text{(HIM constraint)}$$
$$10x_1 + 5x_2 + s_2^- - s_2^+ = 60 \qquad \text{(LIP constraint)}$$
$$5x_1 + 4x_2 + s_3^- - s_3^+ = 35 \qquad \text{(HIW constraint)}$$

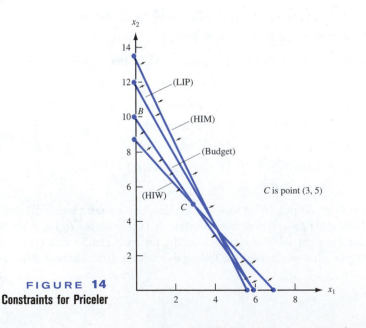

FIGURE 14
Constraints for Priceler

For example, suppose that $x_1 = 5$ and $x_2 = 2$. This advertising schedule yields $7(5) + 3(2) = 41$ million HIM exposures. This exceeds the HIM goal by $41 - 40 = 1$ million exposures, so $s_1^- = 0$ and $s_1^+ = 1$. Also, this schedule yields $10(5) + 5(2) = 60$ million LIP exposures. This exactly meets the LIP requirement, and $s_2^- = s_2^+ = 0$. Finally, this schedule yields $5(5) + 4(2) = 33$ million HIW exposures. We are numerically under the HIW goal by $35 - 33 = 2$ million exposures, so $s_3^- = 2$ and $s_3^+ = 0$.

Suppose Priceler wants to minimize the total penalty from the lost sales. In terms of the deviational variables, the total penalty from lost sales (in thousands of dollars) caused by deviation from the three goals is $200s_1^- + 100s_2^- + 50s_3^-$. The objective function coefficient for the variable associated with goal i is called the **weight** for goal i. The most important goal has the largest weight, and so on. Thus, in the Priceler example, goal 1 (HIM) is most important, goal 2 (LIP) is second most important, and goal 3 (HIW) is least important.

Burnit can minimize the penalty from Priceler's lost sales by solving the following LP:

$$
\begin{aligned}
\min z = 200s_1^- &+ 100s_2^- + 50s_3^- \\
\text{s.t.} \quad 7x_1 + 3x_2 + s_1^- - s_1^+ &= 40 \qquad \text{(HIM constraint)} \\
10x_1 + 5x_2 + s_2^- - s_2^+ &= 60 \qquad \text{(LIP constraint)} \\
5x_1 + 4x_2 + s_3^- - s_3^+ &= 35 \qquad \text{(HIW constraint)} \\
100x_1 + 60x_2 &\le 600 \qquad \text{(Budget constraint)}
\end{aligned}
\tag{22}
$$

All variables nonnegative

The optimal solution to this LP is $z = 250$, $x_1 = 6$, $x_2 = 0$, $s_1^+ = 2$, $s_2^+ = 0$, $s_3^+ = 0$, $s_1^- = 0$, $s_2^- = 0$, $s_3^- = 5$. This meets goal 1 and goal 2 (the goals with the highest costs, or weights, for each unit of deviation from the goal) but fails to meet the least important goal (goal 3).

REMARKS If failure to meet goal i occurs when the attained value of an attribute is numerically smaller than the desired value of goal i, then a term involving s_i^- will appear in the objective function. If failure to meet goal i occurs when the attained value of an attribute is numerically larger than the desired value of goal i, then a term involving s_i^+ will appear in the objective function. Also, if we want to meet a goal exactly and a penalty is assessed for going both over and under a goal, then terms involving both s_i^- and s_i^+ will occur in the objective function.

Suppose we modify the Priceler example by deciding that the budget restriction of $600,000 is a goal. If we decide that a $1 penalty is assessed for each dollar by which this goal is unmet, then the appropriate goal programming formulation would be

$$
\begin{aligned}
\min z = 200s_1^- &+ 100s_2^- + 50s_3^- + s_4^+ \\
\text{s.t.} \quad 7x_1 + 3x_2 + s_1^- - s_1^+ &= 40 \qquad \text{(HIM constraint)} \\
10x_1 + 5x_2 + s_2^- - s_2^+ &= 60 \qquad \text{(LIP constraint)} \\
5x_1 + 4x_2 + s_3^- - s_3^+ &= 35 \qquad \text{(HIW constraint)} \\
100x_1 + 60x_2 + s_4^- - s_4^+ &= 600 \qquad \text{(Budget constraint)}
\end{aligned}
$$

All variables nonnegative

In contrast to our previous optimal solution, the optimal solution to this LP is $z = 33\frac{1}{3}$, $x_1 = 4\frac{1}{3}$, $x_2 = 3\frac{1}{3}$, $s_1^+ = \frac{1}{3}$, $s_2^+ = 0$, $s_3^+ = 0$, $s_4^+ = 33\frac{1}{3}$, $s_1^- = 0$, $s_2^- = 0$, $s_3^- = 0$, $s_4^- = 0$. Thus, when we define the budget restriction to be a goal, the optimal solution is to meet all three advertising goals by going $33\frac{1}{3}$ thousand over budget.

Preemptive Goal Programming

In our LP formulation of the Burnit example, we assumed that Priceler could exactly determine the relative importance of the three goals. For instance, Priceler determined that the HIM goal was $\frac{200}{100} = 2$ times as important as the LIP goal, and the LIP goal was $\frac{100}{50} =$ 2 times as important as the HIW goal. In many situations, however, a decision maker may not be able to determine precisely the relative importance of the goals. When this is the case, *preemptive goal programming* may prove to be a useful tool. To apply preemptive goal programming, the decision maker must rank his or her goals from the most important (goal 1) to least important (goal n). The objective function coefficient for the variable representing goal i will be P_i. We assume that

$$P_1 >>> P_2 >>> P_3 >>> \cdots >>> P_n$$

Thus, the weight for goal 1 is much larger than the weight for goal 2, the weight for goal 2 is much larger than the weight for goal 3, and so on. This definition of the $P_1, P_2, \ldots,$ P_n ensures that the decision maker first tries to satisfy the most important (goal 1) goal. Then, among all points that satisfy goal 1, the decision maker tries to come as close as possible to satisfying goal 2, and so forth. We continue in this fashion until the only way we can come closer to satisfying a goal is to increase the deviation from a higher-priority goal.

For the Priceler problem, the preemptive goal programming formulation is obtained from (22) by replacing (22)'s objective function by $P_1 s_1^- + P_2 s_2^- + P_3 s_3^-$. Thus, the preemptive goal programming formulation of the Priceler problem is

$$
\begin{aligned}
\min z = P_1 s_1^- &+ P_2 s_2^- + P_3 s_3^- \\
\text{s.t.} \quad 7x_1 + 3x_2 + s_1^- - s_1^+ &= 40 \quad &\text{(HIM constraint)} \\
10x_1 + 5x_2 + s_2^- - s_2^+ &= 60 \quad &\text{(LIP constraint)} \\
5x_1 + 4x_2 + s_3^- - s_3^+ &= 35 \quad &\text{(HIW constraint)} \\
100x_1 + 60x_2 \quad &\leq 600 \quad &\text{(Budget constraint)}
\end{aligned}
$$

(23)

All variables non-negative

Assume the decision maker has n goals. To apply preemptive goal programming, we must separate the objective function into n components, where component i consists of the objective function term involving goal i. We define

$$z_i = \text{objective function term involving goal } i$$

For the Priceler example, $z_1 = P_1 s_1^-$, $z_2 = P_2 s_2^-$, and $z_3 = P_3 s_3^-$. Preemptive goal programming problems can be solved by an extension of the simplex known as the **goal programming simplex.** To prepare a problem for solution by the goal programming simplex, we must compute n row 0's, with the ith row 0 corresponding to goal i. Thus, for the Priceler problem, we have

$$\text{Row 0 (goal 1): } z_1 - P_1 s_1^- = 0$$
$$\text{Row 0 (goal 2): } z_2 - P_2 s_2^- = 0$$
$$\text{Row 0 (goal 3): } z_3 - P_3 s_3^- = 0$$

From (23), we find that $BV = \{s_1^-, s_2^-, s_3^-, s_4\}$ (s_4 = slack variable for fourth constraint) is a starting basic feasible solution that could be used to solve (23) via the simplex algorithm (or goal programming simplex algorithm). As with the regular simplex, we must first eliminate all variables in the starting basis from each row 0. Adding P_1 (HIM constraint) to row 0 (goal 1) yields

$$\text{Row 0 (goal 1): } z_1 + 7P_1 x_1 + 3P_1 x_2 - P_1 s_1^+ = 40P_1 \quad \text{(HIM)}$$

Adding P_2 (LIP constraint) to row 0 (goal 2) yields

$$\text{Row 0 (goal 2): } z_2 + 10P_2x_1 + 5P_2x_2 - P_2s_2^+ = 60P_2 \qquad \text{(LIP)}$$

Adding P_3 (HIW constraint) to row 0 (goal 3) yields

$$\text{Row 0 (goal 3): } z_3 + 5P_3x_1 + 4P_3x_2 - P_3s_3^+ = 35P_3 \qquad \text{(HIW)}$$

The Priceler problem can now be solved by the goal programming simplex.

The differences between the goal programming simplex and the ordinary simplex are as follows:

1 The ordinary simplex has a single row 0, whereas the goal programming simplex requires n row 0's (one for each goal).

2 In the goal programming simplex, the following method is used to determine the entering variable: Find the highest-priority goal (goal i') that has not been met (or find the highest-priority goal i' having $z_{i'} > 0$). Find the variable with the most positive coefficient in row 0 (goal i') and enter this variable (subject to the following restriction) into the basis. This will reduce $z_{i'}$ and ensure that we come closer to meeting goal i'. *If, however, a variable has a negative coefficient in row 0 associated with a goal having a higher priority than i', then the variable cannot enter the basis.* Entering such a variable in the basis would increase the deviation from some higher-priority goal. If the variable with the most positive coefficient in row 0 (goal i') cannot be entered into the basis, then try to find another variable with a positive coefficient in row 0 (goal i'). If no variable for row 0 (goal i') can enter the basis, then there is no way to come closer to meeting goal i' without increasing the deviation from some higher-priority goal. In this case, move on to row 0 (goal $i' + 1$) in an attempt to come closer to meeting goal $i' + 1$.

3 When a pivot is performed, row 0 for each goal must be updated.

4 A tableau will yield the optimal solution if all goals are satisfied (that is, $z_1 = z_2 = \cdots = z_n = 0$), or if each variable that can enter the basis and reduce the value of $z_{i'}'$ for an unsatisfied goal i' will increase the deviation from some goal i having a higher priority than goal i'.

We now use the goal programming simplex to solve the Priceler example. In each tableau, the row 0's are listed in order of the goal's priorities (from highest priority to lowest priority). The initial tableau is Table 53. The current bfs is $s_1^- = 40$, $s_2^- = 60$, $s_3^- = 35$, $s_4 = 600$. Because $z_1 = 40P_1$, goal 1 is not satisfied. To reduce the penalty associated with not meeting goal 1, we enter the variable with the most positive coefficient (x_1) in row 0 (HIM). The ratio test indicates that x_1 should enter the basis in the HIM constraint.

After entering x_1 into the basis, we obtain Table 54. The current basic solution is $x_1 = \frac{40}{7}$, $s_2^- = \frac{20}{7}$, $s_3^- = \frac{45}{7}$, $s_4 = \frac{200}{7}$. Because $s_1^- = 0$ and $z_1 = 0$, goal 1 is now satisfied. We now try to satisfy goal 2 (while ensuring that the higher-priority goal 1 is still satisfied). The variable with the most positive coefficient in row 0 (LIP) is s_1^+. Observe that entering s_1^+ into the basis will not increase z_1 [because the coefficient of s_1^+ in row 0 (HIM) is 0]. Thus, after entering s_1^+ into the basis, goal 1 will still be satisfied. The ratio test indicates that s_1^+ could enter the basis in either the LIP or the budget constraint. We arbitrarily choose to enter s_1^+ into the basis in the budget constraint.

After pivoting s_1^+ into the basis, we obtain Table 55. Because $z_1 = z_2 = 0$, goals 1 and 2 are met. Because $z_3 = 5P_3$, however, goal 3 is unmet. The current bfs is $x_1 = 6$, $s_2^- = 0$, $s_3^- = 5$, $s_1^+ = 2$. We now try to come closer to meeting goal 3 (without violating either goal 1 or goal 2). Because x_2 is the only variable with a positive coefficient in row 0 (HIW), the only way to come closer to meeting goal 3 (HIW) is to enter x_2 into the basis. Observe, however, that x_2 has a negative coefficient in row 0 for goal 2 (LIP). Thus,

TABLE 53
Initial Tableau for Preemptive Goal Programming for Priceler

	x_1	x_2	s_1^+	s_2^+	s_3^+	s_1^-	s_2^-	s_3^-	s_4	rhs
Row 0 (HIM)	$7P_1$	$3P_1$	$-P_1$	0	0	0	0	0	0	$z_1 = 40P_1$
Row 0 (LIP)	$10P_2$	$5P_2$	0	$-P_2$	0	0	0	0	0	$z_2 = 60P_2$
Row 0 (HIW)	$5P_3$	$4P_3$	0	0	$-P_3$	0	0	0	0	$z_3 = 35P_3$
HIM	⑦	3	-1	0	0	1	0	0	0	40
LIP	10	5	0	-1	0	0	1	0	0	60
HIW	5	4	0	0	-1	0	0	1	0	35
Budget	100	60	0	0	0	0	0	0	1	600

TABLE 54
First Tableau for Preemptive Goal Programming for Priceler

	x_1	x_2	s_1^+	s_2^+	s_3^+	s_1^-	s_2^-	s_3^-	s_4	rhs
Row 0 (HIM)	0	0	0	0	0	$-P_1$	0	0	0	$z_1 = 0$
Row 0 (LIP)	0	$\frac{5P_2}{7}$	$\frac{10P_2}{7}$	$-P_2$	0	$-\frac{10P_2}{7}$	0	0	0	$z_2 = \frac{20P_2}{7}$
Row 0 (HIW)	0	$\frac{13P_3}{7}$	$\frac{5P_3}{7}$	0	$-P_3$	$-\frac{5P_3}{7}$	0	0	0	$z_3 = \frac{45P_3}{7}$
HIM	1	$\frac{3}{7}$	$-\frac{1}{7}$	0	0	$\frac{1}{7}$	0	0	0	$\frac{40}{7}$
LIP	0	$\frac{5}{7}$	$\frac{10}{7}$	-1	0	$-\frac{10}{7}$	1	0	0	$\frac{20}{7}$
HIW	0	$\frac{13}{7}$	$\frac{5}{7}$	0	-1	$-\frac{5}{7}$	0	1	0	$\frac{45}{7}$
Budget	0	$\frac{120}{7}$	⑩⓪⁄7	0	0	$-\frac{100}{7}$	0	0	1	$\frac{200}{7}$

TABLE 55
Optimal Tableau for Preemptive Goal Programming for Priceler

	x_1	x_2	s_1^+	s_2^+	s_3^+	s_1^-	s_2^-	s_3^-	s_4	rhs
Row 0 (HIM)	0	0	0	0	0	$-P_1$	0	0	0	$z_1 = 0$
Row 0 (LIP)	0	$-P_2$	0	$-P_2$	0	0	0	0	$-\frac{P_2}{10}$	$z_2 = 0$
Row 0 (HIW)	0	P_3	0	0	$-P_3$	0	0	0	$-\frac{P_3}{20}$	$z_3 = 5P_3$
HIM	1	$\frac{3}{5}$	0	0	0	0	0	0	$\frac{1}{100}$	6
LIP	0	-1	0	-1	0	0	1	0	$-\frac{1}{10}$	0
HIW	0	1	0	0	-1	0	0	1	$-\frac{1}{20}$	5
Budget	0	$\frac{6}{5}$	1	0	0	-1	0	0	$\frac{7}{100}$	2

the only way we can come closer to meeting goal 3 (HIW) is to violate a higher-priority goal, goal 2 (LIP). This is therefore an optimal tableau. The preemptive goal programming solution is to purchase 6 minutes of football ads and no soap opera ads. Goals 1 and 2 (HIM and LIP) are met, and Priceler falls 5 million exposures short of meeting goal 3 (HIW).

If the analyst has access to a computerized goal programming code, then by reordering the priorities assigned to the goals, many solutions can be generated. From among these solutions, the decision maker can choose a solution that she feels best fits her preferences. Table 56 lists the solutions found by the preemptive goal programming method for each possible set of priorities. Thus, we see that different ordering of priorities can lead to different advertising strategies.

TABLE 56
Optimal Solutions for Priceler Found by Preemptive Goal Programming

| Priorities | | | Optimal | | | | |
| Highest | Second Highest | Lowest | x_1 Value | x_2 Value | Deviations from | | |
					HIM	LIP	HIW
HIM	LIP	HIW	6	0	0	0	5
HIM	HIW	LIP	5	$\frac{5}{3}$	0	$\frac{5}{3}$	$\frac{10}{3}$
LIP	HIM	HIW	6	0	0	0	5
LIP	HIW	HIM	6	0	0	0	5
HIW	HIM	LIP	3	5	4	5	0
HIW	LIP	HIM	3	5	4	5	0

When a preemptive goal programming problem involves only two decision variables, the optimal solution can be found graphically. For example, suppose HIW is the highest-priority goal, LIP is the second-highest, and HIM is the lowest. From Figure 14, we find that the set of points satisfying the highest-priority goal (HIW) and the budget constraint is bounded by the triangle *ABC*. Among these points, we now try to come as close as we can to satisfying the second-highest-priority goal (LIP). Unfortunately, no point in triangle *ABC* satisfies the LIP goal. We see from the figure, however, that among all points satisfying the highest-priority goal, point *C* (*C* is where the HIW goal is exactly met and the budget constraint is binding) is the unique point that comes the closest to satisfying the LIP goal. Simultaneously solving the equations

$$5x_1 + 4x_2 = 35 \qquad \text{(HIW goal exactly met)}$$
$$100x_1 + 60x_2 = 600 \qquad \text{(Budget constraint binding)}$$

we find that point $C = (3, 5)$. Thus, for this set of priorities, the preemptive goal programming solution is to purchase 3 football game ads and 5 soap opera ads.

Goal programming is not the only approach used to analyze multiple objective decision-making problems under certainty. See Steuer (1985) and Zionts and Wallenius (1976) for other approaches to multiple objective decision making under certainty.

Using LINDO or LINGO to Solve Preemptive Goal Programming Problems

Readers who do not have access to a computer program that will solve preemptive goal programming problems may still use LINDO (or any other LP package) to solve them. To illustrate how LINDO can be used to solve a preemptive goal programming problem, let's look at the Priceler example with our original set of priorities (HIM followed by LIP followed by HIW).

We begin by asking LINDO to minimize the deviation from the highest-priority (HIM) goal by solving the following LP:

$$\min z = s_1^-$$

$$\begin{aligned}
\text{s.t.} \quad & 7x_1 + 3x_2 + s_1^- - s_1^+ = 40 && \text{(HIM constraint)} \\
& 10x_1 + 5x_2 + s_2^- - s_2^+ = 60 && \text{(LIP constraint)} \\
& 5x_1 + 4x_2 + s_3^- - s_3^+ = 35 && \text{(HIW constraint)} \\
& 100x_1 + 60x_2 \leq 600 && \text{(Budget constraint)}
\end{aligned}$$

All variables non-negative

Goal 1 (HIM) can be met, so LINDO reports an optimal z-value of 0. We now want to come as close as possible to meeting goal 2 while ensuring that the deviation from goal 1 remains at its current level (0). Using an objective function of s_2^- (to minimize goal 2) we add the constraint $s_1^- = 0$ (to ensure that goal 1 is still met) and ask LINDO to solve

$$\min z = s_2^-$$

$$
\begin{array}{llll}
\text{s.t.} & 7x_1 + 3x_2 + s_1^- - s_1^+ = 40 & \text{(HIM constraint)} \\
& 10x_1 + 5x_2 + s_2^- - s_2^+ = 60 & \text{(LIP constraint)} \\
& 5x_1 + 4x_2 + s_3^- - s_3^+ = 35 & \text{(HIW constraint)} \\
& 100x_1 + 60x_2 \leq 600 & \text{(Budget constraint)} \\
& s_1^- = 0
\end{array}
$$

All variables non-negative

Because goals 1 and 2 can be simultaneously met, this LP will also yield an optimal z-value of 0. We now come as close as possible to meeting goal 3 (HIW) while keeping the deviations from goals 1 and 2 at their current levels. This requires LINDO to solve the following LP:

$$\min z = s_3^-$$

$$
\begin{array}{llll}
\text{s.t.} & 7x_1 + 3x_2 + s_1^- - s_1^+ = 40 & \text{(HIM constraint)} \\
& 10x_1 + 5x_2 + s_2^- - s_2^+ = 60 & \text{(LIP constraint)} \\
& 5x_1 + 4x_2 + s_3^- - s_3^+ = 35 & \text{(HIW constraint)} \\
& 100x_1 + 60x_2 + s_3^- - s_3^+ \leq 600 & \text{(Budget constraint)} \\
& s_1^- = 0 \\
& s_2^- = 0
\end{array}
$$

All variables non-negative

Of course, the LINDO (or LINGO) full-screen editor makes it easy to go from one step of the goal programming problem to the next. To go from step i to step $i + 1$, simply modify your objective function to minimize the deviation from the $i + 1$ highest-priority goal and add a constraint that ensures that the deviation from the ith highest-priority goal remains at its current level.

REMARKS   **1**   The optimal solution to this LP is $z = 5$, $x_1 = 6$, $x_2 = 0$, $s_1^- = 0$, $s_2^- = 0$, $s_3^- = 5$, $s_1^+ = 2$, $s_2^+ = 0$, $s_3^+ = 0$, which agrees with the solution obtained by the preemptive goal programming method. The z-value of 5 indicates that if goals 1 and 2 are met, then the best that Priceler can do is to come within 5 million exposures of meeting goal 3.
2   By the way, suppose we could only have come within two units of meeting goal 1. When solving our second LP, we would have added the constraint $s_1^- = 2$ (instead of $s_1^- = 0$).
3   The goal programming methodology of this section can be applied without any changes when some or all of the decision variables are restricted to be integer or 0–1 variables (see Problems 11, 12, and 14).
4   Using LINGO, the goal programming methodology of this section can be applied without any changes even if the objective function or some of the constraints are nonlinear.

PROBLEMS

Group A

1   Graphically determine the preemptive goal progamming solution to the Priceler example for the following priorities:

   a   LIP is highest-priority goal, followed by HIW and then HIM.

   b   HIM is highest-priority goal, followed by LIP and then HIW.

   c   HIM is highest-priority goal, followed by HIW and then LIP.

d HIW is highest-priority goal, followed by HIM and then LIP.

2 Fruit Computer Company is ready to make its annual purchase of computer chips. Fruit can purchase chips (in lots of 100) from three suppliers. Each chip is rated as being of excellent, good, or mediocre quality. During the coming year, Fruit will need 5,000 excellent chips, 3,000 good chips, and 1,000 mediocre chips. The characteristics of the chips purchased from each supplier are shown in Table 57. Each year, Fruit has budgeted $28,000 to spend on chips. If Fruit does not obtain enough chips of a given quality, then the company may special-order additional chips at $10 per excellent chip, $6 per good chip, and $4 per mediocre chip. Fruit assesses a penalty of $1 for each dollar by which the amount paid to suppliers 1–3 exceeds the annual budget. Formulate and solve an LP to help Fruit minimize the penalty associated with meeting the annual chip requirements. Also use preemptive goal programming to determine a purchasing strategy. Let the budget constraint have the highest priority, followed in order by the restrictions on excellent, good, and mediocre chips.

3 Highland Appliance must determine how many color TVs and VCRs should be stocked. It costs Highland $300 to purchase a color TV and $200 to purchase a VCR. A color TV requires 3 sq yd of storage space, and a VCR requires 1 sq yd of storage space. The sale of a color TV earns Highland a profit of $150, and the sale of a VCR earns Highland a profit of $100. Highland has set the following goals (listed in order of importance):

Goal 1 A maximum of $20,000 can be spent on purchasing color TVs and VCRs.
Goal 2 Highland should earn at least $11,000 in profits from the sale of color TVs and VCRs.
Goal 3 Color TVs and VCRs should use no more than 200 sq yd of storage space.

Formulate a preemptive goal programming model that Highland could use to determine how many color TVs and VCRs to order. How would the preemptive goal formulation be modified if Highland's goal were to have a profit of exactly $11,000?

4 A company produces two products. Relevant information for each product is shown in Table 58. The company has a goal of $48 in profits and incurs a $1 penalty for each dollar it falls short of this goal. A total of 32 hours of labor are available. A $2 penalty is incurred for each hour of overtime (labor over 32 hours) used, and a $1 penalty is incurred for each hour of available labor that is unused. Marketing considerations require that at least 10 units of product 2 be produced. For each unit (of either product) by which production falls short of demand, a penalty of $5 is assessed.

    a Formulate an LP that can be used to minimize the penalty incurred by the company.

    b Suppose the company sets (in order of importance) the following goals:

Goal 1 Avoid underutilization of labor.
Goal 2 Meet demand for product 1.
Goal 3 Meet demand for product 2.
Goal 4 Do not use any overtime.

Formulate and solve a preemptive goal programming model for this situation.

TABLE 57

Supplier	Characteristics of a Lot of 100 Chips			Price Per 100 Chips ($)
	Excellent	Good	Mediocre	
1	60	20	20	400
2	50	35	15	300
3	40	20	40	250

TABLE 58

	Product 1	Product 2
Labor required	4 hours	2 hours
Contribution to profit	$4	$2

5[†] Deancorp produces sausage by blending together beef head, pork chuck, mutton, and water. The cost per pound, fat per pound, and protein per pound for these ingredients is given in Table 59. Deancorp needs to produce 100 lb of sausage and has set the following goals, listed in order of priority:

Goal 1 Sausage should consist of at least 15% protein.
Goal 2 Sausage should consist of at most 8% fat.
Goal 3 Cost per pound of sausage should not exceed 8¢.

Formulate a preemptive goal programming model for Deancorp.

6[‡] The Touche Young accounting firm must complete three jobs during the next month. Job 1 will require 500 hours of work, job 2 will require 300 hours of work, and job 3 will require 100 hours of work. Currently, the firm consists of 5 partners, 5 senior employees, and 5 junior employees, each of whom can work up to 40 hours per month. The dollar amount (per hour) that the company can bill depends on the type of accountant who is assigned to each job, as shown in Table 60. (The X indicates that a junior employee does not have enough experience to work on job 1.) All jobs must be completed. Touche Young has also set the following goals, listed in order of priority:

Goal 1 Monthly billings should exceed $68,000.
Goal 2 At most, 1 partner should be hired.
Goal 3 At most, 3 senior employees should be hired.
Goal 4 At most, 5 junior employees should be hired.

TABLE 59

	Head	Chuck	Mutton	Moisture
Fat (per lb)	.05	.24	.11	0
Protein (per lb)	.20	.26	.08	0
Cost (in ¢)	.12	9	8	0

[†]Based on Steuer (1984).
[‡]Based on Welling (1977).

TABLE 60

	Job 1	Job 2	Job 3
Partner	160	120	110
Senior employee	120	90	70
Junior employee	X	50	40

Formulate a preemptive goal programming model for this situation.

7 There are four teachers in the Faber College Business School. Each semester, 200 students take each of the following courses: marketing, finance, production, and statistics. The "effectiveness" of each teacher in teaching each class is given in Table 61. Each teacher can teach a total of 200 students during the semester. The dean has set a goal of obtaining an average teaching effectiveness level of about 6 in each course. Deviations from this goal in any course are considered equally important. Formulate a goal programming model that can be used to determine the semester's teaching assignments.

Group B

8[†] Faber College is admitting students for the class of 2001. It has set four goals for this class, listed in order of priority:

Goal 1 Entering class should be at least 5,000 students.
Goal 2 Entering class should have an average SAT score of at least 640.
Goal 3 Entering class should consist of at least 25 percent out-of-state students.
Goal 4 At least 2,000 members of the entering class should not be nerds.

The applicants received by Faber are categorized in Table 62. Formulate a preemptive goal programming model that could determine how many applicants of each type should be admitted. Assume that all applicants who are admitted will decide to attend Faber.

9[‡] During the next four quarters, Wivco faces the following demands for globots: quarter 1—13 globots; quarter 2—14 globots; quarter 3—12 globots; quarter 4—15 globots. Globots may be produced by regular-time labor or by

TABLE 61

Teacher	Marketing	Finance	Production	Statistics
1	7	5	8	2
2	7	8	9	4
3	3	5	7	9
4	5	5	6	7

[†]Based on Lee and Moore, "University Admissions Planning" (1974).
[‡]Based on Lee and Moore, "Production Scheduling" (1974).

TABLE 62

Home State	SAT Score	No. of Nerds	No. of Non-Nerds
In-state	700	1500	400
In-state	600	1300	700
In-state	500	500	500
Out-of-state	700	350	50
Out-of-state	600	400	400
Out-of-state	500	400	600

overtime labor. Production capacity (number of globots) and production costs during the next four quarters are shown in Table 63. Wivco has set the following goals in order of importance:

Goal 1 Meet each quarter's demand on time.
Goal 2 Inventory at the end of each quarter cannot exceed 3 units.
Goal 3 Total production cost should be held below $250.

Formulate a preemptive goal programming model that could be used to determine Wivco's production schedule during the next four quarters. Assume that at the beginning of the first quarter 1 globot is in inventory.

10 Ricky's Record Store now employs five full-time employees and three part-time employees. The normal workload is 40 hours per week for full-time and 20 hours per week for part-time employees. Each full-time employee is paid $6 per hour for work up to 40 hours per week and can sell 5 records per hour. A full-time employee who works overtime is paid $10 per hour. Each part-time employee is paid $3 per hour and can sell 3 records per hour. It costs Ricky $6 to buy a record, and each record sells for $9. Ricky has weekly fixed expenses of $500. He has established the following weekly goals, listed in order of priority:

Goal 1 Sell at least 1,600 records per week.
Goal 2 Earn a profit of at least $2,200 per week.
Goal 3 Full-time employees should work at most 100 hours of overtime.
Goal 4 To increase their sense of job security, the number of hours by which each full-time employee fails to work 40 hours should be minimized.

Formulate a preemptive goal programming model that could be used to determine how many hours per week each employee should work.

TABLE 63

Quarter	Regular-Time Capacity	Regular-Time Cost/Unit	Overtime Capacity	Overtime Cost/Unit
1	9	$4	5	$6
2	10	$4	5	$7
3	11	$5	5	$8
4	12	$6	5	$9

TABLE 66

Site	1	2	3	4	5	6
Golf	31	X	X	X	X	27
Swimming	X	25	21	32	32	X
Gymnasium	X	37	29	28	38	X
Tennis courts	X	20	23	22	20	X

11 Gotham City is trying to determine the type and location of recreational facilities to be built during the next decade. Four types of facilities are under consideration: golf courses, swimming pools, gymnasiams, and tennis courts. Six sites are under consideration. If a golf course is built, it must be built at either site 1 or site 6. Other facilities may be built at sites 2–5. The available land (in thousands of square feet) at each site is given in Table 64.

The cost of building each facility (in thousands of dollars), the annual maintenance cost (in thousands of dollars) for each facility, and the land (in thousands of square feet) required for each facility are given in Table 65.

The number of user days (in thousands) for each type of facility depends on where it is built. The dependence is given in Table 66.

a Consider the following set of priorities:

Priority 1 Limit land use at each site to the land available.
Priority 2 Construction costs should not exceed $1.2 million.
Priority 3 User days should exceed 200,000.
Priority 4 Annual maintenance costs should not exceed $200,000.

For this set of priorities, use preemptive goal programming to determine the type and location of recreation facilities in Gotham City.

b Consider the following set of priorities:

Priority 1 Limit land use at each site to the land available.
Priority 2 User days should exceed 200,000.
Priority 3 Construction costs should not exceed $1.2 million.
Priority 4 Annual maintenance costs should not exceed $200,000.

For this set of priorities, use preemptive goal programming to determine the type and location of recreation facilities in Gotham City.[†]

12 A small aerospace company is considering eight projects:

Project 1 Develop an automated test facility.
Project 2 Barcode all company inventory and machinery.
Project 3 Introduce a CAD/CAM system.
Project 4 Buy a new lathe and deburring system.
Project 5 Institute FMS (flexible manufacturing system).
Project 6 Install a LAN (local area network).
Project 7 Develop AIS (artificial intelligence simulation).
Project 8 Set up a TQM (total quality management) initiative.

Each project has been rated on five attributes: return on investment (ROI), cost, productivity improvement, worker requirements, and degree of technological risk. These ratings are given in Table 67.

The company has set the following five goals (listed in order of priority):

Goal 1 Achieve a return on investment of at least $3,250.
Goal 2 Limit cost to $1,300.
Goal 3 Achieve a productivity improvement of at least 6.
Goal 4 Limit manpower use to 108.
Goal 5 Limit technological risk to a total of 4.

Use preemptive goal programming to determine which projects should be undertaken.

13 The new president has just been elected and has set the following economic goals (listed from highest to lowest priority):

Goal 1 Balance the budget (this means revenues are at least as large as costs).
Goal 2 Cut spending by at most $150 billion.
Goal 3 Raise at most $550 billion in taxes from the upper class.
Goal 4 Raise at most $350 billion in taxes from the lower class.

Currently, the government spends $1 trillion (a trillion = 1,000 billion) per year. Revenue can be raised in two ways: through a gas tax and an income tax. You must determine

G = per gallon tax rate (in cents)
LTR = % tax rate charged on first $30,000 of income
HTR = % tax rate charged on any income earned more than $30,000
C = cut in spending (in billions)

If the government chooses G, LTR, and HTR, then the revenue given in Table 68 (in billions) is raised. Of course, the tax rate on income more than $30,000 must be at least as large as the tax rate on the first $30,000 of income. For-

TABLE 64

	Site			
	2	3	4	5
Land	70	80	95	120

TABLE 65

Site	Construction Cost	Maintenance Cost	Land Required
Golf	340	80	Not relevant
Swimming	300	36	29
Gymnasium	840	50	38
Tennis courts	85	17	45

[†]Based on Taylor and Keown (1984).

TABLE 67

	Project							
	1	2	3	4	5	6	7	8
ROI ($)	2,070	456	670	350	495	380	1,500	480
Cost ($)	900	240	335	700	410	190	500	160
Productivity improvement	3	2	2	0	1	0	3	2
Manpower needed	18	18	27	36	42	6	48	24
Degree of risk	3	2	4	1	1	0	2	3

TABLE 68

	Low Income	High Income
Gas tax	G	.5G
Tax on income up to $30,000	20LTR	5LTR
Tax on income above $30,000	0	15HTR

TABLE 69

Project	NPV (in millions)	Annual Growth Rate	Probability of Success	Cost (in millions)
1	40	20	.75	220
2	30	16	.70	140
3	60	12	.75	280
4	45	8	.90	240
5	55	18	.65	300
6	40	18	.60	200
7	90	19	.65	440

mulate a preemptive goal programming model to help the president meet his goals.

14 HAL computer must determine which of seven research and development (R&D) projects to undertake. For each project four quantities are of interest:

a the net present value (NPV in millions of dollars) of the project

b the annual growth rate in sales generated by the project

c the probability that the project will succeed

d the cost (in millions of dollars) of the project

The relevant information is given in Table 69. HAL has set the following four goals:

Goal 1 The total NPV of all chosen projects should be at least $200 million.

Goal 2 The average probability of success for all projects chosen should be at least .75.

Goal 3 The average growth rate of all projects chosen should be at least 15%.

4.17 Using the Excel Solver to Solve LPs

Excel has the capability to solve linear (and often nonlinear) programming problems. In this section, we show how to use the Excel solver[†] to find the optimal solution to the diet problem of Section 3.4 and the inventory example of Section 3.10.

The key to solving an LP on a spreadsheet is to set up a spreadsheet that tracks everything of interest (costs or profits, resource usage, etc.). Next, identify the cells of interest that can be varied. These are called **Changing Cells.** After defining the Changing Cells, identify the cell that contains your objective function as the **Target Cell.** Next, we identify our constraints and tell the Solver to solve the problem. At this point, the optimal solution to our problem will be placed in the spreadsheet.

[†]To activate the Excel Solver for the first time, select Tools and then select Add-Ins. Checking the Solver Add-in box will cause Excel to open Solver whenever you check Tools and then Solver.

Using the Excel Solver to Solve the Diet Problem

In file Diet1.xls, we set up a spreadsheet model of the diet problem (Example 6 of Chapter 3). To begin (see Figure 15) we enter headings for each type of food in B3:E3. In the range B4:E4, we input trial values for the amount of each food eaten. For example, Figure 15 indicates that we are considering eating three brownies, four scoops of chocolate ice cream, five bottles of cola, and six pieces of pineapple cheesecake. To see if the diet in Figure 15 is an "optimal" diet, we must determine its cost as well as the calories, chocolate, sugar, and fat it provides. In the range B5:E5, we input the per-unit cost for each available food. Then we compute the cost of the diet in cell F5.

We could compute the cost of the diet in cell F5 with the formula

$$=B4 \cdot B5 + C4 \cdot C5 + D4 \cdot D5 + E4 \cdot E5$$

but it is easier to enter the formula

$$=SUMPRODUCT(B\$4:E\$4, B5:E5)$$

The =SUMPRODUCT function requires two ranges as inputs. The first cell in the first range is multiplied by the first cell in the second range; then the second cell in the first range is multiplied by the second cell in the second range; and so on. All of these products are then added. Essentially, the =SUMPRODUCT function duplicates the notion of scalar products of vectors discussed in Section 2.1. Thus, in cell F5 the =SUMPRODUCT function computes total cost as $(3)(50) + 4(20) + 5(30) + 6(80) = 860$ cents.

In the range B6:E6, we enter the calories in each food; in B7:E7, the chocolate content; in B8:E8, the sugar content; and in B9:E9, the fat content. Copying the formula in F5 to the cell range F6:F9 now computes the calories, chocolate, sugar, and fat contained in the diet defined by the values in B4:E4. Note that the =SUMPRODUCT function makes it easy to create many constraints by entering one formula and using the copy command.

In the cell range H6:H9, we have listed the minimum daily requirement for each nutrient. From Figure 15, we see that our current diet is feasible (meets daily requirements for each nutrient) and costs $8.70. We now describe how to use Solver to find the optimal solution to the diet problem.

Step 1 From the Tools menu select Solver. The dialog box in Figure 16 will appear.

Step 2 Move the mouse to the Set Target Cell portion of the dialog box and click (or type in the cell address) on your target cell (total cost in cell F5) and select Min. This tells Solver to minimize total cost.

Step 3 Move the mouse to the By Changing Cells portion of the dialog box and click on the changing cells (B4:E4). This tells Solver it can change the amount eaten of each food.

Step 4 Click on the Add button to add constraints. The screen in Figure 17 will appear. Move to the Cell Reference part of the Add Constraint dialog box and select F6:F9. Then move to the dropdown box and select >=. Finally, click on the constraint portion of the dialog box and select H6:H9. Choose OK because there are no more constraints. If you

	A	B	C	D	E	F	G	H
1			Feasible					
2			solution to Diet Problem					
3		Brownie	Choc IC	Cola	Pine Cheese	Totals		Required
4	Eaten	3	4	5	6			
5	Cost	50	20	30	80	860		
6	Calories	400	200	150	500	5750	>=	500
7	Chocolate	3	2	0	0	17	>=	6
8	Sugar	2	2	4	4	58	>=	10
9	Fat	2	4	1	5	57	>=	8

FIGURE 15

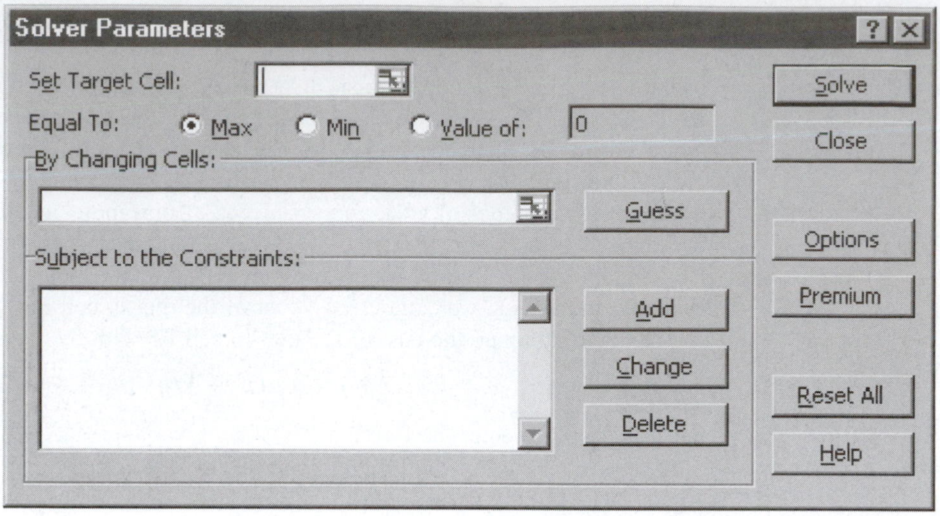

FIGURE 16

need to add more constraints, choose Add. From the main Solver box you may change a constraint by selecting Change or delete a constraint by selecting Delete.

We have now created **four constraints.** Solver will ensure that the changing cells are chosen so F6>=H6, F7>=H7, F8>=H8, and F9>=H9. In short, the diet will be chosen to ensure that enough calories, chocolate, sugar, and fat are eaten.

Our Solver window should now look like Figure 18.

Step 5 Before solving the problem, we need to tell Solver that all Changing Cells must be non-negative. We must also tell Solver that we have a linear model. If we do not tell Solver the model is linear, then Solver will not know it should use the simplex method to solve the problem and Solver may get an incorrect answer. We may accomplish both of these goals by selecting options. The screen in Figure 19 will appear. Checking Assume Non-Negative ensures that all changing cells will be non-negative. Checking the Assume Linear Model box ensures that Solver will use the simplex method to solve our LP. Sometimes in a poorly scaled LP (one with both large and small numbers present in the objective function, right-hand sides, or constraints), the Solver will not recognize an LP as a linear model. Checking the Use Automatic Scaling Box minimizes the chances that a poorly scaled LP will be interpreted as a nonlinear model. By the way, Max Time is the Maximum Time the Solver will run before prompting the user about whether to terminate the solution procedure. Iterations is the maximum number of simplex pivots the Solver will make before asking the user whether the solution procedure will continue. The Precision setting describes how much "error" is tolerated before deciding a constraint is not satisfied. For example, with a precision of .001, a changing cell with a value of $-.0009$ would be deemed to satisfy a non-negativity constraint. The Tolerance and Convergence settings will be discussed in Chapter 8.

FIGURE 17

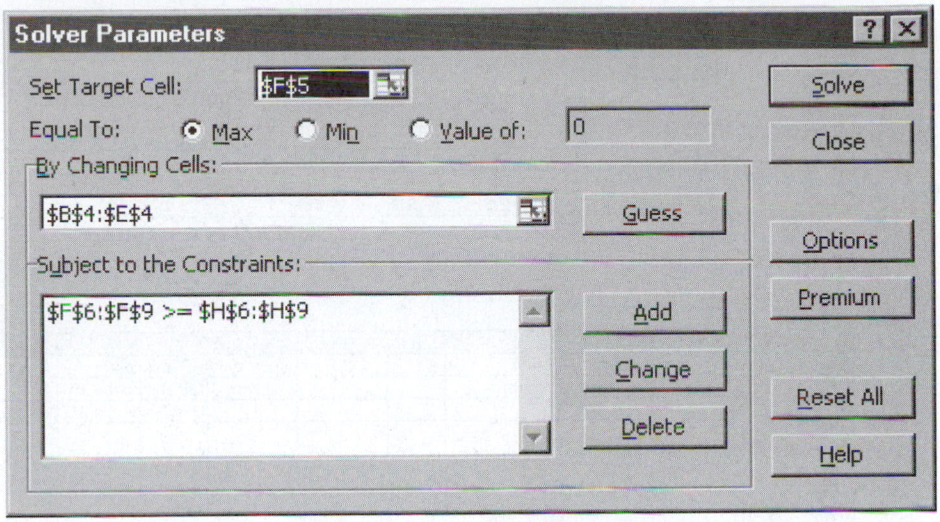

FIGURE 18

Step 6 After choosing OK from the Solver Options box, we then select Solve. Solver yields the optimal solution shown in Figure 20.

Just like LINDO, the Solver says the minimum cost is 90 cents. The minimum cost is obtained by eating no brownies, 3 oz of chocolate ice cream, 1 bottle of cola, and no pineapple cheesecake.

Using the Solver to Solve the Sailco Example

Sailco.xls

We now set up a spreadsheet (Sailco.xls) to solve the Sailco example (Example 12 of Chapter 3). See Figure 21. For each month, we need to keep track of our beginning inventory, ending inventory, and cost. Note that for each month

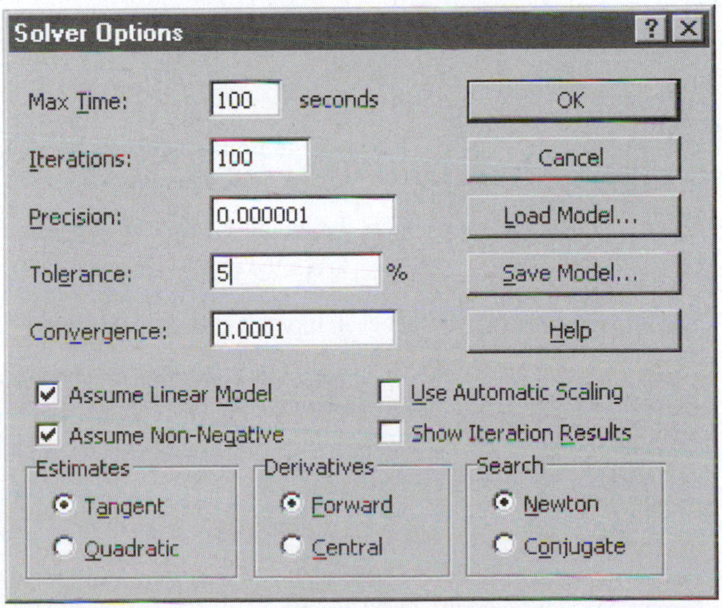

FIGURE 19

	A	B	C	D	E	F	G	H
1			Optimal Solution					
2			to the Diet Problem					
3			Brownie	Choc IC	Cola	Pine Cheese	Totals		Required
4	Eaten	0	3	1	0			
5	Cost	50	20	30	80	90		
6	Calories	400	200	150	500	750	>=	500
7	Chocolate	3	2	0	0	6	>=	6
8	Sugar	2	2	4	4	10	>=	10
9	Fat	2	4	1	5	13	>=	8

FIGURE 20

FIGURE 21

	A	B	C	D	E	F	G	H	I	J	K
1			Optimal solution					RT unit cost	$ 400.00		
2			to Sailco problem					OT unit cost	$ 450.00		
3								Unit Holding cost	$ 20.00		
4	Month	Beg Inventory	OT Production	RT Production		RT Capacity	Demand	Ending Inventory			Monthly Cost
5	1	10	0	40	<=	40	40	10	>=	0	$ 16,200.00
6	2	10	10	40	<=	40	60	0	>=	0	$ 20,500.00
7	3	0	35	40	<=	40	75	0	>=	0	$ 31,750.00
8	4	0	0	25	<=	40	25	0	>=	0	$ 10,000.00
9										Total Cost	$ 78,450.00

Monthly cost = 400(regular-time production) + 450(overtime production) + 20(unit holding cost)

Ending inventory = beginning inventory + monthly production − monthly demand

Step 1 Enter unit costs in I1:I3, regular-time monthly capacities in F5:F8, demands in G5:G8, and beginning month 1 inventory in B5.

Step 2 Enter trial values of each month's overtime and regular-time production in C5:D8.

Step 3 Determine month 1 ending inventory in H5 with the formula

=B5 + C5 + D5 − G5.

This implements the following relationship:

Ending inventory = beginning inventory + monthly production − monthly demand

Step 4 Set month 2 beginning inventory to month 1 ending inventory by entering in cell B6 the formula

=H5.

Step 5 Copying the formula from B5 to B6:B8 computes beginning inventory for months 2–4. Copying the formula from H5 to H6:H8 computes ending inventory for months 2–4.

Step 6 In cell K5, we compute the month 1 cost with the formula

=$I$1*D5 + C5*$I$2 + $I$3*H5.

This implements the fact that each month's cost is given by

Monthly cost = 400(regular-time production) + 450(overtime production) + 20(unit holding cost).

Copying this formula from K5 to K6:K8 computes costs for months 2–4. We compute total cost in cell K9 with the formula

=SUM(K5:K8).

Step 7 We now fill in our Solver dialog box as shown in Figure 22. Our goal is to minimize total cost (cell K9). Our Changing Cells are overtime and regular-time production

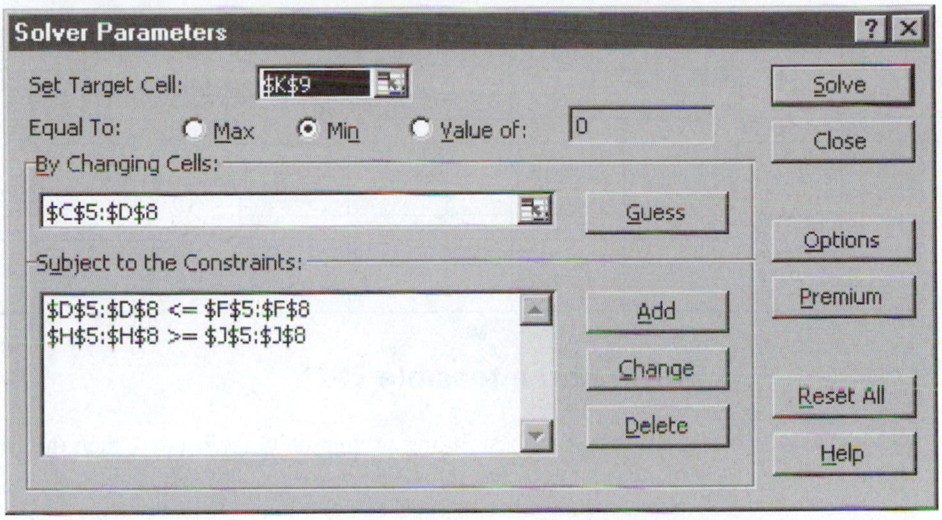

FIGURE 22

(C5:D8). We must ensure that each month's regular-time production is at most 40 (D5:D8 <=F5:F8). Finally, constraining each month's ending inventory to be non-negative (H5:H8 >= J5:J8) ensures that each month's demand is met on time. In Options we check Assume Linear Model, Assume Non-Negative, and use Automatic Scaling. After choosing Solve, we find the optimal solution shown in Figure 21. A minimum cost of $78,450 is achieved by producing 40 units with regular-time production during months 1–3, 25 units of regular-time production during month 4, 10 units of overtime production during month 2, and 35 units of overtime production during month 3.

Using the Value of Option

Recall that in the Sailco problem the minimum cost was $78,450. Suppose that we wanted to find a solution that yielded a cost of exactly, say, $90,000. Then we may use the Solver's value of option. Simply fill in the Solver dialog box as shown in Figure 23 (see the sheet titled Cost of $90,000 in workbook Sailco.xls).

Solver yields the solution in Figure 24. Note that Solver found a feasible solution having a total cost of exactly $90,000.

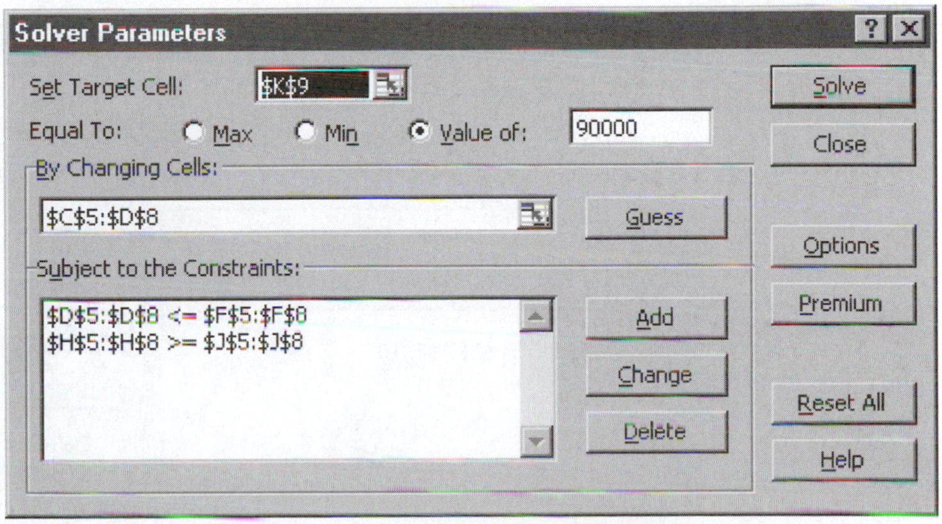

Sailco.xls

FIGURE 23

FIGURE 24

	A	B	C	D	E	F	G	H	I	J	K
1			Optimal solution					RT unit cost	$ 400.00		
2			to Sailco problem					OT unit cost	$ 450.00		
3								Unit Holding cost	$ 20.00		
4	Month	Beg Inventory	OT Production	RT Production		RT Capacity	Demand	Ending Inventory			Monthly Cost
5	1	10	179.090909	0	<=	40	40	149.0909091	>=	0	$ 83,572.73
6	2	149.0909	0	0	<=	40	60	89.09090909	>=	0	$ 1,781.82
7	3	89.09091	0	0	<=	40	75	14.09090909	>=	0	$ 281.82
8	4	14.09091	0	10.909091	<=	40	25	0	>=	0	$ 4,363.64
9										Total Cost	$ 90,000.00

Bevco.xls

Solver and Infeasible LPs

Recall that if at least 36 mg of vitamin C are needed, then the Bevco problem (Example 4 of this chapter) was infeasible. We have set this problem up in Solver in file Bevco.xls. Figure 25 shows the spreadsheet, and Figure 26 shows the Solver window.

When we choose Solve, we obtain the message shown in Figure 27. This indicates that the LP has no feasible solution.

Breadco.xls

Solver and Unbounded LPs

Recall that Example 3 of this chapter was an unbounded LP. The file Breadco.xls (see Figure 28) contains a Solver formulation of this LP. Figure 29 contains the Solver window for the Breadco example. When we choose Solve we obtain the message in Figure 30.

	A	B	C	D	E	F
1						
2	Infeasible LP					
3						
4					Total Cost	
5		Soda	Juice		3 0	
6	Amount	0	10			
7	Unit cost	2	3			
8				Available		Needed
9	Sugar	0.5	0.25	2.5	>=	4
10	Vitamin C	1	3	30	>=	36
11	Total oz.	1	1	10	=	10

FIGURE 25

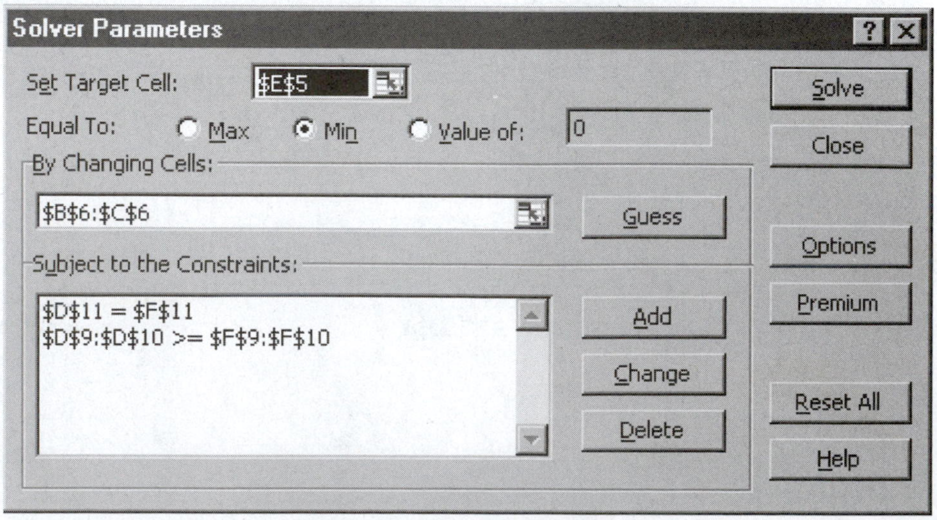

Solver Parameters

Set Target Cell: $E$5

Equal To: ○ Max ● Min ○ Value of: 0

By Changing Cells:

$B$6:$C$6

Subject to the Constraints:

$D$11 = $F$11
$D$9:$D$10 >= $F$9:$F$10

Solve
Close
Guess
Options
Premium
Add
Change
Delete
Reset All
Help

FIGURE 26

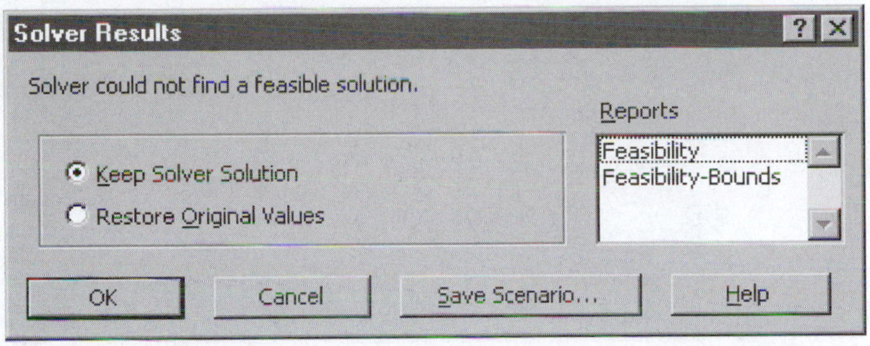

FIGURE 27

	A	B	C	D	E	F	G	H
1	Unbounded LP							
2								
3			FB Baked	SD Baked	Yeast bought	Flour bought		Originally we have
4			5	0	0	20		
5		Price or cost	36	30	3	4		
6		Yeast needed	1	1				5
7		Flour needed	6	5				10
8								
9		Profit	100					
10								
11			Used		Available			
12		Yeast	5	<=	5			
13		Flour	30	<=	30			

FIGURE 28

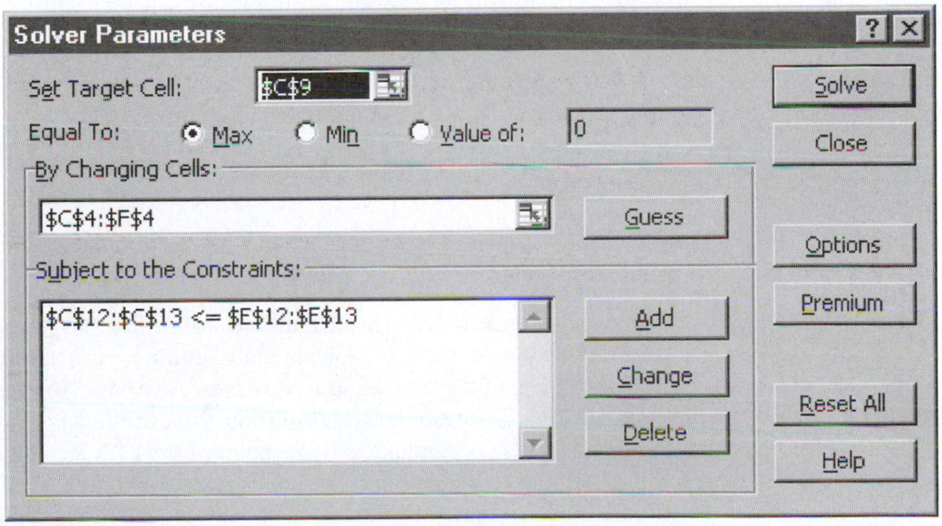

FIGURE 29

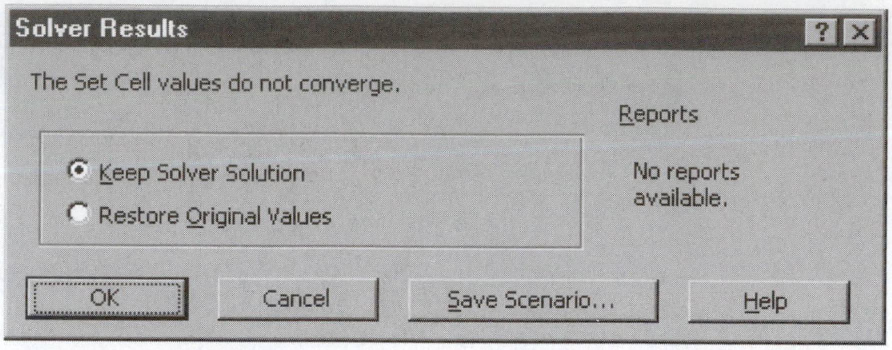

FIGURE 30

The message "Set Cell values do not converge" indicates an unbounded LP; that is, there are values of the changing cells that satisfy all constraints and yield arbitrarily large profit.

PROBLEMS

Group A

Use Excel Solver to find the optimal solution to the following problems:

1 Problem 2 of Section 3.4

2 Example 7 of Chapter 3

3 Example 11 of Chapter 3

4 Problem 3 of Section 3.10

5 Example 14 of Section 3.12

Group B

6 Problem 4 of Section 3.11

7 Problem 5 of Section 3.11

8 Problem 3 of Section 3.12

9 Problem 5 of Section 3.12

SUMMARY   Preparing an LP for Solution by the Simplex

An LP is in **standard form** if all constraints are equality constraints and all variables are non-negative. To place an LP in standard form, we do the following:

Step 1 If the ith constraint is a $\leq$ constraint, then we convert it to an equality constraint by adding a slack variable s_i and the sign restriction $s_i \geq 0$.

Step 2 If the ith constraint is a $\geq$ constraint, then we convert it to an equality constraint by subtracting an excess variable e_i and adding the sign restriction $e_i \geq 0$.

Step 3 If the variable x_i is unrestricted in sign (urs), replace x_i in both the objective function and constraints by $x_i' - x_i''$, where $x_i' \geq 0$ and $x_i'' \geq 0$.

Suppose that once an LP is placed in standard form, it has m constraints and n variables.

A basic solution to $A\mathbf{x} = \mathbf{b}$ is obtained by setting $n - m$ variables equal to 0 and solving for the values of the remaining m variables. Any basic solution in which all variables are non-negative is a **basic feasible solution** (bfs) to the LP.

For any LP, there is a unique extreme point of the LP's feasible region corresponding to each bfs. Also, at least one bfs corresponds to each extreme point of the feasible region.

If an LP has an optimal solution, then there is an extreme point that is optimal. Thus, in searching for an optimal solution to an LP, we may restrict our search to the LP's basic feasible solutions.

The Simplex Algorithm

If the LP is in standard form and a bfs is readily apparent, then the simplex algorithm (for a max problem) proceeds as follows:

Step 1 If all nonbasic variables have non-negative coefficients in row 0, then the current bfs is optimal. If any variables in row 0 have negative coefficients, then choose the variable with the most negative coefficient in row 0 to enter the basis.

Step 2 For each constraint in which the entering variable has a positive coefficient, compute the following ratio:

$$\frac{\text{Right-hand side of constraint}}{\text{Coefficient of entering variable in constraint}}$$

Any constraint attaining the smallest value of this ratio is the winner of the **ratio test.** Use EROs to make the entering variable a basic variable in any constraint that wins the ratio test. Return to Step 1.

If the LP (a max problem) is **unbounded,** then we eventually reach a tableau in which a nonbasic variable has a negative coefficient in row 0 and a nonpositive coefficient in each constraint. Otherwise (barring the extremely rare occurrence of *cycling*), the simplex algorithm will find an optimal solution to an LP.

If a bfs is not readily apparent, then the Big M method or the two-phase simplex method must be used to obtain a bfs.

The Big M Method

Step 1 Modify the constraints so that the right-hand side of each constraint is non-negative.

Step 1′ Identify each constraint that is now (after step 1) an = or ≥ constraint. In step 3, we will add an artificial variable to each of these constraints.

Step 2 Convert each inequality constraint to standard form.

Step 3 If (after step 1 has been completed) constraint i is a ≥ or = constraint, then add an artificial variable a_i and the sign restriction $a_i \geq 0$.

Step 4 Let M denote a very large positive number. If the LP is a min problem, then add (for each artificial variable) Ma_i to the objective function. For a max problem, add $-Ma_i$.

Step 5 Because each artificial variable will be in the starting basis, each must be eliminated from row 0 before beginning the simplex. If all artificial variables are equal to 0 in the optimal solution, then we have found the optimal solution to the original problem. If any artificial variables are positive in the optimal solution, then the original problem is infeasible.

The Two-Phase Method

Step 1 Modify the constraints so that the right-hand side of each constraint is non-negative.

Step 1′ Identify each constraint that is now (after step 1) an = or ≥ constraint. In step 3, we will add an artificial variable to each of these constraints.

Step 2 Convert each inequality constraint to the standard form.

Step 3 If (after step 1′) constraint i is a ≥ or = constraint, then add an artificial variable a_i and the sign restriction $a_i \geq 0$.

Step 4 For now, ignore the original LP's objective function. Instead, solve an LP whose objective function is min $w' =$ (sum of all the artificial variables). This is called the **Phase I LP.**

Because each $a_i \geq 0$, solving the Phase I LP will result in one of the following three cases:

Case 1 The optimal value of w' is greater than zero. In this case, the original LP has no feasible solution.

Case 2 The optimal value of w' is equal to zero, and no artificial variables are in the optimal Phase I basis. In this case, drop all columns in the optimal Phase I tableau that correspond to the artificial variables and combine the original objective function with the constraints from the optimal Phase I tableau. This yields the **Phase II LP.** The optimal solution to the Phase II LP and the original LP are the same.

Case 3 The optimal value of w' is equal to zero, and at least one artificial variable is in the optimal Phase I basis. In this case, we can find the optimal solution to the original LP if, at the end of Phase I, we drop from the optimal Phase I tableau all nonbasic artificial variables and any variable from the original problem that has a negative coefficient in row 0 of the optimal Phase I tableau.

Solving Minimization Problems

To solve a minimization problem by the simplex, choose as the entering variable the nonbasic variable in row 0 with the most positive coefficient. A tableau or canonical form is optimal if each variable in row 0 has a nonpositive coefficient.

Alternative Optimal Solutions

If a nonbasic variable has a zero coefficient in row 0 of an optimal tableau and the nonbasic variable can be pivoted into the basis, the LP may have **alternative optimal solutions.** If two basic feasible solutions are optimal, then any point on the line segment joining the two optimal basic feasible solutions is also an optimal solution to the LP.

Unrestricted-in-Sign Variables

If we replace a urs variable x_i by $x'_i - x''_i$, the LP's optimal solution will have x'_i, x''_i or both x'_i and x''_i equal to zero.

REVIEW PROBLEMS

Group A

1 Use the simplex algorithm to find *two* optimal solutions to the following LP:

$$\max z = 5x_1 + 3x_2 + x_3$$
$$\text{s.t.} \quad x_1 + x_2 + 3x_3 \leq 6$$
$$5x_1 + 3x_2 + 6x_3 \leq 15$$
$$x_3, x_1, x_2 \geq 0$$

2 Use the simplex algorithm to find the optimal solution to the following LP:

$$\min z = -4x_1 + x_2$$
$$\text{s.t.} \quad 3x_1 + x_2 \leq 6$$
$$\text{s.t.} \quad -x_1 + 2x_2 \leq 0$$
$$x_1, x_2 \geq 0$$

3 Use the Big M method and the two-phase method to find the optimal solution to the following LP:

$$\max z = 5x_1 - x_2$$
$$\text{s.t.} \quad 2x_1 + x_2 = 6$$
$$x_1 + x_2 \leq 4$$
$$x_1 + 2x_2 \leq 5$$
$$x_1, x_2 \geq 0$$

4 Use the simplex algorithm to find the optimal solution to the following LP:

$$\max z = 5x_1 - x_2$$
$$\text{s.t.} \quad x_1 - 3x_2 \leq 1$$
$$x_1 - 4x_2 \leq 3$$
$$x_1, x_2 \geq 0$$

5 Use the simplex algorithm to find the optimal solution to the following LP:

$$\min z = -x_1 - 2x_2$$
$$\text{s.t.} \quad 2x_1 + x_2 \leq 5$$
$$x_1 + x_2 \leq 3$$
$$x_1, x_2 \geq 0$$

6 Use the Big M method and the two-phase method to find the optimal solution to the following LP:

$$\max z = x_1 + x_2$$
$$\text{s.t.} \quad 2x_1 + x_2 \geq 3$$
$$3x_1 + x_2 \leq 3.5$$
$$x_1 + x_2 \leq 1$$
$$x_1, x_2 \geq 0$$

7 Use the simplex algorithm to find *two* optimal solutions to the following LP. How many optimal solutions does this LP have? Find a third optimal solution.

$$\max z = 4x_1 + x_2$$
$$\text{s.t.} \quad 2x_1 + 3x_2 \leq 4$$
$$x_1 + x_2 \leq 1$$
$$4x_1 + x_2 \leq 2$$
$$x_1, x_2 \geq 0$$

8 Use the simplex method to find the optimal solution to the following LP:

$$\max z = 5x_1 + x_2$$
$$\text{s.t.} \quad 2x_1 + x_2 \leq 6$$
$$x_1 - x_2 \leq 0$$
$$x_1, x_2 \geq 0$$

9 Use the Big M method and the two-phase method to find the optimal solution to the following LP:

$$\min z = -3x_1 + x_2$$
$$\text{s.t.} \quad x_1 - 2x_2 \geq 2$$
$$-x_1 + x_2 \geq 3$$
$$x_1, x_2 \geq 0$$

10 Suppose that in the Dakota Furniture problem, 10 types of furniture could be manufactured. To obtain an optimal solution, how many types of furniture (at the most) would have to be manufactured?

11 Consider the following LP:

$$\max z = 10x_1 + x_2$$
$$\text{s.t.} \quad x_1 \leq 1$$
$$20x_1 + x_2 \leq 100$$
$$x_1, x_2 \geq 0$$

a Find all the basic feasible solutions for this LP.

b Show that when the simplex is used to solve this LP, every basic feasible solution must be examined before the optimal solution is found.

By generalizing this example, Klee and Minty (1972) constructed (for $n = 2, 3, \ldots$) an LP with n decision variables and n constraints for which the simplex algorithm examines $2^n - 1$ basic feasible solutions before the optimal solution is found. Thus, there exists an LP with 10 variables and 10 constraints for which the simplex requires $2^{10} - 1 = 1,023$ pivots to find the optimal solution. Fortunately, such "pathological" LPs rarely occur in practical applications.

12 Productco produces three products. Each product requires labor, lumber, and paint. The resource requirements, unit price, and variable cost (exclusive of raw materials) for each product are given in Table 70. Currently, 900 labor hours, 1,550 gallons of paint, and 1,600 board feet of lumber are available. Additional labor can be purchased at $6 per hour, additional paint at $2 per gallon, and additional lumber at $3 per board foot. For the following two sets of priorities, use preemptive goal programming to determine an optimal production schedule. For set 1:

Priority 1 Obtain profit of at least $10,500.
Priority 2 Purchase no additional labor.
Priority 3 Purchase no additional paint.
Priority 4 Purchase no additional lumber.

For set 2:

Priority 1 Purchase no additional labor.
Priority 2 Obtain profit of at least $10,500.
Priority 3 Purchase no additional paint.
Priority 4 Purchase no additional lumber.

13 Jobs at Indiana University are rated on three factors:

Factor 1 Complexity of duties
Factor 2 Education required
Factor 3 Mental and or visual demands

For each job at IU, the requirement for each factor has been rated on a scale of 1–4, with a 4 in factor 1 representing high complexity of duty, a 4 in factor 2 representing high educational requirement, and a 4 in factor 3 representing high mental and or visual demands.

TABLE 70

Product	Labor	Lumber	Paint	Price ($)	Variable Cost ($)
1	1.5	2	3	26	10
2	3	3	2	28	6
3	2	4	2	31	7

IU wants to determine a formula for grading each job. To do this, it will assign a point value to the score for each factor that a job requires. For example, suppose level 2 of factor 1 yields a point total of 10, level 3 of factor 2 yields a point total of 20, and level 3 of factor 3 yields a point value of 30. Then a job with these requirements would have a point total of $10 + 20 + 30$. A job's hourly salary equals half its point total.

IU has two goals (listed in order of priority) in setting up the points given to each level of each job factor.

Goal 1 When increasing the level of a factor by one, the points should increase by at least 10. For example, level 2 of factor 1 should earn at least 10 more points than level 1 of factor 1. Goal 1 is to minimize the sum of deviations from this requirement.

Goal 2 For the benchmark jobs in Table 71, the actual point total for each job should come as close as possible to the point total listed in the table. Goal 2 is to minimize the sum of the absolute deviations of the point totals from the desired scores.

Use preemptive goal programming to come up with appropriate point totals. What salary should a job with skill levels of 3 for each factor be paid?

14 A hospital outpatient clinic performs four types of operations. The profit per operation, as well as the minutes of X-ray time and laboratory time used are given in Table 72. The clinic has 500 private rooms and 500 intensive care rooms. Type 1 and Type 2 operations require a patient to stay in an intensive care room for one day while Type 3 and Type 4 operations require a patient to stay in a private room for one day. Each day the hospital is required to perform at least 100 operations of each type. The hospital has set the following goals:

Goal 1 Earn a daily profit of at least $100,000.
Goal 2 Use at most 50 hours daily of X-ray time.
Goal 3 Use at most 40 hours daily of laboratory time.

The cost per unit deviation from each goal is as follows:

Goal 1 Cost of $1 for each dollar by which profit goal is unmet
Goal 2 Cost of $10 for each hour by which X-ray goal is unmet
Goal 3 Cost of $8 for each hour by which laboratory goal is unmet

Formulate a goal programming model to minimize the daily cost incurred due to failing to meet the hospital's goals.

Group B

15 Consider a maximization problem with the optimal tableau in Table 73. The optimal solution to this LP is $z = 10$, $x_3 = 3$, $x_4 = 5$, $x_1 = x_2 = 0$. Determine the second-best bfs to this LP. (*Hint:* Show that the second-best solution must be a bfs that is one pivot away from the optimal solution.)

16 A camper is considering taking two types of items on a camping trip. Item 1 weighs a_1 lb, and item 2 weighs a_2 lb. Each type 1 item earns the camper a benefit of c_1 units, and each type 2 item earns the camper c_2 units. The knapsack can hold items weighing at most b lb.

  a Assuming that the camper can carry a fractional number of items along on the trip, formulate an LP to maximize benefit.

  b Show that if

$$\frac{c_2}{a_2} \geq \frac{c_1}{a_1}$$

then the camper can maximize benefit by filling a knapsack with $\frac{b}{a_2}$ type 2 items.

  c Which of the linear programming assumptions are violated by this formulation of the camper's problem?

17 You are given the tableau shown in Table 74 for a maximization problem. Give conditions on the unknowns a_1, a_2, a_3, b, and c that make the following statements true:

  a The current solution is optimal.

  b The current solution is optimal, and there are alternative optimal solutions.

  c The LP is unbounded (in this part, assume that $b \geq 0$).

18 Suppose we have obtained the tableau in Table 75 for a maximization problem. State conditions on a_1, a_2, a_3, b, c_1, and c_2 that are required to make the following statements true:

  a The current solution is optimal, and there are alternative optimal solutions.

  b The current basic solution is not a basic feasible solution.

TABLE 71

Job	Factor Level 1	2	3	Desired Score
1	4	4	4	105
2	3	3	2	93
3	2	2	2	75
4	1	1	2	68

TABLE 72

	Type of Operation 1	2	3	4
Profit ($)	200	150	100	80
X-ray time (minutes)	6	5	4	3
Laboratory time (minutes)	5	4	3	2

TABLE 73

z	x_1	x_2	x_3	x_4	rhs
1	2	1	0	0	10
0	3	2	1	0	3
0	4	3	0	1	5

TABLE 74

z	x_1	x_2	x_3	x_4	x_5	rhs
1	$-c$	2	0	0	0	10
0	-1	a_1	1	0	0	4
0	a_2	-4	0	1	0	1
0	a_3	3	0	0	1	b

TABLE 75

z	x_1	x_2	x_3	x_4	x_5	x_6	rhs
1	c_1	c_2	0	0	0	0	10
0	4	a_1	1	0	a_2	0	b
0	-1	-5	0	1	-1	0	2
0	a_3	-3	0	0	-4	1	3

c The current basic solution is a degenerate bfs.

d The current basic solution is feasible, but the LP is unbounded.

e The current basic solution is feasible, but the objective function value can be improved by replacing x_6 as a basic variable with x_1.

19 Suppose we are solving a maximization problem and the variable x_r is about to leave the basis.

a What is the coefficient of x_r in the current row 0?

b Show that after the current pivot is performed, the coefficient of x_r in row 0 cannot be less than zero.

c Explain why a variable that has left the basis on a given pivot cannot re-enter the basis on the next pivot.

20 A bus company believes that it will need the following number of bus drivers during each of the next five years: year 1—60 drivers; year 2—70 drivers; year 3—50 drivers; year 4—65 drivers; year 5—75 drivers. At the beginning of each year, the bus company must decide how many drivers should be hired or fired. It costs $4,000 to hire a driver and $2,000 to fire a driver. A driver's salary is $10,000 per year. At the beginning of year 1, the company has 50 drivers. A driver hired at the beginning of a year may be used to meet the current year's requirements and is paid full salary for the current year. Formulate an LP to minimize the bus company's salary, hiring, and firing costs over the next five years.

21 Shoemakers of America forecasts the following demand for each of the next six months: month 1—5,000 pairs; month 2—6,000 pairs; month 3—5,000 pairs; month 4—9,000 pairs; month 5—6,000 pairs; month 6—5,000 pairs. It takes a shoemaker 15 minutes to produce a pair of shoes. Each shoemaker works 150 hours per month plus up to 40 hours per month of overtime. A shoemaker is paid a regular salary of $2,000 per month plus $50 per hour for overtime. At the beginning of each month, Shoemakers can either hire or fire workers. It costs the company $1,500 to hire a worker and $1,900 to fire a worker. The monthly holding cost per pair of shoes is 3% of the cost of producing a pair of shoes with regular-time labor. (The raw materials in a pair of shoes cost $10.) Formulate an LP that minimizes the cost of meeting (on time) the demands of the next six months. At the beginning of month 1, Shoemakers has 13 workers.

22 Monroe County is trying to determine where to place the county fire station. The locations of the county's four major towns are given in Figure 31. Town 1 is at (10, 20); town 2 is at (60, 20); town 3 is at (40, 30); town 4 is at (80, 60). Town 1 averages 20 fires per year; town 2, 30 fires; town 3, 40 fires; and town 4, 25 fires. The county wants to build the fire station in a location that minimizes the average distance that a fire engine must travel to respond to a fire. Since most roads run in either an east–west or a north–south direction, we assume that the fire engine can only do the same. Thus, if the fire station were located at (30, 40) and a fire occurred at town 4, the fire engine would have to travel $(80 - 30) + (60 - 40) = 70$ miles to the fire. Use linear programming to determine where the fire station should be located. (*Hint:* If the fire station is to be located at the point (x, y) and there is a town at the point (a, b), define variables e, w, n, s (east, west, north, south) that satisfy the equations $x - a = w - e$ and $y - b = n - s$. It should now be easy to obtain the correct LP formulation.)

23[†] During the 1972 football season, the games shown in Table 76 were played by the Miami Dolphins, the Buffalo Bills, and the New York Jets. Suppose that on the basis of these games, we want to rate these three teams. Let $M =$ Miami rating, $J =$ Jets rating, and $B =$ Bills rating. Given values of M, J, and B, you would predict that when, for example, the Bills play Miami, Miami is expected to win by $M - B$ points. Thus, for the first Miami–Bills game, your prediction would have been in error by $|M - B - 1|$ points. Show how linear programming can be used to determine ratings for each team that minimize the sum of the prediction errors for all games.

FIGURE 31

TABLE 76

Miami	Bills	Jets
27	—	17
28	—	24
24	23	—
30	16	—
—	24	41
—	3	41

[†]Based on Wagner (1954).

At the conclusion of the season, this method has been used to determine ratings for college football and college basketball. What problems could be foreseen if this method were used to rate teams early in the season?

24 During the next four quarters, Dorian Auto must meet (on time) the following demands for cars: quarter 1—4,000; quarter 2—2,000; quarter 3—5,000; quarter 4—1,000. At the beginning of quarter 1, there are 300 autos in stock, and the company has the capacity to produce at most 3,000 cars per quarter. At the beginning of each quarter, the company can change production capacity by one car. It costs $100 to increase quarterly production capacity. It costs $50 per quarter to maintain one car of production capacity (even if it is unused during the current quarter). The variable cost of producing a car is $2,000. A holding cost of $150 per car is assessed against each quarter's ending inventory. It is required that at the end of quarter 4, plant capacity must be at least 4,000 cars. Formulate an LP to minimize the total cost incurred during the next four quarters.

25 Ghostbusters, Inc., exorcises (gets rid of) ghosts. During each of the next three months, the company will receive the following number of calls from people who want their ghosts exorcised: January, 100 calls; February, 300 calls; March, 200 calls. Ghostbusters is paid $800 for each ghost exorcised during the month in which the customer calls. Calls need not be responded to during the month they are made, but if a call is responded to one month after it is made, then Ghostbusters loses $100 in future goodwill, and if a call is responded to two months after it is made, Ghostbusters loses $200 in goodwill. Each employee of Ghostbusters can exorcise 10 ghosts during a month. Each employee is paid a salary of $4,000 per month. At the beginning of January, the company has 8 workers. Workers can be hired and trained (in 0 time) at a cost of $5,000 per worker. Workers can be fired at a cost of $4,000 per worker. Formulate an LP to maximize Ghostbusters' profit (revenue less costs) over the next three months. Assume that all calls must be handled by the end of March.

26 Carco uses robots to manufacture cars. The following demands for cars must be met (not necessarily on time, but all demands must be met by end of quarter 4): quarter 1—600; quarter 2—800; quarter 3—500; quarter 4—400. At the beginning of the quarter, Carco has two robots. Robots can be purchased at the beginning of each quarter, but a maximum of two per quarter can be purchased. Each robot can build as many as 200 cars per quarter. It costs $5,000 to purchase a robot. Each quarter, a robot incurs $500 in maintenance costs (even if it is not used to build any cars). Robots can also be sold at the beginning of each quarter for $3,000. At the end of each quarter, a holding cost of $200 per car is incurred. If any demand is backlogged, then a cost of $300 per car is incurred for each quarter the demand is backlogged.

At the end of quarter 4, Carco must have at least two robots. Formulate an LP to minimize the total cost incurred in meeting the next four quarters' demands for cars.

27 Suppose we have found an optimal tableau for an LP, and the bfs for that tableau is nondegenerate. Also suppose that there is a nonbasic variable in row 0 with a zero coefficient. Prove that the LP has more than one optimal solution.

28 Suppose the bfs for an optimal tableau is degenerate, and a nonbasic variable in row 0 has a zero coefficient. Show by example that either of the following cases may hold:

Case 1 The LP has more than one optimal solution.
Case 2 The LP has a unique optimal solution.

29 You are the mayor of Gotham City, and you must determine a tax policy for the city. Five types of taxes are used to raise money:

a Property taxes. Let p = property tax percentage rate.

b A sales tax on all items except food, drugs, and durable goods. Let s = sales tax percentage rate.

c A sales tax on durable goods. Let d = durable goods sales tax percentage rate.

d A gasoline sales tax. Let g = gasoline tax sales percentage rate.

e A sales tax on food and drugs. Let f = sales tax on food and drugs.

The city consists of three groups of people: low-income (LI), middle-income (MI), and high-income (HI). The amount of revenue (in millions of dollars) raised from each group by setting a particular tax at a 1% level is given in Table 77.

For example, a 3% tax on durable good sales will raise $360 million from low-income people. Your tax policy must satisfy the following:

Restriction 1 The tax burden on MI people cannot exceed $2.8 billion.
Restriction 2 The tax burden on HI people cannot exceed $2.4 billion.
Restriction 3 The total revenue raised must exceed the current level of $6.5 billion.
Restriction 4 s must be between 1% and 3%.

Given these restrictions, the mayor has set the following three goals:

Goal P Keep the property tax rate less than 3%.
Goal LI Limit the tax burden on LI people to $2 billion.
Goal Suburbs If their tax burden becomes too high, 20% of the LI people, 20% of the MI people, and 40% of the HI people may consider moving to the suburbs. Suppose that this will happen if their total tax burden exceeds $1.5 billion. To discourage this exodus, the suburb goal is to keep the total tax burden on these people below $1.5 billion.

Use goal programming to determine an optimal tax policy if the mayor's goals follow the following set of priorities:

$$LI >>> P >>> Suburbs^\dagger$$

TABLE 77

	p	s	d	g	f
LI	900	300	120	30	90
MI	1,200	400	100	20	60
HI	1,000	250	60	10	40

†Based on Chrisman, Fry, Reeves, Lewis, and Weinstein (1989).

APPENDIX A LINDO Menu Commands and Statements

Menu Commands

LINDO's commands can be accessed from a convenient menu similar to those of other Windows programs. The main menu includes six submenus along the top of the screen that list the various commands. When you click on one of the submenus—File, Edit, Solve, Reports, Window, or Help—a pull-down menu appears with the various commands. You can select commands just like you would in most Window programs—by either clicking on the command with your mouse or pressing the underlined letter in the command name when the appropriate submenu is highlighted. Many commands also have shortcut keys assigned to them (F2, Ctrl+Z, etc.). As an added convenience, some of the most often used commands also may be accessed with icons located in a tool bar at the top of the screen. The following sections briefly describe the various menu commands and list the applicable shortcuts and icons.

File Menu

The File menu commands allow you to manipulate your LINDO data files in various ways. You can use this menu to open, close, save, and print files, as well as perform various tasks unique to LINDO. A description of the File commands follows.

COMMAND		DESCRIPTION
New	F2	Creates a new window for entering input data.
Open	F3	Opens an existing file. Dialog boxes allow you to select from various file types and locations.
View	F4	Opens an existing file for viewing only. No changes can be made to the file.
Save	F5	Saves the window. You can save input data (a model), a Reports window, or a command window. Data can be saved in the following formats: *.LTX, a text format that can be edited with word processing software; *.LPK, for saving compiled models in a "packed" format, but without any special formatting or comments; and *.MPS, the machine-independent industry standard format for transferring LP problems between LINDO and other LP software.
Save As...	F6	Saves the active window with a specified file name. This is useful for renaming a revised file, while keeping the original file intact.
Close	F7	Closes the active window. If the window contains new input data, then you will be asked if you want to save the changes.
Print	F8	Sends the active window to your printer.
Printer Setup...	F9	Selects the printer and various options for print format.
Log Output...	F10	Sends all subsequent screen activity that would normally be sent to the Reports window to a text file. When you have specified a log file location, a check will appear in the File menu by the Log Output line. To disable Log Output, simply select the command again.

COMMAND		DESCRIPTION
Take Commands... F11		"Takes" a LINGO batch file with commands and text for automated operation. A model could be put in memory, solved, and the solution placed in the Reports window and saved to a file. If you use the Batch command before the beginning of the model text, the model and the commands contained in the file, as well as the solution, would be visible in the Reports window.
Basis Read	F12	Retrieves a solution to a model that was saved using the Basis Save command.
Basis Save Shift+F2		Saves the solution for the active model to disk with a specified file name.
Title	Shift+F3	Displays the title of the active model, if one has been included with the optional Title statement in the model.
Date	Shift+F4	Opens a Reports window and displays the current date and time based on your computer's clock.
Elapsed Time Shift+F5		Opens a Reports window and displays the total time elapsed in your current LINDO session.
Exit	Shift+F6	Quits LINDO.

Edit Menu

The Edit menu commands allow you to perform basic editing tasks common to most Windows applications, as well as perform various tasks unique to LINDO. A description of the Edit commands follows.

COMMAND		DESCRIPTION
Undo	Ctrl+Z	Reverses the last action.
Cut	Ctrl+X	Removes any selected text and places it on the clipboard for pasting.
Copy	Ctrl+C	Copies selected text to the clipboard for pasting.
Paste	Ctrl+V	Inserts or pastes clipboard contents at the insertion point.
Clear	Delete	Deletes selected text without placing it on the clipboard.
Find/Replace... Ctrl+F		Searches the active window to find selected text and replaces it with text entered in the "Replace with" box.
Options	Alt+O	Allows viewing and changing of various parameters used in LINDO sessions.
Go To Line...	Ctrl+T	Allows you to move the cursor to any specified line in the active window.
Paste Symbol... Ctrl+P		Allows you to paste variable names and reserved symbols into the active window.
Select All	Ctrl+A	Selects all of the active window for cutting and copying.
Clear All		Deletes the entire contents of the active window.
Choose New Font		Selects a new font for the text in the active window.

Solve Menu

The Solve menu commands are used after you have entered data and are ready to obtain a solution. A description of the Solve command follows.

COMMAND

DESCRIPTION

| Solve | Ctrl+S | Sends the model in the active window to the LINDO solver to obtain the solution. |

Compile Model Ctrl+E — Translates the model into the arithmetic format required by the LINDO solver. Models are also automatically compiled when you use the Solve command.

Debug Ctrl+D — Helps determine problems with infeasible and unbounded models. Sufficient and necessary sets (rows) can be identified, as can crucial constraints—those that make an infeasible model feasible if dropped from the model.

Pivot Ctrl+N — Causes LINDO to perform the next step in the solution process, allowing linear programming problems to be solved step-by-step.

Preemptive Goal Ctrl+G — Performs Lexico optimization (a form of goal programming) on a model.

Reports Menu

The Reports menu commands allow you to specify how LINDO reports are generated. Descriptions of the Reports commands follow.

COMMAND

DESCRIPTION

Solution Alt+0 — Opens the Solution Report Options dialog box, which allows you to specify how you want a solution report to appear.

Range Alt+1 — Creates a range report, or sensitivity analysis, for the active model window.

Parametrics Alt+2 — Performs a parametric analysis on the right-hand side of a constraint.

Statistics Alt+3 — Displays key statistics for the model in the active window.

Peruse Alt+4 — Used to view reports on selected portions of the current model's solution or structure.

Picture Alt+5 — Creates a display of the current model in matrix form. The nonzero coefficients of the matrix may be displayed as either text or graphic.

Basis Picture Alt+6 — Displays a text-format report with a "picture" of the current basis, ordering the rows and columns according to the last inversion or triangularization performed by the solver. The Basis Picture report is sent to the Reports window.

Tableau Alt+7 — Displays the simplex tableau for the active model. This permits observation of the simplex algorithm at each step.

Formulation Alt+8 — Displays all, or selected segments, of your model in the Reports window.

Show Column Alt+9 — Displays a selected column without the rest of the model.

Positive Definite — Checks for a guarantee of global optimality in a quadratic model.

Window Menu

The Window menu commands allow you to adjust active command and status windows, as well as organize the display of multiple windows. Descriptions of the Window commands follow.

COMMAND	DESCRIPTION
Open <u>C</u>ommand Window Alt+C	Provides access to LINDO's command-line interface, where you may enter commands at the colon prompt.
Open <u>S</u>tatus Window	Opens LINDO's Solver Status window, which displays information about the optimizer status, such as number of iterations and elapsed run time. This window also appears when you select Solve from the Solve menu.
Send to <u>B</u>ack Ctrl+B	Sends the frontmost window to the back.
Ca<u>s</u>cade Alt+A	Arranges all open windows in a cascade fashion from upper left to lower right, with the active window on top.
<u>T</u>ile Alt+T	Arranges all open windows so they each occupy equivalent space within the program window.
Close <u>A</u>ll Alt+X 	Closes all active windows.
Arrange <u>I</u>cons Alt+I	Moves icons representing minimized windows so that they are arranged across the bottom of the screen.
List of Windows	At the bottom of the Window menu, a list of the open windows is displayed. The active window is checked.

Help Menu

The Help menu commands provide access to LINDO's online help. Descriptions of the Help commands follow.

COMMAND	DESCRIPTION
<u>C</u>ontents F1	Displays the contents of the help section. The second icon (with the arrow and question mark) enables context-sensitive help, where the cursor indicator will change to a question mark, and help will be provided specifically for a command selected.
<u>S</u>earch for Help on... Alt+F1	Searches the help section for a word or topic.
<u>H</u>ow to Use Help Ctrl+F1	Provides assistance in learning to use the online help system.
<u>A</u>bout LINDO...	Displays the initial startup screen with general information about LINDO.

Optional Modeling Statements

Besides the basic elements of a model, LINDO recognizes several optional statements that may appear after the END statement. These statements provide additional modeling capabilities, such as placing additional limits on variables. Descriptions of these statements follow.

STATEMENT	DESCRIPTION
FREE <Variable>	Removes all bounds on a variable, allowing it to take on any real value—positive or negative.
GIN <Variable>	Restricts a variable to be a general integer (i.e., in the set of non-negative integers).
INT <Variable>	Restricts a variable to be a binary integer (i.e., either 0 or 1).
SLB <Variable> <Value>	Sets a simple lower bound for a variable (i.e., SLB X 10 would required that X be greater than or equal to 10).

SUB \<Variable\> \<Value\>	Sets a simple upper bound for a variable (i.e., SUB X 10 would require that X be less than or equal to 10).
QCP \<Constraint\>	Indicates the first "real" constraints in a quadratic programming model.
TITLE \<Title\>	Allows you to attach a title to your model. The title can then be displayed using the Title command in the File menu.

APPENDIX B   Getting Started with LINGO

Welcome to the LINGO portion of this text. This appendix will give you brief background information on LINGO and help you install the software. Subsequent chapters will describe features of the software and how to apply the software on sample problems.

What Is LINGO?

LINGO is an interactive computer-software package that can be used to solve linear, integer, and nonlinear programming problems. It can be applied in similar situations to those of LINDO, but it offers more flexibility in terms of how models are expressed. Unlike LINDO, LINGO allows parentheses and variables on the right-hand side of an equation. Constraints can therefore be written in original form and do not have to be rewritten with constants on the right-hand side. LINGO is also capable of generating large models with relatively few lines of input. The program also provides a vast library of mathematical, statistical, and probability functions and greater ability to read data from external files and worksheets.

LINGO Fundamentals

Much like LINDO, LINGO can be used to solve problems interactively from the keyboard or solve problems using files created elsewhere—either self-contained or as part of an integrated program containing customized code and LINGO optimization libraries. This appendix will primarily focus on the first method, that of solving problems interactively. More information on the other methods is available from LINDO Systems, Inc.

Entering a model in the Windows version of LINGO is similar to typing in a Windows word-processing format: You simply type in model data much as you would write it if solving a problem manually. The inner window initially labeled "untitled" is provided to accept input data. LINGO also contains basic editing commands for cutting, copying, and pasting text. These tools, and other features, are found in the window commands discussed in Appendix C.

The required elements of LINGO are similar to those of LINDO: LINGO also requires an objective, one or more variables, and one or more constraints. Unlike LINDO, however, LINGO constraints are *not* preceded by any special terms such as SUBJECT TO or SUCH THAT.

LINGO follows a syntax similar to that of LINDO, with the following differences:

- LINGO statements end with semicolons.

- LINGO includes additional mathematical operators, as discussed in Appendix C. An asterisk is required to denote multiplication.

- Parentheses may be included to define the order of mathematical operations if you wish.

- Variable names can be up to 32 characters long.

APPENDIX C LINGO Menu Commands and Functions

Menu Commands

LINGO's commands can be accessed from a convenient menu similar to those of other Windows programs. The main menu includes five submenus along the top of the screen that list the various commands. When you click on one of the submenus—File, Edit, LINGO, Window, or Help—a pull-down menu appears with the various commands. You can select commands just like you would in most Window programs—by either clicking on the command with your mouse or pressing the underlined letter in the command name when the appropriate submenu is highlighted. Many commands also have shortcut keys assigned to them (F2, Ctrl1+Z, etc.). As an additional convenience, some of the most often used commands may also be accessed with icons located in a tool bar at the top of the screen.

File Menu

The File menu commands allow you to manipulate your LINGO data files in various ways. You can use this menu to open, close, save, and print files, as well as perform various tasks unique to LINGO. Descriptions of the File commands follow.

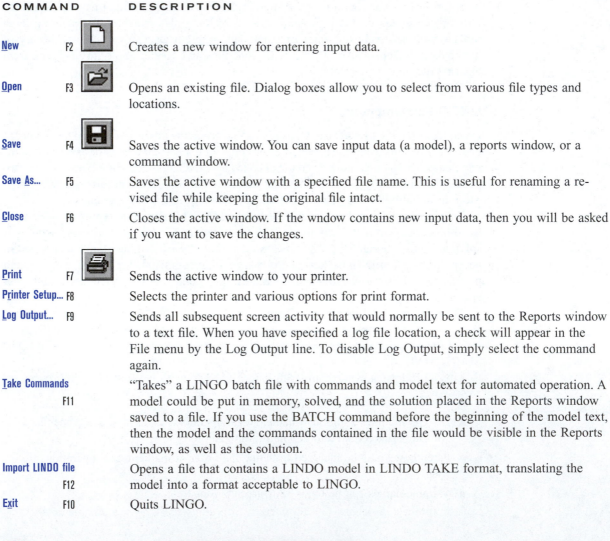

COMMAND		DESCRIPTION
New	F2	Creates a new window for entering input data.
Open	F3	Opens an existing file. Dialog boxes allow you to select from various file types and locations.
Save	F4	Saves the active window. You can save input data (a model), a reports window, or a command window.
Save As...	F5	Saves the active window with a specified file name. This is useful for renaming a revised file while keeping the original file intact.
Close	F6	Closes the active window. If the wndow contains new input data, then you will be asked if you want to save the changes.
Print	F7	Sends the active window to your printer.
Printer Setup...	F8	Selects the printer and various options for print format.
Log Output...	F9	Sends all subsequent screen activity that would normally be sent to the Reports window to a text file. When you have specified a log file location, a check will appear in the File menu by the Log Output line. To disable Log Output, simply select the command again.
Take Commands	F11	"Takes" a LINGO batch file with commands and model text for automated operation. A model could be put in memory, solved, and the solution placed in the Reports window saved to a file. If you use the BATCH command before the beginning of the model text, then the model and the commands contained in the file would be visible in the Reports window, as well as the solution.
Import LINDO file	F12	Opens a file that contains a LINDO model in LINDO TAKE format, translating the model into a format acceptable to LINGO.
Exit	F10	Quits LINGO.

Edit Menu

The Edit menu commands allow you to perform basic editing tasks common to most Windows applications, as well as perform various tasks unique to LINGO. Descriptions of the Edit commands follow.

COMMAND		DESCRIPTION
Undo	Ctrl+Z	Reverses the last action.
Cut	Ctrl+X	Removes any selected text and places it on the clipboard for pasting.
Copy	Ctrl+C	Copies selected text to the clipboard for pasting.
Paste	Ctrl+V	Inserts clipboard contents at the insertion point.
Clear	Delete	Deletes selected text without placing it on the clipboard.
Find/Replace... Ctrl+F		Searches the active window to find selected text and replace it with text entered in the "Replace with" box.
Go To Line... Ctrl+T		Allows you to move the cursor to any specified line in the active window.
Match Parenthesis Ctrl+P		Finds the close parenthesis that corresponds to the selected open parenthesis.
Paste Function		Pastes any of LINGO's built-in functions at the current insertion point. After selecting this command, another submenu appears with the various function categories.
Select All	Ctrl+A	Selects all of the active window for cutting and copying.
Choose New Font		Selects a new font for the text in the active window.

LINGO Menu

The LINGO menu commands are used after you have entered data and are ready to obtain a solution. Descriptions of the LINGO commands follow.

COMMAND		DESCRIPTION
Solve	Ctrl+S	Sends the model in the active window to the LINGO solver.
Solution...	Ctrl+O	Opens the Solution Report Options dialog box, which allows you to specify how you want a solution report to appear.
Range	Ctrl+R	Displays a range report, which shows over what ranges you can change coefficients without changing optimal values.
Look...	Ctrl+L	Displays all or selected lines of a model.
Generate...	Ctrl+S	Creates another version of the current model in algebraic, LINDO, or MPS format. Can be used to number rows and display the model in a more readable format. The GEN command provides a similar capability from the command window.
Export to Spreadsheet Ctrl+E		Exports selected variable values to named ranges in a spreadsheet. A spreadsheet must first be created with ranges sized to accommodate the exported values. The ranges *must* contain numbers. Selecting this command will produce a dialog box that requests the

template and output worksheets (spreadsheet file names), variables to export, and the range to which the values are to be exported. The variables and range are entered in pairs and added to the list of variable and range pairs by clicking the add button.

Options   Alt+O   Allows viewing and changing of various parameters used in LINGO sessions.

Workspace Limit   Allocates memory to LINGO. If you enter "None," LINGO will use all available
Ctrl+S   memory.

Window Menu

The Window menu commands allow you to adjust any open command and status windows, as well as organize the display of multiple windows. Descriptions of the Window commands follow.

COMMAND	DESCRIPTION
Open Command Window	Provides access to LINGO's command-line interface, where you may enter commands at the colon prompt.
Open Status Window	Opens LINGO's Solver Status window, which displays information about the optimizer status, such as number of iterations and elapsed run time. This window also appears when you select Solve from the LINGO menu.

Send to Back   Sends the frontmost window to the back.

Alt+B

Close All   Alt+X   Closes all active windows.

Cascade   Alt+A   Arranges all open windows in a cascade fashion from upper left to lower right, with the active window on top.

Tile   Alt+T   Arranges all open windows so they each occupy equivalent space within the program window.

Arrange Icons   Arranges icons representing minimized windows across the bottom of the screen.
Alt+I

List of Windows   At the bottom of the Window menu, a list of the open windows is displayed. The active window is checked.

Help Menu

The Help menu commands provide access to LINGO's on-line help. Descriptions of the Help commands follow.

COMMAND   DESCRIPTION

Contents   Displays the contents of the help section. The second icon (the arrow with the question mark) enables context-sensitive help; the cursor indicator will change to a question mark, and help will be provided specifically for a command selected.

F1

Search for Help on...   Searches the help section for a word or topic.
Alt+F1

How to Use Help   Provides assistance in learning to use the online help system.
Ctrl+F1

About LINGO...   Displays the initial startup screen with general information about LINGO.

Functions

LINGO has seven main functions—standard operators, file import, financial, mathematical, set-looping, variable-domain, and probability, and an assortment of other functions. Most of these functions are available through the menu commands. The LINGO software includes a detailed description of its functions in online help screens; therefore, only a brief description of LINGO functions is offered here.

Standard Operators

Standard operators include arithmetic operators (i.e., ^, *, /, +, and −), logical operators (#EQ#, #NE#, #GT#, #GE#, #LT#, and #L3#) for determining set membership, and equality–inequality operators (<, =, >, <=, and >=) for specifying whether the left-hand side of an expression should be less than, equal to, or greater than the right-hand side. These operators constitute some of the most basic functions available in LINGO. Note that the "greater than" and "less than" symbols (> and >) are interpreted as "loose" inequalities [i.e., greater than or equal to ($\geq$) and less than or equal to ($\leq$), respectively]. You typically type these operators in at the keyboard rather than access them from a window command.

File Import Functions

File import functions allow you to import text and data from external sources. The @FILE function lets you import text or data from an ASCII file, and the @IMPORT function lets you import data only from a worksheet.

Financial Functions

Financial functions include the @FPA(I,N) function, which returns the present value of an annuity; and the @FPL(I,N) function, which returns the present value of a lump sum of $1 N periods from now if the interest rate is I per period. I is not a percentage but rather a non-negative number representing the interest rate.

Mathematical Functions

Mathematical functions include the following general and trigonometric functions: @ABS(X), @COS(X), @EXP(X), @LGM(X), @LOG(X), @SIGN(X), @SIN(X), @SMAX(list), @SMIN(list), @TAN(X). Combinations of the three basic trigonometric functions (sine, cosine, and tangent) may be used to obtain other trigonometric functions.

Set-Looping Functions

Set-looping functions include @FOR (set_name : constraint_expressions), @MAX (set_name : expression), @MIN (set_name : expression), and @SUM (set_name : expression). These functions operate over an entire set, producing a single result in all cases, except the @FOR function, which generates constraints independently for each element of the set.

Variable Domain Functions

The variable domain functions place additional restrictions on variables and attributes. They include the following: @BND(L, X, U), @BIN(X), @FREE(X), and @GIN(X).

Probability Functions

LINGO provides common statistical capabilities through its probability functions: @PSN(X), @PSL(X), @PPS(A,X), @PPL(A,X), @PBN(P,N,X), @PHG(POP,G,N,X), @PEL(A,X), @PEB(A,X), @PFS(A,X,C), @PFD(N,D,X), @PFD(N,D,X), @PCX(N,X), @PTD(N,X), and @RAND(X).

Other Functions

Other functions provided by LINGO include @IN (set_name, set_element), @SIZE (set_name), @WARN('text', condition), @WRAP(I,N), and @USER. These functions provide a variety of capabilities in addition to those of the categories above.

REFERENCES

There are many fine linear programming texts, including the following books:

Bazaraa, M., and J. Jarvis. *Linear Programming and Network Flows*. New York: Wiley, 1990.

Bersitmas, D., and J. Tsitsiklis. *Introduction to Linear Optimization*. Belmont, Mass.: Athena Publishing, 1997.

Bradley, S., A. Hax, and T. Magnanti. *Applied Mathematical Programming*. Reading, Mass.: Addison-Wesley, 1977.

Chvàtal, V. *Linear Programming*. San Francisco: Freeman, 1983.

Dantzig, G. *Linear Programming and Extensions*. Princeton, N.J.: Princeton University Press, 1963.

Gass, S. *Linear Programming: Methods and Applications,* 5th ed. New York: McGraw-Hill, 1985.

Luenberger, D. *Linear and Nonlinear Programming,* 2d ed. Reading, Mass.: Addison-Wesley, 1984.

Murty, K. *Linear Programming*. New York: Wiley, 1983.

Nash, S., and A. Sofer. *Linear and Nonlinear Programming*. New York: McGraw-Hill, 1995.

Nering, E., and A. Tucker. *Linear Programs and Related Problems*. New York: Academic Press, 1993.

Simmons, D. *Linear Programming for Operations Research*. Englewood Cliffs, N.J.: Prentice Hall, 1972.

Simonnard, M. *Linear Programming*. Englewood Cliffs, N.J.: Prentice Hall, 1966.

Wu, N., and R. Coppins. *Linear Programming and Extensions*. New York: McGraw-Hill, 1981.

Bland, R. "New Finite Pivoting Rules for the Simplex Method," *Mathematics of Operations Research* 2(1977):103–107. Describes simple, elegant approach to prevent cycling.

Dantzig, G., and N. Thapa. *Linear Programming*. New York: Springer-Verlag, 1997.

Karmarkar, N. "A New Polynomial Time Algorithm for Linear Programming," *Combinatorica* 4(1984):373–395. Karmarkar's method for solving LPs.

Klee, V., and G. Minty. "How Good Is the Simplex Algorithm?" In *Inequalities—III*. New York: Academic Press, 1972. Describes LPs for which the simplex method examines every basic feasible solution before finding the optimal solution.

Kotiah, T., and N. Slater. "On Two-Server Poisson Queues with Two Types of Customers," *Operations Research* 21(1973):597–603. Describes an actual application that led to an LP in which cycling occurred.

Love, R., and L. Yerex. "An Application of a Facilities Location Model in the Prestressed Concrete Industry," *Interfaces* 6(no.4, 1976):45–49.

Papadimitriou, C., and K. Steiglitz. *Combinatorial Optimization: Algorithms and Complexity*. Englewood Cliffs, N.J.: Prentice Hall, 1982. More discussion of polynomial time and exponential time algorithms.

Schrage, L. *User's Manual for LINDO*. Palo Alto, Calif.: Scientific Press, 1990. Gives complete details of LINDO.

Schrage, L. *User's Manual for LINGO*. Chicago, Ill.: LINDO Systems Inc., 1991. Gives complete details of LINGO.

Schrage, L. *User's Manual for What's Best*. Chicago, Ill.: LINDO Systems Inc., 1993. Gives complete details of What's Best.

Wagner, H. "Linear Programming Techniques for Regression Analysis," *Journal of the American Statistical Association* 54(1954):206–212.

Sensitivity Analysis: An Applied Approach

In this chapter, we discuss how changes in an LP's parameters affect the optimal solution. This is called *sensitivity analysis*. We also explain how to use the LINDO output to answer questions of managerial interest such as "What is the most money a company would be willing to pay for an extra hour of labor?" We begin with a graphical explanation of sensitivity analysis.

5.1 A Graphical Introduction to Sensitivity Analysis

Sensitivity analysis is concerned with how changes in an LP's parameters affect the optimal solution.

Reconsider the Giapetto problem of Section 3.1:

$$\max z = 3x_1 + 2x_2$$
$$\text{s.t.} \quad 2x_1 + x_2 \le 100 \quad \text{(Finishing constraint)}$$
$$x_1 + x_2 \le 80 \quad \text{(Carpentry constraint)}$$
$$x_1 \le 40 \quad \text{(Demand constraint)}$$
$$x_1, x_2 \ge 0$$

where

$$x_1 = \text{number of soldiers produced per week}$$
$$x_2 = \text{number of trains produced per week}$$

The optimal solution to this problem is $z = 180$, $x_1 = 20$, $x_2 = 60$ (point B in Figure 1), and it has x_1, x_2, and s_3 (the slack variable for the demand constraint) as basic variables. How would changes in the problem's objective function coefficients or right-hand sides change this optimal solution?

Graphical Analysis of the Effect of a Change in an Objective Function Coefficient

If the contribution to profit of a soldier were to increase sufficiently, then it seems reasonable that it would be optimal for Giapetto to produce more soldiers (that is, s_3 would become nonbasic). Similarly, if the contribution to profit of a soldier were to decrease sufficiently, then it would become optimal for Giapetto to produce only trains (x_1 would now be nonbasic). We now show how to determine the values of the contribution to profit for soldiers for which the current optimal basis will remain optimal.

Let c_1 be the contribution to profit by each soldier. For what values of c_1 does the current basis remain optimal?

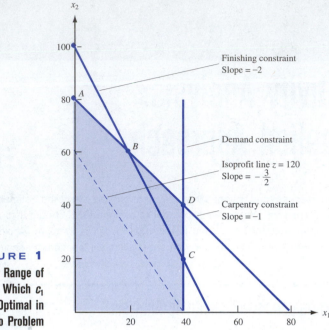

FIGURE 1

Analysis of Range of
Values for Which c_1
Remains Optimal in
Giapetto Problem

Currently, $c_1 = 3$, and each isoprofit line has the form $3x_1 + 2x_2 = $ constant, or

$$x_2 = -\frac{3x_1}{2} + \frac{\text{constant}}{2}$$

and each isoprofit line has a slope of $-\frac{3}{2}$. From Figure 1, we see that if a change in c_1 causes the isoprofit lines to be flatter than the carpentry constraint, then the optimal solution will change from the current optimal solution (point B) to a new optimal solution (point A). If the profit for each soldier is c_1, the slope of each isoprofit line will be $-\frac{c_1}{2}$. Because the slope of the carpentry constraint is -1, the isoprofit lines will be flatter than the carpentry constraint if $-\frac{c_1}{2} > -1$, or $c_1 < 2$, and the current basis will no longer be optimal. The new optimal solution will be $(0, 80)$, point A in Figure 1.

If the isoprofit lines are steeper than the finishing constraint, then the optimal solution will change from point B to point C. The slope of the finishing constraint is -2. If $-\frac{c_1}{2} < -2$, or $c_1 > 4$, then the current basis is no longer optimal and point C, $(40, 20)$, will be optimal. In summary, we have shown that (if all other parameters remain unchanged) the current basis remains optimal for $2 \le c_1 \le 4$, and Giapetto should still manufacture 20 soldiers and 60 trains. Of course, even if $2 \le c_1 \le 4$, Giapetto's profit will change. For instance, if $c_1 = 4$, then Giapetto's profit will now be $4(20) + 2(60) = \$200$ instead of $\$180$.

Graphical Analysis of the Effect of a Change in a Right-Hand Side on the LP's Optimal Solution

A graphical analysis can also be used to determine whether a change in the right-hand side of a constraint will make the current basis no longer optimal. Let b_1 be the number of available finishing hours. Currently, $b_1 = 100$. For what values of b_1 does the current basis remain optimal? From Figure 2, we see that a change in b_1 shifts the finishing constraint parallel to its current position. The current optimal solution (point B in Figure 2)

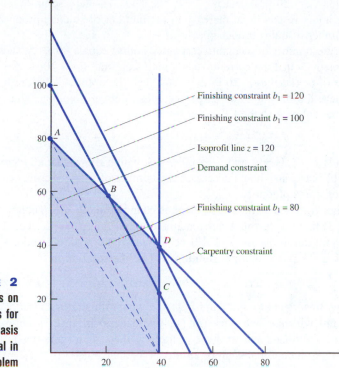

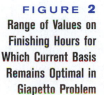

FIGURE 2

Range of Values on
Finishing Hours for
Which Current Basis
Remains Optimal in
Giapetto Problem

is where the carpentry and finishing constraints are binding. If we change the value of b_1, then *as long as the point where the finishing and carpentry constraints are binding remains feasible, the optimal solution will still occur where the finishing and carpentry constraints intersect.* From Figure 2, we see that if $b_1 > 120$, then the point where the finishing and carpentry constraints are both binding will lie on the portion of the carpentry constraint below point D. Note that at point D, $2(40) + 40 = 120$ finishing hours are used. In this region, $x_1 > 40$, and the demand constraint for soldiers is not satisfied. Thus, for $b_1 > 120$, the current basis will no longer be optimal. Similarly, if $b_1 < 80$, then the carpentry and finishing constraints will be binding at an infeasible point having $x_1 < 0$, and the current basis will no longer be optimal. Note that at point A, $0 + 80 = 80$ finishing hours are used. Thus (if all other parameters remain unchanged), the current basis remains optimal if $80 \leq b_1 \leq 120$.

Note that although for $80 \leq b_1 \leq 120$, the current basis remains optimal, *the values of the decision variables and the objective function value change.* For example, if $80 \leq b_1 \leq 100$, then the optimal solution will change from point B to some other point on the line segment AB. Similarly, if $100 \leq b_1 \leq 120$, then the optimal solution will change from point B to some other point on the line BD.

As long as the current basis remains optimal, it is a routine matter to determine how a change in the right-hand side of a constraint changes the values of the decision variables. To illustrate the idea, let $b_1 =$ number of available finishing hours. If we change b_1 to $100 + \Delta$, then we know that the current basis remains optimal for $-20 \leq \Delta \leq 20$. Note that as b_1 changes (as long as $-20 \leq \Delta \leq 20$), the optimal solution to the LP is still the point where the finishing-hour and carpentry-hour constraints are binding. Thus, if $b_1 = 100 + \Delta$, we can find the new values of the decision variables by solving

$$2x_1 + x_2 = 100 + \Delta \quad \text{and} \quad x_1 + x_2 = 80$$

This yields $x_1 = 20 + \Delta$ and $x_2 = 60 - \Delta$. Thus, an increase in the number of available finishing hours results in an increase in the number of soldiers produced and a decrease in the number of trains produced.

If b_2 (the number of available carpentry hours) equals $80 + \Delta$, then it can be shown (see Problem 2) that the current basis remains optimal for $-20 \le \Delta \le 20$. If we change the value of b_2 (keeping $-20 \le \Delta \le 20$), then the optimal solution to the LP is still the point where the finishing and carpentry constraints are binding. Thus, if $b_2 = 80 + \Delta$, the optimal solution to the LP is the solution to

$$2x_1 + x_2 = 100 \qquad \text{and} \qquad x_1 + x_2 = 80 + \Delta$$

This yields $x_1 = 20 - \Delta$ and $x_2 = 60 + 2\Delta$, which shows that an increase in the amount of available carpentry hours decreases the number of soldiers produced and increases the number of trains produced.

Suppose b_3, the demand for soldiers, is changed to $40 + \Delta$. Then it can be shown (see Problem 3) that the current basis remains optimal for $\Delta \ge -20$. For Δ in this range, the optimal solution to the LP will still occur where the finishing and carpentry constraints are binding. Thus, the optimal solution will be the solution to

$$2x_1 + x_2 = 100 \qquad \text{and} \qquad x_1 + x_2 = 80$$

Of course, this yields $x_1 = 20$ and $x_2 = 60$, which illustrates an important fact. Consider a constraint with positive slack (or positive excess) in an LP's optimal solution; if we change the right-hand side of this constraint in the range where the current basis remains optimal, then the optimal solution to the LP is unchanged.

Shadow Prices

As we will see in Sections 5.2 and 5.3, it is often important for managers to determine how a change in a constraint's right-hand side changes the LP's optimal z-value. With this in mind, we define the **shadow price** for the ith constraint of an LP to be the amount by which the optimal z-value is improved—increased in a max problem and decreased in a min problem—if the right-hand side of the ith constraint is increased by 1. This definition applies only if the change in the right-hand side of Constraint i leaves the current basis optimal.

For any two-variable LP, it is a simple matter to determine each constraint's shadow price. For example, we know that if $100 + \Delta$ finishing hours are available (assuming the current basis remains optimal), then the LP's optimal solution is $x_1 = 20 + \Delta$ and $x_2 = 60 - \Delta$. Then the optimal z-value will equal $3x_1 + 2x_2 = 3(20 + \Delta) + 2(60 - \Delta) = 180 + \Delta$. Thus, as long as the current basis remains optimal, a one-unit increase in the number of available finishing hours will increase the optimal z-value by \$1. So the shadow price of the first (finishing hours) constraint is \$1.

For the second (carpentry hours) constraint, we know that if $80 + \Delta$ carpentry hours are available (and the current basis remains optimal), then the optimal solution to the LP is $x_1 = 20 - \Delta$ and $x_2 = 60 + 2\Delta$. Then the new optimal z-value is $3x_1 + 2x_2 = 3(20 - \Delta) + 2(60 + 2\Delta) = 180 + \Delta$. So a one-unit increase in the number of finishing hours will increase the optimal z-value by \$1 (as long as the current basis remains optimal). Thus, the shadow price of the second (carpentry hour) constraint is \$1.

We now find the shadow price of the third (demand) constraint. If the right-hand side is $40 + \Delta$, then (as long as the current basis remains optimal) the optimal values of the decision variables remain unchanged. Then the optimal z-value will also remain unchanged, which shows that the shadow price of the third (demand) constraint is \$0. It turns out that whenever the slack or excess variable for a constraint is positive in an LP's optimal solution, the constraint will have a zero shadow price.

Suppose that the current basis remains optimal as we increase the right-hand side of the ith constraint of an LP by Δb_i. ($\Delta b_i < 0$ means that we are decreasing the right-hand side of the ith constraint.) Then each unit by which Constraint i's right-hand side is increased will increase the optimal z-value (for a max problem) by the shadow price. Thus, the new optimal z-value is given by

(New optimal z-value) = (old optimal z-value) + (Constraint i's shadow price) Δb_i **(1)**

For a minimization problem,

(New optimal z-value) = (old optimal z-value) − (Constraint i's shadow price) Δb_i **(2)**

For example, if 95 carpentry hours are available, then $\Delta b_2 = 15$, and the new z-value is given by

New optimal z-value = $180 + 15(1) = \$195$

We will continue our discussion of shadow prices in Sections 5.2 and 5.3.

Importance of Sensitivity Analysis

Sensitivity analysis is important for several reasons. In many applications, the values of an LP's parameters may change. For example, the prices at which soldiers and trains are sold or the availability of carpentry and finishing hours may change. If a parameter changes, then sensitivity analysis often makes it unnecessary to solve the problem again. For example, if the profit contribution of a soldier increased to $3.50, we would not have to solve the Giapetto problem again, because the current solution remains optimal. Of course, solving the Giapetto problem again would not be much work, but solving an LP with thousands of variables and constraints again would be a chore. A knowledge of sensitivity analysis often enables the analyst to determine from the original solution how changes in an LP's parameters change its optimal solution.

Recall that we may be uncertain about the values of parameters in an LP. For example, we might be uncertain about the weekly demand for soldiers. With the graphical method, it can be shown that if the weekly demand for soldiers is at least 20, then the optimal solution to the Giapetto problem is still (20, 60) (see Problem 3 at the end of this section). Thus, even if Giapetto is uncertain about the demand for soldiers, the company can be fairly confident that it is still optimal to produce 20 soldiers and 60 trains.

PROBLEMS

Group A

1 Show that if the contribution to profit for trains is between $1.50 and $3, the current basis remains optimal. If the contribution to profit for trains is $2.50, then what would be the new optimal solution?

2 Show that if available carpentry hours remain between 60 and 100, the current basis remains optimal. If between 60 and 100 carpentry hours are available, would Giapetto still produce 20 soldiers and 60 trains?

3 Show that if the weekly demand for soldiers is at least 20, then the current basis remains optimal, and Giapetto should still produce 20 soldiers and 60 trains.

4 For the Dorian Auto problem (Example 2 in Chapter 3),

   a Find the range of values on the cost of a comedy ad for which the current basis remains optimal.

   b Find the range of values on the cost of a football ad for which the current basis remains optimal.

   c Find the range of values for required HIW exposures for which the current basis remains optimal. Determine the new optimal solution if $28 + \Delta$ million HIW exposures are required.

   d Find the range of values for required HIM exposures for which the current basis remains optimal. Determine the new optimal solution if $24 + \Delta$ million HIM exposures are required.

e Find the shadow price of each constraint.

f If 26 million HIW exposures are required, determine the new optimal z-value.

5 Radioco manufactures two types of radios. The only scarce resource that is needed to produce radios is labor. At present, the company has two laborers. Laborer 1 is willing to work up to 40 hours per week and is paid $5 per hour. Laborer 2 will work up to 50 hours per week for $6 per hour. The price as well as the resources required to build each type of radio are given in Table 1.

Letting x_i be the number of Type i radios produced each week, Radioco should solve the following LP:

$$\max z = 3x_1 + 2x_2$$
$$\text{s.t.} \quad x_1 + 2x_2 \leq 40$$
$$2x_1 + x_2 \leq 50$$
$$x_1, x_2 \geq 0$$

a For what values of the price of a Type 1 radio would the current basis remain optimal?

b For what values of the price of a Type 2 radio would the current basis remain optimal?

TABLE 1

	Radio 1		Radio 2	
	Price ($)	Resource Required	Price ($)	Resource Required
	25	Laborer 1: 1 hour	22	Laborer 1: 2 hours
		Laborer 2: 2 hours		Laborer 2: 2 hours
		Raw material cost: $5		Raw material cost: $4

c If laborer 1 were willing to work only 30 hours per week, then would the current basis remain optimal? Find the new optimal solution to the LP.

d If laborer 2 were willing to work up to 60 hours per week, then would the current basis remain optimal? Find the new optimal solution to the LP.

e Find the shadow price of each constraint.

5.2 The Computer and Sensitivity Analysis

If an LP has more than two decision variables, the range of values for a right-hand side (or objective function coefficient) for which the current basis remains optimal cannot be determined graphically. These ranges can be computed by hand calculations (see Section 6.3), but this is often tedious, so they are usually determined by packaged computer programs. In this section, we discuss the interpretation of the sensitivity analysis information on the LINDO output.

To obtain a sensitivity report in LINDO, select Yes when asked (after solving LP) whether you want a Range analysis. To obtain sensitivity report in LINGO, go to Options and select Range (after solving LP). If this does not work, then go to Options and choose the General Solver tab. Then go to Dual Computations and select the Ranges and Values option.

EXAMPLE 1   **Winco Products 1**

Winco sells four types of products. The resources needed to produce one unit of each and the sales prices are given in Table 2. Currently, 4,600 units of raw material and 5,000 labor hours are available. To meet customer demands, exactly 950 total units must be produced. Customers also demand that at least 400 units of product 4 be produced. Formulate an LP that can be used to maximize Winco's sales revenue.

Solution   Let x_i = number of units of product i produced by Winco.

$$\max z = 4x_1 + 6x_2 + 7x_3 + 8x_4$$
$$\text{s.t.} \quad x_1 + x_2 + x_3 + x_4 = 950$$
$$x_4 \geq 400$$
$$2x_1 + 3x_2 + 4x_3 + 7x_4 \leq 4,600$$
$$3x_1 + 4x_2 + 5x_3 + 6x_4 \leq 5,000$$
$$x_1, x_2, x_3, x_4 \geq 0$$

TABLE 2
Costs and Resource Requirements for Winco

Resource	Product 1	Product 2	Product 3	Product 4
Raw material	2	3	4	7
Hours of labor	3	4	5	6
Sales price ($)	4	6	7	8

The LINDO output for this LP is given in Figure 3.

When we discuss the interpretation of the LINDO output for minimization problems, we will refer to the following example.

EXAMPLE 2   Tucker Inc.

Tucker Inc. must produce 1,000 Tucker automobiles. The company has four production plants. The cost of producing a Tucker at each plant, along with the raw material and labor needed, is shown in Table 3.

```
MAX        4  X1 + 6  X2  + 7  X3  + 8  X4
SUBJECT TO
       2)     X1  +  X2  +  X3  +  X4  =        950
       3)     X4   >=      400
       4)   2  X1  + 3  X2  + 4  X3 + 7  X4  <=     4600
       5)   3  X1  + 4  X2  + 5  X3 + 6  X4  <=     5000
END

LP OPTIMUM FOUND AT STEP             4
              OBJECTIVE FUNCTION VALUE
              1)   6650.00000

VARIABLE         VALUE          REDUCED COST
   X1           .000000           1.000000
   X2          400.000000          .000000
   X3          150.000000          .000000
   X4          400.000000          .000000


   ROW       SLACK OR SURPLUS    DUAL PRICES
   2)            .000000           3.000000
   3)            .000000          -2.000000
   4)            .000000           1.000000
   5)          250.000000          .000000

NO. ITERATIONS=          4

RANGES IN WHICH THE BASIS IS UNCHANGED:

                      OBJ COEFFICIENT RANGES
VARIABLE    CURRENT        ALLOWABLE        ALLOWABLE
            COEF           INCREASE         DECREASE
   X1     4.000000         1.000000         INFINITY
   X2     6.000000          .666667          .500000
   X3     7.000000         1.000000          .500000
   X4     8.000000         2.000000         INFINITY

                      RIGHTHAND SIDE RANGES
   ROW     CURRENT        ALLOWABLE        ALLOWABLE
            RHS           INCREASE         DECREASE
    2     950.000000       50.000000       100.000000
    3     400.000000       37.000000       125.000000
    4    4600.000000      250.000000       150.000000
    5    5000.000000       INFINITY        250.000000
```

FIGURE 3
LINDO Output
for Winco

TABLE 3
Cost and Requirements for Producing a Tucker

Plant	Cost (in Thousands of Dollars)	Labor	Raw Material
1	15	2	3
2	10	3	4
3	9	4	5
4	7	5	6

The autoworkers' labor union requires that at least 400 cars be produced at plant 3; 3,300 hours of labor and 4,000 units of raw material are available for allocation to the four plants. Formulate an LP whose solution will enable Tucker Inc. to minimize the cost of producing 1,000 cars.

Solution Let x_i = number of cars produced at plant i. Then, expressing the objective function in thousands of dollars, the appropriate LP is

$$\min z = 15x_1 + 10x_2 + 9x_3 + 7x_4$$

$$\text{s.t.} \quad x_1 + x_2 + x_3 + x_4 = 1000$$

$$x_3 \geq 400$$

$$2x_1 + 3x_2 + 4x_3 + 5x_4 \leq 3300$$

$$3x_1 + 4x_2 + 5x_3 + 6x_4 \leq 4000$$

$$x_1, x_2, x_3, x_4 \geq 0$$

The LINDO output for this LP is given in Figure 4.

Objective Function Coefficient Ranges

Recall from Section 5.1 that (at least in a two-variable problem) we can determine the range of values for an objective function coefficient for which the current basis remains optimal. For each objective function coefficient, this range is given in the OBJECTIVE COEFFICIENT RANGES portion of the LINDO output. The ALLOWABLE INCREASE (AI) section indicates the amount by which an objective function coefficient can be increased with the current basis remaining optimal. Similarly, the ALLOWABLE DECREASE (AD) section indicates the amount by which an objective function coefficient can be decreased with the current basis remaining optimal. To illustrate these ideas, let c_i be the objective function coefficient for x_i in Example 1. If c_1 is changed, then the current basis remains optimal if

$$-\infty = 4 - \infty \leq c_1 \leq 4 + 1 = 5$$

If c_2 is changed, then the current basis remains optimal if

$$5.5 = 6 - 0.5 \leq c_2 \leq 6 + 0.666667 = 6.666667$$

We will refer to the range of variables of c_i for which the current basis remains optimal as the **allowable range** for c_i. As discussed in Section 5.1, if c_i remains in its allowable range then the values of the decision variables remain unchanged, although the optimal z-value may change. The following examples illustrate these ideas.

```
MIN        15  X1 + 10  X2  + 9  X3  + 7  X4
SUBJECT TO
        2)      X1  +  X2  +  X3  +  X4  =      1000
        3)      X3  >=      400
        4)     2  X1  +  3  X2  +  4  X3 + 5  X4  <=    3300
        5)     3  X1  +  4  X2  +  5  X3 + 6  X4  <=    4000
END


LP OPTIMUM FOUND AT STEP              3
                OBJECTIVE FUNCTION VALUE
            1)   11600.0000


    VARIABLE            VALUE        REDUCED COST
        X1          400.000000          .000000
        X2          200.000000          .000000
        X3          400.000000          .000000
        X4            .000000          7.000000


        ROW     SLACK OR SURPLUS       DUAL PRICES
        2)           .000000        -30.000000
        3)           .000000         -4.000000
        4)        300.000000          .000000
        5)           .000000         5.000000


    NO. ITERATIONS=          3

    RANGES IN WHICH THE BASIS IS UNCHANGED:

                        OBJ COEFFICIENT RANGES
    VARIABLE    CURRENT        ALLOWABLE        ALLOWABLE
                 COEF          INCREASE         DECREASE
        X1   15.000000        INFINITY         3.500000
        X2   10.000000        2.000000         INFINITY
        X3    9.000000        INFINITY         4.000000
        X4    7.000000        INFINITY         7.000000

                        RIGHTHAND SIDE RANGES
        ROW     CURRENT        ALLOWABLE        ALLOWABLE
                 RHS           INCREASE         DECREASE
        2   1000.000000       66.666660       100.000000
        3    400.000000      100.000000       400.000000
        4   3300.000000        INFINITY       300.000000
        5   4000.000000      300.000000       200.000000
```

FIGURE 4
LINDO Output for Tucker

EXAMPLE 3   **Interpretation of Objective Function Coefficients Sensitivity Analysis**

a Suppose Winco raises the price of product 2 by 50¢ per unit. What is the new optimal solution to the LP?

b Suppose the sales price of product 1 is increased by 60¢ per unit. What is the new optimal solution to the LP?

c Suppose the sales price of product 3 is decreased by 60¢. What is the new optimal solution to the LP?

Solution **a** Because the AI for c_2 is $0.666667, and we are increasing c_2 by only $0.5, the current basis remains optimal. The optimal values of the decision variables remain unchanged ($x_1 = 0$, $x_2 = 400$, $x_3 = 150$, and $x_4 = 400$ is still optimal). The new optimal z-value may be determined in two ways. First, we may simply substitute the optimal values of the decision variables into the new objective function, yielding

$$\text{New optimal } z\text{-value} = 4(0) + 6.5(400) + 7(150) + 8(400) = \$6,850$$

Another way to see that the new optimal z-value is $6,850 is to observe the only difference in sales revenue: Each unit of product 2 brings in 50¢ more in revenue. Thus, total revenue should increase by 400(.50) = $200, so

$$\text{New } z\text{-value} = \text{original } z\text{-value} + 200 = \$6,850$$

b The AI for c_1 is 1, so the current basis remains optimal, and the optimal values of the decision variables remain unchanged. Because the value of x_1 in the optimal solution is 0, the change in the sales price for product 1 will not change the optimal z-value—it will remain $6,650.

c For c_3, AD = .50, so the current basis is no longer optimal. Without resolving the problem by hand or on the computer, we cannot determine the new optimal solution.

Reduced Costs and Sensitivity Analysis

The REDUCED COST portion of the LINDO output gives us information about how changing the objective function coefficient for a nonbasic variable will change the LP's optimal solution. For simplicity, let's assume that the current optimal bfs is nondegenerate (that is, if the LP has m constraints, then the current optimal solution has m variables assuming positive values). For any nonbasic variable x_k, the reduced cost is the amount by which the objective function coefficient of x_k must be improved before the LP will have an optimal solution in which x_k is a basic variable. If the objective function coefficient of a nonbasic variable x_k is improved by its reduced cost, then the LP will have alternative optimal solutions—at least one in which x_k is a basic variable, and at least one in which x_k is not a basic variable. If the objective function coefficient of a nonbasic variable x_k is improved by more than its reduced cost, then (barring degeneracy) any optimal solution to the LP will have x_k as a basic variable and $x_k > 0$. To illustrate these ideas, note that in Example 1 the basic variables associated with the optimal solution are x_2, x_3, x_4, and s_4 (the slack for the labor constraint). The nonbasic variable x_1 has a reduced cost of $1. This implies that if we increase x_1's objective function coefficient (in this case, the sales price per unit of x_1) by exactly $1, then there will be alternative optimal solutions, at least one of which will have x_1 as a basic variable. If we increase x_1's objective function coefficient by more than $1, then (because the current optimal bfs is nondegenerate) any optimal solution to the LP will have x_1 as a basic variable (with $x_1 > 0$). Thus, the reduced cost for x_1 is the amount by which x_1 "misses the optimal basis." We must keep a close watch on x_1's sales price, because a slight increase will change the LP's optimal solution.

Let's now consider Example 2, a minimization problem. Here the basic variables associated with the optimal solution are x_1, x_2, x_3, and s_3 (the slack variable for the labor constraint). Again, the optimal bfs is nondegenerate. The nonbasic variable x_4 has a reduced cost of 7 ($7,000), so we know that if the cost of producing x_4 is decreased by 7, then there will be alternative optimal solutions. In at least one of these optimal solutions, x_4 will be a basic variable. If the cost of producing x_4 is lowered by more than 7, then (because the current optimal solution is nondegenerate) any optimal solution to the LP will have x_4 as a basic variable (with $x_4 > 0$).

Right-Hand Side Ranges

Recall from Section 5.1 that we can determine (at least for a two-variable problem) the range of values for a right-hand side within which the current basis remains optimal. This information is given in the RIGHTHAND SIDE RANGES section of the LINDO output. To illustrate, consider the first constraint in Example 1. Currently, the right-hand side of this constraint (call it b_1) is 950. The current basis remains optimal if b_1 is decreased by up to 100 (the allowable decrease, or AD, for b_1) or increased by up to 50 (the allowable increase, or AI, for b_1). Thus, the current basis remains optimal if

$$850 = 950 - 100 \le b_1 \le 950 + 50 = 1{,}000$$

We call this the allowable range for b_1. Even if a change in the right-hand side of a constraint leaves the current basis optimal, the LINDO output does not provide sufficient information to determine the new values of the decision variables. However, the LINDO output does allow us to determine the LP's new optimal z-value.

Shadow Prices and Dual Prices

In Section 5.1, we defined the shadow price of an LP's ith constraint to be the amount by which the optimal z-value of the LP is improved if the right-hand side is increased by one unit (assuming this change leaves the current basis optimal). If, after a change in a constraint's right-hand side, the current basis is no longer optimal, then the shadow prices of *all* constraints may change. We will discuss this further in Section 5.4. The shadow price for each constraint is found in the DUAL PRICES section of the LINDO output. If we increase the right-hand side of the ith constraint by an amount Δb_i—a decrease in b_i implies that $\Delta b_i < 0$—and the new right-hand side value for Constraint i remains within the allowable range for the right-hand side given in the RIGHTHAND SIDE RANGES section of the output, then formulas (1) and (2) may be used to determine the optimal z-value after a right-hand side is changed. The following example illustrates how shadow prices may be used to determine how a change in a right-hand side affects the optimal z-value.

EXAMPLE 4 **Interpretation of RHS Sensitivity Analysis**

a In Example 1, suppose that a total of 980 units must be produced. Determine the new optimal z-value.

b In Example 1, suppose that 4,500 units of raw material are available. What is the new optimal z-value? What if only 4,400 units of raw material are available?

c In Example 2, suppose that 4,100 units of raw material are available. Find the new optimal z-value.

d In Example 2, suppose that exactly 950 cars must be produced. What will be the new optimal z-value?

Solution **a** $\Delta b_1 = 30$. Because the allowable increase is 50, the current basis remains optimal, and the shadow price of \$3 remains applicable. Then (1) yields

$$\text{New optimal } z\text{-value} = 6{,}650 + 30(3) = \$6{,}740$$

Here we see that (as long as the current basis remains optimal) each additional unit of demand increases revenues by \$3.

b $\Delta b_3 = -100$. Because the allowable decrease is 150, the shadow price of \$1 remains valid. Then (1) yields

$$\text{New optimal } z\text{-value} = 6{,}650 - 100(1) = \$6{,}550$$

Thus (as long as the current basis remains optimal), a decrease in available raw material of one unit decreases revenue by \$1. If only 4,400 units of raw material are available, then $\Delta b_3 = -200$. Because the allowable decrease is 150, we cannot determine the new optimal z-value.

c $\Delta b_4 = 100$. The dual (or shadow) price is 5 (thousand). The current basis remains optimal, so (2) yields

$$\text{New optimal } z\text{-value} = 11{,}600 - 100(5) = 11{,}100 \ (\$11{,}100{,}000)$$

Thus, as long as the current basis remains optimal, each additional unit of raw material decreases costs by $5,000.

d $\Delta b_1 = -50$. The allowable decrease is 100, so the shadow price of -30 (thousand) and (2) yield

New optimal z-value $= 11,600 - (-50)(-30) = 10,100 = \$10,100,000$

Thus, each unit by which demand is reduced (as long as the current basis remains optimal) decreases costs by $30,000.

Let's give an interpretation to the shadow price for each constraint in Examples 1 and 2. Again, all discussions are assuming that we are within the allowable range where the current basis remains optimal. The shadow price of $3 for Constraint 1 in Example 1 implies that each one-unit increase in total demand will increase sales revenues by $3. The shadow price of $-\$2$ for Constraint 2 implies that each unit increase in the requirement for product 4 will decrease revenue by $2. The shadow price of $1 for Constraint 3 implies that an additional unit of raw material given to Winco (for no cost) increases total revenue by $1. Finally, the shadow price of $0 for Constraint 4 implies that an additional unit of labor given to Winco (at no cost) will not increase total revenue. This is reasonable; at present, 250 of the available 5,000 labor hours are not being used, so why should we expect additional labor to raise revenues?

The shadow price of $-\$30$ (thousand) for Constraint 1 of Example 2 means that each extra car that must be produced will decrease costs by $-\$30,000$ (or increase costs by $30,000). The shadow price of $-\$4$ (thousand) for Constraint 2 means that an extra car that the firm is forced to produce at plant 3 will decrease costs by $-\$4,000$ (or increase costs by $4,000). The shadow price of $0 for the third constraint means that an extra hour of labor given to Tucker will decrease costs by $0. Thus, if Tucker is given an additional hour of labor then costs are unchanged. This is reasonable; now 300 hours of available labor are unused. The shadow price for Constraint 4 is $5 (thousand), which means that if Tucker were given an additional unit of raw material, then costs would decrease by $5,000.

Signs of Shadow Prices

A $\geq$ constraint will always have a nonpositive shadow price; a $\leq$ constraint will always have a non-negative shadow price; and an equality constraint may have a positive, negative, or zero shadow price. To see why this is true, observe that adding points to an LP's feasible region can only improve the optimal z-value or leave it the same. Eliminating points from an LP's feasible region can only make the optimal z-value worse or leave it the same. For example, let's look at the shadow price of the raw-material constraint (a $\leq$ constraint) in Example 1. Why must this shadow price be non-negative? The shadow price of the raw-material constraint represents the improvement in the optimal z-value if 4,601 units (instead of 4,600) of raw material are available. Having an additional unit of raw material available adds points to the feasible region—points for which Winco uses $>$ 4,600 but $\leq$ 4,601 units of raw material—so we know that the optimal z-value must increase or stay the same. Thus, the shadow price of this $\leq$ constraint must be nonnegative.

Similarly, let's consider the shadow price of the $x_4 \geq 400$ constraint in Example 1. Increasing the right-hand side of this constraint to 401 eliminates points from the feasible region (points for which Winco produces $\geq$ 400 but $<$ 401 units of product 4). Thus, the optimal z-value must decrease or stay the same, implying that the shadow price of this constraint must be nonpositive. Similar reasoning shows that for a minimization problem, a $\geq$ constraint will have a nonpositive shadow price, and a $\leq$ constraint will have a non-negative shadow price.

An equality constraint's shadow price may be positive, negative, or zero. To see why, consider the following two LPs:

$$\max z = x_1 + x_2$$
$$\text{s.t.} \quad x_1 + x_2 = 1 \qquad \text{(LP 1)}$$
$$x_1, x_2 \geq 0$$

$$\max z = x_1 + x_2$$
$$\text{s.t.} \quad -x_1 - x_2 = -1 \qquad \text{(LP 2)}$$
$$x_1, x_2 \geq 0$$

Both LPs have the same feasible region and set of optimal solutions (the portion of the line segment $x_1 + x_2 = 1$ in the first quadrant). However, LP 1's constraint has a shadow price of $+1$, whereas LP 2's constraint has a shadow price of -1. Thus, the sign of the shadow price for an equality constraint may either be positive, negative, or zero.

Sensitivity Analysis and Slack and Excess Variables

It can be shown (see Section 6.10) that for any inequality constraint, the product of the values of the constraint's slack or excess variable and the constraint's shadow price must equal 0. This implies that any constraint whose slack or excess variable is > 0 will have a zero shadow price. It also implies that any constraint with a nonzero shadow price must be binding (have slack or excess equal to 0). To illustrate these ideas, consider the labor constraint in Example 1. This constraint has positive slack, so its shadow price must be 0. This is reasonable, because slack = 250 for this constraint indicates that 250 hours of currently available labor are unused at present. Thus, an extra hour of labor would not increase revenues. Now consider the raw material constraint of Example 1. Because this constraint has a nonzero shadow price, it must have slack = 0. This is reasonable; the nonzero shadow price means that additional raw material will increase revenue. This can be the case only if all currently available raw material is now being used.

For constraints with nonzero slack or excess, the value of the slack or excess variable is related to the ALLOWABLE INCREASE and ALLOWABLE DECREASE sections of the RIGHTHAND SIDE RANGES portion of the LINDO output. This relationship is detailed in Table 4.

For any constraint having positive slack or excess, the optimal z-value and values of the decision variables remain unchanged within the right-hand side's allowable range. To illustrate these ideas, consider the labor constraint in Example 1. Because slack = 250, we see from Table 4 that AI = ∞ and AD = 250. Thus, the current basis remains optimal

TABLE 4

Allowable Increases and Decreases for Constraints
with Nonzero Slack or Excess

Type of Constraint	AI for rhs	AD for rhs
$\leq$	∞	= Value for slack
$\geq$	= Value of excess	= ∞

for $4{,}750 \le$ available labor $\le \infty$. Within this range, both the optimal z-value and values of the decision variables remain unchanged.

Degeneracy and Sensitivity Analysis

When the optimal solution to an LP is degenerate, caution must be used when interpreting the LINDO output. Recall from Section 4.11 that a bfs is degenerate if at least one basic variable in the optimal solution equals 0. For an LP with m constraints, if the LINDO output indicates that less than m variables are positive, then the optimal solution is a degenerate bfs. To illustrate, consider the following LP:

$$\max z = 6x_1 + 4x_2 + 3x_3 + 2x_4$$

$$\begin{aligned}
\text{s.t.} \quad 2x_1 + 3x_2 + \ x_3 + 2x_4 &\le 400 \\
x_1 + \ x_2 + 2x_3 + \ x_4 &\le 150 \\
2x_1 + \ x_2 + \ x_3 + .5x_4 &\le 200 \\
3x_1 + \ x_2 \qquad \quad x_4 &\le 250 \\
x_1, x_2, x_3, x_4 &\ge 0
\end{aligned}$$

The LINDO output for this LP is in Figure 5. The LP has four constraints and in the optimal solution only two variables are positive, so the optimal solution is a degenerate bfs. By the way, using the **TABLEAU** command indicates that the optimal basis is BV = $\{x_2, x_3, s_3, x_1\}$.

We now discuss three "oddities" that may occur when the optimal solution found by LINDO is degenerate.

Oddity 1   In the RANGES IN WHICH THE BASIS IS UNCHANGED at least one constraint will have a 0 AI or AD. This means that for at least one constraint, the DUAL PRICE can tell us about the new z-value for either an increase or decrease in the right-hand side, but not both.

To understand Oddity 1, consider the first constraint. Its AI is 0. This means that the first constraint's DUAL PRICE of .50 cannot be used to determine a new z-value resulting from any increase in the first constraint's right-hand side.

Oddity 2   For a nonbasic variable to become positive, its objective function coefficient may have to be improved by more than its REDUCED COST.

To understand Oddity 2, consider the nonbasic variable x_4; its REDUCED COST is 1.5. If we increase its objective function coefficient by 2, however, we still find that the new optimal solution has $x_4 = 0$. This oddity occurs because the increase changes the set of basic variables, but not the LP's optimal solution.

Oddity 3   Increasing a variable's objective function coefficient by more than its AI or decreasing it by more than its AD may leave the optimal solution to the LP the same.

Oddity 3 is similar to Oddity 2. To understand it, consider the nonbasic variable x_4. Its AI is 1.5. If we increase its objective function coefficient by 2, however, we still find that the new optimal solution is unchanged. This oddity occurs because the increase changes the set of basic variables, but not the LP's optimal solution.

We close this section by noting that our discussions apply only if one objective function coefficient or one right-hand side is changed. If more than one objective function coefficient or the right-hand side is changed, it is sometimes still possible to use the LINDO output to determine whether the current basis remains optimal. See Section 6.4 for details.

```
MAX       6  X1 + 4  X2  + 3  X3  + 2  X4
SUBJECT TO
        2)     2  X1  +  3 X2  +  X3  +  2  X4  <= 400
        3)        X1  +   X2  + 2 X3  +   X4  <=      150
        4)     2  X1  +   X2  +  X3  + 0.5  X4  <=      200
        5)     3  X1  +   X2  +   X4  <=      250
END

LP OPTIMUM FOUND AT STEP              3
                    OBJECTIVE FUNCTION VALUE
                    1)   700.00000

    VARIABLE           VALUE           REDUCED COST
        X1           50.000000            .000000
        X2          100.000000            .000000
        X3            .000000             .000000
        X4            .000000            1.500000


        ROW      SLACK OR SURPLUS      DUAL PRICES
        2)            .000000            .500000
        3)            .000000           1.250000
        4)            .000000            .000000
        5)            .000000           1.250000

NO. ITERATIONS=           3

RANGES IN WHICH THE BASIS IS UNCHANGED:

                        OBJ COEFFICIENT RANGES
    VARIABLE   CURRENT        ALLOWABLE        ALLOWABLE
                COEF          INCREASE         DECREASE
        X1    6.000000        3.000000         3.000000
        X2    4.000000        5.000000         1.000000
        X3    3.000000        3.000000         2.142857
        X4    2.000000        1.500000         INFINITY

                        RIGHTHAND SIDE RANGES
    ROW      CURRENT        ALLOWABLE        ALLOWABLE
              RHS           INCREASE         DECREASE
     2   400.000000         .000000        200.000000
     3   150.000000         .000000          .000000
     4   200.000000         INFINITY         .000000
     5   250.000000         .000000        120.000000

THE TABLEAU
   ROW   (BASIS)    X1       X2       X3       X4      SLK  2
    1    ART      .000     .000     .000    1.500      .500
    2      X2     .000    1.000     .000     .500      .500
    3      X3     .000     .000    1.000     .167     -.167
    4    SLK  4   .000     .000     .000    -.500      .000
    5      X1    1.000     .000     .000     .167     -.167

   ROW    SLK   3     SLK   4     SLK   5
    1     1.250        .000      1.250     700.000
    2     -.250        .000      -.250     100.000
    3      .583        .000      -.083       .000
    4     -.500       1.000      -.500       .000
    5      .083        .000       .417      50.000
```

FIGURE 5

PROBLEMS

Group A

1 Farmer Leary grows wheat and corn on his 45-acre farm. He can sell at most 140 bushels of wheat and 120 bushels of corn. Each acre planted with wheat yields 5 bushels, and each acre planted with corn yields 4 bushels. Wheat sells for $30 per bushel, and corn sells for $50 per bushel. To harvest an acre of wheat requires 6 hours of labor; 10 hours are needed to harvest an acre of corn. Up to 350 hours of labor can be purchased at $10 per hour. Let A1 = acres planted with wheat; A2 = acres planted with

corn; and L = hours of labor that are purchased. To maximize profits, Leary should solve the following LP:

$$\max z = 150A1 + 200A2 - 10L$$
$$\text{s.t.} \quad A1 + A2 \le 45$$
$$6A1 + 10A2 - L \le 0$$
$$L \le 350$$
$$5A1 \le 140$$
$$4A2 \le 120$$
$$A1, A2, L \ge 0$$

Use the LINDO output in Figure 6 to answer the following questions:

a If only 40 acres of land were available, what would Leary's profit be?

b If the price of wheat dropped to $26, what would be the new optimal solution to Leary's problem?

c Use the SLACK portion of the output to determine the allowable increase and allowable decrease for the amount of wheat that can be sold. If only 130 bushels of wheat could be sold, then would the answer to the problem change?

2 Carco manufactures cars and trucks. Each car contributes $300 to profit, and each truck contributes $400. The resources required to manufacture a car and a truck are shown in Table 5. Each day, Carco can rent up to 98 Type 1 machines at a cost of $50 per machine. The company has 73 Type 2 machines and 260 tons of steel available. Marketing considerations dictate that at least 88 cars and at

TABLE 5

Vehicle	Days on Type 1 Machine	Days on Type 2 Machine	Tons of Steel
Car	0.8	0.6	2
Truck	1	0.7	3

least 26 trucks be produced. Let x_1 = number of cars produced daily; x_2 = number of trucks produced daily; and m_1 = Type 1 machines rented daily.

To maximize profit, Carco should solve the LP in Figure 7. Use the LINDO output to answer the following questions:

a If each car contributed $310 to profit, what would be the new optimal solution to the problem?

FIGURE 6
LINDO Output for Wheat and Corn

```
MAX     150  A1 + 200  A2  -  10 L
SUBJECT TO
    2)     A1 + A2 <=    45
    3)     6 A1 + 10 A2 - L <=   0
    4)     L <=    350
    5)     5 A1 <=    140
    6)     4 A2 <=    120
END

LP OPTIMUM FOUND AT STEP          4

            OBJECTIVE FUNCTION VALUE

        1)   4250.00000

VARIABLE         VALUE          REDUCED COST
    A1         25.000000           .000000
    A2         20.000000           .000000
    L         350.000000           .000000


    ROW      SLACK OR SURPLUS     DUAL PRICES
    2)           .000000          75.000000
    3)           .000000          12.500000
    4)           .000000           2.500000
    5)         15.000000           .000000
    6)         40.000000           .000000

NO. ITERATIONS=        4

RANGES IN WHICH THE BASIS IS UNCHANGED:

              OBJ COEFFICIENT RANGES
VARIABLE   CURRENT     ALLOWABLE      ALLOWABLE
            COEF       INCREASE       DECREASE
    A1   150.000000   10.000000     30.000000
    A2   200.000000   50.000000     10.000000
    L    -10.000000   INFINITY       2.500000

              RIGHTHAND SIDE RANGES
ROW    CURRENT      ALLOWABLE      ALLOWABLE
        RHS         INCREASE       DECREASE
 2    45.000000     1.200000      6.666667
 3     .000000     40.000000     12.000000
 4   350.000000    40.000000     12.000000
 5   140.000000    INFINITY      15.000000
 6   120.000000    INFINITY      40.000000
```

FIGURE 7
LINDO Output for Carco

```
MAX      300  X1 + 400  X2  -  50 M1
SUBJECT TO
    2)    0.8 X1  +    X2  - M1  <=    0
    3)    M1  <=     98
    4)    0.6  X1 +  0.7  X2 <=    73
    5)    2 X1  + 3  X2  <=    260
    6)    X1 >=   88
    7)    X2 >=   26
END

LP OPTIMUM FOUND AT STEP          4

            OBJECTIVE FUNCTION VALUE

        1)   32540.0000

VARIABLE         VALUE          REDUCED COST
    X1         88.000000           .000000
    X2         27.600000           .000000
    M1         98.000000           .000000


    ROW      SLACK OR SURPLUS     DUAL PRICES
    2)           .000000         400.000000
    3)           .000000         350.000000
    4)           .879999           .000000
    5)          1.200003           .000000
    6)           .000000         -20.000000
    7)          1.599999           .000000

NO. ITERATIONS=        4

RANGES IN WHICH THE BASIS IS UNCHANGED:

              OBJ COEFFICIENT RANGES
VARIABLE   CURRENT     ALLOWABLE      ALLOWABLE
            COEF       INCREASE       DECREASE
    X1   300.000000   20.000000      INFINITY
    X2   400.000000    INFINITY     25.000000
    M1   -50.000000    INFINITY    350.000000

              RIGHTHAND SIDE RANGES
ROW    CURRENT      ALLOWABLE      ALLOWABLE
        RHS         INCREASE       DECREASE
 2     .000000      .400001      1.599999
 3    98.000000     .400001      1.599999
 4    73.000000     INFINITY      .879999
 5   260.000000     INFINITY     1.200003
 6    88.000000    1.999999      3.000008
 7    26.000000    1.599999      INFINITY
```

b If Carco were required to produce at least 86 cars, what would Carco's profit become?

3 Consider the diet problem discussed in Section 3.4. Use the LINDO output in Figure 8 to answer the following questions.

a If a Brownie costs 30¢, then what would be the new optimal solution to the problem?

b If a bottle of cola cost 35¢, then what would be the new optimal solution to the problem?

c If at least 8 oz of chocolate were required, then what would be the cost of the optimal diet?

d If at least 600 calories were required, then what would be the cost of the optimal diet?

e If at least 9 oz of sugar were required, then what would be the cost of the optimal diet?

f What would the price of pineapple cheesecake have to be before it would be optimal to eat cheesecake?

g What would the price of a brownie have to be before it would be optimal to eat a brownie?

h Use the SLACK or SURPLUS portion of the LINDO output to determine the allowable increase and allowable decrease for the fat constraint. If 10 oz of fat were required, then would the optimal solution to the problem change?

4 Gepbab Corporation produces three products at two different plants. The cost of producing a unit at each plant is shown in Table 6. Each plant can produce a total of 10,000 units. At least 6,000 units of product 1, at least 8,000 units of product 2, and at least 5,000 units of product 3 must be produced. To minimize the cost of meeting these demands, the following LP should be solved:

$$\min z = 5x_{11} + 6x_{12} + 8x_{13} + 8x_{21} + 7x_{22} + 10x_{23}$$

$$
\begin{aligned}
\text{s.t.} \quad & x_{11} + x_{12} + x_{13} && \leq && 10{,}000 \\
& x_{21} + x_{22} + x_{23} \leq && 10{,}000 \\
& x_{11} && + x_{21} && \geq && 6{,}000 \\
& x_{12} && + x_{22} && \geq && 8{,}000 \\
& x_{13} && + x_{23} \geq && 5{,}000 \\
& \text{All variables} \geq && 0
\end{aligned}
$$

Here, x_{ij} = number of units of product j produced at plant i. Use the LINDO output in Figure 9 to answer the following questions:

a What would the cost of producing product 2 at plant 1 have to be for the firm to make this choice?

b What would total cost be if plant 1 had 9,000 units of capacity?

c If it cost $9 to produce a unit of product 3 at plant 1, then what would be the new optimal solution?

5 Mondo produces motorcycles at three plants. At each plant, the labor, raw material, and production costs (excluding labor cost) required to build a motorcycle are as shown in Table 7. Each plant has sufficient machine capacity to produce up to 750 motorcycles per week. Each of Mondo's workers can work up to 40 hours per week and is paid $12.50 per hour worked. Mondo has a total of 525 workers and now owns 9,400 units of raw material. Each week, at least 1,400 Mondos must be produced. Let x_1 = motorcycles produced at plant 1; x_2 = motorcycles produced at plant 2; and x_3 = motorcycles produced at plant 3.

The LINDO output in Figure 10 enables Mondo to minimize the variable cost (labor + production) of meeting demand. Use the output to answer the following questions:

a What would be the new optimal solution to the problem if the production cost at plant 1 were only $40?

FIGURE 8
LINDO Output for Diet Problem

```
MAX     50 BR + 20 IC + 30 COLA + 80 PC
SUBJECT TO
      2)    400 BR + 200 IC + 150 COLA
                    + 500 PC >=   500
      3)    3  BR  + 2  IC  >=  6
      4)    2  BR  + 2  IC  +   4 COLA
                    + 4  PC >=  10
      5)    2  BR  + 4  IC  +   COLA
                    + 5  PC >=  8
END

LP OPTIMUM FOUND AT STEP          2

            OBJECTIVE FUNCTION VALUE

        1)   90.0000000

VARIABLE        VALUE        REDUCED COST
    BR        .000000         27.500000
    IC       3.000000          .000000
    COLA     1.000000          .000000
    PC        .000000         50.000000


ROW     SLACK OR SURPLUS    DUAL PRICES
   2)    250.000000          .000000
   3)      .000000         -2.500000
   4)      .000000         -7.500000
   5)    5.000000           .000000

NO. ITERATIONS=        2

RANGES IN WHICH THE BASIS IS UNCHANGED:

                OBJ COEFFICIENT RANGES
VARIABLE    CURRENT     ALLOWABLE    ALLOWABLE
             COEF       INCREASE     DECREASE
    BR    50.000000     INFINITY     27.500000
    IC    20.000000    18.333330      5.000000
    COLA  30.000000    10.000000     30.000000
    PC    80.000000     INFINITY     50.000000

                RIGHTHAND SIDE RANGES
ROW     CURRENT     ALLOWABLE    ALLOWABLE
         RHS        INCREASE     DECREASE
  2    500.000000   250.000000    INFINITY
  3      6.000000     4.000000     2.857143
  4     10.000000     INFINITY     4.000000
  5      8.000000     5.000000     INFINITY
```

TABLE 6

Plant	Product ($)		
	1	2	3
1	5	6	8
2	8	7	10

FIGURE 9
LINDO Output for Gepbab

```
MAX      5 X11 + 6 X12 + 8 X13 + 8 X21
                            + 7 X22 + 10 X23
SUBJECT TO
       2)     X11  +  X12  +  X13 <=   10000
       3)     X21  +  X22  +  X23 <=   10000
       4)     X11  +  X21  >=    6000
       5)     X12  +  X22  >=    8000
       6)     X13  +  X23  >=    5000
END

LP OPTIMUM FOUND AT STEP            5

                 OBJECTIVE FUNCTION VALUE

        1)    128000.000

VARIABLE        VALUE          REDUCED COST
   X11      6000.000000          .000000
   X12          .000000         1.000000
   X13      4000.000000          .000000
   X21          .000000         1.000000
   X22      8000.000000          .000000
   X23      1000.000000          .000000


   ROW    SLACK OR SURPLUS    DUAL PRICES
   2)          .000000         2.000000
   3)        1000.000000         .000000
   4)          .000000        -7.000000
   5)          .000000        -7.000000
   6)          .000000       -10.000000

NO. ITERATIONS=        5

RANGES IN WHICH THE BASIS IS UNCHANGED:

                  OBJ COEFFICIENT RANGES
VARIABLE    CURRENT      ALLOWABLE     ALLOWABLE
             COEF        INCREASE      DECREASE
   X11   5.000000       1.000000      7.000000
   X12   6.000000       INFINITY      1.000000
   X13   8.000000       1.000000      1.000000
   X21   8.000000       INFINITY      1.000000
   X22   7.000000       1.000000      7.000000
   X23  10.000000       1.000000      1.000000

                  RIGHTHAND SIDE RANGES
   ROW    CURRENT       ALLOWABLE     ALLOWABLE
            RHS         INCREASE      DECREASE
    2  10000.000000   1000.000000   1000.000000
    3  10000.000000    INFINITY     1000.000000
    4   6000.000000   1000.000000   1000.000000
    5   8000.000000   1000.000000   8000.000000
    6   5000.000000   1000.000000   1000.000000
```

TABLE 7

Plant	Labor Needed (Hours)	Raw Material Needed (Units)	Production Cost ($)
1	20	5	50
2	16	8	80
3	10	7	100

FIGURE 10
LINDO Output for Mondo

```
MAX      300  X1 + 280  X2  +  225  X3
SUBJECT TO
       2)     20 X1  + 16 X2  + 10  X3 <=  21000
       3)      5  X1  + 8 X2  +  7  X3 <=   9400
       4)     X1  <=     750
       5)     X2  <=     750
       6)     X3  <=     750
       7)     X1  +  X2  +  X3  >=    1400
END

LP OPTIMUM FOUND AT STEP             3

                 OBJECTIVE FUNCTION VALUE

        1)    357750.000

VARIABLE        VALUE          REDUCED COST
   X1       350.000000           .000000
   X2       300.000000           .000000
   X3       750.000000           .000000


   ROW    SLACK OR SURPLUS    DUAL PRICES
   2)      1700.000000           .000000
   3)          .000000         6.666668
   4)       400.000000           .000000
   5)       450.000000           .000000
   6)          .000000        61.666660
   7)          .000000      -333.333300

NO. ITERATIONS=        3

RANGES IN WHICH THE BASIS IS UNCHANGED:

                  OBJ COEFFICIENT RANGES
VARIABLE    CURRENT      ALLOWABLE     ALLOWABLE
             COEF        INCREASE      DECREASE
   X1    300.000000      INFINITY     20.000000
   X2    280.000000     20.000010     92.499990
   X3    225.000000     61.666660      INFINITY

                  RIGHTHAND SIDE RANGES
   ROW    CURRENT       ALLOWABLE     ALLOWABLE
            RHS         INCREASE      DECREASE
    2  21000.000000     INFINITY    1700.000000
    3   9400.000000  1050.000000     900.000000
    4    750.000000     INFINITY     400.000000
    5    750.000000     INFINITY     450.000000
    6    750.000000   450.000000     231.818200
    7   1400.000000    63.750000     131.250000
```

b How much money would Mondo save if the capacity of plant 3 were increased by 100 motorcycles?

c By how much would Mondo's cost increase if it had to produce one more motorcycle?

6 Steelco uses coal, iron, and labor to produce three types of steel. The inputs (and sales price) for one ton of each type of steel are shown in Table 8. Up to 200 tons of coal can be purchased at a price of $10 per ton. Up to 60 tons of iron can be purchased at $8 per ton, and up to 100 labor hours can be purchased at $5 per hour. Let x_1 = tons of steel 1 produced; x_2 = tons of steel 2 produced; and x_3 = tons of steel 3 produced.

The LINDO output that yields a maximum profit for the company is given in Figure 11. Use the output to answer the following questions.

a What would profit be if only 40 tons of iron could be purchased?

TABLE 8

Steel	Coal Required (Tons)	Iron Required (Tons)	Labor Required (Hours)	Sales Price ($)
1	3	1	1	51
2	2	0	1	30
3	1	1	1	25

FIGURE 11

LINDO Output for Steelco

```
MAX       8  X1 + 5  X2  + 2  X3
SUBJECT TO
        2)    3  X1  +  2  X2  +  X3 <=  200
        3)       X1  +  X3   <=   60
        4)       X1  +  X2  +  X3 <=  100
END


LP OPTIMUM FOUND AT STEP              2

        OBJECTIVE FUNCTION VALUE

    1)   530.000000

VARIABLE        VALUE        REDUCED COST
    X1        60.000000          .000000
    X2        10.000000          .000000
    X3          .000000         1.000000


    ROW    SLACK OR SURPLUS    DUAL PRICES
    2)          .000000         2.500000
    3)          .000000          .500000
    4)        30.000000          .000000

NO. ITERATIONS=              2

RANGES IN WHICH THE BASIS IS UNCHANGED:

                    OBJ COEFFICIENT RANGES
VARIABLE    CURRENT      ALLOWABLE      ALLOWABLE
            COEF         INCREASE       DECREASE
    X1     8.000000      INFINITY        .500000
    X2     5.000000      .333333        5.000000
    X3     2.000000     1.000000        INFINITY

                   RIGHTHAND SIDE RANGES
    ROW    CURRENT      ALLOWABLE      ALLOWABLE
           RHS         INCREASE       DECREASE
    2   200.000000     60.000000     20.000000
    3    60.000000      6.666667     60.000000
    4   100.000000      INFINITY     30.000000
```

b What is the smallest price per ton for steel 3 that would make it desirable to produce it?

c Find the new optimal solution if steel 1 sold for $55 per ton.

Group B

7 Shoeco must meet (on time) the following demands for pairs of shoes: month 1—300; month 2—500; month 3—100; and month 4—100. At the beginning of month 1, 50 pairs of shoes are on hand, and Shoeco has three workers. A worker is paid $1,500 per month. Each worker can work up to 160 hours per month before receiving overtime. During any month, each worker may be forced to work up to 20

hours of overtime; workers are paid $25 per hour for overtime labor. It takes 4 hours of labor and $5 of raw material to produce each pair of shoes. At the beginning of each month, workers can be hired or fired. Each hired worker costs $1,600, and each fired worker costs $2,000. At the end of each month, a holding cost of $30 per pair of shoes is assessed. Formulate an LP that can be used to minimize the total cost of meeting the next four months' demands. Then use LINDO to solve the LP. Finally, use the LINDO printout to answer the questions that follow these hints (which may help in the formulation.) Let

x_t = Pairs of shoes produced during month t with nonovertime labor

o_t = Pairs of shoes produced during month t with overtime labor

i_t = Inventory of pairs of shoes at end of month t

h_t = Workers hired at beginning of month t

f_t = Workers fired at beginning of month t

w_t = Workers available for month t (after month t hiring and firing)

Four types of constraints will be needed:

Type 1 Inventory equations. For example, during month 1, $i_1 = 50 + x_1 + o_1 - 300$.

Type 2 Relate available workers to hiring and firing. For month 1, for example, the following constraint is needed: $w_1 = 3 + h_1 - f_1$.

Type 3 For each month, the amount of shoes made with nonovertime labor is limited by the number of workers. For example, for month 1, the following constraint is needed: $4x_1 \leq 160w_1$.

Type 4 For each month, the number of overtime labor hours used is limited by the number of workers. For example, for month 1, the following constraint is needed: $4(o_1) \leq 20w_1$.

For the objective function, the following costs must be considered:

1 Workers' salaries

2 Hiring costs

3 Firing costs

4 Holding costs

5 Overtime costs

6 Raw-material costs

a Describe the company's optimal production plan, hiring policy, and firing policy. Assume that it is acceptable to have a fractional number of workers, hirings, or firings.

b If overtime labor during month 1 costs $16 per hour, should any overtime labor be used?

c If the cost of firing workers during month 3 were $1,800, what would be the new optimal solution to the problem?

d If the cost of hiring workers during month 1 were $1,700, what would be the new optimal solution to the problem?

e By how much would total costs be reduced if demand in month 1 were 100 pairs of shoes?

f What would the total cost become if the company had 5 workers at the beginning of month 1 (before month 1's hiring or firing takes place)?

g By how much would costs increase if demand in month 2 were increased by 100 pairs of shoes?

8 Consider the LP:

$$\max \quad 9x_1 + 8x_2 + 5x_3 + 4x_4$$

$$\begin{aligned}
\text{s.t.} \quad x_1 \qquad\qquad\quad + x_4 &\le 200 \\
x_2 + x_3 \qquad &\le 150 \\
x_1 + x_2 + x_3 \qquad &\le 350 \\
2x_1 + x_2 + x_3 + x_4 &\le 550 \\
x_1, x_2, x_3, x_4 &\ge 0
\end{aligned}$$

a Solve this LP with LINDO and use your output to show that the optimal solution is degenerate.

b Use your LINDO output to find an example of Oddities 1–3.

5.3 Managerial Use of Shadow Prices

In this section, we will discuss the managerial significance of shadow prices. In particular, we will learn how shadow prices can often be used to answer the following question: What is the maximum amount that a manager should be willing to pay for an additional unit of a resource? To answer this question, we usually focus our attention on the shadow price of the constraint that describes the availability of the resource. We now discuss four examples of the interpretation of shadow prices.

EXAMPLE 5   Winco Products 2

In Example 1, what is the most that Winco should be willing to pay for an additional unit of raw material? How about an extra hour of labor?

Solution Because the shadow price of the raw-material-availability constraint is 1, an extra unit would increase total revenue by $1. Thus, Winco could pay up to $1 for an extra unit of raw material and be as well off as it is now. This means that Winco should be willing to pay up to $1 for an extra unit of raw material. The labor-availability constraint has a shadow price of 0. This means that an extra hour of labor will not increase revenues, so Winco should not be willing to pay anything for an extra hour of labor. (Note that this discussion is valid because the AIs for the labor and raw-material constraints both exceed 1.)

EXAMPLE 6   Winco Products 3

Let's reconsider Example 1 with the following changes. Suppose as many as 4,600 units of raw material are available, but they must be purchased at a cost of $4 per unit. Also, as many as 5,000 hours of labor are available, but they must be purchased at a cost of $6 per hour. The per-unit sales price of each product is as follows: product 1—$30; product 2—$42; product 3—$53; product 4—$72. A total of 950 units must be produced, of which at least 400 must be product 4. Determine the maximum amount that the firm should be willing to pay for an extra unit of raw material and an extra hour of labor.

Solution The contribution to profit from one unit of each product may be computed as follows:

$$\begin{aligned}
\text{Product 1:} \quad & 30 - 4(2) - 6(3) = \$4 \\
\text{Product 2:} \quad & 42 - 4(3) - 6(4) = \$6 \\
\text{Product 3:} \quad & 53 - 4(4) - 6(5) = \$7 \\
\text{Product 4:} \quad & 72 - 4(7) - 6(6) = \$8
\end{aligned}$$

Thus, Winco's profit is $4x_1 + 6x_2 + 7x_3 + 8x_4$. To maximize profit, Winco should solve the same LP as in Example 1, and the relevant LINDO output is again Figure 3. To determine the most Winco should be willing to pay for an extra unit of raw material, note that the shadow price of the raw material constraint may be interpreted as follows: If Winco has the right to buy one more unit of raw material (at $4 per unit), then profits increase by $1. Thus, paying $4 + $1 = $5 for an extra unit of raw material will increase profits by $1 − $1 = $0. So Winco could pay up to $5 for an extra unit of raw material and still be better off. For the raw-material constraint, the shadow price of $1 represents a *premium* above and beyond the current price Winco is willing to pay for an extra unit of raw material.

The shadow price of the labor-availability constraint is $0, which means that the right to buy an extra hour of labor at $4 an hour will not increase profits. Unfortunately, all this tells us is that at the current price of $4 per hour, Winco should buy no more labor.

EXAMPLE 7   Farmer Leary's Shadow Price

Consider the Farmer Leary problem (Problem 1 in Section 5.2).

a What is the most that Leary should pay for an additional hour of labor?

b What is the most that Leary should pay for an additional acre of land?

Solution   **a** From the $L \leq 350$ constraint's shadow price of 2.5, we see that if 351 hours of labor are available, then (after paying $10 for another hour of labor) profits increase by $2.50. So if Leary pays $10 + $2.50 = $12.50 for an extra hour of labor, profits would increase by $2.50 − $2.50 = $0. This implies that Leary should be willing to pay up to $12.50 for another hour of labor.

To look at it another way, the shadow price of the $6A1 + 10A2 − L \leq 0$ constraint is 12.5. This means that if the constraint $6A1 + 10A2 \leq L$ were replaced by the constraint $6A1 + 10A2 \leq L + 1$, profits would increase by $12.50. So if one extra hour of labor were "given" to Leary (at zero cost), profits would increase by $12.50. Thus, Leary should be willing to pay up to $12.50 for an extra hour of labor.

b If 46 acres of land were available, profits would increase by $75 (the shadow price of the $A1 + A2 \leq 45$ constraint). This includes the cost ($0) of purchasing an additional acre of land. Thus, Leary should be willing to pay up to $75 for an extra acre of land.

We now illustrate some of the managerial insights that can be gained by analyzing the shadow prices for a minimization problem.

EXAMPLE 8   Tucker Inc's Shadow Price

The following questions refer to Example 2.

a What is the most that Tucker should pay for an extra hour of labor?

b What is the most that Tucker should pay for an extra unit of raw material?

c A new customer is willing to purchase 20 cars at a price of $25,000 per vehicle. Should Tucker fill her order?

Solution   **a** Because the shadow price of the labor-availability constraint (row 4) is 0, an extra hour of labor reduces costs by $0. Thus, Tucker should not pay anything for an extra hour of labor.

b Because the shadow price of the raw-material-availability constraint (row 5) is 5 (thousand dollars), an additional unit of raw material reduces costs by $5,000. Thus, Tucker should be willing to pay up to $5,000 for an extra unit of raw material.

c The allowable increase for the constraint $x_1 + x_2 + x_3 + x_4 = 1,000$ is 66.666660. Because the shadow price of this constraint is -30 (thousand dollars), we know that if Tucker fills the order, its costs will increase by $-20(-30,000) = \$600,000$. So Tucker should not fill the order.

In Example 8, the astute reader may notice that each car costs at most \$15,000 to produce. How is it then possible that a unit increase in the number of cars that must be produced increases costs by \$30,000? To see why this is the case, we re-solved Tucker's LP after increasing the number of cars that had to be produced to 1,001. The new optimal solution has $z = 11,630$, $x_1 = 404$, $x_2 = 197$, $x_3 = 400$, $x_4 = 0$. We now see why increasing demand by one car raises costs by \$30,000. To produce one more car, Tucker must produce four more Type 1 cars and three fewer Type 2 cars. This ensures that Tucker still uses only 4,400 units of raw material, but it increases total cost by $4(15,000) - 3(10,000) = \$30,000$!

PROBLEMS

Group A

1 In Problem 2 of Section 5.2, what is the most that Carco should be willing to pay for an extra ton of steel?

2 In Problem 2 of Section 5.2, what is the most that Carco should be willing to pay to rent an additional Type 1 machine for one day?

3 In Problem 3 of Section 5.2, what is the most that one should be willing to pay for an additional ounce of chocolate?

4 In Problem 4 of Section 5.2, how much should Gepbab be willing to pay for another unit of capacity at plant 1?

5 In Problem 5 of Section 5.2, suppose that Mondo could purchase an additional unit of raw material at a cost of \$6. Should the company do it? Explain.

6 In Problem 6 of Section 5.2, what is the most that Steelco should be willing to pay for an extra ton of coal?

7 In Problem 6 of Section 5.2, what is the most that Steelco should be willing to pay for an extra ton of iron?

8 In Problem 6 of Section 5.2, what is the most that Steelco should be willing to pay for an extra hour of labor?

9 In Problem 7 of Section 5.2, suppose that a new customer wishes to buy a pair of shoes during month 1 for \$70. Should Shoeco oblige him?

10 In Problem 7 of Section 5.2, what is the most the company would be willing to pay for having one more worker at the beginning of month 1?

11 In solving part (c) of Example 8, a manager reasons as follows: The average cost of producing a car is \$11,600 up to 1,000 cars. Therefore, if a customer is willing to pay me \$25,000 for a car, I should certainly fill his order. What is wrong with this reasoning?

5.4 What Happens to the Optimal z-Value If the Current Basis Is No Longer Optimal?

In Section 5.2, we used shadow prices to determine the new optimal z-value if the right-hand side of a constraint were changed but remained in the range where the current basis remains optimal. Suppose we change the right-hand side of a constraint to a value where the current basis is no longer optimal. In this situation, the LINDO Parametrics feature can be used to determine how the shadow price of a constraint and the optimal z-value change.

We illustrate the use of the Parametrics feature by varying the amount of raw material available in Example 1. Suppose we want to determine how the optimal z-value and shadow price change as the amount of available raw material varies between 0 and 10,000 units. We first realize that with little raw material available, the LP will be infeasible. To begin, we change the amount of raw material available to 0. We then obtain from the Range and Sensitivity Analysis results that row 4 has an Allowable Decrease of −3,900. This indicates that if at least 3,900 units of raw material are available the problem will be feasible. We therefore change the rhs of the raw material constraint to 3,900 and solve the LP. After finding the optimal solution, select Reports Parametrics. From the dialog box, choose row 4 and set the value to 10,000. We will choose Text output. We obtain the output shown in Figure 12.

From Figure 12 we find that if the amount of available raw material is 3,900, then the shadow price (or dual price) for raw material is now $2, and the optimal z-value is 5,400. The current basis remains optimal until rm = 4,450; between rm = 3,900 and rm = 4,450, each unit increase in rm will increase the optimal z-value by the shadow price of $2. Thus, when rm = 4,450, the optimal z-value will be

$$5{,}400 + 2(4{,}450 - 3{,}900) = \$6{,}500$$

From Figure 12, we see that when rm = 4,450, x_3 enters the basis and x_1 exits. The shadow price of rm is now $1, and each additional unit of rm (up to the next change of basis) will increase the optimal z-value by $1. The next basis change occurs when rm = 4,850. At this point, the new optimal z-value may be computed as (optimal z-value for rm = 6,500) + (4,850 − 4,450)($1) = $6,900. When rm = 4,850, we pivot in SLACK3 (the slack variable for row 3 or constraint 2), and SLACK5 exits. The new shadow price for rm is $0. Thus when rm > 4,850, we see that an additional unit of rm will not increase the optimal z-value. This discussion is summarized in Figure 13, which shows the optimal z-value as a function of the amount of available raw material.

For any LP, a graph of the optimal objective function value as a function of a right-hand side will consist of several straight-line segments of possibly differing slopes. (Such a function is called a *piecewise linear function*.) The slope of each straight-line segment is just the constraint's shadow price. At points where the optimal basis changes (points B, C, and D in Figure 13), the slope of the graph may change. For a ≤ constraint in a maximization problem, the slope of each line segment must be non-negative—more of a re-

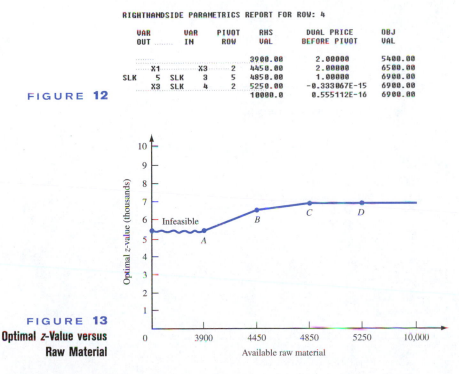

RIGHTHANDSIDE PARAMETRICS REPORT FOR ROW: 4

VAR OUT		VAR IN	PIVOT ROW	RHS VAL	DUAL PRICE BEFORE PIVOT	OBJ VAL
				3900.00	2.00000	5400.00
X1		X3	2	4450.00	2.00000	6500.00
SLK	5	SLK 3	5	4850.00	1.00000	6900.00
		X3 SLK	4 2	5250.00	-0.333067E-15	6900.00
				10000.0	0.555112E-16	6900.00

FIGURE 12

FIGURE 13
Optimal z-Value versus Raw Material

5.4 What Happens to the Optimal z-Value If the Current Basis Is No Longer Optimal? **249**

source can't hurt. In a maximization problem, the slopes of successive line segments for a ≤ constraint will be nonincreasing. This is simply a consequence of diminishing returns; as we obtain more of a resource (and availability of other resources is held constant), the value of an additional unit of the resource cannot increase.

For a ≥ constraint in a maximization problem, the graph of the optimal z-value as a function of the right-hand side will again be a piecewise linear function. The slope of each line segment will be nonpositive (corresponding to the fact that a ≥ constraint has a nonpositive shadow price). The slopes of successive line segments will be nonincreasing. For the $x_4 \geq 400$ constraint in Example 1, plotting the optimal z-value as a function of the constraint's right-hand side yields the graph in Figure 14.

For an equality constraint in a maximization problem, the graph of the optimal z-value as a function of right-hand side will again be piecewise linear. The slopes of each line segment may be positive or negative, but the slopes of successive line segments will again be nonincreasing. For the constraint $x_1 + x_2 + x_3 + x_4 = 950$ in Example 1, we obtain the graph in Figure 15.

For a minimization problem, the plot of the optimal z-value against a constraint's right-hand side is again a piecewise linear function. For all minimization problems, the slopes of successive line segments will be nondecreasing. For a ≤ constraint, the slope of each line segment is nonpositive; for a ≥ constraint, the slope is non-negative; and for an equality constraint, the slope may be positive or negative.

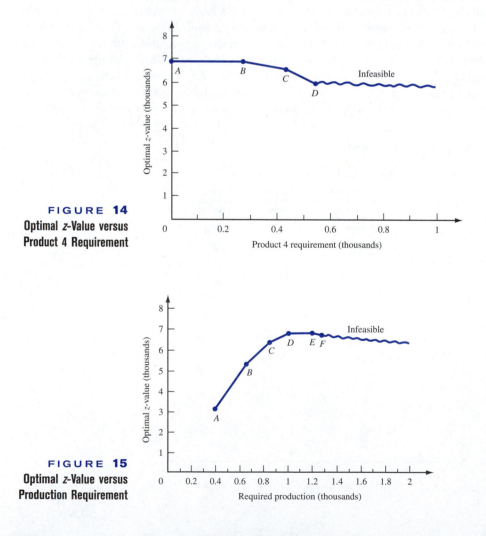

FIGURE 14
Optimal *z*-Value versus Product 4 Requirement

FIGURE 15
Optimal *z*-Value versus Production Requirement

Effect of Change in Objective Function Coefficient on Optimal z-Value

We now discuss how to find the graph of the optimal objective function value as a function of a variable's objective function coefficient. To see how this works, let's reconsider the Giapetto problem.

$$\max z = 3x_1 + 2x_2$$
$$\text{s.t.} \quad 2x_1 + x_2 \leq 100$$
$$x_1 + x_2 \leq 80$$
$$x_1 \quad\quad \leq 40$$
$$x_1, x_2 \geq 0$$

Let c_1 = objective function coefficient for x_1. Currently, we have $c_1 = 3$. We want to determine how the optimal z-value depends on c_1. To determine this relationship, we must find, for each value of c_1, the optimal values of the decision variables. Recall from Figure 1 (p. 228) that point $A = (0, 80)$ is optimal if the isoprofit line is flatter than the carpentry constraint. Also note that point $B = (20, 60)$ is optimal if the slope of the isoprofit line is steeper than the carpentry constraint and flatter than the finishing-hour constraint. Finally, point $C = (40, 20)$ is optimal if the slope of the isoprofit line is steeper than the slope of the finishing-hour constraint. A typical isoprofit line is $c_1x_1 + 2x_2 = k$, so we know that the slope of a typical isoprofit line is $-\frac{c_1}{2}$. This implies that point A is optimal if $-\frac{c_1}{2} \geq -1$ (or $c_1 \leq 2$). We also find that point B is optimal if $-2 \leq -\frac{c_1}{2} \leq -1$ (or $2 \leq c_1 \leq 4$). Finally, point C is optimal if $-\frac{c_1}{2} \leq -2$ (or $c_1 \geq 4$). By substituting the optimal values of the decision variables into the objective function ($c_1x_1 + 2x_2$), we obtain the following information:

Value of c_1	Optimal z-value
$0 \leq c_1 \leq 2$	$c_1(0) + 2(80) = \$160$
$2 \leq c_1 \leq 4$	$c_1(20) + 2(60) = 120 + 20c_1$
$c_1 \geq 4$	$c_1(40) + 2(20) = 40 + 40c_1$

The relationship between c_1 and the optimal z-value is portrayed graphically in Figure 16. As seen in the figure, the graph of the optimal z-value as a function of c_1 is a piecewise linear function. The slope of each line segment in the graph is equal to the value of x_1 in the optimal solution. In a maximization problem, it can be shown (see Problem 5) that as the value of an objective function coefficient increases, the value of the variable in the LP's optimal solution cannot decrease. Thus, the slope of the graph of the optimal z-value as a function of an objective function coefficient will be nondecreasing.

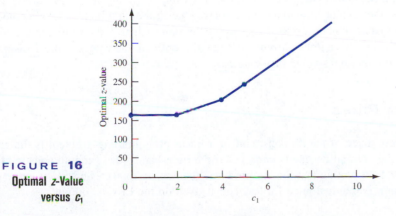

FIGURE 16

Optimal z-Value versus c_1

Similarly, in a minimizing problem, the graph of the optimal z-value as a function of a variable x_i's objective function coefficient c_i is a piecewise linear function. Again, the slope of each line segment is equal to the optimal value of x_i in the bfs corresponding to the line segment. It can be shown (see Problem 6) that the optimal x_i-value is a nonincreasing function of c_i. Thus, in a minimization problem, the graph of the optimal z-value as a function of c_i will be a piecewise linear function having a nonincreasing slope.

PROBLEMS

Group A

In what follows, b_i represents the right-hand side of an LP's ith constraint.

1 Use the LINDO **PARA** command to graph the optimal z-value for Example 1 as a function of b_4.

2 Use the **PARA** command to graph the optimal z-value for Example 2 as a function of b_1. Then answer the same questions for b_2, b_3, and b_4, respectively.

3 For the Giapetto example of Section 3.1, graph the optimal z-value as a function of x_2's objective function coefficient. Also graph the optimal z-value as a function of b_1, b_2, and b_3.

4 For the Dorian Auto example (Example 2 in Chapter 3), let c_1 be the objective function coefficient of x_1. Determine the optimal z-value as a function of c_1.

Group B

5 For Example 1, suppose that we increase the sales price of a product. Show that in the new optimal solution, the amount produced of that product cannot decrease.

6 For Example 2, suppose that we increase the cost of producing a type of car. Show that in the new optimal solution to the LP, the number of cars produced of that type cannot increase.

7 Consider the Sailco problem (Example 12 in Chapter 3). Suppose we want to consider how profit will be affected if we change the number of sailboats that can be produced each month with regular-time labor. How can we use the **PARA** command to answer this question? (*Hint:* Let c = change in number of sailboats that can be produced each month with regular-time labor. Change the right-hand side of some constraints to $40 + c$ and add another constraint to the problem.)

SUMMARY   Graphical Sensitivity Analysis

To determine whether the current basis remains optimal after changing an objective function coefficient, note that changing the objective function coefficient of a variable changes the slope of the isoprofit line. The current basis remains optimal as long as the current optimal solution is the last point in the feasible region to make contact with isoprofit lines as we move in the direction of increasing z (for a max problem). If the current basis remains optimal, the values of the decision variables remain unchanged, but the optimal z-value may change.

To determine if the current basis remains optimal after changing the right-hand side of a constraint, begin by finding the constraints (possibly including sign restrictions) that are binding for the current optimal solution. As we change the right-hand side of a constraint, the current basis remains optimal as long as the point where the constraints are binding remains feasible. Even if the current basis remains optimal, the values of the decision variables and the optimal z-value may change.

Shadow Prices

The **shadow price** of the ith constraint of a linear programming problem is the amount by which the optimal z-value is improved if the right-hand side of the ith constraint is increased by 1 (assuming that the current basis remains optimal). The shadow price of the ith constraint is the dual price for row $i + 1$ given on the LINDO output.

If the right-hand side of the ith constraint is increased by Δb_i, then (assuming the current basis remains optimal) the new optimal z-value for a maximization problem may be found as follows:

(New optimal z-value) = (old optimal z-value) + (Constraint i's shadow price) Δb_i **(1)**

For a minimization problem, the new optimal z-value may be found from

(New optimal z-value) = (old optimal z-value) − (Constraint i's shadow price)Δb_i **(2)**

Objective Function Coefficient Range

The OBJ COEFFICIENT RANGE portion of the LINDO output gives the range of values for an objective function coefficient for which the current basis remains optimal. Within this range, the values of the decision variables remain unchanged, but the optimal z-value may or may not change.

Reduced Cost

For any nonbasic variable, the reduced cost for the variable is the amount by which the nonbasic variable's objective function coefficient must be improved before that variable will become a basic variable in some optimal solution to the LP.

Right-Hand Side Range

If the right-hand side of a constraint remains within the RIGHTHAND SIDE RANGES value given on the LINDO printout, then the current basis remains optimal, and the dual price may be used to determine how a change in the right-hand side changes the optimal z-value. Even if the right-hand side of a constraint remains within the RIGHTHAND SIDE RANGES value on the LINDO output, then the values of the decision variables will probably change.

Signs of Shadow Prices

A $\geq$ constraint will have a nonpositive shadow price; a $\leq$ constraint will have a nonnegative shadow price; and an equality constraint may have a positive, negative, or zero shadow price.

Optimal z-Value as a Function of a Constraint's Right-Hand Side

In all cases, the optimal z-value will be a piecewise linear function of a constraint's right-hand side. The exact form of the function is as shown in Table 9.

Optimal z-Value as a Function of an Objective Function Coefficient

In a maximization problem, the optimal z-value will be a nondecreasing, piecewise linear function of an objective function coefficient. The slope will be a nondecreasing function of the objective function coefficient.

TABLE 9

Type of LP	Type of Constraint	Slopes of Each Piecewise Linear Segment Are
Maximization	≤	Non-negative and nonincreasing
Maximization	≥	Nonpositive and nonincreasing
Maximization	=	Unrestricted in sign and nonincreasing
Minimization	≤	Nonpositive and nondecreasing
Minimization	≥	Non-negative and nondecreasing
Minimization	=	Unrestricted in sign and nondecreasing

In a minimization problem, the optimal z-value will be a nondecreasing, piecewise linear function of an objective function coefficient. The slope will be a nonincreasing function of the objective function coefficient.

REVIEW PROBLEMS

Group A

1 HAL produces two types of computers: PCs and VAXes. The computers are produced in two locations: New York and Los Angeles. New York can produce up to 800 computers and Los Angeles up to 1,000 computers. HAL can sell up to 900 PCs and 900 VAXes. The profit associated with each production site and computer sale is as follows: New York—PC, $600; VAX, $800; Los Angeles—PC, $1,000; VAX, $1,300. The skilled labor required to build each computer at each location is as follows: New York—PC, 2 hours; VAX, 2 hours; Los Angeles—PC, 3 hours; VAX, 4 hours. A total of 4,000 hours of labor are available. Labor is purchased at a cost of $20 per hour. Let

$$XNP = \text{PCs produced in New York}$$
$$XLP = \text{PCs produced in Los Angeles}$$
$$XNV = \text{VAXes produced in New York}$$
$$XLV = \text{VAXes produced in Los Angeles}$$

Use the LINDO printout in Figure 17 to answer the following questions:

a If 3,000 hours of skilled labor were available, what would be HAL's profit?

b Suppose an outside contractor offers to increase the capacity of New York to 850 computers at a cost of $5,000. Should HAL hire the contractor?

c By how much would the profit for a VAX produced in Los Angeles have to increase before HAL would want to produce VAXes in Los Angeles?

d What is the most HAL should pay for an extra hour of labor?

2 Vivian's Gem Company produces two types of gems: Types 1 and 2. Each Type 1 gem contains 2 rubies and 4 diamonds. A Type 1 gem sells for $10 and costs $5 to produce. Each Type 2 gem contains 1 ruby and 1 diamond. A Type 2 gem sells for $6 and costs $4 to produce. A total of 30 rubies and 50 diamonds are available. All gems that are produced can be sold, but marketing considerations dictate that at least 11 Type 1 gems be produced. Let $x_1 =$ number of Type 1 gems produced and $x_2 =$ number of Type 2 gems produced. Assume that Vivian wants to maximize profit. Use the LINDO printout in Figure 18 to answer the following questions:

a What would Vivian's profit be if 46 diamonds were available?

b If Type 2 gems sold for only $5.50, what would be the new optimal solution to the problem?

c What would Vivian's profit be if at least 12 Type 1 gems had to be produced?

3 Wivco produces product 1 and product 2 by processing raw material. Up to 90 lb of raw material may be purchased at a cost of $10/lb. One pound of raw material can be used to produce either 1 lb of product 1 or 0.33 lb of product 2. Using a pound of raw material to produce a pound of product 1 requires 2 hours of labor or 3 hours to produce 0.33 lb of product 2. A total of 200 hours of labor are available, and at most 40 pounds of product 2 can be sold. Product 1 sells for $13/lb and product 2, $40/lb. Let

RM = pounds of raw material processed
P1 = pounds of raw material used to produce product 1
P2 = pounds of raw material used to produce product 2

To maximize profit, Wivco should solve the following LP:

$$\max z = 13P1 + 40(0.33)P2 - 10RM$$
$$\text{s.t.} \quad RM \geq P1 + P2$$
$$2P1 + 3P2 \leq 200$$
$$RM \leq 90$$
$$0.33P2 \leq 40$$
$$P1, P2, RM \geq 0$$

Use the LINDO output in Figure 19 to answer the following questions:

a If only 87 lb of raw material could be purchased, what would be Wivco's profits?

FIGURE **17**
LINDO Output for HAL

```
MAX       600  XNP + 1000 XLP + 800 XNV
                         + 1300 XLV - 20 L

SUBJECT TO
      2)        2 XNP + 3 XLP + 2 XNV
                         + 4 XLV - L <=    0
      3)     XNP  +  XNV   <=   800
      4)     XLP  +  XLV   <=  1000
      5)     XNP  +  XLP   <=   900
      6)     XNV  +  XLV   <=   900
      7)     L    <=   4000
END

LP OPTIMUM FOUND AT STEP          3

            OBJECTIVE FUNCTION VALUE

        1)   1360000.00

VARIABLE        VALUE        REDUCED COST
   XNP         .000000        200.000000
   XLP       800.000000          .000000
   XNV       800.000000          .000000
   XLV          .000000         33.333370
   L        4000.000000          .000000


   ROW     SLACK OR SURPLUS    DUAL PRICES
   2)          .000000        333.333300
   3)          .000000        133.333300
   4)        200.000000          .000000
   5)        100.000000          .000000
   6)        100.000000          .000000
   7)          .000000        313.333300

NO. ITERATIONS=          3

RANGES IN WHICH THE BASIS IS UNCHANGED:

             OBJ COEFFICIENT RANGES
VARIABLE   CURRENT     ALLOWABLE     ALLOWABLE
            COEF       INCREASE      DECREASE
   XNP    600.000000   200.000000    INFINITY
   XLP   1000.000000   200.000000   25.000030
   XNV    800.000000    INFINITY   133.333300
   XLV   1300.000000   33.333370     INFINITY
   L      -20.000000    INFINITY   313.333300

             RIGHTHAND SIDE RANGES
   ROW   CURRENT     ALLOWABLE     ALLOWABLE
          RHS        INCREASE      DECREASE
   2       .000000   300.000000  2400.000000
   3    800.000000   100.000000   150.000000
   4   1000.000000    INFINITY    200.000000
   5    900.000000    INFINITY    100.000000
   6    900.000000    INFINITY    100.000000
   7   4000.000000   300.000000  2400.000000
```

b If product 2 sold for $39.50/lb, what would be the new optimal solution to Wivco's problem?

c What is the most that Wivco should pay for another pound of raw material?

d What is the most that Wivco should pay for another hour of labor?

4 Zales Jewelers uses rubies and sapphires to produce two types of rings. A Type 1 ring requires 2 rubies, 3 sapphires, and 1 hour of jeweler's labor. A Type 2 ring requires 3 rubies, 2 sapphires, and 2 hours of jeweler's labor. Each Type 1 ring sells for $400; type 2 sells for $500. All rings

FIGURE **18**
LINDO Output for Vivian's Gem

```
MAX       5  X1 +  2   X2
SUBJECT TO
      2)      2  X1  +  X2 <=   30
      3)      4  X1  +  X2 <=   50
      4)         X1   >=     11
END

LP OPTIMUM FOUND AT STEP          2

            OBJECTIVE FUNCTION VALUE

        1)   67.0000000

VARIABLE        VALUE        REDUCED COST
   X1        11.000000          .000000
   X2         6.000000          .000000


   ROW     SLACK OR SURPLUS    DUAL PRICES
   2)         2.000000          0.000000
   3)          .000000          2.000000
   4)          .000000         -3.000000

NO. ITERATIONS=          2

RANGES IN WHICH THE BASIS IS UNCHANGED:

             OBJ COEFFICIENT RANGES
VARIABLE   CURRENT     ALLOWABLE     ALLOWABLE
            COEF       INCREASE      DECREASE
   X1     5.000000    3.000000      INFINITY
   X2     2.000000    INFINITY      .750000

             RIGHTHAND SIDE RANGES
   ROW   CURRENT     ALLOWABLE     ALLOWABLE
          RHS        INCREASE      DECREASE
   2    30.000000    INFINITY    2.000000
   3    50.000000    2.000000    6.000000
   4    11.000000    1.500000    1.000000
```

produced by Zales can be sold. At present, Zales has 100 rubies, 120 sapphires, and 70 hours of jeweler's labor. Extra rubies can be purchased at a cost of $100 per ruby. Market demand requires that the company produce at least 20 Type 1 rings and at least 25 Type 2. To maximize profit, Zales should solve the following LP:

$X1$ = Type 1 rings produced

$X2$ = Type 2 rings produced

R = number of rubies purchased

$$\max z = 400X1 + 500X2 - 100R$$

$$\text{s.t.} \quad 2X1 + 3X2 - R \leq 100$$
$$3X1 + 2X2 \leq 120$$
$$X1 + 2X2 \leq 70$$
$$X1 \geq 20$$
$$X2 \geq 25$$
$$X1, X2 \geq 0$$

Use the LINDO output in Figure 20 to answer the following questions:

a Suppose that instead of $100, each ruby costs $190. Would Zales still purchase rubies? What would be the new optimal solution to the problem?

b Suppose that Zales were only required to produce at least 23 Type 2 rings. What would Zales' profit now be?

FIGURE 19
LINDO Output for Wivco

```
MAX        13 P1 + 13.2 P2 - 10 RM
SUBJECT TO
      2) - P1 - P2 + RM >=    0
      3)   2 P1 + 3 P2  <=  200
      4)     RM    <=   90
      5)   0.33 P2  <=   40
END

          LP OPTIMUM FOUND AT STEP       3

                 OBJECTIVE FUNCTION VALUE

      1)            274.000000

VARIABLE         VALUE         REDUCED COST
    P1         70.000000         0.000000
    P2         20.000000         0.000000
    RM         90.000000         0.000000


      ROW    SLACK OR SURPLUS    DUAL PRICES
       2)         0.000000       -12.600000
       3)         0.000000         0.200000
       4)         0.000000         2.600000
       5)        33.400002         0.000000

NO. ITERATIONS=        3

RANGES IN WHICH THE BASIS IS UNCHANGED:

              OBJ COEFFICIENT RANGES
VARIABLE   CURRENT      ALLOWABLE      ALLOWABLE
            COEF        INCREASE       DECREASE
    P1  13.000000      0.200000       0.866667
    P2  13.200000      1.300000       0.200000
    RM -10.000000      INFINITY       2.600000

              RIGHTHAND SIDE RANGES
  ROW    CURRENT      ALLOWABLE      ALLOWABLE
          RHS         INCREASE       DECREASE
   2    0.000000     23.333334      10.000000
   3  200.000000     70.000000      20.000000
   4   90.000000     10.000000      23.333334
   5   40.000000     INFINITY       33.400002
```

c What is the most that Zales would be willing to pay for another hour of jeweler's labor?

d What is the most that Zales would be willing to pay for another sapphire?

5 Beerco manufactures ale and beer from corn, hops, and malt. Currently, 40 lb of corn, 30 lb of hops, and 40 lb of malt are available. A barrel of ale sells for $40 and requires 1 lb of corn, 1 lb of hops, and 2 lb of malt. A barrel of beer sells for $50 and requires 2 lb of corn, 1 lb of hops, and 1 lb of malt. Beerco can sell all ale and beer that is produced. Assume that Beerco's goal is to maximize total sales revenue and solve the following LP:

$$\max z = 40\text{ALE} + 50\text{BEER}$$

s.t.

$$\text{ALE} + 2\text{BEER} \leq 40 \quad \text{(Corn constraint)}$$
$$\text{ALE} + \text{BEER} \leq 30 \quad \text{(Hops constraint)}$$
$$2\text{ALE} + \text{BEER} \leq 40 \quad \text{(Malt constraint)}$$
$$\text{ALE, BEER} \geq 0$$

ALE = barrels of ale produced, and BEER = barrels of beer produced.

FIGURE 20
LINDO Output for Zales

```
MAX        400 X1 + 500 X2 - 100 R
SUBJECT TO
      2)    2 X1 + 3 X2 - R <=   100
      3)    3 X1 + 2 X2 <=  120
      4)      X1 + 2 X2 <=   70
      5)      X1    >=   20
      6)      X2    >=   25
END

          LP OPTIMUM FOUND AT STEP       2

                 OBJECTIVE FUNCTION VALUE

      1)   19000.0000

VARIABLE        VALUE         REDUCED COST
    X1        20.000000         0.000000
    X2        25.000000         0.000000
     R        15.000000         0.000000


   ROW    SLACK OR SURPLUS    DUAL PRICES
    2)         0.000000       100.000000
    3)        10.000000         0.000000
    4)         0.000000       200.000000
    5)         0.000000         0.000000
    6)         0.000000      -200.000000

NO. ITERATIONS=        2

RANGES IN WHICH THE BASIS IS UNCHANGED:

              OBJ COEFFICIENT RANGES
VARIABLE    CURRENT       ALLOWABLE      ALLOWABLE
             COEF         INCREASE       DECREASE
    X1   400.000000       INFINITY      100.000000
    X2   500.000000      200.000000      INFINITY
     R  -100.000000      100.000000     100.000000

              RIGHTHAND SIDE RANGES
  ROW    CURRENT        ALLOWABLE      ALLOWABLE
          RHS           INCREASE       DECREASE
   2   100.000000      15.000000       INFINITY
   3   120.000000      INFINITY        10.000000
   4    70.000000       3.333333        0.000000
   5    20.000000       0.000000        INFINITY
   6    25.000000       0.000000        2.500000
```

a Graphically find the range of values for the price of ale for which the current basis remains optimal.

b Graphically find the range of values for the price of beer for which the current basis remains optimal.

c Graphically find the range of values for the amount of available corn for which the current basis remains optimal. What is the shadow price of the corn constraint?

d Graphically find the range of values for the amount of available hops for which the current basis remains optimal. What is the shadow price of the hops constraint?

e Graphically find the range of values for the amount of available malt for which the current basis remains optimal. What is the shadow price of the malt constraint?

f Find the shadow price of each constraint if the constraints were expressed in ounces instead of pounds.

g Draw a graph of the optimal z-value as a function of the price of ale.

h Draw a graph of the optimal z-value as a function of the amount of available corn.

i Draw a graph of the optimal z-value as a function of the amount of available hops.

j Draw a graph of the optimal z-value as a function of the amount of available malt.

6 Gepbab Production Company uses labor and raw material to produce three products. The resource requirements and sales price for the three products are as shown in Table 10. Currently, 60 units of raw material are available. Up to 90 hours of labor can be purchased at $1 per hour. To maximize Gepbab profits, solve the following LP:

$$\max z = 6X1 + 8X2 + 13X3 - L$$
$$\text{s.t.} \quad 3X1 + 4X2 + 6X3 - L \leq 0$$
$$2X1 + 2X2 + 5X3 \quad\;\; \leq 60$$
$$L \leq 90$$
$$X1, X2, X3, L \geq 0$$

Here, X_i = units of product i produced, and L = number of labor hours purchased. Use the LINDO output in Figure 21 to answer the following questions:

a What is the most the company should pay for another unit of raw material?

b What is the most the company should pay for another hour of labor?

c What would product 1 have to sell for to make it desirable for the company to produce it?

d If 100 hours of labor could be purchased, what would the company's profit be?

e Find the new optimal solution if product 3 sold for $15.

7 Giapetto, Inc., sells wooden soldiers and wooden trains. The resources used to produce a soldier and train are shown in Table 11. A total of 145,000 board feet of lumber and 90,000 hours of labor are available. As many as 50,000 soldiers and 50,000 trains can be sold, with trains selling for $55 and soldiers for $32. In addition to producing trains and soldiers itself, Giapetto can buy (from an outside supplier) extra soldiers at $27 each and extra trains at $50 each. Let

$$SM = \text{thousands of soldiers manufactured}$$
$$SB = \text{thousands of soldiers bought at \$27}$$
$$TM = \text{thousands of trains manufactured}$$
$$TB = \text{thousands of trains bought at \$50}$$

Then Giapetto can maximize profit by solving the LP in the LINDO printout in Figure 22. Use this printout to answer the following questions. (*Hint:* Think about the units of the constraints and objective function.)

a If Giapetto could purchase trains for $48 per train, then what would be the new optimal solution to the LP? Explain.

TABLE 10

Resource	Product 1	Product 2	Product 3
Labor (hours)	3	4	6
Raw material (units)	2	2	5
Sales price ($)	6	8	13

FIGURE 21
LINDO Output for Gepbab

```
MAX        6 X1 + 8 X2 + 13 X3 - L
SUBJECT TO
      2)    3 X1 + 4 X2 + 6 X3 - L  <=    0
      3)    2 X1 + 2 X2 + 5 X3 <=   60
      4)    L     <=    90
END


LP OPTIMUM FOUND AT STEP              3

              OBJECTIVE FUNCTION VALUE

        1)    97.5000000

VARIABLE          VALUE          REDUCED COST
    X1          .000000            .250000
    X2        11.250000            .000000
    X3         7.500000            .000000
    L         90.000000            .000000

    ROW    SLACK OR SURPLUS      DUAL PRICES
     2)         .000000           1.750000
     3)         .000000            .500000
     4)         .000000            .750000

NO. ITERATIONS=          3

RANGES IN WHICH THE BASIS IS UNCHANGED:

                  OBJ COEFFICIENT RANGES
VARIABLE    CURRENT      ALLOWABLE      ALLOWABLE
             COEF        INCREASE       DECREASE
    X1    6.000000        .250000       INFINITY
    X2    8.000000        .666667        .666667
    X3   13.000000       3.000000       1.000000
     L   -1.000000       INFINITY        .750000

                 RIGHTHAND SIDE RANGES
  ROW    CURRENT      ALLOWABLE      ALLOWABLE
           RHS        INCREASE       DECREASE
    2     .000000     30.000000     18.000000
    3   60.000000     15.000000     15.000000
    4   90.000000     30.000000     18.000000
```

TABLE 11

	Soldier	Train
Lumber (board (ft)	3	5
Labor (hours)	2	4

b What is the most Giapetto would be willing to pay for another 100 board feet of lumber? For another 100 hours of labor?

c If 60,000 labor hours are available, what would Giapetto's profit be?

d If only 40,000 trains could be sold, what would Giapetto's profit be?

8 Wivco produces two products: 1 and 2. The relevant data are shown in Table 12. Each week, up to 400 units of raw material can be purchased at a cost of $1.50 per unit. The company employs four workers, who work 40 hours per week. (Their salaries are considered a fixed cost.) Workers are paid $6 per hour to work overtime. Each week, 320 hours of machine time are available.

FIGURE 22

LINDO Output for Giapetto

```
MAX        32 SM + 55 TM + 5 SB + 5  TB
SUBJECT TO
   2)     3 SM + 5 TM    <=    145
   3)     2 SM + 4 TM    <=    90
   4)     SM + SB   <=    50
   5)     TM + TB   <=    50
END

LP OPTIMUM FOUND AT STEP          4

             OBJECTIVE FUNCTION VALUE

          1)   1715.00000

VARIABLE        VALUE        REDUCED COST
   SM         45.000000        .000000
   TM           .000000       4.000000
   SB          5.000000        .000000
   TB         50.000000        .000000

ROW     SLACK OR SURPLUS      DUAL PRICES
  2)        10.000000           .000000
  3)          .000000         13.500000
  4)          .000000          5.000000
  5)          .000000          5.000000

NO. ITERATIONS=          4

RANGES IN WHICH THE BASIS IS UNCHANGED:

             OBJ COEFFICIENT RANGES
VARIABLE    CURRENT      ALLOWABLE     ALLOWABLE
             COEF        INCREASE      DECREASE
   SM     32.000000      INFINITY      2.000000
   TM     55.000000      4.000000      INFINITY
   SB      5.000000      2.000000      5.000000
   TB      5.000000      INFINITY      4.000000

             RIGHTHAND SIDE RANGES
ROW       CURRENT      ALLOWABLE     ALLOWABLE
           RHS         INCREASE      DECREASE
  2    145.000000      INFINITY      10.000000
  3     90.000000      6.666667      90.000000
  4     50.000000      INFINITY      5.000000
  5     50.000000      INFINITY      50.000000
```

TABLE 12

	Product 1	Product 2
Selling price ($)	15	8
Labor required (hours)	0.75	0.50
Machine time required (hours)	1.5	0.80
Raw material required (units)	2	1

In the absence of advertising, 50 units of product 1 and 60 units of product 2 will be demanded each week. Advertising can be used to stimulate demand for each product. Each dollar spent on advertising product 1 increases its demand by 10 units, and each dollar spent for product 2 increases its demand by 15 units. At most $100 can be spent on advertising. Define

$P1$ = number of units of product 1 produced each week

$P2$ = number of units of product 2 produced each week

OT = number of hours of overtime labor used each week

RM = number of units of raw materials purchased each week

$A1$ = dollars spent each week on advertising product 1

$A2$ = dollars spent each week on advertising product 2

Then Wivco should solve the following LP:

$$\max z = 15P1 + 8P2 - 6(OT) - 1.5RM - A1 - A2$$

$$\text{s.t.} \quad P1 - 10A1 \leq 50 \tag{1}$$
$$P2 - 15A2 \leq 60 \tag{2}$$
$$0.75P1 + 0.5P2 \leq 160 + (OT) \tag{3}$$
$$2P1 + P2 \leq RM \tag{4}$$
$$RM \leq 400 \tag{5}$$
$$A1 + A2 \leq 100 \tag{6}$$
$$1.5P1 + 0.8P2 \leq 320 \tag{7}$$

All variables non-negative

Use LINDO to solve this LP. Then use the computer output to answer the following questions:

a If overtime cost only $4 per hour, would Wivco use it?

b If each unit of product 1 sold for $15.50, would the current basis remain optimal? What would be the new optimal solution?

c What is the most that Wivco should be willing to pay for another unit of raw material?

d How much would Wivco be willing to pay for another hour of machine time?

e If each worker were required (as part of the regular workweek) to work 45 hours per week, what would the company's profits be?

f Explain why the shadow price of row (1) is 0.10. (*Hint:* If the right-hand side of (1) were increased from 50 to 51, then in the absence of advertising for product 1, 51 units of product 1 could now be sold each week.)

9 In this problem, we discuss how shadow prices can be interpreted for blending problems (see Section 3.8). To illustrate the ideas, we discuss Problem 2 of Section 3.8. If we define

$$x_{6J} = \text{pounds of grade 6 oranges in juice}$$
$$x_{9J} = \text{pounds of grade 9 oranges in juice}$$
$$x_{6B} = \text{pounds of grade 6 oranges in bags}$$
$$x_{9B} = \text{pounds of grade 9 oranges in bags}$$

then the appropriate formulation is

$$\max z = 0.45(x_{6J} + x_{9J}) + 0.30(x_{6B} + x_{9B})$$

$$\text{s.t.} \quad x_{6J} + x_{6B} \leq 120{,}000 \quad \text{(Grade 6 constraint)}$$

$$x_{9J} + x_{9B} \leq 100{,}000 \quad \text{(Grade 9 constraint)}$$

$$\frac{6x_{6J} + 9x_{9J}}{x_{6J} + x_{9J}} \geq 8 \quad \text{(Orange juice constraint)} \tag{1}$$

$$\frac{6x_{6B} + 9x_{9B}}{x_{6B} + x_{9B}} \geq 7 \quad \text{(Bags constraint)} \tag{2}$$

$$x_{6J}, x_{9J}, x_{6B}, x_{9B} \geq 0$$

Constraints (1) and (2) are examples of blending constraints because they specify the proportion of grade 6 and grade 9 oranges that must be blended to manufacture orange juice and bags of oranges. It would be useful to determine how a slight change in the standards for orange juice and bags of oranges would affect profit. At the end of this problem, we explain how to use the shadow prices of Constraints (1) and (2) to answer the following questions:

a Suppose that the average grade for orange juice is increased to 8.1. Assuming the current basis remains optimal, by how much would profits change?

b Suppose the average grade requirements for bags of oranges is decreased to 6.9. Assuming the current basis remains optimal, by how much would profits change?

The shadow price for both (1) and (2) is -0.15. The optimal solution to the O.J. problem is $x_{6J} = 26,666.67$, $x_{9J} = 53,333.33$, $x_{6B} = 93,333.33$, $x_{9B} = 46,666.67$. To interpret the shadow prices of blending constraints (1) and (2), *we assume that a slight change in the quality standard for a product will not significantly change the quantity of the product that is produced.*

Now note that (1) may be written as

$$6x_{6J} + 9x_{9J} \geq 8(x_{6J} + x_{9J}) \quad \text{or} \quad -2x_{6J} + x_{9J} \geq 0$$

If the quality standard for orange juice is changed to $8 + \Delta$, then (1) can be written as

$$6x_{6J} + 9x_{9J} \geq (8 + \Delta)(x_{6J} + x_{9J})$$

or

$$-2x_{6J} + x_{9J} \geq \Delta(x_{6J} + x_{9J})$$

Because we are assuming that changing orange juice quality from 8 to $8 + \Delta$ does not change the amount of orange juice produced, $x_{6J} + x_{9J}$ will remain equal to 80,000, and (1) will become

$$-2x_{6J} + x_{9J} \geq 80,000\Delta$$

Using the definition of shadow price, answer parts (a) and (b).

10 Use LINDO to solve the Sailco problem of Section 3.10, then use the output to answer the following questions:

a If month 1 demand decreased to 35 sailboats, what would be the total cost of satisfying the demands during the next four months?

b If the cost of producing a sailboat with regular-time labor during month 1 were $420, what would be the new optimal solution to the Sailco problem?

c Suppose a new customer is willing to pay $425 for a sailboat. If his demand must be met during month 1, should Sailco fill the order? How about if his demand must be met during month 4?

11 Autoco has three assembly plants located in various parts of the country. The first plant (built in 1937 and located in Norwood, Ohio) requires 2 hours of labor and 1 hour of machine time to assemble one automobile. The second plant (built in 1958 and located in Bakersfield, California) requires 1.5 hours of labor and 1.5 hours of machine time to assemble one automobile. The third plant (built in 1981 and located in Kingsport, Tennessee) requires 1.1 hours of labor and 2.5 hours of machine time to assemble one automobile.

The firm pays $30 per hour of labor and $10 per hour of machine time at each of its plants. The first plant has a capacity of 1,000 hours of machine time per day; the second, 900 hours; and the third, 2,000 hours. The manufacturer's production target is 1,800 automobiles per day.

The production department sets each plant's schedule by solving a linear programming problem designed to identify the cost-minimizing pattern of assembly across the three plants.

a Use LINDO to determine the cost-minimizing method of meeting Autoco's daily production target.

b The UWA local in Norwood, Ohio, has proposed wage concessions at that plant to raise employment. What is the *smallest* decrease in the wage rate at that plant that would increase employment there?

c What is the cost of assembling an extra automobile given the current output level of 1,800 automobiles? Would your answer be different if the production target were only 1,000 automobiles? Why or why not?

d A team of production specialists has indicated that the auto manufacturer can achieve efficiencies at its Bakersfield plant by reconfiguring the assembly line. The reconfiguration has the effect of increasing the productivity of the labor at this plant from 1.5 hours to 1 hour per automobile. By how much will the firm's costs fall as a result of this change, assuming that it continues to produce 1,800 automobiles?

e If 2,000 autos must be produced, by how much would costs increase?

f If labor costs $32 per hour in Bakersfield, California, what would be the new solution to the problem?

12 Machinco produces four products, requiring time on two machines and two types (skilled and unskilled) of labor. The amount of machine time and labor (in hours) used by each product and the sales prices are given in Table 13. Each month, 700 hours are available on machine 1 and 500 hours on machine 2. Each month, Machinco can purchase up to 600 hours of skilled labor at $8 per hour and up to 650 hours of unskilled labor at $6 per hour. Formulate an LP that will enable Machinco to maximize its monthly profit. Solve this LP and use the output to answer the following questions:

a By how much does the price of product 3 have to increase before it becomes optimal to produce it?

b If product 1 sold for $290, then what would be the new optimal solution to the problem?

c What is the most Machinco would be willing to pay for an extra hour of time on each machine?

d What is the most Machinco would be willing to pay for an extra hour of each type of labor?

e If up to 700 hours of skilled labor could be purchased each month, then what would be Machinco's monthly profits?

13 A company produces tools at two plants and sells them to three customers. The cost of producing 1,000 tools at a

TABLE 13

Product	Machine 1	Machine 2	Skilled	Unskilled	Sales ($)
1	11	4	8	7	300
2	7	6	5	8	260
3	6	5	4	7	220
4	5	4	6	4	180

plant and shipping them to a customer is given in Table 14. Customers 1 and 3 pay $200 per thousand tools; customer 2 pays $150 per thousand tools. To produce 1,000 tools at plant 1, 200 hours of labor are needed, while 300 hours are needed at plant 2. A total of 5,500 hours of labor are available for use at the two plants. Additional labor hours can be purchased at $20 per labor hour. Plant 1 can produce up to 10,000 tools and plant 2, up to 12,000 tools. Demand by each customer is assumed unlimited. If we let X_{ij} = tools (in thousands) produced at plant i and shipped to customer j then the company should solve the LP on the LINDO printout in Figure 23. Use this printout to answer the following questions:

a If it costs $70 to produce 1,000 tools at plant 1 and ship them to customer 1, what would be the new solution to the problem?

b If the price of an additional hour of labor were reduced to $4, would the company purchase any additional labor?

c A consultant offers to increase plant 1's production capacity by 5,000 tools for a cost of $400. Should the company take her offer?

d If the company were given 5 extra hours of labor, what would its profit become?

14 Solve Review Problem 24 of Chapter 3 on LINDO and answer the following questions:

a For which type of DRGs should the hospital seek to increase demand?

b What resources are in excess supply? Which resources should the hospital expand?

c What is the most the hospital should be willing to pay additional nurses?

15 Old Macdonald's 200-acre farm sells wheat, alfalfa, and beef. Wheat sells for $30 per bushel, alfalfa sells for $200 per bushel, and beef sells for $300 per ton. Up to 1,000 bushels of wheat and up to 1,000 bushels of alfalfa can be sold, but demand for beef is unlimited. If an acre of land is devoted to raising wheat, alfalfa, or beef, the yield and the required labor are given in Table 15. As many as 2,000 hours of labor can be purchased at $15 per hour. Each acre devoted to beef requires 5 bushels of alfalfa. The LINDO output in Figure 24 shows how to maximize profit, use it to answer the following questions. The variables are as follows:

$$W = \text{acres devoted to wheat}$$
$$AS = \text{bushels of alfalfa sold}$$
$$A = \text{acres devoted to alfalfa}$$
$$B = \text{acres devoted to beef}$$

TABLE 14

Plant	Customer ($)		
	1	2	3
1	60	30	160
2	130	70	170

FIGURE 23

LINDO Output for Problem 13

```
MAX    140 X11 + 120 X12 + 40 X13
               + 70 X21 + 80 X22 + 30 X23 - 20 L
SUBJECT TO
       2)  X11 + X12 + X13          <=       10
       3)  X21 + X22 + X23          <=       12
       4) 200 X11 + 200 X12 + 200 X13 + 300 X21
               + 300 X22 + 300 X23 - L <= 5500
END

LP OPTIMUM FOUND AT STEP               2

                OBJECTIVE FUNCTION VALUE

          1)   2333.3330

VARIABLE        VALUE         REDUCED COST
   X11       10.000000           .000000
   X12         .000000         20.000000
   X13         .000000        100.000000
   X21         .000000         10.000000
   X22       11.666670           .000000
   X23         .000000         50.000000
   L           .000000         19.733330


ROW      SLACK OR SURPLUS    DUAL PRICES
  2)          .000000         86.666660
  3)          .333333           .000000
  4)          .000000           .266667

NO. ITERATIONS=         2

RANGES IN WHICH THE BASIS IS UNCHANGED:

                OBJ COEFFICIENT RANGES
VARIABLE    CURRENT        ALLOWABLE      ALLOWABLE
             COEF          INCREASE       DECREASE
   X11   140.000000        INFINITY      20.000000
   X12   120.000000       20.000000       INFINITY
   X13    40.000000      100.000000       INFINITY
   X21    70.000000       10.000000       INFINITY
   X22    80.000000      130.000000      10.000000
   X23    30.000000       50.000000       INFINITY
   L     -20.000000       19.733330       INFINITY

                RIGHTHAND SIDE RANGES
ROW     CURRENT        ALLOWABLE      ALLOWABLE
         RHS           INCREASE       DECREASE
  2    10.000000       17.500000        .500000
  3    12.000000        INFINITY        .333333
  4  5500.000000      100.000000     3500.000000
```

TABLE 15

Crop	Yield/Acre	Labor/Acre (Hours)
Wheat	50 bushels	30
Alfalfa	100 bushels	20
Beef	10 tons	50

$$AB = \text{bushels of alfalfa devoted to beef}$$
$$L = \text{hours of labor purchased}$$

a How much must the price of a bushel of wheat increase before it becomes profitable to grow wheat?

b What is the most Old Macdonald should pay for another hour of labor?

FIGURE 24

LINDO Output for Old Macdonald

```
MAX      1500 W + 200 AS + 3000 B - 15  L
SUBJECT TO
     2)   50  W     <=    1000
     3)   AS    <=    1000
     4)   AS + AB - 100  A  =    0
     5)   -  5 B + AB =    0
     6)   W  +  B +  A <= 200
     7)   L  <= 2000
     8)   30 W + 50 B - L + 20  A  <=   0
END

LP OPTIMUM FOUND AT STEP            1

            OBJECTIVE FUNCTION VALUE

        1)   275882.300

VARIABLE        VALUE        REDUCED COST
     W         .000000        264.705800
    AS      1000.000000         .000000
     B        35.294120         .000000
     L      2000.000000         .000000
    AB       176.470600         .000000
     A        11.764710         .000000


   ROW    SLACK OR SURPLUS    DUAL PRICES
    2)     1000.000000          .000000
    3)        .000000         188.235300
    4)        .000000          11.764710
    5)        .000000         -11.764710
    6)      152.941200          .000000
    7)        .000000          43.823530
    8)        .000000          58.823530

NO. ITERATIONS=        1

RANGES IN WHICH THE BASIS IS UNCHANGED:

                  OBJ COEFFICIENT RANGES
VARIABLE   CURRENT      ALLOWABLE       ALLOWABLE
           COEF         INCREASE        DECREASE
     W  1500.000000   264.705800       INFINITY
    AS   200.000000     INFINITY      188.235300
     B  3000.000000 48000.000000      449.999800
     L   -15.000000     INFINITY       43.823530
    AB     .000000   9599.999000       89.999980
     A     .000000     INFINITY      8999.998000

                  RIGHTHAND SIDE RANGES
   ROW   CURRENT       ALLOWABLE       ALLOWABLE
          RHS          INCREASE        DECREASE
    2  1000.000000     INFINITY     1000.000000
    3  1000.000000  8999.999000     1000.000000
    4     .000000   1200.000000     8999.999000
    5     .000000   8999.999000      180.000000
    6   200.000000     INFINITY      152.941200
    7  2000.000000  7428.571000     1800.000000
    8     .000000   7428.571000     1800.000000
```

c What is the most Old Macdonald should pay for another bushel of alfalfa?

d What would be the new optimal solution if alfalfa sold for $20 for bushel?

16 Cornco produces two products: PS and QT. The sales price for each product and the maximum quantity of each that can be sold during each of the next three months are given in Table 16.

Each product must be processed through two assembly lines: 1 and 2. The number of hours required by each product on each assembly line are given in Table 17.

The number of hours available on each assembly line during each month are given in Table 18.

Each unit of PS requires 4 pounds of raw material; each unit of QT requires 3 pounds. As many as 710 units of raw material can be purchased at $3 per pound. At the beginning of month 1, 10 units of PS and 5 units of QT are available. It costs $10 to hold a unit of either product in inventory for a month. Solve this LP on LINDO and use your output to answer the following questions:

a Find the new optimal solution if it costs $11 to hold a unit of PS in inventory at the end of month 1.

b Find the company's new optimal solution if 210 hours on line 1 are available during month 1.

c Find the company's new profit level if 109 hours are available on line 2 during month 3.

d What is the most Cornco should be willing to pay for an extra hour of line 1 time during month 2?

e What is the most Cornco should be willing to pay for an extra pound of raw material?

f What is the most Cornco should be willing to pay for an extra hour of line 1 time during month 3?

g Find the new optimal solution if PS sells for $50 during month 2.

h Find the new optimal solution if QT sells for $50 during month 3.

i Suppose spending $20 on advertising would increase demand for QT in month 2 by 5 units. Should the advertising be done?

TABLE 16

Product	Month 1		Month 2		Month 3	
	Price ($)	Demand	Price ($)	Demand	Price ($)	Demand
PS	40	50	60	45	55	50
QT	35	43	40	50	44	40

TABLE 17

Product	Hours	
	Line 1	Line 2
PS	3	2
QT	2	2

TABLE 18

Line	Month		
	1	2	3
1	1,200	160	190
2	2,140	150	110

Sensitivity Analysis and Duality

Two of the most important topics in linear programming are sensitivity analysis and duality. After studying these important topics, the reader will have an appreciation of the beauty and logic of linear programming and be ready to study advanced linear programming topics such as those discussed in Chapter 10.

In Section 6.1, we illustrate the concept of sensitivity analysis through a graphical example. In Section 6.2, we use our knowledge of matrices to develop some important formulas, which are used in Sections 6.3 and 6.4 to develop the mechanics of sensitivity analysis. The remainder of the chapter presents the important concept of duality. Duality provides many insights into the nature of linear programming, gives us the useful concept of shadow prices, and helps us understand sensitivity analysis. It is a necessary basis for students planning to take advanced topics in linear and nonlinear programming.

6.1 A Graphical Introduction to Sensitivity Analysis

Sensitivity analysis is concerned with how changes in an LP's parameters affect the LP's optimal solution.

Reconsider the Giapetto problem of Section 3.1:

$$\max z = 3x_1 + 2x_2$$
$$\text{s.t.} \quad 2x_1 + x_2 \leq 100 \quad \text{(Finishing constraint)}$$
$$x_1 + x_2 \leq 80 \quad \text{(Carpentry constraint)}$$
$$x_1 \leq 40 \quad \text{(Demand constraint)}$$
$$x_1, x_2 \geq 0$$

where

$$x_1 = \text{number of soldiers produced per week}$$
$$x_2 = \text{number of trains produced per week}$$

The optimal solution to this problem is $z = 180$, $x_1 = 20$, $x_2 = 60$ (point B in Figure 1), and it has x_1, x_2, and s_3 (the slack variable for the demand constraint) as basic variables. How would changes in the problem's objective function coefficients or right-hand sides change this optimal solution?

Graphical Analysis of the Effect of a Change in an Objective Function Coefficient

If the contribution to profit of a soldier were to increase sufficiently, then it would be optimal for Giapetto to produce more soldiers (s_3 would become nonbasic). Similarly, if the

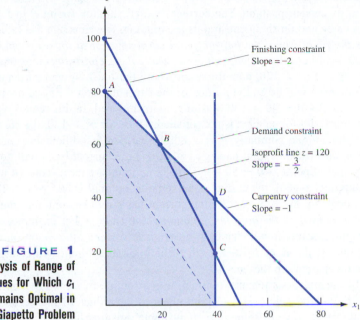

FIGURE 1

Analysis of Range of Values for Which c_1 Remains Optimal in Giapetto Problem

In figure: x_2, 100, 80, 60, 40, 20 ; A, B, D, C ; 20, 40, 60, 80, x_1

Finishing constraint Slope $= -2$

Demand constraint

Isoprofit line $z = 120$ Slope $= -\frac{3}{2}$

Carpentry constraint Slope $= -1$

contribution to profit of a soldier were to decrease sufficiently, it would be optimal for Giapetto to produce only trains (x_1 would now be nonbasic). We now show how to determine the values of the contribution to profit for soldiers for which the current optimal basis will remain optimal.

Let c_1 be the contribution to profit by each soldier. For what values of c_1 does the current basis remain optimal?

At present, $c_1 = 3$, and each isoprofit line has the form $3x_1 + 2x_2 =$ constant, or $x_2 = -\frac{3x}{2} + \frac{\text{constant}}{2}$, and each isoprofit line has a slope of $-\frac{3}{2}$. From Figure 1, we see that if a change in c_1 causes the isoprofit lines to be flatter than the carpentry constraint, then the optimal solution will change from the current optimal solution (point B) to a new optimal solution (point A). If the profit for each soldier is c_1, then the slope of each isoprofit line will be $-\frac{c_1}{2}$. Because the slope of the carpentry constraint is -1, the isoprofit lines will be flatter than the carpentry constraint if $-\frac{c_1}{2} > -1$, or $c_1 < 2$, and the current basis will no longer be optimal. The new optimal solution will be (0, 80), point A in Figure 1.

If the isoprofit lines are steeper than the finishing constraint, then the optimal solution will change from point B to point C. The slope of the finishing constraint is -2. If $-\frac{c_1}{2} < -2$, or $c_1 > 4$, then the current basis is no longer optimal, and point C (40, 20) will be optimal. In summary, we have shown that (if all other parameters remain unchanged) the current basis remains optimal for $2 \leq c_1 \leq 4$, and Giapetto should still manufacture 20 soldiers and 60 trains. Of course, even if $2 \leq c_1 \leq 4$, Giapetto's profit will change. For instance, if $c_1 = 4$, Giapetto's profit will now be $4(20) + 2(60) = \$200$ instead of $\$180$.

Graphical Analysis of the Effect of a Change in a Right-Hand Side on the LP's Optimal Solution

A graphical analysis can also be used to determine whether a change in the right-hand side of a constraint will make the current basis no longer optimal. Let b_1 be the number of available finishing hours. Currently, $b_1 = 100$. For what values of b_1 does the current basis remain optimal? From Figure 2, we see that a change in b_1 shifts the finishing constraint parallel to its current position. The current optimal solution (point B in Figure 2) is where the carpentry and finishing constraints are binding. If we change the value of b_1, then *as long as the point where the finishing and carpentry constraints are binding remains feasible, the optimal solution will still occur where these constraints intersect.* From Figure 2, we see that if $b_1 > 120$, then the point where the finishing and carpentry constraints are both binding will lie on the portion of the carpentry constraint below point D. Note that at point D, $2(40) + 40 = 120$ finishing hours are used. In this region, $x_1 > 40$, and the demand constraint for soldiers is not satisfied. Thus, for $b_1 > 120$, the current basis will no longer be optimal. Similarly, if $b_1 < 80$, the carpentry and finishing constraints will be binding at an infeasible point having $x_1 < 0$, and the current basis will no longer be optimal. Note that at point A, $0 + 80 = 80$ finishing hours are used. Thus (if all other parameters remain unchanged), the current basis remains optimal if $80 \leq b_1 \leq 120$.

Note that although for $80 \leq b_1 \leq 120$, the current basis remains optimal, *the values of the decision variables and the objective function value change.* For example, if $80 \leq b_1 \leq 100$, the optimal solution will change from point B to some other point on the line segment AB. Similarly, if $100 \leq b_1 \leq 120$, then the optimal solution will change from point B to some other point on the line BD.

As long as the current basis remains optimal, it is a routine matter to determine how a change in the right-hand side of a constraint changes the values of the decision variables. To illustrate the idea, let $b_1 =$ number of available finishing hours. If we change b_1 to $100 + \Delta$, we know that the current basis remains optimal for $-20 \leq \Delta \leq 20$. Note that as b_1 changes (as long as $-20 \leq \Delta \leq 20$), the optimal solution to the LP is still the point where the finishing-hour and carpentry-hour constraints are binding. Thus, if $b_1 = 100 + \Delta$, we can find the new values of the decision variables by solving

$$2x_1 + x_2 = 100 + \Delta \quad \text{and} \quad x_1 + x_2 = 80$$

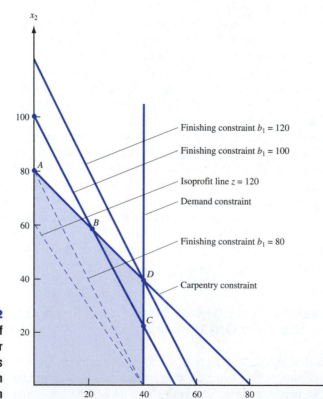

FIGURE 2
Range of Values of
Finishing Hours for
Which Current Basis
Remains Optimal in
Giapetto Problem

This yields $x_1 = 20 + \Delta$ and $x_2 = 60 - \Delta$. Thus, an increase in the number of available finishing hours results in an increase in the number of soldiers produced and a decrease in the number of trains produced.

If b_2 (the number of available carpentry hours) equals $80 + \Delta$, it can be shown (see Problem 2) that the current basis remains optimal for $-20 \leq \Delta \leq 20$. If we change the value of b_2 (keeping $-20 \leq \Delta \leq 20$), then the optimal solution to the LP is still the point where the finishing and carpentry constraints are binding. Thus, if $b_2 = 80 + \Delta$, the optimal solution to the LP is the solution to

$$2x_1 + x_2 = 100 \quad \text{and} \quad x_1 + x_2 = 80 + \Delta$$

This yields $x_1 = 20 - \Delta$ and $x_2 = 60 + 2\Delta$, which shows that an increase in the amount of available carpentry hours decreases the number of soldiers produced and increases the number of trains produced.

Suppose b_3, the demand for soldiers, is changed to $40 + \Delta$. Then it can be shown (see Problem 3) that the current basis remains optimal for $\Delta \geq -20$. For Δ in this range, the optimal solution to the LP will still occur where the finishing and carpentry constraints are binding. Thus, the optimal solution will be the solution to

$$2x_1 + x_2 = 100 \quad \text{and} \quad x_1 + x_2 = 80$$

Of course, this yields $x_1 = 20$ and $x_2 = 60$, which illustrates an important point. In a constraint with positive slack (or positive excess) in an LP's optimal solution, if we change the right-hand side of the constraint to a value in the range where the current basis remains optimal, the optimal solution to the LP is unchanged.

Shadow Prices

As we will see in Section 6.8, it is often important for managers to determine how a change in a constraint's right-hand side changes the LP's optimal z-value. With this in mind, we define the **shadow price** for the ith constraint of an LP to be the amount by which the optimal z-value is improved (improvement means increase in a max problem and decrease in a min problem) if the right-hand side of the ith constraint is increased by 1. This definition applies only if the change in the right-hand side of Constraint i leaves the current basis optimal.

For any two-variable LP, it is a simple matter to determine each constraint's shadow price. For example, we know that if $100 + \Delta$ finishing hours are available (assuming that the current basis remains optimal), then the LP's optimal solution is $x_1 = 20 + \Delta$ and $x_2 = 60 - \Delta$. Then the optimal z-value will equal $3x_1 + 2x_2 = 3(20 + \Delta) + 2(60 - \Delta) = 180 + \Delta$. Thus, as long as the current basis remains optimal, a unit increase in the number of available finishing hours will increase the optimal z-value by \$1. So the shadow price of the first (finishing hour) constraint is \$1.

For the second (carpentry hour) constraint, we know that if $80 + \Delta$ carpentry hours are available (and the current basis remains optimal), then the optimal solution to the LP is $x_1 = 20 - \Delta$ and $x_2 = 60 + 2\Delta$. Then the new optimal z-value is $3x_1 + 2x_2 = 3(20 - \Delta) + 2(60 + 2\Delta) = 180 + \Delta$. Thus, a unit increase in the number of carpentry hours will increase the optimal z-value by \$1 (as long as the current basis remains optimal). So the shadow price of the second (carpentry hour) constraint is \$1.

We now find the shadow price of the third (demand) constraint. If the right-hand side is $40 + \Delta$, then the optimal values of the decision variables remain unchanged, as long as the current basis remains optimal. Then the optimal z-value will also remain unchanged, which shows that the shadow price of the third (demand) constraint is \$0. It turns out that whenever the slack variable or excess variable for a constraint is positive in an LP's optimal solution, the constraint will have a zero shadow price.

Suppose we increase the right-hand side of the ith constraint of an LP by $\Delta b_i(\Delta b_i < 0$ means that we are decreasing the right-hand side) and the current basis remains optimal. Then each unit by which Constraint i's right-hand side is increased will increase the optimal z-value (for a max problem) by the shadow price. Thus, the new optimal z-value is given by

$$\text{(New optimal } z\text{-value)} = \text{(old optimal } z\text{-value)} + \text{(Constraint } i\text{'s shadow price) } \Delta b_i$$

For a minimization problem,

$$\text{(New optimal } z\text{-value)} = \text{(old optimal } z\text{-value)} - \text{(Constraint } i\text{'s shadow price) } \Delta b_i$$

For example, if 95 carpentry hours are available, then $\Delta b_2 = 15$, and the new z-value is given by

$$\text{New optimal } z\text{-value} = 180 + 15(1) = \$195$$

We will continue our discussion of shadow prices in Section 6.8.

Importance of Sensitivity Analysis

Sensitivity analysis is important for several reasons. In many applications, the values of an LP's parameters may change. For example, the prices at which soldiers and trains are sold may change, as may the availability of carpentry and finishing hours. If a parameter changes, sensitivity analysis often makes it unnecessary to solve the problem again. For example, if the profit contribution of a soldier increased to $3.50, we would not have to solve the Giapetto problem again because the current solution remains optimal. Of course, solving the Giapetto problem again would not be much work, but solving an LP with thousands of variables and constraints again would be a chore. A knowledge of sensitivity analysis often enables the analyst to determine from the original solution how changes in an LP's parameters change the optimal solution.

Recall that we may be uncertain about the values of parameters in an LP, for example, the weekly demand for soldiers. With the graphical method, it can be shown that if the weekly demand for soldiers is at least 20, then the optimal solution to the Giapetto problem is still (20, 60) (see Problem 3 at the end of this section). Thus, even if Giapetto is uncertain about the demand for soldiers, the company can still be fairly confident that it is optimal to produce 20 soldiers and 60 trains.

Of course, the graphical approach is not useful for sensitivity analysis on an LP with more than two variables. Before learning how to perform sensitivity analysis on an arbitrary LP, we need to use our knowledge of matrices to express simplex tableaus in matrix form. This is the subject of Section 6.2.

PROBLEMS

Group A

1 Show that if the contribution to profit for trains is between $1.50 and $3, the current basis remains optimal. If the contribution to profit for trains is $2.50, what would be the new optimal solution?

2 Show that if available carpentry hours remain between 60 and 100, the current basis remains optimal. If between 60 and 100 carpentry hours are available, then would Giapetto still produce 20 soldiers and 60 trains?

3 Show that if the weekly demand for soldiers is at least 20, the current basis remains optimal, and Giapetto should still produce 20 soldiers and 60 trains.

4 For the Dorian Auto problem (Example 2 in Chapter 3),
a Find the range of values of the cost of a comedy ad for which the current basis remains optimal.

b Find the range of values of the cost of a football ad for which the current basis remains optimal.

c Find the range of values of required HIW exposures for which the current basis remains optimal. Determine the new optimal solution if $28 + \Delta$ million HIW exposures are required.

d Find the range of values of required HIM exposures for which the current basis remains optimal. Determine the new optimal solution if $24 + \Delta$ million HIM exposures are required.

e Find the shadow price of each constraint.

f If 26 million HIW exposures are required, determine the new optimal z-value.

5 Radioco manufactures two types of radios. The only scarce resource needed to produce radios is labor. The company now has two laborers. Laborer 1 is willing to work as many as 40 hours per week and is paid \$5 per hour. Laborer 2 is willing to work up to 50 hours per week and is paid \$6 per hour. The price as well as the resources required to build each type of radio are given in Table 1.

Letting x_i be the number of type i radios produced each week, show that Radioco should solve the following LP:

$$\max z = 3x_1 + 2x_2$$
$$\text{s.t.} \quad x_1 + 2x_2 \leq 40$$
$$2x_1 + x_2 \leq 50$$
$$x_1, x_2 \geq 0$$

TABLE 1

	Radio 1		Radio 2	
Price (\$)	Resource Required		Price (\$)	Resource Required
25	Laborer 1: 1 hour		22	Laborer 1: 2 hours
	Laborer 2: 2 hours			Laborer 2: 2 hours
	Raw material cost: \$5			Raw material cost: \$4

a For what values of the price of a Type 1 radio would the current basis remain optimal?

b For what values of the price of a Type 2 radio would the current basis remain optimal?

c If laborer 1 were willing to work only 30 hours per week, would the current basis remain optimal? Find the new optimal solution to the LP.

d If laborer 2 were willing to work up to 60 hours per week, would the current basis remain optimal? Find the new optimal solution to the LP.

e Find the shadow price of each constraint.

6.2 Some Important Formulas

In this section, we use our knowledge of matrices to show how an LP's optimal tableau can be expressed in terms of the LP's parameters. The formulas developed in this section are used in our study of sensitivity analysis, duality, and advanced LP topics.

Assume that we are solving a max problem that has been prepared for solution by the Big M method and that at this point, the LP has m constraints and n variables. Although some of these variables may be slack, excess, or artificial, we choose to label them $x_1, x_2, \ldots, x_n$. Then the LP may be written as

$$\max z = c_1 x_1 + c_2 x_2 + \cdots + c_n x_n$$
$$\text{s.t.} \quad a_{11}x_1 + a_{12}x_2 + \cdots + a_{1n}x_n = b_1$$
$$a_{21}x_1 + a_{22}x_2 + \cdots + a_{2n}x_n = b_2$$
$$\vdots \qquad \vdots \qquad \qquad \vdots \qquad \vdots \tag{1}$$
$$a_{m1}x_1 + a_{m2}x_2 + \cdots + a_{mn}x_n = b_m$$
$$x_i \geq 0 \quad (i = 1, 2, \ldots, n)$$

Throughout this chapter, we use the Dakota Furniture problem of Section 4.5 (without the $x_2 \leq 5$ constraint) as an example. For the Dakota problem, the analog of LP (1) is

$$\max z = 60x_1 + 30x_2 + 20x_3 + 0s_1 + 0s_2 + 0s_3$$
$$\text{s.t.} \quad 8x_1 + 6x_2 + x_3 + s_1 = 48$$
$$4x_1 + 2x_2 + 1.5x_3 + s_2 = 20 \tag{1'}$$
$$2x_1 + 1.5x_2 + 0.5x_3 + s_3 = 8$$
$$x_1, x_2, x_3, s_1, s_2, s_3 \geq 0$$

Suppose we have found the optimal solution to (1). Let BV_i be the basic variable for row i of the optimal tableau. Also define $BV = \{BV_1, BV_2, \ldots, BV_m\}$ to be the set of basic variables in the optimal tableau, and define the $m \times 1$ vector

$$\mathbf{x}_{BV} = \begin{bmatrix} x_{BV_1} \\ x_{BV_2} \\ \vdots \\ x_{BV_m} \end{bmatrix}$$

We also define

$NBV =$ the set of nonbasic variables in the optimal tableau

$\mathbf{x}_{NBV} = (n - m) \times 1$ vector listing the nonbasic variables (in any desired order)

To illustrate these definitions, we recall that the optimal tableau for the Dakota problem is

$$
\begin{aligned}
z &+ 5x_2 + &&+ 10s_2 + 10s_3 &= 280 \\
&- 2x_2 + &&+ s_1 + 2s_2 - 8s_3 &= 24 \\
&- 2x_2 + x_3 &&+ 2s_2 - 4s_3 &= 8 \\
x_1 &+ 1.25x_2 &&- 0.5s_2 + 1.5s_3 &= 2
\end{aligned}
$$

(2)

For this optimal tableau, $BV_1 = s_1$, $BV_2 = x_3$, and $BV_3 = x_1$. Then

$$\mathbf{x}_{BV} = \begin{bmatrix} s_1 \\ x_3 \\ x_1 \end{bmatrix}$$

We may choose $NBV = \{x_2, s_2, s_3\}$. Then

$$\mathbf{x}_{NBV} = \begin{bmatrix} x_2 \\ s_2 \\ s_3 \end{bmatrix}$$

Using our knowledge of matrix algebra, we can express the optimal tableau in terms of BV and the original LP (1). Recall that $c_1, c_2, \ldots, c_n$ are the objective function coefficients for the variables $x_1, x_2, \ldots, x_n$ (some of these may be slack, excess, or artificial variables).

DEFINITION ■ $\mathbf{c}_{BV}$ is the $1 \times m$ row vector $[c_{BV_1} \quad c_{BV_2} \quad \cdots \quad c_{BV_m}]$. ■

Thus, the elements of $\mathbf{c}_{BV}$ are the objective function coefficients for the optimal tableau's basic variables. For the Dakota problem, $BV = \{s_1, x_3, x_1\}$. Then from (1′) we find that $\mathbf{c}_{BV} = [0 \quad 20 \quad 60]$.

DEFINITION ■ $\mathbf{c}_{NBV}$ is the $1 \times (n - m)$ row vector whose elements are the coefficients of the nonbasic variables (in the order of NBV). ■

If we choose to list the nonbasic variables for the Dakota problem in the order $NBV = \{x_2, s_2, s_3\}$, then $\mathbf{c}_{NBV} = [30 \quad 0 \quad 0]$.

DEFINITION ■ The $m \times m$ matrix B is the matrix whose jth column is the column for BV_j in (1). ■

For the Dakota problem, the first column of B is the s_1 column in (1′); the second, the x_3 column; and the third, the x_1 column. Thus,

$$B = \begin{bmatrix} 1 & 1 & 8 \\ 0 & 1.5 & 4 \\ 0 & 0.5 & 2 \end{bmatrix}$$

DEFINITION ■ $\mathbf{a}_j$ is the column (in the constraints) for the variable x_j in (1). ■

For example, in the Dakota problem,

$$\mathbf{a}_2 = \begin{bmatrix} 6 \\ 2 \\ 1.5 \end{bmatrix} \quad \text{and} \quad \mathbf{a} \text{ (for } s_1) = \begin{bmatrix} 1 \\ 0 \\ 0 \end{bmatrix}$$

DEFINITION ■ N is the $m \times (n - m)$ matrix whose columns are the columns for the nonbasic variables (in the NBV order) in (1). ■

If for the Dakota problem, we write NBV = $\{x_2, s_2, s_3\}$, then

$$N = \begin{bmatrix} 6 & 0 & 0 \\ 2 & 1 & 0 \\ 1.5 & 0 & 1 \end{bmatrix}$$

DEFINITION ■ The $m \times 1$ column vector $\mathbf{b}$ is the right-hand side of the constraints in (1). ■

For the Dakota problem,

$$\mathbf{b} = \begin{bmatrix} 48 \\ 20 \\ 8 \end{bmatrix}$$

We write b_i for the right-hand side of the ith constraint in the original Dakota problem: $b_2 = 20$.

We can now use matrix algebra to determine how an LP's optimal tableau (with set of basic variables BV) is related to the original LP in the form (1).

Expressing the Constraints in Any Tableau in Terms of B^{-1} and the Original LP

We begin by observing that (1) may be written as

$$\begin{aligned} z &= \mathbf{c}_{BV}\mathbf{x}_{BV} + \mathbf{c}_{NBV}\mathbf{x}_{NBV} \\ \text{s.t.} \quad & B\mathbf{x}_{BV} + N\mathbf{x}_{NBV} = \mathbf{b} \\ & \mathbf{x}_{BV}, \mathbf{x}_{NBV} \geq 0 \end{aligned} \tag{3}$$

Using the format of (3), the Dakota problem can be written as

$$\max z = [0 \quad 20 \quad 60] \begin{bmatrix} s_1 \\ x_3 \\ x_1 \end{bmatrix} + [30 \quad 0 \quad 0] \begin{bmatrix} x_2 \\ s_2 \\ s_3 \end{bmatrix}$$

$$\text{s.t.} \quad \begin{bmatrix} 1 & 1 & 8 \\ 0 & 1.5 & 4 \\ 0 & 0.5 & 2 \end{bmatrix} \begin{bmatrix} s_1 \\ x_3 \\ x_1 \end{bmatrix} + \begin{bmatrix} 6 & 0 & 0 \\ 2 & 1 & 0 \\ 1.5 & 0 & 1 \end{bmatrix} \begin{bmatrix} x_2 \\ s_2 \\ s_3 \end{bmatrix} = \begin{bmatrix} 48 \\ 20 \\ 8 \end{bmatrix}$$

$$
\begin{bmatrix} s_1 \\ x_3 \\ x_1 \end{bmatrix} \geq \begin{bmatrix} 0 \\ 0 \\ 0 \end{bmatrix}, \quad \begin{bmatrix} x_2 \\ s_2 \\ s_3 \end{bmatrix} \geq \begin{bmatrix} 0 \\ 0 \\ 0 \end{bmatrix}
$$

Multiplying the constraints in (3) through by B^{-1}, we obtain

$$
B^{-1}B\mathbf{x}_{\mathrm{BV}} + B^{-1}N\mathbf{x}_{\mathrm{NBV}} = B^{-1}\mathbf{b} \quad \text{or} \quad \mathbf{x}_{\mathrm{BV}} + B^{-1}N\mathbf{x}_{\mathrm{NBV}} = B^{-1}\mathbf{b} \tag{4}
$$

In (4), BV_i occurs with a coefficient of 1 in the ith constraint and a zero coefficient in each other constraint. Thus, BV is the set of the basic variables for (4), and (4) yields the constraints for the optimal tableau.

For the Dakota problem, the Gauss–Jordan method can be used to show that

$$
B^{-1} = \begin{bmatrix} 1 & 2 & -8 \\ 0 & 2 & -4 \\ 0 & -0.5 & 1.5 \end{bmatrix}
$$

Then (4) yields

$$
\begin{bmatrix} s_1 \\ x_3 \\ x_1 \end{bmatrix} + \begin{bmatrix} 1 & 2 & -8 \\ 0 & 2 & -4 \\ 0 & -0.5 & 1.5 \end{bmatrix} \begin{bmatrix} 6 & 0 & 0 \\ 2 & 1 & 0 \\ 1.5 & 0 & 1 \end{bmatrix} \begin{bmatrix} x_2 \\ s_2 \\ s_3 \end{bmatrix} = \begin{bmatrix} 1 & 2 & -8 \\ 0 & 2 & -4 \\ 0 & -0.5 & 1.5 \end{bmatrix} \begin{bmatrix} 48 \\ 20 \\ 8 \end{bmatrix}
$$

or

$$
\begin{bmatrix} s_1 \\ x_3 \\ x_1 \end{bmatrix} + \begin{bmatrix} -2 & 2 & -8 \\ -2 & 2 & -4 \\ 1.25 & -0.5 & 1.5 \end{bmatrix} \begin{bmatrix} x_2 \\ s_2 \\ s_3 \end{bmatrix} = \begin{bmatrix} 24 \\ 8 \\ 2 \end{bmatrix} \tag{4$'$}
$$

Of course, these are the constraints for the Dakota optimal tableau, (2).

From (4), we see that the column of a nonbasic variable x_j in the constraints of the optimal tableau is given by B^{-1} [column for x_j in (1)] $= B^{-1}\mathbf{a}_j$. For example, the x_2 column is B^{-1} (first column of N) $= B^{-1}\mathbf{a}_2$. From (4), we also find that the right-hand side of the constraints is the vector $B^{-1}\mathbf{b}$. The following two equations summarize the preceding discussion:

$$
\text{Column for } x_j \text{ in optimal tableau's constraints} = B^{-1}\mathbf{a}_j \tag{5}
$$

$$
\text{Right-hand side of optimal tableau's constraints} = B^{-1}\mathbf{b} \tag{6}
$$

To illustrate (5), we find:

$$
\begin{array}{l} \text{Column for } x_2 \\ \text{in Dakota optimal tableau} \end{array} = B^{-1}\mathbf{a}_2
$$

$$
= \begin{bmatrix} 1 & 2 & -8 \\ 0 & 2 & -4 \\ 0 & -0.5 & 1.5 \end{bmatrix} \begin{bmatrix} 6 \\ 2 \\ 1.5 \end{bmatrix} = \begin{bmatrix} -2 \\ -2 \\ 1.25 \end{bmatrix}
$$

To illustrate (6), we compute:

$$
\begin{array}{l} \text{Right-hand side of constraints} \\ \text{in Dakota optimal tableau} \end{array} = B^{-1}\mathbf{b}
$$

$$
= \begin{bmatrix} 1 & 2 & -8 \\ 0 & 2 & -4 \\ 0 & -0.5 & 1.5 \end{bmatrix} \begin{bmatrix} 48 \\ 20 \\ 8 \end{bmatrix} = \begin{bmatrix} 24 \\ 8 \\ 2 \end{bmatrix}
$$

Determining the Optimal Tableau's Row 0 in Terms of the Initial LP

We now show how to express row 0 of the optimal tableau in terms of BV and the original LP (1). To begin, we multiply the constraints (expressed in the form $B\mathbf{x}_{BV} + N\mathbf{x}_{NBV} = \mathbf{b}$) through by the vector $\mathbf{c}_{BV}B^{-1}$:

$$\mathbf{c}_{BV}\mathbf{x}_{BV} + \mathbf{c}_{BV}B^{-1}N\mathbf{x}_{NBV} = \mathbf{c}_{BV}B^{-1}\mathbf{b} \tag{7}$$

and rewrite the original objective function, $z = \mathbf{c}_{BV}\mathbf{x}_{BV} + \mathbf{c}_{NBV}\mathbf{x}_{NBV}$, as

$$z - \mathbf{c}_{BV}\mathbf{x}_{BV} - \mathbf{c}_{NBV}\mathbf{x}_{NBV} = 0 \tag{8}$$

By adding (7) to (8), we can eliminate the optimal tableau's basic variables and obtain its row 0:

$$z + (\mathbf{c}_{BV}B^{-1}N - \mathbf{c}_{NBV})\mathbf{x}_{NBV} = \mathbf{c}_{BV}B^{-1}\mathbf{b} \tag{9}$$

From (9), the coefficient of x_j in row 0 is

$$\mathbf{c}_{BV}B^{-1} \text{ (column of } N \text{ for } x_j) - (\text{coefficient for } x_j \text{ in } \mathbf{c}_{NBV}) = \mathbf{c}_{BV}B^{-1}\mathbf{a}_j - c_j$$

and the right-hand side of row 0 is $\mathbf{c}_{BV}B^{-1}\mathbf{b}$.

To help summarize the preceding discussion, we let $\bar{c}_j$ be the coefficient of x_j in the optimal tableau's row 0. Then we have shown that

$$\bar{c}_j = \mathbf{c}_{BV}B^{-1}\mathbf{a}_j - c_j \tag{10}$$

and

$$\text{Right-hand side of optimal tableau's row 0} = \mathbf{c}_{BV}B^{-1}\mathbf{b} \tag{11}$$

To illustrate the use of (10) and (11), we determine row 0 of the Dakota problem's optimal tableau. Recall that

$$\mathbf{c}_{BV} = [0 \quad 20 \quad 60] \quad \text{and} \quad B^{-1} = \begin{bmatrix} 1 & 2 & -8 \\ 0 & 2 & -4 \\ 0 & -0.5 & 1.5 \end{bmatrix}$$

Then $\mathbf{c}_{BV}B^{-1} = [0 \quad 10 \quad 10]$, and from (10) we find that the coefficients of the nonbasic variables in row 0 of the optimal tableau are

$$\bar{c}_2 = \mathbf{c}_{BV}B^{-1}\mathbf{a}_2 - c_2 = [0 \quad 10 \quad 10]\begin{bmatrix} 6 \\ 2 \\ 1.5 \end{bmatrix} - 30 = 20 + 15 - 30 = 5$$

and

$$\text{Coefficient of } s_2 \text{ in optimal row 0} = \mathbf{c}_{BV}B^{-1}\begin{bmatrix} 0 \\ 1 \\ 0 \end{bmatrix} - 0 = 10$$

$$\text{Coefficient of } s_3 \text{ in optimal row 0} = \mathbf{c}_{BV}B^{-1}\begin{bmatrix} 0 \\ 0 \\ 1 \end{bmatrix} - 0 = 10$$

Of course, the optimal tableau's basic variables (x_1, x_3, and s_1) will have zero coefficients in row 0.

From (11), the right-hand side of row 0 is

$$\mathbf{c}_{\text{BV}}B^{-1}\mathbf{b} = \begin{bmatrix} 0 & 10 & 10 \end{bmatrix}\begin{bmatrix} 48 \\ 20 \\ 8 \end{bmatrix} = 280$$

Putting it all together, we see that row 0 is

$$z + 5x_2 + 10s_2 + 10s_3 = 280$$

Of course, this result agrees with (2).

Simplifying Formula (10) for Slack, Excess, and Artificial Variables

Formula (10) can be greatly simplified if x_j is a slack, excess, or artificial variable. For example, if x_j is the slack variable s_i, the coefficient of s_i in the objective function is 0, and the column for s_i in the original tableau has 1 in row i and 0 in all other rows. Then (10) yields

$$\text{Coefficient of } s_i \text{ in optimal row 0} = i\text{th element of } \mathbf{c}_{\text{BV}}B^{-1} - 0 \qquad \text{(10′)}$$
$$= i\text{th element of } \mathbf{c}_{\text{BV}}B^{-1}$$

Similarly, if x_j is the excess variable e_i, then the coefficient of e_i in the objective function is 0 and the column for e_i in the original tableau has -1 in row i and 0 in all other rows. Then (10) reduces to

$$\text{Coefficient of } e_i \text{ in optimal row 0} = -(i\text{th element of } \mathbf{c}_{\text{BV}}B^{-1}) - 0 \qquad \text{(10″)}$$
$$= -(i\text{th element of } \mathbf{c}_{\text{BV}}B^{-1})$$

Finally, if x_j is an artificial variable a_i, then the objective function coefficient of a_i (for a max problem) is $-M$ and the original column for a_i has 1 in row i and 0 in all other rows. Then (10) reduces to

$$\text{Coefficient of } a_i \text{ in optimal row 0} = (i\text{th element of } \mathbf{c}_{\text{BV}}B^{-1}) - (-M) \qquad \text{(10‴)}$$
$$= (i\text{th element of } \mathbf{c}_{\text{BV}}B^{-1}) + (M)$$

The derivations of this section have not been easy. Fortunately, use of (5), (6), (10), and (11) does not require a complete understanding of the derivations. A summary of the formulas derived in this section for computing an optimal tableau from the initial LP follows.

Summary of Formulas for Computing the Optimal Tableau from the Initial LP

$$x_j \text{ column in optimal tableau's constraints} = B^{-1}\mathbf{a}_j \qquad \text{(5)}$$

$$\text{Right-hand side of optimal tableau's constraints} = B^{-1}\mathbf{b} \qquad \text{(6)}$$

$$\bar{c}_j = \mathbf{c}_{\text{BV}}B^{-1}\mathbf{a}_j - c_j \qquad \text{(10)}$$

Coefficient of slack variable s_i in optimal row 0
$$= i\text{th element of } \mathbf{c}_{\text{BV}}B^{-1} \qquad \text{(10′)}$$

Coefficient of excess variable e_i in optimal row 0
$$= -(i\text{th element of } \mathbf{c}_{\text{BV}}B^{-1}) \qquad \text{(10″)}$$

Coefficient of artificial variable a_i in optimal row 0
$$= (i\text{th element of } \mathbf{c}_{\text{BV}}B^{-1}) + M \quad \text{(max problem)} \qquad \text{(10‴)}$$

$$\text{Right-hand side of optimal row 0} = \mathbf{c}_{\text{BV}}B^{-1}\mathbf{b} \qquad \text{(11)}$$

We must first find B^{-1} because it is necessary in order to compute all parts of the optimal tableau. Similarly, we must find $c_{BV}B^{-1}$ to compute the optimal tableau's row 0.

The following example is another illustration of the use of the preceding formulas.

EXAMPLE 1   **Computing Optimal Tableau**

For the following LP, the optimal basis is BV = $\{x_2, s_2\}$. Compute the optimal tableau.

$$\max z = x_1 + 4x_2$$
$$\text{s.t.} \quad x_1 + 2x_2 \leq 6$$
$$2x_1 + x_2 \leq 8$$
$$x_1, x_2 \geq 0$$

Solution   After adding slack variables s_1 and s_2, we obtain the analog of (1):

$$\max z = x_1 + 4x_2$$
$$\text{s.t.} \quad x_1 + 2x_2 + s_1 \qquad = 6$$
$$2x_1 + x_2 \qquad + s_2 = 8$$

First we compute B^{-1}. Because

$$B = \begin{bmatrix} 2 & 0 \\ 1 & 1 \end{bmatrix}$$

we find B^{-1} by applying the Gauss–Jordan method to the following matrix:

$$B|I_2 = \begin{bmatrix} 2 & 0 & 1 & 0 \\ 1 & 1 & 0 & 1 \end{bmatrix}$$

The reader should verify that

$$B^{-1} = \begin{bmatrix} \frac{1}{2} & 0 \\ -\frac{1}{2} & 1 \end{bmatrix}$$

Use (5) and (6) to determine the optimal tableau's constraints. Because

$$\mathbf{a}_1 = \begin{bmatrix} 1 \\ 2 \end{bmatrix}$$

the column for x_1 in the optimal tableau is

$$B^{-1}\mathbf{a}_1 = \begin{bmatrix} \frac{1}{2} & 0 \\ -\frac{1}{2} & 1 \end{bmatrix} \begin{bmatrix} 1 \\ 2 \end{bmatrix} = \begin{bmatrix} \frac{1}{2} \\ \frac{3}{2} \end{bmatrix}$$

The other nonbasic variable is s_1. The column for s_1 in the original problem is

$$\begin{bmatrix} 1 \\ 0 \end{bmatrix}$$

so (5) yields

$$\text{Column for } s_1 \text{ in optimal tableau} = \begin{bmatrix} \frac{1}{2} & 0 \\ -\frac{1}{2} & 1 \end{bmatrix} \begin{bmatrix} 1 \\ 0 \end{bmatrix} = \begin{bmatrix} \frac{1}{2} \\ -\frac{1}{2} \end{bmatrix}$$

Because

$$\mathbf{b} = \begin{bmatrix} 6 \\ 8 \end{bmatrix}$$

(6) yields

$$\text{Right-hand side of optimal tableau} = \begin{bmatrix} \frac{1}{2} & 0 \\ -\frac{1}{2} & 1 \end{bmatrix} \begin{bmatrix} 6 \\ 8 \end{bmatrix} = \begin{bmatrix} 3 \\ 5 \end{bmatrix}$$

Because BV is listed as $\{x_2, s_2\}$, x_2 is the basic variable for row 1, and s_2 is the basic variable for row 2. Thus, the constraints of the optimal tableau are

$$\begin{aligned} \tfrac{1}{2}x_1 + x_2 + \tfrac{1}{2}s_1 & = 3 \\ \tfrac{3}{2}x_1 \quad\quad - \tfrac{1}{2}s_1 + s_2 & = 5 \end{aligned}$$

Because $\mathbf{c}_{BV} = [4 \quad 0]$,

$$\mathbf{c}_{BV}B^{-1} = [4 \quad 0] \begin{bmatrix} \frac{1}{2} & 0 \\ -\frac{1}{2} & 1 \end{bmatrix} = [2 \quad 0]$$

Then (10) yields

$$\text{Coefficient of } x_1 \text{ in row 0 of optimal tableau} = \mathbf{c}_{BV}B^{-1}\mathbf{a}_1 - c_1$$

$$= [2 \quad 0] \begin{bmatrix} 1 \\ 2 \end{bmatrix} - 1 = 1$$

From (10′)

$$\text{Coefficient of } s_1 \text{ in optimal tableau} = \text{First element of } \mathbf{c}_{BV}B^{-1} = 2$$

Because

$$\mathbf{b} = \begin{bmatrix} 6 \\ 8 \end{bmatrix}$$

(11) shows that the right-hand side of the optimal tableau's row 0 is

$$\mathbf{c}_{BV}B^{-1}\mathbf{b} = [2 \quad 0] \begin{bmatrix} 6 \\ 8 \end{bmatrix} = 12$$

Of course, the basic variables x_2 and s_2 will have zero coefficients in row 0. Thus, the optimal tableau's row 0 is $z + x_1 + 2s_1 = 12$, and the complete optimal tableau is

$$\begin{aligned} z + \quad x_1 \quad\quad + 2s_1 \quad\quad & = 12 \\ \tfrac{1}{2}x_1 + x_2 + \tfrac{1}{2}s_1 \quad\quad & = 3 \\ \tfrac{3}{2}x_1 \quad\quad - \tfrac{1}{2}s_1 + s_2 & = 5 \end{aligned}$$

We have used the formulas of this section to create an LP's optimal tableau, but they can also be used to create the tableau for *any* set of basic variables. This observation will be important when we study the revised simplex method in Section 10.1.

PROBLEMS

Group A

1 For the following LP, x_1 and x_2 are basic variables in the optimal tableau. Use the formulas of this section to determine the optimal tableau.

$$\max z = 3x_1 + x_2$$
$$\text{s.t.} \quad 2x_1 - x_2 \le 2$$
$$-x_1 + x_2 \le 4$$
$$x_1, x_2 \ge 0$$

2 For the following LP, x_2 and s_1 are basic variables in the optimal tableau. Use the formulas of this section to determine the optimal tableau.

$$\max z = -x_1 + x_2$$
$$\text{s.t.} \quad 2x_1 + x_2 \leq 4$$
$$x_1 + x_2 \leq 2$$
$$x_1, x_2 \geq 0$$

6.3 Sensitivity Analysis

We now explore how changes in an LP's parameters (objective function coefficients, right-hand sides, and technological coefficients) change the optimal solution. As described in Section 6.1, the study of how an LP's optimal solution depends on its parameters is called *sensitivity analysis*. Our discussion focuses on maximization problems and relies heavily on the formulas of Section 6.2. (The modifications for min problems are straightforward; see Problem 8 at the end of this section.)

As in Section 6.2, we let BV be the set of basic variables in the optimal tableau. Given a change (or changes) in an LP, we want to determine whether BV remains optimal. The mechanics of sensitivity analysis hinge on the following important observation. *From Chapter 4, we know that a simplex tableau (for a max problem) for a set of basic variables BV is optimal if and only if each constraint has a non-negative right-hand side and each variable has a non-negative coefficient in row 0.* This follows, because if each constraint has a non-negative right-hand side, then BV's basic solution is feasible, and if each variable in row 0 has a non-negative coefficient, then there can be no basic feasible solution with a higher z-value than BV. Our observation implies that whether a tableau is feasible and optimal depends only on the right-hand sides of the constraints and on the coefficients of each variable in row 0. For example, if an LP has variables $x_1, x_2, \ldots, x_6$, the following partial tableau would be optimal:

$$z + 2x_2 + x_4 + x_6 = 6$$
$$= 1$$
$$= 2$$
$$= 3$$

This tableau's optimality is not affected by the parts of the tableau that are omitted.

Suppose we have solved an LP and have found that BV is an optimal basis. We can use the following procedure to determine if any change in the LP will cause BV to be no longer optimal.

Step 1 Using the formulas of Section 6.2, determine how changes in the LP's parameters change the right-hand side and row 0 of the optimal tableau (the tableau having BV as the set of basic variables).

Step 2 If each variable in row 0 has a non-negative coefficient and each constraint has a non-negative right-hand side, then BV is still optimal. Otherwise, BV is no longer optimal.

If BV is no longer optimal, then you can find the new optimal solution by using the Section 6.2 formulas to recreate the entire tableau for BV and then continuing the simplex algorithm with the BV tableau as your starting tableau.

There can be two reasons why a change in an LP's parameters causes BV to be no longer optimal. First, a variable (or variables) in row 0 may have a negative coefficient. In this case, a better (larger z-value) bfs can be obtained by pivoting in a nonbasic variable with a negative coefficient in row 0. If this occurs, we say that BV is now a **suboptimal basis.** Second, a constraint (or constraints) may now have a negative right-hand side. In this case, at least one member of BV will now be negative and BV will no longer yield a bfs. If this occurs, we say that BV is now an **infeasible basis.**

We illustrate the mechanics of sensitivity analysis in the Dakota Furniture example. Recall that

$$x_1 = \text{number of desks manufactured}$$
$$x_2 = \text{number of tables manufactured}$$
$$x_3 = \text{number of chairs manufactured}$$

The objective function for the Dakota problem was

$$\max z = 60x_1 + 30x_2 + 20x_3$$

and the initial tableau was

$$
\begin{aligned}
z - 60x_1 - 30x_2 - 20x_3 \qquad\qquad &= 0 \\
8x_1 + 6x_2 + x_3 + s_1 \qquad\qquad &= 48 \qquad \text{(Lumber constraint)} \\
4x_1 + 2x_2 + 1.5x_3 \quad + s_2 \qquad &= 20 \qquad \text{(Finishing constraint)} \\
2x_1 + 1.5x_2 + 0.5x_3 \qquad\quad + s_3 &= 8 \qquad \text{(Carpentry constraint)}
\end{aligned}
$$
(12)

The optimal tableau was

$$
\begin{aligned}
z + 5x_2 \qquad\qquad + 10s_2 + 10s_3 &= 280 \\
- 2x_2 \qquad + s_1 + 2s_2 - 8s_3 &= 24 \\
- 2x_2 + x_3 \qquad + 2s_2 - 4s_3 &= 8 \\
x_1 + 1.25x_2 \qquad\qquad - 0.5s_2 + 1.5s_3 &= 2
\end{aligned}
$$
(13)

Note that BV = $\{s_1, x_3, x_1\}$ and NBV = $\{x_2, s_2, s_3\}$. The optimal bfs is $z = 280$, $s_1 = 24$, $x_3 = 8$, $x_1 = 2$, $x_2 = 0$, $s_2 = 0$, $s_3 = 0$.

We now discuss how six types of changes in an LP's parameters change the optimal solution:

Change 1 Changing the objective function coefficient of a nonbasic variable

Change 2 Changing the objective function coefficient of a basic variable

Change 3 Changing the right-hand side of a constraint

Change 4 Changing the column of a nonbasic variable

Change 5 Adding a new variable or activity

Change 6 Adding a new constraint (see Section 6.11)

Changing the Objective Function Coefficient of a Nonbasic Variable

In the Dakota problem, the only nonbasic decision variable is x_2 (tables). Currently, the objective function coefficient of x_2 is $c_2 = 30$. How would a change in c_2 affect the optimal solution to the Dakota problem? More specifically, for what values of c_2 would BV = $\{s_1, x_3, x_1\}$ remain optimal?

Suppose we change the objective function coefficient of x_2 from 30 to $30 + \Delta$. Then Δ represents the amount by which we have changed c_2 from its current value. For what values of Δ will the current set of basic variables (the current basis) remain optimal? We begin by determining how changing c_2 from 30 to $30 + \Delta$ will change the BV tableau. Note that B^{-1} and **b** are unchanged, and therefore, from (6), the right-hand side of BV's tableau ($B^{-1}\mathbf{b}$) has not changed, so BV is still feasible. Because x_2 is a nonbasic variable, $\mathbf{c}_{BV}$ has not changed. From (10), we can see that the only variable whose row 0 coefficient will be changed by a change in c_2 is x_2. Thus, BV will remain optimal if $\bar{c}_2 \geq 0$, and BV will be suboptimal if $\bar{c}_2 < 0$. In this case, z could be improved by entering x_2 into the basis.

We have

$$\mathbf{a}_2 = \begin{bmatrix} 6 \\ 2 \\ 1.5 \end{bmatrix}$$

and $c_2 = 30 + \Delta$. Also, from Section 6.2, we know that $\mathbf{c}_{BV}B^{-1} = [0 \quad 10 \quad 10]$. Now (10) shows that

$$\bar{c}_2 = [0 \quad 10 \quad 10] \begin{bmatrix} 6 \\ 2 \\ 1.5 \end{bmatrix} - (30 + \Delta) = 35 - 30 - \Delta = 5 - \Delta$$

Thus, $\bar{c}_2 \geq 0$ holds, and BV will remain optimal, if $5 - \Delta \geq 0$, or $\Delta \leq 5$. Similarly, $\bar{c}_2 < 0$ holds if $\Delta > 5$, but then BV is no longer optimal. This means that if the price of tables is decreased or increased by \$5 or less, BV remains optimal. Thus, for $c_2 \leq 30 + 5 = 35$, BV remains optimal.

If BV remains optimal after a change in a nonbasic variable's objective function coefficient, the values of the decision variables and the optimal z-value remain unchanged. This is because a change in the objective function coefficient for a nonbasic variable leaves the right-hand side of row 0 and the constraints unchanged. For example, if the price of tables increases to \$33 ($c_2 = 33$), the optimal solution to the Dakota problem remains unchanged (Dakota should still make 2 desks and 8 chairs, and $z = 280$). On the other hand, if $c_2 > 35$, BV will no longer be optimal, because $\bar{c}_2 < 0$. In this case, we find the new optimal solution by recreating the BV tableau and then using the simplex algorithm. For example, if $c_2 = 40$, we know that the only part of the BV tableau that will change is the coefficient of x_2 in row 0. If $c_2 = 40$, then

$$\bar{c}_2 = [0 \quad 10 \quad 10] \begin{bmatrix} 6 \\ 2 \\ 1.5 \end{bmatrix} - 40 = -5$$

Now the BV "final" tableau is as shown in Table 2. This is not an optimal tableau (it is suboptimal), and we can increase z by making x_2 a basic variable in row 3. The resulting tableau is given in Table 3. This is an optimal tableau. Thus, if $c_2 = 40$, the optimal solution to the Dakota problem changes to $z = 288$, $s_1 = 27.2$, $x_3 = 11.2$, $x_2 = 1.6$, $x_1 = 0$, $s_2 = 0$, $s_3 = 0$. In this case, the increase in the price of tables has made tables sufficiently more attractive to induce Dakota to manufacture them. Note that after changing a nonbasic variable's objective function coefficient, it may, in general, take more than one pivot to find the new optimal solution.

There is a more insightful way to show that the current basis in the Dakota problem remains optimal as long as the price of tables is decreased or increased by \$5 or less. From the optimal row 0 in (13), we see that if $c_2 = 30$, then

$$z = 280 - 10s_2 - 10s_3 - 5x_2$$

This tells us that each table that Dakota manufactures will decrease revenue by \$5 (in other words, the reduced cost for tables is 5). If we increase the price of tables by more than \$5, each table would now increase Dakota's revenue. For example, if $c_2 = 36$, each table would increase revenues by $6 - 5 = \$1$ and Dakota should manufacture tables. Thus, as before, we see that for $\Delta > 5$, the current basis is no longer optimal. This analysis yields another interpretation of the reduced cost of a nonbasic variable: *The reduced cost for a nonbasic variable (in a max problem) is the maximum amount by which the*

TABLE 2
"Final" (Suboptimal) Dakota Tableau ($40/Table)

				Basic Variable	Ratio
$z \quad - \quad 5x_2$		$+ \ 10s_2 + 10s_3 = 280$		$z = 280$	
$- \quad 2x_2$	$+ \ s_1 +$	$2s_2 - \ 8s_3 = 24$		$s_1 = 24$	None
$- \quad 2x_2 + x_3$		$+ \ 2s_2 - \ 4s_3 = 8$		$x_3 = 8$	None
$x_1 + \left(1.25x_2\right)$		$- \ 0.5s_2 + 1.5s_3 = 2$		$x_1 = 2$	1.6*

TABLE 3
Optimal Dakota Tableau ($40/Table)

			Basic Variable
$z + \quad 4x_1$	$+ \quad 8s_2 + 16s_3 = 288$		$z = 288$
$1.6x_1$	$+ \ s_1 + 1.2s_2 - 5.6s_3 = 27.2$		$s_1 = 27.2$
$1.6x_1 \quad + x_3$	$+ \ 1.2s_2 - 1.6s_3 = 11.2$		$x_3 = 11.2$
$0.8x_1 + x_2$	$- \ 0.4s_2 + 1.2s_3 = 1.6$		$x_2 = 1.6$

variable's objective function coefficient can be increased before the current basis becomes suboptimal, and it becomes optimal for the nonbasic variable to enter the basis.

In summary, if the objective function coefficient for a nonbasic variable x_j is changed, the current basis remains optimal if $\bar{c}_j \geq 0$. If $\bar{c}_j < 0$, then the current basis is no longer optimal, and x_j will be a basic variable in the new optimal solution.

Changing the Objective Function Coefficient of a Basic Variable

In the Dakota problem, the decision variables x_1 (desks) and x_3 (chairs) are basic variables. We now explain how a change in the objective function coefficient of a basic variable will affect an LP's optimal solution. We begin by analyzing how this change affects the BV tableau. Because we are not changing B (or therefore B^{-1}) or $\mathbf{b}$, (6) shows that the right-hand side of each constraint will remain unchanged, and BV will remain feasible. Because we are changing $\mathbf{c}_{BV}$, however, so $\mathbf{c}_{BV}B^{-1}$ will change. From (10), we see that a change in $\mathbf{c}_{BV}B^{-1}$ may change more than one coefficient in row 0. To determine whether BV remains optimal, we must use (10) to recompute row 0 for the BV tableau. If each variable in row 0 still has a non-negative coefficient, BV remains optimal. Otherwise, BV is now suboptimal. To illustrate the preceding ideas, we analyze how a change in the objective function coefficient for x_1 (desks) from its current value of $c_1 = 60$ affects the optimal solution to the Dakota problem.

Suppose that c_1 is changed to $60 + \Delta$, changing $\mathbf{c}_{BV}$ to $\mathbf{c}_{BV} = [0 \quad 20 \quad 60 \ + \ \Delta]$. To compute the new row 0, we need to know B^{-1}. We could (as in Section 6.2) use the Gauss–Jordan method to compute B^{-1}. Recall that this method begins by writing down the 3×6 matrix $B|I_3$:

$$B|I_3 = \begin{bmatrix} 1 & 1 & 8 & 1 & 0 & 0 \\ 0 & 1.5 & 4 & 0 & 1 & 0 \\ 0 & 0.5 & 2 & 0 & 0 & 1 \end{bmatrix}$$

Then we use EROs to transform the first three columns of $B|I_3$ to I_3. At this point, the last three columns of the resulting matrix will be B^{-1}.

It turns out that when we solved the Dakota problem by the simplex algorithm, without realizing it, we found B^{-1}. To see why this is the case, note that in going from the initial Dakota tableau (12) to the optimal Dakota tableau (13) we performed a series of EROs on the constraints. These EROs transformed the constraint columns corresponding to the initial basis (s_1, s_2, s_3)

$$
\text{from} \quad
\begin{matrix} s_1 & s_2 & s_3 \end{matrix} \atop
\begin{bmatrix} 1 & 0 & 0 \\ 0 & 1 & 0 \\ 0 & 0 & 1 \end{bmatrix}
\quad \text{to} \quad
\begin{matrix} s_1 & s_2 & s_3 \end{matrix} \atop
\begin{bmatrix} 1 & 2 & -8 \\ 0 & 2 & -4 \\ 0 & -0.5 & 1.5 \end{bmatrix}
$$

These same EROs have transformed the columns corresponding to BV $= \{s_1, x_3, x_1\}$

$$
\text{from} \quad B =
\begin{matrix} s_1 & x_3 & x_1 \end{matrix} \atop
\begin{bmatrix} 1 & 1 & 8 \\ 0 & 1.5 & 4 \\ 0 & 0.5 & 2 \end{bmatrix}
\quad \text{to} \quad
\begin{matrix} s_1 & x_3 & x_1 \end{matrix} \atop
\begin{bmatrix} 1 & 0 & 0 \\ 0 & 1 & 0 \\ 0 & 0 & 1 \end{bmatrix}
$$

This means that in solving the Dakota problem by the simplex algorithm, we have used EROs to transform B to I_3. These same EROs transformed I_3 into

$$
\begin{bmatrix} 1 & 2 & -8 \\ 0 & 2 & -4 \\ 0 & -0.5 & 1.5 \end{bmatrix} = B^{-1}
$$

We have discovered an extremely important fact: *For any simplex tableau, B^{-1} is the $m \times m$ matrix consisting of the columns in the current tableau that correspond to the initial tableau's set of basic variables (taken in the same order).* This means that if the starting basis for an LP consists entirely of slack variables, then B^{-1} for the optimal tableau is simply the columns for the slack variables in the constraints of the optimal tableau. In general, if the starting basic variable for the ith constraint is the artificial variable a_i, then the ith column of B^{-1} will be the column for a_i in the optimal tableau's constraints. Thus, we need not use the Gauss–Jordan method to find the optimal tableau's B^{-1}. We have already found B^{-1} by performing the simplex algorithm.

We can now compute what $\mathbf{c}_{BV}B^{-1}$ will be if $c_1 = 60 + \Delta$:

$$
\mathbf{c}_{BV}B^{-1} = \begin{bmatrix} 0 & 20 & 60 + \Delta \end{bmatrix} \begin{bmatrix} 1 & 2 & -8 \\ 0 & 2 & -4 \\ 0 & -0.5 & 1.5 \end{bmatrix} \tag{14}
$$

$$
= \begin{bmatrix} 0 & 10 - 0.5\Delta & 10 + 1.5\Delta \end{bmatrix}
$$

Observe that for $\Delta = 0$, (14) yields the original $\mathbf{c}_{BV}B^{-1}$. We can now compute the new row 0 corresponding to $c_1 = 60 + \Delta$. After noting that

$$
\mathbf{a}_1 = \begin{bmatrix} 8 \\ 4 \\ 2 \end{bmatrix}, \quad \mathbf{a}_2 = \begin{bmatrix} 6 \\ 2 \\ 1.5 \end{bmatrix}, \quad \mathbf{a}_3 = \begin{bmatrix} 1 \\ 1.5 \\ 0.5 \end{bmatrix}, \quad c_1 = 60 + \Delta, \quad c_2 = 30, \quad c_3 = 20
$$

we can use (10) to compute the new row 0. Because s_1, x_3, and x_1 are basic variables, their coefficients in row 0 must still be 0. The coefficient of each nonbasic variable in the new row 0 is as follows:

$$
\bar{c}_2 = \mathbf{c}_{BV}B^{-1}\mathbf{a}_2 - c_2 = \begin{bmatrix} 0 & 10 - 0.5\Delta & 10 + 1.5\Delta \end{bmatrix} \begin{bmatrix} 6 \\ 2 \\ 1.5 \end{bmatrix} - 30 = 5 + 1.25\Delta
$$

$$\text{Coefficient of } s_2 \text{ in row } 0 = \text{second element of } \mathbf{c}_{BV}B^{-1} = 10 - 0.5\Delta$$
$$\text{Coefficient of } s_3 \text{ in row } 0 = \text{third element of } \mathbf{c}_{BV}B^{-1} = 10 + 1.5\Delta$$

Thus, row 0 of the optimal tableau is now

$$z + (5 + 1.25\Delta)x_2 + (10 - 0.5\Delta)s_2 + (10 + 1.5\Delta)s_3 = ?$$

From the new row 0, we see that BV will remain optimal if and only if the following hold:

$$5 + 1.25\Delta \geq 0 \quad (\text{true iff}^\dagger \ \Delta \geq -4)$$
$$10 - 0.5\Delta \geq 0 \quad (\text{true iff} \ \Delta \geq 20)$$
$$10 + 1.5\Delta \geq 0 \quad (\text{true iff} \ \Delta \geq -(20/3))$$

This means that the current basis remains optimal as long as $\Delta \geq -4$, $\Delta \leq 20$, and $\Delta \geq -\frac{20}{3}$. From Figure 3, we see that the current basis will remain optimal if and only if $-4 \leq \Delta \leq 20$: If c_1 is decreased by \$4 or less or increased by up to \$20, the current basis remains optimal. Thus, as long as $56 = 60 - 4 \leq c_1 \leq 60 + 20 = 80$, the current basis remains optimal. If $c_1 < 56$ or $c_1 > 80$, the current basis is no longer optimal.

If the current basis remains optimal, then the values of the decision variables don't change because $B^{-1}\mathbf{b}$ remains unchanged. The optimal z-value does change, however. To illustrate this, suppose $c_1 = 70$. Because $56 \leq 70 \leq 80$, we know that the current basis remains optimal. Thus, Dakota should still manufacture 2 desks ($x_1 = 2$) and 8 chairs ($x_3 = 8$). However, changing c_1 to 70 changes z to $z = 70x_1 + 30x_2 + 20x_3$. This changes z to $70(2) + 20(8) = \$300$. Another way to see that z is now \$300 is to note that we have increased the revenue from each desk by $70 - 60 = \$10$. Dakota is making 2 desks, so revenue should increase by $2(10) = \$20$, and new revenue $= 280 + 20 = \$300$.

When the Current Basis Is No Longer Optimal

Recall that if $c_1 < 56$ or $c_1 > 80$, then the current basis is no longer optimal. Intuitively, if the price of desks is decreased sufficiently (with all other prices held constant), desks will no longer be worth making. Our analysis shows that this occurs if the price of desks is decreased by more than \$4. The reader should verify (see Problem 2 at the end of this section) that if $c_1 < 56$, x_1 is no longer a basic variable in the new optimal solution. On the other hand, if $c_1 > 80$, desks have become profitable enough to make the current basis suboptimal; desks are now so attractive that we want to make more of them. To do this, we must force another variable out of the basis. Suppose $c_1 = 100$. Because $100 > 80$, we know that the current basis is no longer optimal. How can we determine the new optimal solution? Simply create the optimal tableau for $c_1 = 100$ and proceed with the simplex. If $c_1 = 100$, then $\Delta = 100 - 60 = 40$, and the new row 0 will have

$$\bar{c}_1 = 0, \qquad \bar{c}_2 = 5 + 1.25\Delta = 55, \qquad \bar{c}_3 = 0,$$
$$s_1 \text{ coefficient in row } 0 = 0$$
$$s_2 \text{ coefficient in row } 0 = 10 - 0.5\Delta = -10$$
$$s_3 \text{ coefficient in row } 0 = 10 + 1.5\Delta = 70$$

$$\text{Right-hand side of row } 0 = \mathbf{c}_{BV}B^{-1}\mathbf{b} = \begin{bmatrix} 0 & -10 & 70 \end{bmatrix} \begin{bmatrix} 48 \\ 20 \\ 8 \end{bmatrix} = 360$$

From (6), changing c_1 does not change the constraints in the BV tableau. This means that if $c_1 = 100$, then the BV tableau is as given in Table 4. BV $= \{s_1, x_3, x_1\}$ is now subopti-

---

†"If and only if"

$$-4 \le \Delta \le 20$$

$$\Delta \ge -\frac{20}{3}$$

$$\Delta \ge -4$$

$$\Delta \le 20$$

TABLE 4
"Final" (Suboptimal) Tableau If $c_1 = 100$

				Basic Variable	Ratio
$z + 55x_2 - 10s_2 + 70s_3 = 360$				$z = 360$	
$-2x_2 + s_1 + 2s_2 - 8s_3 = 24$				$s_1 = 24$	12
$-2x_2 + x_3 + (2s_2) - 4s_3 = 8$				$x_3 = 8$	4*
$x_1 + 1.25x_2 - 0.5s_2 + 1.5s_3 = 2$				$x_1 = 2$	None

TABLE 5
Optimal Dakota Tableau If $c_1 = 100$

			Basic Variable
$z + 45x_2 + 5x_3 + 50s_3 = 400$			$z = 400$
$-x_3 + s_1 - 4s_3 = 16$			$s_1 = 16$
$-x_2 + 0.5x_3 + s_2 - 2s_3 = 4$			$s_2 = 4$
$x_1 + 0.75x_2 + 0.25x_3 + 0.5s_3 = 4$			$x_1 = 4$

mal. To find the new Dakota optimal solution, we enter s_2 into the basis in row 2 (Table 5). This is an optimal tableau. If $c_1 = 100$, then the new optimal solution to the Dakota problem is $z = 400$, $s_1 = 16$, $s_2 = 4$, $x_1 = 4$, $x_2 = 0$, $x_3 = 0$. Notice that increasing the profitability of desks has caused Dakota to stop making chairs. The resources that were previously used to make the chairs are now used to make $4 - 2 = 2$ extra desks.

In summary, if the objective function coefficient of a basic variable x_j is changed, then the current basis remains optimal if the coefficient of every variable in row 0 of the BV tableau remains non-negative. If any variable in row 0 has a negative coefficient, then the current basis is no longer optimal.

Interpretation of the Objective Coefficient Ranges Block of the LINDO Output

To obtain a sensitivity report in LINDO, select Yes when asked (after solving LP) whether you want a Range analysis. To obtain a sensitivity report in LINGO, go to Options and select Range (after solving LP). If this does not work, go to Options, choose the General Solver tab, and then go to Dual Computations and select the Ranges and Values option.

In the OBJ COEFFICIENT RANGES block of the LINDO (or LINGO) computer output, we see the amount by which each variable's objective function coefficient may be changed before the current basis becomes suboptimal (assuming all other LP parameters are held constant). Look at the LINDO output for the Dakota problem (Figure 4). For each variable, the CURRENT COEF column gives the current value of the variable's objective function coefficient. For example, the objective function coefficient for DESKS is 60. The ALLOWABLE INCREASE column gives the maximum amount by which the objective

```
MAX          60 DESKS + 30 TABLES + 20 CHAIRS
SUBJECT TO
        2)   8 DESKS + 6 TABLES + CHAIRS <=     48
        3)   4 DESKS + 2 TABLES + 1.5 CHAIRS <=   20
        4)   2 DESKS + 1.5 TABLES + 0.5 CHAIRS <=    8
END

     LP OPTIMUM FOUND  AT STEP         2

          OBJECTIVE FUNCTION VALUE

  1)          280.000000

 VARIABLE         VALUE      REDUCED COST
    DESKS       2.000000       0.000000
   TABLES       0.000000       5.000000
   CHAIRS       8.000000       0.000000

    ROW      SLACK OR SURPLUS      DUAL PRICES
     2)        24.000000            0.000000
     3)         0.000000           10.000000
     4)         0.000000           10.000000

 NO. ITERATIONS=      2

    RANGES IN WHICH THE BASIS IS UNCHANGED

                        OBJ COEFFICIENT RANGES
 VARIABLE        CURRENT      ALLOWABLE       ALLOWABLE
                  COEF        INCREASE        DECREASE
    DESKS       60.000000     20.000000        4.000000
   TABLES       30.000000      5.000000        INFINITY
   CHAIRS       20.000000      2.500000        5.000000

                        RIGHTHAND SIDE RANGES
    ROW         CURRENT       ALLOWABLE       ALLOWABLE
                  RHS         INCREASE        DECREASE
     2          48.000000     INFINITY        24.000000
     3          20.000000      4.000000        4.000000
     4           8.000000      2.000000        1.333333
```

FIGURE 4
LINDO Output for
Dakota Furniture

function coefficient of a variable can be increased with the current basis remaining optimal (assuming all other LP parameters stay constant). For example, if the objective function coefficient for DESKS is increased above $60 + 20 = 80$, then the current basis is no longer optimal. Similarly, the ALLOWABLE DECREASE column gives the maximum amount by which the objective function coefficient of a variable can be decreased with the current basis remaining optimal (assuming all other LP parameters constant). If the objective function coefficient for DESKS drops below $60 - 4 = 56$, the current basis is no longer optimal. In summary, we see from the LINDO output that if the objective function coefficient for DESKS is changed, the current basis remains optimal if

$$56 = 60 - 4 \leq \text{objective coefficient for DESKS} \leq 60 + 20 = 80$$

Of course, this agrees with our earlier computations.

Changing the Right-Hand Side of a Constraint

Effect on the Current Basis

In this section, we examine how the optimal solution to an LP changes if the right-hand side of a constraint is changed. Because **b** does not appear in (10), changing the right-hand side of a constraint will leave row 0 of the optimal tableau unchanged; changing a right-hand side cannot cause the current basis to become suboptimal. From (5) and (6), however, we see that a change in the right-hand side of a constraint will affect the right-hand side of the constraints in the optimal tableau. *As long as the right-hand side of each* *constraint in the optimal tableau remains non-negative, the current basis remains feasible and optimal. If at least one right-hand side in the optimal tableau becomes negative, then the current basis is no longer feasible and therefore no longer optimal.*

Suppose we want to determine how changing the amount of finishing hours (b_2) affects the optimal solution to the Dakota problem. Currently, $b_2 = 20$. If we change b_2 to $20 + \Delta$, then from (6), the right-hand side of the constraints in the optimal tableau will become

$$B^{-1}\begin{bmatrix} 48 \\ 20 + \Delta \\ 8 \end{bmatrix} = \begin{bmatrix} 1 & 2 & -8 \\ 0 & 2 & -4 \\ 0 & -0.5 & 1.5 \end{bmatrix}\begin{bmatrix} 48 \\ 20 + \Delta \\ 8 \end{bmatrix}$$

$$= \begin{bmatrix} 24 + 2\Delta \\ 8 + 2\Delta \\ 2 - 0.5\Delta \end{bmatrix}$$

Of course, for $\Delta = 0$, the right-hand side reduces to the right-hand side of the original optimal tableau. If this does not happen, then an error has been made.

It can be shown (see Problem 9) that if the right-hand side of the ith constraint is increased by Δ, then the right-hand side of the optimal tableau is given by (original right-hand side of the optimal tableau) $+\Delta$(column i of B^{-1}). Because the second column of B^{-1} is

$$\begin{bmatrix} 2 \\ 2 \\ -0.5 \end{bmatrix}$$ and the original right-hand side is $$\begin{bmatrix} 24 \\ 8 \\ 2 \end{bmatrix}$$

we again find that the right-hand side of the constraints in the optimal tableau is

$$\begin{bmatrix} 24 + 2\Delta \\ 8 + 2\Delta \\ 2 - 0.5\Delta \end{bmatrix}$$

For the current basis to remain optimal, we require that the right-hand side of each constraint in the optimal tableau remain non-negative. This means that the current basis will remain optimal if and only if the following hold:

$$24 + 2\Delta \geq 0 \quad \text{(true iff } \Delta \geq -12)$$
$$8 + 2\Delta \geq 0 \quad \text{(true iff } \Delta \geq -4)$$
$$2 - 0.5\Delta \geq 0 \quad \text{(true iff } \Delta \leq 4)$$

As long as $\Delta \geq -12$, $\Delta \geq -4$, and $\Delta \leq 4$, the current basis remains feasible and therefore optimal. From Figure 5, we see that for $-4 \leq \Delta \leq 4$, the current basis remains feasible and therefore optimal. This means that for $20 - 4 \leq b_2 \leq 20 + 4$, or $16 \leq b_2 \leq 24$, the current basis remains optimal: If between 16 and 24 finishing hours are available, BV = $\{s_1, x_3, x_1\}$ remains optimal, and Dakota should still manufacture desks and chairs. If $b_2 > 24$ or if $b_2 < 16$, however, the current basis becomes infeasible and is no longer optimal.

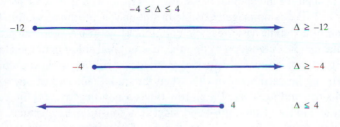

FIGURE 5
Determination of Range of Values on b_2 for Which Current Basis Remains Optimal

Effect on Decision Variables and z

Even if the current basis remains optimal ($16 \leq b_2 \leq 24$), *the values of the decision variables and z change.* This was illustrated in our graphical discussion of sensitivity analysis in Section 6.1. To see how the values of the objective function and decision variables change, recall that the values of the basic variables in the optimal solution are given by $B^{-1}\mathbf{b}$ and the optimal z-value is given by $\mathbf{c}_{BV}B^{-1}\mathbf{b}$. Changing $\mathbf{b}$ will change the values of the basic variables and the optimal z-value. To illustrate this, suppose that 22 finishing hours are available. Because $16 \leq 22 \leq 24$, the current basis remains optimal and, from (6), the new values of the basic variables are as follows (the same basis remains optimal, so the nonbasic variables remain equal to 0):

$$\begin{bmatrix} s_1 \\ x_3 \\ x_1 \end{bmatrix} = B^{-1}\mathbf{b} = \begin{bmatrix} 1 & 2 & -8 \\ 0 & 2 & -4 \\ 0 & -0.5 & 1.5 \end{bmatrix} \begin{bmatrix} 48 \\ 22 \\ 8 \end{bmatrix} = \begin{bmatrix} 28 \\ 12 \\ 1 \end{bmatrix}$$

If 22 finishing hours were available, then Dakota should manufacture 12 chairs and only 1 desk.

To determine how a change in a right-hand side affects the optimal z-value, we may use formula (11). If 22 finishing hours are available, we find that

$$\text{New } z\text{-value} = \mathbf{c}_{BV}B^{-1}(\text{new } \mathbf{b}) = \begin{bmatrix} 0 & 10 & 10 \end{bmatrix} \begin{bmatrix} 48 \\ 22 \\ 8 \end{bmatrix} = 300$$

In Section 6.8, we explain how the important concept of shadow price can be used to determine how changes in a right-hand side change the optimal z-value.

When the Current Basis Is No Longer Optimal

If we change a right-hand side enough that the current basis is no longer optimal, how can we determine the new optimal basis? Suppose we change b_2 to 30. Because $b_2 > 24$, we know that the current basis is no longer optimal. If we re-create the optimal tableau, we see from the formulas of Section 6.2 that the only part of the optimal tableau that will change is the right-hand side of row 0 and the constraints. From (6), the right-hand side of the constraints in the tableau for BV $= \{s_1, x_3, x_1\}$ is

$$B^{-1}\mathbf{b} = \begin{bmatrix} 1 & 2 & -8 \\ 0 & 2 & -4 \\ 0 & -0.5 & 1.5 \end{bmatrix} \begin{bmatrix} 48 \\ 30 \\ 8 \end{bmatrix} = \begin{bmatrix} 44 \\ 28 \\ -3 \end{bmatrix}$$

From (11), the right-hand side of row 0 is now

$$\mathbf{c}_{BV}B^{-1}\mathbf{b} = \begin{bmatrix} 0 & 10 & 10 \end{bmatrix} \begin{bmatrix} 48 \\ 30 \\ 8 \end{bmatrix} = 380$$

The tableau for the optimal basis, BV $= \{s_1, x_3, x_1\}$, is now as shown in Table 6. Because $x_1 = -3$, BV is no longer feasible or optional. Unfortunately, this tableau does not yield a readily apparent basic feasible solution. If we use this as our initial tableau, we can't use the simplex algorithm to find the new optimal solution to the Dakota problem. In Section 6.11, we discuss a different method for solving LPs, the dual simplex algorithm, which can be used to solve LPs when the initial tableau has one or more negative right-hand sides and each variable in row 0 has a non-negative coefficient.

In summary, if the right-hand side of a constraint is changed, then the current basis remains optimal if the right-hand side of each constraint in the tableau remains non-

TABLE 6
Final (Infeasible) Dakota Tableau If $b_2 = 30$

						Basic Variable
z	$+$	$5x_2$		$+ 10s_2 +$	$10s_3 = 380$	$z = 480$
	$-$	$2x_2$	$+ s_1 +$	$2s_2 -$	$8s_3 = 44$	$s_1 = 44$
	$-$	$2x_2 + x_3$		$+ 2s_2 -$	$4s_3 = 28$	$x_3 = 28$
x_1	$+$	$1.25x_2$		$- 0.5s_2 +$	$1.5s_3 = -3$	$x_1 = -3$

negative. If the right-hand side of any constraint is negative, then the current basis is infeasible, and a new optimal solution must be found.

Interpretation of the Right-Hand Side Ranges Block of the LINDO Output

The block of the LINDO (or LINGO) output labeled RIGHTHAND SIDE RANGES (see Figure 4) gives information concerning the amount by which a right-hand side can be changed before the current basis becomes infeasible (all other LP parameters constant). The CURRENT RHS column gives the current right-hand side of each constraint. Thus, for row 3 (the second constraint), the current right-hand side is 20. The ALLOWABLE INCREASE column is the maximum amount by which the right-hand side of the constraint can be increased with the current basis remaining optimal (all other LP parameters constant). For example, if the amount of available finishing hours (second constraint) is increased by up to 4 hours, then the current basis remains optimal. Similarly, the ALLOWABLE DECREASE column gives the maximum amount by which the right-hand side of a constraint can be decreased with the current basis remaining optimal (all other LP parameters constant). If the amount of available finishing hours is decreased by more than 4 hours, then the current basis is no longer optimal. In summary, if the number of finishing hours is changed (all other LP parameters constant), the current basis remains optimal if

$$16 = 20 - 4 \leq \text{available finishing hours} \leq 20 + 4 = 24$$

Changing the Column of a Nonbasic Variable

Currently, 6 board feet of lumber, 2 finishing hours, and 1.5 carpentry hours are required to make a table that can be sold for $30. Also x_2 (the variable for tables) is a nonbasic variable in the optimal solution. This means that Dakota should not manufacture any tables now. Suppose, however, that the price of tables increased to $43 and, because of changes in production technology, a table required 5 board feet of lumber, 2 finishing hours, and 2 carpentry hours. Would this change the optimal solution to the Dakota problem? Here we are changing elements of the column for x_2 in the original problem (including the objective function). Changing the column for a nonbasic variable such as tables leaves B (and B^{-1}) and **b** unchanged. Thus, the right-hand side of the optimal tableau remains unchanged. A glance at (10) also shows that the only part of row 0 that is changed is $\bar{c}_2$; the current basis will remain optimal if and only if $\bar{c}_2 \geq 0$ holds. We now use (10) to compute the new coefficient of x_2 in row 0. This process is called **pricing out** x_2. From (10),

$$\bar{c}_2 = \mathbf{c}_{BV}B^{-1}\mathbf{a}_2 - c_2$$

Note that $c_{BV}B^{-1}$ still equals $[0 \quad 10 \quad 10]$, but a_2 and c_2 have changed to

$$c_2 = 43 \quad \text{and} \quad a_2 = \begin{bmatrix} 5 \\ 2 \\ 2 \end{bmatrix}$$

Now

$$\bar{c}_2 = [0 \quad 10 \quad 10] \begin{bmatrix} 5 \\ 2 \\ 2 \end{bmatrix} - 43 = -3 < 0$$

Because $\bar{c}_2 < 0$, the current basis is no longer optimal. The fact that $\bar{c}_2 = -3$ means that each table that Dakota manufactures now increases revenues by \$3. It is clearly to Dakota's advantage to enter x_2 into the basis. To find the new optimal solution to the Dakota problem, we re-create the tableau for BV $= \{s_1, x_3, x_1\}$ and then apply the simplex algorithm. From (5), the column for x_2 in the constraint portion of the BV tableau is now

$$B^{-1}a_2 = \begin{bmatrix} 1 & 2 & -8 \\ 0 & 2 & -4 \\ 0 & -0.5 & 1.5 \end{bmatrix} \begin{bmatrix} 5 \\ 2 \\ 2 \end{bmatrix} = \begin{bmatrix} -7 \\ -4 \\ 2 \end{bmatrix}$$

The tableau for BV $= \{s_1, x_3, x_1\}$ is now as shown in Table 7. To find the new optimal solution, we enter x_2 into the basis in row 3. This yields the optimal tableau in Table 8. Thus, the new optimal solution to the Dakota problem is $z = 283$, $s_1 = 31$, $x_3 = 12$, $x_2 = 1$, $x_1 = 0$, $s_2 = 0$, $s_3 = 0$. After the column for the nonbasic variable x_2 (tables) has been changed, Dakota should manufacture 12 chairs and 1 table. In summary, if the column of a nonbasic variable x_j is changed, then the current basis remains optimal if $\bar{c}_j \geq 0$. If $\bar{c}_j < 0$, then the current basis is no longer optimal and x_j will be a basic variable in the new optimal solution.

If the column of a basic variable is changed, then it is usually difficult to determine whether the current basis remains optimal. This is because the change may affect both B (and hence B^{-1}) and c_{BV} and thus the entire row 0 and the entire right-hand side of the optimal

TABLE 7
"Final" (Suboptimal) Dakota Tableau for New Method of Making Tables

				Basic Variable
$z \quad - 3x_2$	$+ \ 10s_2 +$	$10s_3 = 280$		$z = 280$
$- 7x_2$	$+ s_1 + \quad 2s_2 -$	$8s_3 = 24$		$s_1 = 24$
$- 4x_2 + x_3$	$+ \quad 2s_2 -$	$4s_3 = 8$		$x_3 = 8$
$x_1 + \textcircled{$2x_2$}$	$- 0.5s_2 + 1.5s_3 = 2$			$x_1 = 2*$

TABLE 8
Optimal Dakota Tableau for New Method of Making Tables

				Basic Variable
$z + 1.5x_1$	$+ 9.25s_2 + 12.25s_3 = 283$			$z = 283$
$3.5x_1$	$+ s_1 + 0.25s_2 - \ 2.75s_3 = 31$			$s_1 = 31$
$2x_1 + x_3$	$+ \quad s_2 - \quad s_3 = 12$			$x_3 = 12$
$0.5x_1 + x_2$	$- 0.25s_2 + \ 0.75s_3 = 1$			$x_1 = 1$

tableau. As always, the current basis would remain optimal if and only if each variable has a non-negative coefficient in row 0 and each constraint has a non-negative right-hand side.

Adding a New Activity

In many situations, opportunities arise to undertake new activities. For example, in the Dakota problem, the company may be presented with the opportunity to manufacture additional types of furniture, such as footstools. If a new activity is available, we can evaluate it by applying the method utilized to determine whether the current basis remains optimal after a change in the column of a nonbasic variable. The following example illustrates the approach.

Suppose that Dakota is considering making footstools. A stool sells for $15 and requires 1 board foot of lumber, 1 finishing hour, and 1 carpentry hour. Should the company manufacture any stools?

To answer this question, define x_4 to be the number of footstools manufactured by Dakota. The initial tableau is now changed by the introduction of the x_4 column. Our new initial tableau is

$$
\begin{aligned}
z - 60x_1 - 30x_2 - 20x_3 - 15x_4 &= 0 \\
8x_1 + 6x_2 + x_3 + x_4 + s_1 &= 48 \\
4x_1 + 2x_2 + 1.5x_3 + x_4 + s_2 &= 20 \\
2x_1 + 1.5x_2 + 0.5x_3 + x_4 + s_3 &= 8
\end{aligned}
\tag{15}
$$

We call the addition of the x_4 column to the problem **adding a new activity.** How will the addition of the new activity change the optimal BV $= \{s_1, x_3, x_1\}$ tableau? From (6), we see that the right-hand sides of all constraints in the optimal tableau will remain unchanged. From (10), we see that the coefficient of each of the old variables in row 0 will remain unchanged. We must, of course, compute $\bar{c}_4$, the coefficient of the new activity in row 0 of the optimal tableau. The right-hand side of each constraint in the optimal tableau is unchanged and the only variable in row 0 that can have a negative coefficient is x_4, so the current basis will remain optimal if $\bar{c}_4 \geq 0$ or become nonoptimal if $\bar{c}_4 < 0$.

To determine whether a new activity causes the current basis to be no longer optimal, price out the new activity. Because

$$
c_4 = 15 \quad \text{and} \quad \mathbf{a}_4 = \begin{bmatrix} 1 \\ 1 \\ 1 \end{bmatrix}
$$

we may use (10) to price out x_4. The result is

$$
\bar{c}_4 = \begin{bmatrix} 0 & 10 & 10 \end{bmatrix} \begin{bmatrix} 1 \\ 1 \\ 1 \end{bmatrix} - 15 = 5
$$

Because $\bar{c}_4 \geq 0$, the current basis is still optimal. Equivalently, the reduced cost of footstools is $5. This means that each stool manufactured will decrease revenues by $5. For this reason, we choose not to manufacture any stools.

In summary, if a new column (corresponding to a variable x_j) is added to an LP, then the current basis remains optimal if $\bar{c}_j \geq 0$. If $\bar{c}_j < 0$, then the current basis is no longer optimal and x_j will be a basic variable in the new optimal solution. Table 9 presents a summary of sensitivity analyses for a maximization problem. When applying the techniques of this section to a minimization problem, just remember that a tableau is optimal if and only if each variable has a *nonpositive* coefficient in row 0 and the right-hand side of each constraint is nonnegative.

TABLE 9
Summary of Sensitivity Analysis (Max Problem)

Change in Initial Problem	Effect on Optimal Tableau	Current Basis Is Still Optimal If:
Changing nonbasic objective function coefficient c_j	Coefficient of x_j in optimal row 0 is changed	Coefficient of x_j in row 0 for current basis is still non-negative
Changing basic objective function coefficient c_j	Entire row 0 may change	Each variable still has a non-negative coefficient in row 0
Changing right-hand side of a constraint	Right-hand side of constraints and row 0 are changed	Right-hand side of each constraint is still non-negative
Changing the column of a nonbasic variable x_j or adding a new variable x_j	Changes the coefficient for x_j in row 0 and x_j's constraint column in optimal tableau	The coefficient of x_j in row 0 is still non-negative

PROBLEMS

Group A

1 In the Dakota problem, show that the current basis remains optimal if c_3, the price of chairs, satisfies $15 \leq c_3 \leq 22.5$. If $c_3 = 21$, find the new optimal solution. Also, if $c_3 = 25$, find the new optimal solution.

2 If $c_1 = 55$ in the Dakota problem, show that the new optimal solution does not produce any desks.

3 In the Dakota problem, show that if the amount of lumber (board ft) available (b_1) satisfies $b_1 \geq 24$, the current basis remains optimal. If $b_1 = 30$, find the new optimal solution.

4 Show that if tables sell for $50 and use 1 board ft of lumber, 3 finishing hours, and 1.5 carpentry hours, the current basis for the Dakota problem will no longer be optimal. Find the new optimal solution.

5 Dakota Furniture is considering manufacturing home computer tables. A home computer table sells for $36 and uses 6 board ft of lumber, 2 finishing hours, and 2 carpentry hours. Should the company manufacture any home computer tables?

6 Sugarco can manufacture three types of candy bar. Each candy bar consists totally of sugar and chocolate. The compositions of each type of candy bar and the profit earned from each candy bar are shown in Table 10. Fifty oz of sugar and 100 oz of chocolate are available. After defining x_i to be the number of Type i candy bars manufactured, Sugarco should solve the following LP:

$$\max z = 3x_1 + 7x_2 + 5x_3$$
$$\text{s.t.} \quad x_1 + x_2 + x_3 \leq 50 \quad \text{(Sugar constraint)}$$
$$2x_1 + 3x_2 + x_3 \leq 100 \quad \text{(Chocolate constraint)}$$
$$x_1, x_2, x_3 \geq 0$$

After adding slack variables s_1 and s_2, the optimal tableau is as shown in Table 11. Using this optimal tableau, answer the following questions:

a For what values of Type 1 candy bar profit does the current basis remain optimal? If the profit for a Type 1 candy bar were 7¢, what would be the new optimal solution to Sugarco's problem?

b For what values of Type 2 candy bar profit would the current basis remain optimal? If the profit for a Type 2 candy bar were 13¢, then what would be the new optimal solution to Sugarco's problem?

c For what amount of available sugar would the current basis remain optimal?

d If 60 oz of sugar were available, what would be Sugarco's profit? How many of each candy bar should the company make? Could these questions be answered if only 30 oz of sugar were available?

e Suppose a Type 1 candy bar used only 0.5 oz of sugar and 0.5 oz of chocolate. Should Sugarco now make Type 1 candy bars?

TABLE 10

Bar	Amount of Sugar (Ounces)	Amount of Chocolate (Ounces)	Profit (Cents)
1	1	2	3
2	1	3	7
3	1	1	5

TABLE 11

z	x_1	x_2	x_3	s_1	s_2	rhs	Basic Variable
1	3	0	0	4	1	300	$z = 300$
0	$\frac{1}{2}$	0	1	$\frac{3}{2}$	$-\frac{1}{2}$	25	$x_3 = 25$
0	$\frac{1}{2}$	1	0	$-\frac{1}{2}$	$\frac{1}{2}$	25	$x_2 = 25$

f Sugarco is considering making Type 4 candy bars. A Type 4 candy bar earns 17¢ profit and requires 3 oz of sugar and 4 oz of chocolate. Should Sugarco manufacture any Type 4 candy bars?

7 The following questions refer to the Giapetto problem (Section 3.1). Giapetto's LP was

$$\max z = 3x_1 + 2x_2$$

s.t.
$$2x_1 + x_2 \le 100 \quad \text{(Finishing constraint)}$$
$$x_1 + x_2 \le 80 \quad \text{(Carpentry constraint)}$$
$$x_1 \le 40 \quad \text{(Limited demand for soldiers)}$$

(x_1 = soldiers and x_2 = trains). After adding slack variables s_1, s_2, and s_3, the optimal tableau is as shown in Table 12. Use this optimal tableau to answer the following questions:

a Show that as long as soldiers (x_1) contribute between $2 and $4 to profit, the current basis remains optimal. If soldiers contribute $3.50 to profit, find the new optimal solution to the Giapetto problem.

b Show that as long as trains (x_2) contribute between $1.50 and $3.00 to profit, the current basis remains optimal.

c Show that if between 80 and 120 finishing hours are available, the current basis remains optimal. Find the new optimal solution to the Giapetto problem if 90 finishing hours are available.

d Show that as long as the demand for soldiers is at least 20, the current basis remains optimal.

e Giapetto is considering manufacturing toy boats. A toy boat uses 2 carpentry hours and 1 finishing hour. Demand for toy boats is unlimited. If a toy boat contributes $3.50 to profit, should Giapetto manufacture any toy boats?

Group B

8 Consider the Dorian Auto problem (Example 2 of Chapter 3):

$$\min z = 50x_1 + 100x_2$$

s.t.
$$7x_1 + 2x_2 \ge 28 \quad \text{(HIW)}$$
$$2x_1 + 12x_2 \ge 24 \quad \text{(HIM)}$$
$$x_1, x_2 \ge 0$$

(x_1 = number of comedy ads, and x_2 = number of football ads). The optimal tableau is given in Table 13. Remember that for a min problem, a tableau is optimal if and only if each variable has a nonpositive coefficient in row 0 and the right-hand side of each constraint is non-negative.

a Find the range of values of the cost of a comedy ad (currently $50,000) for which the current basis remains optimal.

b Find the range of values of the number of required HIW exposures (currently 28 million) for which the current basis remains optimal. If 40 million HIW exposures were required, what would be the new optimal solution?

c Suppose an ad on a news program costs $110,000 and reaches 12 million HIW and 7 million HIM. Should Dorian advertise on the news program?

9 Show that if the right-hand side of the ith constraint is increased by Δ, then the right-hand side of the optimal tableau is given by (original right-hand side of the optimal tableau) + Δ(column i of B^{-1}).

TABLE 12

z	x_1	x_2	s_1	s_2	s_3	rhs	Basic Variable
1	0	0	1	1	0	180	$z = 180$
0	1	0	1	−1	0	20	$x_1 = 20$
0	0	1	−1	2	0	60	$x_2 = 60$
0	0	0	−1	1	1	20	$s_3 = 20$

TABLE 13

z	x_1	x_2	e_1	e_2	a_1	a_2	rhs
1	0	0	−5	−7.5	$5 - M$	$7.5 - M$	320
0	1	0	$-\frac{3}{20}$	$\frac{1}{40}$	$\frac{3}{20}$	$-\frac{1}{40}$	3.6
0	0	1	$\frac{1}{40}$	$-\frac{7}{80}$	$-\frac{1}{40}$	$\frac{7}{80}$	1.4

6.4 Sensitivity Analysis When More Than One Parameter Is Changed: The 100% Rule[†]

In this section, we show how to use the LINDO output to determine whether the current basis remains optimal when more than one objective function coefficient or right-hand side is changed.

The 100% Rule for Changing Objective Function Coefficients

Depending on whether the objective function coefficient of any variable with a zero reduced cost in the optimal tableau is changed, there are two cases to consider:

[†]This section covers topics that may be omitted with no loss of continuity.

Case 1 All variables whose objective function coefficients are changed have nonzero reduced costs in the optimal row 0.

Case 2 At least one variable whose objective function coefficient is changed has a reduced cost of zero.

In Case 1, the current basis remains optimal if and only if the objective function coefficient for each variable remains within the allowable range[†] given on the LINDO printout (see Problem 10 at the end of this section). If the current basis remains optimal, then both the values of the decision variables and objective function remain unchanged. If the objective function coefficient for any variable is outside its allowable range, then the current basis is no longer optimal.

The following two examples of Case 1 refer to the diet problem of Section 3.4. The LINDO printout for this problem is given in Figure 6.

EXAMPLE 2 **100% Rule for Objective Functional Coefficients 1**

Suppose the price of a brownie increases to 60¢ and a piece of pineapple cheesecake decreases to 50¢. Does the current basis remain optimal? What would be the new optimal solution?

Solution Both brownies and pineapple cheesecake have nonzero reduced costs, so we are in Case 1. From Figure 6 and the Case 1 discussion, we see that the current basis remains optimal if and only if

$$22.5 = 50 - 27.5 \leq \text{cost of a brownie} \leq 50 + \infty = \infty$$
$$30 = 80 - 50 \leq \text{cost of a piece of cheesecake} \leq 80 + \infty = \infty$$

Because the new prices satisfy both of these conditions, the current basis remains optimal. Also the optimal z-value and optimal value of the decision variables remain unchanged.

EXAMPLE 3 **100% Rule for Objective Functional Coefficients 2**

If prices drop to 40¢ for a brownie and 25¢ for a piece of pineapple cheesecake, is the current basis still optimal?

Solution From Figure 6, we see that Case 1 again applies. The cost of a brownie remains in its allowable range, but the price of pineapple cheesecake does not. Thus, the current basis is no longer optimal, and the problem must be solved again.

In Case 2, we can often show that the current basis remains optimal by using the **100% Rule.** Let

c_j = original objective function coefficient for x_j

Δc_j = change in c_j

I_j = maximum allowable increase in c_j for which current basis remains optimal (from LINDO output)

D_j = maximum allowable decrease in c_j for which current basis remains optimal (from LINDO output)

[†]The allowable range for c_j is the range of values for which the current basis remains optimal (assuming that only c_j is changed).

```
MIN      50 BR + 20 IC + 30 COLA + 80 PC
SUBJECT TO
        2)    400 BR + 200 IC + 150 COLA + 500 PC >=    500
        3)      3 BR +   2 IC >=    6
        4)      2 BR +   2 IC +   4 COLA +   4 PC >=    10
        5)      2 BR +   4 IC +     COLA +   5 PC >=     8
END

    LP OPTIMUM FOUND   AT STEP      5

        OBJECTIVE FUNCTION VALUE

    1)        90.000000

VARIABLE         VALUE           REDUCED COST
      BR         0.000000          27.500000
      IC         3.000000           0.000000
    COLA         1.000000           0.000000
      PC         0.000000          50.000000

   ROW      SLACK OR SURPLUS      DUAL PRICES
     2)        250.000000           0.000000
     3)          0.000000          -2.500000
     4)          0.000000          -7.500000
     5)          5.000000           0.000000

NO. ITERATIONS=       5

   RANGES IN WHICH THE BASIS IS UNCHANGED

                        OBJ COEFFICIENT RANGES
VARIABLE         CURRENT         ALLOWABLE        ALLOWABLE
                 COEF            INCREASE         DECREASE
      BR         50.000000       INFINITY         27.500000
      IC         20.000000       18.333334         5.000000
    COLA         30.000000       10.000000        30.000000
      PC         80.000000       INFINITY         50.000000

                        RIGHTHAND SIDE RANGES
   ROW           CURRENT         ALLOWABLE        ALLOWABLE
                 RHS             INCREASE         DECREASE
     2          500.000000       250.000000       INFINITY
     3            6.000000         4.000000         2.857143
     4           10.000000       INFINITY          4.000000
     5            8.000000         5.000000        INFINITY
```

FIGURE 6

LINDO Output for Diet Problem

For each variable x_j, we define the ratio r_j:

$$\text{If} \quad \Delta c_j \geq 0, \qquad r_j = \frac{\Delta c_j}{I_j}$$

$$\text{If} \quad \Delta c_j \leq 0, \qquad r_j = \frac{-\Delta c_j}{D_j}$$

If c_j is unchanged, then $r_j = 0$. Thus, r_j measures the ratio of the actual change in c_j to the maximum allowable change in c_j that would keep the current basis optimal. If only one objective function coefficient were being changed, then the current basis would remain optimal if $r_j \leq 1$ (or equivalently, if r_j, expressed as a percentage, were less than or equal to 100%). The 100% Rule for objective function coefficients is a generalization of this idea. It states that if $\Sigma r_j \leq 1$, then we can be sure that the current basis remains optimal. If $\Sigma r_j > 1$, then the current basis may or may not be optimal; we can't be sure. If the current basis does remain optimal, then the values of the decision variables remain unchanged, but the optimal z-value may change. The reader is referred to Bradley, Hax, and Magnanti (1977) for a proof of the 100% Rule. We sketch the proof in Problem 11 at the end of this section.

The following two examples of Case 2 refer to the Dakota Furniture problem and illustrate the use of the 100% Rule.

EXAMPLE 4   Basis No Longer Optimal

Suppose the desk price increases to \$70 and chairs decrease to \$18. Does the current basis remain optimal? What is the new optimal z-value?

Solution   Because both desks and chairs have zero reduced costs (they are basic variables), we must apply the 100% Rule to determine whether the current basis remains optimal. Returning to the notation that x_1 = desks, x_2 = tables, and x_3 = chairs, we may write

$$\Delta c_1 = 70 - 60 = 10, \qquad I_1 = 20, \qquad \text{so} \qquad r_1 = \tfrac{10}{20} = 0.5$$
$$\Delta c_3 = 18 - 20 = -2, \qquad D_3 = 5, \qquad \text{so} \qquad r_3 = \tfrac{2}{5} = 0.4$$
$$\Delta c_2 = 0, \qquad \text{so} \qquad r_2 = 0$$

Because $r_1 + r_2 + r_3 = 0.9 \leq 1$, the current basis remains optimal. Another way of looking at it: We changed c_1 50% of the amount it was "allowed" to change and c_3 40% of the amount it was "allowed" to change. Because 50% + 40% = 90% $\leq$ 100%, the current basis remains optimal.

   The current basis remains optimal, so the values of the decision variables do not change. Note that the revenue from each desk has increased by \$10 and the revenue from each chair has decreased by \$2. Dakota is still producing 2 desks and 8 chairs, so revenue increases by $2(10) - 8(2) = \$4$ and is now $280 + 4 = \$284$.

EXAMPLE 5   100% Rule and Optimal Basis 1

Show that if the price of tables increases to \$33 and desk prices decrease to \$58, the 100% Rule does not tell us whether the current basis is still optimal.

Solution   For this situation,

$$\Delta c_1 = 58 - 60 = -2, \qquad D_1 = 4, \qquad \text{so} \qquad r_1 = \tfrac{2}{4} = 0.5$$
$$\Delta c_2 = 33 - 30 = 3, \qquad I_2 = 5, \qquad \text{so} \qquad r_2 = \tfrac{3}{5} = 0.6$$
$$\Delta c_3 = 0, \qquad \text{so} \qquad r_3 = 0$$

Because $r_1 + r_2 + r_3 = 0.5 + 0.6 + 0 = 1.1 > 1$, the 100% Rule yields no information about whether the current basis is optimal.

The 100% Rule for Changing Right-Hand Sides

Depending on whether any of the constraints whose right-hand sides are being modified are binding constraints, there are two cases to consider:

Case 1   All constraints whose right-hand sides are being modified are nonbinding constraints.

Case 2   At least one of the constraints whose right-hand side is being modified is a binding constraint (that is, has zero slack or zero excess).

In Case 1, the current basis remains optimal if and only if each right-hand side remains within its allowable range.[†] Then the values of the decision variables and optimal objective function remain unchanged. If the right-hand side for any constraint is outside its allowable range, then the current basis is no longer optimal (see Problem 12 at the end of this section). The following examples for the diet problem illustrate the application of Case 1.

[†]The allowable range for a right-hand side b_i is the range of values for which the current basis remains optimal (assuming the other LP parameters remain unchanged).

EXAMPLE 6   **New Optimal Solution**

Suppose the calorie requirement is decreased to 400 calories and the fat requirement is increased to 10 oz. Does the current basis remain optimal? What is the new optimal solution?

Solution   Both the calorie and fat constraints are nonbinding, so Case 1 applies. From Figure 6, we see that the allowable ranges for the calorie and fat constraints are

$$-\infty = 500 - \infty \leq \text{calorie requirement} \leq 500 + 250 = 750$$

$$-\infty = 8 - \infty \leq \text{fat requirement} \leq 8 + 5 = 13$$

The new calorie and fat requirements both remain within their allowable ranges, so the current basis remains optimal. The optimal z-value and the values of the decision variables remain unchanged.

EXAMPLE 7   Basis No Longer Optimal

Suppose the calorie requirement is decreased to 400 calories and the fat requirement is increased to 15 oz. Is the current basis still optimal?

Solution   The fat requirement is no longer in its allowable range, so the current basis is no longer optimal.

In Case 2, we can often show that the current basis remains optimal via another version of the 100% Rule. Let

b_j = current right-hand side of the jth constraint (from row $j + 1$ on LINDO output)

Δb_j = change in b_j

    I_j = maximum allowable increase in b_j for which the current basis remains optimal (from LINDO output)

    D_j = maximum allowable decrease in b_j for which the current basis remains optimal (from LINDO output)

For each constraint, compute the ratio r_j:

$$\text{If} \quad \Delta b_j \geq 0, \qquad r_j = \frac{\Delta b_j}{I_j}$$

$$\text{If} \quad \Delta b_j \leq 0, \qquad r_j = \frac{-\Delta b_j}{I_j}$$

If only the jth right-hand side is changed, then the current basis remains optimal if $r_j \leq 1$. Also note that r_j is the fraction of the maximum allowable change (in the sense that the current basis remains optimal) in b_j that has occurred. The 100% Rule states that if $\Sigma r_j \leq 1$, then the current basis remains optimal. If $\Sigma r_j > 1$, then the current basis may or may not be optimal; we can't be sure (see Problem 13 at the end of this section for a sketch of the proof of this result). The following examples illustrate the use of the 100% Rule for right-hand sides.

EXAMPLE 8   Basis Remains Optimal

In the Dakota problem, suppose 22 finishing hours and 9 carpentry hours are available. Does the current basis remain optimal?

Solution   The finishing and carpentry constraints are binding, so we are in Case 2 and need to use the 100% Rule.

$$\Delta b_1 = 0, \qquad \text{so} \qquad r_1 = 0$$
$$\Delta b_2 = 22 - 20 = 2, \qquad I_2 = 4, \qquad \text{so} \qquad r_2 = \frac{2}{4} = 0.5$$
$$\Delta b_3 = 9 - 8 = 1, \qquad I_3 = 2, \qquad \text{so} \qquad r_3 = \frac{1}{2} = 0.5$$

Because $r_1 + r_2 + r_3 = 1$, the current basis remains optimal.

EXAMPLE 9 | **100% Rule and Optimal Basis 2**

In the diet problem, suppose the chocolate requirement is increased to 8 oz and the sugar requirement is reduced to 7 oz. Does the current basis remain optimal?

Solution The chocolate and sugar constraints are binding, so we are in Case 2 and need to use the 100% Rule.

$$\Delta b_2 = 8 - 6 = 2, \qquad I_2 = 4, \qquad \text{so} \qquad r_2 = \frac{2}{4} = 0.5$$
$$\Delta b_3 = 7 - 10 = -3, \qquad D_3 = 4, \qquad \text{so} \qquad r_3 = \frac{3}{4} = 0.75$$
$$\Delta b_1 = \Delta b_4 = 0, \qquad \text{so} \qquad r_1 = r_4 = 0$$

Because $r_1 + r_2 + r_3 + r_4 = 1.25 > 1$, the 100% Rule yields no information about whether the current basis remains optimal.

PROBLEMS

Group A

The following questions refer to the diet problem:

1 If the cost of a brownie is 70¢ and a piece of cheesecake costs 60¢, does the current basis remain optimal?

2 If the cost of a brownie is 20¢ and a piece of cheesecake is $1, does the current basis remain optimal?

3 If the fat requirement is reduced to 3 oz and the calorie requirement is increased to 800 calories, does the current basis remain optimal?

4 If the fat requirement is 6 oz and the calorie requirement is 600 calories, does the current basis remain optimal?

5 If the price of a bottle of soda is 15¢ and a piece of cheesecake is 60¢, show that the current basis remains optimal. What will be the new optimal solution to the diet problem?

6 If 8 oz of chocolate and 60 calories are required, show that the current basis remains optimal.

The following questions refer to the Dakota problem.

7 Suppose that the price of a desk is $65, a table is $25, and a chair is $18. Show that the current basis remains optimal. What is the new optimal z-value?

8 Suppose that 60 board ft of lumber and 23 finishing hours are available. Show that the current basis remains optimal.

9 Suppose 40 board ft of lumber, 21 finishing hours, and 8.5 carpentry hours are available. Show that the current basis remains optimal.

Group B

10 Prove the Case 1 result for the objective function coefficients.

11 To illustrate the validity of the 100% Rule for objective function coefficients, consider an LP with four decision variables (x_1, x_2, x_3, and x_4) and two constraints in which x_1 and x_2 are basic variables in the optimal basis. Suppose (if only a single objective function coefficient is changed) the current basis is known to be optimal for $L_1 \leq c_1 \leq U_1$ and $L_2 \leq c_2 \leq U_2$. Suppose we change c_1 to $c_1' = c_1 + \Delta c_1$ and c_2 to $c_2' = c_2 + \Delta c_2$, where $\Delta c_1 > 0$ and $\Delta c_2 < 0$. Let

$$\frac{\Delta c_1}{U_1 - c_1} = r_1 \qquad \text{and} \qquad \frac{-\Delta c_2}{c_2 - L_2} = r_2$$

Show that if $r_1 + r_2 \leq 1$, the current basis remains optimal. *Hint:* Any variable x_j prices out to $\mathbf{c}_{BV}B^{-1}\mathbf{a}_j - c_j$. To show that for the new values of c_1 and c_2, all variables still price out non-negative, use the fact that

$$[c_1', c_2'] = r_1[U_1, c_2] + r_2[c_1, L_2] + (1 - r_1 - r_2)[c_1, c_2]$$

12 Prove the Case 1 result for right-hand sides. Use the fact that if a constraint is nonbinding in the optimal solution, then its slack or excess variable is in the optimal basis, and the corresponding column of B^{-1} will have a single 1 and all other elements equal to 0.

13 In this problem, we sketch a proof of the 100% Rule for right-hand sides. Consider an LP with two constraints and right-hand sides b_1 and b_2. Suppose that if only one right-hand side is changed, the current basis remains optimal for $L_1 \leq b_1 \leq U_1$ and $L_2 \leq b_2 \leq U_2$. Suppose we change the right-hand sides to $b_1' = b_1 + \Delta b_1$ and $b_2' = b_2 + \Delta b_2$, where $\Delta b_1 > 0$ and $\Delta b_2 < 0$. Let

$$r_1 = \frac{\Delta b_1}{U_1 - b_1} \quad \text{and} \quad r_2 = \frac{-\Delta b_2}{b_2 - L_2}$$

Show that if $r_1 + r_2 \leq 1$, the current basis remains optimal. (*Hint:* You must show that

$$B^{-1}\begin{bmatrix} b_1' \\ b_2' \end{bmatrix} \geq \begin{bmatrix} 0 \\ 0 \end{bmatrix}$$

Use the fact that

$$[b_1', b_2'] = r_1[U_1, b_2] + r_2[b_1, L_2] + (1 - r_1 - r_2)[b_1, b_2]$$

to show this.)

6.5 Finding the Dual of an LP

Associated with any LP is another LP, called the **dual**. Knowing the relation between an LP and its dual is vital to understanding advanced topics in linear and nonlinear programming. This relation is important because it gives us interesting economic insights. Knowledge of duality will also provide additional insights into sensitivity analysis.

In this section, we explain how to find the dual of any LP; in Section 6.6, we discuss the economic interpretation of the dual; and in Sections 6.7–6.10, we discuss the relation that exists between an LP and its dual.

When taking the dual of a given LP, we refer to the given LP as the **primal.** If the primal is a max problem, then the dual will be a min problem, and vice versa. For convenience, we define the variables for the max problem to be $z, x_1, x_2, \ldots, x_n$ and the variables for the min problem to be $w, y_1, y_2, \ldots, y_m$. We begin by explaining how to find the dual of a max problem in which all variables are required to be non-negative and all constraints are $\leq$ constraints (called a **normal max problem**). A normal max problem may be written as

$$\begin{aligned}
\max z = c_1x_1 &+ c_2x_2 + \cdots + c_nx_n \\
\text{s.t.} \quad a_{11}x_1 &+ a_{12}x_2 + \cdots + a_{1n}x_n \leq b_1 \\
a_{21}x_1 &+ a_{22}x_2 + \cdots + a_{2n}x_n \leq b_2 \\
&\vdots \qquad \vdots \qquad\qquad \vdots \\
a_{m1}x_1 &+ a_{m2}x_2 + \cdots + a_{mn}x_n \leq b_m \\
x_j &\geq 0 \quad (j = 1, 2, \ldots, n)
\end{aligned} \tag{16}$$

The dual of a normal max problem such as (16) is defined to be

$$\begin{aligned}
\min w = b_1y_1 &+ b_2y_2 + \cdots + b_my_m \\
\text{s.t.} \quad a_{11}y_1 &+ a_{21}y_2 + \cdots + a_{m1}y_m \geq c_1 \\
a_{12}y_1 &+ a_{22}y_2 + \cdots + a_{m2}y_m \geq c_2 \\
&\vdots \qquad \vdots \qquad\qquad \vdots \\
a_{1n}y_1 &+ a_{2n}y_2 + \cdots + a_{mn}y_m \geq c_n \\
y_i &\geq 0 \quad (i = 1, 2, \ldots, m)
\end{aligned} \tag{17}$$

A min problem such as (17) that has all $\geq$ constraints and all variables non-negative is called a **normal min problem.** If the primal is a normal min problem such as (17), then we define the dual of (17) to be (16).

Finding the Dual of a Normal Max or Min Problem

A tabular approach makes it easy to find the dual of an LP. If the primal is a normal max problem, then it can be read across (Table 14); the dual is found by reading down. Similarly, if the primal is a normal min problem, we find it by reading down; the dual is found

TABLE 14

Finding the Dual of a Normal Max or Min Problem

min w		max z				
		$(x_1 \geq 0)$	$(x_2 \geq 0)$	$\cdots$	$(x_n \geq 0)$	
		x_1	x_2		x_n	
$(y_1 \geq 0)$	y_1	a_{11}	a_{12}	$\cdots$	a_{1n}	$\leq b_1$
$(y_2 \geq 0)$	y_2	a_{21}	a_{22}	$\cdots$	a_{2n}	$\leq b_2$
$\vdots$	$\vdots$	$\vdots$	$\vdots$		$\vdots$	$\vdots$
$(y_m \geq 0)$	y_m	a_{m1}	a_{m2}	$\cdots$	a_{mn}	$\leq b_m$
		$\geq c_1$	$\geq c_2$		$\geq c_n$	

by reading across in the table. We illustrate the use of the table by finding the dual of the Dakota problem and the dual of the diet problems. The Dakota problem is

$$\max z = 60x_1 + 30x_2 + 20x_3$$

$$\text{s.t.} \quad 8x_1 + 6x_2 + x_3 \leq 48 \qquad \text{(Lumber constraint)}$$

$$4x_1 + 2x_2 + 1.5x_3 \leq 20 \qquad \text{(Finishing constraint)}$$

$$2x_1 + 1.5x_2 + 0.5x_3 \leq 8 \qquad \text{(Carpentry constraint)}$$

$$x_1, x_2, x_3 \geq 0$$

where

$$x_1 = \text{number of desks manufactured}$$

$$x_2 = \text{number of tables manufactured}$$

$$x_3 = \text{number of chairs manufactured}$$

Using the format of Table 14, we read the Dakota problem across in Table 15. Then, reading down, we find the Dakota dual to be

$$\min w = 48y_1 + 20y_2 + 8y_3$$

$$\text{s.t.} \quad 8y_1 + 4y_2 + 2y_3 \geq 60$$

$$6y_1 + 2y_2 + 1.5y_3 \geq 30$$

$$y_1 + 1.5y_2 + 0.5y_3 \geq 20$$

$$y_1, y_2, y_3 \geq 0$$

The tabular method of finding the dual makes it clear that the ith dual *constraint* corresponds to the ith primal *variable* x_i. For example, the first dual constraint corresponds to x_1 (desks), because each number comes from the x_1 (desk) column of the primal. Simi-

TABLE 15

Finding the Dual of the Dakota Problem

min w		max z			
		$(x_1 \geq 0)$	$(x_2 \geq 0)$	$(x_3 \geq 0)$	
		x_1	x_2	x_3	
$(y_1 \geq 0)$	y_1	8	6	1	≤ 48
$(y_2 \geq 0)$	y_2	4	2	1.5	≤ 20
$(y_3 \geq 0)$	y_3	2	1.5	0.5	≤ 8
		≥ 60	≥ 30	≥ 20	

larly, the second dual constraint corresponds to x_2 (tables), and the third dual constraint corresponds to x_3 (chairs). In a similar fashion, dual *variable* y_i is associated with the *i*th primal constraint. For example, y_1 is associated with the first primal constraint (lumber constraint), because each coefficient of y_1 in the dual comes from the lumber constraint, or the availability of lumber. The importance of these correspondences between the primal and the dual will become clear in Section 6.6.

We now find the dual of the diet problem. Because the diet problem is a min problem, we follow the convention of using w to denote the objective function and y_1, y_2, y_3, and y_4 for the variables. Then the diet problem may be written as

$$\min w = 50y_1 + 20y_2 + 30y_3 + 80y_4$$

$$\begin{aligned}
\text{s.t.} \quad 400y_1 + 200y_2 + 150y_3 + 500y_4 &\geq 500 \quad \text{(Calorie constraint)} \\
3y_1 + 2y_2 &\geq 6 \quad \text{(Chocolate constraint)} \\
2y_1 + 2y_2 + 4y_3 + 4y_4 &\geq 10 \quad \text{(Sugar constraint)} \\
2y_1 + 4y_2 + y_3 + 5y_4 &\geq 8 \quad \text{(Fat constraint)} \\
y_1, y_2, y_3, y_4 &\geq 0
\end{aligned}$$

where

y_1 = number of brownies eaten daily

y_2 = number of scoops of chocolate ice cream eaten daily

y_3 = bottles of soda drunk daily

y_4 = pieces of pineapple cheesecake eaten daily

The primal is a normal min problem, so we can read it down, and read its dual across, in Table 16. We find that the dual of the diet problem is

$$\max z = 500x_1 + 6x_2 + 10x_3 + 8x_4$$

$$\begin{aligned}
\text{s.t.} \quad 400x_1 + 3x_2 + 2x_3 + 2x_4 &\leq 50 \\
200x_1 + 2x_2 + 2x_3 + 4x_4 &\leq 20 \\
150x_1 + 4x_3 + x_4 &\leq 30 \\
500x_1 + 4x_3 + 5x_4 &\leq 80 \\
x_1, x_2, x_3, x_4 &\geq 0
\end{aligned}$$

As in the Dakota problem, we see that the *i*th dual constraint corresponds to the *i*th primal variable. For example, the third dual constraint may be thought of as the soda constraint. Also, the *i*th dual variable corresponds to the *i*th primal constraint. For example, x_3 (the third dual variable) may be thought of as the dual sugar variable.

TABLE 16

Finding the Dual of the Diet Problem

min w		max z			
		$(x_1 \geq 0)$	$(x_2 \geq 0)$	$(x_3 \geq 0)$	$(x_4 \geq 0)$
		x_1	x_2	x_3	x_4	
$(y_1 \geq 0)$	y_1	400	3	2	2	≤ 50
$(y_2 \geq 0)$	y_2	200	2	2	4	≤ 20
$(y_3 \geq 0)$	y_3	150	0	4	1	≤ 30
$(y_4 \geq 0)$	y_4	500	0	4	5	≤ 80
		≥ 500	≥ 6	≥ 10	≥ 8	

Finding the Dual of a Nonnormal LP

Unfortunately, many LPs are not normal max or min problems. For example,

$$\max z = 2x_1 + x_2$$
$$\text{s.t.} \quad x_1 + x_2 = 2$$
$$2x_1 - x_2 \geq 3 \qquad\qquad (18)$$
$$x_1 - x_2 \leq 1$$
$$x_1 \geq 0, \ x_2 \text{ urs}$$

is not a normal max problem because it has a $\geq$ constraint, an equality constraint, and an unrestricted-in-sign variable. As another example of a nonnormal LP, consider

$$\min w = 2y_1 + 4y_2 + 6y_3$$
$$\text{s.t.} \quad y_1 + 2y_2 + y_3 \geq 2$$
$$y_1 \qquad\quad - y_3 \geq 1$$
$$y_2 + y_3 = 1 \qquad\qquad (19)$$
$$2y_1 + \ y_2 \qquad\quad \leq 3$$
$$y_1 \text{ urs}, \ y_2, y_3 \geq 0$$

This LP is not a normal min problem because it contains an equality constraint, a $\leq$ constraint, and an unrestricted-in-sign variable.

Fortunately, an LP can be transformed into normal form (either (16) or (17)). To place a max problem into normal form, we proceed as follows:

Step 1 Multiply each $\geq$ constraint by -1, converting it into a $\leq$ constraint. For example, in (18), $2x_1 - x_2 \geq 3$ would be transformed into $-2x_1 + x_2 \leq -3$.

Step 2 Replace each equality constraint by two inequality constraints (a $\leq$ constraint and a $\geq$ constraint). Then convert the $\geq$ constraint to a $\leq$ constraint. For example, in (18), we would replace $x_1 + x_2 = 2$ by the two inequalities $x_1 + x_2 \geq 2$ and $x_1 + x_2 \leq 2$. Then we would convert $x_1 + x_2 \geq 2$ to $-x_1 - x_2 \leq -2$. The net result is that $x_1 + x_2 = 2$ is replaced by the two inequalities $x_1 + x_2 \leq 2$ and $-x_1 - x_2 \leq -2$.

Step 3 As in Section 4.14, replace each urs variable x_i by $x_i = x' - x_i''$, where $x_i' \geq 0$ and $x_i'' \geq 0$. In (18), we would replace x_2 by $x_2' - x_2''$.

After these transformations are complete, (18) has been transformed into the following (equivalent) LP:

$$\max z = 2x_1 + x_2' - x_2''$$
$$\text{s.t.} \quad x_1 + x_2' - x_2'' \leq 2$$
$$-x_1 - x_2' + x_2'' \leq -2$$
$$-2x_1 + x_2' - x_2'' \leq -3 \qquad\qquad (18')$$
$$x_1 - x_2' + x_2'' \leq 1$$
$$x_1, x_2', x_2'' \geq 0$$

Because (18$'$) is a normal max problem, we could use (16) and (17) to find the dual of (18$'$).

If the primal is not a normal min problem, then we can transform it into a normal min problem as follows:

Step 1 Convert each $\leq$ constraint into a $\geq$ constraint by multiplying through by -1. For example, in (19), $2y_1 + y_2 \leq 3$ is transformed into $-2y_1 - y_2 \geq -3$.

Step 2 Replace each equality constraint by a $\leq$ constraint and a $\geq$ constraint. Then transform the $\leq$ constraint into a $\geq$ constraint. For example, in (19), the constraint $y_2 + y_3 = 1$ is equivalent to $y_2 + y_3 \leq 1$ and $y_2 + y_3 \geq 1$. Transforming $y_2 + y_3 \leq 1$ into $-y_2 - y_3 \geq -1$, we see that we can replace the constraint $y_2 + y_3 = 1$ by the two constraints $y_2 + y_3 \geq 1$ and $-y_2 - y_3 \geq -1$.

Step 3 Replace any urs variable y_i by $y_i = y_i' - y_i''$, where $y_i' \geq 0$ and $y_i'' \geq 0$. Applying these steps to (19) yields the following standard min problem:

$$
\begin{aligned}
\min w = 2y_1' - 2y_1'' &+ 4y_2 + 6y_3 \\
\text{s.t.} \quad y_1' - y_1'' + 2y_2 &+ y_3 \geq 2 \\
y_1' - y_1'' \quad\quad &- y_3 \geq 1 \\
y_2 &+ y_3 \geq 1 \\
- y_2 &- 6y_3 \geq -1 \\
-2y_1' + 2y_1'' - y_2 \quad\quad &\geq -3 \\
y_1', y_1'', y_2, y_3 &\geq 0
\end{aligned}
\tag{19$'$}
$$

Because (19$'$) is a normal min problem in standard form, we may use (16) and (17) to find its dual.

We can find the dual of a non-normal LP without going through the transformations that we have described by using the following rules.[†]

Finding the Dual of a Non-normal Max Problem

Step 1 Fill in Table 14 so that the primal can be read across.

Step 2 After making the following changes, the dual can be read down in the usual fashion: (a) If the ith primal constraint is a $\geq$ constraint, then the corresponding dual variable y_i must satisfy $y_i \leq 0$. (b) If the ith primal constraint is an equality constraint, then the dual variable y_i is now unrestricted in sign. (c) If the ith primal variable is urs, then the ith dual constraint will be an equality constraint.

When this method is applied to (18), the Table 14 format yields Table 17. We note with an asterisk (*) the places where the rules must be used to determine part of the dual. For example, x_2 urs causes the second dual constraint to be an equality constraint. Also, the first primal constraint being an equality constraint makes y_1 urs, and the second primal constraint being a $\geq$ constraint makes $y_2 \leq 0$. Filling in the missing information across from the appropriate asterisk yields Table 18. Reading the dual down, we obtain

$$
\begin{aligned}
\min w = 2y_1 &+ 3y_2 + y_3 \\
\text{s.t.} \quad y_1 &+ 2y_2 + y_3 \geq 2 \\
y_1 &- y_2 - y_3 = 1 \\
y_1 \text{ urs}, \; y_2 &\leq 0, \; y_3 \geq 0
\end{aligned}
$$

In Section 6.8, we give an intuitive explanation of why an equality constraint yields an unrestricted-in-sign dual variable and why a $\geq$ constraint yields a negative dual variable.

We can use the following rules to take the dual of a non-normal min problem.

[†]In Problems 5 and 6 at the end of this section, we show that these rules are consistent with taking the dual of the transformed LP via (16) and (17).

TABLE 17
Finding the Dual of LP (18)

min w		max z			
		$(x_1 \geq 0)$	$(x_2$ urs$)^*$		
		x_1	x_2		
	y_1	1	1	$=2^*$	
	y_2	2	-1	$\geq 3^*$	
$(y_3 \geq 0)$	y_3	1	-1	≤ 1	
		≥ 2	$=1$		

TABLE 18
Finding the Dual of LP (18) (Continued)

min w		max z			
		$(x_1 \geq 0)$	$(x_2$ urs$)$		
		x_1	x_2		
$(y_1$ urs$)$	y_1	1	1	$=2$	
$(y_2 \leq 0)$	y_2	2	-1	≥ 3	
$(y_3 \geq 0)$	y_3	1	-1	≤ 1	
		≥ 2	$=1$		

Finding the Dual of a Non-normal Min Problem

Step 1   Write out the primal so it can be read down in Table 14.

Step 2   Except for the following changes, the dual can be read across the table: (a) If the ith primal constraint is a $\leq$ constraint, then the corresponding dual variable x_i must satisfy $x_i \leq 0$. (b) If the ith primal constraint is an equality constraint, then the corresponding dual variable x_i will be urs. (c) If the ith primal variable y_i is urs, then the ith dual constraint is an equality constraint.

When this method is applied to (19), we get Table 19. Asterisks (*) show where the new rules must be used to determine parts of the dual. Because y_1 is urs, the first dual constraint is an equality. The third primal constraint is an equality, so dual variable x_3 is urs. Finally, because the fourth primal constraint is a $\leq$ constraint, the fourth dual variable x_4 must satisfy $x_4 \leq 0$. We can now complete the table (see Table 20). Reading the dual across, we obtain

TABLE 19
Finding the Dual of LP (19)

min w		max z				
		$(x_1 \geq 0)$	$(x_2 \geq 0)$			
		x_1	x_2	x_3	x_4	
$(y_1$ urs$)^*$	y_1	1	1	0	2	2
$(y_2 \geq 0)$	y_2	2	0	1	1	≤ 4
$(y_3 \geq 0)$	y_3	1	-1	1	0	≤ 6
		≥ 2	≥ 1	$=1^*$	$\leq 3^*$	

TABLE 20
Finding the Dual of LP (19) (Continued)

min w		max z			
		$(x_1 \geq 0)$	$(x_2 \geq 0)$	$(x_3 \geq \text{urs})$	$(x_4 \leq 0)$
		x_1	x_2	x_3	x_4
$(y_1 \text{ urs})$	y_1	1	1	0	2	=2
$(y_2 \geq 0)$	y_2	2	0	1	1	≤ 4
$(y_3 \geq 0)$	y_3	1	-1	1	0	≤ 6
		≥ 2	≥ 1	$=1$	≤ 3

$$\max z = 2x_1 + x_2 + x_3 + 3x_4$$
$$\text{s.t.} \quad x_1 + x_2 \qquad\quad + 2x_4 = 2$$
$$2x_1 \qquad + x_3 + x_4 \leq 4$$
$$x_1 - x_2 + x_3 \qquad\quad \leq 6$$
$$x_1, x_2 \geq 0, x_3 \text{ urs}, x_4 \leq 0$$

The reader may verify that with these rules, the dual of the dual is always the primal. This is easily seen from the Table 14 format, because when you take the dual of the dual you are changing the LP back to its original position.

PROBLEMS

Group A

Find the duals of the following LPs:

1 $\max z = 2x_1 + x_2$
  s.t. $-x_1 + x_2 \leq 1$
  $\qquad x_1 + x_2 \leq 3$
  $\qquad x_1 - 2x_2 \leq 4$
  $\qquad\qquad x_1, x_2 \geq 0$

2 $\min w = y_1 - y_2$
  s.t. $2y_1 + y_2 \geq 4$
  $\qquad y_1 + y_2 \geq 1$
  $\qquad y_1 + 2y_2 \geq 3$
  $\qquad\qquad y_1, y_2 \geq 0$

3 $\max z = 4x_1 - x_2 + 2x_3$
  s.t. $x_1 + x_2 \qquad\quad \leq 5$
  $\qquad 2x_1 + x_2 \qquad\quad \leq 7$
  $\qquad\qquad 2x_2 + x_3 \geq 6$
  $\qquad x_1 \qquad\quad + x_3 = 4$
  $\qquad x_1 \geq 0, x_2, x_3 \text{ urs}$

4 $\min w = 4y_1 + 2y_2 - y_3$
  s.t. $y_1 + 2y_2 \qquad\quad \leq 6$
  $\qquad y_1 - y_2 + 2y_3 = 8$
  $\qquad y_1, y_2 \geq 0, y_3 \text{ urs}$

Group B

5 This problem shows why the dual variable for an equality constraint should be urs.

a Use the rules given in the text to find the dual of
  $\max z = x_1 + 2x_2$
  s.t. $3x_1 + x_2 \leq 6$
  $\qquad 2x_1 + x_2 = 5$
  $\qquad\quad x_1, x_2 \geq 0$

b Now transform the LP in part (a) to the normal form. Using (16) and (17), take the dual of the transformed LP. Use y_2' and y_2'' as the dual variables for the two primal constraints derived from $2x_1 + x_2 = 5$.

c Make the substitution $y_2 = y_2' - y_2''$ in the part (b) answer. Now show that the two duals obtained in parts (a) and (b) are equivalent.

6 This problem shows why a dual variable y_i corresponding to a $\geq$ constraint in a max problem must satisfy $y_i \leq 0$.

a Using the rules given in the text, find the dual of
  $\max z = 3x_1 + x_2$
  s.t. $x_1 + x_2 \leq 1$
  $\qquad -x_1 + x_2 \geq 2$
  $\qquad\quad x_1, x_2 \geq 0$

b Transform the LP of part (a) into a normal max problem. Now use (16) and (17) to find the dual of the transformed LP. Let $\bar{y}_2$ be the dual variable corresponding to the second primal constraint.

c Show that, defining $\bar{y}_2 = -y_2$, the dual in part (a) is equivalent to the dual in part (b).

6.6 Economic Interpretation of the Dual Problem

Interpreting the Dual of a Max Problem

The dual of the Dakota problem is

$$\min w = 48y_1 + 20y_2 + 8y_3$$

$$
\begin{aligned}
\text{s.t.} \quad & 8y_1 + 4y_2 + 2y_3 \geq 60 && \text{(Desk constraint)} \\
& 6y_1 + 2y_2 + 1.5y_3 \geq 30 && \text{(Table constraint)} \\
& y_1 + 1.5y_2 + 0.5y_3 \geq 20 && \text{(Chair constraint)} \\
& y_1, y_2, y_3 \geq 0
\end{aligned}
$$

(20)

The first dual constraint is associated with desks, the second with tables, and the third with chairs. Also, y_1 is associated with lumber, y_2 with finishing hours, and y_3 with carpentry hours. The relevant information about the Dakota problem is shown in Table 21.

We are now ready to interpret the Dakota dual (20). Suppose an entrepreneur wants to purchase all of Dakota's resources. Then the entrepreneur must determine the price he or she is willing to pay for a unit of each of Dakota's resources. With this in mind, we define

$$y_1 = \text{price paid for 1 board ft of lumber}$$
$$y_2 = \text{price paid for 1 finishing hour}$$
$$y_3 = \text{price paid for 1 carpentry hour}$$

The resource prices y_1, y_2, and y_3 should be determined by solving the Dakota dual (20). The total price that should be paid for these resources is $48y_1 + 20y_2 + 8y_3$. Because the cost of purchasing the resources is to be minimized,

$$\min w = 48y_1 + 20\,y_2 + 8y_3$$

is the objective function for the Dakota dual.

In setting resource prices, what constraints does the entrepreneur face? Resource prices must be set high enough to induce Dakota to sell. For example, the entrepreneur must offer Dakota at least \$60 for a combination of resources that includes 8 board feet of lumber, 4 finishing hours, and 2 carpentry hours, because Dakota could, if it desires, use these resources to produce a desk that can be sold for \$60. The entrepreneur is offering $8y_1 + 4y_2 + 2y_3$ for the resources used to produce a desk, so he or she must choose y_1, y_2, and y_3 to satisfy

$$8y_1 + 4y_2 + 2y_3 \geq 60$$

But this is just the first (or desk) constraint of the Dakota dual. Similar reasoning shows that at least \$30 must be paid for the resources used to produce a table (6 board feet of lumber, 2 finishing hours, and 1.5 carpentry hours). This means that y_1, y_2, and y_3 must satisfy

$$6y_1 + 2y_2 + 1.5y_3 \geq 30$$

TABLE 21
Relevant Information for Dakota Problem

Resource	Resource/Product			Amount of Resource Available
	Desk	Table	Chair	
Lumber (board ft)	8	6	1	48
Finishing (hours)	4	2	1.5	20
Carpentry (hours)	2	1.5	0.5	8
Selling price ($)	60	30	20	

This is the second (or table) constraint of the Dakota dual.

Similarly, the third (or chair) dual constraint,

$$y_1 + 1.5y_2 + 0.5y_3 \geq 20$$

states that at least \$20 (the price of a chair) must be paid for the resources needed to produce a chair (1 board foot of lumber, 1.5 finishing hours, and 0.5 carpentry hour). The sign restrictions $y_1 \geq 0$, $y_2 \geq 0$, and $y_3 \geq 0$ must also hold. Putting everything together, we see that the solution to the dual of the Dakota problem does yield prices for lumber, finishing hours, and carpentry hours. The preceding discussion also shows that the ith dual variable does indeed correspond in a natural way to the ith primal constraint.

In summary, when the primal is a normal max problem, the dual variables are related to the value of the resources available to the decision maker. For this reason, the dual variables are often referred to as **resource shadow prices.** A more thorough discussion of shadow prices is given in Section 6.8.

Interpreting the Dual of a Min Problem

To interpret the dual of a min problem, we consider the dual of the diet problem of Section 3.4. In Section 6.5, we found that the diet problem dual was

$$\max z = 500x_1 + 6x_2 + 10x_3 + 8x_4$$

$$
\begin{aligned}
\text{s.t.} \quad & 400x_1 + 3x_2 + 2x_3 + 2x_4 \leq 50 && \text{(Brownie constraint)} \\
& 200x_1 + 2x_2 + 2x_3 + 4x_4 \leq 20 && \text{(Ice cream constraint)} \\
& 150x_1 \phantom{{}+ 2x_2} + 4x_3 + x_4 \leq 30 && \text{(Soda constraint)} \\
& 500x_1 \phantom{{}+ 2x_2} + 4x_3 + 5x_4 \leq 80 && \text{(Cheesecake constraint)}
\end{aligned}
$$
(21)

$$x_1, x_2, x_3, x_4 \geq 0$$

The data for the diet problem are shown in Table 22. To interpret (21), suppose Candice is a "nutrient" salesperson who sells calories, chocolate, sugar, and fat. She wants to ensure that a dieter will meet all of his or her daily requirements by purchasing calories, sugar, fat, and chocolate. Then Candice must determine

$$x_1 = \text{price per calorie to charge dieter}$$
$$x_2 = \text{price per ounce of chocolate to charge dieter}$$
$$x_3 = \text{price per ounce of sugar to charge dieter}$$
$$x_4 = \text{price per ounce of fat to charge dieter}$$

Candice wants to maximize her revenue from selling the dieter the daily ration of required nutrients. Because she will receive $500x_1 + 6x_2 + 10x_3 + 8x_4$ cents in revenue from the dieter, her objective is to

$$\max z = 500x_1 + 6x_2 + 10x_3 + 8x_4$$

TABLE 22
Relevant Information for Diet Problem

	Calories	Chocolate (Ounces)	Sugar (Ounces)	Fat (Ounces)	Price (Cents)
Brownie	400	3	2	2	50
Ice cream	200	2	2	4	20
Soda	150	0	4	1	30
Cheesecake	500	0	4	5	80
Requirements	500	6	10	8	

This is the objective function for the dual of the diet problem. But in setting nutrient prices, Candice must set prices low enough so that it will be in the dieter's economic interest to purchase all nutrients from her. For example, by purchasing a brownie for 50¢, the dieter can obtain 400 calories, 3 oz of chocolate, 2 oz of sugar, and 2 oz of fat. So Candice cannot charge more than 50¢ for this combination of nutrients. This leads to the following (brownie) constraint:

$$400x_1 + 3x_2 + 2x_3 + 2x_4 \le 50$$

the first constraint in the diet problem dual. Similar reasoning yields the second dual (ice cream) constraint, the third (soda constraint), and the fourth (cheesecake constraint). Again, the sign restrictions $x_1 \ge 0$, $x_2 \ge 0$, $x_3 \ge 0$, and $x_4 \ge 0$ must be satisfied.

Our discussion shows that the optimal value of x_i may be interpreted as a price for 1 unit of the nutrient associated with the ith dual constraint. Thus, x_1 would be the price for 1 calorie, x_2 would be the price for 1 oz of chocolate, and so on. Again, we see that it is reasonable to associate the ith dual variable (x_i) and the ith primal constraint.

In summary, we have shown that when the primal is a normal max problem or a normal min problem, the dual problem has an intuitive economic interpretation. In Section 6.8, we explain more about the proper interpretation of the dual variables.

PROBLEM

Group A

1 Find the dual of Example 3 in Chapter 3 (an auto company) and give an economic interpretation of the dual problem.

2 Find the dual of Example 2 in Chapter 3 (Dorian Auto) and give an economic interpretation of the dual problem.

6.7 The Dual Theorem and Its Consequences

In this section, we discuss one of the most important results in linear programming: the Dual Theorem. In essence, the Dual Theorem states that the primal and dual have equal optimal objective function values (if the problems have optimal solutions). This result is interesting in its own right, but we will see that in proving the Dual Theorem, we gain many important insights into linear programming.

To simplify the exposition, we assume that the primal is a normal max problem with m constraints and n variables. Then the dual problem will be a normal min problem with m variables and n constraints. In this case, the primal and the dual may be written as follows:

$$\max z = c_1x_1 + c_2x_2 + \cdots + c_nx_n$$

Primal Problem

$$\begin{aligned}
\text{s.t.} \quad & a_{11}x_1 + a_{12}x_2 + \cdots + a_{1n}x_n \le b_1 \\
& a_{21}x_1 + a_{22}x_2 + \cdots + a_{2n}x_n \le b_2 \\
& \quad\vdots \qquad \vdots \qquad\qquad \vdots \qquad \vdots \\
& a_{i1}x_1 + a_{i2}x_2 + \cdots + a_{in}x_n \le b_i \\
& \quad\vdots \qquad \vdots \qquad\qquad \vdots \qquad \vdots \\
& a_{m1}x_1 + a_{m2}x_2 + \cdots + a_{mn}x_n \le b_m \\
& x_j \ge 0 \quad (j = 1, 2, \ldots, n)
\end{aligned}$$

(22)

$$\min w = b_1 y_1 + b_2 y_2 + \cdots + b_m y_m$$

$$\text{s.t.} \quad a_{11} y_1 + a_{21} y_2 + \cdots + a_{m1} y_m \geq c_1$$

$$a_{12} y_1 + a_{22} y_2 + \cdots + a_{m2} y_m \geq c_2$$

$$\vdots \qquad \vdots \qquad\qquad \vdots \qquad \vdots$$

Dual Problem (23)

$$a_{1j} y_1 + a_{2j} y_2 + \cdots + a_{mj} y_m \geq c_j$$

$$\vdots \qquad \vdots \qquad\qquad \vdots \qquad \vdots$$

$$a_{1n} y_1 + a_{2n} y_2 + \cdots + a_{mn} y_m \geq c_n$$

$$y_i \geq 0 \quad (i = 1, 2, \ldots, m)$$

Weak Duality

If we choose any feasible solution to the primal and any feasible solution to the dual, the w-value for the feasible dual solution will be at least as large as the z-value for the feasible primal solution. This result is formally stated in Lemma 1.

LEMMA 1

(Weak Duality). Let

$$\mathbf{x} = \begin{bmatrix} x_1 \\ x_2 \\ \vdots \\ x_n \end{bmatrix}$$

be any feasible solution to the primal and $\mathbf{y} = [y_1 \quad y_2 \quad \cdots \quad y_m]$ be any feasible solution to the dual. Then (z-value for $\mathbf{x}$) $\leq$ (w-value for $\mathbf{y}$).

Proof Because $y_i \geq 0$, multiplying the ith primal constraint in (22) by y_i will yield the following valid inequality:

$$y_i a_{i1} x_1 + y_i a_{i2} x_2 + \cdots + y_i a_{in} x_n \leq b_i y_i \quad (i = 1, 2, \ldots, m) \tag{24}$$

Adding the m inequalities in (24), we find that

$$\sum_{i=1}^{i=m} \sum_{j=1}^{j=n} y_i a_{ij} x_j \leq \sum_{i=1}^{i=m} b_i y_i \tag{25}$$

Because $x_j \geq 0$, multiplying the jth dual constraint in (23) by x_j yields the following valid inequality:

$$x_j a_{1j} y_1 + x_j a_{2j} y_2 + \cdots + x_j a_{mj} y_m \geq c_j x_j \quad (j = 1, 2, \ldots, n) \tag{26}$$

Adding the n inequalities in (26) yields

$$\sum_{i=1}^{i=m} \sum_{j=1}^{j=n} y_i a_{ij} x_j \geq \sum_{j=1}^{j=n} c_j x_j \tag{27}$$

Combining (25) and (27), we obtain

$$\sum_{j=1}^{j=n} c_j x_j \leq \sum_{i=1}^{i=m} \sum_{j=1}^{j=n} y_i a_{ij} x_j \leq \sum_{i=1}^{i=m} b_i y_i$$

which is the desired result.

If a feasible solution to either the primal or the dual is readily available, weak duality can be used to obtain a bound on the optimal objective function value for the other problem. For example, in looking at the Dakota problem, it is easy to see that $x_1 = x_2 = x_3 = 1$ is primal feasible. This solution has a z-value of $60 + 30 + 20 = 110$. Weak duality now implies that any dual feasible solution (y_1, y_2, y_3) must satisfy

$$48y_1 + 20y_2 + 8y_3 \geq 110$$

Because the dual is a min problem, and any dual feasible solution must have $w \geq 110$, this means that the optimal w-value for the dual ≥ 110 (see Figure 7). This shows that weak duality enables us to use any primal feasible solution to bound the optimal value of the dual objective function.

Analogously, we can use any feasible solution to the dual to develop a bound on the optimal value of the primal objective function. For example, looking at the Dakota dual, it can readily be verified that $y_1 = 10$, $y_2 = 10$, $y_3 = 0$ is dual feasible. This dual solution has a dual objective function value of $48(10) + 20(10) + 8(0) = 680$. From weak duality, we see that any primal feasible solution

must satisfy

$$60x_1 + 30x_2 + 20x_3 \leq 680$$

Because the primal is a max problem and every primal feasible solution has $z \leq 680$, we may conclude that the optimal primal objective function value ≤ 680 (see Figure 8).

If we define

$$\mathbf{b} = \begin{bmatrix} b_1 \\ b_2 \\ \vdots \\ b_m \end{bmatrix} \quad \text{and} \quad \mathbf{c} = \begin{bmatrix} c_1 & c_2 & \cdots & c_n \end{bmatrix}$$

then for a point

$$\mathbf{x} = \begin{bmatrix} x_1 \\ x_2 \\ \vdots \\ x_n \end{bmatrix}$$

the primal objective function value may be written as $\mathbf{cx}$, and for a point $\mathbf{y} = \begin{bmatrix} y_1 & y_2 & \cdots & y_m \end{bmatrix}$ the dual objective function value may be written as $\mathbf{yb}$. We now use weak duality to prove the following important result.

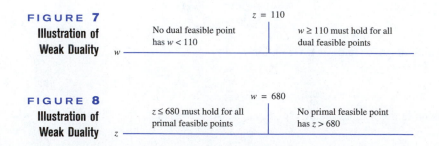

FIGURE 7
Illustration of
Weak Duality

FIGURE 8
Illustration of
Weak Duality

Let

$$\overline{\mathbf{x}} = \begin{bmatrix} \overline{x}_1 \\ \overline{x}_2 \\ \vdots \\ \overline{x}_n \end{bmatrix}$$

be a feasible solution to the primal and $\overline{\mathbf{y}} = [\overline{y}_1 \quad \overline{y}_2 \quad \cdots \quad \overline{y}_m]$ be a feasible solution to the dual. If $\mathbf{c}\overline{\mathbf{x}} = \overline{\mathbf{y}}\mathbf{b}$, then $\overline{\mathbf{x}}$ is optimal for the primal and $\overline{\mathbf{y}}$ is optimal for the dual.

Proof From weak duality we know that for any primal feasible point $\mathbf{x}$,

$$\mathbf{c}\mathbf{x} \leq \overline{\mathbf{y}}\mathbf{b}$$

Thus, any primal feasible point must yield a z-value that does not exceed $\overline{\mathbf{y}}\mathbf{b}$. Because $\overline{\mathbf{x}}$ is primal feasible and has a primal objective function value of $\mathbf{c}\overline{\mathbf{x}} = \overline{\mathbf{y}}\mathbf{b}$, $\overline{\mathbf{x}}$ must be primal optimal. Similarly, because $\overline{\mathbf{x}}$ is primal feasible, weak duality implies that for any dual feasible point $\mathbf{y}$,

$$\mathbf{c}\overline{\mathbf{x}} \leq \mathbf{y}\mathbf{b}$$

Thus, any dual feasible point must yield an objective function value exceeding $\mathbf{c}\overline{\mathbf{x}}$. Because $\overline{\mathbf{y}}$ is dual feasible and has a dual objective function value $\overline{\mathbf{y}}\mathbf{b} = \mathbf{c}\overline{\mathbf{x}}$, $\overline{\mathbf{y}}$ must be an optimal solution for the dual.

We use the Dakota problem to illustrate the use of Lemma 2. The reader may verify that

$$\overline{\mathbf{x}} = \begin{bmatrix} 2 \\ 0 \\ 8 \end{bmatrix}$$

is primal feasible and that $\overline{\mathbf{y}} = [0 \quad 10 \quad 10]$ is dual feasible. Because $\mathbf{c}\overline{\mathbf{x}} = \overline{\mathbf{y}}\mathbf{b} = 280$, Lemma 2 implies that $\overline{\mathbf{x}}$ is optimal for the Dakota primal, and $\overline{\mathbf{y}}$ is optimal for the Dakota dual. Lemma 2 plays an important role in our proof of the Dual Theorem.

The Dual Theorem

Before proceeding with our proof of the Dual Theorem, we note that weak duality can be used to prove the following results.

If the primal is unbounded, then the dual problem is infeasible.

Proof See Problem 7 at the end of this section.

If the dual is unbounded, then the primal is infeasible.

Proof See Problem 8 at the end of this section.

Lemmas 3 and 4 describe the relation between the primal and dual in two relatively unimportant cases.[†]

These cases are of limited interest. We are primarily interested in the relation between the primal and dual when the primal has an optimal solution. In what follows, we let $\bar{z}$ = optimal primal objective function value and $\bar{w}$ = optimal dual objective function value. If the primal has an optimal solution, then the following important result (the Dual Theorem) describes the relation between the primal and the dual.

THEOREM 1

The Dual Theorem

Suppose BV is an optimal basis for the primal. Then $\mathbf{c}_{BV}B^{-1}$ is an optimal solution to the dual. Also, $\bar{z} = \bar{w}$.

Proof The argument used to prove the Dual Theorem includes the following steps:

1 Use the fact that BV is an optimal basis for the primal to show that $\mathbf{c}_{BV}B^{-1}$ is dual feasible.

2 Show that the optimal primal objective function value = the dual objective function for $\mathbf{c}_{BV}B^{-1}$.

3 We have found a primal feasible solution (from BV) and a dual feasible solution ($\mathbf{c}_{BV}B^{-1}$) that have equal objective function values. From Lemma 2, we can now conclude that $\mathbf{c}_{BV}B^{-1}$ is optimal for the dual and $\bar{z} = \bar{w}$.

We now verify step 1 for the case where the primal is a normal maximization problem with n variables and m constraints.[‡] After adding slack variables $s_1, s_2, \ldots, s_m$ to the primal, we write the primal and dual problems as follows:

$$\max z = c_1x_1 + c_2x_2 + \cdots + c_nx_n$$

Primal Problem

$$\begin{aligned}
\text{s.t.} \quad & a_{11}x_1 + a_{12}x_2 + \cdots + a_{1n}x_n + s_1 && = b_1 \\
& a_{21}x_1 + a_{22}x_2 + \cdots + a_{2n}x_n && + s_2 && = b_2 \\
& \quad\vdots \qquad\quad \vdots \qquad\qquad\quad \vdots \qquad\qquad\qquad \vdots \\
& a_{m1}x_1 + a_{m2}x_2 + \cdots + a_{mn}x_n && + s_m = b_m
\end{aligned}$$
(28)

$$x_j \geq 0 \quad (j = 1, 2, \ldots, n); \qquad s_i \geq 0 \quad (i = 1, 2, \ldots, m)$$

$$\min w = b_1y_1 + b_2y_2 + \cdots + b_my_m$$

Dual Problem

$$\begin{aligned}
\text{s.t.} \quad & a_{11}y_1 + a_{21}y_2 + \cdots + a_{m1}y_m \geq c_1 \\
& a_{12}y_1 + a_{22}y_2 + \cdots + a_{m2}y_m \geq c_2 \\
& \quad\vdots \qquad\quad \vdots \qquad\qquad\quad \vdots \qquad\qquad\qquad \vdots \\
& a_{1n}y_1 + a_{2n}y_2 + \cdots + a_{mn}y_m \geq c_n
\end{aligned}$$
(29)

$$y_i \geq 0 \quad (i = 1, 2, \ldots, m)$$

Let BV be an optimal basis for the primal, and define $\mathbf{c}_{BV}B^{-1} = [y_1, y_2, \ldots, y_m]$. Thus, for the optimal basis BV, y_i is the ith element of $\mathbf{c}_{BV}B^{-1}$. BV is optimal for

---

[†]It can happen that both the primal and dual can be infeasible, as in the following example:

$$\begin{array}{llll}
& \max z = x_2 & & \min w = -y_1 + y_2 \\
& \text{s.t.} \quad x_1 \qquad\quad \leq -1 & & \text{s.t.} \quad y_1 \qquad\quad \geq 0 \\
\textbf{Primal} & \qquad\quad -x_2 \leq 1 & \textbf{Dual} & \qquad -y_2 \geq 1 \\
& \qquad x_1, x_2 \geq 0 & & \qquad y_1, y_2 \geq 0
\end{array}$$

[‡]Our proof can easily be modified to handle the situation where the primal is not a normal max problem.

the primal, so the coefficient of each variable in row 0 of BV's primal tableau must be non-negative. From (10), the coefficient of x_j in row 0 of the BV tableau ($\bar{c}_j$) is given by

$$\bar{c}_j = \mathbf{c}_{BV}B^{-1}\mathbf{a}_j - c_j$$

$$= [\,y_1 \quad y_2 \quad \cdots \quad y_m\,] \begin{bmatrix} a_{1j} \\ a_{2j} \\ \vdots \\ a_{mj} \end{bmatrix} - c_j$$

$$= y_1 a_{1j} + y_2 a_{2j} + \cdots + y_m a_{mj} - c_j$$

But we know that $\bar{c}_j \geq 0$, so for $j = 1, 2, \ldots, n$,

$$y_1 a_{1j} + y_2 a_{2j} + \cdots + y_m a_{mj} - c_j \geq 0$$

Thus, $\mathbf{c}_{BV}B^{-1}$ satisfies each of the n dual constraints.

Because BV is an optimal basis for the primal, we also know that each slack variable has a non-negative coefficient in the BV primal tableau. From (10'), we find that the coefficient of s_i in BV's row 0 is y_i, the ith element of $\mathbf{c}_{BV}B^{-1}$. Thus, for $i = 1, 2, \ldots, m$, $y_i \geq 0$. We have shown that $\mathbf{c}_{BV}B^{-1}$ satisfies all n constraints in (29) and that all the elements of $\mathbf{c}_{BV}B^{-1}$ are non-negative. Thus, $\mathbf{c}_{BV}B^{-1}$ is indeed dual feasible.

Step 2 of the Dual Theorem proof requires that we show

Dual objective function value for $\mathbf{c}_{BV}B^{-1}$
$$= \text{primal objective function value for BV} \quad (30)$$

From (11), we know that the primal objective function value for BV is $\mathbf{c}_{BV}B^{-1}\mathbf{b}$. But the dual objective function value for the dual feasible solution $\mathbf{c}_{BV}B^{-1}$ is

$$b_1 y_1 + b_2 y_2 + \cdots + b_m y_m = [\,y_1 \quad y_2 \quad \cdots \quad y_m\,] \begin{bmatrix} b_1 \\ b_2 \\ \vdots \\ b_m \end{bmatrix} = \mathbf{c}_{BV}B^{-1}\mathbf{b}$$

Thus, (30) is valid.

We have shown that steps 1 and 2 of the Dual Theorem proof are valid. Step 3 now completes our proof of the Dual Theorem.

REMARKS **1** In step 1 of the Dual Theorem proof, we showed that a basis BV that is feasible for the primal is optimal if and only if $\mathbf{c}_{BV}B^{-1}$ is dual feasible. In Section 6.9, we use this result to gain useful insights into sensitivity analysis.

2 When we find the optimal solution to the primal by using the simplex algorithm, we have also found the optimal solution to the dual.

To see why Remark 2 is true, suppose that the primal is a normal max problem with m constraints. To use the simplex to solve this problem, we must add a slack variable s_i to the ith primal constraint. Suppose BV is an optimal basis for the primal. Then the Dual Theorem tells us that $\mathbf{c}_{BV}B^{-1} = [\,y_1 \quad y_2 \quad \cdots \quad y_m\,]$ is the optimal solution to the dual. Recall from (10'), however, that y_i is the coefficient of s_i in row 0 of the optimal (BV) primal tableau. Thus, we have shown that *if the primal is a normal max problem, then the optimal value of the ith dual variable is the coefficient of s_i in row 0 of the optimal primal tableau.*

TABLE 23
Optimal Solution to the Dakota Problem

								Basic Variable
z	$+$	$5x_2$		$+$	$10s_2 +$	$10s_3 = 280$		$z_1 = 280$
	$-$	$2x_2$	$+ s_1 +$		$2s_2 -$	$8s_3 = 24$		$s_1 = 24$
	$-$	$2x_2 + x_3$		$+$	$2s_2 -$	$4s_3 = 8$		$x_3 = 8$
x_1	$+$	$1.25x_2$			$- 0.5s_2 +$	$1.5s_3 = 2$		$x_1 = 2$

We use the Dakota problem to illustrate Remark 2. The optimal tableau for the Dakota problem is shown in Table 23. The optimal primal solution is $z = 280$, $s_1 = 24$, $x_3 = 8$, $x_1 = 2$, $x_2 = 0$, $s_2 = 0$, $s_3 = 0$. From the preceding discussion, the optimal dual solution is $y_1 = 0$, $y_2 = 10$, $y_3 = 10$, $w = 48(0) + 20(10) + 8(10) = 280$. Observe that the optimal primal and dual objective function values are equal, as required by the Dual Theorem.

Of course, we may always compute the optimal dual solution directly by solving

$$\mathbf{c}_{\text{BV}}B^{-1} = [0 \quad 20 \quad 60] \begin{bmatrix} 1 & 2 & -8 \\ 0 & 2 & -4 \\ 0 & -0.5 & 1.5 \end{bmatrix} = [0 \quad 10 \quad 10]$$

Of course, the two methods of obtaining the dual solution agree.

If the primal has $\geq$ or equality constraints, then we can still find the optimal dual solution from the optimal primal tableau. To see how this is done, recall that the Dual Theorem tells us that the optimal value of the ith dual variable (y_i) is the ith element of $\mathbf{c}_{\text{BV}}B^{-1}$. From (10″), we see that if the ith constraint of the primal is a $\geq$ constraint, then

Optimal value of ith dual variable $= y_i = -$(coefficient of e_i in the optimal row 0)

The coefficient of e_i in the optimal row 0 must be non-negative, so this shows that if the ith constraint in the primal is a $\geq$ constraint, then $y_i \leq 0$. This agrees with our previous convention (see Section 6.5) that a $\geq$ constraint must have a nonpositive dual variable. From (10‴), we see that if the ith primal constraint is an equality constraint, then

$$y_i = \text{(coefficient of } a_i \text{ in optimal row 0)} - M$$

Although the coefficient of a_i in the optimal row 0 must be non-negative, the fact that M is a large positive number means that $y_i \geq 0$ or $y_i \leq 0$ is possible. This agrees with our previous convention, which stated that the dual variable for an equality constraint is urs.

How to Read the Optimal Dual Solution from Row 0 of the Optimal Tableau If the Primal Is a Max Problem

$$\frac{\text{Optimal value of dual variable } y_i}{\text{if Constraint } i \text{ is a} \leq \text{constraint}} = \text{coefficient of } s_i \text{ in optimal row 0} \qquad \text{(31)}$$

$$\frac{\text{Optimal value of dual variable } y_i}{\text{if Constraint } i \text{ is a} \geq \text{constraint}} = -\text{(coefficient of } e_i \text{ in optimal row 0)} \qquad \text{(31′)}$$

$$\frac{\text{Optimal value of dual variable } y_i}{\text{if Constraint } i \text{ is an equality constraint}} = \text{(coefficient of } a_i \text{ in optimal row 0)} - M \quad \text{(31″)}$$

The following example illustrates how to find the optimal dual solution to a problem with $\leq$, $\geq$, and equality constraints.

EXAMPLE 10   Finding Dual Solution to Non-Normal Max Problem

To solve the following LP,

$$\max z = 3x_1 + 2x_2 + 5x_3$$

$$\text{s.t.} \quad x_1 + 3x_2 + 2x_3 \leq 15$$

$$2x_2 - x_3 \geq 5 \tag{32}$$

$$2x_1 + x_2 - 5x_3 = 10$$

$$x_1, x_2, x_3 \geq 0$$

we add a slack variable s_1, subtract an excess variable e_2, and add two artificial variables a_2 and a_3. The optimal tableau for (32) is given in Table 24. From this tableau, the optimal solution is $z = \frac{565}{23}$, $x_3 = \frac{15}{23}$, $x_2 = \frac{65}{23}$, $x_1 = \frac{120}{23}$, $s_1 = e_2 = a_2 = a_3 = 0$. Use this information to find the optimal solution to the dual of (32).

Solution   Following the steps in Section 6.5, we find the dual of (32) from the tableau in Table 25:

$$\min w = 15y_1 + 5y_2 + 10y_3$$

$$\text{s.t.} \quad y_1 \qquad\quad + 2y_3 \geq 3$$

$$3y_1 + 2y_2 + y_3 \geq 2 \tag{33}$$

$$2y_1 - y_2 - 5y_3 \geq 5$$

$$y_1 \geq 0, \ y_2 \leq 0, \ y_3 \text{ urs}$$

From (31) and the optimal primal tableau, we can find the optimal solution to (33) as follows:

Because the first primal constraint is a $\leq$ constraint, we see from (31) that $y_1 = $ coefficient of s_1 in optimal row $0 = \frac{51}{23}$. The second primal constraint is a $\geq$ constraint, so we see from (31') that $y_2 = -(\text{coefficient of } e_2 \text{ in optimal row } 0) = -\frac{58}{23}$. Because the third constraint is an equality constraint, we see from (31'') that $y_3 = (\text{coefficient of } a_3 \text{ in the optimal row } 0) - M = \frac{9}{23}$.

TABLE 24
Optimal Tableau for LP (32)

z	x_1	x_2	x_3	s_1	e_2	a_2	a_3	rhs	Basic Variable
1	0	0	0	$\frac{51}{23}$	$\frac{58}{23}$	$M - \frac{58}{23}$	$M + \frac{9}{23}$	$\frac{565}{23}$	$z = \frac{565}{23}$
0	0	0	1	$\frac{4}{23}$	$\frac{5}{23}$	$-\frac{5}{23}$	$-\frac{2}{23}$	$\frac{15}{23}$	$x_3 = \frac{15}{23}$
0	0	1	0	$\frac{2}{23}$	$-\frac{9}{23}$	$\frac{9}{23}$	$-\frac{1}{23}$	$\frac{65}{23}$	$x_2 = \frac{65}{23}$
0	1	0	0	$\frac{9}{23}$	$-\frac{17}{23}$	$-\frac{17}{23}$	$\frac{7}{23}$	$\frac{120}{23}$	$x_1 = \frac{120}{23}$

TABLE 25
Finding the Dual of LP (32)

min w		max z			
		$(x_1 \geq 0)$	$(x_2 \geq 0)$	$(x_3 \geq 0)$	
		x_1	x_2	x_3	
$(y_1 \geq 0)$	y_1	1	3	2	≤ 15
$(y_2 \leq 0)$	y_2	0	2	-1	$\geq 5^*$
$(y_3 \text{ urs})$	y_3	2	1	-5	$=10^*$
		≥ 3	≥ 2	≥ 5	

By the Dual Theorem, the optimal dual objective function value w must equal $\frac{565}{23}$. In summary, the optimal dual solution is

$$\overline{w} = \frac{565}{23}, y_1 = \frac{51}{23}, y_2 = -\frac{58}{23}, y_3 = \frac{9}{23}$$

The reader should check that this solution is indeed feasible (all dual constraints are satisfied with equality) and that

$$\overline{w} = 15(\tfrac{51}{23}) + 5(-\tfrac{58}{23}) + 10(\tfrac{9}{23}) = \frac{565}{23}$$

Even if the primal is a min problem, we may read the optimal dual solution from the optimal primal tableau.

How to Read the Optimal Dual Solution from Row 0 of the Optimal Tableau If the Primal Is a Min Problem

Optimal value of dual variable x_i if Constraint i is a $\leq$ constraint = coefficient of s_i in optimal row 0

Optimal value of dual variable x_i if Constraint i is a $\geq$ constraint = $-$(coefficient of e_i in optimal row 0)

Optimal value of dual variable x_i if Constraint i is an equality constraint = (coefficient of a_i in optimal row 0) + M

To illustrate how the optimal solution to the dual of a min problem may be read from the optimal primal tableau, consider

$$\min w = 3y_1 + 2y_2 + y_3$$
$$\text{s.t.} \quad y_1 + y_2 + y_3 \geq 4$$
$$\phantom{\text{s.t.} \quad y_1 + {}} y_2 - y_3 \leq 2$$
$$y_1 + y_2 + 2y_3 = 6$$
$$y_1, y_2, y_3 \geq 0$$

The optimal tableau for this problem is given in Table 26. Thus, the optimal primal solution is $w = 6$, $y_2 = y_3 = 2$, $y_1 = 0$. The dual of the preceding LP is

$$\max z = 4x_1 + 2x_2 + 6x_3$$
$$\text{s.t.} \quad x_1 + x_3 \leq 3$$
$$x_1 + x_2 + x_3 \leq 2$$
$$x_1 - x_2 + 2x_3 \leq 1$$
$$x_1 \geq 0, x_2 \leq 0, x_3 \text{ urs}$$

TABLE 26
Finding the Optimal Solution to the Dual When Primal Is a Min Problem

w	y_1	y_2	y_3	e_1	s_2	a_1	a_3	rhs
1	-1	0	0	-3	0	$3 - M$	$-1 - M$	6
0	1	1	0	-2	0	2	-1	2
0	-1	0	0	3	1	-3	2	2
0	0	0	1	1	0	-1	1	2

From the optimal primal tableau, we find that the optimal dual solution is $z = 6$, $x_1 = 3$, $x_2 = 0$, $x_3 = -1$.

PROBLEMS

Group A

1 The following questions refer to the Giapetto problem (see Problem 7 of Section 6.3).

a Find the dual of the Giapetto problem.

b Use the optimal tableau of the Giapetto problem to determine the optimal dual solution.

c Verify that the Dual Theorem holds in this instance.

2 Consider the following LP:

$$\max z = -2x_1 - x_2 + x_3$$
$$\text{s.t.} \quad x_1 + x_2 + x_3 \leq 3$$
$$x_2 + x_3 \geq 2$$
$$x_1 \quad + x_3 = 1$$
$$x_1, x_2, x_3 \geq 0$$

a Find the dual of this LP.

b After adding a slack variable s_1, subtracting an excess variable e_2, and adding artificial variables a_2 and a_3, row 0 of the LP's optimal tableau is found to be

$$z + 4x_1 + e_2 + (M-1)a_2 + (M+2)a_3 = 0$$

Find the optimal solution to the dual of this LP.

3 For the following LP,

$$\max z = -x_1 + 5x_2$$
$$\text{s.t.} \quad x_1 + 2x_2 \leq 0.5$$
$$-x_1 + 3x_2 \leq 0.5$$
$$x_1, x_2 \geq 0$$

row 0 of the optimal tableau is $z + 0.4s_1 + 1.4s_2 = ?$ Determine the optimal z-value for the given LP.

4 The following questions refer to the Bevco problem of Section 4.10.

a Find the dual of the Bevco problem.

b Use the optimal tableau for the Bevco problem that is given in Section 4.10 to find the optimal solution to the dual. Verify that the Dual Theorem holds in this instance.

5 Consider the following linear programming problem:

$$\max z = 4x_1 + x_2$$
$$\text{s.t.} \quad 3x_1 + 2x_2 \leq 6$$
$$6x_1 + 3x_2 \leq 10$$
$$x_1, x_2 \geq 0$$

Suppose that in solving this problem, row 0 of the optimal tableau is found to be $z + 2x_2 + s_2 = \frac{20}{3}$. Use the Dual Theorem to prove that the computations must be incorrect.

6 Show that (for a max problem) if the ith primal constraint is a $\geq$ constraint, then the optimal value of the ith dual variable may be written as (coefficient of a_i in optimal row 0) $- M$.

Group B

7 In this problem, we use weak duality to prove Lemma 3.

a Show that Lemma 3 is equivalent to the following: If the dual is feasible, then the primal is bounded. (*Hint:* Do you remember, from plane geometry, what the contrapositive is?)

b Use weak duality to show the validity of the form of Lemma 3 given in part (a). (*Hint:* If the dual is feasible, then there must be a dual feasible point having a w-value of, say, w_o. Now use weak duality to show that the primal is bounded.)

8 Following along the lines of Problem 7, use weak duality to prove Lemma 4.

9 Use the information given in Problem 8 of Section 6.3 to determine the dual of the Dorian Auto problem and its optimal solution.

6.8 Shadow Prices

We now return to the concept of shadow price that was discussed in Section 6.1. A more formal definition follows.

DEFINITION ■ The **shadow price** of the ith constraint is the amount by which the optimal z-value is improved (increased in a max problem and decreased in a min problem) if we increase b_i by 1 (from b_i to $b_i + 1$).[†]

[†]This assumes that after the right-hand side of Constraint i has been changed to $b_i + 1$, the current basis remains optimal.

By using the Dual Theorem, we can easily determine the shadow price of the ith constraint. To illustrate, we find the shadow price of the second constraint (finishing hours) of the Dakota problem. Let $\mathbf{c}_{BV}B^{-1} = [y_1 \quad y_2 \quad y_3] = [0 \quad 10 \quad 10]$ be the optimal solution to the dual of the max problem. From the Dual Theorem, we know that

Optimal z-value when rhs of constraints are ($b_1 = 48$, $b_2 = 20$, $b_3 = 8$)

$$= 48y_1 + 20y_2 + 8y_3 \quad \text{(31)}$$

What happens to the optimal z-value for the Dakota problem if b_2 (currently 20 finishing hours) is increased by 1 unit (to 21 hours)? We know that changing a right-hand side may cause the current basis to no longer be optimal (see Section 6.3). For the moment, however, we assume that the current basis remains optimal when we increase b_2 by 1. Then $\mathbf{c}_{BV}$ and B^{-1} remain unchanged, so the optimal solution to the dual of the Dakota problem remains unchanged.

We next find

Optimal z-value when rhs of finishing constraint is $21 = 48y_1 + 21y_2 + 8y_3$ \quad (35)

Subtracting (34) from (35) yields

Change in optimal z-value if finishing hours are increased by 1

= shadow price for finishing constraint 2 \qquad\qquad\qquad (36)

$= y_2 = 10$

This example shows that *the shadow price of the ith constraint of a max problem is the optimal value of the ith dual variable*. The shadow prices are the dual variables, so we know that the shadow price for a $\leq$ constraint will be non-negative; for a $\geq$ constraint, nonpositive; and for an equality constraint, unrestricted in sign. The examples discussed later in this section give intuitive justifications for these sign conventions.

Similar reasoning can be used to show that if (in a maximization problem) the right-hand side of the ith constraint is increased by an amount Δb_i, then (assuming the current basis remains optimal) the new optimal z-value may be found from

New optimal z-value = old optimal z-value $+ \Delta b_i$(Constraint i shadow price) \quad (37)

For a minimization problem, the shadow price of the ith constraint is the amount by which a unit increase in the right-hand side improves, or decreases, the optimal z-value (assuming that the current basis remains optimal). It can be shown that the shadow price of the ith constraint of a min problem $= -$(optimal value of the ith dual variable). If the right-hand side is increased by an amount Δb_i, then (assuming the current basis remains optimal) the new optimal z-value may be found from

New optimal z-value = old optimal z-value $- \Delta b_i$(Constraint i shadow price) \quad (37′)

The following three examples should clarify the shadow price concept.

EXAMPLE 11   **Shadow Prices for Normal Max Problem**

For the Dakota problem:

1 Find and interpret the shadow prices

2 If 18 finishing hours were available, what would be Dakota's revenue? (It can be shown by the methods of Section 6.3 that if $16 \leq$ finishing hours ≤ 24, the current basis remains optimal.)

3 If 9 carpentry hours were available, what would be Dakota's revenue? (For $\frac{20}{3} \leq$ carpentry hours ≤ 10, the current basis remains optimal.)

4 If 30 board feet of lumber were available, what would be Dakota's revenue? (For $24 \leq$ lumber $\leq \infty$, the current basis remains optimal.)

5 If 30 carpentry hours were available, why couldn't the shadow price for the carpentry constraint be used to determine the new z-value?

Solution **1** In Section 6.7, we found the optimal solution to the Dakota dual to be $y_1 = 0$, $y_2 = 10$, $y_3 = 10$. Thus, the shadow price for the lumber constraint is 0; for the finishing constraint, 10; and for the carpentry constraint, 10. The fact that the lumber constraint has a shadow price of 0 means that increasing the amount of available lumber by 1 board foot (or *any* amount) will not increase revenue. This is reasonable because we are currently using only 24 of the available 48 board feet of lumber, so adding any more will not do Dakota any good. Dakota's revenue would increase by $10 if 1 more finishing hour were available. Similarly, 1 more carpentry hour would increase Dakota's revenue by $10. In this problem, the shadow price of the ith constraint may be thought of as the maximum amount that the company would pay for an extra unit of the resource associated with the ith constraint. For example, an extra carpentry hour would raise revenue by $y_3 = \$10$ (see Example 12 for a max problem in which this interpretation is invalid). Thus, Dakota could pay up to $10 for an extra carpentry hour and still be better off. Similarly, the company would be willing to pay nothing ($0) for an extra board foot of lumber and up to $10 for an extra finishing hour. To answer questions 2–4, we apply (37), using the fact that the old z-value $= 280$.

2 $y_2 = 10$, $\Delta b_2 = 18 - 20 = -2$. The current basis is still optimal because $16 \leq 18 \leq 24$. Then (37) yields (new revenue) $= 280 + 10(-2) = \$260$.

3 $y_3 = 10$, $\Delta b_3 = 9 - 8 = 1$. Because $\frac{20}{3} \leq 9 \leq 10$, the current basis remains optimal. Then (37) yields (new revenue) $= 280 + 10(1) = \$290$.

4 $y_1 = 0$, $\Delta b_1 = 30 - 48 = -18$. Because $24 \leq 30 \leq \infty$, the current basis is still optimal. Then (37) yields (new revenue) $= 280 + 0(-18) = \$280$.

5 If $b_3 = 30$, the current basis is no longer optimal, because $30 > 10$. This means that BV (and therefore $\mathbf{c}_{BV}B^{-1}$) changes, and we cannot use the current set of shadow prices to determine the new revenue level.

Intuitive Explanation of the Sign of Shadow Prices

We can now give an intuitive explanation of why (in a max problem) the shadow price of a $\leq$ constraint will always be nonnegative. Consider the following situation: We are given two LP max problems (LP 1 and LP 2) that have the same objective functions. Suppose that every point that is feasible for LP 1 is also feasible for LP 2. This means that LP 2's feasible region contains all the points in LP 1's feasible region and possibly some other points. Then the optimal z-value for LP 2 must be at least as large as the optimal z-value for LP 1. To see this, suppose that point x' (with z-value z') is optimal for LP 1. Because x' is also feasible for LP 2 (which has the same objective function as LP 1), LP 2 can attain a z-value of z' (by using the feasible point x'). It is also possible that by using one of the points feasible for only LP 2 (and not for LP 1), LP 2 might do better than z'. In short, *adding points to the feasible region of a max problem cannot decrease the optimal z-value.*

We can use this observation to show why a $\leq$ constraint must have a nonnegative shadow price. For the Dakota problem, if we increase the right-hand side of the carpentry constraint by 1 (from 8 to 9), we see that all points that were originally feasible remain feasible, and some new points (which use > 8 and ≤ 9 carpentry hours) may be feasible. Thus, the optimal z-value cannot decrease, and the shadow price for the carpentry constraint must be nonnegative.

The purpose of the following example is to show that (contrary to what many books say) the shadow price of a $\leq$ constraint is not always the maximum price you would be willing to pay for an additional unit of a resource.

EXAMPLE 12 | **Shadow Price as a Premium**

Leatherco manufactures belts and shoes. A belt requires 2 square yards of leather and 1 hour of skilled labor. A pair of shoes requires 3 sq yd of leather and 2 hours of skilled labor. As many as 25 sq yd of leather and 15 hours of skilled labor can be purchased at a price of \$5/sq yd of leather and \$10/hour of skilled labor. A belt sells for \$23, and a pair of shoes sells for \$40. Leatherco wants to maximize profits (revenues − costs). Formulate an LP that can be used to maximize Leatherco's profits. Then find and interpret the shadow prices for this LP.

Solution Define

$$x_1 = \text{number of belts produced}$$
$$x_2 = \text{number of pairs of shoes produced}$$

After noting that

$$\text{Cost/belt} = 2(5) + 1(10) = \$20$$
$$\text{Cost/pair of shoes} = 3(5) + 2(10) = \$35$$

we find that Leatherco's objective function is

$$\max z = (23 - 20)x_1 + (40 - 35)x_2 = 3x_1 + 5x_2$$

Leatherco faces the following two constraints:

Constraint 1 Leatherco can use at most 25 sq yd of leather.

Constraint 2 Leatherco can use at most 15 hours of skilled labor.

Constraint 1 is expressed by

$$2x_1 + 3x_2 \leq 25 \qquad \text{(Leather constraint)}$$

while Constraint 2 is expressed by

$$x_1 + 2x_2 \leq 15 \qquad \text{(Skilled-labor constraint)}$$

After adding the sign restrictions $x_1 \geq 0$ and $x_2 \geq 0$, we obtain the following LP:

$$\max z = 3x_1 + 5x_2$$
$$\text{s.t.} \quad 2x_1 + 3x_2 \leq 25 \qquad \text{(Leather constraint)}$$
$$x_1 + 2x_2 \leq 15 \qquad \text{(Skilled-labor constraint)}$$
$$x_1, x_2 \geq 0$$

After adding slack variables s_1 and s_2 to the leather and skilled-labor constraints, respectively, we obtain the optimal tableau shown in Table 27. Thus, the optimal solution to Leatherco's problem is $z = 40$, $x_1 = 5$, $x_2 = 5$. The shadow prices are

$$y_1 = \text{leather shadow price} = \text{coefficient of } s_1 \text{ in optimal row } 0 = 1$$
$$y_2 = \text{skilled-labor shadow price} = \text{coefficient of } s_2 \text{ in optimal row } 0 = 1$$

TABLE 27
Optimal Tableau for Leatherco

		Basic Variable
z $+$ s_1 $+$ $s_2 = 40$		$z = 40$
x_1 $+ 2s_1 - 3s_2 = 5$		$x_1 = 5$
$x_2 - s_1 + 2s_2 = 5$		$x_2 = 5$

The meaning of the leather shadow price is that if one more square yard of leather were available, then Leatherco's objective function (profits) would increase by \$1. Let's look further at what happens if an additional square yard of leather is available. Because s_1 is non-basic, the extra square yard of leather will be purchased. Also, because s_2 is nonbasic, we will still use all available labor. This means that the \$1 increase in profits includes the cost of purchasing an extra square yard of leather. If the availability of an extra square yard of leather increases profits by \$1, then it must be increasing revenue by $1 + 5 = \$6$. Thus, the maximum amount Leatherco should pay for an extra square yard of leather is \$6 (not \$1).

Another way to see this is as follows: If we purchase another square yard of leather at the current price of \$5, profits increase by $y_1 = \$1$. If we purchase another square yard of leather at a price of $\$6 = \$5 + \$1$, then profits increase by $\$1 - \$1 = \$0$. Thus, the most Leatherco would be willing to pay for an extra square yard of leather is \$6.

Similarly, the most Leatherco would be willing to pay for an extra hour of labor is $y_2 +$ (cost of an extra hour of skilled labor) $= 1 + 10 = \$11$. In this problem, we see that the shadow price for a resource represents the *premium* over and above the cost of the resource that Leatherco would be willing to pay for an extra unit of resource.

The two preceding examples show that we must be careful when interpreting the shadow price of a $\leq$ constraint. Remember that the shadow price for a constraint in a max problem is the amount by which the objective function increases if the right-hand side is increased by 1.

The following example illustrates the interpretation of the shadow prices of $\geq$ and equality constraints.

EXAMPLE 13 **Shadow Prices for $\geq$ and $=$ Constraints**

Steelco has received an order for 100 tons of steel. The order must contain at least 3.5 tons of nickel, at most 3 tons of carbon, and exactly 4 tons of manganese. Steelco receives \$20/ton for the order. To fill the order, Steelco can combine four alloys, whose chemical composition is given in Table 28. Steelco wants to maximize the profit (revenues − costs) obtained from filling the order. Formulate the appropriate LP. Also find and interpret the shadow prices for each constraint.

Solution After we define $x_i =$ number of tons of alloy i used to fill the order, Steelco's LP is seen to be

$$\max z = (20 - 12)x_1 + (20 - 10)x_2 + (20 - 8)x_3 + (20 - 6)x_4$$

$$\begin{aligned}
\text{s.t.} \quad & 0.06x_1 + 0.03x_2 + 0.02x_3 + 0.01x_4 \geq 3.5 && \text{(Nickel constraint)} \\
& 0.03x_1 + 0.02x_2 + 0.05x_3 + 0.06x_4 \leq 3 && \text{(Carbon constraint)} \\
& 0.08x_1 + 0.03x_2 + 0.02x_3 + 0.01x_4 = 4 && \text{(Manganese constraint)} \\
& x_1 + x_2 + x_3 + x_4 = 100 && \text{(Order size = 100 tons)} \\
& x_1, x_2, x_3, x_4 \geq 0
\end{aligned}$$

TABLE 28
Relevant Information for Steelco

Cement	Alloy (%)			
	1	2	3	4
Nickel	6	3	2	1
Carbon	3	2	5	6
Manganese	8	3	2	1
Cost/ton ($)	12	10	8	6

After adding a slack variable s_2, subtracting an excess variable e_1, and adding artificial variables a_1, a_3, and a_4, the following optimal solution is obtained: $z = 1{,}000$, $s_2 = 0.25$, $x_1 = 25$, $x_2 = 62.5$, $x_4 = 12.5$, $e_1 = 0$, $x_3 = 0$. The optimal row 0 is

$$z + 400e_1 + (M - 400)a_1 + (M + 200)a_3 + (M + 16)a_4 = 1{,}000$$

Using (31), (31'), and (31''), we obtain

Shadow price of nickel constraint $= -(\text{coefficient of } e_1 \text{ in optimal row 0})$
$$= -400$$

Shadow price of carbon constraint $= \text{coefficient of } s_2 \text{ in optimal row 0}$
$$= 0$$

Shadow price of manganese constraint $= (\text{coefficient of } a_3 \text{ in optimal row 0}) - M$
$$= 200$$

Shadow price of order size constraint $= (\text{coefficient of } a_4 \text{ in optimal row 0}) - M$
$$= 16$$

By the sensitivity analysis procedures of Section 6.3, it can be shown that the current basis remains optimal if $3.46 \leq b_1 \leq 3.6$. As long as the nickel requirement is in this range, increasing the nickel requirement by an amount Δb_1 will increase Steelco's profits by $-400 \, \Delta b_1$. For example, increasing the nickel requirement to 3.55 tons ($\Delta b_1 = 0.05$) would "increase" (actually decrease) profits by $-400(0.05) = \$20$. The nickel constraint has a negative shadow price because increasing the right-hand side of the nickel constraint makes it harder to satisfy the nickel constraint. In fact, an increase in the nickel requirement forces Steelco to use more of the expensive type 1 alloy. This raises costs and lowers profits. As we have already seen, the shadow price of a $\geq$ constraint (in a max problem) will always be nonpositive because increasing the right-hand side of a $\geq$ constraint eliminates points from the feasible region. Thus, the optimal z-value must decrease or remain unchanged.

By the Section 6.3 sensitivity analysis procedures, for $2.75 \leq b_2 \leq \infty$, the current basis remains optimal. As stated before, the carbon constraint has a zero shadow price. This means that if we increase Steelco's carbon requirement, Steelco's profit will not change. Intuitively, this is because our present optimal solution contains only $2.75 < 3$ tons of carbon. Thus, relaxing the carbon requirement won't enable Steelco to reduce costs, so Steelco's profit will remain unchanged.

By the sensitivity analysis procedures, the current basis remains optimal if $3.83 \leq b_3 \leq 4.07$. The shadow price of the third (manganese) constraint is 200, so we know that as long as the manganese requirement remains in the given range, increasing it by an amount of Δb_3 will increase profit by $200\Delta b_3$. For example, if the manganese requirement were 4.05 tons ($\Delta b_3 = 0.05$), then profits would increase by $(0.05)200 = \$10$.

By the sensitivity analysis procedures, the current basis remains optimal if $91.67 \leq b_4 \leq 103.12$. Because the shadow price of the fourth (order size) constraint is 16, increasing the order size by Δb_4 tons (with nickel, carbon, and manganese requirements unchanged) would increase profits by $16\Delta b_4$. For example, the profit from a 103-ton order that required ≥ 3.5 tons of nickel, ≤ 3 tons of carbon, and exactly 4 tons of manganese would be $1,000 + 3(16) = \$1,048$.

In this problem, both equality constraints had positive shadow prices. In general, we know that it is possible for an equality constraint's dual variable (and shadow price) to be negative. If this occurs, then the equality constraint will have a negative shadow price. To illustrate this possibility, suppose that Steelco's customer required exactly 4.5 tons of manganese in the order. Because $4.5 > 4.07$, the current basis is no longer optimal. If we solve Steelco's LP again, it can be shown that the shadow price for the manganese constraint has changed to -54.55. This means that an increase in the manganese requirement will decrease Steelco's profits.

Interpretation of the Dual Prices Column of the LINDO Output

For a max problem, LINDO gives the values of the shadow prices in the DUAL PRICES column of the output. The dual price for row $i + 1$ on the LINDO output is the shadow price for the ith constraint and the optimal value for the ith dual variable. Thus, in Figure 4, we see that for the Dakota problem,

$$y_1 = \text{shadow price for lumber constraint} = \text{row 2 dual price} = 0$$

$$y_2 = \text{shadow price for finishing constraint} = \text{row 3 dual price} = 10$$

$$y_3 = \text{shadow price for carpentry constraint} = \text{row 4 dual price} = 10$$

For a maximization problem, the vector $c_{BV}B^{-1}$ (needed for pricing out new activities) is the same as the vector of dual prices given in the LINDO output. For the Dakota problem, we would price out new activities using $c_{BV}B^{-1} = [0 \quad 10 \quad 10]$.

For a minimization problem, the entry in the DUAL PRICE column for any constraint is the shadow price. Thus, from the LINDO printout in Figure 6, we find that the shadow prices for the constraints in the diet problem are as follows: calorie = 0; chocolate = -2.5ϕ; sugar = -7.5ϕ; and fat = 0. This implies that

1 Increasing the calorie requirement by 1 will leave the cost of the optimal diet unchanged.

2 Increasing the chocolate requirement by 1 oz will decrease the cost of the optimal diet by -2.5ϕ (that is, increase the cost of the optimal diet by 2.5ϕ).

3 Increasing the sugar requirement by 1 oz will decrease the cost of the optimal diet by -7.5ϕ (that is, increase the cost of the optimal diet by 7.5ϕ).

4 Increasing the fat requirement by 1 oz will leave the cost of the optimal diet unchanged.

The entry in the DUAL PRICE column for any constraint is, however, the negative of the constraint's dual variable. Thus, for the diet problem, we see from Figure 6 that the optimal dual solution to the diet problem is given by $c_{BV}B^{-1} = [0 \quad 2.5 \quad 7.5 \quad 0]$. When pricing out a new activity for a minimization problem, use the negative of each dual price as the corresponding element of $\mathbf{c}_{\mathrm{BV}}B^{-1}$.

Remember that for any LP, the dual prices remain valid only as long as the current basis remains optimal. As stated in Section 6.3, the range of right-hand side values for which the current basis remains optimal may be obtained from the RIGHTHAND SIDE RANGES block of the LINDO output.

Degeneracy and Sensitivity Analysis

When the optimal solution to an LP is degenerate, caution must be used when interpreting the LINDO output. Recall from Section 4.11 that a bfs is degenerate if at least one basic variable in the optimal solution equals 0. For an LP with m constraints, if the LINDO output indicates that less than m variables are positive, then the optimal solution is a degenerate bfs. To illustrate, consider the following LP:

$$\max z = 6X_1 + 4X_2 + 3X_3 + 2X_4$$
$$\text{s.t.} \quad 2X_1 + 3X_2 + X_3 + 2X_4 \le 400$$
$$X_1 + X_2 + 2X_3 + X_4 \le 150$$
$$2X_1 + X_2 + X_3 + .5X_4 \le 200$$
$$3X_1 + X_2 + X_4 \le 250$$
$$X_1, X_2, X_3, X_4 \ge 0$$

The LINDO output for this LP is in Figure 9. The LP has four constraints and only two positive variables in the optimal solution, so the bfs is degenerate. By the way, results from using the **TABLEAU** command indicate that the optimal basis is BV = $\{X_2, X_3, S_3, X_1\}$.

We now discuss three "oddities" that may occur when the optimal solution found by LINDO is degenerate.

Oddity 1 At least one constraint's RANGE IN WHICH THE BASIS IS UNCHANGED will have a 0 ALLOWABLE INCREASE or ALLOWABLE DECREASE. This means that for at least one constraint the DUAL PRICE can tell us about the new z-value for either an increase or a decrease in the constraint's right-hand side, but not both.

To understand Oddity 1, consider the first constraint. Its AI is 0. Thus, its DUAL PRICE of .50 cannot be used to determine a new z-value resulting from any increase in the first constraint's right-hand side.

Oddity 2 For a nonbasic variable to become positive, its objective function coefficient may have to be improved by more than its REDUCED COST.

To understand Oddity 2 consider the nonbasic variable X_4; its REDUCED COST is 1.5. If we increase X_4's objective function coefficient by 2, however, we still find that the new optimal solution has $X_4 = 0$ because the change affects the set of basic variables, but not the LP's optimal solution. If we increase X_4's objective function coefficient by 4.5 or more, then we find that X_4 is positive.

Oddity 3 If you increase a variable's objective function coefficient by more than its AI or decrease it by more than its AD, then the optimal solution to the LP may remain the same.

Oddity 3 is similar to Oddity 2. To understand Oddity 3, consider the nonbasic variable X_4; its AI is 1.5. If we increase X_4's objective function coefficient by 2, however, we still find that the new optimal solution is unchanged. This oddity occurs because the change affects the set of basic variables, but not the LP's optimal solution.

```
MAX      6 X1 + 4 X2 + 3 X3 + 2 X4
SUBJECT TO
        2)   2 X1 + 3 X2 + X3 + 2 X4 <=        400
        3)   X1 + X2 + 2 X3 + X4 <=   150
        4)   2 X1 + X2 + X3 + 0.5 X4 <=     200
        5)   3 X1 + X2 + X4 <=   250
END

LP OPTIMUM FOUND AT STEP  3

        OBJECTIVE FUNCTION VALUE

        1)      700.00000

   VARIABLE          VALUE          REDUCED COST
        X1        50.000000            .000000
        X2       100.000000            .000000
        X3          .000000            .000000
        X4          .000000           1.500000


   ROW    SLACK  OR SURPLUS       DUAL PRICES
        2)          .000000            .500000
        3)          .000000           1.250000
        4)          .000000            .000000
        5)          .000000           1.250000

NO. ITERATIONS=         3

RANGES IN WHICH THE BASIS IS UNCHANGED:

                        OBJ COEFFICIENT RANGES
   VARIABLE         CURRENT        ALLOWABLE        ALLOWABLE
                     COEF          INCREASE         DECREASE
        X1        6.000000         3.000000         3.000000
        X2        4.000000         5.000000         1.000000
        X3        3.000000         3.000000         2.142857
        X4        2.000000         1.500000         INFINITY

                      RIGHTHAND SIDE RANGES
   ROW           CURRENT        ALLOWABLE        ALLOWABLE
                  RHS           INCREASE         DECREASE
        2       400.000000        .000000        200.000000
        3       150.000000        .000000          .000000
        4       200.000000        INFINITY         .000000
        5       250.000000        .000000        120.000000


THE TABLEAU
   ROW  (BASIS)        X1        X2        X3        X4     SLK     2
     1 ART           .000      .000      .000     1.500      .500
     2       X2      .000     1.000      .000      .500      .500
     3       X3      .000      .000     1.000      .167     -.167
     4 SLK    4      .000      .000      .000     -.500      .000
     5       X1     1.000      .000      .000      .167     -.167


   ROW   SLK    3  SLK    4  SLK    5
     1    1.250      .000     1.250   700.000
     2    -.250      .000     -.250   100.000
     3     .583      .000     -.083      .000
     4    -.500     1.000     -.500      .000
     5     .083      .000      .417    50.000
```

FIGURE 9

PROBLEMS

Group A

1   Use the Dual Theorem to prove (37).

2   The following questions refer to the Sugarco problem (Problem 6 of Section 6.3):

   a   Find the shadow prices for the Sugarco problem.

   b   If 60 oz of sugar were available, what would be Sugarco's profit?

   c   How about 40 oz of sugar?

   d   How about 30 oz of sugar?

TABLE 29

	Product	
Resource	1	2
Skilled labor (hours)	3	4
Unskilled labor (hours)	2	3
Raw material (units)	1	2

3 Suppose we are working with a min problem and increase the right-hand side of a $\geq$ constraint. What can happen to the optimal z-value?

4 Suppose we are working with a min problem and increase the right-hand side of a $\leq$ constraint. What can happen to the optimal z-value?

5 A company manufactures two products (1 and 2). Each unit of product 1 can be sold for $15, and each unit of product 2 for $25. Each product requires raw material and two types of labor (skilled and unskilled) (see Table 29). Currently, the company has available 100 hours of skilled labor, 70 hours of unskilled labor, and 30 units of raw material. Because of marketing considerations, at least 3 units of product 2 must be produced.

   a Explain why the company's goal is to maximize revenue.

   b The relevant LP is

$$\max z = 15x_1 + 25x_2$$

s.t.
$$3x_1 + 4x_2 \leq 100 \quad \text{(Skilled labor constraint)}$$
$$2x_1 + 3x_2 \leq 70 \quad \text{(Unskilled labor constraint)}$$
$$x_1 + 2x_2 \leq 30 \quad \text{(Raw material constraint)}$$
$$x_2 \geq 3 \quad \text{(Product 2 constraint)}$$
$$x_1, x_2 \geq 0$$

The optimal tableau for this problem has the following row 0:

$$z + 15s_3 + 5e_4 + (M - 5)a_4 = 435$$

The optimal solution to the LP is $z = 435$, $x_1 = 24$, $x_2 = 3$. Find and interpret the shadow price of each constraint. How much would the company be willing to pay for an additional unit of each type of labor? How much would it be willing to pay for an extra unit of raw material?

   c Assuming the current basis remains optimal (it does), what would the company's revenue be if 35 units of raw material were available?

   d With the current basis optimal, what would the company's revenue be if 80 hours of skilled labor were available?

   e With the current basis optimal, what would the company's new revenue be if at least 5 units of product 2 were required? How about if at least 2 units of product 2 were required?

6 Suppose that the company in Problem 5 owns no labor and raw material but can purchase them at the following prices: as many as 100 hours of skilled labor at $3/hour, 70 hours of unskilled labor at $2/hour, and 30 units of raw material at $1 per unit of raw material. If the company's goal is to maximize profit, show that the appropriate LP is

$$\max z = x_1 + 5x_2$$

s.t.
$$3x_1 + 4x_2 \leq 100$$
$$2x_1 + 3x_2 \leq 70$$
$$x_1 + 2x_2 \leq 30$$
$$x_2 \geq 3$$
$$x_1, x_2 \geq 0$$

The optimal row 0 for this LP is

$$z + 1.5x_1 + 2.5s_3 + Ma_4 = 75$$

and the optimal solution is $z = 75$, $x_1 = 0$, $x_2 = 15$. In answering parts (a) and (b), assume that the current basis remains optimal.

   a How much should the company pay for an extra unit of raw material?

   b How much should the company pay for an extra hour of skilled labor? Unskilled labor? (Be careful here!)

7 For the Dorian problem (see Problem 8 of Section 6.3), answer the following questions:

   a What would Dorian's cost be if 40 million HIW exposures were required?

   b What would Dorian's cost be if only 20 million HIM exposures were required?

8 If it seems difficult to believe that the shadow price of an equality constraint should be urs, try this problem. Consider the following two LPs:

$$\max z = x_2$$
(LP 1) s.t. $\quad x_1 + x_2 = 2$
$$x_1, x_2 \geq 0$$

$$\max z = x_2$$
(LP 2) s.t. $\quad -x_1 - x_2 = -2$
$$x_1, x_2 \geq 0$$

In which LP will the constraint have a positive shadow price? Which will have a negative shadow price?

Group B

9 For the Dakota problem, suppose that 22 finishing hours and 9 carpentry hours are available. What would be the new optimal z-value? [*Hint:* Use the 100% Rule to show that the current basis remains optimal, and mimic (34)–(36).]

10 For the diet problem, suppose at least 8 oz of chocolate and at least 9 oz of sugar are required (with other requirements remaining the same). What is the new optimal z-value?

11 Consider the LP:

$$\max z = 9x_1 + 8x_2 + 5x_3 + 4x_4$$

s.t.
$$x_1 \qquad\qquad + x_4 \leq 200$$
$$x_2 + x_3 \qquad \leq 150$$
$$x_1 + x_2 + x_3 \qquad \leq 350$$
$$2x_1 + x_2 + x_3 + x_4 \leq 550$$
$$x_1, x_2, x_3, x_4 \geq 0$$

   a Solve this LP with LINDO and use your output to show that the optimal solution is degenerate.

   b Use your LINDO output to find an example of Oddities 1–3.

6.9 Duality and Sensitivity Analysis

Our proof of the Dual Theorem demonstrated the following result: *Assuming that a set of basic variables BV is feasible, then BV is optimal (that is, each variable in row 0 has a nonnegative coefficient) if and only if the associated dual solution* ($c_{BV}B^{-1}$) *is dual feasible.*

This result can be used for an alternative way of doing the following types of sensitivity analysis (see list of changes at the beginning of Section 6.3).

Change 1 Changing the objective function coefficient of a nonbasic variable

Change 4 Changing the column of a nonbasic variable

Change 5 Adding a new activity

In each case, the change leaves BV feasible. BV will remain optimal if the BV row 0 remains non-negative. Primal optimality and dual feasibility are equivalent, so we see that *the above changes will leave the current basis optimal if and only if the current dual solution* $c_{BV}B^{-1}$ *remains dual feasible.* If the current dual solution is no longer dual feasible, then BV will be suboptimal, and a new optimal solution must be found.

We illustrate the duality-based approach to sensitivity analysis by reworking some of the Section 6.3 illustrations. Recall that these illustrations dealt with the Dakota problem:

$$\max z = 60x_1 + 30x_2 + 20x_3$$

$$
\begin{array}{llll}
\text{s.t.} & 8x_1 + 6x_2 + x_3 \le 48 & \text{(Lumber constraint)} \\
& 4x_1 + 2x_2 + 1.5x_3 \le 20 & \text{(Finishing constraint)} \\
& 2x_1 + 1.5x_2 + 0.5x_3 \le 8 & \text{(Carpentry constraint)} \\
& x_1, x_2, x_3 \ge 0
\end{array}
$$

The optimal solution was $z = 280$, $s_1 = 24$, $x_3 = 8$, $x_1 = 2$, $x_2 = 0$, $s_2 = 0$, $s_3 = 0$. The only nonbasic decision variable in the optimal solution is x_2 (tables). The dual of the Dakota problem is

$$\min w = 48y_1 + 20y_2 + 8y_3$$

$$
\begin{array}{llll}
\text{s.t.} & 8y_1 + 4y_2 + 2y_3 \ge 60 & \text{(Desk constraint)} \\
& 6y_1 + 2y_2 + 1.5y_3 \ge 30 & \text{(Table constraint)} \\
& y_1 + 1.5y_2 + 0.5y_3 \ge 20 & \text{(Chair constraint)} \\
& y_1, y_2, y_3 \ge 0
\end{array}
$$

Recall that the optimal dual solution—and therefore the constraint shadow prices—are $y_1 = 0$, $y_2 = 10$, $y_3 = 10$. We now show how knowledge of duality can be applied to sensitivity analysis.

EXAMPLE 14	Changing Objective Function Coefficient of Nonbasic Variable

We want to change the objective function coefficient of a nonbasic variable. Let c_2 be the coefficient of x_2 (tables) in the Dakota objective function. In other words, c_2 is the price at which a table is sold. For what values of c_2 will the current basis remain optimal?

Solution If $y_1 = 0$, $y_2 = 10$, $y_3 = 10$ remains dual feasible, then the current basis—and the values of all the variables—are unchanged. Note that if the objective function coefficient for x_2 is changed, then the first and third dual constraints remain unchanged, but the second (table) dual constraint is changed to

$$6y_1 + 2y_2 + 1.5y_3 \ge c_2$$

If $y_1 = 0$, $y_2 = 10$, $y_3 = 10$ satisfies this inequality, then dual feasibility (and therefore primal optimality) is maintained. Thus, the current basis remains optimal if c_2 satisfies $6(0) + 2(10) + 1.5(10) \geq c_2$, or $c_2 \leq 35$. This shows that for $c_2 \leq 35$, the current basis remains optimal. Conversely, if $c_2 > 35$, the current basis is no longer optimal. This agrees with the result obtained in Section 6.3.

Using shadow prices, we may give an alternative interpretation of this result. We can use shadow prices to compute the implied value of the resources needed to construct a table (see Table 30). A table uses $35 worth of resources, so the only way producing tables can increase Dakota's revenues is if a table sells for more than $35. Thus, the current basis fails to be optimal if $c_2 > 35$, and the current basis remains optimal if $c_2 \leq 35$.

TABLE 30
Why a Table Is Profitable at > $35/Table

Resource in a Table	Shadow Price of Resource ($)	Amount of Resource Used	Value of Resource Used
Lumber	0	6 board ft	0(6) = $0
Finishing	10	2 hours	10(2) = $20
Carpentry	10	1.5 hours	10(1.5) = $15
			Total: = $35

EXAMPLE 15 **Changing a Nonbasic Variable**

We want to change the column for a nonbasic activity. Suppose a table sells for $43 and uses 5 board feet of lumber, 2 finishing hours, and 2 carpentry hours. Does the current basis remain optimal?

Solution Changing the column for the nonbasic variable "tables" leaves the first and third dual constraints unchanged but changes the second to

$$5y_1 + 2y_2 + 2y_3 \geq 43$$

Because $y_1 = 0$, $y_2 = 10$, $y_3 = 10$ does not satisfy the new second dual constraint, dual feasibility is not maintained, and the current basis is no longer optimal. In terms of shadow prices, this result is reasonable (see Table 31). Each table uses $40 worth of resources and sells for $43, so Dakota can increase its revenue by $43 - 40 = \$3$ for each table that is produced. Thus, the current basis is no longer optimal, and x_2 (tables) will be basic in the new optimal solution.

TABLE 31
Shadow Price Interpretation of Table Production Decision ($40/Table)

Resource in a Table	Shadow Price of Resource ($)	Amount of Resource Used	Value of Resource Used ($)
Lumber	0	5 board ft	0(5) = $0
Finishing	10	2 hours	10(2) = $20
Carpentry	10	2 hours	10(2) = $20
			Total: = $40

EXAMPLE 16   **Adding a New Activity**

We want to add a new activity. Suppose Dakota is considering manufacturing footstools (x_4). A footstool sells for \$15 and uses 1 board foot of lumber, 1 finishing hour, and 1 carpentry hour. Does the current basis remain optimal?

Solution   Introducing the new activity (footstools) leaves the three dual constraints unchanged, but the new variable x_4 adds a new dual constraint (corresponding to footstools). The new dual constraint will be

$$y_1 + y_2 + y_3 \geq 15$$

The current basis remains optimal if $y_1 = 0, y_2 = 10, y_3 = 10$ satisfies the new dual constraint. Because $0 + 10 + 10 \geq 15$, the current basis remains optimal. In terms of shadow prices, a stool utilizes $1(0) = \$0$ worth of lumber, $1(10) = \$10$ worth of finishing hours, and $1(10) = \$10$ worth of carpentry time. A stool uses $0 + 10 + 10 = \$20$ worth of resources and sells for only \$15, so Dakota should not make footstools, and the current basis remains optimal.

PROBLEMS

Group A

1   For the Dakota problem, suppose computer tables sell for \$35 and use 6 board feet of lumber, 2 hours of finishing time, and 1 hour of carpentry time. Is the current basis still optimal? Interpret this result in terms of shadow prices.

2   The following questions refer to the Sugarco problem (Problem 6 of Section 6.3):

 a   For what values of profit on a Type 1 candy bar does the current basis remain optimal?

 b   If a Type 1 candy bar used 0.5 oz of sugar and 0.75 oz of chocolate, would the current basis remain optimal?

 c   A Type 4 candy bar is under consideration. A Type 4 candy bar yields a 10¢ profit and uses 2 oz of sugar and 1 oz of chocolate. Does the current basis remain optimal?

3   Suppose, in the Dakota problem, a desk still sells for \$60 but now uses 8 board ft of lumber, 4 finishing hours, and 15 carpentry hours. Determine whether the current basis remains optimal. What is wrong with the following reasoning?

The change in the column for desks leaves the second and third dual constraints unchanged and changes the first to

$$8y_1 + 4y_2 + 15y_3 \geq 60$$

Because $y_1 = 0, y_2 = 10, y_3 = 10$ satisfies the new dual constraint, the current basis remains optimal.

6.10   Complementary Slackness

The Theorem of Complementary Slackness is an important result that relates the optimal primal and dual solutions. To state this theorem, we assume that the primal is a normal max problem with variables $x_1, x_2, \ldots, x_n$ and $m \leq$ constraints. Let $s_1, s_2, \ldots, s_m$ be the slack variables for the primal. Then the dual is a normal min problem with variables $y_1, y_2, \ldots, y_m$ and $n \geq$ constraints. Let $e_1, e_2, \ldots, e_n$ be the excess variables for the dual. A statement of the Theorem of Complementary Slackness follows.

Let

$$\mathbf{x} = \begin{bmatrix} x_1 \\ x_2 \\ \vdots \\ x_n \end{bmatrix}$$

be a feasible primal solution and $\mathbf{y} = [y_1 \quad y_2 \quad \cdots \quad y_m]$ be a feasible dual solution. Then $\mathbf{x}$ is primal optimal and $\mathbf{y}$ is dual optimal if and only if

$$s_i y_i = 0 \qquad (i = 1, 2, \ldots, m) \tag{38}$$
$$e_j x_j = 0 \qquad (j = 1, 2, \ldots, n) \tag{39}$$

In Problem 4 at the end of this section, we sketch the proof of the Theorem of Complementary Slackness, but first we discuss the intuitive meaning of this theorem.

From (38), it follows that the optimal primal and dual solutions must satisfy

$$i\text{th primal slack} > 0 \text{ implies } i\text{th dual variable} = 0 \tag{40}$$
$$i\text{th dual variable} > 0 \text{ implies } i\text{th primal slack} = 0 \tag{41}$$

From (39), it follows that the optimal primal and dual solutions must satisfy

$$j\text{th dual excess} > 0 \text{ implies } j\text{th primal variable} = 0 \tag{42}$$
$$j\text{th primal variable} > 0 \text{ implies } j\text{th dual excess} = 0 \tag{43}$$

From (40) and (42), we see that if *a constraint in either the primal or dual is non-binding (has either $s_i > 0$ or $e_j > 0$), then the corresponding variable in the other (or complementary) problem must equal* 0. Hence the name **complementary slackness.**

To illustrate the interpretation of the Theorem of Complementary Slackness, we return to the Dakota problem. Recall that the primal is

$$\max z = 60x_1 + 30x_2 + 20x_3$$

$$\begin{array}{lll}
\text{s.t.} & 8x_1 + 6x_2 + x_3 \le 48 & \text{(Lumber constraint)} \\
& 4x_1 + 2x_2 + 1.5x_3 \le 20 & \text{(Finishing constraint)} \\
& 2x_1 + 1.5x_2 + 0.5x_3 \le 8 & \text{(Carpentry constraint)} \\
& x_1, x_2, x_3 \ge 0 &
\end{array}$$

and the dual is

$$\min w = 48y_1 + 20y_2 + 8y_3$$

$$\begin{array}{lll}
\text{s.t.} & 8y_1 + 4y_2 + 2y_3 \ge 60 & \text{(Desk constraint)} \\
& 6y_1 + 2y_2 + 1.5y_3 \ge 30 & \text{(Table constraint)} \\
& y_1 + 1.5y_2 + 0.5y_3 \ge 20 & \text{(Chair constraint)} \\
& y_1, y_2, y_3 \ge 0 &
\end{array}$$

The optimal primal solution is

$$z = 280, \qquad x_1 = 2, \qquad x_2 = 0, \qquad x_3 = 8$$
$$s_1 = 48 - (8(2) + 6(0) + 1(8)) = 24$$
$$s_2 = 20 - (4(2) + 2(0) + 1.5(8)) = 0$$
$$s_3 = 8 - (2(2) + 1.5(0) + 0.5(8)) = 0$$

The optimal dual solution is

$$w = 280, \qquad y_1 = 0, \qquad y_2 = 10, \qquad y_3 = 10$$
$$e_1 = (8(0) + 4(10) + 2(10)) - 60 = 0$$
$$e_2 = (6(0) + 2(10) + 1.5(10)) - 30 = 5$$
$$e_3 = (1(0) + 1.5(10) + 0.5(10)) - 20 = 0$$

For the Dakota problem, (38) reduces to

$$s_1 y_1 = s_2 y_2 = s_3 y_3 = 0$$

which is indeed satisfied by the optimal primal and dual solutions. Also, (39) reduces to

$$e_1 x_1 = e_2 x_2 = e_3 x_3 = 0$$

which is also satisfied by the optimal primal and dual solutions.

We now illustrate the interpretation of (40)–(43). Note that (40) tells us that because the optimal primal solution has $s_1 > 0$, the optimal dual solution must have $y_1 = 0$. In the context of the Dakota problem, this means that positive slack in the lumber constraint implies that lumber must have a zero shadow price. Slack in the lumber constraint means that extra lumber would not be used, so an extra board foot of lumber should indeed be worthless.

Equation (41) tells us that because $y_2 > 0$ in the optimal dual solution, $s_2 = 0$ must hold in the optimal primal solution. This is reasonable because $y_2 > 0$ means that an extra finishing hour has some value. This can only occur if we are at present using all available finishing hours (or equivalently, if $s_2 = 0$).

Observe that (42) tells us that because $e_2 > 0$ in the optimal dual solution, $x_2 = 0$ must hold in the optimal primal solution. This is reasonable because $e_2 = 6y_1 + 2y_2 + 1.5y_3 - 30$. Because y_1, y_2, and y_3 are resource shadow prices, e_2 may be written as

$$e_2 = (\text{value of resources used by table}) - (\text{sales price of a table})$$

Thus, if $e_2 > 0$, tables are selling for a price that is less than the value of the resources used to make 1 unit of x_2 (tables). This means that no tables should be made (or equivalently, that $x_2 = 0$). This shows that $e_2 > 0$ in the optimal dual solution implies that $x_2 = 0$ must hold in the optimal primal solution.

Note that for the Dakota problem, (43) tells us that $x_1 > 0$ for the optimal primal solution implies that $e_1 = 0$. This result simply reflects the following important fact. *For any variable x_j in the optimal primal basis, the marginal revenue obtained from producing a unit of x_j must equal the marginal cost of the resources used to produce a unit of x_j.* This is a consequence of the fact that each basic variable must have a zero coefficient in row 0 of the optimal primal tableau. In short, (43) is simply the LP version of the well-known economic maxim that an optimal production strategy must have marginal revenue equal marginal cost.

To be more specific, observe that $x_1 > 0$ means that desks are in the optimal basis. Then

$$\text{Marginal revenue obtained by manufacturing desk} = \$60$$

To compute the marginal cost of manufacturing a desk (in terms of shadow prices), note that

$$\text{Cost of lumber in desk} = 8(0) = \$0$$
$$\text{Cost of finishing hours used to make a desk} = 4(10) = \$40$$
$$\text{Cost of carpentry hours used to make a desk} = 2(10) = \$20$$
$$\text{Marginal cost of producing a desk} = 0 + 40 + 20 = \$60$$

Thus, for desks, marginal revenue is equal to marginal cost.

Using Complementary Slackness to Solve LPs

If the optimal solution to the primal or dual is known, complementary slackness can sometimes be used to determine the optimal solution to the complementary problem. For example, suppose we were told that the optimal solution to the Dakota problem was $z = 280$, $x_1 = 2$, $x_2 = 0$, $x_3 = 8$, $s_1 = 24$, $s_2 = 0$, $s_3 = 0$. Can we use Theorem 2 to help us find the optimal solution to the Dakota dual? Because $s_1 > 0$, (40) tells us that the optimal dual solution must have $y_1 = 0$. Because $x_1 > 0$ and $x_3 > 0$, (43) implies that the optimal dual solution must have $e_1 = 0$, and $e_3 = 0$. This means that for the optimal dual solution, the first and third constraints must be binding. We know that $y_1 = 0$, so we know that the optimal values of y_2 and y_3 may be found by solving the first and third dual constraints as equalities (with $y_1 = 0$). Thus, the optimal values of y_2 and y_3 must satisfy

$$4y_2 + 2y_3 = 60 \qquad \text{and} \qquad 1.5y_2 + 0.5y_3 = 20$$

Solving these equations simultaneously shows that the optimal dual solution must have $y_2 = 10$ and $y_3 = 10$. Thus, complementary slackness has helped us find the optimal dual solution $y_1 = 0$, $y_2 = 10$, $y_3 = 10$. (From the Dual Theorem, we know, of course, that the optimal dual solution must have $\bar{w} = 280$.)

PROBLEMS

Group A

1 Glassco manufactures glasses: wine, beer, champagne, and whiskey. Each type of glass requires time in the molding shop, time in the packaging shop, and a certain amount of glass. The resources required to make each type of glass are given in Table 32. Currently, 600 minutes of molding time, 400 minutes of packaging time, and 500 oz of glass are available. Assuming that Glassco wants to maximize revenue, the following LP should be solved:

max $z = 6x_1 + 10x_2 + 9x_3 + 20x_4$

s.t. $\quad 4x_1 + 9x_2 + 7x_3 + 10x_4 \le 600$ (Molding constraint)

$\quad\quad x_1 + x_2 + 3x_3 + 40x_4 \le 400$ (Packaging constraint)

$\quad 3x_1 + 4x_2 + 2x_3 + x_4 \le 500$ (Glass constraint)

$\quad\quad x_1, x_2, x_3, x_4 \ge 0$

It can be shown that the optimal solution to this LP is $z = \frac{2800}{3}$, $x_1 = \frac{400}{3}$, $x_4 = \frac{20}{3}$, $x_2 = 0$, $x_3 = 0$, $s_1 = 0$, $s_2 = 0$, $s_3 = \frac{280}{3}$.

a Find the dual of the Glassco problem.

b Using the given optimal primal solution and the Theorem of Complementary Slackness, find the optimal solution to the dual of the Glassco problem.

c Find an example of each of the complementary slackness conditions, (40)–(43). As in the text, interpret each example in terms of shadow prices.

2 Use the Theorem of Complementary Slackness to show that in the LINDO output, the SLACK or SURPLUS and DUAL PRICE entries for any row cannot both be positive.

3 Consider the following LP:

max $z = 5x_1 + 3x_2 + x_3$

s.t. $\quad 2x_1 + x_2 + x_3 \le 6$

$\quad\quad x_1 + 2x_2 + x_3 \le 7$

$\quad\quad x_1, x_2, x_3 \ge 0$

Graphically solve the dual of this LP. Then use complementary slackness to solve the max problem.

TABLE 32

	Glass			
	X_1 Wine	X_2 Beer	X_3 Champagne	X_4 Whiskey
Molding time	4 minutes	9 minutes	7 minutes	10 minutes
Packaging time	1 minute	1 minute	3 minutes	40 minutes
Glass	3 oz	4 oz	2 oz	1 oz
Selling price	$6	$10	$9	$20

4 Let $\mathbf{x} = [x_1 \quad x_2 \quad x_3 \quad s_1 \quad s_2 \quad s_3]$ be a primal feasible point for the Dakota problem and $\mathbf{y} = [y_1 \quad y_2 \quad y_3 \quad e_1 \quad e_2 \quad e_3]$ be a dual feasible point.

   a Multiply the ith constraint (in standard form) of the primal by y_i and sum the resulting constraints.

   b Multiply the jth dual constraint (in standard form) by x_j and sum them.

   c Compute: part (a) answer minus part (b) answer.

   d Use the part (c) answer and the Dual Theorem to show that if $\mathbf{x}$ is primal optimal and $\mathbf{y}$ is dual optimal, then (38) and (39) hold.

   e Use the part (c) answer to show that if (38) and (39) both hold, then $\mathbf{x}$ is primal optimal and $\mathbf{y}$ is dual optimal. (*Hint:* Look at Lemma 2.)

6.11 The Dual Simplex Method

When we use the simplex method to solve a max problem (we will refer to the max problem as a primal), we begin with a primal feasible solution (because each constraint in the initial tableau has a non-negative right-hand side). At least one variable in row 0 of the initial tableau has a negative coefficient, so our initial primal solution is not dual feasible. Through a sequence of simplex pivots, we maintain primal feasibility and obtain an optimal solution when dual feasibility (a non-negative row 0) is attained. In many situations, however, it is easier to solve an LP by beginning with a tableau in which each variable in row 0 has a non-negative coefficient (so the tableau is dual feasible) and at least one constraint has a negative right-hand side (so the tableau is primal infeasible). The dual simplex method maintains a non-negative row 0 (dual feasibility) and eventually obtains a tableau in which each right-hand side is non-negative (primal feasibility). At this point, an optimal tableau has been obtained. Because this technique maintains dual feasibility, it is called the **dual simplex method.**

Dual Simplex Method for a Max Problem

Step 1 Is the right-hand side of each constraint non-negative? If so, an optimal solution has been found; if not, at least one constraint has a negative right-hand side, and we go to step 2.

Step 2 Choose the most negative basic variable as the variable to leave the basis. The row in which the variable is basic will be the pivot row. To select the variable that enters the basis, we compute the following ratio for each variable x_j that has a *negative* coefficient in the pivot row:

$$\frac{\text{Coefficient of } x_j \text{ in row } 0}{\text{Coefficient of } x_j \text{ in pivot row}}$$

Choose the variable with the smallest ratio (absolute value) as the entering variable. This form of the ratio test maintains a dual feasible tableau (all variables in row 0 have non-negative coefficients). Now use EROs to make the entering variable a basic variable in the pivot row.

$-x_1 + x_2 \leq 1$

Step 3 If there is any constraint in which the right-hand side is negative and each variable has a non-negative coefficient, then the LP has no feasible solution. If no constraint indicating infeasibility is found, return to step 1.

To illustrate the case of an infeasible LP, suppose the dual simplex method yielded a constraint such as $x_1 + 2x_2 + x_3 = -5$. Because $x_1 \geq 0$, $2x_2 \geq 0$, and $x_3 \geq 0$, $x_1 + 2x_2 + x_3 \geq 0$, and the constraint $x_1 + 2x_2 + x_3 = -5$ cannot be satisfied. In this case, the original LP must be infeasible.

$x_2 - x_1 = 1$

Three uses of the dual simplex follow:

1 Finding the new optimal solution after a constraint is added to an LP

2 Finding the new optimal solution after changing a right-hand side of an LP

3 Solving a normal min problem

Finding the New Optimal Solution After a Constraint Is Added to an LP

The dual simplex method is often used to find the new optimal solution to an LP after a constraint is added. When a constraint is added, one of the following three cases will occur:

Case 1 The current optimal solution satisfies the new constraint.

Case 2 The current optimal solution does not satisfy the new constraint, but the LP still has a feasible solution.

Case 3 The additional constraint causes the LP to have no feasible solutions.

If Case 1 occurs, then the current optimal solution satisfies the new constraint, and the current solution remains optimal. To illustrate why this is true, suppose we have added the constraint $x_1 + x_2 + x_3 \leq 11$ to the Dakota problem. The current optimal solution ($z = 280, x_1 = 2, x_2 = 0, x_3 = 8$) satisfies this constraint. To see why this solution remains optimal after the constraint $x_1 + x_2 + x_3 \leq 11$ is added, recall that adding a constraint to an LP either leaves the feasible region unchanged or eliminates points from the feasible region. In this case, the Section 6.8 discussion tells us that adding a constraint (to a max problem) either reduces the optimal z-value or leaves it unchanged. This means that if we add the constraint $x_1 + x_2 + x_3 \leq 11$ to the Dakota problem, the new optimal z-value can be at most 280. The current solution is still feasible and has $z = 280$, so it must still be optimal.

If Case 2 occurs, the current solution is no longer feasible, so it can no longer be optimal. The dual simplex method can be used to determine the new optimal solution. Suppose that in the Dakota problem, marketing considerations dictate that at least 1 table be manufactured. This adds the constraint $x_2 \geq 1$. Because the current optimal solution has $x_2 = 0$, it is no longer feasible and cannot be optimal. To find the new optimal solution, we subtract an excess variable e_4 from the constraint $x_2 \geq 1$. This yields the constraint $x_2 - e_4 = 1$. If we multiply this constraint through by -1, we obtain $-x_2 + e_4 = -1$, and we can use e_4 as a basic variable for this constraint. Appending this constraint to the optimal Dakota tableau yields Table 33.

Because we are using the row 0 from an optimal tableau, each variable has a nonnegative coefficient in row 0, and we may proceed with the dual simplex method. The variable $e_4 = -1$ is the most negative basic variable, so e_4 will exit from the basis, and row 4 will be the pivot row. Because x_2 is the only variable with a negative coefficient in row 4, x_2 must enter into the basis (see Table 34).

This is an optimal tableau. Thus, if the constraint $x_2 \geq 1$ is added to the Dakota problem, the optimal solution becomes $z = 275, s_1 = 26, x_3 = 10, x_1 = \frac{3}{4}, x_2 = 1$, which has reduced Dakota's objective function (revenue) by \$5 (the reduced cost for tables).

If we had wanted to, we could simply have added the constraint $x_2 \geq 1$ to the original Dakota initial tableau and used the regular simplex method to solve the problem. This would have entailed adding an artificial variable to the $x_2 \geq 1$ constraint and would probably have required many pivots. When we use the dual simplex to solve a problem again after a constraint has been added, we are taking advantage of the fact that we have already

TABLE 33
"Old" Optimal Dakota Tableau If $x_2 \geq 1$ Is Required

					Basic Variable
z	$+\ 5x_2$		$+\ 10s_2 + 10s_3$	$= 280$	$z = 280$
	$-\ 2x_2$	$+\ s_1 +$	$2s_2 -\ 8s_3$	$= 24$	$s_1 = 24$
	$-\ 2x_2 + x_3$		$+\ 2s_2 -\ 4s_3$	$= 8$	$x_3 = 8$
$x_1 +$	$\frac{5}{4}x_2$		$-\ \frac{1}{2}s_2 + \frac{3}{2}s_3$	$= 2$	$x_1 = 2$
	$-\ \boxed{x_2}$		$+\ e_4$	$= -1$	$e_4 = -1$

TABLE 34
"New" Optimal Dakota Tableau If $x_2 \geq 1$ Is Required

				Basic Variable
z		$+\ 10s_2 + 10s_3 + 5e_4 = 275$		$z = 275$
	$s_1 +$	$2s_2 -\ 8s_3 - 2e_4 = 26$		$s_1 = 26$
	x_3	$+\ 2s_2 -\ 4s_3 - 2e_4 = 10$		$x_3 = 10$
x_1		$-\ \frac{1}{2}s_2 + \frac{3}{4}s_3 + \frac{5}{4}e_4 = \frac{3}{4}$		$x_1 = \frac{3}{4}$
	x_2	$-\ e_4 = 1$		$x_2 = 1$

obtained a non-negative row 0 and that most of our right-hand sides have non-negative coefficients. This is why the dual simplex usually requires relatively few pivots to find a new optimal solution when a constraint is added to an LP.

If Case 3 occurs, step 3 of the dual simplex method allows us to show that the LP is now infeasible. To illustrate the idea, suppose we add the constraint $x_1 + x_2 \geq 12$ to the Dakota problem. After subtracting an excess variable e_4 from this constraint, we obtain

$$x_1 + x_2 - e_4 = 12 \quad \text{or} \quad -x_1 - x_2 + e_4 = -12$$

Appending this constraint to the optimal Dakota tableau yields Table 35.

Because x_1 appears in the new constraint, it seems that x_1 can no longer be used as a basic variable for row 3. To remedy this problem, we eliminate x_1 (and in general all basic variables) from the new constraint by replacing row 4 by row 3 + row 4 (see Table 36). Because $e_4 = -10$ is the most negative basic variable, e_4 will leave the basis and row 4 will be the pivot row. The variable s_2 is the only one with a negative coefficient in row 4, so s_2 enters the basis and becomes a basic variable in row 4 (see Table 37). Now x_3 must leave the basis, and row 2 will be the pivot row. Because x_2 is the only variable in row 2 with a negative coefficient, x_2 now enters the basis (see Table 38). Because $x_1 \geq 0$, $x_3 \geq 0$, $2s_3 \geq 0$, and $3e_4 \geq 0$, the left side of row 3 must be non-negative and cannot equal -20. Hence, the Dakota problem with the additional constraint $x_1 + x_2 \geq 12$ has no feasible solution.

TABLE 35
"Old" Optimal Dakota Tableau If $x_1 + x_2 \geq 12$ Is Required

					Basic Variable
z	$+\quad 5x_2$		$+\ 10s_2 + 10s_3$	$= 280$	$z = 280$
	$-\quad 2x_2$	$+\ s_1 +$	$2s_2 -\ 8s_3$	$= 24$	$s_1 = 24$
	$-\quad 5x_2 + x_3$		$+\ 2s_2 -\ 4s_3$	$= 8$	$x_3 = 8$
	$x_1 + 1.25x_2$		$-\ 0.5s_2 + 1.5s_3$	$= 2$	$x_1 = 2$
	$-\ x_1 -\quad x_2$		$+\ e_4$	$= -12$	$e_4 = -12$

TABLE 36
e_4 Is Now a Basic Variable in Row 4

			Basic Variable
z $+$ $5x_2$ $+ 10s_2 + 10s_3$ $= 280$			$z = 280$
$- 2x_2 + s_1 + 2s_2 - 8s_3 = 24$			$s_1 = 24$
$- 2x_2 + x_3 + 2s_2 - 4s_3 = 8$			$x_3 = 8$
$x_1 + 1.25x_2 - 0.5s_2 + 1.5s_3 = 2$			$x_1 = 2$
$0.25x_2 \;\; \boxed{- 0.5s_2} + 1.5s_3 + e_4 = -10$			$e_4 = -10$

TABLE 37
s_2 Enters the Basis in Row 4

			Basic Variable
z $+ 10x_2$ $+ 40s_3 + 20e_4 = 80$			$z = 80$
$- x_2 + s_1 - 2s_3 + 4e_4 = -16$			$s_1 = -16$
$- \boxed{x_2} + x_3 + 2s_3 + 4e_4 = -32$			$x_3 = -32$
$x_1 + x_2 - e_4 = 12$			$x_2 = 12$
$- 0.5x_2 + s_2 - 3s_3 - 2e_4 = 20$			$s_2 = 20$

TABLE 38
Tableau Indicating Infeasibility of Dakota Example When $x_1 + x_2 \geq 12$ Is Required

			Basic Variable
z $+ 10x_3$ $+ 60s_3 + 60e_4 = -240$			$z = -240$
$- x_3 + s_1 - 4s_3 = 16$			$s_1 = 16$
$x_2 - x_3 - 2s_3 - 4e_4 = 32$			$x_2 = 32$
$x_1 + x_3 + 2s_3 + 3e_4 = -20$			$x_1 = -20$
$- 0.5x_3 + s_2 - 4s_3 - 4e_4 = 36$			$s_2 = 36$

Finding the New Optimal Solution After Changing a Right-Hand Side

If the right-hand side of a constraint is changed and the current basis becomes infeasible, the dual simplex can be used to find the new optimal solution. To illustrate, suppose that 30 finishing hours are now available. In Section 6.3, we showed that this changed the current optimal tableau to that shown in Table 39.

Because each variable in row 0 has a non-negative coefficient, the dual simplex method may be used to find the new optimal solution. The variable x_1 is the most negative one, so x_1 must leave the basis, and row 3 will be the pivot row. Because s_2 has the only negative coefficient in row 3, s_2 will enter the basis (see Table 40).

This is an optimal tableau. If 30 finishing hours are available, the new optimal solution to the Dakota problem is to manufacture 16 chairs, 0 tables, and 0 desks. Of course, if we change the right-hand side of a constraint, it is possible that the LP will be infeasible. Step 3 of the dual simplex algorithm will indicate whether this is the case.

TABLE 39
"Old" Optimal Dakota Tableau If 30 Finishing Hours Are Available

					Basic Variable
z	$+ \quad 5x_2$	$+ \quad 10s_2 + 10s_3 = 380$			$z = 380$
	$- \quad 2x_2$	$+ s_1 + \quad 2s_2 - \quad 8s_3 = 44$			$s_1 = 44$
	$- \quad 2x_2 + x_3$	$+ \quad 2s_2 - \quad 4s_3 = 28$			$x_3 = 28$
$x_1 + 1.25x_2$		$- \,(0.5s_2)\, + 1.5s_3 = -3$			$x_1 = -3$

TABLE 40
"New" Optimal Dakota Tableau If 30 Finishing Hours Are Available

				Basic Variable
$z + 20x_1 + 30x_2$		$+ \; 40s_3 = 320$		$z = 320$
$4x_1 + \quad 3x_2$	$+ s_1$	$- \; 2s_3 = 32$		$s_1 = 32$
$4x_1 + \quad 3x_2 + x_3$		$+ \; 2s_3 = 16$		$x_3 = 16$
$- \; 2x_1 - 2.5x_2$		$+ s_2 - \; 3s_3 = 6$		$x_1 = 6$

Solving a Normal Min Problem

To illustrate how the dual simplex can be used to solve a normal min problem, we solve the following LP:

$$\min z = x_1 + 2x_2$$
$$\text{s.t.} \qquad x_1 - 2x_2 + x_3 \geq 4$$
$$2x_1 + \quad x_2 - x_3 \geq 6$$
$$x_1, x_2, x_3 \geq 0$$

We begin by multiplying z by -1 to convert the LP to a max problem with objective function $z' = -x_1 - 2x_2$. After subtracting excess variables e_1 and e_2 from the two constraints, we obtain the initial tableau in Table 41. Each variable has a non-negative coefficient in row 0, so the dual simplex method can be applied. Before proceeding, we need to find the basic variables for the constraints. If we multiply each constraint through by -1, we can use e_1 and e_2 as basic variables. This yields the tableau in Table 42. At least one constraint has a negative right-hand side, so this is not an optimal tableau, and we proceed to step 2.

We choose the most negative basic variable (e_2) to exit from the basis. Because e_2 is basic in row 2, row 2 will be the pivot row. To determine the entering variable, we find the following ratios:

$$x_1 \text{ ratio} = 1/-2 = -\tfrac{1}{2}$$
$$x_2 \text{ ratio} = 2/-1 = -2$$

The smaller ratio (in absolute value) is the x_1 ratio, so we use EROs to enter x_1 into the basis in row 2 (see Table 43).[†]

There is no constraint indicating infeasibility (step 3), so we return to step 1. The first constraint has a negative right-hand side, so the tableau is not optimal, and we go to step 2. Because $e_1 = -1$ is the most negative basic variable, e_1 will exit from the basis, and

---

[†]The interested reader may verify that if we had made an error in performing the ratio test and had chosen x_2 to enter the basis, then a negative coefficient in row 0 would have resulted, and dual feasibility would have been destroyed.

TABLE 41
Initial Tableau for Solving Normal Min Problem

$$z' + x_1 + 2x_2 \qquad\qquad\qquad = 0$$
$$x_1 - 2x_2 + x_3 - e_1 \qquad\quad = 4$$
$$2x_1 + x_2 - x_3 \qquad\quad - e_2 = 6$$

TABLE 42
Initial Tableau in Canonical Form

			Basic Variable
$z' + x_1 + 2x_1$		$= 0$	$z' = 0$
$- x_1 + 2x_2 - x_3 + e_1$		$= -4$	$e_1 = -4$
$-\boxed{2x_1} - x_2 + x_3$	$+ e_2$	$= -6$	$e_2 = -6$

TABLE 43
First Dual Simplex Tableau

			Basic Variable
$z' + \frac{3}{2}x_2 + \frac{1}{2}x_3$	$+ \frac{1}{2}e_2$	$= -3$	$z' = -3$
$\frac{5}{2}x_2 - \boxed{\frac{3}{2}x_3} + e_1 - \frac{1}{2}e_2$		$= -1$	$e_1 = -1$
$x_1 + \frac{1}{2}x_2 - \frac{1}{2}x_3$	$- \frac{1}{2}e_2$	$= 3$	$x_1 = 3$

row 1 will be the pivot row. The possible entering variables are x_3 and e_2. The relevant ratios are

$$x_3 \text{ ratio} = \frac{\frac{1}{2}}{-\frac{3}{2}} = -\frac{1}{3}$$

$$e_2 \text{ ratio} = \frac{\frac{1}{2}}{-\frac{1}{2}} = -1$$

The smallest ratio (in absolute value) is $-\frac{1}{3}$, so x_3 will enter the basis in row 1. After pivoting in x_3, the new tableau is as shown in Table 44.[†] Each right-hand side is non-negative, so this is an optimal tableau. The original problem was a min problem, so the optimal solution to the original min problem is $z = \frac{10}{3}$, $x_1 = \frac{10}{3}$, $x_3 = \frac{2}{3}$, and $x_2 = 0$.

Observe that each dual simplex tableau (except the optimal dual simplex tableau) has a z'-value exceeding the optimal z'-value. For this reason, we say that the dual simplex tableaus are superoptimal. As the dual simplex proceeds, each pivot brings us closer to a primal feasible solution. Each pivot (barring degeneracy) decreases z', and we are "less superoptimal." Once primal feasibility is obtained, our solution is optimal.

TABLE 44
Optimal Tableau for Dual Simplex Example

			Basic Variable
$z' + \frac{7}{3}x_2$	$+ \frac{1}{3}e_1 + \frac{1}{3}e_2$	$= -\frac{10}{3}$	$z' = -\frac{10}{3}$
$- \frac{5}{3}x_2 + x_3 - \frac{2}{3}e_1 + \frac{1}{3}e_2$		$= \frac{2}{3}$	$x_3 = \frac{2}{3}$
$x_1 - \frac{1}{3}x_2$	$- \frac{1}{3}e_1 - \frac{1}{3}e_2$	$= \frac{10}{3}$	$x_1 = \frac{10}{3}$

[†]If we had chosen to enter into the basis any variable with a positive coefficient in the pivot row, then we would have ended up with some negative entries in row 0. This is why any variable that is entered into the basis must have a negative coefficient in the pivot row.

PROBLEMS

Group A

1 Use the dual simplex method to solve the following LP:

$$\max z = -2x_1 - x_3$$
$$\text{s.t.} \quad x_1 + x_2 - x_3 \geq 5$$
$$x_1 - 2x_2 + 4x_3 \geq 8$$
$$x_1, x_2, x_3 \geq 0$$

2 In solving the following LP, we obtain the optimal tableau shown in Table 45.

$$\max z = 6x_1 + x_2$$
$$\text{s.t.} \quad x_1 + x_2 \leq 5$$
$$2x_1 + x_2 \leq 6$$
$$x_1, x_2 \geq 0$$

a Find the optimal solution to this LP if we add the constraint $3x_1 + x_2 \leq 10$.

b Find the optimal solution if we add the constraint $x_1 - x_2 \geq 6$.

TABLE 45

				Basic Variable
z	$+ 2x_2$	$+ 3s_2 = 18$		$z_1 = 18$
	$0.5x_2 + s_1$	$- 0.5s_2 = 2$		$s_1 = 2$
	$x_1 + 0.5x_2$	$+ 0.5s_2 = 3$		$x_1 = 3$

c Find the optimal solution if we add the constraint $8x_1 + x_2 \leq 12$.

3 Find the new optimal solution to the Dakota problem if only 20 board ft of lumber are available.

4 Find the new optimal solution to the Dakota problem if 15 carpentry hours are available.

6.12 Data Envelopment Analysis[†]

Often we wonder if a university, hospital, restaurant, or other business is operating efficiently. The **Data Envelopment Analysis (DEA) method** can be used to answer this question. Our presentation is based on Callen (1991). To illustrate how DEA works, let's consider a group of three hospitals. To simplify matters, we assume that each hospital "converts" two inputs into three different outputs. The two inputs used by each hospital are

Input 1 = capital (measured by the number of hospital beds)

Input 2 = labor (measured in thousands of labor hours used during a month)

The outputs produced by each hospital are

Output 1 = hundreds of patient-days during month for patients under age 14

Output 2 = hundreds of patient-days during month for patients between 14 and 65

Output 3 = hundreds of patient-days during month for patients over 65

Suppose that the inputs and outputs for the three hospitals are as given in Table 46.

To determine whether a hospital is efficient, let's define t_r = price or value of one unit of output r and w_s = cost of one unit of input s. The *efficiency* of hospital i is defined to be

$$\frac{\text{value of hospital } i\text{'s outputs}}{\text{cost of hospital } i\text{'s inputs}}$$

For the data in Table 46, we find the efficiency of each hospital to be as follows:

$$\text{Hospital 1 efficiency} = \frac{9t_1 + 4t_2 + 16t_3}{5w_1 + 14w_2}$$

$$\text{Hospital 2 efficiency} = \frac{5t_1 + 7t_2 + 10t_3}{8w_1 + 15w_2}$$

$$\text{Hospital 3 efficiency} = \frac{4t_1 + 9t_2 + 13t_3}{7w_1 + 12w_2}$$

[†]This section may be omitted without loss of continuity.

TABLE 46
Inputs and Outputs for Hospitals

Hospital	Inputs		Outputs		
	1	2	1	2	3
1	5	14	9	4	16
2	8	15	5	7	10
3	7	12	4	9	13

The DEA approach uses the following four ideas to determine if a hospital is efficient.

1 No hospital can be more than 100% efficient. Thus, the efficiency of each hospital must be less than or equal to 1. For hospital 1, we find that $(9t_1 + 4t_2 + 16t_3)/(5w_1 + 14w_2) \leq 1$. Multiplying both sides of this inequality by $(5w_1 + 14w_2)$ (this is the trick we used to simplify blending constraints in Section 3.8!) yields the LP constraint $5w_1 + 14w_2 - 9t_1 - 4t_2 - 16t_3 \geq 0$.

2 Suppose we are interested in evaluating the efficiency of hospital i. We attempt to choose output prices (t_1, t_2, and t_3) and input costs (w_1 and w_2) that maximize efficiency. If the efficiency of hospital i equals 1, then it is efficient; if the efficiency is less than 1, then it is inefficient.

3 To simplify computations, we may scale the output prices so that the cost of hospital i's inputs equals 1. Thus, for hospital 2 we add the constraint $8w_1 + 15w_2 = 1$.

4 We must ensure that each input cost and output price is strictly positive. If, for example, $t_i = 0$, then DEA could not detect an inefficiency involving output i; if $w_j = 0$, then DEA could not detect an inefficiency involving input j.

Points (1)–(4) lead to the following LPs for testing the efficiency of each hospital.

Hospital 1 LP

$$\max z = 9t_1 + 4t_2 + 16t_3 \tag{1}$$

$$\text{s.t.} \quad -9t_1 - 4t_2 - 16t_3 + 5w_1 + 14w_2 \geq 0 \tag{2}$$

$$-5t_1 - 7t_2 - 10t_3 + 8w_1 + 15w_2 \geq 0 \tag{3}$$

$$-4t_1 - 9t_2 - 13t_3 + 7w_1 + 12w_2 \geq 0 \tag{4}$$

$$5w_1 + 14w_2 = 1 \tag{5}$$

$$t_1 \geq .0001 \tag{6}$$

$$t_2 \geq .0001 \tag{7}$$

$$t_3 \geq .0001 \tag{8}$$

$$w_1 \geq .0001 \tag{9}$$

$$w_2 \geq .0001 \tag{10}$$

Hospital 2 LP

$$\max z = 5t_1 + 7t_2 + 10t_3 \tag{1}$$

$$\text{s.t.} \quad -9t_1 - 4t_2 - 16t_3 + 5w_1 + 14w_2 \geq 0 \tag{2}$$

$$-5t_1 - 7t_2 - 10t_3 + 8w_1 + 15w_2 \geq 0 \tag{3}$$

$$-4t_1 - 9t_2 - 13t_3 + 7w_1 + 12w_2 \geq 0 \tag{4}$$

$$8w_1 + 15w_2 = 1 \tag{5}$$

$$t_1 \geq .0001 \tag{6}$$

$$t_2 \geq .0001 \quad (7)$$
$$t_3 \geq .0001 \quad (8)$$
$$w_1 \geq .0001 \quad (9)$$
$$w_2 \geq .0001 \quad (10)$$

Hospital 3 LP

$$\max z = 4t_1 + 9t_2 + 13t_3 \qquad (1)$$
$$\text{s.t.} \quad -9t_1 - 4t_2 - 16t_3 + 5w_1 + 14w_2 \geq 0 \qquad (2)$$
$$-5t_1 - 7t_2 - 10t_3 + 8w_1 + 15w_2 \geq 0 \qquad (3)$$
$$-4t_1 - 9t_2 - 13t_3 + 7w_1 + 12w_2 \geq 0 \qquad (4)$$
$$7w_1 + 12w_2 = 1 \qquad (5)$$
$$t_1 \geq .0001 \quad (6)$$
$$t_2 \geq .0001 \quad (7)$$
$$t_3 \geq .0001 \quad (8)$$
$$w_1 \geq .0001 \quad (9)$$
$$w_2 \geq .0001 \quad (10)$$

Let's see how the hospital 1 LP incorporates points (1)–(4). Point (1) maximizes the efficiency of hospital 1. This is because Constraint (5) implies that the total cost of hospital 1's inputs equal 1. Constraints (2)–(4) ensure that no hospital is more than 100% efficient. Constraints (6)–(10) ensure that each input cost and output price is strictly positive (the .0001 right-hand side is arbitrary; any small positive number may be used).

The LINDO output for these LPs is given in Figures 10(a)–(c). From the optimal ob-

```
MAX  9 T1 + 4 T2 + 16 T3
SUBJECT TO
    2)  - 9 T1 - 4 T2 - 16 T3 + 5 W1 + 14 W2 >=    0
    3)  - 5 T1 - 7 T2 - 10 T3 + 8 W1 + 15 W2 >=    0
    4)  - 4 T1 - 9 T2 - 13 T3 + 7 W1 + 12 W2 >=    0
    5)     W1 >=     0.0001
    6)     W2 >=     0.0001
    7)     T1 >=     0.0001
    8)     T2 >=     0.0001
    9)     T3 >=     0.0001
   10)    5 W1 + 14 W2 =      1
END

LP OPTIMUM FOUND AT STEP          6

        OBJECTIVE FUNCTION VALUE

     1)      1.00000000

VARIABLE          VALUE          REDUCED COST
    T1           .110889            .000000
    T2           .000100            .000000
    T3           .000100            .000000
    W1           .000100            .000000
    W2           .071393            .000000


   ROW     SLACK OR SURPLUS       DUAL PRICES
    2)          .000000           -1.000000
    3)          .515548            .000000
    4)          .411659            .000000
    5)          .000000            .000000
    6)          .071293            .000000
    7)          .110789            .000000
    8)          .000000            .000000
    9)          .000000            .000000
   10)          .000000           1.000000

NO. ITERATIONS=          6
```

FIGURE 10(a)

Hospital 1 LP

```
MAX    5 T1 + 7 T2 + 10 T3
SUBJECT TO
      2)  - 9 T1 -  4 T2 - 16 T3 + 5 W1 + 14 W2 >=    0
      3)  - 5 T1 -  7 T2 - 10 T3 + 8 W1 + 15 W2 >=    0
      4)  - 4 T1 -  9 T2 - 13 T3 + 7 W1 + 12 W2 >=    0
      5)    8 W1 + 15 W2 =    1
      6)    W1  >=  0.0001
      7)    W2  >=  0.0001
      8)    T1  >=  0.0001
      9)    T2  >=  0.0001
     10)    T3  >=  0.0001
END

LP OPTIMUM FOUND AT STEP          0

         OBJECTIVE FUNCTION VALUE

     1)     .773030000

VARIABLE         VALUE          REDUCED COST
     T1         .079821           .000000
     T2         .053275           .000000
     T3         .000100           .000000
     W1         .000100           .000000
     W2         .066613           .000000


     ROW     SLACK OR SURPLUS     DUAL PRICES
     2)          .000000          -.261538
     3)          .226970           .000000
     4)          .000000          -.661538
     5)          .000000           .773333
     6)          .000000          -.248206
     7)          .066513           .000000
     8)          .079721           .000000
     9)          .053175           .000000
    10)          .000000         -2.784615

NO. ITERATIONS=         0
```

FIGURE 10(b)

Hospital 2 LP

jective function value to each LP we find that

$$\text{Hospital 1 efficiency} = 1$$

$$\text{Hospital 2 efficiency} = .773$$

$$\text{Hospital 3 efficiency} = 1$$

Thus we find that hospital 2 is inefficient and hospitals 1 and 3 are efficient.

REMARK   1   An easy way to create the hospital 2 LP is to use LINDO to modify the objective function of the hospital 1 LP and the constraint $5w_1 + 14w_2 = 1$. Then it is easy to modify the hospital 2 LP to create the hospital 3 LP.

Using LINGO to Run a DEA

DEA.lng

The following LINGO program (see file DEA.lng) will solve our Hospital DEA problem. When faced with another DEA problem, we begin by changing the numbers of inputs, outputs, and units. Next we change the resource usage and outputs for each unit. Finally, by changing number to say, 1, we can evaluate the efficiency of unit 1. If the optimal objective function value for unit 1 is less than 1, then unit I is inefficient. Otherwise, unit 1 is efficient.

```
SETS:
INPUTS/1..2/:COSTS;
OUTPUTS/1..3/:PRICES;
```

```
        MAX  4 TI + 9 + T2 + 13 T3
        SUBJECT TO
               2) - 9 T1 - 4 T2 - 16 T3 + 5 W1 + 14 W2 >= 0
               3) - 5 T1 - 7 T2 - 10 T3 + 8 W1 + 15 W2 >= 0
               4) - 4 T1 - 9 T2 - 13 T3 + 7 W1 + 12 W2 >= 0
               5)   W1 >=    0.0001
               6)   W2 >=    0.0001
               7)   T1 >=    0.0001
               8)   T2 >=    0.0001
               9)   T3 >=    0.0001
              10)   7 W1 + 12 W2 = 1
        END

        LP OPTIMUM FOUND AT STEP 7

              OBJECTIVE FUNCTION VALUE

           1)    1.00000000

        VARIABLE         VALUE        REDUCED COST
            T1          .099815          .000000
            T2          .066605          .000000
            T3          .000100          .000000
            W1          .000100          .000000
            W2          .083275          .000000


        ROW     SLACK OR SURPLUS     DUAL PRICES
         2)         .000000            .000000
         3)         .283620            .000000
         4)         .000000          -1.000000
         5)         .000000            .000000
         6)         .083175            .000000
         7)         .099715            .000000
         8)         .066505            .000000
         9)         .000000            .000000
        10)         .000000           1.000000
```

FIGURE 10(c)

Hospital 3 LP   NO. ITERATIONS=   7

```
UNITS/1..3/;
UNIN(UNITS,INPUTS):USED;
UNOUT(UNITS,OUTPUTS):PRODUCED;
ENDSETS
NUMBER=2;
@FOR(UNITS(J)|j#EQ#NUMBER:MAX=@SUM(OUTPUTS(I):PRICES(I)*PRODUCED(J,I)));
@FOR(UNITS(J)|J#EQ#NUMBER:@SUM(INPUTS(I):COSTS(I)*USED(J,I))=1);
@FOR(INPUTS(I):COSTS(I)>=.0001);
@FOR(OUTPUTS(I):PRICES(I)>=.0001);
@FOR(UNITS(I):@SUM(INPUTS(J):COSTS(J)*USED(I,J))>=@SUM(OUTPUTS(J):PRICES(J)*PRODUCED(I,J))
);
DATA:
USED=5,14,
     8,15,
     7,12;
PRODUCED=9,4,16,
         5,7,10,
         4,9,13;
ENDDATA
END
```

Dual Prices and DEA

The DUAL PRICES section of the LINDO output gives us great insight into Hospital 2's (or any organization found inefficient by DEA) inefficiency. Consider all hospitals whose efficiency constraints have nonzero dual prices in the hospital 2 LP (Figure 10b). (In our example, hospitals 1 and 3 have nonzero dual prices.) If we average the output vectors and input vectors for these hospitals (using the absolute value of the dual price for each hospital as the weight) we obtain the following:

Averaged Output Vector

$$.261538 \begin{bmatrix} 9 \\ 4 \\ 16 \end{bmatrix} + .661538 \begin{bmatrix} 4 \\ 9 \\ 13 \end{bmatrix} = \begin{bmatrix} 5 \\ 7 \\ 12.785 \end{bmatrix}$$

Averaged Input Vector

$$.261538 \begin{bmatrix} 5 \\ 14 \end{bmatrix} + .661538 \begin{bmatrix} 7 \\ 12 \end{bmatrix} = \begin{bmatrix} 5.938 \\ 11.6 \end{bmatrix}$$

Suppose we create a composite hospital by combining .261538 of hospital 1 with .661538 of hospital 3. The averaged output vector tells us that the composite hospital produces the same amount of outputs 1 and 2 as hospital 2, but the composite hospital produces $12.785 - 10 = 2.785$ more of output 3 (patient days for more than 65 patients). From the averaged input vector for the composite hospital, we find that the composite hospital uses less of each input than does hospital 2. We now see exactly where hospital 2 is inefficient!

By the way, the objective function value of .7730 for the hospital 2 LP implies that the more efficient composite hospital produces its superior outputs by using at most 77.30% as much of each input. Note that

Input 1 used by composite hospital $<.7730 *$ (Input 1 used by hospital 2) $= 6.2186$

and

Input 2 used by composite hospital $= .7730 *$ (Input 2 used by hospital 2) $= 11.6$

An explanation of why the dual prices are needed to find a composite hospital that is superior to an inefficient hospital is given in Problems 5–7.

PROBLEMS

Group A

1 The Salem Board of Education wants to evaluate the efficiency of the town's four elementary schools. The three outputs of the schools are defined to be

> Output 1 = average reading score
> Output 2 = average mathematics score
> Output 3 = average self-esteem score

The three inputs to the schools are defined to be

Input 1 = average educational level of mothers (defined by highest grade completed—12 = high school graduate; 16 = college graduate, and so on).

Input 2 = number of parent visits to school (per child)

Input 3 = teacher to student ratio

The relevant information for the four schools is given in Table 47. Determine which (if any) schools are inefficient. For any inefficient school, determine the nature of the inefficiency.

2 Pine Valley Bank has three branches. You have been assigned to evaluate the efficiency of each. The following inputs and outputs are to be used for the study.

TABLE 47

School	Inputs			Outputs		
	1	2	3	1	2	3
1	13	4	.05	9	7	6
2	14	5	.05	10	8	7
3	11	6	.06	11	7	8
4	15	8	.08	9	9	9

Input 1 = labor hours used (hundreds per month)
Input 2 = space used (in hundreds of square feet)
Input 3 = supplies used per month (in dollars)
Output 1 = loan applications per month
Output 2 = deposits processed per month (in thousands)
Output 3 = checks processed per month (in thousands)

The relevant information is given in Table 48. Use this data to determine if any bank branches are inefficient. If any

TABLE 48

Bank	Inputs 1	2	3	Outputs 1	2	3
1	15	20	50	200	15	35
2	14	23	51	220	18	45
3	16	19	51	210	17	20

TABLE 49

Precinct	Inputs 1	2	Outputs 1	2
1	200	60	6	8
2	300	90	8	9.5
3	400	120	10	11

bank branches are inefficient, determine the nature of the inefficiency.

3 You have been assigned to evaluate the efficiency of the Port Charles Police Department. Three precincts are to be evaluated. The inputs and outputs for each precinct are as follows:

Input 1 = number of police officers

Input 2 = number of vehicles used

Output 1 = number of patrol units responding to service requests (thousands per year)

Output 2 = number of convictions obtained each year (in hundreds)

You are given the data in Table 49. Use this information to determine which precincts, if any, are inefficient. For any inefficient precincts, determine the nature of the inefficiency.

4 You have been assigned by Indiana University to evaluate the relative efficiency of four degree-granting units: Business; Education; Arts and Sciences; and Health, Physical Education, and Recreation (HPER). You are given the

information in Table 50. Use DEA to find all inefficient units. Comment on the nature of the inefficiencies you found.

Group B

5 Explain why the amount of each output produced by the composite hospital obtained by averaging hospitals 1 and 3 (with the absolute value of the dual prices as weights) is at least as large as the amount of the corresponding output produced by hospital 2. (*Hint:* Price out variables t_1, t_2, and t_3, and use the fact that the coefficient of these variables in row 0 of the optimal tableau must equal 0.)

6 Explain why the dual price for the $8w_1 + 15w_2 = 1$ constraint must equal the optimal z-value for the hospital 2 LP.

7 a Explain why the amount of each input used by the composite hospital is at most (efficiency of hospital 2) * (the amount of the corresponding input used by hospital 2). (*Hint:* Price out w_1 and w_2 and use Problem 6.)

b Explain why the amount of each input used by the composite hospital is no larger than the amount of the corresponding input used by hospital 2.

TABLE 50

	Faculty	Support Staff	Supply Budget (in Millions)	Credit Hours (in Thousands)	Research Publications
Business	150	70	5	15	225
Education	60	20	3	5.4	70
Arts and Sciences	800	140	20	56	1,300
HPER	30	15	1	2.1	40

SUMMARY   Graphical Sensitivity Analysis

To determine whether the current basis remains optimal after changing an objective function coefficient, note that the change affects the slope of the isoprofit line. The current basis remains optimal as long as the current optimal solution is the last point in the feasible region to make contact with isoprofit lines as we move in the direction of increasing z (for a max problem). If the current basis remains optimal, then the values of the decision variables remain unchanged, but the optimal z-value may change.

To determine whether the current basis remains optimal after changing the right-hand side of a constraint, find the constraints (possibly including sign restrictions) that are binding for the current optimal solution. As we change the right-hand side of a constraint, the current basis remains optimal as long as the point where the constraints are binding remains feasible. Even if the current basis remains optimal, the values of the decision variables and the optimal z-value may change.

Shadow Prices

The **shadow price** of the ith constraint of a linear programming problem is the amount by which the optimal z-value is improved if the right-hand side is increased by 1. The shadow price of the ith constraint is the DUAL PRICE for row $i + 1$ in the LINDO output.

Notation

BV_i = basic variable for ith constraint in the optimal tableau

$\mathbf{c}_{BV}$ = row vector whose ith element is the objective function coefficient for BV_i in the LP

$\mathbf{a}_j$ = column for variable x_j in constraints of original LP

$\mathbf{b}$ = right-hand side vector for original LP

$\bar{c}_j$ = coefficient of x_j in row 0 of the optimal tableau

How to Compute Optimal Tableau from Initial LP

$$\text{Column for } x_j \text{ in optimal tableau's constraints} = B^{-1}\mathbf{a}_j \tag{5}$$

$$\text{Right-hand side of optimal tableau's constraints} = B^{-1}\mathbf{b} \tag{6}$$

$$\bar{c}_j = \mathbf{c}_{BV}B^{-1}\mathbf{a}_j - c_j \tag{10}$$

$$\text{Coefficient of slack variable } s_i \text{ in optimal row 0} = i\text{th element of } \mathbf{c}_{BV}B^{-1} \tag{10'}$$

$$\text{Coefficient of excess variable } e_i \text{ in optimal row 0} = -(i\text{th element of } \mathbf{c}_{BV}B^{-1}) \tag{10''}$$

$$\text{Coefficient of artificial variable } a_i \text{ in optimal row 0} = (i\text{th element of } \mathbf{c}_{BV}B^{-1}) + M \tag{10'''}$$

$$\text{Right-hand side of optimal row 0} = \mathbf{c}_{BV}B^{-1}\mathbf{b} \tag{11}$$

Sensitivity Analysis

For a max problem, a tableau is optimal if and only if each variable has a non-negative coefficient in row 0 and each constraint has a non-negative right-hand side. For a min problem, a tableau is optimal if and only if each variable has a nonpositive coefficient in row 0 and each constraint has a non-negative right-hand side.

If the current basis remains optimal after changing the objective function coefficient of a nonbasic variable, the values of the decision variables and the optimal z-value remain unchanged. With a basic variable, the values of the decision variables remain unchanged, but the optimal z-value may change. Both the values of the decision variables and the optimal z-value may change after changing a right-hand side. The new values of the decision variables may be found by computing B^{-1} (new right-hand side vector). The new optimal z-value may be determined by using shadow prices or Equation (11).

Objective Function Coefficient Range

The OBJ COEFFICIENT RANGES section of the LINDO output gives the range of values for an objective function coefficient for which the current basis remains optimal. Within this range, the values of the decision variables remain unchanged, but the optimal z-value may or may not change.

Reduced Cost

For any nonbasic variable, the reduced cost for the variable is the amount by which its objective function coefficient must be improved before that variable will be a basic variable in some optimal solution to the LP.

Right-Hand Side Range

If the right-hand side of a constraint remains within the RIGHTHAND SIDE RANGE of the LINDO printout, the current basis remains optimal, and the LINDO listing for the constraint's dual price may be used to determine how the change affects the optimal z-value. Even if the right-hand side of a constraint remains within the range, the values of the decision variables will probably change.

Finding the Dual of an LP

For a normal (all $\leq$ constraints and all variables non-negative) max problem or a normal min (all $\geq$ constraints and all variables non-negative) problem, we find the dual as follows:

If we read the primal across in Table 14, we read the dual down. If we read the primal down in Table 14, we read the dual across. We use x_i's and z as variables for a maximization problem and y_j's and w as variables for a minimization problem.

To find the dual of a non-normal max problem:

Step 1 Fill in Table 14 so that the primal can be read across.

Step 2 After making the following changes, the dual can be read down in the usual fashion: (a) If the ith primal constraint is a $\geq$ constraint, the corresponding dual variable y_i must satisfy $y_i \leq 0$. (b) If the ith primal constraint is an equality, then the dual variable y_i is now urs. (c) If the ith primal variable is urs, then the ith dual constraint will be an equality.

To find the dual of a non-normal min problem:

Step 1 Write out the primal so it can be read down in Table 14.

Step 2 Except for the following changes, the dual can be read across the table: (a) If the ith primal constraint is a $\leq$ constraint, then the corresponding dual variable x_i must satisfy $x_i \leq 0$. (b) If the ith primal constraint is an equality, then the corresponding dual variable x_i will be urs. (c) If the ith primal variable y_i is urs, then the ith dual constraint is an equality.

The Dual Theorem

Suppose BV is an optimal basis for the primal. Then $c_{BV}B^{-1}$ is an optimal solution to the dual. Also, $\bar{z} = \bar{w}$.

Finding the Optimal Solution to the Dual of an LP

If the primal is a max problem, then the optimal dual solution may be read from row 0 of the optimal tableau by using the following rules:

$$\text{Optimal value of dual variable } y_i \text{ if Constraint } i \text{ is a} \leq \text{constraint} = \text{coefficient of } s_i \text{ in optimal row 0} \qquad \text{(31)}$$

$$\text{Optimal value of dual variable } y_i \text{ if Constraint } i \text{ is a} \geq \text{constraint} = -(\text{coefficient of } e_i \text{ in optimal row 0}) \qquad \text{(31')}$$

$$\text{Optimal value of dual variable } y_i \text{ if Constraint } i \text{ is an equality constraint} = (\text{coefficient of } a_i \text{ in optimal row 0}) - M \qquad \text{(31'')}$$

If the primal is a min problem, then the optimal dual solution may be read from row 0 of the optimal tableau by using the following rules:

$$\text{Optimal value of dual variable } x_i \text{ if Constraint } i \text{ is a} \leq \text{constraint} = \text{coefficient of } s_i \text{ in optimal row 0}$$

$$\text{Optimal value of dual variable } x_i \text{ if Constraint } i \text{ is a} \geq \text{constraint} = -(\text{coefficient of } e_i \text{ in optimal row 0})$$

$$\text{Optimal value of dual variable } x_i \text{ if Constraint } i \text{ is an equality constraint} = (\text{coefficient of } a_i \text{ in optimal row 0}) + M$$

Shadow Prices (Again)

For a maximization LP, the shadow price of the ith constraint is the value of the ith dual variable in the optimal dual solution. For a minimization LP, the shadow price of the ith constraint $= -(i$th dual variable in the optimal dual solution). The shadow price of the ith constraint is found in row $i + 1$ of the DUAL PRICES portion of the LINDO printout.

$$\text{New optimal } z\text{-value} = (\text{old optimal } z\text{-value}) + (\text{Constraint } i \text{ shadow price}) \, \Delta b_i \quad (\text{max problem}) \qquad \text{(37)}$$

$$\text{New optimal } z\text{-value} = (\text{old optimal } z\text{-value}) - (\text{Constraint } i \text{ shadow price}) \, \Delta b_i \quad (\text{min problem}) \qquad \text{(37')}$$

A $\geq$ constraint will have a nonpositive shadow price; a $\leq$ constraint will have a nonnegative shadow price; and an equality constraint may have a positive, negative, or zero shadow price.

Duality and Sensitivity Analysis

Our proof of the Dual Theorem showed that if a set of basic variables BV is feasible, then BV is optimal (that is, each variable in row 0 has a non-negative coefficient) if and only if the associated dual solution, $\mathbf{c}_{\mathrm{BV}}B^{-1}$, is dual feasible.

This result can be used to yield an alternative way of doing the following types of sensitivity analysis:

Change 1   Changing the objective function coefficient of a nonbasic variable

Change 4   Changing the column of a nonbasic variable

Change 5   Adding a new activity

In each case, simply determine whether a change in the original LP maintains dual feasibility. If dual feasibility is maintained, then the current basis remains optimal. If dual feasibility is not maintained, then the current basis is no longer optimal.

Complementary Slackness

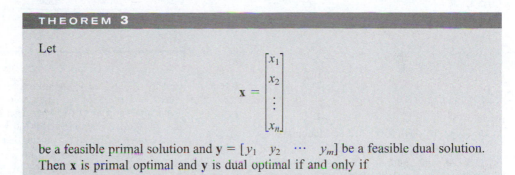

THEOREM 3

Let

$$\mathbf{x} = \begin{bmatrix} x_1 \\ x_2 \\ \vdots \\ x_n \end{bmatrix}$$

be a feasible primal solution and $\mathbf{y} = [y_1 \quad y_2 \quad \cdots \quad y_m]$ be a feasible dual solution. Then $\mathbf{x}$ is primal optimal and $\mathbf{y}$ is dual optimal if and only if

$$s_i y_i = 0 \quad (i = 1, 2, \ldots, m) \tag{38}$$

$$e_j x_j = 0 \quad (j = 1, 2, \ldots, n) \tag{39}$$

The Dual Simplex Method

The dual simplex method can be applied (to a max problem) whenever there is a basic solution in which each variable has a non-negative coefficient in row 0. If we have found such a basic solution, then the dual simplex method proceeds as follows:

Step 1 If the right-hand side of each constraint is non-negative, then an optimal solution has been found; if not, then at least one constraint has a negative right-hand side and we go to step 2.

Step 2 Choose the most negative basic variable as the variable to leave the basis. The row in which this variable is basic will be the pivot row. To select the variable that enters the basis, compute the following ratio for each variable x_j that has a *negative* coefficient in the pivot row:

$$\frac{\text{Coefficient of } x_j \text{ in row 0}}{\text{Coefficient of } x_j \text{ in the pivot row}}$$

Choose the variable that has the smallest ratio (absolute value) as the entering variable. Use EROs to make the entering variable a basic variation in the pivot row.

Step 3 If there is any constraint in which the right-hand side is negative and each variable has a non-negative coefficient, then the LP has no feasible solution. Infeasibility would be indicated by the presence (after possibly several pivots) of a constraint such as $x_1 + 2x_2 + x_3 = -5$. If no constraint indicating infeasibility is found, return to step 1.

The dual simplex method is often used in the following situations:

1 Finding the new optimal solution after a constraint is added to an LP

2 Finding the new optimal solution after changing an LP's right-hand side

3 Solving a normal min problem

REVIEW PROBLEMS

All problems from Sections 5.2 and 5.3 are relevant, along with Chapter 5 Review Problems 1, 2, 6, and 7.

Group A

1 Consider the following LP and its optimal tableau (Table 51):

$$\max z = 4x_1 + x_2$$
$$\text{s.t.} \quad x_1 + 2x_2 = 6$$
$$x_1 - x_2 \geq 3$$
$$2x_1 + x_2 \leq 10$$
$$x_1, x_2 \geq 0$$

a Find the dual of this LP and its optimal solution.

b Find the range of values of b_3 for which the current basis remains optimal. If $b_3 = 11$, what would be the new optimal solution?

2 For the LP in Problem 1, graphically determine the range of values on c_1 for which the current basis remains optimal. (*Hint:* The feasible region is a line segment.)

3 Consider the following LP and its optimal tableau (Table 52):

$$\max z = 5x_1 + x_2 + 2x_3$$
$$\text{s.t.} \quad x_1 + x_2 + x_3 \leq 6$$
$$6x_1 + x_3 \leq 8$$
$$x_2 + x_3 \leq 2$$
$$x_1, x_2, x_3 \geq 0$$

a Find the dual to this LP and its optimal solution.

b Find the range of values of c_1 for which the current basis remains optimal.

c Find the range of values of c_2 for which the current basis remains optimal.

4 Carco manufactures cars and trucks. Each car contributes $300 to profit and each truck, $400. The

TABLE 51

z	x_1	x_2	e_2	s_3	a_1	a_2	rhs
1	0	0	0	$\frac{7}{3}$	$M - \frac{2}{3}$	M	$\frac{58}{3}$
0	0	1	0	$-\frac{1}{3}$	$\frac{2}{3}$	0	$\frac{2}{3}$
0	1	0	0	$\frac{2}{3}$	$-\frac{1}{3}$	0	$\frac{14}{3}$
0	0	0	1	1	-1	-1	1

TABLE 52

z	x_1	x_2	x_3	s_1	s_2	s_3	rhs
1	0	$\frac{1}{6}$	0	0	$\frac{5}{6}$	$\frac{7}{6}$	9
0	0	$\frac{1}{6}$	0	1	$-\frac{1}{6}$	$-\frac{5}{6}$	3
0	1	$-\frac{1}{6}$	0	0	$\frac{1}{6}$	$-\frac{1}{6}$	1
0	0	1	1	0	0	1	2

TABLE 53

Vehicle	Days on Type 1 Machine	Days on Type 2 Machine	Tons of Steel
Car	0.8	0.6	2
Truck	1	0.7	3

resources required to manufacture a car and a truck are shown in Table 53. Each day, Carco can rent up to 98 Type 1 machines at a cost of $50 per machine. The company now has 73 Type 2 machines and 260 tons of steel available. Marketing considerations dictate that at least 88 cars and at least 26 trucks be produced. Let

$$X1 = \text{number of cars produced daily}$$
$$X2 = \text{number of trucks produced daily}$$
$$M1 = \text{type 1 machines rented daily}$$

To maximize profit, Carco should solve the LP given in Figure 11. Use the LINDO output to answer the following questions:

a If cars contributed $310 to profit, what would be the new optimal solution to the problem?

b What is the most that Carco should be willing to pay to rent an additional Type 1 machine for 1 day?

c What is the most that Carco should be willing to pay for an extra ton of steel?

d If Carco were required to produce at least 86 cars, what would Carco's profit become?

e Carco is considering producing jeeps. A jeep contributes $600 to profit and requires 1.2 days on machine 1, 2 days on machine 2, and 4 tons of steel. Should Carco produce any jeeps?

5 The following LP has the optimal tableau shown in Table 54.

$$\max z = 4x_1 + x_2$$
$$\text{s.t.} \quad 3x_1 + x_2 \geq 6$$
$$2x_1 + x_2 \geq 4$$
$$x_1 + x_2 = 3$$
$$x_1, x_2 \geq 0$$

a Find the dual of this LP and its optimal solution.

b Find the range of values of the objective function coefficient of x_2 for which the current basis remains optimal.

c Find the range of values of the objective function coefficient of x_1 for which the current basis remains optimal.

6 Consider the following LP and its optimal tableau (Table 55):

$$\max z = 3x_1 + x_2 - x_3$$
$$\text{s.t.} \quad 2x_1 + x_2 + x_3 \leq 8$$
$$4x_1 + x_2 - x_3 \leq 10$$
$$x_1, x_2, x_3 \geq 0$$

FIGURE 11
LINDO Output for Carco (Problem 4)

```
MAX          300 X1 + 400 X2 - 50 M1
SUBJECT TO
     2)   0.8 X1 + X2 - M1 <=         0
     3)    M1 <=   98
     4)   0.6 X1 + 0.7 X2 <=    73
     5)    2 X1 + 3 X2 <=    260
     6)     X1 >=    88
     7)     X2 >=    26
END

     LP OPTIMUM FOUND   AT STEP     1

          OBJECTIVE FUNCTION VALUE

     1)        32540.0000

   VARIABLE          VALUE         REDUCED COST
        X1        88.000000          0.000000
        X2        27.599998          0.000000
        M1        98.000000          0.000000

     ROW      SLACK OR SURPLUS     DUAL PRICES
      2)         0.000000          400.000000
      3)         0.000000          350.000000
      4)         0.879999            0.000000
      5)         1.200003            0.000000
      6)         0.000000          -20.000000
      7)         1.599999            0.000000

NO. ITERATIONS=          1

   RANGES IN WHICH THE BASIS IS UNCHANGED

                      OBJ COEFFICIENT RANGES
   VARIABLE      CURRENT        ALLOWABLE      ALLOWABLE
                  COEF          INCREASE       DECREASE
        X1     300.000000      20.000000       INFINITY
        X2     400.000000      INFINITY       25.000000
        M1     -50.000000      INFINITY      350.000000

                      RIGHTHAND SIDE RANGES
     ROW       CURRENT        ALLOWABLE       ALLOWABLE
                RHS           INCREASE        DECREASE
      2       0.000000       0.400001        1.599999
      3      98.000000       0.400001        1.599999
      4      73.000000       INFINITY        0.879999
      5     260.000000       INFINITY        1.200003
      6      88.000000       1.999999        3.000008
      7      26.000000       1.599999        INFINITY
```

TABLE 54

z	x_1	x_2	e_1	e_2	a_1	a_2	a_3	rhs
1	0	3	0	0	M	M	$M + 4$	12
0	1	1	0	0	0	0	1	3
0	0	2	1	0	-1	0	3	3
0	0	1	0	1	0	-1	2	2

TABLE 55

z	x_1	x_2	x_3	s_1	s_2	rhs
1	0	0	1	$\frac{1}{2}$	$\frac{1}{2}$	9
0	0	1	3	2	-1	6
0	1	0	-1	$-\frac{1}{2}$	$\frac{1}{2}$	1

a Find the dual of this LP and its optimal solution.

b Find the range of values of b_2 for which the current basis remains optimal. If $b_2 = 12$, what is the new optimal solution?

7 Consider the following LP:

$$\max z = 3x_1 + 4x_2$$
$$\text{s.t.} \quad 2x_1 + x_2 \le 8$$
$$4x_1 + x_2 \le 10$$
$$x_1, x_2 \ge 0$$

The optimal solution to this LP is $z = 32$, $x_1 = 0$, $x_2 = 8$, $s_1 = 0$, $s_2 = 2$. Graphically find the range of values of c_1 for which the current basis remains optimal.

8 Wivco produces product 1 and product 2 by processing raw material. As much as 90 lb of raw material may be purchased at a cost of \$10/lb. One pound of raw material can be used to produce either 1 lb of product 1 or 0.33 lb of product 2. Using a pound of raw material to produce a pound of product 1 requires 2 hours of labor or 3 hours to produce 0.33 lb of product 2. A total of 200 hours of labor are available, and at most 40 pounds of product 2 can be sold. Product 1 sells for $13/lb, and product 2 sells for $40/lb. Let

RM = pounds of raw material processed

$P1$ = pounds of raw material used to produce product 1

$P2$ = pounds of raw material used to produce product 2

To maximize profit, Wivco should solve the following LP:

$$\max z = 13P1 + 40(0.33)P2 - 10RM$$

$$\text{s.t.} \quad RM \geq P1 + P2$$
$$2P1 + 3P2 \leq 200$$
$$RM \leq 90$$
$$0.33P2 \leq 40$$
$$P1, P2, RM \geq 0$$

Use the LINDO output in Figure 12 to answer the following questions:

a If only 87 lb of raw material could be purchased, what would be Wivco's profits?

b If product 2 sold for $39.50/lb, what would be the new optimal solution?

c What is the most that Wivco should pay for another pound of raw material?

d What is the most that Wivco should pay for another hour of labor?

e Suppose that 1 lb of raw material could also be used to produce 0.8 lb of product 3, which sells for $24/lb. Processing 1 lb of raw material into 0.8 lb of product 3 requires 7 hours of labor. Should Wivco produce any of product 3?

9 Consider the following LP and its optimal tableau (Table 56):

$$\max z = 3x_1 + 4x_2 + x_3$$

$$\text{s.t.} \quad x_1 + x_2 + x_3 \leq 50$$
$$2x_1 - x_2 + x_3 \geq 15$$
$$x_1 + x_2 = 10$$
$$x_1, x_2, x_3 \geq 0$$

a Find the dual of this LP and its optimal solution.

b Find the range of values of the objective function coefficient of x_1 for which the current basis remains optimal.

c Find the range of values of the objective function coefficient for x_2 for which the current basis remains optimal.

FIGURE 12
LINDO Output for Wivco (Problem 8)

```
MAX      13 P1 + 13.2 P2 - 10 RM
SUBJECT TO
       2)  - P1 - P2 + RM >=  0
       3)    2 P1 + 3 P2 <=   200
       4)    RM <=   90
       5)    0.33 P2 <=   40
END

    LP OPTIMUM FOUND  AT STEP     3

           OBJECTIVE FUNCTION VALUE

   1)        274.000000

 VARIABLE         VALUE          REDUCED COST
      P1        70.000000          0.000000
      P2        20.000000          0.000000
      RM        90.000000          0.000000

    ROW    SLACK OR SURPLUS      DUAL PRICES
      2)        0.000000         -12.600000
      3)        0.000000           0.200000
      4)        0.000000           2.600000
      5)       33.400002           0.000000

 NO. ITERATIONS=       3

   RANGES IN WHICH THE BASIS IS UNCHANGED

                        OBJ COEFFICIENT RANGES
 VARIABLE         CURRENT        ALLOWABLE        ALLOWABLE
                    COEF         INCREASE         DECREASE
      P1        13.000000         0.200000         0.866667
      P2        13.200000         1.300000         0.200000
      RM       -10.000000         INFINITY         2.600000

                        RIGHTHAND SIDE RANGES
     ROW         CURRENT        ALLOWABLE        ALLOWABLE
                   RHS          INCREASE         DECREASE
      2          0.000000        23.333334        10.000000
      3        200.000000        70.000000        20.000000
      4         90.000000        10.000000        23.333334
      5         40.000000        INFINITY         33.400002
```

TABLE 56

z	x_1	x_2	x_3	s_1	e_2	a_2	a_3	rhs
1	1	0	0	1	0	M	$M+3$	80
0	-3	0	0	1	1	-1	-2	15
0	0	0	1	1	0	0	-2	40
0	1	1	0	0	0	0	1	10

TABLE 57

z	x_1	x_2	s_1	s_2	rhs
1	0	0	0	1	10
0	0	$\frac{1}{3}$	1	$-\frac{2}{3}$	$\frac{4}{3}$
0	1	$\frac{7}{3}$	0	$\frac{1}{3}$	$\frac{10}{3}$

10 Consider the following LP and its optimal tableau (Table 57):

$$\max z = 3x_1 + 2x_2$$
$$\text{s.t.} \quad 2x_1 + 5x_2 \leq 8$$
$$3x_1 + 7x_2 \leq 10$$
$$x_1, x_2 \geq 0$$

a Find the dual of this LP and its optimal solution.

b Find the range of values of b_2 for which the current basis remains optimal. Also find the new optimal solution if $b_2 = 5$.

11 Consider the following LP:

$$\max z = 3x_1 + x_2$$
$$\text{s.t.} \quad 2x_1 + x_2 \leq 8$$
$$4x_1 + x_2 \leq 10$$
$$x_1, x_2 \geq 0$$

The optimal solution to this LP is $z = 9$, $x_1 = 1$, $x_2 = 6$. Graphically find the range of values of b_2 for which the current basis remains optimal.

12 Farmer Leary grows wheat and corn on his 45-acre farm. He can sell at most 140 bushels of wheat and 120 bushels of corn. Each planted acre yields either 5 bushels of wheat or 4 bushels of corn. Wheat sells for $30 per bushel, and corn sells for $50 per bushel. Six hours of labor are needed to harvest an acre of wheat, and 10 hours are needed to harvest an acre of corn. As many as 350 hours of labor can be purchased at $10 per hour. Let

A1 = acres planted with wheat

A2 = acres planted with corn

L = hours of labor that are purchased

To maximize profits, farmer Leary should solve the following LP:

$$\max z = 150A1 + 200A2 - 10L$$
$$\text{s.t.} \quad A1 + A2 \leq 45$$
$$6A1 + 10A2 - L \leq 0$$
$$L \leq 350$$
$$5A1 \leq 140$$
$$4A2 \leq 120$$
$$A1, A2, L \geq 0$$

Use the LINDO output in Figure 13 to answer the following questions:

a What is the most that Leary should pay for an additional hour of labor?

b What is the most that Leary should pay for an additional acre of land?

c If only 40 acres of land were available, what would be Leary's profit?

d If the price of wheat dropped to $26, what would be the new optimal solution?

e Farmer Leary is considering growing barley. Demand for barley is unlimited. An acre yields 4 bushels of barley and requires 3 hours of labor. If barley sells for $30 per bushel, should Leary produce any barley?

13 Consider the following LP and its optimal tableau (Table 58):

$$\max z = 4x_1 + x_2 + 2x_3$$
$$\text{s.t.} \quad 8x_1 + 3x_2 + x_3 \leq 2$$
$$6x_1 + x_2 + x_3 \leq 8$$
$$x_1, x_2, x_3 \geq 0$$

a Find the dual to this LP and its optimal solution.

b Find the range of values of the objective function coefficient of x_3 for which the current basis remains optimal.

c Find the range of values of the objective function coefficient of x_1 for which the current basis remains optimal.

14 Consider the following LP and its optimal tableau (Table 59):

$$\max z = 3x_1 + x_2$$
$$\text{s.t.} \quad 2x_1 + x_2 \leq 4$$
$$3x_1 + 2x_2 \geq 6$$
$$4x_1 + 2x_2 = 7$$
$$x_1 \geq 0, x_2 \geq 0$$

a Find the dual to this LP and its optimal solution.

b Find the range of values of the right-hand side of the third constraint for which the current basis remains optimal. Also find the new optimal solution if the right-hand side of the third constraint were $\frac{15}{2}$.

15 Consider the following LP:

$$\max z = 3x_1 + x_2$$
$$\text{s.t.} \quad 4x_1 + x_2 \leq 7$$
$$5x_1 + 2x_2 \leq 10$$
$$x_1, x_2 \geq 0$$

The optimal solution to this LP is $z = \frac{17}{3}$, $x_1 = \frac{4}{3}$, $x_2 = \frac{5}{3}$. Use the graphical approach to determine the range of values for the right-hand side of the second constraint for which the current basis remains optimal.

16 Zales Jewelers uses rubies and sapphires to produce two types of rings. A Type 1 ring requires 2 rubies, 3 sapphires, and 1 hour of jeweler's labor. A Type 2 ring requires 3 rubies, 2 sapphires, and 2 hours of jeweler's labor. Each Type 1 ring sells for $400, and each Type 2 ring sells for $500. All rings produced by Zales can be sold. Zales now has 100 rubies, 120 sapphires, and 70 hours of jeweler's

FIGURE **13**

LINDO Output for Wheat/Corn (Problem 12)

```
MAX 150A1+200A2-10L
ST
A1+A2<45
6A1+10A2-L<0
L<350
5A1<140
4A2<120
END
```

LP OPTIMUM FOUND AT STEP   4

OBJECTIVE FUNCTION VALUE

1)   4250.000

VARIABLE	VALUE	REDUCED COST
A1	25.000000	0.000000
A2	20.000000	0.000000
L	350.000000	0.000000

ROW	SLACK OR SURPLUS	DUAL PRICES
2)	0.000000	75.000000
3)	0.000000	12.500000
4)	0.000000	2.500000
5)	15.000000	0.000000
6)	40.000000	0.000000

NO. ITERATIONS=   4

RANGES IN WHICH THE BASIS IS UNCHANGED:

OBJ COEFFICIENT RANGES

VARIABLE	CURRENT COEF	ALLOWABLE INCREASE	ALLOWABLE DECREASE
A1	150.000000	10.000000	30.000000
A2	200.000000	50.000000	10.000000
L	-10.000000	INFINITY	2.500000

RIGHTHAND SIDE RANGES

ROW	CURRENT RHS	ALLOWABLE INCREASE	ALLOWABLE DECREASE
2	45.000000	1.200000	6.666667
3	0.000000	40.000000	12.000000
4	350.000000	40.000000	12.000000
5	140.000000	INFINITY	15.000000
6	120.000000	INFINITY	40.000000

TABLE **58**

z	x_1	x_2	x_3	s_1	s_2	rhs
1	8	1	0	0	2	16
0	2	2	0	1	−1	4
0	6	1	1	0	1	8

TABLE **59**

z	x_1	x_2	s_1	e_2	a_2	a_3	rhs
1	0	0	0	1	$M-1$	$M+\frac{3}{2}$	$\frac{9}{2}$
0	0	0	1	0	0	$-\frac{1}{2}$	$\frac{1}{2}$
0	0	1	0	−2	2	$-\frac{3}{2}$	$\frac{3}{2}$
0	1	0	0	1	−1	1	1

labor. Extra rubies can be purchased at a cost of $100 per ruby. Market demand requires that the company produce at least 20 Type 1 rings and at least 25 Type 2 rings. To maximize profit, Zales should solve the following LP:

$X1$ = Type 1 rings produced

$X2$ = Type 2 rings produced

R = number of rubies purchased

$$\max z = 400X1 + 500X2 - 100R$$

$$
\begin{array}{ll}
\text{s.t.} & 2X1 + 3X2 - R \le 100 \\
& 3X1 + 2X2 \quad\;\; \le 120 \\
& X1 + 2X2 \quad\;\; \le 70 \\
& X1 \quad\qquad\quad\; \ge 20 \\
& \qquad\quad X2 \quad\;\; \ge 25 \\
& X1, X2 \ge 0
\end{array}
$$

Use the LINDO output in Figure 14 to answer the following questions:

a Suppose that instead of $100, each ruby costs $190. Would Zales still purchase rubies? What would be the new optimal solution to the problem?

FIGURE **14**
LINDO Output for Jewelry (Problem 16)

```
MAX      400 X1 + 500 X2 - 100 R
SUBJECT TO
      2)    2 X1 + 3 X2 - R <= 100
      3)    3 X1 + 2 X2 <= 120
      4)     X1 + 2 X2 <=  70
      5)     X1 >=   20
      6)     X2 >=   25
END

    LP OPTIMUM FOUND   AT STEP  2

          OBJECTIVE FUNCTION VALUE

  1)          19000.0000

 VARIABLE        VALUE        REDUCED COST
      X1        20.000000        0.000000
      X2        25.000000        0.000000
       R        15.000000        0.000000

   ROW     SLACK OR SURPLUS      DUAL PRICES
      2)        0.000000        100.000000
      3)       10.000000          0.000000
      4)        0.000000        200.000000
      5)        0.000000          0.000000
      6)        0.000000       -200.000000

 NO. ITERATIONS=      2

  RANGES IN WHICH THE BASIS IS UNCHANGED

                        OBJ COEFFICIENT RANGES
 VARIABLE       CURRENT        ALLOWABLE        ALLOWABLE
                 COEF          INCREASE         DECREASE
      X1       400.000000      INFINITY        100.000000
      X2       500.000000      200.000000      INFINITY
       R      -100.000000      100.000000      100.000000

                        RIGHTHAND SIDE RANGES
   ROW        CURRENT         ALLOWABLE        ALLOWABLE
                 RHS           INCREASE         DECREASE
      2       100.000000      15.000000        INFINITY
      3       120.000000      INFINITY         10.000000
      4        70.000000       3.333333         0.000000
      5        20.000000       0.000000        INFINITY
      6        25.000000       0.000000         2.500000
```

b Suppose that Zales were only required to produce at least 23 Type 2 rings. What would Zales' profit now be?

c What is the most that Zales would be willing to pay for another hour of jeweler's labor?

d What is the most that Zales would be willing to pay for another sapphire?

e Zales is considering producing Type 3 rings. Each Type 3 ring can be sold for $550 and requires 4 rubies, 2 sapphires, and 1 hour of jeweler's labor. Should Zales produce any Type 3 rings?

17 Use the dual simplex method to solve the following LP:

$$\max z = -2x_1 - x_2$$
$$\text{s.t.} \quad x_1 + x_2 \geq 5$$
$$x_1 - 2x_2 \geq 8$$
$$x_1, x_2 \geq 0$$

18 Consider the following LP:

$$\max z = -4x_1 - x_2$$
$$\text{s.t.} \quad 4x_1 + 3x_2 \geq 6$$
$$x_1 + 2x_2 \leq 3$$
$$3x_1 + x_2 = 3$$
$$x_1, x_2 \geq 0$$

After subtracting an excess variable e_1 from the first constraint, adding a slack variable s_2 to the second constraint, and adding artificial variables a_1 and a_3 to the first and third constraints, the optimal tableau for this LP is as shown in Table 60.

a Find the dual to this LP and its optimal solution.

b If we changed this LP to

$$\max z = -4x_1 - x_2 - x_3$$
$$\text{s.t.} \quad 4x_1 + 3x_2 + x_3 \geq 6$$
$$x_1 + 2x_2 + x_3 \leq 3$$
$$3x_1 + x_2 + x_3 = 3$$
$$x_1, x_2, x_3 \geq 0$$

would the current optimal solution remain optimal?

TABLE 60

z	x_1	x_2	e_1	s_2	a_1	a_3	rhs
1	0	0	0	$\frac{1}{5}$	M	$M - \frac{7}{5}$	$-\frac{18}{5}$
0	0	1	0	$\frac{3}{5}$	0	$-\frac{1}{5}$	$\frac{6}{5}$
0	1	0	0	$-\frac{1}{5}$	0	$\frac{2}{5}$	$\frac{3}{5}$
0	0	0	1	1	-1	1	0

19 Consider the following LP:

$$\max z = -2x_1 + 6x_2$$
$$\text{s.t.} \quad x_1 + x_2 \geq 2$$
$$-x_1 + x_2 \leq 1$$
$$x_1, x_2 \geq 0$$

This LP is unbounded. Use this fact to show that the following LP has no feasible solution:

$$\min 2y_1 + y_2$$
$$\text{s.t.} \quad y_1 - y_2 \geq -2$$
$$y_1 + y_2 \geq 6$$
$$y_1 \leq 0, y_2 \geq 0$$

20 Use the Theorem of Complementary Slackness to find the optimal solution to the following LP and its dual:

$$\max z = 3x_1 + 4x_2 + x_3 + 5x_4$$
$$\text{s.t.} \quad x_1 + 2x_2 + x_3 + 2x_4 \leq 5$$
$$2x_1 + 3x_2 + x_3 + 3x_4 \leq 8$$
$$x_1, x_2, x_3, x_4 \geq 0$$

21 $z = 8$, $x_1 = 2$, $x_2 = 0$ is the optimal solution to the following LP:

$$\max z = 4x_1 + x_2$$
$$\text{s.t.} \quad 3x_1 + x_2 \leq 6$$
$$5x_1 + 3x_2 \leq 15$$
$$x_1, x_2 \geq 0$$

Use the graphical approach to answer the following questions:

a Determine the range of values of c_1 for which the current basis remains optimal.

b Determine the range of values of c_2 for which the current basis remains optimal.

c Determine the range of values of b_1 for which the current basis remains optimal.

d Determine the range of values of b_2 for which the current basis remains optimal.

22 Radioco manufactures two types of radios. The only scarce resource that is needed to produce radios is labor. The company now has two laborers. Laborer 1 is willing to work up to 40 hours per week and is paid $5 per hour. Laborer 2 is willing to work up to 50 hours per week and is paid $6 per hour. The price as well as the resources required to build each type of radio are given in Table 61.

a Letting x_i be the number of type i radios produced each week, show that Radioco should solve the following LP (its optimal tableau is given in Table 62):

TABLE 61

Radio 1		Radio 2	
Price ($)	Resource Required	Price ($)	Resource Required
25	Laborer 1: 1 hour	22	Laborer 1: 2 hours
	Laborer 2: 2 hours		Laborer 2: 2 hours
	Raw material cost: $5		Raw material cost: $4

TABLE 62

z	x_1	x_2	s_1	s_2	rhs
1	0	0	$\frac{1}{3}$	$\frac{4}{3}$	80
0	1	0	$-\frac{1}{3}$	$\frac{2}{3}$	20
0	0	1	$\frac{2}{3}$	$-\frac{1}{3}$	10

$$\max z = 3x_1 + 2x_2$$
$$\text{s.t.} \quad x_1 + 2x_2 \leq 40$$
$$2x_1 + x_2 \leq 50$$
$$x_1, x_2 \geq 0$$

b For what values of the price of a Type 1 radio would the current basis remain optimal?

c For what values of the price of a Type 2 radio would the current basis remain optimal?

d If laborer 1 were willing to work only 30 hours per week, would the current basis remain optimal?

e If laborer 2 were willing to work as many as 60 hours per week, would the current basis remain optimal?

f If laborer 1 were willing to work an additional hour, what is the most that Radioco should pay?

g If laborer 2 were willing to work only 48 hours, what would Radioco's profits be? Verify your answer by determining the number of radios of each type that would be produced.

h A Type 3 radio is under consideration for production. The specifications of a Type 3 radio are as follows: price, $30; 2 hours from laborer 1; 2 hours from laborer 2; cost of raw materials, $3. Should Radioco manufacture any Type 3 radios?

23 Beerco manufactures ale and beer from corn, hops, and malt. Currently, 40 lb of corn, 30 lb of hops, and 40 lb of malt are available. A barrel of ale sells for $40 and requires 1 lb of corn, 1 lb of hops, and 2 lb of malt. A barrel of beer sells for $50 and requires 2 lb of corn, 1 lb of hops, and 1 lb of malt. Beerco can sell all ale and beer that is produced. To maximize total sales revenue, Beerco should solve the following LP:

$$\max z = 40\text{ALE} + 50\text{BEER}$$
$$\text{s.t.} \quad \text{ALE} + 2\text{BEER} \leq 40 \quad \text{(Corn constraint)}$$
$$\text{ALE} + \text{BEER} \leq 30 \quad \text{(Hops constraint)}$$

TABLE 63

z	Ale	Beer	s_1	s_2	s_3	rhs
1	0	0	20	0	10	1,200
0	0	1	$\frac{2}{3}$	0	$-\frac{1}{3}$	$\frac{40}{3}$
0	0	0	$-\frac{1}{3}$	1	$-\frac{1}{3}$	$\frac{10}{3}$
0	1	0	$-\frac{1}{3}$	0	$\frac{2}{3}$	$\frac{40}{3}$

TABLE 64

	Product 1	Product 2
Selling price	$15	$8
Labor required	0.75 hour	0.50 hour
Machine time required	1.5 hours	0.80 hour
Raw material required	2 units	1 unit

$$2\text{ALE} + \text{BEER} \le 40 \qquad \text{(Malt constraint)}$$
$$\text{ALE, BEER} \ge 0$$

ALE = barrels of ale produced, and BEER = barrels of beer produced. An optimal tableau for this LP is shown in Table 63.

a Write down the dual to Beerco's LP and find its optimal solution.

b Find the range of values of the price of ale for which the current basis remains optimal.

c Find the range of values of the price of beer for which the current basis remains optimal.

d Find the range of values of the amount of available corn for which the current basis remains optimal.

e Find the range of values of the amount of available hops for which the current basis remains optimal.

f Find the range of values of the amount of available malt for which the current basis remains optimal.

g Suppose Beerco is considering manufacturing malt liquor. A barrel of malt liquor requires 0.5 lb of corn, 3 lb of hops, and 3 lb of malt and sells for $50. Should Beerco manufacture any malt liquor?

h Suppose we express the Beerco constraints in ounces. Write down the new LP and its dual.

i What is the optimal solution to the dual of the new LP? (*Hint:* Think about what happens to $c_{BV}B^{-1}$. Use the idea of shadow prices to explain why the dual to the original LP (pounds) and the dual to the new LP (ounces) should have different optimal solutions.)

Group B

24 Consider the following LP:

$$\max z = -3x_1 + x_2 + 2x_3$$
$$\text{s.t.} \qquad x_2 + 2x_3 \le 3$$
$$-x_1 \qquad + 3x_3 \le -1$$
$$-2x_1 - 3x_2 \qquad \le -2$$
$$x_1, x_2, x_3 \ge 0$$

a Find the dual to this LP and show that it has the same feasible region as the original LP.

b Use weak duality to show that the optimal objective function value for the LP (and its dual) must be 0.

25 Consider the following LP:

$$\max z = 2x_1 + x_2 + x_3$$
$$\text{s.t.} \quad x_1 \qquad + x_3 \le 1$$
$$x_2 + x_3 \le 2$$
$$x_1 + x_2 \qquad \le 3$$
$$x_1, x_2, x_3 \ge 0$$

It is given that

$$\begin{bmatrix} 1 & 0 & 1 \\ 0 & 1 & 1 \\ 1 & 1 & 0 \end{bmatrix}^{-1} = \begin{bmatrix} \frac{1}{2} & -\frac{1}{2} & \frac{1}{2} \\ -\frac{1}{2} & \frac{1}{2} & \frac{1}{2} \\ \frac{1}{2} & \frac{1}{2} & -\frac{1}{2} \end{bmatrix}$$

a Show that the basic solution with basic variables x_1, x_2, and x_3 is optimal. Find the optimal solution.

b Write down the dual to this LP and find its optimal solution.

c Show that if we multiply the right-hand side of each constraint by a non-negative constant k, then the new optimal solution is obtained simply by multiplying the value of each variable in the original optimal solution by k.

26 Wivco produces two products: 1 and 2. The relevant data are shown in Table 64. Each week, as many as 400 units of raw material can be purchased at a cost of $1.50 per unit. The company employs four workers, who work 40 hours per week (their salaries are considered a fixed cost). Workers can be asked to work overtime and are paid $6 per hour for overtime work. Each week, 320 hours of machine time are available.

In the absence of advertising, 50 units of product 1 and 60 units of product 2 will be demanded each week. Advertising can be used to stimulate demand for each product. Each dollar spent on advertising product 1 increases its demand by 10 units; each dollar spent for product 2 increases its demand by 15 units. At most $100 can be spent on advertising. Define

P1 = number of units of product 1 produced each week
P2 = number of units of product 2 produced each week
OT = number of hours of overtime labor used each week
RM = number of units of raw material purchased each week
A1 = dollars spent each week on advertising product 1
A2 = dollars spent each week on advertising product 2

Then Wivco should solve the following LP:

$$\max z = 15P1 + 8P2 - 6(OT) - 1.5RM$$
$$- A1 - A2$$

$$\text{s.t.} \qquad P1 - 10A1 \le 50 \qquad (1)$$
$$P2 - 15A2 \le 60 \qquad (2)$$
$$0.75P1 + 0.5P2 \le 160 + (OT) \qquad (3)$$
$$2P1 + P2 \le RM \qquad (4)$$
$$RM \le 400 \qquad (5)$$
$$A1 + A2 \le 100 \qquad (6)$$
$$1.5P1 + 0.8P2 \le 320 \qquad (7)$$

All variables non-negative

Use LINDO to solve this LP. Then use the computer output to answer the following questions:

a If overtime were only $4 per hour, would Wivco use it?

b If each unit of product 1 sold for $15.50, would the current basis remain optimal? What would be the new optimal solution?

c What is the most that Wivco should be willing to pay for another unit of raw material?

d How much would Wivco be willing to pay for another hour of machine time?

e If each worker were required (as part of the regular workweek) to work 45 hours per week, what would the company's profits be?

f Explain why the shadow price of row (1) is 0.10. (*Hint:* If the right-hand side of (1) were increased from 50 to 51, then in the absence of advertising for product 1, 51 units could now be sold each week.)

g Wivco is considering producing a new product (product 3). Each unit sells for $17 and requires 2 hours of labor, 1 unit of raw material, and 2 hours of machine time. Should Wivco produce any of product 3?

h If each unit of product 2 sold for $10, would the current basis remain optimal?

27 The following question concerns the Rylon example discussed in Section 3.9. After defining

RB = ounces of Regular Brute produced annually
LB = ounces of Luxury Brute produced annually
RC = ounces of Regular Chanelle produced annually
LC = ounces of Luxury Chanelle produced annually
RM = pounds of raw material purchased annually

the LINDO output in Figure 15 was obtained for this problem. Use this output to answer the following questions:

a Interpret the shadow price of each constraint.

FIGURE 15
LINDO Output for Brute/Chanelle (Problem 27)

```
MAX      7 RB + 14 LB + 6 RC + 10 LC - 3 RM
SUBJECT TO
        2)     RM <=    4000
        3)    3 LB + 2 LC +  RM <=    6000
        4)     RM + LB - 3 RM =      0
        5)     RC + LC - 4 RM =      0
END

   LP OPTIMUM FOUND   AT STEP      6

        OBJECTIVE FUNCTIONS VALUE

  1)        172666.672

VARIABLE        VALUE          REDUCED COST
     RB     11333.333008         0.000000
     LB       666.666687         0.000000
     RC     16000.000000         0.000000
     LC         0.000000         0.666667
     RM      4000.000000         0.000000

   ROW   SLACK OR SURPLUS      DUAL PRICES
     2)        0.000000        39.666668
     3)        0.000000         2.333333
     4)        0.000000         7.000000
     5)        0.000000         6.000000

NO.  ITERATIONS=    6

   RANGES IN WHICH THE BASIS IS UNCHANGED

                        OBJ COEFFICIENT RANGES
VARIABLE        CURRENT        ALLOWABLE        ALLOWABLE
                 COEF          INCREASE         DECREASE
     RB        7.000000        1.000000        11.900001
     LB       14.000000      119.000000         1.000000
     RC        6.000000        INFINITY         0.666667
     LC       10.000000        0.666667         INFINITY
     RM       -3.000000        INFINITY        39.666668

                        RIGHTHAND SIDE RANGES
   ROW        CURRENT        ALLOWABLE        ALLOWABLE
                RHS          INCREASE         DECREASE
     2      4000.000000     2000.000000      3400.000000
     3      6000.000000    33999.996094      2000.000000
     4         0.000000        INFINITY     11333.333008
     5         0.000000        INFINITY     16000.000000
```

b If the price of RB were to increase by 50¢, what would be the new optimal solution to the Rylon problem?

c If 8,000 laboratory hours were available each year, but only 2,000 lb of raw material were available each year, would Rylon's profits increase or decrease? [*Hint:* Use the 100% Rule to show that the current basis remains optimal. Then use reasoning analogous to (34)–(37) to determine the new objective function value.]

d Rylon is considering expanding its laboratory capacity. Two options are under consideration:

Option 1 For a cost of $10,000 (incurred now), annual laboratory capacity can be increased by 1,000 hours.

Option 2 For a cost of $200,000 (incurred now), annual laboratory capacity can be increased by 10,000 hours.

Suppose that all other aspects of the problem remain unchanged and that future profits are discounted, with the interest rate being $11\frac{1}{9}\%$ per year. Which option, if any, should Rylon choose?

e Rylon is considering purchasing a new type of raw material. Unlimited quantities can be purchased at $8/lb. It requires 3 laboratory hours to process a pound of the new raw material. Each processed pound yields 2 oz of RB and 1 oz of RC. Should Rylon purchase any of the new material?

28 Consider the following two LPs:

$$\max z = c_1 x_1 + c_2 x_2$$
$$\text{s.t.} \quad a_{11}x_1 + a_{12}x_2 \leq b_1 \qquad \textbf{(LP 1)}$$
$$a_{21}x_1 + a_{22}x_2 \leq b_2$$
$$x_1, x_2 \geq 0$$

$$\max z = 100c_1 x_1 + 100c_2 x_2$$
$$\text{s.t.} \quad 100a_{11}x_1 + 100a_{12}x_2 \leq b_1 \qquad \textbf{(LP 2)}$$
$$100a_{21}x_1 + 100a_{22}x_2 \leq b_2$$
$$x_1, x_2 \geq 0$$

Suppose that BV = $\{x_1, x_2\}$ is an optimal basis for both LPs, and the optimal solution to LP 1 is $x_1 = 50$, $x_2 = 500$, $z = 550$. Also suppose that for LP 1, the shadow price of both Constraint 1 and Constraint 2 is $\frac{100}{3}$. Find the optimal solution to LP 2 and the optimal solution to the dual of LP 2. (*Hint:* If we multiply each number in a matrix by 100, what happens to B^{-1}?)

29 The following questions pertain to the Star Oil capital budgeting example of Section 3.6. The LINDO output for this problem is shown in Figure 16.

a Find and interpret the shadow price for each constraint.

b If the NPV of investment 1 were $5 million, would the optimal solution to the problem change?

c If the NPV of investment 2 and investment 4 were each decreased by 25%, would the optimal solution to the problem change? (This part requires knowledge of the 100% Rule.)

d Suppose that Star Oil's investment budget were changed to $50 million at time 0 and $15 million at time 1. Would Star be better off? (This part requires knowledge of the 100% Rule.)

e Suppose a new investment (investment 6) is available. Investment 6 yields an NPV of $10 million and re-

quires a cash outflow of $5 million at time 0 and $10 million at time 1. Should Star Oil invest any money in investment 6?

30 The following questions pertain to the Finco investment example of Section 3.11. The LINDO output for this problem is shown in Figure 17.

a If Finco has $2,000 more on hand at time 0, by how much would their time 3 cash increase?

b Observe that if Finco were given a dollar at time 1, the cash available for investment at time 1 would now be $0.5A + 1.2C + 1.08S_0 + 1$. Use this fact and the shadow price of Constraint 2 to determine by how much Finco's time 3 cash position would increase if an extra dollar were available at time 1.

c By how much would Finco's time 3 cash on hand change if Finco were given an extra dollar at time 2?

d If investment D yielded $1.80 at time 3, would the current basis remain optimal?

e Suppose that a super money market fund yielded 25% for the period between time 0 and time 1. Should Finco invest in this fund at time 0?

f Show that if the investment limitations of $75,000 on investments A, B, C, and D were all eliminated, the current basis would remain optimal. (Knowledge of the 100% Rule is required for this part.) What would be the new optimal z-value?

g A new investment (investment F) is under consideration. One dollar invested in investment F generates the following cash flows: time 0, −$1.00; time 1, + $1.10; time 2, + $0.20; time 3, + $0.10. Should Finco invest in investment F?

31 In this problem, we discuss how shadow prices can be interpreted for blending problems (see Section 3.8). To illustrate the ideas, we discuss Problem 2 of Section 3.8. If we define

$$x_{6J} = \text{pounds of grade 6 oranges in juice}$$
$$x_{9J} = \text{pounds of grade 9 oranges in juice}$$
$$x_{6B} = \text{pounds of grade 6 oranges in bags}$$
$$x_{9B} = \text{pounds of grade 9 oranges in bags}$$

then the appropriate formulation is

$$\max z = 0.45(x_{6J} + x_{9J}) + 0.30(x_{6B} + x_{9B})$$

s.t.	x_{6J}	$+ x_{6B}$	$\leq 120,000$	(Grade 6 constraint)
	x_{9J}	$+ x_{9B}$	$\leq 100,000$	(Grade 9 constraint)
(1)	$\dfrac{6x_{6J} + 9x_{9J}}{x_{6J} + x_{9J}}$		≥ 8	(Orange Juice constraint)
(2)	$\dfrac{6x_{6B} + 9x_{9B}}{x_{6B} + x_{9B}}$		≥ 7	(Bags constraint)

$$x_{6J}, x_{9J}, x_{6B}, x_{9B} \geq 0$$

Constraints (1) and (2) are examples of blending constraints, because they specify the proportion of grade 6 and grade 9 oranges that must be blended to manufacture orange juice and bags of oranges. It would be useful to determine how a slight change in the standards for orange juice and bags of

FIGURE 16
LINDO Output for Star Oil (Problem 29)

```
MAX     13 X1 + 16 X2 + 16 X3 + 14 X4 + 39 X5
SUBJECT TO
       2)   11 X1 + 53 X2 + 5 X3 + 5 X4 + 29 X5 <= 40
       3)    3 X1 + 6 X2 + 5 X3 + X4 + 34 X5 <= 20
       4)   X1 <=   1
       5)   X2 <=   1
       6)   X3 <=   1
       7)   X4 <=   1
       8)   X5 <=   1
END

     LP OPTIMUM FOUND    AT STEP   5

          OBJECTIVE FUNCTION VALUE

 1)        57.4490166

 VARIABLE          VALUE            REDUCED COST
      X1         1.000000             0.000000
      X2         0.200860             0.000000
      X3         1.000000             0.000000
      X4         1.000000             0.000000
      X5         0.288084             0.000000

   ROW     SLACK OR SURPLUS      DUAL PRICES
      2)        0.000000           0.190418
      3)        0.000000           0.984644
      4)        0.000000           7.951474
      5)        0.799140           0.000000
      6)        0.000000          10.124693
      7)        0.000000          12.063268
      8)        0.711916           0.000000

 NO. ITERATIONS=      5

     RANGES IN WHICH THE BASIS IS UNCHANGED

                    OBJ COEFFICIENT RANGES
 VARIABLE        CURRENT        ALLOWABLE        ALLOWABLE
                 COEF           INCREASE         DECREASE
      X1       13.000000        INFINITY          7.951474
      X2       16.000000       45.104530          9.117648
      X3       16.000000        INFINITY         10.124693
      X4       14.000000        INFINITY         12.063268
      X5       39.000000       51.666668         30.245283

                    RIGHTHAND SIDE RANGES
   ROW          CURRENT        ALLOWABLE        ALLOWABLE
                RHS            INCREASE         DECREASE
      2        40.000000       38.264709          9.617647
      3        20.000000       11.275863          8.849057
      4         1.000000        1.139373          1.000000
      5         1.000000        INFINITY          0.799140
      6         1.000000        1.995745          1.000000
      7         1.000000        2.319149          1.000000
      8         1.000000        INFINITY          0.711916
```

oranges would affect profit. At the end of this problem, we explain how to use the shadow prices of Constraints (1) and (2) to answer the following questions:

a Suppose that the average grade for orange juice is increased to 8.1. Assuming the current basis remains optimal, by how much would profits change?

b Suppose the average grade requirement for bags of oranges is decreased to 6.9. Assuming the current basis remains optimal, by how much would profits change?

The shadow price for both (1) and (2) is -0.15. The optimal solution is $x_{6J} = 26,666.67$, $x_{9J} = 53,333.33$, $x_{6B} = $ 93,333.33, $x_{9B} = 46,666.67$. To interpret the shadow prices of blending Constraints (1) and (2), *we assume that a slight change in the quality standard for a product will not significantly change the quantity of the product that is produced.*

Now note that (1) may be written as

$$6x_{6J} + 9x_{9J} \geq 8(x_{6J} + x_{9J}), \qquad \text{or} \qquad -2x_{6J} + x_{9J} \geq 0$$

If the quality standard for orange juice is changed to $8 + \Delta$, then (1) can be written as

$$6x_{6J} + 9x_{9J} \geq (8 + \Delta)(x_{6J} + x_{9J})$$

or

FIGURE **17**

LINDO Output for Finco (Problem 30)

```
MAX    B + 1.9 D + 1.5 E + 1.08 S2
SUBJECT TO
      2)   D + A + C  + SO =       100000
      3) - B + 0.5 A + 1.2 C + 1.08 SO -  S1 =   0
      4)  0.5 B - E- S2 +  A + 1.08 S1 =   0
      5)   A <=  75000
      6)   B <=  75000
      7)   C <=  75000
      8)   D <=  75000
      9)   E <=  75000
END

   LP OPTIMUM FOUND  AT STEP    8

        OBJECTIVE FUNCTION VALUE

 1)       218500.000

VARIABLE        VALUE          REDUCED COST
        B    30000.000000        0.000000
        D    40000.000000        0.000000
        E    75000.000000        0.000000
       S2        0.000000        0.040000
        A    60000.000000        0.000000
        C        0.000000        0.028000
       SO        0.000000        0.215200
       S1        0.000000        0.350400


   ROW    SLACK OR SURPLUS     DUAL PRICES
      2)        0.000000         1.900000
      3)        0.000000        -1.560000
      4)        0.000000        -1.120000
      5)    15000.000000         0.000000
      6)    45000.000000         0.000000
      7)    75000.000000         0.000000
      8)    35000.000000         0.000000
      9)        0.000000         0.380000

NO. ITERATIONS=        8

   RANGES IN WHICH THE BASIS IS UNCHANGED

                   OBJ COEFFICIENT RANGES
VARIABLE       CURRENT      ALLOWABLE      ALLOWABLE
               COEF         INCREASE       DECREASE
        B      1.000000     0.029167       0.284416
        D      1.900000     0.475000       0.050000
        E      1.500000     INFINITY       0.380000
       S2      1.080000     0.040000       INFINITY
        A      0.000000     0.050000       0.058333
        C      0.000000     0.028000       INFINITY
       SO      0.000000     0.215200       INFINITY
       S1      0.000000     0.350400       INFINITY

                   RIGHTHAND SIDE RANGES
ROW          CURRENT        ALLOWABLE      ALLOWABLE
             RHS            INCREASE       DECREASE
      2   100000.000000    35000.000000   40000.000000
      3        0.000000    37500.000000   56250.000000
      4        0.000000    18750.000000   43750.000000
      5    75000.000000    INFINITY       15000.000000
      6    75000.000000    INFINITY       45000.000000
      7    75000.000000    INFINITY       75000.000000
      8    75000.000000    INFINITY       35000.000000
      9    75000.000000    18750.000000   43750.000000
```

$$-2x_{6J} + x_{9J} \geq \Delta(x_{6J} + x_{9J})$$

Because we are assuming that changing orange juice quality from 8 to $8 + \Delta$ does not change the amount produced, $x_{6J} + x_{9J}$ will remain equal to 80,000, and (1) will become

$$-2x_{6J} + x_{9J} \geq 80,000\Delta$$

Using the definition of shadow price, now answer parts (a) and (b).

32 Ballco manufactures large softballs, regular softballs, and hardballs. Each type of ball requires time in three departments: cutting, sewing, and packaging, as shown in

TABLE 65

Balls	Cutting Time	Sewing Time	Packaging Time
Regular softballs	15	15	3
Large softballs	10	15	4
Hardballs	8	4	2

Table 65 (in minutes). Because of marketing considerations, at least 1,000 regular softballs must be produced. Each regular softball can be sold for $3, each large softball, for $5; and each hardball, for $4. A total of 18,000 minutes of cutting time, 18,000 minutes of sewing time, and 9,000 minutes of packaging time are available. Ballco wants to maximize sales revenue. If we define

RS = number of regular softballs produced
LS = number of large softballs produced
HB = number of hardballs produced

then the appropriate LP is

max $z = 3RS + 5LS + 4HB$

s.t. $15RS + 10LS + 8HB \leq 18{,}000$ (Cutting constraint)

$15RS + 15LS + 4HB \leq 18{,}000$ (Sewing constraint)

$3RS + 4LS + 2HB \leq 9{,}000$ (Packaging constraint)

$RS \geq 1{,}000$ (Demand constraint)

RS, LS, HB ≥ 0

The optimal tableau for this LP is shown in Table 66.

a Find the dual of the Ballco problem and its optimal solution.

b Show that the Ballco problem has an alternative optimal solution. Find it. How many minutes of sewing time are used by the alternative optimal solution?

c By how much would an increase of 1 minute in the amount of available sewing time increase Ballco's revenue? How can this answer be reconciled with the fact that the sewing constraint is binding? (*Hint:* Look at the answer to part (b).)

d Assuming the current basis remains optimal, how would an increase of 100 in the regular softball requirement affect Ballco's revenue?

33 Consider the following LP:

max $z = c_1 x_1 + c_2 x_2$
s.t. $3x_1 + 4x_2 \leq 6$
$2x_1 + 3x_2 \leq 4$
$x_1, x_2 \geq 0$

The optimal tableau for this LP is

$z \quad + s_1 + 2s_2 = 14$
$x_1 \quad + 3s_1 - 4s_2 = 2$
$x_2 - 2s_1 + 3s_2 = 0$

Without doing any pivots, determine c_1 and c_2.

34 Consider the following LP and its partial optimal tableau (Table 67):

max $z = 20x_1 + 10x_2$
s.t. $x_1 + x_2 = 150$
$x_1 \leq 40$
$x_2 \geq 20$
$x_1, x_2 \geq 0$

a Complete the optimal tableau.

b Find the dual to this LP and its optimal solution.

35 Consider the following LP and its optimal tableau (Table 68):

max $z = c_1 x_1 + c_2 x_2$
s.t. $a_{11} x_1 + a_{12} x_2 \leq b_1$
$a_{21} x_1 + a_{22} x_2 \leq b_2$
$x_1, x_2 \geq 0$

Determine c_1, c_2, b_1, b_2, a_{11}, a_{12}, a_{21}, and a_{22}.

36 Consider an LP with three $\leq$ constraints. The right-hand sides are 10, 15, and 20, respectively. In the optimal tableau, s_2 is a basic variable in the second constraint, which has a right-hand side of 12. Determine the range of values of b_2 for which the current basis remains optimal. (*Hint:* If rhs of Constraint 2 is $15 + \Delta$, this should help in finding the rhs of the optimal tableau.)

37 Use LINDO to solve the Sailco problem of Section 3.10. Then use the output to answer the following questions:

a If month 1 demand decreased to 35 sailboats, what would be the total cost of satisfying the demands during the next four months?

b If the cost of producing a sailboat with regular-time labor during month 1 were $420, what would be the new optimal solution?

c Suppose a new customer is willing to pay $425 for a sailboat. If his demand must be met during month 1, should Sailco fill the order? How about if his demand must be met during month 4?

TABLE 66

z	RS	LS	HB	s_1	s_2	s_3	e_4	a_4	rhs
1	0	0	0	0.5	0	0	4.5	$M - 4.5$	4,500
0	0	0	1	0.19	-0.125	0	0.94	-0.94	187.5
0	0	1	0	-0.05	0.10	0	0.75	-0.75	150
0	0	0	0	-0.17	-0.15	1	-1.88	1.88	5,025
0	1	0	0	0	0	0	-1	1	1,000

TABLE 67

z	x_1	x_2	s_2	e_3	a_1	a_3	rhs
1	0	0		0			1,900
0	0	0	−1	1	1	−1	90
0	1	0	1	0	0	0	40
0	0	1	−1	0	1	0	110

TABLE 68

z	x_1	x_2	s_1	s_2	b
1	0	0	2	3	$\frac{5}{2}$
0	1	0	3	2	$\frac{5}{2}$
0	1	1	1	1	1

REFERENCES

The following texts contain extensive discussions of sensitivity analysis and duality:

Bazaraa, M., and J. Jarvis. *Linear Programming and Network Flows.* New York: Wiley, 1990.

Bersitmas, D., and Tsitsiklis, J. *Introduction to Linear Optimization.* Belmont, Mass.: Athena, 1997.

Bradley, S., A. Hax, and T. Magnanti. *Applied Mathematical Programming.* Reading, Mass.: Addison-Wesley, 1977.

Dantzig, G. *Linear Programming and Extensions.* Princeton, N.J.: Princeton University Press, 1963.

Dantzig, G., and Thapa, N. *Linear Programming.* New York: Springer-Verlag, 1997.

Gass, S. *Linear Programming: Methods and Applications,* 5th ed. New York: McGraw-Hill, 1985.

Luenberger, D. *Linear and Nonlinear Programming,* 2d ed. Reading, Mass.: Addison-Wesley, 1984.

Murty, K. *Linear Programming.* New York: Wiley, 1983.

Nash, S., and Sofer, A. *Linear and Nonlinear Programming.* New York: McGraw-Hill, 1995.

Nering, E., and Tucker, A. *Linear Programs and Related Problems.* New York: Academic Press, 1993.

Simmons, D. *Linear Programming for Operations Research.* Englewood Cliffs, N.J.: Prentice Hall, 1972.

Simonnard, M. *Linear Programming.* Englewood Cliffs, N.J.: Prentice Hall, 1966.

Wu, N., and R. Coppins. *Linear Programming and Extensions.* New York: McGraw-Hill, 1981.

The following contains a lucid discussion of DEA:

Callen, J. "Data Envelopment Analysis: Practical Survey and Managerial Accounting Applications," *Journal of Management Accounting Research* 3(1991):35–57.

Transportation, Assignment, and Transshipment Problems

In this chapter, we discuss three special types of linear programming problems: transportation, assignment, and transshipment. Each of these can be solved by the simplex algorithm, but specialized algorithms for each type of problem are much more efficient.

7.1 Formulating Transportation Problems

We begin our discussion of transportation problems by formulating a linear programming model of the following situation.

EXAMPLE 1 — Powerco Formulation

Powerco has three electric power plants that supply the needs of four cities.[†] Each power plant can supply the following numbers of kilowatt-hours (kwh) of electricity: plant 1—35 million; plant 2—50 million; plant 3—40 million (see Table 1). The peak power demands in these cities, which occur at the same time (2 P.M.), are as follows (in kwh): city 1—45 million; city 2—20 million; city 3—30 million; city 4—30 million. The costs of sending 1 million kwh of electricity from plant to city depend on the distance the electricity must travel. Formulate an LP to minimize the cost of meeting each city's peak power demand.

Solution   To formulate Powerco's problem as an LP, we begin by defining a variable for each decision that Powerco must make. Because Powerco must determine how much power is sent from each plant to each city, we define (for $i = 1, 2, 3$ and $j = 1, 2, 3, 4$)

$$x_{ij} = \text{number of (million) kwh produced at plant } i \text{ and sent to city } j$$

In terms of these variables, the total cost of supplying the peak power demands to cities 1–4 may be written as

$$
\begin{aligned}
& 8x_{11} + 6x_{12} + 10x_{13} + 9x_{14} && \text{(Cost of shipping power from plant 1)} \\
+\; & 9x_{21} + 12x_{22} + 13x_{23} + 7x_{24} && \text{(Cost of shipping power from plant 2)} \\
+\; & 14x_{31} + 9x_{32} + 16x_{33} + 5x_{34} && \text{(Cost of shipping power from plant 3)}
\end{aligned}
$$

Powerco faces two types of constraints. First, the total power supplied by each plant cannot exceed the plant's capacity. For example, the total amount of power sent from plant

[†]This example is based on Aarvik and Randolph (1975).

TABLE 1

Shipping Costs, Supply, and Demand for Powerco

From	To				Supply (million kwh)
	City 1	City 2	City 3	City 4	
Plant 1	$8	$6	$10	$9	35
Plant 2	$9	$12	$13	$7	50
Plant 3	$14	$9	$16	$5	40
Demand (million kwh)	45	20	30	30	

1 to the four cities cannot exceed 35 million kwh. Each variable with first subscript 1 represents a shipment of power from plant 1, so we may express this restriction by the LP constraint

$$x_{11} + x_{12} + x_{13} + x_{14} \leq 35$$

In a similar fashion, we can find constraints that reflect plant 2's and plant 3's capacities. Because power is supplied by the power plants, each is a **supply point.** Analogously, a constraint that ensures that the total quantity shipped from a plant does not exceed plant capacity is a **supply constraint.** The LP formulation of Powerco's problem contains the following three supply constraints:

$$x_{11} + x_{12} + x_{13} + x_{14} \leq 35 \quad \text{(Plant 1 supply constraint)}$$
$$x_{21} + x_{22} + x_{23} + x_{24} \leq 50 \quad \text{(Plant 2 supply constraint)}$$
$$x_{31} + x_{32} + x_{33} + x_{34} \leq 40 \quad \text{(Plant 3 supply constraint)}$$

Second, we need constraints that ensure that each city will receive sufficient power to meet its peak demand. Each city demands power, so each is a **demand point.** For example, city 1 must receive at least 45 million kwh. Each variable with second subscript 1 represents a shipment of power to city 1, so we obtain the following constraint:

$$x_{11} + x_{21} + x_{31} \geq 45$$

Similarly, we obtain a constraint for each of cities 2, 3, and 4. A constraint that ensures that a location receives its demand is a **demand constraint.** Powerco must satisfy the following four demand constraints:

$$x_{11} + x_{21} + x_{31} \geq 45 \quad \text{(City 1 demand constraint)}$$
$$x_{12} + x_{22} + x_{32} \geq 20 \quad \text{(City 2 demand constraint)}$$
$$x_{13} + x_{23} + x_{33} \geq 30 \quad \text{(City 3 demand constraint)}$$
$$x_{14} + x_{24} + x_{34} \geq 30 \quad \text{(City 4 demand constraint)}$$

Because all the x_{ij}'s must be non-negative, we add the sign restrictions $x_{ij} \geq 0$ ($i = 1, 2, 3; j = 1, 2, 3, 4$).

Combining the objective function, supply constraints, demand constraints, and sign restrictions yields the following LP formulation of Powerco's problem:

$$\min z = 8x_{11} + 6x_{12} + 10x_{13} + 9x_{14} + 9x_{21} + 12x_{22} + 13x_{23} + 7x_{24}$$
$$+ 14x_{31} + 9x_{32} + 16x_{33} + 5x_{34}$$
$$\text{s.t.} \quad x_{11} + x_{12} + x_{13} + x_{14} \leq 35 \quad \text{(Supply constraints)}$$
$$x_{21} + x_{22} + x_{23} + x_{24} \leq 50$$
$$x_{31} + x_{32} + x_{33} + x_{34} \leq 40$$

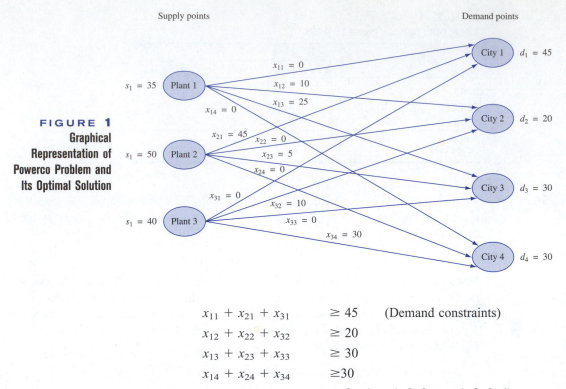

Supply points

Demand points

$s_1 = 35$ Plant 1

$s_1 = 50$ Plant 2

$s_1 = 40$ Plant 3

City 1 $d_1 = 45$

City 2 $d_2 = 20$

City 3 $d_3 = 30$

City 4 $d_4 = 30$

$x_{11} = 0$
$x_{12} = 10$
$x_{13} = 25$
$x_{14} = 0$
$x_{21} = 45$
$x_{22} = 0$
$x_{23} = 5$
$x_{24} = 0$
$x_{31} = 0$
$x_{32} = 10$
$x_{33} = 0$
$x_{34} = 30$

$$
\begin{aligned}
x_{11} + x_{21} + x_{31} &\geq 45 \qquad \text{(Demand constraints)} \\
x_{12} + x_{22} + x_{32} &\geq 20 \\
x_{13} + x_{23} + x_{33} &\geq 30 \\
x_{14} + x_{24} + x_{34} &\geq 30 \\
x_{ij} &\geq 0 \quad (i = 1, 2, 3; j = 1, 2, 3, 4)
\end{aligned}
$$

In Section 7.3, we will find that the optimal solution to this LP is $z = 1020$, $x_{12} = 10$, $x_{13} = 25$, $x_{21} = 45$, $x_{23} = 5$, $x_{32} = 10$, $x_{34} = 30$. Figure 1 is a graphical representation of the Powerco problem and its optimal solution. The variable x_{ij} is represented by a line, or arc, joining the ith supply point (plant i) and the jth demand point (city j).

General Description of a Transportation Problem

In general, a transportation problem is specified by the following information:

1 A set of m *supply points* from which a good is shipped. Supply point i can supply at most s_i units. In the Powerco example, $m = 3$, $s_1 = 35$, $s_2 = 50$, and $s_3 = 40$.

2 A set of n *demand points* to which the good is shipped. Demand point j must receive at least d_j units of the shipped good. In the Powerco example, $n = 4$, $d_1 = 45$, $d_2 = 20$, $d_3 = 30$, and $d_4 = 30$.

3 Each unit produced at supply point i and shipped to demand point j incurs a *variable cost* of c_{ij}. In the Powerco example, $c_{12} = 6$.

Let

x_{ij} = number of units shipped from supply point i to demand point j

then the general formulation of a transportation problem is

$$
\min \sum_{i=1}^{i=m} \sum_{j=1}^{j=n} c_{ij} x_{ij}
$$

$$\text{s.t.} \quad \sum_{j=1}^{j=n} x_{ij} \leq s_i \quad (i = 1, 2, \ldots, m) \qquad \text{(Supply constraints)}$$

(1)

$$\sum_{i=1}^{i=m} x_{ij} \geq d_j \quad (j = 1, 2, \ldots, n) \qquad \text{(Demand constraints)}$$

$$x_{ij} \geq 0 \quad (i = 1, 2, \ldots, m; j = 1, 2, \ldots, n)$$

If a problem has the constraints given in (1) and is a *maximization* problem, then it is still a transportation problem (see Problem 7 at the end of this section). If

$$\sum_{i=1}^{i=m} s_i = \sum_{j=1}^{j=n} d_j$$

then total supply equals total demand, and the problem is said to be a **balanced transportation problem.**

For the Powerco problem, total supply and total demand both equal 125, so this is a balanced transportation problem. In a balanced transportation problem, all the constraints must be binding. For example, in the Powerco problem, if any supply constraint were nonbinding, then the remaining available power would not be sufficient to meet the needs of all four cities. For a balanced transportation problem, (1) may be written as

$$\min \sum_{i=1}^{i=m} \sum_{j=1}^{j=n} c_{ij} x_{ij}$$

$$\text{s.t.} \quad \sum_{j=1}^{j=n} x_{ij} = s_i \quad (i = 1, 2, \ldots, m) \qquad \text{(Supply constraints)}$$

(2)

$$\sum_{i=1}^{i=m} x_{ij} = d_j \quad (j = 1, 2, \ldots, n) \qquad \text{(Demand constraints)}$$

$$x_{ij} \geq 0 \quad (i = 1, 2, \ldots, m; j = 1, 2, \ldots, n)$$

Later in this chapter, we will see that it is relatively simple to find a basic feasible solution for a balanced transportation problem. Also, simplex pivots for these problems do not involve multiplication and reduce to additions and subtractions. For these reasons, it is desirable to formulate a transportation problem as a balanced transportation problem.

Balancing a Transportation Problem If Total Supply Exceeds Total Demand

If total supply exceeds total demand, we can balance a transportation problem by creating a **dummy demand point** that has a demand equal to the amount of excess supply. Because shipments to the dummy demand point are not real shipments, they are assigned a cost of zero. Shipments to the dummy demand point indicate unused supply capacity. To understand the use of a dummy demand point, suppose that in the Powerco problem, the demand for city 1 were reduced to 40 million kwh. To balance the Powerco problem, we would add a dummy demand point (point 5) with a demand of $125 - 120 = 5$ million kwh. From each plant, the cost of shipping 1 million kwh to the dummy is 0. The optimal solution to this balanced transportation problem is $z = 975$, $x_{13} = 20$, $x_{12} = 15$, $x_{21} = 40$, $x_{23} = 10$, $x_{32} = 5$, $x_{34} = 30$, and $x_{35} = 5$. Because $x_{35} = 5$, 5 million kwh of plant 3 capacity will be unused (see Figure 2).

A transportation problem is specified by the supply, the demand, and the shipping costs, so the relevant data can be summarized in a **transportation tableau** (see Table 2). The square, or **cell,** in row i and column j of a transportation tableau corresponds to the

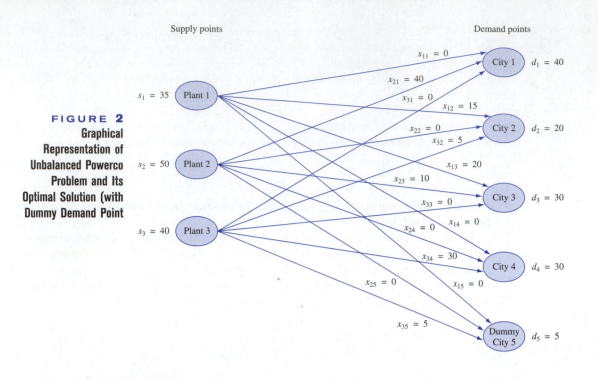

FIGURE 2
Graphical
Representation of
Unbalanced Powerco
Problem and Its
Optimal Solution (with
Dummy Demand Point

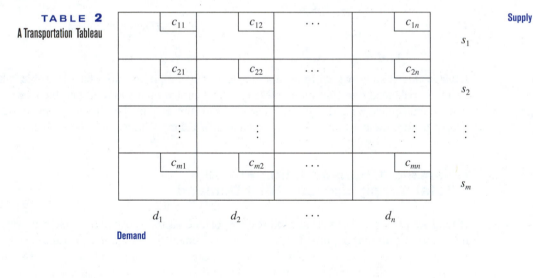

TABLE 2
A Transportation Tableau

				Supply
c_{11}	c_{12}	$\cdots$	c_{1n}	s_1
c_{21}	c_{22}	$\cdots$	c_{2n}	s_2
$\vdots$	$\vdots$		$\vdots$	$\vdots$
c_{m1}	c_{m2}	$\cdots$	c_{mn}	s_m
d_1	d_2	$\cdots$	d_n	

Demand

TABLE 3
Transportation Tableau
for Powerco

	City 1	City 2	City 3	City 4	Supply
Plant 1	8	6 10	10 25	9	35
Plant 2	9 45	12	13 5	7	50
Plant 3	14	9 10	16	5 30	40
Demand	45	20	30	30	

variable x_{ij}. If x_{ij} is a basic variable, its value is placed in the lower left-hand corner of the ijth cell of the tableau. For example, the balanced Powerco problem and its optimal solution could be displayed as shown in Table 3. The tableau format implicitly expresses the supply and demand constraints through the fact that the sum of the variables in row i must equal s_i and the sum of the variables in column j must equal d_j.

Balancing a Transportation Problem If Total Supply Is Less Than Total Demand

If a transportation problem has a total supply that is strictly less than total demand, then the problem has no feasible solution. For example, if plant 1 had only 30 million kwh of capacity, then a total of only 120 million kwh would be available. This amount of power would be insufficient to meet the total demand of 125 million kwh, and the Powerco problem would no longer have a feasible solution.

When total supply is less than total demand, it is sometimes desirable to allow the possibility of leaving some demand unmet. In such a situation, a penalty is often associated with unmet demand. Example 2 illustrates how such a situation can yield a balanced transportation problem.

EXAMPLE 2 **Handling Shortages**

Two reservoirs are available to supply the water needs of three cities. Each reservoir can supply up to 50 million gallons of water per day. Each city would like to receive 40 million gallons per day. For each million gallons per day of unmet demand, there is a penalty. At city 1, the penalty is $20; at city 2, the penalty is $22; and at city 3, the penalty is $23. The cost of transporting 1 million gallons of water from each reservoir to each city is shown in Table 4. Formulate a balanced transportation problem that can be used to minimize the sum of shortage and transport costs.

Solution In this problem,

$$\text{Daily supply} = 50 + 50 = 100 \text{ million gallons per day}$$
$$\text{Daily demand} = 40 + 40 + 40 = 120 \text{ million gallons per day}$$

To balance the problem, we add a dummy (or shortage) *supply point* having a supply of $120 - 100 = 20$ million gallons per day. The cost of shipping 1 million gallons from the dummy supply point to a city is just the shortage cost per million gallons for that city. Table 5 shows the balanced transportation problem and its optimal solution. Reservoir 1 should send 20 million gallons per day to city 1 and 30 million gallons per day to city 2, whereas reservoir 2 should send 10 million gallons per day to city 2 and 40 million gallons per day to city 3. Twenty million gallons per day of city 1's demand will be unsatisfied.

TABLE 4
Shipping Costs for Reservoir

From	To		
	City 1	City 2	City 3
Reservoir 1	$7	$8	$10
Reservoir 2	$9	$7	$8

TABLE 5
Transportation Tableau
for Reservoir

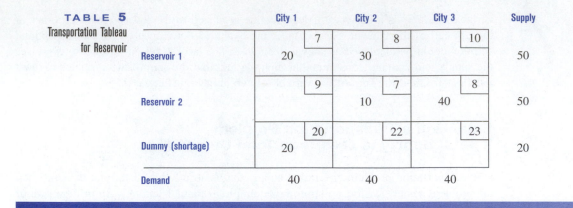

	City 1	City 2	City 3	Supply
Reservoir 1	7 20	8 30	10	50
Reservoir 2	9	7 10	8 40	50
Dummy (shortage)	20 20	22	23	20
Demand	40	40	40	

Modeling Inventory Problems as Transportation Problems

Many inventory planning problems can be modeled as balanced transportation problems. To illustrate, we formulate a balanced transportation model of the Sailco problem of Section 3.10.

EXAMPLE 3   Setting Up an Inventory Problem as a Transportation Problem

Sailco Corporation must determine how many sailboats should be produced during each of the next four quarters (one quarter is three months). Demand is as follows: first quarter, 40 sailboats; second quarter, 60 sailboats; third quarter, 75 sailboats; fourth quarter, 25 sailboats. Sailco must meet demand on time. At the beginning of the first quarter, Sailco has an inventory of 10 sailboats. At the beginning of each quarter, Sailco must decide how many sailboats should be produced during the current quarter. For simplicity, we assume that sailboats manufactured during a quarter can be used to meet demand for the current quarter. During each quarter, Sailco can produce up to 40 sailboats at a cost of $400 per sailboat. By having employees work overtime during a quarter, Sailco can produce additional sailboats at a cost of $450 per sailboat. At the end of each quarter (after production has occurred and the current quarter's demand has been satisfied), a carrying or holding cost of $20 per sailboat is incurred. Formulate a balanced transportation problem to minimize the sum of production and inventory costs during the next four quarters.

Solution   We define supply and demand points as follows:

$$Point\ 1 = initial\ inventory\qquad (s_1 = 10)$$
$$Point\ 2 = quarter\ 1\ regular\text{-}time\ (RT)\ production\qquad (s_2 = 40)$$
$$Point\ 3 = quarter\ 1\ overtime\ (OT)\ production\qquad (s_3 = 150)$$
$$Point\ 4 = quarter\ 2\ RT\ production\qquad (s_4 = 40)$$

Supply Points
$$Point\ 5 = quarter\ 2\ OT\ production\qquad (s_5 = 150)$$
$$Point\ 6 = quarter\ 3\ RT\ production\qquad (s_6 = 40)$$
$$Point\ 7 = quarter\ 3\ OT\ production\qquad (s_7 = 150)$$
$$Point\ 8 = quarter\ 4\ RT\ production\qquad (s_8 = 40)$$
$$Point\ 9 = quarter\ 4\ OT\ production\qquad (s_9 = 150)$$

There is a supply point corresponding to each source from which demand for sailboats can be met:

$$\text{Point 1} = \text{quarter 1 demand} \quad (d_1 = 40)$$
$$\text{Point 2} = \text{quarter 2 demand} \quad (d_2 = 60)$$

Demand Points $\quad$ Point 3 = quarter 3 demand $\quad (d_3 = 75)$

$$\text{Point 4} = \text{quarter 4 demand} \quad (d_4 = 25)$$
$$\text{Point 5} = \text{dummy demand point} \quad (d_5 = 770 - 200 = 570)$$

A shipment from, say, quarter 1 RT to quarter 3 demand means producing 1 unit on regular time during quarter 1 that is used to meet 1 unit of quarter 3's demand. To determine, say, c_{13}, observe that producing 1 unit during quarter 1 RT and using that unit to meet quarter 3 demand incurs a cost equal to the cost of producing 1 unit on quarter 1 RT plus the cost of holding a unit in inventory for $3 - 1 = 2$ quarters. Thus, $c_{13} = 400 + 2(20) = 440$.

Because there is no limit on the overtime production during any quarter, it is not clear what value should be chosen for the supply at each overtime production point. Total demand = 200, so at most $200 - 10 = 190$ (-10 is for initial inventory) units will be produced during any quarter. Because 40 units must be produced on regular time before any units are produced on overtime, overtime production during any quarter will never exceed $190 - 40 = 150$ units. Any unused overtime capacity will be "shipped" to the dummy demand point. To ensure that no sailboats are used to meet demand during a quarter prior to their production, a cost of M (M is a large positive number) is assigned to any cell that corresponds to using production to meet demand for an earlier quarter.

TABLE 6
Transportation Tableau for Sailco

	1	2	3	4	Dummy	Supply
Initial	0 — 10	20	40	60	0	10
Qtr 1 RT	400 — 30	420 — 10	440	460	0	40
Qtr 1 OT	450	470	490	510	0 — 150	150
Qtr 2 RT	M	400 — 40	420	440	0	40
Qtr 2 OT	M	450 — 10	470	490	0 — 140	150
Qtr 3 RT	M	M	400 — 40	420	0	40
Qtr 3 OT	M	M	450 — 35	470	0 — 115	150
Qtr 4 RT	M	M	M	400 — 25	0 — 15	40
Qtr 4 OT	M	M	M	450	0 — 150	150
Demand	40	60	75	25	570	

Total supply = 770 and total demand = 200, so we must add a dummy demand point with a demand of 770 − 200 = 570 to balance the problem. The cost of shipping a unit from any supply point to the dummy demand point is 0.

Combining these observations yields the balanced transportation problem and its optimal solution shown in Table 6. Thus, Sailco should meet quarter 1 demand with 10 units of initial inventory and 30 units of quarter 1 RT production; quarter 2 demand with 10 units of quarter 1 RT, 40 units of quarter 2 RT, and 10 units of quarter 2 OT production; quarter 3 demand with 40 units of quarter 3 RT and 35 units of quarter 3 OT production; and finally, quarter 4 demand with 25 units of quarter 4 RT production.

In Problem 12 at the end of this section, we show how this formulation can be modified to incorporate other aspects of inventory problems (backlogged demand, perishable inventory, and so on).

Solving Transportation Problems on the Computer

To solve a transportation problem with LINDO, type in the objective function, supply constraints, and demand constraints. Other menu-driven programs are available that accept the shipping costs, supply values, and demand values. From these values, the program can generate the objective function and constraints.

LINGO can be used to easily solve any transportation problem. The following LINGO model can be used to solve the Powerco example (file Trans.lng).

file Trans.lng

```
MODEL:
  1]SETS:
  2]PLANTS/P1,P2,P3/:CAP;
  3]CITIES/C1,C2,C3,C4/:DEM;
  4]LINKS(PLANTS,CITIES):COST,SHIP;
  5]ENDSETS
  6]MIN=@SUM(LINKS:COST*SHIP);
  7]@FOR(CITIES(J):
  8]@SUM(PLANTS(I):SHIP(I,J))>DEM(J));
  9]@FOR(PLANTS(I):
 10]@SUM(CITIES(J):SHIP(I,J))<CAP(I));
 11]DATA:
 12]CAP=35,50,40;
 13]DEM=45,20,30,30;
 14]COST=8,6,10,9,
 15]9,12,13,7,
 16]14,9,16,5;
 17]ENDDATA
END
```

Lines 1–5 define the **SETS** needed to generate the objective function and constraints. In line 2, we create the three power plants (the supply points) and specify that each has a capacity (given in the **DATA** section). In line 3, we create the four cities (the demand points) and specify that each has a demand (given in the **DATA** section). The **LINK** statement in line 4 creates a LINK(I,J) as I runs over all PLANTS and J runs over all CITIES. Thus, objects LINK(1,1), LINK (1,2), LINK(1,3), LINK(1,4), LINK(2,1), LINK (2,2), LINK(2,3), LINK(2,4), LINK(3,1), LINK (3,2), LINK(3,3), LINK(3,4) are created and stored in this order. Attributes with multiple subscripts are stored so that the rightmost subscripts advance most rapidly. Each LINK has two attributes: a per-unit shipping cost [(COST), given in the **DATA** section] and the amount shipped (SHIP), for which LINGO will solve.

Line 6 creates the objective function. We sum over all links the product of the unit shipping cost and the amount shipped. Using the **@FOR** and **@SUM** operators, lines 7–8 generate all demand constraints. They ensure that for each city, the sum of the amount shipped into the city will be at least as large as the city's demand. Note that the extra parenthesis after SHIP(I,J) in line 8 is to close the **@SUM** operator, and the extra parenthesis after DEM(J) is to close the **@FOR** operator. Using the **@FOR** and **@SUM** operators, lines 9–10 generate all supply constraints. They ensure that for each plant, the total shipped out of the plant will not exceed the plant's capacity.

Lines 11–17 contain the data needed for the problem. Line 12 defines each plant's capacity and line 13 defines each city's demand. Lines 14–16 contain the unit shipping cost from each plant to each city. These costs correspond to the ordering of the links described previously. **ENDDATA** ends the data section, and **END** ends the program. Typing **GO** will solve the problem.

This program can be used to solve any transportation problem. If, for example, we wanted to solve a problem with 15 supply points and 10 demand points, we would change line 2 to create 15 supply points and line 3 to create 10 demand points. Moving to line 12, we would type in the 15 plant capacities. In line 13, we would type in the demands for the 10 demand points. Then in line 14, we would type in the 150 shipping costs. Observe that the part of the program (lines 6–10) that generates the objective function and constraints remains unchanged! Notice also that our LINGO formulation does not require that the transportation problem be balanced.

Obtaining LINGO Data from an Excel Spreadsheet

Powerco.xls

Often it is easier to obtain data for a LINGO model from a spreadsheet. For example, shipping costs for a transportation problem may be the end result of many computations. As an example, suppose we have created the capacities, demands, and shipping costs for the Powerco model in the file Powerco.xls (see Figure 3). We have created capacities in the cell range F9:F11 and named the range Cap. As you probably know, you can name a range of cells in Excel by selecting the range and clicking in the name box in the upper left-hand corner of your spreadsheet. Then type the range name and hit the Enter key. In a similar fashion, name the city demands (in cells B12:E12) with the name Demand and the unit shipping costs (in cells B4:E6) with the name Costs.

FIGURE 3

	A	B	C	D	E	F	G	H
1		OPTIMAL SOLUTION	FOR	POWERCO		COSTS		
2	COSTS		CITY			1020		
3	PLANT	1	2	3	4			
4	1	8	6	10	9			
5	2	9	12	13	7			
6	3	14	9	16	5			
7	SHIPMENTS		CITY			SHIPPED		SUPPLIES
8	PLANT	1	2	3	4			
9	1	0	10	25	0	35	<=	35
10	2	45	0	5	0	50	<=	50
11	3	0	10	0	30	40	<=	40
12	RECEIVED	45	20	30	30			
13		>=	>=	>=	>=			
14	DEMANDS	45	20	30	30			

Using an **@OLE** statement, LINGO can read from a spreadsheet the values of data that are defined in the Sets portion of a program. The LINGO program (see file Transpspread.lng) needed to read our input data from the Powerco.xls file is shown below.

```
MODEL:
SETS:
PLANTS/P1,P2,P3/:CAP;
CITIES/C1,C2,C3,C4/:DEM;
LINKS(PLANTS,CITIES):COST,SHIP;
ENDSETS
MIN=@SUM(LINKS:COST*SHIP);
@FOR(CITIES(J):
@SUM(PLANTS(I):SHIP(I,J))>DEM(J));
@FOR(PLANTS(I);
@SUM(CITIES(J):SHIP(I,J))<CAP(I));
DATA:
CAP, DEM, COST=@OLE('C:\MPROG\POWERCO.XLS','Cap','Demand','Costs');
ENDDATA
  END
```

The key statement is

```
CAP, DEM, COST=@OLE('C:\MPROG\POWERCO.XLS','Cap','Demand','Costs');.
```

This statement reads the defined data sets CAP, DEM, and COSTS from the Powerco.xls spreadsheet. Note that the full path location of our Excel file (enclosed in single quotes) must be given first followed by the spreadsheet range names that contain the needed data. The range names are paired with the data sets in the order listed. Therefore, CAP values are found in range Cap and so on. The **@OLE** statement is very powerful, because a spreadsheet will usually greatly simplify the creation of data for a LINGO program.

Spreadsheet Solution of Transportation Problems

In the file Powerco.xls, we show how easy it is to use the EXCEL Solver to find the optimal solution to a transportation problem. After entering the plant capacities, city demands, and unit shipping costs as shown, we enter trial values of the units shipped from each plant to each city in the range B9:E11. Then we proceed as follows:

Step 1 Compute the total amount shipped out of each city by copying from F9 to F10:F11 the formula

$$=SUM(B9:E9).$$

Step 2 Compute the total received by each city by copying from B12 to C12:E12 the formula

$$=SUM(B9:B11).$$

Step 3 Compute the total shipping cost in cell F2 with the formula

$$=SUMPRODUCT(B9:E11,Costs).$$

Note the =SUMPRODUCT function works on rectangles as well as rows or columns of numbers. Also, we have named the range of unit shipping costs (B4:E6) as COSTS.

Step 4 We now fill in the Solver window shown in Figure 4. We minimize total shipping costs (F2) by changing units shipped from each plant to each city (B9:E11). We constrain amount received by each city (B12:E12) to be at least each city's demand (range name Demand). We constrain the amount shipped out of each plant (F9:F11) to be at most each plant's capacity (range name Cap). After checking the Assume Nonnegative option and Assume Linear Model option, we obtain the optimal solution shown in Figure 3. Note, of course, that the objective function of the optimal solution found by Excel equals the ob-

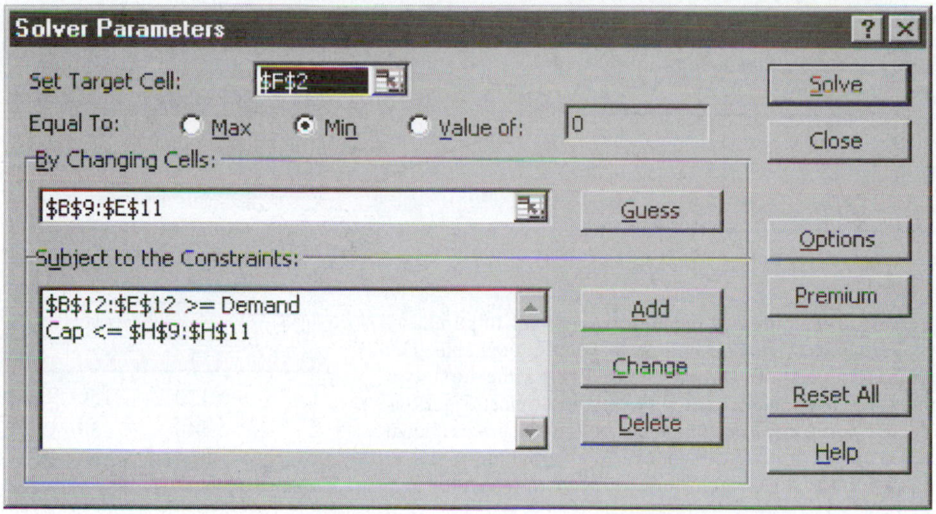

FIGURE 4

jective function value found by LINGO and our hand solution. If the problem had multiple optimal solutions, then it is possible that the values of the shipments found by LINGO, Excel, and our hand solution might be different.

PROBLEMS

Group A

1 A company supplies goods to three customers, who each require 30 units. The company has two warehouses. Warehouse 1 has 40 units available, and warehouse 2 has 30 units available. The costs of shipping 1 unit from warehouse to customer are shown in Table 7. There is a penalty for each unmet customer unit of demand: With customer 1, a penalty cost of $90 is incurred; with customer 2, $80; and with customer 3, $110. Formulate a balanced transportation problem to minimize the sum of shortage and shipping costs.

2 Referring to Problem 1, suppose that extra units could be purchased and shipped to either warehouse for a total cost of $100 per unit and that all customer demand must be met. Formulate a balanced transportation problem to minimize the sum of purchasing and shipping costs.

3 A shoe company forecasts the following demands during the next six months: month 1—200; month 2—260; month 3—240; month 4—340; month 5—190; month 6—150. It costs $7 to produce a pair of shoes with regular-time labor (RT) and $11 with overtime labor (OT). During each month, regular production is limited to 200 pairs of shoes, and

overtime production is limited to 100 pairs. It costs $1 per month to hold a pair of shoes in inventory. Formulate a balanced transportation problem to minimize the total cost of meeting the next six months of demand on time.

4 Steelco manufactures three types of steel at different plants. The time required to manufacture 1 ton of steel (regardless of type) and the costs at each plant are shown in Table 8. Each week, 100 tons of each type of steel (1, 2, and 3) must be produced. Each plant is open 40 hours per week.

a Formulate a balanced transportation problem to minimize the cost of meeting Steelco's weekly requirements.

b Suppose the time required to produce 1 ton of steel depends on the type of steel as well as on the plant at which it is produced (see Table 9, page 372). Could a transportation problem still be formulated?

5 A hospital needs to purchase 3 gallons of a perishable medicine for use during the current month and 4 gallons for use during the next month. Because the medicine is

TABLE 7

From	To		
	Customer 1	Customer 2	Customer 3
Warehouse 1	$15	$35	$25
Warehouse 2	$10	$50	$40

TABLE 8

Plant	Cost ($)			Time (minutes)
	Steel 1	Steel 2	Steel 3	
1	60	40	28	20
2	50	30	30	16
3	43	20	20	15

TABLE 9

	Time (minutes)		
Plant	Steel 1	Steel 2	Steel 3
1	15	12	15
2	15	15	20
3	10	10	15

TABLE 12

	To ($)	
From ($)	England	Japan
Field 1	1	2
Field 2	2	1

TABLE 13

	Project ($)		
Auditor	1	2	3
1	120	150	190
2	140	130	120
3	160	140	150

perishable, it can only be used during the month of purchase. Two companies (Daisy and Laroach) sell the medicine. The medicine is in short supply. Thus, during the next two months, the hospital is limited to buying at most 5 gallons from each company. The companies charge the prices shown in Table 10. Formulate a balanced transportation model to minimize the cost of purchasing the needed medicine.

6 A bank has two sites at which checks are processed. Site 1 can process 10,000 checks per day, and site 2 can process 6,000 checks per day. The bank processes three types of checks: vendor, salary, and personal. The processing cost per check depends on the site (see Table 11). Each day, 5,000 checks of each type must be processed. Formulate a balanced transportation problem to minimize the daily cost of processing checks.

7[†] The U.S. government is auctioning off oil leases at two sites: 1 and 2. At each site, 100,000 acres of land are to be auctioned. Cliff Ewing, Blake Barnes, and Alexis Pickens are bidding for the oil. Government rules state that no bidder can receive more than 40% of the land being auctioned. Cliff has bid $1,000/acre for site 1 land and $2,000/acre for site 2 land. Blake has bid $900/acre for site 1 land and $2,200/acre for site 2 land. Alexis has bid $1,100/acre for site 1 land and $1,900/acre for site 2 land. Formulate a balanced transportation model to maximize the government's revenue.

TABLE 10

Company	Current Month's Price per Gallon ($)	Next Month's Price per Gallon ($)
Daisy	800	720
Laroach	710	750

TABLE 11

	Site (¢)	
Checks	1	2
Vendor	5	3
Salary	4	4
Personal	2	5

†This problem is based on Jackson (1980).

8 The Ayatola Oil Company controls two oil fields. Field 1 can produce up to 40 million barrels of oil per day, and field 2 can produce up to 50 million barrels of oil per day. At field 1, it costs $3 to extract and refine a barrel of oil; at field 2, the cost is $2. Ayatola sells oil to two countries: England and Japan. The shipping cost per barrel is shown in Table 12. Each day, England is willing to buy up to 40 million barrels (at $6 per barrel), and Japan is willing to buy up to 30 million barrels (at $6.50 per barrel). Formulate a balanced transportation problem to maximize Ayatola's profits.

9 For the examples and problems of this section, discuss whether it is reasonable to assume that the proportionality assumption holds for the objective function.

10 Touche Young has three auditors. Each can work as many as 160 hours during the next month, during which time three projects must be completed. Project 1 will take 130 hours; project 2, 140 hours; and project 3, 160 hours. The amount per hour that can be billed for assigning each auditor to each project is given in Table 13. Formulate a balanced transportation problem to maximize total billings during the next month.

Group B

11[‡] Paperco recycles newsprint, uncoated paper, and coated paper into recycled newsprint, recycled uncoated paper, and recycled coated paper. Recycled newsprint can be produced by processing newsprint or uncoated paper. Recycled coated paper can be produced by recycling any type of paper. Recycled uncoated paper can be produced by processing uncoated paper or coated paper. The process used to produce recycled newsprint removes 20% of the input's pulp, leaving 80% of the input's pulp for recycled paper. The process used to produce recycled coated paper removes 10% of the input's pulp. The process used to produce recycled uncoated paper removes 15% of the input's pulp. The purchasing costs, processing costs, and availability of each type of paper are shown in Table 14. To meet demand,

‡This problem is based on Glassey and Gupta (1974).

TABLE 14

	Purchase Cost per Ton of Pulp ($)	Processing Cost per Ton of Input ($)	Availability
Newsprint	10		500
Coated paper	9		300
Uncoated paper	8		200
NP used for RNP		3	
NP used for RCP		4	
UCP used for RNP		4	
UCP used for RUP		1	
UCP used for RCP		6	
CP used for RUP		5	
CP used for RCP		3	

Paperco must produce at least 250 tons of recycled newsprint pulp, at least 300 tons of recycled uncoated paper pulp, and at least 150 tons of recycled coated paper pulp. Formulate a balanced transportation problem that can be used to minimize the cost of meeting Paperco's demands.

12 Explain how each of the following would modify the formulation of the Sailco problem as a balanced transportation problem:

a Suppose demand could be backlogged at a cost of $30/sailboat/month. (*Hint:* Now it is permissible to ship from, say, month 2 production to month 1 demand.)

b If demand for a sailboat is not met on time, the sale is lost and an opportunity cost of $450 is incurred.

c Sailboats can be held in inventory for a maximum of two months.

d At a cost of $440/sailboat, Sailco can purchase up to 10 sailboats/month from a subcontractor.

7.2 Finding Basic Feasible Solutions for Transportation Problems

Consider a balanced transportation problem with m supply points and n demand points. From (2), we see that such a problem contains $m + n$ equality constraints. From our experience with the Big M method and the two-phase simplex method, we know it is difficult to find a bfs if all of an LP's constraints are equalities. Fortunately, the special structure of a balanced transportation problem makes it easy for us to find a bfs.

Before describing three methods commonly used to find a bfs to a balanced transportation problem, we need to make the following important observation. *If a set of values for the x_{ij}'s satisfies all but one of the constraints of a balanced transportation problem, then the values for the x_{ij}'s will automatically satisfy the other constraint.* For example, in the Powerco problem, suppose a set of values for the x_{ij}'s is known to satisfy all the constraints with the exception of the first supply constraint. Then this set of x_{ij}'s must supply $d_1 + d_2 + d_3 + d_4 = 125$ million kwh to cities 1–4 and supply $s_2 + s_3 = 125 - s_1 = 90$ million kwh from plants 2 and 3. Thus, plant 1 must supply $125 - (125 - s_1) = 35$ million kwh, so the x_{ij}'s must also satisfy the first supply constraint.

The preceding discussion shows that when we solve a balanced transportation problem, we may omit from consideration any one of the problem's constraints and solve an LP having $m + n - 1$ constraints. We (arbitrarily) assume that the first supply constraint is omitted from consideration.

In trying to find a bfs to the remaining $m + n - 1$ constraints, you might think that any collection of $m + n - 1$ variables would yield a basic solution. Unfortunately, this is not the case. For example, consider (3), a balanced transportation problem. (We omit the costs because they are not needed to find a bfs.)

(3)

In matrix form, the constraints for this balanced transportation problem may be written as

$$\begin{bmatrix} 1 & 1 & 1 & 0 & 0 & 0 \\ 0 & 0 & 0 & 1 & 1 & 1 \\ 1 & 0 & 0 & 1 & 0 & 0 \\ 0 & 1 & 0 & 0 & 1 & 0 \\ 0 & 0 & 1 & 0 & 0 & 1 \end{bmatrix} \begin{bmatrix} x_{11} \\ x_{12} \\ x_{13} \\ x_{21} \\ x_{22} \\ x_{23} \end{bmatrix} = \begin{bmatrix} 4 \\ 5 \\ 3 \\ 2 \\ 4 \end{bmatrix} \tag{3'}$$

After dropping the first supply constraint, we obtain the following linear system:

$$\begin{bmatrix} 0 & 0 & 0 & 1 & 1 & 1 \\ 1 & 0 & 0 & 1 & 0 & 0 \\ 0 & 1 & 0 & 0 & 1 & 0 \\ 0 & 0 & 1 & 0 & 0 & 1 \end{bmatrix} \begin{bmatrix} x_{11} \\ x_{12} \\ x_{13} \\ x_{21} \\ x_{22} \\ x_{23} \end{bmatrix} = \begin{bmatrix} 5 \\ 3 \\ 2 \\ 4 \end{bmatrix} \tag{3''}$$

A basic solution to (3″) must have four basic variables. Suppose we try BV = $\{x_{11}, x_{12}, x_{21}, x_{22}\}$. Then

$$B = \begin{bmatrix} 0 & 0 & 1 & 1 \\ 1 & 0 & 1 & 0 \\ 0 & 1 & 0 & 1 \\ 0 & 0 & 0 & 0 \end{bmatrix}$$

For $\{x_{11}, x_{12}, x_{21}, x_{22}\}$ to yield a basic solution, it must be possible to use EROs to transform B to I_4. Because rank $B = 3$ and EROs do not change the rank of a matrix, there is no way that EROs can be used to transform B into I_4. Thus, BV = $\{x_{11}, x_{12}, x_{21}, x_{22}\}$ cannot yield a basic solution to (3″). Fortunately, the simple concept of a loop may be used to determine whether an arbitrary set of $m + n - 1$ variables yields a basic solution to a balanced transportation problem.

DEFINITION ■ An ordered sequence of at least four different cells is called a **loop** if

1 Any two consecutive cells lie in either the same row or same column

2 No three consecutive cells lie in the same row or column

3 The last cell in the sequence has a row or column in common with the first cell in the sequence ■

In the definition of a loop, the first cell is considered to follow the last cell, so the loop may be thought of as a closed path. Here are some examples of the preceding definition: Figure 5 represents the loop (2, 1)–(2, 4)–(4, 4)–(4, 1). Figure 6 represents the loop (1, 1)–(1, 2)–(2, 2)–(2, 3)–(4, 3)–(4, 5)–(3, 5)–(3, 1). In Figure 7, the path (1, 1)–(1, 2)–(2, 3)–(2, 1) does not represent a loop, because (1, 2) and (2, 3) do not lie in the same row or column. In Figure 8, the path (1, 2)–(1, 3)–(1, 4)–(2, 4)–(2, 2) does not represent a loop, because (1, 2), (1, 3), and (1, 4) all lie in the same row.

Theorem 1 (which we state without proof) shows why the concept of a loop is important.

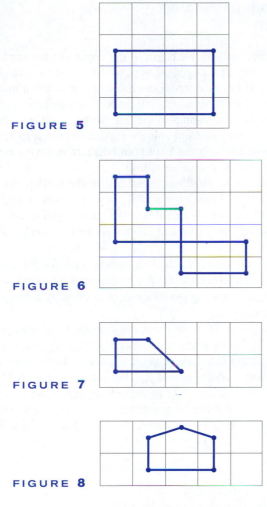

FIGURE 5

FIGURE 6

FIGURE 7

FIGURE 8

THEOREM 1

In a balanced transportation problem with m supply points and n demand points, the cells corresponding to a set of $m + n - 1$ variables contain no loop if and only if the $m + n - 1$ variables yield a basic solution.

Theorem 1 follows from the fact that a set of $m + n - 1$ cells contains no loop if and only if the $m + n - 1$ columns corresponding to these cells are linearly independent. Because $(1, 1)$–$(1, 2)$–$(2, 2)$–$(2, 1)$ is a loop, Theorem 1 tells us that $\{x_{11}, x_{12}, x_{22}, x_{21}\}$ cannot yield a basic solution for $(3'')$. On the other hand, no loop can be formed with the cells $(1, 1)$–$(1, 2)$–$(1, 3)$–$(2, 1)$, so $\{x_{11}, x_{12}, x_{13}, x_{21}\}$ will yield a basic solution to $(3'')$.

We are now ready to discuss three methods that can be used to find a basic feasible solution for a balanced transportation problem:

1 northwest corner method

2 minimum-cost method

3 Vogel's method

Northwest Corner Method for Finding a Basic Feasible Solution

To find a bfs by the northwest corner method, we begin in the upper left (or northwest) corner of the transportation tableau and set x_{11} as large as possible. Clearly, x_{11} can be no larger than the smaller of s_1 and d_1. If $x_{11} = s_1$, cross out the first row of the transportation tableau; this indicates that no more basic variables will come from row 1. Also change d_1 to $d_1 - s_1$. If $x_{11} = d_1$, cross out the first column of the transportation tableau; this indicates that no more basic variables will come from column 1. Also change s_1 to $s_1 - d_1$. If $x_{11} = s_1 = d_1$, cross out either row 1 or column 1 (but not both). If you cross out row 1, change d_1 to 0; if you cross out column 1, change s_1 to 0.

Continue applying this procedure to the most northwest cell in the tableau that does not lie in a crossed-out row or column. Eventually, you will come to a point where there is only one cell that can be assigned a value. Assign this cell a value equal to its row or column demand, and cross out both the cell's row and column. A basic feasible solution has now been obtained.

We illustrate the use of the northwest corner method by finding a bfs for the balanced transportation problem in Table 15. (We do not list the costs because they are not needed to apply the algorithm.) We indicate the crossing out of a row or column by placing an $\times$ by the row's supply or column's demand.

To begin, we set $x_{11} = \min\{5, 2\} = 2$. Then we cross out column 1 and change s_1 to $5 - 2 = 3$. This yields Table 16. The most northwest remaining variable is x_{12}. We set $x_{12} = \min\{3, 4\} = 3$. Then we cross out row 1 and change d_2 to $4 - 3 = 1$. This yields Table 17. The most northwest available variable is now x_{22}. We set $x_{22} = \min\{1, 1\} = 1$. Because both the supply and demand corresponding to the cell are equal, we may cross out either row 2 or column 2 (but not both). For no particular reason, we choose to cross out row 2. Then d_2 must be changed to $1 - 1 = 0$. The resulting tableau is Table 18. At the next step, this will lead to a *degenerate* bfs.

TABLE 15

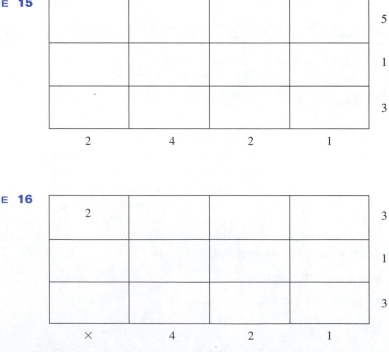

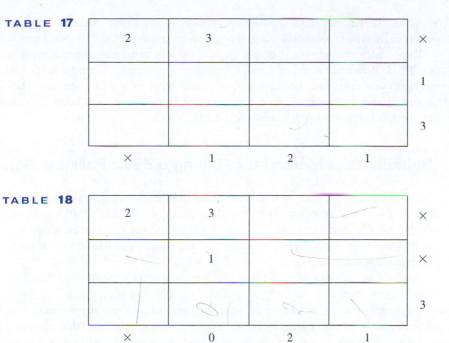

TABLE 17

TABLE 18

The most northwest available cell is now x_{32}, so we set $x_{32} = \min\{3, 0\} = 0$. Then we cross out column 2 and change s_3 to $3 - 0 = 3$. The resulting tableau is Table 19. We now set $x_{33} = \min\{3, 2\} = 2$. Then we cross out column 3 and reduce s_3 to $3 - 2 = 1$. The resulting tableau is Table 20. The only available cell is x_{34}. We set $x_{34} = \min\{1, 1\} = 1$. Then we cross out row 3 and column 4. No cells are available, so we are finished. We have obtained the bfs $x_{11} = 2$, $x_{12} = 3$, $x_{22} = 1$, $x_{32} = 0$, $x_{33} = 2$, $x_{34} = 1$.

Why does the northwest corner method yield a bfs? The method ensures that no basic variable will be assigned a negative value (because no right-hand side ever becomes nega-

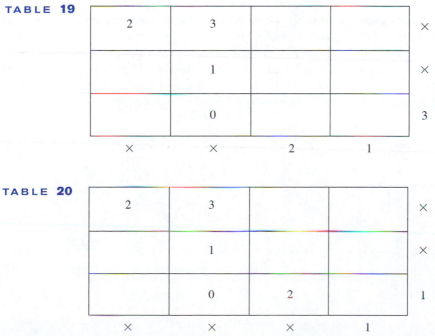

TABLE 19

TABLE 20

tive) and also that each supply and demand constraint is satisfied (because every row and column is eventually crossed out). Thus, the northwest corner method yields a feasible solution.

To complete the northwest corner method, $m + n$ rows and columns must be crossed out. The last variable assigned a value results in a row and column being crossed out, so the northwest corner method will assign values to $m + n - 1$ variables. The variables chosen by the northwest corner method cannot form a loop, so Theorem 1 implies that the northwest corner method must yield a bfs.

Minimum-Cost Method for Finding a Basic Feasible Solution

The northwest corner method does not utilize shipping costs, so it can yield an initial bfs that has a very high shipping cost. Then determining an optimal solution may require several pivots. The minimum-cost method uses the shipping costs in an effort to produce a bfs that has a lower total cost. Hopefully, fewer pivots will then be required to find the problem's optimal solution.

To begin the minimum-cost method, find the variable with the smallest shipping cost (call it x_{ij}). Then assign x_{ij} its largest possible value, $\min\{s_i, d_j\}$. As in the northwest corner method, cross out row i or column j and reduce the supply or demand of the noncrossed-out row or column by the value of x_{ij}. Then choose from the cells that do not lie in a crossed-out row or column the cell with the minimum shipping cost and repeat the procedure. Continue until there is only one cell that can be chosen. In this case, cross out both the cell's row and column. Remember that (with the exception of the last variable) if a variable satisfies both a supply and demand constraint, only cross out a row or column, not both.

To illustrate the minimum cost method, we find a bfs for the balanced transportation problem in Table 21. The variable with the minimum shipping cost is x_{22}. We set $x_{22} = \min\{10, 8\} = 8$. Then we cross out column 2 and reduce s_2 to $10 - 8 = 2$ (Table 22). We could now choose either x_{11} or x_{21} (both having shipping costs of 2). We arbitrarily choose x_{21} and set $x_{21} = \min\{2, 12\} = 2$. Then we cross out row 2 and change d_1 to $12 - 2 = 10$ (Table 23). Now we set $x_{11} = \min\{5, 10\} = 5$, cross out row 1, and change d_1 to $10 - 5 = 5$ (Table 24). The minimum cost that does not lie in a crossed-out row or column is x_{31}. We set $x_{31} = \min\{15, 5\} = 5$, cross out column 1, and reduce s_3 to $15 - 5 = 10$ (Table 25). Now we set $x_{33} = \min\{10, 4\} = 4$, cross out column 3, and reduce s_3 to $10 - 4 = 6$ (Table 26). The only cell that we can choose is x_{34}. We set $x_{34} = \min\{6, 6\}$ and cross out both row 3 and column 4. We have now obtained the bfs: $x_{11} = 5$, $x_{21} = 2$, $x_{22} = 8$, $x_{31} = 5$, $x_{33} = 4$, and $x_{34} = 6$.

Because the minimum-cost method chooses variables with small shipping costs to be basic variables, you might think that this method would always yield a bfs with a relatively low total shipping cost. The following problem shows how the minimum-cost method can be fooled into choosing a relatively high-cost bfs.

 TABLE 21

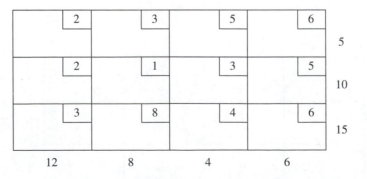

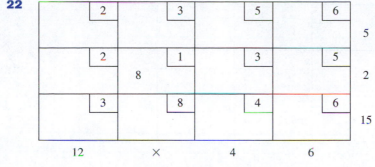

TABLE 22

2	3	5	6	5
2 8	1	3	5	2
3	8	4	6	15
12	×	4	6	

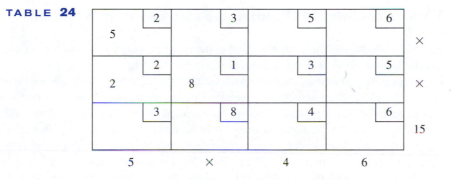

TABLE 23

2	3	5	6	5
2 8	1	3	5	×
3	8	4	6	15
10	×	4	6	

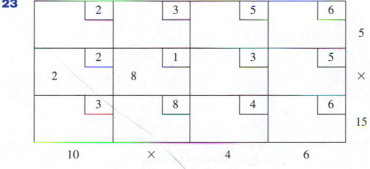

TABLE 24

5 2	3	5	6	×
2 8	1	3	5	×
3	8	4	6	15
5	×	4	6	

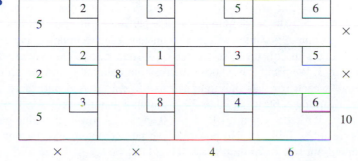

TABLE 25

5 2	3	5	6	×
2 8	1	3	5	×
5 3	8	4	6	10
×	×	4	6	

TABLE 26

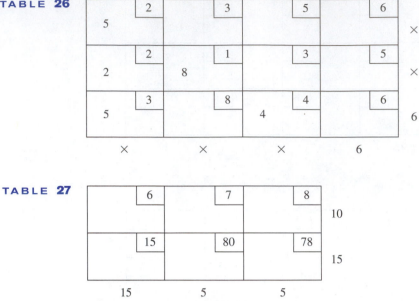

TABLE 27

6	7	8	10
15	80	78	15
15	5	5	

If we apply the minimum-cost method to Table 27, we set $x_{11} = 10$ and cross out row 1. This forces us to make x_{22} and x_{23} basic variables, thereby incurring their high shipping costs. Thus, the minimum-cost method will yield a costly bfs. Vogel's method for finding a bfs usually avoids extremely high shipping costs.

Vogel's Method for Finding a Basic Feasible Solution

Begin by computing for each row (and column) a "penalty" equal to the difference between the two smallest costs in the row (column). Next find the row or column with the largest penalty. Choose as the first basic variable the variable in this row or column that has the smallest shipping cost. As described in the northwest corner and minimum-cost methods, make this variable as large as possible, cross out a row or column, and change the supply or demand associated with the basic variable. Now recompute new penalties (using only cells that do not lie in a crossed-out row or column), and repeat the procedure until only one uncrossed cell remains. Set this variable equal to the supply or demand associated with the variable, and cross out the variable's row and column. A bfs has now been obtained.

We illustrate Vogel's method by finding a bfs to Table 28. Column 2 has the largest penalty, so we set $x_{12} = \min\{10, 5\} = 5$. Then we cross out column 2 and reduce s_1 to $10 - 5 = 5$. After recomputing the new penalties (observe that after a column is crossed out, the column penalties will remain unchanged), we obtain Table 29. The largest penalty now occurs in column 3, so we set $x_{13} = \min\{5, 5\}$. We may cross out either row 1 or column 3. We arbitrarily choose to cross out column 3, and we reduce s_1 to $5 - 5 = 0$. Because each row has only one cell that is not crossed out, there are no row penalties. The resulting tableau is Table 30. Column 1 has the only (and, of course, the largest) penalty. We set $x_{11} = \min\{0, 15\} = 0$, cross out row 1, and change d_1 to $15 - 0 = 15$. The result is Table 31. No penalties can be computed, and the only cell that is not in a crossed-out row or column is x_{21}. Therefore, we set $x_{21} = 15$ and cross out both column 1 and row 2. Our application of Vogel's method is complete, and we have obtained the bfs: $x_{11} = 0$, $x_{12} = 5$, $x_{13} = 5$, and $x_{21} = 15$ (see Table 32).

TABLE 28

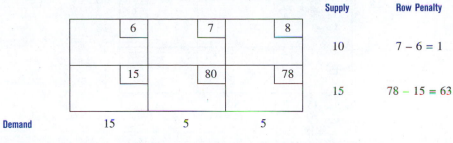

	6	7	8	Supply	Row Penalty
				10	$7 - 6 = 1$
	15	80	78	15	$78 - 15 = 63$
Demand	15	5	5		
Column Penalty	$15 - 6 = 9$	$80 - 7 = 73$	$78 - 8 = 70$		

TABLE 29

	6	7	8	Supply	Row Penalty
	5			5	$8 - 6 = 2$
	15	80	78	15	$78 - 15 = 63$
Demand	15	×	5		
Column Penalty	9	—	70		

TABLE 30

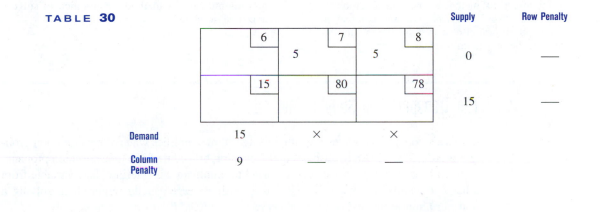

	6	7	8	Supply	Row Penalty
	5	5		0	—
	15	80	78	15	—
Demand	15	×	×		
Column Penalty	9	—	—		

TABLE 31

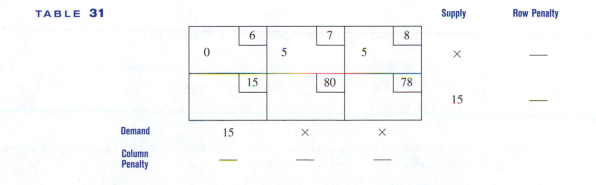

	6	7	8	Supply	Row Penalty
	0	5	5	×	—
	15	80	78	15	—
Demand	15	×	×		
Column Penalty	—	—	—		

TABLE 32

	6		7		8	
0		5		5		10
	15		80		78	
15						15
15		5		5		

Observe that Vogel's method avoids the costly shipments associated with x_{22} and x_{23}. This is because the high shipping costs resulted in large penalties that caused Vogel's method to choose other variables to satisfy the second and third demand constraints.

Of the three methods we have discussed for finding a bfs, the northwest corner method requires the least effort and Vogel's method requires the most effort. Extensive research [Glover et al. (1974)] has shown, however, that when Vogel's method is used to find an initial bfs, it usually takes substantially fewer pivots than if the other two methods had been used. For this reason, the northwest corner and minimum-cost methods are rarely used to find a basic feasible solution to a large transportation problem.

PROBLEMS

Group A

1 Use the northwest corner method to find a bfs for Problems 1, 2, and 3 of Section 7.1.

2 Use the minimum cost method to find a bfs for Problems 4, 7, and 8 of Section 7.1. (*Hint:* For a maximization problem, call the minimum-cost method the maximum-profit method or the maximum-revenue method.)

3 Use Vogel's method to find a bfs for Problems 5 and 6 of Section 7.1.

4 How should Vogel's method be modified to solve a maximization problem?

7.3 The Transportation Simplex Method

In this section, we show how the simplex algorithm simplifies when a transportation problem is solved. We begin by discussing the pivoting procedure for a transportation problem.

Recall that when the pivot row was used to eliminate the entering basic variable from other constraints and row 0, many multiplications were usually required. In solving a transportation problem, however, *pivots require only additions and subtractions.*

How to Pivot in a Transportation Problem

By using the following procedure, the pivots for a transportation problem may be performed within the confines of the transportation tableau:

Step 1 Determine (by a criterion to be developed shortly) the variable that should enter the basis.

Step 2 Find the loop (it can be shown that there is only one loop) involving the entering variable and some of the basic variables.

Step 3 Counting *only cells in the loop,* label those found in Step 2 that are an even number (0, 2, 4, and so on) of cells away from the entering variable as *even* cells. Also label those that are an odd number of cells away from the entering variable as *odd* cells.

Step 4 Find the odd cell whose variable assumes the smallest value. Call this value θ. The variable corresponding to this odd cell will leave the basis. To perform the pivot, decrease the value of each odd cell by θ and increase the value of each even cell by θ. The values of variables not in the loop remain unchanged. The pivot is now complete. If $\theta = 0$, then the entering variable will equal 0, and an odd variable that has a current value of 0 will leave the basis. In this case, a degenerate bfs existed before and will result after the pivot. If more than one odd cell in the loop equals θ, you may arbitrarily choose one of these odd cells to leave the basis; again, a degenerate bfs will result.

We illustrate the pivoting procedure on the Powerco example. When the northwest corner method is applied to the Powerco example, the bfs in Table 33 is found. For this bfs, the basic variables are $x_{11} = 35$, $x_{21} = 10$, $x_{22} = 20$, $x_{23} = 20$, $x_{33} = 10$, and $x_{34} = 30$.

Suppose we want to find the bfs that would result if x_{14} were entered into the basis. The loop involving x_{14} and some of the basic variables is

$$
\begin{array}{cccccc}
\text{E} & \text{O} & \text{E} & \text{O} & \text{E} & \text{O} \\
(1, 4)\text{--}(3, 4)\text{--}(3, 3)\text{--}(2, 3)\text{--}(2, 1)\text{--}(1, 1)
\end{array}
$$

In this loop, $(1, 4)$, $(3, 3)$, and $(2, 1)$ are the even cells, and $(1, 1)$, $(3, 4)$, and $(2, 3)$ are the odd cells. The odd cell with the smallest value is $x_{23} = 20$. Thus, after the pivot, x_{23} will have left the basis. We now add 20 to each of the even cells and subtract 20 from each of the odd cells. The bfs in Table 34 results. Because each row and column has as many $+20$s as -20s, the new solution will satisfy each supply and demand constraint. By choosing the smallest odd variable (x_{23}) to leave the basis, we have ensured that all variables will remain non-negative. Thus, the new solution is feasible. There is no loop involving the cells $(1, 1)$, $(1, 4)$, $(2, 1)$, $(2, 2)$, $(3, 3)$, and $(3, 4)$, so the new solution is a bfs. After the pivot, the new bfs is $x_{11} = 15$, $x_{14} = 20$, $x_{21} = 30$, $x_{22} = 20$, $x_{33} = 30$, and $x_{34} = 10$, and all other variables equal 0.

TABLE 33
Northwest Corner Basic Feasible Solution for Powerco

35				35
10	20	20		50
		10	30	40
45	20	30	30	

TABLE 34
New Basic Feasible Solution After x_{14} Is Pivoted into Basis

35 − 20			0 + 20	35
10 + 20	20	20 − 20 (nonbasic)		50
		10 + 20	30 − 20	40
45	20	30	30	

The preceding illustration of the pivoting procedure makes it clear that each pivot in a transportation problem involves only additions and subtractions. Using this fact, we can show that *if all the supplies and demands for a transportation problem are integers, then the transportation problem will have an optimal solution in which all the variables are integers.* Begin by observing that, by the northwest corner method, we can find a bfs in which each variable is an integer. Each pivot involves only additions and subtractions, so each bfs obtained by performing the simplex algorithm (including the optimal solution) will assign all variables integer values. The fact that a transportation problem with integer supplies and demands has an optimal integer solution is useful, because it ensures that we need not worry about whether the Divisibility Assumption is justified.

Pricing Out Nonbasic Variables (Based on Chapter 6)

To complete our discussion of the transportation simplex, we now show how to compute row 0 for any bfs. From Section 6.2, we know that for a bfs in which the set of basic variables is BV, the coefficient of the variable x_{ij} (call it $\bar{c}_{ij}$) in the tableau's row 0 is given by

$$\bar{c}_{ij} = \mathbf{c}_{BV}B^{-1}\mathbf{a}_{ij} - c_{ij}$$

where c_{ij} is the objective function coefficient for x_{ij} and $\mathbf{a}_{ij}$ is the column for x_{ij} in the original LP (we are assuming that the first supply constraint has been dropped).

Because we are solving a minimization problem, the current bfs will be optimal if all the $\bar{c}_{ij}$'s are nonpositive; otherwise, we enter into the basis the variable with the most positive $\bar{c}_{ij}$.

After determining $\mathbf{c}_{BV}B^{-1}$, we can easily determine $\bar{c}_{ij}$. Because the first constraint has been dropped, $\mathbf{c}_{BV}B^{-1}$ will have $m + n - 1$ elements. We write

$$\mathbf{c}_{BV}B^{-1} = [u_2 \quad u_3 \quad \cdots \quad u_m \quad v_1 \quad v_2 \quad \cdots \quad v_n]$$

where $u_2, u_3, \ldots, u_m$ are the elements of $\mathbf{c}_{BV}B^{-1}$ corresponding to the $m - 1$ supply constraints, and $v_1, v_2, \ldots, v_n$ are the elements of $\mathbf{c}_{BV}B^{-1}$ corresponding to the n demand constraints.

To determine $\mathbf{c}_{BV}B^{-1}$, we use the fact that in any tableau, each basic variable x_{ij} must have $\bar{c}_{ij} = 0$. Thus, for each of the $m + n - 1$ variables in BV,

$$\mathbf{c}_{BV}B^{-1}\mathbf{a}_{ij} - c_{ij} = 0 \tag{4}$$

For a transportation problem, the equations in (4) are very easy to solve. To illustrate the solution of (4), we find $\mathbf{c}_{BV}B^{-1}$ for (5), by applying the northwest corner method bfs to the Powerco problem.

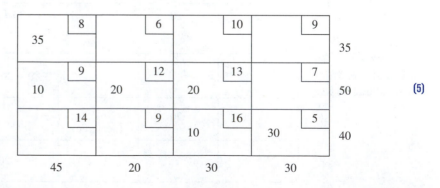

For this bfs, BV $= \{x_{11}, x_{21}, x_{22}, x_{23}, x_{33}, x_{34}\}$. Applying (4) we obtain

$$\bar{c}_{11} = [u_2 \quad u_3 \quad v_1 \quad v_2 \quad v_3 \quad v_4] \begin{bmatrix} 0 \\ 0 \\ 1 \\ 0 \\ 0 \\ 0 \end{bmatrix} - 8 = v_1 - 8 = 0$$

$$\bar{c}_{21} = [u_2 \quad u_3 \quad v_1 \quad v_2 \quad v_3 \quad v_4] \begin{bmatrix} 1 \\ 0 \\ 1 \\ 0 \\ 0 \\ 0 \end{bmatrix} - 9 = u_2 + v_1 - 9 = 0$$

$$\bar{c}_{22} = [u_2 \quad u_3 \quad v_1 \quad v_2 \quad v_3 \quad v_4] \begin{bmatrix} 1 \\ 0 \\ 0 \\ 1 \\ 0 \\ 0 \end{bmatrix} - 12 = u_2 + v_2 - 12 = 0$$

$$\bar{c}_{23} = [u_2 \quad u_3 \quad v_1 \quad v_2 \quad v_3 \quad v_4] \begin{bmatrix} 1 \\ 0 \\ 0 \\ 0 \\ 1 \\ 0 \end{bmatrix} - 13 = u_2 + v_3 - 13 = 0$$

$$\bar{c}_{33} = [u_2 \quad u_3 \quad v_1 \quad v_2 \quad v_3 \quad v_4] \begin{bmatrix} 0 \\ 1 \\ 0 \\ 0 \\ 1 \\ 0 \end{bmatrix} - 16 = u_3 + v_3 - 16 = 0$$

$$\bar{c}_{34} = [u_2 \quad u_3 \quad v_1 \quad v_2 \quad v_3 \quad v_4] \begin{bmatrix} 0 \\ 1 \\ 0 \\ 0 \\ 0 \\ 1 \end{bmatrix} - 5 = u_3 + v_4 - 5 = 0$$

For each basic variable x_{ij} (except those having $i = 1$), we see that (4) reduces to $u_i + v_j = c_{ij}$. If we define $u_1 = 0$, we see that (4) reduces to $u_i + v_j = c_{ij}$ for all basic variables. Thus, to solve for $\mathbf{c}_{BV}B^{-1}$, we must solve the following system of $m + n$ equations: $u_1 = 0$, $u_i + v_j = c_{ij}$ for all basic variables.

For (5), we find $\mathbf{c}_{BV}B^{-1}$ by solving

$$u_1 = 0 \tag{6}$$
$$u_1 + v_1 = 8 \tag{7}$$
$$u_2 + v_1 = 9 \tag{8}$$
$$u_2 + v_2 = 12 \tag{9}$$
$$u_2 + v_3 = 13 \tag{10}$$
$$u_3 + v_3 = 16 \tag{11}$$
$$u_3 + v_4 = 5 \tag{12}$$

From (7), $v_1 = 8$. From (8), $u_2 = 1$. Then (9) yields $v_2 = 11$, and (10) yields $v_3 = 12$. From (11), $u_3 = 4$. Finally, (12) yields $v_4 = 1$. For each nonbasic variable, we now compute $\bar{c}_{ij} = u_i + v_j - c_{ij}$. We obtain

$$\bar{c}_{12} = 0 + 11 - 6 = 5 \qquad \bar{c}_{13} = 0 + 12 - 10 = 2$$
$$\bar{c}_{14} = 0 + 1 - 9 = -8 \qquad \bar{c}_{24} = 1 + 1 - 7 = -5$$
$$\bar{c}_{31} = 4 + 8 - 14 = -2 \qquad \bar{c}_{32} = 4 + 11 - 9 = 6$$

Because $\bar{c}_{32}$ is the most positive $\bar{c}_{ij}$, we would next enter x_{32} into the basis. Each unit of x_{32} that is entered into the basis will decrease Powerco's cost by \$6.

How to Determine the Entering Nonbasic Variable (Based on Chapter 5)

For readers who have not covered Chapter 6, we now discuss how to determine whether a bfs is optimal, and, if it is not, how to determine which nonbasic variable should enter the basis. Let $-u_i$ ($i = 1, 2, \ldots, m$) be the shadow price of the ith supply constraint, and let $-v_j$ ($j = 1, 2, \ldots, n$) be the shadow price of the jth demand constraint. We assume that the first supply constraint has been dropped, so we may set $-u_1 = 0$. From the definition of shadow price, if we were to increase the right-hand side of the ith supply and jth demand constraint by 1, the optimal z-value would decrease by $-u_i - v_j$. Equivalently, if we were to decrease the right-hand side of the ith supply and jth demand constraint by 1, the optimal z-value would increase by $-u_i - v_j$. Now suppose x_{ij} is a nonbasic variable. Should we enter x_{ij} into the basis? Observe that if we increase x_{ij} by 1, costs directly increase by c_{ij}. Also, increasing x_{ij} by 1 means that one less unit will be shipped from supply point i and one less unit will be shipped to demand point j. This is equivalent to reducing the right-hand sides of the ith supply constraint and jth demand constraint by 1. This will increase z by $-u_i - v_j$. Thus, increasing x_{ij} by 1 will increase z by a total of $c_{ij} - u_i - v_j$. So if $c_{ij} - u_i - v_j \geq 0$ (or $u_i + v_j - c_{ij} \leq 0$) for all nonbasic variables, the current bfs will be optimal. If, however, a nonbasic variable x_{ij} has $c_{ij} - u_i - v_j < 0$ (or $u_i + v_j - c_{ij} > 0$), then z can be decreased by $u_i + v_j - c_{ij}$ per unit of x_{ij} by entering x_{ij} into the basis. Thus, we may conclude that if $u_i + v_j - c_{ij} \leq 0$ for all nonbasic variables, then the current bfs is optimal. Otherwise, the nonbasic variable with the most positive value of $u_i + v_j - c_{ij}$ should enter the basis. How do we find the u_i's and v_j's? The coefficient of a nonbasic variable x_{ij} in row 0 of any tableau is the amount by which a unit increase in x_{ij} will decrease z, so we can conclude that the coefficient of any nonbasic variable (and, it turns out, any basic variable) in row 0 is $u_i + v_j - c_{ij}$. So we may solve for the u_i's and v_j's by solving the following system of equations: $u_1 = 0$ and $u_i + v_j - c_{ij} = 0$ for all basic variables.

To illustrate the previous discussion, consider the bfs for the Powerco problem shown in (5).

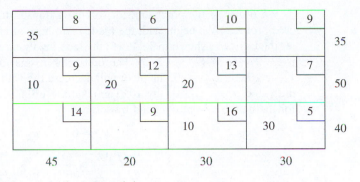

We find the u_i's and v_j's by solving

$$u_1 = 0 \tag{6}$$
$$u_1 + v_1 = 8 \tag{7}$$
$$u_2 + v_1 = 9 \tag{8}$$
$$u_2 + v_2 = 12 \tag{9}$$
$$u_2 + v_3 = 13 \tag{10}$$
$$u_3 + v_3 = 16 \tag{11}$$
$$u_3 + v_4 = 5 \tag{12}$$

From (7), $v_1 = 8$. From (8), $u_2 = 1$. Then (9) yields $v_2 = 11$, and (10) yields $v_3 = 12$. From (11), $u_3 = 4$. Finally, (12) yields $v_4 = 1$. For each nonbasic variable, we now compute $\bar{c}_{ij} = u_i + v_j - c_{ij}$. We obtain

$$\bar{c}_{12} = 0 + 11 - 6 = 5 \qquad \bar{c}_{13} = 0 + 12 - 10 = 2$$
$$\bar{c}_{14} = 0 + 1 - 9 = -8 \qquad \bar{c}_{24} = 1 + 1 - 7 = -5$$
$$\bar{c}_{31} = 4 + 8 - 14 = -2 \qquad \bar{c}_{32} = 4 + 11 - 9 = 6$$

Because $\bar{c}_{32}$ is the most positive $\bar{c}_{ij}$, we would next enter x_{32} into the basis. Each unit of x_{32} that is entered into the basis will decrease Powerco's cost by \$6.

We can now summarize the procedure for using the transportation simplex to solve a transportation (min) problem.

Summary and Illustration of the Transportation Simplex Method

Step 1 If the problem is unbalanced, balance it.

Step 2 Use one of the methods described in Section 7.2 to find a bfs.

Step 3 Use the fact that $u_1 = 0$ and $u_i + v_j = c_{ij}$ for all basic variables to find the $[u_1 \quad u_2 \quad \ldots \quad u_m \quad v_1 \quad v_2 \quad \ldots \quad v_n]$ for the current bfs.

Step 4 If $u_i + v_j - c_{ij} \leq 0$ for all nonbasic variables, then the current bfs is optimal. If this is not the case, then we enter the variable with the most positive $u_i + v_j - c_{ij}$ into the basis using the pivoting procedure. This yields a new bfs.

Step 5 Using the new bfs, return to steps 3 and 4.

For a maximization problem, proceed as stated, but replace step 4 by step 4′.

Step 4′ If $u_i + v_j - c_{ij} \geq 0$ for all nonbasic variables, then the current bfs is optimal. Otherwise, enter the variable with the most negative $u_i + v_j - c_{ij}$ into the basis using the pivoting procedure described earlier.

We illustrate the procedure for solving a transportation problem by solving the Powerco problem. We begin with the bfs (5). We have already determined that x_{32} should enter the basis. As shown in Table 35, the loop involving x_{32} and some of the basic variables is $(3, 2)$–$(3, 3)$–$(2, 3)$–$(2, 2)$. The odd cells in this loop are $(3, 3)$ and $(2, 2)$. Because $x_{33} = 10$ and $x_{22} = 20$, the pivot will decrease the value of x_{33} and x_{22} by 10 and increase the value of x_{32} and x_{23} by 10. The resulting bfs is shown in Table 36. The u_i's and v_j's for the new bfs were obtained by solving

$$u_1 = 0 \qquad u_2 + v_3 = 13$$
$$u_2 + v_2 = 12 \qquad u_2 + v_1 = 9$$
$$u_3 + v_4 = 5 \qquad u_3 + v_2 = 9$$
$$u_1 + v_1 = 8$$

In computing $\bar{c}_{ij} = u_i + v_j - c_{ij}$ for each nonbasic variable, we find that $\bar{c}_{12} = 5$, $\bar{c}_{24} = 1$, and $\bar{c}_{13} = 2$ are the only positive $\bar{c}_{ij}$'s. Thus, we next enter x_{12} into the basis. The loop involving x_{12} and some of the basic variables is $(1, 2)$–$(2, 2)$–$(2, 1)$–$(1, 1)$. The odd cells are $(2, 2)$ and $(1, 1)$. Because $x_{22} = 10$ is the smallest entry in an odd cell, we decrease x_{22} and x_{11} by 10 and increase x_{12} and x_{21} by 10. The resulting bfs is shown in Table 37. For this bfs, the u_i's and v_j's were determined by solving

$$u_1 = 0 \qquad u_1 + v_2 = 6$$
$$u_2 + v_1 = 9 \qquad u_3 + v_2 = 9$$
$$u_1 + v_1 = 8 \qquad u_3 + v_4 = 5$$
$$u_2 + v_3 = 13$$

In computing $\bar{c}_{ij}$ for each nonbasic variable, we find that the only positive $\bar{c}_{ij}$ is $\bar{c}_{13} = 2$. Thus, x_{13} enters the basis. The loop involving x_{13} and some of the basic variables is

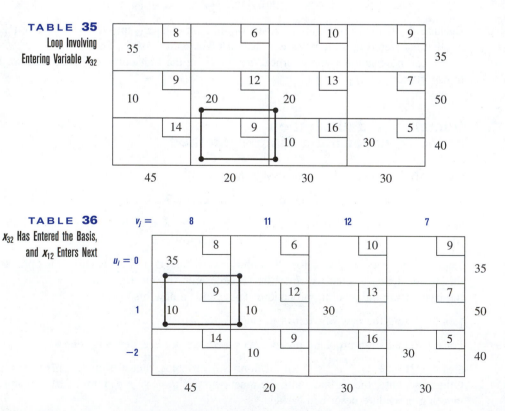

TABLE 35
Loop Involving
Entering Variable x_{32}

TABLE 36
x_{32} Has Entered the Basis,
and x_{12} Enters Next

TABLE 37
x_{12} Has Entered the Basis, and x_{13} Enters Next

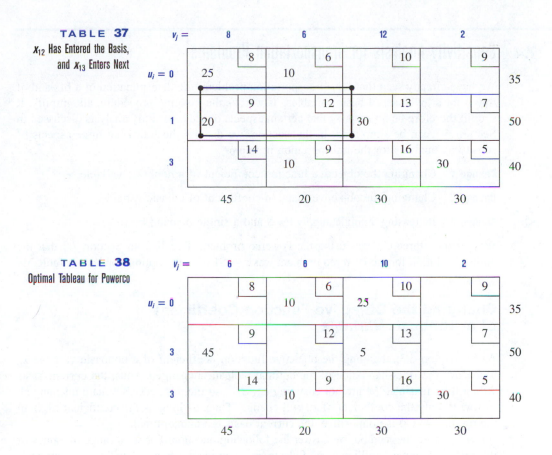

$v_j =$	8	6	12	2	
$u_i = 0$	8 25	6 10	10	9	35
1	9 20	12 30	13	7	50
3	14 10	9	16 30	5	40
	45	20	30	30	

TABLE 38
Optimal Tableau for Powerco

$v_j =$	6	8	10	2	
$u_i = 0$	8	6 10 25	10	9	35
3	9 45	12	13 5	7	50
3	14 10	9	16 30	5	40
	45	20	30	30	

(1, 3)–(2, 3)–(2, 1)–(1, 1). The odd cells are x_{23} and x_{11}. Because $x_{11} = 25$ is the smallest entry in an odd cell, we decrease x_{23} and x_{11} by 25 and increase x_{13} and x_{21} by 25. The resulting bfs is shown in Table 38. For this bfs, the u_i's and v_j's were obtained by solving

$$u_1 = 0 \qquad u_2 + v_3 = 13$$
$$u_2 + v_1 = 9 \qquad u_1 + v_3 = 10$$
$$u_3 + v_4 = 5 \qquad u_3 + v_2 = 9$$
$$u_1 + v_2 = 6$$

The reader should check that for this bfs, all $\bar{c}_{ij} \leq 0$, so an optimal solution has been obtained. Thus, the optimal solution to the Powerco problem is $x_{12} = 10$, $x_{13} = 25$, $x_{21} = 45$, $x_{23} = 5$, $x_{32} = 10$, $x_{34} = 30$, and

$$z = 6(10) + 10(25) + 9(45) + 13(5) + 9(10) + 5(30) = \$1,020$$

PROBLEMS

Group A

Use the transportation simplex to solve Problems 1–8 in Section 7.1. Begin with the bfs found in Section 7.2.

7.4 Sensitivity Analysis for Transportation Problems[†]

We have already seen that for a transportation problem, the determination of a bfs and of row 0 for a given set of basic variables, as well as the pivoting procedure, all simplify. It should therefore be no surprise that certain aspects of the sensitivity analysis discussed in Section 6.3 can be simplified. In this section, we discuss the following three aspects of sensitivity analysis for the transportation problem:

Change 1 Changing the objective function coefficient of a nonbasic variable.

Change 2 Changing the objective function coefficient of a basic variable.

Change 3 Increasing a single supply by Δ and a single demand by Δ.

We illustrate three changes using the Powerco problem. Recall from Section 7.3 that the optimal solution for the Powerco problem was $z = \$1,020$; the optimal tableau is Table 39.

Changing the Objective Function Coefficient of a Nonbasic Variable

As in Section 6.3, changing the objective function coefficient of a nonbasic variable x_{ij} will leave the right-hand side of the optimal tableau unchanged. Thus, the current basis will still be feasible. We are not changing $c_{BV}B^{-1}$, so the u_i's and v_j's remain unchanged. In row 0, only the coefficient of x_{ij} will change. Thus, as long as the coefficient of x_{ij} in the optimal row 0 is nonpositive, the current basis remains optimal.

To illustrate the method, we answer the following question: For what range of values of the cost of shipping 1 million kwh of electricity from plant 1 to city 1 will the current basis remain optimal? Suppose we change c_{11} from 8 to $8 + \Delta$. For what values of Δ will the current basis remain optimal? Now $\bar{c}_{11} = u_1 + v_1 - c_{11} = 0 + 6 - (8 + \Delta) = -2 - \Delta$. Thus, the current basis remains optimal for $-2 - \Delta \le 0$, or $\Delta \ge -2$, and $c_{11} \ge 8 - 2 = 6$.

Changing the Objective Function Coefficient of a Basic Variable

Because we are changing $c_{BV}B^{-1}$, the coefficient of each nonbasic variable in row 0 may change, and to determine whether the current basis remains optimal, we must find the new u_i's and v_j's and use these values to price out all nonbasic variables. The current basis remains optimal as long as all nonbasic variables price out nonpositive. To illustrate the idea, we determine for the Powerco problem the range of values of the cost of shipping 1 million kwh from plant 1 to city 3 for which the current basis remains optimal.

Suppose we change c_{13} from 10 to $10 + \Delta$. Then the equation $\bar{c}_{13} = 0$ changes from $u_1 + v_3 = 10$ to $u_1 + v_3 = 10 + \Delta$. Thus, to find the u_i's and v_j's, we must solve the following equations:

$$u_1 = 0 \qquad u_3 + v_2 = 9$$
$$u_2 + v_1 = 9 \qquad u_1 + v_3 = 10 + \Delta$$
$$u_1 + v_2 = 6 \qquad u_3 + v_4 = 5$$
$$u_2 + v_3 = 13$$

[†]This section covers topics that may be omitted with no loss of continuity.

TABLE 39
Optimal Tableau for Powerco

		City 1	City 2	City 3	City 4	Supply
	$v_j =$	6	6	10	2	
Plant 1	$u_i = 0$	8	6 / 10	10 / 25	9	35
Plant 2	3	9 / 45	12	13 / 5	7	50
Plant 3	3	14	9 / 10	16	5 / 30	40
Demand		45	20	30	30	

Solving these equations, we obtain $u_1 = 0$, $v_2 = 6$, $v_3 = 10 + \Delta$, $v_1 = 6 + \Delta$, $u_2 = 3 - \Delta$, $u_3 = 3$, and $v_4 = 2$.

We now price out each nonbasic variable. The current basis will remain optimal as long as each nonbasic variable has a nonpositive coefficient in row 0.

$$\bar{c}_{11} = u_1 + v_1 - 8 = \Delta - 2 \le 0 \qquad \text{for } \Delta \le 2$$
$$\bar{c}_{14} = u_1 + v_4 - 9 = -7$$
$$\bar{c}_{22} = u_2 + v_2 - 12 = -3 - \Delta \le 0 \qquad \text{for } \Delta \ge -3$$
$$\bar{c}_{24} = u_2 + v_4 - 7 = -2 - \Delta \le 0 \qquad \text{for } \Delta \ge -2$$
$$\bar{c}_{31} = u_3 + v_1 - 14 = -5 + \Delta \le 0 \qquad \text{for } \Delta \le 5$$
$$\bar{c}_{33} = u_3 + v_3 - 16 = \Delta - 3 \le 0 \qquad \text{for } \Delta \le 3$$

Thus, the current basis remains optimal for $-2 \le \Delta \le 2$, or $8 = 10 - 2 \le c_{13} \le 10 + 2 = 12$.

Increasing Both Supply s_i and Demand d_j by Δ

Observe that this change maintains a balanced transportation problem. Because the u_i's and v_j's may be thought of as the negative of each constraint's shadow prices, we know from (37') of Chapter 6 that if the current basis remains optimal,

$$\text{New } z\text{-value} = \text{old } z\text{-value} + \Delta u_i + \Delta v_j$$

For example, if we increase plant 1's supply and city 2's demand by 1 unit, then (new cost) $= 1,020 + 1(0) + 1(6) = \$1,026$.

We may also find the new values of the decision variables as follows:

1 If x_{ij} is a basic variable in the optimal solution, then increase x_{ij} by Δ.

2 If x_{ij} is a nonbasic variable in the optimal solution, then find the loop involving x_{ij} and some of the basic variables. Find an odd cell in the loop that is in row i. Increase the value of this odd cell by Δ and go around the loop, alternately increasing and then decreasing current basic variables in the loop by Δ.

To illustrate the first situation, suppose we increase s_1 and d_2 by 2. Because x_{12} is a basic variable in the optimal solution, the new optimal solution will be the one shown in Table 40. The new optimal z-value is $1,020 + 2u_1 + 2v_2 = \$1,032$. To illustrate the second situation, suppose we increase both s_1 and d_1 by 1. Because x_{11} is a nonbasic variable in the current optimal solution, we must find the loop involving x_{11} and some of the

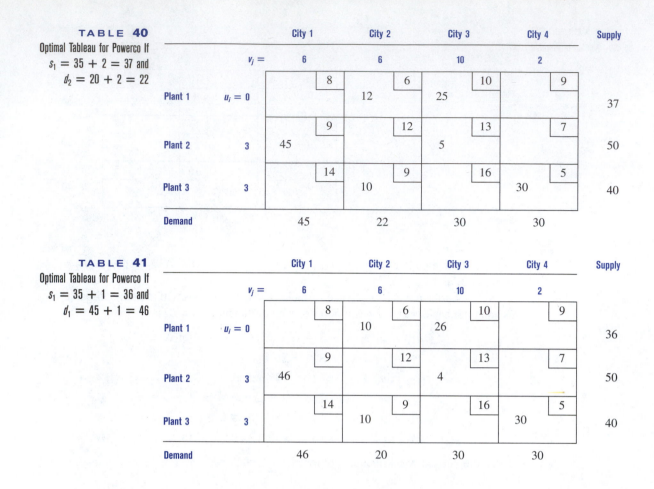

TABLE 40
Optimal Tableau for Powerco If $s_1 = 35 + 2 = 37$ and $d_2 = 20 + 2 = 22$

		City 1	City 2	City 3	City 4	Supply
	$v_j =$	6	6	10	2	
Plant 1	$u_i = 0$	8	6 / 12	10 / 25	9	37
Plant 2	3	9 / 45	12	13 / 5	7	50
Plant 3	3	14	9 / 10	16	5 / 30	40
Demand		45	22	30	30	

TABLE 41
Optimal Tableau for Powerco If $s_1 = 35 + 1 = 36$ and $d_1 = 45 + 1 = 46$

		City 1	City 2	City 3	City 4	Supply
	$v_j =$	6	6	10	2	
Plant 1	$u_i = 0$	8	6 / 10	10 / 26	9	36
Plant 2	3	9 / 46	12	13 / 4	7	50
Plant 3	3	14	9 / 10	16	5 / 30	40
Demand		46	20	30	30	

basic variables. The loop is (1, 1)–(1, 3)–(2, 3)–(2, 1). The odd cell in the loop and row 1 is x_{13}. Thus, the new optimal solution will be obtained by increasing both x_{13} and x_{21} by 1 and decreasing x_{23} by 1. This yields the optimal solution shown in Table 41. The new optimal z-value is found from (new z-value) $= 1,020 + v_1 + v_1 = \$1,026$. Observe that if both s_1 and d_1 were increased by 6, the current basis would be infeasible. (Why?)

PROBLEMS

Group A

The following problems refer to the Powerco example.

1 Determine the range of values of c_{14} for which the current basis remains optimal.

2 Determine the range of values of c_{34} for which the current basis remains optimal.

3 If s_2 and d_3 are both increased by 3, what is the new optimal solution?

4 If s_3 and d_3 are both decreased by 2, what is the new optimal solution?

5 Two plants supply three customers with medical supplies. The unit costs of shipping from the plants to the customers, along with the supplies and demands, are given in Table 42.

a The company's goal is to minimize the cost of meeting customers' demands. Find two optimal bfs for this transportation problem.

b Suppose that customer 2's demand increased by one unit. By how much would costs increase?

TABLE 42

From	To			Supply
	Customer 1	Customer 2	Customer 3	
Plant 1	$55	$65	$80	35
Plant 2	$10	$15	$25	50
Demand	10	10	10	

7.5 Assignment Problems

Although the transportation simplex appears to be very efficient, there is a certain class of transportation problems, called assignment problems, for which the transportation simplex is often very inefficient. In this section, we define assignment problems and discuss an efficient method that can be used to solve them.

EXAMPLE 4   Machine Assignment Problem

Machineco has four machines and four jobs to be completed. Each machine must be assigned to complete one job. The time required to set up each machine for completing each job is shown in Table 43. Machineco wants to minimize the total setup time needed to complete the four jobs. Use linear programming to solve this problem.

Solution   Machineco must determine which machine should be assigned to each job. We define (for $i, j = 1, 2, 3, 4$)

$$x_{ij} = 1 \text{ if machine } i \text{ is assigned to meet the demands of job } j$$
$$x_{ij} = 0 \text{ if machine } i \text{ is not assigned to meet the demands of job } j$$

Then Machineco's problem may be formulated as

$$\min z = 14x_{11} + 5x_{12} + 8x_{13} + 7x_{14} + 2x_{21} + 12x_{22} + 6x_{23} + 5x_{24}$$
$$+ 7x_{31} + 8x_{32} + 3x_{33} + 9x_{34} + 2x_{41} + 4x_{42} + 6x_{43} + 10x_{44}$$

$$\text{s.t.} \quad x_{11} + x_{12} + x_{13} + x_{14} = 1 \quad \text{(Machine constraints)}$$
$$x_{21} + x_{22} + x_{23} + x_{24} = 1$$
$$x_{31} + x_{32} + x_{33} + x_{34} = 1$$
$$x_{41} + x_{42} + x_{43} + x_{44} = 1 \qquad\qquad\qquad\qquad\text{(13)}$$
$$x_{11} + x_{21} + x_{31} + x_{41} = 1 \quad \text{(Job constraints)}$$
$$x_{12} + x_{22} + x_{32} + x_{42} = 1$$
$$x_{13} + x_{23} + x_{33} + x_{43} = 1$$
$$x_{14} + x_{24} + x_{34} + x_{44} = 1$$
$$x_{ij} = 0 \quad \text{or} \quad x_{ij} = 1$$

The first four constraints in (13) ensure that each machine is assigned to a job, and the last four ensure that each job is completed. If $x_{ij} = 1$, then the objective function will pick up the time required to set up machine i for job j; if $x_{ij} = 0$, then the objective function will not pick up the time required.

Ignoring for the moment the $x_{ij} = 0$ or $x_{ij} = 1$ restrictions, we see that Machineco faces a balanced transportation problem in which each supply point has a supply of 1 and each

TABLE **43**
Setup Times for Machineco

Machine	Time (Hours)			
	Job 1	Job 2	Job 3	Job 4
1	14	5	8	7
2	2	12	6	5
3	7	8	3	9
4	2	4	6	10

demand point has a demand of 1. In general, an **assignment problem** is a balanced transportation problem in which all supplies and demands are equal to 1. Thus, an assignment problem is characterized by knowledge of the cost of assigning each supply point to each demand point. The assignment problem's matrix of costs is its **cost matrix.**

All the supplies and demands for the Machineco problem (and for any assignment problem) are integers, so our discussion in Section 7.3 implies that all variables in Machineco's optimal solution must be integers. Because the right-hand side of each constraint is equal to 1, each x_{ij} must be a non-negative integer that is no larger than 1, so each x_{ij} must equal 0 or 1. This means that we can ignore the restrictions that $x_{ij} = 0$ or 1 and solve (13) as a balanced transportation problem. By the minimum cost method, we obtain the bfs in Table 44. The current bfs is highly degenerate. (In any bfs to an $m \times m$ assignment problem, there will always be m basic variables that equal 1 and $m - 1$ basic variables that equal 0.)

We find that $\bar{c}_{43} = 1$ is the only positive $\bar{c}_{ij}$. We therefore enter x_{43} into the basis. The loop involving x_{43} and some of the basic variables is (4, 3)–(1, 3)–(1, 2)–(4, 2). The odd variables in the loop are x_{13} and x_{42}. Because $x_{13} = x_{42} = 0$, either x_{13} or x_{42} will leave

TABLE 44
Basic Feasible Solution for Machineco

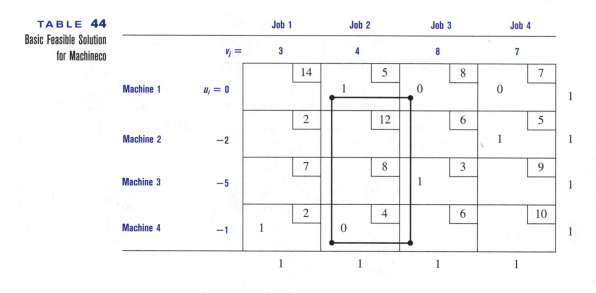

TABLE 45
x_{43} Has Entered the Basis

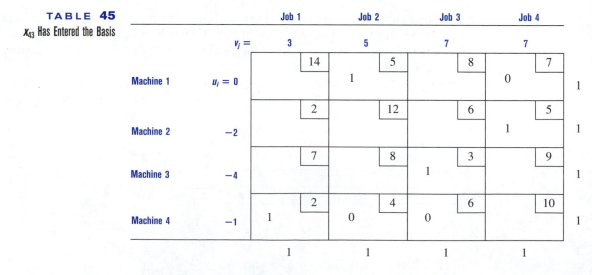

the basis. We arbitrarily choose x_{13} to leave the basis. After performing the pivot, we obtain the bfs in Table 45. All $\bar{c}_{ij}$'s are now nonpositive, so we have obtained an optimal assignment: $x_{12} = 1$, $x_{24} = 1$, $x_{33} = 1$, and $x_{41} = 1$. Thus, machine 1 is assigned to job 2, machine 2 is assigned to job 4, machine 3 is assigned to job 3, and machine 4 is assigned to job 1. A total setup time of $5 + 5 + 3 + 2 = 15$ hours is required.

The Hungarian Method

Looking back at our initial bfs, we see that it was an optimal solution. We did not know that it was optimal, however, until performing one iteration of the transportation simplex. This suggests that the high degree of degeneracy in an assignment problem may cause the transportation simplex to be an inefficient way of solving assignment problems. For this reason (and the fact that the algorithm is even simpler than the transportation simplex), the Hungarian method is usually used to solve assignment (min) problems:

Step 1 Find the minimum element in each row of the $m \times m$ cost matrix. Construct a new matrix by subtracting from each cost the minimum cost in its row. For this new matrix, find the minimum cost in each column. Construct a new matrix (called the reduced cost matrix) by subtracting from each cost the minimum cost in its column.

Step 2 Draw the minimum number of lines (horizontal, vertical, or both) that are needed to cover all the zeros in the reduced cost matrix. If m lines are required, then an optimal solution is available among the covered zeros in the matrix. If fewer than m lines are needed, then proceed to step 3.

Step 3 Find the smallest nonzero element (call its value k) in the reduced cost matrix that is uncovered by the lines drawn in step 2. Now subtract k from each uncovered element of the reduced cost matrix and add k to each element that is covered by two lines. Return to step 2.

REMARKS
1 To solve an assignment problem in which the goal is to maximize the objective function, multiply the profits matrix through by -1 and solve the problem as a minimization problem.
2 If the number of rows and columns in the cost matrix are unequal, then the assignment problem is **unbalanced.** The Hungarian method may yield an incorrect solution if the problem is unbalanced. Thus, any assignment problem should be balanced (by the addition of one or more dummy points) before it is solved by the Hungarian method.
3 In a large problem, it may not be easy to find the minimum number of lines needed to cover all zeros in the current cost matrix. For a discussion of how to find the minimum number of lines needed, see Gillett (1976). It can be shown that if j lines are required, then only j "jobs" can be assigned to zero costs in the current matrix. This explains why the algorithm terminates when m lines are required.

Solution of Machineco Example by the Hungarian Method

We illustrate the Hungarian method by solving the Machineco problem (see Table 46).

Step 1 For each row, we subtract the row minimum from each element in the row, obtaining Table 47. We now subtract 2 from each cost in column 4, obtaining Table 48.

Step 2 As shown, lines through row 1, row 3, and column 1 cover all the zeros in the reduced cost matrix. From remark 3, it follows that only three jobs can be assigned to zero costs in the current cost matrix. Fewer than four lines are required to cover all the zeros, so we proceed to step 3.

TABLE 46
Cost Matrix for Machineco

				Row Minimum
14	5	8	7	5
2	12	6	5	2
7	8	3	9	3
2	4	6	10	2

TABLE 47
Cost Matrix After Row
Minimums Are Subtracted

9	0	3	2
0	10	4	3
4	5	0	6
0	2	4	8

Column Minimum	0	0	2

TABLE 48
Cost Matrix After Column
Minimums Are Subtracted

9	0	3	0
0	10	4	1
4	5	0	4
0	2	4	6

Step 3 The smallest uncovered element equals 1, so we now subtract 1 from each uncovered element in the reduced cost matrix and add 1 to each twice-covered element. The resulting matrix is Table 49. Four lines are now required to cover all the zeros. Thus, an optimal solution is available. To find an optimal assignment, observe that the only covered 0 in column 3 is x_{33}, so we must have $x_{33} = 1$. Also, the only available covered zero in column 2 is x_{12}, so we set $x_{12} = 1$ and observe that neither row 1 nor column 2 can be used again. Now the only available covered zero in column 4 is x_{24}. Thus, we choose $x_{24} = 1$ (which now excludes both row 2 and column 4 from further use). Finally, we choose $x_{41} = 1$.

TABLE 49
Four Lines Required; Optimal
Solution Is Available

10	0	3	0
0	9	3	0
5	5	0	4
0	1	3	5

Thus, we have found the optimal assignment $x_{12} = 1$, $x_{24} = 1$, $x_{33} = 1$, and $x_{41} = 1$. Of course, this agrees with the result obtained by the transportation simplex.

Intuitive Justification of the Hungarian Method

To give an intuitive explanation of why the Hungarian algorithm works, we need to discuss the following result: *If a constant is added to each cost in a row (or column) of a balanced transportation problem, then the optimal solution to the problem is unchanged.* To show why the result is true, suppose we add k to each cost in the first row of the Machineco problem. Then

New objective function = old objective function + $k(x_{11} + x_{12} + x_{13} + x_{14})$

Because any feasible solution to the Machineco problem must have $x_{11} + x_{12} + x_{13} + x_{14} = 1$,

New objective function = old objective function + k

Thus, the optimal solution to the Machineco problem remains unchanged if a constant k is added to each cost in the first row. A similar argument applies to any other row or column.

Step 1 of the Hungarian method consists (for each row and column) of subtracting a constant from each element in the row or column. Thus, step 1 creates a new cost matrix having the same optimal solution as the original problem. Step 3 of the Hungarian method is equivalent (see Problem 7 at the end of this section) to adding k to each cost that lies in a covered row and subtracting k from each cost that lies in an uncovered column (or vice versa). Thus, step 3 creates a new cost matrix with the same optimal solution as the initial assignment problem. Each time step 3 is performed, at least one new zero is created in the cost matrix.

Steps 1 and 3 also ensure that all costs remain non-negative. Thus, the net effect of steps 1 and 3 of the Hungarian method is to create a sequence of assignment problems (with non-negative costs) that all have the same optimal solution as the original assignment problem. Now consider an assignment problem in which all costs are non-negative. Any feasible assignment in which all the x_{ij}'s that equal 1 have zero costs must be optimal for such an assignment problem. Thus, when step 2 indicates that m lines are required to cover all the zeros in the cost matrix, an optimal solution to the original problem has been found.

Computer Solution of Assignment Problems

To solve assignment problems in LINDO, type in the objective function and constraints. Also, many menu-driven programs require the user to input only a list of supply and demand points (such as jobs and machines, respectively) and a cost matrix. LINGO can also be used to easily solve assignment problems, including the following model to solve the Machineco example (file Assign.lng).

```
MODEL:
 1]SETS:
 2]MACHINES/1..4/;
 3]JOBS/1..4/;
 4]LINKS(MACHINES,JOBS):COST,ASSIGN;
 5]ENDSETS
 6]MIN=@SUM(LINKS:COST*ASSIGN);
 7]@FOR(MACHINES(I):
 8]@SUM(JOBS(J):ASSIGN(I,J))<1);
 9]@FOR(JOBS(J):
10]@SUM(MACHINES(I):ASSIGN(I,J))>1);
11]DATA:
12]COST = 14,5,8,7,
13]2,12,6,5,
14]7,8,3,9,
15]2,4,6,10;
16]ENDDATA
END
```

Line 2 defines the four supply points (machines), and line 3 defines the four demand points (jobs). In line 4, we define each possible combination of jobs and machines (16 in all) and associate with each combination an assignment cost [for example COST(1, 2) = 5] and a variable ASSIGN(I,J). ASSIGN(I,J) equals 1 if machine i is used to perform job j; it equals 0 otherwise. Line 5 ends the definition of sets.

Line 6 expresses the objective function by summing over all possible (I,J) combinations the product of the assignment cost and ASSIGN(I,J). Lines 7–8 limit each MACHINE to performing at most one job by forcing (for each machine) the sum of ASSIGN(I,J) over all JOBS to be at most 1. Lines 9–10 require that each JOB be completed by forcing (for each job) the sum of ASSIGN(I,J) over all MACHINES to be at least 1.

Lines 12–16 input the cost matrix.

Observe that this LINGO program can (with simple editing) be used to solve any assignment problem (even if it is not balanced!). For example, if you had 10 machines available to perform 8 jobs, you would edit line 2 to indicate that there are 10 machines (replace 1..4 with 1..10). Then edit line 3 to indicate that there are 8 jobs. Finally, in line 12, you would type the 80 entries of your cost matrix, following "COST=" and you would be ready to roll!

REMARK   1   From our discussion of the Machineco example, it is unnecessary to force the ASSIGN(I,J) to equal 0 or 1; this will happen automatically!

PROBLEMS

Group A

1 Five employees are available to perform four jobs. The time it takes each person to perform each job is given in Table 50. Determine the assignment of employees to jobs that minimizes the total time required to perform the four jobs.

2[†] Doc Councillman is putting together a relay team for the 400-meter relay. Each swimmer must swim 100 meters of breaststroke, backstroke, butterfly, or freestyle. Doc believes that each swimmer will attain the times given in

[†]This problem is based on Machol (1970).

TABLE 50

Person	Time (hours)			
	Job 1	Job 2	Job 3	Job 4
1	22	18	30	18
2	18	—	27	22
3	26	20	28	28
4	16	22	—	14
5	21	—	25	28

Note: Dashes indicate person cannot do that particular job.

Table 51. To minimize the team's time for the race, which swimmer should swim which stroke?

3 Tom Cruise, Freddy Prinze Jr., Harrison Ford, and Matt LeBlanc are marooned on a desert island with Jennifer Aniston, Courteney Cox, Gwyneth Paltrow, and Julia Roberts. The "compatibility measures" in Table 52 indicate how much happiness each couple would experience if they spent all their time together. The happiness earned by a couple is proportional to the fraction of time they spend together. For example, if Freddie and Gwyneth spend half their time together, they earn happiness of $\frac{1}{2}(9) = 4.5$.

a Let x_{ij} be the fraction of time that the ith man spends with the jth woman. The goal of the eight people is to maximize the total happiness of the people on the island. Formulate an LP whose optimal solution will yield the optimal values of the x_{ij}'s.

b Explain why the optimal solution in part (a) will have four $x_{ij} = 1$ and twelve $x_{ij} = 0$. The optimal solution requires that each person spend all his or her time with one person of the opposite sex, so this result is often referred to as the Marriage Theorem.

c Determine the marriage partner for each person.

d Do you think the Proportionality Assumption of linear programming is valid in this situation?

4 A company is taking bids on four construction jobs. Three people have placed bids on the jobs. Their bid (in thousands of dollars) are given in Table 53 (a * indicates that the person

did not bid on the given job). Person 1 can do only one job, but persons 2 and 3 can each do as many as two jobs. Determine the minimum cost assignment of persons to jobs.

5 Greydog Bus Company operates buses between Boston and Washington, D.C. A bus trip between these two cities takes 6 hours. Federal law requires that a driver rest for four or more hours between trips. A driver's workday consists of two trips: one from Boston to Washington and one from Washington to Boston. Table 54 gives the departure times for the buses. Greydog's goal is to minimize the total downtime for all drivers. How should Greydog assign crews to trips? *Note:* It is permissible for a driver's "day" to overlap midnight. For example, a Washington-based driver can be assigned to the Washington–Boston 3 P.M. trip and the Boston–Washington 6 A.M. trip.

6 Five male characters (Billie, John, Fish, Glen, and Larry) and five female characters (Ally, Georgia, Jane, Rene, and Nell) from *Ally McBeal* are marooned on a desert island. The problem is to determine what percentage of time each woman on the island should spend with each man. For example, Ally could spend 100% of her time with John or she could "play the field" by spending 20% of her time with each man. Table 55 shows a "happiness index" for each potential pairing of a man and woman. For example, if Larry and Rene spend all their time together, they earn 8 units of happiness for the island.

a Play matchmaker and determine an allocation of each man and woman's time that earns the maximum total happiness for the island. Assume that happiness earned by a couple is proportional to the amount of time they spend together.

b Explain why the optimal solution to this problem will, for any matrix of "happiness indices," always involve each woman spending all her time with one man.

TABLE 51

	Time (seconds)			
Swimmer	Free	Breast	Fly	Back
Gary Hall	54	54	51	53
Mark Spitz	51	57	52	52
Jim Montgomery	50	53	54	56
Chet Jastremski	56	54	55	53

TABLE 52

	JA	CC	GP	JR
TC	7	5	8	2
FP	7	8	9	4
HF	3	5	7	9
ML	5	5	6	7

TABLE 53

	Job			
Person	1	2	3	4
1	50	46	42	40
2	51	48	44	*
3	*	47	45	45

TABLE 54

Trip	Departure Time	Trip	Departure Time
Boston 1	6 A.M.	Washington 1	5:30 A.M.
Boston 2	7:30 A.M.	Washington 2	9 A.M.
Boston 3	11.30 A.M.	Washington 3	3 P.M.
Boston 4	7 P.M.	Washington 4	6:30 P.M.
Boston 5	12:30 A.M.	Washington 5	12 midnight

TABLE 55

	Ally	Georgia	Jane	Rene	Nell
Billie	8	6	4	7	5
John	5	7	6	4	9
Fish	10	6	5	2	10
Glen	1	0	0	0	0
Larry	5	7	9	8	6

c What assumption made in the problem is needed for the Marriage Theorem to hold?

Group B

7 Any transportation problem can be formulated as an assignment problem. To illustrate the idea, determine an assignment problem that could be used to find the optimal solution to the transportation problem in Table 56. (*Hint:* You will need five supply and five demand points).

8 The Chicago board of education is taking bids on the city's four school bus routes. Four companies have made the bids in Table 57.

    a Suppose each bidder can be assigned only one route. Use the assignment method to minimize Chicago's cost of running the four bus routes.

TABLE 56

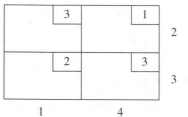

TABLE 57

Company	Route 1	Route 2	Route 3	Route 4
1	$4,000	$5,000	—	—
2		$4,000	—	$4,000
3	$3,000		$2,000	
4	—	—	$4,000	$5,000

    b Suppose that each company can be assigned two routes. Use the assignment method to minimize Chicago's cost of running the four bus routes. (*Hint:* Two supply points will be needed for each company.)

9 Show that step 3 of the Hungarian method is equivalent to performing the following operations: (1) Add k to each cost that lies in a covered row. (2) Subtract k from each cost that lies in an uncovered column.

10 Suppose c_{ij} is the smallest cost in row i and column j of an assignment problem. Must $x_{ij} = 1$ in any optimal assignment?

7.6 Transshipment Problems

A transportation problem allows only shipments that go directly from a supply point to a demand point. In many situations, shipments are allowed between supply points or between demand points. Sometimes there may also be points (called *transshipment points*) through which goods can be transshipped on their journey from a supply point to a demand point. Shipping problems with any or all of these characteristics are transshipment problems. Fortunately, the optimal solution to a transshipment problem can be found by solving a transportation problem.

    In what follows, we define a **supply point** to be a point that can send goods to another point but cannot receive goods from any other point. Similarly, a **demand point** is a point that can receive goods from other points but cannot send goods to any other point. A **transshipment point** is a point that can both receive goods from other points and send goods to other points. The following example illustrates these definitions ("—" indicates that a shipment is impossible).

EXAMPLE 5   **Transshipment**

Widgetco manufactures widgets at two factories, one in Memphis and one in Denver. The Memphis factory can produce as many as 150 widgets per day, and the Denver factory can produce as many as 200 widgets per day. Widgets are shipped by air to customers in Los Angeles and Boston. The customers in each city require 130 widgets per day. Because of the deregulation of airfares, Widgetco believes that it may be cheaper to first fly some widgets to New York or Chicago and then fly them to their final destinations. The costs of flying a widget are shown in Table 58. Widgetco wants to minimize the total cost of shipping the required widgets to its customers.

TABLE 58
Shipping Costs for Transshipments

From	Memphis	Denver	N.Y.	Chicago	L.A.	Boston
			To ($)			
Memphis	0	—	8	13	25	28
Denver	—	0	15	12	26	25
N.Y.	—	—	0	6	16	17
Chicago	—	—	6	0	14	16
L.A.	—	—	—	—	0	—
Boston	—	—	—	—	—	0

In this problem, Memphis and Denver are supply points, with supplies of 150 and 200 widgets per day, respectively. New York and Chicago are transshipment points. Los Angeles and Boston are demand points, each with a demand of 130 widgets per day. A graphical representation of possible shipments is given in Figure 9.

We now describe how the optimal solution to a transshipment problem can be found by solving a transportation problem. Given a transshipment problem, we create a balanced transportation problem by the following procedure (assume that total supply exceeds total demand):

Step 1 If necessary, add a dummy demand point (with a supply of 0 and a demand equal to the problem's excess supply) to balance the problem. Shipments to the dummy and from a point to itself will, of course, have a zero shipping cost. Let s = total available supply.

Step 2 Construct a transportation tableau as follows: A row in the tableau will be needed for each supply point and transshipment point, and a column will be needed for each demand point and transshipment point. Each supply point will have a supply equal to its original supply, and each demand point will have a demand equal to its original demand. Let s = total available supply. Then each transshipment point will have a supply equal to (point's original supply) + s and a demand equal to (point's original demand) + s. This ensures that any transshipment point that is a net supplier will have a net outflow equal to the point's original supply, and, similarly, a net demander will have a net inflow equal to the point's original demand. Although we don't know how much will be shipped through each transshipment point, we can be sure that the total amount will not exceed s. This explains why we add s to the supply and demand at each transshipment point. By adding the same amounts to the supply and demand, we ensure that the net outflow at each transshipment point will be correct, and we also maintain a balanced transportation tableau.

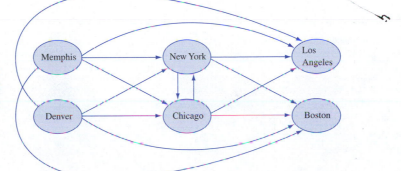

FIGURE 9
A Transshipment Problem

For the Widgetco example, this procedure yields the transportation tableau and its optimal solution given in Table 59. Because s = (total supply) = $150 + 200 = 350$ and (total demand) = $130 + 130 = 260$, the dummy demand point has a demand of $350 - 260 = 90$. The other supplies and demands in the transportation tableau are obtained by adding $s = 350$ to each transshipment point's supply and demand.

In interpreting the solution to the transportation problem created from a transshipment problem, we simply ignore the shipments to the dummy and from a point to itself. From Table 59, we find that Widgetco should produce 130 widgets at Memphis, ship them to New York, and transship them from New York to Los Angeles. The 130 widgets produced at Denver should be shipped directly to Boston. The net outflow from each city is

$$
\begin{array}{lll}
\text{Memphis:} & 130 + 20 & = 150 \\
\text{Denver:} & 130 + 70 & = 200 \\
\text{N.Y.:} & 220 + 130 - 130 - 220 & = 0 \\
\text{Chicago:} & 350 - 350 & = 0 \\
\text{L.A.:} & -130 & \\
\text{Boston:} & -130 & \\
\text{Dummy:} & -20 - 70 & = -90
\end{array}
$$

A negative net outflow represents an inflow. Observe that each transshipment point (New York and Chicago) has a net outflow of 0; whatever flows into the transshipment point must leave the transshipment point. A graphical representation of the optimal solution to the Widgetco example is given in Figure 10.

Suppose that we modify the Widgetco example and allow shipments between Memphis and Denver. This would make Memphis and Denver transshipment points and would add columns for Memphis and Denver to the Table 59 tableau. The Memphis row in the tableau would now have a supply of $150 + 350 = 500$, and the Denver row would have

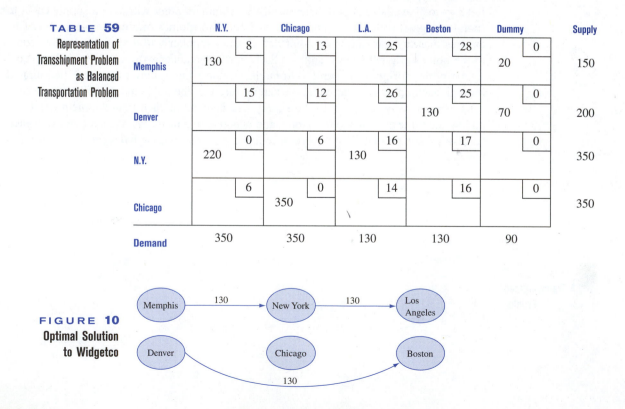

TABLE 59
Representation of Transshipment Problem as Balanced Transportation Problem

	N.Y.	Chicago	L.A.	Boston	Dummy	Supply
Memphis	8 — 130	13	25	28	0 — 20	150
Denver	15	12	26	25 — 130	0 — 70	200
N.Y.	0 — 220	6	16 — 130	17	0	350
Chicago	6	0 — 350	14	16	0	350
Demand	350	350	130	130	90	

FIGURE 10
Optimal Solution to Widgetco

CHAPTER **7** Transportation, Assignment, and Transshipment Problems

a supply of 200 + 350 = 550. The new Memphis column would have a demand of 0 + 350 = 350, and the new Denver column would have a demand of 0 + 350 = 350. Finally, suppose that shipments between demand points L.A. and Boston were allowed. This would make L.A. and Boston transshipment points and add rows for L.A. and Boston. The supply for both the L.A. and Boston rows would be 0 + 350 = 350. The demand for both the L.A. and Boston columns would now be 130 + 350 = 480.

PROBLEMS

Group A

1 General Ford produces cars at L.A. and Detroit and has a warehouse in Atlanta; the company supplies cars to customers in Houston and Tampa. The cost of shipping a car between points is given in Table 60 ("—" means that a shipment is not allowed). L.A. can produce as many as 1,100 cars, and Detroit can produce as many as 2,900 cars. Houston must receive 2,400 cars, and Tampa must receive 1,500 cars.

   a Formulate a balanced transportation problem that can be used to minimize the shipping costs incurred in meeting demands at Houston and Tampa.

   b Modify the answer to part (a) if shipments between L.A. and Detroit are not allowed.

   c Modify the answer to part (a) if shipments between Houston and Tampa are allowed at a cost of $5.

2 Sunco Oil produces oil at two wells. Well 1 can produce as many as 150,000 barrels per day, and well 2 can produce as many as 200,000 barrels per day. It is possible to ship oil directly from the wells to Sunco's customers in Los Angeles and New York. Alternatively, Sunco could transport oil to the ports of Mobile and Galveston and then ship it by tanker to New York or Los Angeles. Los Angeles requires 160,000 barrels per day, and New York requires 140,000 barrels per day. The costs of shipping 1,000 barrels between two points are shown in Table 61. Formulate a transshipment model (and equivalent transportation model) that could be used to minimize the transport costs in meeting the oil demands of Los Angeles and New York.

3 In Problem 2, assume that before being shipped to Los Angeles or New York, all oil produced at the wells must be refined at either Galveston or Mobile. To refine 1,000 barrels of oil costs $12 at Mobile and $10 at Galveston. Assuming that both Mobile and Galveston have infinite refinery capacity,

TABLE 60

From	To ($)				
	L.A.	Detroit	Atlanta	Houston	Tampa
L.A.	0	140	100	90	225
Detroit	145	0	111	110	119
Atlanta	105	115	0	113	78
Houston	89	109	121	0	—
Tampa	210	117	82	—	0

TABLE 61

From	To ($)					
	Well 1	Well 2	Mobile	Galveston	N.Y.	L.A.
Well 1	0	—	10	13	25	28
Well 2	—	0	15	12	26	25
Mobile	—	—	0	6	16	17
Galveston	—	—	6	0	14	16
N.Y.	—	—	—	—	0	15
L.A.	—	—	—	—	15	0

Note: Dashes indicate shipments that are not allowed.

formulate a transshipment and balanced transportation model to minimize the daily cost of transporting and refining the oil requirements of Los Angeles and New York.

4 Rework Problem 3 under the assumption that Galveston has a refinery capacity of 150,000 barrels per day and Mobile has one of 180,000 barrels per day. (*Hint:* Modify the method used to determine the supply and demand at each transshipment point to incorporate the refinery capacity restrictions, but make sure to keep the problem balanced.)

5 General Ford has two plants, two warehouses, and three customers. The locations of these are as follows:

   Plants: Detroit and Atlanta
   Warehouses: Denver and New York
   Customers: Los Angeles, Chicago, and Philadelphia

Cars are produced at plants, then shipped to warehouses, and finally shipped to customers. Detroit can produce 150 cars per week and Atlanta can produce 100 cars per week. Los Angeles requires 80 cars per week; Chicago, 70; and Philadelphia, 60. It costs $10,000 to produce a car at each plant, and the cost of shipping a car between two cities is given in Table 62. Determine how to meet General Ford's weekly demands at minimum cost.

Group B

6[†] A company must meet the following demands for cash at the beginning of each of the next six months: month 1,

[†]Based on Srinivasan (1974).

TABLE **62**

TABLE 62

	To ($)	
From	**Denver**	**New York**
Detroit	1,253	637
Atlanta	1,398	841

	To ($)		
From	**Los Angeles**	**Chicago**	**Philadelphia**
Denver	1,059	996	1,691
New York	2,786	802	100

TABLE 63

	Month of Sale					
Bond	**1**	**2**	**3**	**4**	**5**	**6**
1	$0.21	$0.19	$0.17	$0.13	$0.09	$0.05
2	$0.50	$0.50	$0.50	$0.33	$0	$0
3	$1.00	$1.00	$1.00	$1.00	$1.00	$0

$200; month 2, $100; month 3, $50; month 4, $80; month 5, $160; month 6, $140. At the beginning of month 1, the company has $150 in cash and $200 worth of bond 1, $100 worth of bond 2, and $400 worth of bond 3. The company will have to sell some bonds to meet demands, but a penalty will be charged for any bonds sold before the end of month 6. The penalties for selling $1 worth of each bond are as shown in Table 63.

a Assuming that all bills must be paid on time, formulate a balanced transportation problem that can be used to minimize the cost of meeting the cash demands for the next six months.

b Assume that payment of bills can be made after they are due, but a penalty of 5¢ per month is assessed for each dollar of cash demands that is postponed for one month. Assuming all bills must be paid by the end of month 6, develop a transshipment model that can be used to minimize the cost of paying the next six months' bills. (*Hint:* Transshipment points are needed, in the form Ct = cash available at beginning of month t after bonds for month t have been sold, but before month t demand is met. Shipments into Ct occur from bond sales and $Ct - 1$. Shipments out of Ct occur to $Ct + 1$ and demands for months $1, 2, \ldots t$.)

SUMMARY   Notation

$$m = \text{number of supply points}$$

$$n = \text{number of demand points}$$

$$x_{ij} = \text{number of units shipped from supply point } i \text{ to demand point } j$$

$$c_{ij} = \text{cost of shipping 1 unit from supply point } i \text{ to demand point } j$$

$$s_i = \text{supply at supply point } i$$

$$d_j = \text{demand at demand point } j$$

$$\bar{c}_{ij} = \text{coefficient of } x_{ij} \text{ in row 0 of a given tableau}$$

$$\mathbf{a}_{ij} = \text{column for } x_{ij} \text{ in transportation constraints}$$

A transportation problem is **balanced** if total supply equals total demand. To use the methods of this chapter to solve a transportation problem, the problem must first be balanced by use of a dummy supply or a dummy demand point. A balanced transportation problem may be written as

$$\min \sum_{i=1}^{i=m} \sum_{j=1}^{j=n} c_{ij}x_{ij}$$

$$\text{s.t.} \quad \sum_{j=1}^{j=n} x_{ij} = s_i \quad (i = 1, 2, \ldots, m) \quad \text{(Supply constraints)}$$

$$\sum_{i=1}^{i=m} x_{ij} = d_j \quad (j = 1, 2, \ldots, n) \quad \text{(Demand constraints)}$$

$$x_{ij} \geq 0 \quad (i = 1, 2, \ldots, m; j = 1, 2, \ldots, n)$$

Finding Basic Feasible Solutions for Balanced Transportation Problems

We can find a bfs for a balanced transportation problem by the northwest corner method, the minimum-cost method, or Vogel's method. To find a bfs by the northwest corner method, begin in the upper left-hand (or northwest) corner of the transportation tableau and set x_{11} as large as possible. Clearly, x_{11} can be no larger than the smaller of s_1 and d_1. If $x_{11} = s_1$, then cross out the first row of the transportation tableau; this indicates that no more basic variables will come from row 1 of the tableau. Also change d_1 to $d_1 - s_1$. If $x_{11} = d_1$, then cross out the first column of the transportation tableau and change s_1 to $s_1 - d_1$. If $x_{11} = s_1 = d_1$, cross out either row 1 or column 1 (but not both) of the transportation tableau. If you cross out row 1, change d_1 to 0; if you cross out column 1, change s_1 to 0. Continue applying this procedure to the most northwest cell in the tableau that does not lie in a crossed-out row or column. Eventually, you will come to a point where there is only one cell that can be assigned a value. Assign this cell a value equal to its row or column demand, and cross out both the cell's row and its column. A basic feasible solution has now been obtained.

Finding the Optimal Solution for a Transportation Problem

Step 1 If the problem is unbalanced, balance it.

Step 2 Use one of the methods described in Section 7.2 to find a bfs.

Step 3 Use the fact that $u_1 = 0$ and $u_i + v_j = c_{ij}$ for all basic variables to find the $[u_1 \quad u_2 \ldots u_m \quad v_1 \quad v_2 \ldots v_n]$ for the current bfs.

Step 4 If $u_i + v_j - c_{ij} \leq 0$ for all nonbasic variables, then the current bfs is optimal. If this is not the case, then we enter the variable with the most positive $u_i + v_j - c_{ij}$ into the basis. To do this, find the loop. Then, *counting only cells in the loop,* label the even cells. Also label the odd cells. Now find the odd cell whose variable assumes the smallest value, θ. The variable corresponding to this odd cell will leave the basis. To perform the pivot, decrease the value of each odd cell by θ and increase the value of each even cell by θ. The values of variables not in the loop remain unchanged. The pivot is now complete. If $\theta = 0$, then the entering variable will equal 0, and an odd variable that has a current value of 0 will leave the basis. In this case, a degenerate bfs will result. If more than one odd cell in the loop equals θ, you may arbitrarily choose one of these odd cells to leave the basis; again, a degenerate bfs will result. The pivoting yields a new bfs.

Step 5 Using the new bfs, return to steps 3 and 4.

For a maximization problem, proceed as stated, but replace step 4 by step 4'.

Step 4' If $u_i + v_j - c_{ij} \geq 0$ for all nonbasic variables, the current bfs is optimal. Otherwise, enter the variable with the most negative $u_i + v_j - c_{ij}$ into the basis using the pivoting procedure.

Assignment Problems

An **assignment problem** is a balanced transportation problem in which all supplies and demands equal 1. An $m \times m$ assignment problem may be efficiently solved by the Hungarian method:

Step 1   Find the minimum element in each row of the cost matrix. Construct a new matrix by subtracting from each cost the minimum cost in its row. For this new matrix, find the minimum cost in each column. Construct a new matrix (reduced cost matrix) by subtracting from each cost the minimum cost in its column.

Step 2   Cover all the zeros in the reduced cost matrix using the minimum number of lines needed. If m lines are required, then an optimal solution is available among the covered zeros in the matrix. If fewer than m lines are needed, then proceed to step 3.

Step 3   Find the smallest nonzero element (k) in the reduced cost matrix that is uncovered by the lines drawn in step 2. Now subtract k from each uncovered element and add k to each element that is covered by two lines. Return to step 2.

REMARKS   1 To solve an assignment problem in which the goal is to maximize the objective function, multiply the profits matrix through by -1 and solve it as a minimization problem.
2 If the number of rows and columns in the cost matrix are unequal, then the problem is **unbalanced.** The Hungarian method may yield an incorrect solution if the problem is unbalanced. Thus, any assignment problem should be balanced (by the addition of one or more dummy points) before it is solved by the Hungarian method.

Transshipment Problems

A transshipment problem allows shipment between supply points and between demand points, and it may also contain transshipment points through which goods may be shipped on their way from a supply point to a demand point. Using the following method, a transshipment problem may be transformed into a balanced transportation problem.

Step 1   If necessary, add a dummy demand point (with a supply of 0 and a demand equal to the problem's excess supply) to balance the problem. Shipments to the dummy and from a point to itself will, of course, have a zero shipping cost. Let s = total available supply.

Step 2   Construct a transportation tableau creating a row for each supply point and transshipment point, and a column for each demand point and transshipment point. Each supply point will have a supply equal to its original supply, and each demand point will have a demand equal to its original demand. Let s = total available supply. Then each transshipment point will have a supply equal to (point's original supply) + s and a demand equal to (point's original demand) + s.

Sensitivity Analysis for Transportation Problems

Following the discussion of sensitivity analysis in Chapter 6, we can analyze how a change in a transportation problem affects the problem's optimal solution.

Change 1   Changing the objective function coefficient of a nonbasic variable. As long as the coefficient of x_{ij} in the optimal row 0 is nonpositive, the current basis remains optimal.

Change 2   Changing the objective function coefficient of a basic variable. To see whether the current basis remains optimal, find the new u_i's and v_j's and use these values to price out all nonbasic variables. The current basis remains optimal as long as all nonbasic variables have a nonpositive coefficient in row 0.

Change 3   Increasing both supply s_i and demand d_j by Δ.

$$\text{New } z\text{-value} = \text{old } z\text{-value} + \Delta u_i + \Delta v_j$$

We may find the new values of the decision variables as follows:

1 If x_{ij} is a basic variable in the optimal solution, then increase x_{ij} by Δ.

2 If x_{ij} is a nonbasic variable in the optimal solution, find the loop involving x_{ij} and some of the basic variables. Find an odd cell in the loop that is in row i. Increase the value of this odd cell by Δ and go around the loop, alternately increasing and then decreasing current basic variables in the loop by Δ.

REVIEW PROBLEMS

Group A

1 Televco produces TV picture tubes at three plants. Plant 1 can produce 50 tubes per week; plant 2, 100 tubes per week; and plant 3, 50 tubes per week. Tubes are shipped to three customers. The profit earned per tube depends on the site where the tube was produced and on the customer who purchases the tube (see Table 64). Customer 1 is willing to purchase as many as 80 tubes per week; customer 2, as many as 90; and customer 3, as many as 100. Televco wants to find a shipping and production plan that will maximize profits.

a Formulate a balanced transportation problem that can be used to maximize Televco's profits.

b Use the northwest corner method to find a bfs to the problem.

c Use the transportation simplex to find an optimal solution to the problem.

2 Five workers are available to perform four jobs. The time it takes each worker to perform each job is given in Table 65. The goal is to assign workers to jobs so as to minimize the total time required to perform the four jobs. Use the Hungarian method to solve the problem.

3 A company must meet the following demands for a product: January, 30 units; February, 30 units; March, 20 units. Demand may be backlogged at a cost of $5/unit/month. All demand must be met by the end of March. Thus, if 1 unit of January demand is met during March, a backlogging cost of $5(2) = \$10$ is incurred. Monthly production capacity and unit production cost during each month are given in Table 66. A holding cost of $20/unit is assessed on the inventory at the end of each month.

a Formulate a balanced transportation problem that could be used to determine how to minimize the total cost (including backlogging, holding, and production costs) of meeting demand.

b Use Vogel's method to find a basic feasible solution.

c Use the transportation simplex to determine how to meet each month's demand. Make sure to give an interpretation of your optimal solution (for example, 20 units of month 2 demand is met from month 1 production).

4 Appletree Cleaning has five maids. To complete cleaning my house, they must vacuum, clean the kitchen, clean the bathroom, and do general straightening up. The time it takes each maid to do each job is shown in Table 67. Each maid

TABLE 64

From	To ($)		
	Customer 1	Customer 2	Customer 3
Plant 1	75	60	69
Plant 2	79	73	68
Plant 3	85	76	70

TABLE 66

Month	Production Capacity	Unit Production Cost
January	35	$400
February	30	$420
March	35	$410

TABLE 65

Worker	Time (Hours)			
	Job 1	Job 2	Job 3	Job 4
1	10	15	10	15
2	12	8	20	16
3	12	9	12	18
4	6	12	15	18
5	16	12	8	12

TABLE 67

Maid	Time (Hours)			
	Vacuum	Clean Kitchen	Clean Bathroom	Straighten Up
1	6	5	2	1
2	9	8	7	3
3	8	5	9	4
4	7	7	8	3
5	5	5	6	4

is assigned one job. Use the Hungarian method to determine assignments that minimize the total number of maid-hours needed to clean my house.

5[†] Currently, State University can store 200 files on hard disk, 100 files in computer memory, and 300 files on tape. Users want to store 300 word-processing files, 100 packaged-program files, and 100 data files. Each month a typical word-processing file is accessed eight times; a typical packaged-program file, four times; and a typical data file, two times. When a file is accessed, the time it takes for the file to be retrieved depends on the type of file and on the storage medium (see Table 68).

a If the goal is to minimize the total time per month that users spend accessing their files, formulate a balanced transportation problem that can be used to determine where files should be stored.

b Use the minimum cost method to find a bfs.

c Use the transportation simplex to find an optimal solution.

6 The Gotham City police have just received three calls for police. Five cars are available. The distance (in city blocks) of each car from each call is given in Table 69. Gotham City wants to minimize the total distance cars must travel to respond to the three police calls. Use the Hungarian method to determine which car should respond to which call.

7 There are three school districts in the town of Busville. The number of black and white students in each district are shown in Table 70. The Supreme Court requires the schools in Busville to be racially balanced. Thus, each school must have exactly 300 students, and each school must have the same number of black students. The distances between districts are shown in Table 70.

TABLE 68

Storage Medium	Time (Minutes)		
	Word Processing	Packaged Program	Data
Hard disk	5	4	4
Memory	2	1	1
Tape	10	8	6

TABLE 69

Car	Distance (Blocks)		
	Call 1	Call 2	Call 3
1	10	11	18
2	6	7	7
3	7	8	5
4	5	6	4
5	9	4	7

[†]This problem is based on Evans (1984).

TABLE 70

District	No. of Students		Distance to (Miles)	
	Whites	Blacks	District 2	District 3
1	210	120	3	5
2	210	30	—	4
3	180	150	—	—

Formulate a balanced transportation problem that can be used to determine the minimum total distance that students must be bussed while still satisfying the Supreme Court's requirements. Assume that a student who remains in his or her own district will not be bussed.

8 Using the northwest corner method to find a bfs, find (via the transportation simplex) an optimal solution to the transportation (minimization) problem shown in Table 71.

9 Solve the following LP:

$$\min z = 2x_1 + 3x_2 + 4x_3 + 3x_4$$
$$\text{s.t.} \quad x_1 + x_2 \qquad\qquad \leq 4$$
$$x_3 + x_4 \leq 5$$
$$x_1 \qquad + x_3 \qquad \geq 3$$
$$x_2 \qquad + x_4 \geq 6$$
$$\min x_j \geq 0 \quad (j = 1, 2, 3, 4)$$

10 Find the optimal solution to the balanced transportation problem in Table 72 (minimization).

11 In Problem 10, suppose we increase s_i to 16 and d_3 to 11. The problem is still balanced, and because 31 units (instead of 30 units) must be shipped, one would think that the total shipping costs would be increased. Show that the total shipping cost has actually decreased by $2, however. This is called the "more for less" paradox. Explain why increasing both the supply and the demand has decreased cost. Using the theory of shadow prices, explain how one could have predicted that increasing s_1 and d_3 by 1 would decrease total cost by $2.

12 Use the northwest corner method, the minimum-cost method, and Vogel's method to find basic feasible solutions to the transportation problem in Table 73.

13 Find the optimal solution to Problem 12.

TABLE 71

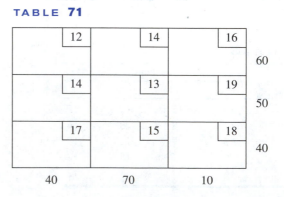

12	14	16	60
14	13	19	50
17	15	18	40
40	70	10	

TABLE 72

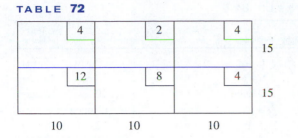

TABLE 73

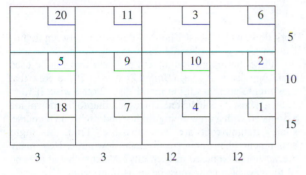

TABLE 74

From	To ($)			
	Dallas	**Houston**	**N.Y.**	**Chicago**
L.A.	300	110	—	—
San Diego	420	100	—	—
Dallas	—	—	450	550
Houston	—	—	470	530

TABLE 75

Plant	Cars Available
Atlanta	5,000
Boston	6,000
Chicago	4,000
L.A.	3,000

TABLE 76

Warehouse	Cars Required
Memphis	6,000
Milwaukee	4,000
N.Y.	4,000
Denver	2,000
San Francisco	2,000

TABLE 77

	Memphis	**Milwaukee**	**N.Y.**	**Denver**	**S.F.**
Atlanta	371	761	841	1,398	2,496
Boston	1,296	1,050	206	1,949	3,095
Chicago	530	87	802	996	2,142
L.A.	1,817	2,012	2,786	1,059	379

14 Oilco has oil fields in San Diego and Los Angeles. The San Diego field can produce 500,000 barrels per day, and the Los Angeles field can produce 400,000 barrels per day. Oil is sent from the fields to a refinery, either in Dallas or in Houston (assume that each refinery has unlimited capacity). It costs $700 to refine 100,000 barrels of oil at Dallas and $900 at Houston. Refined oil is shipped to customers in Chicago and New York. Chicago customers require 400,000 barrels per day of refined oil; New York customers require 300,000. The costs of shipping 100,000 barrels of oil (refined or unrefined) between cities are given in Table 74. Formulate a balanced transportation model of this situation.

15 For the Powerco problem, find the range of values of c_{24} for which the current basis remains optimal.

16 For the Powerco problem, find the range of values of c_{23} for which the current basis remains optimal.

17 A company produces cars in Atlanta, Boston, Chicago, and Los Angeles. The cars are then shipped to warehouses in Memphis, Milwaukee, New York City, Denver, and San Francisco. The number of cars available at each plant is given in Table 75.

Each warehouse needs to have available the number of cars given in Table 76.

The distance (in miles) between the cities is given in Table 77.

a Assuming that the cost (in dollars) of shipping a car equals the distance between two cities, determine an optimal shipping schedule.

b Assuming that the cost (in dollars) of shipping a car equals the square root of the distance between two cities, determine an optimal shipping schedule.

18 During the next three quarters, Airco faces the following demands for air conditioner compressors: quarter 1—200; quarter 2—300; quarter 3—100. As many as 240 air compressors can be produced during each quarter. Production costs/compressor during each quarter are given in Table 78. The cost of holding an air compressor in inventory is $100/quarter. Demand may be backlogged (as long as it is met by the end of quarter 3) at a cost of $60/compressor/quarter. Formulate the tableau for a balanced transportation problem whose solution tells Airco how to minimize the total cost of meeting the demands for quarters 1–3.

19 A company is considering hiring people for four types of jobs. It would like to hire the number of people in Table 79 for each type of job.

Four types of people can be hired by the company. Each type is qualified to perform two types of jobs according to

TABLE 78

Quarter 1	Quarter 2	Quarter 3
$200	$180	$240

TABLE 79

	Job			
	1	2	3	4
Number of people	30	30	40	20

TABLE 80

	Type of Person			
	1	2	3	4
Jobs qualified for	1 and 3	2 and 3	3 and 4	1 and 4

Table 80. A total of 20 Type 1, 30 Type 2, 40 Type 3, and 20 Type 4 people have applied for jobs. Formulate a balanced transportation problem whose solution will tell the company how to maximize the number of employees assigned to suitable jobs. (*Note:* Each person can be assigned to at most one job.)

20 During each of the next two months you can produce as many as 50 units/month of a product at a cost of $12/unit during month 1 and $15/unit during month 2. The customer is willing to buy as many as 60 units/month during each of the next two months. The customer will pay $20/unit during month 1, and $16/unit during month 2. It costs $1/unit to hold a unit in inventory for a month. Formulate a balanced transportation problem whose solution will tell you how to maximize profit.

Group B

21[†] The Carter Caterer Company must have the following number of clean napkins available at the beginning of each of the next four days: day 1—15; day 2—12; day 3—18; day 4—6. After being used, a napkin can be cleaned by one of two methods: fast service or slow service. Fast service costs 10¢ per napkin, and a napkin cleaned via fast service is available for use the day after it is last used. Slow service costs 6¢ per napkin, and these napkins can be reused two days after they are last used. New napkins can be purchased for a cost of 20¢ per napkin. Formulate a balanced transportation problem to minimize the cost of meeting the demand for napkins during the next four days.

22 Braneast Airlines must staff the daily flights between New York and Chicago shown in Table 81. Each of Braneast's crews lives in either New York or Chicago. Each day a crew must fly one New York–Chicago and one Chicago–New

[†]This problem is based on Jacobs (1954).

TABLE 81

Flight	Leave Chicago	Arrive New York	Flight	Leave New York	Arrive Chicago
1	6 A.M.	10 A.M.	1	7 A.M.	9 A.M.
2	9 A.M.	1 P.M.	2	8 A.M.	10 A.M.
3	12 noon	4 P.M.	3	10 A.M.	12 noon
4	3 P.M.	7 P.M.	4	12 noon	2 P.M.
5	5 P.M.	9 P.M.	5	2 P.M.	4 P.M.
6	7 P.M.	11 P.M.	6	4 P.M.	6 P.M.
7	8 P.M.	12 midnight	7	6 P.M.	8 P.M.

York flight with at least 1 hour of downtime between flights. Braneast wants to schedule the crews to minimize the total downtime. Set up an assignment problem that can be used to accomplish this goal. (*Hint:* Let $x_{ij} = 1$ if the crew that flies flight i also flies flight j, and $x_{ij} = 0$ otherwise. If $x_{ij} = 1$, then a cost c_{ij} is incurred, corresponding to the downtime associated with a crew flying flight i and flight j.) Of course, some assignments are not possible. Find the flight assignments that minimize the total downtime. How many crews should be based in each city? Assume that at the end of the day, each crew must be in its home city.

23 A firm producing a single product has three plants and four customers. The three plants will produce 3,000, 5,000, and 5,000 units, respectively, during the next time period. The firm has made a commitment to sell 4,000 units to customer 1, 3,000 units to customer 2, and at least 3,000 units to customer 3. Both customers 3 and 4 also want to buy as many of the remaining units as possible. The profit associated with shipping a unit from plant i to customer j is given in Table 82. Formulate a balanced transportation problem that can be used to maximize the company's profit.

24 A company can produce as many as 35 units/month. The demands of its primary customers must be met on time each month; if it wishes, the company may also sell units to secondary customers each month. A $1/unit holding cost is assessed against each month's ending inventory. The relevant data are shown in Table 83. Formulate a balanced transportation problem that can be used to maximize profits earned during the next three months.

25 My home has four valuable paintings that are up for sale. Four customers are bidding for the paintings. Customer 1 is willing to buy two paintings, but each other customer is willing to purchase at most one painting. The prices that each customer is willing to pay are given in Table 84. Use

TABLE 82

From	To Customer ($)			
	1	2	3	4
Plant 1	65	63	62	64
Plant 2	68	67	65	62
Plant 3	63	60	59	60

TABLE 83

Month	Production Cost/Unit ($)	Primary Demand	Available for Secondary Demand	Sales Price/Unit ($)
1	13	20	15	15
2	12	15	20	14
3	13	25	15	16

TABLE 84

Customer	Bid for ($)			
	Painting 1	Painting 2	Painting 3	Painting 4
1	8	11	—	—
2	9	13	12	7
3	9	—	11	—
4	—	—	12	9

the Hungarian method to determine how to maximize the total revenue received from the sale of the paintings.

26 Powerhouse produces capacitors at three locations: Los Angeles, Chicago, and New York. Capacitors are shipped from these locations to public utilities in five regions of the country: northeast (NE), northwest (NW), midwest (MW), southeast (SE), and southwest (SW). The cost of producing and shipping a capacitor from each plant to each region of the country is given in Table 85. Each plant has an annual production capacity of 100,000 capacitors. Each year, each region of the country must receive the following number of capacitors: NE, 55,000; NW, 50,000; MW, 60,000; SE, 60,000; SW, 45,000. Powerhouse feels shipping costs are too high, and the company is therefore considering building one or two more production plants. Possible sites are Atlanta and Houston. The costs of producing a capacitor and shipping it to each region of the country are given in Table 86. It costs $3 million (in current dollars) to build a new plant, and operating each plant incurs a fixed cost (in addition to variable shipping and production costs) of $50,000 per year. A plant at Atlanta or Houston will have the capacity to produce 100,000 capacitors per year.

Assume that future demand patterns and production costs will remain unchanged. If costs are discounted at a rate of $11\frac{1}{9}\%$ per year, how can Powerhouse minimize the present value of all costs associated with meeting current and future demands?

TABLE 85

From	To ($)				
	NE	NW	MW	SE	SW
L.A.	27.86	4.00	20.54	21.52	13.87
Chicago	8.02	20.54	2.00	6.74	10.67
N.Y.	2.00	27.86	8.02	8.41	15.20

TABLE 86

From	To ($)				
	NE	NW	MW	SE	SW
Atlanta	8.41	21.52	6.74	3.00	7.89
Houston	15.20	13.87	10.67	7.89	3.00

27[†] During the month of July, Pittsburgh resident B. Fly must make four round-trip flights between Pittsburgh and Chicago. The dates of the trips are as shown in Table 87. B. Fly must purchase four round-trip tickets. Without a discounted fare, a round-trip ticket between Pittsburgh and Chicago costs $500. If Fly's stay in a city includes a weekend, then he gets a 20% discount on the round-trip fare. If his stay in a city is at least 21 days, then he receives a 35% discount; and if his stay is more than 10 days, then he receives a 30% discount. Of course, only one discount can be applied toward the purchase of any ticket. Formulate and solve an assignment problem that minimizes the total cost of purchasing the four round-trip tickets. (*Hint:* Let $x_{ij} = 1$ if a round-trip ticket is purchased for use on the ith flight out of Pittsburgh and the jth flight out of Chicago. Also think about where Fly should buy a ticket if, for example, $x_{21} = 1$.)

28 Three professors must be assigned to teach six sections of finance. Each professor must teach two sections of finance, and each has ranked the six time periods during which finance is taught, as shown in Table 88. A ranking of 10 means that the professor wants to teach that time, and a ranking of 1 means that he or she does not want to teach at that time. Determine an assignment of professors to sections that will maximize the total satisfaction of the professors.

29[‡] Three fires have just broken out in New York. Fires 1 and 2 each require two fire engines, and fire 3 requires three fire engines. The "cost" of responding to each fire depends on the time at which the fire engines arrive. Let t_{ij} be the time (in minutes) when the jth engine arrives at fire i. Then the cost of responding to each fire is as follows:

$$\text{Fire 1:} \quad 6t_{11} + 4t_{12}$$
$$\text{Fire 2:} \quad 7t_{21} + 3t_{22}$$
$$\text{Fire 3:} \quad 9t_{31} + 8t_{32} + 5t_{33}$$

Three fire companies can respond to the three fires. Company 1 has three engines available, and companies 2

TABLE 87

Leave Pittsburgh	Leave Chicago
Monday, July 1	Friday, July 5
Tuesday, July 9	Thursday, July 11
Monday, July 15	Friday, July 19
Wednesday, July 24	Thursday, July 25

[†]Based on Hansen and Wendell (1982).
[‡]Based on Denardo, Rothblum, and Swersey (1988).

TABLE 88

TABLE 89

Professor	9 A.M.	10 A.M.	11 A.M.	1 P.M.	2 P.M.	3 P.M.
1	8	7	6	5	7	6
2	9	9	8	8	4	4
3	7	6	9	6	9	9

Company	Fire 1	Fire 2	Fire 3
1	6	7	9
2	5	8	11
3	6	9	10

and 3 each have two engines available. The time (in minutes) it takes an engine to travel from each company to each fire is shown in Table 89.

a Formulate and solve a transportation problem that can be used to minimize the cost associated with as-

signing the fire engines. (*Hint:* Seven demand points will be needed.)

b Would the formulation in part (a) still be valid if the cost of fire 1 were $4t_{11} + 6t_{12}$?

REFERENCES

The following six texts discuss transportation, assignment, and transshipment problems:

Bazaraa, M., and J. Jarvis. *Linear Programming and Network Flows.* New York: Wiley, 1990.

Bradley, S., A. Hax, and T. Magnanti. *Applied Mathematical Programming.* Reading, Mass.: Addison-Wesley, 1977.

Dantzig, G. *Linear Programming and Extensions.* Princeton, N.J.: Princeton University Press, 1963.

Gass, S. *Linear Programming: Methods and Applications,* 5th ed. New York: McGraw-Hill, 1985.

Murty, K. *Linear Programming.* New York: Wiley, 1983.

Wu, N., and R. Coppins. *Linear Programming and Extensions.* New York: McGraw-Hill, 1981.

Aarvik, O., and P. Randolph. "The Application of Linear Programming to the Determination of Transmission Line Fees in an Electrical Power Network," *Interfaces* 6(1975):17–31.

Denardo, E., U. Rothblum, and A. Swersey. "Transportation Problem in Which Costs Depend on Order of Arrival," *Management Science* 34(1988):774–784.

Evans, J. "The Factored Transportation Problem," *Management Science* 30(1984):1021–1024.

Gillett, B. *Introduction to Operations Research: A Computer-*

Oriented Algorithmic Approach. New York: McGraw-Hill, 1976.

Glassey, R., and V. Gupta. "A Linear Programming Analysis of Paper Recycling," *Management Science* 21(1974): 392–408.

Glover, F., et al. "A Computational Study on Starting Procedures, Basis Change Criteria and Solution Algorithms for Transportation Problems," *Management Science* 20(1974):793–813. This article discusses the computational efficiency of various methods used to find basic feasible solutions for transportation problems.

Hansen, P., and R. Wendell. "A Note on Airline Commuting," *Interfaces* 11(no. 12, 1982):85–87.

Jackson, B. "Using LP for Crude Oil Sales at Elk Hills: A Case Study," *Interfaces* 10(1980):65–70.

Jacobs, W. "The Caterer Problem," *Naval Logistics Research Quarterly* 1(1954):154–165.

Machol, R. "An Application of the Assignment Problem," *Operations Research* 18(1970):745–746.

Srinivasan, P. "A Transshipment Model for Cash Management Decisions," *Management Science* 20(1974): 1350–1363.

Wagner, H., and D. Rubin. "Shadow Prices: Tips and Traps for Managers and Instructors," *Interfaces* 20(no. 4, 1990):150–157.

8

Network Models

8.2.1,5.8

Many important optimization problems can best be analyzed by means of a graphical or network representation. In this chapter, we consider four specific network models—shortest path problems, maximum-flow problems, CPM–PERT project-scheduling models, and minimum-spanning tree problems—for which efficient solution procedures exist. We also discuss minimum-cost network flow problems (MCNFPs), of which transportation, assignment, transshipment, shortest path, and maximum flow problems and the CPM project-scheduling models are all special cases. Finally, we discuss a generalization of the transportation simplex, the network simplex, which can be used to solve MCNFPs. We begin the chapter with some basic terms used to describe graphs and networks.

8.1 Basic Definitions

A **graph,** or **network,** is defined by two sets of symbols: nodes and arcs. First, we define a set (call it V) of points, or **vertices.** The vertices of a graph or network are also called **nodes.**

We also define a set of arcs A.

DEFINITION ■ An **arc** consists of an ordered pair of vertices and represents a possible direction of motion that may occur between vertices. ■

For our purposes, if a network contains an arc (j, k), then motion is possible from node j to node k. Suppose nodes 1, 2, 3, and 4 of Figure 1 represent cities, and each arc represents a (one-way) road linking two cities. For this network, $V = \{1, 2, 3, 4\}$ and $A = \{(1, 2), (2, 3), (3, 4), (4, 3), (4, 1)\}$. For the arc (j, k), node j is the **initial node,** and node k is the **terminal node.** The arc (j, k) is said to go from node j to node k. Thus, the arc $(2, 3)$ has initial node 2 and terminal node 3, and it goes from node 2 to node 3. The arc $(2, 3)$ may be thought of as a (one-way) road on which we may travel from city 2 to city 3. In Figure 1, the arcs show that travel is allowed from city 3 to city 4, and from city 4 to city 3, but that travel between the other cities may be one way only.

Later, we often discuss a group or collection of arcs. The following definitions are convenient ways to describe certain groups or collections of arcs.

DEFINITION ■ A sequence of arcs such that every arc has exactly one vertex in common with the previous arc is called a **chain.** ■

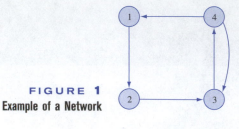

FIGURE 1
Example of a Network

DEFINITION ■ A **path** is a chain in which the terminal node of each arc is identical to the initial node of the next arc. ■

For example, in Figure 1, (1, 2)–(2, 3)–(4, 3) is a chain but not a path; (1, 2)–(2, 3)–(3, 4) is a chain *and* a path. The path (1, 2)–(2, 3)–(3, 4) represents a way to travel from node 1 to node 4.

8.2 Shortest Path Problems

In this section, we assume that each arc in the network has a length associated with it. Suppose we start at a particular node (say, node 1). The problem of finding the shortest path (path of minimum length) from node 1 to any other node in the network is called a **shortest path problem.** Examples 1 and 2 are shortest path problems.

EXAMPLE 1   Shortest Path

Let us consider the Powerco example (Figure 2). Suppose that when power is sent from plant 1 (node 1) to city 1 (node 6), it must pass through relay substations (nodes 2–5). For any pair of nodes between which power can be transported, Figure 2 gives the distance (in miles) between the nodes. Thus, substations 2 and 4 are 3 miles apart, and power cannot be sent between substations 4 and 5. Powerco wants the power sent from plant 1 to city 1 to travel the minimum possible distance, so it must find the shortest path in Figure 2 that joins node 1 to node 6.

If the cost of shipping power were proportional to the distance the power travels, then knowing the shortest path between plant 1 and city 1 in Figure 2 (and the shortest path between plant *i* and city *j* in similar diagrams) would be necessary to determine the shipping costs for the transportation version of the Powerco problem discussed in Chapter 7.

FIGURE 2
Network for Powerco

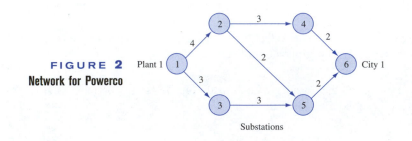

Substations

EXAMPLE 2   **Equipment Replacement**

I have just purchased (at time 0) a new car for $12,000. The cost of maintaining a car during a year depends on its age at the beginning of the year, as given in Table 1. To avoid the high maintenance costs associated with an older car, I may trade in my car and purchase a new car. The price I receive on a trade-in depends on the age of the car at the time of trade-in (see Table 2). To simplify the computations, we assume that at any time, it costs $12,000 to purchase a new car. My goal is to minimize the net cost (purchasing costs + maintenance costs − money received in trade-ins) incurred during the next five years. Formulate this problem as a shortest path problem.

Solution   Our network will have six nodes (1, 2, 3, 4, 5, and 6). Node i is the beginning of year i. For $i < j$, an arc (i, j) corresponds to purchasing a new car at the beginning of year i and keeping it until the beginning of year j. The length of arc (i, j) (call it c_{ij}) is the total net cost incurred in owning and operating a car from the beginning of year i to the beginning of year j if a new car is purchased at the beginning of year i and this car is traded in for a new car at the beginning of year j. Thus,

$$c_{ij} = \text{maintenance cost incurred during years } i, i + 1, \ldots, j - 1$$
$$+ \text{ cost of purchasing car at beginning of year } i$$
$$- \text{ trade-in value received at beginning of year } j$$

Applying this formula to the information in the problem yields (all costs are in thousands)

$c_{12} = 2 + 12 - 7 = 7$ $c_{16} = 2 + 4 + 5 + 9 + 12 + 12 - 0 = 44$

$c_{13} = 2 + 4 + 12 - 6 = 12$ $c_{23} = 2 + 12 - 7 = 7$

$c_{14} = 2 + 4 + 5 + 12 - 2 = 21$ $c_{24} = 2 + 4 + 12 - 6 = 12$

$c_{15} = 2 + 4 + 5 + 9 + 12 - 1 = 31$ $c_{25} = 2 + 4 + 5 + 12 - 2 = 21$

TABLE 1
Car Maintenance Costs

Age of Car (Years)	Annual Maintenance Cost ($)
0	2,000
1	4,000
2	5,000
3	9,000
4	12,000

TABLE 2
Car Trade-in Prices

Age of Car (Years)	Trade-in Price
1	7,000
2	6,000
3	2,000
4	1,000
5	0

FIGURE **3**
Network for Minimizing
Car Costs

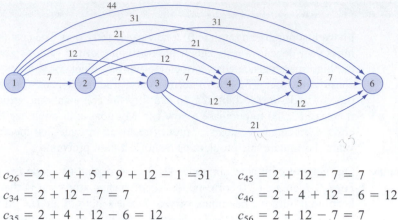

$$c_{26} = 2 + 4 + 5 + 9 + 12 - 1 = 31 \qquad c_{45} = 2 + 12 - 7 = 7$$
$$c_{34} = 2 + 12 - 7 = 7 \qquad c_{46} = 2 + 4 + 12 - 6 = 12$$
$$c_{35} = 2 + 4 + 12 - 6 = 12 \qquad c_{56} = 2 + 12 - 7 = 7$$
$$c_{36} = 2 + 4 + 5 + 12 - 2 = 21$$

We now see that the length of any path from node 1 to node 6 is the net cost incurred during the next five years corresponding to a particular trade-in strategy. For example, suppose I trade in the car at the beginning of year 3 and next trade in the car at the end of year 5 (the beginning of year 6). This strategy corresponds to the path 1–3–6 in Figure 3. The length of this path ($c_{13} + c_{36}$) is the total net cost incurred during the next five years if I trade in the car at the beginning of year 3 and at the beginning of year 6. Thus, the length of the shortest path from node 1 to node 6 in Figure 3 is the minimum net cost that can be incurred in operating a car during the next five years.

Dijkstra's Algorithm

Assuming that all arc lengths are non-negative, the following method, known as **Dijkstra's algorithm,** can be used to find the shortest path from a node (say, node 1) to all other nodes. To begin, we label node 1 with a permanent label of 0. Then we label each node i that is connected to node 1 by a single arc with a "temporary" label equal to the length of the arc joining node 1 to node i. Each other node (except, of course, for node 1) will have a temporary label of ∞. Choose the node with the smallest temporary label and make this label permanent.

Now suppose that node i has just become the $(k + 1)$th node to be given a permanent label. Then node i is the kth closest node to node 1. At this point, the temporary label of any node (say, node i') is the length of the shortest path from node 1 to node i' that passes only through nodes contained in the $k - 1$ closest nodes to node 1. For each node j that now has a temporary label and is connected to node i by an arc, we replace node j's temporary label by

$$\min \begin{cases} \text{node } j\text{'s current temporary label} \\ \text{node } i\text{'s permanent label} + \text{length of arc } (i, j) \end{cases}$$

(Here, $\min\{a, b\}$ is the smaller of a and b.) The new temporary label for node j is the length of the shortest path from node 1 to node j that passes only through nodes contained in the k closest nodes to node 1. We now make the smallest temporary label a permanent label. The node with this new permanent label is the $(k + 1)$th closest node to node 1. Continue this process until all nodes have a permanent label. To find the shortest path from node 1 to node j, work backward from node j by finding nodes having labels differing by exactly the length of the connecting arc. Of course, if we want the shortest path from node 1 to node j, we can stop the labeling process as soon as node j receives a permanent label.

To illustrate Dijkstra's algorithm, we find the shortest path from node 1 to node 6 in Figure 2. We begin with the following labels (a * represents a permanent label, and the ith number is the label of the node i): [0* 4 3 ∞ ∞ ∞]. Node 3 now has the smallest temporary label. We therefore make node 3's label permanent and obtain the following labels:

$$[0^* \quad 4 \quad 3^* \quad \infty \quad \infty \quad \infty]$$

We now know that node 3 is the closest node to node 1. We compute new temporary labels for all nodes that are connected to node 3 by a single arc. In Figure 2 that is node 5.

$$\text{New node 5 temporary label} = \min\{\infty, 3 + 3\} = 6$$

Node 2 now has the smallest temporary label; we now make node 2's label permanent. We now know that node 2 is the second closest node to node 1. Our new set of labels is

$$[0^* \quad 4^* \quad 3^* \quad \infty \quad 6 \quad \infty]$$

Because nodes 4 and 5 are connected to the newly permanently labeled node 2, we must change the temporary labels of nodes 4 and 5. Node 4's new temporary label is min $\{\infty, 4 + 3\} = 7$ and node 5's new temporary label is min $\{6, 4 + 2\} = 6$. Node 5 now has the smallest temporary label, so we make node 5's label permanent. We now know that node 5 is the third closest node to node 1. Our new labels are

$$[0^* \quad 4^* \quad 3^* \quad 7 \quad 6^* \quad \infty]$$

Only node 6 is connected to node 5, so node 6's temporary label will change to min $\{\infty, 6 + 2\} = 8$. Node 4 now has the smallest temporary label, so we make node 4's label permanent. We now know that node 4 is the fourth closest node to node 1. Our new labels are

$$[0^* \quad 4^* \quad 3^* \quad 7^* \quad 6^* \quad 8]$$

Because node 6 is connected to the newly permanently labeled node 4, we must change node 6's temporary label to min $\{8, 7 + 2\} = 8$. We can now make node 6's label permanent. Our final set of labels is [0* 4* 3* 7* 6* 8*]. We can now work backward and find the shortest path from node 1 to node 6. The difference between node 6's and node 5's permanent labels is 2 = length of arc (5, 6), so we go back to node 5. The difference between node 5's and node 2's permanent labels is 2 = length of arc (2, 5), so we may go back to node 2. Then, of course, we must go back to node 1. Thus, 1–2–5–6 is a shortest path (of length 8) from node 1 to node 6. Observe that when we were at node 5, we could also have worked backward to node 3 and obtained the shortest path 1–3–5–6.

The Shortest Path Problem as a Transshipment Problem

Finding the shortest path between node i and node j in a network may be viewed as a transshipment problem. Simply try to minimize the cost of sending one unit from node i to node j (with all other nodes in the network being transshipment points), where the cost of sending one unit from node k to node k' is the length of arc (k, k') if such an arc exists and is M (a large positive number) if such an arc does not exist. As in Section 7.6, the cost of shipping one unit from a node to itself is zero. Following the method described in Section 7.6, this transshipment problem may be transformed into a balanced transportation problem.

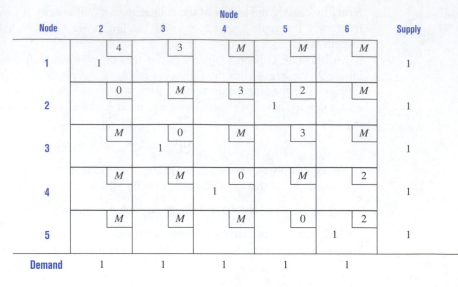

Node	2	3	4	5	6	Supply
1	4 [1]	3	M	M	M	1
2	0	M	3	2 [1]	M	1
3	M	0 [1]	M	3	M	1
4	M	M	0 [1]	M	2	1
5	M	M	M	0	2 [1]	1
Demand	1	1	1	1	1	

To illustrate the preceding ideas, we formulate the balanced transportation problem associated with finding the shortest path from node 1 to node 6 in Figure 2. We want to send one unit from node 1 to node 6. Node 1 is a supply point, node 6 is a demand point, and nodes 2, 3, 4, and 5 will be transshipment points. Using $s = 1$, we obtain the balanced transportation problem shown in Table 3. This transportation problem has two optimal solutions:

1 $z = 4 + 2 + 2 = 8$, $x_{12} = x_{25} = x_{56} = x_{33} = x_{44} = 1$ (all other variables equal 0). This solution corresponds to the path 1–2–5–6.

2 $z = 3 + 3 + 2 = 8$, $x_{13} = x_{35} = x_{56} = x_{22} = x_{44} = 1$ (all other variables equal 0). This solution corresponds to the path 1–3–5–6.

REMARK After formulating a shortest path problem as a transshipment problem, the problem may be solved easily by using LINGO or a spreadsheet optimizer. See Section 7.1 for details.

PROBLEMS

Group A

1 Find the shortest path from node 1 to node 6 in Figure 3.

2 Find the shortest path from node 1 to node 5 in Figure 4.

3 Formulate Problem 2 as a transshipment problem.

4 Use Dijkstra's algorithm to find the shortest path from node 1 to node 4 in Figure 5. Why does Dijkstra's algorithm fail to obtain the correct answer?

FIGURE **4**
Network for Problem 2

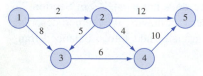

FIGURE **5**
Network for Problem 4

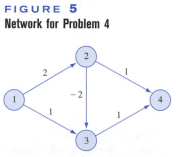

5 Suppose it costs $10,000 to purchase a new car. The annual operating cost and resale value of a used car are shown in Table 4. Assuming that one now has a new car, determine a replacement policy that minimizes the net costs of owning and operating a car for the next six years.

TABLE 4

Age of Car (Years)	Resale Value ($)	Operating Cost ($)
1	7,000	300 (year 1)
2	6,000	500 (year 2)
3	4,000	800 (year 3)
4	3,000	1,200 (year 4)
5	2,000	1,600 (year 5)
6	1,000	2,200 (year 6)

TABLE 6

Year	Purchase Cost ($)
1	170,000
2	190,000
3	210,000
4	250,000
5	300,000

6 It costs $40 to buy a telephone from the department store. Assume that I can keep a telephone for at most five years and that the estimated maintenance cost each year of operation is as follows: year 1, $20; year 2, $30; year 3, $40; year 4, $60; year 5, $70. I have just purchased a new telephone. Assuming that a telephone has no salvage value, determine how to minimize the total cost of purchasing and operating a telephone for the next six years.

7 At the beginning of year 1, a new machine must be purchased. The cost of maintaining a machine i years old is given in Table 5.

The cost of purchasing a machine at the beginning of each year is given in Table 6.

There is no trade-in value when a machine is replaced. Your goal is to minimize the total cost (purchase plus maintenance) of having a machine for five years. Determine the years in which a new machine should be purchased.

Group B

8[†] A library must build shelving to shelve 200 4-inch high books, 100 8-inch high books, and 80 12-inch high books.

TABLE 5

Age at Beginning of Year	Maintenance Cost for Next Year ($)
0	38,000
1	50,000
2	97,000
3	182,000
4	304,000

[†]Based on Ravindran (1971).

Each book is 0.5 inch thick. The library has several ways to store the books. For example, an 8-inch high shelf may be built to store all books of height less than or equal to 8 inches, and a 12-inch high shelf may be built for the 12-inch books. Alternatively, a 12-inch high shelf might be built to store all books. The library believes it costs $2,300 to build a shelf and that a cost of $5 per square inch is incurred for book storage. (Assume that the area required to store a book is given by height of storage area times book's thickness.)

Formulate and solve a shortest path problem that could be used to help the library determine how to shelve the books at minimum cost. (*Hint:* Have nodes 0, 4, 8, and 12, with c_{ij} being the total cost of shelving all books of height $> i$ and $\leq j$ on a single shelf.)

9 A company sells seven types of boxes, ranging in volume from 17 to 33 cubic feet. The demand and size of each box is given in Table 7. The variable cost (in dollars) of producing each box is equal to the box's volume. A fixed cost of $1,000 is incurred to produce any of a particular box. If the company desires, demand for a box may be satisfied by a box of larger size. Formulate and solve a shortest path problem whose solution will minimize the cost of meeting the demand for boxes.

10 Explain how by solving a single transshipment problem you can find the shortest path from node 1 in a network to *each other node* in the network.

TABLE 7

	Box						
	1	2	3	4	5	6	7
Size	33	30	26	24	19	18	17
Demand	400	300	500	700	200	400	200

8.3 Maximum-Flow Problems

Many situations can be modeled by a network in which the arcs may be thought of as having a capacity that limits the quantity of a product that may be shipped through the arc. In these situations, it is often desired to transport the maximum amount of flow from a starting point (called the **source**) to a terminal point (called the **sink**). Such problems are called **maximum-flow problems.** Several specialized algorithms exist to solve maximum-flow problems. In this section, we begin by showing how linear programming can be used to solve a maximum-flow problem. Then we discuss the Ford–Fulkerson (1962) method for solving maximum-flow problems.

LP Solution of Maximum-Flow Problems

EXAMPLE 3 | **Maximum Flow**

Sunco Oil wants to ship the maximum possible amount of oil (per hour) via pipeline from node *so* to node *si* in Figure 6. On its way from node *so* to node *si*, oil must pass through some or all of stations 1, 2, and 3. The various arcs represent pipelines of different diameters. The maximum number of barrels of oil (millions of barrels per hour) that can be pumped through each arc is shown in Table 8. Each number is called an **arc capacity.** Formulate an LP that can be used to determine the maximum number of barrels of oil per hour that can be sent from *so* to *si*.

Solution   Node *so* is called the source node because oil flows out of it but no oil flows into it. Analogously, node *si* is called the sink node because oil flows into it and no oil flows out of it. For reasons that will soon become clear, we have added an artificial arc a_0 from the sink to the source. The flow through a_0 is not actually oil, hence the term **artificial arc.**

To formulate an LP that will yield the maximum flow from node *so* to *si*, we observe that Sunco must determine how much oil (per hour) should be sent through arc (i, j). Thus, we define

x_{ij} = millions of barrels of oil per hour that will pass through arc (i,j) of pipeline

As an example of a possible flow (termed a *feasible flow*), consider the flow indentified by the numbers in parentheses in Figure 6.

$$x_{so,1} = 2, \quad x_{13} = 0, \quad x_{12} = 2, \quad x_{3,si} = 0, \quad x_{2,si} = 2, \quad x_{si,so} = 2, \quad x_{so,2} = 0$$

FIGURE 6
Network for Sunco Oil

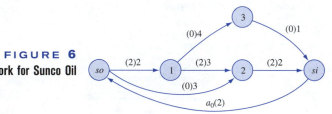

TABLE 8
Arc Capacities for
Sunco Oil

Arc	Capacity
(*so*, 1)	2
(*so*, 2)	3
(1, 2)	3
(1, 3)	4
(3, *si*)	1
(2, *si*)	2

For a flow to be feasible, it must have two characteristics:

$$0 \leq \text{flow through each arc} \leq \text{arc capacity} \qquad (1)$$

and

$$\text{Flow into node } i = \text{flow out of node } i \qquad (2)$$

We assume that no oil gets lost while being pumped through the network, so at each node, a feasible flow must satisfy (2), the *conservation-of-flow* constraint. The introduction of the artificial arc a_0 allows us to write the conservation-of-flow constraint for the source and sink.

If we let x_0 be the flow through the artificial arc, then conservation of flow implies that $x_0 =$ total amount of oil entering the sink. Thus, Sunco's goal is to maximize x_0 subject to (1) and (2):

$$\max z = x_0$$

s.t.
$$x_{so,1} \leq 2 \qquad \text{(Arc capacity constraints)}$$
$$x_{so,2} \leq 3$$
$$x_{12} \leq 3$$
$$x_{2,si} \leq 2$$
$$x_{13} \leq 4$$
$$x_{3,si} \leq 1$$
$$x_0 = x_{so,1} + x_{so,2} \qquad \text{(Node } so \text{ flow constraint)}$$
$$x_{so,1} = x_{12} + x_{13} \qquad \text{(Node 1 flow constraint)}$$
$$x_{so,2} + x_{12} = x_{2,si} \qquad \text{(Node 2 flow constraint)}$$
$$x_{13} = x_{3,si} \qquad \text{(Node 3 flow constraint)}$$
$$x_{3,si} + x_{2,si} = x_0 \qquad \text{(Node } si \text{ flow constraint)}$$
$$x_{ij} \geq 0$$

One optimal solution to this LP is $z = 3$, $x_{so,1} = 2$, $x_{13} = 1$, $x_{12} = 1$, $x_{so,2} = 1$, $x_{3,si} = 1$, $x_{2,si} = 2$, $x_0 = 3$. Thus, the maximum possible flow of oil from node so to si is 3 million barrels per hour, with 1 million barrels each sent via the following paths: so–1–2–si, so–1–3–si, and so–2–si.

The linear programming formulation of maximum flow problems is a special case of the minimum-cost network flow problem (MCNFP) discussed in Section 8.5. A generalization of the transportation simplex (known as the *network simplex*) can be used to solve MCNFPs.

Before discussing the Ford–Fulkerson method for solving maximum-flow problems, we give two examples for situations in which a maximum-flow problem might arise.

EXAMPLE 4 **Airline Maximum-Flow**

Fly-by-Night Airlines must determine how many connecting flights daily can be arranged between Juneau, Alaska, and Dallas, Texas. Connecting flights must stop in Seattle and then stop in Los Angeles or Denver. Because of limited landing space, Fly-by-Night is limited to making the number of daily flights between pairs of cities shown in Table 9. Set up a maximum-flow problem whose solution will tell the airline how to maximize the number of connecting flights daily from Juneau to Dallas.

TABLE 9
TABLE 9

Arc Capacities for Fly-by-Night Airlines

Cities	Maximum Number of Daily Flights
Juneau–Seattle (*J, S*)	3
Seattle–L.A. (*S, L*)	2
Seattle–Denver (*S, De*)	3
L.A.–Dallas (*L, D*)	1
Denver–Dallas (*De, D*)	2

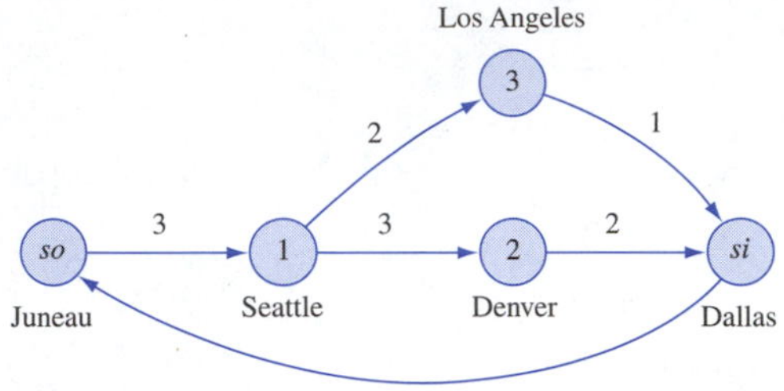

FIGURE 7

Network for Fly-by-Night Airlines

Solution   The appropriate network is given in Figure 7. Here the capacity of arc (i, j) is the maximum number of daily flights between city i and city j. The optimal solution to this maximum flow problem is $z = x_0 = 3$, $x_{J,S} = 3$, $x_{S,L} = 1$, $x_{S,De} = 2$, $x_{L,D} = 1$, $x_{De,D} = 2$. Thus, Fly-by-Night can send three flights daily connecting Juneau and Dallas. One flight connects via Juneau–Seattle–L.A.–Dallas, and two flights connect via Juneau–Seattle–Denver–Dallas.

EXAMPLE 5   Matchmaking

Five male and five female entertainers are at a dance. The goal of the matchmaker is to match each woman with a man in a way that maximizes the number of people who are matched with compatible mates. Table 10 describes the compatibility of the entertainers. Draw a network that makes it possible to represent the problem of maximizing the number of compatible pairings as a maximum-flow problem.

Solution   Figure 8 is the appropriate network. In Figure 8, there is an arc with capacity 1 joining the source to each man, an arc with capacity 1 joining each pair of compatible mates, and an arc with capacity 1 joining each woman to the sink. The maximum flow in this network is the number of compatible couples that can be created by the matchmaker. For ex-

TABLE 10

Compatibilities for Matching

	Loni Anderson	Meryl Streep	Katharine Hepburn	Linda Evans	Victoria Principal
Kevin Costner	—	C	—	—	—
Burt Reynolds	C	—	—	—	—
Tom Selleck	C	C	—	—	—
Michael Jackson	C	C	—	—	C
Tom Cruise	—	—	C	C	C

Note: C indicates compatibility.

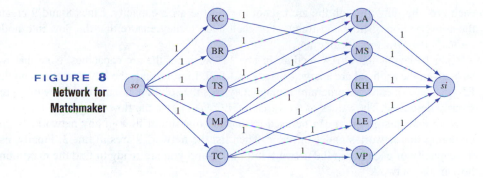

FIGURE 8
**Network for
Matchmaker**

ample, if the matchmaker pairs KC and MS, BR and LA, MJ and VP, and TC and KH, a flow of 4 from source to sink would be obtained. (This turns out to be a maximum flow for the network.)

To see why our network representation correctly models the matchmaker's problem, note that because the arc joining each woman to the sink has a capacity of 1, conservation of flow ensures that each woman will be matched with at most one man. Similarly, because each arc from the source to a man has a capacity of 1, each man can be paired with at most one woman. Because arcs do not exist between noncompatible mates, we can be sure that a flow of k units from source to sink represents an assignment of men to women in which k compatible couples are created.

Solving Maximum-Flow Problems with LINGO

The maximum flow in a network can be found using LINDO, but LINGO greatly lessens the effort needed to communicate the necessary information to the computer. The following LINGO program (in the file Maxflow.lng) can be used to find the maximum flow from source to sink in Figure 6.

Maxflow.lng

```
MODEL:
 1]SETS:
 2]NODES/1..5/;
 3]ARCS(NODES,NODES)/1,2  1,3  2,3  2,4  3,5  4,5  5,1/
 4]:CAP,FLOW;
 5]ENDSETS
 6]MAX=FLOW  (5,1);
 7]@FOR(ARCS(I,J):FLOW(I,J)<CAP(I,J));
 8]@FOR(NODES(I):@SUM(ARCS(J,I):FLOW(J,I))
 9]=@SUM(ARCS(I,J):FLOW(I,J)));
10]DATA:
11]CAP=2,3,3,4,2,1,1000;
12]ENDDATA
END
```

If some nodes are identified by numbers, then LINGO will not allow you to identify other nodes with names involving letters. Thus, we have identified node 1 in line 2 with node *so* in Figure 6 and node 5 in line 2 with node *si*. Also nodes 1, 2, and 3 in Figure 6 correspond to nodes 2, 3, and 4, respectively, in line 2 of our LINGO program. Thus, line 2 defines the nodes of the flow network. In line 3, we define the arcs of the network by listing them (separated by spaces). For example, 1, 2 represents the arc from the source to node 1 in Figure 6 and 5,1 is the artificial arc. In line 4, we indicate that an arc capacity and a flow are associated with each arc. Line 5 ends the definition of the relevant sets.

In line 6, we indicate that our objective is to maximize the flow through the artificial arc (this equals the flow into the sink). Line 7 specifies the arc capacity constraints; for each arc, the flow through the arc cannot exceed the arc's capacity. Lines 8 and 9 create the conservation of flow constraints. For each node I, they ensure that the flow into node I equals the flow out of node I.

Line 10 begins the DATA section. In line 11, we input the arc capacities. Note that we have given the artificial arc a large capacity of 1,000. Line 12 ends the DATA section and the **END** statement ends the program. Typing GO yields the solution, a maximum flow of 3 previously described. The values of the variable FLOW(I,J) give the flow through each arc.

Note that this program can be used to find the maximum flow in any network. Begin by listing the network's nodes in line 2. Then list the network's arcs in line 3. Finally, list the capacity of each arc in the network in line 11, and you are ready to find the maximum flow in the network!

The Ford–Fulkerson Method
for Solving Maximum-Flow Problems

We assume that a feasible flow has been found (letting the flow in each arc equal zero gives a feasible flow), and we turn our attention to the following important questions:

Question 1   Given a feasible flow, how can we tell if it is an optimal flow (that is, maximizes x_0)?

Question 2   If a feasible flow is nonoptimal, how can we modify the flow to obtain a new feasible flow that has a larger flow from the source to the sink?

First, we answer question 2. We determine which of the following properties is possessed by each arc in the network:

Property 1   The flow through arc (i, j) is below the capacity of arc (i, j). In this case, the flow through arc (i, j) can be increased. For this reason, we let I represent the set of arcs with this property.

Property 2   The flow in arc (i, j) is positive. In this case, the flow through arc (i, j) can be reduced. For this reason, we let R be the set of arcs with this property.

As an illustration of the definitions of I and R, consider the network in Figure 9. The arcs in this figure may be classified as follows: $(so, 1)$ is in I and R; $(so, 2)$ is in I; $(1, si)$ is in R; $(2, si)$ is in I; and $(2, 1)$ is in I.

We can now describe the Ford–Fulkerson labeling procedure used to modify a feasible flow in an effort to increase the flow from the source to the sink.

Step 1   Label the source.

Step 2   Label nodes and arcs (except for arc a_0) according to the following rules: (1) If node x is labeled, then node y is unlabeled and arc (x, y) is a member of I; then label node y and arc (x, y). In this case, arc (x, y) is called a **forward arc.** (2) If node y is unlabeled, node x is labeled and arc (y, x) is a member of R; label node y and arc (y, x). In this case, (y, x) is called a **backward arc.**

FIGURE 9
Illustration of *I* and
***R* arcs**

Step 3 Continue this labeling process until the sink has been labeled or until no more vertices can be labeled.

If the labeling process results in the sink being labeled, then there will be a chain of labeled arcs (call it C) leading from the source to the sink. By adjusting the flow of the arcs in C, we can maintain a feasible flow and increase the total flow from source to sink. To see this, observe that C must consist of one of the following:

Case 1 C consists entirely of forward arcs.

Case 2 C contains both forward and backward arcs.[†]

In each case, we can obtain a new feasible flow that has a larger flow from source to sink than the current feasible flow. In Case 1, the chain C consists entirely of forward arcs. For each forward arc in C, let $i(x, y)$ be the amount by which the flow in arc (x, y) can be increased without violating the capacity constraint for arc (x, y) . Let

$$k = \min_{(x,\ y) \in C} i(x,\ y)$$

Then $k > 0$. To create a new flow, increase the flow through each arc in C by k units. No capacity constraints are violated, and conservation of flow is still maintained. Thus, the new flow is feasible, and the new feasible flow will transport k more units from source to sink than does the current feasible flow.

We use Figure 10 to illustrate Case 1. Currently, 2 units are being transported from source to sink. The labeling procedure results in the sink being labeled by the chain $C = (so, 1) - (1, 2) - (2, si)$. Each arc is in I, and $i(so, 1) = 5 - 2 = 3$; $i(1, 2) = 3 - 2 = 1$; and $i(2, si) = 4 - 2 = 2$. Hence, $k = \min(3, 1, 2) = 1$. Thus, an improved feasible flow can be obtained by increasing the flow on each arc in C by 1 unit. The resulting flow transports 3 units from source to sink (see Figure 11).

In Case 2, the chain C leading from the source to the sink contains both backward and forward arcs. For each backward arc in C, let $r(x, y)$ be the amount by which the flow through arc (x, y) can be reduced. Also define

$$k_1 = \min_{x,\ y \in C \cap R} r(x,\ y) \quad \text{and} \quad k_2 = \min_{x,\ y \in C \cap I} i(x,\ y)$$

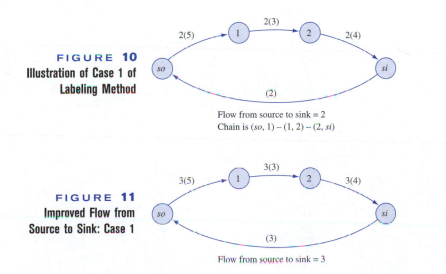

FIGURE 10
Illustration of Case 1 of Labeling Method

Flow from source to sink = 2
Chain is $(so, 1) - (1, 2) - (2, si)$

FIGURE 11
Improved Flow from Source to Sink: Case 1

Flow from source to sink = 3

[†]Because we exclude arc a_0 from the labeling procedure, no chain made entirely of backward arcs can lead from source to sink.

Of course, both k_1 and k_2 and min (k_1, k_2) are > 0. To increase the flow from source to sink (while maintaining a feasible flow), decrease the flow in all of C's backward arcs by min (k_1, k_2) and increase the flow in all of C's forward arcs by min(k_1, k_2). This will maintain conservation of flow and ensure that no arc capacity constraints are violated. Because the last arc in C is a forward arc leading into the sink, we have found a new feasible flow and have increased the total flow into the sink by min(k_1, k_2). We now adjust the flow in the arc a_0 to maintain conservation of flow. To illustrate Case 2, suppose we have found the feasible flow in Figure 12. For this flow, $(so, 1) \in R$; $(so, 2) \in I$; $(1, 3) \in I$; $(1, 2) \in I$ and R; $(2, si) \in R$; and $(3, si) \in I$.

We begin by labeling arc $(so, 2)$ and node 2 (thus $(so, 2)$ is a forward arc). Then we label arc $(1, 2)$ and node 1. Arc $(1, 2)$ is a backward arc, because node 1 was unlabeled before we labeled arc $(1, 2)$, and arc $(1, 2)$ is in R. Nodes so, 1, and 2 are labeled, so we can label arc $(1, 3)$ and node 3. [Arc $(1, 3)$ is a forward arc, because node 3 has not yet been labeled.] Finally we label arc $(3, si)$ and node si. Arc $(3, si)$ is a forward arc, because node si has not yet been labeled. We have now labeled the sink via the chain $C = (so, 2) - (1, 2) - (1, 3) - (3, si)$. With the exception of arc $(1, 2)$, all arcs in the chain are forward arcs. Because $i(so, 2) = 3$; $i(1, 3) = 4$; $i(3, si) = 1$; and $r(1, 2) = 2$, we have

$$\min_{(x, y) \in C \cap R} r(x, y) = 2 \quad \text{and} \quad \min_{(x, y) \in C \cap I} i(x, y) = 1$$

Thus, we can increase the flow on all forward arcs in C by 1 and decrease the flow in all backward arcs by 1. The new result, pictured in Figure 13, has increased the flow from source to sink by 1 unit (from 2 to 3). We accomplish this by diverting 1 unit that was transported through the arc $(1, 2)$ to the path $1-3-si$. This enabled us to transport an extra unit from source to sink via the path $so-2-si$. Observe that the concept of a backward arc was needed to find this improved flow.

If the sink cannot be labeled, then the current flow is optimal. The proof of this fact relies on the concept of a cut for a network.

DEFINITION ■ Choose any set of nodes V' that contains the sink but does not contain the source. Then the set of arcs (i, j) with i not in V' and j a member of V' is a **cut** for the network. ■

FIGURE 12
Illustration of Case 2 of
Labeling Method

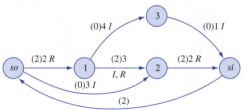

Flow from source to sink = 2
Chain is $(so, 2) - (1, 2) - (1, 3) - (3, si)$

FIGURE 13
Improved Flow from
Source to Sink: Case 2

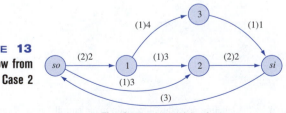

Flow from source to sink = 3

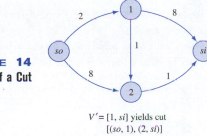

FIGURE 14
Example of a Cut

$V' = [1, si]$ yields cut
$[(so, 1), (2, si)]$

The capacity of a cut is the sum of the capacities of the arcs in the cut. ■

In short, a cut is a set of arcs whose removal from the network makes it impossible to travel from the source to the sink. A network may have many cuts. For example, in the network in Figure 14, $V' = \{1, si\}$ yields the cut containing the arcs $(so, 1)$ and $(2, si)$, which has capacity $2 + 1 = 3$. The set $V' = \{1, 2, si\}$ yields the cut containing the arcs $(so, 1)$ and $(so, 2)$, which has capacity $2 + 8 = 10$.

Lemma 1 and Lemma 2 indicate the connection between cuts and maximum flows.

LEMMA 1

The flow from source to sink for any feasible flow is less than or equal to the capacity of *any* cut.

Proof   Consider an arbitrary cut specified by a set of nodes V' that contains the sink but does not contain the source. Let V be all other nodes in the network. Also let x_{ij} be the flow in arc (i, j) for any feasible flow and f be the flow from source to sink for this feasible flow. Summing the flow balance equations (flow out of node i − flow into node $i = 0$) over all nodes i in V, we find that the terms involving arcs (i, j) having i and j both members of V will cancel, and we obtain

$$\sum_{\substack{i \in V; \\ j \in V'}} x_{ij} - \sum_{\substack{i \in V'; \\ j \in V}} x_{ij} = f \tag{3}$$

Now the first sum in (3) is less than or equal to the capacity of the cut. Each x_{ij} is non-negative, so we see that $f \leq$ capacity of the cut, which is the desired result.

Lemma 1 is analogous to the weak duality result discussed in Chapter 6. From Lemma 1, we see that the capacity of any cut is an upper bound for the maximum flow from source to sink. Thus, if we can find a feasible flow and a cut for which the flow from source to sink equals the capacity of the cut, then we have found the maximum flow from source to sink.

Suppose that we find a feasible flow and cannot label the sink. Let CUT be the cut corresponding to the set of unlabeled nodes.

LEMMA 2

If the sink cannot be labeled, then

Capacity of CUT = current flow from source to sink

Proof   Let V' be the set of unlabeled nodes and V be the set of labeled nodes. Consider an arc (i, j) such that i is in V and j is in V'. Then we know that $x_{ij} =$ capacity of arc (i, j) must hold; otherwise, we could label node j (via a forward arc) and node j would not be in V'. Now consider an arc (i, j) such that i is in V' and j is in V. Then $x_{ij} = 0$ must hold; otherwise, we could label node i (via a backward arc) and node i would not be in V'. Now (3) shows that the current flow must satisfy

$$\text{Capacity of CUT} = \text{current flow from source to sink}$$

which is the desired result.

From the remarks following Lemma 1, when the sink cannot be labeled, the maximum flow from source to sink has been obtained.

Summary and Illustration of the Ford–Fulkerson Method

Step 1 Find a feasible flow (setting each arc's flow to zero will do).

Step 2 Using the labeling procedure, try to label the sink. If the sink cannot be labeled, then the current feasible flow is a maximum flow; if the sink is labeled, then go on to step 3.

Step 3 Using the method previously described, adjust the feasible flow and increase the flow from the source to the sink. Return to step 2.

To illustrate the Ford–Fulkerson method, we find the maximum flow from source to sink for Sunco Oil, Example 3 (see Figure 6). We begin by letting the flow in each arc equal zero. We then try to label the sink—label the source, and then arc $(so, 1)$ and node 1; then label arc $(1, 2)$ and node 2; finally, label arc $(2, si)$ and node si. Thus, $C =$ $(so, 1)$–$(1, 2)$–$(2, si)$. Each arc in C is a forward arc, so we can increase the flow through each arc in C by min $(2, 3, 2) = 2$ units. The resulting flow is pictured in Figure 15.

As we saw previously (Figure 12), we can label the sink by using the chain $C =$ $(so, 2)$–$(1, 2)$–$(1, 3)$–$(3, si)$. We can increase the flow through the forward arcs $(so, 2)$, $(1, 3)$, and $(3, si)$ by 1 unit and decrease the flow through the backward arc $(1, 2)$ by 1 unit. The resulting flow is pictured in Figure 16. It is now impossible to label the sink. Any attempt to label the sink must begin by labeling arc $(so, 2)$ and node 2; then we could label arc $(1, 2)$ and arc $(1, 3)$. But there is no way to label the sink.

We can verify that the current flow is maximal by finding the capacity of the cut corresponding to the set of unlabeled vertices (in this case, si). The cut corresponding to si is the set of arcs $(2, si)$ and $(3, si)$, with capacity $2 + 1 = 3$. Thus, Lemma 1 implies that any feasible flow can transport at most 3 units from source to sink. Our current flow transports 3 units from source to sink, so it must be an optimal flow.

Another example of the Ford–Fulkerson method is given in Figure 17. Note that without the concept of a backward arc, we could not have obtained the maximum flow of 7

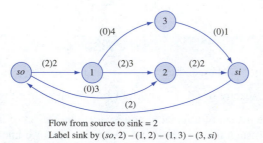

FIGURE 15
Network for Sunco Oil
(Increased Flow)

Flow from source to sink = 2
Label sink by $(so, 2) - (1, 2) - (1, 3) - (3, si)$

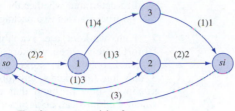

FIGURE 16
Network for Sunco Oil
(Optimal Flow)

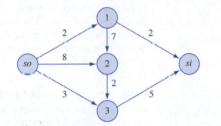

Flow from source to sink = 3
Since sink cannot be labeled, this is an optimal flow

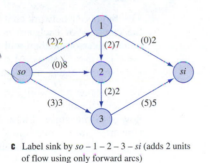

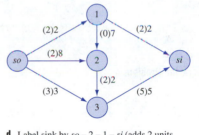

a Original network

b Label sink by $so - 3 - si$ (adds 3 units of flow using only forward arcs)

FIGURE 17
Example of
Ford–Fulkerson Method

c Label sink by $so - 1 - 2 - 3 - si$ (adds 2 units of flow using only forward arcs)

d Label sink by $so - 2 - 1 - si$ (adds 2 units of flow using backward arc (1, 2); maximum flow of 7 has been obtained)

units from source to sink. The minimum cut (with capacity 7, of course) corresponds to nodes 1, 3, and si and consists of arcs $(so, 1)$, $(so, 3)$ and $(2, 3)$.

PROBLEMS

Group A

1–3 Figures 18–20 show the networks for Problems 1–3. Find the maximum flow from source to sink in each network. Find a cut in the network whose capacity equals the maximum flow in the network. Also, set up an LP that could be used to determine the maximum flow in the network.

FIGURE 18
Network for Problem 1

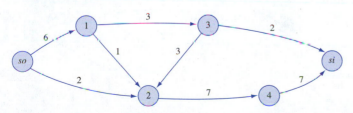

FIGURE 19

Network for Problem 2

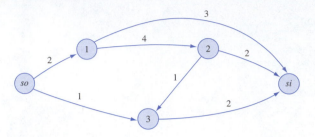

FIGURE 20

Nework for Problem 3

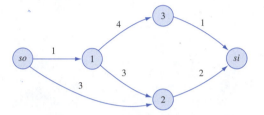

FIGURE 21

Network for Problem 4

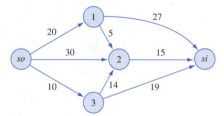

FIGURE 22

Network for Problem 5

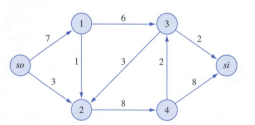

4–5 For the networks in Figures 21 and 22, find the maximum flow from source to sink. Also find a cut whose capacity equals the maximum flow in the network.

6 Seven types of packages are to be delivered by five trucks. There are three packages of each type, and the capacities of the five trucks are 6, 4, 5, 4, and 3 packages, respectively. Set up a maximum-flow problem that can be used to determine whether the packages can be loaded so that no truck carries two packages of the same type.

7 Four workers are available to perform jobs 1–4. Unfortunately, three workers can do only certain jobs: worker 1, only job 1; worker 2, only jobs 1 and 2; worker 3, only job 2; worker 4, any job. Draw the network for the maximum-flow problem that can be used to determine whether all jobs can be assigned to a suitable worker.

8 The Hatfields, Montagues, McCoys, and Capulets are going on their annual family picnic. Four cars are available to transport the families to the picnic. The cars can carry the following number of people: car 1, four; car 2, three; car 3, three; and car 4, four. There are four people in each family, and no car can carry more than two people from any one family. Formulate the problem of transporting the maximum possible number of people to the picnic as a maximum-flow problem.

9–10 For the networks in Figures 23 and 24, find the maximum flow from source to sink. Also find a cut whose capacity equals the maximum flow in the network.

Group B

11 Suppose a network contains a finite number of arcs and the capacity of each arc is an integer. Explain why the Ford–Fulkerson method will find the maximum flow in the finite number of steps. Also show that the maximum flow from source to sink will be an integer.

12 Consider a network flow problem with several sources and several sinks in which the goal is to maximize the total flow into the sinks. Show how such a problem can be converted into a maximum flow problem having only a single source and a single sink.

FIGURE 23

FIGURE 24

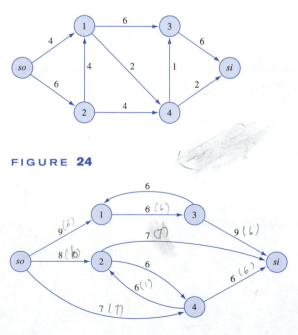

13 Suppose the total flow into a node of a network is restricted to 10 units or less. How can we represent this restriction via an arc capacity constraint? (This still allows us to use the Ford–Fulkerson method to find the maximum flow.)

14 Suppose as many as 300 cars per hour can travel between any two of the cities 1, 2, 3, and 4. Set up a maximum-flow problem that can be used to determine how many cars can be sent in the next two hours from city 1 to city 4. (*Hint:* Have portions of the network represent $t = 0$, $t = 1$, and $t = 2$.)

15 Fly-by-Night Airlines is considering flying three flights. The revenue from each flight and the airports used by each flight are shown in Table 11. When Fly-by-Night uses an airport, the company must pay the following landing fees (independent of the number of flights using the airport): airport 1, \$300; airport 2, \$700; airport 3, \$500. Thus, if flights 1 and 3 are flown, a profit of $900 + 800 - 300 - 700 - 500 = \$200$ will be earned. Show that for the network in Figure 25 (maximum profit) = (total revenue from all flights) − (capacity of minimal cut). Explain how this result can be used to help Fly-by-Night maximize profit (even if it has hundreds of possible flights). (*Hint:* Consider any set of flights F (say flights 1 and 3). Consider the cut corresponding

TABLE 11

Flight	Revenue ($)	Airport Used
1	900	1 and 2
2	600	2
3	800	2 and 3

to the sink, the nodes associated with the flights not in F, and the nodes associated with the airports not used by F. Show that (capacity of this cut) = (revenue from flights not in F) + (costs associated with airports used by F).)

16 During the next four months, a construction firm must complete three projects. Project 1 must be completed within three months and requires 8 months of labor. Project 2 must be completed within four months and requires 10 months of labor. Project 3 must be completed at the end of two months and requires 12 months of labor. Each month, 8 workers are available. During a given month, no more than 6 workers can work on a single job. Formulate a maximum-flow problem that could be used to determine whether all three projects can be completed on time. (*Hint:* If the maximum flow in the network is 30, then all projects can be completed on time.)

FIGURE 25
Network for Problem 15

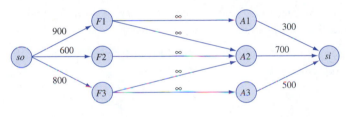

8.4 CPM and PERT

Network models can be used as an aid in scheduling large complex projects that consist of many activities. If the duration of each activity is known with certainty, then the **critical path method (CPM)** can be used to determine the length of time required to complete a project. CPM also can be used to determine how long each activity in the project can be delayed without delaying the completion of the project. CPM was developed in the late 1950s by researchers at DuPont and Sperry Rand.

If the duration of the activities is not known with certainty, the Program Evaluation and Review Technique (PERT) can be used to estimate the probability that the project will be completed by a given deadline. PERT was developed in the late 1950s by consultants working on the development of the Polaris missile. CPM and PERT were given a major share of the credit for the fact that the Polaris missile was operational two years ahead of schedule.

CPM and PERT have been successfully used in many applications, including:

1 Scheduling construction projects such as office buildings, highways, and swimming pools

2 Scheduling the movement of a 400-bed hospital from Portland, Oregon, to a suburban location

3 Developing a countdown and "hold" procedure for the launching of space flights

4 Installing a new computer system

5 Designing and marketing a new product

6 Completing a corporate merger

7 Building a ship

To apply CPM and PERT, we need a list of the activities that make up the project. The project is considered to be completed when all the activities have been completed. For each activity, there is a set of activities (called the **predecessors** of the activity) that must be completed before the activity begins. A project network is used to represent the precedence relationships between activities. In our discussion, activities will be represented by directed arcs, and nodes will be used to represent the completion of a set of activities. (For this reason, we often refer to the nodes in our project network as **events.**) This type of project network is called an **AOA (activity on arc)** network.[†]

To understand how an AOA network represents precedence relationships, suppose that activity A is a predecessor of activity B. Each node in an AOA network represents the completion of one or more activities. Thus, node 2 in Figure 26 represents the completion of activity A and the beginning of activity B. Suppose activities A and B must be completed before activity C can begin. In Figure 27, node 3 represents the event that activities A and B are completed. Figure 28 shows activity A as a predecessor of both activities B and C.

Given a list of activities and predecessors, an AOA representation of a project (called a **project network** or **project diagram**) can be constructed by using the following rules:

1 Node 1 represents the start of the project. An arc should lead from node 1 to represent each activity that has no predecessors.

FIGURE 26
Activity A Must Be Completed Before Activity B Can Begin

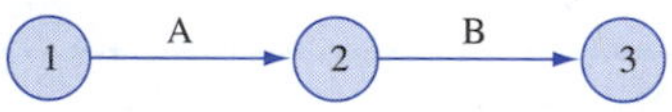

FIGURE 27
Activities A and B Must Be Completed Before Activity C Can Begin

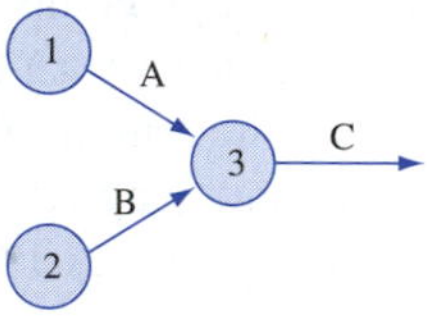

FIGURE 28
Activity A Must Be Completed Before Activities B and C Can Begin

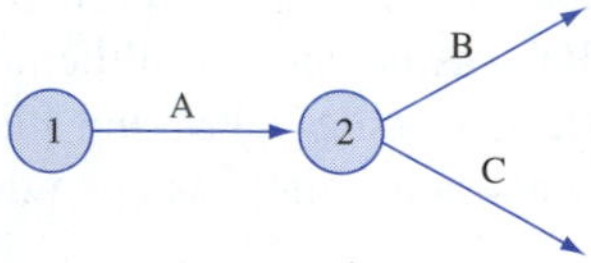

[†]In an AON (activity on node) project network, the nodes of the network are used to represent activities. See Wiest and Levy (1977) for details.

FIGURE **29**
Violation of Rule 5

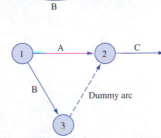

FIGURE **30**
Use of Dummy Activity

2 A node (called the **finish node**) representing the completion of the project should be included in the network.

3 Number the nodes in the network so that the node representing the completion of an activity always has a larger number than the node representing the beginning of an activity (there may be more than one numbering scheme that satisfies rule 3).

4 An activity should not be represented by more than one arc in the network.

5 Two nodes can be connected by at most one arc.

To avoid violating rules 4 and 5, it is sometimes necessary to utilize a **dummy activity** that takes zero time. For example, suppose activities A and B are both predecessors of activity C and can begin at the same time. In the absence of rule 5, we could represent this by Figure 29. However, because nodes 1 and 2 are connected by more than one arc, Figure 29 violates rule 5. By using a dummy activity (indicated by a dotted arc), as in Figure 30, we may represent the fact that A and B are both predecessors of C. Figure 30 ensures that activity C cannot begin until both A and B are completed, but it does not violate rule 5. Problem 10 at the end of this section illustrates how dummy activities may be needed to avoid violating rule 4.

Example 6 illustrates a project network.

EXAMPLE 6 | **Drawing a Project Network**

Widgetco is about to introduce a new product (product 3). One unit of product 3 is produced by assembling 1 unit of product 1 and 1 unit of product 2. Before production begins on either product 1 or 2, raw materials must be purchased and workers must be trained. Before products 1 and 2 can be assembled into product 3, the finished product 2 must be inspected. A list of activities and their predecessors and of the duration of each activity is given in Table 12. Draw a project diagram for this project.

TABLE 12
Duration of Activities and Predecessor Relationships for Widgetco

Activity	Predecessors	Duration (Days)
A = train workers	—	6
B = purchase raw materials	—	9
C = produce product 1	A, B	8
D = produce product 2	A, B	7
E = test product 2	D	10
F = assemble products 1 and 2	C, E	12

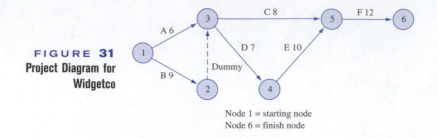

FIGURE 31
Project Diagram for
Widgetco

Node 1 = starting node
Node 6 = finish node

Solution   Observe that although we list only C and E as predecessors of F, it is actually true that activities A, B, and D must also be completed before F begins. C cannot begin until A and B are completed, and E cannot begin until D is completed, however, so it is redundant to state that A, B, and D are predecessors of F. Thus, in drawing the project network, we need only be concerned with the immediate predecessors of each activity.

The AOA network for this project is given in Figure 31 (the number above each arc represents activity duration in days). Node 1 is the beginning of the project, and node 6 is the finish node representing completion of the project. The dummy arc $(2, 3)$ is needed to ensure that rule 5 is not violated.

The two key building blocks in CPM are the concepts of early event time (ET) and late event time (LT) for an event.

DEFINITION ■   The **early event time** for node i, represented by $ET(i)$, is the earliest time at which the event corresponding to node i can occur.   ■

The **late event time** for node i, represented by $LT(i)$, is the latest time at which the event corresponding to node i can occur without delaying the completion of the project.   ■

Computation of Early Event Time

To find the early event time for each node in the project network, we begin by noting that because node 1 represents the start of the project, $ET(1) = 0$. We then compute $ET(2)$, $ET(3)$, and so on, stopping when ET(finish node) has been calculated. To illustrate how $ET(i)$ is calculated, suppose that for the segment of a project network in Figure 32, we have already determined that $ET(3) = 6$, $ET(4) = 8$, and $ET(5) = 10$. To determine $ET(6)$, observe that the earliest time that node 6 can occur is when the activities corresponding to arc $(3, 6)$, $(4, 6)$, and $(5, 6)$ have *all* been completed.

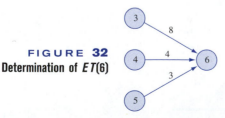

FIGURE 32
Determination of $ET(6)$

$$ET(6) = \max \begin{cases} ET(3) + 8 = 14 \\ ET(4) + 4 = 12 \\ ET(5) + 3 = 13 \end{cases}$$

Thus, the earliest time that node 6 can occur is 14, and $ET(6) = 14$.

From this example, it is clear that computation of $ET(i)$ requires (for $j < i$) knowledge of one or more of the $ET(j)$'s. This explains why we begin by computing the predecessor ETs. In general, if $ET(1), ET(2), \ldots, ET(i - 1)$ have been determined, then we compute $ET(i)$ as follows:

Step 1 Find each prior event to node i that is connected by an arc to node i. These events are the **immediate predecessors** of node i.

Step 2 To the ET for each immediate predecessor of the node i add the duration of the activity connecting the immediate predecessor to node i.

Step 3 $ET(i)$ equals the maximum of the sums computed in Step 2.

We now compute the $ET(i)$'s for Example 6. We begin by observing that $ET(1) = 0$. Node 1 is the only immediate predecessor of node 2, so $ET(2) = ET(1) + 9 = 9$. The immediate predecessors of node 3 are nodes 1 and 2. Thus,

$$ET(3) = \max \begin{cases} ET(1) + 6 = 6 \\ ET(2) + 0 = 9 \end{cases} = 9$$

Node 4's only immediate predecessor is node 3. Thus, $ET(4) = ET(3) + 7 = 16$. Node 5's immediate predecessors are nodes 3 and 4. Thus,

$$ET(5) = \max \begin{cases} ET(3) + 8 = 17 \\ ET(4) + 10 = 26 \end{cases} = 26$$

Finally, node 5 is the only immediate predecessor of node 6. Thus, $ET(6) = ET(5) + 12 = 38$. Because node 6 represents the completion of the project, we see that the earliest time that product 3 can be assembled is 38 days from now.

It can be shown that $ET(i)$ is the length of the longest path in the project network from node 1 to node i.

Computation of Late Event Time

To compute the $LT(i)$'s, we begin with the finish node and work backward (in descending numerical order) until we determine $LT(1)$. The project in Example 6 can be completed in 38 days, so we know that $LT(6) = 38$. To illustrate how $LT(i)$ is computed for nodes other than the finish node, suppose we are working with a network (Figure 33) for which we have already determined that $LT(5) = 24$, $LT(6) = 26$, and $LT(7) = 28$. In this situation, how can we compute $LT(4)$? If the event corresponding to node 4 occurs after $LT(5) - 3$, node 5 will occur after $LT(5)$, and the completion of the project will be delayed.

FIGURE 33
Computation of $LT(4)$

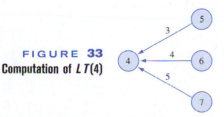

Similarly, if node 4 occurs after $LT(6) - 4$ or if node 4 occurs after $LT(7) - 5$, the completion of the project will be delayed. Thus,

$$LT(4) = \min \begin{cases} LT(5) - 3 = 21 \\ LT(6) - 4 = 22 = 21 \\ LT(7) - 5 = 23 \end{cases}$$

In general, if $LT(j)$ is known for $j > i$, we can find $LT(i)$ as follows:

Step 1   Find each node that occurs after node i and is connected to node i by an arc. These events are the **immediate successors** of node i.

Step 2   From the LT for each immediate successor to node i, subtract the duration of the activity joining the successor the node i.

Step 3 $LT(i)$ is the smallest of the differences determined in step 2.

We now compute the $LT(i)$'s for Example 6. Recall that $LT(6) = 38$. Because node 6 is the only immediate successor of node 5, $LT(5) = LT(6) - 12 = 26$. Node 4's only immediate successor is node 5. Thus, $LT(4) = LT(5) - 10 = 16$. Nodes 4 and 5 are immediate successors of node 3. Thus,

$$LT(3) = \min \begin{cases} LT(4) - 7 = 9 \\ LT(5) - 8 = 18 \end{cases}$$

Node 3 is the only immediate successor of node 2. Thus, $LT(2) = LT(3) - 0 = 9$. Finally, node 1 has nodes 2 and 3 as immediate successors. Thus,

$$LT(1) = \min \begin{cases} LT(3) - 6 = 3 \\ LT(2) - 9 = 0 \end{cases}$$

Table 13 summarizes our computations for Example 6. If $LT(i) = ET(i)$, any delay in the occurrence of node i will delay the completion of the project. For example, because $LT(4) = ET(4)$, any delay in the occurrence of node 4 will delay the completion of the project.

Total Float

Before the project is begun, the duration of an activity is unknown, and the duration of each activity used to construct the project network is just an estimate of the activity's actual completion time. The concept of total float of an activity can be used as a measure of how important it is to keep each activity's duration from greatly exceeding our estimate of its completion time.

TABLE 13
ET and *LT* for Widgetco

Node	$ET(i)$	$LT(i)$
1	0	0
2	9	9
3	9	9
4	16	16
5	26	26
6	38	38

DEFINITION ■ For an arbitrary arc representing activity (i, j), the **total float,** represented by $TF(i, j)$, of the activity represented by (i, j) is the amount by which the starting time of activity (i, j) could be delayed beyond its earliest possible starting time without delaying the completion of the project (assuming no other activities are delayed). ■

Equivalently, the total float of an activity is the amount by which the duration of the activity can be increased without delaying the completion of the project.

If we define t_{ij} to be the duration of activity (i, j), then $TF(i, j)$ can easily be expressed in terms of $LT(j)$ and $ET(i)$. Activity (i, j) begins at node i. If the occurrence of node i, or the duration of activity (i, j), is delayed by k time units, then activity (i, j) will be completed at time $ET(i) + k + t_{ij}$. Thus, the completion of the project will not be delayed if

$$ET(i) + k + t_{ij} \leq LT(j) \qquad \text{or} \qquad k \leq LT(j) - ET(i) - t_{ij}$$

Therefore,

$$TF(i, j) = LT(j) - ET(i) - t_{ij}$$

For Example 6, the $TF(i, j)$ are as follows:

$$
\begin{aligned}
\text{Activity B:} &\quad TF(1, 2) = LT(2) - ET(1) - 9 = 0 \\
\text{Activity A:} &\quad TF(1, 3) = LT(3) - ET(1) - 6 = 3 \\
\text{Activity D:} &\quad TF(3, 4) = LT(4) - ET(3) - 7 = 0 \\
\text{Activity C:} &\quad TF(3, 5) = LT(5) - ET(3) - 8 = 9 \\
\text{Activity E:} &\quad TF(4, 5) = LT(5) - ET(4) - 10 = 0 \\
\text{Activity F:} &\quad TF(5, 6) = LT(6) - ET(5) - 12 = 0 \\
\text{Dummy activity:} &\quad TF(2, 3) = LT(3) - ET(2) - 0 = 0
\end{aligned}
$$

Finding a Critical Path

If an activity has a total float of zero, then any delay in the start of the activity (or the duration of the activity) will delay the completion of the project. In fact, increasing the duration of an activity by Δ days will increase the length of the project by Δ days. Such an activity is critical to the completion of the project on time.

DEFINITION ■ Any activity with a total float of zero is a **critical activity.** ■

A path from node 1 to the finish node that consists entirely of critical activities is called a **critical path.** ■

In Figure 31, activities B, D, E, F, and the dummy activity are critical activities and the path 1–2–3–4–5–6 is the critical path (it is possible for a network to have more than one critical path). A critical path in any project network is the longest path from the start node to the finish node (see Problem 2 in Section 8.5).

Any delay in the duration of a critical activity will delay the completion of the project, so it is advisable to monitor closely the completion of critical activities.

Free Float

As we have seen, the total float of an activity can be used as a measure of the flexibility in the duration of an activity. For example, activity A can take up to 3 days longer than its scheduled duration of 6 days without delaying the completion of the project. Another measure of the flexibility available in the duration of an activity is free float.

DEFINITION ■ The **free float** of the activity corresponding to arc (i, j), denoted by $FF(i, j)$, is the amount by which the starting time of the activity corresponding to arc (i, j) (or the duration of the activity) can be delayed without delaying the start of any later activity beyond its earliest possible starting time. ■

Suppose the occurrence of node i, or the duration of activity (i, j), is delayed by k units. Then the earliest that node j can occur is $ET(i) + t_{ij} + k$. Thus, if $ET(i) + t_{ij} + k \leq ET(j)$, or $k \leq ET(j) - ET(i) - t_{ij}$, then node j will not be delayed. If node j is not delayed, then no other activities will be delayed beyond their earliest possible starting times. Therefore,

$$FF(i, j) = ET(j) - ET(i) - t_{ij}$$

For Example 6, the $FF(i, j)$ are as follows:

$$\begin{aligned}
\text{Activity B:} \quad & FF(1, 2) = 9 - 0 - 9 = 0 \\
\text{Activity A:} \quad & FF(1, 3) = 9 - 0 - 6 = 3 \\
\text{Activity D:} \quad & FF(3, 4) = 16 - 9 - 7 = 0 \\
\text{Activity C:} \quad & FF(3, 5) = 26 - 9 - 8 = 9 \\
\text{Activity E:} \quad & FF(4, 5) = 26 - 16 - 10 = 0 \\
\text{Activity F:} \quad & FF(5, 6) = 38 - 26 - 12 = 0
\end{aligned}$$

For example, because the free float for activity C is 9 days, a delay in the start of activity C (or in the occurrence of node 3) or a delay in the duration of activity C of more than 9 days will delay the start of some later activity (in this case, activity F).

Using Linear Programming to Find a Critical Path

Although the previously described method for finding a critical path in a project network is easily programmed on a computer, linear programming can also be used to determine the length of the critical path. Define

$$x_j = \text{the time that the event corresponding to node } j \text{ occurs}$$

For each activity (i, j), we know that before node j occurs, node i must occur and activity (i, j) must be completed. This implies that for each arc (i, j) in the project network, $x_j \geq x_i + t_{ij}$. Let F be the node that represents completion of the project. Our goal is to minimize the time required to complete the project, so we use an objective function of $z = x_F - x_1$.

To illustrate how linear programming can be used to find the length of the critical path, we apply the preceding approach to Example 6. The appropriate LP is

$$\begin{aligned}
\min z = x_6 - x_1 \\
\text{s.t.} \quad x_3 \geq x_1 + 6 \quad & \text{(Arc } (1, 3) \text{ constraint)} \\
x_2 \geq x_1 + 9 \quad & \text{(Arc } (1, 2) \text{ constraint)} \\
x_5 \geq x_3 + 8 \quad & \text{(Arc } (3, 5) \text{ constraint)} \\
x_4 \geq x_3 + 7 \quad & \text{(Arc } (3, 4) \text{ constraint)}
\end{aligned}$$

$$x_5 \geq x_4 + 10 \qquad \text{(Arc (4, 5) constraint)}$$
$$x_6 \geq x_5 + 12 \qquad \text{(Arc (5, 6) constraint)}$$
$$x_3 \geq x_2 \qquad \text{(Arc (2, 3) constraint)}$$

All variables urs

An optimal solution to this LP is $z = 38$, $x_1 = 0$, $x_2 = 9$, $x_3 = 9$, $x_4 = 16$, $x_5 = 26$, and $x_6 = 38$. This indicates that the project can be completed in 38 days.

This LP has many alternative optimal solutions. In general, the value of x_i in any optimal solution may assume any value between $ET(i)$ and $LT(i)$. All optimal solutions to this LP, however, will indicate that the length of any critical path is 38 days.

A critical path for this project network consists of a path from the start of the project to the finish in which each arc in the path corresponds to a constraint having a dual price of -1. From the LINDO output in Figure 34, we find, as before, that 1–2–3–4–5–6 is a critical path. For each constraint with a dual price of -1, increasing the duration of the activity corresponding to that constraint by Δ days will increase the duration of the project by Δ days. For example, an increase of Δ days in the duration of activity B will increase the duration of the project by Δ days. This assumes that the current basis remains optimal.

Crashing the Project

In many situations, the project manager must complete the project in a time that is less than the length of the critical path. For instance, suppose Widgetco believes that to have any chance of being a success, product 3 must be available for sale before the competitor's product hits the market. Widgetco knows that the competitor's product is scheduled to hit the market 26 days from now, so Widgetco must introduce product 3 within 25 days. Because the critical path in Example 6 has a length of 38 days, Widgetco will have to expend additional resources to meet the 25-day project deadline. In such a situation, linear programming can often be used to determine the allocation of resources that minimizes the cost of meeting the project deadline.

Suppose that by allocating additional resources to an activity, Widgetco can reduce the duration of any activity by as many as 5 days. The cost per day of reducing the duration of an activity is shown in Table 14. To find the minimum cost of completing the project by the 25-day deadline, define variables $A, B, C, D, E,$ and F as follows:

$$A = \text{number of days by which duration of activity } A \text{ is reduced}$$
$$\vdots \qquad\qquad\qquad \vdots$$
$$F = \text{number of days by which duration of activity } F \text{ is reduced}$$
$$x_j = \text{time that the event corresponding to node } j \text{ occurs}$$

Then Widgetco should solve the following LP:

$$\min z = 10A + 20B + 3C + 30D + 40E + 50F$$
$$\text{s.t.} \quad A \leq 5$$
$$B \leq 5$$
$$C \leq 5$$
$$D \leq 5$$
$$E \leq 5$$
$$F \leq 5$$

```
MIN     X6 - X1
SUBJECT TO
        2) - X1 + X3 >=  6
        3) - X1 + X2 >=  9
        4) - X3 + X5 >=  8
        5) - X3 + X4 >=  7
        6)   X5 - X4 >= 10
        7)   X6 - X5 >= 12
        8)   X3 - X2 >=  0
END


        LP OPTIMUM FOUND  AT STEP      7

             OBJECTIVE FUNCTION VALUE

    1)         38.0000000

    VARIABLE          VALUE        REDUCED COST
        X6         38.000000        0.000000
        X1          0.000000        0.000000
        X3          9.000000        0.000000
        X2          9.000000        0.000000
        X5         26.000000        0.000000
        X4         16.000000        0.000000

       ROW     SLACK OR SURPLUS    DUAL PRICES
        2)          3.000000        0.000000
        3)          0.000000       -1.000000
        4)          9.000000        0.000000
        5)          0.000000       -1.000000
        6)          0.000000       -1.000000
        7)          0.000000       -1.000000
        8)          0.000000       -1.000000

    NO. ITERATIONS=         7


    RANGES IN WHICH THE BASIS IS UNCHANGED

                      OBJ COEFFICIENT RANGES
    VARIABLE        CURRENT      ALLOWABLE      ALLOWABLE
                    COEF         INCREASE       DECREASE
        X6        1.000000       INFINITY       0.000000
        X1       -1.000000       INFINITY       0.000000
        X3        1.000000       INFINITY       0.000000
        X2        1.000000       INFINITY       0.000000
        X5        1.000000       INFINITY       0.000000
        X4        1.000000       INFINITY       0.000000

                      RIGHTHAND SIDE RANGES
       ROW         CURRENT      ALLOWABLE      ALLOWABLE
                   RHS          INCREASE       DECREASE
        2         6.000000      3.000000       INFINITY
        3         9.000000      INFINITY       3.000000
        4         8.000000      9.000000       INFINITY
        5         7.000000      INFINITY       9.000000
        6        10.000000      INFINITY       9.000000
        7        12.000000      INFINITY      38.000000
        8         0.000000      INFINITY       3.000000
```

FIGURE 34
LINDO Output
for Widgetco

TABLE 14

A	B	C	D	E	F
$10	$20	$3	$30	$40	$50

$$x_2 \geq x_1 + 9 - B \qquad \text{(Arc (1, 2) constraint)}$$
$$x_3 \geq x_1 + 6 - A \qquad \text{(Arc (1, 3) constraint)}$$
$$x_5 \geq x_3 + 8 - C \qquad \text{(Arc (3, 5) constraint)}$$
$$x_4 \geq x_3 + 7 - D \qquad \text{(Arc (3, 4) constraint)}$$
$$x_5 \geq x_4 + 10 - E \qquad \text{(Arc (4, 5) constraint)}$$
$$x_6 \geq x_5 + 12 - F \qquad \text{(Arc (5, 6) constraint)}$$
$$x_3 \geq x_2 + 0 \qquad \text{(Arc (2, 3) constraint)}$$
$$x_6 - x_1 \leq 25$$
$$A, B, C, D, E, F \geq 0, x_j \text{urs}$$

The first six constraints stipulate that the duration of each activity can be reduced by at most 5 days. As before, the next seven constraints ensure that event j cannot occur until after node i occurs and activity (i, j) is completed. For example, activity B (arc (1, 2)) now has a duration of $9 - B$. Thus, we need the constraint $x_2 \geq x_1 + (9 - B)$. The constraint $x_6 - x_1 \leq 25$ ensures that the project is completed within the 25-day deadline. The objective function is the total cost incurred in reducing the duration of the activities. An optimal solution to this LP is $z = \$390$, $x_1 = 0$, $x_2 = 4$, $x_3 = 4$, $x_4 = 6$, $x_5 = 13$, $x_6 = 25$, $A = 2$, $B = 5$, $C = 0$, $D = 5$, $E = 3$, $F = 0$. After reducing the durations of projects B, A, D, and E by the given amounts, we obtain the project network pictured in Figure 35. The reader should verify that A, B, D, E, and F are critical activities and that 1–2–3–4–5–6 and 1–3–4–5–6 are both critical paths (each having length 25). Thus, the project deadline of 25 days can be met for a cost of $390.

Using LINGO to Determine the Critical Path

Many computer packages (such as Microsoft Project) enable the user to determine (among other things!) the critical path(s) and critical activities in a project network. You can always find a critical path and critical activities using LINDO, but LINGO makes it very easy to communicate the necessary information to the computer. The following LINGO program (file Widget1.lng) generates the objective function and constraints needed to find the critical path for the project network of Example 6 via linear programming.

Widget1.lng

```
MODEL:
  1]SETS:
  2]NODES/1..6/:TIME;
  3]ARCS(NODES,NODES)/
  4]1,2   1,3   2,3   3,4   3,5   4,5   5,6/:DUR;
  5]ENDSETS
  6]MIN=TIME(6)-TIME(1);
  7]@FOR(ARCS(I,J):TIME(J)>TIME(I)+DUR(I,J));
  8]DATA:
  9]DUR=9,6,0,7,8,10,12;
 10]ENDDATA
END
```

Line 1 begins the SETS portion of the program. In line 2, we define the six nodes of the project network and associate with each node a time that the events corresponding to

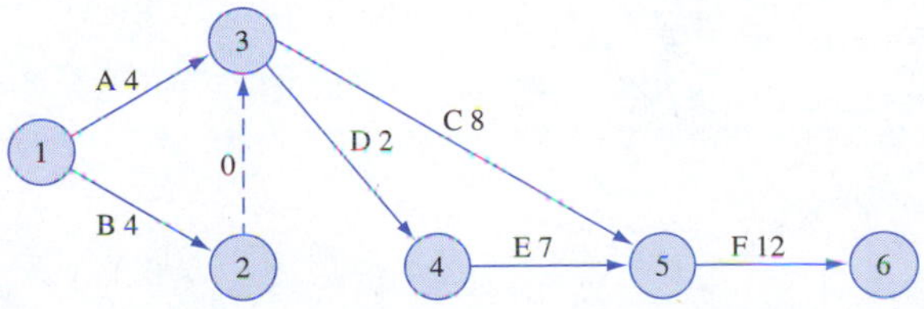

FIGURE 35
Duration of Activities
After Crashing

```
MIN      -ET(1 + ET(6
SUBJECT TO
 2) - ET(1 + ET(2 >=   9
 3) - ET(1 + ET(3 >=   6
 4) - ET(2 + ET(3 >=   0
 5) - ET(3 + ET(4 >=   7
 6) - ET(3 + ET(5 >=   8
 7) - ET(4 + ET(5 >=  10
 8) - ET(5 + ET(6 >=  12
END

LP OPTIMUM FOUND AT STEP      6
OBJECTIVE VALUE =   38.0000000

              VARIABLE        VALUE            REDUCED COST
                ET( 1)      0.0000000E+00       0.0000000E+00
                ET( 2)      9.000000            0.0000000E+00
                ET( 3)      9.000000            0.0000000E+00
                ET( 4)      16.00000            0.0000000E+00
                ET( 5)      26.00000            0.0000000E+00
                ET( 6)      38.00000            0.0000000E+00
             DUR( 1, 2)     9.000000            0.0000000E+00
             DUR( 1, 3)     6.000000            0.0000000E+00
             DUR( 2, 3)     0.0000000E+00       0.0000000E+00
             DUR( 3, 4)     7.000000            0.0000000E+00
             DUR( 3, 5)     8.000000            0.0000000E+00
             DUR( 4, 5)     10.00000            0.0000000E+00
             DUR( 5, 6)     12.00000            0.0000000E+00

                ROW      SLACK OR SURPLUS      DUAL PRICE
                 1          38.00000            1.000000
                 2          0.0000000E+00      -1.000000
                 3          3.000000            0.0000000E+00
                 4          0.0000000E+00      -1.000000
                 5          0.0000000E+00      -1.000000
                 6          9.000000            0.0000000E+00
                 7          0.0000000E+00      -1.000000
                 8          0.0000000E+00      -1.000000
```

FIGURE 36

the node occurs. For example, TIME(3) represents the time when activities A and B have just been completed. In line 3, we generate the arcs in the project network by listing them (separated by spaces). For example, arc (3, 4) represents activity D. In line 4, we associate a duration (DUR) of each activity with each arc. Line 5 ends the SETS section of the program.

Line 6 specifies the objective, to minimize the time it takes to complete the project. For each arc defined in line 3, line 7 creates a constraint analagous to $x_j \geq x_i + t_{ij}$.

Line 8 begins the DATA section of the program. In line 9, we list the duration of each activity. Line 10 concludes the data entry and the **END** statement concludes the program. The output from this LINGO model is given in Figure 36, where by following the arcs corresponding to constraints having dual prices of -1, we find the critical path to be 1–2–3–4–5–6.

To find the critical path in any network we would begin by listing the nodes, arcs, and activity durations in our program. Then we would modify the objective function created by line 6 to reflect the number of nodes in the network. For example, if there were 10 nodes in the project network, we would change line 6 to **MIN**=TIME(10)–TIME(1); and we would be ready to go!

Widget2.lng The following LINGO program (file Widget2.lng) enables the user to determine the critical path and total float at each node for Example 6 without using linear programming.

```
MODEL:
  1]MODEL:
  2]SETS:
  3]NODES/1..6/:ET,LT;
  4]ARCS(NODES,NODES)/1,2  1,3  2,3  3,4  3,5  4,5  5,6/:DUR,TFLOAT;
  5]ENDSETS
  6]DATA:
  7]DUR = 9,6,0,7,8,10,12;
```

```
 8]ENDDATA
 9]ET(1)=0;
10]@FOR(NODES(J) | J#GT#1:
11]ET(J)  = @MAX(ARCS(I,J): ET(I)+DUR(I,J)););
12]LNODE=@SIZE(NODES);
13]LT(LNODE)  = ET(LNODE);
14]@FOR(NODES(I) | I#LT#LNODE:
15]LT(I)  = @MIN(ARCS(I,J): LT(J) - DUR(I,J)););
16]@FOR(ARCS(I,J):TFLOAT(I,J)=LT(J)-ET(I)-DUR(I,J));
END
```

In line 3, we define the nodes of the project network and associate an early event time (ET) and late event time (LT) with each node. We define the arcs of the project network by listing them in line 4. With each arc we associate the duration of the arc's activity and the total float of the activity. In line 7, we input the duration of each activity.

To begin the computation of the ET(J)'s for each node, we set ET(1) = 0 in line 9. In lines 10–11, we compute ET(J) for all other nodes. For J > 1 ET(J) is the maximum value of ET(I) + DUR(I, J) for all (I, J) such that (I, J) is an arc in the network. By using the **@SIZE** function, which returns the number of elements in a set, we identify the finish node in the network in line 12. Thus, line 12 defines node 6 as the last node. In line 13, we set LT(6) = ET(6). Lines 14–15 work backward from node 6 toward node 1 to compute the LT(I)'s. For every node I other than the last node (6), LT(I) is the minimum of LT(J) − DUR (I, J), where the minimum is taken over all (I, J) such that (I, J) is an arc in the project network.

Finally, line 16 computes the total float for each activity (I, J) from total float for activity (I, J) = LT(Node J) − ET(Node I) − Duration (I, J). All activities whose total float equals 0 are critical activities.

After inputting a list of nodes, arcs, and activity durations we can use this program to analyze any project network (without changing any of lines 9–16). It is also easy to write a LINGO program that can be used to crash the network (see Problem 14).

PERT: Program Evaluation and Review Technique

CPM assumes that the duration of each activity is known with certainty. For many projects, this is clearly not applicable. PERT is an attempt to correct this shortcoming of CPM by modeling the duration of each activity as a random variable. For each activity, PERT requires that the project manager estimate the following three quantities:

a = estimate of the activity's duration
under the most favorable conditions

b = estimate of the activity's duration
under the least favorable conditions

m = most likely value for the activity's duration

Let $\mathbf{T}_{ij}$ (random variables are printed in boldface) be the duration of activity (i, j). PERT requires the assumption that $\mathbf{T}_{ij}$ follows a beta distribution. The specific definition of a beta distribution need not concern us, but it is important to realize that it can approximate a wide range of random variables, including many positively skewed, negatively skewed, and symmetric random variables. If $\mathbf{T}_{ij}$ follows a beta distribution, then it can be shown that the mean and variance of $\mathbf{T}_{ij}$ may be approximated by

$$E(\mathbf{T}_{ij}) = \frac{a + 4m + b}{6} \tag{4}$$

$$\text{var}\,\mathbf{T}_{ij} = \frac{(b - a)^2}{36} \tag{5}$$

PERT requires the assumption that the durations of all activities are independent. Then for any path in the project network, the mean and variance of the time required to complete the activities on the path are given by

$$\sum_{(i,\ j)\in\text{path}} E(\mathbf{T}_{ij}) = \text{expected duration of activities on any path} \tag{6}$$

$$\sum_{(i,\ j)\in\text{path}} \text{var}\mathbf{T}_{ij} = \text{variance of duration of activities on any path} \tag{7}$$

Let **CP** be the random variable denoting the total duration of the activities on a critical path found by CPM. PERT assumes that the critical path found by CPM contains enough activities to allow us to invoke the Central Limit Theorem and conclude that

$$\mathbf{CP} = \sum_{(i,\ j)\in\text{critical path}} \mathbf{T}_{ij}$$

is normally distributed. With this assumption, (4)–(7) can be used to answer questions concerning the probability that the project will be completed by a given date. For example, suppose that for Example 6, a, b, and m for each activity are shown in Table 15. Now (4) and (5) yield

$$E(\mathbf{T}_{12}) = \frac{\{5 + 13 + 36\}}{6} = 9 \qquad \text{var}\mathbf{T}_{12} = \frac{(13 - 5)^2}{36} = 1.78$$

$$E(\mathbf{T}_{13}) = \frac{\{2 + 10 + 24\}}{6} = 6 \qquad \text{var}\mathbf{T}_{13} = \frac{(10 - 2)^2}{36} = 1.78$$

$$E(\mathbf{T}_{35}) = \frac{\{3 + 13 + 32\}}{6} = 8 \qquad \text{var}\mathbf{T}_{35} = \frac{(13 - 3)^2}{36} = 2.78$$

$$E(\mathbf{T}_{34}) = \frac{\{1 + 13 + 28\}}{6} = 7 \qquad \text{var}\mathbf{T}_{34} = \frac{(13 - 1)^2}{36} = 4$$

$$E(\mathbf{T}_{45}) = \frac{\{8 + 12 + 40\}}{6} = 10 \qquad \text{var}\mathbf{T}_{45} = \frac{(12 - 8)^2}{36} = 0.44$$

$$E(\mathbf{T}_{56}) = \frac{\{9 + 15 + 48\}}{6} = 12 \qquad \text{var}\mathbf{T}_{56} = \frac{(15 - 9)^2}{36} = 1$$

Of course, the fact that arc (2, 3) is a dummy arc yields

$$E(\mathbf{T}_{23}) = \text{var } \mathbf{T}_{23} = 0$$

Recall that the critical path for Example 6 was 1–2–3–4–5–6. From Equations (6) and (7),

$$E(\mathbf{CP}) = 9 + 0 + 7 + 10 + 12 = 38$$
$$\text{var}\mathbf{CP} = 1.78 + 0 + 4 + 0.44 + 1 = 7.22$$

Then the standard deviation for **CP** is $(7.22)^{1/2} = 2.69$.

TABLE 15

a, *b*, and *m* for Activities in Widgeto

Activity	a	b	m
(1, 2)	5	13	9
(1, 3)	2	10	6
(3, 5)	3	13	8
(3, 4)	1	13	7
(4, 5)	8	12	10
(5, 6)	9	15	12

Applying the assumption that **CP** is normally distributed, we can answer questions such as the following: What is the probability that the project will be completed within 35 days? To answer this question, we must also make the following assumption: *No matter what the durations of the project's activities turn out to be, 1–2–3–4–5–6 will be a critical path.* This assumption implies that the probability that the project will be completed within 35 days is just $P(\mathbf{CP} \le 35)$. Standardizing and applying the assumption that **CP** is normally distributed, we find that **Z** is a standardized normal random variable with mean 0 and variance 1. The cumulative distribution function for a normal random variable is tabulated in Table 16. For example, $P(\mathbf{Z} \le -1) = 0.1587$ and $P(\mathbf{Z} \le 2) = 0.9772$. Thus,

$$P(\mathbf{CP} \le 35) = P\left(\frac{\mathbf{CP} - 38}{2.69} \le \frac{35 - 38}{2.69} \right) = P(\mathbf{Z} \le -1.12) = .13$$

where $F(-1.12) = .13$ may be obtained using the NORMSDIST function in Excel. Entering the formula =NORMSDIST(x) returns the probability that a standard normal random variable with mean 0 and standard deviation 1 is less than or equal to x. For example =NORMDIST(-1.12) yields .1313.

Difficulties with PERT

There are several difficulties with PERT:

1 The assumption that the activity durations are independent is difficult to justify.

2 Activity durations may not follow a beta distribution.

3 The assumption that the critical path found by CPM will always be the critical path for the project may not be justified.

The last difficulty is the most serious. For example, in our analysis of Example 6, we assumed that 1–2–3–4–5–6 would always be the critical path. If, however, activity A were significantly delayed and activity B were completed ahead of schedule, then the critical path might be 1–3–4–5–6.

Here is a more concrete example of the fact that (because of the uncertain duration of activities) the critical path found by CPM may not actually be the path that determines the completion date of the project. Consider the simple project network in Figure 37. As-

TABLE 16
_a_, _b_, and _m_ for Figure 37

Activity	a	b	m
A	1	9	5
B	6	14	10
C	5	7	6
D	7	9	8

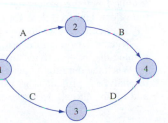

FIGURE 37
Project Network to Illustrate Difficulties with PERT

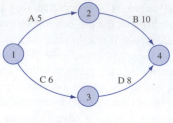

TABLE 17
Probability That Each Arc
Is on a Critical Path

Activity	Probability
A	$\frac{17}{27}$
B	$\frac{17}{27}$
C	$\frac{12}{27}$
D	$\frac{12}{27}$

sume that for each activity in Table 16, *a*, *b*, and *m* each occur with probability $\frac{1}{3}$. If CPM were applied (using the expected duration of each activity as the duration of the activity), then we would obtain the network in Figure 38. For this network, the critical path is 1–2–4. In actuality, however, the critical path could be 1–3–4. For example, if the optimistic duration of B (6 days) occurred and all other activities had a duration *m*, then 1–3–4 would be the critical path in the network. If we assume that the durations of the four activities are independent random variables, then using elementary probability (see Problem 11 at the end of this section), it can be shown that there is a $\frac{10}{27}$ probability that 1–3–4 is the critical path, a $\frac{15}{27}$ chance that 1–2–4 is the critical path, and a $\frac{2}{27}$ chance that 1–2–4 and 1–3–4 will both be critical paths. This example shows that one must be cautious in designating an activity as critical. In this situation, the probability that each activity is actually a critical activity is shown in Table 17.

PROBLEMS

Group A

1 What problem would arise if the network in Figure 39 were a portion of a project network?

2 A company is planning to manufacture a product that consists of three parts (A, B, and C). The company anticipates that it will take 5 weeks to design the three parts and to determine the way in which these parts must be assembled to make the final product. Then the company estimates that it will take 4 weeks to make part A, 5 weeks to make part B, and 3 weeks to make part C. The company must test part A after it is completed (this takes 2 weeks). The assembly line process will then proceed as follows: assemble parts A and B (2 weeks) and then attach part C (1 week). Then the final product must undergo 1 week of testing. Draw the project network and find the critical path, total float, and free float for each activity. Also set up the LP that could be used to find the critical path.

When determining the critical path in Problems 3 and 4, assume that *m* = activity duration.

When the duration of activities is uncertain, the best way to analyze a project is to use a Monte Carlo simulation add-in for Excel. In *Stochastic Models in Operations*

FIGURE 39

Network for Problem 1

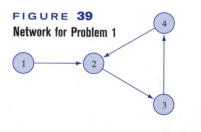

Research: Application and Algorithms, we will show how to use the Excel add-in @RISK to perform Monte Carlo simulations. With @RISK, we can easily determine the probability a project is completed on time and determine the probability that each activity is critical.

3 Consider the project network in Figure 40. For each activity, you are given the estimates of *a*, *b*, and *m* in Table 18. Determine the critical path for this network, the total float for each activity, the free float for each activity, and the probability that the project is completed within 40 days. Also set up the LP that could be used to find the critical path.

4 The promoter of a rock concert in Indianapolis must perform the tasks shown in Table 19 before the concert can be held (all durations are in days).

  a Draw the project network.

  b Determine the critical path.

  c If the advance promoter wants to have a 99% chance of completing all preparations by June 30, when should work begin on finding a concert site?

  d Set up the LP that could be used to find the project's critical path.

5 Consider the (simplified) list of activities and predecessors that are involved in building a house (Table 20).

FIGURE 40
Network for Problem 3

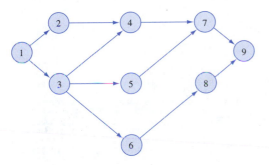

TABLE 18

Activity	a	b	m
(1, 2)	4	8	6
(1, 3)	2	8	4
(2, 4)	1	7	3
(3, 4)	6	12	9
(3, 5)	5	15	10
(3, 6)	7	18	12
(4, 7)	5	12	9
(5, 7)	1	3	2
(6, 8)	2	6	3
(7, 9)	10	20	15
(8, 9)	6	11	9

TABLE 19

Activity	Description	Immediate Predecessors	a	b	m
A	Find site	—	2	4	3
B	Find engineers	A	1	3	2
C	Hire opening act	A	2	10	6
D	Set radio and TV ads	C	1	3	2
E	Set up ticket agents	A	1	5	3
F	Prepare electronics	B	2	4	3
G	Print advertising	C	3	7	5
H	Set up transportation	C	0.5	1.5	1
I	Rehearsals	F, H	1	2	1.5
J	Last-minute details	I	1	3	2

TABLE 20

Activity	Description	Immediate Predecessors	Duration (Days)
A	Build foundation	—	5
B	Build walls and ceilings	A	8
C	Build roof	B	10
D	Do electrical wiring	B	5
E	Put in windows	B	4
F	Put on siding	E	6
G	Paint house	C, F	3

  a Draw a project network, determine the critical path, find the total float for each activity, and find the free float for each activity.

  b Suppose that by hiring additional workers, the duration of each activity can be reduced. The costs per day of reducing the duration of the activities are given in Table 21. Write down the LP to be solved to minimize the total cost of completing the project within 20 days.

6 Horizon Cable is about to expand its cable TV offerings in Smalltown by adding MTV and other exciting stations. The activities in Table 22 must be completed before the service expansion is completed.

  a Draw the project network and determine the critical path for the network, the total float for each activity, and the free float for each activity.

  b Set up the LP that can be used to find the project's critical path.

7 When an accounting firm audits a corporation, the first phase of the audit involves obtaining "knowledge of the business." This phase of the audit requires the activities in Table 23.

  a Draw the project network and determine the critical path for the network, the total float for each activity, and the free float for each activity. Also set up the LP that can be used to find the project's critical path.

TABLE 21

Activity	Cost per Day of Reducing Duration of Activity ($)	Maximum Possible Reduction in Duration of Activity (Days)
Foundation	30	2
Walls and ceiling	15	3
Roof	20	1
Electrical wiring	40	2
Windows	20	2
Siding	30	3
Paint	40	1

TABLE 24

Activity	Cost per Day of Reducing Duration of Activity ($)	Maximum Possible Reduction in Duration of Activity (Days)
A	100	3
B	80	4
C	60	5
D	70	2
E	30	4
F	20	4
G	50	4

TABLE 22

Activity	Description	Immediate Predecessors	Duration (Weeks)
A	Choose stations	—	2
B	Get town council to approve expansion	A	4
C	Order converters needed to expand service	B	3
D	Install new dish to receive new stations	B	2
E	Install converters	C, D	10
F	Change billing system	B	4

FIGURE 41
LINDO Output for Problem 8

```
MIN      X6 - X1
   SUBJECT TO
         2)  - X1 + X2 >=   5
         3)  - X2 + X3 >=   8
         4)  - X3 + X4 >=   4
         5)  - X3 + X5 >=  10
         6)  - X4 + X5 >=   6
         7)    X6 - X3 >=   5
         8)    X6 - X5 >=   3
   END


      LP OPTIMUM FOUND  AT STEP     6

          OBJECTIVE FUNCTION VALUE

    1)        26.0000000

     VARIABLE        VALUE        REDUCED COST
         X6        26.000000         0.000000
         X1         0.000000         0.000000
         X2         5.000000         0.000000
         X3        13.000000         0.000000
         X4        17.000000         0.000000
         X5        23.000000         0.000000


        ROW      SLACK OR SURPLUS   DUAL PRICES
         2)         0.000000        -1.000000
         3)         0.000000        -1.000000
         4)         0.000000        -1.000000
         5)         0.000000         0.000000
         6)         0.000000        -1.000000
         7)         8.000000         0.000000
         8)         0.000000        -1.000000

     NO. ITERATIONS=        6


   RANGES IN WHICH THE BASIS IS UNCHANGED

                      OBJ COEFFICIENT RANGES
   VARIABLE      CURRENT      ALLOWABLE      ALLOWABLE
                  COEF        INCREASE       DECREASE
       X6       1.000000      INFINITY       0.000000
       X1      -1.000000      INFINITY       0.000000
       X2       0.000000      INFINITY       0.000000
       X3       0.000000      INFINITY       0.000000
       X4       0.000000      INFINITY       0.000000
       X5       0.000000      INFINITY       0.000000

                     RIGHTHAND SIDE RANGES
   ROW         CURRENT       ALLOWABLE      ALLOWABLE
                RHS          INCREASE       DECREASE
     2       5.000000        INFINITY       5.000000
     3       8.000000        INFINITY      13.000000
     4       4.000000        0.000000       8.000000
     5      10.000000        INFINITY       0.000000
     6       6.000000        0.000000       8.000000
     7       5.000000        8.000000       INFINITY
     8       3.000000        INFINITY       8.000000
```

TABLE 23

Activity	Description	Immediate Predecessors	Duration (Days)
A	Determining terms of engagement	—	3
B	Appraisal of auditability risk and materiality	A	6
C	Identification of types of transactions and possible errors	A	14
D	Systems description	C	8
E	Verification of systems description	D	4
F	Evaluation of internal controls	B, E	8
G	Design of audit approach	F	9

b Assume that the project must be completed in 30 days. The duration of each activity can be reduced by incurring the costs shown in Table 24. Formulate an LP that can be used to minimize the cost of meeting the project deadline.

8 The LINDO output in Figure 41 can be used to determine the critical path for Problem 5. Use this output to do the following:

a Draw the project diagram.

b Determine the length of the critical path and the critical activities for this project.

9 Explain why an activity's free float can never exceed the activity's total float.

10 A project is complete when activities A–E are completed. The predecessors of each activity are shown in Table 25. Draw the appropriate project diagram. (*Hint:* Don't violate rule 4.)

11 Determine the probabilities that 1–2–4 and 1–3–4 are critical paths for Figure 37.

12 Given the information in Table 26, **(a)** draw the appropriate project network, and **(b)** find the critical path.

13 The government is going to build a high-speed computer in Austin, Texas. Once the computer is designed (D), we can select the exact site (S), the building contractor (C), and the operating personnel (P). Once the site is selected, we can begin erecting the building (B). We can start manufacturing the computer (COM) and preparing the operations manual (M) only after contractor is selected. We can begin training the computer operators (T) when the operating manual and personnel selection are completed. When the computer and the building are both finished, the computer may be installed (I). Then the computer is considered operational. Draw a project network that could be used to determine when the project is operational.

14 Write a LINGO program that can be used to crash the project network of Example 6 with the crashing costs given in Table 14.

15 Consider the project diagram in Figure 42. This project must be completed in 90 days. The time required to complete each activity can be reduced by up to five days at the costs given in Table 27.

Formulate an LP whose solution will enable us to minimize the cost of completing the project in 90 days.

16–17 Find the critical path, total float, and free float for each activity in the project networks of Figures 43 and 44.

TABLE 25

Activity	Predecessors
A	—
B	A
C	A
D	B
E	B, C

TABLE 26

Activity	Immediate Predecessors	Duration (Days)
A	—	3
B	—	3
C	—	1
D	A, B	3
E	A, B	3
F	B, C	2
G	D, E	4
H	E	3

FIGURE 42

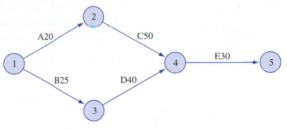

TABLE 27

Activity	Cost of Reducing Activities Duration by 1 Day ($)
A	300
B	200
C	350
D	260
E	320

FIGURE 43

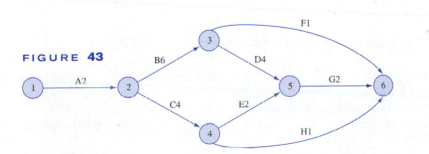

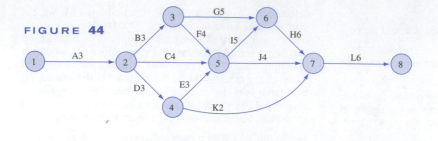

FIGURE 44

8.5 Minimum-Cost Network Flow Problems

The transportation, assignment, transshipment, shortest path, maximum flow, and CPM problems are all special cases of the minimum-cost network flow problem (MCNFP). Any MCNFP can be solved by a generalization of the transportation simplex called the **network simplex.**

To define an MCNFP, let

x_{ij} = number of units of flow sent from node i to node j through arc (i, j)

b_i = net supply (outflow − inflow) at node i

c_{ij} = cost of transporting 1 unit of flow from node i to node j via arc (i, j)

L_{ij} = lower bound on flow through arc (i, j)
(if there is no lower bound, let $L_{ij} = 0$)

U_{ij} = upper bound on flow through arc (i, j)
(if there is no upper bound, let $U_{ij} = \infty$)

Then the MCNFP may be written as

$$\min \sum_{\text{all arcs}} c_{ij} x_{ij}$$

$$\text{s.t.} \quad \sum_j x_{ij} - \sum_k x_{ki} = b_i \qquad \text{(for each node } i \text{ in the network)} \tag{8}$$

$$L_{ij} \leq x_{ij} \leq U_{ij} \qquad \text{(for each arc in the network)} \tag{9}$$

Constraints (8) stipulate that the net flow out of node i must equal b_i. Constraints (8) are referred to as the **flow balance equations** for the network. Constraints (9) ensure that the flow through each arc satisfies the arc capacity restrictions. In all our previous examples, we have set $L_{ij} = 0$.

Let us show that transportation and maximum-flow problems are special cases of the minimum-cost network flow problem.

Formulating a Transportation Problem as an MCNFP

Consider the transportation problem in Table 28. Nodes 1 and 2 are the two supply points, and nodes 3 and 4 are the two demand points. Then $b_1 = 4$, $b_2 = 5$, $b_3 = -6$, and $b_4 = -3$. The network corresponding to this transportation problem contains arcs $(1, 3)$, $(1, 4)$, $(2, 3)$, and $(2, 4)$ (see Figure 45). The LP for this transportation problem may be written as shown in Table 29.

The first two constraints are the supply constraints, and the last two constraints are (after being multiplied by −1) the demand constraints. Because this transportation problem

TABLE 28

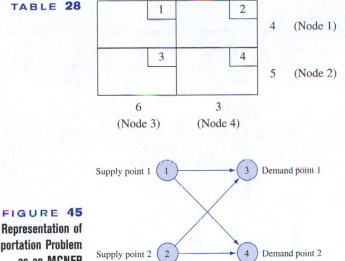

1	2	4 (Node 1)
3	4	5 (Node 2)
6	3	
(Node 3)	(Node 4)	

FIGURE 45
Representation of Transportation Problem as an MCNFP

Supply point 1 (1) → (3) Demand point 1

Supply point 2 (2) → (4) Demand point 2

TABLE 29
MCNFP Representation of Transportation Problem

min $z = x_{13} + 2x_{14} + 3x_{23} + 4x_{24}$						
x_{13}	x_{14}	x_{23}	x_{24}		rhs	Constraint
1	1	0	0	=	4	Node 1
0	0	1	1	=	5	Node 2
−1	0	−1	0	=	−6	Node 3
1	−1	0	−1	=	−3	Node 4
All variables non-negative						

has no arc capacity restrictions, the flow balance equations are the only constraints. We note that if the problem had not been balanced, we could not have formulated the problem as an MCNFP. This is because if total supply exceeded total demand, we would not know with certainty the net outflow at each supply point. Thus, to formulate a transportation (or a transshipment) problem as an MCNFP, it may be necessary to add a dummy point.

Formulating a Maximum-Flow Problem as an MCNFP

To see how a maximum-flow problem fits into the minimum-cost network flow context, consider the problem of finding the maximum flow from source to sink in the network of Figure 6. After creating an arc a_0 joining the sink to the source, we have $b_{so} = b_1 = b_2 = b_3 = b_{si} = 0$. Then the LP constraints for finding the maximum flow in Figure 6 may be written as shown in Table 30.

The first five constraints are the flow balance equations for the nodes of the network, and the last six constraints are the arc capacity constraints. Because there is no upper limit on the flow through the artificial arc, there is no arc capacity constraint for a_0.

The flow balance equations in any MCNFP have the following important property: *Each variable x_{ij} has a coefficient of $+1$ in the node i flow balance equation, a coefficient of -1 in the node j flow balance equation, and a coefficient of 0 in all other flow balance equations.* For example, in a transportation problem, the variable x_{ij} will have a coeffi-

TABLE 30
MCNFP Representation of Maximum-Flow Problem

			min $z = x_0$						
$x_{so,1}$	$x_{so,2}$	x_{13}	x_{12}	$x_{3,si}$	$x_{2,si}$	x_0		rhs	Constraint
1	1	0	0	0	0	-1	$=$	0	Node so
-1	0	1	1	0	0	0	$=$	0	Node 1
0	-1	0	-1	0	1	0	$=$	0	Node 2
0	0	-1	0	1	0	0	$=$	0	Node 3
0	0	0	0	-1	-1	1	$=$	0	Node si
1	0	0	0	0	0	0	$\leq$	2	Arc $(so, 1)$
0	1	0	0	0	0	0	$\leq$	3	Arc $(so, 2)$
0	0	1	0	0	0	0	$\leq$	4	Arc $(1, 3)$
0	0	0	1	0	0	0	$\leq$	3	Arc $(1, 2)$
0	0	0	0	1	0	0	$\leq$	1	Arc $(3, si)$
0	0	0	0	0	1	0	$\leq$	2	Arc $(2, si)$

All variables non-negative

cient of $+1$ in the flow balance equation for supply point i, a coefficient of -1 in the flow balance equation for demand point j, and a coefficient of 0 in all other flow balance equations. Even if the constraints of an LP do not appear to contain the flow balance equations of a network, clever transformation of an LP's constraints can often show that an LP is equivalent to an MCNFP (see Problem 6 at the end of this section).

An MCNFP can be solved by a generalization of the transportation simplex known as the *network simplex algorithm* (see Section 8.7). As with the transportation simplex, the pivots in the network simplex involve only additions and subtractions. This fact can be used to prove that if all the b_i's and arc capacities are integers, then in the optimal solution to an MCNFP, all the variables will be integers. Computer codes that use the network simplex can quickly solve even extremely large network problems. For example, MCNFPs with 5,000 nodes and 600,000 arcs have been solved in under 10 minutes. To use a network simplex computer code, the user need only input a list of the network's nodes and arcs, the c_{ij}'s and arc capacity for each arc, and the b_i's for each node. The network simplex is efficient and easy to use, so it is extremely important to formulate an LP, if at all possible, as an MCNFP.

To close this section, we formulate a simple traffic assignment problem as an MCNFP.

EXAMPLE 7   **Traffic MCNFP**

Each hour, an average of 900 cars enter the network in Figure 46 at node 1 and seek to travel to node 6. The time it takes a car to traverse each arc is shown in Table 31. In Figure 46, the number above each arc is the maximum number of cars that can pass by any point on the arc during a one-hour period. Formulate an MCNFP that minimizes the total time required for all cars to travel from node 1 to node 6.

Solution   Let

$$x_{ij} = \text{number of cars per hour that traverse the arc from node } i \text{ to node } j$$

Then we want to minimize

$$z = 10x_{12} + 50x_{13} + 70x_{25} + 30x_{24} + 30x_{56} + 30x_{45} + 60x_{46} + 60x_{35} + 10x_{34}$$

We are given that $b_1 = 900$, $b_2 = b_3 = b_4 = b_5 = 0$, and $b_6 = -900$ (we will not introduce the artificial arc connecting node 6 to node 1). The constraints for this MCNFP are shown in Table 32.

FIGURE **46**

Representation of
Traffic Example as
MCNFP

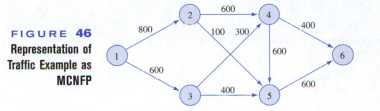

TABLE **31**

Travel Times for Traffic
Example

Arc	Time (Minutes)
(1, 2)	10
(1, 3)	50
(2, 5)	70
(2, 4)	30
(5, 6)	30
(4, 5)	30
(4, 6)	60
(3, 5)	60
(3, 4)	10

TABLE **32**

MCNFP Representation of
Traffic Example

X_{12}	X_{13}	X_{24}	X_{25}	X_{34}	X_{35}	X_{45}	X_{46}	X_{56}		rhs	Constraint
1	1	0	0	0	0	0	0	0	=	900	Node 1
−1	0	1	1	0	0	0	0	0	=	0	Node 2
0	−1	0	0	1	1	0	0	0	=	0	Node 3
0	0	−1	0	−1	0	1	1	0	=	0	Node 4
0	0	0	−1	0	−1	−1	0	1	=	0	Node 5
0	0	0	0	0	0	0	−1	−1	=	−900	Node 6
1	0	0	0	0	0	0	0	0	≤	800	Arc (1, 2)
0	1	0	0	0	0	0	0	0	≤	600	Arc (1, 3)
0	0	1	0	0	0	0	0	0	≤	600	Arc (2, 4)
0	0	0	1	0	0	0	0	0	≤	100	Arc (2, 5)
0	0	0	0	1	0	0	0	0	≤	300	Arc (3, 4)
0	0	0	0	0	1	0	0	0	≤	400	Arc (3, 5)
0	0	0	0	0	0	1	0	0	≤	600	Arc (4, 5)
0	0	0	0	0	0	0	1	0	≤	400	Arc (4, 6)
0	0	0	0	0	0	0	0	1	≤	600	Arc (5, 6)

All variables non-negative

Solving an MCNFP with LINGO

Traffic.lng

The following LINGO program (file Traffic.lng) can be used to find the optimal solution
to Example 7 (or any MCNFP).

```
MODEL:
 1] SETS:
 2] NODES/1..6/:SUPP;
 3] ARCS(NODES,NODES)/1,2  1,3  2,4  2,5  3,4  3,5  4,5  4,6  5,6/
 4] :CAP,FLOW,COST;
 5] ENDSETS
 6] MIN=@SUM(ARCS:COST*FLOW);
 7] @FOR(ARCS(I,J):FLOW(I,J)<CAP(I,J));
 8] @FOR(NODES(I):-@SUM(ARCS)(J,I):FLOW(J,I))
 9] +@SUM(ARCS(I,J):FLOW(I,J))=SUPP(I));
10] DATA:
11] COST=10,50,30,70,10,60,30,60,30;
12] SUPP=900,0,0,0,0,-900;
13] CAP=800,600,600,100,300,400,600,400,600;
14] ENDDATA
END      .
```

In line 2, we define the network's nodes and associate a net supply (flow out−flow in) with each node. The supplies data are entered in line 12. In line 3, we define, by listing, the arcs in the network and in line 4 associate a capacity (CAP), a flow (FLOW), and a cost-per-unit-shipped (COST) with each arc. The unit shipping costs data are entered in line 11. Line 6 generates the objective function by summing over all arcs (unit cost for arc)*(flow through arc). Line 7 generates each arc's capacity constraint (arc capacities data are entered in line 13). For each node, lines 8–9 generate the conservation-of-flow constraint. They imply that for each node I, −(flow into node I) + (flow out of node I) = (supply of node I). When solved on LINGO, we find that the solution to Example 7 is $z = 95{,}000$ minutes, $x_{12} = 700$, $x_{13} = 200$, $x_{24} = 600$, $x_{25} = 100$, $x_{34} = 200$, $x_{45} = 400$, $x_{46} = 400$, $x_{56} = 500$.

Our LINGO program can be used to solve any MCNFP. Just input the set of nodes, supplies, arcs, and unit shipping cost; hit **GO** and you are done!

PROBLEMS

Note: To formulate a problem as an MCNFP, you should draw the appropriate network and determine the c_{ij}'s, the b_i's, and the arc capacities.

Group A

1 Formulate the problem of finding the shortest path from node 1 to node 6 in Figure 2 as an MCNFP. (*Hint:* Think of finding the shortest path as the problem of minimizing the total cost of sending 1 unit of flow from node 1 to node 6.)

2 a Find the dual of the LP that was used to find the length of the critical path for Example 6 of Section 8.4.

b Show that the answer in part (a) is an MCNFP.

c Explain why the optimal objective function value for the LP found in part (a) is the longest path in the project network from node 1 to node 6. Why does this justify our earlier claim that the critical path in a project network is the longest path from the start node to the finish node?

3 Fordco produces cars in Detroit and Dallas. The Detroit plant can produce as many as 6,500 cars, and the Dallas plant can produce as many as 6,000 cars. Producing a car costs $2,000 in Detroit and $1,800 in Dallas. Cars must be shipped to three cities. City 1 must receive 5,000 cars, city 2 must receive 4,000 cars, and city 3 must receive 3,000 cars. The cost of shipping a car from each plant to each city is given in Table 33. At most, 2,200 cars may be sent from a given plant to a given city. Formulate an MCNFP that can be used to minimize the cost of meeting demand.

4 Each year, Data Corporal produces as many as 400 computers in Boston and 300 computers in Raleigh. Los Angeles customers must receive 400 computers, and 300 computers must be supplied to Austin customers. Producing a computer costs $800 in Boston and $900 in Raleigh. Computers are transported by plane and may be sent through Chicago. The costs of sending a computer between pairs of cities are shown in Table 34.

a Formulate an MCNFP that can be used to minimize the total (production + distribution) cost of meeting Data Corporal's annual demand.

TABLE 33

From	To ($)		
	City 1	City 2	City 3
Detroit	800	600	300
Dallas	500	200	200

TABLE 34

From	To ($)		
	Chicago	Austin	Los Angeles
Boston	80	220	280
Raleigh	100	140	170
Chicago	—	40	50

b How would you modify the part (a) formulation if at most 200 units could be shipped through Chicago? [*Hint:* Add an additional node and arc to this part (a) network.]

5 Oilco has oil fields in San Diego and Los Angeles. The San Diego field can produce 500,000 barrels per day, and the Los Angeles field can produce 400,000 barrels per day. Oil is sent from the fields to a refinery, in either Dallas or Houston (assume each refinery has unlimited capacity). To refine 100,000 barrels costs $700 at Dallas and $900 at Houston. Refined oil is shipped to customers in Chicago and New York. Chicago customers require 400,000 barrels per day, and New York customers require 300,000 barrels per day. The costs of shipping 100,000 barrels of oil (refined or unrefined) between cities are shown in Table 35.

a Formulate an MCNFP that can be used to determine how to minimize the total cost of meeting all demands.

b If each refinery had a capacity of 500,000 barrels per day, how would the part (a) answer be modified?

Group B

6 Workco must have the following number of workers available during the next three months: month 1, 20; month 2, 16; month 3, 25. At the beginning of month 1, Workco has no workers. It costs Workco $100 to hire a worker and $50 to fire a worker. Each worker is paid a salary of $140/month. We will show that the problem of determining a hiring and firing strategy that minimizes the total cost incurred during the next three (or in general, the next n) months can be formulated as an MCNFP.

a Let

x_{ij} = number of workers hired at beginning of month i and fired after working till end of month $j - 1$

(if $j = 4$, the worker is never fired). Explain why the following LP will yield a minimum-cost hiring and firing strategy:

TABLE 35

From	To ($)			
	Dallas	Houston	New York	Chicago
Los Angeles	300	110	—	—
San Diego	420	100	—	—
Dallas	—	—	450	550
Houston	—	—	470	530

$$\min z = 50(x_{12} + x_{13} + x_{23})$$
$$+ 100(x_{12} + x_{13} + x_{14} + x_{23} + x_{24} + x_{34})$$
$$+ 140(x_{12} + x_{23} + x_{34})$$
$$+ 280(x_{13} + x_{24}) + 420x_{14}$$

s.t.   (1) $x_{12} + x_{13} + x_{14} \quad\quad\quad - e_1 = 20$
                             (Month 1 constraint)

         (2) $x_{13} + x_{14} + x_{23} + x_{24} - e_2 = 16$
                             (Month 2 constraint)

         (3) $x_{14} + x_{24} + x_{34} \quad\quad - e_3 = 25$
                             (Month 3 constraint)

$$x_{ij} \geq 0$$

b To obtain an MCNFP, replace the constraints in part (a) by

     i   Constraint (1);
     ii   Constraint (2) − Constraint (1);
     iii   Constraint (3) − Constraint (2);
     iv   − (Constraint (3)).

Explain why an LP with Constraints (i)–(iv) is an MCNFP.

c Draw the network corresponding to the MCNFP obtained in answering part (b).

7[†] Braneast Airlines must determine how many airplanes should serve the Boston–New York–Washington air corridor and which flights to fly. Braneast may fly any of the daily flights shown in Table 36. The fixed cost of operating an airplane is $800/day. Formulate an MCNFP that can be used to maximize Braneast's daily profits. (*Hint:* Each node in the network represents a city and a time. In addition to arcs representing flights, we must allow for the possibility that an airplane will stay put for an hour or more. We must ensure that the model includes the fixed cost of operating a plane. To include this cost, the following three arcs might be included in the network: from Boston 7 P.M. to Boston 9 A.M.; from New York 7 P.M. to New York 9 A.M.; and from Washington 7 P.M. to Washington 9 A.M.)

8 Daisymay Van Line moves people between New York, Philadelphia, and Washington, D.C. It takes a van one day to travel between any two of these cities. The company incurs costs of $1,000 per day for a van that is fully loaded and traveling, $800 per day for an empty van that travels, $700 per day for a fully loaded van that stays in a city, and $400 per day for an empty van that remains in a city. Each day of the week, the loads described in Table 37 must be shipped. On Monday, for example, two trucks must be sent from Philadelphia to New York (arriving on Tuesday). Also, two trucks must be sent from Philadelphia to Washington on Friday (assume that Friday shipments must arrive on Monday). Formulate an MCNFP that can be used to minimize the cost of meeting weekly requirements. To simplify the formulation, assume that the requirements repeat each week. Then it seems plausible to assume that any of the company's trucks will begin each week in the same city in which it began the previous week.

[†]This problem is based on Glover et al. (1982).

TABLE 36

Leaves		Arrives		Flight Revenue	Variable Cost of Flight (S)
City	Time	City	Time		
N.Y.	9 A.M.	Wash.	10 A.M.	$900	400
N.Y.	2 P.M.	Wash.	3 P.M.	$600	350
N.Y.	10 A.M.	Bos.	11 A.M.	$800	400
N.Y.	4 P.M.	Bos.	5 P.M.	$1,200	450
Wash.	9 A.M.	N.Y.	10 A.M.	$1,100	400
Wash.	3 P.M.	N.Y.	4 P.M.	$900	350
Wash.	10 A.M.	Bos.	12 noon	$1,500	700
Wash.	5 P.M.	Bos.	7 P.M.	$1,800	900
Bos.	10 A.M.	N.Y.	11 A.M.	$900	500
Bos.	2 P.M.	N.Y.	3 P.M.	$800	450
Bos.	11 A.M.	Wash.	1 P.M.	$1,100	600
Bos.	3 P.M.	Wash.	5 P.M.	$1,200	650

TABLE 37

Trip	Monday	Tuesday	Wednesday	Thursday	Friday
Phil.–N.Y.	2	—	—	—	—
Phil.–Wash.	—	2	—	—	2
N.Y.–Phil.	3	2	—	—	—
N.Y.–Wash.	—	—	2	2	—
N.Y.–Phil.	1	—	—	—	—
Wash.–N.Y.	—	—	1	—	1

8.6 Minimum Spanning Tree Problems

Suppose that each arc (i, j) in a network has a length associated with it and that arc (i, j) represents a way of connecting node i to node j. For example, if each node in a network represents a computer at State University, then arc (i, j) might represent an underground cable that connects computer i with computer j. In many applications, we want to determine the set of arcs in a network that connect all nodes such that the sum of the length of the arcs is minimized. Clearly, such a group of arcs should contain no loop. (A loop is often called a *closed path* or *cycle.*) For example, in Figure 47, the sequence of arcs $(1, 2)$–$(2, 3)$–$(3, 1)$ is a loop.

DEFINITION ■ For a network with n nodes, a **spanning tree** is a group of $n - 1$ arcs that connects all nodes of the network and contains no loops. ■

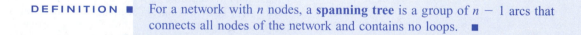

FIGURE 47
Illustration of Loop and Minimum Spanning Tree

$(1, 2)$–$(2, 3)$–$(3, 1)$
is a loop
$(1, 3)$, $(2, 3)$ is the
minimum spanning tree

In Figure 47, there are three spanning trees:

1 Arcs (1, 2) and (2, 3)

2 Arcs (1, 2) and (1, 3)

3 Arcs (1, 3) and (2, 3)

A spanning tree of minimum length in a network is a **minimum spanning tree (MST).** In Figure 47, the spanning tree consisting of arcs (1, 3) and (2, 3) is the unique minimum spanning tree.

The following method (MST algorithm) may be used to find a minimum spanning tree.

Step 1 Begin at any node i, and join node i to the node in the network (call it node j) that is closest to node i. The two nodes i and j now form a connected set of nodes $C = \{i, j\}$, and arc (i, j) will be in the minimum spanning tree. The remaining nodes in the network (call them C') are referred to as the *unconnected* set of nodes.

Step 2 Now choose a member of C' (call it n) that is closest to some node in C. Let m represent the node in C that is closest to n. Then the arc (m, n) will be in the minimum spanning tree. Now update C and C'. Because n is now connected to $\{i, j\}$, C now equals $\{i, j, n\}$ and we must eliminate node n from C'.

Step 3 Repeat this process until a minimum spanning tree is found. Ties for closest node and arc to be included in the minimum spanning tree may be broken arbitrarily.

At each step the algorithm chooses the shortest arc that can be used to expand C, so the algorithm is often referred to as a "greedy" algorithm. It is remarkable that the act of being "greedy" at each step of the algorithm can never force us later to follow a "bad arc." In Example 1 of Chapter 9 we will see that for some types of problems, a greedy algorithm may not yield an optimal solution! A justification of the MST algorithm is given in Problem 3 at the end of this section. Example 8 illustrates the algorithm.

EXAMPLE 8   **MST Algorithm**

The State University campus has five minicomputers. The distance between each pair of computers (in city blocks) is given in Figure 48. The computers must be interconnected by underground cable. What is the minimum length of cable required? Note that if no arc is drawn connecting a pair of nodes, this means that (because of underground rock formations) no cable can be laid between these two computers.

Solution   We want to find the minimum spanning tree for Figure 48.

Iteration 1   Following the MST algorithm, we arbitrarily choose to begin at node 1. The closest node to node 1 is node 2. Now $C = \{1, 2\}$, $C' = \{3, 4, 5\}$, and arc (1, 2) will be in the minimum spanning tree (see Figure 49a).

Iteration 2   Node 5 is closest (two blocks distant) to C. Because node 5 is two blocks from node 1 and from node 2, we may include either arc (2, 5) or arc (1, 5) in the minimum spanning tree. We arbitrarily choose to include arc (2, 5). Then $C = \{1, 2, 5\}$ and $C' = \{3, 4\}$ (see Figure 49b).

Iteration 3   Node 3 is two blocks from node 5, so we may include arc (5, 3) in the minimum spanning tree. Now $C = \{1, 2, 3, 5\}$ and $C' = 4$ (see Figure 49c).

Iteration 4   Node 5 is the closest node to node 4, so we add arc (5, 4) to the minimum spanning tree (see Figure 49d).

We have now obtained the minimum spanning tree consisting of arcs (1, 2), (2, 5), (5, 3), and (5, 4). The length of the minimum spanning tree is $1 + 2 + 2 + 4 = 9$ blocks.

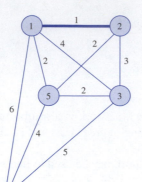

FIGURE 48
Distances between
State University
Computers

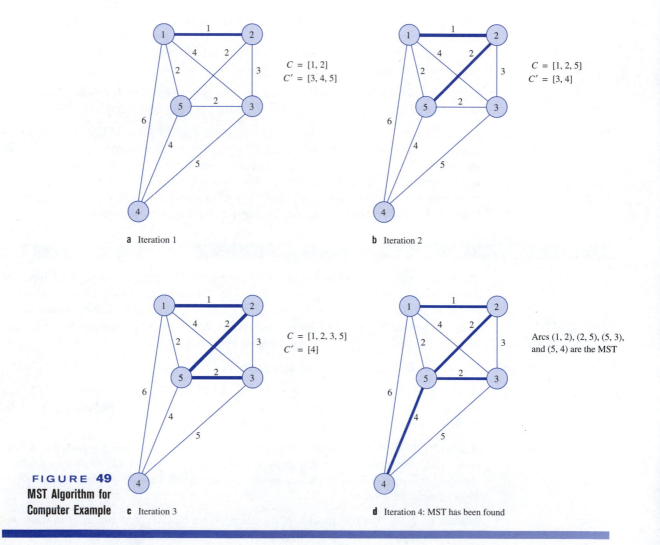

a Iteration 1

$C = [1, 2]$
$C' = [3, 4, 5]$

b Iteration 2

$C = [1, 2, 5]$
$C' = [3, 4]$

c Iteration 3

$C = [1, 2, 3, 5]$
$C' = [4]$

d Iteration 4: MST has been found

Arcs (1, 2), (2, 5), (5, 3),
and (5, 4) are the MST

FIGURE 49
MST Algorithm for
Computer Example

PROBLEMS

Group A

1 The distances (in miles) between the Indiana cities of Gary, Fort Wayne, Evansville, Terre Haute, and South Bend are shown in Table 38. It is necessary to build a state road system that connects all these cities. Assume that for political reasons no road can be built connecting Gary and Fort Wayne, and no road can be built connecting South Bend and Evansville. What is the minimum length of road required?

2 The city of Smalltown consists of five subdivisions. Mayor John Lion wants to build telephone lines to ensure that all the subdivisions can communicate with each other. The distances between the subdivisions are given in Figure 50. What is the minimum length of telephone line required? Assume that no telephone line can be built between subdivisions 1 and 4.

Group B

3 In this problem, we explain why the MST algorithm works. Define

S = minimum spanning tree

C_t = nodes connected after iteration t of MST algorithm has been completed

C_t' = nodes not connected after iteration t of MST algorithm has been completed

A_t = set of arcs in minimum spanning tree after t iterations of MST algorithm have been completed

TABLE 38

	Gary	Fort Wayne	Evansville	Terre Haute	South Bend
Gary	—	132	217	164	58
Fort Wayne	132	—	290	201	79
Evansville	217	290	—	113	303
Terre Haute	164	201	113	—	196
South Bend	58	79	303	196	—

FIGURE 50
Network for Problem 2

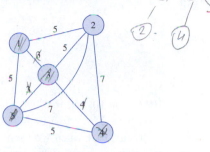

Suppose the MST algorithm does not yield a minimum spanning tree. Then, for some t, it must be the case that all arcs in A_{t-1} are in S, but the arc chosen at iteration t (call it a_t) of the MST algorithm is not in S. Then S must contain some arc a_t' that leads from a node in C_{t-1} to a node in C_{t-1}'. Show that by replacing arc a_t' with arc a_t, we can obtain a shorter spanning tree than S. This contradiction proves that all arcs chosen by the MST algorithm must be in S. Thus, the MST algorithm does indeed find a minimum spanning tree.

4 a Three cities are at the vertices of an equilateral triangle of unit length. Flying Lion Airlines needs to supply connecting service between these three cities. What is the minimum length of the two routes needed to supply the connecting service?

b Now suppose Flying Lion Airlines adds a hub at the "center" of the equilateral triangle. Show that the length of the routes needed to connect the three cities has decreased by 13%. *(Note:* It has been shown that no matter how many "hubs" you add and no matter how many points must be connected, you can never save more than 13% of the total distance needed to "span" all the original points by adding hubs.)[†]

8.7 The Network Simplex Method[‡]

In this section, we describe how the simplex algorithm simplifies for MCNFPs. To simplify our presentation, we assume that for each arc, $L_{ij} = 0$. Then the information needed to describe an MCNFP of the form (8)–(9) may be summarized graphically as in Figure 51. We will denote the c_{ij} for each arc by the symbol \$, and the other number on each arc will represent the arc's upper bound (U_{ij}). The b_i for any node with nonzero outflow will be listed in parentheses. Thus, Figure 51 represents an MCNFP with $c_{12} = 5$, $c_{25} = 2$, $c_{13} = 4$, $c_{35} = 8$,

---

[†]Based on Peterson (1990).
[‡]This section covers topics that may be omitted with no loss of continuity.

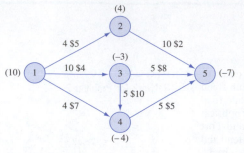

FIGURE 51
Graphical
Representation of
an MCNFP

$c_{14} = 7$, $c_{34} = 10$, $c_{45} = 5$, $b_1 = 10$, $b_2 = 4$, $b_3 = -3$, $b_4 = -4$, $b_5 = -7$, $U_{12} = 4$, $U_{25} = 10$, $U_{13} = 10$, $U_{35} = 5$, $U_{14} = 4$, $U_{34} = 5$, $U_{45} = 5$. For the network simplex to be used, we must have $\Sigma b_i = 0$; usually this can be ensured by adding a dummy node.

Recall that when we used the simplex method to solve a transportation problem, the following aspects of the simplex algorithm simplified: finding a basic feasible solution, computing the coefficient of a nonbasic variable in row 0, and pivoting. We now describe how these aspects of the simplex algorithm simplify when we are solving an MCNFP.

Basic Feasible Solutions for MCNFPs

How can we determine whether a feasible solution to an MCNFP is a bfs? Begin by observing that any bfs to an MCNFP will contain three types of variables:

1 Basic variables: In the absence of degeneracy, each basic variable x_{ij} will satisfy $L_{ij} < x_{ij} < U_{ij}$; with degeneracy, it is possible for a basic variable x_{ij} to equal arc (i, j)'s upper or lower bound.

2 Nonbasic variables x_{ij}: These equal arc (i, j)'s upper bound U_{ij}.

3 Nonbasic variables x_{ij}: These equal arc (i, j)'s lower bound L_{ij}.

Suppose we are solving an MCNFP with n nodes. In solving an MCNFP, we consider the n conservation-of-flow constraints and ignore the upper- and lower-bound constraints (for reasons that will soon become apparent). As in the transportation problem, any solution satisfying $n - 1$ of the conservation-of-flow constraints will automatically satisfy the last conservation-of-flow constraint, so we may drop one such constraint. This means that a bfs to an n-node MCNFP will have $n - 1$ basic variables. Suppose we choose a set of $n - 1$ variables (or arcs). How can we determine whether this set of $n - 1$ variables yields a basic feasible solution? A set of $n - 1$ variables will yield a bfs if and only if the arcs corresponding to the basic variables form a spanning tree for the network. For example, consider the MCNFP in Figure 52. In Figure 53, we give a bfs for this MCNFP. The basic variables are x_{13}, x_{35}, x_{25}, and x_{45}. The variables $x_{12} = 5$ and $x_{14} = 4$ are nonbasic vari-

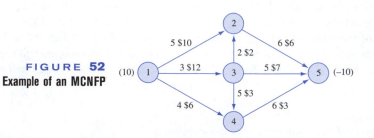

FIGURE 52
Example of an MCNFP

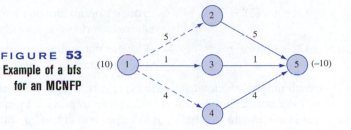

FIGURE 53
Example of a bfs
for an MCNFP

ables at their upper bound. (Such variables will be indicated by dashed arcs.) Because the arcs $(1, 3)$, $(3, 5)$, $(2, 5)$, and $(4, 5)$ form a spanning tree (they connect all nodes of the graph and do not contain any cycles), we know that this is a bfs. As will soon become clear, a bfs for small problems can often be obtained by trial and error.

Computing Row 0 for Any bfs

For any given bfs, how do we determine the objective function coefficient for a nonbasic variable? Suppose we arbitrarily choose to drop the conservation-of-flow constraint for node 1. For a given bfs, let $c_{BV}B^{-1} = [y_2 \quad y_3 \quad \cdots \quad y_n]$. Each variable x_{ij} will have a $+1$ coefficient in the node i flow constraint and a -1 coefficient in the node j constraint. If we define $y_1 = 0$, then the coefficient of x_{ij} in row 0 of a given tableau may be written as $\bar{c}_{ij} = y_i - y_j - c_{ij}$. Each basic variable must have $\bar{c}_{ij} = 0$, so we can find $y_1, y_2, \ldots, y_n$ by solving the following system of linear equations:

$$y_1 = 0, \quad y_i - y_j = c_{ij} \quad \text{for each basic variable}$$

The $y_1, y_2, \ldots, y_n$ corresponding to a bfs are often called the **simplex multipliers** for the bfs.

How can we determine whether a bfs is optimal? For a bfs to be optimal, it must be possible to improve (decrease) the value of z by changing the value of a nonbasic variable. Note that $\bar{c}_{ij} \leq 0$ if and only if increasing x_{ij} cannot decrease z. Also note that $\bar{c}_{ij} \geq 0$ if and only if decreasing x_{ij} cannot decrease z. These observations can be used to show that a bfs is optimal if and only if the following conditions are met:

1 If a variable $x_{ij} = L_{ij}$, then an increase in x_{ij} cannot result in a decrease in z. Thus, if $x_{ij} = L_{ij}$ and the bfs is optimal, then $\bar{c}_{ij} \leq 0$ must hold.

2 If a variable $x_{ij} = U_{ij}$, then a decrease in x_{ij} cannot result in a decrease in z. Thus, if $x_{ij} = U_{ij}$ and the bfs is optimal, then $\bar{c}_{ij} \geq 0$ must hold.

If conditions 1 and 2 are not met, then z can be improved (barring degeneracy) by pivoting into the basis any nonbasic variable violating either condition. To illustrate, let's determine the objective function coefficient for each nonbasic variable in the simplex tableau corresponding to the bfs in Figure 53. To find y_1, y_2, y_3, y_4, and y_5, we solve the following set of equations:

$$y_1 = 0, \quad y_1 - y_3 = 12, \quad y_2 - y_5 = 6, \quad y_3 - y_5 = 7, \quad y_4 - y_5 = 3$$

The solutions to these equations are $y_1 = 0$, $y_2 = -13$, $y_3 = -12$, $y_4 = -16$, and $y_5 = -19$. We now "price out" each nonbasic variable and obtain

$\bar{c}_{12} = y_1 - y_2 - c_{12} = 0 - (-13) - 10 = 3$ (Satisfies optimality condition for nonbasic variable at upper bound)

$\bar{c}_{14} = y_1 - y_4 - c_{14} = 0 - (-16) - 6 = 10$ (Satisfies optimality condition for nonbasic variable at upper bound)

$$\bar{c}_{32} = y_3 - y_2 - c_{32} = -12 - (-13) - 2 = -1 \qquad \text{(Satisfies optimality condition for nonbasic variable at lower bound)}$$

$$\bar{c}_{34} = y_3 - y_4 - c_{34} = -12 - (-16) - 3 = 1 \qquad \text{(Violates optimality condition for nonbasic variable at lower bound)}$$

Because $\bar{c}_{34} = 1 > 0$, each unit by which we increase x_{34} (x_{34} is at its lower bound, so it's okay to increase it) will decrease z by one unit. Thus, we can improve z by entering x_{34} into the basis. Note that if a nonbasic variable x_{ij} at its upper bound had $\bar{c}_{ij} < 0$, then we could decrease z by entering x_{ij} into the basis and decreasing x_{ij}. We now show that when solving an MCNFP, the pivot step may be performed almost by inspection.

Pivoting in the Network Simplex

As we have just shown, for the bfs in Figure 53, we want to enter x_{34} into the basis. To do this, note that if we add the arc (3, 4) to the set of arcs corresponding to the current set of basic variables, a cycle (or loop) will be formed. To enter x_{34} into the basis, note that $x_{34} = 0$ is at its lower bound, we want to increase x_{34}. Suppose we try to increase x_{34} by θ. The values of all variables after x_{34} is entered into the basis may be found by invoking the conservation-of-flow constraints. In Figure 54, we find that arc (3, 4), (4, 5), and (3, 5) form a cycle. After the pivot, all variables corresponding to arcs not in the cycle will remain unchanged, but when we set $x_{34} = \theta$, the values of the variables corresponding to arcs in the cycle will change. Setting $x_{34} = \theta$ increases the flow into node 4 by θ, so the flow out of node 4 must increase by θ. This requires $x_{45} = 4 + \theta$. Because the flow into node 5 has now increased by θ, conservation of flow requires that $x_{35} = 1 - \theta$. The pivot leaves all other variables unchanged. To find the new values of the variables, observe that we want to increase x_{34} by as much as possible. We can increase x_{34} to the point where a basic variable first attains its upper or lower bound. Thus, arc (3, 4) implies that $\theta \leq 5$; arc (3, 5) requires $1 - \theta \geq 0$ or $\theta \leq 1$; arc (4, 5) requires $4 + \theta \leq 6$ or $\theta \leq 2$. So the best we can do is set $\theta = 1$. The basic variable that first hits its upper or lower bound as θ is increased is chosen to exit the basis (in case of a tie, we can choose the exiting variable arbitrarily). Now x_{35} exits the basis, and the new bfs is shown in Figure 55. The spanning tree corresponding to the current set of basic variables is (1, 3), (3, 4), (4, 5), and (2, 5). We now compute the coefficient of each nonbasic variable in row 0. To begin, we solve the following set of equations:

$$y_1 = 0, \qquad y_1 - y_3 = 12, \qquad y_3 - y_4 = 3, \qquad y_2 - y_5 = 6, \qquad y_4 - y_5 = 3$$

This yields $y_1 = 0$, $y_2 = -12$, $y_3 = -12$, $y_4 = -15$, and $y_5 = -18$.

The nonbasic variables that currently equal their upper bounds will have row 0 coefficients of

$$\bar{c}_{12} = 0 - (-12) - 10 = 2 \qquad \text{and} \qquad \bar{c}_{14} = 0 - (-15) - 6 = 9$$

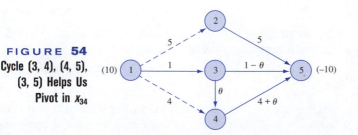

FIGURE 54
Cycle (3, 4), (4, 5), (3, 5) Helps Us Pivot in x_{34}

FIGURE 55
New bfs ($\theta = 1$) After
x_{34} Enters and x_{35} Exits

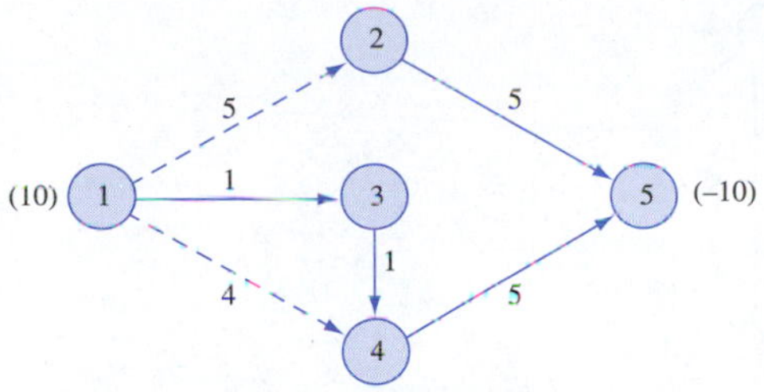

The nonbasic variables that currently equal their lower bounds will have row 0 coefficients of

$$\bar{c}_{32} = -12 - (-12) - 2 = -2 \quad \text{and} \quad \bar{c}_{35} = -12 - (-18) - 7 = -1$$

Because each nonbasic variable at its upper bound has $\bar{c}_{ij} \geq 0$, and each nonbasic variable at its lower bound has $\bar{c}_{ij} \leq 0$, the current bfs is optimal. Thus, the optimal solution to the MCNFP in Figure 52 is

$$\text{Upper bounded variables:} \quad x_{12} = 5, \quad x_{14} = 4$$

$$\text{Lower bounded variables:} \quad x_{32} = x_{35} = 0$$

$$\text{Basic variables:} \quad x_{13} = 1, x_{34} = 1, x_{25} = 5, x_{45} = 5$$

Summary of the Network Simplex Method

Step 1 Determine a starting bfs. The $n - 1$ basic variables will correspond to a spanning tree. Indicate nonbasic variables at their upper bound by dashed arcs.

Step 2 Compute $y_1, y_2, \ldots y_n$ (often called the simplex multipliers) by solving $y_1 = 0$, $y_i - y_j = c_{ij}$ for all basic variables x_{ij}. For all nonbasic variables, determine the row 0 coefficient $\bar{c}_{ij}$ from $\bar{c}_{ij} = y_i - y_j - c_{ij}$. The current bfs is optimal if $\bar{c}_{ij} \leq 0$ for all $x_{ij} = L_{ij}$ and $\bar{c}_{ij} \geq 0$ for all $x_{ij} = U_{ij}$. If the bfs is not optimal, choose the nonbasic variable that most violates the optimality conditions as the entering basic variable.

Step 3 Identify the cycle (there will be exactly one!) created by adding the arc corresponding to the entering variable to the current spanning tree of the current bfs. Use conservation of flow to determine the new values of the variables in the cycle. The variable that exits the basis will be the variable that first hits its upper or lower bound as the value of the entering basic variable is changed.

Step 4 Find the new bfs by changing the flows of the arcs in the cycle found in step 3. Now go to step 2.

Example 9 illustrates the network simplex.

EXAMPLE 9 | **Network Simplex Solution to MCNFP**

Use the network simplex to solve the MCNFP in Figure 56.

Solution A bfs requires that we find a spanning tree (three arcs that connect nodes 1, 2, 3, and 4 and do not form a cycle). Any arcs not in the spanning tree may be set equal to their upper or lower bound. By trial and error, we find the bfs in Figure 57 involving the spanning tree (1, 2), (1, 3), and (2, 4).

To find y_1, y_2, y_3, and y_4 we solve

$$y_1 = 0, \quad y_1 - y_2 = 4, \quad y_2 - y_4 = 3, \quad y_1 - y_3 = 3$$

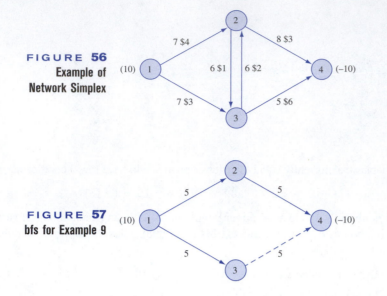

FIGURE 56
Example of
Network Simplex

FIGURE 57
bfs for Example 9

This yields $y_1 = 0$, $y_2 = -4$, $y_3 = -3$, and $y_4 = -7$. The row 0 coefficients for each nonbasic variable are

$$\bar{c}_{34} = -3 - (-7) - 6 = -2 \quad \text{(Violates optimality condition)}$$
$$\bar{c}_{23} = -4 - (-3) - 1 = -2 \quad \text{(Satisfies optimality condition)}$$
$$\bar{c}_{32} = -3 - (-4) - 2 = -1 \quad \text{(Satisfies optimality condition)}$$

Thus, x_{34} enters the basis. We set $x_{34} = 5 - \theta$ and obtain the cycle in Figure 58. From arc $(1, 2)$, we find $5 + \theta \leq 7$ or $\theta \leq 2$. From arc $(1, 3)$, we find $5 - \theta \geq 0$ or $\theta \leq 5$. From arc $(2, 4)$, we find $5 + \theta \leq 8$ or $\theta \leq 3$. From arc $(3, 4)$, we find $5 - \theta \geq 0$ or $\theta \leq 5$. Thus, we can set $\theta = 2$. Now x_{12} exits the basis at its upper bound, and x_{34} enters, yielding the bfs in Figure 59.

The new bfs is associated with the spanning tree $(1, 3)$, $(2, 4)$, and $(3, 4)$. Solving for the new values of the simplex multipliers, we obtain

$$y_1 = 0, \quad y_1 - y_3 = 3, \quad y_3 - y_4 = 6, \quad y_2 - y_4 = 3$$

This yields $y_1 = 0$, $y_2 = -6$, $y_3 = -3$, $y_4 = -9$. The coefficient of each nonbasic variable in row 0 is given by

$$\bar{c}_{12} = 0 - (-6) - 4 = 2 \qquad \text{(Satisfies optimality condition)}$$
$$\bar{c}_{23} = -6 - (-3) - 1 = -4 \qquad \text{(Satisfies optimality condition)}$$
$$\bar{c}_{32} = -3 - (-6) - 2 = 1 \qquad \text{(Violates optimality condition)}$$

Now x_{32} enters the basis, yielding the cycle in Figure 60. From arc $(2, 4)$, we find $7 + \theta \leq 8$ or $\theta \leq 1$); from arc $(3, 4)$, we find $3 - \theta \geq 0$ or $\theta \leq 3$. From arc $(3, 2)$, we find $\theta \leq 6$. So we now set $\theta = 1$ and have x_{24} exit from the basis at its upper bound. The new bfs is given in Figure 61.

The current set of basic values corresponds to the spanning tree $(1, 3)$, $(3, 2)$, and $(3, 4)$. The new values of the simplex multipliers are found by solving

$$y_1 = 0, \quad y_1 - y_3 = 3, \quad y_3 - y_2 = 2, \quad y_3 - y_4 = 6$$

which yields $y_1 = 0$, $y_2 = -5$, $y_3 = -3$, $y_4 = -9$. The coefficient of each nonbasic variable in row 0 is now

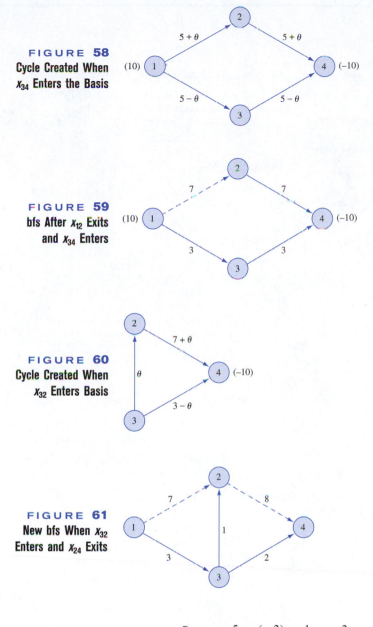

FIGURE 58
Cycle Created When x_{34} Enters the Basis

FIGURE 59
bfs After x_{12} Exits and x_{34} Enters

FIGURE 60
Cycle Created When x_{32} Enters Basis

FIGURE 61
New bfs When x_{32} Enters and x_{24} Exits

$$\bar{c}_{23} = -5 - (-3) - 1 = -3 \qquad \text{(Satisfies optimality condition)}$$
$$\bar{c}_{12} = 0 - (-5) - 4 = 1 \qquad \text{(Satisfies optimality condition)}$$
$$\bar{c}_{24} = -5 - (-9) - 3 = 1 \qquad \text{(Satisfies optimality condition)}$$

Thus, the current bfs is optimal. The optimal solution to the MCNFP is

Basic variables: $\quad x_{13} = 3, \qquad x_{32} = 1, \qquad x_{34} = 2$

Nonbasic variables at their upper bound: $\quad x_{12} = 7, \qquad x_{24} = 8$

Nonbasic variable at lower bound: $\quad x_{23} = 0$

The optimal z-value is obtained from

$$z = 7(4) + 3(3) + 1(2) + 8(3) + 2(6) = \$75$$

PROBLEMS

Group A

1 Consider the problem of finding the shortest path from node 1 to node 6 in Figure 2.

   a Formulate this problem as an MCNFP.

   b Find a bfs in which x_{12}, x_{24}, and x_{46} are positive. (*Hint:* A degenerate bfs will be obtained.)

   c Use the network simplex to find the shortest path from node 1 to node 6.

2 For the MCNFP in Figure 62, find a bfs.

3 Find the optimal solution to the MCNFP in Figure 63 using the bfs in Figure 64 as a starting basis.

4 Find a bfs for the network in Figure 65.

5 Find the optimal solution to the MCNFP in Figure 66 using the bfs in Figure 67 as a starting basis.

FIGURE 65

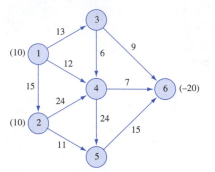

FIGURE 62

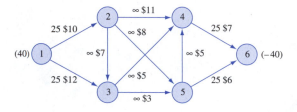

FIGURE 63

FIGURE 66

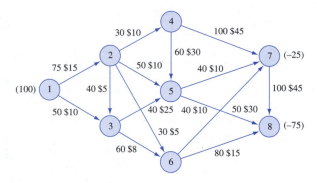

FIGURE 64

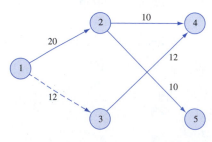

FIGURE 67

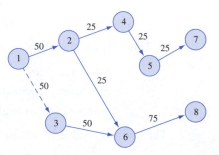

Shortest Path Problems

Suppose we want to find the shortest path from node 1 to node j in a network in which all arcs have non-negative lengths.

Dijkstra's Algorithm

1 Label node 1 with a permanent label of 0. Then label each arc connected to node 1 by a single arc with a "temporary" label equal to the length of the arc joining node 1 and node i. Remaining nodes will have a temporary label of ∞. Choose the node with the smallest temporary label and make this label permanent.

2 Suppose that node i is the $(k + 1)$th node to be given a permanent label. For each node j that now has a temporary label and is connected to node i by an arc, replace node j's temporary label by min {node j's current temporary label, (node i's permanent label) + length of arc (i, j)}. Make the smallest temporary label a permanent label. Continue this process until all nodes have permanent labels. To find the shortest path from node 1 to node j, work backward from node j by finding nodes having labels differing by exactly the length of the connecting arc. If the shortest path from node 1 to node j is desired, stop the labeling process as soon as node j receives a permanent label.

The Shortest Path Problem as a Transshipment Problem

To find the shortest path from node 1 to node j, try to minimize the cost of sending one unit from node 1 to node j (with all other nodes in the network being transshipment points), where the cost of sending one unit from node k to node k' is the length of arc (k, k') if such an arc exists and is M (a large positive number) if such an arc does not exist. As in Section 7.6, the cost of shipping one unit from a node to itself is zero.

Maximum-Flow Problems

We can find the maximum flow from source to sink in a network by linear programming or by the Ford–Fulkerson method.

Finding Maximum Flow by Linear Programming

Let

$$x_0 = \text{flow through artificial arc going from sink to source}$$

Then to find the maximum flow from source to sink, maximize x_0 subject to the following two sets of constraints:

1 The flow through each arc must be non-negative and cannot exceed the arc capacity.

2 Flow into node i = flow out of node i (Conservation of flow)

Finding Maximum Flow by the Ford–Fulkerson Method

Let

$$I = \text{set of arcs in which flow may be increased}$$
$$R = \text{set of arcs in which flow may be reduced}$$

Step 1   Find a feasible flow (setting each arc's flow to zero will do).

Step 2   Using the following procedure, try to find a chain of labeled arcs and nodes that can be used to label the sink. Label the source. Then label vertices and arcs (except for arc a_0) according to the following rules: (1) If vertex x is labeled, then vertex y is unlabeled and arc (x, y) is a member of I; then label vertex y and arc (x, y). Arc (x, y) is called a **forward arc.** (2) If vertex y is unlabeled, then vertex x is labeled and arc (y, x) is a member of R; then label vertex y and arc (y, x). Arc (y, x) is called a **backward arc.**

If the sink cannot be labeled, the current feasible flow is a maximum flow; if the sink is labeled, go on to step 3.

Step 3   If the chain used to label the sink consists entirely of forward arcs, the flow through each of the forward arcs in the chain may be increased, thereby increasing the flow from source to sink. If the chain used to label the sink consists of both forward and backward arcs, increase the flow in each forward arc in the chain and decrease the flow in each backward arc in the chain. Again, this will increase the flow from source to sink. Return to step 2.

Critical Path Method

Assuming the duration of each activity is known, the critical path method (CPM) may be used to find the duration of a project.

Rules for Constructing an AOA Project Diagram

1   Node 1 represents the start of the project. An arc should lead from node 1 to represent each activity that has no predecessors.

2   A node (called the finish node) representing the completion of the project should be included in the network.

3   Number the nodes in the network so that the node representing the completion of an activity always has a larger number than the node representing the beginning of an activity (there may be more than one numbering scheme that satisfies rule 3).

4   An activity should not be represented by more than one arc in the network.

5   Two nodes can be connected by at most one arc.

To avoid violating rules 4 and 5, it is sometimes necessary to utilize a **dummy activity** that takes zero time.

Computation of Early Event Time

The early event time for node i, denoted $ET(i)$, is the earliest time at which the event corresponding to node i can occur. We compute $ET(i)$ as follows:

Step 1   Find each prior event to node i that is connected by an arc to node i. These events are the **immediate predecessors** of node i.

Step 2   To the ET for each immediate predecessor of node i, add the duration of the activity connecting the immediate predecessor to node i.

Step 3 $ET(i)$ equals the maximum of the sums computed in step 2.

Computation of Late Event Time

The late event time for node i, denoted $LT(i)$, is the latest time at which the event corresponding to node i can occur without delaying the completion of the project. We compute $LT(i)$ as follows:

Step 1   Find each node that occurs after node i and is connected to node i by an arc. These events are the **immediate successors** of node i.

Step 2   From the LT for each immediate successor to node i, subtract the duration of the activity joining the successor to node i.

Step 3 $LT(i)$ is the smallest of the differences determined in step 2.

Total Float

For an arbitrary arc representing activity (i, j), the total float (denoted $TF(i, j)$ of the activity represented by (i, j) is the amount by which the starting time of activity (i, j) could be delayed beyond its earliest possible starting time without delaying the completion of the project (assuming no other activities are delayed):

$$TF(i, j) = LT(j) - ET(i) - t_{ij} \quad [t_{ij} = \text{duration of activity represented by arc } (i, j)]$$

Any activity with a total float of zero is a **critical activity.** A path from node 1 to the finish node that consists entirely of critical activities is called a **critical path.** Any critical path (there may be more than one in a project network) is the longest path in the network from the start node (node 1) to the finish node. If the start of a critical activity is delayed, or if the duration of a critical activity is longer than expected, then the completion of the project will be delayed.

Free Float

The free float of the activity corresponding to arc (i, j), denoted by $FF(i, j)$, is the amount by which the starting time of the activity corresponding to arc (i, j) (or the duration of the activity) can be delayed without delaying the start of any later activity beyond its earliest possible starting time:

$$FF(i, j) = ET(j) - ET(i) - t_{ij}$$

Linear programming can be used to find a critical path and the duration of the project. Let

$$x_j = \text{time at which node } j \text{ in project network occurs}$$

$$F = \text{node representing finish or completion of the project}$$

To find a critical path, minimize $z = x_F - x_1$ subject to

$$x_j \geq x_i + t_{ij} \quad \text{or} \quad x_j - x_i \geq t_{ij} \quad \text{for each arc}$$

$$x_j \text{ urs}$$

The optimal objective function value is the length of any critical path (or time to project completion). To find a critical path, simply find a path from node 1 to node F for which each arc in the path is represented by an arc (i, j) whose constraint $(x_j - x_i \geq t_{ij})$ has a dual price of -1.

Linear programming can also be used to determine the minimum-cost method of reducing the duration of activities (crashing) to meet a project completion deadline.

PERT

If the durations of the project's activities are not known with certainty, then PERT may be used to estimate the probability that the project will be completed in a specified amount of time. PERT requires that for each activity the following three numbers be specified:

a = estimate of the activity's duration under the most favorable conditions

b = estimate of the activity's duration under the least favorable conditions

m = most likely value for the activity's duration

If the estimates a, b, and m refer to the activity represented by arc (i, j), then $\mathbf{T}_{ij}$ is the random variable representing the duration of the activity represented by arc (i, j). $\mathbf{T}_{ij}$ has (approximately) the following properties:

$$E(\mathbf{T}_{ij}) = \frac{a + 4m + b}{6}$$

$$\text{var}\mathbf{T}_{ij} = \frac{(b - a)^2}{36}$$

Then

$$\sum_{(i, j)\in\text{path}} E(\mathbf{T}_{ij}) = \text{expected duration of activities on any path}$$

$$\sum_{(i, j)\in\text{path}} \text{var}\mathbf{T}_{ij} = \text{variance of duration of activities on any path}$$

Assuming (sometimes incorrectly) that the critical path found by CPM is the critical path, and assuming that the duration of the critical path is normally distributed, the preceding equations may be used to estimate the probability that the project will be completed within any specified length of time.

Minimum-Cost Network Flow Problems

The transportation, assignment, transshipment, shortest path, maximum-flow, and critical path problems are all special cases of the minimum-cost network flow problem (MCNFP).

x_{ij} = number of units of flow sent from node i to node j through arc (i, j)

b_i = net supply (outflow − inflow) at node i

c_{ij} = cost of transporting one unit of flow from node i to node j via arc (i, j)

L_{ij} = lower bound on flow through arc (i, j) (if there is no lower bound, let $L_{ij} = 0$)

U_{ij} = upper bound on flow through arc (i, j) (if there is no upper bound, let $U_{ij} = \infty$)

Then an MCNFP may be written as

$$\min \sum_{\text{all arcs}} c_{ij}x_{ij}$$

$$\text{s.t.} \quad \sum_{j} x_{ij} - \sum_{k} x_{ki} = b_i \quad \text{(for each node } i \text{ in the network)}$$

$$L_{ij} \leq x_{ij} \leq U_{ij} \quad \text{(for each arc in the network)}$$

The first set of constraints are the **flow balance equations,** and the second set of constraints express limitations on arc capacities.

Any MCNFP may be solved by a computer code using the **network simplex;** the user need only input the nodes and arcs in the network, the c_{ij}'s and arc capacity for each arc, and the b_i's for each node. Formulation of a problem as an MCNFP may require adding a dummy point to the problem.

Minimum Spanning Tree Problems

The following method (MST algorithm) may be used to find a minimum spanning tree for a network:

Step 1 Begin at any node i, and join node i to the node in the network (node j) that is closest to node i. The two nodes i and j now form a connected set of nodes $C = \{i, j\}$ and arc (i, j) will be in the minimum spanning tree. The remaining nodes in the network (C') are the unconnected set of nodes.

Step 2 Choose a member of $C'(n)$ that is closest to some node in C. Let m represent the node in C that is closest to n. Then the arc (m, n) will be in the minimum spanning tree. Update C and C'. Because n is now connected to $\{i, j\}$, C now equals $\{i, j, n\}$, and we must eliminate node n from C'.

Step 3 Repeat this process until a minimum spanning tree is found. Ties for closest node and arc may be broken arbitrarily.

Network Simplex Method

Step 1 Determine a starting bfs. The $n - 1$ basic variables will correspond to a spanning tree. Indicate nonbasic variables at their upper bound by dashed arcs.

Step 2 Compute $y_1, y_2, \ldots y_n$ (often called the *simplex multipliers*) by solving $y_1 = 0$, $y_i - y_j = c_{ij}$ for all basic variables x_{ij}. For all nonbasic variables, determine the row 0 coefficient $\bar{c}_{ij}$ from $\bar{c}_{ij} = y_i - y_j - c_{ij}$. The current bfs is optimal if $\bar{c}_{ij} \leq 0$ for all $x_{ij} = L_{ij}$ and $\bar{c}_{ij} \geq 0$ for all $x_{ij} = U_{ij}$. If the bfs is not optimal, then choose the nonbasic variable that most violates the optimality conditions as the entering basic variable.

Step 3 Identify the cycle (there will be exactly one!) created by adding the arc corresponding to the entering variable to the current spanning tree of the current bfs. Use conservation of flow to determine the new values of the variables in the cycle. The variable that first hits its upper or lower bound as the value of the entering basic variable is changed exits the basis.

Step 4 Find the new bfs by changing the flows of the arcs in the cycle found in step 3. Go to step 2.

REVIEW PROBLEMS

Group A

1 A truck must travel from New York to Los Angeles. As shown in Figure 68, a variety of routes are available. The number associated with each arc is the number of gallons of fuel required by the truck to traverse the arc.

    a Use Dijkstra's algorithm to find the route from New York to Los Angeles that uses the minimum amount of gas.

b Formulate a balanced transportation problem that could be used to find the route from New York to Los Angeles that uses the minimum amount of gas.

c Formulate as an MCNFP the problem of finding the New York to Los Angeles route that uses the minimum amount of gas.

FIGURE **68**

Network for Problem 1

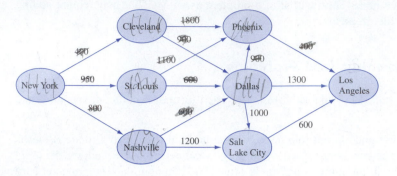

2 Telephone calls from New York to Los Angeles are transported as follows: The call is sent first to either Chicago or Memphis, then routed through either Denver or Dallas, and finally sent to Los Angeles. The number of phone lines joining each pair of cities is shown in Table 39.

    a Formulate an LP that can be used to determine the maximum number of calls that can be sent from New York to Los Angeles at any given time.

    b Use the Ford–Fulkerson method to determine the maximum number of calls that can be sent from New York to Los Angeles at any given time.

TABLE **39**

Cities	No. of Telephone Lines
N.Y.–Chicago	500
N.Y.–Memphis	400
Chicago–Denver	300
Chicago–Dallas	250
Memphis–Denver	200
Memphis–Dallas	150
Denver–L.A.	400
Dallas–L.A.	350

3 Before a new product can be introduced, the activities in Table 40 must be completed (all times are in weeks).

    a Draw the project diagram.

    b Determine all critical paths and critical activities.

    c Determine the total float and free float for each activity.

    d Set up an LP that can be used to determine the critical path.

    e Formulate an MCNFP that can be used to find the critical path.

    f It is now 12 weeks before Christmas. What is the probability that the product will be in the stores before Christmas?

    g The duration of each activity can be reduced by up to 2 weeks at the following cost per week: A, $80; B, $60; C, $30; D, $60; E, $40; F, $30; G, $20. Assuming that the duration of each activity is known with certainty, formulate an LP that will minimize the cost of getting the product into the stores by Christmas.

4 During the next three months, Shoemakers, Inc. must meet (on time) the following demands for shoes: month 1, 1,000 pairs; month 2, 1,500 pairs; month 3, 1,800 pairs. It takes 1 hour of labor to produce a pair of shoes. During each of the next three months, the following number of regular-time labor hours are available: month 1, 1,000 hours; month 2, 1,200 hours; month 3, 1,200 hours. Each month, the company can require workers to put in up to 400 hours of overtime. Workers

TABLE **40**

Activity	Description	Predecessors	Duration	a	b	m
A	Design the product	—	6	2	10	6
B	Survey the market	—	5	4	6	5
C	Place orders for raw materials	A	3	2	4	3
D	Receive raw materials	C	2	1	3	2
E	Build prototype of product	A, D	3	1	5	3
F	Develop ad campaign	B	2	3	5	4
G	Set up plan for mass production	E	4	2	6	4
H	Deliver product to stores	G, F	2	0	4	2

are paid only for the hours they work, and a worker receives $4 per hour for regular-time work and $6 per hour for overtime work. At the end of each month, a holding cost of $1.50 per pair of shoes is incurred. Formulate an MCNFP that can be used to minimize the total cost incurred in meeting the demands of the next three months. A formulation requires drawing the appropriate network and determining the c_{ij}'s, b_i's, and arc capacities. How would you modify your answer if demand could be backlogged (all demand must still be met by the end of month 3) at a cost of $20/pair/month?

5 Find a minimum spanning tree for the network in Figure 68.

6 A company produces a product at two plants, 1 and 2. The unit production cost and production capacity during each period are given in Table 41. The product is instantaneously shipped to the company's only customer according to the unit shipping costs given in Table 42. If a unit is produced and shipped during period 1, it can still be used to meet a period 2 demand, but a holding cost of $13 per unit in inventory is assessed. At the end of period 1, at most six units may be held in inventory. Demands are as follows: period 1, 9; period 2, 11. Formulate an MCNFP that can be used to minimize the cost of meeting all demands on time. Draw the network and determine the net outflow at each node, the arc capacities, and shipping costs.

7 A project is considered completed when activities A–F have all been completed. The duration and predecessors of each activity are given in Table 43. The LINDO output in Figure 69 can be used to determine the critical path for this project.

a Use the LINDO output to draw the project network. Indicate the activity represented by each arc.

b Determine a critical path in the network. What is the earliest the project can be completed?

8[†] State University has three professors who each teach four courses per year. Each year, four sections of marketing, finance, and production must be offered. At least one section of each class must be offered during each semester (fall and spring). Each professor's time preference and preference for teaching various courses are given in Table 44.

TABLE 41

	Unit Production Cost ($)	Capacity
Plant 1 (period 1)	33	7
Plant 1 (period 2)	43	4
Plant 2 (period 1)	30	9
Plant 2 (period 2)	41	9

TABLE 42

	Period 1	Period 2
Plant 1 to customer	$51	$60
Plant 2 to customer	$42	$71

[†]Based on Mulvey (1979).

FIGURE 69

```
MIN      X6 - X1
SUBJECT TO
       2)  - X1 + X3 >=    3
       3)    X4 - X2 >=    1
       4)  - X3 + X4 >=    0
       5)  - X4 + X5 >=    7
       6)  - X3 + X5 >=    5
       7)    X6 - X5 >=    5
       8)    X3 - X2 >=    0
       9)  - X1 + X2 >=    2
END

        LP OPTIMUM FOUND AT STEP    3

              OBJECTIVE FUNCTION VALUE

      1)          15.0000000

      VARIABLE        VALUE          REDUCED COST
         X6        15.000000           0.000000
         X1         0.000000           0.000000
         X3         3.000000           0.000000
         X4         3.000000           0.000000
         X2         2.000000           0.000000
         X5        10.000000           0.000000

      ROW        SLACK OR SURPLUS     DUAL PRICES
        2)          0.000000          -1.000000
        3)          0.000000           0.000000
        4)          0.000000          -1.000000
        5)          0.000000          -1.000000
        6)          2.000000           0.000000
        7)          0.000000          -1.000000
        8)          1.000000           0.000000
        9)          0.000000           0.000000

      NO. ITERATIONS=       3
```

TABLE 43

Activity	Duration	Immediate Predecessors
A	2	—
B	3	—
C	1	A
D	5	A, B
E	7	B, C
F	5	D, E

The total satisfaction a professor earns teaching a class is the sum of the semester satisfaction and the course satisfaction. Thus, professor 1 derives a satisfaction of $3 + 6 = 9$ from teaching marketing during the fall semester. Formulate an MCNFP that can be used to assign professors to courses so as to maximize the total satisfaction of the three professors.

Group B

9[†] During the next two months, Machineco must meet (on time) the demands for three types of products shown in Table 45. Two machines are available to produce these

[†]This problem is based on Brown, Geoffrion, and Bradley (1981).

TABLE 44

	Professor 1	Professor 2	Professor 3
Fall Preference	3	5	4
Spring Preference	4	3	4
Marketing	6	4	5
Finance	5	6	4
Production	4	5	6

TABLE 45

Month	Product 1	Product 2	Product 3
1	50 units	70 units	80 units
2	60 units	90 units	120 units

products. Machine 1 can only produce products 1 and 2, and machine 2 can only produce products 2 and 3. Each machine can be used for up to 40 hours per month. Table 46 shows the time required to produce one unit of each product (independent of the type of machine); the cost of producing one unit of each product on each type of machine; and the cost of holding one unit of each product in inventory for one month. Formulate an MCNFP that could be used to minimize the total cost of meeting all demands on time.

TABLE 46

Product	Production Time (minutes)	Production Cost ($) Machine 1	Production Cost ($) Machine 2	Holding Cost ($)
1	30	40	—	15
2	20	45	60	10
3	15	—	55	5

REFERENCES

Brown, G., A. Geoffrion, and G. Bradley. "Production and Sales Planning with Limited Shared Tooling at the Key Operation," *Management Science* 27(1981):247–259.

Glover, F., et al. "The Passenger-Mix Problem in the Scheduled Airlines," *Interfaces* 12(1982):73–80.

Mulvey, M. "Strategies in Modeling: A Personnel Example," *Interfaces* 9(no. 3, 1979):66–75.

Peterson, I. "Proven Path for Limiting Shortest Shortcut," *Science News* December 22, 1990: 389.

Ravidran, A. "On Compact Book Storage in Libraries," *Opsearch* 8(1971).

The following three texts contain an overview of networks at an elementary level:

Chachra, V., P. Ghare, and J. Moore. *Applications of Graph Theory Algorithms.* New York: North-Holland, 1979.

Mandl, C. *Applied Network Optimization.* Orlando, Fla.: Academic Press, 1979.

Phillips, D., and A. Diaz. *Fundamentals of Network Analysis.* Englewood Cliffs, N.J.: Prentice Hall, 1981.

The two best comprehensive references on network models are:

Ahuja, R., Magnanti, T., and Orlin, J. *Network Flows: Theory Algorithms and Applications.* Englewood-Cliffs, N.J.: Prentice-Hall, 1993.

Bersetkas, D. *Linear Network Optimization: Algorithms and Codes.* Cambridge, Mass. MIT Press, 1991.

Detailed discussion of methods for solving shortest path problems can be found in the following three texts:

Denardo, E. *Dynamic Programming: Theory and Applications.* Englewood Cliffs, N.J.: Prentice Hall, 1982.

Evans, T., and E. Minieka. *Optimization Algorithms for Networks and Graphs.* New York: Dekker, 1992. Also discusses minimum spanning tree algorithms.

Hu, T. *Combinatorial Algorithms.* Reading, Mass.: Addison-Wesley, 1982. Also discusses minimum spanning tree algorithms.

Evans and Minieka (1992) and Hu (1982) discuss the maximum-flow problem in detail, as do the following three texts:

Ford, L., and D. Fulkerson. *Flows in Networks.* Princeton, N.J.: Princeton University Press, 1962.

Jensen, P., and W. Barnes. *Network Flow Programming.* New York: Wiley, 1980.

Lawler, E. *Combinatorial Optimization: Networks and Matroids.* Chicago: Holt, Rinehart & Winston, 1976.

Excellent discussions of CPM and PERT are contained in:

Hax, A., and D. Candea. *Production and Inventory Management.* Englewood Cliffs, N.J.: Prentice Hall, 1984.

Wiest, J., and F. Levy. *A Management Guide to PERT/CPM,* 2d ed. Englewood Cliffs, N.J.: Prentice Hall, 1977.

Jensen and Barnes (1980) and the following references each contain a detailed discussion of the network simplex method used to solve an MCNFP.

Chvàtal, V. *Linear Programming.* San Francisco: Freeman, 1983.

Shapiro, J. *Mathematical Programming: Structures and Algorithms.* New York: Wiley, 1979.

Wu, N., and R. Coppins. *Linear Programming and Extensions.* New York: McGraw-Hill, 1981.

An excellent discussion of applications of MCNFPs is contained in the following:

Glover, F., D. Klingman, and N. Phillips. *Network Models and Their Applications in Practice.* New York: Wiley, 1992.

9

Integer Programming

Recall that we defined integer programming problems in our discussion of the Divisibility Assumption in Section 3.1. Simply stated, an *integer programming problem* (IP) is an LP in which some or all of the variables are required to be non-negative integers.[†]

In this chapter (as for LPs in Chapter 3), we find that many real-life situations may be formulated as IPs. Unfortunately, we will also see that IPs are usually much harder to solve than LPs.

In Section 9.1, we begin with necessary definitions and some introductory comments about IPs. In Section 9.2, we explain how to formulate integer programming models. We also discuss how to solve IPs on the computer with LINDO, LINGO, and Excel Solver. In Sections 9.3–9.8, we discuss other methods used to solve IPs.

9.1 Introduction to Integer Programming

An IP in which all variables are required to be integers is called a **pure integer programming problem.** For example,

$$\max z = 3x_1 + 2x_2$$
$$\text{s.t.} \quad x_1 + x_2 \leq 6 \tag{1}$$
$$x_1, x_2 \geq 0, x_1, x_2 \text{ integer}$$

is a pure integer programming problem.

An IP in which only some of the variables are required to be integers is called a **mixed integer programming problem.** For example,

$$\max z = 3x_1 + 2x_2$$
$$\text{s.t.} \quad x_1 + x_2 \leq 6$$
$$x_1, x_2 \geq 0, x_1 \text{ integer}$$

is a mixed integer programming problem (x_2 is not required to be an integer).

An integer programming problem in which all the variables must equal 0 or 1 is called a 0–1 IP. In Section 9.2, we see that 0–1 IPs occur in surprisingly many situations.[‡] The following is an example of a 0–1 IP:

$$\max z = x_1 - x_2$$
$$\text{s.t.} \quad x_1 + 2x_2 \leq 2$$
$$2x_1 - x_2 \leq 1 \tag{2}$$
$$x_1, x_2 = 0 \text{ or } 1$$

Solution procedures especially designed for 0–1 IPs are discussed in Section 9.7.

[†]A nonlinear integer programming problem is an optimization problem in which either the objective function or the left-hand side of some of the constraints are nonlinear functions and some or all of the variables must be integers. Such problems may be solved with LINGO or Excel Solver.
[‡]Actually, any pure IP can be reformulated as an equivalent 0–1 IP (Section 9.7).

The concept of LP relaxation of an integer programming problem plays a key role in the solution of IPs.

DEFINITION ■ The LP obtained by omitting all integer or 0–1 constraints on variables is called the **LP relaxation** of the IP. ■

For example, the LP relaxation of (1) is

$$\max z = 3x_1 + 2x_2$$
$$\text{s.t.} \quad x_1 + x_2 \leq 6 \tag{1'}$$
$$x_1, x_2 \geq 0$$

and the LP relaxation of (2) is

$$\max z = x_1 - x_2$$
$$\text{s.t.} \quad x_1 + 2x_2 \leq 2$$
$$2x_1 - x_2 \leq 1 \tag{2'}$$
$$x_1, x_2 \geq 0$$

Any IP may be viewed as the LP relaxation plus additional constraints (the constraints that state which variables must be integers or be 0 or 1). Hence, the LP relaxation is a less constrained, or more relaxed, version of the IP. This means that *the feasible region for any IP must be contained in the feasible region for the corresponding LP relaxation.* For any IP that is a max problem, this implies that

$$\text{Optimal } z\text{-value for LP relaxation} \geq \text{optimal } z\text{-value for IP} \tag{3}$$

This result plays a key role when we discuss the solution of IPs.

To shed more light on the properties of integer programming problems, we consider the following simple IP:

$$\max z = 21x_1 + 11x_2$$
$$\text{s.t.} \quad 7x_1 + 4x_2 \leq 13 \tag{4}$$
$$x_1, x_2 \geq 0; \ x_1, x_2 \text{ integer}$$

From Figure 1, we see that the feasible region for this problem consists of the following set of points: $S = \{(0, 0), (0, 1), (0, 2), (0, 3), (1, 0), (1, 1)\}$. Unlike the feasible region for any LP, the one for (4) is not a convex set. By simply computing and comparing the z-values for each of the six points in the feasible region, we find the optimal solution to (4) is $z = 33$, $x_1 = 0$, $x_2 = 3$.

If the feasible region for a pure IP's LP relaxation is bounded, as in (4), then the feasible region for the IP will consist of a finite number of points. In theory, such an IP could be solved (as described in the previous paragraph) by enumerating the z-values for each feasible point and determining the feasible point having the largest z-value. The problem with this approach is that most actual IPs have feasible regions consisting of billions of feasible points. In such cases, a complete enumeration of all feasible points would require a large amount of computer time. As we explain in Section 9.3, IPs often are solved by cleverly enumerating all the points in the IP's feasible region.

Further study of (4) sheds light on other interesting properties of IPs. Suppose that a naive analyst suggests the following approach for solving an IP: First solve the LP relaxation; then round off (to the nearest integer) each variable that is required to be an integer and that assumes a fractional value in the optimal solution to the LP relaxation.

Applying this approach to (4), we first find the optimal solution to the LP relaxation: $x_1 = \frac{13}{7}$, $x_2 = 0$. Rounding this solution yields the solution $x_1 = 2$, $x_2 = 0$ as a possible

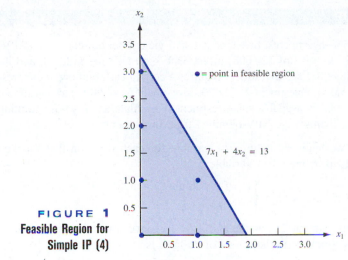

FIGURE 1

Feasible Region for Simple IP (4)

optimal solution to (4). But $x_1 = 2$, $x_2 = 0$ is infeasible for (4), so it cannot possibly be the optimal solution to (4). Even if we round x_1 downward (yielding the candidate solution $x_1 = 1$, $x_2 = 0$), we do not obtain the optimal solution ($x_1 = 0$, $x_2 = 3$ is the optimal solution).

For some IPs, it can even turn out that every roundoff of the optimal solution to the LP relaxation is infeasible. To see this, consider the following IP:

$$\max z = 4x_1 + x_2$$
$$\text{s.t.} \quad 2x_1 + x_2 \leq 5$$
$$2x_1 + 3x_2 = 5$$
$$x_1, x_2 \geq 0; \ x_1, x_2 \text{ integer}$$

The optimal solution to the LP relaxation for this IP is $z = 10$, $x_1 = \frac{5}{2}$, $x_2 = 0$. Rounding off this solution, we obtain either the candidate $x_1 = 2$, $x_2 = 0$ or the candidate $x_1 = 3$, $x_2 = 0$. Neither candidate is a feasible solution to the IP.

Recall from Chapter 4 that the simplex algorithm allowed us to solve LPs by going from one basic feasible solution to a better one. Also recall that in most cases, the simplex algorithm examines only a small fraction of all basic feasible solutions before the optimal solution is obtained. This property of the simplex algorithm enables us to solve relatively large LPs by expending a surprisingly small amount of computational effort. Analogously, one would hope that an IP could be solved via an algorithm that proceeded from one feasible integer solution to a better feasible integer solution. Unfortunately, no such algorithm is known.

In summary, even though the feasible region for an IP is a subset of the feasible region for the IP's LP relaxation, the IP is usually much more difficult to solve than the IP's LP relaxation.

9.2 Formulating Integer Programming Problems

In this section, we show how practical solutions can be formulated as IPs. After completing this section, the reader should have a good grasp of the art of developing integer programming formulations. We begin with some simple problems and gradually build to more complicated formulations. Our first example is a capital budgeting problem reminiscent of the Star Oil problem of Section 3.6.

EXAMPLE 1   Capital Budgeting IP

Stockco is considering four investments. Investment 1 will yield a net present value (NPV) of $16,000; investment 2, an NPV of $22,000; investment 3, an NPV of $12,000; and investment 4, an NPV of $8,000. Each investment requires a certain cash outflow at the present time: investment 1, $5,000; investment 2, $7,000; investment 3, $4,000; and investment 4, $3,000. Currently, $14,000 is available for investment. Formulate an IP whose solution will tell Stockco how to maximize the NPV obtained from investments 1–4.

Solution   As in LP formulations, we begin by defining a variable for each decision that Stockco must make. This leads us to define a 0–1 variable:

$$x_j(j = 1, 2, 3, 4) = \begin{cases} 1 & \text{if investment } j \text{ is made} \\ 0 & \text{otherwise} \end{cases}$$

For example, $x_2 = 1$ if investment 2 is made, and $x_2 = 0$ if investment 2 is not made.

The NPV obtained by Stockco (in thousands of dollars) is

$$\text{Total NPV obtained by Stockco} = 16x_1 + 22x_2 + 12x_3 + 8x_4 \tag{5}$$

To see this, note that if $x_j = 1$, then (5) includes the NPV of investment j, and if $x_j = 0$, (5) does not include the NPV of investment j. This means that whatever combination of investments is undertaken, (5) gives the NPV of that combination of projects. For example, if Stockco invests in investments 1 and 4, then an NPV of $16,000 + 8,000 = \$24,000$ is obtained. This combination of investments corresponds to $x_1 = x_4 = 1$, $x_2 = x_3 = 0$, so (5) indicates that the NPV for this investment combination is $16(1) + 22(0) + 12(0) + 8(1) = \$24$ (thousand). This reasoning implies that Stockco's objective function is

$$\max z = 16x_1 + 22x_2 + 12x_3 + 8x_4 \tag{6}$$

Stockco faces the constraint that at most $14,000 can be invested. By the same reasoning used to develop (5), we can show that

$$\text{Total amount invested (in thousands of dollars)} = 5x_1 + 7x_2 + 4x_3 + 3x_4 \tag{7}$$

For example, if $x_1 = 0$, $x_2 = x_3 = x_4 = 1$, then Stockco makes investments 2, 3, and 4. In this case, Stockco must invest $7 + 4 + 3 = \$14$ (thousand). Equation (7) yields a total amount invested of $5(0) + 7(1) + 4(1) + 3(1) = \$14$ (thousand). Because at most $14,000 can be invested, x_1, x_2, x_3, and x_4 must satisfy

$$5x_1 + 7x_2 + 4x_3 + 3x_4 \leq 14 \tag{8}$$

Combining (6) and (8) with the constraints $x_j = 0$ or 1 ($j = 1, 2, 3, 4$) yields the following 0–1 IP:

$$\max z = 16x_1 + 22x_2 + 12x_3 + 8x_4$$
$$\text{s.t.} \quad 5x_1 + 7x_2 + 4x_3 + 3x_4 \leq 14 \tag{9}$$
$$x_j = 0 \text{ or } 1 \quad (j = 1, 2, 3, 4)$$

REMARKS   1   In Section 9.5, we show that the optimal solution to (9) is $x_1 = 0$, $x_2 = x_3 = x_4 = 1$, $z = \$42,000$. Hence, Stockco should make investments 2, 3, and 4, but not 1. Investment 1 yields a higher NPV per dollar invested than any of the others (investment 1 yields $3.20 per dollar invested, investment 2, $3.14; investment 3, $3; and investment 4, $2.67), so it may seem surprising that investment 1 is not undertaken. To see why the optimal solution to (9) does not involve making the "best" investment, note that any investment combination that includes investment 1 cannot use more than $12,000. This means that using investment 1 forces Stockco to forgo investing $2,000. On the other hand, the optimal investment combination uses all $14,000 of the investment budget. This en-

TABLE 1
Weights and Benefits for
Items in Josie's Knapsack

Item	Weight (Pounds)	Benefit
1	5	16
2	7	22
3	4	12
4	3	8

ables the optimal combination to obtain a higher NPV than any combination that includes investment 1. If, as in Chapter 3, fractional investments were allowed, the optimal solution to (9) would be $x_1 - x_2 = 1, x_3 = 0.50, x_4 = 0, z = \$44,000$, and investment 1 would be used. This simple example shows that the choice of modeling a capital budgeting problem as a linear programming or as an integer programming problem can significantly affect the optimal solution to the problem.

2 Any IP, such as (9), that has only one constraint is referred to as a **knapsack problem.** Suppose that Josie Camper is going on an overnight hike. There are four items Josie is considering taking along on the trip. The weight of each item and the benefit Josie feels she would obtain from each item are listed in Table 1.

Suppose Josie's knapsack can hold up to 14 lb of items. For $j = 1, 2, 3, 4$, define

$$x_j = \begin{cases} 1 \text{ if Josie takes item } j \text{ on the hike} \\ 0 \text{ otherwise} \end{cases}$$

Then Josie can maximize the total benefit by solving (9).

In the following example, we show how the Stockco formulation can be modified to handle additional constraints.

EXAMPLE 2 | **Capital Budgeting (Continued)**

Modify the Stockco formulation to account for each of the following requirements:

1 Stockco can invest in at most two investments.

2 If Stockco invests in investment 2, they must also invest in investment 1.

3 If Stockco invests in investment 2, they cannot invest in investment 4.

Solution **1** Simply add the constraint

$$x_1 + x_2 + x_3 + x_4 \leq 2 \tag{10}$$

to (9). Because any choice of three or four investments will have $x_1 + x_2 + x_3 + x_4 \geq 3$, (10) excludes from consideration all investment combinations involving three or more investments. Thus, (10) eliminates from consideration exactly those combinations of investments that do not satisfy the first requirement.

2 In terms of x_1 and x_2, this requirement states that if $x_2 = 1$, then x_1 must also equal 1. If we add the constraint

$$x_2 \leq x_1 \quad \text{or} \quad x_2 - x_1 \leq 0 \tag{11}$$

to (9), then we will have taken care of the second requirement. To show that (11) is equivalent to requirement 2, we consider two possibilities: either $x_2 = 1$ or $x_2 = 0$.

Case 1 $x_2 = 1$. If $x_2 = 1$, then the (11) implies that $x_1 \geq 1$. Because x_1 must equal 0 or 1, this implies that $x_1 = 1$, as required by 2.

Case 2 $x_2 = 0$. In this case, (11) reduces to $x_1 \geq 0$, which allows $x_1 = 0$ or $x_1 = 1$. In short, if $x_2 = 0$, (11) does not restrict the value of x_1. This is also consistent with requirement 2.

In summary, for any value of x_2, (11) is equivalent to requirement 2.

3 Simply add the constraint

$$x_2 + x_4 \leq 1 \tag{12}$$

to (9). We now show that for the two cases $x_2 = 1$ and $x_2 = 0$, (12) is equivalent to the third requirement.

Case 1 $x_2 = 1$. In this case, we are investing in investment 2, and requirement 3 implies that Stockco cannot invest in investment 4 (that is, x_4 must equal 0). Note that if $x_2 = 1$, then (12) does imply $1 + x_4 \leq 1$, or $x_4 \leq 0$. Thus, if $x_2 = 1$, then (12) is consistent with requirement 3.

Case 2 $x_2 = 0$. In this case, requirement 3 does not restrict the value of x_4. Note that if $x_2 = 0$, then (12) reduces to $x_4 \leq 1$, which also leaves x_4 free to equal 0 or 1.

Fixed-Charge Problems

Example 3 illustrates an important trick that can be used to formulate many location and production problems as IPs.

EXAMPLE 3 **Fixed-Charge IP**

Gandhi Cloth Company is capable of manufacturing three types of clothing: shirts, shorts, and pants. The manufacture of each type of clothing requires that Gandhi have the appropriate type of machinery available. The machinery needed to manufacture each type of clothing must be rented at the following rates: shirt machinery, $200 per week; shorts machinery, $150 per week; pants machinery, $100 per week. The manufacture of each type of clothing also requires the amounts of cloth and labor shown in Table 2. Each week, 150 hours of labor and 160 sq yd of cloth are available. The variable unit cost and selling price for each type of clothing are shown in Table 3. Formulate an IP whose solution will maximize Gandhi's weekly profits.

Solution As in LP formulations, we define a decision variable for each decision that Gandhi must make. Clearly, Gandhi must decide how many of each type of clothing should be manufactured each week, so we define

$$x_1 = \text{number of shirts produced each week}$$
$$x_2 = \text{number of shorts produced each week}$$
$$x_3 = \text{number of pants produced each week}$$

TABLE 2
Resource Requirements for Gandhi

Clothing Type	Labor (Hours)	Cloth (Square Yards)
Shirt	3	4
Shorts	2	3
Pants	6	4

TABLE 3

Revenue and Cost Information for Gandhi

Clothing Type	Sales Price ($)	Variable Cost ($)
Shirt	12	6
Shorts	8	4
Pants	15	8

Note that the cost of renting machinery depends only on the types of clothing produced, not on the amount of each type of clothing. This enables us to express the cost of renting machinery by using the following variables:

$$y_1 = \begin{cases} 1 & \text{if any shirts are manufactured} \\ 0 & \text{otherwise} \end{cases}$$

$$y_2 = \begin{cases} 1 & \text{if any shorts are manufactured} \\ 0 & \text{otherwise} \end{cases}$$

$$y_3 = \begin{cases} 1 & \text{if any pants are manufactured} \\ 0 & \text{otherwise} \end{cases}$$

In short, if $x_j > 0$, then $y_j = 1$, and if $x_j = 0$, then $y_j = 0$. Thus, Gandhi's weekly profits = (weekly sales revenue) − (weekly variable costs) − (weekly costs of renting machinery).

Also,

$$\text{Weekly cost of renting machinery} = 200y_1 + 150y_2 + 100y_3 \qquad \text{(13)}$$

To justify (13), note that it picks up the rental costs only for the machines needed to manufacture those products that Gandhi is actually manufacturing. For example, suppose that shirts and pants are manufactured. Then $y_1 = y_3 = 1$ and $y_2 = 0$, and the total weekly rental cost will be $200 + 100 = \$300$.

Because the cost of renting, say, shirt machinery does not depend on the number of shirts produced, the cost of renting each type of machinery is called a **fixed charge**. A fixed charge for an activity is a cost that is assessed whenever the activity is undertaken at a nonzero level. The presence of fixed charges will make the formulation of the Gandhi problem much more difficult.

We can now express Gandhi's weekly profits as

$$\text{Weekly profit} = (12x_1 + 8x_2 + 15x_3) - (6x_1 + 4x_2 + 8x_3)$$
$$- (200y_1 + 150y_2 + 100y_3)$$
$$= 6x_1 + 4x_2 + 7x_3 - 200y_1 - 150y_2 - 100y_3$$

Thus, Gandhi wants to maximize

$$z = 6x_1 + 4x_2 + 7x_3 - 200y_1 - 150y_2 - 100y_3$$

Because its supply of labor and cloth is limited, Gandhi faces the following two constraints:

Constraint 1 At most, 150 hours of labor can be used each week.

Constraint 2 At most, 160 sq yd of cloth can be used each week.

Constraint 1 is expressed by

$$3x_1 + 2x_2 + 6x_3 \leq 150 \qquad \text{(Labor constraint)} \qquad \text{(14)}$$

Constraint 2 is expressed by

$$4x_1 + 3x_2 + 4x_3 \leq 160 \qquad \text{(Cloth constraint)} \tag{15}$$

Observe that $x_j > 0$ and x_j integer ($j = 1, 2, 3$) must hold along with $y_j = 0$ or 1 ($j = 1, 2, 3$). Combining (14) and (15) with these restrictions and the objective function yields the following IP:

$$\begin{aligned}
\max z = 6x_1 &+ 4x_2 + 7x_3 - 200y_1 - 150y_2 - 100y_3 \\
\text{s.t.} \quad 3x_1 &+ 2x_2 + 6x_3 \leq 150 \\
4x_1 &+ 3x_2 + 4x_3 \leq 160 \\
&x_1, x_2, x_3 \geq 0; \ x_1, x_2, x_3 \text{ integer} \\
&y_1, y_2, y_3 = 0 \text{ or } 1
\end{aligned} \tag{IP 1}$$

The optimal solution to this problem is found to be $x_1 = 30$, $x_3 = 10$, $x_2 = y_1 = y_2 = y_3 = 0$. This cannot be the optimal solution to Gandhi's problem because it indicates that Gandhi can manufacture shirts and pants without incurring the cost of renting the needed machinery. The current formulation is incorrect because the variables y_1, y_2, and y_3 are not present in the constraints. This means that there is nothing to stop us from setting $y_1 = y_2 = y_3 = 0$. Setting $y_i = 0$ is certainly less costly than setting $y_i = 1$, so a minimum cost solution to (IP 1) will always set $y_i = 0$. Somehow we must modify (IP 1) so that whenever $x_i > 0$, $y_i = 1$ must hold. The following trick will accomplish this goal. Let M_1, M_2, and M_3 be three large positive numbers, and add the following constraints to (IP 1):

$$x_1 \leq M_1 y_1 \tag{16}$$

$$x_2 \leq M_2 y_2 \tag{17}$$

$$x_3 \leq M_3 y_3 \tag{18}$$

Adding (16)–(18) to IP 1 will ensure that if $x_i > 0$, then $y_i = 1$. To illustrate, let us show that (16) ensures that if $x_1 > 0$, then $y_1 = 1$. If $x_1 > 0$, then y_1 cannot be 0. For if $y_1 = 0$, then (16) would imply $x_1 \leq 0$ or $x_1 = 0$. Thus, if $x_1 > 0$, $y_1 = 1$ must hold. If any shirts are produced ($x_1 > 0$), (16) ensures that $y_1 = 1$, and the objective function will include the cost of the machinery needed to manufacture shirts. Note that if $y_1 = 1$, then (16) becomes $x_1 \leq M_1$, which does not unnecessarily restrict the value of x_1. If M_1 were not chosen large, however (say, $M_1 = 10$), then (16) would unnecessarily restrict the value of x_1. In general, M_i should be set equal to the maximum value that x_i can attain. In the current problem, at most 40 shirts can be produced (if Gandhi produced more than 40 shirts, the company would run out of cloth), so we can safely choose $M_1 = 40$. The reader should verify that we can choose $M_2 = 53$ and $M_3 = 25$.

If $x_1 = 0$, (16) becomes $0 \leq M_1 y_1$. This allows either $y_1 = 0$ or $y_1 = 1$. Because $y_1 = 0$ is less costly than $y_1 = 1$, the optimal solution will choose $y_1 = 0$ if $x_1 = 0$. In summary, we have shown that if (16)–(18) are added to (IP 1), then $x_i > 0$ will imply $y_i = 1$, and $x_i = 0$ will imply $y_i = 0$.

The optimal solution to the Gandhi problem is $z = \$75$, $x_3 = 25$, $y_3 = 1$. Thus, Gandhi should produce 25 pants each week.

The Gandhi problem is an example of a **fixed-charge problem.** In a fixed-charge problem, there is a cost associated with performing an activity at a nonzero level that does not depend on the level of the activity. Thus, in the Gandhi problem, if we make any shirts at all (no matter how many we make), we must pay the fixed charge of $200 to rent a shirt machine. Problems in which a decision maker must choose where to locate facilities are often fixed-charge problems. The decision maker must choose where to locate various facilities (such as plants, warehouses, or business offices), and a fixed charge is often associated with building or operating a facility. Example 4 is a typical location problem involving the idea of a fixed charge.

EXAMPLE 4   **The Lockbox Problem**

J. C. Nickles receives credit card payments from four regions of the country (West, Midwest, East, and South). The average daily value of payments mailed by customers from each region is as follows: the West, $70,000; the Midwest, $50,000; the East, $60,000; the South, $40,000. Nickles must decide where customers should mail their payments. Because Nickles can earn 20% annual interest by investing these revenues, it would like to receive payments as quickly as possible. Nickles is considering setting up operations to process payments (often referred to as lockboxes) in four different cities: Los Angeles, Chicago, New York, and Atlanta. The average number of days (from time payment is sent) until a check clears and Nickles can deposit the money depends on the city to which the payment is mailed, as shown in Table 4. For example, if a check is mailed from the West to Atlanta, it would take an average of 8 days before Nickles could earn interest on the check. The annual cost of running a lockbox in any city is $50,000. Formulate an IP that Nickles can use to minimize the sum of costs due to lost interest and lockbox operations. Assume that each region must send all its money to a single city and that there is no limit on the amount of money that each lockbox can handle.

Solution   Nickles must make two types of decisions. First, Nickles must decide where to operate lockboxes. We define, for $j = 1, 2, 3, 4$,

$$y_j = \begin{cases} 1 & \text{if a lockbox is operated in city } j \\ 0 & \text{otherwise} \end{cases}$$

Thus, $y_2 = 1$ if a lockbox is operated in Chicago, and $y_3 = 0$ if no lockbox is operated in New York. Second, Nickles must determine where each region of the country should send payments. We define (for $i, j = 1, 2, 3, 4$)

$$x_{ij} = \begin{cases} 1 & \text{if region } i \text{ sends payments to city } j \\ 0 & \text{otherwise} \end{cases}$$

For example, $x_{12} = 1$ if the West sends payments to Chicago, and $x_{23} = 0$ if the Midwest does not send payments to New York.

Nickles wants to minimize (total annual cost) = (annual cost of operating lockboxes) + (annual lost interest cost). To determine how much interest Nickles loses annually, we must determine how much revenue would be lost if payments from region i were sent to region j. For example, how much in annual interest would Nickles lose if customers from the West region sent payments to New York? On any given day, 8 days' worth, or $8(70,000) = \$560,000$ of West payments will be in the mail and will not be earning in-

TABLE 4
Average Number of Days from Mailing of Payment Until Payment Clears

| From | To | | | |
	City 1 (Los Angeles)	City 2 (Chicago)	City 3 (New York)	City 4 (Atlanta)
Region 1 West	2	6	8	8
Region 2 Midwest	6	2	5	5
Region 3 East	8	5	2	5
Region 4 South	8	5	5	2

terest. Because Nickles can earn 20% annually, each year West funds will result in $0.20(560,000) = \$112,000$ in lost interest. Similar calculations for the annual cost of lost interest for each possible assignment of a region to a city yield the results shown in Table 5. The lost interest cost from sending region i's payments to city j is only incurred if $x_{ij} = 1$, so Nickles's annual lost interest costs (in thousands) are

$$
\begin{aligned}
\text{Annual lost interest costs} = {}& 28x_{11} + 84x_{12} + 112x_{13} + 112x_{14} \\
& + 60x_{21} + 20x_{22} + 50x_{23} + 50x_{24} \\
& + 96x_{31} + 60x_{32} + 24x_{33} + 60x_{34} \\
& + 64x_{41} + 40x_{42} + 40x_{43} + 16x_{44}
\end{aligned}
$$

The cost of operating a lockbox in city i is incurred if and only if $y_i = 1$, so the annual lockbox operating costs (in thousands) are given by

$$
\text{Total annual lockbox operating cost} = 50y_1 + 50y_2 + 50y_3 + 50y_4
$$

Thus, Nickles's objective function may be written as

$$
\begin{aligned}
\min z = {}& 28x_{11} + 84x_{12} + 112x_{13} + 112x_{14} \\
& + 60x_{21} + 20x_{22} + 50x_{23} + 50x_{24} \\
& + 96x_{31} + 60x_{32} + 24x_{33} + 60x_{34} \\
& + 64x_{41} + 40x_{42} + 40x_{43} + 16x_{44} \\
& + 50y_1 + 50y_2 + 50y_3 + 50y_4
\end{aligned}
\tag{19}
$$

Nickles faces two types of constraints.

Type 1 Constraint   Each region must send its payments to a single city.

Type 2 Constraint   If a region is assigned to send its payments to a city, that city must have a lockbox.

TABLE 5
Calculation of Annual Lost Interest

Assignment	Annual Lost Interest Cost ($)
West to L.A.	$0.20(70,000)2 = 28,000$
West to Chicago	$0.20(70,000)6 = 84,000$
West to N.Y.	$0.20(70,000)8 = 112,000$
West to Atlanta	$0.20(70,000)8 = 112,000$
Midwest to L.A.	$0.20(50,000)6 = 60,000$
Midwest to Chicago	$0.20(50,000)2 = 20,000$
Midwest to N.Y.	$0.20(50,000)5 = 50,000$
Midwest to Atlanta	$0.20(50,000)5 = 50,000$
East to L.A.	$0.20(60,000)8 = 96,000$
East to Chicago	$0.20(60,000)5 = 60,000$
East to N.Y.	$0.20(60,000)2 = 24,000$
East to Atlanta	$0.20(60,000)5 = 60,000$
South to L.A.	$0.20(40,000)8 = 64,000$
South to Chicago	$0.20(40,000)5 = 40,000$
South to N.Y	$0.20(40,000)5 = 40,000$
South to Atlanta	$0.20(40,000)2 = 16,000$

The type 1 constraints state that for region i ($i = 1, 2, 3, 4$) exactly one of x_{i1}, x_{i2}, x_{i3}, and x_{i4} must equal 1 and the others must equal 0. This can be accomplished by including the following four constraints:

$$x_{11} + x_{12} + x_{13} + x_{14} = 1 \qquad \text{(West region constraint)} \qquad \textbf{(20)}$$
$$x_{21} + x_{22} + x_{23} + x_{24} = 1 \qquad \text{(Midwest region constraint)} \qquad \textbf{(21)}$$
$$x_{31} + x_{32} + x_{33} + x_{34} = 1 \qquad \text{(East region constraint)} \qquad \textbf{(22)}$$
$$x_{41} + x_{42} + x_{43} + x_{44} = 1 \qquad \text{(South region constraint)} \qquad \textbf{(23)}$$

The type 2 constraints state that if

$$x_{ij} = 1 \qquad \text{(that is, customers in region } i \text{ send payments to city } j) \qquad \textbf{(24)}$$

then y_j must equal 1. For example, suppose $x_{12} = 1$. Then there must be a lockbox at city 2, so $y_2 = 1$ must hold. This can be ensured by adding 16 constraints of the form

$$x_{ij} \leq y_j \qquad (i = 1, 2, 3, 4; j = 1, 2, 3, 4) \qquad \textbf{(25)}$$

If $x_{ij} = 1$, then (25) ensures that $y_j = 1$, as desired. Also, if $x_{1j} = x_{2j} = x_{3j} = x_{4j} = 0$, then (25) allows $y_j = 0$ or $y_j = 1$. As in the fixed-charge example, the act of minimizing costs will result in $y_j = 0$. In summary, the constraints in (25) ensure that Nickles pays for a lockbox at city i if it uses a lockbox at city i.

Combining (19)–(23) with the $4(4) = 16$ constraints in (25) and the 0–1 restrictions on the variables yields the following formulation:

$$\min z = 28x_{11} + 84x_{12} + 112x_{13} + 112x_{14} + 60x_{21} + 20x_{22} + 50x_{23} + 50x_{24}$$
$$+ 96x_{31} + 60x_{32} + 24x_{33} + 60x_{34} + 64x_{41} + 40x_{42} + 40x_{43} + 16x_{44}$$
$$+ 50y_1 + 50y_2 + 50y_3 + 50y_4$$

s.t.
$$x_{11} + x_{12} + x_{13} + x_{14} = 1 \qquad \text{(West region constraint)}$$
$$x_{21} + x_{22} + x_{23} + x_{24} = 1 \qquad \text{(Midwest region constraint)}$$
$$x_{31} + x_{32} + x_{33} + x_{34} = 1 \qquad \text{(East region constraint)}$$
$$x_{41} + x_{42} + x_{43} + x_{44} = 1 \qquad \text{(South region constraint)}$$
$$x_{11} \leq y_1, x_{21} \leq y_1, x_{31} \leq y_1, x_{41} \leq y_1, x_{12} \leq y_2, x_{22} \leq y_2, x_{32} \leq y_2, x_{42} \leq y_2,$$
$$x_{13} \leq y_3, x_{23} \leq y_3, x_{33} \leq y_3, x_{43} \leq y_3, x_{14} \leq y_4, x_{24} \leq y_4, x_{34} \leq y_4, x_{44} \leq y_4$$
$$\text{All } x_{ij} \text{ and } y_j = 0 \text{ or } 1$$

The optimal solution is $z = 242$, $y_1 = 1$, $y_3 = 1$, $x_{11} = 1$, $x_{23} = 1$, $x_{33} = 1$, $x_{43} = 1$. Thus, Nickles should have a lockbox operation in Los Angeles and New York. West customers should send payments to Los Angeles, and all other customers should send payments to New York.

There is an alternative way of modeling the Type 2 constraints. Instead of the 16 constraints of the form $x_{ij} \leq y_j$, we may include the following four constraints:

$$x_{11} + x_{21} + x_{31} + x_{41} \leq 4y_1 \qquad \text{(Los Angeles constraint)}$$
$$x_{12} + x_{22} + x_{32} + x_{42} \leq 4y_2 \qquad \text{(Chicago constraint)}$$
$$x_{13} + x_{23} + x_{33} + x_{43} \leq 4y_3 \qquad \text{(New York constraint)}$$
$$x_{14} + x_{24} + x_{34} + x_{44} \leq 4y_4 \qquad \text{(Atlanta constraint)}$$

For the given city, each constraint ensures that if the lockbox is used, then Nickles must pay for it. For example, consider $x_{14} + x_{24} + x_{34} + x_{44} \leq 4y_4$. The lockbox in Atlanta is used if $x_{14} = 1$, $x_{24} = 1$, $x_{34} = 1$, or $x_{44} = 1$. If any of these variables equals 1, then the Atlanta constraint ensures that $y_4 = 1$, and Nickles must pay for the lockbox. If all these variables are 0, then the act of minimizing costs will cause $y_4 = 0$, and the cost of the Atlanta lockbox will not be incurred. Why does the right-hand side of each constraint equal 4? This ensures that for each city, it is possible to send money from all four regions to the city. In Section 9.3, we discuss which of the two alternative formulations of the lockbox problem is easier for a computer to solve. The answer may surprise you!

Set-Covering Problems

The following example is typical of an important class of IPs known as set-covering problems.

EXAMPLE 5   Facility-Location Set-Covering Problem

There are six cities (cities 1–6) in Kilroy County. The county must determine where to build fire stations. The county wants to build the minimum number of fire stations needed to ensure that at least one fire station is within 15 minutes (driving time) of each city. The times (in minutes) required to drive between the cities in Kilroy County are shown in Table 6. Formulate an IP that will tell Kilroy how many fire stations should be built and where they should be located.

Solution For each city, Kilroy must determine whether to build a fire station there. We define the 0–1 variables x_1, x_2, x_3, x_4, x_5, and x_6 by

$$x_i = \begin{cases} 1 & \text{if a fire station is built in city } i \\ 0 & \text{otherwise} \end{cases}$$

Then the total number of fire stations that are built is given by $x_1 + x_2 + x_3 + x_4 + x_5 + x_6$, and Kilroy's objective function is to minimize

$$z = x_1 + x_2 + x_3 + x_4 + x_5 + x_6$$

What are Kilroy's constraints? Kilroy must ensure that there is a fire station within 15 minutes of each city. Table 7 indicates which locations can reach the city in 15 minutes or less. To ensure that at least one fire station is within 15 minutes of city 1, we add the constraint

$$x_1 + x_2 \geq 1 \qquad \text{(City 1 constraint)}$$

This constraint ensures that $x_1 = x_2 = 0$ is impossible, so at least one fire station will be built within 15 minutes of city 1. Similarly the constraint

$$x_1 + x_2 + x_6 \geq 1 \qquad \text{(City 2 constraint)}$$

ensures that at least one fire station will be located within 15 minutes of city 2. In a similar fashion, we obtain constraints for cities 3–6. Combining these six constraints with the

TABLE 6
Time Required to Travel between Cities in Kilroy County

From	To					
	City 1	City 2	City 3	City 4	City 5	City 6
City 1	0	10	20	30	30	20
City 2	10	0	25	35	20	10
City 3	20	25	0	15	30	20
City 4	30	35	15	0	15	25
City 5	30	20	30	15	0	14
City 6	20	10	20	25	14	0

TABLE 7
Cities within 15 Minutes of
Given City

City	Within 15 Minutes
1	1, 2
2	1, 2, 6
3	3, 4
4	3, 4, 5
5	4, 5, 6
6	2, 5, 6

objective function (and with the fact that each variable must equal 0 or 1), we obtain the following 0–1 IP:

$$\min z = x_1 + x_2 + x_3 + x_4 + x_5 + x_6$$

$$
\begin{array}{lll}
\text{s.t.} & x_1 + x_2 \geq 1 & \text{(City 1 constraint)} \\
& x_1 + x_2 \qquad\quad + x_6 \geq 1 & \text{(City 2 constraint)} \\
& x_3 + x_4 \geq 1 & \text{(City 3 constraint)} \\
& x_3 + x_4 + x_5 \geq 1 & \text{(City 4 constraint)} \\
& x_4 + x_5 + x_6 \geq 1 & \text{(City 5 constraint)} \\
& x_2 \qquad\quad + x_5 + x_6 \geq 1 & \text{(City 6 constraint)} \\
& x_i = 0 \text{ or } 1 \quad (i = 1, 2, 3, 4, 5, 6)
\end{array}
$$

One optimal solution to this IP is $z = 2$, $x_2 = x_4 = 1$, $x_1 = x_3 = x_5 = x_6 = 0$. Thus, Kilroy County can build two fire stations: one in city 2 and one in city 4.

As noted, Example 5 represents a class of IPs known as **set-covering problems.** In a set-covering problem, each member of a given set (call it set 1) must be "covered" by an acceptable member of some set (call it set 2). The objective in a set-covering problem is to minimize the number of elements in set 2 that are required to cover all the elements in set 1. In Example 5, set 1 is the cities in Kilroy County, and set 2 is the set of fire stations. The station in city 2 covers cities 1, 2, and 6, and the station in city 4 covers cities 3, 4, and 5. Set-covering problems have many applications in areas such as airline crew scheduling, political districting, airline scheduling, and truck routing.

Either–Or Constraints

The following situation commonly occurs in mathematical programming problems. We are given two constraints of the form

$$f(x_1, x_2, \ldots, x_n) \leq 0 \tag{26}$$

$$g(x_1, x_2, \ldots, x_n) \leq 0 \tag{27}$$

We want to ensure that at least one of (26) and (27) is satisfied, often called **either–or constraints.** Adding the two constraints (26') and (27') to the formulation will ensure that at least one of (26) and (27) is satisfied:

$$f(x_1, x_2, \ldots, x_n) \leq My \tag{26'}$$

$$g(x_1, x_2, \ldots, x_n) \leq M(1 - y) \tag{27'}$$

In (26′) and (27′), y is a 0–1 variable, and M is a number chosen large enough to ensure that $f(x_1, x_2, \ldots, x_n) \leq M$ and $g(x_1, x_2, \ldots, x_n) \leq M$ are satisfied for all values of $x_1, x_2, \ldots, x_n$ that satisfy the other constraints in the problem.

Let us show that the inclusion of constraints (26′) and (27′) is equivalent to at least one of (26) and (27) being satisfied. Either $y = 0$ or $y = 1$. If $y = 0$, then (26′) and (27′) become $f \leq 0$ and $g \leq M$. Thus, if $y = 0$, then (26) (and possibly (27)) must be satisfied. Similarly, if $y = 1$, then (26′) and (27′) become $f \leq M$ and $g \leq 0$. Thus, if $y = 1$, then (27) (and possibly (26)) must be satisfied. Therefore, whether $y = 0$ or $y = 1$, (26′) and (27′) ensure that at least one of (26) and (27) is satisfied.

The following example illustrates the use of either–or constraints.

EXAMPLE 6   Either–Or Constraint

Dorian Auto is considering manufacturing three types of autos: compact, midsize, and large. The resources required for, and the profits yielded by, each type of car are shown in Table 8. Currently, 6,000 tons of steel and 60,000 hours of labor are available. For production of a type of car to be economically feasible, at least 1,000 cars of that type must be produced. Formulate an IP to maximize Dorian's profit.

Solution   Because Dorian must determine how many cars of each type should be built, we define

$$x_1 = \text{number of compact cars produced}$$
$$x_2 = \text{number of midsize cars produced}$$
$$x_3 = \text{number of large cars produced}$$

Then contribution to profit (in thousands of dollars) is $2x_1 + 3x_2 + 4x_3$, and Dorian's objective function is

$$\max z = 2x_1 + 3x_2 + 4x_3$$

We know that if any cars of a given type are produced, then at least 1,000 cars of that type must be produced. Thus, for $i = 1, 2, 3$, we must have $x_i \leq 0$ or $x_i \geq 1,000$. Steel and labor are limited, so Dorian must satisfy the following five constraints:

Constraint 1 $x_1 \leq 0$ or $x_1 \geq 1,000$.

Constraint 2 $x_2 \leq 0$ or $x_2 \geq 1,000$.

Constraint 3 $x_3 \leq 0$ or $x_3 \geq 1,000$.

Constraint 4   The cars produced can use at most 6,000 tons of steel.

Constraint 5   The cars produced can use at most 60,000 hours of labor.

TABLE 8
Resources and Profits for Three Types of Cars

Resource	Car Type		
	Compact	Midsize	Large
Steel required	1.5 tons	3 tons	5 tons
Labor required	30 hours	25 hours	40 hours
Profit yielded ($)	2,000	3,000	4,000

From our previous discussion, we see that if we define $f(x_1, x_2, x_3) = x_1$ and $g(x_1, x_2, x_3) = 1{,}000 - x_1$, we can replace Constraint 1 by the following pair of constraints:

$$x_1 \leq M_1 y_1$$
$$1{,}000 - x_1 \leq M_1(1 - y_1)$$
$$y_1 = 0 \text{ or } 1$$

To ensure that both x_1 and $1{,}000 - x_1$ will never exceed M_1, it suffices to choose M_1 large enough so that M_1 exceeds 1,000 and x_1 is always less than M_1. Building $\frac{60{,}000}{30} = 2{,}000$ compacts would use all available labor (and still leave some steel), so at most 2,000 compacts can be built. Thus, we may choose $M_1 = 2{,}000$. Similarly, Constraint 2 may be replaced by the following pair of constraints:

$$x_2 \leq M_2 y_2$$
$$1{,}000 - x_2 \leq M_2(1 - y_2)$$
$$y_2 = 0 \text{ or } 1$$

You should verify that $M_2 = 2{,}000$ is satisfactory. Similarly, Constraint 3 may be replaced by

$$x_3 \leq M_3 y_3$$
$$1{,}000 - x_3 \leq M_3(1 - y_3)$$
$$y_3 = 0 \text{ or } 1$$

Again, you should verify that $M_3 = 1{,}200$ is satisfactory. Constraint 4 is a straightforward resource constraint that reduces to

$$1.5x_1 + 3x_2 + 5x_3 \leq 6{,}000 \qquad \text{(Steel constraint)}$$

Constraint 5 is a straightforward resource usage constraint that reduces to

$$30x_1 + 25x_2 + 40x_3 \leq 60{,}000 \qquad \text{(Labor constraint)}$$

After noting that $x_i \geq 0$ and that x_i must be an integer, we obtain the following IP:

$$\max z = 2x_1 + 3x_2 + 4x_3$$
$$\text{s.t.} \qquad x_1 \leq 2{,}000y_1$$
$$1{,}000 - x_1 \leq 2{,}000(1 - y_1)$$
$$x_2 \leq 2{,}000y_2$$
$$1{,}000 - x_2 \leq 2{,}000(1 - y_2)$$
$$x_3 \leq 1{,}200y_3$$
$$1{,}000 - x_3 \leq 1{,}200(1 - y_3)$$
$$1.5x_1 + 3x_2 + 5x_3 \leq 6{,}000 \qquad \text{(Steel constraint)}$$
$$30x_1 + 25x_2 + 40x_3 \leq 60{,}000 \qquad \text{(Labor constraint)}$$
$$x_1, x_2, x_3 \geq 0; \; x_1, x_2, x_3 \text{ integer}$$
$$y_1, y_2, y_3 = 0 \text{ or } 1$$

The optimal solution to the IP is $z = 6{,}000$, $x_2 = 2{,}000$, $y_2 = 1$, $y_1 = y_3 = x_1 = x_3 = 0$. Thus, Dorian should produce 2,000 midsize cars. If Dorian had not been required to manufacture at least 1,000 cars of each type, then the optimal solution would have been to produce 570 compacts and 1,715 midsize cars.

If–Then Constraints

In many applications, the following situation occurs: We want to ensure that if a constraint $f(x_1, x_2, \ldots, x_n) > 0$ is satisfied, then the constraint $g(x_1, x_2, \ldots, x_n) \geq 0$ must be satisfied, while if $f(x_1, x_2, \ldots, x_n) > 0$ is not satisfied, then $g(x_1, x_2, \ldots, x_n) \geq 0$ may or may not be satisfied. In short, we want to ensure that $f(x_1, x_2, \ldots, x_n) > 0$ implies $g(x_1, x_2, \ldots, x_n) \geq 0$.

To ensure this, we include the following constraints in the formulation:

$$-g(x_1, x_2, \ldots, x_n) \leq My \tag{28}$$

$$f(x_1, x_2, \ldots, x_n) \leq M(1 - y) \tag{29}$$

$$y = 0 \text{ or } 1$$

As usual, M is a large positive number. (M must be chosen large enough so that $f \leq M$ and $-g \leq M$ hold for all values of $x_1, x_2, \ldots, x_n$ that satisfy the other constraints in the problem.) Observe that if $f > 0$, then (29) can be satisfied only if $y = 0$. Then (28) implies $-g \leq 0$, or $g \geq 0$, which is the desired result. Thus, if $f > 0$, then (28) and (29) ensure that $g \geq 0$. Also, if $f > 0$ is not satisfied, then (29) allows $y = 0$ or $y = 1$. By choosing $y = 1$, (28) is automatically satisfied. Thus, if $f > 0$ is not satisfied, then the values of $x_1, x_2, \ldots, x_n$ are unrestricted and $g < 0$ or $g \geq 0$ are both possible.

To illustrate the use of this idea, suppose we add the following constraint to the Nickles lockbox problem: If customers in region 1 send their payments to city 1, then no other customers may send their payments to city 1. Mathematically, this restriction may be expressed by

$$\text{If } x_{11} = 1, \quad \text{then} \quad x_{21} = x_{31} = x_{41} = 0 \tag{30}$$

Because all x_{ij} must equal 0 or 1, (30) may be written as

If $x_{11} > 0$,   then $x_{21} + x_{31} + x_{41} \leq 0$,   or $-x_{21} - x_{31} - x_{41} \geq 0$ (30′)

If we define $f = x_{11}$ and $g = -x_{21} - x_{31} - x_{41}$, we can use (28) and (29) to express (30′) [and therefore (30)] by the following two constraints:

$$x_{21} + x_{31} + x_{41} \leq My$$

$$x_{11} \leq M(1 - y)$$

$$y = 0 \text{ or } 1$$

Because $-g$ and f can never exceed 3, we can choose $M = 3$ and add the following constraints to the original lockbox formulation:

$$x_{21} + x_{31} + x_{41} \leq 3y$$

$$x_{11} \leq 3(1 - y)$$

$$y = 0 \text{ or } 1$$

Integer Programming and Piecewise Linear Functions[†]

The next example shows how 0–1 variables can be used to model optimization problems involving piecewise linear functions. A **piecewise linear function** consists of several straight-line segments. The piecewise linear function in Figure 2 is made of four straight-line segments. The points where the slope of the piecewise linear function changes (or the range of definition of the function ends) are called the **break points** of the function. Thus, 0, 10, 30, 40, and 50 are the break points of the function pictured in Figure 2.

[†]This section covers topics that may be omitted with no loss of continuity.

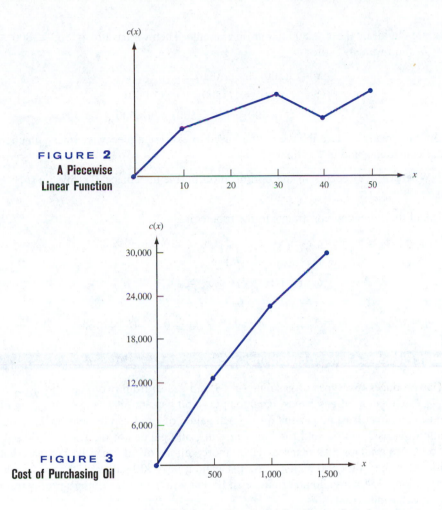

FIGURE 2
A Piecewise
Linear Function

FIGURE 3
Cost of Purchasing Oil

To illustrate why piecewise linear functions can occur in applications, suppose we manufacture gasoline from oil. In purchasing oil from our supplier, we receive a quantity discount. The first 500 gallons of oil purchased cost 25¢ per gallon; the next 500 gallons cost 20¢ per gallon; and the next 500 gallons cost 15¢ per gallon. At most, 1,500 gallons of oil can be purchased. Let x be the number of gallons of oil purchased and $c(x)$ be the cost (in cents) of purchasing x gallons of oil. For $x \leq 0$, $c(x) = 0$. Then for $0 \leq x \leq 500$, $c(x) = 25x$. For $500 \leq x \leq 1,000$, $c(x) =$ (cost of purchasing first 500 gallons at 25¢ per gallon) + (cost of purchasing next $x - 500$ gallons at 20¢ per gallon) $= 25(500) + 20(x - 500) = 20x + 2,500$. For $1,000 \leq x \leq 1,500$, $c(x) =$ (cost of purchasing first 1,000 gallons) + (cost of purchasing next $x - 1,000$ gallons at 15¢ per gallon) $= c(1,000) + 15(x - 1,000) = 7,500 + 15x$. Thus, $c(x)$ has break points 0, 500, 1,000, and 1,500 and is graphed in Figure 3.

A piecewise linear function is not a linear function, so one might think that linear programming could not be used to solve optimization problems involving these functions. By using 0–1 variables, however, piecewise linear functions can be represented in linear form. Suppose that a piecewise linear function $f(x)$ has break points $b_1, b_2, \ldots, b_n$. For some k ($k = 1, 2, \ldots, n - 1$), $b_k \leq x \leq b_{k+1}$. Then, for some number z_k ($0 \leq z_k \leq 1$), x may be written as

$$x = z_k b_k + (1 - z_k)b_{k+1}$$

Because $f(x)$ is linear for $b_k \leq x \leq b_{k+1}$, we may write

$$f(x) = z_k f(b_k) + (1 - z_k)f(b_{k+1})$$

To illustrate the idea, take $x = 800$ in our oil example. Then we have $b_2 = 500 \le 800 \le 1,000 = b_3$, and we may write

$$x = \tfrac{2}{5}(500) + \tfrac{3}{5}(1,000)$$
$$f(x) = f(800) = \tfrac{2}{5}f(500) + \tfrac{3}{5}f(1,000)$$
$$= \tfrac{2}{5}(12,500) + \tfrac{3}{5}(22,500) = 18,500$$

We are now ready to describe the method used to express a piecewise linear function via linear constraints and 0–1 variables:

Step 1 Wherever $f(x)$ occurs in the optimization problem, replace $f(x)$ by $z_1 f(b_1) + z_2 f(b_2) + \cdots + z_n f(b_n)$.

Step 2 Add the following constraints to the problem:

$$z_1 \le y_1, z_2 \le y_1 + y_2, z_3 \le y_2 + y_3, \ldots, z_{n-1} \le y_{n-2} + y_{n-1}, z_n \le y_{n-1}$$
$$y_1 + y_2 + \cdots + y_{n-1} = 1$$
$$z_1 + z_2 + \cdots + z_n = 1$$
$$x = z_1 b_1 + z_2 b_2 + \cdots + z_n b_n$$
$$y_i = 0 \text{ or } 1 \quad (i = 1, 2, \ldots, n-1); \qquad z_i \ge 0 \quad (i = 1, 2, \ldots, n)$$

EXAMPLE 7 **IP with Piecewise Linear Functions**

Euing Gas produces two types of gasoline (gas 1 and gas 2) from two types of oil (oil 1 and oil 2). Each gallon of gas 1 must contain at least 50 percent oil 1, and each gallon of gas 2 must contain at least 60 percent oil 1. Each gallon of gas 1 can be sold for 12¢, and each gallon of gas 2 can be sold for 14¢. Currently, 500 gallons of oil 1 and 1,000 gallons of oil 2 are available. As many as 1,500 more gallons of oil 1 can be purchased at the following prices: first 500 gallons, 25¢ per gallon; next 500 gallons, 20¢ per gallon; next 500 gallons, 15¢ per gallon. Formulate an IP that will maximize Euing's profits (revenues − purchasing costs).

Solution Except for the fact that the cost of purchasing additional oil 1 is a piecewise linear function, this is a straightforward blending problem. With this in mind, we define

$$x = \text{amount of oil 1 purchased}$$
$$x_{ij} = \text{amount of oil } i \text{ used to produce gas } j \quad (i, j = 1, 2)$$

Then (in cents)

$$\text{Total revenue} - \text{cost of purchasing oil 1} = 12(x_{11} + x_{21}) + 14(x_{12} + x_{22}) - c(x)$$

As we have seen previously,

$$c(x) = \begin{cases} 25x & (0 \le x \le 500) \\ 20x + 2,500 & (500 \le x \le 1,000) \\ 15x + 7,500 & (1,000 \le x \le 1,500) \end{cases}$$

Thus, Euing's objective function is to maximize

$$z = 12x_{11} + 12x_{21} + 14x_{12} + 14x_{22} - c(x)$$

Euing faces the following constraints:

Constraint 1 Euing can use at most $x + 500$ gallons of oil 1.

Constraint 2 Euing can use at most 1,000 gallons of oil 2.

Constraint 3 The oil mixed to make gas 1 must be at least 50% oil 1.

Constraint 4 The oil mixed to make gas 2 must be at least 60% oil 1.

Constraint 1 yields

$$x_{11} + x_{12} \leq x + 500$$

Constraint 2 yields

$$x_{21} + x_{22} \leq 1{,}000$$

Constraint 3 yields

$$\frac{x_{11}}{x_{11} + x_{21}} \geq 0.5 \quad \text{or} \quad 0.5x_{11} - 0.5x_{21} \geq 0$$

Constraint 4 yields

$$\frac{x_{12}}{x_{12} + x_{22}} \geq 0.6 \quad \text{or} \quad 0.4x_{12} - 0.6x_{22} \geq 0$$

Also all variables must be non-negative. Thus, Euing Gas must solve the following optimization problem:

$$
\begin{aligned}
\max z = {} & 12x_{11} + 12x_{21} + 14x_{12} + 14x_{22} - c(x) \\
\text{s.t.} \quad & x_{11} + x_{12} \leq x + 500 \\
& x_{21} + x_{22} \leq 1{,}000 \\
& 0.5x_{11} - 0.5x_{21} \geq 0 \\
& 0.4x_{12} - 0.6x_{22} \geq 0 \\
& x_{ij} \geq 0, \ 0 \leq x \leq 1{,}500
\end{aligned}
$$

Because $c(x)$ is a piecewise linear function, the objective function is not a linear function of x, and this optimization is not an LP. By using the method described earlier, however, we can transform this problem into an IP. After recalling that the break points for $c(x)$ are 0, 500, 1,000, and 1,500, we proceed as follows:

Step 1 Replace $c(x)$ by $c(x) = z_1 c(0) + z_2 c(500) + z_3 c(1{,}000) + z_4 c(1{,}500)$.

Step 2 Add the following constraints:

$$x = 0z_1 + 500z_2 + 1{,}000z_3 + 1{,}500z_4$$
$$z_1 \leq y_1, z_2 \leq y_1 + y_2, z_3 \leq y_2 + y_3, z_4 \leq y_3$$
$$z_1 + z_2 + z_3 + z_4 = 1, \quad y_1 + y_2 + y_3 = 1$$
$$y_i = 0 \text{ or } 1 \ (i = 1, 2, 3); z_i \geq 0 \ (i = 1, 2, 3, 4)$$

Our new formulation is the following IP:

$$
\begin{aligned}
\max z = {} & 12x_{11} + 12x_{21} + 14x_{12} + 14x_{22} - z_1 c(0) - z_2 c(500) \\
& - z_3 c(1{,}000) - z_4 c(1{,}500) \\
\text{s.t.} \quad & x_{11} + x_{12} \leq x + 500 \\
& x_{21} + x_{22} \leq 1{,}000 \\
& 0.5x_{11} - 0.5x_{21} \geq 0 \\
& 0.4x_{12} - 0.6x_{22} \geq 0
\end{aligned}
$$

$$x = 0z_1 + 500z_2 + 1{,}000z_3 + 1{,}500z_4 \qquad \text{(31)}$$
$$z_1 \leq y_1 \qquad \text{(32)}$$
$$z_2 \leq y_1 + y_2 \qquad \text{(33)}$$
$$z_3 \leq y_2 + y_3 \qquad \text{(34)}$$

$$z_4 \leq y_3 \tag{35}$$

$$y_1 + y_2 + y_3 = 1 \tag{36}$$

$$z_1 + z_2 + z_3 + z_4 = 1 \tag{37}$$

$$y_i = 0 \text{ or } 1 \quad (i = 1, 2, 3); z_i \geq 0 \quad (i = 1, 2, 3, 4)$$

$$x_{ij} \geq 0$$

To see why this formulation works, observe that because $y_1 + y_2 + y_3 = 1$ and $y_i = 0$ or 1, exactly one of the y_i's will equal 1, and the others will equal 0. Now, (32)–(37) imply that if $y_i = 1$, then z_i and z_{i+1} may be positive, but all the other z_i's must equal 0. For instance, if $y_2 = 1$, then $y_1 = y_3 = 0$. Then (32)–(35) become $z_1 \leq 0$, $z_2 \leq 1$, $z_3 \leq 1$, and $z_4 \leq 0$. These constraints force $z_1 = z_4 = 0$ and allow z_2 and z_3 to be any non-negative number less than or equal to 1. We can now show that (31)–(37) correctly represent the piecewise linear function $c(x)$. Choose any value of x, say $x = 800$. Note that $b_2 = 500 \leq 800 \leq 1,000 = b_3$. For $x = 800$, what values do our constraints assign to y_1, y_2, and y_3? The value $y_1 = 1$ is impossible, because if $y_1 = 1$, then $y_2 = y_3 = 0$. Then (34)–(35) force $z_3 = z_4 = 0$. Then (31) reduces to $800 = x = 500z_2$, which cannot be satisfied by $z_2 \leq 1$. Similarly, $y_3 = 1$ is impossible. If we try $y_2 = 1$ (32) and (35) force $z_1 = z_4 = 0$. Then (33) and (34) imply $z_2 \leq 1$ and $z_3 \leq 1$. Now (31) becomes $800 = x = 500z_2 + 1,000z_3$. Because $z_2 + z_3 = 1$, we obtain $z_2 = \frac{2}{5}$ and $z_3 = \frac{3}{5}$. Now the objective function reduces to

$$12x_{11} + 12x_{21} + 14x_{21} + 14x_{22} - \frac{2c(500)}{5} - \frac{3c(1,000)}{5}$$

Because

$$c(800) = \frac{2c(500)}{5} + \frac{3c(1,000)}{5}$$

our objective function yields the correct value of Euing's profits!

The optimal solution to Euing's problem is $z = 12,500$, $x = 1,000$, $x_{12} = 1,500$, $x_{22} = 1,000$, $y_3 = z_3 = 1$. Thus, Euing should purchase 1,000 gallons of oil 1 and produce 2,500 gallons of gas 2.

In general, constraints of the form (31)–(37) ensure that if $b_i \leq x \leq b_{i+1}$, then $y_i = 1$ and only z_i and z_{i+1} can be positive. Because $c(x)$ is linear for $b_i \leq x \leq b_{i+1}$, the objective function will assign the correct value to $c(x)$.

If a piecewise linear function $f(x)$ involved in a formulation has the property that the slope of $f(x)$ becomes less favorable to the decision maker as x increases, then the tedious IP formulation we have just described is unnecessary.

EXAMPLE 8　Media Selection with Piecewise Linear Functions

Dorian Auto has a $20,000 advertising budget. Dorian can purchase full-page ads in two magazines: *Inside Jocks* (IJ) and *Family Square* (FS). An exposure occurs when a person reads a Dorian Auto ad for the first time. The number of exposures generated by each ad in IJ is as follows: ads 1–6, 10,000 exposures; ads 7–10, 3,000 exposures; ads 11–15, 2,500 exposures; ads 16+, 0 exposures. For example, 8 ads in IJ would generate $6(10,000) + 2(3,000) = 66,000$ exposures. The number of exposures generated by each ad in FS is as follows: ads 1–4, 8,000 exposures; ads 5–12, 6,000 exposures; ads 13–15, 2,000 exposures; ads 16+, 0 exposures. Thus, 13 ads in FS would generate $4(8,000) +$

$8(6{,}000) + 1(2{,}000) = 82{,}000$ exposures. Each full-page ad in either magazine costs \$1,000. Assume there is no overlap in the readership of the two magazines. Formulate an IP to maximize the number of exposures that Dorian can obtain with limited advertising funds.

Solution If we define

$$x_1 = \text{number of IJ ads yielding 10,000 exposures}$$
$$x_2 = \text{number of IJ ads yielding 3,000 exposures}$$
$$x_3 = \text{number of IJ ads yielding 2,500 exposures}$$
$$y_1 = \text{number of FS ads yielding 8,000 exposures}$$
$$y_2 = \text{number of FS ads yielding 6,000 exposures}$$
$$y_3 = \text{number of FS ads yielding 2,000 exposures}$$

then the total number of exposures (in thousands) is given by

$$10x_1 + 3x_2 + 2.5x_3 + 8y_1 + 6y_2 + 2y_3$$

Thus, Dorian wants to maximize

$$z = 10x_1 + 3x_2 + 2.5x_3 + 8y_1 + 6y_2 + 2y_3$$

Because the total amount spent (in thousands) is just the toal number of ads placed in both magazines, Dorian's budget constraint may be written as

$$x_1 + x_2 + x_3 + y_1 + y_2 + y_3 \leq 20$$

The statement of the problem implies that $x_1 \leq 6$, $x_2 \leq 4$, $x_3 \leq 5$, $y_1 \leq 4$, $y_2 \leq 8$, and $y_3 \leq 3$ all must hold. Adding the sign restrictions on each variable and noting that each variable must be an integer, we obtain the following IP:

$$\max z = 10x_1 + 3x_2 + 2.5x_3 + 8y_1 + 6y_2 + 2y_3$$
$$\text{s.t.} \quad x_1 + x_2 + x_3 + y_1 + y_2 + y_3 \leq 20$$
$$x_1 \qquad\qquad\qquad\qquad \leq 6$$
$$x_2 \qquad\qquad\qquad \leq 4$$
$$x_3 \qquad\qquad \leq 5$$
$$y_1 \qquad\quad \leq 4$$
$$y_2 \quad \leq 8$$
$$y_3 \leq 3$$
$$x_i, y_i \text{ integer} \quad (i = 1, 2, 3)$$
$$x_i, y_i \geq 0 \quad (i = 1, 2, 3)$$

Observe that the statement of the problem implies that x_2 cannot be positive unless x_1 assumes its maximum value of 6. Similarly, x_3 cannot be positive unless x_2 assumes its maximum value of 4. Because x_1 ads generate more exposures than x_2 ads, however, the act of maximizing ensures that x_2 will be positive only if x_1 has been made as large as possible. Similarly, because x_3 ads generate fewer exposures than x_2 ads, x_3 will be positive only if x_2 assumes its maximum value. (Also, y_2 will be positive only if $y_1 = 4$, and y_3 will be positive only if $y_2 = 8$.)

The optimal solution to Dorian's IP is $z = 146{,}000$, $x_1 = 6$, $x_2 = 2$, $y_1 = 4$, $y_2 = 8$, $x_3 = 0$, $y_3 = 0$. Thus, Dorian will place $x_1 + x_2 = 8$ ads in IJ and $y_1 + y_2 = 12$ ads in FS.

In Example 8, additional advertising in a magazine yielded diminishing returns. This ensured that x_i (y_i) would be positive only if x_{i-1} (y_{i-1}) assumed its maximum value. If additional advertising generated increasing returns, then this formulation would not yield the correct solution. For example, suppose that the number of exposures generated by each IJ ad was as follows: ads 1–6, 2,500 exposures; ads 7–10, 3,000 exposures; ads 11–15, 10,000 exposures. Suppose also that the number of exposures generated by each FS is as follows: ads 1–4, 2,000 exposures; ads 5–12, 6,000 exposures; ads 13–15, 8,000 exposures.

If we define

$$x_1 = \text{number of IJ ads generating 2,500 exposures}$$
$$x_2 = \text{number of IJ ads generating 3,000 exposures}$$
$$x_3 = \text{number of IJ ads generating 10,000 exposures}$$
$$y_1 = \text{number of FS ads generating 2,000 exposures}$$
$$y_2 = \text{number of FS ads generating 6,000 exposures}$$
$$y_3 = \text{number of FS ads generating 8,000 exposures}$$

the reasoning used in the previous example would lead to the following formulation:

$$\max z = 2.5x_1 + 3x_2 + 10x_3 + 2y_1 + 6y_2 + 8y_3$$
$$\text{s.t.} \quad x_1 + x_2 + x_3 + y_1 + y_2 + y_3 \leq 20$$
$$x_1 \qquad\qquad\qquad \leq 6$$
$$x_2 \qquad\qquad \leq 4$$
$$x_3 \qquad\quad \leq 5$$
$$y_1 \quad\quad \leq 4$$
$$y_2 \quad \leq 8$$
$$y_3 \leq 3$$
$$x_i, y_i \text{ integer} \quad (i = 1, 2, 3)$$
$$x_i, y_i \leq 0 \quad (i = 1, 2, 3)$$

The optimal solution to this IP is $x_3 = 5$, $y_3 = 3$, $y_2 = 8$, $x_2 = 4$, $x_1 = 0$, $y_1 = 0$, which cannot be correct. According to this solution, $x_1 + x_2 + x_3 = 9$ ads should be placed in IJ. If 9 ads were placed in IJ, however, then it must be that $x_1 = 6$ and $x_2 = 3$. Therefore, we see that the type of formulation used in the Dorian Auto example is correct only if the piecewise linear objective function has a less favorable slope for larger values of x. In our second example, the effectiveness of an ad increased as the number of ads in a magazine increased, and the act of maximizing will not ensure that x_i can be positive only if x_{i-1} assumes its maximum value. In this case, the approach used in the Euing Gas example would yield a correct formulation (see Problem 8).

Solving IPs with LINDO

LINDO can be used to solve pure or mixed IPs. In addition to the optimal solution, the LINDO output for an IP gives shadow prices and reduced costs. Unfortunately, the shadow prices and reduced costs refer to subproblems generated during the branch-and-bound solution—*not* to the IP. Unlike linear programming, there is no well-developed theory of sensitivity analysis for integer programming. The reader interested in a discussion of sensitivity analysis for IPs should consult Williams (1985).

To use LINDO to solve an IP, begin by entering the problem as if it were an LP. After typing in the **END** statement (to designate the end of the LP constraints), type for each 0–1 variable x the following statement:

$$\text{INTE } x$$

Thus, for an IP in which x and y are 0–1 variables, the following statements would be typed after the **END** statement:

$$\text{INTE } x$$
$$\text{INTE } y$$

A variable (say, w) that can assume any non-negative integer value is indicated by the **GIN** statement. Thus, if w may assume the values $0, 1, 2, \ldots$, we would type the following statement after the **END** statement:

$$\text{GIN } w$$

To tell LINDO that the first n variables appearing in the formulation must be 0–1 variables, use the command **INT** n.

To tell LINDO that the first n variables appearing in the formulation may assume any non-negative integer value, use the command **GIN** n.

To illustrate how to use LINDO to solve IPs, we show how to solve Example 3 with LINDO. We typed the following input (file Gandhi):

```
MAX        6 X1 + 4 X2 + 7 X3 - 200 Y1 - 150 Y2 - 100 Y3
SUBJECT TO
       2)    3 X1 + 2 X2 + 6 X3 <= 150
       3)    4 X1 + 3 X2 + 4 X3 <= 160
       4)    X1 - 40 Y1 <= 0
       5)    X2 - 53 Y2 <= 0
       6)    X3 - 25 Y3 <= 0
END
GIN        X1
GIN        X2
GIN        X3
INTE       Y1
INTE       Y2
INTE       Y3
```

Thus we see that X1, X2, and X3 can be any non-negative integer, while Y1, Y2, and Y3 must equal 0 or 1. By the way, we could have typed GIN 3 to ensure that X1, X2, and X3 must be non-negative integers. The optimal solution found by LINDO is given in Figure 4.

Solving IPs with LINGO

LINGO can also be used to solve IPs. To indicate that a variable must equal 0 or 1 use the **@BIN** operator (see the following example). To indicate that a variable must equal a non-negative integer, use the **@GIN** operator. We illustrate how LINGO is used to solve IPs with Example 4 (the Lockbox Problem). The following LINGO program (file Lock.lng) can be used to solve Example 4 (or any reasonably sized lockbox program).

```
MODEL:
  1]SETS:
  2]REGIONS/W,MW,E,S/:DEMAND;
  3]CITIES/LA,CHIC,NY,ATL/:Y;
  4]LINKS(REGIONS,CITIES):DAYS,COST,ASSIGN;
  5]ENDSETS
  6]MIN=@SUM(CITIES:50000*Y)+@SUM(LINKS:COST*ASSIGN);
  7]@FOR(LINKS(I,J):ASSIGN(I,J) < Y(J));
  8]@FOR(REGIONS(I):
  9]@SUM(CITIES(J):ASSIGN(I,J))=1);
 10]@FOR(CITIES(I):@BIN(Y(I)););
```

```
MAX      6 X1 + 4 X2 + 7 X3 - 200 Y1 - 150 Y2 - 100 Y3
SUBJECT TO
        2)    3 X1 + 2 X2 + 6 X3 <=    150
        3)    4 X1 + 3 X2 + 4 X3 <=    160
        4)    X1 - 40 Y1 <=    0
        5)    X2 - 53 Y2 <=    0
        6)    X3 - 25 Y3 <=    0
END
GIN      X1
GIN      X2
GIN      X3
INTE     Y1
INTE     Y2
INTE     Y3


        OBJECTIVE FUNCTION VALUE

    1)      75.000000

VARIABLE        VALUE          REDUCED COST
    X1          .000000         -6.000000
    X2          .000000         -4.000000
    X3        25.000000         -7.000000
    Y1          .000000        200.000000
    Y2          .000000        150.000000
    Y3         1.000000        100.000000


   ROW    SLACK OR SURPLUS    DUAL PRICES
    2)          .000000          .000000
    3)        60.000000          .000000
    4)          .000000          .000000
    5)          .000000          .000000
    6)          .000000          .000000

NO. ITERATIONS=       11
BRANCHES=      1 DETERM.=  1.000E     0
```

FIGURE 4

```
11]@FOR(LINKS(I,J):@BIN(ASSIGN(I,J)););
12]@FOR(LINKS(I,J):COST(I,J)=.20*DEMAND(I)*DAYS(I,J));
13]DATA:
14]DAYS=2,6,8,8,
15]6,2,5,5,
16]8,5,2,5,
17]8,5,5,2;
18]DEMAND=70000,50000,60000,40000;
19]ENDDATA
END
```

In line 2, we define the four regions of the country and associate a daily demand for cash payments from each region. Line 3 specifies the four cities where a lockbox may be built. With each city I, we associate a 0–1 variable (Y(I)) that equals 1 if a lockbox is built in the city or 0 otherwise. In line 4, we create a "link" (LINK(I,J)) between each region of the country and each potential lockbox site. Associated with each link are the following quantities:

1   The average number of days (DAYS) it takes a check to clear when mailed from region I to city J. This information is given in the DATA section.

2   The annual lost interest cost for funds sent from region i (COST) incurred if region I sends its money to city J.

3   A 0–1 variable ASSIGN(I,J) which equals 1 if region I sends its money to city J and 0 otherwise.

In line 6, we compute the total cost by summing 50000*Y(I) over all cities. This computes the total annual cost of running lockboxes. Then we sum COST*ASSIGN over all links. This picks up the total annual lost interest cost. The line 7 constraints ensure that (for all combinations of I and J) if region I sends its money to city J, then Y(J) = 1. This forces us to pay for lockboxes we use. Lines 8–9 ensure that each region of the country sends its money to some city. Line 10 ensures that each Y(I) equals 0 or 1. Line 11 ensures that each ASSIGN(I,J) equals 0 or 1 (actually we do not need this statement; see Problem 44). We compute the lost annual interest cost if region I sends its money to city J in line 12. This duplicates the calculations in Table 5. Note that an * is needed to ensure that multiplications are performed.

In lines 14–17, we input the average number of days required for a check to clear when it is sent from region I to city J. In line 18, we input the daily demand for each region.

Note that to obtain the objective function and constraints we selected the Model window and then chose LINDO, Generate, Display Model. See Figure 8.

Using the Excel Solver to Solve IP Problems

It is easy to use the Excel Solver to solve integer programming problems. The file Gandhi.xls contains a spreadsheet solution to Example 3. See Figure 7 for the optimal solution. In our spreadsheet, the changing cells J4:J6 (the number of each product produced) must be integers. To tell the Solver that these changing cells must be integers, just select Add Constraint and point to the cells J4:J6. Then select int from the drop-down arrow in the middle.

The changing cells K4:K6 are the binary fixed charge variables. To tell the Solver that these changing cells must be binary, select Add Constraint and point to cells K4:K6. Then select bin from the drop-down arrow. See Figure 6.

From Figure 7, we find that the optimal solution (as found with LINDO) is to make 25 pairs of pants.

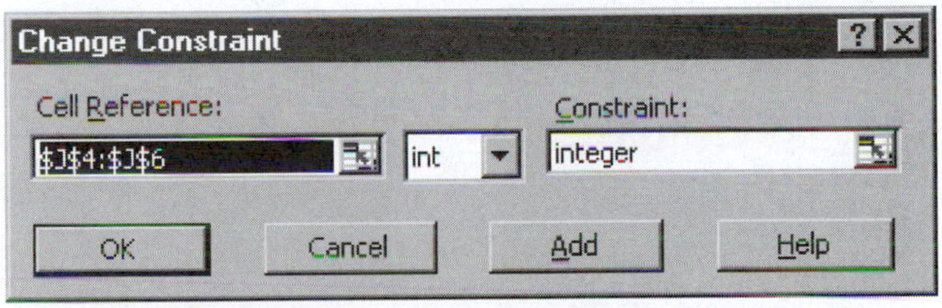

FIGURE 5

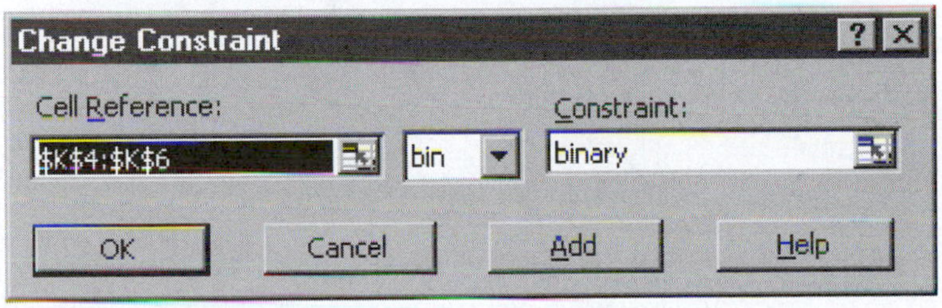

FIGURE 6

FIGURE 7

	A	B	C	D	E	F	G	H	I	J	K
1	Gandhi										
2											
3				Labor hours used	Cloth yards used	Unit price	Unit cost	Unit profit	Fixed Cost	Number Made	Binary variable
4			Shirt	3	4	$ 12.00	$ 6.00	$ 6.00	$ 200.00	0	0
5			Shorts	2	3	$ 8.00	$ 4.00	$ 4.00	$ 150.00	0	0
6			Pants	6	4	$ 15.00	$ 8.00	$ 7.00	$ 100.00	25	1
7		Resource Constraints									
8			Used		Available				Fixed charge	$ 100.00	
9		Labor	150	<=	150				Variable cost	$ 200.00	
10		Cloth	100	<=	160				Revenue	$ 375.00	
11									Profit	$ 75.00	
12		Fixed Charge Constraints	Number Made		Logical Upper Bound	Max possible to make					
13		Shirts	0	<=	0	40					
14		Shorts	0	<=	0	53.33333					
15		Pants	25	<=	25	25					

```
MIN     50000 Y(ATL + 50000 Y(NY + 50000 Y(CHIC + 50000 Y(LA + 16000 ASSIGNSA
        + 40000 ASSIGNSN + 40000 ASSIGNSC + 64000 ASSIGNSL + 60000 ASSIGNEA
        + 24000 ASSIGNEN + 60000 ASSIGNEC + 96000 ASSIGNEL + 50000 ASSIGNMW
        + 50000 ASSIGNMW + 20000 ASSIGNMW + 60000 ASSIGNMW + 112000 ASSIGNWA
        + 112000 ASSIGNWN + 84000 ASSIGNWC + 28000 ASSIGNWL
SUBJECT TO
2)- Y(LA + ASSIGNWL <=    0
3)- Y(CHIC + ASSIGNWC <=    0
4)- Y(NY + ASSIGNWN <=    0
5)- Y(ATL + ASSIGNWA <=    0
6)- Y(LA + ASSIGNMW <=    0
7)- Y(CHIC + ASSIGNMW <=    0
8)- Y(NY + ASSIGNMW <=    0
9)- Y(ATL + ASSIGNMW <=    0
10)- Y(LA + ASSIGNEL <=    0
11)- Y(CHIC + ASSIGNEC <=    0
12)- Y(NY + ASSIGNEN <=    0
13)- Y(ATL + ASSIGNEA <=    0
14)- Y(LA + ASSIGNSL <=    0
15)- Y(CHIC + ASSIGNSC <=    0
16)- Y(NY + ASSIGNSN <=    0
17)- Y(ATL + ASSIGNSA <=    0
18)   ASSIGNWA + ASSIGNWN + ASSIGNWC + ASSIGNWL =    1
19)   ASSIGNMW + ASSIGNMW + ASSIGNMW + ASSIGNMW =    1
20)   ASSIGNEA + ASSIGNEN + ASSIGNEC + ASSIGNEL =    1
21)   ASSIGNSA + ASSIGNSN + ASSIGNSC + ASSIGNSL =    1
END
INTE    20


[ERROR CODE: 96]
WARNING: SEVERAL LINGO NAMES MAY HAVE BEEN TRANSFORMED INTO A
SINGLE LINDO NAME.

LP OPTIMUM FOUND AT STEP     14
OBJECTIVE VALUE =    242000.000
ENUMERATION COMPLETE. BRANCHES=      0 PIVOTS=      14
```

FIGURE 8

LAST INTEGER SOLUTION IS THE BEST FOUND
RE-INSTALLING BEST SOLUTION...

VARIABLE	VALUE	REDUCED COST
DEMAND(W)	70000.00	0.0000000E+00
DEMAND(MW)	50000.00	0.0000000E+00
DEMAND(E)	60000.00	0.0000000E+00
DEMAND(S)	40000.00	0.0000000E+00
Y(LA)	1.000000	50000.00
Y(CHIC)	0.0000000E+00	50000.00
Y(NY)	1.000000	50000.00
Y(ATL)	0.0000000E+00	50000.00
DAYS(W, LA)	2.000000	0.0000000E+00
DAYS(W, CHIC)	6.000000	0.0000000E+00
DAYS(W, NY)	8.000000	0.0000000E+00
DAYS(W, ATL)	8.000000	0.0000000E+00
DAYS(MW, LA)	6.000000	0.0000000E+00
DAYS(MW, CHIC)	2.000000	0.0000000E+00
DAYS(MW, NY)	5.000000	0.0000000E+00
DAYS(MW, ATL)	5.000000	0.0000000E+00
DAYS(E, LA)	8.000000	0.0000000E+00
DAYS(E, CHIC)	5.000000	0.0000000E+00
DAYS(E, NY)	2.000000	0.0000000E+00
DAYS(E, ATL)	5.000000	0.0000000E+00
DAYS(S, LA)	8.000000	0.0000000E+00
DAYS(S, CHIC)	5.000000	0.0000000E+00
DAYS(S, NY)	5.000000	0.0000000E+00
DAYS(S, ATL)	2.000000	0.0000000E+00
COST(W, LA)	28000.00	0.0000000E+00
COST(W, CHIC)	84000.00	0.0000000E+00
COST(W, NY)	112000.0	0.0000000E+00
COST(W, ATL)	112000.0	0.0000000E+00
COST(MW, LA)	60000.00	0.0000000E+00
COST(MW, CHIC)	20000.00	0.0000000E+00
COST(MW, NY)	50000.00	0.0000000E+00
COST(MW, ATL)	50000.00	0.0000000E+00
COST(E, LA)	96000.00	0.0000000E+00
COST(E, CHIC)	60000.00	0.0000000E+00
COST(E, NY)	24000.00	0.0000000E+00
COST(E, ATL)	60000.00	0.0000000E+00
COST(S, LA)	64000.00	0.0000000E+00
COST(S, CHIC)	40000.00	0.0000000E+00
COST(S, NY)	40000.00	0.0000000E+00
COST(S, ATL)	16000.00	0.0000000E+00
ASSIGN(W, LA)	1.000000	28000.00
ASSIGN(W, CHIC)	0.0000000E+00	84000.00
ASSIGN(W, NY)	0.0000000E+00	112000.0
ASSIGN(W, ATL)	0.0000000E+00	112000.0
ASSIGN(MW, LA)	0.0000000E+00	60000.00
ASSIGN(MW, CHIC)	0.0000000E+00	20000.00
ASSIGN(MW, NY)	1.000000	50000.00
ASSIGN(MW, ATL)	0.0000000E+00	50000.00
ASSIGN(E, LA)	0.0000000E+00	96000.00
ASSIGN(E, CHIC)	0.0000000E+00	60000.00
ASSIGN(E, NY)	1.000000	24000.00
ASSIGN(E, ATL)	0.0000000E+00	60000.00
ASSIGN(S, LA)	0.0000000E+00	64000.00
ASSIGN(S, CHIC)	0.0000000E+00	40000.00
ASSIGN(S, NY)	1.000000	40000.00
ASSIGN(S, ATL)	0.0000000E+00	16000.00

FIGURE 8
(Continued)

ROW	SLACK OR SURPLUS	DUAL PRICE
1	242000.0	-1.000000
2	0.0000000E+00	0.0000000E+00
3	0.0000000E+00	0.0000000E+00
4	1.000000	0.0000000E+00
5	0.0000000E+00	0.0000000E+00
6	1.000000	0.0000000E+00
7	0.0000000E+00	0.0000000E+00
8	0.0000000E+00	0.0000000E+00
9	0.0000000E+00	0.0000000E+00
10	1.000000	0.0000000E+00
11	0.0000000E+00	0.0000000E+00
12	0.0000000E+00	0.0000000E+00
13	0.0000000E+00	0.0000000E+00
14	1.000000	0.0000000E+00
15	0.0000000E+00	0.0000000E+00
16	0.0000000E+00	0.0000000E+00
17	0.0000000E+00	0.0000000E+00
18	0.0000000E+00	0.0000000E+00
19	0.0000000E+00	0.0000000E+00
20	0.0000000E+00	0.0000000E+00
21	0.0000000E+00	0.0000000E+00
22	0.0000000E+00	-1.000000
23	0.0000000E+00	0.0000000E+00
24	0.0000000E+00	0.0000000E+00
25	0.0000000E+00	0.0000000E+00
26	0.0000000E+00	0.0000000E+00
27	0.0000000E+00	0.0000000E+00
28	0.0000000E+00	-1.000000
29	0.0000000E+00	0.0000000E+00
30	0.0000000E+00	0.0000000E+00
31	0.0000000E+00	0.0000000E+00
32	0.0000000E+00	-1.000000
33	0.0000000E+00	0.0000000E+00
34	0.0000000E+00	0.0000000E+00
35	0.0000000E+00	0.0000000E+00
36	0.0000000E+00	-1.000000
37	0.0000000E+00	0.0000000E+00

FIGURE 8
(Continued)

PROBLEMS

Group A

1 Coach Night is trying to choose the starting lineup for the basketball team. The team consists of seven players who have been rated (on a scale of 1 = poor to 3 = excellent) according to their ball-handling, shooting, rebounding, and defensive abilities. The positions that each player is allowed to play and the player's abilities are listed in Table 9.

The five-player starting lineup must satisfy the following restrictions:

1 At least 4 members must be able to play guard, at least 2 members must be able to play forward, and at least 1 member must be able to play center.

2 The average ball-handling, shooting, and rebounding level of the starting lineup must be at least 2.

3 If player 3 starts, then player 6 cannot start.

4 If player 1 starts, then players 4 and 5 must both start.

5 Either player 2 or player 3 must start.

Given these constraints, Coach Night wants to maximize the total defensive ability of the starting team. Formulate an IP that will help him choose his starting team.

2 Because of excessive pollution on the Momiss River, the state of Momiss is going to build pollution control stations. Three sites (1, 2, and 3) are under consideration. Momiss is

TABLE 9

Player	Position	Ball-Handling	Shooting	Rebounding	Defense
1	G	3	3	1	3
2	C	2	1	3	2
3	G-F	2	3	2	2
4	F-C	1	3	3	1
5	G-F	3	3	3	3
6	F-C	3	1	2	3
7	G-F	3	2	2	1

interested in controlling the pollution levels of two pollutants (1 and 2). The state legislature requires that at least 80,000 tons of pollutant 1 and at least 50,000 tons of pollutant 2 be removed from the river. The relevant data for this problem are shown in Table 10. Formulate an IP to minimize the cost of meeting the state legislature's goals.

3 A manufacturer can sell product 1 at a profit of $2/unit and product 2 at a profit of $5/unit. Three units of raw material are needed to manufacture 1 unit of product 1, and

TABLE 10

Site	Cost of Building Station ($)	Cost of Treating 1 Ton Water ($)	Amount Removed per Ton of Water	
			Pollutant 1	Pollutant 2
1	100,000	20	0.40	0.30
2	60,000	30	0.25	0.20
3	40,000	40	0.20	0.25

TABLE 11

From	To ($)		
	Region 1	Region 2	Region 3
New York	20	40	50
Los Angeles	48	15	26
Chicago	26	35	18
Atlanta	24	50	35

6 units of raw material are needed to manufacture 1 unit of product 2. A total of 120 units of raw material are available. If any of product 1 is produced, a setup cost of $10 is incurred, and if any of product 2 is produced, a setup cost of $20 is incurred. Formulate an IP to maximize profits.

4 Suppose we add the following restriction to Example 1 (Stockco): If investments 2 and 3 are chosen, then investment 4 must be chosen. What constraints would be added to the formulation given in the text?

5 How would the following restrictions modify the formulation of Example 6 (Dorian car sizes)? (Do each part separately.)

  a If midsize cars are produced, then compacts must also be produced.

  b Either compacts or large cars must be manufactured.

6 To graduate from Basketweavers University with a major in operations research, a student must complete at least two math courses, at least two OR courses, and at least two computer courses. Some courses can be used to fulfill more than one requirement: Calculus can fulfill the math requirement; operations research, math and OR requirements; data structures, computer and math requirements; business statistics, math and OR requirements; computer simulation, OR and computer requirements; introduction to computer programming, computer requirement; and forecasting, OR and math requirements.

Some courses are prerequisites for others: Calculus is a prerequisite for business statistics; introduction to computer programming is a prerequisite for computer simulation and for data structures; and business statistics is a prerequisite for forecasting. Formulate an IP that minimizes the number of courses needed to satisfy the major requirements.

7 In Example 7 (Euing Gas), suppose that $x = 300$. What would be the values of $y_1, y_2, y_3, z_1, z_2, z_3,$ and z_4? How about if $x = 1,200$?

8 Formulate an IP to solve the Dorian Auto problem for the advertising data that exhibit increasing returns as more ads are placed in a magazine (pages 495–496).

9 How can integer programming be used to ensure that the variable x can assume only the values 1, 2, 3, and 4?

10 If x and y are integers, how could you ensure that $x + y \leq 3$, $2x + 5y \leq 12$, or both are satisfied by x and y?

11 If x and y are both integers, how would you ensure that whenever $x \leq 2$, then $y \leq 3$?

12 A company is considering opening warehouses in four cities: New York, Los Angeles, Chicago, and Atlanta. Each warehouse can ship 100 units per week. The weekly fixed cost of keeping each warehouse open is $400 for New York, $500 for Los Angeles, $300 for Chicago, and $150 for Atlanta. Region 1 of the country requires 80 units per week, region 2 requires 70 units per week, and region 3 requires 40 units per week. The costs (including production and shipping costs) of sending one unit from a plant to a region are shown in Table 11. We want to meet weekly demands at minimum cost, subject to the preceding information and the following restrictions:

  1 If the New York warehouse is opened, then the Los Angeles warehouse must be opened.

  2 At most two warehouses can be opened.

  3 Either the Atlanta or the Los Angeles warehouse must be opened.

Formulate an IP that can be used to minimize the weekly costs of meeting demand.

13 Glueco produces three types of glue on two different production lines. Each line can be utilized by up to seven workers at a time. Workers are paid $500 per week on production line 1, and $900 per week on production line 2. A week of production costs $1,000 to set up production line 1 and $2,000 to set up production line 2. During a week on a production line, each worker produces the number of units of glue shown in Table 12. Each week, at least 120 units of glue 1, at least 150 units of glue 2, and at least 200 units of glue 3 must be produced. Formulate an IP to minimize the total cost of meeting weekly demands.

14[†] The manager of State University's DED computer wants to be able to access five different files. These files are scattered on 10 disks as shown in Table 13. The amount of storage required by each disk is as follows: disk 1, 3K; disk 2, 5K; disk 3, 1K; disk 4, 2K; disk 5, 1K; disk 6, 4K; disk 7, 3K; disk 8, 1K; disk 9, 2K; disk 10, 2K.

  a Formulate an IP that determines a set of disks requiring the minimum amount of storage such that each

TABLE 12

Production Line	Glue		
	1	2	3
1	20	30	40
2	50	35	45

[†]Based on Day (1965).

TABLE 13

File	Disk 1	2	3	4	5	6	7	8	9	10
1	x	x		x	x			x	x	
2	x		x							
3		x			x		x			x
4			x			x		x		
5	x	x		x		x	x		x	x

file is on at least one of the disks. For a given disk, we must either store the entire disk or store none of the disk; we cannot store part of a disk.

b Modify your formulation so that if disk 3 or disk 5 is used, then disk 2 must also be used.

15 Fruit Computer produces two types of computers: Pear computers and Apricot computers. Relevant data are given in Table 14. A total of 3,000 chips and 1,200 hours of labor are available. Formulate an IP to help Fruit maximize profits.

16 The Lotus Point Condo Project will contain both homes and apartments. The site can accommodate up to 10,000 dwelling units. The project must contain a recreation project: either a swimming–tennis complex or a sailboat marina, but not both. If a marina is built, then the number of homes in the project must be at least triple the number of apartments in the project. A marina will cost $1.2 million, and a swimming–tennis complex will cost $2.8 million. The developers believe that each apartment will yield revenues with an NPV of $48,000, and each home will yield revenues with an NPV of $46,000. Each home (or apartment) costs $40,000 to build. Formulate an IP to help Lotus Point maximize profits.

17 A product can be produced on four different machines. Each machine has a fixed setup cost, variable production costs per-unit-processed, and a production capacity given in Table 15. A total of 2,000 units of the product must be produced. Formulate an IP whose solution will tell us how to minimize total costs.

TABLE 14

Computer	Labor	Chips	Equipment Costs ($)	Selling Price ($)
Pear	1 hour	2	5,000	400
Apricot	2 hours	5	7,000	900

TABLE 15

Machine	Fixed Cost ($)	Variable Cost per Unit ($)	Capacity
1	1,000	20	900
2	920	24	1,000
3	800	16	1,200
4	700	28	1,600

TABLE 16

	Book 1	2	3	4	5
Maximum Demand	5,000	4,000	3,000	4,000	3,000
Variable Cost ($)	25	20	15	18	22
Sales Price ($)	50	40	38	32	40
Fixed Cost ($ Thousands)	80	50	60	30	40

18 Use LINDO, LINGO, or Excel Solver to find the optimal solution to the following IP:

Bookco Publishers is considering publishing five textbooks. The maximum number of copies of each textbook that can be sold, the variable cost of producing each textbook, the sales price of each textbook, and the fixed cost of a production run for each book are given in Table 16. Thus, for example, producing 2,000 copies of book 1 brings in a revenue of 2,000(50) = $100,000 but costs 80,000 + 25(2,000) = $130,000. Bookco can produce at most 10,000 books if it wants to maximize profit.

19 Comquat owns four production plants at which personal computers are produced. Comquat can sell up to 20,000 computers per year at a price of $3,500 per computer. For each plant the production capacity, the production cost per computer, and the fixed cost of operating a plant for a year are given in Table 17. Determine how Comquat can maximize its yearly profit from computer production.

20 WSP Publishing sells textbooks to college students. WSP has two sales reps available to assign to the A–G state area. The number of college students (in thousands) in each state is given in Figure 9. Each sales rep must be assigned to two adjacent states. For example, a sales rep could be assigned to A and B, but not A and D. WSP's goal is to

TABLE 17

Plant	Production Capacity	Plant Fixed Cost ($ Million)	Cost per Computer ($)
1	10,000	9	1,000
2	8,000	5	1,700
3	9,000	3	2,300
4	6,000	1	2,900

FIGURE 9

maximize the number of total students in the states assigned to the sales reps. Formulate an IP whose solution will tell you where to assign the sales reps. Then use LINDO to solve your IP.

21 Eastinghouse sells air conditioners. The annual demand for air conditioners in each region of the country is as follows: East, 100,000; South, 150,000; Midwest, 110,000; West, 90,000. Eastinghouse is considering building the air conditioners in four different cities: New York, Atlanta, Chicago, and Los Angeles. The cost of producing an air conditioner in a city and shipping it to a region of the country is given in Table 18. Any factory can produce as many as 150,000 air conditioners per year. The annual fixed cost of operating a factory in each city is given in Table 19. At least 50,000 units of the Midwest demand for air conditioners must come from New York, or at least 50,000 units of the Midwest demand must come from Atlanta. Formulate an IP whose solution will tell Eastinghouse how to minimize the annual cost of meeting demand for air conditioners.

22 Consider the following puzzle. You are to pick out 4 three-letter "words" from the following list:

   DBA DEG ADI FFD GHI BCD FDF BAI

For each word, you earn a score equal to the position that the word's third letter appears in the alphabet. For example, DBA earns a score of 1, DEG earns a score of 7, and so on. Your goal is to choose the four words that maximize your total score, subject to the following constraint: The sum of the positions in the alphabet for the first letter of each word chosen must be at least as large as the sum of the positions in the alphabet for the second letter of each word chosen. Formulate an IP to solve this problem.

23 At a machine tool plant, five jobs must be completed each day. The time it takes to do each job depends on the machine used to do the job. If a machine is used at all, there is a setup time required. The relevant times are given in Table 20. The company's goal is to minimize the sum of the setup and machine operation times needed to complete all

TABLE 18

City	Price by Region ($)			
	East	South	Midwest	West
New York	206	225	230	290
Atlanta	225	206	221	270
Chicago	230	221	208	262
Los Angeles	290	270	262	215

TABLE 19

City	Annual Fixed Cost ($ Million)
New York	6
Atlanta	5.5
Chicago	5.8
Los Angeles	6.2

TABLE 20

Machine	Job					Machine Setup Time (Minutes)
	1	2	3	4	5	
1	42	70	93	X	X	30
2	X	85	45	X	X	40
3	58	X	X	37	X	50
4	58	X	55	X	38	60
5	X	60	X	54	X	20

jobs. Formulate and solve (with LINDO, LINGO, or Excel Solver) an IP whose solution will do this.

Group B

24[†] Breadco Bakeries is a new bakery chain that sells bread to customers throughout the state of Indiana. Breadco is considering building bakeries in three locations: Evansville, Indianapolis, and South Bend. Each bakery can bake as many as 900,000 loaves of bread each year. The cost of building a bakery at each site is $5 million in Evansville, $4 million in Indianapolis, and $4.5 million in South Bend. To simplify the problem, we assume that Breadco has only three customers, whose demands each year are 700,000 loaves (customer 1); 400,000 loaves (customer 2); and 300,000 loaves (customer 3). The total cost of baking and shipping a loaf of bread to a customer is given in Table 21.

Assume that future shipping and production costs are discounted at a rate of $11\frac{1}{9}\%$ per year. Assume that once built, a bakery lasts forever. Formulate an IP to minimize Breadco's total cost of meeting demand (present and future). (*Hint:* You will need the fact that for $x < 1$, $a + ax + ax^2 + ax^3 + \cdots = a/(1 - x)$.) How would you modify the formulation if either Evansville or South Bend must produce at least 800,000 loaves per year?

25[‡] Speaker's Clearinghouse must disburse sweepstakes checks to winners in four different regions of the country: Southeast (SE), Northeast (NE), Far West (FW), and Midwest (MW). The average daily amount of the checks written to winners in each region of the country is as follows: SE, $40,000; NE, $60,000; FW, $30,000; MW, $50,000. Speaker's must issue the checks the day they find out a customer has won. They can delay winners from quickly cashing their checks by giving a winner a check drawn on an out-of-the-way bank (this will cause the check to clear

TABLE 21

From	To		
	Customer 1	Customer 2	Customer 3
Evansville	16¢	34¢	26¢
Indianapolis	40¢	30¢	35¢
South Bend	45¢	45¢	23¢

[†]Based on Efroymson and Ray (1966).
[‡]Based on Shanker and Zoltners (1972).

slowly). Four bank sites are under consideration: Frosbite Falls, Montana (FF), Redville, South Carolina (R), Painted Forest, Arizona (PF), and Beanville, Maine (B). The annual cost of maintaining an account at each bank is as follows: FF, $50,000; R, $40,000; PF, $30,000; B, $20,000. Each bank has a requirement that the average daily amount of checks written cannot exceed $90,000. The average number of days it takes a check to clear is given in Table 22. Assuming that money invested by Speaker's earns 15% per year, where should the company have bank accounts, and from which bank should a given customer's check be written?

26[†] Governor Blue of the state of Berry is attempting to get the state legislature to gerrymander Berry's congressional districts. The state consists of 10 cities, and the numbers of registered Republicans and Democrats (in thousands) in each city are shown in Table 23. Berry has five congressional representatives. To form congressional districts, cities must be grouped according to the following restrictions:

1 All voters in a city must be in the same district.

2 Each district must contain between 150,000 and 250,000 voters (there are no independent voters).

Governor Blue is a Democrat. Assume that each voter always votes a straight party ticket. Formulate an IP to help Governor Blue maximize the number of Democrats who will win congressional seats.

27[‡] The Father Domino Company sells copying machines. A major factor in making a sale is Domino's quick service. Domino sells copiers in six cities: Boston, New York,

Philadelphia, Washington, Providence, and Atlantic City. The annual sales of copiers projected depend on whether a service representative is within 150 miles of a city (see Table 24).

Each copier costs $500 to produce and sells for $1,000. The annual cost per service representative is $80,000. Domino must determine in which of its markets to base a service representative. Only Boston, New York, Philadelphia, and Washington are under consideration as bases for service representative. The distance (in miles) between the cities is shown in Table 25. Formulate an IP that will help Domino maximize annual profits.

28[§] Thailand inducts naval draftees at three drafting centers. Then the draftees must each be sent to one of three naval bases for training. The cost of transporting a draftee from a drafting center to a base is given in Table 26. Each year, 1,000 men are inducted at center 1; 600 at center 2; and 700 at center 3. Base 1 can train 1,000 men a year, base 2, 800 men; and base 3, 700 men. After the inductees are trained, they are sent to Thailand's main naval base (B). They may be transported on either a small ship or a large ship. It costs $5,000 plus $2 per mile to use a small ship. A small ship can transport up to 200 men to the main base and may visit up to two bases on its way to the main base. Seven small and five large ships are available. It costs $10,000 plus $3 per mile to use a large ship. A large ship may visit up to

TABLE 22

Region	FF	R	PF	B
SE	7	2	6	5
NE	8	4	5	3
FW	4	8	2	11
MW	5	4	7	5

TABLE 23

City	Republicans	Democrats
1	80	34
2	60	44
3	40	44
4	20	24
5	40	114
6	40	64
7	70	14
8	50	44
9	70	54
10	70	64

TABLE 24

Representative Within 150 Miles?	Sales					
	Boston	N.Y.	Phila.	Wash.	Prov.	Atl. City
Yes	700	1,000	900	800	400	450
No	500	750	700	450	200	300

TABLE 25

	Boston	N.Y.	Phila.	Wash.
Boston	0	222	310	441
New York	222	0	89	241
Philadelphia	310	89	0	146
Washington	441	241	146	0
Providence	47	186	255	376
Atlantic City	350	123	82	178

TABLE 26

From	To ($)		
	Base 1	Base 2	Base 3
Center 1	200	200	300
Center 2	300	400	220
Center 3	300	400	250

[†]Based on Garfinkel and Nemhauser (1970).
[‡]Based on Gelb and Khumawala (1984).

[§]Based on Choypeng, Puakpong, and Rosenthal (1986).

three bases on its way to the main base and may transport up to 500 men. The possible "tours" for each type of ship are given in Table 27.

Assume that the assignment of draftees to training bases is done using the transportation method. Then formulate an IP that will minimize the total cost incurred in sending the men from the training bases to the main base. (*Hint:* Let y_{ij} = number of men sent by tour i from base j to main base (B) on a small ship, x_{ij} = number of men sent by tour i from base j to B on a large ship, S_i = number of times tour i is used by a small ship, and L_i = number of times tour i is used by a large ship.)

29 You have been assigned to arrange the songs on the cassette version of Madonna's latest album. A cassette tape has two sides (1 and 2). The songs on each side of the cassette must total between 14 and 16 minutes in length. The length and type of each song are given in Table 28. The assignment of songs to the tape must satisfy the following conditions:

1 Each side must have exactly two ballads.

2 Side 1 must have at least three hit songs.

3 Either song 5 or song 6 must be on side 1.

4 If songs 2 and 4 are on side 1, then song 5 must be on side 2.

Explain how you could use an integer programming formulation to determine whether there is an arrangement of songs satisfying these restrictions.

30 Cousin Bruzie of radio station WABC schedules radio commercials in 60-second blocks. This hour, the station has sold commercial time for commercials of 15, 16, 20, 25, 30, 35, 40, and 50 seconds. Formulate an integer programming model that can be used to determine the minimum number of 60-second blocks of commercials that must be scheduled to fit in all the current hour's commercials. (*Hint:* Certainly no more than eight blocks of time are needed. Let $y_i = 1$ if block i is used and $y_i = 0$ otherwise).

31[†] A Sunco oil delivery truck contains five compartments, holding up to 2,700, 2,800, 1,100, 1,800, and 3,400 gallons of fuel, respectively. The company must deliver three types of fuel (super, regular, and unleaded) to a customer. The demands, penalty per gallon short, and the maximum allowed shortage are given in Table 29. Each compartment of the truck can carry only one type of gasoline. Formulate an IP whose solution will tell Sunco how to load the truck in a way that minimizes shortage costs.

32[‡] Simon's Mall has 10,000 sq ft of space to rent and wants to determine the types of stores that should occupy the mall. The minimum number and maximum number of each type of store (along with the square footage of each type) is given in Table 30. The annual profit made by each type of store will, of course, depend on how many stores of that type are in the mall. This dependence is given in Table 31 (all profits are in units of $10,000). Thus, if there are two department stores in the mall, each department store earns $210,000 profit per year. Each store pays 5% of its annual profit as rent to Simon's. Formulate an IP whose solution will tell Simon's how to maximize rental income from the mall.

33[§] Boris Milkem's financial firm owns six assets. The expected sales price (in millions of dollars) for each asset is given in Table 32. If asset 1 is sold in year 2, the firm receives $20 million. To maintain a regular cash flow, Milkem must sell at least $20 million of assets during year 1, at least $30 million worth during year 2, and at least $35 million worth during year 3. Set up an IP that Milkem can

TABLE 27

Tour Number	Locations Visited	Miles Traveled
1	B–1–B	370
2	B–1–2–B	515
3	B–2–3–B	665
4	B–2–B	460
5	B–3–B	600
6	B–1–3–B	640
7	B–1–2–3–B	720

TABLE 28

Song	Type	Length (in minutes)
1	Ballad	4
2	Hit	5
3	Ballad	3
4	Hit	2
5	Ballad	4
6	Hit	3
7		5
8	Ballad and hit	4

TABLE 29

Type of Gasoline	Demand	Cost per Gallon Short ($)	Maximum Allowed Shortage
Super	2,900	10	500
Regular	4,000	8	500
Unleaded	4,900	6	500

TABLE 30

Store Type	Square Footage	Minimum	Maximum
Jewelry	500	1	3
Shoe	600	1	3
Department	1,500	1	3
Book	700	0	3
Clothing	900	1	3

TABLE 31

	Number of Stores		
Type of Store	1	2	3
Jewelry	9	8	7
Shoe	10	9	5
Department	27	21	20
Book	16	9	7
Clothing	17	13	10

TABLE 32

	Sold In		
Asset	Year 1	Year 2	Year 3
1	15	20	24
2	16	18	21
3	22	30	36
4	10	20	30
5	17	19	22
6	19	25	29

TABLE 33

Alarm Box	Two Closest Ladder Companies
1	2, 3
2	3, 4
3	1, 5
4	2, 6
5	3, 6
6	4, 7
7	5, 7

TABLE 34

Boiler Number	Minimum Steam	Maximum Steam	Cost/Ton ($)
1	500	1,000	10
2	300	900	8
3	400	800	6

TABLE 35

Turbine Number	Minimum	Maximum	Kwh per Ton of Steam	Processing Cost per Ton ($)
1	300	600	4	2
2	500	800	5	3
3	600	900	6	4

use to determine how to maximize total revenue from assets sold during the next three years. In implementing this model, how could the idea of a rolling planning horizon be used?

34[†] The Smalltown Fire Department currently has seven conventional ladder companies and seven alarm boxes. The two closest ladder companies to each alarm box are given in Table 33. The city fathers want to maximize the number of conventional ladder companies that can be replaced with tower ladder companies. Unfortunately, political considerations dictate that a conventional company can be replaced only if, after replacement, at least one of the two closest companies to each alarm box is still a conventional company.

a Formulate an IP that can be used to maximize the number of conventional companies that can be replaced by tower companies.

b Suppose $y_k = 1$ if conventional company k is replaced. Show that if we let $z_k = 1 - y_k$, the answer in part (a) is equivalent to a set-covering problem.

35[‡] A power plant has three boilers. If a given boiler is operated, it can be used to produce a quantity of steam (in tons) between the minimum and maximum given in Table 34. The cost of producing a ton of steam on each boiler is also given. Steam from the boilers is used to produce power on three turbines. If operated, each turbine can process an amount of steam (in tons) between the minimum and maximum given in Table 35. The cost of processing a ton of steam and the power produced by each turbine is also given. Formulate an IP that can be used to minimize the cost of producing 8,000 kwh of power.

36[§] An Ohio company, Clevcinn, consists of three subsidiaries. Each has the respective average payroll, unemployment reserve fund, and estimated payroll given in Table 36. (All figures are in millions of dollars.) Any employer in the state of Ohio whose reserve/average payroll ratio is less than 1 must pay 20% of its estimated payroll in unemployment insurance premiums or 10% if the ratio is at least 1. Clevcinn can aggregate its subsidiaries and label them as separate employers. For instance, if subsidiaries 2 and 3 are aggregated, they must pay 20% of their combined payroll in unemployment insurance premiums. Formulate an IP that can be used to determine which subsidiaries should be aggregated.

37 The Indiana University Business School has two rooms that each seat 50 students, one room that seats 100 students, and one room that seats 150 students. Classes are held five hours a day. The four types of requests for rooms are listed in Table 37. The business school must decide how many requests of each type should be assigned to each type of room. Penalties for each type of assignment are given in Table 38. An X means that a request must be satisfied by a room of adequate size. Formulate an IP whose solution will tell the business school how to assign classes to rooms in a way that minimizes total penalties.

[†]Based on Walker (1974).
[‡]Based on Cavalieri, Roversi, and Ruggeri (1971).

[§]Based on Salkin (1979).

TABLE 36

Subsidiary	Average Payroll	Reserve	Estimated Payroll
1	300	400	350
2	600	510	400
3	800	600	500

TABLE 37

Type	Size Room Requested (Seats)	Hours Requested	Number of Requests
1	50	2, 3, 4	3
2	150	1, 2, 3	1
3	100	5	1
4	50	1, 2	2

TABLE 38

Size Requested	Sizes Used to Satisfy Request			Penalty
	50	100	150	
50	0	2	4	100* (Hours requested)
100	X	0	1	100* (Hours requested)
150	X	X	0	100* (Hours requested)

38 A company sells seven types of boxes, ranging in volume from 17 to 33 cubic feet. The demand and size of each box are given in Table 39. The variable cost (in dollars) of producing each box is equal to the box's volume. A fixed cost of $1,000 is incurred to produce any of a particular box. If the company desires, demand for a box may be satisfied by a box of larger size. Formulate and solve (with LINDO, LINGO, or Excel Solver) an IP whose solution will minimize the cost of meeting the demand for boxes.

39 Huntco produces tomato sauce at five different plants. The capacity (in tons) of each plant is given in Table 40. The tomato sauce is stored at one of three warehouses. The per-ton cost (in hundreds of dollars) of producing tomato sauce at each plant and shipping it to each warehouse is given in Table 41. Huntco has four customers. The cost of shipping a ton of sauce from each warehouse to each customer is as given in Table 42. Each customer must be delivered the amount (in tons) of sauce given in Table 43.

TABLE 39

	Box						
	1	2	3	4	5	6	7
Size	33	30	26	24	19	18	17
Demand	400	300	500	700	200	400	200

TABLE 40

	Plant				
	1	2	3	4	5
Tons	300	200	300	200	400

TABLE 41

From	To		
	Warehouse 1	Warehouse 2	Warehouse 3
Plant 1	8	10	12
Plant 2	7	5	7
Plant 3	8	6	5
Plant 4	5	6	7
Plant 5	7	6	5

TABLE 42

From	To			
	Customer 1	Customer 2	Customer 3	Customer 4
Warehouse 1	40	80	90	50
Warehouse 2	70	70	60	80
Warehouse 3	80	30	50	60

TABLE 43

	Customer			
	1	2	3	4
Demand	200	300	150	250

a Formulate a balanced transportation problem whose solution will tell us how to minimize the cost of meeting the customer demands.

b Modify this problem if these are annual demands and there is a fixed annual cost of operating each plant and warehouse. These costs (in thousands) are given in Table 44.

40 To satisfy telecommunication needs for the next 20 years, Telstar Corporation estimates that the number of circuits required between the United States and Germany, France, Switzerland, and the United Kingdom will be as given in Table 45.

Two types of circuits may be created: cable and satellite. Two types of cable circuits (TA7 and TA8) are available. The fixed cost of building each type of cable and the circuit capacity of each type are as given in Table 46.

TA7 and TA8 cable go underseas from the United States to the English Channel. Thus, it costs an additional amount to extend these circuits to other European countries. The annual variable cost per circuit is given in Table 47.

TABLE 44[†]

Facility	Fixed Annual Cost (in Thousands) $
Plant 1	35
Plant 2	45
Plant 3	40
Plant 4	42
Plant 5	40
Warehouse 1	30
Warehouse 2	40
Warehouse 3	30

[†]Based on Geoffrion and Graves (1974).

TABLE 45

Country	Required Circuits
France	20,000
Germany	60,000
Switzerland	16,000
United Kingdom	60,000

TABLE 46

Cable Type	Fixed Operating Cost ($ Billion)	Capacity
TA7	1.6	8,500
TA8	2.3	37,800

TABLE 47

Country	Variable Cost per Circuit ($)
France	0
Germany	310
Switzerland	290
United Kingdom	0

To create and use a satellite circuit Telstar must launch a satellite and each country using the satellite must have an earth station(s) to receive the signal. It costs $3 billion to launch a satellite. Each launched satellite can handle up to 140,000 circuits. All earth stations have a maximum capacity of 190 circuits and cost $6,000 per year to operate. Formulate an integer programming model to help determine how to supply the needed circuits and minimize total cost incurred during the next 20 years.

Then use LINDO (or LINGO) to find a near optimal solution. LINDO after 300 pivots did not think it had an optimal solution! By the way, do not require that the number of cable or satellite circuits in a country be integers, or your

model will never get solved! For some variables, however, the integer requirement is vital![†]

41 A large drug company must determine how many sales representatives to assign to each of four sales districts. The cost of having n representatives in a district is ($88,000 + $80,000$n$) per year. If a rep is based in a given district, the time it takes to complete a call on a doctor is given in Table 48 (times are in hours).

Each sales rep can work up to 160 hours per month. Each month the number of calls given in Table 49 must be made in each district. A fractional number of representatives in a district is not permissible. Determine how many representatives should be assigned to each district.

42[‡] In this assignment, we will use integer programming and the concept of bond duration to show how Wall Street firms can select an optimal bond portfolio. The *duration* of a bond (or any stream of payments) is defined as follows: Let $C(t)$ be the payment of the bond at time t ($t = 1, 2, \ldots, n$). Let r = market interest rate. If the time-weighted average of the bond's payments is given by:

$$\sum_{t=1}^{t=n} tC(t)/(1 + r)^t$$

and the market price P of the bond is given by:

$$\sum_{t=1}^{t=n} C(t)/(1 + r)^t$$

then the duration of the bond D is given by:

$$D = (1/P) \sum_{t=1}^{n} \frac{tC(t)}{(1 + r)^t}$$

Thus, the duration of a bond measures the "average" time (in years) at which a randomly chosen $1 of NPV is received. Suppose an insurance company needs to make payments of $20,000 every six months for the next 10 years. If the market

TABLE 48

Rep's Base District	Actual Sales Call District			
	1	2	3	4
1	1	4	5	7
2	4	1	3	5
3	5	3	1	2
4	7	5	2	1

TABLE 49

District	Number of Calls
1	50
2	80
3	100
4	60

[†]Based on Calloway, Cummins, and Freeland (1990).
[‡]Based on Strong (1989).

rate of interest is 10% per year, then this stream of payments has an NPV of $251,780 and a duration of 4.47 years. If we want to minimize the sensitivity of our bond portfolio to interest risk and still meet our payment obligations, then it has been shown that we should invest $251,780 at the beginning of year 1 in a bond portfolio having a duration equal to the duration of the payment stream.

Suppose the only cost of owning a bond portfolio is the transaction cost associated with the cost of purchasing the bonds. Let's suppose six bonds are available. The payment streams for these six bonds are given in Table 50. The transaction cost of purchasing any units of bond i equals $500 + $5 per bond purchased. Thus, purchasing one unit of bond 1 costs $505 and purchasing 10 units of bond 1 costs $550. Assume that a fractional number of bond i unit purchases is permissible, but in the interests of diversification at most 100 units of any bond can be purchased. Treasury bonds may also be purchased (with no transaction cost). A treasury bond costs $980 and has a duration of .25 year (90 days).

After computing the price and duration for each bond, use integer programming to determine the immunized bond portfolio that incurs the smallest transaction costs. You may assume the duration of your portfolio is a weighted average of the durations of the bonds included in the portfolio, where the weight associated with each bond is equal to the money invested in that bond.

43 Ford has four automobile plants. Each is capable of producing the Taurus, Lincoln, or Escort, but it can only produce one of these cars. The fixed cost of operating each plant for a year and the variable cost of producing a car of each type at each plant are given in Table 51.

Ford faces the following restrictions:

a Each plant can produce only one type of car.

b The total production of each type of car must be at a single plant; that is, for example, if any Tauruses are made at plant 1, then all Tauruses must be made there.

c If plants 3 and 4 are used, then plant 1 must also be used.

TABLE 50

			Available Bonds			
Year	Bond 1	Bond 2	Bond 3	Bond 4	Bond 5	Bond 6
1	50	100	130	20	100	120
2	60	90	130	20	100	100
3	70	80	130	20	100	80
4	80	70	130	20	100	140
5	90	60	130	20	100	100
6	100	50	130	80	100	90
7	110	40	130	40	100	110
8	120	30	130	150	100	130
9	130	20	130	200	100	180
10	1,010	1,040	1,130	1,200	1,100	950

TABLE 51

Plant	Fixed Cost ($)	Variable Cost ($)		
		Taurus	Lincoln	Escort
1	7 billion	12,000	16,000	9,000
2	6 billion	15,000	18,000	11,000
3	4 billion	17,000	19,000	12,000
4	2 billion	19,000	22,000	14,000

Each year, Ford must produce 500,000 of each type of car. Formulate an IP whose solution will tell Ford how to minimize the annual cost of producing cars.

44 Venture capital firm JD is trying to determine in which of 10 projects it should invest. It knows how much money is available for investment each of the next N years, the NPV of each project, and the cash required by each project during each of the next N years (see Table 52).

a Write a LINGO program to determine the projects in which JD should invest.

b Use your LINGO program to determine which of the 10 projects should be selected. Each project requires cash investment during the next three years. During year 1, $80 million is available for investment. During year 2, $60 million is available for investment. During year 3, $70 million is available for investment. (All figures are in millions of dollars.)

45 Write a LINGO program that can solve a fixed charge problem of the type described in Example 3. Assume there is a limited demand for each product. Then use your program to solve a four-product, three-resource fixed charge problem with the parameters shown in Tables 53, 54, and 55.

TABLE 52

Investment ($ Million)	Project									
	1	2	3	4	5	6	7	8	9	10
Year 1	6	9	12	15	18	21	24	27	30	35
Year 2	3	5	7	9	11	13	15	17	19	21
Year 3	5	7	9	12	12	14	16	11	20	24
NPV	20	30	40	50	60	70	80	90	100	130

TABLE 53

Resource	Resource Availability
1	40
2	60
3	80

TABLE **54**

Product	Demand	Unit Profit Contribution ($)	Fixed Charge ($)
1	40	2	30
2	60	5	40
3	65	6	50
4	70	7	60

TABLE **55**

Resource Usage	Product			
	1	2	3	4
1	1	2	3.5	4
2	5	6	7	9
3	3	4	5	6

9.3 The Branch-and-Bound Method for Solving Pure Integer Programming Problems

In practice, most IPs are solved by using the technique of branch-and-bound. Branch-and-bound methods find the optimal solution to an IP by efficiently enumerating the points in a subproblem's feasible region. Before explaining how branch-and-bound works, we need to make the following elementary but important observation: *If you solve the LP relaxation of a pure IP and obtain a solution in which all variables are integers, then the optimal solution to the LP relaxation is also the optimal solution to the IP.*

To see why this observation is true, consider the following IP:

$$\max z = 3x_1 + 2x_2$$
$$\text{s.t.} \quad 2x_1 + x_2 \le 6$$
$$x_1, x_2 \ge 0; x_1, x_2 \text{ integer}$$

The optimal solution to the LP relaxation of this pure IP is $x_1 = 0$, $x_2 = 6$, $z = 12$. Because this solution gives integer values to all variables, the preceding observation implies that $x_1 = 0$, $x_2 = 6$, $z = 12$ is also the optimal solution to the IP. Observe that the feasible region for the IP is a subset of the points in the LP relaxation's feasible region (see Figure 10). Thus, the optimal z-value for the IP cannot be larger than the optimal z-value for the LP relaxation. This means that the optimal z-value for the IP must be ≤ 12. But the point $x_1 = 0$, $x_2 = 6$, $z = 12$ is feasible for the IP and has $z = 12$. Thus, $x_1 = 0$, $x_2 = 6$, $z = 12$ must be optimal for the IP.

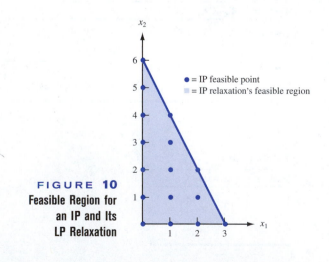

FIGURE 10
Feasible Region for an IP and Its LP Relaxation

EXAMPLE 9   **Branch-and-Bound Method**

The Telfa Corporation manufactures tables and chairs. A table requires 1 hour of labor and 9 square board feet of wood, and a chair requires 1 hour of labor and 5 square board feet of wood. Currently, 6 hours of labor and 45 square board feet of wood are available. Each table contributes $8 to profit, and each chair contributes $5 to profit. Formulate and solve an IP to maximize Telfa's profit.

Solution   Let

$$x_1 = \text{number of tables manufactured}$$

$$x_2 = \text{number of chairs manufactured}$$

Because x_1 and x_2 must be integers, Telfa wants to solve the following IP:

$$\max z = 8x_1 + 5x_2$$

s.t. $x_1 + x_2 \le 6$ (Labor constraint)

$9x_1 + 5x_2 \le 45$ (Wood constraint)

$x_1, x_2 \ge 0; x_1, x_2 \text{ integer}$

The branch-and-bound method begins by solving the LP relaxation of the IP. If all the decision variables assume integer values in the optimal solution to the LP relaxation, then the optimal solution to the LP relaxation will be the optimal solution to the IP. We call the LP relaxation subproblem 1. Unfortunately, the optimal solution to the LP relaxation is $z = \frac{165}{4}$, $x_1 = \frac{15}{4}$, $x_2 = \frac{9}{4}$ (see Figure 11). From Section 9.1, we know that (optimal z-value for IP) $\le$ (optimal z-value for LP relaxation). This implies that the optimal z-value for the IP cannot exceed $\frac{165}{4}$. Thus, the optimal z-value for the LP relaxation is an **upper bound** for Telfa's profit.

Our next step is to partition the feasible region for the LP relaxation in an attempt to find out more about the location of the IP's optimal solution. We arbitrarily choose a variable that is fractional in the optimal solution to the LP relaxation—say, x_1. Now observe that every point in the feasible region for the IP must have either $x_1 \le 3$ or $x_1 \ge 4$. (Why can't a feasible solution to the IP have $3 < x_1 < 4$?) With this in mind, we "branch" on the variable x_1 and create the following two additional subproblems:

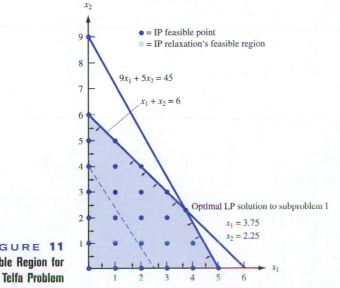

FIGURE 11
Feasible Region for Telfa Problem

9x_1 + 5x_2 = 45

x_1 + x_2 = 6

• = IP feasible point
■ = IP relaxation's feasible region

Optimal LP solution to subproblem 1
$x_1 = 3.75$
$x_2 = 2.25$

Subproblem 2 Subproblem 1 + Constraint $x_1 \geq 4$.

Subproblem 3 Subproblem 1 + Constraint $x_1 \leq 3$.

Observe that neither subproblem 2 nor subproblem 3 includes any points with $x_1 = \frac{15}{4}$. This means that the optimal solution to the LP relaxation cannot recur when we solve subproblem 2 or subproblem 3.

From Figure 12, we see that every point in the feasible region for the Telfa IP is included in the feasible region for subproblem 2 or subproblem 3. Also, the feasible regions for subproblems 2 and 3 have no points in common. Because subproblems 2 and 3 were created by adding constraints involving x_1, we say that subproblems 2 and 3 were created by **branching** on x_1.

We now choose any subproblem that has not yet been solved as an LP. We arbitrarily choose to solve subproblem 2. From Figure 12, we see that the optimal solution to subproblem 2 is $z = 41$, $x_1 = 4$, $x_2 = \frac{9}{5}$ (point C). Our accomplishments to date are summarized in Figure 13.

A display of all subproblems that have been created is called a **tree**. Each subproblem is referred to as a **node** of the tree, and each line connecting two nodes of the tree is called an **arc**. The constraints associated with any node of the tree are the constraints for the LP relaxation plus the constraints associated with the arcs leading from subproblem 1 to the node. The label t indicates the chronological order in which the subproblems are solved.

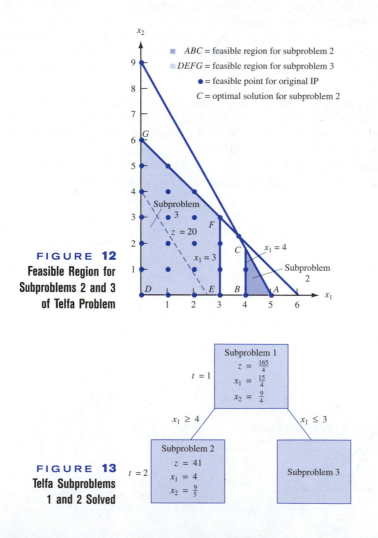

FIGURE 12
Feasible Region for Subproblems 2 and 3 of Telfa Problem

ABC = feasible region for subproblem 2
$DEFG$ = feasible region for subproblem 3
● = feasible point for original IP
C = optimal solution for subproblem 2

FIGURE 13
Telfa Subproblems 1 and 2 Solved

The optimal solution to subproblem 2 did not yield an all-integer solution, so we choose to use subproblem 2 to create two new subproblems. We choose a fractional-valued variable in the optimal solution to subproblem 2 and then branch on that variable. Because x_2 is the only fractional variable in the optimal solution to subproblem 2, we branch on x_2. We partition the feasible region for subproblem 2 into those points having $x_2 \geq 2$ and $x_2 \leq 1$. This creates the following two subproblems:

Subproblem 4 Subproblem 1 + Constraints $x_1 \geq 4$ and $x_2 \geq 2$ = subproblem 2 + Constraint $x_2 \geq 2$.

Subproblem 5 Subproblem 1 + Constraints $x_1 \geq 4$ and $x_2 \leq 1$ = subproblem 2 + Constraint $x_2 \leq 1$.

The feasible regions for subproblems 4 and 5 are displayed in Figure 14. The set of unsolved subproblems consists of subproblems 3, 4, and 5. We now choose a subproblem to solve. For reasons that are discussed later, we choose to solve the most recently created subproblem. (This is called the LIFO, or last-in-first-out, rule.) The LIFO rule implies that we should next solve subproblem 4 or subproblem 5. We arbitrarily choose to solve subproblem 4. From Figure 14 we see that subproblem 4 is infeasible. Thus, subproblem 4 cannot yield the optimal solution to the IP. To indicate this fact, we place an × by subproblem 4 (see Figure 15). Because any branches emanating from subproblem 4 will yield no useful information, it is fruitless to create them. When further branching on a subproblem cannot yield any useful information, we say that the subproblem (or node) is **fathomed.** Our results to date are displayed in Figure 15.

Now the only unsolved subproblems are subproblems 3 and 5. The LIFO rule implies that subproblem 5 should be solved next. From Figure 14, we see that the optimal solution to subproblem 5 is point I in Figure 14: $z = \frac{365}{9}$, $x_1 = \frac{40}{9}$, $x_2 = 1$. This solution does not yield any immediately useful information, so we choose to partition subproblem 5's feasible region by branching on the fractional-valued variable x_1. This yields two new subproblems (see Figure 16).

Subproblem 6 Subproblem 5 + Constraint $x_1 \geq 5$.

Subproblem 7 Subproblem 5 + Constraint $x_1 \leq 4$.

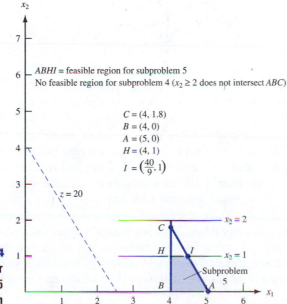

$ABHI$ = feasible region for subproblem 5
No feasible region for subproblem 4 ($x_2 \geq 2$ does not intersect ABC)

$C = (4, 1.8)$
$B = (4, 0)$
$A = (5, 0)$
$H = (4, 1)$
$I = \left(\frac{40}{9}, 1\right)$

$z = 20$

$x_2 = 2$

$x_2 = 1$

Subproblem 5

FIGURE 14

Feasible Regions for Subproblems 4 and 5 of Telfa Problem

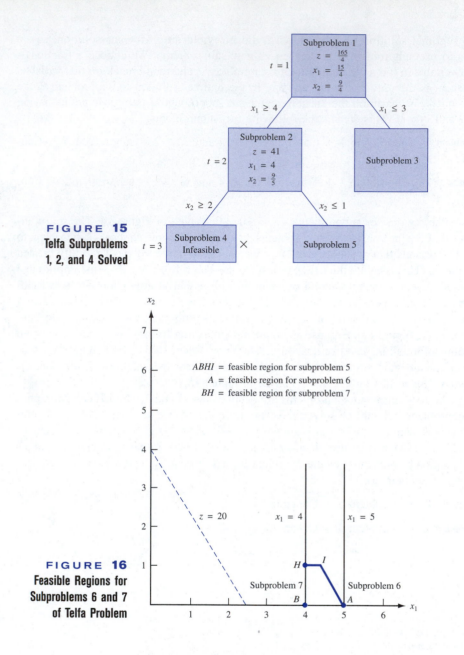

FIGURE 15
Telfa Subproblems 1, 2, and 4 Solved

Subproblem 1
$z = \frac{165}{4}$
$x_1 = \frac{15}{4}$
$x_2 = \frac{9}{4}$
$t = 1$

$x_1 \geq 4$ $x_1 \leq 3$

Subproblem 2
$z = 41$
$x_1 = 4$
$x_2 = \frac{9}{5}$
$t = 2$

Subproblem 3

$x_2 \geq 2$ $x_2 \leq 1$

Subproblem 4
Infeasible
$t = 3$ $\times$

Subproblem 5

FIGURE 16
Feasible Regions for Subproblems 6 and 7 of Telfa Problem

$ABHI$ = feasible region for subproblem 5
A = feasible region for subproblem 6
BH = feasible region for subproblem 7

$z = 20$ $x_1 = 4$ $x_1 = 5$

H I

Subproblem 7   Subproblem 6

B A

Together, subproblems 6 and 7 include all integer points that were included in the feasible region for subproblem 5. Also, no point having $x_1 = \frac{40}{9}$ can be in the feasible region for subproblem 6 or subproblem 7. Thus, the optimal solution to subproblem 5 will not recur when we solve subproblems 6 and 7. Our tree now looks as shown in Figure 17.

Subproblems 3, 6, and 7 are now unsolved. The LIFO rule implies that we next solve subproblem 6 or subproblem 7. We arbitrarily choose to solve subproblem 7. From Figure 16, we see that the optimal solution to subproblem 7 is point H: $z = 37$, $x_1 = 4$, $x_2 = 1$. Both x_1 and x_2 assume integer values, so this solution is feasible for the original IP. We now know that subproblem 7 yields a feasible integer solution with $z = 37$. We also know that subproblem 7 cannot yield a feasible integer solution having $z > 37$. Thus, further branching on subproblem 7 will yield no new information about the optimal solution to the IP, and subproblem has been fathomed. The tree to date is pictured in Figure 18.

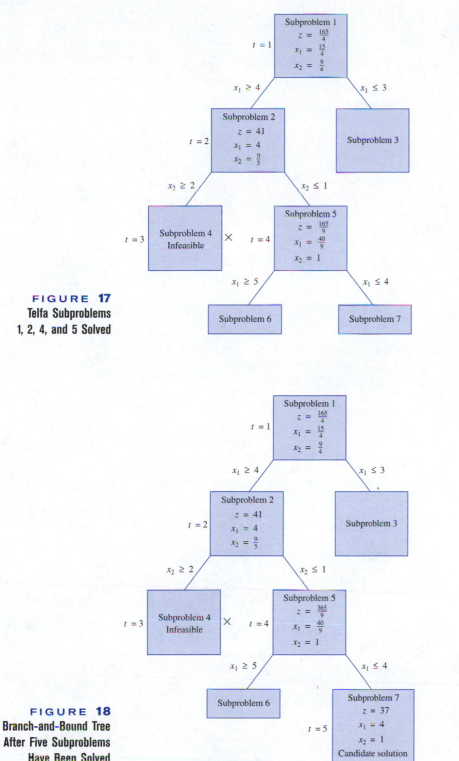

FIGURE 17
Telfa Subproblems
1, 2, 4, and 5 Solved

Subproblem 1
$t = 1$
$z = \frac{165}{4}$
$x_1 = \frac{15}{4}$
$x_2 = \frac{9}{4}$

$x_1 \geq 4$ $x_1 \leq 3$

Subproblem 2
$t = 2$
$z = 41$
$x_1 = 4$
$x_2 = \frac{9}{5}$

Subproblem 3

$x_2 \geq 2$ $x_2 \leq 1$

$t = 3$ Subproblem 4
Infeasible $\times$

$t = 4$ Subproblem 5
$z = \frac{165}{9}$
$x_1 = \frac{40}{9}$
$x_2 = 1$

$x_1 \geq 5$ $x_1 \leq 4$

Subproblem 6   Subproblem 7

FIGURE 18
Branch-and-Bound Tree
After Five Subproblems
Have Been Solved

Subproblem 1
$t = 1$
$z = \frac{165}{4}$
$x_1 = \frac{15}{4}$
$x_2 = \frac{9}{4}$

$x_1 \geq 4$ $x_1 \leq 3$

Subproblem 2
$t = 2$
$z = 41$
$x_1 = 4$
$x_2 = \frac{9}{5}$

Subproblem 3

$x_2 \geq 2$ $x_2 \leq 1$

$t = 3$ Subproblem 4
Infeasible $\times$

$t = 4$ Subproblem 5
$z = \frac{365}{9}$
$x_1 = \frac{40}{9}$
$x_2 = 1$

$x_1 \geq 5$ $x_1 \leq 4$

Subproblem 6   Subproblem 7
$z = 37$
$x_1 = 4$
$t = 5$ $x_2 = 1$
Candidate solution

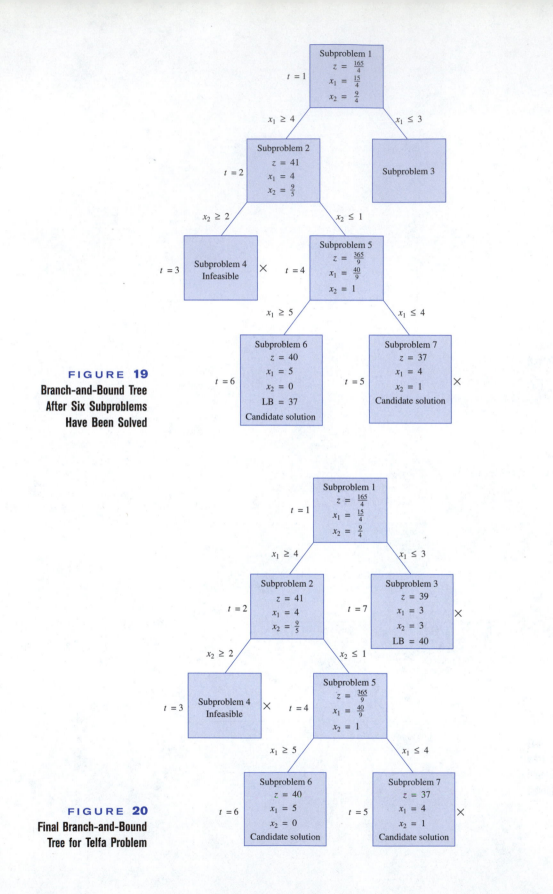

FIGURE 19
Branch-and-Bound Tree
After Six Subproblems
Have Been Solved

FIGURE 20
Final Branch-and-Bound
Tree for Telfa Problem

A solution obtained by solving a subproblem in which all variables have integer values is a **candidate solution.** Because the candidate solution may be optimal, we must keep a candidate solution until a better feasible solution to the IP (if any exists) is found. We have a feasible solution to the original IP with $z = 37$, so we may conclude that the optimal z-value for the IP ≥ 37. Thus, the z-value for the candidate solution is a **lower bound** on the optimal z-value for the original IP. We note this by placing the notation LB = 37 in the box corresponding to the *next* solved subproblem (see Figure 19).

The only remaining unsolved subproblems are 6 and 3. Following the LIFO rule, we next solve subproblem 6. From Figure 16, we find that the optimal solution to subproblem 6 is point A: $z = 40$, $x_1 = 5$, $x_2 = 0$. All decision variables have integer values, so this is a candidate solution. Its z-value of 40 is larger than the z-value of the best previous candidate (candidate 7 with $z = 37$). Thus, subproblem 7 cannot yield the optimal solution of the IP (we denote this fact by placing an $\times$ by subproblem 7). We also update our LB to 40. Our progress to date is summarized in Figure 20.

Subproblem 3 is the only remaining unsolved problem. From Figure 12, we find that the optimal solution to subproblem 3 is point F: $z = 39$, $x_1 = x_2 = 3$. Subproblem 3 cannot yield a z-value exceeding the current lower bound of 40, so it cannot yield the optimal solution to the original IP. Therefore, we place an $\times$ by it in Figure 20. From Figure 20, we see that there are no remaining unsolved subproblems, and that only subproblem 6 can yield the optimal solution to the IP. Thus, the optimal solution to the IP is for Telfa to manufacture 5 tables and 0 chairs. This solution will contribute $40 to profits.

In using the branch-and-bound method to solve the Telfa problem, we have implicitly enumerated all points in the IP's feasible region. Eventually, all such points (except for the optimal solution) are eliminated from consideration, and the branch-and-bound procedure is complete. To show that the branch-and-bound procedure actually does consider all points in the IP's feasible region, we examine several possible solutions to the Telfa problem and show how the procedure found these points to be nonoptimal. For example, how do we know that $x_1 = 2$, $x_2 = 3$ is not optimal? This point is in the feasible region for subproblem 3, and we know that all points in the feasible region for subproblem 3 have $z \leq 39$. Thus, our analysis of subproblem 3 shows that $x_1 = 2$, $x_2 = 3$ cannot beat $z = 40$ and cannot be optimal. As another example, why isn't $x_1 = 4$, $x_2 = 2$ optimal? Following the branches of the tree, we find that $x_1 = 4$, $x_2 = 2$ is associated with subproblem 4. Because no point associated with subproblem 4 is feasible, $x_1 = 4$, $x_2 = 2$ must fail to satisfy the constraints for the original IP and thus cannot be optimal for the Telfa problem. In a similar fashion, the branch-and-bound analysis has eliminated all points x_1, x_2 (except for the optimal solution) from consideration.

For the simple Telfa problem, the use of the branch-and-bound method may seem like using a cannon to kill a fly, but for an IP in which the feasible region contains a large number of integer points, the procedure can be very efficient for eliminating nonoptimal points from consideration. For example, suppose we are applying the branch-and-bound method and our current LB = 42. Suppose we solve a subproblem that contains 1 million feasible points for the IP. If the optimal solution to this subproblem has $z < 42$, then we have eliminated 1 million nonoptimal points by solving a single LP!

The key aspects of the branch-and-bound method for solving pure IPs (mixed IPs are considered in the next section) may be summarized as follows:

Step 1   If it is unnecessary to branch on a subproblem, then it is fathomed. The following three situations result in a subproblem being fathomed: (1) The subproblem is infeasible; (2) the subproblem yields an optimal solution in which all variables have integer values; and (3) the optimal z-value for the subproblem does not exceed (in a max problem) the current LB.

Step 2 A subproblem may be eliminated from consideration in the following situations: (1) The subproblem is infeasible (in the Telfa problem, subproblem 4 was eliminated for this reason); (2) the LB (representing the z-value of the best candidate to date) is at least as large as the z-value for the subproblem (in the Telfa problem, subproblems 3 and 7 were eliminated for this reason).

Recall that in solving the Telfa problem by the branch-and-bound procedure, many seemingly arbitrary choices were made. For example, when x_1 and x_2 were both fractional in the optimal solution to subproblem 1, how did we determine the branching variable? Or how did we determine which subproblem should next be solved? The manner in which these questions are answered can result in trees that differ greatly in size and in the computer time required to find an optimal solution. Through experience and ingenuity, practitioners of the procedure have developed guidelines on how to make the necessary decisions.

Two general approaches are commonly used to determine which subproblems should be solved next. The most widely used is the LIFO rule, which chooses to solve the most recently created subproblem.[†] LIFO leads us down one side of the branch-and-bound tree (as in the Telfa problem) and quickly finds a candidate solution. Then we backtrack our way up to the top of the other side of the tree. For this reason, the LIFO approach is often called **backtracking.**

The second commonly used method is **jumptracking.** When branching on a node, the jumptracking approach solves all the problems created by the branching. Then it branches again on the node with the best z-value. Jumptracking often jumps from one side of the tree to the other. It usually creates more subproblems and requires more computer storage than backtracking. The idea behind jumptracking is that moving toward the subproblems with good z-values should lead us more quickly to the best z-value.

If two or more variables are fractional in a subproblem's optimal solution, then on which variable should we branch? Branching on the fractional-valued variable that has the greatest economic importance is often the best strategy. In the Nickles example, suppose the optimal solution to a subproblem had y_1 and x_{12} fractional. Our rule would say to branch on y_1 because y_1 represents the decision to operate (or not operate) a lockbox in city 1, and this is presumably a more important decision than whether region 1 payments should be sent to city 2. When more than one variable is fractional in a subproblem solution, many computer codes will branch on the lowest-numbered fractional variable. Thus, if an integer programming computer code requires that variables be numbered, they should be numbered in order of their economic importance (1 = most important).

REMARKS **1** For some IP's, the optimal solution to the LP relaxation will also be the optimal solution to the IP. Suppose the constraints of the IP are written as $A\mathbf{x} = \mathbf{b}$. If the determinant[‡] of every square submatrix of A is $+1$, -1, or 0, we say that the matrix A is **unimodular.** If A is unimodular and each element of $\mathbf{b}$ is an integer, then the optimal solution to the LP relaxation will assign all variables integer values [see Shapiro (1979) for a proof] and will therefore be the optimal solution to the IP. It can be shown that the constraint matrix of any MCNFP is unimodular. Thus, as was discussed in Chapter 8, any MCNFP in which each node's net outflow and each arc's capacity are integers will have an integer-valued solution.
2 As a general rule, the more an IP looks like an MCNFP, the easier the problem is to solve by branch-and-bound methods. Thus, in formulating an IP, it is good to choose a formulation in which as many variables as possible have coefficients of $+1$, -1, and 0. To illustrate this idea, recall that the formulation of the Nickles (lockbox) problem given in Section 9.2 contained 16 constraints of the following form:

Formulation 1 $$x_{ij} \le y_j \ (i = 1, 2, 3, 4; j = 1, 2, 3, 4) \tag{25}$$

[†]For two subproblems created at the same time, many sophisticated methods have been developed to determine which one should be solved first. See Taha (1975) for details.
[‡]The determinant of a matrix is defined in Section 2.6.

As we have already seen in Section 9.2, if the 16 constraints in (25) are replaced by the following 4 constraints, then an equivalent formulation results:

Formulation 2

$$x_{11} + x_{21} + x_{31} + x_{41} \leq 4y_1$$
$$x_{12} + x_{22} + x_{32} + x_{42} \leq 4y_2$$
$$x_{13} + x_{23} + x_{33} + x_{43} \leq 4y_3$$
$$x_{14} + x_{24} + x_{34} + x_{44} \leq 4y_4$$

Because formulation 2 has $16 - 4 = 12$ fewer constraints than formulation 1, one might think that formulation 2 would require less computer time to find the optimal solution. This turns out to be untrue. To see why, recall that the branch-and-bound method begins by solving the LP relaxation of the IP. The feasible region of the LP relaxation of formulation 2 contains many more noninteger points than the feasible region of formulation 1. For example, the point $y_1 = y_2 = y_3 = y_4 = \frac{1}{4}$, $x_{11} = x_{22} = x_{33} = x_{44} = 1$ (all other x_{ij}'s equal 0) is in the feasible region for the LP relaxation of formulation 2, but not for formulation 1. The branch-and-bound method must eliminate all noninteger points before obtaining the optimal solution to the IP, so it seems reasonable that formulation 2 will require more computer time than formulation 1. Indeed, when the LINDO package was used to find the optimal solution to formulation 1, the LP relaxation yielded the optimal solution. But, 17 subproblems were solved before the optimal solution was found for formulation 2. Note that formulation 2 contains the terms $4y_1$, $4y_2$, $4y_3$, and $4y_4$. These terms "disturb" the network-like structure of the lockbox problem and cause the branch-and-bound method to be less efficient.

3 When solving an IP in the real world, we are usually happy with a near-optimal solution. For example, suppose that we are solving a lockbox problem and the LP relaxation yields a cost of $200,000. This means that the optimal solution to the lockbox IP will certainly have a cost of at least $200,000. If we find a candidate solution during the course of the branch-and-bound procedure that has a cost of, say, $205,000, why bother to continue with the branch-and-bound procedure? Even if we found the optimal solution to the IP, it could not save more than $5,000 in costs over the candidate solution with $z = 205,000$. It might even cost more than $5,000 in computer time to find the optimal lockbox solution. For this reason, the branch-and-bound procedure is often terminated when a candidate solution is found with a z-value close to the z-value of the LP relaxation.

4 Subproblems for branch-and-bound problems are often solved using some variant of the dual simplex algorithm. To illustrate this, we return to the Telfa example. The optimal tableau for the LP relaxation of the Telfa problem is

$$z \quad + 1.25s_1 + 0.75s_2 = 41.25$$
$$x_2 + 2.25s_1 - 0.25s_2 = \quad 2.25$$
$$x_1 \quad - 1.25s_1 + 0.25s_2 = \quad 3.75$$

After solving the LP relaxation, we solved subproblem 2, which is just subproblem 1 plus the constraint $x_1 \geq 4$. Recall that the dual simplex is an efficient method for finding the new optimal solution to an LP when we know the optimal tableau and a new constraint is added to the LP. We have added the constraint $x_1 \geq 4$ (which may be written as $x_1 - e_3 = 4$). To utilize the dual simplex, we must eliminate the basic variable x_1 from this constraint and use e_3 as a basic variable for $x_1 - e_3 = 4$. Adding $-$(second row of optimal tableau) to the constraint $x_1 - e_3 = 4$, we obtain the constraint $1.25s_1 - 0.25s_2 - e_3 = 0.25$. Multiplying this constraint through by -1, we obtain $-1.25s_1 + 0.25s_2 + e_3 = -0.25$. After adding this constraint to subproblem 1's optimal tableau, we obtain the tableau in Table 56. The dual simplex method states that we should enter a variable from row 3 into the basis. Because s_1 is the only variable with a negative coefficient in row 3, s_1 will enter the basis in row 3. After the pivot, we obtain the (optimal) tableau in Table 57. Thus, the optimal solution to subproblem 2 is $z = 41$, $x_2 = 1.8$, $x_1 = 4$, $s_1 = 0.20$.

TABLE 56
Initial Tableau for Solving Subproblem 2 by Dual Simplex

			Basic Variable
z	$+ 1.25s_1 + 0.75s_2$	$= 41.25$	$z = 41.25$
	$x_2 + 2.25s_1 - 0.25s_2$	$= 2.25$	$x_2 = 2.25$
x_1	$- 1.25s_1 + 0.25s_2$	$= 3.75$	$x_1 = 3.75$
	$- 1.25s_1 + 0.25s_2 + e_3$	$= -0.25$	$e_3 = -0.25$

TABLE **57**

Optimal Tableau for Solving Subproblem 2 by Dual Simplex

			Basic Variable
z	$+\quad s_2 +\quad e_3 = 41$		$z = 41$
	$x_2\quad + 0.20s_2 +\quad 1.8e_3 = 1.8$		$x_2 = 1.8$
	$x_1\qquad\qquad -\qquad e_3 = 4$		$x_1 = 4$
	$s_1 - 0.20s_2 - 0.80e_3 = 0.20$		$s_1 = 0.20$

5 In Problem 8, we show that if we create two subproblems by adding the constraints $x_k \leq i$ and $x_k \geq i + 1$, then the optimal solution to the first subproblem will have $x_k = i$ and the optimal solution to the second subproblem will have $x_k = i + 1$. This observation is very helpful when we graphically solve subproblems. For example, we know the optimal solution to subproblem 5 of Example 9 will have $x_2 = 1$. Then we can find the value of x_1 that solves subproblem 5 by choosing x_1 to be the largest integer satisfying all constraints when $x_2 = 1$.

Solver Tolerance Option for Solving IPs

When solving integer programming problems with the Excel Solver, you may go to Options and set a tolerance. A tolerance value of, say, .20, causes the Excel Solver to stop when a feasible solution is found that has an objective function value within 20% of the optimal z-value for the problem's LP relaxation. For instance, in Example 9, the optimal z-value for the LP relaxation was 41.25. With a tolerance of .20, the Solver would stop whenever a feasible integer solution is found with a z-value exceeding $(1 - .2)(41.25) = 33$. Thus, if we solved Example 9 with the Excel Solver and found a feasible integer solution having $z = 35$, then the Solver would stop because this solution would be within 20% of the LP relaxation bound.

Why set a nonzero tolerance? For many large IP problems, it might take a long time (weeks or months!) to find an optimal solution. It might take much less time to find a near optimal solution (say, within 5% of the optimal LP relaxation). In this case, we would be much better off with a near-optimal solution, and use of the tolerance option might be appropriate.

PROBLEMS

Group A

Use branch-and-bound to solve the following IPs:

1
$$\max z = 5x_1 + 2x_2$$
$$\text{s.t.} \quad 3x_1 + x_2 \leq 12$$
$$x_1 + x_2 \leq 5$$
$$x_1, x_2 \geq 0; x_1, x_2 \text{ integer}$$

2 The Dorian Auto example of Section 3.2.

3
$$\max z = 2x_1 + 3x_2$$
$$\text{s.t.} \quad x_1 + 2x_2 \leq 10$$
$$3x_1 + 4x_2 \leq 25$$
$$x_1, x_2 \geq 0; x_1, x_2 \text{ integer}$$

4
$$\max z = 4x_1 + 3x_2$$
$$\text{s.t.} \quad 4x_1 + 9x_2 \leq 26$$
$$8x_1 + 5x_2 \leq 17$$
$$x_1, x_2 \geq 0; x_1, x_2 \text{ integer}$$

5
$$\max z = 4x_1 + 5x_2$$
$$\text{s.t.} \quad x_1 + 4x_2 \geq 5$$
$$3x_1 + 2x_2 \geq 7$$
$$x_1, x_2 \geq 0; x_1, x_2 \text{ integer}$$

6
$$\max z = 4x_1 + 5x_2$$
$$\text{s.t.} \quad 3x_1 + 2x_2 \leq 10$$
$$x_1 + 4x_2 \leq 11$$
$$3x_1 + 3x_2 \leq 13$$
$$x_1, x_2 \geq 0; x_1, x_2 \text{ integer}$$

7 Use the branch-and-bound method to find the optimal solution to the following IP:

$$\max z = 7x_1 + 3x_2$$
$$\text{s.t.} \quad 2x_1 + x_2 \leq 9$$
$$3x_1 + 2x_2 \leq 13$$
$$x_1, x_2 \geq 0; x_1, x_2 \text{ integer}$$

8 Suppose we have branched on a subproblem (call it subproblem 0, having optimal solution SOL0) and have obtained the following two subproblems:

Subproblem 1 Subproblem 0 + Constraint $x_1 \leq i$.
Subproblem 2 Subproblem 0 + Constraint $x_1 \geq i + 1$ (i is some integer).

Prove that there will exist at least one optimal solution to subproblem 1 having $x_1 = i$ and at least one optimal solution to subproblem 2 having $x_1 = i + 1$. [*Hint:* Suppose an optimal solution to subproblem 1 (call it SOL1) has $x_1 = \bar{x}_1$, where $\bar{x}_1 < i$. For some number c ($0 < c < 1$), $c(\text{SOL0}) + (1 - c)\text{SOL1}$ will have the following three properties:

a The value of x_1 in $c(\text{SOL0}) + (1 - c)\text{SOL1}$ will equal i.

b $c(\text{SOL0}) + (1 - c)\text{SOL1}$ will be feasible in subproblem 1.

c The z-value for $c(\text{SOL0}) + (1 - c)\text{SOL1}$ will be at least as good as the z-value for SOL1.

Explain how this result can help when we graphically solve branch-and-bound problems.]

9 During the next five periods, the demands in Table 58 must be met on time. At the beginning of period 1, the

TABLE 58

	Period				
	1	2	3	4	5
Demand	220	280	360	140	270

inventory level is 0. Each period that production occurs a setup cost of $250 and a per-unit production cost of $2 are incurred. At the end of each period a per-unit holding cost of $1 is incurred.

a Solve for the cost-minimizing production schedule using the following decision variables: x_t = units produced during month t and $y_t = 1$ if any units are produced during period t, $y_t = 0$ otherwise.

b Solve for the cost-minimizing production schedule using the following variables: y_t's defined in part (a) and x_{it} = number of units produced during period i to satisfy period t demand.

c Which formulation took LINDO or LINGO less time to solve?

d Give an intuitive explanation of why the part (b) formulation is solved faster than the part (a) formulation.

9.4 The Branch-and-Bound Method for Solving Mixed Integer Programming Problems

Recall that, in a mixed IP, some variables are required to be integers and others are allowed to be either integers or nonintegers. To solve a mixed IP by the branch-and-bound method, modify the method described in Section 9.3 by branching only on variables that are required to be integers. Also, for a solution to a subproblem to be a candidate solution, it need only assign integer values to those variables that are required to be integers. To illustrate, let us solve the following mixed IP:

$$\max z = 2x_1 + x_2$$
$$\text{s.t.} \quad 5x_1 + 2x_2 \leq 8$$
$$x_1 + x_2 \leq 3$$
$$x_1, x_2 \geq 0; x_1 \text{ integer}$$

As before, we begin by solving the LP relaxation of the IP. The optimal solution of the LP relaxation is $z = \frac{11}{3}$, $x_1 = \frac{2}{3}$, $x_2 = \frac{7}{3}$. Because x_2 is allowed to be fractional, we do not branch on x_2; if we did so, we would be excluding points having x_2 values between 2 and 3, and we don't want to do that. Thus, we must branch on x_1. This yields subproblems 2 and 3 in Figure 21.

We next choose to solve subproblem 2. The optimal solution to subproblem 2 is the candidate solution $z = 3$, $x_1 = 0$, $x_2 = 3$. We now solve subproblem 3 and obtain the candidate solution $z = \frac{7}{2}$, $x_1 = 1$, $x_2 = \frac{3}{2}$. The z-value from the subproblem 3 candidate exceeds the z-value for the subproblem 2 candidate, so subproblem 2 can be eliminated from consideration, and the subproblem 3 candidate ($z = \frac{7}{2}$, $x_1 = 1$, $x_2 = \frac{3}{2}$) is the optimal solution to the mixed IP.

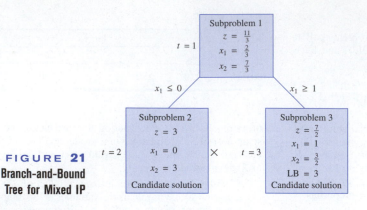

FIGURE **21**
Branch-and-Bound
Tree for Mixed IP

PROBLEMS

Group A

Use the branch-and-bound method to solve the following IPs:

1
$$\max z = 3x_1 + x_2$$
$$\text{s.t.} \quad 5x_1 + 2x_2 \leq 10$$
$$4x_1 + x_2 \leq 7$$
$$x_1, x_2 \geq 0; x_2 \text{ integer}$$

2
$$\min z = 3x_1 + x_2$$
$$\text{s.t.} \quad x_1 + 5x_2 \geq 8$$
$$x_1 + 2x_2 \geq 4$$
$$x_1, x_2 \geq 0; x_1 \text{ integer}$$

3
$$\max z = 4x_1 + 3x_2 + x_3$$
$$\text{s.t.} \quad 3x_1 + 2x_2 + x_3 \leq 7$$
$$2x_1 + x_2 + 2x_3 \leq 11$$
$$x_2, x_3 \text{ integer}, x_1, x_2, x_3 \geq 0$$

9.5 Solving Knapsack Problems by the Branch-and-Bound Method

In Section 9.2, we learned that a knapsack problem is an IP with a single constraint. In this section, we discuss knapsack problems in which each variable must equal 0 or 1 (see Problem 1 at the end of this section for an explanation of how any knapsack problem can be reformulated so that each variable must equal 0 or 1). A knapsack problem in which each variable must equal 0 or 1 may be written as

$$\max z = c_1x_1 + c_2x_2 + \cdots + c_nx_n$$
$$\text{s.t.} \quad a_1x_1 + a_2x_2 + \cdots + a_nx_n \leq b \tag{38}$$
$$x_i = 0 \text{ or } 1 \quad (i = 1, 2, \ldots, n)$$

Recall that c_i is the benefit obtained if item i is chosen, b is the amount of an available resource, and a_i is the amount of the available resource used by item i.

When knapsack problems are solved by the branch-and-bound method, two aspects of the method greatly simplify. Because each variable must equal 0 or 1, branching on x_i will yield an $x_i = 0$ and an $x_i = 1$ branch. Also, the LP relaxation (and other subproblems) may be solved by inspection. To see this, observe that $\frac{c_i}{a_i}$ may be interpreted as the benefit item i earns for each unit of the resource used by item i. Thus, the best items have the largest values of $\frac{c_i}{a_i}$ and the worst items have the smallest values of $\frac{c_i}{a_i}$. To solve any subproblem resulting from a knapsack problem, compute all the ratios $\frac{c_i}{a_i}$. Then put the best item in the knapsack. Then put the second-best item in the knapsack. Continue in this fashion until the best remaining item will overfill the knapsack. Then fill the knapsack with as much of this item as possible.

To illustrate, we solve the LP relaxation of

$$\max z = 40x_1 + 80x_2 + 10x_3 + 10x_4 + 4x_5 + 20x_6 + 60x_7$$

$$\text{s.t.} \quad 40x_1 + 50x_2 + 30x_3 + 10x_4 + 10x_5 + 40x_6 + 30x_7 \le 100 \tag{39}$$

$$x_i = 0 \text{ or } 1 \quad (i = 1, 2, \ldots, 7)$$

We begin by computing the $\frac{c_i}{a_i}$ ratios and ordering the variables from best to worst (see Table 59). To solve the LP relaxation of (39), we first choose item 7 ($x_7 = 1$). Then $100 - 30 = 70$ units of the resource remain. Now we include the second-best item (item 2) in the knapsack by setting $x_2 = 1$. Now $70 - 50 = 20$ units of the resource remain. Item 4 and item 1 have the same $\frac{c_i}{a_i}$ ratio, so we can next choose either of these items. We arbitrarily choose to set $x_4 = 1$. Then $20 - 10 = 10$ units of the resource remain. The best remaining item is item 1. We now fill the knapsack with as much of item 1 as we can. Because only 10 units of the resource remain, we set $x_1 = \frac{10}{40} = \frac{1}{4}$. Thus an optimal solution to the LP relaxation of (39) is $z = 80 + 60 + 10 + (\frac{1}{4})(40) = 160$, $x_2 = x_7 = x_4 = 1$, $x_1 = \frac{1}{4}$, $x_3 = x_5 = x_6 = 0$.

To show how the branch-and-bound method can be used to solve a knapsack problem, let us find the optimal solution to the Stockco capital budgeting problem (Example 1). Recall that this problem was

$$\max z = 16x_1 + 22x_2 + 12x_3 + 8x_4$$

$$\text{s.t.} \quad 5x_1 + 7x_2 + 4x_3 + 3x_4 \le 14$$

$$x_j = 0 \text{ or } 1$$

[handwritten: $X_1 : 3.2$ $X_2 : 3.14$ $X_3 = 3$ $X_4 = 2.9$]

The branch-and-bound tree for this problem is shown in Figure 22. From the tree, we find that the optimal solution to Example 1 is $z = 42$, $x_1 = 0$, $x_2 = x_3 = x_4 = 1$. Thus, we should invest in investments 2, 3, and 4 and earn an NPV of $42,000. As discussed in Section 9.2, the "best" investment is not used.

REMARKS   The method we used in traversing the tree of Figure 22 is as follows:

1   We used the LIFO approach to determine which subproblem should be solved.

2   We arbitrarily chose to solve subproblem 3 before subproblem 2. To solve subproblem 3, we first set $x_3 = 1$ and then solved the resulting knapsack problem. After setting $x_3 = 1$, $14 - 4 = \$10$ million was still available for investment. Applying the technique used to solve the LP relaxation of a knapsack problem yielded the following optimal solution to subproblem 3: $x_3 = 1$, $x_1 = 1$, $x_2 = \frac{5}{7}$, $x_4 = 0$, $z = 16 + (\frac{5}{7})(22) + 12 = \frac{306}{7}$. Other subproblems were solved similarly; of course, if a subproblem specified $x_i = 0$, the optimal solution to that subproblem could not use investment i.

TABLE 59
Ordering Items from Best to Worst in a Knapsack Problem

Item	$\dfrac{c_i}{a_i}$	Ranking (1 = best, 7 = worst)
1	1	3.5 (tie for third or fourth)
2	$\frac{8}{5}$	2
3	$\frac{1}{3}$	7
4	1	3.5
5	$\frac{4}{10}$	6
6	$\frac{1}{2}$	5
7	2	1

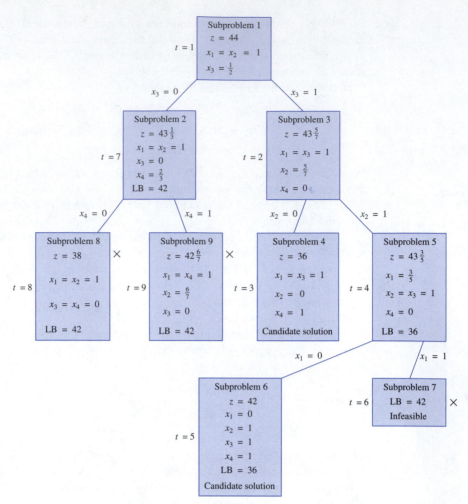

FIGURE 22
Branch-and-Bound Tree for Stockco Knapsack Problem

3 Subproblem 4 yielded the candidate solution $x_1 = x_3 = x_4 = 1$, $z = 36$. We then set LB = 36.
4 Subproblem 6 yielded a candidate solution with $z = 42$. Thus, subproblem 4 was eliminated from consideration, and the LB was updated to 42.
5 Subproblem 7 was infeasible because it required $x_1 = x_2 = x_3 = 1$, and such a solution requires at least $16 million.
6 Subproblem 8 was eliminated because its z-value ($z = 38$) did not exceed the current LB of 42.
7 Subproblem 9 had a z-value of $42\frac{6}{7}$. Because the z-value for any all-integer solution must also be an integer, this meant that branching on subproblem 9 could never yield a z-value larger than 42. Thus, further branching on subproblem 9 could not beat the current LB of 42, and subproblem 9 was eliminated from consideration.

In Chapter 13, we show how dynamic programming can be used to solve knapsack problems.

PROBLEMS

Group A

1 Show how the following problem can be expressed as a knapsack problem in which all variables must equal 0 or 1. NASA is determining how many of three types of objects should be brought on board the space shuttle. The weight and benefit of each of the items are given in Table 60. If the space shuttle can carry a maximum of 26 lb of items 1–3, which items should be taken on the space shuttle?

TABLE 60

Item	Benefit	Weight (Pounds)
1	10	3
2	15	4
3	17	5

TABLE 61

Item	Value ($)	Volume (Cubic Feet)
Bedroom set	60	800
Dining room set	48	600
Stereo	14	300
Sofa	31	400
TV set	10	200

2 I am moving from New Jersey to Indiana and have rented a truck that can haul up to 1,100 cu ft of furniture. The volume and value of each item I am considering moving on the truck are given in Table 61. Which items should I bring to Indiana? To solve this problem as a knapsack problem, what unrealistic assumptions must we make?

3 Four projects are available for investment. The projects require the cash flows and yield the net present values (NPV) (in millions) shown in Table 62. If $6 million is available for investment at time 0, find the investment plan that maximizes NPV.

TABLE 62

Project	Cash Outflow at Time 0 ($)	NPV ($)
1	3	5
2	5	8
3	2	3
4	4	7

9.6 Solving Combinatorial Optimization Problems by the Branch-and-Bound Method

Loosely speaking, a **combinatorial optimization problem** is any optimization problem that has a finite number of feasible solutions. A branch-and-bound approach is often the most efficient way to solve them. Three examples of combinatorial optimization problems follow:

1 Ten jobs must be processed on a single machine. You know the time it takes to complete each job and the time at which each job must be completed (the job's due date). What ordering of the jobs minimizes the total delay of the 10 jobs?

2 A salesperson must visit each of 10 cities once before returning to his home. What ordering of the cities minimizes the total distance the salesperson must travel before returning home? Not surprisingly, this problem is called the *traveling salesperson problem* (TSP).

3 Determine how to place eight queens on a chessboard so that no queen can capture any other queen (see Problem 7 at the end of this section).

In each of these problems, many possible solutions must be considered. For instance, in Problem 1, the first job to be processed can be one of 10 jobs, the next job can be one of 9 jobs, and so on. Thus, even for this relatively small problem there are $10(9)(8) \cdots (1) = 10! = 3,628,000$ possible ways to schedule the jobs. A combinatorial optimization problem may have many feasible solutions, so it can require a great deal of computer time to enumerate all possible solutions explicitly. For this reason, branch-and-bound methods are often used for *implicit* enumeration of all possible solutions to a combinatorial optimization problem. As we will see, the branch-and-bound method should take advantage of the structure of the particular problem that is being solved.

To illustrate how branch-and-bound methods are used to solve combinatorial optimization problems, we show how the approach can be used to solve problems 1 and 2 of the preceding list.

Branch-and-Bound Approach for Machine-Scheduling Problem

Example 10 illustrates how a branch-and-bound approach may be used to schedule jobs on a single machine. See Baker (1974) and Hax and Candea (1984) for a discussion of other branch-and-bound approaches to machine-scheduling problems.

EXAMPLE 10   **Branch-and-Bound Machine Scheduling**

Four jobs must be processed on a single machine. The time required to process each job and the date the job is due are shown in Table 63. The delay of a job is the number of days after the due date that a job is completed (if a job is completed on time or early, the job's delay is zero). In what order should the jobs be processed to minimize the total delay of the four jobs?

Solution   Suppose the jobs are processed in the following order: job 1, job 2, job 3, and job 4. Then the delays shown in Table 64 would occur. For this sequence, total delay = 0 + 6 + 3 + 7 = 16 days. We now describe a branch-and-bound approach for solving this type of machine-scheduling problem.

Because a possible solution to the problem must specify the order in which the jobs are processed, we define

$$x_{ij} = \begin{cases} 1 & \text{if job } i \text{ is the } j\text{th job to be processed} \\ 0 & \text{otherwise} \end{cases}$$

The branch-and-bound approach begins by partitioning all solutions according to the job that is *last* processed. Any sequence of jobs must process some job last, so each sequence of jobs must have $x_{14} = 1$, $x_{24} = 1$, $x_{34} = 1$, or $x_{44} = 1$. This yields four branches with nodes 1–4 in Figure 23. After we create a node by branching, we obtain a lower bound on the total delay (D) associated with the node. For example, if $x_{44} = 1$, we know that job 4 is the last job to be processed. In this case, job 4 will be completed at the end of day $6 + 4 + 5 + 8 = 23$ and will be $23 - 16 = 7$ days late. Thus, any schedule having

---

TABLE 63
Durations and Due Date of Jobs

Job	Days Required to Complete Job	Due Date
1	6	End of day 8
2	4	End of day 4
3	5	End of day 12
4	8	End of day 16

TABLE 64
Delays Incurred If Jobs Are Processed in the Order 1-2-3-4

Job	Completion Time of Job	Delay of Job
1	6	0
2	$6 + 4 = 10$	$10 - 4 = 6$
3	$6 + 4 + 5 = 15$	$15 - 12 = 3$
4	$6 + 4 + 5 + 8 = 23$	$23 - 16 = 7$

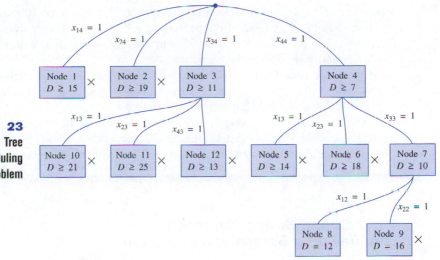

FIGURE 23
Branch-and-Bound Tree
for Machine-Scheduling
Problem

$x_{44} = 1$ must have $D \geq 7$. Thus, we write $D \geq 7$ inside node 4 of Figure 23. Similar reasoning shows that any sequence of jobs having $x_{34} = 1$ will have $D \geq 11$, $x_{24} = 1$ will have $D \geq 19$, and $x_{14} = 1$ will have $D \geq 15$. We have no reason to exclude any of nodes 1–4 from consideration as part of the optimal job sequence, so we choose to branch on a node. We use the jumptracking approach and branch on the node that has the smallest bound on D: node 4. Any job sequence associated with node 4 must have $x_{13} = 1$, $x_{23} = 1$, or $x_{33} = 1$. Branching on node 4 yields nodes 5–7 in Figure 23. For each new node, we need a lower bound for the total delay. For example, at node 7, we know from our analysis of node 1 that job 4 will be processed last and will be delayed by 7 days. For node 7, we know that job 3 will be the third job processed. Thus, job 3 will be completed after $6 + 4 + 5 = 15$ days and will be $15 - 12 = 3$ days late. Any sequence associated with node 7 must have $D \geq 7 + 3 = 10$ days. Similar reasoning shows that node 5 must have $D \geq 14$, and node 6 must have $D \geq 18$. We still do not have any reason to eliminate any of nodes 1–7 from consideration, so we again branch on a node. The jumptracking approach directs us to branch on node 7. Any job sequence associated with node 7 must have either job 1 or job 2 as the second job processed. Thus, any job sequence associated with node 7 must have $x_{12} = 1$ or $x_{22} = 1$. Branching on node 7 yields nodes 8 and 9 in Figure 23.

Node 9 corresponds to processing the jobs in the order 1–2–3–4. This sequence yields a total delay of 7 (for job 4) + 3 (for job 3) + (6 + 4 − 4) (for job 2) + 0 (for job 1) = 16 days. Node 9 is a feasible sequence and may be considered a candidate solution having $D = 16$. We now know that any node that cannot have a total delay of less than 16 days can be eliminated.

Node 8 corresponds to the sequence 2–1–3–4. This sequence has a total delay of 7 (for job 4) + 3 (for job 3) + (4 + 6 − 8) (for job 1) + 0 (for job 2) = 12 days. Node 8 is a feasible sequence and may be viewed as a candidate solution with $D = 12$. Because node 8 is better than node 9, node 9 may be eliminated from consideration.

Similarly, node 5 (having $D \geq 14$), node 6 (having $D \geq 18$), node 1 (having $D \geq 15$), and node 2 (having $D \geq 19$) can be eliminated. Node 3 cannot yet be eliminated, because it is still possible for node 3 to yield a sequence having $D = 11$. Thus, we now branch on node 3. Any job sequence associated with node 3 must have $x_{13} = 1$, $x_{23} = 1$, or $x_{43} = 1$, so we obtain nodes 10–12.

For node 10, $D \geq$ (delay from processing job 3 last) + (delay from processing job 1 third) = 11 + (6 + 4 + 8 − 8) = 21. Because any sequence associated with node 10 must have $D \geq 21$ and we have a candidate with $D = 12$, node 10 may be eliminated.

For node 11, $D \geq$ (delay from processing job 3 last) + (delay from processing job 2 third) = $11 + (6 + 4 + 8 - 4) = 25$. Any sequence associated with node 11 must have $D \geq 25$, and node 11 may be eliminated.

Finally, for node 12, $D \geq$ (delay from processing job 3 last) + (delay from processing job 4 third) = $11 + (6 + 4 + 8 - 16) = 13$. Any sequence associated with node 12 must have $D \geq 13$, and node 12 may be eliminated.

With the exception of node 8, every node in Figure 23 has been eliminated from consideration. Node 8 yields the delay-minimizing sequence $x_{44} = x_{33} = x_{12} = x_{21} = 1$. Thus, the jobs should be processed in the order 2–1–3–4 with a total delay of 12 days resulting.

Branch-and-Bound Approach for Traveling Salesperson Problem

EXAMPLE 11   Traveling Salesperson Problem

Joe State lives in Gary, Indiana. He owns insurance agencies in Gary, Fort Wayne, Evansville, Terre Haute, and South Bend. Each December, he visits each of his insurance agencies. The distance between each agency (in miles) is shown in Table 65. What order of visiting his agencies will minimize the total distance traveled?

Solution   Joe must determine the order of visiting the five cities that minimizes the total distance traveled. For example, Joe could choose to visit the cities in the order 1–3–4–5–2–1. Then he would travel a total of $217 + 113 + 196 + 79 + 132 = 737$ miles.

To tackle the traveling salesperson problem, define

$$x_{ij} = \begin{cases} 1 & \text{if Joe leaves city } i \text{ and travels next to city } j \\ 0 & \text{otherwise} \end{cases}$$

Also, for $i \neq j$,

$$c_{ij} = \text{distance between cities } i \text{ and } j$$
$$c_{ii} = M, \text{ where } M \text{ is a large positive number}$$

It seems reasonable that we might be able to find the answer to Joe's problem by solving an assignment problem having a cost matrix whose ijth element is c_{ij}. For instance, suppose we solved this assignment problem and obtained the solution $x_{12} = x_{24} = x_{45} = x_{53} = x_{31} = 1$. Then Joe should go from Gary to Fort Wayne, from Fort Wayne to Terre Haute, from Terre Haute to South Bend, from South Bend to Evansville, and from Evansville to Gary. This solution can be written as 1–2–4–5–3–1. An itinerary that begins and ends at the same city and visits each city once is called a **tour.**

TABLE 65
Distance between Cities in Traveling Salesperson Problem

Day	Gary	Fort Wayne	Evansville	Terre Haute	South Bend
City 1 Gary	0	132	217	164	58
City 2 Fort Wayne	132	0	290	201	79
City 3 Evansville	217	290	0	113	303
City 4 Terre Haute	164	201	113	0	196
City 5 South Bend	58	79	303	196	0

If the solution to the preceding assignment problem yields a tour, then it is the optimal solution to the traveling salesperson problem. (Why?) Unfortunately, the optimal solution to the assignment problem need not be a tour. For example, the optimal solution to the assignment problem might be $x_{15} = x_{21} = x_{34} = x_{43} = x_{52} = 1$. This solution suggests going from Gary to South Bend, then to Fort Wayne, and then back to Gary. This solution also suggests that if Joe is in Evansville he should go to Terre Haute and then to Evansville (see Figure 24). Of course, if Joe begins in Gary, this solution will never get him to Evansville or Terre Haute. This is because the optimal solution to the assignment problem contains two **subtours.** A subtour is a round trip that does not pass through all cities. The current assignment contains the two subtours 1–5–2–1 and 3–4–3. If we could exclude all feasible solutions that contain subtours and then solve the assignment problem, we would obtain the optimal solution to the traveling salesperson problem. This is not easy to do, however. In most cases, a branch-and-bound approach is the most efficient approach for solving a TSP.

Several branch-and-bound approaches have been developed for solving TSPs [see Wagner (1975)]. We describe an approach here in which the subproblems reduce to assignment problems. To begin, we solve the preceding assignment problem, in which, for $i \ne j$, the cost c_{ij} is the distance between cities i and j and $c_{ii} = M$ (this prevents a person in a city from being assigned to visit that city itself). Because this assignment problem contains no provisions to prevent subtours, it is a relaxation (or less constrained problem) of the original traveling salesperson problem. Thus, if the optimal solution to the assignment problem is feasible for the traveling salesperson problem (that is, if the assignment solution contains no subtours), then it is also optimal for the traveling salesperson problem. The results of the branch-and-bound procedure are given in Figure 25.

We first solve the assignment problem in Table 66 (referred to as subproblem 1). The optimal solution is $x_{15} = x_{21} = x_{34} = x_{43} = x_{52} = 1$, $z = 495$. This solution contains two subtours (1–5–2–1 and 3–4–3) and cannot be the optimal solution to Joe's problem.

We now branch on subproblem 1 in a way that will prevent one of subproblem 1's subtours from recurring in solutions to subsequent subproblems. We choose to exclude the subtour 3–4–3. Observe that the optimal solution to Joe's problem must have either $x_{34} = 0$ or $x_{43} = 0$ (if $x_{34} = x_{43} = 1$, the optimal solution would have the subtour 3–4–3). Thus, we can branch on subproblem 1 by adding the following two subproblems:

Subproblem 2 Subproblem 1 + ($x_{34} = 0$, or $c_{34} = M$).

Subproblem 3 Subproblem 1 + ($x_{43} = 0$, or $c_{43} = M$).

We now arbitrarily choose subproblem 2 to solve, applying the Hungarian method to the cost matrix as shown in Table 67. The optimal solution to subproblem 2 is $z = 652$, $x_{14} = x_{25} = x_{31} = x_{43} = x_{52} = 1$. This solution includes the subtours 1–4–3–1 and 2–5–2, so this cannot be the optimal solution to Joe's problem.

We now branch on subproblem 2 in an effort to exclude the subtour 2–5–2. We must ensure that either x_{25} or x_{52} equals zero. Thus, we add the following two subproblems:

Subproblem 4 Subproblem 2 + ($x_{25} = 0$, or $c_{25} = M$).

Subproblem 5 Subproblem 2 + ($x_{52} = 0$, or $c_{52} = M$).

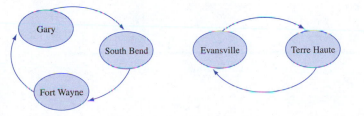

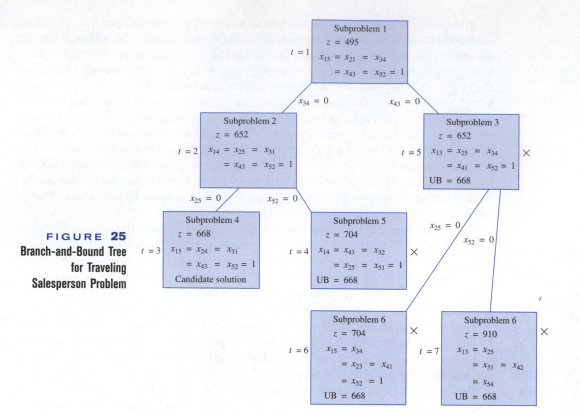

FIGURE 25
Branch-and-Bound Tree for Traveling Salesperson Problem

Subproblem 1
$t = 1$
$z = 495$
$x_{15} = x_{21} = x_{34}$
$= x_{43} = x_{52} = 1$

$x_{34} = 0$ $x_{43} = 0$

Subproblem 2
$t = 2$
$z = 652$
$x_{14} = x_{25} = x_{31}$
$= x_{43} = x_{52} = 1$

Subproblem 3
$t = 5$
$z = 652$
$x_{13} = x_{25} = x_{34}$
$= x_{41} = x_{52} = 1$
UB = 668   ✕

$x_{25} = 0$ $x_{52} = 0$

Subproblem 4
$t = 3$
$z = 668$
$x_{15} = x_{24} = x_{31}$
$= x_{43} = x_{52} = 1$
Candidate solution

Subproblem 5
$t = 4$
$z = 704$
$x_{14} = x_{43} = x_{32}$
$= x_{25} = x_{51} = 1$
UB = 668   ✕

$x_{25} = 0$
$x_{52} = 0$

Subproblem 6
$t = 6$
$z = 704$
$x_{15} = x_{34}$
$= x_{23} = x_{41}$
$= x_{52} = 1$
UB = 668   ✕

Subproblem 6
$t = 7$
$z = 910$
$x_{13} = x_{25}$
$= x_{31} = x_{42}$
$= x_{54}$
UB = 668   ✕

TABLE 66
Cost Matrix for Subproblem 1

	City 1	City 2	City 3	City 4	City 5
City 1	M	132	217	164	58
City 2	132	M	290	201	79
City 3	217	290	M	113	303
City 4	164	201	113	M	196
City 5	58	79	303	196	M

TABLE 67
Cost Matrix for Subproblem 2

	City 1	City 2	City 3	City 4	City 5
City 1	M	132	217	164	58
City 2	132	M	290	201	79
City 3	217	290	M	M	303
City 4	164	201	113	M	196
City 5	58	79	303	196	M

Following the LIFO approach, we should next solve subproblem 4 or subproblem 5. We arbitrarily choose to solve subproblem 4. Applying the Hungarian method to the cost matrix shown in Table 68, we obtain the optimal solution $z = 668$, $x_{15} = x_{24} = x_{31} = x_{43} = x_{52} = 1$. This solution contains no subtours and yields the tour 1–5–2–4–3–1. Thus, subproblem 4 yields a candidate solution with $z = 668$. Any node that cannot yield a z-value < 668 may be eliminated from consideration.

Following the LIFO rule, we next solve subproblem 5, applying the Hungarian method to the matrix in Table 69. The optimal solution to subproblem 5 is $z = 704$, $x_{14} = x_{43} = x_{32} = x_{25} = x_{51} = 1$. This solution is a tour, but $z = 704$ is not as good as the subproblem 4 candidate's $z = 668$. Thus, subproblem 5 may be eliminated from consideration.

Only subproblem 3 remains. We find the optimal solution to the assignment problem in Table 70, $x_{13} = x_{25} = x_{34} = x_{41} = x_{52} = 1$, $z = 652$. This solution contains the subtours 1–3–4–1 and 2–5–2. Because $652 < 668$, however, it is still possible for subproblem 3 to yield a solution with no subtours that beats $z = 668$. Thus, we now branch on subproblem 3 in an effort to exclude the subtours. Any feasible solution to the traveling salesperson problem that emanates from subproblem 3 must have either $x_{25} = 0$ or $x_{52} = 0$ (why?), so we create subproblems 6 and 7.

Subproblem 6   Subproblem 3 + ($x_{25} = 0$, or $c_{25} = M$).

Subproblem 7   Subproblem 3 + ($x_{52} = 0$, or $c_{52} = M$).

TABLE **68**
Cost Matrix for Subproblem 4

	City 1	City 2	City 3	City 4	City 5
City 1	M	132	217	164	58
City 2	132	M	290	201	M
City 3	217	290	M	M	303
City 4	164	201	113	M	196
City 5	58	79	303	196	M

TABLE **69**
Cost Matrix for Subproblem 5

	City 1	City 2	City 3	City 4	City 5
City 1	M	132	217	164	58
City 2	132	M	290	201	79
City 3	217	290	M	M	303
City 4	164	201	113	M	196
City 5	58	M	303	196	M

TABLE **70**
Cost Matrix for Subproblem 3

	City 1	City 2	City 3	City 4	City 5
City 1	M	132	217	164	58
City 2	132	M	290	201	79
City 3	217	290	M	113	303
City 4	164	201	M	M	196
City 5	58	79	303	196	M

We next choose to solve subproblem 6. The optimal solution to subproblem 6 is $x_{15} = x_{34} = x_{23} = x_{41} = x_{52} = 1$, $z = 704$. This solution contains no subtours, but its z-value of 704 is inferior to the candidate solution from subproblem 4, so subproblem 6 cannot yield the optimal solution to the problem.

The only remaining subproblem is subproblem 7. The optimal solution to subproblem 7 is $x_{13} = x_{25} = x_{31} = x_{42} = x_{54} = 1$, $z = 910$. Again, $z = 910$ is inferior to $z = 668$, so subproblem 7 cannot yield the optimal solution.

Subproblem 4 thus yields the optimal solution: Joe should travel from Gary to South Bend, from South Bend to Fort Wayne, from Fort Wayne to Terre Haute, from Terre Haute to Evansville, and from Evansville to Gary. Joe will travel a total distance of 668 miles.

Heuristics for TSPs

When using branch-and-bound methods to solve TSPs with many cities, large amounts of computer time may be required. For this reason, **heuristic methods,** or **heuristics,** which quickly lead to a good (but not necessarily optimal) solution to a TSP, are often used. A heuristic is a method used to solve a problem by trial and error when an algorithmic approach is impractical. Heuristics often have an intuitive justification. We now discuss two heuristics for the TSP: the nearest-neighbor and the cheapest-insertion heuristics.

To apply the nearest-neighbor heuristic (NNH), we begin at any city and then "visit" the nearest city. Then we go to the unvisited city closest to the city we have most recently visited. Continue in this fashion until a tour is obtained. We now apply the NNH to Example 11. We arbitrarily choose to begin at city 1. City 5 is the closest city to city 1, so we have now generated the arc 1–5. Of cities 2, 3, and 4, city 2 is closest to city 5, so we have now generated the arcs 1–5–2. Of cities 3 and 4, city 4 is closest to city 2. We now have generated the arcs 1–5–2–4. Of course, we must next visit city 3 and then return to city 1; this yields the tour 1–5–2–4–3–1. In this case, the NNH yields an optimal tour. If we had begun at city 3, however, the reader should verify that the tour 3–4–1–5–2–3 would be obtained. This tour has length $113 + 164 + 58 + 79 + 290 = 704$ miles and is not optimal. Thus, the NNH need not yield an optimal tour. A popular heuristic is to apply the NNH beginning at each city and then take the best tour obtained.

In the cheapest-insertion heuristic (CIH), we begin at any city and find its closest neighbor. Then we create a subtour joining those two cities. Next, we replace an arc in the subtour [say, arc (i, j)] by the combination of two arcs—(i, k) and (k, j), where k is not in the current subtour—that will increase the length of the subtour by the smallest (or cheapest) amount. Let c_{ij} be the length of arc (i, j). Note that if arc (i, j) is replaced by arcs (i, k) and (k, j), then a length $c_{ik} + c_{kj} - c_{ij}$ is added to the subtour. Then we continue with this procedure until a tour is obtained. Suppose we begin the CIH at city 1. City 5 is closest to city 1, so we begin with the subtour $(1, 5)$–$(5, 1)$. Then we could replace $(1, 5)$ by $(1, 2)$–$(2, 5)$, $(1, 3)$–$(3, 5)$, or $(1, 4)$–$(4, 5)$. We could also replace arc $(5, 1)$ by $(5, 2)$–$(2, 1)$, $(5, 3)$–$(3, 1)$, or $(5, 4)$–$(4, 1)$. The calculations used to determine which arc of $(1,5)$–$(5,1)$ should be replaced are given in Table 71 (* indicates the correct replacement). As seen in the table, we may replace either $(1, 5)$ or $(5, 1)$. We arbitrarily choose to replace arc $(1, 5)$ by arcs $(1, 2)$ and $(2, 5)$. We currently have the subtour $(1, 2)$–$(2, 5)$–$(5, 1)$. We must now replace an arc (i, j) of this subtour by the arcs (i, k) and (k, j), where $k = 3$ or 4. The relevant computations are shown in Table 72.

We now replace $(1, 2)$ by arcs $(1, 4)$ and $(4, 2)$. This yields the subtour $(1, 4)$–$(4, 2)$–$(2, 5)$–$(5, 1)$. An arc (i, j) in this subtour must now be replaced by arcs $(i, 3)$ and $(3, j)$. The relevant computations are shown in Table 73. We now replace arc $(1, 4)$ by arcs $(1, 3)$ and $(3, 4)$. This yields the tour $(1, 3)$–$(3, 4)$–$(4, 2)$–$(2, 5)$–$(5, 1)$. In this example, the CIH yields an optimal tour—but, in general, the CIH does not necessarily do so.

TABLE 71
Determining Which Arc of (1, 5)–(5, 1) Is Replaced

Arc Replaced	Arcs Added to Subtour	Added Length
(1, 5)*	(1, 2)–(2, 5)	$c_{12} + c_{25} - c_{15} = 153$
(1, 5)	(1, 3)–(3, 5)	$c_{13} + c_{35} - c_{15} = 462$
(1, 5)	(1, 4)–(4, 5)	$c_{14} + c_{45} - c_{15} = 302$
(5, 1)*	(5, 2)–(2, 1)	$c_{52} + c_{21} - c_{51} = 153$
(5, 1)	(5, 3)–(3, 1)	$c_{53} + c_{31} - c_{51} = 462$
(5, 1)	(5, 4)–(4, 1)	$c_{54} + c_{41} - c_{51} = 302$

TABLE 72
Determining Which Arc of (1, 2)–(2, 5)–(5, 1) Is Replaced

Arc Replaced	Arcs Added	Added Length
(1, 2)	(1, 3)–(3, 2)	$c_{13} + c_{32} - c_{12} = 375$
(1, 2)*	(1, 4)–(4, 2)	$c_{14} + c_{42} - c_{12} = 233$
(2, 5)	(2, 3)–(3, 5)	$c_{23} + c_{35} - c_{25} = 514$
(2, 5)	(2, 4)–(4, 5)	$c_{24} + c_{45} - c_{25} = 318$
(5, 1)	(5, 3)–(3, 1)	$c_{53} + c_{31} - c_{51} = 462$
(5, 1)	(5, 4)–(4, 1)	$c_{54} + c_{41} - c_{51} = 302$

TABLE 73
Determining Which Arc of (1, 4)–(4, 2)–(2, 5)–(5, 1) Is Replaced

Arc Replaced	Arcs Added	Added Length
(1, 4)*	(1, 3)–(3, 4)	$c_{13} + c_{34} - c_{14} = 166$
(4, 2)	(4, 3)–(3, 2)	$c_{43} + c_{32} - c_{42} = 202$
(2, 5)	(2, 3)–(3, 5)	$c_{23} + c_{35} - c_{25} = 514$
(5, 1)	(5, 3)–(3, 1)	$c_{53} + c_{31} - c_{51} = 462$

Evaluation of Heuristics

The following three methods have been suggested for evaluating heuristics:

1 Performance guarantees

2 Probabilistic analysis

3 Empirical analysis

A performance guarantee for a heuristic gives a worst-case bound on how far away from optimality a tour constructed by the heuristic can be. For the NNH, it can be shown that for any number r, a TSP can be constructed such that the NNH yields a tour that is r times as long as the optimal tour. Thus, in a worst-case scenario, the NNH fares poorly. For a symmetric TSP satisfying the triangle inequality (that is, for which $c_{ij} = c_{ji}$ and $c_{lk} \leq c_{ij} + c_{jk}$ for all $i, j,$ and k), it has been shown that the length of the tour obtained by the CIH cannot exceed twice the length of the optimal tour.

In probabilistic analysis, a heuristic is evaluated by assuming that the location of cities follows some known probability distribution. For example, we might assume that the cities are independent random variables that are uniformly distributed on a cube of unit length, width, and height. Then, for each heuristic, we would compute the following ratio:

$$\frac{\text{Expected length of the path found by the heuristic}}{\text{Expected length of an optimal tour}}$$

The closer the ratio is to 1, the better the heuristic.

For empirical analysis, heuristics are compared to the optimal solution for a number of problems for which the optimal tour is known. As an illustration, for five 100-city TSPs, Golden, Bodin, Doyle, and Stewart (1980) found that the NNH—taking the best of all solutions found when the NNH was applied beginning at each city—produced tours that averaged 15% longer than the optimal tour. For the same set of problems, it was found that the CIH (again applying the best solution obtained by applying CIH to all cities) produced tours that also averaged 15% longer than the optimal tour.

REMARKS **1** Golden, Bodin, Doyle, and Stewart (1980) describe a heuristic that regularly comes within 2–3% of the optimal tour.
2 It is also important to compare heuristics with regard to computer running time and ease of implementation.
3 For an excellent discussion of heuristics, see Chapters 5–7 of Lawler (1985).

An Integer Programming Formulation of the TSP

We now discuss how to formulate an IP whose solution will solve a TSP. We note, however, that the formulation of this section becomes unwieldy and inefficient for large TSPs. Suppose the TSP consists of cities $1, 2, 3, \ldots, N$. For $i \neq j$ let $c_{ij} =$ distance from city i to city j and let $c_{ii} = M$, where M is a very large number (relative to the actual distances in the problem). Setting $c_{ii} = M$ ensures that we will not go to city i immediately after leaving city i. Also define

$$x_{ij} = \begin{cases} 1 & \text{if the solution to TSP goes from city } i \text{ to city } j \\ 0 & \text{otherwise} \end{cases}$$

Then the solution to a TSP can be found by solving

$$\min z = \sum_i \sum_j c_{ij} x_{ij} \tag{40}$$

$$\text{s.t.} \quad \sum_{i=1}^{i=N} x_{ij} = 1 \quad (\text{for } j = 1, 2, \ldots, N) \tag{41}$$

$$\sum_{j=1}^{j=N} x_{ij} = 1 \quad (\text{for } i = 1, 2, \ldots, N) \tag{42}$$

$$u_i - u_j + N x_{ij} \leq N - 1 \quad (\text{for } i \neq j; i = 2, 3, \ldots, N; j = 2, 3, \ldots, N) \tag{43}$$

$$\text{All } x_{ij} = 0 \text{ or } 1, \text{ All } u_j \geq 0$$

The objective function (40) gives the total length of the arcs included in a tour. The constraints in (41) ensure that we arrive once at each city. The constraints in (42) ensure that we leave each city once. The constraints in (43) are the key to the formulation. They ensure the following:

1 Any set of x_{ij}'s containing a subtour will be infeasible [that is, they violate (43)].

2 Any set of x_{ij}'s that forms a tour will be feasible [there will exist a set of u_j's that satisfy (43)].

To illustrate that any set of x_{ij}'s containing a subtour will violate (43), consider the subtour illustration given in Figure 24. Here $x_{15} = x_{21} = x_{43} = x_{43} = x_{52} = 1$. This assignment contains the two subtours 1–5–2–1 and 3–4–3. Choose the subtour that does *not* contain city 1 (3–4–3) and write down the constraints in (43) corresponding to the arcs in this subtour. We obtain $u_3 - u_4 + 5x_{34} \leq 4$ and $u_4 - u_3 + 5x_{43} \leq 4$. Adding these constraints yields $5(x_{34} + x_{43}) \leq 8$. Clearly, this rules out the possibility that $x_{43} = x_{34} = 1$, so the subtour 3–4–3 (and any other subtour!) is ruled out by the constraints in (43).

We now show that for any set of x_{ij}'s that does not contain a subtour, there exist values of the u_j's that will satisfy all constraints in (43). Assume that city 1 is the first city visited (we visit all cities eventually, so this is okay). Let $t_i =$ the position in the tour where city i is visited. Then setting $u_i = t_i$ will satisfy all constraints in (43). To illustrate, consider the tour 1–3–4–5–2–1. Then we choose $u_1 = 1$, $u_2 = 5$, $u_3 = 2$, $u_4 = 3$, $u_5 = 4$. We now show that with this choice of the u_i's all constraints in (43) are satisfied. First, consider any constraint corresponding to an arc having $x_{ij} = 1$. For example, the constraint corresponding to x_{52} is $u_5 - u_2 + 5x_{52} \leq 4$. Because city 2 immediately follows city 5, $u_5 - u_2 = -1$. Then the constraint for x_{52} in (43) reduces to $-1 + 5 \leq 4$, which is true. Now consider a constraint corresponding to an x_{ij} (say, x_{32}) satisfying $x_{ij} = 0$. For x_{32}, we obtain the constraint $u_3 - u_2 + 5x_{32} \leq 4$. This reduces to $u_3 - u_2 \leq 4$. Because $u_3 \leq 5$ and $u_2 > 1$, $u_3 - u_2$ cannot exceed $5 - 2$.

This shows that the formulation defined by (40)–(43) eliminates from consideration all sequences of N cities that begin in city 1 and include a subtour. We have also shown that this formulation does not eliminate from consideration any sequence of N cities beginning in city 1 that does not include a subtour. Thus, (40)–(43) will (if solved) yield the optimal solution to the TSP.

Using LINGO to Solve TSPs

The IP described in (40)–(43) can easily be implemented with the following LINGO program (file TSP.lng).

```
MODEL:
  1]SETS:
  2]CITY/1..5/:U;
  3]LINK(CITY,CITY):DIST,X;
  4]ENDSETS
  5]DATA:
  6]DIST= 50000 132 217 164 58
  7]132 50000 290 201 79
  8]217 290 50000 113 303
  9]164 201 113 50000 196
 10]58 79 303 196 5000;
 11]ENDDATA
 12]N=@SIZE(CITY);
 13]MIN=@SUM(LINK:DIST*X);
 14]@FOR(CITY(K):@SUM(CITY(I):X(I,K))=1;);
 15]@FOR(CITY(K):@SUM(CITY(J):X(K,J))=1;);
 16]@FOR(CITY(K):@FOR(CITY(J)|J#GT#1#AND#K#GT#1:
 17]U(J)-U(K)+N*X(J,K)<N-1;));
 18]@FOR(LINK:@BIN(X););
END
```

In line 2, we define our five cities and associate a U(J) with city J. In line 3, we create the arcs joining each combination of cities. With the arc from city I to city J, we associate the distance between city I and J and a 0–1 variable X(I,J), which equals 1 if city J immediately follows city I in a tour.

In lines 6–10, we input the distance between the cities given in Example 11. Note that the distance between city I and itself is assigned a large number, to ensure that city I does not follow itself.

In line 12, we use **@SIZE** to compute the number of cities (we use this in line 17). In line 13, we create the objective function by summing over each link (I,J) the product of the distance between cities I and J and X(I,J). Line 14 ensures that for each city we enter the city exactly once. Line 15 ensures that for each city we leave the city exactly once. Lines 16–17 create the constraints in (43). Note that we only create these constraints for combinations J,K where J > 1 and K > 1. This agrees with (43). Note that when J = K line 17 generates constraints of the form N*X(J,J) ≤ N − 1, which imply that all X(J,J) = 0. In line 18, we ensure that each X(I,J) = 0 or 1. We need not constrain the U(J)'s, because LINGO assumes they are non-negative. *Note:* Even for small TSPs this formulation will exceed the capacity of student LINGO.

PROBLEMS

Group A

1 Four jobs must be processed on a single machine. The time required to perform each job and the due date for each job are shown in Table 74. Use the branch-and-bound method to determine the order of performing the jobs that minimizes the total time the jobs are delayed.

2 Each day, Sunco manufactures four types of gasoline: lead-free premium (LFP), lead-free regular (LFR), leaded premium (LP), and leaded regular (LR). Because of cleaning and resetting of machinery, the time required to produce a batch of gasoline depends on the type of gasoline last produced. For example, it takes longer to switch between a lead-free gasoline and a leaded gasoline than it does to switch between two lead-free gasolines. The time (in minutes) required to manufacture each day's gasoline requirements are shown in Table 75. Use a branch-and-bound approach to determine the order in which the gasolines should be produced each day.

3 A Hamiltonian path in a network is a closed path that passes exactly once through each node in the network before

returning to its starting point. Taking a four-city TSP as an example, explain why solving a TSP is equivalent to finding the shortest Hamiltonian path in a network.

4 There are four pins on a printed circuit. The distance between each pair of pins (in inches) is given in Table 76.

a Suppose we want to place three wires between the pins in a way that connects all the wires and uses the minimum amount of wire. Solve this problem by using one of the techniques discussed in Chapter 8.

b Now suppose that we again want to place three wires between the pins in a way that connects all the wires and uses the minimum amount of wire. Also suppose that if more than two wires touch a pin, a short circuit will occur. Now set up a traveling-salesperson problem that can be used to solve this problem. (*Hint:* Add a pin 0 such that the distance between pin 0 and any other pin is 0.)

5 a Use the NNH to find a solution to the TSP in Problem 2. Begin with LFR.

b Use the CIH to find a solution to the TSP in Problem 2. Begin with the subtour LFR–LFP–LFR.

6 LL Pea stores clothes at five different locations. Several times a day it sends an "order picker" out to each location to pick up orders. Then the order picker must return to the packaging area. Describe a TSP that could be used to minimize the time needed to pick up orders and return to the packaging area.

Group B

7 Use branch-and-bound to determine a way (if any exists to place four queens on a 4 × 4 chessboard so that no queen can capture another queen. (*Hint:* Let $x_{ij} = 1$ if a queen is placed in row i and column j of the chessboard and $x_{ij} = 0$ otherwise. Then branch as in the machine-delay problem.

TABLE 74

Job	Time to Perform Job (Minutes)	Due Date of Job
1	7	End of minute 14
2	5	End of minute 13
3	9	End of minute 18
4	11	End of minute 15

TABLE 75

Last-Produced Gasoline	Gas to Be Next Produced			
	LFR	LFP	LR	LP
LFR	—	50	120	140
LFP	60	—	140	110
LR	90	130	—	60
LP	130	120	80	—

Note: Assume that the last gas produced yesterday precedes the first gas produced today.

TABLE 76

	1	2	3	4
1	0	1	2	2
2	1	0	3	2.9
3	2	3	0	3
4	2	2.9	3	0

Many nodes may be eliminated from consideration because they are infeasible. For example, the node associated with the arcs $x_{11} = x_{22} = 1$ is infeasible, because the two queens can capture each other.)

8 Although the Hungarian method is an efficient method for solving an assignment problem, the branch-and-bound method can also be used to solve an assignment problem. Suppose a company has five factories and five warehouses. Each factory's requirements must be met by a single warehouse, and each warehouse can be assigned to only one factory. The costs of assigning a warehouse to meet a factory's demand (in thousands) are shown in Table 77.

Let $x_{ij} = 1$ if warehouse i is assigned to factory j and 0 otherwise. Begin by branching on the warehouse assigned to factory 1. This creates the following five branches: $x_{11} = 1$, $x_{21} = 1$, $x_{31} = 1$, $x_{41} = 1$, and $x_{51} = 1$. How can we obtain a lower bound on the total cost associated with a branch? Examine the branch $x_{21} = 1$. If $x_{21} = 1$, no further assignments can come from row 2 or column 1 of the cost matrix. In determining the factory to which each of the unassigned warehouses (1, 3, 4, and 5) is assigned, we cannot do better than assign each to the smallest cost in the warehouse's row (excluding the factory 1 column). Thus, the minimum-cost assignment having $x_{21} = 1$ must have a total cost of at least $10 + 10 + 9 + 5 + 5 = 39$.

Similarly, in determining the warehouse to which each of the unassigned factories (2, 3, 4, and 5) is assigned, we cannot do better than to assign each to the smallest cost in the factory's column (excluding the warehouse 2 row). Thus, the minimum-cost assignment having $x_{21} = 1$ must have a total cost of at least $10 + 9 + 5 + 5 + 7 = 36$. Thus, the total cost of any assignment having $x_{21} = 1$ must be at least $\max(36, 39) = 39$. So, if branching ever leads to a candidate solution having a total cost of 39 or less, the $x_{21} = 1$ branch may be eliminated from consideration. Use this idea to solve the problem by branch-and-bound.

9[†] Consider a long roll of wallpaper that repeats its pattern every yard. Four sheets of wallpaper must be cut from the roll. With reference to the beginning (point 0) of the wallpaper, the beginning and end of each sheet are located as shown in Table 78. Thus, sheet 1 begins 0.3 yd from the beginning of the roll (and 1.3 yd from the beginning of the roll) and sheet 1 ends 0.7 yd from the beginning of the roll (and 1.7 yd from the beginning of the roll). Assume we are

TABLE 78

Sheet	Beginning (Yards)	End (Yards)
1	0.3	0.7
2	0.4	0.8
3	0.2	0.5
4	0.7	0.9

at the beginning of the roll. In what order should the sheets be cut to minimize the total amount of wasted paper? Assume that a final cut is made to bring the roll back to the beginning of the pattern.

10[‡] A manufacturer of printed circuit boards uses programmable drill machines to drill six holes in each board. The x and y coordinates of each hole are given in Table 79. The time (in seconds) it takes the drill machine to move from one hole to the next is equal to the distance between the points. What drilling order minimizes the total time that the drill machine spends moving between holes?

11 Four jobs must be processed on a single machine. The time required to perform each job, the due date, and the penalty (in dollars) per day the job is late are given in Table 80.

Use branch-and-bound to determine the order of performing the jobs that will minimize the total penalty costs due to delayed jobs.

TABLE 79

x	y	Hole
1	2	1
3	1	2
5	3	3
7	2	4
8	3	5

TABLE 77

Warehouse	Factory ($)				
	1	2	3	4	5
1	5	15	20	25	10
2	10	12	5	15	19
3	5	17	18	9	11
4	8	9	10	5	12
5	9	10	5	11	7

TABLE 80

Job	Time (Days)	Due Date	Penalty
1	4	Day 4	4
2	5	Day 2	5
3	2	Day 13	7
4	3	Day 8	2

[†]Based on Garfinkle (1977).

[‡]Based on Magirou (1986).

9.7 Implicit Enumeration

The method of implicit enumeration is often used to solve 0–1 IPs. Implicit enumeration uses the fact that each variable must equal 0 or 1 to simplify both the branching and bounding components of the branch-and-bound process and to determine efficiently when a node is infeasible.

Before discussing implicit enumeration, we show how any pure IP may be expressed as a 0–1 IP: Simply express each variable in the original IP as the sum of powers of 2. For example, suppose the variable x_i is required to be an integer. Let n be the smallest integer such that we can be sure that $x_i < 2^{n+1}$. Then x_i may be (uniquely) expressed as the sum of $2^0, 2^1, \ldots, 2^{n-1}, 2^n$, and

$$x_i = u_n 2^n + u_{n-1} 2^{n-1} + \cdots + u_2 2^2 + 2u_1 + u_0 \tag{44}$$

where $u_i = 0$ or 1 $(i = 0, 1, \ldots, n)$.

To convert the original IP to a 0–1 IP, replace each occurrence of x_i by the right side of (44). For example, suppose we know that $x_i \le 100$. Then $x_i < 2^{6+1} = 128$. Then (44) yields

$$x_i = 64u_6 + 32u_5 + 16u_4 + 8u_3 + 4u_2 + 2u_1 + u_0 \tag{45}$$

where $u_i = 0$ or 1 $(i = 0, 1, \ldots, 6)$. Then replace each occurrence of x_i by the right side of (45). How can we find the values of the u's corresponding to a given value of x_i? Suppose $x_i = 93$. Then u_6 will be the largest multiple of $2^6 = 64$ that is contained in 93. This yields $u_6 = 1$; then the rest of the right side of (45) must equal $93 - 64 = 29$. Then u_5 will be the largest multiple of $2^5 = 32$ contained in 29. This yields $u_5 = 0$. Then u_4 will be the largest multiple of $2^4 = 16$ contained in 29. This yields $u_4 = 1$. Continuing in this fashion, we obtain $u_3 = 1$, $u_2 = 1$, $u_1 = 0$, and $u_0 = 1$. Thus $93 = 2^6 + 2^4 + 2^3 + 2^2 + 2^0$.

We will soon discover that 0–1 IP's are generally easier to solve than other pure IP's. Why, then, don't we transform every pure IP into a 0–1 IP? Simply because transforming a pure IP into a 0–1 IP greatly increases the number of variables. However, many situations (such as lockbox and knapsack problems) naturally yield 0–1 problems. Thus, it is certainly worthwhile to learn how to solve 0–1 IPs.

The tree used in the implicit enumeration method is similar to those used to solve 0–1 knapsack problems in Section 9.5. Each branch of the tree will specify, for some variable x_i, that $x_i = 0$ or $x_i = 1$. At each node, the values of some of the variables are specified. For instance, suppose a 0–1 problem has variables $x_1, x_2, x_3, x_4, x_5, x_6$, and part of the tree looks like Figure 26. At node 4, the values of x_3, x_4, and x_2 are specified. These variables are referred to as **fixed variables.** All variables whose values are unspecified at a node are called **free variables.** Thus, at node 4, x_1, x_5, and x_6 are free variables. For any node, a

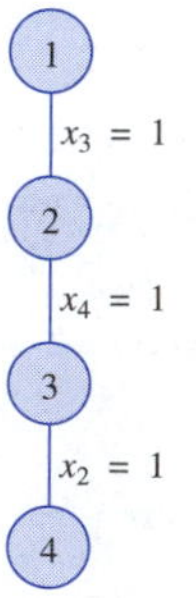

FIGURE 26
Illustration of Free and
Fixed Variables

specification of the values of all the free variables is called a **completion** of the node. Thus $x_1 = 1$, $x_5 = 1$, $x_6 = 0$ is a completion of node 4.

We are now ready to outline the three main ideas used in implicit enumeration.

1 Suppose we are at any node. Given the values of the fixed variables at that node, is there an easy way to find a good completion of that node that is feasible in the original 0–1 IP? To answer this question, we complete the node by setting each free variable equal to the value (0 or 1) that makes the objective function largest (in a max problem) or smallest (in a min problem). If this completion of the node is feasible, then it is certainly the best feasible completion of the node, and further branching of the node is unnecessary. Suppose we are solving

$$\max z = 4x_1 + 2x_2 - x_3 + 2x_4$$

$$\text{s.t.} \quad x_1 + 3x_2 - x_3 - 2x_4 \geq 1$$

$$x_i = 0 \text{ or } 1 \quad (i = 1, 2, 3, 4)$$

If we are at a node (call it node 4) where $x_1 = 0$ and $x_2 = 1$ are fixed, then the best we can do is set $x_3 = 0$ and $x_4 = 1$. Because $x_1 = 0$, $x_2 = 1$, $x_3 = 0$, and $x_4 = 1$ is feasible in the original problem, we have found the best feasible completion of node 4. Thus, node 4 is fathomed and $x_1 = 0$, $x_2 = 1$, $x_3 = 0$, $x_4 = 1$ (along with its z-value of 4) may be used as a candidate solution.

2 Even if the best completion of a node is not feasible, the best completion gives us a bound on the best objective function value that can be obtained via a feasible completion of the node. This bound can often be used to eliminate a node from consideration. For example, suppose we have previously found a candidate solution with $z = 6$, and our objective is to maximize

$$z = 4x_1 + 2x_2 + x_3 - x_4 + 2x_5$$

Also suppose that we are at a node where the fixed variables are $x_1 = 0$, $x_2 = 1$, and $x_3 = 1$. Then the best completion of this node is $x_4 = 0$ and $x_5 = 1$. This yields a z-value of $2 + 1 + 2 = 5$. Because $z = 5$ cannot beat the candidate with $z = 6$, we can immediately eliminate this node from consideration (whether or not the completion is feasible is irrelevant).

3 At any node, is there an easy way to determine if all completions of the node are infeasible? Suppose we are at node 4 of Figure 26 and one of the constraints is

$$-2x_1 + 3x_2 + 2x_3 - 3x_4 - x_5 + 2x_6 \leq -5 \tag{46}$$

Is there any completion of node 4 that can satisfy this constraint? We assign values to the free variables that make the left side of (46) as small as possible. If this completion of node 4 won't satisfy (46), then certainly no completion of node 4 can. Thus, we set $x_1 = 1$, $x_5 = 1$, and $x_6 = 0$. Substituting these values and the values of the fixed variables, we obtain $-2 + 3 + 2 - 3 - 1 \leq -5$. This inequality does not hold, so no completion of node 4 can satisfy (46). No completion of node 4 can be feasible for the original problem, and node 4 may be eliminated from consideration.

In general, we check whether a node has a feasible completion by looking at each constraint and assigning each free variable the best value (as described in Table 81) for satisfying the constraint.[†] If even one constraint is not satisfied by its most feasible completion, then we know that the node has no feasible completion. In this case, the node cannot yield the optimal solution to the original IP.

[†]Each equality constraint should be replaced by a $\leq$ and a $\geq$ constraint.

TABLE **81**
How to Determine Whether a Node Has a Completion
Satisfying a Given Constraint

Type of Constraint	Sign of Free Variable's Coefficient in Constraint	Value Assigned to Free Variable in Feasibility Check
≤	+	0
≤	−	1
≥	+	1
≥	−	0

We note, however, that even if a node has no feasible completion, our crude infeasibility check may not reveal that the node has no feasible completion until we have moved further down the tree to a node where there are more fixed variables. If we have failed to obtain any information about a node, we now branch on a free variable x_i and add two new nodes: one with $x_i = 1$ and another with $x_i = 0$.

EXAMPLE 12 **Implicit Enumeration**

Use implicit enumeration to solve the following 0–1 IP:

$$\max z = -7x_1 - 3x_2 - 2x_3 - x_4 - 2x_5$$
$$\text{s.t.} \quad -4x_1 - 2x_2 + x_3 - 2x_4 - x_5 \le -3 \tag{47}$$
$$-4x_1 - 2x_2 - 4x_3 + x_4 + 2x_5 \le -7 \tag{48}$$
$$x_i = 0 \text{ or } 1 \quad (i = 1, 2, 3, 4, 5)$$

Solution   At the beginning (node 1), all variables are free. We first check whether the best completion of node 1 is feasible. The best completion of node 1 is $x_1 = 0$, $x_2 = 0$, $x_3 = 0$, $x_4 = 0$, $x_5 = 0$, which is not feasible (it violates both constraints). We now check to see whether node 1 has no feasible completion. Checking (47) for feasibility, we set $x_1 = 1$, $x_2 = 1$, $x_3 = 0$, $x_4 = 1$, $x_5 = 1$. This satisfies (47) (it yields $-9 \le -3$). We now check (48) for feasibility by setting $x_1 = 1$, $x_2 = 1$, $x_3 = 1$, $x_4 = 0$, $x_5 = 0$. This completion of node 1 satisfies (48) (it yields $-10 \le -7$). Thus, node 1 has a feasible completion satisfying (48). Therefore, our infeasibility check does not allow us to classify node 1 as having no feasible completion. We now choose to branch on a free variable: arbitrarily, x_1. This yields two new nodes: node 2 with the constraint $x_1 = 1$ and node 3 with the constraint $x_1 = 0$ (see Figure 27).

We now choose to analyze node 2. The best completion of node 2 is $x_1 = 1$, $x_2 = 0$, $x_3 = 0$, $x_4 = 0$, and $x_5 = 0$. Unfortunately, this completion is not feasible. We now try to determine whether node 2 has a feasible completion. We check whether $x_1 = 1$, $x_2 = 1$, $x_3 = 0$, $x_4 = 1$, $x_5 = 1$ satisfies (47) (this yields $-9 \le -3$). Then we check whether $x_1 = 1$, $x_2 = 1$, $x_3 = 1$, $x_4 = 0$, $x_5 = 0$ satisfies (48) (this yields $-10 \le -7$). Thus, our infeasibility check has yielded no information about whether node 2 has a feasible completion.

We now choose to branch on node 2, arbitrarily, on the free variable x_2. This yields nodes 4 and 5 in Figure 28. Using the LIFO rule, we choose to next analyze node 5. The best completion of node 5 is $x_1 = 1$, $x_2 = 0$, $x_3 = 0$, $x_4 = 0$, $x_5 = 0$. Again, this completion is infeasible. We now perform a feasibility check on node 5. We determine whether $x_1 = 1$, $x_2 = 0$, $x_3 = 0$, $x_4 = 1$, $x_5 = 1$ satisfies (47) (this yields $-7 \le -3$). Then we check whether $x_1 = 1$, $x_2 = 0$, $x_3 = 1$, $x_4 = 0$, $x_5 = 0$ satisfies (48) (this yields $-8 \le -7$). Again our feasibility check has yielded no information. Thus, we branch on node 5, arbitrarily choosing the free variable x_3. This adds nodes 6 and 7 in Figure 29.

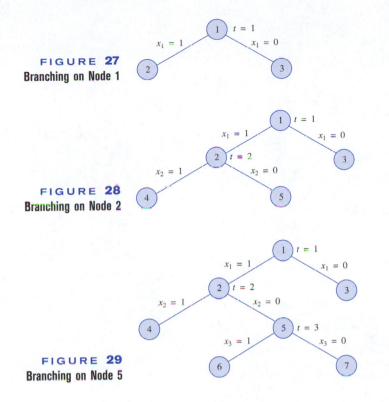

FIGURE 27
Branching on Node 1

FIGURE 28
Branching on Node 2

FIGURE 29
Branching on Node 5

Applying the LIFO rule, we next choose to analyze node 6. The best completion of node 6 is $x_1 = 1, x_2 = 0, x_3 = 1, x_4 = 0, x_5 = 0, z = -9$. This point is feasible, so we have found a candidate solution with $z = -9$. Using the LIFO rule, we next analyze node 7. The best completion of node 7 is $x_1 = 1, x_2 = 0, x_3 = 0, x_4 = 0, x_5 = 0, z = -7$. Because $z = -7$ is better than $z = -9$, it is possible for node 7 to beat the current candidate. Thus, we must check node 7 to see whether it has any feasible completion. We see whether $x_1 = 1, x_2 = 0, x_3 = 0, x_4 = 1, x_5 = 1$ satisfies (47) (this yields $-7 \le -3$). Then we see whether $x_1 = 1, x_2 = 0, x_3 = 0, x_4 = 0, x_5 = 0$ satisfies (48) (this yields $-4 \le -7$). This means that no completion of node 7 can satisfy (48). Thus, node 7 has no feasible completion, and it may be eliminated from consideration (indicated by an $\times$ in Figure 30).

The LIFO rule now indicates that we should analyze node 4. The best completion of node 4 is $x_1 = 1, x_2 = 1, x_3 = 0, x_4 = 0, x_5 = 0$. This solution has $z = -10$. Thus, node 4 cannot beat the previous candidate solution from node 6 (having $z = -9$), and node 4 may be eliminated from consideration.

We are now facing the tree in Figure 31, where only node 3 remains to be analyzed. The best completion of node 3 is $x_1 = 0, x_2 = 0, x_3 = 0, x_4 = 0, x_5 = 0$. This point is infeasible. This point has $z = 0$, however, so it is possible that node 3 can yield a feasible solution that is better than our current candidate (with $z = -9$). We now check whether node 3 has any feasible completion: Does $x_1 = 0, x_2 = 1, x_3 = 1, x_4 = 1, x_5 = 1$ satisfy (47)? This yields $-5 \le -3$, so node 3 does have a completion satisfying (47). Then we see whether node 3 has any completion satisfying (48): Does $x_1 = 0, x_2 = 1, x_3 = 1, x_4 = 0, x_5 = 0$ satisfy (48)? This yields $-6 \le -7$, which is untrue. Thus, node 3 has no completion satisfying (48), and node 3 may be eliminated from consideration. We now have the tree in Figure 32.

Because there are no nodes left to analyze, the node 6 candidate with $z = -9$ must be optimal. Thus, $x_1 = 1, x_2 = 0, x_3 = 1, x_4 = 0, x_5 = 0, z = -9$ is the optimal solution to

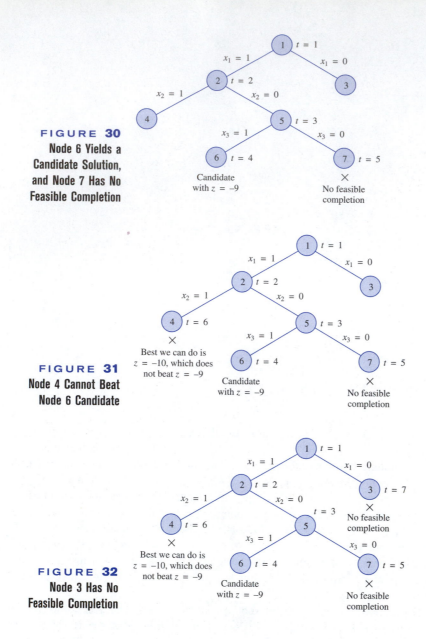

FIGURE 30
Node 6 Yields a
Candidate Solution,
and Node 7 Has No
Feasible Completion

FIGURE 31
Node 4 Cannot Beat
Node 6 Candidate

FIGURE 32
Node 3 Has No
Feasible Completion

the 0–1 IP. Note that every possible point $(x_1, x_2, x_3, x_4, x_5)$ where $x_i = 0$ or 1 has been implicitly considered, and all but the optimal solution have been eliminated. For example, for the point $x_1 = 1$, $x_2 = 1$, $x_3 = 1$, $x_4 = 1$, $x_5 = 0$, the analysis of node 4 shows that this point cannot be optimal because it cannot have a z-value of better than -9. As another example, the point $x_1 = 0$, $x_2 = 1$, $x_3 = 1$, $x_4 = 1$, $x_5 = 1$ cannot be optimal, because our analysis of node 3 shows that no completion can be feasible.

The use of subtler infeasibility tests (called **surrogate constraints**) can often reduce the number of nodes that must be examined before an optimal solution is found. For example, consider a 0–1 IP with the following constraints:

$$x_1 + x_2 + x_3 + x_4 + x_5 \le 2 \tag{49}$$

$$x_1 - x_2 + x_3 - x_4 - x_5 \ge 1 \tag{50}$$

Suppose we are at a node where $x_1 = x_2 = 1$. To check whether this node has a feasible completion, we would first see whether $x_1 = 1, x_2 = 1, x_3 = 0, x_4 = 0, x_5 = 0$ satisfies (49) (it does). Then we would see whether $x_1 = 1, x_2 = 1, x_3 = 1, x_4 = 0, x_5 = 0$ (50) (it does). In this situation, our crude infeasibility tests do not yet indicate that this node is infeasible. Observe, however, that because $x_1 = x_2 = 1$, the only way to satisfy (49) is by choosing $x_3 = x_4 = x_5 = 0$, but this completion of the $x_1 = x_2 = 1$ node fails to satisfy (50). Thus, the node with $x_1 = x_2 = 1$ will have no feasible completion. Eventually, our crude infeasibility test would have indicated this fact, but we might have been forced to examine several more nodes before we found that the node with $x_1 = x_2 = 1$ had no feasible completion. In a more complex problem, a subtler infeasibility test that combined information from both constraints might have enabled us to examine fewer nodes. Of course, a subtler infeasibility test would require more computation, so it might not be worth the effort. For a discussion of surrogate constraints, see Salkin (1975), Taha (1975), and Nemhauser and Wolsey (1988).

As with any branch-and-bound algorithm, many arbitrary choices determine the efficiency of the implicit enumeration algorithm. See Salkin, Taha, and Nemhauser and Wolsey for further discussion of implicit enumeration techniques.

PROBLEMS

Group A

Use implicit enumeration to solve the following 0–1 IPs:

1 max $z = 3x_1 + x_2 + 2x_3 - x_4 + x_5$
s.t. $2x_1 + x_2 \qquad - 3x_4 \qquad \leq 1$
$\quad x_1 + 2x_2 - 3x_3 - x_4 + 2x_5 \geq 2$
$\qquad\qquad x_i = 0$ or 1

2 max $z = 2x_1 - x_2 + x_3$
s.t. $x_1 + 2x_2 - x_3 \leq 1$
$\quad x_1 + x_2 + x_3 \leq 2$
$\qquad x_i = 0$ or 1

TABLE 82

Project	Time 0 Cash Outflow ($)	NPV ($)
1	4	5
2	6	9
3	5	6
4	4	3
5	3	2

3 Finco is considering investing in five projects. Each requires a cash outflow at time 0 and yields an NPV as described in Table 82 (all dollars in millions). At time 0, $10 million is available for investment. Projects 1 and 2 are mutually exclusive (that is, Finco cannot undertake both). Similarly, projects 3 and 4 are mutually exclusive. Also, project 2 cannot be undertaken unless project 5 is undertaken. Use implicit enumeration to determine which projects should be undertaken to maximize NPV.

4 Use implicit enumeration to find the optimal solution to Example 5 (the set-covering problem).

5 Use implicit enumeration to solve Problem 1 of Section 9.2.

Group B

6 Why are the values of $u_0, u_1, \ldots, u_n$ in (44) unique?

9.8 The Cutting Plane Algorithm[†]

In previous portions of this chapter, we have described in some detail how branch-and-bound methods can be used to solve IPs. In this section, we discuss an alternative method, **the cutting plane algorithm.** We illustrate the cutting plane algorithm by solving the Telfa Corporation problem (Example 9). Recall from Section 9.3 that this problem was

[†]This section covers topics that may be omitted with no loss of continuity.

TABLE 83
Optimal Tableau for LP Relaxation of Telfa

z	x_1	x_2	s_1	s_2	rhs
1	0	0	1.25	0.75	41.25
0	0	1	2.25	−0.25	2.25
0	1	0	−1.25	0.25	3.75

$$\max z = 8x_1 + 5x_2$$
$$\text{s.t.} \quad x_1 + x_2 \leq 6$$
$$9x_1 + 5x_2 \leq 45 \tag{51}$$
$$x_1, x_2 \geq 0; \ x_1, x_2 \text{ integer}$$

After adding slack variables s_1 and s_2, we found the optimal tableau for the LP relaxation of the Telfa example to be as shown in Table 83.

To apply the cutting plane method, we begin by choosing any constraint in the LP relaxation's optimal tableau in which a basic variable is fractional. We arbitrarily choose the second constraint, which is

$$x_1 - 1.25s_1 + 0.25s_2 = 3.75 \tag{52}$$

We now define $[x]$ to be the largest integer less than or equal to x. For example, $[3.75] = 3$ and $[-1.25] = -2$. Any number x can be written in the form $[x] + f$, where $0 \leq f < 1$. We call f the fractional part of x. For example, $3.75 = 3 + 0.75$, and $-1.25 = -2 + 0.75$. In (51)'s optimal tableau, we now write each variable's coefficient and the constraint's right-hand side in the form $[x] + f$, where $0 \leq f < 1$. Now (52) may be written as

$$x_1 - 2s_1 + 0.75s_1 + 0s_2 + 0.25s_2 = 3 + 0.75 \tag{53}$$

Putting all terms with integer coefficients on the left side and all terms with fractional coefficients on the right side yields

$$x_1 - 2s_1 + 0s_2 - 3 = 0.75 - 0.75s_1 - 0.25s_2 \tag{54}$$

The cutting plane algorithm now suggests adding the following constraint to the LP relaxation's optimal tableau:

$$\text{Right-hand side of (54)} \leq 0$$

or

$$0.75 - 0.75s_1 - 0.25s_2 \leq 0 \tag{55}$$

This constraint is called (for reasons that will soon become apparent) a **cut.** We now show that a cut generated by this method has two properties:

1 Any feasible point for the IP will satisfy the cut.

2 The current optimal solution to the LP relaxation will not satisfy the cut.

Thus, a cut "cuts off" the current optimal solution to the LP relaxation, but not any feasible solutions to the IP. When the cut to the LP relaxation is added, we hope we will obtain a solution where all variables are integer-valued. If so, we have found the optimal solution to the original IP. If our new optimal solution (to the LP relaxation plus the cut) has some fractional-valued variables, then we generate another cut and continue the process. Gomory (1958) showed that this process will yield an optimal solution to the IP after a finite number of cuts. Before finding the optimal solution to the IP (51), we show why the cut (55) satisfies properties 1 and 2.

We now show that any feasible solution to the IP (51) will satisfy the cut (55). Consider any point that is feasible for the IP. For such a point, x_1 and x_2 take on integer values, and the point must be feasible in the LP relaxation of (51). Because (54) is just a rearrangement of the optimal tableau's second constraint, any feasible point for the IP must satisfy (54). Any feasible solution to the IP must have $s_1 \geq 0$ and $s_2 \geq 0$. Because $0.75 < 1$, any feasible solution to the IP will make the right-hand side of (54) less than 1. Also note that for any point that is feasible for the IP, the left-hand side of (54) will be an integer. Thus, for any feasible point to the IP, the right-hand side must be an integer that is less than 1. This implies that any point that is feasible for the IP satisfies (55), so our cut does not eliminate any feasible integer points from consideration!

We now show that the current optimal solution to the LP relaxation cannot satisfy the cut (55). The current optimal solution to the LP relaxation has $s_1 = s_2 = 0$. Thus, it cannot satisfy (55). This argument works because 0.75 (the fractional part of the right-hand side of the second constraint) is greater than 0. Thus, if we choose any constraint whose right-hand side in the optimal tableau is fractional, we can cut off the LP relaxation's optimal solution.

The effect of the cut (55) can be seen in Figure 33; all points feasible for the IP (51) satisfy the cut (55), but the current optimal solution to the LP relaxation ($x_1 = 3.75$ and $x_2 = 2.25$) does not. To obtain the graph of the cut, we replaced s_1 by $6 - x_1 - x_2$ and s_2 by $45 - 9x_1 - 5x_2$. This enabled us to rewrite the cut as $3x_1 + 2x_2 \leq 15$.

We now add (55) to the LP relaxation's optimal tableau and use the dual simplex to solve the resulting LP. Cut (55) may be written as $-0.75s_1 - 0.25s_2 \leq -0.75$. After adding a slack variable s_3 to this constraint, we obtain the tableau shown in Table 84.

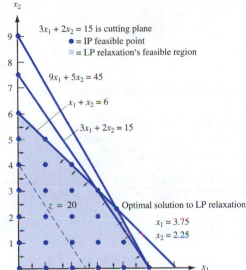

FIGURE 33

Example of Cutting Plane

$3x_1 + 2x_2 = 15$ is cutting plane
● = IP feasible point
■ = LP relaxation's feasible region
$9x_1 + 5x_2 = 45$
$x_1 + x_2 = 6$
$3x_1 + 2x_2 = 15$
$z = 20$
Optimal solution to LP relaxation
$x_1 = 3.75$
$x_2 = 2.25$

TABLE 84

Cutting Plane Tableau After Adding Cut (55)

z	x_1	x_2	s_1	s_2	s_3	rhs
1	0	0	1.25	0.75	0	41.25
0	0	1	2.25	-0.25	0	2.25
0	1	0	-1.25	0.25	0	3.75
0	0	0	-0.75	-0.25	1	-0.75

TABLE **85**
Optimal Tableau for Cutting Plane

z	x_1	x_2	s_1	s_2	s_3	rhs
1	0	0	0	0.33	1.67	40
0	0	1	0	-1	3	0
0	1	0	0	0.67	-1.67	5
0	0	0	1	0.33	-1.33	1

The dual simplex ratio test indicates that s_1 should enter the basis in the third constraint. The resulting tableau is given in Table 85, which yields the optimal solution $z = 40$, $x_1 = 5$, $x_2 = 0$.

Recall that a cut does not eliminate any points that are feasible for the IP. This means that whenever we solve the LP relaxation to an IP with several cuts as additional constraints and find an optimal solution in which all variables are integers, we have solved our original IP. Because x_1 and x_2 are integers in our current optimal solution, this point must be optimal for (51). Of course, if the first cut had not yielded the optimal solution to the IP, we would have kept adding cuts until we obtained an optimal tableau in which all variables were integers.

REMARKS **1** The algorithm requires that all coefficients of variables in the constraints and all right-hand sides of constraints be integers. This is to ensure that if the original decision variables are integers, then the slack and excess variables will also be integers. Thus, a constraint such as $x_1 + 0.5x_2 \leq 3.6$ must be replaced by $10x_1 + 5x_2 \leq 36$.
2 If at any stage of the algorithm, two or more constraints have fractional right-hand sides, then best results are often obtained if the next cut is generated by using the constraint whose right-hand side has the fractional part closest to $\frac{1}{2}$.

Summary of the Cutting Plane Algorithm

Step 1 Find the optimal tableau for the IP's linear programming relaxation. If all variables in the optimal solution assume integer values, then we have found an optimal solution to the IP; otherwise, proceed to step 2.

Step 2 Pick a constraint in the LP relaxation optimal tableau whose right-hand side has the fractional part closest to $\frac{1}{2}$. This constraint will be used to generate a cut.

Step 2a For the constraint identified in step 2, write its right-hand side and each variables's coefficient in the form $[x] + f$, where $0 \leq f < 1$.

Step 2b Rewrite the constraint used to generate the cut as

All terms with integer coefficients = all terms with fractional coefficients

Then the cut is

All terms with fractional coefficients ≤ 0

Step 3 Use the dual simplex to find the optimal solution to the LP relaxation, with the cut as an additional constraint. If all variables assume integer values in the optimal solution, we have found an optimal solution to the IP. Otherwise, pick the constraint with the most fractional right-hand side and use it to generate another cut, which is added to the tableau. We continue this process until we obtain a solution in which all variables are integers. This will be an optimal solution to the IP.

PROBLEMS

Group A

1 Consider the following IP:

$$\max z = 14x_1 + 18x_2$$
$$\text{s.t.} \quad -x_1 + 3x_2 \le 6$$
$$7x_1 + x_2 \le 35$$
$$x_1, x_2 \ge 0; \; x_1, x_2 \text{ integer}$$

The optimal tableau for this IP's linear programming relaxation is given in Table 86. Use the cutting plane algorithm to solve this IP.

2 Consider the following IP:

$$\min z = 6x_1 + 8x_2$$
$$\text{s.t.} \quad 3x_1 + x_2 \ge 4$$
$$x_1 + 2x_2 \ge 4$$
$$x_1, x_2 \ge 0; \; x_1, x_2 \text{ integer}$$

The optimal tableau for this IP's linear programming relaxation is given in Table 87. Use the cutting plane algorithm to find the optimal solution.

3 Consider the following IP:

$$\max z = 2x_1 - 4x_2$$
$$\text{s.t.} \quad 2x_1 + x_2 \le 5$$
$$-4x_1 + 4x_2 \le 5$$
$$x_1, x_2 \ge 0; \; x_1, x_2 \text{ integer}$$

The optimal tableau for this IP's linear programming relaxation is given in Table 88. Use the cutting plane algorithm to find the optimal solution.

TABLE 86

z	x_1	x_2	s_1	s_2	rhs
1	0	0	$\frac{56}{11}$	$\frac{30}{11}$	126
0	0	1	$\frac{7}{22}$	$\frac{1}{22}$	$\frac{7}{2}$
0	1	0	$-\frac{1}{22}$	$\frac{3}{22}$	$\frac{9}{2}$

TABLE 87

z	x_1	x_2	e_1	e_2	rhs
1	0	0	$-\frac{4}{5}$	$-\frac{18}{5}$	$\frac{88}{5}$
0	1	0	$-\frac{2}{5}$	$\frac{1}{5}$	$\frac{4}{5}$
0	0	1	$\frac{1}{5}$	$-\frac{3}{5}$	$\frac{8}{5}$

TABLE 88

z	x_1	x_2	s_1	s_2	rhs
1	0	0	$-\frac{2}{3}$	$-\frac{5}{6}$	$-\frac{15}{2}$
0	1	0	$\frac{1}{3}$	$-\frac{1}{12}$	$\frac{5}{4}$
0	0	1	$\frac{1}{3}$	$\frac{1}{6}$	$\frac{5}{2}$

SUMMARY

Integer programming problems (IP's) are usually much harder to solve than linear programming problems.

Integer Programming Formulations

Most integer programming formulations involve **0–1 variables.**

Fixed-Charge Problems

Suppose activity i incurs a fixed charge if undertaken at any positive level. Let

$$x_i = \text{level of activity } i$$

$$y_i = \begin{cases} 1 & \text{if activity } i \text{ is undertaken at positive level } (x_i > 0) \\ 0 & \text{if } x_i = 0 \end{cases}$$

Then a constraint of the form $x_i \le M_i y_i$ must be added to the formulation. Here, M_i must be large enough to ensure that x_i will be less than or equal to M_i.

Either–Or Constraints

Suppose we want to ensure that at least one of the following two constraints (and possibly both) are satisfied:

$$f(x_1, x_2, \ldots, x_n) \le 0 \tag{26}$$

$$g(x_1, x_2, \ldots, x_n) \le 0 \tag{27}$$

Adding the following two constraints to the formulation will ensure that at least one of (26) and (27) is satisfied:

$$f(x_1, x_2, \ldots, x_n) \le My \tag{26$'$}$$

$$g(x_1, x_2, \ldots, x_n) \le M(1 - y) \tag{27$'$}$$

In (26$'$) and (27$'$), y is a 0–1 variable, and M is a number chosen large enough to ensure that $f(x_1, x_2, \ldots, x_n) \le M$ and $g(x_1, x_2, \ldots, x_n) \le M$ are satisfied for all values of $x_1, x_2,$ $\ldots, x_n$ that satisfy the other constraints in the problem.

If–Then Constraints

Suppose we want to ensure that $f(x_1, x_2, \ldots, x_n) > 0$ implies $g(x_1, x_2, \ldots, x_n) \ge 0$. Then we include the following constraints in the formulation:

$$-g(x_1, x_2, \ldots, x_n) \le My \tag{28$'$}$$

$$f(x_1, x_2, \ldots, x_n) \le M(1 - y) \tag{29}$$

$$y = 0 \text{ or } 1$$

Here, M is a large positive number, chosen large enough so that $f \le M$ and $-g \le M$ hold for all values of $x_1, x_2, \ldots, x_n$ that satisfy the other constraints in the problem.

How to Model a Piecewise Linear Function $f(x)$ with 0–1 Variables

Suppose the piecewise linear function $f(x)$ has break points $b_1, b_2, \ldots, b_n$.

Step 1   Wherever $f(x)$ occurs in the optimization problem, replace $f(x)$ by $z_1 f(b_1) + z_2 f(b_2) + \cdots + z_n f(b_n)$.

Step 2   Add the following constraints to the problem:

$$z_1 \le y_1, z_2 \le y_1 + y_2, z_3 \le y_2 + y_3, \ldots, z_{n-1} \le y_{n-2} + y_{n-1}, z_n \le y_{n-1}$$

$$y_1 + y_2 + \cdots + y_{n-1} = 1$$

$$z_1 + z_2 + \cdots + z_n = 1$$

$$x = z_1 b_1 + z_2 b_2 + \cdots + z_n b_n$$

$$y_i = 0 \text{ or } 1 \ (i = 1, 2, \ldots, n - 1); z_i \ge 0 \ (i = 1, 2, \ldots, n)$$

Branch-and-Bound Method

Usually, IPs are solved by some version of the **branch-and-bound** procedure. Branch-and-bound methods implicitly enumerate all possible solutions to an IP. By solving a single **subproblem,** many possible solutions may be eliminated from consideration.

Branch-and-Bound for Pure IP's

Subproblems are generated by branching on an appropriately chosen fractional-valued variable x_i. Suppose that in a given subproblem (call it old subproblem), x_i assumes a fractional value between the integers i and $i + 1$. Then the two newly generated subproblems are

New Subproblem 1 Old subproblem + Constraint $x_i \leq i$.

New Subproblem 2 Old subproblem + Constraint $x_i \geq i + 1$.

If it is unnecessary to branch on a subproblem, then we say it is **fathomed.** The following three situations (for a max problem) result in a subproblem being fathomed: (1) The subproblem is infeasible, thus it cannot yield the optimal solution to the IP. (2) The subproblem yields an optimal solution in which all variables have integer values. If this optimal solution has a better z-value than any previously obtained solution that is feasible in the IP, then it becomes a **candidate solution,** and its z-value becomes the current lower bound (LB) on the optimal z-value for the IP. In this case, the current subproblem may yield the optimal solution to the IP. (3) The optimal z-value for the subproblem does not exceed (in a max problem) the current LB, so it may be eliminated from consideration.

Branch-and-Bound for Mixed IPs

When branching on a fractional variable, only branch on those required to be integers.

Branch-and-Bound for Knapsack Problems

Subproblems may easily be solved by first putting the best (in terms of benefit per-unit weight) item in the knapsack, then the next best, and so on, until a fraction of an item is used to completely fill the knapsack.

Branch-and-Bound to Minimize Delay on a Single Machine

Begin the branching by determining which job should be processed last. Suppose there are n jobs. At a node where the jth job to be processed, $(j + 1)$th job to be processed, . . . , nth job to be processed are fixed, a lower bound on the total delay is given by (delay of jth job to be processed) + (delay of $(j + 1)$th job to be processed) + $\cdots$ + (delay of nth job to be processed).

Branch-and-Bound for Traveling Salesperson Problem

Subproblems are assignment problems. If the optimal solution to a subproblem contains no subtours, then it is a feasible solution to the traveling salesperson problem. Create new subproblems by branching to exclude a subtour. Eliminate a subproblem if its optimal z-value is inferior to the best previously found feasible solution.

Heuristics for the TSP

To apply the nearest-neighbor heuristic (NNH), we begin at any city and then "visit" the nearest city. Then we go to the unvisited city closest to the city we have most recently visited. We continue in this fashion until a tour is obtained. After applying this procedure beginning at each city, we take the best tour found.

In the cheapest-insertion heuristic (CIH), we begin at any city and find its closest neighbor. Then we create a subtour joining those two cities. Next, we replace an arc in the subtour [say, arc (i, j)] by the combination of two arcs—(i, k) and (k, j), where k is not in the current subtour—that will increase the length of the subtour by the smallest (or cheapest) amount. We continue with this procedure until a tour is obtained. After applying this procedure beginning with each city, we take the best tour found.

Implicit Enumeration

In a 0–1 IP, implicit enumeration may be used to find an optimal solution. When branching at a node, create two new subproblems by (for some free variable x_i) adding constraints $x_i = 0$ and $x_i = 1$. If the best completion of a node is feasible, then we need not branch on the node. If the best completion is feasible and better than the current candidate solution, then the current node yields a new LB (in a max problem) and may be optimal. If the best completion is feasible and is not better than the current candidate solution, then the current node may be eliminated from consideration. If at a given node, there is at least one constraint that is not satisfied by any completion of the node, then the node cannot yield a feasible solution nor an optimal solution to the IP.

Cutting Plane Algorithm

Step 1   Find the optimal tableau for the IP's linear programming relaxation. If all variables in the optimal solution assume integer values, we have found an optimal solution to the IP; otherwise, proceed to step 2.

Step 2   Pick a constraint in the LP relaxation optimal tableau whose right-hand side has the fractional part closest to $\frac{1}{2}$. This constraint will be used to generate a cut.

Step 2a   For the constraint identified in step 2, write its right-hand side and each variable's coefficient in the form $[x] + f$, where $0 \leq f < 1$.

Step 2b   Rewrite the constraint used to generate the cut as

All terms with integer coefficients = all terms with fractional coefficients

Then the cut is

All terms with fractional coefficients ≤ 0

Step 3   Use the dual simplex to find the optimal solution to the LP relaxation, with the cut as an additional constraint. If all variables assume integer values in the optimal solution, then we have found an optimal solution to the IP. Otherwise, pick the constraint with the most fractional right-hand side and use it to generate another cut, which is added to the tableau. We continue this process until we obtain a solution in which all variables are integers. This will be an optimal solution to the IP.

REVIEW PROBLEMS

Group A

1   In the Sailco problem of Section 3.10, suppose that a fixed cost of $200 is incurred during each quarter that production takes place. Formulate an IP to minimize Sailco's total cost of meeting the demands for the four quarters.

2   Explain how you would use integer programming and piecewise linear functions to solve the following optimization problem. (*Hint:* Approximate x^2 and y^2 by piecewise linear functions.)

$$\max z = 3x^2 + y^2$$
$$\text{s.t.} \quad x + y \le 1$$
$$x, y \ge 0$$

3[†] The Transylvania Olympic Gymnastics Team consists of six people. Transylvania must choose three people to enter both the balance beam and floor exercises. They must also enter a total of four people in each event. The score that each individual gymnast can attain in each event is shown in Table 89. Formulate an IP to maximize the total score attained by the Transylvania gymnasts.

4[‡] A court decision has stated that the enrollment of each high school in Metropolis must be at least 20 percent black. The numbers of black and white high school students in each of the city's five school districts are shown in Table 90. The distance (in miles) that a student in each district must travel to each high school is shown in Table 91. School board policy requires that all the students in a given district attend the same school. Assuming that each school must have an enrollment of at least 150 students, formulate an IP that will minimize the total distance that Metropolis students must travel to high school.

5 The Cubs are trying to determine which of the following free agent pitchers should be signed: Rick Sutcliffe (RS), Bruce Sutter (BS), Dennis Eckersley (DE), Steve Trout (ST), Tim Stoddard (TS). The cost of signing each pitcher and the number of victories each pitcher will add to the Cubs are shown in Table 92. Subject to the following restrictions, the Cubs want to sign the pitchers who will add the most victories to the team.

TABLE 89

Gymnast	Balance Beam	Floor Exercise
1	8.8	7.9
2	9.4	8.3
3	9.2	8.5
4	7.5	8.7
5	8.7	8.1
6	9.1	8.6

TABLE 90

District	Whites	Blacks
1	80	30
2	70	5
3	90	10
4	50	40
5	60	30

[†]Based on Ellis and Corn (1984).
[‡]Based on Liggett (1973).

TABLE 91

District	High School 1	High School 2
1	1	2
2	0.5	1.7
3	0.8	0.8
4	1.3	0.4
5	1.5	0.6

TABLE 92

Pitcher	Cost of Signing Pitcher ($) Millions	Victories Added to Cubs
RS	6	6 (righty)
BS	4	5 (righty)
DE	3	3 (righty)
ST	2	3 (lefty)
TS	2	2 (righty)

a At most, $12 million can be spent.
b If DE and ST are signed, then BS cannot be signed.
c At most two right-handed pitchers can be signed.
d The Cubs cannot sign both BS and RS.

Formulate an IP to help the Cubs determine who they should sign.

6 State University must purchase 1,100 computers from three vendors. Vendor 1 charges $500 per computer plus a delivery charge of $5,000. Vendor 2 charges $350 per computer plus a delivery charge of $4,000. Vendor 3 charges $250 per computer plus a delivery charge of $6,000. Vendor 1 will sell the university at most 500 computers; vendor 2, at most 900; and vendor 3, at most 400. Formulate an IP to minimize the cost of purchasing the needed computers.

7 Use the branch-and-bound method to solve the following IP:

$$\max z = 3x_1 + x_2$$
$$\text{s.t.} \quad 5x_1 + x_2 \le 12$$
$$2x_1 + x_2 \le 8$$
$$x_1, x_2 \ge 0; \ x_1, x_2 \text{ integer}$$

8 Use the branch-and-bound method to solve the following IP:

$$\min z = 3x_1 + x_2$$
$$\text{s.t.} \quad 2x_1 - x_2 \le 6$$
$$x_1 + x_2 \le 4$$
$$x_1, x_2 \ge 0; \ x_1 \text{ integer}$$

9 Use the branch-and-bound method to solve the following IP:

$$\max z = x_1 + 2x_2$$
$$\text{s.t.} \quad x_1 + x_2 \le 10$$
$$2x_1 + 5x_2 \le 30$$
$$x_1, x_2 \ge 0; \ x_1, x_2 \text{ integer}$$

10 Consider a country where there are 1¢, 5¢, 10¢, 20¢, 25¢, and 50¢ pieces. You work at the Two-Twelve Convenience Store and must give a customer 91¢ in change. Formulate an IP that can be used to minimize the number of coins needed to give the correct change. Use what you know about knapsack problems to solve the IP by the branch-and-bound method. (*Hint:* We need only solve a 90¢ problem.)

11 Use the branch-and-bound approach to find the optimal solution to the traveling salesperson problem shown in Table 93.

12 Use the implicit enumeration method to find the optimal solution to Problem 5.

13 Use the implicit enumeration method to find the optimal solution to the following 0–1 IP:

$$\max z = 5x_1 - 7x_2 + 10x_3 + 3x_4 - x_5$$
$$\text{s.t.} \quad -x_1 - 3x_2 + 3x_3 - x_4 - 2x_5 \leq 0$$
$$2x_1 - 5x_2 + 3x_3 - 2x_4 - 2x_5 \leq 3$$
$$- x_2 + x_3 + x_4 - x_5 \geq 2$$
$$\text{All variables 0 or 1}$$

14 A soda delivery truck starts at location 1 and must deliver soda to locations 2, 3, 4, and 5 before returning to location 1. The distance between these locations is given in Table 94. The soda truck wants to minimize the total distance traveled. In what order should the delivery truck make its deliveries?

15 At Blair General Hospital, six types of surgical operations are performed. The types of operations each surgeon is qualified to perform (indicated by an X) are given in Table 95. Suppose that surgeon 1 and surgeon 2 dislike each other and cannot be on duty at the same time. Formulate

TABLE 95

Surgeon	Operation 1	2	3	4	5	6
1	X	X		X		
2			X		X	X
3			X		X	
4	X					X
5		X				
6					X	X

an IP whose solution will determine the minimum number of surgeons required so that the hospital can perform all types of surgery.

16 Eastinghouse ships 12,000 capacitors per month to their customers. The capacitors may be produced at three different plants. The production capacity, fixed monthly cost of operation, and variable cost of producing a capacitor at each plant are given in Table 96. The fixed cost for a plant is incurred only if the plant is used to make any capacitors. Formulate an integer programming model whose solution will tell Eastinghouse how to minimize their monthly costs of meeting their customers' demands.

17[†] Newcor's steel mill has received an order for 25 tons of steel. The steel must be 5% carbon and 5% molybdenum by weight. The steel is manufactured by combining three types of metal: steel ingots, scrap steel, and alloys. Four steel ingots are available for purchase. The weight (in tons), cost per ton, carbon and molybdenum content of each ingot are given in Table 97.

Three types of alloys can be purchased. The cost per ton and chemical makeup of each alloy are given in Table 98.

TABLE 93

City	City 1	2	3	4	5
1	—	3	1	7	2
2	3	—	4	4	2
3	1	4	—	4	2
4	7	4	4	—	7
5	2	2	2	7	—

TABLE 96

Plant	Fixed Cost (in $ Thousands)	Variable Cost ($)	Production Capacity
1	80	20	6,000
2	40	25	7,000
3	30	30	6,000

TABLE 94

Location	Location 1	2	3	4	5
1	0	20	4	10	25
2	20	0	5	30	10
3	4	5	0	6	6
4	10	25	6	0	20
5	35	10	6	20	0

TABLE 97

Ingot	Weight	Cost per Ton ($)	Carbon %	Molybdenum %
1	5	350	5	3
2	3	330	4	3
3	4	310	5	4
4	6	280	3	4

[†]Based on Westerberg, Bjorklund, and Hultman (1977).

TABLE 98

Alloy	Cost per Ton ($)	Carbon %	Molybdenum %
1	500	8	6
2	450	7	7
3	400	6	

Steel scrap may be purchased at a cost of $100 per ton. Steel scrap contains 3% carbon and 9% molybdenum. Formulate a mixed integer programming problem whose solution will tell Newcor how to minimize the cost of filling their order.

18[†] Monsanto annually produces 359 million lb of the chemical maleic anhydride. A total of four reactors are available to produce maleic anhydride. Each reactor can be run on one of three settings. The cost (in thousands of dollars) and pounds produced (in millions) annually for each reactor and each setting are given in Table 99. A reactor can only be run on one setting for the entire year. Set up an IP whose solution will tell Monsanto the minimum-cost method to meet its annual demand for maleic anhydride.

19[‡] Hallco runs a day shift and a night shift. No matter how many units are produced, the only production cost during a shift is a setup cost. It costs $8,000 to run the day shift and $4,500 to run the night shift. Demand for the next two days is as follows: day 1, 2,000; night 1, 3,000; day 2, 2,000; night 2, 3,000. It costs $1 per unit to hold a unit in inventory for a shift. Determine a production schedule that minimizes the sum of setup and inventory costs. All demand must be met on time.

20[‡] After listening to a seminar on the virtues of the Japanese theory of production, Hallco has cut its day shift setup cost to $1,000 per shift and its night shift setup cost to $3,500 per shift. Determine a production schedule that minimizes the sum of setup and inventory costs. All demand must be met on time. Show that the decrease in setup costs has actually *raised* the average inventory level!

Group B

21[§] Gotham City has been divided into eight districts. The time (in minutes) it takes an ambulance to travel from one district to another is shown in Table 100. The population of each district (in thousands) is as follows: district 1, 40; district 2, 30; district 3, 35; district 4, 20; district 5, 15; district 6, 50; district 7, 45; district 8, 60. The city has only two ambulances and wants to locate them to maximize the number of people who live within 2 minutes of an ambulance. Formulate an IP to accomplish this goal.

22 A company must complete three jobs. The amounts of processing time (in minutes) required are shown in Table 101. A job cannot be processed on machine j unless for all $i < j$ the job has completed its processing on machine i. Once a job begins its processing on machine j, the job cannot be preempted on machine j. The flow time for a job is the difference between its completion time and the time at which the job begins its first stage of processing. Formulate an IP whose solution can be used to minimize the average flow time of the three jobs. (*Hint:* Two types of constraints will be needed: Constraint type 1 ensures that a job cannot begin to be processed on a machine until all earlier portions of the job are completed. You will need five constraints of this type. Constraint type 2 ensures that only one job will occupy a machine at any given time. For example, on machine 1 either job 1 is completed before job 2 begins, or job 2 is completed before job 1 begins.)

TABLE 99

Reactor	Setting	Cost ($ Thousands)	Pounds
1	1	50	80
1	2	80	140
1	3	100	170
2	1	65	100
2	2	90	140
2	3	120	215
3	1	70	112
3	2	90	153
3	3	110	195
4	1	40	65
4	2	60	105
4	3	70	130

TABLE 100

District	District 1	2	3	4	5	6	7	8
1	10	3	4	6	8	9	8	10
2	3	0	5	4	8	6	12	9
3	4	5	0	2	2	3	5	7
4	6	4	2	0	3	2	5	4
5	8	8	2	3	0	2	2	4
6	9	6	3	2	2	0	3	2
7	8	12	5	5	2	3	0	2
8	10	9	7	4	4	2	2	0

TABLE 101

Job	Machine 1	2	3	4
1	20	—	25	30
2	15	20	—	18
3	—	35	28	—

[†]Based on Boykin (1985).
[‡]Based on Zangwill (1992).

[§]Based on Eaton et al. (1985).

23 Arthur Ross, Inc., must complete many corporate tax returns during the period February 15–April 15. This year the company must begin and complete the five jobs shown in Table 102 during this eight-week period. Arthur Ross employs four full-time accountants who normally work 40 hours per week. If necessary, however, they will work up to 20 hours of overtime per week for which they are paid $100 per hour. Use integer programming to determine how Arthur Ross can minimize the overtime cost incurred in completing all jobs by April 15.

24[†] PSI believes it will need the amounts of generating capacity shown in Table 103 during the next five years. The company has a choice of building (and then operating) power plants with the specifications shown in Table 104. Formulate an IP to minimize the total costs of meeting the generating capacity requirements of the next five years.

25[†] Reconsider Problem 24. Suppose that at the beginning of year 1, power plants 1–4 have been constructed and are in operation. At the beginning of each year, PSI may shut down a plant that is operating or reopen a shut-down plant.

The costs associated with reopening or shutting down a plant are shown in Table 105. Formulate an IP to minimize the total cost of meeting the demands of the next five years. (*Hint:* Let

$X_{it} = 1$ if plant i is operated during year t

$Y_{it} = 1$ if plant i is shut down at end of year t

$Z_{it} = 1$ if plant i is reopened at beginning of year t

You must ensure that if $X_{it} = 1$ and $X_{i,t+1} = 0$, then $Y_{it} = 1$. You must also ensure that if $X_{i,t-1} = 0$ and $X_{it} = 1$, then $Z_{it} = 1$.)

26[‡] Houseco Developers is considering erecting three office buildings. The time required to complete each and the number of workers required to be on the job at all times are shown in Table 106. Once a building is completed, it brings in the following amount of rent per year: building 1, $50,000; building 2, $30,000; building 3, $40,000. Houseco faces the following constraints:

a During each year, 60 workers are available.

b At most, one building can be started during any year.

c Building 2 must be completed by the end of year 4.

Formulate an IP that will maximize the total rent earned by Houseco through the end of year 4.

27 Four trucks are available to deliver milk to five groceries. The capacity and daily operating cost of each truck are shown in Table 107. The demand of each grocery store can be supplied by only one truck, but a truck may deliver to more than one grocery. The daily demands of each grocery are as follows: grocery 1, 100 gallons; grocery 2, 200 gallons; grocery 3, 300 gallons; grocery 4, 500 gallons; grocery 5, 800 gallons. Formulate an IP that can be used to minimize the daily cost of meeting the demands of the four groceries.

TABLE 102

Job	Duration (Weeks)	Accountant Hours Needed per Week
1	3	120
2	4	160
3	3	80
4	2	80
5	4	100

TABLE 103

Year	Generating Capacity (Million kwh)
1	80
2	100
3	120
4	140
5	160

TABLE 104

Plant	Generating Capacity (Million kwh)	Construction Cost ($ Millions)	Annual Operating Cost ($ Millions)
1	70	20	1.5
2	50	16	0.8
3	60	18	1.3
4	40	14	0.6

TABLE 105

Plant	Reopening Cost ($ Million)	Shutdown Cost ($ Millions)
1	1.9	1.7
2	1.5	1.2
3	1.6	1.3
4	1.1	0.8

TABLE 106

Building	Duration of Project (Years)	Number of Workers Required
1	2	30
2	2	20
3	3	20

[†]Based on Muckstadt and Wilson (1968).

[‡]Based on Peiser and Andrus (1983).

TABLE **107**

Truck	Capacity (Gallons)	Daily Operating Cost ($)
1	400	45
2	500	50
3	600	55
4	1,100	60

TABLE **108**

	Auditor Cost ($)			
	Northeast	Midwest	West	South
New York	1,100	1,400	1,900	1,400
Chicago	1,200	1,000	1,500	1,200
Los Angeles	1,900	1,700	1,100	1,400
Atlanta	1,300	1,400	1,500	1,050

TABLE **109**

Project	Required Workers	Revenue ($)
1	1,4,5,8	10,000
2	2,3,7,10	15,000
3	1,6,8,9	6,000
4	2,3,5,10	8,000
5	1,6,7,9	12,000
6	2,4,8,10	9,000

28[†] The State of Texas frequently does tax audits of companies doing business in Texas. These companies often have headquarters located outside the state, so auditors must be sent to out-of-state locations. Each year, auditors must make 500 trips to cities in the Northeast, 400 trips to cities in the Midwest, 300 trips to cities in the West, and 400 trips to cities in the South. Texas is considering basing auditors in Chicago, New York, Atlanta, and Los Angeles. The annual cost of basing auditors in any city is $100,000. The cost of

TABLE **110**

	Worker									
	1	2	3	4	5	6	7	8	9	10
Retainer ($)	800	500	600	700	800	600	400	500	400	500

TABLE **111**

	Project					
	1	2	3	4	5	6
Fee ($)	250	300	250	300	175	180

TABLE **112**

District	Coordinates x	Coordinates y	Tons	Cost ($ Millions) Fixed	Cost ($ Millions) Variable
1	4	3	49	2	310
2	2	5	874	1	40
3	10	8	555	1	51
4	2	8	352	1	341
5	5	3	381	3	131
6	4	5	428	2	182
7	10	5	985	1	20
8	5	1	105	2	40
9	5	8	258	4	177
10	1	7	210	2	75

sending an auditor from any of these cities to a given region of the country is given in Table 108. Formulate an IP whose solution will minimize the annual cost of conducting out-of-state audits.

29 A consulting company has 10 employees, each of whom can work on at most two team projects. Six projects are under consideration. Each project requires 4 of our 10 workers. The required workers and the revenue earned from each project are shown in Table 109.

Each worker who is used on *any project* must be paid the retainer shown in Table 110.

Finally, each worker on a project is paid the project fee shown in Table 111.

How can we maximize our profit?

30 New York City has 10 trash districts and is trying to determine which of the districts should be a site for dumping trash. It costs $1,000 to haul one ton of trash one mile. The location of each district, the number of tons of trash produced per year by the district, the annual fixed cost (in millions of dollars) of running a dumping site, and the variable cost (per ton) of processing a ton of trash at a site are shown in Table 112.

[†]Based on Fitzsimmons and Allen (1983).

TABLE 113

City	Calls Required
San Antonio	2
Phoenix	3
Los Angeles	6
Seattle	3
Detroit	4
Minneapolis	2
Chicago	7
Atlanta	5
New York	9
Boston	5
Philadelphia	4

For example, district 3 is located at coordinates (10,8). District 3 produces 555 tons of trash a year, and it costs $1 million per year in fixed costs to operate a dump site in district 3. Each ton of trash processed at site 3 incurs $51 in variable costs. Each dump site can handle at most 1,500 tons of trash. Each district must send all its trash to a single site. Determine how to locate the dump sites in order to minimize total cost per year.

31 You are the sales manager for Eli Lilly. You want to have sales headquarters located in four of the cities in Table 113. The number of sales calls (in thousands) that must be made in each city are given in Table 113. For example, San Antonio requires 2,000 calls and is 602 miles from Phoenix. The distance between each pair of cities is given below and in file Test1.xls. Where should the headquarters be located to minimize the total distance that must be traveled to make the needed calls?

32 Alcoa produces 100-, 200-, and 300-foot-long aluminum ingots for customers. This week's demand for ingots is shown in Table 115.

Alcoa has 4 furnaces in which ingots can be produced. During a week, each furnace can be operated for 50 hours. Because ingots are produced by cutting long strips of aluminum, longer ingots take less time to produce than shorter ingots. If a furnace is devoted completely to producing one type of ingot; the number it can produce in a week is shown in Table 116.

For example, furnace 1 could produce 350 300-foot ingots per week. The material in an ingot costs $10 per foot. If a customer wants a 100- or 200-foot ingot, then she will accept an ingot of that length or longer. How can Alcoa minimize the material costs incurred in meeting required weekly demands?

33[‡] In treating a brain tumor with radiation, physicians want the maximum amount of radiation possible to bombard

the tissue containing the tumors. The constraint is, however, that there is a maximum amount of radiation that normal tissue can handle without suffering tissue damage. Physicians must therefore decide how to aim the radiation so as to maximize the radiation that hits the tumor tissue subject to the constraint of not damaging the normal tissue. As a simple example of this situation, suppose six types of radiation beams (beams differ in where they are aimed and their intensity) can be aimed at a tumor. The region containing the tumor has been divided into six regions: three regions contain tumors and three contain normal tissue. The amount of radiation delivered to each region by each type of beam is shown in Table 117.

If each region of normal tissue can handle at most 60 units of radiation, then which beams should be used to maximize the total amount of radiation received by the tumors?

34 It is currently the beginning of 2003. Gotham City is trying to sell municipal bonds to support improvements in recreational facilities and highways. The face value and due date at which principal comes due of the bonds are in Table 118.

Gold and Silver (GS) wants to underwrite Gotham City's bonds. A proposal to Gotham for underwriting this issue consists of the following:

- An interest rate (3%, 4%, 5%, 6%, or 7%) for each bond. Coupons are paid annually
- An up-front premium paid by GS to Gotham City

GS has determined the fair prices (in thousands) for possible bonds as shown in Table 119.

For example, if GS underwrites the bond maturing in 2006 at 5%, then it would charge Gotham City $444,000 for that bond. GS is constrained to use at most three different interest rates. GS wants to make a profit of at least $46,000. GS profit is given by

(Sales price of bonds) − (Face value of bonds)

− (Premium)

To maximize the chances that GS will get Gotham City's business, GS wants to minimize the total cost of the bond issue to Gotham City. The total cost of the bond issue to Gotham City is given by

(Total interests on bonds) − (Premium)

For example, if the year 2005 bond is issued at a 4% rate, then Gotham City must pay 2 years of coupon interest or 2*(.04)*($700,000) = $56,000 of interest.

What assignment of interest rates to each bond and up-front premium ensures that GS makes the desired profit (if it gets the contract) and maximizes the chances of GS getting Gotham City's business?

35 When you lease 800-phone numbers from AT&T for telemarketing, AT&T uses a Solver model to tell you where you should locate calling centers to minimize your operating costs over a 10-year horizon. To illustrate the model, suppose you are considering 7 calling center locations: Boston, New York, Charlotte, Dallas, Chicago, L.A., and Omaha. We know the average cost (in dollars) incurred if a telemarketing call is made from any of these cities to any region of the country. We also know the hourly wage that we must pay workers in each city (see Table 120).

We assume that an average call requires 4 minutes. We make calls 250 days per year, and the average number of

[‡]Based on "Radiotherapy Design Using Mathematical Programming Models," by D. Sonderman and P. Abrahamson, *Operations Research*, Vol. 33, No. 4 (1985):705–725.

TABLE 114

	San Antonio	Phoenix	Los Angeles	Seattle	Detroit	Minneapolis	Chicago	Atlanta	New York	Boston	Philadelphia
San Antonio	—	602	1,376	1,780	1,262	1,140	1,060	935	1,848	2,000	1,668
Phoenix	602	—	851	1,193	1,321	1,026	1,127	1,290	2,065	2,201	1,891
Los Angeles	1,376	851	—	971	2,088	1,727	1,914	2,140	2,870	2,995	2,702
Seattle	1,780	1,193	971	—	1,834	1,432	1,734	2,178	2,620	2,707	2,486
Detroit	1,262	1,321	2,088	1,834	—	403	205	655	801	912	654
Minneapolis	1,140	1,026	1,727	1,432	403	—	328	876	1,200	1,304	1,057
Chicago	1,060	1,127	1,914	1,734	205	328	—	564	957	1,082	794
Atlanta	935	1,290	2,140	2,178	655	876	564	—	940	1,096	765
New York	1,848	2,065	2,870	2,620	801	1,200	957	940	—	156	180
Boston	2,000	2,201	2,995	2,707	912	1,304	1,082	1,096	156	—	333
Philadelphia	1,668	1,891	2,702	2,486	654	1,057	794	765	180	333	—

TABLE 115

Ingot (ft)	Demand
100	700
200	300
300	150

TABLE 116

	Ingot Length		
Furnace	100'	200'	300'
1	230	340	350
2	230	260	280
3	240	300	310
4	200	280	300

TABLE 117

Normal			Tumor			
1	2	3	1	2	3	Beam
16	12	8	20	12	6	1
12	10	6	18	15	8	2
9	8	13	13	10	17	3
4	12	12	6	18	16	4
9	4	11	13	5	14	5
8	7	7	10	10	10	6

TABLE 118

Due Date	Principal ($ Thousands)
2005	700
2006	450
2007	250
2008	600
2009	300

TABLE 119

	Amount at Maturity ($ Thousands)				
Interest Rate (%)	2005	2006	2007	2008	2009
3	695	427	233	504	248
4	701	433	235	522	256
5	715	444	247	548	268
6	731	460	255	575	288
7	750	478	269	605	307

calls made per day to each region of the country is shown in Table 121.

The cost of building a calling center in each possible location is in Table 122.

Each calling center can make as many as 5,000 calls per day. Given this information, how can we minimize the discounted cost (at 10% per year) of running the telemarketing operation for 10 years? Assume all wage and calling costs are paid at the end of each year.

36 Cook County needs to build two hospitals. There are nine cities where the hospitals can be built. The number of hospital visits made annually by the inhabitants of each city and the x and y coordinates of each city are as shown in Table 123.

To minimize the total distance patients must travel to hospitals, where should the hospitals be located? (*Hint:* Use Lookup functions to generate the distances between each pair of cities.)

TABLE 120

Cost Call	New England	Middle Atlantic	Southeast	Southwest	Great Lakes	Plains	Rocky Mountains	Pacific	Hourly Wage ($)
Boston	1.2	1.4	1.1	2.6	2	2.2	2.8	2.2	14
New York	1.3	1	1.3	2.2	1.8	1.9	2.5	2.8	16
Charlotte	1.5	1.4	0.9	1.9	2.1	2.3	2.6	3.3	11
Dallas	2	1.8	1.2	1	1.7	2.2	1.8	2.7	12
Chicago	2.1	1.9	2.3	1.5	0.9	1.3	1.2	2.2	13
LA	2.5	2.1	1.9	1.2	1.7	1.5	1.4	1	18
Omaha	2.2	2.1	2	1.3	1.4	0.6	0.9	1.5	10

TABLE 121

Region	Daily Calls
New England	1,000
Middle Atlantic	2,000
Southeast	2,000
Southwest	2,000
Great Lakes	3,000
Plains	1,000
Rocky Mountain	2,000
Pacific	4,000

TABLE 123

City	x	y	Visits
1	0	0	3,000
2	10	3	4,000
3	12	15	5,000
4	14	13	6,000
5	16	9	4,000
6	18	6	3,000
7	8	12	2,000
8	6	10	4,000
9	4	8	1,200

TABLE 122[†]

City	Building Cost ($ Millions)
Boston	2.7
New York	3
Charlotte	2.1
Dallas	2.1
Chicago	2.4
LA	3.6
Omaha	2.1

[†]Based on Spencer, T., Brigandi, A., Dargon D., and Sheehan, M., "AT&T's Telemarketing Site Selection System Offers Customer Support," *Interfaces,* Vol. 20, no. 1, 1990.

REFERENCES

The following eight texts offer a more advanced discussion of integer programming:

Garfinkel, R., and G. Nemhauser. *Integer Programming.* New York: Wiley, 1972.

Nemhauser, G., and L. Wolsey. *Integer and Combinatorial Optimization.* New York: Wiley, 1999.

Parker, G., and R. Rardin. *Discrete Optimization.* San Diego: Academic Press, 1988.

Salkin, H. *Integer Programming.* Reading, Mass.: Addison-Wesley, 1975.

Schrijver, A. *Theory of Linear and Integer Programming.* New York: Wiley, 1998.

Shapiro, J. *Mathematical Programming: Structures and Algorithms.* New York: Wiley, 1979.

Taha, H. *Integer Programming: Theory, Applications, and Computations.* Orlando, Fla.: Academic Press, 1975. Also details branch-and-bound methods for traveling salesperson problem.

Wolsey, L. *Integer Programming*. New York: Wiley, 1998.

The following three texts contain extensive discussion of the art of formulating integer programming problems:

Plane, D., and C. McMillan. *Discrete Optimization: Integer Programming and Network Analysis for Management Decisions.* Englewood Cliffs, N.J.: Prentice Hall, 1971.

Wagner, H. *Principles of Operations Research,* 2d ed. Englewood Cliffs, N.J.: Prentice Hall, 1975. Also details branch-and-bound methods for traveling salesperson problem.

Williams, H. *Model Building in Mathematical Programming,* 4th ed. New York: Wiley, 1999.

Recently, the techniques of Lagrangian Relaxation and Benders' Decomposition have been used to solve many large integer programming problems. Discussion of these techniques is beyond the scope of the text. The reader interested in Lagrangian Relaxation should read Shapiro (1979), Nemhauser and Wolsey (1988), or

Fisher, M. "An Applications-Oriented Guide to Lagrangian Relaxation," *Interfaces* 15(no. 2, 1985):10–21.

Geoffrion, A. "Lagrangian Relaxation for Integer Programming," in *Mathematical Programming Study 2: Approaches to Integer Programming,* ed. M. Balinski. New York: North-Holland, 1974, pp. 82–114.

The reader interested in Benders' Decomposition should read Shapiro (1979), Taha (1975), Nemhauser and Wolsey (1988), or the following reference:

Geoffrion, A., and G. Graves. "Multicommodity Distribution System Design by Benders' Decomposition," *Management Science* 20(1974):822–844.

Baker, K. *Introduction to Sequencing and Scheduling.* New York: Wiley, 1974. Discusses branch-and-bound methods for traveling salesperson and machine-scheduling problems.

Bean, J., C. Noon, and J. Salton. "Asset Divestiture at Homart Development Company," *Interfaces* 17(no. 1, 1987):48–65.

Bean, J., et al. "Selecting Tenants in a Shopping Mall," *Interfaces* 18(no. 2, 1988):1–10.

Boykin, R. "Optimizing Chemical Production at Monsanto," *Interfaces* 15(no. 1, 1985):88–95.

Brown, G., et al. "Real-Time Wide Area Dispatch of Mobil Tank Trucks," *Interfaces* 17(no. 1, 1987):107–120.

Calloway, R., M. Cummins, and J. Freeland, "Solving Spreadsheet-Based Integer Programming Models: An Example from International Telecommunications," *Decision Sciences* 21(1990):808–824.

Cavalieri, F., A. Roversi, and R. Ruggeri. "Use of Mixed Integer Programming to Investigate Optimal Planning Policy for a Thermal Power Station and Extension to Capacity," *Operational Research Quarterly* 22(1971):221–236.

Choypeng, P., P. Puakpong, and R. Rosenthal. "Optimal Ship Routing and Personnel Assignment for Naval Recruitment in Thailand," *Interfaces* 16(no. 4, 1986):47–52.

Day, R. "On Optimal Extracting from a Multiple File Data Storage System: An Application of Integer Programming," *Operations Research* 13(1965):482–494.

Eaton, D., et al. "Determining Emergency Medical Service Vehicle Deployment in Austin, Texas," *Interfaces* 15(1985):96–108.

Efroymson, M., and T. Ray. "A Branch-Bound Algorithm for Plant Location," *Operations Research* 14(1966):361–368.

Ellis, P., and R. Corn, "Using Bivalent Integer Programming to Select Teams for Intercollegiate Women's Gymnastics Competition," *Interfaces* 14(1984):41–46.

Fitzsimmons, J., and L. Allen. "A Warehouse Location Model Helps Texas Comptroller Select Out-of-State Audit Offices," *Interfaces* 13 (no. 5, 1983):40–46.

Garfinkel, R. "Minimizing Wallpaper Waste I: A Class of Traveling Salesperson Problems," *Operations Research* 25(1977):741–751.

Garfinkel, R., and G. Nemhauser. "Optimal Political Districting by Implicit Enumeration Techniques," *Management Science* 16(1970):B495–B508.

Gelb, B., and B. Khumawala. "Reconfiguration of an Insurance Company's Sales Regions," *Interfaces* 14(1984):87–94.

Golden, B., L. Bodin, T. Doyle, and W. Stewart. "Approximate Traveling Salesmen Algorithms," *Operations Research* 28(1980):694–712. Contains an excellent discussion of heuristics for the TSP.

Gomory, R. "Outline of an Algorithm for Integer Solutions to Linear Programs," *Bulletin of the American Mathematical Society* 64(1958):275–278.

Hax, A., and D. Candea. *Production and Inventory Management.* Englewood Cliffs, N.J.: Prentice Hall, 1984. Branch-and-bound methods for machine-scheduling problems.

Lawler, L., et al. *The Traveling Salesman Problem.* New York: Wiley, 1985. Everything you ever wanted to know about this problem.

Liggett, R. "The Application of an Implicit Enumeration Algorithm to the School Desegregation Problem," *Management Science* 20(1973):159–168.

Magirou, V.F. "The Efficient Drilling of Printed Circuit Boards," *Interfaces* 16(no. 4, 1984):13–23.

Muckstadt, J., and R. Wilson. "An Application of Mixed Integer Programming Duality to Scheduling Thermal Generating Systems," *IEEE Transactions on Power Apparatus and Systems* (1968):1968–1978.

Peiser, R., and S. Andrus. "Phasing of Income-Producing Real Estate," *Interfaces* 13(1983):1–11.

Salkin, H., and C. Lin. "Aggregation of Subsidiary Firms for Minimal Unemployment Compensation Payments via Integer Programming," *Management Science* 25(1979):405–408.

Shanker, R., and A. Zoltners. "The Corporate Payments Problem," *Journal of Bank Research* (1972):47–53.

Strong, R. "LP Solves Problem: Eases Duration Matching Process," *Pension and Investment Age* 17(no. 26, 1989):21.

Walker, W. "Using the Set Covering Problem to Assign Fire Companies to Firehouses," *Operations Research* 22(1974):275–277.

Westerberg, C., B. Bjorklund, and E. Hultman. "An Application of Mixed Integer Programming in a Swedish Steel Mill," *Interfaces* 7(no. 2, 1977):39–43.

Zangwill, W. "The Limits of Japanese Production Theory," *Interfaces* 22(no. 5, 1992):14–25.

10

Advanced Topics in Linear Programming[†]

In this chapter, we discuss six advanced linear programming topics: the revised simplex method, the product form of the inverse, column generation, the Dantzig–Wolfe decomposition algorithm, the simplex method for upper-bounded variables, and Karmarkar's method for solving LPs. The techniques discussed are often utilized to solve large linear programming problems. The results of Section 6.2 play a key role throughout this chapter.

10.1 The Revised Simplex Algorithm

In Section 6.2, we demonstrated how to create an optimal tableau from an initial tableau, given an optimal set of basic variables. Actually, the results of Section 6.2 can be used to create a tableau corresponding to *any set of basic variables*. To show how to create a tableau for any set of basic variables BV, we first describe the following notation (assume the LP has m constraints):

BV = any set of basic variables (the first element of BV is the basic variable in the first constraint, the second variable in BV is the basic variable in the second constraint, and so on; thus, BV_j is the basic variable for constraint j in the desired tableau)

$\mathbf{b}$ = right-hand-side vector of the original tableau's constraints

$\mathbf{a}_j$ = column for x_j in the constraints of the original problem

B = $m \times m$ matrix whose jth column is the column for BV_j in the original constraints

c_j = coefficients of x_j in the objective function

$\mathbf{c}_{BV}$ = $1 \times m$ row vector whose jth element is the objective function coefficient for BV_j

$\mathbf{u}_i$ = $m \times 1$ column vector with ith element 1 and all other elements equal to zero

Summarizing the formulas of Section 6.2, we write:

$$B^{-1}\mathbf{a}_j = \text{column for } x_j \text{ in BV tableau} \tag{1}$$

$$\mathbf{c}_{BV}B^{-1}\mathbf{a}_j - c_j = \text{coefficient of } x_j \text{ in row 0} \tag{2}$$

$$B^{-1}\mathbf{b} = \text{right-hand side of constraints in BV tableau} \tag{3}$$

$$\mathbf{c}_{BV}B^{-1}\mathbf{u}_i = \text{coefficient of slack variable } s_i \text{ in BV in row 0} \tag{4}$$

[†]This chapter covers topics that may be omitted with no loss of continuity.

$$\mathbf{c}_{BV}B^{-1}(-\mathbf{u}_i) = \text{coefficient of excess variable } e_i \text{ in BV row 0} \qquad (5)$$

$$M + \mathbf{c}_{BV}B^{-1}\mathbf{u}_i = \text{coefficient of artificial variable } a_i \text{ in BV row 0} \qquad (6)$$
$$\text{(in a max problem)}$$

$$\mathbf{c}_{BV}B^{-1}\mathbf{b} = \text{right-hand side of BV row 0} \qquad (7)$$

If we know BV, B^{-1}, and the original tableau, formulas (1)–(7) enable us to compute any part of the simplex tableau for any set of basic variables BV. This means that if a computer is programmed to perform the simplex algorithm, then all the computer needs to store on any pivot is the current set of basic variables, B^{-1}, and the initial tableau. Then (1)–(7) can be used to generate any portion of the simplex tableau. This idea is the basis of the revised simplex algorithm.

We illustrate the revised simplex algorithm by using it to solve the Dakota problem of Chapter 6. Recall that after adding slack variables s_1, s_2, and s_3, the initial tableau (tableau 0) for the Dakota problem is

$$\max z = 60x_1 + 30x_2 + 20x_3$$
$$\text{s.t.} \quad 8x_1 + 6x_2 + x_3 + s_1 \qquad\qquad = 48$$
$$4x_1 + 2x_2 + 1.5x_3 \qquad + s_2 \qquad = 20$$
$$2x_1 + 1.5x_2 + 0.5x_3 \qquad\qquad + s_3 = 8$$

No matter how many pivots have been completed, B^{-1} for the current tableau will simply be the 3×3 matrix whose jth column is the column for s_j in the current tableau. Thus, for the original tableau BV(0), the set of basic variables is given by

$$\text{BV}(0) = \{s_1, s_2, s_3\}$$
$$\text{NBV}(0) = \{x_1, x_2, x_3\}$$

We let B_i be the columns in the original LP that correspond to the basic variables for tableau i. Then

$$B_0^{-1} = B_0 = \begin{bmatrix} 1 & 0 & 0 \\ 0 & 1 & 0 \\ 0 & 0 & 1 \end{bmatrix}$$

We can now determine which nonbasic variable should enter the basis by computing the coefficient of each nonbasic variable in the current row 0. This procedure is often referred to as **pricing out** the nonbasic variable. From (2)–(5), we see that we can't price out the nonbasic variables until we have determined $\mathbf{c}_{BV}B_0^{-1}$. Because $\mathbf{c}_{BV} = [0 \quad 0 \quad 0]$, we have

$$\mathbf{c}_{BV}B_0^{-1} = [0 \quad 0 \quad 0]\begin{bmatrix} 1 & 0 & 0 \\ 0 & 1 & 0 \\ 0 & 0 & 1 \end{bmatrix} = [0 \quad 0 \quad 0]$$

We now use (2) to price out each nonbasic variable:

$$\bar{c}_1 = [0 \quad 0 \quad 0]\begin{bmatrix} 8 \\ 4 \\ 2 \end{bmatrix} - 60 = -60$$

$$\bar{c}_2 = [0 \quad 0 \quad 0] \begin{bmatrix} 6 \\ 2 \\ 1.5 \end{bmatrix} - 30 = -30$$

$$\bar{c}_3 = [0 \quad 0 \quad 0] \begin{bmatrix} 1 \\ 1.5 \\ 0.5 \end{bmatrix} - 20 = -20$$

Because x_1 has the most negative coefficient in the current row 0, x_1 should enter the basis. To continue the simplex, all we need to know about the new tableau is the new set of basic variables, BV(1), and the corresponding B_1^{-1}. To determine BV(1), we find the row in which x_1 enters the basis. We compute the column for x_1 in the current tableau and the right-hand side of the current tableau.

From(1),

$$\text{Column for } x_1 \text{ in current tableau} = \begin{bmatrix} 1 & 0 & 0 \\ 0 & 1 & 0 \\ 0 & 0 & 1 \end{bmatrix} \begin{bmatrix} 8 \\ 4 \\ 2 \end{bmatrix} = \begin{bmatrix} 8 \\ 4 \\ 2 \end{bmatrix}$$

From (3),

$$\text{Right-hand side of current tableau} = \begin{bmatrix} 1 & 0 & 0 \\ 0 & 1 & 0 \\ 0 & 0 & 1 \end{bmatrix} \begin{bmatrix} 48 \\ 20 \\ 8 \end{bmatrix} = \begin{bmatrix} 48 \\ 20 \\ 8 \end{bmatrix}$$

We now use the ratio test to determine the row in which x_1 should enter the basis. The appropriate ratios are row 1, $\frac{48}{8} = 6$; row 2, $\frac{20}{4} = 5$; and row 3, $\frac{8}{2} = 4$. Thus, x_1 should enter the basis in row 3. This means that our new tableau (tableau 1) will have BV(1) = $\{s_1, s_2, x_1\}$ and NBV(1) = $\{s_3, x_2, x_3\}$.

The new B^{-1} will be the columns of s_1, s_2, and s_3 in the new tableau. To determine the new B^{-1}, look at the column in tableau 0 for the entering variable x_1. From this column, we see that in going from tableau 0 to tableau 1, we must perform the following EROs:

1 Multiply row 3 of tableau 0 by $\frac{1}{2}$.

2 Replace row 1 of tableau 0 by -4(row 3 of tableau 0) + row 1 of tableau 0.

3 Replace row 2 of tableau 0 by -2(row 3 of tableau 0) + row 2 of tableau 0.

Applying these EROs to B_0^{-1} yields

$$B_1^{-1} = \begin{bmatrix} 1 & 0 & -4 \\ 0 & 1 & -2 \\ 0 & 0 & \frac{1}{2} \end{bmatrix}$$

We can now price out all the nonbasic variables for the new tableau. First we compute

$$\mathbf{c}_{BV}B_1^{-1} = [0 \quad 0 \quad 60] \begin{bmatrix} 1 & 0 & -4 \\ 0 & 1 & -2 \\ 0 & 0 & \frac{1}{2} \end{bmatrix} = [0 \quad 0 \quad 30]$$

Then use (2) and (4) to price out tableau 1's nonbasic variables:

$$\bar{c}_2 = [0 \quad 0 \quad 30] \begin{bmatrix} 6 \\ 2 \\ 1.5 \end{bmatrix} - 30 = 15$$

$$\bar{c}_3 = [0 \quad 0 \quad 30] \begin{bmatrix} 1 \\ 1.5 \\ 0.5 \end{bmatrix} - 20 = -5$$

$$\text{Coefficient of } s_3 \text{ in row } 0 = [0 \quad 0 \quad 30] \begin{bmatrix} 0 \\ 0 \\ 1 \end{bmatrix} - 0 = 30$$

Because x_3 is the only variable with a negative coefficient in row 0 of tableau 1, we enter x_3 into the basis. To determine the new set of basic variables, BV(2), and the corresponding B_2^{-1}, we find the row in which x_3 enters the basis and compute

$$x_3 \text{ column in tableau } 1 = B_1^{-1}\mathbf{a}_3 = \begin{bmatrix} 1 & 0 & -4 \\ 0 & 1 & -2 \\ 0 & 0 & 0.5 \end{bmatrix} \begin{bmatrix} 1 \\ 1.5 \\ 0.5 \end{bmatrix} = \begin{bmatrix} -1 \\ 0.5 \\ 0.25 \end{bmatrix}$$

$$\text{Right-hand side of tableau } 1 = B_1^{-1}\mathbf{b} = \begin{bmatrix} 1 & 0 & -4 \\ 0 & 1 & -2 \\ 0 & 0 & 0.5 \end{bmatrix} \begin{bmatrix} 48 \\ 20 \\ 8 \end{bmatrix} = \begin{bmatrix} 16 \\ 4 \\ 4 \end{bmatrix}$$

The appropriate ratios for determining where x_3 should enter the basis are row 1, none; row 2, $\frac{4}{0.5} = 8$; and row 3, $\frac{4}{0.25} = 16$. Hence, x_3 should enter the basis in row 2. Then tableau 2 will have BV(2) = $\{s_1, x_3, x_1\}$ and NBV(2) = $\{s_2, s_3, x_2\}$.

To compute B_2^{-1}, note that to make x_3 a basic variable in row 2, we must perform the following EROs on tableau 1:

1   Replace row 2 of tableau 1 by 2(row 2 of tableau 1).

2   Replace row 1 of tableau 1 by 2(row 2 of tableau 1) + row 1 of tableau 1.

3   Replace row 3 of tableau 1 by $-\frac{1}{2}$(row 2 of tableau 1) + row 3 of tableau 1.

Applying these EROs to B_1^{-1}, we obtain

$$B_2^{-1} = \begin{bmatrix} 1 & 2 & -8 \\ 0 & 2 & -4 \\ 0 & -0.5 & 1.5 \end{bmatrix}$$

We now price out the nonbasic variables in tableau 2. First we compute

$$\mathbf{c}_{BV}B_2^{-1} = [0 \quad 20 \quad 60] \begin{bmatrix} 1 & 2 & -8 \\ 0 & 2 & -4 \\ 0 & -0.5 & 1.5 \end{bmatrix} = [0 \quad 10 \quad 10]$$

Then we price out the nonbasic variables x_2, s_2, and s_3:

$$\bar{c}_2 = [0 \quad 10 \quad 10] \begin{bmatrix} 6 \\ 2 \\ 1.5 \end{bmatrix} - 30 = 5$$

$$\text{Coefficient of } s_2 \text{ in row } 0 = [0 \quad 10 \quad 10] \begin{bmatrix} 0 \\ 1 \\ 0 \end{bmatrix} - 0 = 10$$

$$\text{Coefficient of } s_3 \text{ in row } 0 = [0 \quad 10 \quad 10] \begin{bmatrix} 0 \\ 0 \\ 1 \end{bmatrix} - 0 = 10$$

Each nonbasic variable has a non-negative coefficient in row 0, so tableau 2 is an optimal tableau. To find the optimal solution, we find the right-hand side of tableau 2. From (3), we obtain

$$\text{Right-hand side of tableau } 2 = \begin{bmatrix} 1 & 2 & -8 \\ 0 & 2 & -4 \\ 0 & -0.5 & 1.5 \end{bmatrix} \begin{bmatrix} 48 \\ 20 \\ 8 \end{bmatrix} = \begin{bmatrix} 24 \\ 8 \\ 2 \end{bmatrix}$$

Because $\text{BV}(2) = \{s_1, x_3, x_1\}$, the optimal solution to the Dakota problem is

$$\begin{bmatrix} s_1 \\ x_3 \\ x_1 \end{bmatrix} = \begin{bmatrix} 24 \\ 8 \\ 2 \end{bmatrix}$$

or $s_1 = 24, x_3 = 8, x_1 = 2, x_2 = s_2 = s_3 = 0$. The optimal z-value may be found from (7):

$$\mathbf{c}_{\text{BV}} B_2^{-1} \mathbf{b} = [0 \quad 10 \quad 10] \begin{bmatrix} 48 \\ 20 \\ 8 \end{bmatrix} = 280$$

A summary of the revised simplex method (for a max problem) follows:

Step 0   Note the columns from which the current B^{-1} will be read. Initially, $B^{-1} = I$.

Step 1   For the current tableau, compute $\mathbf{c}_{\text{BV}} B^{-1}$.

Step 2   Price out all nonbasic variables in the current tableau. If each nonbasic variable prices out to be non-negative, then the current basis is optimal. If the current basis is not optimal, then enter into the basis the nonbasic variable with the most negative coefficient in row 0. Call this variable x_k.

Step 3   To determine the row in which x_k enters the basis, compute x_k's column in the current tableau $(B^{-1} \mathbf{a}_k)$ and compute the right-hand side of the current tableau $(B^{-1} \mathbf{b})$. Then use the ratio test to determine the row in which x_k should enter the basis. We now know the set of basic variables (BV) for the new tableau.

Step 4   Use the column for x_k in the current tableau to determine the EROs needed to enter x_k into the basis. Perform these EROs on the current B^{-1}. This will yield the new B^{-1}. Return to step 1.

Most linear programming computer codes use some version of the revised simplex to solve LPs. Knowing the current tableau's B^{-1} and the initial tableau is all that is needed to obtain the next tableau, so the computational effort required to solve an LP by the revised simplex depends primarily on the size of B^{-1}. Suppose the LP being solved has m constraints and n variables. Then each B^{-1} will be an $m \times m$ matrix, and the effort required to solve an LP will depend primarily on the number of constraints (not the number of variables). This fact has important computational implications. For example, if we are solving an LP that has 500 constraints and 10 variables, the LP's dual will have 10 constraints and 500 variables. Then all the B^{-1}'s for the dual will be 10×10 matrices, and all the B^{-1}'s for the primal will be 500×500. Thus, it will be much easier to solve the dual than to solve the primal. In this situation, computation can be greatly reduced by solving the dual and reading the optimal primal solution from the SHADOW PRICE or DUAL VARIABLE section of a computer printout.

PROBLEMS

Group A

Use the revised simplex method to solve the following LPs:

1
$$\max z = 3x_1 + x_2 + x_3$$
$$\text{s.t.} \quad x_1 + x_2 + x_3 \le 6$$
$$2x_1 \quad\quad - x_3 \le 4$$
$$x_2 + x_3 \le 2$$
$$x_1, x_2, x_3 \ge 0$$

2
$$\max z = 4x_1 + x_2$$
$$\text{s.t.} \quad x_1 + x_2 \le 4$$
$$2x_1 + x_2 \ge 6$$
$$3x_2 \ge 6$$
$$x_1, x_2, x_3 \ge 0$$

Remember that B^{-1} is always found under the columns corresponding to the starting basis.)

3
$$\min z = 3x_1 + x_2 - 3x_3$$
$$\text{s.t.} \quad x_1 - x_2 + x_3 \le 4$$
$$x_1 \quad\quad + x_3 \le 6$$
$$2x_2 - x_3 \le 5$$
$$x_1, x_2, x_3 \ge 0$$

10.2 The Product Form of the Inverse

Much of the computation in the revised simplex algorithm is concerned with updating B^{-1} from one tableau to the next. In this section, we develop an efficient method to update B^{-1}.

Suppose we are solving an LP with m constraints. Assume that we have found that x_k should enter the basis, in row r. Let the column for x_k in the current tableau be

Define the $m \times m$ matrix E:

$$
E =
\begin{bmatrix}
1 & 0 & \cdots & -\dfrac{\bar{a}_{1k}}{\bar{a}_{rk}} & \cdots & 0 & 0 \\
0 & 1 & \cdots & -\dfrac{\bar{a}_{2k}}{\bar{a}_{rk}} & \cdots & 0 & 0 \\
\vdots & \vdots & & \vdots & & \vdots & \vdots \\
0 & 0 & \cdots & \dfrac{1}{\bar{a}_{rk}} & \cdots & 0 & 0 \\
\vdots & \vdots & & \vdots & & \vdots & \vdots \\
0 & 0 & \cdots & -\dfrac{\bar{a}_{m-1,k}}{\bar{a}_{rk}} & \cdots & 1 & 0 \\
0 & 0 & \cdots & -\dfrac{\bar{a}_{mk}}{\bar{a}_{rk}} & \cdots & 0 & 1
\end{bmatrix}
\quad \text{(row } r\text{)}
$$

(column r)

In short, E is simply I_m with column r replaced by the column vector

$$\begin{bmatrix} -\dfrac{\bar{a}_{1k}}{\bar{a}_{rk}} \\[2mm] -\dfrac{\bar{a}_{2k}}{\bar{a}_{rk}} \\[2mm] \vdots \\[2mm] \dfrac{1}{\bar{a}_{rk}} \\[2mm] \vdots \\[2mm] -\dfrac{\bar{a}_{m-1,k}}{\bar{a}_{rk}} \\[2mm] -\dfrac{\bar{a}_{mk}}{\bar{a}_{rk}} \end{bmatrix}$$

DEFINITION ■ A matrix (such as E) that differs from the identity matrix in only one column is called an **elementary matrix.** ■

We now show that

$$B^{-1} \text{ for new tableau} = E(B^{-1} \text{ for current tableau}) \tag{8}$$

To see why this is true, note that the EROs used to go from the current tableau to the new tableau boil down to

$$\text{Row } r \text{ of new } B^{-1} = \left(\frac{1}{\bar{a}_{rk}}\right)(\text{row } r \text{ of current } B^{-1}) \tag{9}$$

and for $i \neq r$,

$$\text{Row } i \text{ of new } B^{-1}$$
$$= (\text{row } i \text{ of current } B^{-1}) - \left(\frac{\bar{a}_{ik}}{\bar{a}_{rk}}\right)(\text{row } r \text{ of current } B^{-1}) \tag{10}$$

Recall from Section 2.1 that

$$\text{Row } i \text{ of } E(\text{current } B^{-1}) = (\text{row } i \text{ of } E)(\text{current } B^{-1}) \tag{11}$$

Combining (11) with the definition of E, we find that

$$\text{Row } r \text{ of } E(\text{current } B^{-1}) = \left(\frac{1}{\bar{a}_{rk}}\right)(\text{row } r \text{ of current } B^{-1})$$

and for $i \neq r$,

$$\text{Row } i \text{ of } E(\text{current } B^{-1})$$
$$= (\text{row } i \text{ of current } B^{-1}) - \left(\frac{\bar{a}_{ik}}{\bar{a}_{rk}}\right)(\text{row } r \text{ of current } B^{-1})$$

Hence, (8) does agree with (9) and (10). Thus, we can use (8) to find the new B^{-1} from the current B^{-1}.

Define the initial tableau to be tableau 0, and let E_i be the elementary matrix E associated with the ith simplex tableau. Recall that $B_0^{-1} = I_m$. We now write

$$B_1^{-1} = E_0 B_0^{-1} = E_0$$

Then

$$B_2^{-1} = E_1 B_1^{-1} = E_1 E_0$$

and, in general,

$$B_k^{-1} = E_{k-1} E_{k-2} \cdots E_1 E_0 \qquad \text{(12)}$$

Equation (12) is called the **product form of the inverse.** Most linear programming computer codes utilize the revised simplex method and compute successive B^{-1}'s by using the product form of the inverse.

EXAMPLE 1   **Product Form of the Inverse**

Use the product form of the inverse to compute B_1^{-1} and B_2^{-1} for the Dakota problem that was solved by the revised simplex in Section 10.1.

Solution   Recall that in tableau 0, x_1 entered the basis in row 3. Hence, for tableau 0, $r = 3$ and $k = 1$. For tableau 0,

$$\begin{bmatrix} \bar{a}_{11} \\ \bar{a}_{21} \\ \bar{a}_{31} \end{bmatrix} = \begin{bmatrix} 8 \\ 4 \\ 2 \end{bmatrix}$$

Then

$$E_0 = \begin{bmatrix} 1 & 0 & -\frac{8}{2} \\ 0 & 1 & -\frac{4}{2} \\ 0 & 0 & \frac{1}{2} \end{bmatrix} = \begin{bmatrix} 1 & 0 & -4 \\ 0 & 1 & -2 \\ 0 & 0 & \frac{1}{2} \end{bmatrix}$$

$$B_1^{-1} = \begin{bmatrix} 1 & 0 & -4 \\ 0 & 1 & -2 \\ 0 & 0 & \frac{1}{2} \end{bmatrix} \begin{bmatrix} 1 & 0 & 0 \\ 0 & 1 & 0 \\ 0 & 0 & 1 \end{bmatrix} = \begin{bmatrix} 1 & 0 & -4 \\ 0 & 1 & -2 \\ 0 & 0 & \frac{1}{2} \end{bmatrix}$$

As we proceeded from tableau 1 to tableau 2, x_3 entered the basis in row 2. Hence, in computing E_1, we set $r = 2$ and $k = 3$. To compute E_1, we need to find the column for the entering variable (x_3) in tableau 1:

$$\begin{bmatrix} \bar{a}_{13} \\ \bar{a}_{23} \\ \bar{a}_{33} \end{bmatrix} = B_1^{-1} \mathbf{a}_3 = \begin{bmatrix} 1 & 0 & -4 \\ 0 & 1 & -2 \\ 0 & 0 & 0.5 \end{bmatrix} \begin{bmatrix} 1 \\ 1.5 \\ 0.5 \end{bmatrix} = \begin{bmatrix} -1 \\ 0.5 \\ 0.25 \end{bmatrix}$$

As before, x_3 enters the basis in row 2. Then

$$E_1 = \begin{bmatrix} 1 & -(-\frac{1}{0.5}) & 0 \\ 0 & \frac{1}{0.5} & 0 \\ 0 & -\frac{0.25}{0.50} & 1 \end{bmatrix} = \begin{bmatrix} 1 & 2 & 0 \\ 0 & 2 & 0 \\ 0 & -0.5 & 1 \end{bmatrix}$$

and (as before)

$$B_2^{-1} = E_1 B_1^{-1} = \begin{bmatrix} 1 & 2 & 0 \\ 0 & 2 & 0 \\ 0 & -0.5 & 1 \end{bmatrix} \begin{bmatrix} 1 & 0 & -4 \\ 0 & 1 & -2 \\ 0 & 0 & 0.5 \end{bmatrix} = \begin{bmatrix} 1 & 2 & -8 \\ 0 & 2 & -4 \\ 0 & -0.5 & 1.5 \end{bmatrix}$$

In the next two sections, we use the product form of the inverse in our study of column generation and of the Dantzig–Wolfe decomposition algorithm.

PROBLEM

Group A

For the problems of Section 10.1, use the product form of the inverse to perform the revised simplex method.

10.3 Using Column Generation to Solve Large-Scale LPs

We have already seen that the revised simplex algorithm requires less computation than the simplex algorithm of Chapter 4. In this section, we discuss the method of column generation, devised by Gilmore and Gomory (1961). For LPs that have many variables, column generation can be used to increase the efficiency of the revised simplex algorithm. Column generation is also a very important component of the Dantzig–Wolfe decomposition algorithm, which is discussed in Section 10.4. To explain the idea of column generation, we solve a simple version of the classic *cutting stock problem*.

EXAMPLE 2   **Odds and Evens**

Woodco sells 3-ft, 5-ft, and 9-ft pieces of lumber. Woodco's customers demand 25 3-ft boards, 20 5-ft boards, and 15 9-ft boards. Woodco, who must meet its demands by cutting up 17-ft boards, wants to minimize the waste incurred. Formulate an LP to help Woodco accomplish its goal, and solve the LP by column generation.

Solution   Woodco must decide how each 17-ft board should be cut. Hence, each decision corresponds to a way in which a 17-ft board can be cut. For example, one decision variable would correspond to a board being cut into three 5-ft boards, which would incur waste of $17 - 15 = 2$ ft. Many possible ways of cutting a board need not be considered. For example, it would be foolish to cut a board into one 9-ft and one 5-ft piece; we could just as easily cut the board into a 9-ft piece, a 5-ft piece, *and* a 3-ft piece. In general, any cutting pattern that leaves 3 ft or more of waste need not be considered because we could use the waste to obtain one or more 3-ft boards. Table 1 lists the sensible ways to cut a 17-ft board.

TABLE 1

Ways to Cut a Board in the Cutting Stock Problem

Combination	Number of 3-ft Boards	Number of 5-ft Boards	Number of 9-ft Boards	Waste (Feet)
1	5	0	0	2
2	4	1	0	0
3	2	2	0	1
4	2	0	1	2
5	1	1	1	0
6	0	3	0	2

We now define

$$x_i = \text{number of 17-ft boards cut according to combination } i$$

and formulate Woodco's LP:

$$\text{Woodco's waste} + \text{total customer demand} = \text{total length of board cut}$$

Because

$$\text{Total customer demand} = 25(3) + 20(5) + 15(9) = 310 \text{ ft}$$
$$\text{Total length of boards cut} = 17(x_1 + x_2 + x_3 + x_4 + x_5 + x_6)$$

we write

$$\text{Woodco's waste (in feet)} = 17x_1 + 17x_2 + 17x_3 + 17x_4 + 17x_5 + 17x_6 - 310$$

Then Woodco's objective function is to minimize

$$\min z = 17x_1 + 17x_2 + 17x_3 + 17x_4 + 17x_5 + 17x_6 - 310$$

This is equivalent to minimizing

$$17(x_1 + x_2 + x_3 + x_4 + x_5 + x_6)$$

which is equivalent to minimizing

$$x_1 + x_2 + x_3 + x_4 + x_5 + x_6$$

Hence, Woodco's objective function is

$$\min z = x_1 + x_2 + x_3 + x_4 + x_5 + x_6 \tag{13}$$

This means that Woodco can minimize its total waste by minimizing the number of 17-ft boards that are cut.

Woodco faces the following three constraints:

Constraint 1   At least 25 3-ft boards must be cut.

Constraint 2   A t least 20 5-ft boards must be cut.

Constraint 3   At least 15 9-ft boards must be cut.

Because the total number of 3-ft boards that are cut is given by $5x_1 + 4x_2 + 2x_3 + 2x_4 + x_5$, Constraint 1 becomes

$$5x_1 + 4x_2 + 2x_3 + 2x_4 + x_5 \geq 25 \tag{14}$$

Similarly, Constraint 2 becomes

$$x_2 + 2x_3 + x_5 + 3x_6 \geq 20 \tag{15}$$

and Constraint 3 becomes

$$x_4 + x_5 \geq 15 \tag{16}$$

Note that the coefficient of x_i in the constraint for k-ft boards is just the number of k-ft boards yielded if a board is cut according to combination i.

It is clear that the x_i should be required to assume integer values. Despite this fact, in problems with large demands, a near-optimal solution can be obtained by solving the cutting stock problem as an LP and then rounding all fractional variables upward. This procedure may not yield the best possible integer solution, but it usually yields a near-optimal integer solution. For this reason, we concentrate on the LP version of the cutting stock problem. Combining the sign restrictions with (13)–(16), we obtain the following LP:

$$\min z = x_1 + x_2 + x_3 + x_4 + x_5 + x_6$$

$$\text{s.t.} \quad 5x_1 + 4x_2 + 2x_3 + 2x_4 + x_5 \qquad\qquad \geq 25 \quad \text{(3-ft constraint)}$$

$$x_2 + 2x_3 \qquad + x_5 + 3x_6 \geq 20 \quad \text{(5-ft constraint)} \qquad \text{(17)}$$

$$x_4 + x_5 \qquad\qquad \geq 15 \quad \text{(9-ft constraint)}$$

$$x_1, x_2, x_3, x_4, x_5, x_6 \geq 0$$

Note that x_1 only occurs in the 3-ft constraint (because combination 1 yields only 3-ft boards), and x_6 occurs in the 5-ft constraint (because combination 6 yields only 5-ft boards). This means that x_1 and x_6 can be used as starting basic variables for the 3-ft and 5-ft constraints. Unfortunately, none of combinations 1–6 yields only 9-ft boards, so the 9-ft constraint has no obvious basic variable. To avoid having to add an artificial variable to the 9-ft constraint, we define combination 7 to be the cutting combination that yields only one 9-ft board. Also, define x_7 to be the number of boards cut according to combination 7. Clearly, x_7 will be equal to zero in the optimal solution, but inserting x_7 in the starting basis allows us to avoid using the Big M or the two-phase simplex method. Note that the column for x_7 in the LP constraints will be

$$\begin{bmatrix} 0 \\ 0 \\ 1 \end{bmatrix}$$

and a term x_7 will be added to the objective function. We can now use BV $= \{x_1, x_6, x_7\}$ as a starting basis for LP (17). If we let the tableau for this basis be tableau 0, then we have

$$B_0 = \begin{bmatrix} 5 & 0 & 0 \\ 0 & 3 & 0 \\ 0 & 0 & 1 \end{bmatrix}$$

$$B_0^{-1} = \begin{bmatrix} \frac{1}{5} & 0 & 0 \\ 0 & \frac{1}{3} & 0 \\ 0 & 0 & 1 \end{bmatrix}$$

Then

$$\mathbf{c}_{BV} B_0^{-1} = \begin{bmatrix} 1 & 1 & 1 \end{bmatrix} \begin{bmatrix} \frac{1}{5} & 0 & 0 \\ 0 & \frac{1}{3} & 0 \\ 0 & 0 & 1 \end{bmatrix} = \begin{bmatrix} \frac{1}{5} & \frac{1}{3} & 1 \end{bmatrix}$$

If we now priced out each nonbasic variable it would tell us which variable should enter the basis. However, in a large-scale cutting stock problem there may be thousands of variables, so pricing out each nonbasic variable would be an extremely tedious chore. This is the type of situation in which column generation comes into play. Because we are solving a minimization problem, we want to find a column that will price out positive (have a positive coefficient in row 0). In the cutting stock problem, each column, or variable, represents a combination for cutting up a board: A variable is specified by three numbers: a_3, a_5, and a_9, where a_i is the number of i-ft boards yielded by cutting one 17-ft board according to the given combination. For example, the variable x_2 is specified by $a_3 = 4$, $a_5 = 1$, and $a_9 = 0$. The idea of column generation is to search efficiently for a column that will price out favorably (positive in a min problem and negative in a max problem). For our current basis, a combination specified by a_3, a_5, and a_9 will price out as

$$\mathbf{c}_{BV}B_0^{-1}\begin{bmatrix} a_3 \\ a_5 \\ a_9 \end{bmatrix} - 1 = \frac{1}{5}a_3 + \frac{1}{3}a_5 + a_9 - 1$$

Note that a_3, a_5, and a_9 must be chosen so they don't use more than 17 ft of wood. We also know that a_3, a_5, and a_9 must be non-negative integers. In short, for any combination, a_3, a_5, and a_9 must satisfy

$$3a_3 + 5a_5 + 9a_9 \le 17 \quad (a_3 \ge 0, a_5 \ge 0, a_9 \ge 0; a_3, a_5, a_9 \text{ integer}) \qquad \text{(18)}$$

We can now find the combination that prices out most favorably by solving the following knapsack problem:

$$\begin{aligned} \max z &= \tfrac{1}{5}a_3 + \tfrac{1}{3}a_5 + a_9 - 1 \\ \text{s.t.} \quad 3a_3 &+ 5a_5 + 9a_9 \le 17 \\ a_3, a_5, a_9 &\ge 0; a_3, a_5, a_9 \text{ integer} \end{aligned} \qquad \text{(19)}$$

Because (19) is a knapsack problem (without 0–1 restrictions on the variables), it can easily be solved by using the branch-and-bound procedure outlined in Section 9.5.

The resulting branch-and-bound tree is given in Figure 1. For example, to solve Problem 6 in Figure 1, we first set $a_5 = 1$ (because $a_5 \ge 1$ is necessary). Then we have 12 ft left in the knapsack, and we choose to make a_9 (the best item) as large as possible. Because $a_9 \ge 1$, we set $a_9 = 1$. This leaves 3 ft, so we set $a_3 = 1$ to fill the knapsack. From Figure 1, we find that the optimal solution to LP (19) is $z = \frac{8}{15}$, $a_3 = a_5 = a_9 = 1$. This corresponds to combination 5 and variable x_5. Hence, x_5 prices out $\frac{8}{15}$, and entering x_5 into the basis will decrease Woodco's waste. To enter x_5 into the basis, we create the right-hand side of the current tableau and the x_5 column of the current tableau.

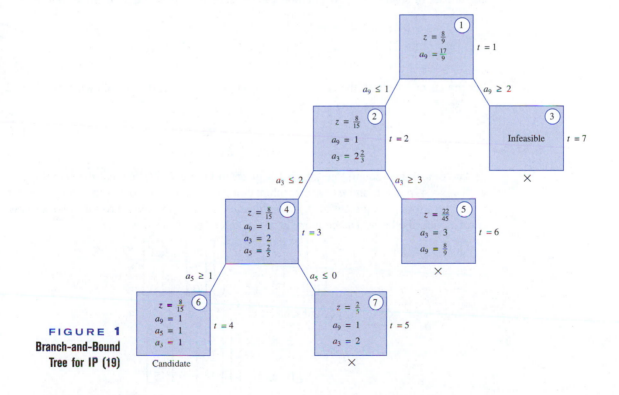

FIGURE 1
Branch-and-Bound Tree for IP (19)

$$x_5 \text{ column in current tableau} = B_0^{-1}\begin{bmatrix} 1 \\ 1 \\ 1 \end{bmatrix} = \begin{bmatrix} \frac{1}{5} & 0 & 0 \\ 0 & \frac{1}{3} & 0 \\ 0 & 0 & 1 \end{bmatrix}\begin{bmatrix} 1 \\ 1 \\ 1 \end{bmatrix} = \begin{bmatrix} \frac{1}{5} \\ \frac{1}{3} \\ 1 \end{bmatrix}$$

$$\text{Right-hand side of current tableau} = B_0^{-1}\mathbf{b} = \begin{bmatrix} \frac{1}{5} & 0 & 0 \\ 0 & \frac{1}{3} & 0 \\ 0 & 0 & 1 \end{bmatrix}\begin{bmatrix} 25 \\ 20 \\ 15 \end{bmatrix} = \begin{bmatrix} 5 \\ \frac{20}{3} \\ 15 \end{bmatrix}$$

The ratio test indicates that x_5 should enter the basis in row 3. This yields BV(1) = $\{x_1, x_6, x_5\}$. Using the product form of the inverse, we obtain

$$B_1^{-1} = E_0 B_0^{-1} = \begin{bmatrix} 1 & 0 & -\frac{1}{5} \\ 0 & 1 & -\frac{1}{3} \\ 0 & 0 & 1 \end{bmatrix}\begin{bmatrix} \frac{1}{5} & 0 & 0 \\ 0 & \frac{1}{3} & 0 \\ 0 & 0 & 1 \end{bmatrix}$$

$$= \begin{bmatrix} \frac{1}{5} & 0 & -\frac{1}{5} \\ 0 & \frac{1}{3} & -\frac{1}{3} \\ 0 & 0 & 1 \end{bmatrix}$$

Now

$$\mathbf{c}_{BV}B_1^{-1} = [1 \quad 1 \quad 1]\begin{bmatrix} \frac{1}{5} & 0 & -\frac{1}{5} \\ 0 & \frac{1}{3} & -\frac{1}{3} \\ 0 & 0 & 1 \end{bmatrix} = [\tfrac{1}{5} \quad \tfrac{1}{3} \quad \tfrac{7}{15}]$$

With our new set of shadow prices ($\mathbf{c}_{BV}B_1^{-1}$), we can again use column generation to determine whether there is any combination that should be entered into the basis. For the current set of shadow prices, a combination specified by a_3, a_5, and a_9 prices out to

$$\begin{bmatrix} \frac{1}{5} & \frac{1}{3} & \frac{7}{15} \end{bmatrix}\begin{bmatrix} a_3 \\ a_5 \\ a_9 \end{bmatrix} - 1 = \frac{1}{5}a_3 + \frac{1}{3}a_5 + \frac{7}{15}a_9 - 1$$

For the current tableau, the column generation procedure yields the following problem:

$$\max z = \tfrac{1}{5}a_3 + \tfrac{1}{3}a_5 + \tfrac{7}{15}a_9 - 1$$
$$\text{s.t.} \quad 3a_3 + 5a_5 + 9a_9 \leq 17 \tag{20}$$
$$a_3, a_5, a_9 \geq 0; \ a_3, a_5, a_9 \text{ integer}$$

The branch-and-bound tree for (20) is given in Figure 2. We see that the combination with $a_3 = 4$, $a_5 = 1$, and $a_9 = 0$ (combination 2) will price out better than any other (it will have a row 0 coefficient of $\frac{2}{15}$). Combination 2 prices out most favorably, so we now enter x_2 into the basis. The column for x_2 in the current tableau is

$$B_1^{-1}\begin{bmatrix} 4 \\ 1 \\ 0 \end{bmatrix} = \begin{bmatrix} \frac{1}{5} & 0 & -\frac{1}{5} \\ 0 & \frac{1}{3} & -\frac{1}{3} \\ 0 & 0 & 1 \end{bmatrix}\begin{bmatrix} 4 \\ 1 \\ 0 \end{bmatrix} = \begin{bmatrix} \frac{4}{5} \\ \frac{1}{3} \\ 0 \end{bmatrix}$$

The right-hand side of the current tableau is

$$B_1^{-1}\mathbf{b} = \begin{bmatrix} \frac{1}{5} & 0 & -\frac{1}{5} \\ 0 & \frac{1}{3} & -\frac{1}{3} \\ 0 & 0 & 1 \end{bmatrix}\begin{bmatrix} 25 \\ 20 \\ 15 \end{bmatrix} = \begin{bmatrix} 2 \\ \frac{5}{3} \\ 15 \end{bmatrix}$$

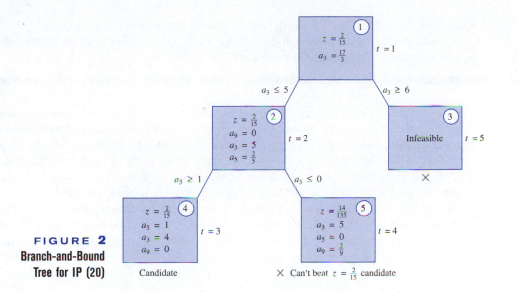

FIGURE 2
Branch-and-Bound
Tree for IP (20)

Candidate   ✕ Can't beat $z = \frac{2}{15}$ candidate

The ratio test indicates the x_2 should enter the basis in row 1. Hence, BV(2) = $\{x_2, x_6, x_5\}$. Using the product form of the inverse, we find that

$$E_1 = \begin{bmatrix} \frac{5}{4} & 0 & 0 \\ -\frac{5}{12} & 1 & 0 \\ 0 & 0 & 1 \end{bmatrix}$$

Then

$$B_2^{-1} = E_1 B_1^{-1} = \begin{bmatrix} \frac{5}{4} & 0 & 0 \\ -\frac{5}{12} & 1 & 0 \\ 0 & 0 & 1 \end{bmatrix} \begin{bmatrix} \frac{1}{5} & 0 & -\frac{1}{5} \\ 0 & \frac{1}{3} & -\frac{1}{3} \\ 0 & 0 & 1 \end{bmatrix} = \begin{bmatrix} \frac{1}{4} & 0 & -\frac{1}{4} \\ -\frac{1}{12} & \frac{1}{3} & -\frac{1}{4} \\ 0 & 0 & 1 \end{bmatrix}$$

The new set of shadow prices is given by

$$\mathbf{c}_{BV} B_2^{-1} = \begin{bmatrix} 1 & 1 & 1 \end{bmatrix} \begin{bmatrix} \frac{1}{4} & 0 & -\frac{1}{4} \\ -\frac{1}{12} & \frac{1}{3} & -\frac{1}{4} \\ 0 & 0 & 1 \end{bmatrix} = \begin{bmatrix} \frac{1}{6} & \frac{1}{3} & \frac{1}{2} \end{bmatrix}$$

For this set of shadow prices, a combination specified by a_3, a_5, and a_9 will price out to $\frac{1}{6}a_3 + \frac{1}{3}a_5 + \frac{1}{2}a_9 - 1$. Thus, the column-generation procedure requires us to solve the following problem:

$$\max z = \tfrac{1}{6}a_3 + \tfrac{1}{3}a_5 + \tfrac{1}{2}a_9 - 1$$
$$\text{s.t.} \quad 3a_3 + 5a_5 + 9a_9 \le 17 \tag{21}$$
$$a_3, a_5, a_9 \ge 0; \; a_3, a_5, a_9 \text{ integer}$$

The branch-and-bound tree for IP (21) is left as an exercise (see Problem 1 at the end of this section). The optimal z-value for (21) is found to be $z = 0$. This means that no combination can price out favorably. Hence, our current basic solution must be an optimal solution. To find the values of the basic variables in the optimal solution, we find the right-hand side of the current tableau:

$$B_2^{-1}\mathbf{b} = \begin{bmatrix} \frac{1}{4} & 0 & -\frac{1}{4} \\ -\frac{1}{12} & \frac{1}{3} & -\frac{1}{4} \\ 0 & 0 & 1 \end{bmatrix} \begin{bmatrix} 25 \\ 20 \\ 15 \end{bmatrix} = \begin{bmatrix} \frac{5}{2} \\ \frac{5}{6} \\ 15 \end{bmatrix}$$

Therefore, the optimal solution to Woodco's cutting stock problem is given by $x_{2'} = \frac{5}{2}$, $x_6 = \frac{5}{6}$, $x_5 = 15$. If desired, we could obtain a "reasonable" integer solution by rounding x_2 and x_6 upward. This yields the integer solution $x_2 = 3$, $x_6 = 1$, $x_5 = 15$.

If we have a starting bfs for a cutting stock problem, we need not list all possible ways in which a board may be cut. At each iteration, a good combination (one that will improve the z-value when entered into the basis) is generated by solving a branch-and-bound problem. The fact that we don't have to list all the ways a board can be cut is very helpful; a cutting stock problem that was solved in Gilmore and Gomory (1961) for which customers demanded boards of 40 different lengths involved more than 100 million possible ways a board could be cut. At the last stage of the column-generation procedure for this problem, solving a single branch-and-bound problem indicated that none of the 100 million (nonbasic) ways would price out favorably. This method is certainly more pleasant than using $\mathbf{c}_{BV}B^{-1}$ to price out all 100 million variables!

PROBLEMS

Group A

1 Show that the optimal solution to IP (21) has $z = 0$.

2 Use column generation to solve a cutting stock problem in which 15-ft boards are cut to satisfy the following requirements: 10 3-ft boards, 20 5-ft boards, and 15 8-ft boards.

3 Use column generation to solve a cutting stock problem in which 15-ft boards are cut to meet the folowing requirements: 80 4-ft boards, 50 6-ft boards, and 100 7-ft boards.

10.4 The Dantzig–Wolfe Decomposition Algorithm

In many LPs, the constraints and variables may be decomposed in the following manner:

Constraints in set 1 only involve variables in Variable set 1.
Constraints in set 2 only involve variables in Variable set 2.

$$\vdots$$

Constraints in set k only involve variables in Variable set k.

Constraints in set $k + 1$ may involve any variable. The constraints in set $k + 1$ are referred to as the **central constraints.** LPs that can be decomposed in this fashion can often be solved efficiently by the Dantzig–Wolfe decomposition algorithm.

EXAMPLE 3 **Decomposition**

Steelco manufactures two types of steel (steel 1 and steel 2) at two locations (plants 1 and 2). Three resources are needed to manufacture a ton of steel: iron, coal, and blast furnace time. The two plants have different types of furnaces, so the resources needed to manufacture a ton of steel depend on the location (see Table 2). Each plant has its own coal mine. Each day, 12 tons of coal are available at plant 1 and 15 tons at plant 2. Coal cannot be shipped between plants. Each day, plant 1 has 10 hours of blast furnace time available, and plant 2 has 4 hours available. Iron ore is mined in a mine located midway between the two plants; 80 tons of iron are available each day. Each ton of steel 1 can be sold for $170/ton, and each ton of steel 2 can be sold for $160/ton. All steel that is sold is shipped to a single customer. It costs $80 to ship a ton of steel from plant 1, and $100

TABLE 2
Resource Requirements for Steelco

Product (1 Ton)	Iron Required (Tons)	Coal Required (Tons)	Blast Furnace Time Requested (Hours)
Steel 1 at plant 1	8	3	2
Steel 2 at plant 1	6	1	1
Steel 1 at plant 2	7	3	1
Steel 2 at plant 2	5	2	1

a ton from plant 2. Assuming that the only variable cost is the shipping cost, formulate and solve an LP to maximize Steelco's revenues less shipping costs.

Solution Define

$$x_1 = \text{tons of steel 1 produced daily at plant 1}$$
$$x_2 = \text{tons of steel 2 produced daily at plant 1}$$
$$x_3 = \text{tons of steel 1 produced daily at plant 2}$$
$$x_4 = \text{tons of steel 2 produced daily at plant 2}$$

Steelco's revenue is given by $170(x_1 + x_3) + 160(x_2 + x_4)$, and Steelco's shipping cost is $80(x_1 + x_2) + 100(x_3 + x_4)$. Therefore, Steelco wants to maximize

$$z = (170 - 80)x_1 + (160 - 80)x_2 + (170 - 100)x_3 + (160 - 100)x_4$$
$$= 90x_1 + 80x_2 + 70x_3 + 60x_4$$

Steelco faces the following five constraints:

Constraint 1   At plant 1, no more than 12 tons of coal can be used daily.

Constraint 2   At plant 1, no more than 10 hours of blast furnace time can be used daily.

Constraint 3   At plant 2, no more than 15 tons of coal can be used daily.

Constraint 4   At plant 2, no more than 4 hours of blast furnace time can be used daily.

Constraint 5   At most, 80 tons of iron ore can be used daily.

Constraints 1–5 lead to the following five LP constraints:

$3x_1 + x_2 \leq 12$	(Plant 1 coal constraint)	(23)
$2x_1 + x_2 \leq 10$	(Plant 1 furnace constraint)	(24)
$3x_3 + 2x_4 \leq 15$	(Plant 2 coal constraint)	(25)
$x_3 + x_4 \leq 4$	(Plant 2 furnace constraint)	(26)
$8x_1 + 6x_2 + 7x_3 + 5x_4 \leq 80$	(Iron ore constraint)	(27)

We also need the sign restrictions $x_i \geq 0$. Putting it all together, we write Steelco's LP as

$$\max z = 90x_1 + 80x_2 + 70x_3 + 60x_4$$

s.t.		
$3x_1 + x_2 \leq 12$	(Plant 1 coal constraint)	(22)
$2x_1 + x_2 \leq 10$	(Plant 1 furnace constraint)	(23)
$3x_3 + 2x_4 \leq 15$	(Plant 2 coal constraint)	(24)
$x_3 + x_4 \leq 4$	(Plant 2 furnace constraint)	(25)
$8x_1 + 6x_2 + 7x_3 + 5x_4 \leq 80$	(Iron ore constraint)	(26)
$x_1, x_2, x_3, x_4 \geq 0$		

Using our definition of decomposition, we may decompose the Steelco LP in the following manner:

Variable set 1 x_1 and x_2 (plant 1 variables).

Variable set 2 x_3 and x_4 (plant 2 variables).

Constraint 1 (22) and (23) (plant 1 constraints).

Constraint 2 (24) and (25) (plant 2 constraints).

Constraint 3 (26).

Constraint set 1 and Variable set 1 involve activities at plant 1 and do not involve x_3 and x_4 (which represent plant 2 activities). Constraint set 2 and Variable set 2 involve activities at plant 2 and do not involve x_1 and x_2 (plant 1 activities). Constraint set 3 may be thought of as a centralized constraint that interrelates the two sets of variables. (Solution to be continued.)

Problems in which several plants manufacture several products can easily be decomposed along the lines of Example 3.

To efficiently solve LPs that decompose along the lines of Example 3, Dantzig and Wolfe developed the Dantzig–Wolfe decomposition algorithm. To simplify our discussion of this algorithm, we assume we are solving an LP in which each subproblem has a bounded feasible region.[†] The decomposition algorithm depends on the results in Theorem 1.

THEOREM 1

Suppose the feasible region for an LP is bounded and the extreme points (or basic feasible solutions) of the LP's feasible region are $P_1, P_2, \ldots, P_k$. Then any point x in the LP's feasible region may be written as a linear combination of $P_1, P_2, \ldots, P_k$. In other words, there exist weights $\mu_1, \mu_2, \ldots, \mu_k$ satisfying

$$x = \mu_1 P_1 + \mu_2 P_2 + \cdots + \mu_k P_k \tag{27}$$

Moreover, the weights $\mu_1, \mu_2, \ldots, \mu_k$ in (27) may be chosen such that

$$\mu_1 + \mu_2 + \cdots + \mu_k = 1 \quad \text{and} \quad \mu_i \geq 0 \quad \text{for } i = 1, 2, \ldots, k \tag{28}$$

Any linear combination of vectors for which the weights satisfy (28) is called a **convex combination.** Thus, Theorem 1 states that if an LP's feasible region is bounded, then any point within may be written as a convex combination of the extreme points of the LP's feasible region.

We illustrate Theorem 1 by showing how it applies to the LPs defined by Constraint set 1 and Constraint set 2 of Example 3. To begin, we look at the feasible region defined by the sign restrictions $x_1 \geq 0$ and $x_2 \geq 0$ and Constraint set 1 (consisting of (22) and (23)). This feasible region is the interior and the boundary of the shaded quadrilateral $P_1P_2P_3P_4$ in Figure 3. The extreme points are $P_1 = [0 \quad 0]$, $P_2 = [4 \quad 0]$, $P_3 = [2 \quad 6]$, and $P_4 = [0 \quad 10]$. For this feasible region, Theorem 1 states that any point

$$\begin{bmatrix} x_1 \\ x_2 \end{bmatrix}$$

[†]See Bradley, Hax, and Magnanti (1977) for a discussion of decomposition that includes the case where at least one subproblem has an unbounded feasible region.

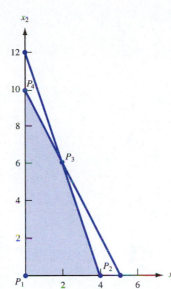

FIGURE 3
Feasible Region for
Constraint Set 1

in the feasible region for Constraint set 1 may be written as

$$\begin{bmatrix} x_1 \\ x_2 \end{bmatrix} = \mu_1 \begin{bmatrix} 0 \\ 0 \end{bmatrix} + \mu_2 \begin{bmatrix} 4 \\ 0 \end{bmatrix} + \mu_3 \begin{bmatrix} 2 \\ 6 \end{bmatrix} + \mu_4 \begin{bmatrix} 0 \\ 10 \end{bmatrix} = \begin{bmatrix} 4\mu_2 + 2\mu_3 \\ 6\mu_3 + 10\mu_4 \end{bmatrix}$$

where $\mu_i \geq 0 (i = 1, 2, 3, 4)$ and $\mu_1 + \mu_2 + \mu_3 + \mu_4 = 1$. For example, the point

$$\begin{bmatrix} 2 \\ 2 \end{bmatrix}$$

is in the feasible region $P_1 P_2 P_3 P_4$. A glance at Figure 3 shows that

$$\begin{bmatrix} 2 \\ 2 \end{bmatrix}$$

may be written as a linear combination of $\mathbf{P}_1, \mathbf{P}_2$, and $\mathbf{P}_3$. A little algebra shows that

$$\begin{bmatrix} 2 \\ 2 \end{bmatrix} = \frac{1}{3} \begin{bmatrix} 0 \\ 0 \end{bmatrix} + \frac{1}{3} \begin{bmatrix} 4 \\ 0 \end{bmatrix} + \frac{1}{3} \begin{bmatrix} 2 \\ 6 \end{bmatrix}$$

As another illustration of Theorem 1, consider the feasible region defined by the sign restrictions $x_3 \geq 0$ and $x_4 \geq 0$ and Constraint set 2 [(24) and (25)]. The feasible region for this LP is the shaded area $Q_1 Q_2 Q_3$ in Figure 4. The extreme points are $\mathbf{Q}_1 = (0, 0)$, $\mathbf{Q}_2 = (4, 0)$, and $\mathbf{Q}_3 = (0, 4)$. Theorem 1 tells us that any point

$$\begin{bmatrix} x_3 \\ x_4 \end{bmatrix}$$

that is in the feasible region for Constraint set 2 may be written as

$$\begin{bmatrix} x_3 \\ x_4 \end{bmatrix} = \mu_1 \begin{bmatrix} 0 \\ 0 \end{bmatrix} + \mu_2 \begin{bmatrix} 4 \\ 0 \end{bmatrix} + \mu_3 \begin{bmatrix} 0 \\ 4 \end{bmatrix}$$

where $\mu_i \geq 0$ and $\mu_1 + \mu_2 + \mu_3 = 1$. For example, the feasible point

$$\begin{bmatrix} 2 \\ 1 \end{bmatrix}$$

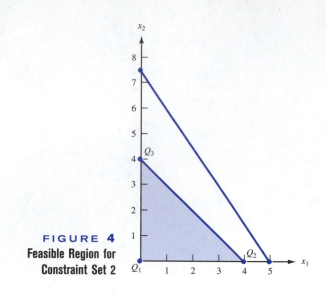

FIGURE 4
Feasible Region for
Constraint Set 2

may be written as

$$\begin{bmatrix} 2 \\ 1 \end{bmatrix} = \frac{1}{4}\begin{bmatrix} 0 \\ 0 \end{bmatrix} + \frac{1}{2}\begin{bmatrix} 4 \\ 0 \end{bmatrix} + \frac{1}{4}\begin{bmatrix} 0 \\ 4 \end{bmatrix}$$

For our purposes, it is not important to know how to determine the set of weights corresponding to a particular feasible point. The decomposition algorithm does not require us to be able to find the weights for an arbitrary point.

To explain the basic ideas of the decomposition algorithm, we assume that the set of variables has been decomposed into set 1 and set 2. The reader should have no trouble generalizing to a situation where the set of variables is decomposed into more than two sets of variables.

The Dantzig–Wolfe decomposition algorithm proceeds as follows:

Step 1   Let the variables in Variable set 1 be $x_1, x_2, \ldots, x_{n_1}$. Express the variables as a convex combination (see Theorem 1) of the extreme points of the feasible region for Constraint set 1 (the constraints that only involve the variables in Variable set 1). If we let $\mathbf{P}_1, \mathbf{P}_2, \ldots, \mathbf{P}_k$ be the extreme points of this feasible region, then any point

$$\begin{bmatrix} x_1 \\ x_2 \\ \vdots \\ x_{n_1} \end{bmatrix}$$

in the feasible region for Constraint set 1 may be written in the form

$$\begin{bmatrix} x_1 \\ x_2 \\ \vdots \\ x_{n_1} \end{bmatrix} = \mu_1\mathbf{P}_1 + \mu_2\mathbf{P}_2 + \cdots + \mu_k\mathbf{P}_k \tag{29}$$

where $\mu_1 + \mu_2 + \cdots + \mu_k = 1$ and $\mu_i \geq 0$ $(i = 1, 2, \ldots, k)$.

Step 2   Express the variables in Variable set 2, $x_{n_1+1}, x_{n_1+2}, \ldots, x_n$, as a convex combination of the extreme points of Constraint set 2's feasible region. If we let the extreme points of the feasible region be $Q_1, Q_2, \ldots, Q_m$, then any point in Constraint set 2's feasible region may be written as

$$\begin{bmatrix} x_{n_1+1} \\ x_{n_1+2} \\ \vdots \\ x_n \end{bmatrix} = \lambda_1 Q_1 + \lambda_2 Q_2 + \cdots + \lambda_m Q_m \tag{30}$$

where $\lambda_i \geq 0$ $(i = 1, 2, \ldots, m)$ and $\lambda_1 + \lambda_2 + \cdots + \lambda_m = 1$.

Step 3 Using (29) and (30), express the LP's objective function and centralized constraints in terms of the μ_i's and the λ_i's. After adding the constraints (called convexity constraints) $\mu_1 + \mu_2 + \cdots + \mu_k = 1$ and $\lambda_1 + \lambda_2 + \cdots + \lambda_m = 1$ and the sign restrictions $\mu_i \geq 0$ $(i = 1, 2, \ldots, k)$ and $\lambda_i \geq 0$ $(i = 1, 2, \ldots, m)$, we obtain the following LP, which is referred to as the **restricted master:**

$$\max \text{ (or min) [objective function in terms of } \mu_i\text{'s and } \lambda_i\text{'s]}$$

s.t. [central constraints in terms of μ_i's and λ_i's]

$$\mu_1 + \mu_2 + \cdots + \mu_k = 1 \qquad \text{(Convexity constraints)}$$
$$\lambda_1 + \lambda_2 + \cdots + \lambda_m = 1$$
$$\mu_i \geq 0 \quad (i = 1, 2, \ldots, k) \qquad \text{(Sign restrictions)}$$
$$\lambda_i \geq 0 \quad (i = 1, 2, \ldots, m)$$

In many large-scale LPs, the restricted master may have millions of variables (corresponding to the many basic feasible solutions of extreme points for each constraint set). Fortunately, however, we rarely have to write down the entire restricted master; all we need is to generate the column in the restricted master that corresponds to a specific μ_i or λ_i.

Step 4 Assume that a basic feasible solution for the restricted master is readily available.[†] Then use the column generation method of Section 10.3 to solve the restricted master.

Step 5 Substitute the optimal values of the μ_i's and λ_i's found in Step 4 into (29) and (30). This will yield the optimal values of $x_1, x_2, \ldots, x_n$.

Solution **Example 3 (Continued)** For Example 3, we have already seen that

$$\text{Variable set } 1 = \{x_1, x_2\} \tag{22}$$

$$\text{Constraint set } 1 = \begin{cases} 3x_1 + x_2 \leq 12 \\ 2x_1 + x_2 \leq 10 \end{cases} \tag{23}$$

We have also seen that the feasible region for Constraint set 1 has four extreme points, and any feasible point

$$\begin{bmatrix} x_1 \\ x_2 \end{bmatrix}$$

for Constraint set 1 may be written as

$$\begin{bmatrix} x_1 \\ x_2 \end{bmatrix} = \mu_1 \begin{bmatrix} 0 \\ 0 \end{bmatrix} + \mu_2 \begin{bmatrix} 4 \\ 0 \end{bmatrix} + \mu_3 \begin{bmatrix} 2 \\ 6 \end{bmatrix} + \mu_4 \begin{bmatrix} 0 \\ 10 \end{bmatrix} = \begin{bmatrix} 4\mu_2 + 2\mu_3 \\ 6\mu_3 + 10\mu_4 \end{bmatrix} \tag{29'}$$

[†]If this is not the case, then the two-phase simplex method must be used. See Bradley, Hax, and Magnanti (1977) for details.

where $\mu_1 + \mu_2 + \mu_3 + \mu_4 = 1$ and $\mu_i \geq 0$.

$$\text{Variable set } 2 = x_3 \text{ and } x_4 \qquad (24)$$

$$\text{Constraint set } 2 = \begin{cases} 3x_3 + 2x_4 \leq 15 \\ x_3 + x_4 \leq 4 \end{cases} \qquad (25)$$

Any point

$$\begin{bmatrix} x_3 \\ x_4 \end{bmatrix}$$

in the feasible region for Constraint set 2 may be written as

$$\begin{bmatrix} x_3 \\ x_4 \end{bmatrix} = \lambda_1 \begin{bmatrix} 0 \\ 0 \end{bmatrix} + \lambda_2 \begin{bmatrix} 4 \\ 0 \end{bmatrix} + \lambda_3 \begin{bmatrix} 0 \\ 4 \end{bmatrix} = \begin{bmatrix} 4\lambda_2 \\ 4\lambda_3 \end{bmatrix} \qquad (30')$$

where $\lambda_1 + \lambda_2 + \lambda_3 = 1$ and $\lambda_i \geq 0$ $(i = 1, 2, 3)$.

We now obtain the restricted master by substituting (29') and (30') into the objective function and the centralized constraint. The objective function for (21) becomes

$$90x_1 + 80x_2 + 70x_3 + 60x_4 = 90(4\mu_2 + 2\mu_3) + 80(6\mu_3 + 10\mu_4) + 70(4\lambda_2) + 60(4\lambda_3)$$
$$= 360\mu_2 + 660\mu_3 + 800\mu_4 + 280\lambda_2 + 240\lambda_3$$

The centralized constraint becomes

$$8(4\mu_2 + 2\mu_3) + 6(6\mu_3 + 10\mu_4) + 7(4\lambda_2) + 5(4\lambda_3) \leq 80$$

or

$$32\mu_2 + 52\mu_3 + 60\mu_4 + 28\lambda_2 + 20\lambda_3 \leq 80$$

After adding a slack variable s_1 to this constraint and writing down the convexity constraints and the sign restrictions, we obtain the following restricted master program:

$$\max z = 360\mu_2 + 660\mu_3 + 800\mu_4 + 280\lambda_2 + 240\lambda_3$$
$$\text{s.t.} \quad 32\mu_2 + 52\mu_3 + 60\mu_4 + 28\lambda_2 + 20\lambda_3 + s_1 = 80$$
$$\mu_1 + \mu_2 + \mu_3 + \mu_4 \qquad\qquad\qquad = 1$$
$$\lambda_1 + \lambda_2 + \lambda_3 \qquad = 1$$
$$\mu_i, \lambda_i \geq 0$$

There is a more insightful way to obtain the column for a variable in the restricted master. Recall that each variable in the restricted master corresponds to an extreme point for the feasible region of Constraint set 1 or Constraint set 2. As an example, let's focus on how to find the column in the restricted master for a variable μ_i, which corresponds to an extreme point

$$\begin{bmatrix} x_1 \\ x_2 \end{bmatrix}$$

for Constraint set 1. Because x_1 and x_2 correspond to activity at plant 1, we may consider any specification of x_1 and x_2 as a "proposal" from plant 1. For example, the point

$$\begin{bmatrix} 2 \\ 6 \end{bmatrix}$$

corresponds to plant 1 proposing to produce 2 tons of type 1 steel and 6 tons of type 2 steel. Then the weight μ_i may be thought of as a fraction of the proposal corresponding to extreme point $\mathbf{P}_i$ that is included in the actual production schedule. For example, because

$$\begin{bmatrix} 2 \\ 2 \end{bmatrix} = \frac{1}{3}\mathbf{P}_1 + \frac{1}{3}\mathbf{P}_2 + \frac{1}{3}\mathbf{P}_3$$

we may think of

$$\begin{bmatrix} 2 \\ 2 \end{bmatrix}$$

as consisting of one-third of plant 1 proposal $\mathbf{P}_1$, one-third of plant 1 proposal $\mathbf{P}_2$, and one-third of plant 1 proposal $\mathbf{P}_3$.

We can now describe an easy method to determine the column for any variable in the restricted master. Suppose we want to determine the column for the extreme point

$$\begin{bmatrix} x_1 \\ x_2 \end{bmatrix}$$

corresponding to the weight μ_i. If we include a fraction μ_i of the extreme point

$$\begin{bmatrix} x_1 \\ x_2 \end{bmatrix}$$

what will this contribute to the objective function? If $\mu_i = 1$, then

$$\mu_i \begin{bmatrix} x_1 \\ x_2 \end{bmatrix}$$

will contribute $90x_1 + 80x_2$ to the objective function. By the Proportionality Assumption, if we use a fraction μ_i of the extreme point

$$\begin{bmatrix} x_1 \\ x_2 \end{bmatrix}$$

then it will contribute $\mu_i(90x_1 + 80x_2)$ to the objective function. Similarly, if $\mu_i = 1$, then

$$\mu_i \begin{bmatrix} x_1 \\ x_2 \end{bmatrix}$$

will contribute $8x_1 + 6x_2$ of iron usage. Thus, for an arbitrary value of μ_i,

$$\mu_i \begin{bmatrix} x_1 \\ x_2 \end{bmatrix}$$

will contribute an amount $\mu_i(8x_1 + 6x_2)$ to the left-hand side of the iron ore usage constraint.

To be more specific, let's use the reasoning we have just described to determine the column in the restricted master for the weight μ_3 corresponding to the extreme point

$$\begin{bmatrix} 2 \\ 6 \end{bmatrix}$$

Our logic shows that the left-hand side of the objective function involving μ_3 is μ_3 $[90(2) + 80(6)] = 660\mu_3$. Similarly, the term involving μ_3 on the left-hand side of the iron ore constraint will be $\mu_3[8(2) + 6(6)] = 52\mu_3$. Also, μ_3 will have a coefficient of 1 in the first convexity constraint and a zero coefficient in the other convexity constraint. (If the reader understood how we obtained the μ_3 column, there should be little trouble with what follows; readers who are confused should reread the last two pages before continuing.)

We now solve the restricted master by using the revised simplex method and column generation. We refer to our initial tableau as tableau 0. Then $BV(0) = \{s_1, \mu_1, \lambda_1\}$. Also,

$$B_0 = \begin{bmatrix} 1 & 0 & 0 \\ 0 & 1 & 0 \\ 0 & 0 & 1 \end{bmatrix}, \quad \text{so} \quad B_0^{-1} = \begin{bmatrix} 1 & 0 & 0 \\ 0 & 1 & 0 \\ 0 & 0 & 1 \end{bmatrix}$$

Because s_1, μ_1, and λ_1 don't appear in the objective function of the restricted master, we have $c_{BV} = [0 \quad 0 \quad 0]$, and the tableau 0 shadow prices are given by

$$c_{BV}B_0^{-1} = [0 \quad 0 \quad 0] \begin{bmatrix} 1 & 0 & 0 \\ 0 & 1 & 0 \\ 0 & 0 & 1 \end{bmatrix} = [0 \quad 0 \quad 0]$$

We now apply the idea of column generation in two stages. First, we determine whether there is any weight μ_i associated with Constraint set 1 that prices out favorably (because we are solving a max problem, a negative coefficient in row 0 is favorable). A weight μ_i associated with an extreme point

$$\begin{bmatrix} x_1 \\ x_2 \end{bmatrix}$$

of Constraint set 1 will have the following column in the restricted master:

$$\text{Objective function coefficient for } \mu_i = 90x_1 + 80x_2$$

$$\text{Column in constraints for } \mu_i = \begin{bmatrix} 8x_1 + 6x_2 \\ 1 \\ 0 \end{bmatrix}$$

From this information, we see that in tableau 0, the column for the weight μ_i corresponding to

$$\begin{bmatrix} x_1 \\ x_2 \end{bmatrix}$$

will price out to

$$c_{BV}B_0^{-1} \begin{bmatrix} 8x_1 + 6x_2 \\ 1 \\ 0 \end{bmatrix} - (90x_1 + 80x_2) = -90x_1 - 80x_2$$

Since

$$\begin{bmatrix} x_1 \\ x_2 \end{bmatrix}$$

must satisfy Constraint set 1 (or the plant 1 constraints), the weight μ_i that prices out most negatively will be the weight associated with the extreme point that is the optimal solution to the following LP:

Tableau 0
Plant 1 Subproblem

$$\min z = -90x_1 - 80x_2$$
$$\text{s.t.} \quad 3x_1 + x_2 \le 12$$
$$2x_1 + x_2 \le 10$$
$$x_1, x_2 \ge 0$$

Solving the plant 1 subproblem graphically, we obtain the solution $z = -800$, $x_1 = 0$, $x_2 = 10$. This means that the weight μ_i associated with the extreme point

$$\begin{bmatrix} 0 \\ 10 \end{bmatrix}$$

will price out most negatively. Recall that

$$\mathbf{P}_4 = \begin{bmatrix} 0 \\ 10 \end{bmatrix}$$

This means that μ_4 will price out with a coefficient of -800 in the restricted master.

We now look at the weights associated with Constraint set 2 and try to determine the weight λ_i that will price out most negatively. The λ_i corresponding to an extreme point

$$\begin{bmatrix} x_3 \\ x_4 \end{bmatrix}$$

of Constraint set 2 will have the following column in the restricted master:

$$\text{Objective function coefficient for } \lambda_i = 70x_3 + 60x_4$$

$$\text{Column in constraints for } \lambda_i = \begin{bmatrix} 7x_3 + 5x_4 \\ 0 \\ 1 \end{bmatrix}$$

This means that the λ_i corresponding to the extreme point

$$\begin{bmatrix} x_3 \\ x_4 \end{bmatrix}$$

will price out to

$$\mathbf{c}_{\text{BV}} B_0^{-1} \begin{bmatrix} 7x_3 + 5x_4 \\ 0 \\ 1 \end{bmatrix} - (70x_3 + 60x_4) = -70x_3 - 60x_4$$

Note that

$$\begin{bmatrix} x_3 \\ x_4 \end{bmatrix}$$

must satisfy Constraint set 2. Thus, the extreme point whose weight λ_i prices out most favorably will be the solution to the following LP:

Tableau 0
Plant 2 Subproblem

$$\min z = -70x_3 - 60x_4$$
$$\text{s.t.} \quad 3x_3 + 2x_4 \leq 15$$
$$\quad\quad x_3 + x_4 \leq 4$$
$$\quad\quad x_3, x_4 \geq 0$$

The optimal solution to this LP is $z = -280$, $x_3 = 4$, $x_4 = 0$. Because

$$\begin{bmatrix} 4 \\ 0 \end{bmatrix} = \mathbf{Q}_2$$

λ_2 prices out the most negatively of all the λ_i's. But μ_4 prices out more negatively than λ_2, so we enter μ_4 into the basis (by using the revised simplex procedure). To do this, we need to find the column for μ_4 in tableau 0 and also find the right-hand side of tableau 0. The column for μ_4 in tableau 0 is

$$B_0^{-1} \begin{bmatrix} 8(0) + 6(10) \\ 1 \\ 0 \end{bmatrix} = \begin{bmatrix} 60 \\ 1 \\ 0 \end{bmatrix}$$

and the right-hand side of tableau 0 is

$$B_0^{-1}\mathbf{b} = \begin{bmatrix} 1 & 0 & 0 \\ 0 & 1 & 0 \\ 0 & 0 & 1 \end{bmatrix} \begin{bmatrix} 80 \\ 1 \\ 1 \end{bmatrix} = \begin{bmatrix} 80 \\ 1 \\ 1 \end{bmatrix}$$

The ratio test now indicates that μ_4 should enter the basis in the second constraint. Then $BV(1) = \{s_1, \mu_4, \lambda_1\}$. Because

$$E_0 = \begin{bmatrix} 1 & -60 & 0 \\ 0 & 1 & 0 \\ 0 & 0 & 1 \end{bmatrix}$$

$$B_1^{-1} = E_0 B_0^{-1} = \begin{bmatrix} 1 & -60 & 0 \\ 0 & 1 & 0 \\ 0 & 0 & 1 \end{bmatrix}$$

The objective function coefficient for μ_4 is $90(0) + 80(10) = 800$, so the new set of shadow prices may be found from

$$\mathbf{c}_{BV}B_1^{-1} = \begin{bmatrix} 0 & 800 & 0 \end{bmatrix} \begin{bmatrix} 1 & -60 & 0 \\ 0 & 1 & 0 \\ 0 & 0 & 1 \end{bmatrix} = \begin{bmatrix} 0 & 800 & 0 \end{bmatrix}$$

We now try to find the weight that prices out most negatively in the current tableau. As before, we solve the current tableau's plant 1 and plant 2 subproblems. Also, as before, a weight μ_i that corresponds to a Constraint 1 extreme point

$$\begin{bmatrix} x_1 \\ x_2 \end{bmatrix}$$

will price out to

$$\mathbf{c}_{BV}B_1^{-1} \begin{bmatrix} 8x_1 + 6x_2 \\ 1 \\ 0 \end{bmatrix} - (90x_1 + 80x_2)$$

$$= \begin{bmatrix} 0 & 800 & 0 \end{bmatrix} \begin{bmatrix} 8x_1 + 6x_2 \\ 1 \\ 0 \end{bmatrix} - (90x_1 + 80x_2) = 800 - 90x_1 - 80x_2$$

Because

$$\begin{bmatrix} x_1 \\ x_2 \end{bmatrix}$$

must satisfy Constraint set 1, the μ_i that prices out most favorably will correspond to the point

$$\begin{bmatrix} x_1 \\ x_2 \end{bmatrix}$$

that solves the following LP:

Tableau 1
Plant 1 Subproblem

$$\min z = 800 - 90x_1 - 80x_2$$
$$\text{s.t.} \quad 3x_1 + x_2 \leq 12$$
$$2x_1 + x_2 \leq 10$$
$$x_1, x_2 \geq 0$$

The optimal solution to this LP is $z = 0$, $x_1 = 0$, $x_2 = 10$. This means that no μ_i can price out favorably. We now solve the plant 2 subproblem in an effort to find a λ_i that prices out favorably. A λ_i corresponding to an extreme point

$$\begin{bmatrix} x_3 \\ x_4 \end{bmatrix}$$

of Constraint set 2 will price out to

$$\mathbf{c}_{BV}B_1^{-1} \begin{bmatrix} 7x_3 + 5x_4 \\ 0 \\ 1 \end{bmatrix} - (70x_3 + 60x_4) = -70x_3 - 60x_4$$

Because

$$\begin{bmatrix} x_3 \\ x_4 \end{bmatrix}$$

must satisfy the plant 2 constraints, the λ_i that will price out most negatively will correspond to the extreme point

$$\begin{bmatrix} x_3 \\ x_4 \end{bmatrix}$$

that solves the plant 2 subproblem for tableau 1:

Tableau 1
Plant 2 Subproblem

$$\min z = -70x_3 - 60x_4$$
$$\text{s.t.} \quad 3x_3 + 2x_4 \leq 15$$
$$x_3 + x_4 \leq 4$$
$$x_3, x_4 \geq 0$$

The optimal solution to this LP is $x_3 = 4$, $x_4 = 0$, $z = -280$. This means that the λ_i corresponding to

$$\begin{bmatrix} 4 \\ 0 \end{bmatrix}$$

prices out to -280. Because

$$\begin{bmatrix} 4 \\ 0 \end{bmatrix} = \mathbf{Q}_2$$

λ_2 prices out to -280. No μ_i has priced out negatively, so the best we can do is to enter λ_2 into the basis. To enter λ_2 into the basis, we need the column for λ_2 in tableau 1 and the right-hand side for tableau 1. The column for λ_2 in tableau 1 is given by

$$B_1^{-1} \begin{bmatrix} 7(4) + 5(0) \\ 0 \\ 1 \end{bmatrix} = \begin{bmatrix} 1 & -60 & 0 \\ 0 & 1 & 0 \\ 0 & 0 & 1 \end{bmatrix} \begin{bmatrix} 28 \\ 0 \\ 1 \end{bmatrix} = \begin{bmatrix} 28 \\ 0 \\ 1 \end{bmatrix}$$

and the right-hand side of tableau 1 is

$$B_1^{-1}\mathbf{b} = \begin{bmatrix} 1 & -60 & 0 \\ 0 & 1 & 0 \\ 0 & 0 & 1 \end{bmatrix} \begin{bmatrix} 80 \\ 1 \\ 1 \end{bmatrix} = \begin{bmatrix} 20 \\ 1 \\ 1 \end{bmatrix}$$

The ratio test indicates that λ_2 should enter the basis in row 1. Thus, BV(2) = $\{\lambda_2, \mu_4, \lambda_1\}$. Because

$$E_1 = \begin{bmatrix} \frac{1}{28} & 0 & 0 \\ 0 & 1 & 0 \\ -\frac{1}{28} & 0 & 1 \end{bmatrix}$$

$$B_2^{-1} = E_1 B_1^{-1} = \begin{bmatrix} \frac{1}{28} & 0 & 0 \\ 0 & 1 & 0 \\ -\frac{1}{28} & 0 & 1 \end{bmatrix} \begin{bmatrix} 1 & -60 & 0 \\ 0 & 1 & 0 \\ 0 & 0 & 1 \end{bmatrix} = \begin{bmatrix} \frac{1}{28} & -\frac{60}{28} & 0 \\ 0 & 1 & 0 \\ -\frac{1}{28} & \frac{60}{28} & 1 \end{bmatrix}$$

To compute $\mathbf{c}_{BV}B_2^{-1}$, note that λ_2 has a coefficient of $70x_3 + 60x_4 = 70(4) + 60(0) = 280$ in the objective function of the restricted master. Recall that μ_4 has an objective coefficient of 800 in the restricted master objective function, and λ_1 has an objective function coefficient of 0 in the restricted master. Then the new set of shadow prices is

$$\mathbf{c}_{BV}B_2^{-1} = \begin{bmatrix} 280 & 800 & 0 \end{bmatrix} \begin{bmatrix} \frac{1}{28} & -\frac{60}{28} & 0 \\ 0 & 1 & 0 \\ -\frac{1}{28} & \frac{60}{28} & 1 \end{bmatrix} = \begin{bmatrix} 10 & 200 & 0 \end{bmatrix}$$

By solving the plant 1 subproblem for tableau 2, we can determine whether any μ_i prices out favorably. The μ_i corresponding to

$$\begin{bmatrix} x_1 \\ x_2 \end{bmatrix}$$

prices out to

$$\mathbf{c}_{BV}B_2^{-1} \begin{bmatrix} 8x_1 + 6x_2 \\ 1 \\ 0 \end{bmatrix} - (90x_1 + 80x_2)$$

$$= \begin{bmatrix} 10 & 200 & 0 \end{bmatrix} \begin{bmatrix} 8x_1 + 6x_2 \\ 1 \\ 0 \end{bmatrix} - (90x_1 + 80x_2) = 200 - 10x_1 - 20x_2$$

Thus, we have

Tableau 2
Plant 1 Subproblem

$$\min z = 200 - 10x_1 - 20x_2$$
$$\text{s.t.} \quad 3x_1 + x_2 \le 12$$
$$2x_1 + x_2 \le 10$$
$$x_1, x_2 \ge 0$$

The optimal solution to this LP is $z = 0$, $x_1 = 0$, $x_2 = 10$. As before, this means that no μ_i can price out favorably.

To determine whether the λ_i corresponding to the extreme point

$$\begin{bmatrix} x_3 \\ x_4 \end{bmatrix}$$

should be entered into the basis, observe that it prices out to

$$\begin{bmatrix} 10 & 200 & 0 \end{bmatrix} \begin{bmatrix} 7x_3 + 5x_4 \\ 0 \\ 1 \end{bmatrix} - (70x_3 + 60x_4) = -10x_4$$

Because

$$\begin{bmatrix} x_3 \\ x_4 \end{bmatrix}$$

must satisfy Constraint set 2, the λ_i that prices out most favorably will be the λ_i associated with the point

$$\begin{bmatrix} x_3 \\ x_4 \end{bmatrix}$$

that solves the following LP:

Tableau 2
Plant 2 Subproblem

$$\min z = -10x_4$$
$$\text{s.t.} \quad 3x_3 + 2x_4 \leq 15$$
$$x_3 + x_4 \leq 4$$
$$x_3, x_4 \geq 0$$

This LP has the solution $z = -40$, $x_3 = 0$, $x_4 = 4$. Thus, the λ_i corresponding to

$$\begin{bmatrix} 0 \\ 4 \end{bmatrix} = Q_3$$

should enter the basis, and λ_3 should be entered into the basis. The λ_3 column in tableau 2 is

$$B_2^{-1} \begin{bmatrix} 7(0) + 5(4) \\ 0 \\ 1 \end{bmatrix} = \begin{bmatrix} \frac{1}{28} & -\frac{60}{28} & 0 \\ 0 & 1 & 0 \\ -\frac{1}{28} & \frac{60}{28} & 1 \end{bmatrix} \begin{bmatrix} 20 \\ 0 \\ 1 \end{bmatrix} = \begin{bmatrix} \frac{20}{28} \\ 0 \\ \frac{8}{28} \end{bmatrix}$$

Tableau 2's right-hand side is

$$B_2^{-1} b = \begin{bmatrix} \frac{1}{28} & -\frac{60}{28} & 0 \\ 0 & 1 & 0 \\ -\frac{1}{28} & \frac{60}{28} & 1 \end{bmatrix} \begin{bmatrix} 80 \\ 1 \\ 1 \end{bmatrix} = \begin{bmatrix} \frac{20}{28} \\ 1 \\ \frac{8}{28} \end{bmatrix}$$

The ratio test indicates that λ_3 should enter the basis in Constraint 1 or Constraint 3; we arbitrarily choose Constraint 1. Thus, $BV(3) = \{\lambda_3, \mu_4, \lambda_1\}$. Because

$$E_2 = \begin{bmatrix} \frac{28}{20} & 0 & 0 \\ 0 & 1 & 0 \\ -\frac{2}{5} & 0 & 1 \end{bmatrix}$$

$$B_3^{-1} = E_2 B_2^{-1} = \begin{bmatrix} \frac{28}{20} & 0 & 0 \\ 0 & 1 & 0 \\ -\frac{2}{5} & 0 & 1 \end{bmatrix} \begin{bmatrix} \frac{1}{28} & -\frac{60}{28} & 0 \\ 0 & 1 & 0 \\ -\frac{1}{28} & \frac{60}{28} & 1 \end{bmatrix} = \begin{bmatrix} \frac{1}{20} & -3 & 0 \\ 0 & 1 & 0 \\ -\frac{1}{20} & 3 & 1 \end{bmatrix}$$

λ_3 corresponds to

$$\begin{bmatrix} 0 \\ 4 \end{bmatrix}$$

so the coefficient of λ_3 in the objective function of the restricted master is $70x_3 + 60x_4 = 70(0) + 60(4) = 240$. The μ_4 and λ_1 coefficients in the objective function have already been found to be 800 and 0, respectively, so we have $\mathbf{c}_{BV} = [240 \quad 800 \quad 0]$, and the new set of shadow prices is given by

$$\mathbf{c}_{BV}B_3^{-1} = [240 \quad 800 \quad 0] \begin{bmatrix} \frac{1}{20} & -3 & 0 \\ 0 & 1 & 0 \\ -\frac{1}{20} & 3 & 1 \end{bmatrix} = [12 \quad 80 \quad 0]$$

With these shadow prices, the μ_i corresponding to the extreme point

$$\begin{bmatrix} x_1 \\ x_2 \end{bmatrix}$$

will price out to

$$[12 \quad 80 \quad 0] \begin{bmatrix} 8x_1 + 6x_2 \\ 1 \\ 0 \end{bmatrix} - (90x_1 + 80x_2) = 80 + 6x_1 - 8x_2$$

Then we have

Tableau 3
Plant 1 Subproblem

$$\min z = 80 + 6x_1 - 8x_2$$

s.t. $\quad 3x_1 + x_2 \le 12$

$\quad\quad\ 2x_1 + x_2 \le 10$

$\quad\quad\quad\quad x_1, x_2 \ge 0$

The optimal solution to this LP is $z = 0$, $x_1 = 0$, $x_2 = 10$. Again, this means that no μ_i prices out favorably.

Using the new shadow prices, we now determine whether any λ_i will price out favorably. If no λ_i prices out favorably, then we will have found an optimal tableau. The λ_i corresponding to

$$\begin{bmatrix} x_3 \\ x_4 \end{bmatrix}$$

will price out to

$$[12 \quad 80 \quad 0] \begin{bmatrix} 7x_3 + 5x_4 \\ 0 \\ 1 \end{bmatrix} - (70x_3 + 60x_4) = 14x_3$$

Then we have

Tableau 3
Plant 2 Subproblem

$$\min z = 14x_3$$

s.t. $\quad 3x_3 + 2x_4 \le 15$

$\quad\quad\ x_3 + \ x_4 \le 4$

$\quad\quad\quad\ x_3, x_4 \ge 0$

The optimal solution to this LP is $z = 0$, $x_3 = x_4 = 0$. This means that no λ_i can price out favorably. Because no μ_i or λ_i prices out favorably for tableau 3, tableau 3 must be an optimal tableau for the restricted master. Recall that BV(3) = $\{\lambda_3, \mu_4, \lambda_1\}$. Thus,

$$\begin{bmatrix} \lambda_3 \\ \mu_4 \\ \lambda_1 \end{bmatrix} = B_3^{-1} = \begin{bmatrix} \frac{1}{20} & -3 & 0 \\ 0 & 1 & 0 \\ -\frac{1}{20} & 3 & 1 \end{bmatrix} \begin{bmatrix} 80 \\ 1 \\ 1 \end{bmatrix} = \begin{bmatrix} 1 \\ 1 \\ 0 \end{bmatrix}$$

Thus the optimal solution to the restricted master is $\lambda_3 = 1$, $\mu_4 = 1$, $\lambda_1 = 0$, and all other weights equal 0.

We can now use the representation of the Constraint set 1 feasible region as a convex combination of its extreme points to determine that the optimal value of

$$\begin{bmatrix} x_1 \\ x_2 \end{bmatrix}$$

is given by

$$\begin{bmatrix} x_1 \\ x_2 \end{bmatrix} = 0\mathbf{P}_1 + 0\mathbf{P}_2 + 0\mathbf{P}_3 + \mathbf{P}_4 = \begin{bmatrix} 0 \\ 10 \end{bmatrix}$$

Similarly, we can use the representation of the Constraint set 2 feasible region as a convex combination of its extreme points to determine that the optimal value of

$$\begin{bmatrix} x_3 \\ x_4 \end{bmatrix}$$

is given by

$$\begin{bmatrix} x_3 \\ x_4 \end{bmatrix} = 0\mathbf{Q}_1 + 0\mathbf{Q}_2 + \mathbf{Q}_3 = \begin{bmatrix} 0 \\ 4 \end{bmatrix}$$

Then the optimal solution to Steelco's problem is $x_2 = 10$, $x_4 = 4$, $x_1 = x_3 = 0$, $z = 1040$. Thus, Steelco can maximize its net profit by manufacturing 10 tons of steel 2 at plant 1 and 4 tons of steel 2 at plant 2.

REMARKS

1 If there are k sets of variables, then the restricted master will contain the central constraints and k convexity constraints (one convexity constraint for each set of variables). For each tableau, there will also be k subproblems that must be solved (one for the weights associated with the extreme points of the constraint set corresponding to each set of variables). After solving these subproblems, use the revised simplex algorithm to enter into the basis the weight that prices out most favorably.

2 A major virtue of decomposition is that solving several relatively small LPs is often much easier than solving one large LP. For example, consider an analog of Example 3 in which there are five plants and each plant has 50 constraints. Also suppose that there are 40 central constraints. Then the master problem will involve a $45 \times 45\ B^{-1}$, and each subproblem will involve a $50 \times 50\ B^{-1}$. Solving the original LP would involve a $290 \times 290\ B^{-1}$. Clearly, storing a 290×290 matrix requires more computer memory than storing five 50×50 matrices and a 45×45 matrix. This illustrates how decomposition greatly reduces storage requirements.

3 Decomposition has an interesting economic interpretation. What is the meaning of the shadow prices for the restricted master of Example 3? For each tableau, the shadow price for the central constraint (reflecting the limited amount of iron ore) is the amount by which an extra unit of iron would increase profits. It can be shown that for any tableau, the shadow price for the plant i ($i = 1, 2$) convexity constraint is the profit obtained from the current mix of extreme points being used at plant i less the value of the centralized resource (calculated via the centralized shadow price) required by the current mix of extreme points that is being used at plant i. For example, in tableau 3, the shadow price for the plant 1 convexity constraint is 80. Currently, plant 1 is utilizing the mix $x_1 = 0$ and $x_2 = 10$. This mix yields a profit of $80(10) = \$800$, and it uses $6(10) = 60$ tons of iron worth $60(12) = \$720$. Thus, the plant 1 convexity constraint has a shadow price of $800 - 720 = \$80$. This means that if Δ of the plant 1 weight were taken away, profits would be reduced by 80Δ.

We can now give an economic interpretation of the pricing-out procedure that we use to generate our subproblems. If we are at tableau 3, what are the benefits and costs if we try to introduce the μ_i associated with the extreme point

$$\begin{bmatrix} x_1 \\ x_2 \end{bmatrix}$$

into the basis? Recall that for tableau 3, the iron shadow price is 12 and the plant 1 convexity constraint has a shadow price of 80. In determining whether μ_i should enter the basis, we must balance

$$\text{Increased profits for } \mu_i = \text{profits earned by } \mu_i \begin{bmatrix} x_1 \\ x_2 \end{bmatrix}$$

$$= 90(\mu_i x_1) + 80(\mu_i x_2)$$

against the costs incurred if μ_i is entered into the basis.

If we enter μ_i into the basis, we incur two costs: first, \$12 for each ton of iron used. This amounts to a cost of $12[8(\mu_i x_1) + 6(\mu_i x_2)]$. By entering μ_i into the basis, we are also diverting a fraction μ_i of the available plant 1 weights away from the current mix. This incurs an opportunity cost of $80\mu_i$. Hence,

$$\text{Increase in cost from entering } \mu_i \text{ into basis} = 96\mu_i x_1 + 72\mu_i x_2 + 80\mu_i$$

This means that entering μ_i into the basis can increase profits if and only if

$$90\mu_i x_1 + 80\mu_i x_2 > 96\mu_i x_1 + 72\mu_i x_2 + 80\mu_i$$

Canceling the μ_i's from both sides, we see that μ_i will price out favorably if

$$90x_1 + 80x_2 > 96x_1 + 72x_2 + 80 \qquad \text{or} \qquad 0 > 80 + 6x_1 - 8x_2$$

Thus, the best μ_i will be the μ_i associated with the extreme point

$$\begin{bmatrix} x_1 \\ x_2 \end{bmatrix}$$

that minimizes $80 + 6x_1 - 8x_2$. This is indeed the objective function for the plant 1 tableau 3 subproblem.

This discussion shows that the Dantzig–Wolfe decomposition algorithm combines centralized information (from the shadow prices of the centralized constraints) with local information (the shadow price of each plant's convexity constraint) in an effort to determine which weights should be entered into the basis (or equivalently, which extreme points from each plant should be used).

PROBLEMS

Group A

Use the Dantzig–Wolfe decomposition algorithm to solve the following problems:

1

$$\max z = 7x_1 + 5x_2 + 3x_3$$
$$\text{s.t.} \quad x_1 + 2x_2 + x_3 \leq 10$$
$$x_2 + x_3 \leq 5$$
$$x_1 \qquad\qquad \leq 3$$
$$2x_2 + x_3 \leq 8$$
$$x_1, x_2, x_3 \geq 0$$

2

$$\max z = 4x_1 + 2x_2 + 3x_3 + 4x_4 + 2x_5$$
$$\text{s.t.} \quad x_1 + 2x_2 + x_3 \qquad\qquad \leq 8$$
$$x_1 + 2x_2 + 2x_3 \qquad\qquad \leq 8$$
$$x_4 + x_5 \leq 3$$
$$x_1, x_2, x_3, x_4, x_5 \geq 0$$

3

$$\max z = 3x_1 + 6x_2 + 5x_3$$
$$\text{s.t.} \quad x_1 + 2x_2 + x_3 \leq 4$$
$$2x_1 + 3x_2 + 2x_3 \leq 6$$
$$x_1 + x_2 \qquad \leq 2$$
$$2x_1 + x_2 \qquad \leq 3$$
$$x_1, x_2, x_3 \geq 0$$

(*Hint:* There is no law against having only one set of variables and one subproblem.)

4 Give an economic interpretation to explain why λ_3 priced out favorably in the plant 2 tableau 2 subproblem.

5 Give an example to show why Theorem 1 does not hold for an LP with an unbounded feasible region.

10.5 The Simplex Method for Upper-Bounded Variables

Often, LPs contain many constraints of the form $x_i \leq u_i$ (where u_i is a constant). For example, in a production-scheduling problem, there may be many constraints of the type $x_i \leq u_i$, where

$$x_i = \text{period } i \text{ production}$$

$$u_i = \text{period } i \text{ production capacity}$$

Because a constraint of the form $x_i \leq u_i$ provides an upper bound on x_i, it is called an **upper-bound constraint.** Because $x_i \leq u_i$ is a legal LP constraint, we can clearly use the ordinary simplex method to solve an LP that has upper-bound constraints. However, if an LP contains several upper-bound constraints, then the procedure described in this section (called the simplex method for upper-bounded variables) is much more efficient than the ordinary simplex algorithm.

To efficiently solve an LP with upper-bound constraints, we allow the variable x_i to be nonbasic if $x_i = 0$ (the usual criterion for a nonbasic variable) or if $x_i = u_i$. To accomplish this, we use the following gimmick: For each variable x_i that has an upper-bound constraint $x_i \leq u_i$, we define a new variable x_i' by the relationship $x_i + x_i' = u_i$, or $x_i = u_i - x_i'$. Note that if $x_i = 0$, then $x_i' = u_i$, whereas if $x_i = u_i$, then $x_i' = 0$. Whenever we want x_i to equal its upper bound of u_i, we simply replace x_i by $u_i - x_i'$. This is called an **upper-bound substitution.**

We are now ready to describe the simplex method for upper-bounded variables. We assume that a basic solution is available and that we are solving a max problem. As usual, at each iteration, we choose to increase the variable x_i that has the most negative coefficient in row 0. Three possible occurrences, or bottlenecks, can restrict the amount by which we increase x_i:

Bottleneck 1 x_i cannot exceed its upper bound of u_i.

Bottleneck 2 x_i increases to a point where it causes one of the current basic variables to become negative. The smallest value of x_i that will cause one of the current basic variables to become negative may be found by expressing each basic variable in terms of x_i (recall that we used this idea in Chapter 4, in discussing the simplex algorithm).

Bottleneck 3 x_i increases to a point where it causes one of the current basic variables to exceed its upper bound. As in bottleneck 2, the smallest value of x_i for which this bottleneck occurs can be found by expressing each basic variable in terms of x_i.

Let BN_k ($k = 1, 2, 3$) be the value of x_i where bottleneck k occurs. Then x_i can be increased only to a value of $\min\{BN_1, BN_2, BN_3\}$. The smallest of BN_1, BN_2, and BN_3 is called the winning bottleneck. If the winning bottleneck is BN_1, then we make an upper-bound substitution on x_i by replacing x_i by $u_i - x_i'$. If the winning bottleneck is BN_2, then we enter x_i into the basis in the row corresponding to the basic variable that caused BN_2 to occur. If the winning bottleneck is BN_3, then we make an upper-bound substitution of the variable x_j (by replacing x_j by $u_j - x_j'$) that reaches its upper bound when $x_i = BN_3$. Then we enter x_i into the basis in the row for which x_j was a basic variable.

After following this procedure, we examine the new row 0. If each variable has a nonnegative coefficient in row 0, then we have obtained an optimal tableau. Otherwise, we try to increase the variable with the most negative coefficient in row 0. Our procedure ensures (through BN_1 and BN_3) that no upper-bound constraint is ever violated and (through BN_2) that all of the non-negativity constraints are satisfied.

EXAMPLE 4 Simplex with Upper Bounds 1

Solve the following LP:

$$\max z = 4x_1 + 2x_2 + 3x_3$$

$$\text{s.t.} \quad 2x_1 + x_2 + x_3 \leq 10$$

$$x_1 + \tfrac{1}{2}x_2 + \tfrac{1}{2}x_3 \leq 6$$

$$2x_1 + 2x_2 + 4x_3 \leq 20$$

$$x_1 \leq 4$$

$$x_2 \leq 3$$

$$x_3 \leq 1$$

$$x_1, x_2, x_3 \geq 0$$

Solution   The initial tableau for this problem is given in Table 3. Because x_1 has the most negative coefficient in row 0, we try to increase x_1 as much as we can. The three bottlenecks for x_1 are computed as follows: x_1 cannot exceed its upper bound of 4, so $BN_1 = 4$. To compute BN_2, we solve for the current set of basic variables in terms of x_1:

$$s_1 = 10 - 2x_1 \qquad (s_1 \geq 0 \text{ iff } x_1 \leq 5)$$

$$s_2 = 6 - x_1 \qquad (s_2 \geq 0 \text{ iff } x_1 \leq 6)$$

$$s_3 = 20 - 2x_1 \qquad (s_3 \geq 0 \text{ iff } x_1 \leq 10)$$

Hence, $BN_2 = \min\{5, 6, 10\} = 5$. The current basic variables ($\{s_1, s_2, s_3\}$) have no upper bounds, so there is no value of BN_3. Then the winning bottleneck is $\min\{4, 5\} = 4 = BN_1$. Thus, we must make an upper-bound substitution on x_1 by replacing x_1 by $4 - x_1'$. The resulting tableau is Table 4.

Because x_3 has the most negative coefficient in row 0, we try to increase x_3 as much as possible. The x_3 bottlenecks are computed as follows: x_3 cannot exceed its upper bound of 1, so $BN_1 = 1$. For BN_2, we solve for the current set of basic variables in terms of x_3:

$$s_1 = 2 - x_3 \qquad (s_1 \geq 0 \text{ iff } x_3 \leq 2)$$

$$s_2 = 2 - \tfrac{1}{2}x_3 \qquad (s_2 \geq 0 \text{ iff } x_3 \leq 4)$$

$$s_3 = 12 - 4x_3 \qquad (s_3 \geq 0 \text{ iff } x_3 \leq 3)$$

Thus, $BN_2 = \min\{2, 4, 3\} = 2$. Because s_1, s_2, and s_3 do not have an upper bound, there is no BN_3. The winning bottleneck is $\min\{1, 2\} = BN_1 = 1$, so we make an upper-bound substitution on x_3 by replacing x_3 by $1 - x_3'$. The resulting tableau is Table 5.

Because x_2 now has the most negative coefficient in row 0, we try to increase x_2. The computation of the bottlenecks follows: For BN_1, x_2 cannot exceed its upper bound of 3, so $BN_1 = 3$. For BN_2,

TABLE 3
Initial Tableau for Example 5

		Basic Variable
$z - 4x_1 - 2x_2 - 3x_3$	$= 0$	$z = 0$
$2x_1 + x_2 + x_3 + s_1$	$= 10$	$s_1 = 10$
$x_1 + \tfrac{1}{2}x_2 + \tfrac{1}{2}x_3 \quad + s_2$	$= 6$	$s_2 = 6$
$2x_1 + 2x_2 + 4x_3 \quad\quad + s_3$	$= 20$	$s_3 = 20$

TABLE 4
Replacing x_1 by $4 - x_1'$

		Basic Variable
$z + 4x_1' - 2x_2 - 3x_3$	$= 16$	$z = 16$
$-2x_1' + x_2 + x_3 + s_1$	$= 2$	$s_1 = 2$
$-x_1' + \frac{1}{2}x_2 + \frac{1}{2}x_3 + s_2$	$= 2$	$s_2 = 2$
$-2x_1' + 2x_2 + 4x_3 + s_3$	$= 12$	$s_3 = 12$

TABLE 5
Replacing x_3 by $1 - x_3'$

		Basic Variable
$z + 4x_1' - 2x_2 + 3x_3'$	$= 19$	$z = 19$
$-2x_1' + x_2 - x_3' + s_1$	$= 1$	$s_1 = 1$
$-x_1' + \frac{1}{2}x_2 - \frac{1}{2}x_3' + s_2$	$= \frac{3}{2}$	$s_2 = \frac{3}{2}$
$-2x_1' + 2x_2 - 4x_3' + s_3$	$= 8$	$s_3 = 8$

TABLE 6
Optimal Tableau for Example 4

		Basic Variable
$z \quad + x_3' + 2s_1$	$= 21$	$z = 21$
$-2x_1' + x_2 - x_3' + s_1$	$= 1$	$x_2 = 1$
$-\frac{1}{2}s_1 + s_2$	$= 1$	$s_2 = 1$
$2x_1' - 2x_3' - 2s_1 + s_3$	$= 6$	$s_3 = 6$

$$s_1 = 1 - x_2 \qquad (s_1 \geq 0 \text{ iff } x_2 \leq 1)$$
$$s_2 = \frac{3}{2} - \frac{1}{2}x_2 \qquad (s_2 \geq 0 \text{ iff } x_2 \leq 3)$$
$$s_3 = 8 - 2x_2 \qquad (s_3 \geq 0 \text{ iff } x_2 \leq 4)$$

Thus, $BN_2 = \min\{1, 3, 4\} = 1$. Note that BN_2 occurs because s_1 is forced to zero. None of the basic variables in the current set has an upper-bound constraint, so there is no BN_3. The winning bottleneck is $\min\{3, 1\} = 1 = BN_2$, so x_2 will enter the basis in the row in which s_1 was a basic variable (row 1). After the pivot is performed, the new tableau is Table 6. Because each variable has a non-negative coefficient in row 0, this is an optimal tableau. Thus, the optimal solution to the LP is $z = 21$, $s_2 = 1$, $x_2 = 1$, $s_3 = 6$, $x_1' = 0$, $s_1 = 0$, $x_3' = 0$. Because $x_1' = 4 - x_1$ and $x_3' = 1 - x_3$, we also have $x_1 = 4$ and $x_3 = 1$.

EXAMPLE 5 **Simplex with Upper Bounds 2**

Solve the following LP:

$$\max z = 6x_3$$
$$\text{s.t.} \quad x_1 - x_3 = 6$$
$$x_2 + 2x_3 = 8$$
$$x_1 \leq 8, x_2 \leq 10, x_3 \leq 5; \; x_1, x_2, x_3 \geq 0$$

TABLE 7
Initial Tableau for Example 6

		Basic Variable
$z \quad - 6x_3 = 0$	$z = 0$	
$x_1 \quad - \; x_3 = 6$	$x_1 = 6$	
$x_2 + 2x_3 = 8$	$x_2 = 8$	

TABLE 8
Replacing x_1 by $8 - x_1'$

		Basic Variable
$z \quad - 6x_3 = 0$	$z = 0$	
$x_1' \quad + \; \boxed{x_3} = 2$	$x_1' = 6$	
$x_2 + 2x_3 = 8$	$x_2 = 8$	

TABLE 9
Optimal Tableau for Example 5

		Basic Variable
$z + 6x_1' \quad\quad = 12$	$z = 12$	
$x_1' \quad + x_3 = 2$	$x_3 = 2$	
$-2x_1' + x_2 \quad\quad = 4$	$x_2 = 4$	

Solution After putting the objective function in our standard row 0 format, we obtain the tableau in Table 7. Fortunately, the basic feasible solution $z = 0$, $x_1 = 6$, $x_2 = 8$, $x_3 = 0$ is readily apparent. We can now proceed with the simplex method for upper-bounded variables. Because x_3 has the most negative coefficient in row 0, we try to increase x_3. Because x_3 cannot exceed its upper bound of 5, $BN_1 = 5$. To compute BN_2,

$$x_1 = 6 + x_3 \qquad (x_1 \geq 0 \text{ iff } x_3 \geq -6)$$
$$x_2 = 8 - 2x_3 \qquad (x_2 \geq 0 \text{ iff } x_3 \leq 4)$$

Thus, all the current basic variables will remain non-negative as long as $x_3 \leq 4$. Hence, $BN_2 = 4$. For BN_3, note that $x_1 \leq 8$ will hold iff $6 + x_3 \leq 8$, or $x_3 \leq 2$. Also, $x_2 \leq 10$ will hold iff $8 - 2x_3 \leq 10$, or $x_3 \geq -1$. Thus, for $x_3 \leq 2$, each basic variable remains less than or equal to its upper bound, so $BN_3 = 2$. Note that BN_3 occurs when the basic variable x_1 attains its upper bound. The winning bottleneck is $\min\{5, 4, 2\} = 2 = BN_3$, so the largest that we can make x_3 is 2, and the bottleneck occurs because x_1 attains its upper bound of 8. Thus, we make an upper-bound substitution on x_1 by replacing x_1 by $8 - x_1'$. The resulting tableau is

$$z \quad\quad\quad - 6x_3 = 0$$
$$-x_1' \quad\quad - \; x_3 = -2$$
$$x_2 + 2x_3 = 8$$

After rewriting $-x_1' - x_3 = -2$ as $x_1' + x_3 = 2$, we obtain the tableau in Table 8.

Because x_1, the variable that caused BN_3, was basic in row 1, we now make x_3 a basic variable in row 1. After the pivot, we obtain the tableau in Table 9, which is optima. Thus, the optimal solution to the LP is $z = 12$, $x_3 = 2$, $x_2 = 4$, $x_1' = 0$. Because $x_1' = 0$, $x_1 = 8 - x_1' = 8$.

To illustrate the efficiencies obtained by using the simplex algorithm with upper bounds, suppose we are solving an LP (call it LP 1) with 100 variables, each having an upper-bound constraint, with five other constraints. If we were to solve LP 1 by the revised simplex method, the B^{-1} for each tableau would be a 105×105 matrix. If we were to use the simplex method for upper-bounded variables, however, the B^{-1} for each tableau would be only a 5×5 matrix. Although the computation of the winning bottleneck in each iteration is more complicated than the ordinary ratio test, solving LP 1 by the simplex method for upper-bounded variables would still be much more efficient than by the ordinary revised simplex.

PROBLEMS

Use the upper-bounded simplex algorithm to solve the following LPs:

Group A

1 max $z = 4x_1 + 3x_2 + 5x_3$
  s.t. $2x_1 + 2x_2 + x_3 + x_4 \qquad \leq 9$
    $4x_1 - x_2 - x_3 \qquad + x_5 \leq 6$
      $2x_2 + x_3 \qquad \leq 5$
    $x_1 \qquad\qquad\qquad \leq 2$
      $x_2 \qquad\qquad \leq 3$
        $x_3 \qquad \leq 4$
          $x_4 \qquad \leq 5$
            $x_5 \leq 7$
    $x_1, x_2, x_3, x_4, x_5 \geq 0$

2 min $z = -4x_1 - 9x_2$
  s.t. $3x_1 + 5x_2 \leq 6$
    $5x_1 + 6x_2 \leq 10$
    $2x_1 - 3x_2 \leq 4$
    $x_1 \qquad \leq 2$
      $x_2 \leq 1$
    $x_1, x_2 \geq 0$

3 max $z = 4x_1 + 3x_2$
  s.t. $2x_1 - x_2 \leq 1$
    $x_1 + 6x_2 \leq 6$
      $x_2 \leq 5$
    $x_1, x_2 \geq 0$

4 Suppose an LP contained lower-bound constraints of the following form: $x_j \geq L_j$. Suggest an algorithm that could be used to solve such a problem efficiently.

10.6 Karmarkar's Method for Solving LPs

As discussed in Section 4.13, Karmarkar's method for solving LPs is a polynomial time algorithm. This is in contrast to the simplex algorithm, an exponential time algorithm. Unlike the ellipsoid method (another polynomial time algorithm), Karmarkar's method appears to solve many LPs faster than does the simplex algorithm. In this section, we give a description of the basic concepts underlying Karmarkar's method. Note that several versions of Karmarkar's method are computationally more efficient than the version we describe; our goal is simply to introduce the reader to the exciting ideas used in Karmarkar's method. For a more detailed description of Karmarkar's method, see Hooker (1986), Parker and Rardin (1988), and Murty (1989).

Karmarkar's method is applied to an LP in the following form:

$$\min z = \mathbf{c}\mathbf{x}$$
$$\text{s.t.} \quad A\mathbf{x} = \mathbf{0}$$
$$x_1 + x_2 + \cdots + x_n = 1 \qquad (31)$$
$$\mathbf{x} \geq \mathbf{0}$$

In (31), $\mathbf{x} = [x_1 \quad x_2 \quad \cdots \quad x_n]^T$, A is an $m \times n$ matrix, $\mathbf{c} = [c_1 \quad c_2 \quad \cdots \quad c_n]$ and $\mathbf{0}$ is an n-dimensional column vector of zeros. The LP must also satisfy

$$[\tfrac{1}{n} \quad \tfrac{1}{n} \quad \cdots \quad \tfrac{1}{n}]^T \qquad \text{is feasible} \tag{32}$$

$$\text{Optimal } z\text{-value} = 0 \tag{33}$$

Although it may seem unlikely that an LP would have the form (31) and satisfy (32)–(33), it is easy to show that any LP may be put in a form such that (31)–(33) are satisfied. We will demonstrate this at the end of this section.

The following three concepts play a key role in Karmarkar's method:

1 Projection of a vector onto the set of $\mathbf{x}$ satisfying $A\mathbf{x} = \mathbf{0}$

2 Karmarkar's centering transformation

3 Karmarkar's potential function

We now discuss the first two concepts, leaving a discussion of Karmarkar's potential function to the end of the section. Before discussing the ideas just listed, we need a definition.

DEFINITION ■ The ***n*-dimensional unit simplex** S is the set of points $[x_1 \quad x_2 \quad \cdots \quad x_n]^T$ satisfying $x_1 + x_2 + \cdots + x_n = 1$ and $x_j \geq 0, j = 1, 2, \ldots, n$. ■

Projection

Suppose we are given a point $\mathbf{x}^0$ that is feasible for (31), and we want to move from $\mathbf{x}^0$ to another feasible point (call it $\mathbf{x}^1$) that, for some fixed vector $\mathbf{v}$, will have a larger value of $\mathbf{vx}$. Suppose that we find $\mathbf{x}^1$ by moving away from $\mathbf{x}^0$ in a direction $\mathbf{d} = [d_1 \quad d_2 \quad \cdots \quad d_n]$. For $\mathbf{x}^1$ to be feasible, $\mathbf{d}$ must satisfy $A\mathbf{d} = \mathbf{0}$ and $d_1 + d_2 + \cdots + d_n = 0$. If we choose the direction $\mathbf{d}$ that solves the optimization problem

$$\max \mathbf{vd}$$
$$\text{s.t.} \qquad A\mathbf{d} = \mathbf{0}$$
$$d_1 + d_2 + \cdots + d_n = 0$$
$$\|\mathbf{d}\| = 1$$

then we will be moving in the "feasible" direction that maximizes the increase in $\mathbf{vx}$ per unit of length moved. The direction $\mathbf{d}$ that solves this optimization problem is given by the **projection** of $\mathbf{v}$ onto the set of $\mathbf{x} = [x_1 \quad x_2 \quad \cdots \quad x_n]^T$ satisfying $A\mathbf{x} = \mathbf{0}$ and $x_1 + x_2 + \cdots + x_n = 0$. The projection of $\mathbf{v}$ onto the set of $\mathbf{x}$ satisfying $A\mathbf{x} = \mathbf{0}$ and $x_1 + x_2 + \cdots + x_n = 0$ is given by $[I - B^T(BB^T)^{-1}B]\mathbf{v}$, where B is the $(m + 1) \times n$ matrix whose first m rows are A and whose last row is a vector of 1's.

Geometrically, what does it mean to project a vector $\mathbf{v}$ onto the set of $\mathbf{x}$ satisfying $A\mathbf{x} = \mathbf{0}$? It can be shown that any vector $\mathbf{v}$ may be written (uniquely) in the form $\mathbf{v} = \mathbf{p} + \mathbf{w}$, where $\mathbf{p}$ satisfies $A\mathbf{p} = \mathbf{0}$ and $\mathbf{w}$ is perpendicular to all vectors $\mathbf{x}$ satisfying $A\mathbf{x} = \mathbf{0}$. Then $\mathbf{p}$ is the projection of $\mathbf{v}$ onto the set of $\mathbf{x}$ satisfying $A\mathbf{x} = \mathbf{0}$. An example of this idea is given in Figure 5, where $\mathbf{v} = [-2 \quad -1 \quad 7]$ is projected onto the set of three-dimensional vectors satisfying $x_3 = 0$ (the x_1–x_2-plane). In this case, we decompose $\mathbf{v}$ as $\mathbf{v} = [-2 \quad -1 \quad 0] + [0 \quad 0 \quad 7]$. Thus, $\mathbf{p} = [-2 \quad -1 \quad 0]$. It is easy to show that $\mathbf{p}$ is the vector in the set of $\mathbf{x}$ satisfying $A\mathbf{x} = \mathbf{0}$ that is "closest" to $\mathbf{v}$. This is apparent from Figure 5.

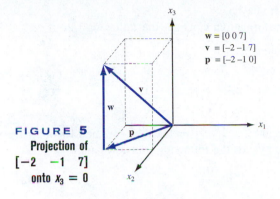

FIGURE 5
Projection of
$[-2 \quad -1 \quad 7]$
onto $x_3 = 0$

$\mathbf{w} = [0 \ 0 \ 7]$
$\mathbf{v} = [-2 \ -1 \ 7]$
$\mathbf{p} = [-2 \ -1 \ 0]$

Karmarkar's Centering Transformation

Given a feasible point (in (31)) $\mathbf{x}^k = [x_1^k \quad x_2^k \quad \cdots \quad x_n^k]$ in S having $x_j^k > 0, j = 1, 2, \ldots, n$, we write the **centering transformation** associated with the point $\mathbf{x}^k$ as $f([x_1 \quad x_2 \quad \cdots \quad x_n] \| \mathbf{x}^k)$. If $\mathbf{x}^k$ is a point in S, then $f([x_1 \quad x_2 \quad \cdots \quad x_n] | \mathbf{x}^k)$ transforms a point $[x_1 \quad x_2 \quad \cdots \quad x_n]^T$ in S into a point $[y_1 \quad y_2 \quad \cdots \quad y_n]^T$ in S, where

$$y_j = \frac{\dfrac{x_j}{x_j^k}}{\displaystyle\sum_{r=1}^{r=n} \dfrac{x_r}{x_r^k}} \tag{34}$$

Let $\text{Diag}(\mathbf{x}^k)$ be the $n \times n$ matrix with all off-diagonal entries equal to 0 and $\text{Diag}(\mathbf{x}^k)_{ii} = x_i^k$. The centering transformation specified by (34) can be shown to have the properties listed in Lemma 1.

LEMMA 1

Karmarkar's centering transformation has the following properties:

$$f(\mathbf{x}^k \mid \mathbf{x}^k) = [\tfrac{1}{n} \quad \tfrac{1}{n} \quad \cdots \quad \tfrac{1}{n}]^T. \tag{35}$$

For $\mathbf{x} \neq \mathbf{x}'$, $\quad f(\mathbf{x} \mid \mathbf{x}^k) \neq f(\mathbf{x}' \mid \mathbf{x}^k) \tag{36}$

$f(\mathbf{x} \mid \mathbf{x}^k) \in S \tag{37}$

For any point $[y_1 \quad y_2 \quad \cdots \quad y_n]^T$ in S, there is a unique point $\tag{38}$

$[x_1 \quad x_2 \quad \cdots \quad x_n]^T$ in S satisfying

$f([x_1 \quad x_2 \quad \cdots \quad x_n]^T \mid \mathbf{x}^k) = [y_1 \quad y_2 \quad \cdots \quad y_n]^T \tag{38$'$}$

The point $[x_1 \quad x_2 \quad \cdots \quad x_n]^T$ is given by

$$x_j = \frac{x_j^k y_j}{\displaystyle\sum_{r=1}^{r=n} x_r^k y_r}$$

If $[x_1 \quad x_2 \quad \cdots \quad x_n]^T$ and $[y_1 \quad y_2 \quad \cdots \quad y_n]^T$ satisfy (38$'$), we write $f^{-1}([y_1 \quad y_2 \quad \cdots \quad y_n]^T \mid \mathbf{x}^k) = [x_1 \quad x_2 \quad \cdots \quad x_n]^T$.

A point $\mathbf{x}$ in S will satisfy $A\mathbf{x} = \mathbf{0}$ $\quad$ if $\quad A[\text{Diag}(\mathbf{x}^k)] f(\mathbf{x} \mid \mathbf{x}^k) = \mathbf{0} \tag{39}$

(See Problem 5 for a proof of Lemma 1.)

To illustrate the centering transformation, consider the following LP:

$$\min z = x_1 + 3x_2 - 3x_3$$
$$\text{s.t.} \qquad x_2 - x_3 = 0 \tag{40}$$
$$x_1 + x_2 + x_3 = 1$$
$$x_i \geq 0$$

This LP is of the form (31); the point $[\frac{1}{3} \quad \frac{1}{3} \quad \frac{1}{3}]^T$ is feasible, and the LP's optimal z-value is 0. The feasible point $[\frac{1}{4} \quad \frac{3}{8} \quad \frac{3}{8}]$ yields the following transformation:

$$f([x_1 \quad x_2 \quad x_3] \mid [\tfrac{1}{4} \quad \tfrac{3}{8} \quad \tfrac{3}{8}])$$

$$= \left[\frac{4x_1}{4x_1 + \dfrac{8x_2}{3} + \dfrac{8x_3}{3}} \quad \frac{\dfrac{8x_2}{3}}{4x_1 + \dfrac{8x_2}{3} + \dfrac{8x_3}{3}} \quad \frac{\dfrac{8x_3}{3}}{4x_1 + \dfrac{8x_2}{3} + \dfrac{8x_3}{3}} \right]$$

For example,

$$f([\tfrac{1}{3} \quad \tfrac{1}{3} \quad \tfrac{1}{3}] \mid [\tfrac{1}{4} \quad \tfrac{3}{8} \quad \tfrac{3}{8}]) = [\tfrac{12}{28} \quad \tfrac{8}{28} \quad \tfrac{8}{28}]$$

We now refer to the variables $x_1, x_2, \ldots, x_n$ as being the *original* space and the variables $y_1, y_2, \ldots, y_n$ as being the *transformed* space. The unit simplex involving variables $y_1, y_2, \ldots, y_n$ will be called the transformed unit simplex. We now discuss the intuitive meaning of (35)–(39). Equation (35) implies that $f(\cdot \mid \mathbf{x}^k)$ maps $\mathbf{x}^k$ into the "center" of the transformed unit simplex. Equations (36)–(37) imply that any point in S is transformed into a point in the transformed unit simplex, and no two points in S can yield the same point in the transformed unit simplex (that is, f is a one-to-one mapping). Equation (38) implies that for any point $\mathbf{y}$ in the transformed unit simplex, there is a point $\mathbf{x}$ in S that is transformed into $\mathbf{y}$. The formula for the $\mathbf{x}$ that is transformed into $\mathbf{y}$ is also given. Thus, (36)–(38) imply that f is a one-to-one and an onto mapping from S to S. Finally, (39) states that feasible points in the original problem correspond to points $\mathbf{y}$ in the transformed unit simplex that satisfy $A[\text{Diag}(\mathbf{x}^k)]\mathbf{y} = 0$.

Description and Example of Karmarkar's Method

We assume that we will be satisfied with a feasible point having an optimal z-value $< \epsilon$ (for some small ϵ). Karmarkar's method proceeds as follows:

Step 1 Begin at the feasible point $\mathbf{x}^0 = [\frac{1}{n} \quad \frac{1}{n} \quad \cdots \quad \frac{1}{n}]^T$ and set $k = 0$.

Step 2 Stop if $\mathbf{c}\mathbf{x}^k < \epsilon$. If not, go to step 3.

Step 3 Find the new point $\mathbf{y}^{k+1} = [y_1^{k+1} \quad y_2^{k+1} \quad \cdots \quad y_n^{k+1}]^T$ in the transformed unit simplex given by

$$\mathbf{y}^{k+1} = [\tfrac{1}{n} \quad \tfrac{1}{n} \quad \cdots \quad \tfrac{1}{n}]^T - \frac{\theta(I - P^T(PP^T)^{-1}P)[\text{Diag}(\mathbf{x}^k)]\mathbf{c}^T}{\|\mathbf{c}_p\|\sqrt{n(n-1)}}$$

Here, $\|\mathbf{c}_p\| = $ the length of $(I - P^T(PP^T)^{-1}P)[\text{Diag}(\mathbf{x}^k)]\mathbf{c}^T$, P is the $(m+1) \times n$ matrix whose first m rows are $A[\text{Diag}(\mathbf{x}^k)]$ and whose last row is a vector of 1's, and $0 < \theta < 1$ is chosen to ensure convergence of the algorithm. $\theta = \frac{1}{4}$ is known to ensure convergence.

Now obtain a new point $\mathbf{x}^{k+1}$ in the original space by using the centering transformation to determine the point corresponding to $\mathbf{y}^{k+1}$. That is, $\mathbf{x}^{k+1} = f^{-1}(\mathbf{y}^{k+1} \mid \mathbf{x}^k)$. Increase k by 1 and return to Step 2.

REMARKS

1 In step 3, we move from the "center" of the transformed unit simplex in a direction opposite to the projection of $\text{Diag}(\mathbf{x}^k)\mathbf{c}^T$ onto the transformation of the feasible region (the set of $\mathbf{y}$ satisfying $A[\text{Diag}(\mathbf{x}^k)]\mathbf{y} = \mathbf{0}$). From our discussion of the projection, this ensures that we maintain feasibility (in the transformed space) and move in a direction that maximizes the rate of decrease of $[\text{Diag}(\mathbf{x}^k)]\mathbf{c}^T$.

2 By moving a distance

$$\frac{\theta}{\sqrt{n(n-1)}}$$

from the center of the transformed unit simplex, we ensure that $\mathbf{y}^{k+1}$ will remain in the interior of the transformed unit simplex.

3 When we use the inverse of Karmarkar's centering transformation to transform $\mathbf{y}^{k+1}$ back into $\mathbf{x}^{k+1}$, the definition of projection and (39) imply that $\mathbf{x}^{k+1}$ will be feasible for the original LP (see Problem 6).

4 Why do we project $[\text{Diag}(\mathbf{x}^k)]\mathbf{c}^T$ rather than $\mathbf{c}^T$ onto the transformed feasible region? The answer to this question must await our discussion of Karmarkar's potential function. Problem 7 provides another explanation of why we project $[\text{Diag}(\mathbf{x}^k)]\mathbf{c}^T$ rather than $\mathbf{c}^T$.

We now work out the first iteration of Karmarkar's method when applied to (40), choosing $\epsilon = 0.10$.

First Iteration of Karmarkar's Method

Step 1 $\mathbf{x}^0 = [\frac{1}{3} \quad \frac{1}{3} \quad \frac{1}{3}]^T$ and $k = 0$.

Step 2 $\mathbf{x}^0$ yields $z = \frac{1}{3} > 0.10$, so we must proceed to step 3.

Step 3

$$A = [0 \quad 1 \quad -1], \qquad \text{Diag}(\mathbf{x}^k) = \begin{bmatrix} \frac{1}{3} & 0 & 0 \\ 0 & \frac{1}{3} & 0 \\ 0 & 0 & \frac{1}{3} \end{bmatrix}$$

$$A[\text{Diag}(\mathbf{x}^k)] = [0 \quad \frac{1}{3} \quad -\frac{1}{3}], \qquad P = \begin{bmatrix} 0 & \frac{1}{3} & -\frac{1}{3} \\ 1 & 1 & 1 \end{bmatrix}$$

$$PP^T = \begin{bmatrix} \frac{2}{9} & 0 \\ 0 & 3 \end{bmatrix}, \qquad (PP^T)^{-1} = \begin{bmatrix} \frac{9}{2} & 0 \\ 0 & \frac{1}{3} \end{bmatrix}$$

$$(I - P^T(PP^T)^{-1}P) = \begin{bmatrix} \frac{2}{3} & -\frac{1}{3} & -\frac{1}{3} \\ -\frac{1}{3} & \frac{1}{6} & \frac{1}{6} \\ -\frac{1}{3} & \frac{1}{6} & \frac{1}{6} \end{bmatrix}, \qquad \mathbf{c} = [1 \quad 3 \quad -3]$$

$$[\text{Diag } \mathbf{x}^k]\mathbf{c}^T = \begin{bmatrix} \frac{1}{3} \\ 1 \\ -1 \end{bmatrix}$$

$$(I - P^T(PP^T)^{-1}P)[\text{Diag } \mathbf{x}^k]\mathbf{c}^T = [\frac{2}{9} \quad -\frac{1}{9} \quad -\frac{1}{9}]$$

Now, (using $\theta = 0.25$), we obtain

$$\mathbf{y}^1 = [\frac{1}{3} \quad \frac{1}{3} \quad \frac{1}{3}]^T - \frac{0.25[\frac{2}{9} \quad -\frac{1}{9} \quad -\frac{1}{9}]^T}{\sqrt{3(2)}\|[\frac{2}{9} \quad -\frac{1}{9} \quad -\frac{1}{9}]\|}$$

Because

$$\|[\frac{2}{9} \quad -\frac{1}{9} \quad -\frac{1}{9}]\|^T = \sqrt{(\tfrac{2}{9})^2 + (-\tfrac{1}{9})^2 + (-\tfrac{1}{9})^2}$$

$$= \frac{\sqrt{6}}{9}$$

we obtain

$$\mathbf{y}^1 = [\tfrac{1}{3} \quad \tfrac{1}{3} \quad \tfrac{1}{3}]^T - [\tfrac{6}{72} \quad -\tfrac{3}{72} \quad -\tfrac{3}{72}]^T = [\tfrac{1}{4} \quad \tfrac{3}{8} \quad \tfrac{3}{8}]^T$$

Using (38'), we now obtain $\mathbf{x}^1 = [x_1^1 \quad x_2^1 \quad x_3^1]^T$ from

$$x_1^1 = \frac{\tfrac{1}{3}(\tfrac{1}{4})}{\tfrac{1}{3}(\tfrac{1}{4}) + \tfrac{1}{3}(\tfrac{3}{8}) + \tfrac{1}{3}(\tfrac{3}{8})} = \tfrac{1}{4}$$

$$x_2^1 = \frac{\tfrac{1}{3}(\tfrac{3}{8})}{\tfrac{1}{3}(\tfrac{1}{4}) + \tfrac{1}{3}(\tfrac{3}{8}) + \tfrac{1}{3}(\tfrac{3}{8})} = \tfrac{3}{8}$$

$$x_3^1 = \frac{\tfrac{1}{3}(\tfrac{3}{8})}{\tfrac{1}{3}(\tfrac{1}{4}) + \tfrac{1}{3}(\tfrac{3}{8}) + \tfrac{1}{3}(\tfrac{3}{8})} = \tfrac{3}{8}$$

Thus, $\mathbf{x}^1 = [\tfrac{1}{4} \quad \tfrac{3}{8} \quad \tfrac{3}{8}]^T$. It will always be the case (see Problem 3) that $\mathbf{x}^1 = \mathbf{y}^1$, but for $k > 1$, $\mathbf{x}^k$ need not equal $\mathbf{y}^k$. Note that for $\mathbf{x}^1$, we have $z = \tfrac{1}{4} + 3(\tfrac{3}{8}) - 3(\tfrac{3}{8}) = \tfrac{1}{4} < \tfrac{1}{3}$ (the z-value for $\mathbf{x}^0$).

Potential Function

Because we are projecting $[\text{Diag}(\mathbf{x}^k)]\mathbf{c}^T$ rather than $\mathbf{c}^T$, we cannot be sure that each iteration of Karmarkar's method will decrease z. In fact, it is possible for $\mathbf{cx}^{k+1} > \mathbf{cx}^k$ to occur. To explain why Karmarkar projects $[\text{Diag}(\mathbf{x}^k)]\mathbf{c}^T$, we need to discuss Karmarkar's potential function. For $\mathbf{x} = [x_1 \quad x_2 \quad \cdots \quad x_n]^T$, we define the potential function $f(\mathbf{x})$ by

$$f(\mathbf{x}) = \sum_{j=1}^{j=n} \ln\left(\frac{\mathbf{cx}^T}{x_j}\right)$$

Karmarkar showed that if we project $[\text{Diag}(\mathbf{x}^k)]\mathbf{c}^T$ (not $\mathbf{c}^T$) onto the feasible region in the transformed space, then for some $\delta > 0$, it will be true that for $k = 0, 1, 2, \ldots$,

$$f(\mathbf{x}^k) - f(\mathbf{x}^{k+1}) \geq \delta \qquad \text{(41)}$$

Inequality (41) states that each iteration of Karmarkar's method decreases the potential function by an amount bounded away from 0. Karmarkar shows that if the potential function evaluated at $\mathbf{x}^k$ is small enough, then $\mathbf{z} = \mathbf{cx}^k$ will be near 0. Because $f(\mathbf{x}^k)$ is decreased by at least δ per iteration, it follows that by choosing k sufficiently large, we can ensure that the z-value for $\mathbf{x}^k$ is less than ϵ.

Putting an LP in Standard Form for Karmarkar's Method

We now show how to convert any LP to the form defined by (31)–(33). To illustrate, we show how to transform the following LP

$$\max z = 3x_1 + x_2$$
$$\text{s.t.} \quad 2x_1 - x_2 \leq 2$$
$$x_1 + 2x_2 \leq 5 \qquad \text{(42)}$$
$$x_1, x_2 \geq 0$$

into the form defined by (31)–(33).

We begin by finding the dual of (42).

$$\min w = 2y_1 + 5y_2$$
$$\text{s.t.} \quad 2y_1 + y_2 \geq 3$$
$$-y_1 + 2y_2 \geq 1 \tag{42'}$$
$$y_1, y_2 \geq 0$$

From the Dual Theorem (Theorem 1 of Chapter 6), we know that if (x_1, x_2) is feasible in (42), (y_1, y_2) is feasible in (42'), and the z-value for (x_1, x_2) in (42) equals the w-value for (y_1, y_2) in (42'), then (x_1, x_2) is optimal for (42). This means that any feasible solution to the following set of constraints will yield the optimal solution to (42):

$$3x_1 + x_2 - 2y_1 - 5y_2 = 0$$
$$2x_1 - x_2 \leq 2$$
$$x_1 + 2x_2 \leq 5$$
$$2y_1 + y_2 \geq 3 \tag{43}$$
$$-y_1 + 2y_2 \geq 1$$
$$\text{All variables} \geq 0$$

Inserting slack and excess variables into (43) yields

$$3x_1 + x_2 - 2y_1 - 5y_2 = 0$$
$$2x_1 - x_2 + s_1 = 2$$
$$x_1 + 2x_2 + s_2 = 5$$
$$2y_1 + y_2 - e_1 = 3 \tag{44}$$
$$-y_1 + 2y_2 - e_2 = 1$$
$$\text{All variables} \geq 0$$

We now find a number M such that any feasible solution to (44) will satisfy

$$\text{sum of all variables in (44)} \leq M \tag{45}$$

and add constraint (45) to (44). Being conservative, we can see that any values of the variables that yield an optimal primal solution to (42) and an optimal dual solution to (42') will have no variable exceeding 10. This would yield $M = 10(8) = 80$. We then add a slack variable (dummy variable d_1) to (45). Our new goal is then to find a feasible solution to

$$3x_1 + x_2 - 2y_1 - 5y_2 = 0$$
$$2x_1 - x_2 + s_1 = 2$$
$$x_1 + 2x_2 + s_2 = 5$$
$$2y_1 + y_2 - e_1 = 3 \tag{46}$$
$$-y_1 + 2y_2 - e_2 = 1$$
$$x_1 + x_2 + y_1 + y_2 + s_1 + s_2 + e_1 + e_2 + d_1 = 80$$
$$\text{All variables} \geq 0$$

We now define a new dummy variable d_2; $d_2 = 1$. We can use this new variable to "homogenize" the constraints in (46), which have nonzero right-hand sides. To do this, we add the appropriate multiple of the constraint $d_2 = 1$ to each constraint in (46) (except the last constraint) having a nonzero right-hand side. For example we add $-2(d_2 = 1)$ to the constraint $2x_1 - x_2 + s_1 = 2$. We also replace the last constraint in (46) by the following two constraints:

(a) Add $d_2 = 1$ to the last constraint

(b) Subtract M times $(d_2 = 1)$ from (46)

Together (a) and (b) are equivalent to $d_2 = 1$ and the last constraint in (46).
We now seek a feasible solution to

$$
\begin{aligned}
3x_1 + x_2 - 2y_1 - 5y_2 &= 0 \\
2x_1 - x_2 \quad\quad + s_1 - 2d_2 &= 0 \\
x_1 + 2x_2 \quad\quad + s_2 - 5d_2 &= 0 \\
2y_1 + y_2 - e_1 - 3d_2 &= 0 \\
- y_1 + 2y_2 - e_2 - d_2 &= 0 \\
x_1 + x_2 + y_1 + y_2 + s_1 + s_2 + e_1 + e_2 + d_1 - 80d_2 &= 0 \\
x_1 + x_2 + y_1 + y_2 + s_1 + s_2 + e_1 + e_2 + d_1 + d_2 &= 81
\end{aligned}
\tag{47}
$$

$$\text{All variables} \geq 0$$

Now we make the following change of variables in (47):

$$
x_j = (M + 1)x_j', \; y_j = (M + 1)y_j', \; s_j = (M + 1)s_j', \; e_j = (M + 1)e_j',
$$
$$
d_j = (M + 1)d_j' \; (j = 1, 2)
$$

This yields

$$
\begin{aligned}
3x_1' + x_2' - 2y_1' - 5y_2' &= 0 \\
2x_1' - x_2' \quad\quad + s_1' - 2d_2' &= 0 \\
x_1' + 2x_2' \quad\quad + s_2' - 5d_2' &= 0 \\
2y_1' + y_2' - e_1' - 3d_2' &= 0 \\
- y_1' + 2y_2' - e_2' - d_2' &= 0 \\
x_1' + x_2' + y_1' + y_2' + s_1' + s_2' + e_1' + e_2' + d_1' - 80d_2' &= 0 \\
x_1' + x_2' + y_1' + y_2' + s_1' + s_2' + e_1' + e_2' + d_1' + d_2' &= 1
\end{aligned}
\tag{48}
$$

$$\text{All variables} \geq 0$$

We now ensure that a point that sets all variables equal is feasible in (48). (Recall that this is requirement (33) for Karmarkar's method.) To accomplish this, we first add a dummy variable d_3' to the last constraint in (48) and then add a multiple of d_3' to each of the other constraints. This multiple is chosen so that the sum of the coefficients of all variables in each constraint (except the last) will equal 0. This yields LP (49).

$$
\begin{aligned}
\min z = d_3' \\
\text{s.t.} \quad 3x_1' + x_2' - 2y_1' - 5y_2' + 3d_3' &= 0 \\
2x_1' - x_2' \quad\quad + s_1' - 2d_2' + d_3' &= 0 \\
x_1' + 2x_2' \quad\quad + s_2' - 5d_2' + d_3' &= 0 \\
2y_1' + y_2' - e_1' - 3d_2' + d_3' &= 0 \\
- y_1' + 2y_2' - e_2' - d_2' + d_3' &= 0 \\
x_1' + x_2' + y_1' + y_2' + s_1' + s_2' + e_1' + e_2' + d_1' - 80d_2' + 71d_3' &= 0 \\
x_1' + x_2' + y_1' + y_2' + s_1' + s_2' + e_1' + e_2' + d_1' + d_2' + d_3' &= 1
\end{aligned}
\tag{49}
$$

$$\text{All variables} \geq 0$$

In (49) the point $x_1' = x_2' = y_1' = y_2' = s_1' = s_2' = e_1' = e_2' = d_1' = d_2' = d_3' = 1/11$ is feasible. Because d_3' should equal 0 in a feasible solution to (48), we need to have (49) minimize d_3'. If (48) is feasible, then the minimum value of d_3', in (49) will equal 0, and the values of the remaining variables in an optimal solution to (49) will yield a feasible solution to (48). The values of x_1 and x_2 in the optimal solution to (49) will yield an optimal solution to our original LP (42). The LP in (49) satisfies (31)–(33) and is ready for solution by Karmarkar's method.

PROBLEMS

Group A

1 Perform one iteration of Karmarkar's method for the following LP:

$$\min z = x_1 + 2x_2 - x_3$$
$$\text{s.t.} \quad x_1 \qquad - x_3 = 0$$
$$x_1 + x_2 + x_3 = 1$$
$$x_1, x_2, x_3 \geq 0$$

2 Perform one iteration of Karmarkar's method for the following LP:

$$\min z = x_1 - x_2 + 6x_3$$
$$\text{s.t.} \quad x_1 - x_2 \qquad = 0$$
$$x_1 + x_2 + x_3 = 1$$
$$x_1, x_2, x_3 \geq 0$$

3 Prove that in Karmarkar's method, $\mathbf{x}^1 = \mathbf{y}^1$.

4 Perform two iterations of Karmarkar's method for the following LP:

$$\min z = 2x_2$$
$$\text{s.t.} \quad x_1 + x_2 - 2x_3 = 0$$
$$x_1 + x_2 + x_3 = 1$$
$$x_1, x_2, x_3 \geq 0$$

Group B

5 Prove Lemma 1.

6 Show that the point $\mathbf{x}^k$ in Karmarkar's method is feasible for the original LP.

7 Given a point $\mathbf{y}^k$ in Karmarkar's method, express the LP's original objective function as a function of $\mathbf{y}^k$. Use the answer to this question to give a reason why $[\text{Diag}(\mathbf{x}^k)]\mathbf{c}^T$ is projected, rather than $\mathbf{c}^T$.

SUMMARY

The Revised Simplex Method and the Product Form of the Inverse

Step 0 Note the columns from which the current B^{-1} will be read. Initially $B^{-1} = I$.

Step 1 For the current tableau, compute $\mathbf{c}_{\text{BV}}B^{-1}$.

Step 2 Price out all nonbasic variables in the current tableau. If (for a max problem) each nonbasic variable prices out non-negative, the current basis is optimal. If the current basis is not optimal, enter into the basis the nonbasic variable with the most negative coefficient in row 0. Call this variable x_k.

Step 3 To determine the row in which x_k enters the basis, compute x_k's column in the current tableau $(B^{-1}\mathbf{a}_k)$ and compute the right-hand side of the current tableau $(B^{-1}\mathbf{b})$. Then use the ratio test to determine the row in which x_k should enter the basis. We now know the set of basic variables (BV) for the new tableau.

Step 4 Use the column for x_k in the current tableau to determine the EROs needed to enter x_k into the basis. Perform these EROs on the current B^{-1} to yield the new B^{-1}. Return to Step 1.

Alternatively, we may use the product form of the inverse to update B^{-1}. Suppose we have found that x_k should enter the basis in row r. Let the column for x_k in the current tableau be

$$\begin{bmatrix} \bar{a}_{1k} \\ \bar{a}_{2k} \\ \vdots \\ \bar{a}_{mk} \end{bmatrix}$$

Define the $m \times m$ matrix E by

$$
\text{(column } r \text{)}
$$

$$
E = \begin{bmatrix}
1 & 0 & \cdots & -\dfrac{\bar{a}_{1k}}{\bar{a}_{rk}} & \cdots & 0 & 0 \\
0 & 1 & \cdots & -\dfrac{\bar{a}_{2k}}{\bar{a}_{rk}} & \cdots & 0 & 0 \\
\vdots & \vdots & & \vdots & & \vdots & \vdots \\
0 & 0 & \cdots & \dfrac{1}{\bar{a}_{rk}} & \cdots & 0 & 0 \\
\vdots & \vdots & & \vdots & & \vdots & \vdots \\
0 & 0 & \cdots & -\dfrac{\bar{a}_{m-1,k}}{\bar{a}_{rk}} & \cdots & 1 & 0 \\
0 & 0 & \cdots & -\dfrac{\bar{a}_{mk}}{\bar{a}_{rk}} & \cdots & 0 & 1
\end{bmatrix} \quad \text{(row } r \text{)}
$$

Then

$$
B^{-1} \text{ for new tableau} = E(B^{-1} \text{ for current tableau})
$$

Return to step 1.

Column Generation

When an LP has many variables, it is very time-consuming to price out each nonbasic variable individually. The column generation approach lets us determine the nonbasic variable that prices out most favorably by solving a subproblem (such as the branch-and-bound problems in the cutting stock problem).

Dantzig–Wolfe Decomposition Method

In many LPs, the constraints and variables may be decomposed in the following manner:

Constraints in set 1 only involve variables in Variable set 1.

Constraints in set 2 only involve variables in Variable set 2.

$$\vdots$$

Constraints in set k only involve variables in Variable set k.

Constraints in set $k + 1$ may involve any variable. The constraints in set $k + 1$ are referred to as the **central constraints.**

LPs that can be decomposed in this fashion can often be efficiently solved by the Dantzig–Wolfe decomposition algorithm. The following explanation assumes that $k = 2$.

Step 1 Let the variables in Variable set 1 be $x_1, x_2, \ldots, x_{n_1}$. Express the variables in Variable set 1 as a convex combination of the extreme points of the feasible region for Constraint set 1 (the constraints that involve only the variables in Variable set 1). If we let $\mathbf{P}_1$, $\mathbf{P}_2, \ldots, \mathbf{P}_k$ be the extreme points of this feasible region, then any point

$$\begin{bmatrix} x_1 \\ x_2 \\ \vdots \\ x_{n_1} \end{bmatrix}$$

in the feasible region for Constraint set 1 may be written in the form

$$\begin{bmatrix} x_1 \\ x_2 \\ \vdots \\ x_{n_1} \end{bmatrix} = \mu_1 \mathbf{P}_1 + \mu_2 \mathbf{P}_2 + \cdots + \mu_k \mathbf{P}_k \tag{29}$$

where $\mu_1 + \mu_2 + \cdots + \mu_k = 1$ and $\mu_i \geq 0$ $(i = 1, 2, \ldots, k)$.

Step 2 Express the variables in Variable set 2, $x_{n_1+1}, x_{n_1+2}, \ldots, x_n$, as a convex combination of the extreme points of Constraint set 2's feasible region. If we let the extreme points of the feasible region be $\mathbf{Q}_1, \mathbf{Q}_2, \ldots, \mathbf{Q}_m$, then any point in Constraint set 2's feasible region may be written as

$$\begin{bmatrix} x_{n_1+1} \\ x_{n_1+2} \\ \vdots \\ x_n \end{bmatrix} = \lambda_1 \mathbf{Q}_1 + \lambda_2 \mathbf{Q}_2 + \cdots + \lambda_m \mathbf{Q}_m \tag{30}$$

where $\lambda_i \geq 0$ $(i = 1, 2, \ldots, m)$ and $\lambda_1 + \lambda_2 + \cdots + \lambda_m = 1$.

Step 3 Using (29) and (30), express the LP's objective function and centralized constraints in terms of the μ_i's and the λ_i's. After adding the constraints (called convexity constraints), $\mu_1 + \mu_2 + \cdots + \mu_k = 1$ and $\lambda_1 + \lambda_2 + \cdots + \lambda_m = 1$ and the sign restrictions $\mu_i \geq 0$ $(i = 1, 2, \ldots, k)$ and $\lambda_i \geq 0$ $(i = 1, 2, \ldots, m)$, we obtain the following LP, which is referred to as the **restricted master**:

$$\max \text{ (or min) } [\text{objective function in terms of } \mu_i\text{'s and } \lambda_i\text{'s}]$$

s.t.   [central constraints in terms of μ_i's and λ_i's]

$$\mu_1 + \mu_2 + \cdots + \mu_k = 1 \quad \text{(Convexity constraints)}$$
$$\lambda_1 + \lambda_2 + \cdots + \lambda_m = 1$$
$$\mu_i \geq 0 \quad (i = 1, 2, \ldots, k) \quad \text{(Sign restrictions)}$$
$$\lambda_i \geq 0 \quad (i = 1, 2, \ldots, m)$$

Step 4 Assume that a basic feasible solution for the restricted master is readily available. Then use the column generation method of Section 10.3 to determine whether there is any μ_i or λ_i that can improve the z-value for the restricted master. If so, use the revised simplex method to enter that variable into the basis. Otherwise, the current tableau is optimal for the restricted master. If the current tableau is not optimal, continue with column generation until an optimal solution is found.

Step 5 Substitute the optimal values of the μ_i's and λ_i's found in step 4 into (29) and (30). This will yield the optimal values of $x_1, x_2, \ldots, x_n$.

The Simplex Method for Upper-Bounded Variables

For each variable x_i that has an upper-bound constraint $x_i \leq u_i$, we define a new variable x_i' by the relationship $x_i + x_i' = u_i$, or $x_i = u_i - x_i'$.

At each iteration, we choose (for a max problem) to increase the variable x_i with the most negative coefficient in row 0. Three possible occurrences, or bottlenecks, can restrict the amount by which we increase x_i:

Bottleneck 1 x_i cannot exceed its upper bound of u_i.

Bottleneck 2 x_i increases to a point where it causes one of the current basic variables to become negative.

Bottleneck 3 x_i increases to a point where it causes one of the current basic variables to exceed its upper bound.

Let BN_k ($k = 1, 2, 3$) be the value of x_i where bottleneck k occurs. Then x_i can only be increased to a value of $\min\{BN_1, BN_2, BN_3\}$, the winning bottleneck. If the winning bottleneck is BN_1, then we make an upper-bound substitution on x_i by replacing x_i by $u_i - x_i'$. If the winning bottleneck is BN_2, then we enter x_i into the basis in the row corresponding to the basic variable that caused BN_2 to occur. If the winning bottleneck is BN_3, we make an upper-bound substitution on the variable x_j (by replacing x_j by $u_j - x_j'$) that reaches its upper bound when $x_i = BN_3$. Then enter x_i into the basis in the row for which x_j was a basic variable.

After following this procedure, examine the new row 0. If each variable has a nonnegative coefficient, we have obtained an optimal tableau. Otherwise, we try to increase the variable with the most negative coefficient in row 0.

Karmarkar's Method

Step 1 Begin at the feasible point $\mathbf{x}^0 = [\frac{1}{n} \quad \frac{1}{n} \quad \cdots \quad \frac{1}{n}]^T$ and set $k = 0$.

Step 2 Stop if $\mathbf{c}\mathbf{x}^k < \epsilon$. If not, go to step 3.

Step 3 Find the new point $\mathbf{y}^{k+1} = [y_1^{k+1} \quad y_2^{k+1} \quad \cdots \quad y_n^{k+1}]^T$ in the transformed unit simplex given by

$$\mathbf{y}^{k+1} = [\tfrac{1}{n} \quad \tfrac{1}{n} \quad \cdots \quad \tfrac{1}{n}]^T - \frac{\theta(I - P^T(PP^T)^{-1}P)[\mathrm{Diag}(\mathbf{x}^k)]\mathbf{c}^T}{\|\mathbf{c}_p\|\sqrt{n(n-1)}}$$

Here, $\|\mathbf{c}_p\|$ = the length of $(I - P^T(PP^T)^{-1}P)[\mathrm{Diag}(\mathbf{x}^k)]\mathbf{c}^T$, P is the $(m + 1) \times n$ matrix whose first m rows are $A[\mathrm{Diag}(\mathbf{x}^k)]$ and whose last row is a vector of 1's, and $0 < \theta < 1$ is chosen to ensure convergence of the algorithm. $\theta = \frac{1}{4}$ is known to ensure convergence.

Now obtain a new point $\mathbf{x}^{k+1}$ in the original space by using the centering transformation to determine the point corresponding to $\mathbf{y}^{k+1}$. That is, $\mathbf{x}^{k+1} = f^{-1}(\mathbf{y}^{k+1}|\mathbf{x}^k)$. Increase k by 1 and return to step 2.

REVIEW PROBLEMS

Group A

1 Use the revised simplex with the product form of the inverse to solve the following LP:

$$\max z = 4x_1 + 3x_2 + x_3$$
$$\text{s.t.} \quad 3x_1 + 2x_2 + x_3 \leq 6$$
$$x_2 + x_3 \leq 3$$
$$x_1 \quad\quad + x_3 \leq 2$$
$$x_1, x_2, x_3 \geq 0$$

2 Use the column generation technique to solve a cutting stock problem in which a customer demands 20 3-ft boards, 25 4-ft boards, and 30 5-ft boards, and demand is met by cutting up 14-ft boards.

3 Use the Dantzig–Wolfe decompostion method to solve the following LP:

$$\min z = 2x_1 - x_2 + x_3 - x_4$$
$$\text{s.t.} \quad x_1 + 2x_2 \quad\quad\quad \leq 4$$

$$x_1 - x_2 \leq 1$$
$$x_3 - 3x_4 \leq 7$$
$$2x_3 + x_4 \leq 10$$
$$x_1 + 3x_2 - x_3 - 2x_4 \leq 10$$
$$x_i \geq 0 \ (i = 1, 2, 3, 4)$$

4 Consider the following situation:

a Two types of cars are produced at three production plants and are demanded by three customers.

b You are given the cost of producing each type of car at each plant and the cost of shipping each type of car from each plant to each customer.

c You are given the production capacity of each plant (for each type of car).

d You are also told that at most one half of the total number of cars demanded by customer 1 can be met from plant 1 production.

Explain how you would use decomposition to minimize the cost of meeting the customers' demands.

5 A company manufactures products 1 and 2. A total of 100 hours of production time is available at each plant. The times required to produce a unit of each product at each plant are shown in Table 10, and the profits earned for a unit of each product produced at each plant are shown in Table 11. At most, 35 units of each product can be sold. Use decomposition to determine how the company can maximize profits.

TABLE 10

Plant	Hours	
	Product 1	Product 2
1	2	3
2	3	4

TABLE 11

Plant	Profit per Product ($)	
	Product 1	Product 2
1	8	6
2	10	8

REFERENCES

The following three references are classic works that detail methods used to solve large LPs:

Beale, E. *Mathematical Programming in Practice*. Pittman, 1968.

Lasdon, L. *Optimization Theory for Large Systems*. New York: Macmillan, 1970.

Orchard-Hays, W. *Advanced LP Computing Techniques*. New York: McGraw-Hill, 1968.

The following three references contain excellent discussions of Dantzig–Wolfe decomposition:

Bradley, S., A. Hax, and T. Magnanti. *Applied Mathematical Programming*. Reading, Mass.: Addison-Wesley, 1977.

Chvàtal, V. *Linear Programming*. San Francisco: Freeman, 1983.

Shapiro, J. *Mathematical Programming: Structures and Algorithms*. New York: Wiley, 1979.

The following two references discuss column generation and the cutting stock problem:

Gilmore, P., and R. Gomory. "A Linear Programming Approach to the Cutting Stock Problem," *Operations Research* 9(1961):849–859.

———. "A Linear Programming Approach to the Cutting Stock Problem: Part II," *Operations Research* 11(1963):863–888.

The following three references contain lucid discussions of Karmarkar's method:

Hooker, J. N. "Karmarkar's Linear Programming Algorithm," *Interfaces* 16(no. 4, 1986)75–90.

Murty, K. G. *Linear Complementarity, Linear and Nonlinear Programming*. Berlin, Germany: Heldermann Verlag, 1989.

Parker, G., and R. Rardin. *Discrete Optimization*. San Diego: Academic Press, 1988.

Game Theory

In previous chapters, we have encountered many situations in which a *single* decision maker chooses an optimal decision without reference to the effect that the decision has on other decision makers (and without reference to the effect that the decisions of others have on him or her). In many business situations, however, two or more decision makers simultaneously choose an action, and the action chosen by each player affects the rewards earned by the other players. For example, each fast-food company must determine an advertising and pricing policy for its product, and each company's decision will affect the revenues and profits of other fast-food companies.

Game theory is useful for making decisions in cases where two or more decision makers have conflicting interests. Most of our study of game theory deals with situations where there are only two decision makers (or players), but we briefly study n-person (where $n > 2$) game theory also. We begin our study of game theory with a discussion of two-player games in which the players have no common interest.

11.1 Two-Person Zero-Sum and Constant-Sum Games: Saddle Points

Characteristics of Two-Person Zero-Sum Games

1 There are two players (called the row player and column player).

2 The row player must choose 1 of m strategies. Simultaneously, the column player must choose 1 of n strategies.

3 If the row player chooses her ith strategy and the column player chooses her jth strategy, then the row player receives a reward of a_{ij} and the column player loses an amount a_{ij}. Thus, we may think of the row player's reward of a_{ij} as coming from the column player.

Such a game is called a **two-person zero-sum game,** which is represented by the matrix in Table 1 (the game's **reward matrix**). As previously stated, a_{ij} is the row player's

TABLE 1
Example of Two-Person Zero-Sum Game

Row Player's Strategy	Column Player's Strategy			
	Column 1	Column 2	$\cdots$	Column n
Row 1	a_{11}	a_{12}	$\cdots$	a_{1n}
Row 2	a_{21}	a_{22}	$\cdots$	a_{2n}
$\vdots$	$\vdots$	$\vdots$		$\vdots$
Row m	a_{m1}	a_{m2}	$\cdots$	a_{mn}

reward (and the column player's loss) if the row player chooses her ith strategy and the column player chooses her jth column strategy.

For example, in the two-person zero-sum game in Table 2, the row player would receive two units (and the column player would lose two units) if the row player chose her second strategy and the column player chose her first strategy.

A two-person zero-sum game has the property that for any choice of strategies, the sum of the rewards to the players is zero. In a zero-sum game, every dollar that one player wins comes out of the other player's pocket, so the two players have totally conflicting interests. Thus, cooperation between the two players would not occur.

John von Neumann and Oskar Morgenstern developed a theory of how two-person zero-sum games should be played, based on the following assumption.

Basic Assumption of Two-Person Zero-Sum Game Theory

Each player chooses a strategy that enables him to do the best he can, given that his opponent *knows the strategy he is following*. Let's use this assumption to determine how the row and column players should play the two-person zero-sum game in Table 3.

How should the row player play this game? If he chooses row 1, then the assumption implies that the column player will choose column 1 or column 2 and hold the row player to a reward of four units (the smallest number in row 1 of the game matrix). Similarly, if the row player chooses row 2, then the assumption implies that the column player will choose column 3 and hold the row player's reward to one unit (the smallest or minimum number in the second row of the game matrix). If the row player chooses row 3, then he will be held to the smallest number in the third row (5). Thus, the assumption implies that the row player should choose the row having the largest minimum. Because max (4, 1, 5) = 5, the row player should choose row 3. By choosing row 3, the row player can ensure that he will win at least max (row minimum) = five units.

From the column player's viewpoint, if he chooses column 1, then the row player will choose the strategy that makes the column player's losses as large as possible (and the row player's winnings as large as possible). Thus, if the column player chooses column 1, then the row player will choose row 3 (because the largest number in the first column is the 6 in the third row). Similarly, if the column player chooses column 2, then the row player will again choose row 3, because 5 = max (4, 3, 5). Finally, if the column player chooses column 3, the row player will choose row 1, causing the column player to lose 10 = max

TABLE 2

$$\begin{array}{cccc} 1 & 2 & 3 & -1 \\ 2 & 1 & -2 & 0 \end{array}$$

TABLE 3
A Game with a Saddle Point

Row Player's Strategy	Column Player's Strategy			Row Minimum
	Column 1	Column 2	Column 3	
Row 1	4	4	10	4
Row 2	2	3	1	1
Row 3	6	5	7	5
Column Maximum	6	5	10	

(10, 1, 7) units. Thus, the column player can hold his losses to min (column maximum) = min (6, 5, 10) = 5 by choosing column 2.

We have shown that the row player can ensure that he will win at least five units and the column player can hold the row player's winnings to at most five units. Thus, the only rational outcome of this game is for the row player to win exactly five units; the row player cannot expect to win more than five units, because the column player (by choosing column 2) can hold the row player's winnings to five units.

The game matrix we have just analyzed has the property of satisfying the **saddle point condition:**

$$\max_{\substack{\text{all} \\ \text{rows}} } (\text{row minimum}) = \min_{\substack{\text{all} \\ \text{columns}} } (\text{column maximum}) \tag{1}$$

Any two-person zero-sum game satisfying (1) is said to have a **saddle point.** If a two-person zero-sum game has a saddle point, then the row player should choose any strategy (row) attaining the maximum on the left side of (1). The column player should choose any strategy (column) attaining the minimum on the right side of (1). Thus, for the game we have just analyzed, a saddle point occurred where the row player chose row 3 and the column player chose column 2. The row player could make sure of receiving a reward of at least five units (by choosing the optimal strategy of row 3), and the column player could ensure that the row player would receive a reward of at most five units (by choosing the optimal strategy of column 2). If a game has a saddle point, then we call the common value of both sides of (1) the **value** (v) of the game to the row player. Thus, this game has a value of 5.

An easy way to spot a saddle point is to observe that the reward for a saddle point must be the smallest number in its row and the largest number in its column (see Problem 4 at the end of this section). Thus, like the center point of a horse's saddle, a saddle point for a two-person zero-sum game is a local minimum in one direction (looking across the row) and a local maximum in another direction (looking up and down the column).

A saddle point can also be thought of as an **equilibrium point** in that neither player can benefit from a unilateral change in strategy. For example, if the row player were to change from the optimal strategy of row 3 (to either row 1 or row 2), his reward would decrease, while if the column player changed from his optimal strategy of column 2 (to either column 1 or column 3), the row player's reward (and the column player's losses) would increase. Thus, a saddle point is stable in that neither player has an incentive to move away from it.

Many two-person zero-sum games do not have saddle points. For example, the game in Table 4 does not have a saddle point, because

$$\max (\text{row minimum}) = -1 < \min (\text{column maximum}) = +1$$

In Sections 11.2 and 11.3, we explain how to find the value and the optimal strategies for two-person zero-sum games that do not have saddle points.

TABLE 4
A Game with No Saddle Point

Row Player's Strategy	Column Player's Strategy		Row Minimum
	Column 1	Column 2	
Row 1	-1	$+1$	-1
Row 2	$+1$	-1	-1
Column Maximum	$+1$	$+1$	

Two-Person Constant-Sum Games

Even if a two-person game is not zero-sum, two players can still be in total conflict. To illustrate this, we now consider two-person constant-sum games.

Of course, a two-person zero-sum game is just a two-person constant-sum game with $c = 0$. A two-person constant-sum game maintains the feature that the row and column players are in total conflict, because a unit increase in the row player's reward will always result in a unit decrease in the column player's reward. In general, the optimal strategies and value for a two-person constant-sum game may be found by the same methods used to find the optimal strategies and value for a two-person zero-sum game.

EXAMPLE 1   Constant Sum TV Game

During the 8 to 9 P.M. time slot, two networks are vying for an audience of 100 million viewers. The networks must simultaneously announce the type of show they will air in that time slot. The possible choices for each network and the number of network 1 viewers (in millions) for each choice are shown in Table 5. For example, if both networks choose a western, the matrix indicates that 35 million people will watch network 1 and $100 - 35 = 65$ million people will watch network 2. Thus, we have a two-person constant-sum game with $c = 100$ (million). Does this game have a saddle point? What is the value of the game to network 1?

Solution   Looking at the row minima, we find that by choosing a soap opera, network 1 can be sure of at least max (15, 45, 14) = 45 million viewers. Looking at the column maxima, we find that by choosing a western, network 2 can hold network 1 to at most min (45, 58, 70) = 45 million viewers. Because

$$\text{max (row minimum)} = \text{min (column maximum)} = 45$$

we find that Equation (1) is satisfied. Thus, network 1's choosing a soap opera and network 2's choosing a western yield a saddle point; neither side will do better if it unilaterally changes strategy (check this). Thus, the value of the game to network 1 is 45 million viewers, and the value of the game to network 2 is $100 - 45 = 55$ million viewers. The optimal strategy for network 1 is to choose a soap opera, and the optimal strategy for network 2 is to choose a western.

TABLE 5
A Constant-Sum Game

Network 1	Network 2			Row Minimum
	Western	Soap Opera	Comedy	
Western	35	15	60	15
Soap Opera	45	58	50	45
Comedy	38	14	70	14
Column Maximum	45	58	70	

PROBLEMS

Group A

1 Find the value and optimal strategy for the game in Table 6.

2 Find the value and the optimal strategies for the two-person zero-sum game in Table 7.

Group B

3 Mad Max wants to travel from New York to Dallas by the shortest possible route. He may travel over the routes shown in Table 8. Unfortunately, the Wicked Witch can block one road leading out of Atlanta and one road leading out of Nashville. Mad Max will not know which roads have been blocked until he arrives at Atlanta or Nashville. Should Mad Max start toward Atlanta or Nashville? Which routes should the Wicked Witch block?

Group C

4 Explain why the reward for a saddle point must be the smallest number in its row and the largest number in its column. Suppose a reward is the smallest in its row and the largest in its column. Must that reward yield a saddle point? (*Hint:* Think about the idea of weak duality discussed in Chapter 6.)

TABLE 8

Route	Length of Route (Miles)
New York–Atlanta	800
New York–Nashville	900
Nashville–St. Louis	400
Nashville–New Orleans	200
Atlanta–St. Louis	300
Atlanta–New Orleans	600
St. Louis–Dallas	500
New Orleans–Dallas	300

TABLE 6

$$\begin{array}{cc} 2 & 2 \\ 1 & 3 \end{array}$$

TABLE 7

$$\begin{array}{cccc} 4 & 5 & 5 & 8 \\ 6 & 7 & 6 & 9 \\ 5 & 7 & 5 & 4 \\ 6 & 6 & 5 & 5 \end{array}$$

11.2 Two-Person Zero-Sum Games: Randomized Strategies, Domination, and Graphical Solution

In the previous section, we found that not all two-person zero-sum games have saddle points. We now discuss how to find the value and optimal strategies for a two-person zero-sum game that does not have a saddle point. We begin with the simple game of Odds and Evens.

EXAMPLE 2   Odds and Evens

Two players (called Odd and Even) simultaneously choose the number of fingers (1 or 2) to put out. If the sum of the fingers put out by both players is odd, then Odd wins $1 from Even. If the sum of the fingers is even, then Even wins $1 from Odd. We consider the row player to be Odd and the column player to be Even. Determine whether this game has a saddle point.

Solution   This is a zero-sum game, with the reward matrix shown in Table 9. Because max (row minimum) = -1 and min (column maximum) = $+1$, Equation (1) is not satisfied, and this game has no saddle point. All we know is that Odd can be sure of a reward of at least

TABLE 9

Reward Matrix for Odds and Evens

Row Player (Odd)	Column Player (Even)		Row Minimum
	1 Finger	2 Fingers	
1 Finger	-1	$+1$	-1
2 Fingers	$+1$	-1	-1
Column Maximum	$+1$	$+1$	

-1, and Even can hold Odd to a reward of at most $+1$. Thus, it is unclear how to determine the value of the game and the optimal strategies. Observe that for any choice of strategies by both players, there is a player who can benefit by unilaterally changing his or her strategy. For example, if both players put out one finger, then Odd could have increased her reward from -1 to $+1$ by putting out two fingers. Thus, no choice of strategies by the player is stable. We now determine optimal strategies and the value for this game.

Randomized or Mixed Strategies

To progress further with the analysis of Example 2 (and other games without saddle points), we must expand the set of allowable strategies for each player to include **randomized strategies.** Until now, we have assumed that each time a player plays a game, the player will choose the same strategy. Why not allow each player to select a probability of playing each strategy? For Example 2, we might define

$$x_1 = \text{probability that Odd puts out one finger}$$
$$x_2 = \text{probability that Odd puts out two fingers}$$
$$y_1 = \text{probability that Even puts out one finger}$$
$$y_2 = \text{probability that Even puts out two fingers}$$

If $x_1 \geq 0$, $x_2 \geq 0$, and $x_1 + x_2 = 1$, then (x_1, x_2) is a randomized, or mixed, strategy for Odd. For example, the mixed strategy $(\frac{1}{2}, \frac{1}{2})$ could be realized by Odd if she tossed a coin before each play of the game and put out one finger for heads and two fingers for tails. Similarly, if $y_1 \geq 0$, $y_2 \geq 0$, and $y_1 + y_2 = 1$, then (y_1, y_2) is a mixed strategy for Even.

Any mixed strategy $(x_1, x_2, \ldots, x_m)$ for the row player is a **pure strategy** if any of the x_i equals 1. Similarly, any mixed strategy $(y_1, y_2, \ldots, y_n)$ for the column player is a pure strategy if any of the y_i equals 1. A pure strategy is a special case of a mixed strategy in which a player always chooses the same action. Recall from Section 11.1 that the game in Table 10 had a value of 5 (corresponding to a saddle point), so the row player's optimal strategy could be represented as the pure strategy $(0, 0, 1)$, and the column player's optimal strategy could be represented as the pure strategy $(0, 1, 0)$.

We continue to assume that both players will play two-person zero-sum games in accordance with the basic assumption of Section 11.1. In the context of randomized strate-

TABLE 10

4	4	10
2	3	1
6	5	7

gies, the assumption (from the standpoint of Odd) may be stated as follows: Odd should choose x_1 and x_2 to maximize her expected reward under the assumption that Even knows the value of x_1 and x_2.

It is important to realize that even though we assume that Even knows the values of x_1 and x_2, on a particular play of the game, he is not assumed to know Odd's actual strategy choice until the instant the game is played.

Graphical Solution of Odds and Evens

Finding Odd's Optimal Strategy

With this version of the basic assumption, we can determine the optimal strategy for Odd. Because $x_1 + x_2 = 1$, we know that $x_2 = 1 - x_1$. Thus, any mixed strategy may be written as $(x_1, 1 - x_1)$, and it suffices to determine the value of x_1. Suppose Odd chooses a particular mixed strategy $(x_1, 1 - x_1)$. What is Odd's expected reward against each of Even's strategies? If Even puts out one finger, then Odd will receive a reward of -1 with probability x_1 and a reward of $+1$ with probability $x_2 = 1 - x_1$. Thus, if Even puts out one finger and Odd chooses the mixed strategy $(x_1, 1 - x_1)$, then Odd's expected reward is

$$(-1)x_1 + (+1)(1 - x_1) = 1 - 2x_1$$

As a function of x_1, this expected reward is drawn as line segment AC in Figure 1. Similarly, if Even puts out two fingers and Odd chooses the mixed strategy $(x_1, 1 - x_1)$, Odd's expected reward is

$$(+1)(x_1) + (-1)(1 - x_1) = 2x_1 - 1$$

which is line segment DE in Figure 1.

Suppose Odd chooses the mixed strategy $(x_1, 1 - x_1)$. Because Even is assumed to know the value of x_1, for any value of x_1 Even will choose the strategy (putting out one or two fingers) that yields a smaller expected reward for Odd. From Figure 1, we see that, as a function of x_1, Odd's expected reward will be given by the y coordinate in DBC. Odd wants to maximize her expected reward, so she should choose the value of x_1 corresponding to point B. Point B occurs where the line segments AC and DE intersect, or

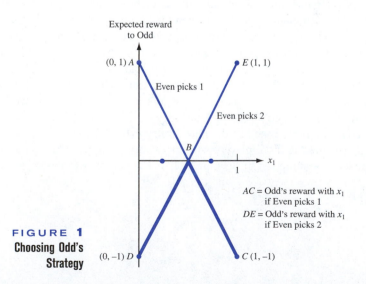

FIGURE 1
Choosing Odd's Strategy

where $1 - 2x_1 = 2x_1 - 1$. Solving this equation, we obtain $x_1 = \frac{1}{2}$. Thus, Odd should choose the mixed strategy $(\frac{1}{2}, \frac{1}{2})$. The reader should verify that against each of Even's strategies, $(\frac{1}{2}, \frac{1}{2})$ yields an expected reward of zero. Thus, zero is a **floor** on Odd's expected reward, because by choosing the mixed strategy $(\frac{1}{2}, \frac{1}{2})$, Odd can be sure that (for any choice of Even's strategy) her expected reward will always be at least zero.

Finding Even's Optimal Strategy

We now consider how Even should choose a mixed strategy (y_1, y_2). Again, because $y_2 = 1 - y_1$, we may ask how Even should choose a mixed strategy $(y_1, 1 - y_1)$. The basic assumption implies that Even should choose y_1 to minimize his expected losses (or, equivalently, minimize Odd's expected reward) under the assumption that Odd knows the value of y_1. Suppose Even chooses the mixed strategy $(y_1, 1 - y_1)$. What will Odd do? If Odd puts out one finger, then her expected reward is

$$(-1)y_1 + (+1)(1 - y_1) = 1 - 2y_1$$

which is line segment AC in Figure 2. If Odd puts out two fingers, then her expected reward is

$$(+1)(y_1) + (-1)(1 - y_1) = 2y_1 - 1$$

which is line segment DE in Figure 2. Because Odd is assumed to know the value of y_1, she will put out the number of fingers corresponding to max $(1 - 2y_1, 2y_1 - 1)$. Thus, for a given value of y_1, Odd's expected reward (and Even's expected loss) will be given by the y-coordinate on the piecewise linear curve ABE.

Now Even chooses the mixed strategy $(y_1, 1 - y_1)$ that will make Odd's expected reward as small as possible. Thus, Even should choose the value of y_1 corresponding to the lowest point on ABE (point B). Point B is where the line segments AC and DE intersect, or where $1 - 2y_1 = 2y_1 - 1$, or $y_1 = \frac{1}{2}$. The basic assumption implies that Even should choose the mixed strategy $(\frac{1}{2}, \frac{1}{2})$. For this mixed strategy, Even's expected loss (and Odd's expected reward) is zero. We say that zero is a **ceiling** on Even's expected loss

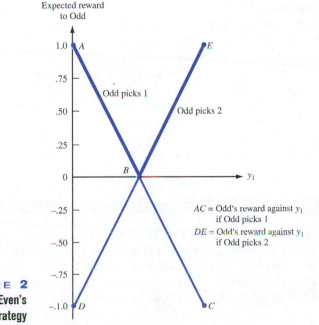

FIGURE 2
Choosing Even's Strategy

TABLE 11

How to Make a Nonoptimal Strategy Pay the Price

Odd's Mixed Strategy	Even Can Choose	Odd's Expected Reward (Even's expected losses)
$x_1 < \frac{1}{2}$	2 fingers	< 0 (on BD in Figure 1)
$x_1 > \frac{1}{2}$	1 finger	< 0 (on BC in Figure 1)
Even's Mixed Strategy	**Odd Can Choose**	**Odd's Expected Reward (Even's expected losses)**
$y_1 < \frac{1}{2}$	1 finger	> 0 (on AB in Figure 2)
$y_1 > \frac{1}{2}$	2 fingers	> 0 (on BE in Figure 2)

(or Odd's expected reward), because by choosing the mixed strategy $(\frac{1}{2}, \frac{1}{2})$, Even can ensure that his expected loss (for any choice of strategies by Odd) will not exceed zero.

More on the Idea of Value and Optimal Strategies

For the game of Odds and Evens, the row player's *floor* and the column player's *ceiling* are equal. This is not a coincidence. When each player is allowed to choose mixed strategies, the row player's floor will always equal the column player's ceiling. In Section 11.3, we use the Dual Theorem of Chapter 6 to prove this interesting result. We call the common value of the floor and ceiling the **value** of the game to the row player. Any mixed strategy for the row player that guarantees that the row player gets an expected reward at least equal to the value of the game is an **optimal strategy** for the row player. Similarly, any mixed strategy for the column player that guarantees that the column player's expected loss is no more than the value of the game is an optimal strategy for the column player. Thus, for Example 2, we have shown that the value of the game is zero, the row player's optimal strategy is $(\frac{1}{2}, \frac{1}{2})$, and the column player's optimal strategy is $(\frac{1}{2}, \frac{1}{2})$.

Example 2 illustrates that by allowing mixed strategies, we have enabled each player to find an optimal strategy in that *if the row player departs from her optimal strategy, the column player may have a strategy that reduces the row player's expected reward below the value of the game, and if the column player departs from his optimal strategy, the row player may have a strategy that increases her expected reward above the value of the game.* Table 11 illustrates this idea for the game of Odds and Evens.

For example, suppose that Odd chooses a nonoptimal mixed strategy with $x_1 < \frac{1}{2}$. Then, by choosing two fingers, Even ensures that Odd's expected reward can be read from BD in Figure 1. This means that if Odd chooses a mixed strategy having $x_1 < \frac{1}{2}$, then her expected reward can be negative (less than the value of the game).

To close this section, we find the value and optimal strategies for a more complicated game.

EXAMPLE 3 **Coin Toss Game with Bluffing**

A fair coin is tossed, and the result is shown to player 1. Player 1 must then decide whether to pass or bet. If player 1 passes, then he must pay player 2 $1. If player 1 bets, then player 2 (who does not know the result of the coin toss) may either fold or call the bet. If player 2 folds, then she pays player 1 $1. If player 2 calls and the coin comes up heads, then she pays player 1 $2; if player 2 calls and the coin comes up tails, then player 1 must pay her

$2. Formulate this as a two-person zero-sum game. Then graphically determine the value of the game and each player's optimal strategy.

Solution Player 1's strategies may be represented as follows: **PP,** pass on heads and pass on tails; **PB,** pass on heads and bet on tails; **BP,** bet on heads and pass on tails; and **BB,** bet on heads and bet on tails. Player 2 simply has the two strategies call and fold. For each choice of strategies, player 1's expected reward is as shown in Table 12.

To illustrate these computations, suppose player 1 chooses **BP** and player 2 calls. Then with probability $\frac{1}{2}$, heads is tossed. Then player 1 bets, is called, and wins $2 from player 2. With probability $\frac{1}{2}$, tails is tossed. In this case, player 1 passes and pays player 2 $1. Thus, if player 1 chooses **BP** and player 2 calls, then player 1's expected reward is $(\frac{1}{2})(2) + (\frac{1}{2})(-1) = \$0.50$. For each line in Table 12, the first term in the expectation corresponds to heads being tossed, and the second term corresponds to tails being tossed.

Example 3 may be described as the two-person zero-sum game represented by the reward matrix in Table 13. Because max (row minimum) $= 0 <$ min (column maximum) $= \frac{1}{2}$, this game does not have a saddle point. Observe that player 1 would be unwise ever to choose the strategy **PP,** because (for each of player 2's strategies) player 1 could do better than **PP** by choosing either **BP** or **BB**. In general, a strategy i for a given player is **dominated** by a strategy i' if, for each of the other player's possible strategies, the given player does at least as well with strategy i' as he or she does with strategy i, and if for at least one of the other player's strategies, strategy i' is superior to strategy i. A player may eliminate all dominated strategies from consideration. We have just shown that for player 1, **BP**

TABLE 12
Computation of Reward Matrix for Example 3

Player 1's Expected Reward	
PP vs. call	$(\frac{1}{2})(-1) + (\frac{1}{2})(-1) = -\$1$
PP vs. fold	$(\frac{1}{2})(-1) + (\frac{1}{2})(-1) = -\$1$
PB vs. call	$(\frac{1}{2})(-1) + (\frac{1}{2})(-2) = -\$1.50$
PB vs. fold	$(\frac{1}{2})(-1) + (\frac{1}{2})(1) \quad= \quad\$0$
BP vs. call	$(\frac{1}{2})(2) \quad+ (\frac{1}{2})(-1) = \quad\$0.50$
BP vs. fold	$(\frac{1}{2})(1) \quad+ (\frac{1}{2})(-1) = \quad\$0$
BB vs. call	$(\frac{1}{2})(2) \quad+ (\frac{1}{2})(-2) = \quad\$0$
BB vs. fold	$(\frac{1}{2})(1) \quad+ (\frac{1}{2})(1) \quad= \quad\$1$

TABLE 13
Reward Matrix for Example 3

	Player 2		
Player 1	**Call**	**Fold**	**Row Minimum**
PP	-1	-1	-1
PB	$-\frac{3}{2}$	0	$-\frac{3}{2}$
BP	$\frac{1}{2}$	0	0
BB	0	1	0
Column Maximum	$\frac{1}{2}$	1	

or **BB** dominates **PP.** Similarly, the reader should be able to show that player 1's **PB** strategy is dominated by **BP** or **BB.** After eliminating the dominated strategies **PP** and **PB,** we are left with the game matrix shown in Table 14.

As with Odds and Evens, this game has no saddle point, and we proceed with a graphical solution. Let

$$x_1 = \text{probability that player 1 chooses } \textbf{BP}$$
$$x_2 = 1 - x_1 = \text{probability that player 1 chooses } \textbf{BB}$$
$$y_1 = \text{probability that player 2 chooses call}$$
$$y_2 = 1 - y_1 = \text{probability that player 2 chooses fold}$$

To determine the optimal strategy for player 1, observe that for any value of x_1, his expected reward against calling is

$$(\tfrac{1}{2})(x_1) + 0(1 - x_1) = \frac{x_1}{2}$$

which is line segment AB in Figure 3. Against folding, player 1's expected reward is

$$0(x_1) + 1(1 - x_1) = 1 - x_1$$

which is line segment CD in Figure 3. Player 2 is assumed to know the value of x_1, so player 1's expected reward (as a function of x_1) is given by the piecewise linear curve AED

TABLE 14
Reward Matrix for Example 3 after Dominated Strategies Have Been Eliminated

Player 1	Player 2		Row Minimum
	Call	Fold	
BP	$\frac{1}{2}$	0	0
BB	0	1	0
Column Maximum	$\frac{1}{2}$	1	

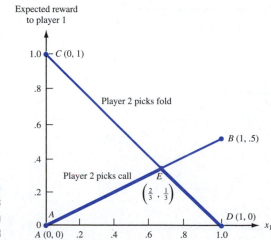

FIGURE 3
How Player 1 Chooses Optimal Strategy in Example 3

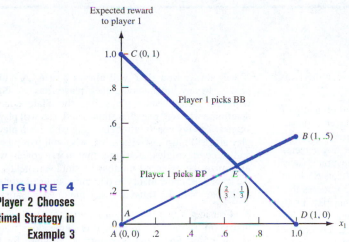

FIGURE 4
How Player 2 Chooses Optimal Strategy in Example 3

in Figure 3. Thus, to maximize his expected reward, player 1 should choose the value of x_1 corresponding to point E, which solves $x_1/2 = 1 - x_1$, or $x_1 = \frac{2}{3}$. Then $x_2 = 1 - \frac{2}{3} = \frac{1}{3}$, and player 1's expected reward against either of player 2's strategies is $\frac{x_1}{2}$ (or $1 - x_1$) $= \frac{1}{3}$.

How should player 2 choose y_1? (Remember, $y_2 = 1 - y_1$.) For a given value of y_1, suppose player 1 chooses **BP.** Then his expected reward is

$$(\tfrac{1}{2})(y_1) + 0(1 - y_1) = \frac{y_1}{2}$$

which is line segment AB in Figure 4. For a given value of y_1, suppose player 1 chooses **BB.** Then his expected reward is

$$0(y_1) + 1(1 - y_1) = 1 - y_1$$

which is line segment CD in Figure 4. Thus, for a given value of y_1, player 1 will choose a strategy that causes his expected reward to be given by the piecewise linear curve CEB in Figure 4. Knowing this, player 2 should choose the value of y_1 corresponding to point E in Figure 4. The value of y_1 at point E is the solution to $\frac{y_1}{2} = 1 - y_1$, or $y_1 = \frac{2}{3}$ (and $y_2 = \frac{1}{3}$). You should check that no matter what player 1 does, player 2's mixed strategy $(\frac{2}{3}, \frac{1}{3})$ ensures that player 1 earns an expected reward of $\frac{1}{3}$.

In summary, the value of the game is $\frac{1}{3}$ to player 1; the optimal mixed strategy for player 1 is $(\frac{2}{3}, \frac{1}{3})$; and the optimal strategy for player 2 is also $(\frac{2}{3}, \frac{1}{3})$.

REMARKS **1** Observe that player 1 should bet $\frac{1}{3}$ of the time that he has a losing coin. Thus, our simple model indicates that player 1's optimal strategy includes bluffing.
2 In Problem 4 at the end of this section, it will be shown that if player 1 deviates from his optimal strategy, player 2 can hold him to an expected reward that is less than the value ($\frac{1}{3}$) of the game. Similarly, Problem 5 will show that if player 2 deviates from her optimal strategy, player 1 can earn an expected reward in excess of the value ($\frac{1}{3}$) of the game.
3 Although we have only applied the graphical method to games in which each player (after dominated strategies have been eliminated) has only two strategies, the graphical approach can be used to solve two-person zero-sum games in which only one player has two strategies (games in which the reward matrix is $2 \times n$ or $m \times 2$). We choose, however, to solve all non-2×2 two-person games by the linear programming method outlined in the next section.

PROBLEMS

Group A

1 Find the value and the optimal strategies for the two-person zero-sum game in Table 15.

2 Player 1 writes an integer between 1 and 20 on a slip of paper. Without showing this slip of paper to player 2, player 1 tells player 2 what he has written. Player 1 may lie or tell the truth. Player 2 must then guess whether or not player 1 has told the truth. If caught in a lie, player 1 must pay player 2 $10; if falsely accused of lying, player 1 collects $5 from player 2. If player 1 tells the truth and player 2 guesses that player 1 has told the truth, then player 1 must pay $1 to player 2. If player 1 lies and player 2 does not guess that player 1 has lied, player 1 wins $5 from player 2. Determine the value of this game and each player's optimal strategy.

3 Find the value and optimal strategies for the two-person zero-sum game in Table 16.

4 For Example 3, show that if player 1 deviates from his optimal strategy, then player 2 can ensure that player 1 earns an expected reward that is less than the value $(\frac{1}{3})$ of the game.

5 For Example 3, show that if player 2 deviates from her optimal strategy, then player 1 can ensure that he earns an expected reward that is more than the value $(\frac{1}{3})$ of the game.

6 Two competing firms must simultaneously determine how much of a product to produce. The total profit earned by the two firms is always $1,000. If firm 1's production level is low and firm 2's is also low, then firm 1 earns a profit of $500; if firm 1's level is low and 2's is high, then firm 1's profit is $400. If firm 1's production level is high and so is firm 2's, then firm 1's profit is $600; but if firm 1's level is high while firm 2's level is low, then firm 1's profit is only $300. Find the value and optimal strategies for this constant-sum game.

7 Mo and Bo each have a quarter and a penny. Simultaneously, they each display a coin. If the coins match, then Mo wins both coins; if they don't match, then Bo wins both coins. Determine optimal strategies for this game.

Group B

8 State University is about to play Ivy College for the state tennis championship. The State team has two players (A and B), and the Ivy team has three players (X, Y, and Z). The following facts are known about the players' relative abilities:

X will always beat B; Y will always beat A; A will always beat Z. In any other match, each player has a $\frac{1}{2}$ chance of winning. Before State plays Ivy, the State coach must determine who will play first singles and who will play second singles. The Ivy coach (after choosing which two players will play singles) must also determine who will play first singles and second singles. Assume that each coach wants to maximize the expected number of singles matches won by the team. Use game theory to determine optimal strategies for each coach and the value of the game to each team.

9 Consider a two-person zero-sum game with the reward matrix in Table 17. Suppose this game does not have a saddle point. Show that the optimal strategy for the row player is to play the first row a fraction $(d - c)/(a + d - b - c)$ of the time and the optimal strategy for the column player is to play the first column a fraction $(d - b)/(a + d - b - c)$ of the time.

10 Consider the following simplified version of football. On each play the offense chooses to run or pass. At the same time, the defense chooses to play a run defense or pass defense. The number of yards gained on each play is determined by the reward matrix in Table 18. The offense's goal is to maximize the average yards gained per play.

a Use Problem 9 to show that the offense should run 10/17 of the time.

b Suppose that the effectiveness of a pass against the run defense improves. Use the results of Problem 9 to show that the offense should pass less! Can you give an explanation for this strange phenomenon?

11 Use the idea of dominated strategies to determine optimal strategies for the reward matrix in Table 19.

TABLE 17

a	b
c	d

TABLE 18

	Defense	
Offense	Run	Pass
Run	1	8
Pass	10	0

TABLE 15

1	2	3
2	0	3

TABLE 16

2	1	3
4	3	2

TABLE 19

−5	−10	−1	−10	2	−1
−1	2	−10	7	−5	20
2	7	−5	−10	−10	7
7	20	−1	−1	−1	2
20	7	−10	7	−1	−10

11.3 Linear Programming and Zero-Sum Games

Linear programming can be used to find the value and optimal strategies (for the row and column players) for any two-person zero-sum game. To illustrate the main ideas, we consider the well-known game Stone, Paper, Scissors.

EXAMPLE 4 **Stone, Paper, Scissors**

Two players simultaneously utter one of the three words *stone, paper,* or *scissors* and show corresponding hand signs. If both players utter the same word, then the game is a draw. Otherwise, one player wins $1 from the other player according to the following: Scissors defeats (cuts) paper, paper defeats (covers) stone, and stone defeats (breaks) scissors. Find the value and optimal strategies for this two-person zero-sum game. The solution is given later in this section.

The reward matrix is shown in Table 20. Observe that no strategies are dominated and that the game does not have a saddle point. To determine optimal mixed strategies for the row and the column player, define

$$x_1 = \text{probability that row player chooses stone}$$
$$x_2 = \text{probability that row player chooses paper}$$
$$x_3 = \text{probability that row player chooses scissors}$$
$$y_1 = \text{probability that column player chooses stone}$$
$$y_2 = \text{probability that column player chooses paper}$$
$$y_3 = \text{probability that column player chooses scissors}$$

The Row Player's LP

If the row player chooses the mixed strategy (x_1, x_2, x_3), then her expected reward against each of the column player's strategies is as shown in Table 21. Suppose the row player chooses the mixed strategy (x_1, x_2, x_3). By the basic assumption, the column player will choose a strategy that makes the row player's expected reward equal to $\min (x_2 - x_3, -x_1 + x_3, x_1 - x_2)$. Then the row player should choose (x_1, x_2, x_3) to make $\min (x_2 - x_3, -x_1 + x_3, x_1 - x_2)$ as *large* as possible. To obtain an LP formulation (called the row player's LP) that will yield the row player's optimal strategy, observe that for any values of x_1, x_2, and x_3, $\min (x_2 - x_3, -x_1 + x_3, x_1 - x_2)$ is just the largest number (call it v) that is simulta-

TABLE 20
Reward Matrix for Stone, Paper, Scissors

Row Player	Column Player			Row Minimum
	Stone	Paper	Scissors	
Stone	0	-1	$+1$	-1
Paper	$+1$	0	-1	-1
Scissors	-1	$+1$	0	-1
Column Maximum	$+1$	$+1$	$+1$	

TABLE **21**

Expected Reward to Row Player in Stone, Paper, Scissors

Column Player Chooses	Row Player's Expected Reward If Row Player Chooses (x_1, x_2, x_3)
Stone	$x_2 - x_3$
Paper	$-x_1 + x_3$
Scissors	$x_1 - x_2$

neously less than or equal to $x_2 - x_3$, $-x_1 + x_3$, and $x_1 - x_2$. After noting that x_1, x_2, and x_3 must satisfy $x_1 \geq 0$, $x_2 \geq 0$, $x_3 \geq 0$, and $x_1 + x_2 + x_3 = 1$, we see that the row player's optimal strategy can be found by solving the following LP:

$$\max z = v$$
$$\text{s.t.} \quad v \leq x_2 - x_3 \qquad \text{(Stone constraint)}$$
$$v \leq -x_1 + x_3 \qquad \text{(Paper constraint)}$$
$$v \leq x_1 - x_2 \qquad \text{(Scissors constraint)} \tag{2}$$
$$x_1 + x_2 + x_3 = 1$$
$$x_1, x_2, x_3 \geq 0; \; v \text{ urs}$$

Note that there is a constraint in (2) for each of the column player's strategies. The value of v in the optimal solution to (2) is the row player's *floor*, because no matter what strategy (pure or mixed) is chosen by the column player, the row player is sure to receive an expected reward of at least v.

The Column Player's LP

How should the column player choose an optimal mixed strategy (y_1, y_2, y_3)? Suppose the column player has chosen the mixed strategy (y_1, y_2, y_3). For each of the row player's strategies, we may compute the row player's expected reward if the column player chooses (y_1, y_2, y_3) (see Table 22). The row player is assumed to know (y_1, y_2, y_3), the row player will choose a strategy to ensure that she obtains an expected reward of max $(-y_2 + y_3, y_1 - y_3, -y_1 + y_2)$. Thus, the column player should choose (y_1, y_2, y_3) to make max $(-y_2 + y_3, y_1 - y_3, -y_1 + y_2)$ as *small* as possible. To obtain an LP formulation that will yield the column player's optimal strategies, observe that for any choice of (y_1, y_2, y_3), max $(-y_2 + y_3, y_1 - y_3, -y_1 + y_2)$ will equal the smallest number that is simultaneously greater than or equal to $-y_2 + y_3$, $y_1 - y_3$, and $-y_1 + y_2$ (call this number w). Also note that for (y_1, y_2, y_3) to be a mixed strategy, (y_1, y_2, y_3) must satisfy $y_1 + y_2 + y_3 = 1$, $y_1 \geq 0$, $y_2 \geq 0$, and $y_3 \geq 0$. Thus, the column player may find his optimal strategy by solving the following LP:

$$\min z = w$$
$$\text{s.t.} \quad w \geq -y_2 + y_3 \qquad \text{(Stone constraint)}$$
$$w \geq y_1 - y_3 \qquad \text{(Paper constraint)}$$
$$w \geq -y_1 + y_2 \qquad \text{(Scissors constraint)} \tag{3}$$
$$y_1 + y_2 + y_3 = 1$$
$$y_1, y_2, y_3 \geq 0; \; w \text{ urs}$$

Observe that (3) contains a constraint corresponding to each of the row player's strategies. Also, the optimal objective function w for (3) is a *ceiling* on the column player's expected

TABLE **22**
Expected Reward to Row Player in
Stone, Paper, Scissors

Row Player Chooses	Row Player's Expected Reward If Column Player Chooses (y_1, y_2, y_3)
Stone	$-y_2 + y_3$
Paper	$y_1 - y_3$
Scissors	$-y_1 + y_2$

losses (or the row player's expected reward), because by choosing a mixed strategy (y_1, y_2, y_3) that solves (3), the column player can ensure that his expected losses will be (against any of the row player's strategies) at most w.

Relation Between the Row and the Column Player's LPs

It is easy to show that the column player's LP is the dual of the row player's LP. Begin by rewriting the row player's LP (2) as

$$\max z = v$$
$$\text{s.t.} \quad -x_2 + x_3 + v \leq 0$$
$$x_1 \quad - x_3 + v \leq 0$$
$$-x_1 + x_2 \quad + v \leq 0 \tag{4}$$
$$x_1 + x_2 + x_3 \quad = 1$$
$$x_1, x_2, x_3 \geq 0; \; v \text{ urs}$$

Let the dual variables for the constraints in (4) be y_1, y_2, y_3, and w, respectively. We can now show that the dual of the row player's LP is the column player's LP. As in Section 6.5, we read the row player's LP across in Table 23 and find the dual of the row player's LP by reading down. Recall that the dual constraint corresponding to the variable v will be an equality constraint (because v is urs), and the dual variable w corresponding to the primal constraint $x_1 + x_2 + x_3 = 1$ will be urs (because $x_1 + x_2 + x_3 = 1$ is an equality constraint). Reading down in Table 23, we find the dual of the row player's LP (4) to be

$$\min z = w$$
$$\text{s.t.} \quad y_2 - y_3 + w \geq 0$$
$$-y_1 \quad + y_3 + w \geq 0$$
$$y_1 - y_2 \quad + w \geq 0$$
$$y_1 + y_2 + y_3 \quad = 1$$
$$y_1, y_2, y_3 \geq 0; \; w \text{ urs}$$

After transposing all terms involving y_1, y_2, and y_3 in the first three constraints to the right-hand side, we see that the last LP is the same as the column player's LP (3). Thus, the dual of the row player's LP is the column player's LP. (Of course, the dual of the column player's LP would be the row player's LP.)

It is easy to show that both the row player's LP (2) and the column player's LP (3) have an optimal solution (that is, neither LP can be infeasible or unbounded). Then the Dual Theorem of Section 6.7 implies that v, the optimal objective function value for the row player's LP, and w, the optimal objective function value for the column player's LP, are equal. Thus, the row player's floor equals the column player's ceiling. This result is often

TABLE 23
Dual or Row Player's LP

Min		Max			
	x_1	x_2	x_3	v	
y_1 $(y_1 \geq 0)$	0	-1	1	1	≤ 0
y_2 $(y_2 \geq 0)$	1	0	-1	1	≤ 0
y_3 $(y_3 \geq 0)$	-1	1	0	1	≤ 0
w (urs)	1	1	1	0	$= 1$
	≥ 0	≥ 0	≥ 0	$= 1$	

known as the **Minimax Theorem.** We call the common value of v and w the **value** of the game to the row player. As in Sections 11.1 and 11.2, the row player can (by playing an optimal strategy) guarantee that her expected reward will at least equal the value of the game. Similarly, the column player can (by playing an optimal strategy) guarantee that his expected losses will not exceed the value of the game. It can also be shown (see Problem 6 at the end of this section) that the optimal strategies obtained via linear programming represent a stable equilibrium, because neither player can improve his or her situation by a unilateral change in strategy.

For the Stone, Paper, Scissors game, the optimal solution to the row player's LP (2) is $w = 0$, $x_1 = \frac{1}{3}$, $x_2 = \frac{1}{3}$, $x_3 = \frac{1}{3}$, and the optimal solution to the column player's LP (3) is $v = 0$, $y_1 = \frac{1}{3}$, $y_2 = \frac{1}{3}$, $y_3 = \frac{1}{3}$. Note that the first solution is feasible in (2), and the second solution is feasible in (3). Each solution yields an objective function value of zero, so Lemma 2 in Chapter 6 shows that $x_1 = \frac{1}{3}$, $x_2 = \frac{1}{3}$, $x_3 = \frac{1}{3}$ is optimal for the row player's LP, and $y_1 = \frac{1}{3}$, $y_2 = \frac{1}{3}$, $y_3 = \frac{1}{3}$ is optimal for the column player's LP.

The complementary slackness theory of linear programming (discussed in Section 6.10) could have been used to find the optimal strategies and value for Stone, Paper, Scissors (as well as other games). Before showing how, we state the row and the column player's LPs for a general two-person zero-sum game.

Consider a two-person zero-sum game with the reward matrix shown in Table 24. The reasoning used to derive (2) and (3) yields the following LPs:

$$\max z = v$$

Row Player's LP

$$\text{s.t.} \quad v \leq a_{11}x_1 + a_{21}x_2 + \cdots + a_{m1}x_m \quad \text{(Column 1 constraint)}$$
$$v \leq a_{12}x_1 + a_{22}x_2 + \cdots + a_{m2}x_m \quad \text{(Column 2 constraint)}$$
$$\vdots$$
$$v \leq a_{1n}x_1 + a_{2n}x_2 + \cdots + a_{mn}x_m \quad \text{(Column } n \text{ constraint)}$$
$$x_1 + x_2 + \cdots + x_m = 1$$
$$x_i \geq 0 \quad (i = 1, 2, \ldots, m); \ v \text{ urs}$$

(5)

$$\min z = w$$

Column Player's LP

$$\text{s.t.} \quad w \geq a_{11}y_1 + a_{12}y_2 + \cdots + a_{1n}y_n \quad \text{(Row 1 constraint)}$$
$$w \geq a_{21}y_1 + a_{22}y_2 + \cdots + a_{2n}y_n \quad \text{(Row 2 constraint)}$$
$$\vdots$$
$$w \geq a_{m1}y_1 + a_{m2}y_2 + \cdots + a_{mn}y_n \quad \text{(Row } m \text{ constraint)}$$
$$y_1 + y_2 + \cdots + y_n = 1$$
$$y_j \geq 0 \quad (j = 1, 2, \ldots, n); \ w \text{ urs}$$

(6)

TABLE 24
A General Two-Person Zero-Sum Game

	Column Player			
Row Player	Strategy 1	Strategy 2	$\cdots$	Strategy n
Strategy 1	a_{11}	a_{12}	$\cdots$	a_{1n}
Strategy 2	a_{21}	a_{22}	$\cdots$	a_{2n}
$\vdots$	$\vdots$	$\vdots$		$\vdots$
Strategy m	a_{m1}	a_{m2}	$\cdots$	a_{mn}

In (5), x_i = probability that the row player chooses row i, and in (6), y_j = probability that the column player chooses column j. The jth constraint ($j = 1, 2, \ldots, n$) in the row player's LP implies that her expected reward against column j must at least equal v; otherwise, the column player could hold the row player's expected reward below v by choosing column j. Similarly, the ith ($i = 1, 2, \ldots, m$) constraint in the column player's LP implies that if the row player chooses row i, then the column player's expected losses cannot exceed w; if this were not the case, the row player could obtain an expected reward that exceeded w by choosing row i.

How to Solve the Row and the Column Players' LPs

It is easy to show (see Problem 9 at the end of this section) that if we add a constant c to each entry in a game's reward matrix, the optimal strategies for each player remain unchanged, but the optimal values of w and v (and thus the value of the game) are both increased by c. Let A be the original reward matrix. Suppose we add $c = |$most negative entry in reward matrix$|$ to each element of A. Call the new reward matrix A'. A' is a two-person constant-sum game. (Why?) Let $\bar{v}$ and $\bar{w}$ be the optimal objective function values for the row and the column players' LP's for A, and let $\bar{v}'$ and $\bar{w}'$ denote the same quantities for the game A'. Because A' will have no negative rewards, $\bar{v}' \geq 0$ and $\bar{w}' \geq 0$ must hold. Thus, when solving the row and the column players' LP's for A', we may assume that $v' \geq 0$ and $w' \geq 0$ and ignore v urs and w urs. Then the optimal strategies for A' will be identical to the optimal strategies for A, and the value of $A' =$ (value of A) $+ c$, or value of $A =$ (value of A') $- c$.

In solving small games by hand, it is often helpful to use the constraint $x_1 + x_2 + \cdots + x_m = 1$ to eliminate one of the x_i's from the row player's LP and to use the constraint $y_1 + y_2 + \cdots + y_n = 1$ to eliminate one of the y_i's from the column player's LP. Then (as illustrated by Examples 5 and 6, which follow), the complementary slackness results of Section 6.10 can often be used to solve the row and the column player's LP's simultaneously.

EXAMPLE 4   **Stone, Paper, Scissors (Continued)**

Solution   The most negative element in the Stone, Paper, Scissors reward matrix is -1. Therefore, we add $|-1| = 1$ to each element of the reward matrix. This yields the constant-sum game shown in Table 25. The row player's LP is as follows:

$$\max v'$$
$$\text{s.t.} \quad v' \leq x_1 + 2x_2$$
$$v' \leq x_2 + 2x_3$$
$$v' \leq 2x_1 + x_3$$
$$x_1 + x_2 + x_3 = 1$$
$$x_1, x_2, x_3, v' \geq 0$$

TABLE 25
Modified Reward Matrix for Stone, Paper, Scissors

Row Player	Column Player		
	Stone	Paper	Scissors
Stone	1	0	2
Paper	2	1	0
Scissors	0	2	1

Substituting $x_3 = 1 - x_1 - x_2$ transforms the row player's LP into the following LP:

$$\max v'$$

s.t.   (a) $v' - x_1 - 2x_2 \leq 0$ (y_1, or column 1, constraint)

      (b) $v' + 2x_1 + x_2 \leq 2$ (y_2, or column 2, constraint)

      (c) $v' - x_1 + x_2 \leq 1$ (y_3, or column 3, constraint)   (7)

$$x_1, x_2, v' \geq 0$$

The column player's LP is as follows:

$$\min w'$$

s.t. $w' \geq y_1 + 2y_3$ (x_1, or row 1, constraint)

      $w' \geq 2y_1 + y_2$ (x_2, or row 2, constraint)

      $w' \geq 2y_2 + y_3$ (x_3, or row 3, constraint)

$$y_1 + y_2 + y_3 = 1$$

$$y_1, y_2, y_3, w' \geq 0$$

Substituting $y_3 = 1 - y_1 - y_2$ transforms the column player's LP into the following LP:

$$\min w'$$

s.t.   (a) $w' + y_1 + 2y_2 \geq 2$ (x_1, or row 1, constraint)

      (b) $w' - 2y_1 - y_2 \geq 0$ (x_2, or row 2, constraint)   (8)

      (c) $w' + y_1 - y_2 \geq 1$ (x_3, or row 3, constraint)

$$y_1, y_2, w' \geq 0$$

Stone, Paper, Scissors appears to be a fair game, so we might conjecture that $v = w = 0$. This would make $v' = w' = 0 + 1 = 1$. Let's try this and conjecture that constraints (7a) and (7b) are binding in the optimal solution to (7). If this is the case, then solving (7a) and (7b) simultaneously (with $v' = 1$) yields $x_1 = x_2 = \frac{1}{3}$. Because $x_1 = \frac{1}{3}, x_2 = \frac{1}{3}, w' = 1$ satisfies (7c) with equality, we have obtained a feasible solution to the row player's LP. Suppose this solution is optimal for the row player's LP. Then by complementary slackness (see Section 6.10), $x_1 > 0$ and $x_2 > 0$ would imply that the first two dual constraints in (8) must be binding in the optimal solution to (8). Solving (8a) and (8b) simultaneously (using $w' = 1$) yields $y_1 = y_2 = \frac{1}{3}, w' = 1$. This solution is dual feasible. Thus, we have found a primal feasible and a dual feasible solution, both of which have the same objective function value and are optimal. Thus:

1   The value of Stone, Paper, Scissors is $v' - 1 = 0$.

2   The optimal strategy for the row player is $(\frac{1}{3}, \frac{1}{3}, \frac{1}{3})$.

3   The optimal strategy for the column player is $(\frac{1}{3}, \frac{1}{3}, \frac{1}{3})$.

Suppose we had not been able to conjecture that $v' = w' = 1$. Then the row player's LP (7) would have had three unknowns (x_1, x_2, and v'), and we might have hoped that the optimal solution to (7) occurred where all three constraints (7a)–(7c) were binding. Solving (7a)–(7c) simultaneously yields $v' = 1$, $x_1 = x_2 = \frac{1}{3}$, $x_3 = 1 - \frac{2}{3} = \frac{1}{3}$. If this is the optimal solution to the row player's LP, then complementary slackness implies that constraints (8a)–(8c) must all be binding. Simultaneously solving (8a)–(8c) yields $w' = 1$, $y_1 = y_2 = \frac{1}{3}$, $y_3 = 1 - \frac{2}{3} = \frac{1}{3}$. Again we have obtained a primal feasible point and a dual feasible point having the same objective function value, and both solutions must be optimal.

EXAMPLE 5 Using Complementary Slackness to Solve a Two-Person Zero-Sum Game

Find the value and optimal strategies for the two-person zero-sum game in Table 26.

Solution The game has no saddle point and no dominated strategies, so we set up the row and the column players' LP's. All entries in the reward matrix are non-negative, so we are sure that the value of the game is non-negative. The row and the column players' LP's for this game are as follows:

$$\max v$$
$$\text{s.t.} \quad v \le 30x_1 + 60x_2$$
$$v \le 40x_1 + 10x_2$$
$$v \le 36x_1 + 36x_2 \tag{9}$$
$$x_1 + x_2 = 1$$
$$x_1, x_2, v \ge 0$$

Substituting $x_2 = 1 - x_1$ into the row player's LP yields

$$\max v$$
$$\text{s.t.} \quad \text{(a)} \ v + 30x_1 \le 60 \quad (y_1, \text{ or column 1, constraint})$$
$$\text{(b)} \ v - 30x_1 \le 10 \quad (y_2, \text{ or column 2, constraint}) \tag{9'}$$
$$\text{(c)} \ v \quad\quad\ \le 36 \quad (y_3, \text{ or column 3, constraint})$$
$$x_1, v \ge 0$$

Similarly, we find

$$\min w$$
$$\text{s.t.} \quad \text{(a)} \quad w \ge 30y_1 + 40y_2 + 36y_3$$
$$\text{(b)} \quad w \ge 60y_1 + 10y_2 + 36y_3 \tag{10}$$
$$y_1 + y_2 + y_3 = 1$$
$$y_1, y_2, y_3, w \ge 0$$

and substituting $y_3 = 1 - y_1 - y_2$ into the column player's LP yields

$$\min w$$
$$\text{s.t.} \quad \text{(a)} \quad w + 6y_1 - 4y_2 \ge 36 \quad (x_1, \text{ or row 1, constraint})$$
$$\text{(b)} \quad w - 24y_1 + 26y_2 \ge 36 \quad (x_2, \text{ or row 2, constraint}) \tag{10'}$$
$$y_1, y_2, w \ge 0$$

TABLE 26
Reward Matrix for Example 5

	Column Player			Row Minimum
Row Player	30	40	36	30
	60	10	36	10
Column Maximum	60	40	36	

When using complementary slackness to solve an LP and its dual, it is usually easier to first examine the LP with the smaller number of variables. Thus, we first examine (9′). We assume that (9′a) and (9′b) are both binding in the optimal solution to the row player's LP. Then $v = 35$, $x_1 = \frac{5}{6}$, $x_2 = \frac{1}{6}$ would be the optimal solution to the row player's LP. This solution is feasible in the row player's LP and makes constraint (9′c) nonbinding. If this solution is optimal for the row player's LP, then complementary slackness implies that (10′a) and (10′b) must both be binding and $y_3 = 0$ must hold. This implies that $y_1 + y_2 = 1$, or $y_2 = 1 - y_1$. Trying $w = 35$ and substituting $y_2 = 1 - y_1$ in (10′a) and (10′b) yields $y_1 = y_2 = \frac{1}{2}$. Thus, we have found a feasible solution ($v = 35$, $x_1 = \frac{5}{6}$, $x_2 = \frac{1}{6}$) to the row player's LP and a feasible solution ($w = 35$, $y_1 = \frac{1}{2}$, $y_2 = \frac{1}{2}$, $y_3 = 0$) to the column player's LP, both of which have the same objective function value. We have therefore found the value of the game and the optimal strategy for each player.

In closing, we note that while the column player's third strategy is not dominated by column 1 or column 2, he should still never choose column 3. Why?

In the following two-person zero-sum game, the complementary slackness method does not yield optimal strategies.

EXAMPLE 6   Two-Finger Morra

Two players in the game of Two-Finger Morra simultaneously put out either one or two fingers. Each player must also announce the number of fingers that he believes his opponent has put out. If neither or both players correctly guess the number of fingers put out by the opponent, the game is a draw. Otherwise, the player who guesses correctly wins (from the other player) the sum (in dollars) of the fingers put out by the two players. If we let (i, j) represent the strategy of putting out i fingers and guessing the opponent has put out j fingers, the appropriate reward matrix is as shown in Table 27.

Solution   Again, this game has no saddle point and no dominated strategies. To ensure that the value of the game is non-negative, we add 4 to each entry in the reward matrix. This yields the reward matrix in Table 28. For this game, the row player's and the column player's LP's are as follows (recall that the value for the original Two-Finger Morra game $= v' - 4$):

$$\max v'$$

	s.t.	(a) $v' \leq 4x_1 + 2x_2 + 7x_3 + 4(1 - x_1 - x_2 - x_3)$	(y_1 constraint)
Row		(b) $v' \leq 6x_1 + 4x_2 + 4x_3 + (1 - x_1 - x_2 - x_3)$	(y_2 constraint)
Player's		(c) $v' \leq x_1 + 4x_2 + 4x_3 + 8(1 - x_1 - x_2 - x_3)$	(y_3 constraint)
LP		(d) $v' \leq 4x_1 + 7x_2 + \ 4(1 - x_1 - x_2 - x_3)$	(y_4 constraint)

$$x_1, x_2, x_3, v' \geq 0$$

$$\max w'$$

	s.t.	(a) $w' \geq 4y_1 + 6y_2 + y_3 + 4(1 - y_1 - y_2 - y_3)$	(x_1 constraint)
Column		(b) $w' \geq 2y_1 + 4y_2 + 4y_3 + 7(1 - y_1 - y_2 - y_3)$	(x_2 constraint)
Player's		(c) $w' \geq 7y_1 + 4y_2 + 4y_3$	(x_3 constraint)
LP		(d) $w' \geq 4y_1 + y_2 + 8y_3 + 4(1 - y_1 - y_2 - y_3)$	(x_4 constraint)

$$y_1, y_2, y_3, w' \geq 0$$

An attempt to use complementary slackness to solve the row and the column players' LP's fails, because the optimal strategies for both players are degenerate. (Try complementary

TABLE **27**
Reward Matrix for Two-Finger Morra

TABLE 27
Reward Matrix for Two-Finger Morra

Row Player		Column Player			Row Minimum
	(1, 1)	(1, 2)	(2, 1)	(2, 2)	
(1, 1)	0	2	−3	0	−3
(1, 2)	−2	0	0	3	−2
(2, 1)	3	0	0	−4	−4
(2, 2)	0	−3	4	0	−3
Column Maximum	3	2	4	3	

TABLE 28
Transformed Reward Matrix for Two-Finger Morra

Row Player	Column Player			
	(1, 1)	(1, 2)	(2, 1)	(2, 2)
(1, 1)	4	6	1	4
(1, 2)	2	4	4	7
(2, 1)	7	4	4	0
(2, 2)	4	1	8	4

slackness and see what happens.) Using LINDO (or the simplex) to solve the LP's yields the following solutions: For the row player's problem, $v' = 4$, $x_1 = 0$, $x_2 = \frac{3}{5}$, $x_3 = \frac{2}{5}$, $x_4 = 0$ or $v' = 4$, $x_1 = 0$, $x_2 = \frac{4}{7}$, $x_3 = \frac{3}{7}$, $x_4 = 0$; for the column player's problem, $w' = 4$, $y_1 = 0$, $y_2 = \frac{3}{5}$, $y_3 = \frac{2}{5}$, $y_4 = 0$ or $w' = 4$, $y_1 = 0$, $y_2 = \frac{4}{7}$, $y_3 = \frac{3}{7}$, $y_4 = 0$. Each player's LP has alternative optimal solutions, so each player actually has an infinite number of optimal strategies. For example, for any c satisfying $0 \le c \le 1$, $x_1 = 0$, $x_2 = \frac{3c}{5} + \frac{4(1-c)}{7}$, $x_3 = \frac{2c}{5} + \frac{3(1-c)}{7}$, $x_4 = 0$ would be an optimal strategy for the row player. Of course, the value (to the row player) of the original Two-Finger Morra game is $v' - 4 = 0$.

REMARKS **1** Observe that both players have the same optimal strategies. (This is no accident; see Problem 5 at the end of this section.)

2 Also note that if each player utilizes her optimal strategy, then neither player will ever lose or win any money. This illustrates the fact that if both players follow the basic assumption of two-person zero-sum game theory, then conservative play will generally result.

3 Finally, we see that if each player uses his or her optimal strategy, then a player never guesses the same number of fingers that he or she has actually put out. This fact is explained in Table 29.

Now suppose the row player chooses the optimal strategy $(0, \frac{3}{5}, \frac{2}{5}, 0)$. Then the column player only breaks even by playing (1, 1) and loses an average of $\frac{1}{5}$ per play when he plays (2, 2). Similarly, if the row player chooses the optimal strategy $(0, \frac{4}{7}, \frac{3}{7}, 0)$, the column player breaks even with (2, 2)

TABLE 29
Expected Reward to Row Player

Row Plays Optimal Strategy	Expected Reward to Row	
	Column Plays	Column Plays
	(1, 1)	(2, 2)
$(0, \frac{3}{5}, \frac{2}{5}, 0)$	$-2(\frac{3}{5}) + 3(\frac{2}{5}) = 0$	$3(\frac{3}{5}) - 4(\frac{2}{5}) = \frac{1}{5}$
$(0, \frac{4}{7}, \frac{3}{7}, 0)$	$-2(\frac{4}{7}) + 3(\frac{3}{7}) = \frac{1}{7}$	$3(\frac{4}{7}) - 4(\frac{3}{7}) = 0$

TABLE 30
Expected Reward to Row Player If Column Player Plays $(\frac{1}{4}, \frac{1}{4}, \frac{1}{4}, \frac{1}{4})$

Row Chooses	Column Chooses	Reward to Row	Probability of Occurrence
(1, 2)	(1, 1)	−2	$(\frac{3}{5})(\frac{1}{4}) = \frac{3}{20}$
(1, 2)	(1, 2)	0	$(\frac{3}{5})(\frac{1}{4}) = \frac{3}{20}$
(1, 2)	(2, 1)	0	$(\frac{3}{5})(\frac{1}{4}) = \frac{3}{20}$
(1, 2)	(2, 2)	3	$(\frac{3}{5})(\frac{1}{4}) = \frac{3}{20}$
(2, 1)	(1, 1)	3	$(\frac{2}{5})(\frac{1}{4}) = \frac{2}{20}$
(2, 1)	(1, 2)	0	$(\frac{2}{5})(\frac{1}{4}) = \frac{2}{20}$
(2, 1)	(2, 1)	0	$(\frac{2}{5})(\frac{1}{4}) = \frac{2}{20}$
(2, 1)	(2, 2)	−4	$(\frac{2}{5})(\frac{1}{4}) = \frac{2}{20}$

and loses an average of $\frac{1}{7}$ per play when he plays (1, 1). Thus, putting out the same number of fingers as you guess cannot have a positive expected reward against the other player's optimal strategy.

The preceding discussion explains why the seemingly reasonable strategy $(\frac{1}{4}, \frac{1}{4}, \frac{1}{4}, \frac{1}{4})$ is not optimal for either player. For instance, if the column player chooses the strategy $(\frac{1}{4}, \frac{1}{4}, \frac{1}{4}, \frac{1}{4})$ and the row player plays the optimal strategy $(0, \frac{3}{5}, \frac{2}{5}, 0)$, the row player's expected reward may be computed as in Table 30. In this situation, the expected reward received by the row player is $-2(\frac{3}{20}) + 0(\frac{3}{20}) + 0(\frac{3}{20}) + 3(\frac{3}{20}) + 3(\frac{2}{20}) + 0(\frac{2}{20}) + 0(\frac{2}{20}) - 4(\frac{2}{20}) = \frac{1}{20}$. Another way to see this: Each time the column player chooses (1, 1), (1, 2), or (2, 1), the players break even, but on the plays for which the column player chooses (2, 2), the row player wins an average of $\frac{1}{5}$ unit. Thus, the row player's expected reward is $(\frac{1}{4})(\frac{1}{5}) = \frac{1}{20}$ unit.

4 In Odds and Evens and in Stone, Paper, Scissors, the optimal strategies may have been intuitively obvious, but the game of Two-Finger Morra shows that game theory can often yield subtle insights into how a two-person zero-sum game should be played.

Using LINDO or LINGO to Solve Two-Person Zero-Sum Games

To use LINDO to solve for the value and optimal strategies in a two-person zero-sum game, simply type in either the row or column player's problem. If, for example you type in the row player's problem your optimal z-value is the value of the game; your optimal values of the decision variables are the row player's optimal strategies; and the absolute value of the dual prices are the column player's optimal strategies. By the way, because v is unrestricted in sign, you should use the command **FREE** v after the **END** statement.

Game.lng

The following LINGO model (file Game.lng) can be used to solve for the value and optimal strategies for Two-Finger Morra (or any two-person zero-sum game).

```
MODEL:
1]SETS:
2]ROWS/1..4/:X;
3]COLS/1..4/;
4]MATRIX(ROWS,COLS):REW;
5]ENDSETS
6]@FOR(COLS(J):@SUM(ROWS(I):REW(I,J)*X(I))>V;);
7]@SUM(ROWS(I):X(I))=1;
8]MAX=V;
9]@FREE(V);
10]DATA:
11]REW=0,2,-3,0,
12]-2,0,0,3,
13]3,0,0,-4,
14]0,-3,4,0;
15]ENDDATA
16]END
```

In line 2, we define the rows of our reward matrix, associating row i with $X(I)$ = probability that the row player plays row i. In line 3, we define the columns of the reward matrix. In line 4, we create the reward matrix itself and define the reward REW(I,J) to the row player when row i and column j are played. For each column j, line 6 creates the constraint that $\sum \text{REW}(I,J)*X(I) \geq V$. In line 7, we ensure that the row player's probabilities sum to 1. Row 8 creates the objective function of max $z = v$. Row 9 uses the **@FREE** statement to allow v to be negative. In rows 11 through 14 we input the reward matrix.

To use this model to solve for optimal strategies in any two-person zero-sum game, change the number of rows and columns and change the entries in the reward matrix. Remember that the dual prices yield the column player's optimal strategies.

Summary of How to Solve a Two-Person Zero-Sum Game

To close our discussion of two-person zero-sum games, we summarize a procedure that can be used to find the value and optimal strategies for any two-person zero-sum (or constant-sum) game.

Step 1   Check for a saddle point. If the game has no saddle point, then go on to step 2.

Step 2   Eliminate any of the row player's dominated strategies. Looking at the reduced matrix (dominated rows crossed out), eliminate any of the column player's dominated strategies. Now eliminate any of the row player's dominated strategies. Continue in this fashion until no more dominated strategies can be found. Now proceed to step 3.

Step 3   If the game matrix is now 2×2, solve the game graphically. Otherwise, solve the game by using the linear programming methods of this section.

PROBLEMS

Group A

1   A soldier can hide in one of five foxholes (1, 2, 3, 4, or 5) (see Figure 5). A gunner has a single shot and may fire at any of the four spots A, B, C, or D. A shot will kill a soldier if the soldier is in a foxhole adjacent to the spot where the shot was fired. For example, a shot fired at spot B will kill the soldier if he is in foxhole 2 or 3, while a shot fired at spot D will kill the soldier if he is in foxhole 4 or 5. Suppose the gunner receives a reward of 1 if the soldier is killed and a reward of 0 if the soldier survives the shot.

    a   Assuming this to be a zero-sum game, construct the reward matrix.

    b   Find and eliminate all dominated strategies.

    c   We are given that an optimal strategy for the soldier is to hide $\frac{1}{3}$ of the time in foxholes 1, 3, and 5. We are also told that for the gunner, an optimal strategy is to shoot $\frac{1}{3}$ of the time at A, $\frac{1}{3}$ of the time at D, and $\frac{1}{3}$ of the time at B or C. Determine the value of the game to the gunner.

    d   Suppose the soldier chooses the following nonoptimal strategy: $\frac{1}{2}$ of the time, hide in foxhole 1; $\frac{1}{4}$ of the time, hide in foxhole 3; and $\frac{1}{4}$ of the time, hide in fox-

hole 5. Find a strategy for the gunner that ensures that his expected reward will exceed the value of the game.

    e   Write down each player's LP and verify that the strategies given in part (c) are optimal strategies.

2   Find each player's optimal strategy and the value of the two-person zero-sum game in Table 31.

3   Find each player's optimal strategy and the value of the two-person zero-sum game in Table 32.

4   Two armies are advancing on two cities. The first army is commanded by General Custard and has four regiments;

TABLE 31

4	5	1	4
2	1	6	3
1	0	0	2

TABLE 32

2	4	6
3	1	5

FIGURE 5

① A ② B ③ C ④ D ⑤

the second army is commanded by General Peabody and has three regiments. At each city, the army that sends more regiments to the city captures both the city and the opposing army's regiments. If both armies send the same number of regiments to a city, then the battle at the city is a draw. Each army scores 1 point per city captured and 1 point per captured regiment. Assume that each army wants to maximize the difference between its reward and its opponent's reward. Formulate this situation as a two-person zero-sum game and solve for the value of the game and each player's optimal strategies.

Group B

5 A two-person zero-sum game with an $n \times n$ reward matrix A is a **symmetric** game if $A = -A^T$.

a Explain why a game having $A = -A^T$ is called a symmetric game.

b Show that a symmetric game must have a value of zero.

c Show that if $(\bar{x}_1, \bar{x}_2, \ldots, \bar{x}_n)$ is an optimal strategy for the row player, then $(\bar{x}_1, \bar{x}_2, \ldots, \bar{x}_n)$ is also an optimal strategy for the column player.

d What examples discussed in this chapter are symmetric games? How could the results of this problem make it easier to solve for the value and optimal strategies of a symmetric game?

6 For a two-person zero-sum game with an $m \times n$ reward matrix, let $\bar{x} = (\bar{x}_1, \bar{x}_2, \ldots, \bar{x}_m)$ be a solution to the row player's LP and $\bar{y} = (\bar{y}_1, \bar{y}_2, \ldots, \bar{y}_n)$ be a solution to the column player's LP. Show that if the row player departs from his optimal strategy, he cannot increase his expected reward against $\bar{y}$.

7 Interpret the complementary slackness conditions for the row and the column players' LP's.

8 Wivco has observed the daily production and the daily variable production costs of widgets at the New York City

plant. The data in Table 33 have been collected. Wivco believes that daily production and daily variable production costs are related as follows: For some numbers a and b,

Daily production cost $= a + b$(daily production)

Wivco wants to find estimates of a and b ($\hat{a}$ and $\hat{b}$) that minimize the maximum error (in absolute value) incurred in estimating daily production costs. For example, if Wivco chooses $\hat{a} = 3$ and $\hat{b} = 2$, then the predicted daily costs are shown in Table 34. In this case, the maximum error would be $3,000. Formulate an LP that can be used to find the optimal estimates $\hat{a}$ and $\hat{b}$.

9 Suppose we add a constant c to every element in a reward matrix A. Call the new game matrix A'. Show that A and A' have the same optimal strategies and that value of $A' =$ (value of A) $+ c$.

TABLE 33

Day	Production	Variable Production Cost ($)
1	4,000	9,000
2	6,000	12,000
3	7,000	14,000
4	1,000	5,000
5	3,000	8,000

TABLE 34

Day	Predicted Cost ($)	Absolute Error ($)
1	11,000	2,000
2	15,000	3,000
3	17,000	3,000
4	5,000	0
5	9,000	1,000

11.4 Two-Person Nonconstant-Sum Games

Most game-theoretic models of business situations are not constant-sum games, because it is unusual for business competitors to be in total conflict.

In this section, we briefly discuss the analysis of two-person nonconstant-sum games in which cooperation between the players is not allowed. We begin with a discussion of the famous Prisoner's Dilemma.

EXAMPLE 7 | **Prisoner's Dilemma**

Two prisoners who escaped and participated in a robbery have been recaptured and are awaiting trial for their new crime. Although they are both guilty, the Gotham City

TABLE 35

Reward Matrix for Prisoner's Dilemma

| | Prisoner 2 | |
Prisoner 1	Confess	Don't Confess
Confess	$(-5, -5)$	$(0, -20)$
Don't confess	$(-20, 0)$	$(-1, -1)$

district attorney is not sure he has enough evidence to convict them. To entice them to testify against each other, the district attorney tells each prisoner the following: "If only one of you confesses and testifies against your partner, the person who confesses will go free while the person who does not confess will surely be convicted and given a 20-year jail sentence. If both of you confess, then you will both be convicted and sent to prison for 5 years. Finally, if neither of you confesses, I can convict you both of a misdemeanor and you will each get 1 year in prison." What should each prisoner do?

Solution If we assume that the prisoners cannot communicate with each other, the strategies and rewards for each are as shown in Table 35. The first number in each cell of this matrix is the reward (negative, because years in prison is undesirable) to prisoner 1, and the second matrix in each cell is the reward to prisoner 2. Note that the sum of the rewards in each cell varies from a high of -2 ($-1 - 1$) to a low of -20 ($-20 + 0$). Thus, this is not a constant-sum two-player game.

Suppose each prisoner seeks to eliminate any dominated strategies from consideration. For each prisoner, the "confess" strategy dominates the "don't confess" strategy. If each prisoner follows his undominated ("confess") strategy, however, each prisoner will spend 5 years in jail. On the other hand, if each prisoner chooses the dominated "don't confess" strategy, then each prisoner will spend only 1 year in prison. Thus, if each prisoner chooses his dominated strategy, both are better off than if each prisoner chooses his undominated strategy.

DEFINITION ■ As in a two-person zero-sum game, a choice of strategy by each player (prisoner) is an **equilibrium point** if neither player can benefit from a unilateral change in strategy. ■

Thus, $(-5, -5)$ is an equilibrium point, because if either prisoner changes his strategy, then his reward decreases (from -5 to -20). Clearly, however, each prisoner is better off at the point $(-1, -1)$. To see that the outcome $(-1, -1)$ may not occur, observe that $(-1, -1)$ is not an equilibrium point, because if we are currently at the outcome $(-1, -1)$, either prisoner can increase his reward (from -1 to 0) by changing his strategy from "don't confess" to "confess" (that is, each prisoner can benefit from double-crossing his opponent). This illustrates an important aspect of the Prisoner's Dilemma type of game: If the players are cooperating (if each prisoner chooses "don't confess"), then each player can gain by double-crossing his opponent (assuming his opponent's strategy remains unchanged). If both players double-cross each other, however, then both will be worse off than if they had both chosen their cooperative strategy. This anomaly cannot occur in a two-person constant-sum game. (Why not?)

TABLE 36

A General Prisoner's Dilemma
Reward Matrix

Player 1	Player 2	
	NC	C
NC	(P, P)	(T, S)
C	(S, T)	(R, R)

More formally, a Prisoner's Dilemma game may be described as in Table 36, where

NC = noncooperative action

C = cooperative action

P = punishment for not cooperating

S = payoff to person who is double-crossed

R = reward for cooperating if both players cooperate

T = temptation for double-crossing opponent

In a Prisoner's Dilemma game, (P, P) is an equilibrium point. This requires $P > S$. For (R, R) not to be an equilibrium point requires $T > R$. (This gives each player a temptation to double-cross his opponent.) The game is reasonable only if $R > P$. Thus, for Table 36 to represent a Prisoner's Dilemma game, we require that $T > R > P > S$. The Prisoner's Dilemma game is of interest because it explains why two adversaries often fail to cooperate with each other. This is illustrated by Examples 8 and 9.

EXAMPLE 8 **Advertising Prisoner's Dilemma Game**

Competing restaurants Hot Dog King and Hot Dog Chef are attempting to determine their advertising budgets for next year. The two restaurants will have combined sales of $240 million and can spend either $6 million or $10 million on advertising. If one restaurant spends more money than the other, then the restaurant that spends more money will have sales of $190 million. If both companies spend the same amount on advertising, then they will have equal sales. Each dollar of sales yields 10¢ of profit. Suppose each restaurant is interested in maximizing (contribution of sales to profit) − (advertising costs). Find an equilibrium point for this game.

Solution The appropriate reward matrix is shown in Table 37. If we identify spending $10 million on advertising as the noncooperative action and spending $6 million as the cooperative action, then (2, 2) (corresponding to heavy advertising by both restaurants) is an equilibrium point. Although both restaurants are better off at (6, 6) than at (2, 2), (6, 6) is unstable because either restaurant may gain by changing its strategy. Thus, to protect its market share, each restaurant must spend heavily on advertising.

TABLE 37

Reward Matrix for Advertising Game

Hot Dog King	Hot Dog Chef	
	Spend $10 Million	Spend $6 Million
Spend $10 million	(2, 2)	(9, −1)
Spend $6 million	(−1, 9)	(6, 6)

EXAMPLE 9 **Arms Race Prisoner's Dilemma**

The Vulcans and the Klingons are engaged in an arms race in which each nation is assumed to have two possible strategies: develop a new missile or maintain the status quo. The reward matrix is assumed to be as shown in Table 38. This reward matrix is based on the assumption that if only one nation develops a new missile, the nation with the new missile will conquer the other nation. In this case, the conquering nation earns a reward of 20 units and the conquered nation loses 100 units. It is also assumed that the cost of developing a new missile is 10 units. Identify an equilibrium point for this game.

Solution Identifying "develop" as the noncooperative action and "maintain" as the cooperative action, we see that $(-10, -10)$ (both nations choosing their noncooperative action) is an equilibrium point. Although $(0, 0)$ leaves both nations better off than $(-10, -10)$, we see that in this situation, each nation can gain from a double-cross. Thus, $(0, 0)$ is not stable. This example shows how maintaining the balance of power may lead to an arms race.

TABLE 38
Reward Matrix for Arms Race Game

Vulcans	Klingons	
	Develop New Missile	**Maintain Status Quo**
Develop new missile	$(-10, -10)$	$(10, -100)$
Maintain status quo	$(-100, 10)$	$(0, 0)$

The following two-person nonconstant-sum game is not a Prisoner's Dilemma game.

EXAMPLE 10 **"Chicken" Game**

Angry Max drives toward James Bound on a deserted road. Each person has two strategies: swerve or don't swerve. The reward matrix in Table 39 needs no explanation! Find the equilibrium point(s) for this game.

Solution For both $(5, -5)$ and $(-5, 5)$, neither player can gain by a unilateral change in strategy. Thus, $(5, -5)$ and $(-5, 5)$ are both equilibrium points.

TABLE 39
Reward Matrix for Swerve Game

Angry Max	James Bound	
	Swerve	**Don't Swerve**
Swerve	$(0, 0)$	$(-5, 5)$
Don't swerve	$(5, -5)$	$(-100, -100)$

Like constant-sum games, a nonconstant-sum game may fail to have an equilibrium point in pure strategies. It can be shown that if mixed strategies are allowed, then in any two-person nonconstant-sum game, each player has an equilibrium strategy (in that if one player plays her equilibrium strategy, the other player cannot benefit by deviating from her equilibrium strategy) [see Owen (1982, p. 127)]. For example, consider the two-

TABLE **40**

A Game with No Equilibrium in Pure Strategies

Player 1	Player 2	
	Strategy 1	Strategy 2
Strategy 1	(2, −1)	(−2, 1)
Strategy 2	(−2, 1)	(2, −1)

person nonconstant-sum game in Table 40. For this game, the reader should verify that there is no equilibrium in pure strategies and also that each player's choice of the mixed strategy $(\frac{1}{2}, \frac{1}{2})$ is an equilibrium because neither player can benefit from a unilateral change in strategy (see Problem 4 at the end of this section). Owen (1982, Chapter 7) discusses two-person nonconstant-sum games in which the players are allowed to cooperate.

PROBLEMS

Group A

1 Find an equilibrium point (if one exists in pure strategies) for the two-person nonconstant-sum game in Table 41.

2 Find an equilibrium point in pure strategies (if any exists) for the two-person nonconstant-sum game in Table 42.

3 The New York City Council is ready to vote on two bills that authorize the construction of new roads in Manhattan and Brooklyn. If the two boroughs join forces, they can pass both bills, but neither borough by itself has enough power to pass a bill. If a bill is passed, then it will cost the taxpayers of each borough $1 million, but if roads are built in a borough, the benefits to the borough are estimated to be $10 million. The council votes on both bills simultaneously, and each councilperson must vote on the bills without knowing how anybody else will vote. Assuming that each borough supports its own bill, determine whether this game has any equilibrium points. Is this game analogous to the Prisoner's Dilemma? Explain why or why not.

Group B

4 Given that each player's goal is to maximize her expected reward, show that for the game in Table 43 each player's choice of the mixed strategy $(\frac{1}{2}, \frac{1}{2})$ is an equilibrium point.

TABLE 41

(9, −1)	(−2, −3)
(8, 7)	(−9, 11)

TABLE 42

(9, 9)	(−10, 10)
(10, −10)	(−1, 1)

TABLE 43

Player 1	Player 2	
	Strategy 1	Strategy 2
Strategy 1	(2, −1)	(−2, 1)
Strategy 2	(−2, 1)	(2, −1)

5[†] A Japanese electronics company and an American electronics company are both considering working on developing a superconductor. If both companies work on the superconductor, they will have to share the market, and each company will lose $10 billion. If only one company works on the superconductor, that company will earn $100 billion in profits. Of course, if neither company works on the superconductor, then each company earns profits of $0.

a Formulate this situation as a two-person nonconstant-sum game. Does the game have any equilibrium points?

b Now suppose the Japanese government offers the Japanese electronics company a $15 billion subsidy to work on the superconductor. Formulate the reward matrix for this game. Does this game have any equilibrium points?

c Businesspeople have often said that a protectionist attitude toward trade can increase exports, but economists have usually argued that it will reduce exports. Whose viewpoint does this problem support?

[†]Based on "Protectionism Gets Clever" (1988).

11.5 Introduction to *n*-Person Game Theory

In many competitive situations, there are more than two competitors. With this in mind, we now turn our attention to games with three or more players. Let $N = \{1, 2, \ldots, n\}$ be the set of players. Any game with n players is an **n-person game.** For our purposes, an *n*-person game is specified by the game's characteristic function.

DEFINITION ■ For each subset S of N, the **characteristic function** v of a game gives the amount $v(S)$ that the members of S can be sure of receiving if they act together and form a coalition. ■

Thus, $v(S)$ can be determined by calculating the amount that members of S can get without any help from players who are not in S.

EXAMPLE 11   The Drug Game

Joe Willie has invented a new drug. Joe cannot manufacture the drug himself, but he can sell the drug's formula to company 2 or company 3. The lucky company will split a $1 million profit with Joe Willie. Find the characteristic function for this game.

Solution   Letting Joe Willie be player 1, company 2 be player 2, and company 3 be player 3, we find the characteristic function for this game to be:

$$v(\{\ \}) = v(\{1\}) = v(\{2\}) = v(\{3\}) = v(\{2, 3\}) = 0$$
$$v(\{1, 2\}) = v(\{1, 3\}) = v(\{1, 2, 3\}) = \$1{,}000{,}000$$

EXAMPLE 12   The Garbage Game

Each of four property owners has one bag of garbage and must dump it on somebody's property. If b bags of garbage are dumped on the coalition of property owners, then the coalition receives a reward of $-b$. Find the characteristic function for this game.

Solution   The best that the members of any coalition can do is to dump all of their garbage on the property of owners who are not in S. Thus, the characteristic function for the garbage game ($|S|$ is the number of players in S) is given by

$$v(\{S\}) = -(4 - |S|) \qquad (\text{if } |S| < 4) \tag{11}$$
$$v(\{1, 2, 3, 4\}) = -4 \qquad (\text{if } |S| = 4) \tag{11.1}$$

Equation (11.1) follows because if players are in S, they must dump their garbage on members of S.

EXAMPLE 13   The Land Development Game

Player 1 owns a piece of land and values the land at $10,000. Player 2 is a subdivider who can develop the land and increase its worth to $20,000. Player 3 is a subdivider who can develop the land and increase its worth to $30,000. There are no other prospective buyers. Find the characteristic function for this game.

Solution   Note that any coalition that does not contain player 1 has a worth or value of $0. Any other coalition has a value equal to the maximum value that a member of the coalition places on the piece of land. Thus, we obtain the following characteristic function:

$$v(\{1\}) = \$10{,}000, \quad v(\{\ \}) = v(\{2\}) = v(\{3\}) = \$0, \quad v(\{1, 2\}) = \$20{,}000,$$
$$v(\{1, 3\}) = \$30{,}000, \quad v(\{2, 3\}) = \$0, \quad v(\{1, 2, 3\}) = \$30{,}000$$

Consider any two subsets of sets A and B such that A and B have no players in common ($A \cap B = \emptyset$). Then for each of our examples (and any n-person game), the characteristic function must satisfy the following inequality:

$$v(A \cup B) \geq v(A) + v(B) \tag{12}$$

This property of the characteristic function is called **superadditivity.** Equation (12) is reasonable, because if the players in $A \cup B$ band together, one of their options (but not their only option) is to let the players in A fend for themselves and let the players in B fend for themselves. This would result in the coalition receiving an amount $v(A) + v(B)$. Thus, $v(A \cup B)$ must be at least as large as $v(A) + v(B)$.

There are many solution concepts for n-person games. A solution concept should indicate the reward that each player will receive. More formally, let $\mathbf{x} = \{x_1, x_2, \ldots, x_n\}$ be a vector such that player i receives a reward x_i. We call such a vector a **reward vector.** A reward vector $\mathbf{x} = (x_1, x_2, \ldots, x_n)$ is not a reasonable candidate for a solution unless $\mathbf{x}$ satisfies

$$v(N) = \sum_{i=1}^{i=n} x_i \qquad \text{(Group rationality)} \tag{13}$$

$$x_i \geq v(\{i\}) \quad \text{(for each } i \in N) \qquad \text{(Individual rationality)} \tag{14}$$

If $\mathbf{x}$ satisfies both (13) and (14), we say that $\mathbf{x}$ is an **imputation.** Equation (13) states that any reasonable reward vector must give all the players an amount that equals the amount that can be attained by the supercoalition consisting of all players. Equation (14) implies that player i must receive a reward at least as large as what he can get for himself ($v\{i\}$).

To illustrate the idea of an imputation, consider the payoff vectors for Example 13, shown in Table 44. Any solution concept for n-person games chooses some subset of the set of imputations (possibly empty) as the solution to the n-person game. In Sections 11.6 and 11.7, we discuss two solution concepts, the core and the Shapley value. See Owen (1999) for a discussion of other solution concepts for n-person games. The problems involving n-person game theory are at the end of Section 11.7.

TABLE 44

Examples of Imputation

x	Is x an Imputation?
($10,000, $10,000, $10,000)	Yes
($5,000, $2,000, $5,000)	No, $x_1 < v(\{1\})$, so (14) is violated
($12,000, $19,000, −$1000)	No, (14) is violated
($11,000, $11,000, $11,000)	No, (13) is violated

11.6 The Core of an *n*-Person Game

An important solution concept for an *n*-person game is the core. Before defining this, we must define the concept of **domination**. Given an imputation $\mathbf{x} = (x_1, x_2, \ldots, x_n)$, we say that the imputation $\mathbf{y} = (y_1, y_2, \ldots, y_n)$ *dominates* $\mathbf{x}$ through a coalition S (written $\mathbf{y} > {}^s\mathbf{x}$) if

$$\sum_{i \in S} y_i \leq v(S) \qquad \text{and for all } i \in S, \qquad y_i > x_i \tag{15}$$

If $\mathbf{y} > {}^s\mathbf{x}$, then both the following must be true:

1 Each member of S prefers $\mathbf{y}$ to $\mathbf{x}$.

2 Because $\Sigma_{i \in S} y_i \leq v(S)$, the members of S can attain the rewards given by $\mathbf{y}$.

Thus, if $\mathbf{y} > {}^s\mathbf{x}$, then $\mathbf{x}$ should not be considered a possible solution to the game, because the players in S can object to the rewards given by $\mathbf{x}$ and enforce their objection by banding together and thereby receiving the rewards given by $\mathbf{y}$ [because members of S can surely receive an amount equal to $v(S)$].

The founders of game theory, John von Neumann and Oskar Morgenstern, argued that a reasonable solution concept for an *n*-person game was the set of all undominated imputations.

DEFINITION ■   The **core** of an *n*-person game is the set of all undominated imputations.   ■

Examples 14 and 15 illustrate the concept of domination.

EXAMPLE 14   Dominance

Consider a three-person game with the following characteristic function:

$$v(\{\ \}) = v(\{1\}) = v(\{2\}) = v(\{3\}) = 0$$
$$v(\{1, 2\}) = 0.1, \quad v(\{1, 3\}) = 0.2, \quad v(\{2, 3\}) = 0.2, \quad v(\{1, 2, 3\}) = 1$$

Let $\mathbf{x} = (0.05, 0.90, 0.05)$ and $\mathbf{y} = (0.10, 0.80, 0.10)$. Show that $\mathbf{y} > {}^{\{1,3\}}\mathbf{x}$.

Solution   First, note that both $\mathbf{x}$ and $\mathbf{y}$ are imputations. Next, observe that with the imputation $\mathbf{y}$, players 1 and 3 both receive more than they receive with $\mathbf{x}$. Also, $\mathbf{y}$ gives the players in $\{1, 3\}$ a total of $0.10 + 0.10 = 0.20$. Because 0.20 does not exceed $v(\{1, 3\}) = 0.20$, it is reasonable to assume that players 1 and 3 can band together and receive a total reward of 0.20. Thus, players 1 and 3 will never allow the rewards given by $\mathbf{x}$ to occur.

EXAMPLE 15   Dominance in Land Development Game

For the land development game (Example 13), let $\mathbf{x} = (\$19,000, \$1,000, \$10,000)$ and $\mathbf{y} = (\$19,800, \$100, \$10,100)$. Show that $\mathbf{y} > {}^{\{1,3\}}\mathbf{x}$.

Solution   We need only observe that players 1 and 3 both receive more from $\mathbf{y}$ than they receive from $\mathbf{x}$, and the total received by players 1 and 3 from $\mathbf{y}$ ($\$29,900$) does not exceed $v(\{1, 3\})$. If $\mathbf{x}$ were proposed as a solution to the land development game, player 1 would sell the land to player 3 and $\mathbf{y}$ (or some other imputation that dominates $\mathbf{x}$) would result. The important point is that $\mathbf{x}$ cannot occur, because players 1 and 3 will never allow $\mathbf{x}$ to occur.

We are now ready to show how to determine the core of an n-person game, for which Theorem 1 is often useful.

THEOREM 1

An imputation $\mathbf{x} = \{x_1, x_2, \ldots, x_n\}$ is in the core of an n-person game if and only if for each subset S of N,

$$\sum_{i \in S} x_i \geq v(S)$$

Theorem 1 states that an imputation $\mathbf{x}$ is in the core (that $\mathbf{x}$ is undominated) if and only if for every coalition S, the total of the rewards received by the players in S (according to $\mathbf{x}$) is at least as large as $v(S)$.

To illustrate the use of Theorem 1, we find the core of the three games discussed in Section 11.5.

EXAMPLE 11   **The Drug Game (Continued)**

Find the core of the drug game.

Solution   For this game, $\mathbf{x} = (x_1, x_2, x_3)$ will be an imputation if and only if

$$x_1 \geq 0 \tag{16}$$

$$x_2 \geq 0 \tag{17}$$

$$x_3 \geq 0 \tag{18}$$

$$x_1 + x_2 + x_3 = \$1,000,000 \tag{19}$$

Theorem 1 shows that $\mathbf{x} = (x_1, x_2, x_3)$ will be in the core if and only if $x_1, x_2,$ and x_3 satisfy (16)–(19) and the following inequalities:

$$x_1 + x_2 \geq \$1,000,000 \tag{20}$$

$$x_1 + x_3 \geq \$1,000,000 \tag{21}$$

$$x_2 + x_3 \geq \$0 \tag{22}$$

$$x_1 + x_2 + x_3 \geq \$1,000,000 \tag{23}$$

To determine the core, note that if $\mathbf{x} = (x_1, x_2, x_3)$ is in the core, then $x_1, x_2,$ and x_3 must satisfy the inequality generated by adding together inequalities (20)–(22). Adding (20)–(22) yields $2(x_1 + x_2 + x_3) \geq \$2,000,000$, or

$$x_1 + x_2 + x_3 \geq \$1,000,000 \tag{24}$$

By (19), $x_1 + x_2 + x_3 = \$1,000,000$. Thus, (20)–(22) must all be binding.[†] Simultaneously solving (20)–(22) as equalities yields $x_1 = \$1,000,000$, $x_2 = \$0$, $x_3 = \$0$. A quick check shows that ($\$1,000,000, \$0, \$0$) does satisfy (16)–(23). In summary, the core of this game is the imputation ($\$1,000,000, \$0, \$0$). Thus, the core emphasizes the importance of player 1.

REMARKS   1   In Section 11.7, we show that for this game, an alternative solution concept, the Shapley value, gives player 1 less than $\$1,000,000$ and gives both player 2 and player 3 some money.

[†]If (20), (21), or (22) were nonbinding, then for any point in the core, the sum of (20)–(22) would also be nonbinding. Because we know that (24) must be binding, this implies that for any point in the core, (20), (21), and (22) must all be binding.

2 For the drug game, if we choose an imputation that is not in the core, then we can show how it is dominated. Consider the imputation $\mathbf{x} = (\$900,000, \$50,000, \$50,000)$. If we let $\mathbf{y} = (\$925,000, \$75,000, \$0)$, then $\mathbf{y} >^{\{1,2\}} \mathbf{x}$.

EXAMPLE 12 **The Garbage Game (Continued)**

Determine the core of the garbage game.

Solution Note that $\mathbf{x} = (x_1, x_2, x_3, x_4)$ will be an imputation if and only if x_1, x_2, x_3, and x_4 satisfy the following inequalities:

$$x_1 \geq -3 \tag{25}$$
$$x_2 \geq -3 \tag{26}$$
$$x_3 \geq -3 \tag{27}$$
$$x_4 \geq -3 \tag{28}$$
$$x_1 + x_2 + x_3 + x_4 = -4 \tag{29}$$

Applying Theorem 1 to all three-player coalitions, we find that for $\mathbf{x} = \{x_1, x_2, x_3, x_4\}$ to be in the core, it is necessary that x_1, x_2, x_3, and x_4 satisfy the following inequalities:

$$x_1 + x_2 + x_3 \geq -1 \tag{30}$$
$$x_1 + x_2 + x_4 \geq -1 \tag{31}$$
$$x_1 + x_3 + x_4 \geq -1 \tag{32}$$
$$x_2 + x_3 + x_4 \geq -1 \tag{33}$$

We now show that no imputation $\mathbf{x} = (x_1, x_2, x_3, x_4)$ can satisfy (30)–(33) and that the garbage game has an empty core. Consider an imputation $\mathbf{x} = (x_1, x_2, x_3, x_4)$. If $\mathbf{x}$ is to be in the core of the garbage game, $\mathbf{x}$ must satisfy the inequality generated by adding together (30)–(33):

$$3(x_1 + x_2 + x_3 + x_4) \geq -4 \tag{34}$$

Equation (29) implies that any imputation $\mathbf{x} = (x_1, x_2, x_3, x_4)$ must satisfy $x_1 + x_2 + x_3 + x_4 = -4$. Thus, (34) cannot hold. This means that no imputation $\mathbf{x} = (x_1, x_2, x_3, x_4)$ can satisfy (30)–(33) and the core of the garbage game is empty.

To understand why the garbage game has an empty core, consider the imputation $\mathbf{x} = (-2, -1, -1, 0)$, which treats players 1 and 2 unfairly. By joining, players 1 and 2 could ensure that the imputation $\mathbf{y} = (-1.5, -0.5, -1, -1)$ occurred. Thus, $\mathbf{y} >^{\{1,2\}} \mathbf{x}$. In a similar fashion, any imputation can be dominated by another imputation. We note that for a two-player version of the garbage game, the core consists of the imputation $(-1, -1)$, and for $n > 2$, the n-player garbage game has an empty core (see Problems 4 and 5 at the end of Section 11.7).

EXAMPLE 13 **The Land Development Game (Continued)**

Find the core of the land development game.

Solution For the land development game, any imputation $\mathbf{x} = (x_1, x_2, x_3)$ must satisfy

$$x_1 \geq \$10,000 \tag{35}$$
$$x_2 \geq \$0 \tag{36}$$
$$x_3 \geq \$0 \tag{37}$$
$$x_1 + x_2 + x_3 = \$30,000 \tag{38}$$

An imputation $\mathbf{x} = (x_1, x_2, x_3)$ is in the core if and only if it satisfies the following inequalities:

$$x_1 + x_2 \geq \$20{,}000 \tag{39}$$

$$x_1 + x_3 \geq \$30{,}000 \tag{40}$$

$$x_2 + x_3 \geq \$0 \tag{41}$$

$$x_1 + x_2 + x_3 \geq \$30{,}000 \tag{42}$$

Adding (36) and (40), we find that if $\mathbf{x} = (x_1, x_2, x_3)$ is in the core, then x_1, x_2, and x_3 must satisfy $x_1 + x_2 + x_3 \geq \$30{,}000$. From (38), $x_1 + x_2 + x_3 = \$30{,}000$. Thus, (36) and (40) must be binding. This argument shows that for $\mathbf{x} = (x_1, x_2, x_3)$ to be in the core, x_1, x_2, and x_3 must satisfy

$$x_2 = \$0 \quad \text{and} \quad x_1 + x_3 = \$30{,}000 \tag{43}$$

Now (39) implies that

$$x_1 \geq \$20{,}000 \tag{44}$$

Thus, for $\mathbf{x} = (x_1, x_2, x_3)$ to be in the core, (43) and (44) must both be satisfied. Any vector in the core must also satisfy $x_3 \geq 0$ and $x_1 \leq \$30{,}000$, and any vector $\mathbf{x} = (x_1, x_2, x_3)$ satisfying (43), (44), $x_3 \geq \$0$, and $x_1 \leq \$30{,}000$ will be in the core of the land development game. Thus, if $\$20{,}000 \leq x_1 \leq \$30{,}000$, then any vector of the form $(x_1, \$0, \$30{,}000 - x_1)$ will be in the core of the land development game. The interpretation of the core is as follows: Player 3 outbids player 2 and purchases the land from player 1 for a price x_1 ($\$20{,}000 \leq x_1 \leq \$30{,}000$). Then player 1 receives a reward of x_1 dollars, and player 3 receives a reward of $\$30{,}000 - x_1$. Player 2 is shut out and receives nothing. In this example, the core contains an infinite number of points.

The problems involving n-person game theory are at the end of Section 11.7.

11.7 The Shapley Value[†]

In Section 11.6, we found that the core of the drug game gave all benefits or rewards to the game's most important player (the inventor of the drug). Now we discuss an alternative solution concept for n-person games, the **Shapley value,** which in general gives more equitable solutions than the core does.[‡]

For any characteristic function, Lloyd Shapley showed there is a unique reward vector $\mathbf{x} = (x_1, x_2, \ldots, x_n)$ satisfying the following axioms:

Axiom 1 Relabeling of players interchanges the players' rewards. Suppose the Shapley value for a three-person game is $\mathbf{x} = (10, 15, 20)$. If we interchange the roles of player 1 and player 3 [for example, if originally $v(\{1\}) = 10$ and $v(\{3\}) = 15$, we would make $v(\{1\}) = 15$ and $v(\{3\}) = 10$], then the Shapley value for the new game would be $\mathbf{x} = (20, 15, 10)$.

Axiom 2 $\sum_{i=1}^{i=n} x_i = v(N)$. This is simply group rationality.

Axiom 3 If $v(S - \{i\}) = v(S)$ holds for all coalitions S, then the Shapley value has $x_i = 0$. If player i adds no value to any coalition, then player i receives a reward of zero from the Shapley value.

Before stating Axiom 4, we define the sum of two n-person games. Let v and $\bar{v}$ be two characteristic functions for games with identical players. Define the game $(v + \bar{v})$ to be

[†]This section covers topics that can be omitted with no loss of continuity.
[‡]See Owen (1982) for an excellent discussion of the Shapley value. See also Shapley (1953).

the game with the characteristic function $(v + \bar{v})$ given by $(v + \bar{v})(S) = v(S) + \bar{v}(S)$. For example, if $v(\{1, 2\}) = 10$ and $\bar{v}(\{1, 2\}) = -3$, then in the game $(v + \bar{v})$ the coalition $\{1, 2\}$ would have $(v + \bar{v})(\{1, 2\}) = 10 - 3 = 7$.

Axiom 4 Let **x** be the Shapley value vector for game v, and let **y** be the Shapley value vector for game $\bar{v}$. Then the Shapley value vector for the game $(v + \bar{v})$ is the vector **x** + **y**.

The validity of this axiom has often been questioned, because adding rewards from two different games may be like adding apples and oranges. If Axioms 1–4 are assumed to be valid, however, Shapley proved the remarkable result in Theorem 2.

THEOREM 2

Given any n-person game with the characteristic function v, there is a unique reward vector $\mathbf{x} = (x_1, x_2, \ldots, x_n)$ satisfying Axioms 1–4. The reward of the ith player (x_i) is given by

$$x_i = \sum_{\substack{\text{all } S \text{ for which} \\ i \text{ is not in } S}} p_n(S)[v(S \cup \{i\}) - v(S)] \tag{45}$$

In (45),

$$p_n(S) = \frac{|S|!(n - |S| - 1)!}{n!} \tag{46}$$

where $|S|$ is the number of players in S, and for $n \geq 1$, $n! = n(n-1) \cdots 2(1)$ $(0! = 1)$.

Although (45) seems complex, the equation has a simple interpretation. Suppose that players $1, 2, \ldots, n$ arrive in a random order. That is, any of the $n!$ permutations of $1, 2, \ldots, n$ has a $\frac{1}{n!}$ chance of being the order in which the players arrive. For example, if $n = 3$, then there is a $\frac{1}{3!} = \frac{1}{6}$ probability that the players arrive in any one of the following sequences:

$$1, 2, 3 \quad 2, 3, 1$$
$$1, 3, 2 \quad 3, 1, 2$$
$$2, 1, 3 \quad 3, 2, 1$$

Suppose that when player i arrives, he finds that the players in the set S have already arrived. If player i forms a coalition with the players who are present when he arrives, then player i adds $v(S \cup \{i\}) - v(S)$ to the coalition S. The probability that when player i arrives the players in the coalition S are present is $p_n(S)$. Then (45) implies that *player i's reward should be the expected amount that player i adds to the coalition made up of the players who are present when he or she arrives.*

We now show that $p_n(S)$ [as given by (46)] is the probability that when player i arrives, the players in the subset S will be present. Observe that the number of permutations of $1, 2, \ldots, n$ that result in player i's arriving when the players in the coalition S are present is given by

$$\underbrace{|S|(|S| - 1)(|S| - 2) \cdots + (2)(1)}_{S \text{ arrives}} \underbrace{(1)}_{i \text{ arrives}} \underbrace{(n - |S| - 1)(n - |S| - 2) \cdots (2)(1)}_{\text{Players not in } S \cup \{i\} \text{ arrive}}$$

$$= |S|!(n - |S| - 1)!$$

Because there are a total of $n!$ permutations of $1, 2, \ldots, n$, the probability that player i will arrive and see the players in S is

$$\frac{|S|!(n - |S| - 1)!}{n!} = p_n(S)$$

We now compute the Shapley value for the drug game.

EXAMPLE 11   **The Drug Game (Continued)**

Find the Shapley value for the drug game.

Solution   To compute x_1, the reward that player 1 should receive, we list all coalitions S for which player 1 is not a member. For each of these coalitions, we compute $v(S \cup \{i\}) - v(S)$ and $p_3(S)$ (see Table 45). Because player 1 adds (on the average)

$$(\tfrac{2}{6})(0) + (\tfrac{1}{6})(1{,}000{,}000) + (\tfrac{2}{6})(1{,}000{,}000) + (\tfrac{1}{6})(1{,}000{,}000) = \tfrac{\$4{,}000{,}000}{6}$$

the Shapley value concept recommends that player 1 receive a reward of $\frac{\$4{,}000{,}000}{6}$.

To compute the Shapley value for player 2, we require the information in Table 46. Thus, the Shapley value recommends a reward of

$$(\tfrac{1}{6})(1{,}000{,}000) = \tfrac{\$1{,}000{,}000}{6}$$

for player 2. The Shapley value must allocate a total of $v(\{1, 2, 3\}) = \$1{,}000{,}000$ to the players, so the Shapley value will recommend that player 3 receive $\$1{,}000{,}000 - x_1 - x_2 = \frac{\$1{,}000{,}000}{6}$.

TABLE 45
Computation of Shapley Value for Player 1
(Joe Willie)

S	$p_3(S)$	$v(S \cup \{1\}) - v(S)$
{ }	$\frac{2}{6}$	\$0
{2}	$\frac{1}{6}$	\$1,000,000
{2, 3}	$\frac{2}{6}$	\$1,000,000
{3}	$\frac{1}{6}$	\$1,000,000

TABLE 46
Computation of Shapley Value for Player 2

S	$p_3(S)$	$v(S \cup \{2\}) - v(S)$
{ }	$\frac{2}{6}$	\$0
{1}	$\frac{1}{6}$	\$1,000,000
{3}	$\frac{1}{6}$	\$0
{1, 3}	$\frac{2}{6}$	\$0

REMARKS   **1** Recall that the core of this game assigned \$1,000,000 to player 1 and no money to players 2 and 3. Thus, the Shapley value treats players 2 and 3 more fairly than the core. In general, the Shapley value provides more equitable solutions than the core.
2 For a game with few players, it may be easier to compute each player's Shapley value by using the fact that player i should receive the expected amount that she adds to the coalition present when she arrives. For Example 11, this method yields the computations in Table 47. Each of the six or-

TABLE 47
Alternative Method for Determining Shapley Value

Order of Arrival	Amount Added by Player's Arrival ($)		
	Player 1	Player 2	Player 3
1, 2, 3	0	1,000,000	0
1, 3, 2	0	0	1,000,000
2, 1, 3	1,000,000	0	0
2, 3, 1	1,000,000	0	0
3, 1, 2	1,000,000	0	0
3, 2, 1	1,000,000	0	0

derings of the arrivals of the players is equally likely, so we find that the Shapley value to each player is as follows:

$$x_1 = \frac{\$4,000,000}{6}, \qquad x_2 = \frac{\$1,000,000}{6}, \qquad x_3 = \frac{\$1,000,000}{6}$$

3   The Shapley value can be used as a measure of the power of individual members of a political or business organization. For example, the UN Security Council consists of five permanent members (who have veto power over any resolution) and ten nonpermanent members. For a resolution to pass the Security Council, it must receive at least nine votes, including the votes of all permanent members. Assigning a value of 1 to all coalitions that can pass a resolution and a value of 0 to all coalitions that cannot pass a resolution defines a characteristic function. For this characteristic function, it can be shown that the Shapley value of each permanent member is 0.1963 and of each nonpermanent member is 0.001865, giving $5(0.1963) + 10(0.001865) = 1$. Thus, the Shapley value indicates that $5(0.1963) = 98.15\%$ of the power in the Security Council resides with the permanent members.

As a final application of the Shapley value, we discuss how it can be used to determine a pricing schedule for landing fees at an airport.

EXAMPLE 16   **Airport Pricing**

Suppose three types of planes (Piper Cubs, DC-10s, and 707s) use an airport. A Piper Cub requires a 100-yd runway, a DC-10 requires a 150-yd runway, and a 707 requires a 400-yd runway. Suppose the cost (in dollars) of maintaining a runway for one year is equal to the length of the runway. Because 707s land at the airport, the airport will have a 400-yd runway. For simplicity, suppose that each year only one plane of each type lands at the airport. How much of the $400 annual maintenance cost should be charged to each plane?

Solution   Let player 1 = Piper Cub, player 2 = DC-10, and player 3 = 707. We can now define a three-player game in which the value to a coalition is the cost associated with the runway length needed to service the largest plane in the coalition. Thus, the characteristic function for this game (we list a cost as a negative revenue) would be

$$v(\{\ \}) = \$0, \quad v(\{1\}) = -\$100, \quad v(\{1, 2\}) = v(\{2\}) = -\$150,$$
$$v(\{3\}) = v(\{2, 3\}) = v(\{1, 3\}) = v(\{1, 2, 3\}) = -\$400$$

To find the Shapley value (cost) to each player, we assume that the three planes land in a random order, and we determine how much cost (on the average) each plane adds to the cost incurred by the planes that are already present (see Table 48). The Shapley cost for each player is as follows:

$$\text{Player 1 cost} = (\tfrac{1}{6})(100 + 100) = \tfrac{\$200}{6}$$
$$\text{Player 2 cost} = (\tfrac{1}{6})(50 + 150 + 150) = \tfrac{\$350}{6}$$
$$\text{Player 3 cost} = (\tfrac{1}{6})(250 + 300 + 250 + 250 + 400 + 400) = \tfrac{\$1850}{6}$$

TABLE **48**
Computation of Shapley Value for Airport Game

Order of Arrival	Probability of Order	Cost Added by Player's Arrival ($)		
		Player 1	Player 2	Player 3
1, 2, 3	$\frac{1}{6}$	100	50	250
1, 3, 2	$\frac{1}{6}$	100	0	300
2, 1, 3	$\frac{1}{6}$	0	150	250
2, 3, 1	$\frac{1}{6}$	0	150	250
3, 1, 2	$\frac{1}{6}$	0	0	400
3, 2, 1	$\frac{1}{6}$	0	0	400

Thus, the Shapley value concept suggests that the Piper Cub pay $33.33, the DC-10 pay $58.33, and the 707 pay $308.33.

In general, even if more than one plane of each type lands, it has been shown that the Shapley value for the airport problem allocates runway operating cost as follows: All planes that use a portion of the runway should divide equally the cost of that portion of the runway (see Littlechild and Owen (1973)). Thus, all planes should cover the cost of the first 100 yd of runway, the DC-10s and 707s should pay for the next $150 - 100 = 50$ yd of runway, and the 707s should pay for the last $400 - 150 = 250$ yd of runway. If there were ten Piper Cub landings, five DC-10 landings, and two 707 landings, the Shapley value concept would recommend that each Piper Cub pay $\frac{100}{10+5+2} = \$5.88$ in landing fees, each DC-10 pay $5.88 + \frac{150-100}{5+2} = \$13.03$, and each 707 pay $13.03 + \frac{400-150}{2} = \$138.03$.

PROBLEMS

Group A

1 Consider the four-player game with the following characteristic function:
$$v(\{1, 2, 3\}) = v(\{1, 2, 4\}) = v(\{1, 3, 4\})$$
$$= v(\{2, 3, 4\}) = 75$$
$$v(\{1, 2, 3, 4\}) = 100$$
$$v(\{3, 4\}) = 60$$
$$v(S) = 0 \text{ for all other coalitions}$$
Show that this game has an empty core.

2 Show that if $v(\{3, 4\})$ in Problem 1 were changed to 50, then the game's core would consist of a single point.

3 The game of Odd Man Out is a three-player coin toss game in which each player must choose heads or tails. If all the players make the same choice, the house pays each player $1; otherwise, the odd man out pays each of the other players $1.

    a Find the characteristic function for this game.

    b Find the core of this game.

    c Find the Shapley value for this game.

4 Show that for $n = 2$, the core of the garbage game is the imputation $(-1, -1)$.

5 Show that for $n > 2$, the n-player garbage game has an empty core.

6 For the four-player garbage game, find an imputation that dominates $(-1, -1, -1, -1)$.

7 The Gotham City airport runway is 5,000 ft long and costs $100,000 per year to maintain. Last year there were 2,000 landings at the airport. Four types of planes landed. The length of runway required by each type of plane and the number of landings of each type are shown in Table 49. Assuming that the cost of operating a length of runway is proportional to the length of the runway, how much per landing should be paid by each type of plane?

TABLE **49**

Type of Plane	Number of Landings	Length of Runway (ft)
1	600	2,000
2	700	3,000
3	500	4,000
4	200	5,000

8 Consider the following three-person game:

$$v(\{\ \}) = 0, \qquad v(\{1\}) = 0.2,$$
$$v(\{2\}) = v(\{3\}) = 0, \quad v(\{1, 2\}) = 1.5,$$
$$v(\{1, 3\}) = 1.6, \qquad v(\{2, 3\}) = 1.8,$$
$$v(\{1, 2, 3\}) = 2$$

a Find the core of this game.

b Find the Shapley value for this game.

c Find an imputation dominating the imputation $(1, \frac{1}{2}, \frac{1}{2})$.

9 Howard Whose has left an estate of $200,000 to support his three ex-wives. Unfortunately, Howard's attorney has determined that each ex-wife needs the following amount of money to take care of Howard's children: wife 1—$100,000; wife 2—$200,000; wife 3—$300,000. Howard's attorney must determine how to divide the money among the three wives. He defines the value of a coalition S of ex-wives to be the maximum amount of money left for the ex-wives in S after all ex-wives not in S receive what they need. Using this definition, construct a characteristic function for this problem. Then determine the core and Shapley value for this game.

10 Indiana University leases WATS lines and is charged according to the following rules: $400 per month for each of the first five lines; $300 per month for each of the next five lines; $100 per month for each additional line. The College of Arts and Sciences makes 150 calls per hour, the School of Business makes 120 calls per hour, and the rest

of the university makes 30 calls per hour. Assume that each line can handle 30 calls per hour. Thus, the university will rent 10 WATS lines. The university wants to determine how much each part of the university should pay for long-distance phone service.

a Set up a characteristic function representation of the problem.

b Use the Shapley value to allocate the university's long-distance phone costs.

11 Three doctors have banded together to form a joint practice: the Port Charles Trio. The overhead for the practice is $40,000 per year. Each doctor brings in annual revenues and incurs annual variable costs as follows: doctor 1—$155,000 in revenue, $40,000 in variable cost; doctor 2—$160,000 in revenue, $35,000 in variable cost; and doctor 3—$140,000 in revenue, $38,000 in variable cost.

The Port Charles Trio wants to use game theory to determine how much each doctor should be paid. Determine the relevant characteristic function and show that the core of the game consists of an infinite number of points. Also determine the Shapley value of the game. Does the Shapley value give a reasonable division of the practice's profits?

Group B

12 Consider an n-person game in which the only winning coalitions are those containing player 1 and at least one other player. If a winning coalition receives a reward of $1, find the Shapley value to each player.

SUMMARY Two-Person Zero-Sum and Constant-Sum Games

John von Neumann and Oskar Morgenstern suggested that two-person zero-sum and constant-sum games be played according to the following basic assumption of two-person zero-sum game theory: Each player chooses a strategy that enables him to do the best he can, given that his opponent *knows the strategy he is following*.

A two-person zero-sum game has a saddle point if and only if

$$\max_{\substack{\text{all} \\ \text{rows}}} (\text{row minimum}) = \min_{\substack{\text{all} \\ \text{columns}}} (\text{column maximum}) \tag{1}$$

If a two-person zero-sum or constant-sum game has a saddle point, then the row player should choose any strategy (row) attaining the maximum on the left side of (1). The column player should choose any strategy (column) attaining the minimum on the right side of (1).

In general, we may use the following method to find the optimal strategies and the value of a two-person zero-sum or constant-sum game:

Step 1 Check for a saddle point. If the game has none, go on to step 2.

Step 2 Eliminate any of the row player's dominated strategies. Looking at the reduced matrix (dominated rows crossed out), eliminate any of the column player's dominated strategies and then those of the row player. Continue until no more dominated strategies can be found. Then proceed to step 3.

Step 3 If the game matrix is now 2×2, solve the game graphically. Otherwise, solve by using the linear programming method in Table 24 (page 627).

The value of the game and the optimal strategies for the row and column players in the Table 24 reward matrix may be found by solving the row player's LP and the column player's LP, respectively (page 630).

The dual of the row (column) player's LP is the column (row) player's LP. The optimal objective function value for either the row or the column player's LP is the value of the game to the row player. If the row player departs from her optimal strategy, then she may receive an expected reward that is less than the value of the game. If the column player departs from his optimal strategy, then he may incur an expected loss that exceeds the value of the game. Complementary slackness may be used to simultaneously solve the row and the column players' LPs.

Two-Person Nonconstant-Sum Games

As in a two-person zero-sum game, a choice of strategy by each player is an **equilibrium point** if neither player can benefit from a unilateral change in strategy.

A two-person nonconstant-sum game of particular interest is Prisoner's Dilemma. If $T > R > P > S$, a reward matrix like the one in Table 36 (page 636) will be a Prisoner's Dilemma game. For such a game, (NC, NC) (both players choosing a noncooperative action) is an equilibrium point.

n-Person Games

When more than two players are involved, the structure of a competitive situation may be summarized by the **characteristic function.** For each set of players S, the characteristic function v of a game gives the amount $v(S)$ that the members of S can be sure of receiving if they act together and form a coalition.

Let $\mathbf{x} = (x_1, x_2, \ldots, x_n)$ be a vector such that player i receives a reward x_i. We call such a vector a **reward vector.** A reward vector $\mathbf{x} = (x_1, x_2, \ldots, x_n)$ is an **imputation** if and only if

$$v(N) = \sum_{i=1}^{i=n} x_i \qquad \text{(Group rationality)} \qquad (13)$$

$$x_i \geq v(\{i\}) \quad \text{(for each } i \in N) \qquad \text{(Individual rationality)} \qquad (14)$$

The imputation $\mathbf{y} = (y_1, y_2, \ldots, y_n)$ **dominates** $\mathbf{x}$ through a coalition S (written $\mathbf{y} > {}^s\mathbf{x}$) if

$$\sum_{i \in S} y_i \leq v(S) \qquad \text{and for all } i \in S, \quad y_i > x_i \qquad (15)$$

The **core** and the **Shapley value** are two alternative solution concepts for n-person games. The *core* of an n-person game is the set of all undominated imputations. An imputation $\mathbf{x} = (x_1, x_2, \ldots, x_n)$ is in the core of an n-person game if and only if for each subset S of $N = \{1, 2, \ldots, n\}$

$$\sum_{i \in S} x_i \geq v(S)$$

The Shapley value gives a reward x_i to the ith player, where x_i is given by

$$x_i = \sum_{\substack{\text{all } S \text{ for which} \\ i \text{ is not in } S}} p_n(S)[v(S \cup \{i\}) - v(S)] \qquad (45)$$

In (45),

$$p_n(S) = \frac{|S|!(n - |S| - 1)!}{n!} \qquad (46)$$

Equation (45) implies that player i's reward should be the expected amount that player i adds to the coalition made up of the players who are present when player i arrives.

REVIEW PROBLEMS

Group A

1 Two competing firms are deciding whether to locate a new store at point A, B, or C. There are 52 prospective customers for the two stores. Twenty customers live in village A, 20 customers live in village B, and 12 customers live in village C (see Figure 6). Each customer will shop at the nearer store. If a customer is equidistant from both stores, then assume there is a $\frac{1}{2}$ chance that he or she will shop at either store. Each firm wants to maximize the expected number of customers that will shop at its store. Where should each firm locate its store? ($AB = BC = 10$ miles.)

2 A total of 90,000 customers frequent the Ruby and the Swamp supermarkets. To induce customers to enter, each store gives away a free item. Each week, the giveaway item is announced in the Monday newspaper. Of course, neither store knows which item the other store will choose to give away this week. Ruby's is considering giving away a carton of soda or a half gallon of milk. Swamp's is considering giving away a pound of butter or a half gallon of orange juice. For each possible choice of items, the number of customers who will stop at Ruby's during the current week is shown in Table 50. Each store wants to maximize its expected number of customers during the current week. Use game theory to determine an optimal strategy for each store and the value of the game. Interpret the value of the game.

3 Consider the two-person zero-sum game in Table 51.

a Write down each player's LP.

b We are told that player 1's optimal strategy has $x_1 > 0$, $x_2 > 0$, and $x_3 > 0$. Find the value of the game and each player's optimal strategies.

c Suppose the column player plays the nonoptimal strategy $(\frac{1}{2}, \frac{1}{2}, 0)$. Show how the row player can earn an expected reward that exceeds the value of the game.

4 Find optimal strategies for each player and the value of the two-person zero-sum game in Table 52.

FIGURE 6

20 customers	20 customers	12 customers
A	B	C

TABLE 50

Ruby Chooses	Swamp Chooses	
	Butter	Orange Juice
Soda	40,000	50,000
Milk	60,000	30,000

TABLE 51

$\frac{1}{2}$	-1	-1
-1	$\frac{1}{2}$	-1
-1	-1	1

TABLE 52

20	1	2
12	10	4
24	8	-2

5 Airway (a Midwestern department store chain) and Corvett (an Eastern department store chain) are determining whether to expand their geographical bases. The only viable manner by which expansion might be carried out is for a chain to open stores in the other's area. If neither chain expands, then Airway's profits will be $3 million and Corvett's will be $2 million. If Airway expands and Corvett does not, then Airway's profits will be $5 million, and Corvett will lose $2 million. If Airway does not expand and Corvett does, Airway will lose $1 million, and Corvett will earn $4 million. Finally, if both chains expand, Airway will earn $1 million and Corvett will earn $500,000 in profits. Determine the equilibrium points, if any, for this game.

6 The stock in Alden Corporation is held by three people. Person 1 owns 1%, person 2 owns 49%, and person 3 owns 50%. To pass a resolution at the annual stockholders' meeting, 51% of the stock is needed. A coalition receives a reward of 1 if it can pass a resolution and a reward of 0 if it cannot pass a resolution.

a Find the characteristic function for this game.

b Find the core of this game.

c Find the Shapley value for this game.

d Because $(\frac{1}{3}, \frac{1}{3}, \frac{1}{3})$ is not in the core, there must be an imputation dominating $(\frac{1}{3}, \frac{1}{3}, \frac{1}{3})$. Find one.

Group B

7 In addition to the core and the Shapley value, the stable set is an alternative solution concept for n-person games. A set I of imputations is called a **stable set** if each imputation in I is undominated and every imputation that is not in I is dominated by some member of I. Consider the three-person game in which all zero- and one-member coalitions have a characteristic function value of 0, and each two- and three-player coalition has a value of 1. Show that for this game $I = \{(\frac{1}{2}, \frac{1}{2}, 0), (0, \frac{1}{2}, \frac{1}{2}), (\frac{1}{2}, 0, \frac{1}{2})\}$ is a stable set.

REFERENCES

The following books take an elementary, applications-oriented approach to game theory:

Davis, M. *Game Theory: An Introduction*. New York: Basic Books, 1983.

Dixit, A., and B. Nalebuff, *Thinking Strategically*. New York: Norton, 1991.

McMillian, J. *Games, Strategies, and Managers*. New York: Oxford, 1992.

Poundstone, W. *The Prisoner's Dilemma*. New York: Doubleday, 1992.

Rapoport, A. *Two-Person Game Theory*. Ann Arbor, Mich.: University of Michigan Press, 1973.

The following classics are still worth reading:

Luce, R., and H. Raiffa. *Games and Decisions*. New York: Wiley, 1957.

Von Neumann, J., and O. Morgenstern. *Theory of Games and Economic Behavior*. Princeton, N.J.: Princeton University Press, 1944.

For the more mathematically inclined reader, the next eleven books are recommended:

Dutta, P. *Strategies and Games: Theory and Practice*. Cambridge, Mass.: MIT Press, 1999.

Friedman, J. *Game Theory with Applications to Economics*. New York: Oxford Press, 1990.

Fudenberg, D., and J. Tirole. *Game Theory*. Cambridge, Mass: MIT Press, 1991.

Gibbons, R. *Game Theory for Applied Economists*. Princeton, N.J.: Princeton University Press, 1992.

Gitnis, H. *Game Theory Evolving*. Princeton, N.J.: Princeton University Press, 2000.

Hargreaves, S., and Varoufakis, Y. *Game Theory: A Critical Introduction*. New York: Routledge, 1995.

Osborne, M., and Rubenstein, A. *A Course in Game Theory*. Cambridge, Mass.: MIT Press, 1994.

Owen, G. *Game Theory*. Orlando, Fla.: Academic Press, 1999.

Shubik, M. *Game Theory in the Social Sciences: Concepts and Solutions*. Cambridge, Mass.: MIT Press, 1982.

———. *A Game-Theoretic Approach to Political Economy*. Cambridge, Mass.: MIT Press, 1984.

Thomas, L. C. *Games, Theory and Applications*. Chichester, England: Ellis Horwood, 1986.

Vorobev, N. *Game Theory Lectures for Economists and Social Sciences*. New York: Springer-Verlag, 1977.

Littlechild, S., and G. Owen. "A Simple Expression for the Shapley Value in a Special Case," *Management Science* 20(1973): 370–372. Discusses applications of the Shapley value to airport landings.

"Protectionism Gets Clever," *The Economist* (November 21, 1988): 78.

Shapley, L. "Quota Solutions of *n*-Person Games." In *Contributions to the Theory of Games II,* ed. H. Kuhn and A. Tucker. Princeton, N.J.: Princeton University Press, 1953.

12

Nonlinear Programming

In previous chapters, we have studied linear programming problems. For an LP, our goal was to maximize or minimize a linear function subject to linear constraints. But in many interesting maximization and minimization problems, the objective function may not be a linear function, or some of the constraints may not be linear constraints. Such an optimization problem is called a *nonlinear programming* problem (NLP). In this chapter, we discuss techniques used to solve NLPs.

We begin with a review of material from differential calculus, which will be needed for our study of nonlinear programming.

12.1 Review of Differential Calculus

Limits

This idea of a limit is one of the most basic ideas in calculus.

DEFINITION ■ The equation

$$\lim_{x \to a} f(x) = c$$

means that as x gets closer to a (but not equal to a), the value of $f(x)$ gets arbitrarily close to c. ■

It is also possible that $\lim_{x \to a} f(x)$ may not exist.

EXAMPLE 1 **Limits**

1 Show that $\lim_{x \to 2} x^2 - 2x = 2^2 - 2(2) = 0$.

2 Show that $\lim_{x \to 0} \frac{1}{x}$ does not exist.

Solution 1 To verify this result, evaluate $x^2 - 2x$ for values of x close to, but not equal to, 2.

2 To verify this result, observe that as x gets near 0, $\frac{1}{x}$ becomes either a very large positive number or a very large negative number. Thus, as x approaches 0, $\frac{1}{x}$ will not approach any single number.

Continuity

A function $f(x)$ is **continuous** at a point a if

$$\lim_{x \to a} f(x) = f(a)$$

If $f(x)$ is not continuous at $x = a$, we say that $f(x)$ is **discontinuous** (or has a discontinuity) at a. ■

EXAMPLE 2   **Continuous Functions**

Bakeco orders sugar from Sugarco. The per-pound purchase price of the sugar depends on the size of the order (see Table 1). Let

$$x = \text{number of pounds of sugar purchased by Bakeco}$$
$$f(x) = \text{cost of ordering } x \text{ pounds of sugar}$$

Then

$$f(x) = 25x \text{ for } 0 \le x < 100$$
$$f(x) = 20x \text{ for } 100 \le x \le 200$$
$$f(x) = 15x \text{ for } x > 200$$

For all values of x, determine if x is continuous or discontinuous.

Solution   From Figure 1, it is clear that

$$\lim_{x \to 100} f(x) \qquad \text{and} \qquad \lim_{x \to 200} f(x)$$

do not exist. Thus, $f(x)$ is discontinuous at $x = 100$ and $x = 200$ and is continuous for all other values of x satisfying $x \ge 0$.

TABLE 1
Price of Sugar Paid by Bakeco

Size of Order	Price per Pound (c)
$0 \le x < 100$	25
$100 \le x \le 200$	20
$x > 200$	15

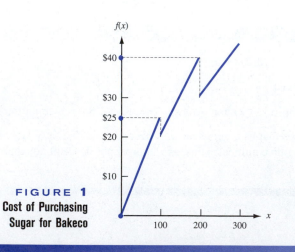

FIGURE 1
Cost of Purchasing Sugar for Bakeco

TABLE 2
Rules for Finding the Derivative of a Function

Function	Derivative of Function
a	0
x	1
$af(x)$	$af'(x)$
$f(x) + g(x)$	$f'(x) + g'(x)$
x^n	nx^{n-1}
e^x	e^x
a^x	$a^x \ln a$
$\ln x$	$\frac{1}{x}$
$[f(x)]^n$	$n [f(x)]^{n-1} f'(x)$
$e^{f(x)}$	$e^{f(x)} f'(x)$
$a^{f(x)}$	$a^{f(x)} f'(x) \ln a$
$\ln f(x)$	$\dfrac{f'(x)}{fx}$
$f(x)g(x)$	$f(x)g'(x) + f'(x)g(x)$
$\dfrac{f(x)}{g(x)}$	$\dfrac{g(x)f'(x) - f(x)g'(x)}{g(x)^2}$

Differentiation

DEFINITION ■ The **derivative** of a function $f(x)$ at $x = a$ [written $f'(a)$] is defined to be

$$\lim_{\Delta x \to 0} \frac{f(a + \Delta x) - f(a)}{\Delta x} \quad ■$$

If this limit does not exist, then $f(x)$ has no derivative at $x = a$.

We may think of $f'(a)$ as the slope of $f(x)$ at $x = a$. Thus, if we begin at $x = a$ and increase x by a small amount Δ (Δ may be positive or negative), then $f(x)$ will increase by an amount approximately equal to $\Delta f'(a)$. If $f'(a) > 0$, then $f(x)$ is increasing at $x = a$, whereas if $f'(a) < 0$, then $f(x)$ is decreasing at $x = a$. The derivatives of many functions can be found via application of the rules in Table 2 (a represents an arbitrary constant). Example 3 illustrates the use and interpretation of the derivative.

EXAMPLE 3 **Product Profitability**

If a company charges a price p for a product, then it can sell $3e^{-p}$ thousand units of the product. Then, $f(p) = 3,000pe^{-p}$ is the company's revenue if it charges a price p.

1 For what values of p is $f(p)$ decreasing? For what values of p is $f(p)$ increasing?

2 Suppose the current price is \$4 and the company increases the price by 5¢. By approximately how much would the company's revenue change?

Solution We have

$$f'(p) = -3,000pe^{-p} + 3,000e^{-p} = 3,000e^{-p}(1 - p)$$

1 For $p < 1, f'(p) > 0$ and $f(p)$ is increasing, whereas for $p > 1, f'(p) < 0$ and $f(p)$ is decreasing.

2 Using the interpretation of $f'(4)$ as the slope of $f(p)$ at $p = 4$ (with $\Delta p = 0.05$), we see that the company's revenue would increase by approximately

$$0.05(3{,}000e^{-4})(1 - 4) = -8.24$$

In actuality, of course, the company's revenue would increase by

$$f(4.05) - f(4) = 3{,}000(4.05)e^{-4.05} - 3{,}000(4)e^{-4}$$
$$= 211.68 - 219.79 = -8.11$$

Higher Derivatives

We define $f^{(2)}(a) = f''(a)$ to be the derivative of the function $f'(x)$ at $x = a$. Similarly, we can define (if it exists) $f^{(n)}(a)$ to be the derivative of $f^{(n-1)}(x)$ at $x = a$. Thus, for Example 3,

$$f''(p) = 3{,}000e^{-p}(-1) - 3{,}000e^{-p}(1 - p)$$

Taylor Series Expansion

In the Taylor series expansion of a function $f(x)$, given that $f^{(n+1)}(x)$ exists for every point on the interval $[a, b]$, we can write for any h satisfying $0 \leq h \leq b - a$,

$$f(a + h) = f(a) + \sum_{i=1}^{i=n} \frac{f^{(i)}(a)}{i!} h^i + \frac{f^{(n+1)}(p)}{(n + 1)!} h^{n+1} \tag{1}$$

where (1) will hold for some number p between a and $a + h$. Equation (1) is the **nth-order Taylor series expansion** of $f(x)$ about a.

EXAMPLE 4 — Taylor Series Expansion

Find the first-order Taylor series expansion of e^{-x} about $x = 0$.

Solution Because $f'(x) = -e^{-x}$ and $f''(x) = e^{-x}$, we know that (1) will hold on any interval $[0, b]$. Also, $f(0) = 1$, $f'(0) = -1$, and $f''(x) = e^{-x}$. Then (1) yields the following first-order Taylor series expansion for e^{-x} about $x = 0$:

$$e^{-h} = f(h) = 1 - h + \frac{h^2 e^{-p}}{2}$$

This equation holds for some p between 0 and h.

Partial Derivatives

We now consider a function f of $n > 1$ variables $(x_1, x_2, \ldots, x_n)$, using the notation $f(x_1, x_2, \ldots, x_n)$ to denote such a function.

The **partial derivative** of $f(x_1, x_2, \ldots, x_n)$ with respect to the variable x_i is written $\dfrac{\partial f}{\partial x_i}$, where

$$\frac{\partial f}{\partial x_i} = \lim_{\Delta x_i \to 0} \frac{f(x_1, \ldots, x_i + \Delta x_i, \ldots, x_n) - f(x_1, \ldots, x_i, \ldots, x_n)}{\Delta x_i} \qquad ■$$

Intuitively, if x_i is increased by Δ (and all other variables are held constant), then for small values of Δ, the value of $f(x_1, x_2, \ldots, x_n)$ will increase by approximately $\Delta \dfrac{\partial f}{\partial x_i}$. We find $\dfrac{\partial f}{\partial x_i}$ by treating all variables other than x_i as constants and finding the derivatives of $f(x_1, x_2, \ldots, x_n)$. More generally, suppose that for each i, we increase x_i by a small amount Δx_i. Then the value of f will increase by approximately

$$\sum_{i=1}^{i=n} \frac{\partial f}{\partial x_i} \Delta x_i$$

EXAMPLE 5 When Is a Function Increasing?

The demand $f(p, a) = 30{,}000p^{-2}a^{1/6}$ for a product depends on $p =$ product price (in dollars) and $a =$ dollars spent advertising the product. Is demand an increasing or decreasing function of price? Is demand an increasing or decreasing function of advertising expenditure? If $p = 10$ and $a = 1{,}000{,}000$, then by how much (approximately) will a \$1 cut in price increase demand?

Solution

$$\frac{\partial f}{\partial p} = 30{,}000(-2p^{-3})a^{1/6} = -60{,}000p^{-3}a^{1/6} < 0$$

$$\frac{\partial f}{\partial a} = 30{,}000p^{-2}\left(\frac{a^{-5/6}}{6}\right) = 5{,}000p^{-2}a^{-5/6} > 0$$

Thus, an increase in price (with advertising held constant) will decrease demand, while an increase in advertising (with price held constant) will increase demand. Because

$$\frac{\partial f}{\partial p}(10, 1{,}000{,}000) = -60{,}000\left(\frac{1}{1{,}000}\right)(1{,}000{,}000)^{1/6} = -600$$

a \$1 price cut will increase demand by approximately $(-1)(-600)$, or 600 units.

We will also use *second-order partial derivatives* extensively. We use the notation $\dfrac{\partial^2}{\partial x_i \partial x_j}$ to denote a second-order partial derivative. To find $\dfrac{\partial^2}{\partial x_i \partial x_j}$, we first find $\dfrac{\partial f}{\partial x_i}$ and then take its partial derivative with respect to x_j. If the second-order partials exist and are everywhere continuous, then

$$\frac{\partial^2 f}{\partial x_i \partial x_j} = \frac{\partial^2 f}{\partial x_j \partial x_i}$$

EXAMPLE 6 Second-Order Partial Derivatives

For $f(p, a) = 30{,}000p^{-2}a^{1/6}$, find all second-order partial derivatives.

Solution

$$\frac{\partial^2 f}{\partial p^2} = -60{,}000(-3p^{-4})a^{1/6} = \frac{180{,}000a^{1/6}}{p^4}$$

$$\frac{\partial^2 f}{\partial a^2} = 5{,}000p^{-2}\left(\frac{-5a^{-11/6}}{6}\right) = -\frac{25{,}000p^{-2}a^{-11/6}}{6}$$

$$\frac{\partial^2 f}{\partial a \partial p} = 5{,}000(-2p^{-3})a^{-5/6} = -10{,}000p^{-3}a^{-5/6}$$

$$\frac{\partial^2 f}{\partial p \partial a} = -60{,}000p^{-3}\left(\frac{a^{-5/6}}{6}\right) = -10{,}000p^{-3}a^{-5/6}$$

Observe that for $p \neq 0$ and $a \neq 0$,

$$\frac{\partial^2 f}{\partial a \partial p} = \frac{\partial^2 f}{\partial p \partial a}$$

PROBLEMS

Group A

1 Find $\lim_{h \to 0} \dfrac{3h + h^2}{h}$.

2 It costs Sugarco 25¢/lb to purchase the first 100 lb of sugar, 20¢/lb to purchase the next 100 lb, and 15¢ to buy each additional pound. Let $f(x)$ be the cost of purchasing x pounds of sugar. Is $f(x)$ continuous at all points? Are there any points where $f(x)$ has no derivative?

3 Find $f'(x)$ for each of the following functions:

a xe^{-x}

b $\dfrac{x^2}{x^2 + 1}$

c e^{3x}

d $(3x + 2)^{-2}$

e $\ln x^3$

4 Find all first- and second-order partial derivatives for $f(x_1, x_2) = x_1^2 e^{x_2}$.

5 Find the second-order Taylor series expansion of $\ln x$ about $x = 1$.

Group B

6 Let $q = f(p)$ be the demand for a product when the price is p. For a given price p, the price elasticity E of the product is defined by

$$E = \frac{\text{percentage change in demand}}{\text{percentage change in price}}$$

If the change in price (Δp) is small, this formula reduces to

$$E = \frac{\frac{\Delta q}{q}}{\frac{\Delta p}{p}} = \left(\frac{p}{q}\right)\left(\frac{dq}{dp}\right)$$

a Would you expect $f(p)$ to be positive or negative?

b Show that if $E < -1$, a small decrease in price will increase the firm's total revenue (in this case, we say that demand is elastic).

c Show that if $-1 < E < 0$, a small price decrease will decrease total revenue (in this case, we say demand is inelastic).

7 Suppose that if x dollars are spent on advertising during a given year, $k(1 - e^{-cx})$ customers will purchase a product ($c > 0$).

a As x grows large, the number of customers purchasing the product approaches a limit. Find this limit.

b Can you give an interpretation for k?

c Show that the sales response from a dollar of advertising is proportional to the number of potential customers who are not purchasing the product at present.

8 Let the total cost of producing x units, $c(x)$, be given by $c(x) = kx^{1-b}$ ($0 < b < 1$). This cost curve is called the **learning**, or **experience cost curve.**

a Show that the cost of producing a unit is a decreasing function of the number of units that have been produced.

b Suppose that each time the number of units produced is doubled, the per-unit product cost drops to $r\%$ of its previous value (because workers learn how to perform their jobs better). Show that $r = 100(2^{-b})$.

9 If a company has m hours of machine time and w hours of labor, it can produce $3m^{1/3}w^{2/3}$ units of a product. Currently, the company has 216 hours of machine time and 1,000 hours of labor. An extra hour of machine time costs \$100, and an extra hour of labor costs \$50. If the company has \$100 to invest in purchasing additional labor and machine time, would it be better off buying 1 hour of machine time or 2 hours of labor?

12.2 Introductory Concepts

DEFINITION ■ A general **nonlinear programming problem** (NLP) can be expressed as follows: Find the values of decision variables $x_1, x_2, \ldots, x_n$ that

$$\max \quad (\text{or min}) \ z = f(x_1, x_2, \ldots, x_n)$$

$$\text{s.t.} \quad g_1(x_1, x_2, \ldots, x_n) \ (\leq, =, \text{ or } \geq) \ b_1$$

$$\text{s.t.} \quad g_2(x_1, x_2, \ldots, x_n) \ (\leq, =, \text{ or } \geq) \ b_2 \qquad (2)$$

$$\vdots$$

$$g_m(x_1, x_2, \ldots, x_n) \ (\leq, =, \text{ or } \geq) \ b_m \quad ■$$

As in linear programming, $f(x_1, x_2, \ldots, x_n)$ is the NLP's **objective function,** and $g_1(x_1, x_2, \ldots, x_n) \ (\leq, =, \text{ or } \geq) \ b_1, \ldots, g_m(x_1, x_2, \ldots, x_n) \ (\leq, =, \text{ or } \geq) \ b_m$ are the NLP's **constraints.** An NLP with no constraints is an **unconstrained NLP.**

The set of all points $(x_1, x_2, \ldots, x_n)$ such that x_i is a real number is R^n. Thus, R^1 is the set of all real numbers. The following subsets of R^1 (called intervals) will be of particular interest:

$$[a, b] = \text{all } x \text{ satisfying } a \leq x \leq b$$

$$[a, b) = \text{all } x \text{ satisfying } a \leq x < b$$

$$(a, b] = \text{all } x \text{ satisfying } a < x \leq b$$

$$(a, b) = \text{all } x \text{ satisfying } a < x < b$$

$$[a, \infty) = \text{all } x \text{ satisfying } x \geq a$$

$$(-\infty, b] = \text{all } x \text{ satisfying } x \leq b$$

The following definitions are analogous to the corresponding definitions for LPs given in Section 3.1.

DEFINITION ■ The **feasible region** for NLP (2) is the set of points $(x_1, x_2, \ldots, x_n)$ that satisfy the m constraints in (2). A point in the feasible region is a *feasible point,* and a point that is not in the feasible region is an *infeasible point.* ■

Suppose (2) is a maximization problem.

DEFINITION ■ Any point $\bar{x}$ in the feasible region for which $f(\bar{x}) \geq f(x)$ holds for all points x in the feasible region is an **optimal solution** to the NLP. [For a minimization problem, $\bar{x}$ is the optimal solution if $f(\bar{x}) \leq f(x)$ for all feasible x.] ■

Of course, if $f, g_1, g_2, \ldots, g_m$ are all linear functions, then (2) is a linear programming problem and may be solved by the simplex algorithm.

Examples of NLPs

EXAMPLE 7   Profit Maximization

It costs a company c dollars per unit to manufacture a product. If the company charges p dollars per unit for the product, customers demand $D(p)$ units. To maximize profits, what price should the firm charge?

Solution   The firm's decision variable is p. Since the firm's profit is $(p - c)D(p)$, the firm wants to solve the following unconstrained maximization problem: $\max(p - c)D(p)$.

EXAMPLE 8   **Production Maximization**

If K units of capital and L units of labor are used, a company can produce KL units of a manufactured good. Capital can be purchased at \$4/unit and labor can be purchased at \$1/unit. A total of \$8 is available to purchase capital and labor. How can the firm maximize the quantity of the good that can be manufactured?

Solution   Let K = units of capital purchased and L = units of labor purchased. Then K and L must satisfy $4K + L \leq 8$, $K \geq 0$, and $L \geq 0$. Thus, the firm wants to solve the following constrained maximization problem:

$$\max z = KL$$
$$\text{s.t.} \quad 4K + L \leq 8$$
$$K, L \geq 0$$

Solving NLPs with LINGO

Cap.lng

LINGO may be used to solve NLPs on a PC. Figure 2 (file Cap.lng) contains the LINGO formulation and output for Example 8. From the Value column, we see that LINGO has found the solution K = 1 and L = 4, which has an objective function value of 4. As we shall soon see, this is indeed the optimal solution to Example 8. However, in general, there is no guarantee that the solution found by LINGO is an optimal solution. Throughout this chapter, we will detail the circumstances in which you can be sure that LINGO will find the optimal solution to an NLP.

Note that the ^ symbol is used to indicate raising to a power and * indicates multiplication. LINGO has several built-in functions including

- ABS(X) = absolute value of X
- EXP(X) = e^x
- LOG(X) = natural logarithm of X

In Sections 12.9 and 12.10 we will discuss the Price column of the LINGO output. We will not discuss the Reduced Cost column.

Differences Between NLPs and LPs

Recall from Chapter 3 that the feasible region for any LP is a convex set (that is, if A and B are feasible for an LP, then the entire line segment joining A and B is also feasible). Also recall that if an LP has an optimal solution, then there is an extreme point of the feasible region that is optimal. We will soon see, however, that even if the feasible region for an NLP is a convex set, the optimal solution (unlike the optimal solution for an LP) need not be an extreme point of the NLP's feasible region. The previous example illustrates this idea. Figure 3 shows graphically the feasible region (bounded by triangle ABC) for the example and the isoprofit curves $KL = 1$, $KL = 2$, and $KL = 4$. We see that the optimal solution to the example occurs where an isoprofit curve is tangent to the boundary of the feasible region. Thus, the optimal solution to the example is $z = 4$, $K = 1$, $L = 4$ (point D). Of course, point D is not an extreme point of the NLP's feasible region. For this ex-

FIGURE 2

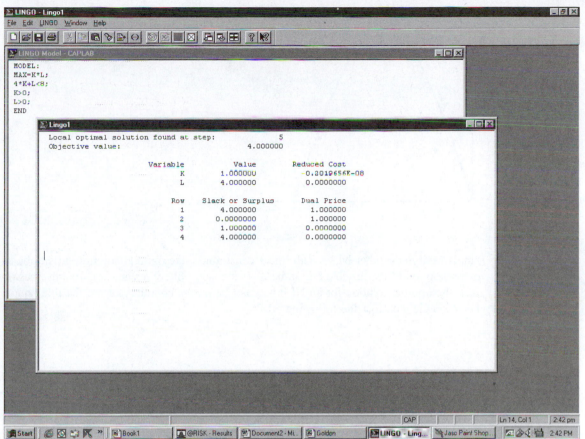

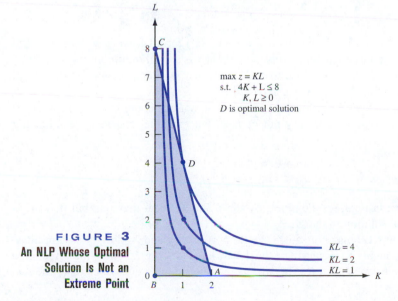

FIGURE 3
An NLP Whose Optimal Solution Is Not an Extreme Point

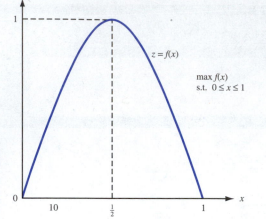

FIGURE 4
An NLP Whose Optimal
Solution Is Not on
Boundary of Feasible
Region

ample (and many other NLPs with linear constraints), the optimal solution fails to be an extreme point of the feasible region because the isoprofit curves are not straight lines. In fact, the optimal solution for an NLP may not be on the boundary of the feasible region. For example, consider the following NLP:

$$\max z = f(x)$$
$$\text{s.t.} \quad 0 \le x \le 1$$

where $f(x)$ is pictured in Figure 4. The optimal solution for this NLP is $z = 1$, $x = \frac{1}{2}$. Of course, $x = \frac{1}{2}$ is not on the boundary of the feasible region.

Local Extremum

DEFINITION ■ For any NLP (maximization), a feasible point $x = (x_1, x_2, \ldots, x_n)$ is a **local maximum** if for sufficiently small ϵ, any feasible point $x' = (x'_1, x'_2, \ldots, x'_n)$ having $|x_i - x'_i| < \epsilon$ $(i = 1, 2, \ldots, n)$ satisfies $f(x) \ge f(x')$. ■

In short, a point x is a local maximum if $f(x) \ge f(x')$ for all feasible x' that are close to x. Analogously, for a minimization problem, a point x is a local minimum if $f(x) \le f(x')$ holds for all feasible x' that are close to x. A point that is a local maximum or a local minimum is called a **local,** or **relative, extremum.**

For an LP (max problem), any local maximum is an optimal solution to the LP. (Why?) For a general NLP, however, this may not be true. For example, consider the following NLP:

$$\max z = f(x)$$
$$\text{s.t.} \quad 0 \le x \le 10$$

where $f(x)$ is given in Figure 5. Points A, B, and C are all local maxima, but point C is the unique optimal solution to the NLP.

Unlike an LP, an NLP may not satisfy the Proportionality and Additivity assumptions. For instance, in Example 8, increasing L by 1 will increase z by K. Thus, the effect on z of increasing L by 1 depends on K. This means that the example does not satisfy the Additivity Assumption.

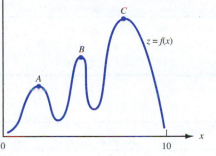

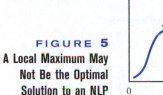

FIGURE 5
A Local Maximum May Not Be the Optimal Solution to an NLP

The NLP

$$\max z = x^{1/3} + y^{1/3}$$
$$\text{s.t.} \quad x + y = 1$$
$$x, y \geq 0$$

does not satisfy the Proportionality Assumption because doubling the value of x does not double the contribution of x to the objective function.

More Examples of NLP Formulations

We now give three more examples of nonlinear programming formulations.

EXAMPLE 9 **Oilco NLP**

Oilco produces three types of gasoline: regular, unleaded, and premium. All three are produced by combining lead and crude oil brought in from Alaska and Texas. The required sulphur content, octane levels, minimum daily demand (in gallons), and sales price per gallon of each type of gasoline are given in Table 3. The crude brought in from Alaska is made by blending two types of crude: Alaska1 and Alaska2. The Alaska crude is blended in Alaska and shipped via pipeline to Oilco's Texas refinery. At most, 10,000 gallons of crude per day can be shipped from Alaska. The sulphur content, octane level, daily maximum amount available (in gallons) and purchase cost (per gallon) for each type of Alaska crude, Texas crude, and lead are given in Table 4. Of course, unleaded gasoline can contain no lead. Formulate an NLP to help Oilco maximize the daily profit obtained from selling gasoline.[†]

Solution After defining the following decision variables:

R = gallons of regular gasoline produced daily

U = gallons of unleaded gasoline produced daily

P = gallons of premium gasoline produced daily

$A1$ = gallons of Alaska1 crude purchased daily

$A2$ = gallons of Alaska2 crude purchased daily

T = gallons of Texas crude purchased daily

L = gallons of lead purchased daily

[†]Based on Haverly (1978).

TABLE 3

Type of Gasoline	Sulphur Content (%)	Octane Level	Minimum Daily Demand (Gallons)	Sales Price ($)
Regular	≤ 3	≥ 90	5,000	.86
Unleaded	≤ 3	≥ 88	5,000	.93
Premium	≤ 2.8	≥ 94	5,000	1.06

TABLE 4

Type of Input	Sulphur Content (%)	Octane Level	Maximum Availability (Gallons)	Cost (per Gallon) ($)
Alaska 1	4	91	0	.78
Alaska 2	1	97	0	.88
Texas	2	83	11,000	.75
Lead	0	800	6,000	1.30

SA = sulphur content of crude purchased from Alaska

OA = octane level of crude purchased from Alaska

A = total gallons of crude purchased from Alaska

LP = gallons of lead used daily to make premium gasoline

TP = gallons of Texas crude used daily to make premium gasoline

AP = gallons of Alaska crude used daily to make premium gasoline

TU = Texas crude used daily to make unleaded gasoline

AU = Alaska crude used daily to make unleaded gasoline

AR = Alaska crude used daily to make regular gasoline

TR = Texas crude used daily to make regular gasoline

LR = gallons of lead used daily to make regular gasoline

Alas.lng

we find the appropriate formulation in the LINGO printout given in Figure 6 (file Alas.lng).

The objective function maximizes daily revenues (86 * R + 93 * U + 106 * P) less the daily costs of purchasing crude (78 * A1 + 88 * A2 + 75 * T + 130 * L). Rows 2–4 specify that the amount of each input cannot exceed its daily availability. Rows 5–7 ensure that the minimum demand requirements for each gasoline are met.

The percentage sulphur content (as a decimal) of Alaska crude in terms of the amount of each type of Alaska crude purchased is defined in row 8. Similarly, the octane level of Alaska crude in terms of the amount of each type of Alaska crude purchased is defined in row 9. Row 10 defines the total amount of Alaska crude purchased as the sum of the amount of Alaska1 and Alaska2 purchased. Similarly, row 11 expresses the amount of premium produced as the sum of its lead, Texas crude, and Alaska crude inputs, and row 12 expresses the amount of unleaded gasoline produced as the sum of its inputs. Rows 13–15 tell us that inputs are fully consumed in production—all lead is used to produce premium or regular gasoline; all Alaskan crude is used to make premium, unleaded, or regular gasoline; and all Texas crude is used to make premium, unleaded, or regular gasoline.

Row 16 requires that the average octane level of the inputs used to produce regular gasoline is at least 90. Notice that this is not a linear constraint because of the presence of the term AR * OA. Similarly, row 17 (again, not a linear constraint) ensures that the

```
MODEL:
   1) MAX= 86 * R + 93 * U + 106 * P - 78 * A1 - 88 * A2 - 75 * T - 130 *
      L ;
   2) A < 10000 ;
   3) T < 11000 ;
   4) L < 6000 ;
   5) R > 5000 ;
   6) U > 5000 ;
   7) P > 5000 ;
   8) SA = ( .04 * A1 + .01 * A2 ) / A ;
   9) OA = ( 91 * A1 + 97 * A2 ) / A ;
  10) A = A1 + A2 ;
  11) P = LP + TP + AP ;
  12) U = TU + AU ;
  13) L = LP + LR ;
  14) A = AP + AU + AR ;
  15) T = TP + TU + TR ;
  16) ( AR * OA + 83 * TR + 800 * LR ) / R > 90 ;
  17) ( AP * OA + 83 * TP + 800 * LP ) / P > 94 ;
  18) ( AU * OA + TU * 83 ) / U > 88 ;
  19) ( SA * AR + .02 * TR ) / R < .03 ;
  20) ( SA * AP + .02 * TP ) / P < .028 ;
  21) ( SA * AU + .02 * TU ) / U < .03 ;
  22) LP > 0 ;
  23) TP > 0 ;
  24) AP > 0 ;
  25) TU > 0 ;
  26) AU > 0 ;
  27) LR > 0 ;
  28) TR > 0 ;
  29) AR > 0 ;
  30) R = TR + AR + LR ;
END
```

SOLUTION STATUS: OPTIMAL TO TOLERANCES. DUAL CONDITIONS: SATISFIED.

OBJECTIVE FUNCTION VALUE

          1)   443237.052541

VARIABLE	VALUE	REDUCED COST
R	5000.000000	.000000
U	5000.000000	.000000
P	11134.965633	.000000
A1	9047.622772	.000000
A2	952.377228	.000000
T	11000.000000	.000000
L	134.965633	.000000
A	10000.000000	.000000
SA	.037143	.000000
OA	91.571426	.000000
LP	121.210474	.000000
TP	6863.136139	.000000
AP	4150.619020	.000000
TU	2083.333333	.000000
AU	2916.666667	.000000
LR	13.755159	.000000
AR	2932.714313	.000000
TR	2053.530528	.000000

ROW	SLACK OR SURPLUS	PRICE
2)	.000000	26.965066
3)	.000000	30.626062
4)	5865.034367	.000000
5)	.000000	-19.864023
6)	.000000	-12.796034
7)	6134.965633	.000000
8)	.000000	-2332388.904850
9)	.000000	5004.725331
10)	.000000	41.786554
11)	.000000	102.804532
12)	.000000	-13.948311
13)	.000000	-130.000000

(continued)

FIGURE 6
Oilco Problem and Solution

```
14)         .000000      -105.917442
15)         .000000      -105.626062
16)        -.000001      -169.971719
17)         .000001      -378.525763
18)        -.000001     -8166.740734
19)         .000000         .000000
20)         .001828         .000000
21)         .000000    3998380.979742
22)      121.210474         .000000
23)     6863.136139         .000000
24)     4150.619020         .000000
25)     2083.333333         .000000
26)     2916.666667         .000000
27)       13.755159         .000000
28)     2053.530528         .000000
29)     2932.714313         .000000
30)         .000000       102.804532
```

FIGURE 6
(Continued)

average octane level of the inputs used to produce premium gasoline is at least 94, and row 18 (again, not a linear constraint) that the octane level of the unleaded gasoline inputs is at least 88.

Row 19 (again a nonlinear constraint due to the presence of the term SA * AR) ensures that regular gasoline contains at most 3% sulphur; row 20, that premium gasoline contains at most 2.8% sulphur; and row 21, that unleaded gasoline contains at most 3% sulphur.

The amount of each input used to produce each output must be non-negative, required by rows 22–29. Row 30 specifies that the amount of regular gasoline sold must equal the sum of the inputs used to produce regular gasoline.

When solved on LINGO, we obtain a solution with a profit of $4,432.37 (remember the objective function is in cents) earned by producing 5,000 gallons of regular gasoline (with 13.76 gallons of lead, 2,932.71 gallons of Alaska crude, and 2,053.53 gallons of Texas crude); 5,000 gallons of unleaded gasoline (with 2,916.67 gallons of Alaska crude and 2,083.33 gallons of Texas crude); and 11,134.97 gallons of premium gasoline (using 121.21 gallons of lead, 6,863.14 gallons of Texas crude, and 4,150.62 gallons of Alaska crude). The mix of 10,000 gallons of Alaska crude was 90.48% Alaska1 and 9.52% Alaska2.

In Section 12.10 we will discuss how we can be sure that the solution found by LINGO is optimal.

REMARK By using a nonlinear blending model to optimize production of its gasoline products Texaco saves at least $30 million per year. See Dewitt et al. (1989) for details.

EXAMPLE 10   Warehouse Location

Truckco is trying to determine where it should locate a single warehouse. The positions in the x–y plane (in miles) of four customers and the number of shipments made annually to each customer are given in Table 5. Truckco wants to locate the warehouse to minimize the total distance trucks must travel annually from the warehouse to the four customers.

Solution Define

$$X = x\text{-coordinate of warehouse}$$

$$Y = y\text{-coordinate of warehouse}$$

$$Di = \text{Distance from customer } i \text{ to warehouse.}$$

Ware.lng

The appropriate NLP is given in the LINGO printout in Figure 7 (file Ware.lng). The objective function minimizes the total distance trucks travel each year from the warehouse to the four customers. Rows 2–5 define the distance from each customer to the warehouse in

TABLE 5

Customer	Coordinate X	Coordinate Y	Number of Shipments
1	5	10	200
2	10	5	150
3	0	12	200
4	12	0	300

```
MODEL:
  1) MIN= 200 * D1 + 150 * D2 + 200 * D3 + 300 * D4 ;
  2) D1 = ( ( X - 5 ) ^ 2 + ( Y - 10 ) ^ 2 ) ^ .5 ;
  3) D2 = ( ( X - 10 ) ^ 2 + ( Y - 5 ) ^ 2 ) ^ .5 ;
  4) D3 = ( X ^ 2 + ( Y - 12 ) ^ 2 ) ^ .5 ;
  5) D4 = ( ( X - 12 ) ^ 2 + Y ^ 2 ) ^ .5 ;
END
```

```
SOLUTION STATUS:  OPTIMAL TO TOLERANCES.  DUAL CONDITIONS:  UNSATISFIED.

              OBJECTIVE FUNCTION VALUE

       1)      5456.539688

   VARIABLE           VALUE        REDUCED COST
         D1        6.582238            .000000
         D2         .686433            .000000
         D3       11.634119            .000000
         D4        5.701011            .000000
          X        9.314167            .000176
          Y        5.028701            .000167

   ROW    SLACK OR SURPLUS              PRICE
     2)            .000000       -200.000000
     3)            .000000       -150.000000
     4)            .000000       -200.000000
     5)            .000000       -300.000000
```

FIGURE 7

Truckco Problem and Solution

terms of the warehouse location. LINGO located the warehouse at $X = 9.31$ and $Y = 5.03$. Each year the truck will travel a total of 5,456.54 miles from the warehouse to the customers.

EXAMPLE 11   Tire Production

Firerock produces rubber used for tires by combining three ingredients: rubber, oil, and carbon black. The cost in cents per pound of each ingredient is given in Table 6.

The rubber used in automobile tires must have a hardness of between 25 and 35, an elasticity of at least 16, and a tensile strength of at least 12. To manufacture a set of four automobile tires, 100 pounds of product is needed. The rubber used to make a set of four tires must contain between 25 and 60 pounds of rubber and at least 50 pounds of carbon black. If we define

R = pounds of rubber in mixture used to produce four tires

O = pounds of oil in mixture used to produce four tires

C = pounds of carbon black used to produce four tires

then statistical analysis has shown that the hardness, elasticity, and tensile strength of a 100-pound mixture of rubber, oil, and carbon black is as follows:

Tensile strength = $12.5 - .10(O) - .001\,(O)^2$

Elasticity = $17 + .35R - .04(O) - .002(R)^2$

Hardness = $34 + .10R + .06(O) - .3(C) + .001(R)(O) + .005(O)^2 + .001C^2$

TABLE 6

Product	Cost (Cents/Pound)
Rubber	4
Oil	1
Carbon black	7

Formulate an NLP whose solution will tell Firerock how to minimize the cost of producing the rubber product needed to manufacture a set of automobile tires.†

Solution After defining

$$TS = \text{Tensile strength of mixture}$$
$$E = \text{Elasticity of mixture}$$
$$H = \text{Hardness of mixture}$$

Rubber.lng

the LINGO program in Figure 8 (file Rubber.lng) gives the correct formulation. Row 1 minimizes the cost of producing the needed rubber product. Rows 2–4 express the tensile strength, elasticity, and hardness, respectively, of the mixture in terms of its component ingredients. Observe that tensile strength, elasticity, and hardness are each nonlinear functions of R, O, and C. Row 5 requires that we combine 100 pounds of inputs to produce

FIGURE 8

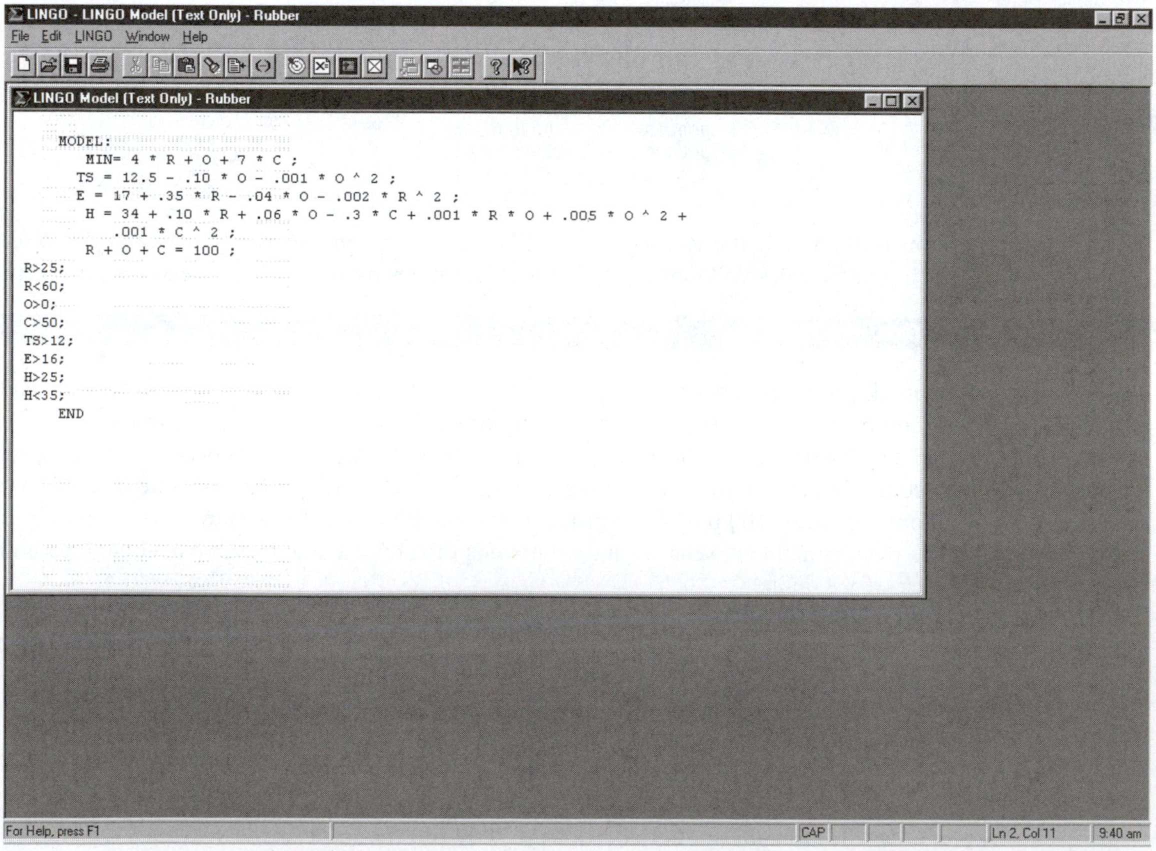
LINGO - LINGO Model (Text Only) - Rubber
File Edit LINGO Window Help
LINGO Model (Text Only) - Rubber
MODEL:
MIN= 4 * R + O + 7 * C ;
TS = 12.5 - .10 * O - .001 * O ^ 2 ;
E = 17 + .35 * R - .04 * O - .002 * R ^ 2 ;
H = 34 + .10 * R + .06 * O - .3 * C + .001 * R * O + .005 * O ^ 2 +
.001 * C ^ 2 ;
R + O + C = 100 ;
R>25;
R<60;
O>0;
C>50;
TS>12;
E>16;
H>25;
H<35;
END
For Help, press F1
CAP
Ln 2, Col 11
9:40 am

†Based on Nicholson (1971).

FIGURE 9

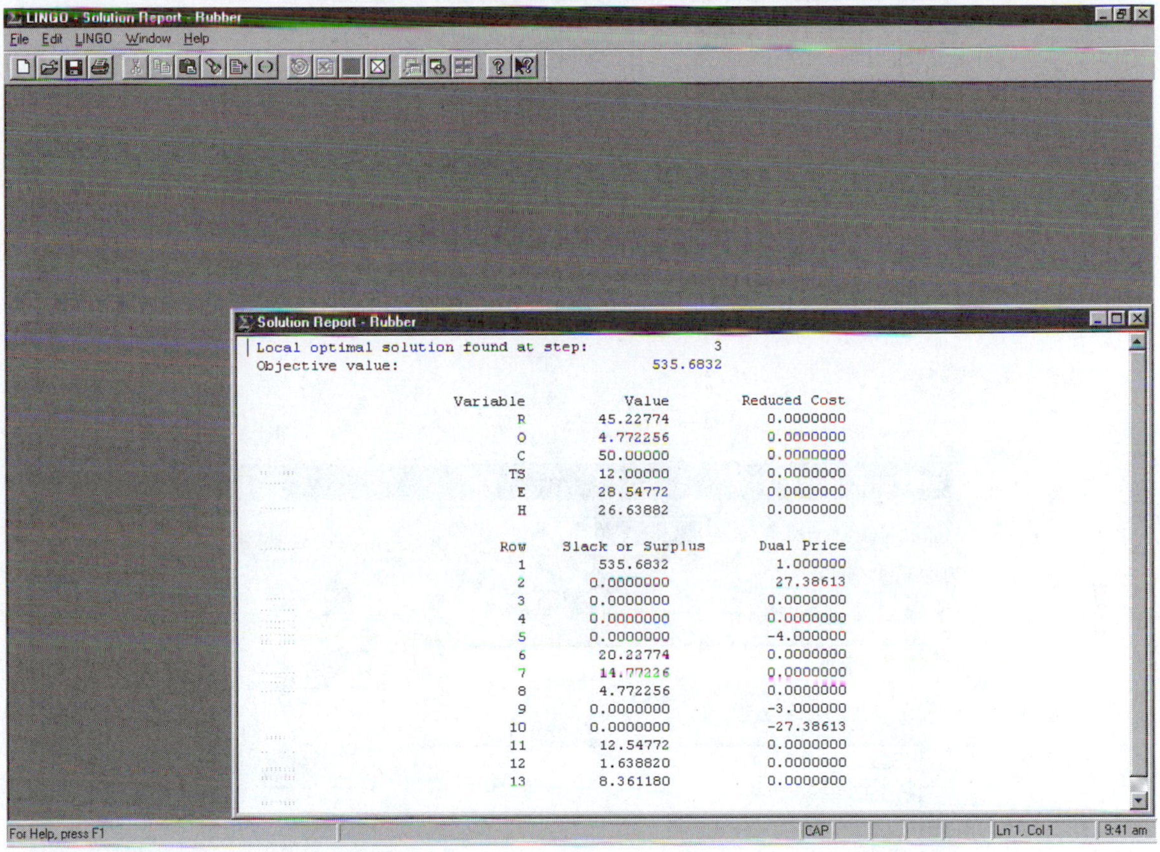

Local optimal solution found at step: 3
Objective value: 535.6832

Variable	Value	Reduced Cost
R	45.22774	0.0000000
O	4.772256	0.0000000
C	50.00000	0.0000000
TS	12.00000	0.0000000
E	28.54772	0.0000000
H	26.63882	0.0000000

Row	Slack or Surplus	Dual Price
1	535.6832	1.000000
2	0.0000000	27.38613
3	0.0000000	0.0000000
4	0.0000000	0.0000000
5	0.0000000	-4.000000
6	20.22774	0.0000000
7	14.77226	0.0000000
8	4.772256	0.0000000
9	0.0000000	-3.000000
10	0.0000000	-27.38613
11	12.54772	0.0000000
12	1.638820	0.0000000
13	8.361180	0.0000000

our final rubber product. The solution found by LINGO (Figure 9) is 45.23 pounds of rubber, 4.77 pounds of oil, and 50 pounds of carbon black. The total cost of the 100-pound mixture is \$5.36.

In Section 12.9 we will discuss whether or not this solution is optimal.

Solving NLPs with Excel

Caplabor.xls

It is easy to use the Excel Solver to solve NLPs. You proceed as you would with a linear model but do not select the Linear Model option. To illustrate, we solve Example 8 in the file Caplabor.xls (see Figure 10).

Our changing cells are the capital and labor purchased (cells C5 and D5, respectively). Our target cell is the total number of units produced (computed in cell C8). Our constraint is that the total spent (in cell B11) is less than or equal to \$8. Of course, the quantity of capital and labor purchased must be non-negative. Our Solver window is shown in Figure 11. Note we find that the optimal solution is $K = 1$, $L = 4$, and $z = 8$.

For NLPs having multiple local optimal solutions, the Excel Solver may fail to find the optimal solution because it may pick a local extremum that is not a global extremum. To illustrate, consider the following NLP:

	A	B	C	D
3				
4			Capital	Labor
5		Purchased	1	4
6		Cost	$ 4.00	$ 1.00
7				
8		Units produced	4	
9				
10		Total spent		Available
11		$ 8.00	<=	$ 8.00

FIGURE 10

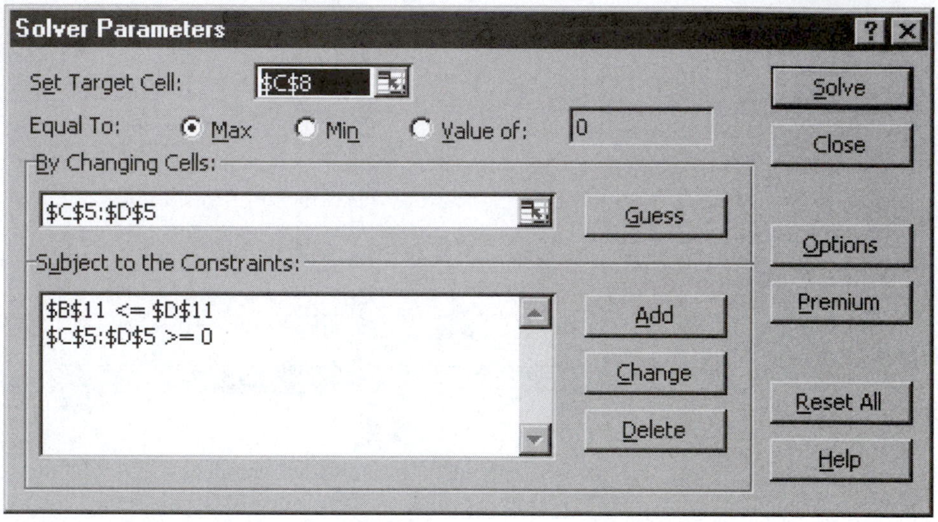

FIGURE 11

$$\max z = (x - 1)(x - 2)(x - 3)(x - 4)(x - 5)$$
$$\text{s.t.} \quad x \geq 1$$
$$x \leq 5$$

Multiple.xls

The graph of this function is shown in Figure 12. Note there are two local maxima for this problem. In workbook Multiple.xls, we solve this problem twice. The first time we begin with $x = 2$ and we find the optimal solution, which is $x = 1.36$ and $z = 3.63$ (see Figure 13).

The second time, we begin with $x = 3.5$ and we find the other local maximum, $x = 3.54$ and $z = 1.42$ (see Figure 14). The reason for this is that when we start with $x = 3.5$, the Solver soon hits $x = 3.54$ and finds that the objective function cannot be improved by small moves in either direction. Both LINGO and Solver use calculus-based methods (to be described later in this chapter) to solve NLPs. Any calculus-based approach to solving NLPs runs the risk of finding a local extremum that is not a global extremum. Evolutionary algorithms (to be discussed in Chapter 14) do not have this drawback.

CHAPTER 1 2 Nonlinear Programming

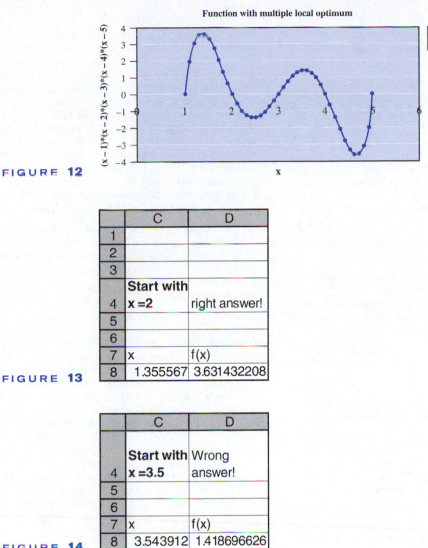

Function with multiple local optimum

Series 1

FIGURE **12**

	C	D
1		
2		
3		
4	**Start with** **x =2**	right answer!
5		
6		
7	x	f(x)
8	1.355567	3.631432208

FIGURE **13**

	C	D
4	**Start with** **x =3.5**	Wrong answer!
5		
6		
7	x	f(x)
8	3.543912	1.418696626

FIGURE **14**

PROBLEMS

Group A

1 Q & H Company advertises on soap operas and football games. Each soap opera ad costs $50,000, and each football game ad costs $100,000. Giving all figures in millions of viewers, if S soap opera ads are bought, they will be seen by $5\sqrt{S}$ men and $20\sqrt{S}$ women. If F football ads are bought, they will be seen by $17\sqrt{F}$ men and $7\sqrt{F}$ women. Q & H wants at least 40 million men and at least 60 million women to see its ads.

a Formulate an NLP that will minimize Q & H's cost of reaching sufficient viewers.

b Does the NLP violate the Proportionality and Additivity Assumptions?

c Suppose that the number of women reached by F football ads and S soap opera ads is $7\sqrt{F} + 20\sqrt{S} - 0.2\sqrt{FS}$. Why might this be a more realistic representation of the number of women viewers seeing Q & H's ads?

2 The area of a triangle with sides of length a, b, and c is $\sqrt{s(s - a)(s - b)(s - c)}$, where s is half the perimeter of the triangle. We have 60 ft of fence and want to fence a triangular-shaped area. Formulate an NLP that will enable us to maximize the fenced area.

3 The energy used in compressing a gas (in three stages) from an initial pressure I to a final pressure F is given by

$$K\left\{\sqrt{\frac{p_1}{I}} + \sqrt{\frac{p_2}{p_1}} + \sqrt{\frac{F}{p_2}} - 3\right\}$$

Formulate an NLP whose solution describes how to minimize the energy used in compressing the gas.

4 Use LINGO to solve Problem 1.

5 Use LINGO to solve Problem 2.

6 Use LINGO to solve Problem 3. Use $I = 64$ and $F = 1,000$.

7 For Example 6 of Chapter 8, let A = number of days duration of A is reduced, B = number of days duration of B is reduced, and so on. Suppose that the cost of crashing each activity is as follows:

A, $5A^2$; B, $20B^2$; C, $2C^2$; D, $20D^2$; E, $10E^2$; F, $15F^2$

and that each activity may be "crashed" to a duration of 0 days, if desired. Formulate an NLP that will minimize the cost of finishing the project in 25 days or less.

8 Beerco has $100,000 to spend on advertising in four markets. The sales revenue (in thousands of dollars) that can be created in each market by spending x_i thousand dollars in market i is given in Table 7. To maximize sales revenue, how much money should be spent in each market?

9 Widgetco produces widgets at plant 1 and plant 2. It costs $20x^{1/2}$ to produce x units at plant 1 and $40x^{1/3}$ to produce x units at plant 2. Each plant can produce as many as 70 units. Each unit produced can be sold for $10. At most, 120 widgets can be sold. Formulate an NLP whose solution will tell Widgetco how to maximize profit.

10 Three cities are located at the vertices of an equilateral triangle. An airport is to be built at a location that minimizes the total distance from the airport to the three cities. Formulate an NLP whose solution will tell us where to build the airport. Then solve your NLP on LINGO.

11 The yield of a chemical process depends on the length of time T (in minutes) that the process is run and the temperature TEMP (in degrees centigrade) at which the process is operated. This dependence is described by the equation

$\text{YIELD} = 87 - 1.4T' + .4\text{TEMP}' - 2.2T'^2 -$
$$3.2\text{TEMP}'^2 - 4.9(T')(\text{TEMP}')$$

where $T' = (T - 90)/10$ and $\text{TEMP}' = (\text{TEMP} - 150)/5$. T must be between 60 and 120 minutes, while TEMP must be between 100 and 200 degrees. Set up an NLP that could be used to maximize the yield of the process. Use LINGO to solve your NLP.

TABLE 7

Market	Sales Revenue
1	$10x_1^{.4}$
2	$8x_2^{.5}$
3	$12x_3^{.3}$
4	$16x_4^{.6}$

Group B

12 Consider Problem 5 of Section 3.8 with the following modification: Suppose that we can add a chemical called Superquality (SQ) to improve the quality level of gasoline and heating oil. If we add an amount x of SQ to each barrel of gasoline we improve its quality level by $x^{.5}$ over what it would have been. If we add an amount x of SQ to each barrel of heating oil we improve its quality level by $.6x^{.6}$ over what it would have been. The amount of SQ added to heating oil cannot exceed (by weight) 5% of the oils used to make heating oil. Similarly, the amount of SQ added to gasoline cannot exceed (by weight) 5% of the oils used to make gasoline. SQ may be purchased at a cost of $20 per pound. Formulate (and solve with LINGO) an NLP that will help CEO Adam Chandler maximize his profits.

13 A salesperson for Fuller Brush has three options: quit, put forth a low-effort level, or put forth a high-effort level. Suppose for simplicity that each salesperson will either sell $0, $5,000, or $50,000 worth of brushes. The probability of each sales amount depends on the effort level in the manner described in Table 8.

If the salesperson is paid $w, then he or she earns a benefit $w^{1/2}$. Low effort costs the salesperson 0 benefit units, while high effort costs 50 benefit units. If the salesperson were to quit Fuller and work elsewhere, then he or she could earn a benefit of 20. Fuller wants all salespeople to put forth a high-effort level. The question is how to minimize the cost of doing it. The company cannot observe the level of effort put forth by a salesperson, but they can observe the size of his or her sale. Thus, the wage is completely determined by the size of the sale. Fuller must then determine w_0 = wage paid for $0 in sale, w_{5000} = wage paid for $5,000 in sales, and $w_{50,000}$ = wage paid for $50,000 in sales. These wages must be set so that the salespeople value the expected benefit from high effort more than quitting and more than low effort. Formulate (and solve on LINGO) an NLP that can be used to ensure that all salespeople put forth high effort. This problem is an example of *agency theory*.[†]

TABLE 8

Size of Sale ($)	Effort Level	
	Low	High
0	.6	.3
5,000	.3	.2
50,000	.1	.5

12.3 Convex and Concave Functions

Convex and concave functions play an extremely important role in the study of nonlinear programming problems.

Let $f(x_1, x_2, \ldots, x_n)$ be a function that is defined for all points $(x_1, x_2, \ldots, x_n)$ in a convex set S.[‡]

DEFINITION ■ A function $f(x_1, x_2, \ldots, x_n)$ is a **convex function** on a convex set S if for any $x' \in S$ and $x'' \in S$

$$f[cx' + (1 - c)x''] \leq cf(x') + (1 - c)f(x'') \qquad (3)$$

holds for $0 \leq c \leq 1$. ■

DEFINITION ■ A function $f(x_1, x_2, \ldots, x_n)$ is a **concave function** on a convex set S if for any $x' \in S$ and $x'' \in S$

$$f[cx' + (1 - c)x''] \geq cf(x') + (1 - c)f(x'') \qquad (4)$$

holds for $0 \leq c \leq 1$. ■

From (3) and (4), we see that $f(x_1, x_2, \ldots, x_n)$ is a convex function if and only if $-f(x_1, x_2, \ldots, x_n)$ is a concave function, and conversely.

To gain some insights into these definitions, let $f(x)$ be a function of a single variable. From Figure 15 and inequality (3), we find that $f(x)$ is convex if and only if the line segment joining any two points on the curve $y = f(x)$ is never below the curve $y = f(x)$. Similarly, Figure 16 and inequality (4) show that $f(x)$ is a concave function if and only if the straight line joining any two points on the curve $y = f(x)$ is never above the curve $y = f(x)$.

FIGURE 15
A Convex Function

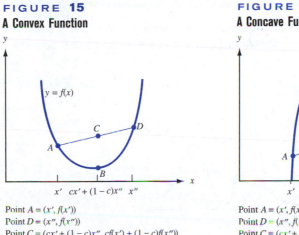

Point $A = (x', f(x'))$
Point $D = (x'', f(x''))$
Point $C = (cx' + (1 - c)x'', cf(x') + (1 - c)f(x''))$
Point $B = (cx' + (1 - c)x'', f(cx' + (1 - c)x''))$
From figure: $f(cx' + (1 - c)x'') \leq cf(x') + (1 - c)f(x'')$

FIGURE 16
A Concave Function

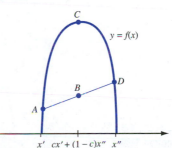

Point $A = (x', f(x'))$
Point $D = (x'', f(x''))$
Point $C = (cx' + (1 - c)x'', f(cx' + (1 - c)x''))$
Point $B = (cx' + (1 - c)x'', cf(x') + (1 - c)f(x''))$
From figure: $f(cx' + (1 - c)x'') \geq cf(x') + (1 - c)f(x'')$

[†]Based on Grossman and Hart (1983).
[‡]Recall from Chapter 3 that a set S is convex if $x' \in S$ and $x'' \in S$ imply that all points on the line segment joining x' and x'' are members of S. This ensures that $cx' + (1 - c)x''$ will be a member of S.

EXAMPLE **12**   Convex and Concave Functions

For $x \geq 0$, $f(x) = x^2$ and $f(x) = e^x$ are convex functions and $f(x) = x^{1/2}$ is a concave function. These facts are evident from Figure 17.

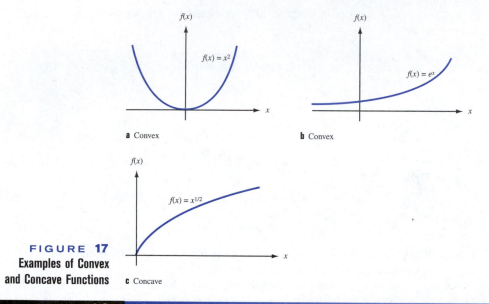

a Convex   **b** Convex

c Concave

FIGURE 17
Examples of Convex and Concave Functions

EXAMPLE **13**   Sum of Convex Functions

It can be shown (see Problem 12 at the end of this section) that the sum of two convex functions is convex and the sum of two concave functions is concave. Thus, $f(x) = x^2 + e^x$ is a convex function.

EXAMPLE **14**   Neither Convex nor Concave Function

Because the line segment AB lies below $y = f(x)$ and the line segment BC lies above $y = f(x)$, $f(x)$ as pictured in Figure 18 is not a convex or a concave function.

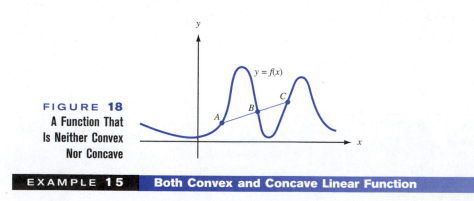

FIGURE 18
A Function That Is Neither Convex Nor Concave

EXAMPLE **15**   Both Convex and Concave Linear Function

A linear function of the form $f(x) = ax + b$ is both a convex and a concave function. This follows from

$$f[cx' + (1 - c)x''] = a[cx' + (1 - c)x''] + b$$
$$= c(ax' + b) + (1 - c)(ax'' + b)$$
$$= cf(x') + (1 - c)f(x'')$$

Both (3) and (4) hold with equality, so $f(x) = ax + b$ is both a convex and a concave function.

Before discussing how to determine whether a given function is convex or concave, we prove a result that illustrates the importance of convex and concave functions.

THEOREM 1

Consider NLP (2) and assume it is a maximization problem. Suppose the feasible region S for NLP (2) is a convex set. If $f(x)$ is concave on S, then any local maximum for NLP (2) is an optimal solution to this NLP.

Proof If Theorem 1 is false, then there must be a local maximum $\bar{x}$ that is not an optimal solution to NLP (2). Let S be the feasible region for NLP (2) (we have assumed that S is a convex set). Then, for some $x \in S, f(x) > f(\bar{x})$. The inequality (4) implies that for any c satisfying $0 < c < 1$,

$$f[c\bar{x} + (1 - c)x] \geq cf(\bar{x}) + (1 - c)f(x)$$
$$> cf(\bar{x}) + (1 - c)f(\bar{x}) \qquad [\text{from } f(x) > f(\bar{x})]$$
$$= f(\bar{x})$$

Now observe that for c arbitrarily near 1, $c\bar{x} + (1 - c)x$ is feasible (because S is convex) and is near $\bar{x}$. Thus, $\bar{x}$ cannot be a local maximum. This contradiction proves Theorem 1.

Similar reasoning can be used to prove Theorem 1$'$ (see Problem 11 at the end of this section).

THEOREM 1$'$

Consider NLP (2) and assume it is a minimization problem. Suppose the feasible region S for NLP (2) is a convex set. If $f(x)$ is convex on S, then any local minimum for NLP (2) is an optimal solution to this NLP.

Theorems 1 and 1$'$ demonstrate that if we are maximizing a concave function (or minimizing a convex function) over a convex feasible region S, then any local maximum (or local minimum) will solve NLP (2). As we solve NLPs, we will repeatedly apply Theorems 1 and 1$'$.

We now explain how to determine if a function $f(x)$ of a single variable is convex or concave. Recall that if $f(x)$ is a convex function of a single variable, the line joining any two points on $y = f(x)$ is never below the curve $y = f(x)$. From Figures 9 and 10, we see that $f(x)$ convex implies that the slope of $f(x)$ must be nondecreasing for all values of x.

THEOREM 2

Suppose $f''(x)$ exists for all x in a convex set S. Then $f(x)$ is a convex function on S if and only if $f''(x) \geq 0$ for all x in S.

Because $f(x)$ is convex if and only if $-f(x)$ is concave, Theorem 2' must also be true.

THEOREM 2'

Suppose $f''(x)$ exists for all x in a convex set S. Then $f(x)$ is a concave function on S if and only if $f''(x) \leq 0$ for all x in S.

EXAMPLE 16   **Determining If a Function Is Convex or Concave**

1   Show that $f(x) = x^2$ is a convex function on $S = R^1$.

2   Show that $f(x) = e^x$ is a convex function on $S = R^1$.

3   Show that $f(x) = x^{1/2}$ is a concave function on $S = (0, \infty)$.

4   Show that $f(x) = ax + b$ is both a convex and a concave function on $S = R^1$.

Solution   1 $f''(x) = 2 \geq 0$, so $f(x)$ is convex on $S = R^1$.

2   $f''(x) = e^x \geq 0$, so $f(x)$ is convex on $S = R^1$.

3   $f''(x) = -x^{-3/2}/4 \leq 0$, so $f(x)$ is a concave function on $S(0, \infty)$.

4   $f''(x) = 0$, so $f(x)$ is both convex and concave on $S = R^1$.

How can we determine whether a function $f(x_1, x_2, \ldots, x_n)$ of n variables is convex or concave on a set $S \subset R^n$? We assume that $f(x_1, x_2, \ldots, x_n)$ has continous second-order partial derivatives. Before stating the criterion used to determine whether $f(x_1, x_2, \ldots, x_n)$ is convex or concave, we require three definitions.

DEFINITION ■ The **Hessian** of $f(x_1, x_2, \ldots, x_n)$ is the $n \times n$ matrix whose ijth entry is

$$\frac{\partial^2 f}{\partial x_i \partial x_j} \quad ■$$

We let $H(x_1, x_2, \ldots, x_n)$ denote the value of the Hessian at $(x_1, x_2, \ldots, x_n)$. For example, if $f(x_1, x_2) = x_1^3 + 2x_1x_2 + x_2^2$, then

$$H(x_1, x_2) = \begin{bmatrix} 6x_1 & 2 \\ 2 & 2 \end{bmatrix}$$

DEFINITION ■ An ***i*th principal minor** of an $n \times n$ matrix is the determinant of any $i \times i$ matrix obtained by deleting $n - i$ rows and the corresponding $n - i$ columns of the matrix. ■

Thus, for the matrix

$$\begin{bmatrix} -2 & -1 \\ -1 & -4 \end{bmatrix}$$

the first principal minors are -2 and -4, and the second principal minor is $-2(-4) - (-1)(-1) = 7$. For any matrix, the first principal minors are just the diagonal entries of the matrix.

The **kth leading principal minor** of an $n \times n$ matrix is the determinant of the $k \times k$ matrix obtained by deleting the last $n - k$ rows and columns of the matrix. ■

We let $H_k(x_1, x_2, \ldots, x_n)$ be the kth leading principal minor of the Hessian matrix evaluated at the point $(x_1, x_2, \ldots, x_n)$. Thus, if $f(x_1, x_2) = x_1^3 + 2x_1x_2 + x_2^2$, then $H_1(x_1, x_2) = 6x_1$, and $H_2(x_1, x_2) = 6x_1(2) - 2(2) = 12x_1 - 4$.

By applying Theorems 3 and 3' (stated below, without proof), the Hessian matrix can be used to determine whether $f(x_1, x_2, \ldots, x_n)$ is a convex or a concave (or neither) function on a convex set $S \subset R^n$. [See Bazaraa and Shetty pages 91–93 (1993) for proof of Theorems 3 and 3'.]

THEOREM 3

Suppose $f(x_1, x_2, \ldots, x_n)$ has continuous second-order partial derivatives for each point $x = (x_1, x_2, \ldots, x_n) \in S$. Then $f(x_1, x_2, \ldots, x_n)$ is a convex function on S if and only if for each $x \in S$, all principal minors of H are non-negative.

EXAMPLE 17   Using the Hessian to Ascertain Convexity or Concavity 1

Show that $f(x_1, x_2) = x_1^2 + 2x_1x_2 + x_2^2$ is a convex function on $S = R^2$.

Solution   We find that

$$H(x_1, x_2) = \begin{bmatrix} 2 & 2 \\ 2 & 2 \end{bmatrix}$$

The first principal minors of the Hessian are the diagonal entries (both equal $2 \geq 0$). The second principal minor is $2(2) - 2(2) = 0 \geq 0$. For any point, all principal minors of H are non-negative, so Theorem 3 shows that $f(x_1, x_2)$ is a convex function on R^2.

THEOREM 3'

Suppose $f(x_1, x_2, \ldots, x_n)$ has continuous second-order partial derivatives for each point $x = (x_1, x_2, \ldots, x_n) \in S$. Then $f(x_1, x_2, \ldots, x_n)$ is a concave function on S if and only if for each $x \in S$ and $k = 1, 2, \ldots, n$, all nonzero principal minors have the same sign as $(-1)^k$.

EXAMPLE 18   Using the Hessian to Ascertain Convexity or Concavity 2

Show that $f(x_1, x_2) = -x_1^2 - x_1x_2 - 2x_2^2$ is a concave function on R^2.

Solution   We find that

$$H(x_1, x_2) = \begin{bmatrix} -2 & -1 \\ -1 & -4 \end{bmatrix}$$

The first principal minors are the diagonal entries of the Hessian (-2 and -4). These are both nonpositive. The second principal minor is the determinant of $H(x_1, x_2)$ and equals $-2(-4) - (-1)(-1) = 7 > 0$. Thus, $f(x_1, x_2)$ is a concave function on R^2.

EXAMPLE 19   Using the Hessian to Ascertain Convexity or Concavity 3

Show that for $S = R^2$, $f(x_1, x_2) = x_1^2 - 3x_1x_2 + 2x_2^2$ is not a convex or a concave function.

Solution   We have

$$H(x_1, x_2) = \begin{bmatrix} 2 & -3 \\ -3 & 4 \end{bmatrix}$$

The first principal minors of the Hessian are 2 and 4. Because both the first principal minors are positive, $f(x_1, x_2)$ cannot be concave. The second principal minor is $2(4) - (-3)(-3) = -1 < 0$. Thus, $f(x_1, x_2)$ cannot be convex. Together, these facts show that $f(x_1, x_2)$ cannot be a convex or a concave function.

EXAMPLE 20   Using the Hessian to Ascertain Convexity or Concavity 4

Show that for $S = R^3$, $f(x_1, x_2, x_3) = x_1^2 + x_2^2 + 2x_3^2 - x_1x_2 - x_2x_3 - x_1x_3$ is a convex function.

Solution   The Hessian is given by

$$H(x_1, x_2, x_3) = \begin{bmatrix} 2 & -1 & -1 \\ -1 & 2 & -1 \\ -1 & -1 & 4 \end{bmatrix}$$

By deleting rows (and columns) 1 and 2 of Hessian, we obtain the first-order principal minor $4 > 0$. By deleting rows (and columns) 1 and 3 of Hessian, we obtain the first-order principal minor $2 > 0$. By deleting rows (and columns) 2 and 3 of Hessian, we obtain the first-order principal minor $2 > 0$.

By deleting row 1 and column 1 of Hessian, we find the second-order principal minor

$$\det \begin{bmatrix} 2 & -1 \\ -1 & 4 \end{bmatrix} = 7 > 0.$$

By deleting row 2 and column 2 of Hessian, we find the second-order principal minor

$$\det \begin{bmatrix} 2 & -1 \\ -1 & 4 \end{bmatrix} = 7 > 0$$

By deleting row 3 and column 3 of Hessian, we find the second-order principal minor

$$\det \begin{bmatrix} 2 & -1 \\ -1 & 2 \end{bmatrix} = 3 > 0.$$

The third-order principal minor is simply the determinant of the Hessian itself. Expanding by row 1 cofactors we find the third-order principal minor

$$2[(2)(4) - (-1)(-1)] - (-1)[(-1)(4) - (-1)(-1)]$$
$$+(-1)[(-1)(-1) - (-1)(2)] = 14 - 5 - 3 = 6 > 0.$$

Because for all (x_1, x_2, x_3) all principal minors of the Hessian are non-negative, we have shown that $f(x_1, x_2, x_3)$ is a convex function on R^3.

PROBLEMS

Group A

On the given set S, determine whether each function is convex, concave, or neither.

1 $f(x) = x^3$; $S = [0, \infty)$

2 $f(x) = x^3$; $S = R^1$

3 $f(x) = \frac{1}{x}$; $S = (0, \infty)$

4 $f(x) = x^a$ ($0 \leq a \leq 1$); $S = (0, \infty)$

5 $f(x) = \ln x$; $S = (0, \infty)$

6 $f(x_1, x_2) = x_1^3 + 3x_1x_2 + x_2^2$; $S = R^2$

7 $f(x_1, x_2) = x_1^2 + x_2^2$; $S = R^2$

8 $f(x_1, x_2) = -x_1^2 - x_1x_2 - 2x_2^2$; $S = R^2$

9 $f(x_1, x_2, x_3) = -x_1^2 - x_2^2 - 2x_3^2 + .5x_1x_2$; $S = R^3$

10 For what values of a, b, and c will $ax_1^2 + bx_1x_2 + cx_2^2$ be a convex function on R^2? A concave function on R^2?

Group B

11 Prove Theorem $1'$.

12 Show that if $f(x_1, x_2, \ldots, x_n)$ and $g(x_1, x_2, \ldots, x_n)$ are convex functions on a convex set S, then $h(x_1, x_2, \ldots, x_n) = f(x_1, x_2, \ldots, x_n) = g(x_1, x_2, \ldots, x_n)$ is a convex function on S.

13 If $f(x_1, x_2, \ldots, x_n)$ is a convex function on a convex set S, show that for $c \geq 0$, $g(x, x_2, \ldots, x_n) = cf(x_1, x_2, \ldots, x_n)$ is a convex function on S, and for $c \leq 0$, $g(x_1, x_2, \ldots, x_n) = cf(x_1, x_2, \ldots, x_n)$ is a concave function on S.

14 Show that if $y = f(x)$ is a concave function on R^1, then $z = \frac{1}{f(x)}$ is a convex function [assume that $f(x) > 0$].

15 A function $f(x_1, x_2, \ldots, x_n)$ is *quasi-concave* on a convex set $S \subset R^n$ if $x' \in S$, $x'' \in S$, and $0 \leq c \leq 1$ implies

$$f[cx' + (1 - c)x''] \geq \min[f(x'), f(x'')]$$

Show that if f is concave on R^1, then f is quasi-concave. Which of the functions in Figure 19 is quasi-concave? Is a quasi-concave function necessarily a concave function?

16 From Problem 12, it follows that the sum of concave functions is concave. Is the sum of quasi-concave functions necessarily quasi-concave?

17 Suppose a function's Hessian has both positive and negative entries on its diagonal. Show that the function is neither concave nor convex.

18 Show that if $f(x)$ is a non-negative, increasing concave function, then $\ln [f(x)]$ is also a concave function.

19 Show that if a function $f(x_1, x_2, \ldots, x_n)$ is quasi-concave on a convex set S, then for any number a the set $S_a = $ all points satisfying $f(x_1, x_2, \ldots, x_n) \geq a$ is a convex set.

20 Show that Theorem 1 is untrue if f is a quasi-concave function.

21 Suppose the constraints of an NLP are of the form $g_i(x_1, x_2, \ldots, x_n) \leq b_i (i = 1, 2, \ldots m)$. Show that if each of the g_i is a convex function, then the NLP's feasible region is convex.

Group C

22 If $f(x_1, x_2)$ is a concave function on R^2, show that for any number a, the set of (x_1, x_2) satisfying $f(x_1, x_2) \geq a$ is a convex set.

23 Let $\mathbf{Z}$ be a $N(0, 1)$ random variable, and let $F(x)$ be the cumulative distribution function for Z. Show that on $S = (-\infty, 0]$, $F(x)$ is an increasing convex function, and on $S = [0, \infty)$, $F(x)$ is an increasing concave function.

24 Recall the Dakota LP discussed in Chapter 6. Let $v(L, FH, CH)$ be the maximum revenue that can be earned when L sq board ft of lumber, FH finishing hours, and CH carpentry hours are available.

a Show that $v(L, FH, CH)$ is a concave function.

b Explain why this result shows that the value of each additional available unit of a resource must be a nonincreasing function of the amount of the resource that is available.

FIGURE 19

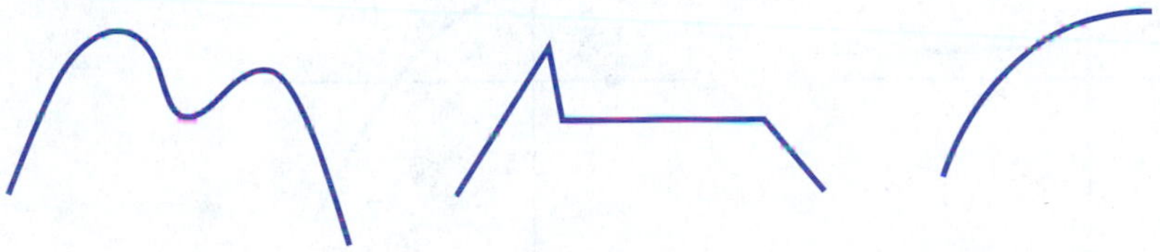

a   b   c

12.4 Solving NLPs with One Variable

In this section, we explain how to solve the NLP

$$\text{max (or min) } f(x)$$

$$\text{s.t.} \quad x \in [a, b] \tag{5}$$

[If $b = \infty$, then the feasible region for NLP (5) is $x \geq a$, and if $a = -\infty$, then the feasible region for (5) is $x \leq b$.]

To find the optimal solution to (5), we find all local maxima (or minima). A point that is a local maximum or a local minimum for (5) is called a local extremum. Then the optimal solution to (5) is the local maximum (or minimum) having the largest (or smallest) value of $f(x)$. Of course, if $a = -\infty$ or $b = \infty$, then (5) may have no optimal solution (see Figure 20).

There are three types of points for which (5) can have a local maximum or minimum (these points are often called *extremum candidates*):

Case 1   Points where $a < x < b$, and $f'(x) = 0$ [called a stationary point of $f(x)$].

Case 2   Points where $f'(x)$ does not exist.

Case 3   Endpoints a and b of the interval $[a, b]$.

Case 1. Points Where $a < x < b$ and $f'(x) = 0$

Suppose $a < x < b$, and $f'(x_0)$ exists. If x_0 is a local maximum or a local minimum, then $f'(x_0) = 0$. To see this, look at Figures 21a and 21b. From Figure 21a, we see that if $f'(x_0) > 0$, then there are points x_1 and x_2 near x_0 where $f(x_1) < f(x_0)$ and $f(x_2) > f(x_0)$. Thus, if $f'(x_0) > 0$, x_0 cannot be a local maximum or a local minimum. Similarly, Figure 21b shows that if $f'(x_0) < 0$, then x_0 cannot be a local maximum or a local minimum. From Figures 21c and 21d, however, we see $f'(x_0) = 0$, then x_0 may be a local maximum or a local minimum. Unfortunately, Figure 21e shows that $f'(x_0)$ can equal zero without x_0 being a local maximum or a local minimum. From Figure 21c, we see that if $f'(x)$ changes from positive to negative as we pass through x_0, then x_0 is a local maximum. Thus, if $f''(x_0) < 0$, x_0 is a local maximum. Similarly, from Figure 21d, we see that if $f'(x)$ changes from negative to positive as we pass through x_0, x_0 is a local minimum. Thus, if $f''(x_0) > 0$, x_0 is a local minimum.

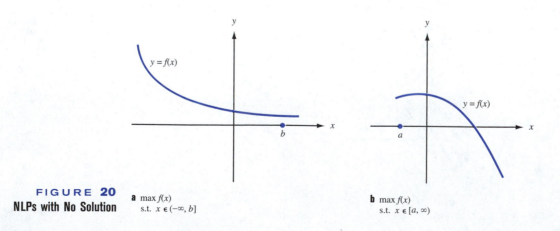

FIGURE 20
NLPs with No Solution

a max $f(x)$
   s.t. $x \in (-\infty, b]$

b max $f(x)$
   s.t. $x \in [a, \infty)$

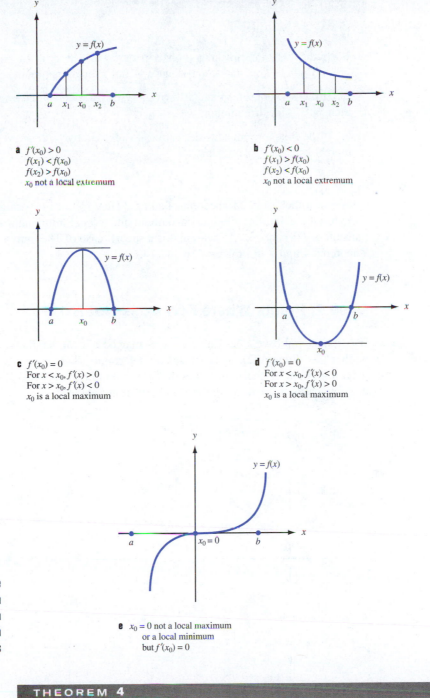

a $f'(x_0) > 0$
$f(x_1) < f(x_0)$
$f(x_2) > f(x_0)$
x_0 not a local extremum

b $f'(x_0) < 0$
$f(x_1) > f(x_0)$
$f(x_2) < f(x_0)$
x_0 not a local extremum

c $f'(x_0) = 0$
For $x < x_0, f'(x) > 0$
For $x > x_0, f'(x) < 0$
x_0 is a local maximum

d $f'(x_0) = 0$
For $x < x_0, f'(x) < 0$
For $x > x_0, f'(x) > 0$
x_0 is a local maximum

e $x_0 = 0$ not a local maximum
or a local minimum
but $f'(x_0) = 0$

FIGURE 21
How to Determine Whether x_0 Is a Local Maximum or a Local Minimum When $f'(x_0)$ Exists

---

THEOREM 4

If $f'(x_0) = 0$ and $f''(x_0) < 0$, then x_0 is a local maximum. If $f'(x_0) = 0$ and $f'''(x_0) > 0$, then x_0 is a local minimum.

What happens if $f'(x_0) = 0$ and $f''(x_0) = 0$ (this is the case in Figure 21e)? In this case, we determine whether x_0 is a local maximum or a local minimum by applying Theorem 5.

THEOREM 5

If $f'(x_0) = 0$, and

1 If the first nonvanishing (nonzero) derivative at x_0 is an odd-order derivative $[f^{(3)}(x_0), f^{(5)}(x_0),$ and so on], then x_0 is not a local maximum or a local minimum.

2 If the first nonvanishing derivative at x_0 is positive and an even-order derivative, then x_0 is a local minimum.

3 If the first nonvanishing derivative at x_0 is negative and an even-order derivative, then x_0 is a local maximum.

We omit the proofs of Theorems 4 and 5. [They follow in a straightforward fashion by applying the definition of a local maximum and a local minimum to the Taylor series expansion of $f(x)$ about x_0.] Theorem 4 is a special case of Theorem 5. We ask you to prove Theorems 4 and 5 in Problems 16 and 17.

Case 2. Points Where $f'(x)$ Does Not Exist

If $f(x)$ does not have a derivative at x_0, x_0 may be a local maximum, a local minimum, or neither (see Figure 22). In this case, we determine whether x_0 is a local maximum or a local minimum by checking values of $f(x)$ at points $x_1 < x_0$ and $x_2 > x_0$ near x_0. The four possible cases that can occur are summarized in Table 9.

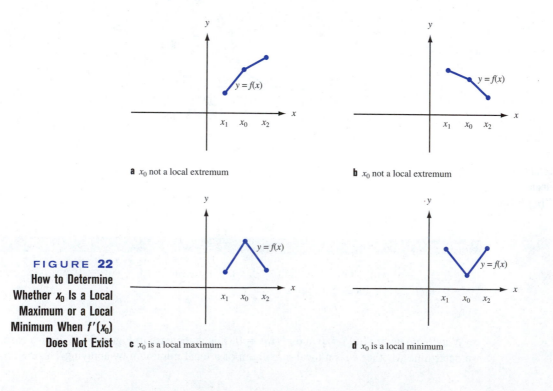

a x_0 not a local extremum

b x_0 not a local extremum

c x_0 is a local maximum

d x_0 is a local minimum

FIGURE 22
How to Determine Whether x_0 Is a Local Maximum or a Local Minimum When $f'(x_0)$ Does Not Exist

TABLE 9
How to Determine Whether a Point Where $f'(x)$ Does Not Exist Is a Local Maximum
or a Local Minimum

Relationship Between $f(x_0)$, $f(x_1)$, and $f(x_2)$	x_0	Figure
$f(x_0) > f(x_1)$; $f(x_0) < f(x_2)$	Not local extremum	16a
$f(x_0) < f(x_1)$; $f(x_0) > f(x_2)$	Not local extremum	16b
$f(x_0) \geq f(x_1)$; $f(x_0) \geq f(x_2)$	Local maximum	16c
$f(x_0) \leq f(x_1)$; $f(x_0) \leq f(x_2)$	Local minimum	16d

Case 3. Endpoints *a* and *b* of [*a*, *b*]

From Figure 23, we see that

If $f'(a) > 0$, then a is a local minimum.

If $f'(a) < 0$, then a is a local maximum.

If $f'(b) > 0$, then b is a local maximum.

If $f'(b) < 0$, then b is a local minimum.

If $f'(a) = 0$ or $f'(b) = 0$, draw a sketch like Figure 22 to determine whether a or b is a local extremum.

The following examples illustrate how these ideas can be applied to solve NLPs of the form (5).

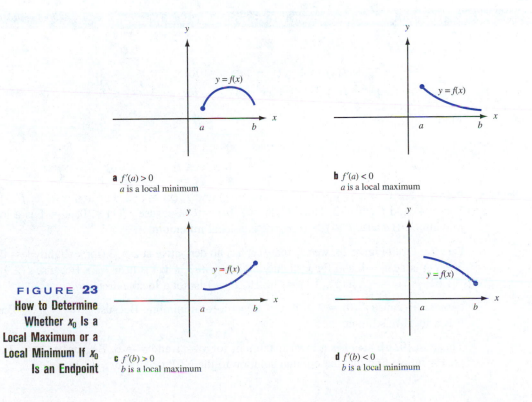

a $f'(a) > 0$
  a is a local minimum

b $f'(a) < 0$
  a is a local maximum

c $f'(b) > 0$
  b is a local maximum

d $f'(b) < 0$
  b is a local minimum

FIGURE 23
How to Determine
Whether x_0 Is a
Local Maximum or a
Local Minimum If x_0
Is an Endpoint

EXAMPLE 21 — Profit Maximization by Monopolist

It costs a monopolist \$5/unit to produce a product. If he produces x units of the product, then each can be sold for $10 - x$ dollars ($0 \le x \le 10$). To maximize profit, how much should the monopolist produce?

Solution Let $P(x)$ be the monopolist's profit if he produces x units. Then

$$P(x) = x(10 - x) - 5x = 5x - x^2 \qquad (0 \le x \le 10)$$

Thus, the monopolist wants to solve the following NLP:

$$\max P(x)$$
$$\text{s.t.} \qquad 0 \le x \le 10$$

We now classify all extremum candidates:

Case 1 $P'(x) = 5 - 2x$, so $P'(2.5) = 0$. Because $P''(x) = -2$, $x = 2.5$ is a local maximum yielding a profit of $P(2.5) = 6.25$.

Case 2 $P'(x)$ exists for all points in $[0, 10]$, so there are no Case 2 candidates.

Case 3 $a = 0$ has $P'(0) = 5 > 0$, so $a = 0$ is a local minimum; $b = 10$ has $P'(10) = -15 < 0$, so $b = 10$ is a local minimum.

Thus, $x = 2.5$ is the only local maximum. This means that the monopolist's profits are maximized by choosing $x = 2.5$.

Observe that $P''(x) = -2$ for all values of x. This shows that $P(x)$ is a concave function. Any local maximum for $P(x)$ must be the optimal solution to the NLP. Thus, Theorem 1 implies that once we have determined that $x = 2.5$ is a local maximum, we know that it is the optimal solution to the NLP.

EXAMPLE 22 — Finding Global Maximum When Endpoint Is a Maximum

Let

$$f(x) = 2 - (x - 1)^2 \qquad \text{for} \qquad 0 \le x < 3$$
$$f(x) = -3 + (x - 4)^2 \qquad \text{for} \qquad 3 \le x \le 6$$

Find

$$\max f(x)$$
$$\text{s.t.} \qquad 0 \le x \le 6$$

Solution **Case 1** For $0 \le x < 3$, $f'(x) = -2(x - 1)$ and $f''(x) = -2$. For $3 < x \le 6$, $f'(x) = 2(x - 4)$ and $f''(x) = 2$. Thus, $f'(1) = f'(4) = 0$. Because $f''(1) < 0$, $x = 1$ is a local maximum. Because $f''(4) > 0$, $x = 4$ is a local minimum.

Case 2 From Figure 24, we see that $f(x)$ has no derivative at $x = 3$ (for x slightly less than 3, $f'(x)$ is near -4, and for x slightly bigger than 3, $f'(x)$ is near -2). Because $f(2.9) = -1.61$, $f(3) = -2$, and $f(3.1) = -2.19$, $x = 3$ is not a local extremum.

Case 3 Because $f'(0) = 2 > 0$, $x = 0$ is a local minimum. Because $f'(6) = 4 > 0$, $x = 6$ is a local maximum.

Thus, on $[0, 6]$, $f(x)$ has a local maximum for $x = 1$ and $x = 6$. Because $f(1) = 2$ and $f(6) = 1$, we find that the optimal solution to the NLP occurs for $x = 1$.

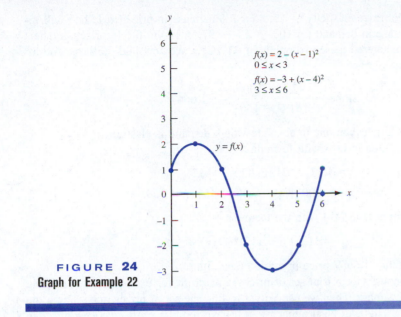

$f(x) = 2 - (x-1)^2$
$0 \leq x < 3$

$f(x) = -3 + (x-4)^2$
$3 \leq x \leq 6$

$y = f(x)$

FIGURE 24
Graph for Example 22

Pricing and Nonlinear Optimization

An important business decision is the determination of the profit-maximizing price that should be charged for a product. Demand for a product is often modeled as a linear function of price

$$\text{Demand} = a - b(\text{Price})$$

where a and b are constants. If the linear demand function is relevant, then profit from a product with a unit cost of c is given by

$$(\text{Price} - c) * [a - b(\text{Price})].$$

This implies that profit is a concave function of price and Solver should find the profit-maximizing price. In this section, we give two Solver models (based on Dolan and Simon, 1997) that can be used to determine optimal prices.

Our first model tackles the following problem: As exchange rates fluctuate, how should a U.S. company change the overseas price of its product? To be more specific, suppose Eli Daisy is selling a drug in Germany. Its goal is to maximize its profit in dollars, but when the drug is sold in Germany it receives marks. To maximize Daisy's dollar profit, how should the price in marks vary with the exchange rate? To illustrate the ideas involved consider the following example.

EXAMPLE 23   **Pricing When Exchange Rates Change**

The drug taxoprol costs $60 to produce. Currently, the exchange rate is .667 $/mark, and we are charging 150 marks for taxoprol. Current demand for taxoprol is 100 units, and it is estimated that the elasticity for taxoprol is 2.5. Assuming a linear demand curve, determine how the price (in marks) for taxoprol should vary with the exchange rate.

Solution   We begin by determining the linear demand curve that relates demand to the price in marks. Currently, demand is 100 and the price is 150 marks. Recall from economics that the price elasticity of a product is the percentage decrease in sales that will result from a

1% increase in price. Price elasticity is 2.5, so a 1% increase in price (to 151.5) will result in a 2.5% decrease in demand (to $100 - 2.5 = 97.5$). In the file Intprice.xls (sheet Linear Demand), we entered these two points in B12:C13. We now find the slope and intercept of the demand curve.

$$\text{Slope} = \frac{97.5 - 100}{151.5 - 150} = -1.6667$$

(because demand is larger than one in absolute value, demand is elastic).

The slope is computed in D13 with formula

$$=(B13 - B12)/(A13 - A12).$$
$$\text{Intercept} = 100 + (-150)(-1.6667) = 350.$$

The intercept is computed in D14 with the formula

$$=B12 + (-A12)*(D13).$$

Thus, demand $= 350 - 1.6667$(price in marks) (see Figure 25).

We can now compute (for a trial set of prices) our profit for exchange rates ranging from .4 $/mark to 1 $/mark. Then we use Solver to find the set of prices maximizing the *sum* of these profits. This will ensure that we will have found a profit-maximizing price for a variety of exchange rates.

Step 1   Enter trial values for the exchange rate ($/mark) in the cell range B4:J4.

Step 2   In B5:J5, we enter the unit cost in dollars ($60).

Step 3   In B6:J6, we enter trial prices (in marks) for taxoprol.

Step 4   Observe that demand for each exchange rate is given by $350 - 1.66667*$(price in marks)

In cells B7:J7, we determine the demand for each exchange rate. In B7, we find the demand for the exchange rate of .6667 $/mark with the formula

$$=\$D\$14 + \$D\$13*B6.$$

Copying this formula to the range C7:J7 computes the demand for all other exchange rates.

Step 5   Observe that profit in dollars is given by

$$[(\$/\text{mark})*\text{price in marks} - \text{cost in dollars}]*(\text{demand}).$$

FIGURE 25

	A	B	C	D	E	F	G	H	I	J	K
1	Price dependence										
2	on exchange rate										
3											
4	Current $/DM	0.666667	0.4	0.5	0.6	0.666667	0.7	0.8	0.9	1	
5	Unit Cost US $	60	60	60	60	60	60	60	60	60	
6	Current price DM	149.9999	179.9999	164.9999	154.9999	149.9999	147.8571	142.4999	138.3333	134.9999	
7	Current demand	100.0002	50.00015	75.00014	91.6668	100.0002	103.5716	112.5001	119.4446	125.0001	
8	Current profit US$	4000.005	600	1687.5	3025	4000.005	4505.357	6075	7704.167	9375	Total Profit
9	Elasticity	2.5	2.5	2.5	2.5	2.5	2.5	2.5	2.5	2.5	36972.03
10											
11	Price DM	demand									
12	150	100									
13	151.5	97.5	slope	-1.666667							
14	Demand = 350-(5/3)*price		intercept	350							

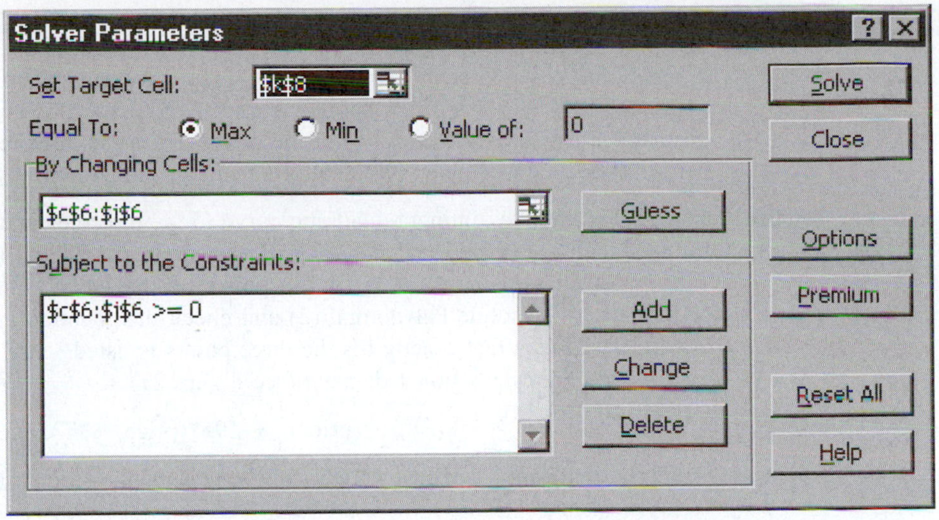

FIGURE 26

In cells B8:J8, we compute the dollar profit (for the trial prices) for each exchange rate. In B8, we find the profit for our current exchange rate (.66667 \$/mark) and current price (150 marks) with the formula

$$=(B4*B6 - B5)*B7.$$

Copying this formula to the cell range C8:J8 computes profits for all other exchange rates.

Step 6   In cell K8, we add the profit for all exchange rates with the formula

$$= SUM(C8:J8).$$

Step 7   We now use the solver to determine the profit-maximizing price for each exchange rate. By changing the non-negative prices for each exchange rate (C6:J6), we can maximize the sum of the profits (K8). Because each price only affects the profit for the exchange rate in its own column, this ensures that we find the profit-maximizing price for each exchange rate. For example, for .66667 \$/mark the optimal price is 150 marks. Note that if the mark drops in value by 25% to .5 \$/mark, we only raise the cost in marks by 10% ($\frac{165 - 150}{150}$ = .10). Because of the elastic demand, profit maximization does not call for making the German customers absorb all of the loss in dollars due to depreciation of the mark. Our Solver window is as shown in Figure 26.

How can we use Solver to determine a profit-maximizing price? One way is to derive a demand curve by breaking the market into segments and identifying a low price, a medium price, and a high price. For each of these prices and market segments, ask company experts to estimate product demand. Then we can use EXCEL's trend-curve-fitting capabilities to fit a quadratic function that can be used to estimate each segment's demand for different prices. Finally, we can add the segment demand curves to derive an aggregate demand curve and use the Solver to determine the profit-maximizing price. The procedure is illustrated in Example 24 (based on Dolan and Simon (1996).

EXAMPLE 24   Pricing a Candy Bar

A candy bar costs 55 cents to produce. We are considering charging a price of between $1.10 and $1.50 for this candy bar. For a price of $1.10, $1.30, and $1.50, the marketing department estimates the demand for the candy bar in the three regions where the candy bar will be sold (see Table 10). What price will maximize profit?

Solution

Expdemand.xls

Step 1   We begin by fitting a quadratic curve to the three demands specified in Table 10 for each region. See file Expdemand.xls. For example, for Region 1 we use the X-Y Chart Wizard option to plot D4:E6. Click the points on the graph until they turn yellow and choose Insert Trendline Polynomial (2) and check the equation option to make sure the quadratic equation that exactly fits the three points is listed.

Thus we estimate region 1 demand (see Figure 27):

$$= -87.5*(\text{price})^2 + 195*(\text{price}) - 73.625.$$

Similarly, in regions 2 and 3 we find the following demand equations (see Figures 28 and 29):

$$\text{Region 2 demand} = -75*(\text{price})^2 + 155*(\text{price}) - 47.75.$$
$$\text{Region 3 demand} = -12.5*(\text{price})^2 - 5*(\text{price}) + 44.625.$$

Step 2   We now enter a trial price in cell H4 and determine in cells I4:K4 the demand (in thousands of units) for that price in each region:

$$\text{Region 1 demand (cell I4)} = -87.5*\text{H4}^2 + 195*\text{H4} - 73.625$$
$$\text{Region 2 demand (cell J4)} = -75*(\text{H4})^2 + 155*\text{H4} - 47.75$$
$$\text{Region 3 demand (cell K4)} = -12.5*(\text{H4})^2 - 5*\text{H4} + 44.625$$

Step 3   In cell L4, compute the total demand (in thousands of units) with the formula

$$= \text{SUM(I4:K4)}.$$

Step 4   In cell I6, compute our profit (in thousands of dollars):

$$= (\text{H4} - \text{I2})*\text{L4}.$$

Step 5   We are now ready to invoke the Solver to find the profit-maximizing price. We simply maximize profit (cell I6), with price (H4) being a changing cell. Because

TABLE 10

Price ($) (Unit cost: 0.55)	Demand (in Thousands)		
	Region 1	Region 2	Region 3
Low (1.10)	35	32	24
Medium (1.30)	32	27	17
High (1.50)	22	16	9

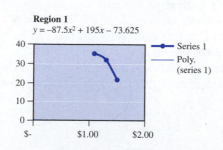

Region 1
$y = -87.5x^2 + 195x - 73.625$

FIGURE 27

our demand curves are, in theory, only valid for prices between \$1.10 and \$1.50, we add the constraints H4>=1.10 and H4<=1.50 (see Figure 30 for the Solver window).

Why is the model nonlinear? As Figure 31 shows, we find the profit-maximizing price to be \$1.29.

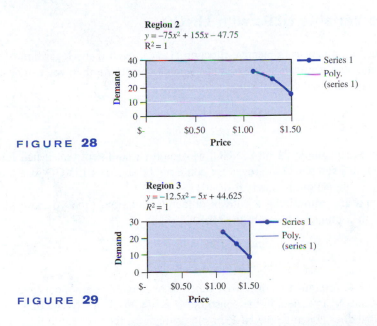

Region 2
$y = -75x^2 + 155x - 47.75$
$R^2 = 1$

FIGURE 28

Region 3
$y = -12.5x^2 - 5x + 44.625$
$R^2 = 1$

FIGURE 29

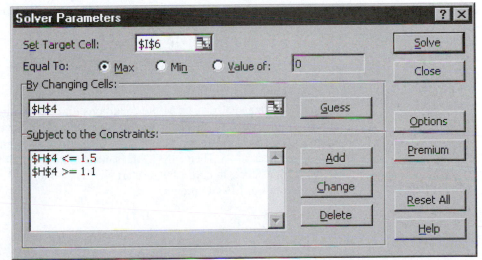

FIGURE 30

	H	I	J	K	L
2	Variable cost	0.55			
3	Price	Region 1 demand	Region 2 demand	Region 3 demand	Total demand
4	1.286325018	32.42807	27.53297	17.51047	77.47152
5					
6	Profit	57.04422			
7	(000's)				

FIGURE 31

If we are trying to maximize a function $f(x)$ that is a product of several functions, then it is often easier to maximize $\ln[f(x)]$. Because ln is an increasing function, we know that any x solving max $z' = \ln[f(x)]$ subject to $x \in S$ will also solve max $z = f(x)$ over $x \in S$. See Problem 4 for application of this idea.

Solving One Variable NLPs with LINGO

If you are maximizing a concave objective function $f(x)$ (or even if the logarithm of the objective function in a maximization problem is a concave function), then you can be certain that LINGO will find the optimal solution to the NLP

$$\text{max } z = f(x)$$

$$\text{s.t.} \quad a \leq x \leq b$$

Thus, if we solved Example 21 on LINGO, we would be confident that it had found the correct answer. In Example 22, however, we could not be sure that LINGO would find the maximum value of $f(x)$ on the interval $[0, 6]$.

Similarly, if you are minimizing a convex objective function, then you know that LINGO will find the optimal solution to the NLP

$$\text{min } z = f(x)$$

$$\text{s.t.} \quad a \leq x \leq b$$

If you are trying to minimize a nonconvex function or maximize a nonconcave function of a one-variable NLP subject to the constraint $a \leq x \leq b$, then LINGO may find a local extremum that does not solve the NLP. In such situations, the user can influence the solution found by LINGO by inputting a starting value for x with the INIT command. For example, if we direct LINGO to solve

$$\text{min } z = x \sin(\pi x)$$

$$\text{s.t.} \quad 0 \leq x \leq 6$$

LINGO may find the local minimum $x = 1.564$. This is because as a default, LINGO first guesses that $x = 0$, and at $x = 1.564$ the conditions for a local minimum $[f'(x) = 0$ and $f''(x) > 0]$ are satisfied. A sketch of the function $x \sin(\pi x)$ reveals that another local minimum occurs for x between 5 and 6. By using the INIT command, we may direct LINGO to start near $x = 5$. Then LINGO does indeed find the optimal solution ($x = 5.52$) to the NLP. For instance, to have LINGO start with $x1=2$ and $x2=3$, we would add the following section to our LINGO program.

INIT:

$x1=2$;

$x2=3$;

ENDINIT

PROBLEMS

Group A

1 It costs a company $100 in variable costs to produce an air conditioner, plus a fixed cost of $5,000 if any air conditioners are produced. If the company spends x dollars on advertising, then it can sell $x^{1/2}$ air conditioners at $300 each. How can the company maximize its profit? If the fixed cost of producing any air conditioners were $20,000, what should the company do?

2 If a monopolist produces q units, she can charge $100 - 4q$ dollars/unit. The fixed cost of production is $50, and the variable per-unit cost is $2. How can the monopolist maximize profits? If a sales tax of $2/unit must be paid by the monopolist, then would she increase or decrease production?

3 Show that for all x, $e^x \geq x + 1$. [Hint: Let $f(x) = e^x - x - 1$. Show that

$$\min f(x)$$
$$\text{s.t.} \quad x \in R$$

occurs for $x = 0$.]

4 Suppose that in n "at bats," a baseball player gets x hits. Suppose we want to estimate the player's probability (p) of getting a hit on each "at bat." The method of maximum likelihood estimates p by $\hat{p}$, where $\hat{p}$ maximizes the probability of observing x hits in n "at bats." Show that the method of maximum likelihood would choose $\hat{p} = \frac{x}{n}$.

5 Find the optimal solution to

$$\max x^3$$
$$\text{s.t.} \quad -1 \leq x \leq 1$$

6 Find the optimal solution to

$$\min x^3 - 3x^2 + 2x - 1$$
$$\text{s.t.} \quad -2 \leq x \leq 4$$

7 During the Reagan administration, economist Arthur Laffer became famous for his Laffer curve, which implied that an increase in the tax rate might decrease tax revenues, while a decrease in the tax rate might increase tax revenues. This problem illustrates the idea behind the Laffer curve. Suppose that if an individual puts in a degree of effort e, he or she earns a revenue of $10e^{1/2}$. Also suppose that an individual associates a cost of e with an effort level of e. Suppose further that the tax rate is T. This means that each individual gets to keep a fraction $1 - T$ of before tax revenue. Show that $T = .5$ maximizes the government's tax revenues. Thus, if the tax rate were 60%, then a cut in the tax rate would increase revenues.

8 The cost per day of running a hospital is $200,000 + .002x^2$ dollars, where $x =$ patients served per day. What size hospital minimizes the per-patient cost of running the hospital?

9 Each morning during rush hour, 10,000 people want to travel from New Jersey to New York City. If a person takes the subway, the trip lasts 40 minutes. If x thousand people per morning drive to New York, it takes $20 + 5x$ minutes to make the trip. This problem illustrates a basic fact of life: If people are left to their own devices, they will cause more congestion than need actually occur!

   a Show that if people are left to their own devices, an average of 4,000 people will travel by road from New Jersey to New York. Here you should assume that people will divide up between the subways and roads in a way that makes the average travel time by road = average travel time by subway. When this "equilibrium" occurs, nobody has an incentive to switch from road to subway or subway to road.

   b Show that the average travel time per person is minimized if 2,000 people travel by road.

10 Currently, the exchange rate is 100 yen per dollar. In Japan, we sell a product that costs $5 to produce for 700 yen. The product has an elasticity of 3. For exchange rates varying from 70 to 130 yen per dollar, determine the optimal product price in Japan and the profit in US dollars. Assume a linear demand curve. Current demand is assumed to equal 100.

11 It costs $250 to produce an X-Box. We are trying to determine the selling price for the X-Box. Prices between $200 and $400 are under consideration, with demand for prices of $200, $250, $350, and $400 given below. Suppose MSFT earns $10 in profit for each game that an X-Box owner purchases. Determine the optimal price and associated profit for the case in which an average X-Box owner buys 10 games.

Console Price ($)	Demand
200	2.00E+06
250	1.20E+06
350	6.00E+05
400	2.00E+05
Unit cost	$250

12 You are the publisher of a new magazine. The variable cost of printing and distributing each weekly copy of the magazine is $0.25. You are thinking of charging between $0.50 and $1.30 per week for the magazine. The estimated numbers of subscribers (in millions) for weekly prices of $0.50, $0.80, and $1.30 are as follows:

Price	Demand (Millions)
0.5	2.00
0.8	1.20
1.3	0.30

What price will maximize weekly profit from the magazine?

Group B

13 It costs a company $c(x)$ dollars to produce x units. The curve $y = c'(x)$ is called the firm's marginal cost curve. (Why?) The firm's average cost curve is given by $z = \frac{c(x)}{x}$. Let x^* be the production level that minimizes the company's average cost. Give conditions under which the marginal cost curve intersects the average cost curve at x^*.

14 When a machine is t years old, it earns revenue at a rate of e^{-t} dollars per year. After t years of use, the machine can be sold for $\frac{1}{t+1}$ dollars.

   a When should the machine be sold to maximize total revenue?

b If revenue is discounted continuously (so that $1 of revenue received t years from now is equivalent to e^{-rt} dollars of revenue received now, how would the answer in part (a) change?

15[†] Suppose a company must service customers lying in an area of A sq mi with n warehouses. Kolesar and Blum have shown that the average distance between a warehouse and a customer is

$$\sqrt{\frac{A}{n}}$$

Assume that it costs the company $60,000 per year to maintain a warehouse and $400,000 to build a warehouse. (Assume that a $400,000 cost is equivalent to forever incurring a cost of $40,000 per year.) The company fills 160,000 orders per year, and the shipping cost per order is $1 per mile. If the company serves an area of 100 sq mi, then how many warehouses should it have?

16 Prove Theorem 4.

17 Prove Theorem 5.

12.5 Golden Section Search

Consider a function $f(x)$. [For some x, $f'(x)$ may not exist.] Suppose we want to solve the following NLP:

$$\max f(x)$$
$$\text{s.t.} \quad a \le x \le b \tag{6}$$

It may be that $f'(x)$ does not exist, or it may be difficult to solve the equation $f'(x) = 0$. In either case, it may be difficult to use the methods of the previous section to solve this NLP. In this section, we discuss how (6) can be solved if $f(x)$ is a special type of function (a unimodal function).

DEFINITION ■ A function $f(x)$ is **unimodal** on $[a, b]$ if for some point $\bar{x}$ on $[a, b]$, $f(x)$ is strictly increasing on $[a, \bar{x}]$ and strictly decreasing on $[\bar{x}, b]$. ■

If $f(x)$ is unimodal on $[a, b]$, then $f(x)$ will have only one local maximum ($\bar{x}$) on $[a, b]$ and that local maximum will solve (6) (see Figure 32). Let $\bar{x}$ denote the optimal solution to (6).

Without any further information, all we can say is that the optimal solution to (6) is some point on the interval $[a, b]$. By evaluating $f(x)$ at two points x_1 and x_2 (assume $x_1 < x_2$) on $[a, b]$, we may reduce the size of the interval in which the solution to (6) must

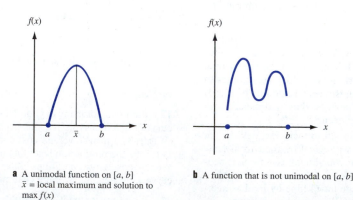

FIGURE 32
Definition of a
Unimodal Function

a A unimodal function on $[a, b]$
$\bar{x}$ = local maximum and solution to
max $f(x)$
s.t. $a \le x \le b$

b A function that is not unimodal on $[a, b]$

[†]Based on Kolesar and Blum (1973).

lie. After evaluating $f(x_1)$ and $f(x_2)$, one of three cases must occur. In each case, we can show that the optimal solution to (6) will lie in a subset of $[a, b]$.

Case 1 $f(x_1) < f(x_2)$. Because $f(x)$ is increasing for at least part of the interval $[x_1, x_2]$, the fact that $f(x)$ is unimodal shows that the optimal solution to (6) cannot occur on $[a, x_1]$. Thus, in Case 1, $\bar{x} \in (x_1, b]$ (see Figure 33).

Case 2 $f(x_1) = f(x_2)$. For some part of the interval $[x_1, x_2]$, $f(x)$ must be decreasing, and the optimal solution to (6) must occur for some $\bar{x} < x_2$. Thus, in Case 2, $\bar{x} \in [a, x_2]$ (see Figure 34).

Case 3 $f(x_1) > f(x_2)$. In this case, $f(x)$ begins decreasing before x reaches x_2. Thus, $\bar{x} \in [a, x_2)$ (see Figure 35).

The interval in which $\bar{x}$ must lie—either $[a, x_2)$ or $(x_1, b]$—is called the **interval of uncertainty.**

Many search algorithms use these ideas to reduce the interval of uncertainty [see Bazaraa and Shetty (1993, Section 8.1)]. Most of these algorithms proceed as follows:

Step 1 Begin with the region of uncertainty for x being $[a, b]$. Evaluate $f(x)$ at two judiciously chosen points x_1 and x_2.

Step 2 Determine which of Cases 1–3 holds, and find a reduced interval of uncertainty.

Step 3 Evaluate $f(x)$ at two new points (the algorithm specifies how the two new points are chosen). Return to step 2 unless the length of the interval of uncertainty is sufficiently small.

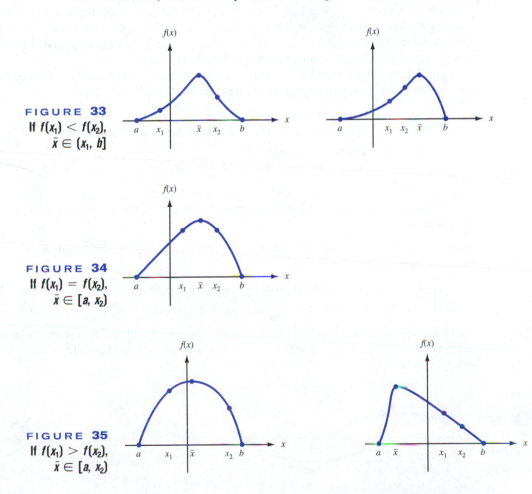

FIGURE 33
If $f(x_1) < f(x_2)$, $\bar{x} \in (x_1, b]$

FIGURE 34
If $f(x_1) = f(x_2)$, $\bar{x} \in [a, x_2)$

FIGURE 35
If $f(x_1) > f(x_2)$, $\bar{x} \in [a, x_2)$

We discuss in detail one such search algorithm: Golden Section Search. In using this algorithm to solve (6) for a unimodal function $f(x)$, we will see that when we choose two new points at step 3, one will always coincide with a point at which we have previously evaluated $f(x)$.

Let r be the unique positive root of the quadratic equation $r^2 + r = 1$. Then the quadratic formula yields that

$$r = \frac{5^{1/2} - 1}{2} = 0.618$$

(See Problem 3 at the end of this section for an explanation of why r is referred to as the Golden Section.) Golden Section Search begins by evaluating $f(x)$ at points x_1 and x_2, where $x_1 = b - r(b - a)$, and $x_2 = a + r(b - a)$ (see Figure 36). From this figure, we see that to find x_1, we move a fraction r of the interval from the right endpoint of the interval; to find x_2, we move a fraction r of the interval from the left endpoint. Then Golden Section Search generates two new points, at which $f(x)$ should again be evaluated with the following moves:

New Left-Hand Point   Move a distance equal to a fraction r of the current interval of uncertainty from the right endpoint of the interval of uncertainty.

New Right-Hand Point   Move a distance equal to a fraction r of the current interval of uncertainty from the left endpoint of the interval.

From our discussion of Cases 1–3, we know that if $f(x_1) < f(x_2)$, then $\bar{x} \in (x_1, b]$, whereas if $f(x_1) \geq f(x_2)$, then $\bar{x} \in [a, x_2)$. If $f(x_1) < f(x_2)$, then the reduced interval of uncertainty has length $b - x_1 = r(b - a)$, and if $f(x_1) \geq f(x_2)$, then the reduced interval of uncertainty has a length $x_2 - a = r(b - a)$. Thus, after evaluating $f(x_1)$ and $f(x_2)$, we have reduced the interval of uncertainty to a length $r(b - a)$.

Each time $f(x)$ is evaluated at two points and the interval of uncertainty is reduced, we say that an iteration of Golden Section Search has been completed. Define

L_k = length of the interval of uncertainty
after k iterations of the algorithm have been completed

I_k = interval of uncertainty
after k iterations have been completed

Then we see that $L_1 = r(b - a)$, and $I_1 = [a, x_2)$ or $I_1 = (x_1, b]$.

Following this procedure, we generate two new points, x_3 and x_4, at which $f(x)$ must be evaluated.

Case 1 $f(x_1) < f(x_2)$. The new interval of uncertainty, $(x_1, b]$, has length $b - x_1 = r(b - a)$. Then (see Figure 37a)

$$x_3 = \text{new left-hand point} = b - r(b - x_1) = b - r^2(b - a)$$
$$x_4 = \text{new right-hand point} = x_1 + r(b - x_1)$$

The new left-hand point, x_3, will equal the old right-hand point, x_2. To see this, use the fact that $r^2 = 1 - r$ to conclude that $x_3 = b - r^2(b - a) = b - (1 - r)(b - a) = a + r(b - a) = x_2$.

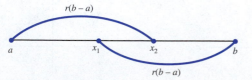

FIGURE 36
Location of x_1
and x_2 for Golden
Section Search

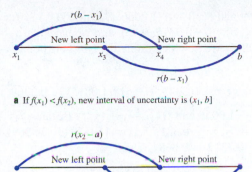

a If $f(x_1) < f(x_2)$, new interval of uncertainty is $(x_1, b]$

b If $f(x_1) \geq f(x_2)$, new interval of uncertainty is $[a, x_2)$

FIGURE **37**
How to Generate New
Points in Golden
Section Search

Case 2 $f(x_1) \geq f(x_2)$. The new interval of uncertainty, $[a, x_2)$, has length $x_2 - a = r(b - a)$. Then (see Figure 37b)

$$x_3 = \text{new left-hand point} = x_2 - r(x_2 - a)$$
$$x_4 = \text{new right-hand point} = a + r(x_2 - a) = a + r^2(b - a)$$

The new right-hand point, x_4, will equal the old left-hand point, x_1. To see this, use the fact that $r^2 = 1 - r$ to conclude that $x_4 = a + r^2(b - a) = a + (1 - r)(b - a) = b - r(b - a) = x_1$.

Now the values of $f(x_3)$ and $f(x_4)$ can be used to further reduce the length of the interval of uncertainty. At this point, two iterations of Golden Section Search have been completed.

We have shown that at each iteration of Golden Section Search, $f(x)$ must be evaluated at only one of the new points. It is easy to see that $L_2 = rL_1 = r^2(b - a)$ and, in general, $L_k = rL_{k-1}$ yields that $L_k = r^k(b - a)$. Thus, if we want our final interval of uncertainty to have a length $< \epsilon$, we must perform k iterations of Golden Section Search, where $r^k(b - a) < \epsilon$.

EXAMPLE 25   Golden Section Search

Use Golden Section Search to find

$$\max \ -x^2 - 1$$
$$\text{s.t.} \quad -1 \leq x \leq 0.75$$

with the final interval of uncertainty having a length less than $\frac{1}{4}$.

Solution   Here $a = -1$, $b = 0.75$, and $b - a = 1.75$. To determine the number k of iterations of Golden Section Search that must be performed, we solve for k using $1.75(0.618^k) < 0.25$, or $0.618^k < \frac{1}{7}$. Taking logarithms to base e of both sides, we obtain

$$k \ln 0.618 < \ln \tfrac{1}{7}$$
$$k(-0.48) < -1.95$$
$$k > \tfrac{1.95}{0.48} = 4.06$$

Thus, five iterations of Golden Section Search must be performed. We first determine x_1 and x_2:

$$x_1 = 0.75 - (0.618)(1.75) = -0.3315$$
$$x_2 = -1 + (0.618)(1.75) = 0.0815$$

Then $f(x_1) = -1.1099$ and $f(x_2) = -1.0066$. Because $f(x_1) < f(x_2)$, the new interval of uncertainty is $I_1 = (x_1, b] = (-0.3315, 0.75]$, and we have that $x_3 = x_2$. Of course, $L_1 = 0.75 + 0.3315 = 1.0815$. We now determine the two new points x_3 and x_4:

$$x_3 = x_2 = 0.0815$$
$$x_4 = -0.3315 + 0.618(1.0815) = 0.3369$$

Now $f(x_3) = f(x_2) = -1.0066$ and $f(x_4) = -1.1135$. Because $f(x_3) > f(x_4)$, the new interval of uncertainty is $I_2 = [-0.3315, x_4) = [-0.3315, 0.3369)$, and x_6 will equal x_3. Also, $L_2 = 0.3369 + 0.3315 = 0.6684$. Then

$$x_5 = 0.3369 - 0.618(0.6684) = -0.0762$$
$$x_6 = x_3 = 0.0815$$

Note that $f(x_5) = -1.0058$ and $f(x_6) = f(x_3) = -1.0066$. Because $f(x_5) > f(x_6)$, the new interval of uncertainty is $I_3 = [-0.3315, x_6) = [-0.3315, 0.0815)$ and $L_3 = 0.0815 + 0.3315 = 0.4130$. Because $f(x_6) < f(x_5)$, we have that $x_5 = x_8$ and $f(x_8) = -1.0058$. Now

$$x_7 = 0.0815 - 0.618(0.413) = -0.1737$$
$$x_8 = x_5 = -0.0762$$

and $f(x_7) = -1.0302$. Because $f(x_8) > f(x_7)$, the new interval of uncertainty is $I_4 = (x_7, 0.0815] = (-0.1737, 0.0815]$, and $L_4 = 0.0815 + 0.1737 = 0.2552$. Also, $x_9 = x_8$ will hold. Finally,

$$x_9 = x_8 = -0.0762$$
$$x_{10} = -0.1737 + 0.618(0.2552) = -0.016$$

Now $f(x_9) = f(x_8) = -1.0058$ and $f(x_{10}) = -1.0003$. Because $f(x_{10}) > f(x_9)$, the new interval of uncertainty is $I_5 = (x_9, 0.0815] = (-0.0762, 0.0815]$ and $L_5 = 0.0815 + 0.0762 = 0.1577 < 0.25$ (as desired).

Thus, we have determined that

$$\max \quad -x^2 - 1$$
$$\text{s.t.} \quad -1 \leq x \leq 0.75$$

must lie within the interval $(-0.0762, 0.0815]$. (Of course, the actual maximum occurs for $\bar{x} = 0$.)

Golden Section Search can be applied to a minimization problem by multiplying the objective function by -1. This assumes that the modified objective function is unimodal.

Using Spreadsheets to Conduct Golden Section Search

Golden.xls

Figure 38 (file Golden.xls) displays an implementation of Golden Section Search on Lotus 1-2-3. We begin by entering the left-hand and right-hand endpoints ($a = -1, b = .75$) of the interval of uncertainty for Example 25 in cells A2 and B2. We compute r by entering the formula (5^.5–1)/2 into G2. Then, we name the cell G2 as the range R (with the **INSERT NAME CREATE** sequence of commands). In all subsequent formulas R refers to the range R and assumes the value of r computed in G2. We compute the initial left-hand point x_1 by entering the formula =B2–R*(B2–A2) in C2 and the initial right-hand point x_2 by entering the formula =A2+R*(B2–A2) in D2. In effect, the formulas in C2 and D2 implement Figure 36. We evaluate $f(x_1)$ by entering –(C2^.2–1) in E2 and $f(x_2)$ by entering –(D2)^2–1 in F2.

A	A	B	C	D	E	F	G
1	LEFTPTUNC	RIGTPTUNC	LEFTPT	RIGHTPT	F(LEFTPT)	F(RIGHTPT)	R
2	.1	0.75	-0.33156	0.081559	-1.10993169	-1.00665195	0.618034
3	-0.33155948	0.75	0.081559	0.336881	-1.00665195	-1.11348883	
4	-0.33155948	0.336881039	-0.07624	0.081559	-1.00581222	-1.00665195	FIGURE
5	-0.33155948	0.08155948	-0.17376	-0.07624	-1.03019326	-1.00581222	25
6	-0.17376208	0.08155948	-0.07624	-0.01596	-1.00581222	-1.00025487	GOLDEN
7	-0.07623792	0.08155948	-0.01596	0.021286	-1.00025487	-1.0004531	SECTION
8							SEARCH

FIGURE 38

Golden Section Search for Example 25

In A3, we determine the new left point of the interval of uncertainty by entering the formula $=\textbf{IF}(\text{E2}<\text{F2, C2, A2})$. This ensures that if $f(x_1) < f(x_2)$, then the new left point of the interval of uncertainty equals the last left-hand point where the function is evaluated (x_1); while if $f(x_1) \geq f(x_2)$, then the new left-hand point of uncertainty equals the old left-hand endpoint (a). Similarly, in B3 we determine the new right-hand endpoint of the interval of uncertainty. In C3, we compute the new left-hand point (x_3) where the function is evaluated by entering the formula $=\textbf{IF}(\text{E2}<\text{F2,D2,D2}-\text{R*(D2}-\text{A2}))$. If $f(x_1) < f(x_2)$, then this formula ensures that the new left-hand point (x_3) will equal the old right-hand point (x_2); if $f(x_1) \geq f(x_2)$, then the new left-hand point (x_3) will equal $x_2 - r(x_2 - a)$ [this equals D2–R*(D2–A2)]. In D3, we compute the new right-hand point (x_4) by entering the formula $=\textbf{IF}(\text{E2}<\text{F2,C2}+\text{R*(B2}-\text{C2)), C2})$. If $f(x_1) < f(x_2)$, then the new right-hand point (x_4) will equal $x_1 + r(b - x_1)$ [this equals C2+R*(B2–C2)]; if $f(x_1) \geq f(x_2)$, then the new right-hand point will equal the old left-hand point (x_1) (which equals C2). In E3, we evaluate the function at the new left-hand point by entering $-(\text{C4})^2-1$, and in F3 we evaluate the function at the new right-hand endpoint by entering $-(\text{D4})^2-1$.

Now copying the formulas from the range A3:F3 to the range A3:F7 will generate four more iterations of Golden Section Search.

PROBLEMS

Group A

1 Use Golden Section Search to determine (within an interval of 0.8) the optimal solution to

$$\max x^2 + 2x$$
$$\text{s.t.} \quad -3 \leq x \leq 5$$

2 Use Golden Section Search to determine (within an interval of 0.6) the optimal solution to

$$\max x - e^x$$
$$\text{s.t.} \quad -1 \leq x \leq 3$$

3 Consider a line segment [0, 1] that is divided into two parts (Figure 39). The line segment is said to be divided into the Golden Section if

$$\frac{\text{Length of whole line}}{\text{Length of larger part of line}}$$
$$= \frac{\text{length of larger part of line}}{\text{length of smaller part of line}}$$

Show that for the line segment to be divided into the Golden Section,

$$r = \frac{5^{1/2} - 1}{2}$$

4 Hughesco is interested in determining how cutting fluid jet pressure (p) affects the useful life of a machine tool (t), using the data in Table 11. Pressure p is constrained to be between 0 and 600 pounds per square inch (psi). Use Golden Section Search to estimate (within 50 units) the value of p that maximizes useful tool life. Assume that t is a unimodal function of p.

FIGURE 39

0   r   1

TABLE 11

p (Pounds per Square Inch)	t (Minutes)
229	39
371	81
458	82
513	79
425	84
404	85
392	84

12.6 Unconstrained Maximization and Minimization with Several Variables

We now discuss how to find an optimal solution (if it exists) or a local extremum for the following unconstrained NLP:

$$\max \text{ (or min) } f(x_1, x_2, \ldots, x_n)$$
$$\text{s.t.} \quad (x_1, x_2, \ldots, x_n) \in R^n \tag{7}$$

We assume that the first and second partial derivatives of $f(x_1, x_2, \ldots, x_n)$ exist and are continuous at all points. Let

$$\frac{\partial f(\bar{x})}{\partial x_i}$$

be the partial derivative of $f(x_1, x_2, \ldots, x_n)$ with respect to x_i, evaluated at $\bar{x}$. A necessary condition for $\bar{x} = (\bar{x}_1, \bar{x}_2, \ldots, \bar{x}_n)$ to be a local extremum for NLP (7) is given in Theorem 6.

THEOREM 6

If $\bar{x}$ is a local extremum for (6), then $\dfrac{\partial f(\bar{x})}{\partial x_i} = 0$.

To see why Theorem 6 holds, suppose $\bar{x}$ is a local extremum for (7)—say, a local maximum. If $\dfrac{\partial f(\bar{x})}{\partial x_i} > 0$ holds for any i, then by slightly increasing x_i (and holding all other variables constant), we can find a point x' near $\bar{x}$ with $f(x') > f(\bar{x})$. This would contradict the fact that $\bar{x}$ is a local maximum. Similarly, if $\bar{x}$ is a local maximum for (7) and $\dfrac{\partial f(\bar{x})}{\partial x_i} < 0$, then by slightly decreasing x_i (and holding all other variables constant), we can find a point x'' near $\bar{x}$ with $f(x'') > f(\bar{x})$. Thus, if $\bar{x}$ is a local maximum for (7), then $\dfrac{\partial f(\bar{x})}{\partial x_i} = 0$ must hold for $i = 1, 2, \ldots, n$. A similar argument shows that if $\bar{x}$ is a local minimum, then $\dfrac{\partial f(\bar{x})}{\partial x_i} = 0$ must hold for $i = 1, 2, \ldots, n$.

DEFINITION ■ A point $\bar{x}$ having $\dfrac{\partial f(\bar{x})}{\partial x_i} = 0$ for $i = 1, 2, \ldots, n$ is called a **stationary point** of f. ■

The following three theorems give conditions (involving the Hessian of f) under which a stationary point is a local minimum, a local maximum, or not a local extremum.

THEOREM 7

If $H_k(\bar{x}) > 0$, $k = 1, 2, \ldots, n$, then a stationary point $\bar{x}$ is a local minimum for NLP (7).

THEOREM 7'

If, for $k = 1, 2, \ldots, n$, $H_k(\bar{x})$ is nonzero and has the same sign as $(-1)^k$, then a stationary point $\bar{x}$ is a local maximum for NLP (7).

If $H_n(\bar{x}) \neq 0$ and the conditions of Theorems 7 and 7′ do not hold, then a stationary point $\bar{x}$ is not a local extremum.

If a stationary point $\bar{x}$ is not a local extremum, then it is called a **saddle point**. If $H_n(\bar{x}) = 0$ for a stationary point $\bar{x}$, then $\bar{x}$ may be a local minimum, a local maximum, or a saddle point, and the preceding tests are inconclusive.

From Theorems 1 and 7′, we know that if $f(x_1, x_2, \ldots, x_n)$ is a concave function (and NLP (7) is a max problem), then any stationary point for (7) is an optimal solution to (7). From Theorems 1′ and 7, we know that if $f(x_1, x_2, \ldots, x_n)$ is a convex function [and NLP (7) is a min problem], then any stationary point for (7) is an optimal solution to (7).

EXAMPLE 26 **Monopolistic Pricing with Multiple Customer Types**

A monopolist producing a single product has two types of customers. If q_1 units are produced for customer 1, then customer 1 is willing to pay a price of $70 - 4q_1$ dollars. If q_2 units are produced for customer 2, then customer 2 is willing to pay a price of $150 - 15q_2$ dollars. For $q > 0$, the cost of manufacturing q units is $100 + 15q$ dollars. To maximize profit, how much should the monopolist sell to each customer?

Solution Let $f(q_1, q_2)$ be the monopolist's profit if she produces q_i units for customer i. Then (assuming some production takes place)

$$f(q_1, q_2) = q_1(70 - 4q_1) + q_2(150 - 15q_2) - 100 - 15q_1 - 15q_2$$

To find the stationary point(s) for $f(q_1, q_2)$, we set

$$\frac{\partial f}{\partial q_1} = 70 - 8q_1 - 15 = 0 \qquad (\text{for } q_1 = \tfrac{55}{8})$$

$$\frac{\partial f}{\partial q_2} = 150 - 30q_2 - 15 = 0 \qquad (\text{for } q_2 = \tfrac{9}{2})$$

Thus, the only stationary point of $f(q_1, q_2)$ is $(\tfrac{55}{8}, \tfrac{9}{2})$. Next we find the Hessian for $f(q_1, q_2)$.

$$H(q_1, q_2) = \begin{bmatrix} -8 & 0 \\ 0 & -30 \end{bmatrix}$$

Since the first leading principal minor of H is $-8 < 0$, and the second leading principal minor of H is $(-8)(-30) = 240 > 0$, Theorem 7′ shows that $(\tfrac{55}{8}, \tfrac{9}{2})$ is a local maximum. Also, Theorem 3′ implies that $f(q_1, q_2)$ is a concave function [on the set of points S of (q_1, q_2) satisfying $q_1 \geq 0$, $q_2 \geq 0$, and $q_1 + q_2 > 0$]. Thus, Theorem 1 implies that $(\tfrac{55}{8}, \tfrac{9}{2})$ maximizes profit among all production possibilities (with the possible exception of no production). Then $(\tfrac{55}{8}, \tfrac{9}{2})$ yields a profit of

$$f(q_1, q_2) = \tfrac{55}{8}(70 - \tfrac{220}{8}) + \tfrac{9}{2}[150 - 15(\tfrac{9}{2})] - 100 - 15(\tfrac{55}{8} + \tfrac{9}{2}) = \$392.81$$

The profit from producing $(\tfrac{55}{8}, \tfrac{9}{2})$ exceeds the profit of $\$0$ that is obtained by producing nothing, so $(\tfrac{55}{8}, \tfrac{9}{2})$ solves the NLP; the monopolist should sell $\tfrac{55}{8}$ units to customer 1 and $\tfrac{9}{2}$ units to customer 2.

EXAMPLE 27   **Least Squares Estimation**

Suppose the grade-point average (GPA) for a student can be accurately predicted from the student's score on the GMAT (Graduate Management Admissions Test). More specifically, suppose that the ith student observed has a GPA of y_i and a GMAT score of x_i. How can we use the **least squares method** to estimate a hypothesized relation of the form $y_i = a + bx_i$?

Solution   Let $\hat{a}$ be our estimate of a and $\hat{b}$ our estimate of b. Given that for students $i = 1, 2, \ldots,$ n we have observed $(x_1, y_1), (x_2, y_2), \ldots, (x_n, y_n)$, $\hat{e}_i = y_i - (\hat{a} + \hat{b}x_i)$ is our error in estimating the GPA of student i. The least squares method chooses $\hat{a}$ and $\hat{b}$ to minimize

$$f(a, b) = \sum_{i=1}^{i=n} \hat{e}_i^2 = \sum_{i=1}^{i=n} (y_i - a - bx_i)^2$$

Since

$$\frac{\partial f}{\partial a} = -2 \sum_{i=1}^{i=n} (y_i - a - bx_i) \quad \text{and} \quad \frac{\partial f}{\partial b} = -2 \sum_{i=1}^{i=n} (y_i - a - bx_i)x_i$$

$\dfrac{\partial f}{\partial a} = \dfrac{\partial f}{\partial b} = 0$ will hold for the point $(\hat{a}, \hat{b})$ satisfying

$$\sum_{i=1}^{i=n} (y_i - a - bx_i) = 0 \quad \text{or} \quad \sum_{i=1}^{i=n} y_i = na + b \sum_{i=1}^{i=n} x_i$$

and

$$\sum_{i=1}^{i=n} x_i(y_i - a - bx_i) = 0 \quad \text{or} \quad \sum_{i=1}^{i=n} x_iy_i = a \sum_{i=1}^{i=n} x_i + b \sum_{i=1}^{i=n} x_i^2$$

These are the well-known **normal equations**. Does the solution $(\hat{a}, \hat{b})$ to the normal equations minimize $f(a, b)$? To answer this question, we must compute the Hessian for $f(a, b)$:

$$\frac{\partial^2 f}{\partial a^2} = 2n, \quad \frac{\partial^2 f}{\partial b^2} = 2 \sum_{i=1}^{i=n} x_i^2, \quad \frac{\partial^2 f}{\partial a \partial b} = \frac{\partial^2 f}{\partial b \partial a} = 2 \sum_{i=1}^{i=n} x_i$$

Thus,

$$H = \begin{bmatrix} 2n & 2 \sum_{i=1}^{i=n} x_i \\ 2 \sum_{i=1}^{i=n} x_i & 2 \sum_{i=1}^{i=n} x_i^2 \end{bmatrix}$$

Since $H_1(\hat{a}, \hat{b}) = 2n > 0$, $(\hat{a}, \hat{b})$ will be a local minimum if

$$H_2(\hat{a}, \hat{b}) = 4n \sum_{i=1}^{i=n} x_i^2 - 4\left(\sum_{i=1}^{i=n} x_i \right)^2 > 0$$

In Example 31 of Section 12.8, we show that

$$n \sum_{i=1}^{i=n} x_i^2 \geq \left(\sum_{i=1}^{i=n} x_i \right)^2$$

with equality holding if and only if $x_1 = x_2 = \cdots = x_n$. Thus, if at least two of the x_i's are different, Theorem 7' implies that $(\hat{a}, \hat{b})$ will be a local minimum. $H(a, b)$ does not depend on the values of a and b, so this reasoning (and Theorem 3) shows that if at least two of the x_i's are different, then $f(a, b)$ is a convex function. If at least two of the x_i's are different, then Theorem 1' shows that $(\hat{a}, \hat{b})$ minimizes $f(a, b)$.

EXAMPLE 28   Finding Maxima, Minima, and Saddle Points

Find all local maxima, local minima, and saddle points for $f(x_1, x_2) = x_1^2 x_2 + x_2^3 x_1 - x_1 x_2$.

Solution   We have

$$\frac{\partial f}{\partial x_1} = 2x_1 x_2 + x_2^3 - x_2, \qquad \frac{\partial f}{\partial x_2} = x_1^2 + 3x_2^2 x_1 - x_1$$

Thus, $\dfrac{\partial f}{\partial x_1} = \dfrac{\partial f}{\partial x_2} = 0$ requires

$$2x_1 x_2 + x_2^3 - x_2 = 0 \qquad \text{or} \qquad x_2(2x_1 + x_2^2 - 1) = 0 \tag{8}$$
$$x_1^2 + 3x_2^2 x_1 - x_1 = 0 \qquad \text{or} \qquad x_1(x_1 + 3x_2^2 - 1) = 0 \tag{9}$$

For (8) to hold, either (i) $x_2 = 0$ or (ii) $2x_1 + x_2^2 - 1 = 0$ must hold. For (9) to hold, either (iii) $x_1 = 0$ or (iv) $x_1 + 3x_2^2 - 1 = 0$ must hold.

Thus, for (x_1, x_2) to be a stationary point, we must have:

(i) and (iii) hold. This is only true at $(0, 0)$.

(i) and (iv) hold. This is only true at $(1, 0)$.

(ii) and (iii) hold. This is only true at $(0, 1)$ and $(0, -1)$.

(ii) and (iv) hold. This requires that $x_2^2 = 1 - 2x_1$ and $x_1 + 3(1 - 2x_1) - 1 = 0$ hold.

Then

$$x_1 = \frac{2}{5} \qquad \text{and} \qquad x_2 = \frac{5^{1/2}}{5} \qquad \text{or} \qquad -\frac{5^{1/2}}{5}$$

Thus, $f(x_1, x_2)$ has the following stationary points:

$$(0, 0), \ (1, 0), \ (0, 1), \ (0, -1), \left(\frac{2}{5}, \frac{5^{1/2}}{5}\right) \qquad \text{and} \qquad \left(\frac{2}{5}, -\frac{5^{1/2}}{5}\right)$$

Also,

$$H(x_1, x_2) = \begin{bmatrix} 2x_2 & 2x_1 + 3(x_2)^2 - 1 \\ 2x_1 + 3(x_2)^2 - 1 & 6x_1 x_2 \end{bmatrix}$$

$$H(0, 0) = \begin{bmatrix} 0 & -1 \\ -1 & 0 \end{bmatrix}$$

Because $H_1(0, 0) = 0$, the conditions of Theorems 7 and 7' cannot be satisfied. Because $H_2(0, 0) = -1 \neq 0$, Theorem 7'' now implies that $(0, 0)$ is a saddle point.

$$H(1, 0) = \begin{bmatrix} 0 & 1 \\ 1 & 0 \end{bmatrix}$$

Then $H_1(1, 0) = 0$ and $H_2(1, 0) = -1$, so by Theorem 7'' $(1, 0)$ is also a saddle point. Since

$$H(0, 1) = \begin{bmatrix} 2 & 2 \\ 2 & 0 \end{bmatrix}$$

we have $H_1(0, 1) = 2 > 0$ (so the hypotheses of Theorem 7' cannot be satisfied) and $H_2(0, 1) = -4$ (so the hypothesis of Theorem 7 cannot be satisfied). Because $H_2(0, 1) \neq 0$, $(0, 1)$ is a saddle point.

For $\left(\dfrac{2}{5}, -\dfrac{5^{1/2}}{5}\right)$, we have

$$H\left(\frac{2}{5}, -\frac{5^{1/2}}{5}\right) = \begin{bmatrix} -\dfrac{2}{5^{1/2}} & \dfrac{2}{5} \\[2mm] \dfrac{2}{5} & -\dfrac{12}{5(5)^{1/2}} \end{bmatrix}$$

Thus,

$$H_1\left(\frac{2}{5}, -\frac{5^{1/2}}{5}\right) = -\frac{2}{5^{1/2}} < 0 \quad \text{and} \quad H_2\left(\frac{2}{5}, -\frac{5^{1/2}}{5}\right) = \frac{20}{25} > 0$$

Thus, Theorem 7' shows that $\left(\dfrac{2}{5}, -\dfrac{5^{1/2}}{5}\right)$ is a local maximum. Finally,

$$H\left(\frac{2}{5}, \frac{5^{1/2}}{5}\right) = \begin{bmatrix} \dfrac{2}{5^{1/2}} & \dfrac{2}{5} \\[2mm] \dfrac{2}{5} & \dfrac{12}{5(5)^{1/2}} \end{bmatrix}$$

Since $H_1\left(\dfrac{2}{5}, \dfrac{5^{1/2}}{5}\right) = \dfrac{2}{5^{1/2}} > 0$ and $H_2\left(\dfrac{2}{5}, \dfrac{5^{1/2}}{5}\right) = \dfrac{20}{25} > 0$, Theorem 7 shows that $\left(\dfrac{2}{5}, \dfrac{5^{1/2}}{5}\right)$ is a local minimum.

When Does LINGO Find the Optimal Solution to an Unconstrained NLP?

If you are maximizing a concave function (with no constraints) or minimizing a convex function (with no constraints), you can be sure that any solution found by LINGO is the optimal solution to your problem. In Example 27, for instance, our work shows that $f(a, b)$ is a convex function, so we know that LINGO would correctly find the least squares line fitting a set of points.

PROBLEMS

Group A

1 A company has n factories. Factory i is located at point (x_i, y_i), in the x–y plane. The company wants to locate a warehouse at a point (x, y) that minimizes

$$\sum_{i=1}^{i=n} (\text{distance from factory } i \text{ to warehouse})^2$$

Where should the warehouse be located?

2 A company can sell all it produces of a given output for \$2/unit. The output is produced by combining two inputs. If q_1 units of input 1 and q_2 units of input 2 are used, then the

company can produce $q_1^{1/3} + q_2^{2/3}$ units of the output. If it costs \$1 to purchase a unit of input 1 and \$1.50 to purchase a unit of input 2, then how can the company maximize its profit?

3 (Collusive Duopoly Model) There are two firms producing widgets. It costs the first firm q_1 dollars to produce q_1 widgets and the second firm $0.5q_2^2$ dollars to produce q_2 widgets. If a total of q widgets are produced, consumers will pay \$200 $- q$ for each widget. If the two manufacturers want to collude in an attempt to maximize the sum of their profits, how many widgets should each company produce?

4 It costs a company \$6/unit to produce a product. If it charges a price p and spends a dollars on advertising, it can sell $10,000p^{-2}a^{1/6}$ units of the product. Find the price and advertising level that will maximize the company's profits.

5 A company manufactures two products. If it charges a price p_i for product i, it can sell q_i units of product i, where $q_1 = 60 - 3p_1 + p_2$ and $q_2 = 80 - 2p_2 + p_1$. It costs \$25 to produce a unit of product 1 and \$72 to produce a unit of product 2. How many units of each product should be produced to maximize profits?

6 Find all local maxima, local minima, and saddle points for $f(x_1, x_2) = x_1^3 - 3x_1x_2^2 + x_2^4$.

7 Find all local maxima, local minima, and saddle points for $f(x_1, x_2) = x_1x_2 + x_2x_3 + x_1x_3$.

Group B

8[†] (Cournot Duopoly Model) Let's reconsider Problem 3. The Cournot solution to this situation is obtained as follows: Firm i will produce $\bar{q}_i$, where if firm 1 changes its production level from $\bar{q}_1$ (and firm 2 still produces $\bar{q}_2$), then firm 1's profit will decrease. Also, if firm 2 changes its production level from $\bar{q}_2$ (and firm 1 still produces $\bar{q}_1$), then firm 2's profit will decrease. If firm i produces $\bar{q}_i$, this solution is stable, because if either firm changes its production level, it will do worse. Find $\bar{q}_1$ and $\bar{q}_2$.

9 In the Bloomington Girls Club basketball league, the following games have been played: team A beat team B by 7 points, team C beat team A by 8 points, team B beat team C by 6 points, and team B beat team C by 9 points. Let A, B, and C represent "ratings" for each team in the sense that if, say, team A plays team B, then we predict that team A will defeat team B by $A - B$ points. Determine values of A, B, and C that best fit (in the least squares sense) these results. To obtain a unique set of ratings, it may be helpful to add the constraint $A + B + C = 0$. This ensures that an "average" team will have a rating of 0.

12.7 The Method of Steepest Ascent

Suppose we want to solve the following unconstrained NLP:

$$\max z = f(x_1, x_2, \ldots, x_n)$$
$$\text{s.t.} \quad (x_1, x_2, \ldots, x_n) \in R^n \tag{10}$$

Our discussion in Section 12.6 shows that if $f(x_1, x_2, \ldots, x_n)$ is a concave function, then the optimal solution to (10) (if there is one) will occur at a stationary point $\bar{x}$ having

$$\frac{\partial f(\bar{x})}{\partial x_1} = \frac{\partial f(\bar{x})}{\partial x_2} = \cdots = \frac{\partial f(\bar{x})}{\partial x_n} = 0$$

In Examples 26 and 28, it was easy to find a stationary point, but in many problems, it may be difficult. In this section, we discuss the *method of steepest ascent*, which can be used to approximate a function's stationary point.

DEFINITION ■ Given a vector $\mathbf{x} = (x_1, x_2, \ldots, x_n) \in R^n$, the **length** of $\mathbf{x}$ (written $\|\mathbf{x}\|$) is

$$\|\mathbf{x}\| = (x_1^2 + x_2^2 + \cdots + x_n^2)^{1/2} \quad ■$$

Recall from Section 2.1 that any n-dimensional vector represents a direction in R^n. Unfortunately, for any direction, there are an infinite number of vectors representing that direction. For example, the vectors $(1, 1)$, $(2, 2)$, and $(3, 3)$ all represent the same direction (moving at a positive 45° angle) in R^2. For any vector $\mathbf{x}$, the vector $\mathbf{x}/\|\mathbf{x}\|$ will have a length of 1 and will define the same direction as $\mathbf{x}$ (see Problem 1 at the end of this section). Thus, with any direction in R^n, we may associate a vector of length 1 (called a unit vector). For example, because $\mathbf{x} = (1, 1)$ has $\|\mathbf{x}\| = 2^{1/2}$, the direction defined by $\mathbf{x} = (1, 1)$ is associated with the unit vector $(1/2^{1/2}, 1/2^{1/2})$. For any vector $\mathbf{x}$, the unit vector

[†]Based on Cournot (1897).

$x/\|\mathbf{x}\|$ is called the **normalized** version of $\mathbf{x}$. Henceforth, any direction in R^n will be described by the normalized vector defining that direction. Thus, the direction in R^2 defined by $(1, 1), (2, 2), (3, 3), \ldots$ will be described by the normalized vector

$$\left(\frac{1}{2^{1/2}}, \frac{1}{2^{1/2}}\right)$$

Consider a function $f(x_1, x_2, \ldots, x_n)$, all of whose partial derivatives exist at every point.

DEFINITION ■ The **gradient vector** for $f(x_1, x_2, \ldots, x_n)$, written $\nabla f(\mathbf{x})$, is given by

$$\nabla f(\mathbf{x}) = \left[\frac{\partial f(\mathbf{x})}{\partial x_1}, \frac{\partial f(\mathbf{x})}{\partial x_2}, \ldots, \frac{\partial f(\mathbf{x})}{\partial x_n}\right] \quad \blacksquare$$

$\nabla f(\mathbf{x})$ defines the direction

$$\frac{\nabla f(\mathbf{x})}{\|\nabla f(\mathbf{x})\|}$$

For example, if $f(x_1, x_2) = x_1^2 + x_2^2$, then $\nabla f(x_1, x_2) = (2x_1, 2x_2)$. Thus, $\nabla f(3, 4) = (6, 8)$. Because $\|\nabla f(3, 4)\| = 10$, $\nabla f(3, 4)$ defines the direction $(\frac{6}{10}, \frac{8}{10}) = (0.6, 0.8)$.

At any point $\bar{\mathbf{x}}$ that lies on the curve $f(x_1, x_2, \ldots, x_n) = f(\bar{\mathbf{x}})$, the vector

$$\frac{\nabla f(\bar{\mathbf{x}})}{\|\nabla f(\bar{\mathbf{x}})\|}$$

will be perpendicular to the curve $f(x_1, x_2, \ldots, x_n) = f(\bar{x})$ (see Problem 5 at the end of this section). For example, let $f(x_1, x_2) = x_1^2 + x_2^2$. Then at $(3, 4)$,

$$\frac{\nabla f(3, 4)}{\|\nabla f(3, 4)\|} = (0.6, 0.8)$$

is perpendicular to $x_1^2 + x_2^2 = 25$ (see Figure 40).

From the definition of $\frac{\partial f(\mathbf{x})}{\partial x_i}$, it follows that if the value of x_i is increased by a small amount δ, the value of $f(\mathbf{x})$ will increase by approximately $\delta\frac{\partial f(\mathbf{x})}{\partial x_i}$. Suppose we move from a point $\mathbf{x}$ a small length δ in a direction defined by a normalized column vector $\mathbf{d}$. By how much does $f(\mathbf{x})$ increase? The answer is that $f(\mathbf{x})$ increases by δ times the scalar product of $\frac{\nabla f(\mathbf{x})}{\|\nabla f(\mathbf{x})\|}$ and $\mathbf{d}$ $\left(\text{written } \frac{\delta\,\nabla f(\mathbf{x})\cdot\mathbf{d}}{\|\nabla f(\mathbf{x})\|}\right)$. Thus, if $\frac{\nabla f(\mathbf{x})\cdot\mathbf{d}}{\|\nabla f(\mathbf{x})\|} > 0$, moving

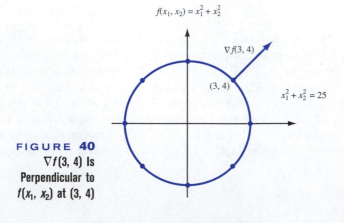

FIGURE 40
$\nabla f(3, 4)$ Is
Perpendicular to
$f(x_1, x_2)$ at $(3, 4)$

in a direction $\mathbf{d}$ away from $\mathbf{x}$ will increase the value of $f(\mathbf{x})$, and if $\dfrac{\nabla f(\mathbf{x}) \cdot \mathbf{d}}{\|\nabla f(\mathbf{x})\|} < 0$, moving in a direction $\mathbf{d}$ away from $\mathbf{x}$ will decrease $f(\mathbf{x})$. For example, suppose $f(x_1, x_2) = x_1^2 + x_2^2$ and we move a length δ in a $45°$ direction away from the point $(3, 4)$. By how much will the value of $f(x_1, x_2)$ change? A $45°$ direction is represented by the vector $\left(\dfrac{1}{2^{1/2}}, \dfrac{1}{2^{1/2}}\right)$ and $\dfrac{\nabla f(3, 4)}{\|\nabla f(3, 4)\|} = (0.6, 0.8)$, so the value of $f(x_1, x_2)$ will increase by approximately

$$\delta[0.6 \quad 0.8] \begin{bmatrix} \dfrac{1}{2^{1/2}} \\[2mm] \dfrac{1}{2^{1/2}} \end{bmatrix} = 0.99\delta$$

Recall from Section 12.6 that the optimal solution $\bar{\mathbf{v}}$ to (10) must satisfy $\nabla f(\bar{\mathbf{v}}) = 0$. Now suppose that we are at a point $\mathbf{v}_0$ and want to find a point $\bar{\mathbf{v}}$ that solves (10). In an attempt to find $\bar{\mathbf{v}}$, it seems reasonable to move away from $\mathbf{v}_0$ in a direction that maximizes the rate (at least locally) at which $f(x_1, x_2, \ldots, x_n)$ increases. Lemma 1 proves useful here (see Review Problem 22).

LEMMA 1

Suppose we are at a point $\mathbf{v}$ and we move from $\mathbf{v}$ a small distance δ in a direction $\mathbf{d}$. Then for a given δ, the maximal increase in the value of $f(x_1, x_2, \ldots, x_n)$ will occur if we choose

$$\mathbf{d} = \frac{\nabla f(\mathbf{x})}{\|\nabla f(\mathbf{x})\|}$$

In short, if we move a small distance away from $\mathbf{v}$ and we want $f(x_1, x_2, \ldots, x_n)$ to increase as quickly as possible, then we should move in the direction of $\nabla f(\mathbf{v})$.

We are now ready to describe the method of steepest ascent. Begin at any point $\mathbf{v}_0$. Moving in the direction of $\nabla f(\mathbf{v}_0)$ will result in a maximum rate of increase for f, so we begin by moving away from $\mathbf{v}_0$ in the direction of $\nabla f(\mathbf{v}_0)$. For some non-negative value of t, we move to a point $\mathbf{v}_1 = \mathbf{v}_0 + t\nabla f(\mathbf{v}_0)$. The maximum possible improvement in the value of f (for a max problem) that can be attained by moving away from $\mathbf{v}_0$ in the direction of $\nabla f(\mathbf{v}_0)$ results from moving to $\mathbf{v}_1 = \mathbf{v}_0 + t_0\nabla f(\mathbf{v}_0)$, where t_0 solves the following one-dimensional optimization problem:

$$\max f(\mathbf{v}_0 + t_0\nabla f(\mathbf{v}_0)) \tag{11}$$
$$\text{s.t.} \quad t_0 \geq 0$$

NLP (11) may be solved by the methods of Section 12.4 or, if necessary, by a search procedure such as the Golden Section Search.

If $\|\nabla f(\mathbf{v}_1)\|$ is small (say, less than 0.01), we may terminate the algorithm with the knowledge that $\mathbf{v}_1$ is near a stationary point $\bar{\mathbf{v}}$ having $\nabla f(\bar{\mathbf{v}}) = 0$. If $\|\nabla f(\mathbf{v}_1)\|$ is not sufficiently small, then we move away from $\mathbf{v}_1$ a distance t_1 in the direction of $\|\nabla f(\mathbf{v}_1)\|$. As before, we choose t_1 by solving

$$\max f(\mathbf{v}_1 + t_1\nabla f(\mathbf{v}_1))$$
$$\text{s.t.} \quad t_1 \geq 0$$

We are now at the point $\mathbf{v}_2 = \mathbf{v}_1 + t_1 \nabla f(\mathbf{v}_1)$. If $\| \nabla f(\mathbf{v}_2) \|$ is sufficiently small, then we terminate the algorithm and choose $\mathbf{v}_2$ as our approximation to a stationary point of $f(x_1, x_2, \ldots, x_n)$. Otherwise, we continue in this fashion until we reach a point $\mathbf{v}_n$ having $\| \nabla f(\mathbf{v}_n) \|$ sufficiently small. Then we choose $\mathbf{v}_n$ as our approximation to a stationary point of $f(x_1, x_2, \ldots, x_n)$.

This algorithm is called the **method of steepest ascent** because to generate points, we always move in the direction that maximizes the rate at which f increases (at least locally).

EXAMPLE 29   **Steepest Ascent Example**

Use the method of steepest ascent to approximate the solution to

$$\max z = -(x_1 - 3)^2 - (x_2 - 2)^2 = f(x_1, x_2)$$
$$\text{s.t.} \quad (x_1, x_2) \in R^2$$

Solution   We arbitrarily choose to begin at the point $\mathbf{v}_0 = (1, 1)$. Because $\nabla f(x_1, x_2) = (-2(x_1 - 3), -2(x_2 - 2))$, we have $\nabla f(1, 1) = (4, 2)$. Thus, we must choose t_0 to maximize

$$f(t_0) = f[(1, 1) + t_0(4, 2)] = f(1 + 4t_0, 1 + 2t_0) = -(-2 + 4t_0)^2 - (-1 + 2t_0)^2$$

Setting $f'(t_0) = 0$, we obtain

$$-8(-2 + 4t_0) - 4(-1 + 2t_0) = 0$$
$$20 - 40t_0 = 0$$
$$t_0 = 0.5$$

Our new point is $\mathbf{v}_1 = (1, 1) + 0.5(4, 2) = (3, 2)$. Now $\nabla f(3, 2) = (0, 0)$, and we terminate the algorithm. Because $f(x_1, x_2)$ is a concave function, we have found the optimal solution to the NLP.

PROBLEMS

Group A

1 For any vector $\mathbf{x}$, show that the vector $\mathbf{x}/\| \mathbf{x} \|$ has unit length.

2 Use the method of steepest ascent to approximate the optimal solution to the following problem: $\max z = -(x_1 - 2)^2 - x_1 - x_2^2$. Begin at the point $(2.5, 1.5)$.

3 Use steepest ascent to approximate the optimal solution to the following problem: $\max z = 2x_1x_2 + 2x_2 - x_1^2 - 2x_2^2$. Begin at the point $(0.5, 0.5)$. Note that at later iterations, successive points are very close together. Variations of steepest ascent have been developed to deal with this problem [see Bazaraa and Shetty (1993, Section 8.6)].

Group B

4 How would you modify the method of steepest ascent if each variable x_1 were constrained to lie in an interval $[a_i, b_i]$?

Group C

5 Show that at any point $\bar{\mathbf{x}} = (\bar{x}_1, \bar{x}_2)$, $\nabla f(\bar{\mathbf{x}})$ is perpendicular to the curve $f(x_1, x_2) = f(\bar{x}_1, \bar{x}_2)$. (*Hint:* Two vectors are perpendicular if their scalar product equals zero.)

12.8 Lagrange Multipliers

Lagrange multipliers can be used to solve NLPs in which all the constraints are equality constraints. We consider NLPs of the following type:

$$\text{max (or min) } z = f(x_1, x_2, \ldots, x_n)$$

$$\text{s.t.} \quad g_1(x_1, x_2, \ldots, x_n) = b_1$$

$$g_2(x_1, x_2, \ldots, x_n) = b_2 \qquad \text{(12)}$$

$$\vdots$$

$$g_m(x_1, x_2, \ldots, x_n) = b_m$$

To solve (12), we associate a **multiplier** λ_i with the ith constraint in (12) and form the **Lagrangian**

$$L(x_1, x_2, \ldots, x_n, \lambda_1, \lambda_2, \ldots, \lambda_m) = f(x_1, x_2, \ldots, x_n)$$

$$+ \sum_{i=1}^{i=m} \lambda_i [b_i - g_i(x_1, x_2, \ldots, x_n)] \qquad \text{(13)}$$

Then we attempt to find a point $(\bar{x}_1, \bar{x}_2, \ldots, \bar{x}_n, \bar{\lambda}_1, \bar{\lambda}_2, \ldots, \bar{\lambda}_m)$ that maximizes (or minimizes) $L(x_1, x_2, \ldots, x_n, \lambda_1, \lambda_2, \ldots, \lambda_m)$. In many situations, $(\bar{x}_1, \bar{x}_2, \ldots, \bar{x}_n)$ will solve (12). Suppose that (12) is a maximization problem. If $(\bar{x}_1, \bar{x}_2, \ldots, \bar{x}_n, \bar{\lambda}_1, \bar{\lambda}_2, \ldots, \bar{\lambda}_m)$ maximizes L, then at $(\bar{x}_1, \bar{x}_2, \ldots, \bar{x}_n, \bar{\lambda}_1, \bar{\lambda}_2, \ldots, \bar{\lambda}_n)$

$$\frac{\partial L}{\partial \lambda_i} = b_i - g_i(x_1, x_2, \ldots, x_n) = 0$$

Here $\dfrac{\partial L}{\partial \lambda_i}$ is the partial derivative of L with respect to λ_i. This shows that $(\bar{x}_1, \bar{x}_2, \ldots, \bar{x}_n)$ will satisfy the constraints in (12). To show that $(\bar{x}_1, \bar{x}_2, \ldots, \bar{x}_n)$ solves (12), let $(x'_1, x'_2, \ldots, x'_n)$ be any point that is in (12)'s feasible region. Since $(\bar{x}_1, \bar{x}_2, \ldots, \bar{x}_n, \bar{\lambda}_1, \bar{\lambda}_2, \ldots, \bar{\lambda}_m)$ maximizes L, for any numbers $\lambda'_1, \lambda'_2, \ldots, \lambda'_m$ we have

$$L(\bar{x}_1, \bar{x}_2, \ldots, \bar{x}_n, \bar{\lambda}_1, \bar{\lambda}_2, \ldots, \bar{\lambda}_m) \geq L(x'_1, x'_2, \ldots, x'_n, \lambda'_1, \lambda'_2, \ldots \lambda'_m) \qquad \text{(14)}$$

Since $(\bar{x}_1, \bar{x}_2, \ldots, \bar{x}_n)$ and $(x'_1, x'_2, \ldots, x'_n)$ are both feasible in (12), the terms in (13) involving the λ's are all zero, and (14) becomes $f(\bar{x}_1, \bar{x}_2, \ldots, \bar{x}_n) \geq f(x'_1, x'_2, \ldots, x'_n)$. Thus, $(\bar{x}_1, \bar{x}_2, \ldots, \bar{x}_n)$ does solve (12). In short, if $(\bar{x}_1, \bar{x}_2 \ldots, \bar{x}_n, \bar{\lambda}_1, \bar{\lambda}_2, \ldots, \bar{\lambda}_m)$ solves the unconstrained maximization problem

$$\text{max } L(x_1, x_2, \ldots, x_n, \lambda_1, \lambda_2, \ldots, \lambda_m) \qquad \text{(15)}$$

then $(\bar{x}_1, \bar{x}_2, \ldots, \bar{x}_n)$ solves (12).

From Section 12.6, we know that for $(\bar{x}_1, \bar{x}_2, \ldots, \bar{x}_n, \bar{\lambda}_1, \bar{\lambda}_2, \ldots, \bar{\lambda}_m)$ to solve (15), it is necessary that at $(\bar{x}_1, \bar{x}_2, \ldots, \bar{x}_n, \bar{\lambda}_1, \bar{\lambda}_2, \ldots, \bar{\lambda}_m)$,

$$\frac{\partial L}{\partial x_1} = \frac{\partial L}{\partial x_2} = \cdots = \frac{\partial L}{\partial x_n} = \frac{\partial L}{\partial \lambda_1} = \frac{\partial L}{\partial \lambda_2} = \cdots = \frac{\partial L}{\partial \lambda_m} = 0 \qquad \text{(16)}$$

Theorem 8 gives conditions implying that any point $(\bar{x}_1, \bar{x}_2, \ldots, \bar{x}_n, \bar{\lambda}_1, \bar{\lambda}_2, \ldots, \bar{\lambda}_m)$ that satisfies (16) will yield an optimal solution $(\bar{x}_1, \bar{x}_2, \ldots, \bar{x}_n)$ to (12).

THEOREM 8

Suppose (12) is a maximization problem. If $f(x_1, x_2, \ldots, x_n)$ is a concave function and each $g_i(x_1, x_2, \ldots, x_n)$ is a linear function, then any point $(\bar{x}_1, \bar{x}_2, \ldots, \bar{x}_n, \bar{\lambda}_1, \bar{\lambda}_2, \ldots, \bar{\lambda}_m)$ satisfying (16) will yield an optimal solution $(\bar{x}_1, \bar{x}_2, \ldots, \bar{x}_n)$ to (12).

Suppose (12) is a minimization problem. If $f(x_1, x_2, \ldots, x_n)$ is a convex function and each $g_i(x_1, x_2, \ldots, x_n)$ is a linear function, then any point $(\bar{x}_1, \bar{x}_2, \ldots, \bar{x}_n, \bar{\lambda}_1, \bar{\lambda}_2, \ldots, \bar{\lambda}_m)$ satisfying (16) will yield an optimal solution $(\bar{x}_1, \bar{x}_2, \ldots, \bar{x}_n)$ to (12).

Even if the hypotheses of these theorems fail to hold, it is possible that any point satisfying (16) will solve (12). See the appendix of Henderson and Quandt (1980) for details.

Geometrical Interpretation of Lagrange Multipliers

From (16) we know that for the point $\bar{x} = (\bar{x}_1, \bar{x}_2, \ldots, \bar{x}_n)$ to solve (12) it is necessary that at $\bar{x}$

$$\frac{\partial L}{\partial x_j} = 0 \quad \text{for } j = 1, 2, \ldots, n$$

This is equivalent to saying that there exist numbers $\lambda_1, \lambda_2, \ldots \lambda_m$ such that at the point $\bar{x}$

$$\nabla f = \sum_{i=1}^{i=m} \lambda_i \nabla g_i \tag{17}$$

To see why this is so, note that the jth component of the left-hand side of (17) is

$$\frac{\partial f}{\partial x_j}$$

and the jth component of the right-hand side is

$$\sum_{i=1}^{i=m} \lambda_i \frac{\partial g_i}{\partial x_j}$$

Thus, (17) implies that for $j = 1, 2, \ldots, n$

$$\frac{\partial f}{\partial x_j} - \sum_{i=1}^{i=m} \lambda_i \frac{\partial g_i}{\partial x_j} = 0 \quad \text{or} \quad \frac{\partial L}{\partial x_j} = 0$$

Another way to look at (17) is as follows: For $\bar{x}$ to solve (12), it is necessary that at $\bar{x}$, ∇f is a linear combination of the constraint gradients.

For an optimization problem with one constraint it is easy to see why (17) must hold at a solution to (12). If (12) has one constraint, then (17) is equivalent to the statement that the gradient of the objective function and the constraint are parallel. The necessity of this condition is illustrated in Figure 41. Here $z = 3$ is the optimal z-value when we try to maximize $f(x_1, x_2)$, subject to $g(x_1, x_2) = 0$. At the optimal point in Figure 41, $\nabla f = \lambda \nabla g$, where $\lambda < 0$.

To see why (17) must hold for an optimal solution to (12), let's consider the following NLP:

$$\max z = f(x_1, x_2, x_3)$$
$$\text{s.t.} \quad g_1(x_1, x_2, x_3) = 0 \tag{18}$$
$$g_2(x_1, x_2, x_3) = 0$$

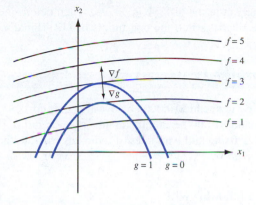

FIGURE 41
One-Constraint
Example of (17)

Suppose $\bar{x} = (\bar{x}_1, \bar{x}_2, \bar{x}_3)$ is an optimal solution to (18). We claim that for any $c \neq 0$, the following system of equations can have no solution (all gradients are evaluated at $\bar{x}$).

$$\begin{bmatrix} \nabla g_1 \\ \nabla g_2 \\ \nabla f \end{bmatrix} \begin{bmatrix} d_1 \\ d_2 \\ d_3 \end{bmatrix} = \begin{bmatrix} 0 \\ 0 \\ c \end{bmatrix} \tag{19}$$

To see why (19) can have no solution, suppose that it has a solution for some $c > 0$. [If (19) has a solution with $c < 0$, then a similar argument holds.] This solution defines a direction $\mathbf{d}$ in three dimensions. If we move in the direction $\mathbf{d}$ a small distance ϵ away from $\bar{x}$ we can find a feasible point $\bar{x} + \epsilon\mathbf{d}$ for (18) that has a larger z-value than $\bar{x}$. This would contradict the optimality of $\bar{x}$. To see that $\bar{x} + \epsilon\mathbf{d}$ is feasible in (18), note that for $i = 1, 2$ (19) implies that $g_i(\bar{x} + \epsilon\mathbf{d})$ is approximately equal to

$$g_i(\bar{x}) + \sum_{j=1}^{j=3} \frac{\partial g_i(\bar{x})}{\partial x_j} \left(\frac{\epsilon d_j}{\| \mathbf{d} \|} \right) = g_i(\bar{x}) = 0$$

Also $f(\bar{x} + \epsilon\mathbf{d})$ is approximately equal to

$$f(\bar{x}) + \sum_{j=1}^{j=3} \frac{\partial f}{\partial x_j} \left(\frac{\epsilon d_j}{\| \mathbf{d} \|} \right) = f(\bar{x}) + c\epsilon / \| \mathbf{d} \| > f(\bar{x})$$

This means that if $\bar{x}$ solves (18), then (19) can have no solution for $c \neq 0$. From Section 2.4, we know that (19) can have no solution if and only if the rank of the matrix on the left side of (19) is less than or equal to 2. This means that ∇f, ∇g_1, ∇g_2 at $\bar{x}$ are linearly dependent vectors. Thus, a nontrivial linear combination of ∇f, ∇g_1, and ∇g_2 must add up to the zero vector. If we assume that ∇g_1 and ∇g_2 are linearly independent (the usual case), then (17) must hold.

Lagrange Multipliers and Sensitivity Analysis

The Lagrange multipliers λ_i can be used in sensitivity analysis. If the right-hand side of the ith constraint is increased by a small amount Δb_i (in either a maximization or minimization problem), then the optimal z-value for (12) will increase by approximately $\sum_{i=1}^{i=m} (\Delta b_i)\lambda_i$. This result is proven in Problem 9 of this section. In particular, if we increase the right-hand side of only constraint i by Δb_i, then the optimal z-value of (12) will increase by $(\Delta b_i)\lambda_i$.

The two examples that follow illustrate the use of Lagrange multipliers. In most cases, the easiest way to find a point $(\bar{x}_1, \bar{x}_2, \ldots, \bar{x}_n, \bar{\lambda}_1, \bar{\lambda}_2, \ldots, \bar{\lambda}_m)$ satisfying (16) is to first solve for $\bar{x}_1, \bar{x}_2, \ldots, \bar{x}_n$ in terms of $\bar{\lambda}_1, \bar{\lambda}_2, \ldots, \bar{\lambda}_m$. Then determine the values of the $\bar{\lambda}_i$'s by substituting these relations into the constraints of (12). Finally, use the values of the $\bar{\lambda}_i$'s to determine $\bar{x}_1, \bar{x}_2, \ldots, \bar{x}_n$.

EXAMPLE 30   Lagrange Multiplier in Advertising

A company is planning to spend $10,000 on advertising. It costs $3,000 per minute to advertise on television and $1,000 per minute to advertise on radio. If the firm buys x minutes of television advertising and y minutes of radio advertising, then its revenue in thousands of dollars is given by $f(x, y) = -2x^2 - y^2 + xy + 8x + 3y$. How can the firm maximize its revenue?

Solution   We want to solve the following NLP:

$$\max z = -2x^2 - y^2 + xy + 8x + 3y$$
$$\text{s.t.} \quad 3x + y = 10$$

Then $L(x, y, \lambda) = -2x^2 - y^2 + xy + 8x + 3y + \lambda(10 - 3x - y)$. We set

$$\frac{\partial L}{\partial x} = \frac{\partial L}{\partial y} = \frac{\partial L}{\partial \lambda} = 0$$

This yields

$$\frac{\partial L}{\partial x} = -4x + y + 8 - 3\lambda = 0 \tag{20}$$

$$\frac{\partial L}{\partial y} = -2y + x + 3 - \lambda = 0 \tag{21}$$

$$\frac{\partial L}{\partial \lambda} = 10 - 3x - y = 0 \tag{22}$$

Observe that $10 - 3x - y = 0$ reduces to the constraint $3x + y = 10$. Equation (20) yields $y = 3\lambda - 8 + 4x$, and (21) yields $x = \lambda - 3 + 2y$. Thus, $y = 3\lambda - 8 + 4(\lambda - 3 + 2y) = 7\lambda - 20 + 8y$, or

$$y = \frac{20}{7} - \lambda \tag{23}$$
$$x = \lambda - 3 + 2(\frac{20}{7} - \lambda) = \frac{19}{7} - \lambda \tag{24}$$

Substituting (23) and (24) into (22) yields $10 - 3(\frac{19}{7} - \lambda) - (\frac{20}{7} - \lambda) = 0$, or $4\lambda - 1 = 0$, or $\lambda = \frac{1}{4}$. Then (23) and (24) yield

$$\bar{y} = \frac{20}{7} - \frac{1}{4} = \frac{73}{28}$$
$$\bar{x} = \frac{19}{7} - \frac{1}{4} = \frac{69}{28}$$

The Hessian for $f(x, y)$ is

$$H(x, y) = \begin{bmatrix} -4 & 1 \\ 1 & -2 \end{bmatrix}$$

Since each first-order principal minor is negative, and $H_2(x, y) = 7 > 0$, $f(x, y)$ is concave. The constraint is linear, so Theorem 8 shows that the Lagrange multiplier method does yield the optimal solution to the NLP.

Thus, the firm should purchase $\frac{69}{28}$ minutes of television time and $\frac{73}{28}$ minutes of radio time. Since $\lambda = \frac{1}{4}$, spending an extra Δ (thousands) (for small Δ) would increase the firm's revenues by approximately 0.25Δ (thousands).

In general, if the firm had a dollars to spend on advertising, then it could be shown that $\lambda = \frac{11-a}{4}$ (see Problem 1 at the end of this section). We see that as more money is spent on advertising, the increase to revenue for each additional advertising dollar becomes smaller.

EXAMPLE 31 **Lagrange Multiplier and Optimal Solution**

Given numbers $x_1, x_2, \ldots, x_n$, show that

$$n \sum_{i=1}^{i=n} x_i^2 \geq \left(\sum_{i=1}^{i=n} x_i \right)^2$$

with equality holding only if $x_1 = x_2 = \cdots = x_n$.

Solution Suppose that $x_1 + x_2 + \cdots + x_n = c$. Consider the NLP

$$\min z = \sum_{i=1}^{i=n} x_i^2$$

$$\text{s.t.} \quad \sum_{i=1}^{i=n} x_i = c \tag{25}$$

To solve (25), we form

$$L(x_1, x_2, \ldots, x_n, \lambda) = x_1^2 + x_2^2 + \cdots + x_n^2 + \lambda(c - x_1 - x_2 - \cdots - x_n)$$

Then to solve (25) we need to find $(x_1, x_2, \ldots, x_n, \lambda)$ that satisfy

$$\frac{\partial L}{\partial x_i} = 2x_i - \lambda = 0 \quad (i = 1, 2, \ldots, n) \quad \text{and}$$

$$\frac{\partial L}{\partial \lambda} = c - x_1 - x_2 - \cdots - x_n = 0$$

From $\frac{\partial L}{\partial x_i} = 0$, we obtain $2\bar{x}_1 = 2\bar{x}_2 = \cdots = 2\bar{x}_n = \bar{\lambda}$, or $x_i = \frac{\bar{\lambda}}{2}$. From $\frac{\partial L}{\partial \lambda} = 0$, we obtain $c - \frac{n\bar{\lambda}}{2} = 0$, or $\bar{\lambda} = \frac{2c}{n}$. The objective function is convex (it is the sum of n convex functions), and the constraint is linear. Thus, Theorem 8′ shows that the Lagrange multiplier method does yield an optimal solution to (25); it has

$$\bar{x}_i = \frac{\left(\dfrac{2c}{n} \right)}{2} = \frac{c}{n} \quad \text{and} \quad z = n \left(\frac{c^2}{n^2} \right) = \frac{c^2}{n}$$

Thus, if

$$\sum_{i=1}^{i=n} x_i = c$$

then

$$n \sum_{i=1}^{i=n} x_i^2 \geq n \left(\frac{c^2}{n} \right) = \left(\sum_{i=1}^{i=n} x_i \right)^2$$

with equality holding if and only if $x_1 = x_2 = \cdots = x_n$.

If we are trying to maximize a function $f(x_1, x_2, \ldots, x_n)$ that is a product of several functions, then it is often easier to maximize $\ln [f(x_1, x_2, \ldots, x_n)]$. Since $\ln$ is an increasing function, we know that any x^* maximizing $\ln [f(x_1, x_2, \ldots, x_n)]$ over any set of possible values for $(x_1, x_2, \ldots, x_n)$ will also maximize $f(x_1, x_2, \ldots, x_n)$ over the same set of possible values for $(x_1, x_2, \ldots, x_n)$. See Problem 2 for an application of this idea.

Solving NLP with Equality Constraints on LINGO

Adv.lng

If the hypotheses of Theorem 8 or Theorem 8′ hold for a problem, LINGO will find the optimal solution to the NLP. You will receive the messages OPTIMAL TO TOLERANCES and DUAL CONDITIONS: SATISFIED. "Optimal to Tolerances" means that LINGO is sure that it has found a local extremum. "Dual Conditions: Satisfied" means that LINGO is sure that the point it has found satisfies (16). Figure 42 (file Adv.lng) contains the LINGO printout for Example 28.

Interpretation of the LINGO Price Column

For a maximization problem, the LINGO PRICE column yields the Lagrange multiplier for each constraint. Thus, if the right-hand side of Constraint i in a maximization problem is increased by a small amount Δ, then the optimal z-value is increased by approximately Δ (PRICE for Constraint i). The PRICE column in Figure 42 implies that in Example 30 spending an extra Δ thousand dollars on advertising will increase revenues by approximately $\$0.25\Delta$ (thousands).

For a minimization problem, the LINGO PRICE column yields the negative of the Lagrange multiplier for each constraint. Thus, if the right-hand side of Constraint i in a minimization problem is increased by a small amount Δ, then the optimal z-value will increase by approximately $\Delta(-\text{PRICE for Constraint } i)$.

```
MODEL:
  1) MAX= - 2 * X ^ 2 - Y ^ 2 + X * Y + 8 * X + 3 * Y ;
  2) 3 * X + Y = 10 ;
  3) X > 0 ;
  4) Y > 0 ;
END


SOLUTION STATUS:  OPTIMAL TO TOLERANCES.  DUAL CONDITIONS:  SATISFIED.

            OBJECTIVE FUNCTION VALUE

        1)         15.017855

    VARIABLE         VALUE          REDUCED COST
        X          2.464283            .000000
        Y          2.607140            .000003

    ROW      SLACK OR SURPLUS          PRICE
     2)           -.000010            .249996
     3)           2.464283            .000000
     4)           2.607140            .000000
```

FIGURE 42
Optimal Solution for Example 28

PROBLEMS

Group A

1 For Example 30, show that if a dollars are available for advertising, then an extra dollar spent on advertising will increase revenues by approximately $\frac{11-a}{4}$.

2 It costs me $2 to purchase an hour of labor and $1 to purchase a unit of capital. If L hours of labor and K units of capital are available, then $L^{2/3}K^{1/3}$ machines can be produced. If I have $10 to purchase labor and capital, what is the maximum number of machines that can be produced?

3 In Problem 2, what is the minimum cost method of producing 6 machines?

4 A beer company has divided Bloomington into two territories. If x_1 dollars are spent on promotion in territory 1, then $6x_1^{1/2}$ cases of beer can be sold there; and if x_2 dollars are spent on promotion in territory 2, then $4x_2^{1/2}$ cases of beer can be sold there. Each case of beer sold in territory 1 sells for $10 and incurs $5 in shipping and production costs.

Each case of beer sold in territory 2 sells for $9 and incurs $4 in shipping and production costs. A total of $100 is available for promotion. How can the beer company maximize profits? If an extra dollar could be spent on promotion, by approximately how much would profits increase? By how much would revenues increase?

Group B

5 We must invest all our money in two stocks: x and y. The variance of the annual return on one share of stock x is var x, and the variance of the annual return on one share of stock y is var y. Assume that the covariance between the annual return for one share of x and one share of y is cov(x, y). If we invest a% of our money in stock x and b% in stock y, then the variance of our return is given by a^2var $x + b^2$var $y + 2ab$ cov(x, y). We want to minimize the variance of the return on our invested money. What percentage of the money should be invested in each stock?

6 As in Problem 5, assume that we must determine the percentage of our money that is invested in stocks x and y. A choice of a and b is called a *portfolio*. A portfolio is efficient if there exists no other portfolio whose return has a higher mean return and lower variance, or a higher mean return and the same variance, or a lower variance with the same mean return. Let $\bar{x}$ be the mean return on stock x and $\bar{y}$ be the mean return on stock y. Consider the following NLP:

$$\max z = c[a\bar{x} + b\bar{y}]$$
$$- (1 - c)[a^2\text{var } x + b^2\text{var } y$$
$$+ 2ab\text{cov}(x, y)]$$
s.t. $\quad a + b = 1$
$$a, b \geq 0$$

Suppose that $1 > c > 0$. Show that any solution to this NLP is an efficient portfolio.

7 Suppose product i ($i = 1, 2$) costs $\$c_i$ per unit. If $x_i(i = 1, 2)$ units of products 1 and 2 are purchased, then a utility $x_1^a x_2^{1-a}$ ($0 < a < 1$) is received.

   a If $\$d$ are available to purchase products 1 and 2, how many of each type should be purchased?

   b Show that an increase in the cost of product i decreases the number of units of product i that should be purchased.

   c Show that an increase in the cost of product i does not change the number of units of the other product that should be purchased.

8 Suppose that a cylindrical soda can must have a volume of 26 cu in. If the soda company wants to minimize the surface area of the soda can, what should be the ratio of the height of the can to the radius of the can? (*Hint:* The volume of a right circular cylinder is $\pi r^2 h$, and the surface area of a right circular cylinder is $2\pi r^2 + 2\pi rh$, where $r =$ the radius of the cylinder and $h =$ the height of the cylinder.)

9 Show that if the right-hand side of the ith constraint is increased by a small amount Δb_i (in either a maximization or minimization problem), then the optimal z-value for (11) will increase by approximately $\sum_{i=1}^{i=m}(\Delta b_i)\lambda_i$.

12.9 The Kuhn–Tucker Conditions

In this section, we discuss necessary and sufficient conditions for $\bar{x} = (\bar{x}_1, \bar{x}_2, \ldots, \bar{x}_n)$ to be an optimal solution for the following NLP:

$$\max \text{ (or min) } f(x_1, x_2, \ldots, x_n)$$
$$\text{s.t.} \quad g_1(x_1, x_2, \ldots, x_n) \leq b_1$$
$$g_2(x_1, x_2, \ldots, x_n) \leq b_2 \quad \quad \text{(26)}$$
$$\vdots$$
$$g_m(x_1, x_2, \ldots, x_n) \leq b_m$$

To apply the results of this section, all the NLP's constraints must be $\leq$ constraints. A constraint of the form $h(x_1, x_2, \ldots, x_n) \geq b$ must be rewritten as $-h(x_1, x_2, \ldots, x_n) \leq -b$. For example, the constraint $2x_1 + x_2 \geq 2$ should be rewritten as $-2x_1 - x_2 \leq -2$. A constraint of the form $h(x_1, x_2, \ldots, x_n) = b$ must be replaced by $h(x_1, x_2, \ldots, x_n) \leq b$ and $-h(x_1, x_2, \ldots, x_n) \leq -b$. For example, $2x_1 + x_2 = 2$ would be replaced by $2x_1 + x_2 \leq 2$ and $-2x_1 - x_2 \leq -2$.

Theorems 9 and 9′ give conditions (the **Kuhn–Tucker**, or **KT**, **conditions**) that are necessary for a point $\bar{x} = (\bar{x}_1, \bar{x}_2, \ldots, \bar{x}_n)$ to solve (26). The partial derivative of a function f with respect to a variable x_j evaluated at $\bar{x}$ is written

$$\frac{\partial f(\bar{x})}{\partial x_j}$$

For the theorems of this section to hold, the functions $g_1, g_2, \ldots, g_m$ must satisfy certain regularity conditions (usually called **constraint qualifications**). We will briefly discuss one constraint qualification at the end of the section. [For a detailed discussion of constraint qualifications we refer the reader to Chapter 5 of Bazaraa and Shetty (1993).]

When the constraints are linear, these regularity assumptions are always satisfied. In other situations (particularly when some of the constraints are equality constraints), the regularity conditions may not be satisfied. We assume that all problems we consider satisfy these regularity conditions.

THEOREM 9

Suppose (26) is a maximization problem. If $\bar{x} = (\bar{x}_1, \bar{x}_2, \ldots, \bar{x}_n)$ is an optimal solution to (26), then $\bar{x} = (\bar{x}_1, \bar{x}_2, \ldots, \bar{x}_n)$ must satisfy the m constraints in (26), and there must exist multipliers $\lambda_1, \lambda_2, \ldots, \lambda_m$ satisfying

$$\frac{\partial f(\bar{x})}{\partial x_j} - \sum_{i=1}^{i=m} \bar{\lambda}_i \frac{\partial g_i(\bar{x})}{\partial x_j} = 0 \qquad (j = 1, 2, \ldots, n) \tag{27}$$

$$\bar{\lambda}_i[b_i - g_i(\bar{x})] = 0 \qquad (i = 1, 2, \ldots, m) \tag{28}$$

$$\bar{\lambda}_i \geq 0 \qquad (i = 1, 2, \ldots, m) \tag{29}$$

THEOREM 9'

Suppose (26) is a minimization problem. If $\bar{x} = (\bar{x}_1, \bar{x}_2, \ldots, \bar{x}_n)$ is an optimal solution to (26), then $\bar{x} = (\bar{x}_1, \bar{x}_2, \ldots, \bar{x}_n)$ must satisfy the m constraints in (26), and there must exist multipliers $\lambda_1, \lambda_2, \ldots, \lambda_m$ satisfying

$$\frac{\partial f(\bar{x})}{\partial x_j} + \sum_{i=1}^{i=m} \bar{\lambda}_i \frac{\partial g_i(\bar{x})}{\partial x_j} = 0 \qquad (j = 1, 2, \ldots, n)$$

$$\bar{\lambda}_i[b_i - g_i(\bar{x})] = 0 \qquad (i = 1, 2, \ldots, m)$$

$$\bar{\lambda}_i \geq 0 \qquad (i = 1, 2, \ldots, m)$$

Like the Lagrange multipliers of the preceding section, the multiplier $\bar{\lambda}_i$ associated with the K–T conditions may be thought of as the shadow price for the ith constraint in (26). Suppose (26) is a maximization problem. If the right-hand side of the ith constraint is increased from b_i to $b_i + \Delta$ (for Δ small), the optimal objective function value will increase by approximately $\Delta\lambda_i$. Suppose (26) is a minimization problem. If the right-hand side of the ith constraint is increased from b_i to $b_i + \Delta$ (for Δ small), then the optimal objective function value is decreased by $\Delta\bar{\lambda}_i$.

Bearing in mind this interpretation of the multipliers as shadow prices, we may interpret (27)–(29) for a max problem. Suppose we consider each constraint in (26) to be a resource-usage constraint. That is, at $\bar{x} = (\bar{x}_1, \bar{x}_2, \ldots, \bar{x}_n)$ we use $g_i(\bar{x}_1, \bar{x}_2, \ldots, \bar{x}_n)$ units of resource i, and b_i units of resource i are available. If we increase the value of x_j by a small amount Δ, then the value of the objective function increases by

$$\frac{\partial f(\bar{x})}{\partial x_j} \Delta$$

Changing the value of x_j to $\bar{x}_j + \Delta$ also changes the ith constraint to

$$g_i(\bar{x}) + \frac{\partial g_i(\bar{x})}{\partial x_j} \Delta \leq b_i \quad \text{or} \quad g_i(\bar{x}) \leq b_i - \frac{\partial g_i(\bar{x})}{\partial x_j} \Delta$$

Thus, increasing x_j by Δ has the effect of increasing the right-hand side of the ith constraint by

$$-\frac{\partial g_i(\bar{x})}{\partial x_j} \Delta$$

These changes in the right-hand sides of the constraints will increase the value of z by approximately

$$-\Delta \sum_{i=1}^{i=m} \bar{\lambda}_i \frac{\partial g_i(\bar{x})}{\partial x_j}$$

In total, the approximate change in z due to increasing x_j by Δ is

$$\Delta \left[\frac{\partial f(\bar{x})}{\partial x_j} - \sum_{i=1}^{i=m} \bar{\lambda}_i \frac{\partial g_i(\bar{x})}{\partial x_j} \right]$$

If the term in brackets is larger than zero, we can increase f by choosing $\Delta > 0$. On the other hand, if this term is smaller than zero, we can increase f by choosing $\Delta < 0$. Thus, for $\bar{x}$ to be optimal, (27) must hold.

Condition (28) is a generalization of the complementary slackness conditions for LPs discussed in Section 6.10. Condition (28) implies that

$$\text{If } \bar{\lambda}_i > 0, \quad \text{then} \quad g_i(\bar{x}) = b_i \quad (i\text{th constraint binding}) \tag{28'}$$

$$\text{If } g_i(\bar{x}) < b_i, \quad \text{then} \quad \bar{\lambda}_i = 0 \tag{28''}$$

Suppose the constraint $g_i(x_1, x_2, \ldots, x_n) \leq b_i$ is a resource-usage constraint representing the fact that at most b_i units of the ith resource can be used. Then (28') states that if an additional unit of the resource associated with the ith constraint is to have any value, then the current optimal solution must use all b_i units of the ith resource currently available. On the other hand, (28'') states that if some of the ith resource currently available is unused, then additional amounts of the ith resource have no value.

If for $\Delta > 0$, we increase the right-hand side of the ith constraint from b_i to $b_i + \Delta$, then the optimal objective function value must increase or stay the same, because the increase adds points to the problem's feasible region. Increasing the right-hand side of the ith constraint by Δ increases the optimal objective function value by $\Delta \bar{\lambda}_i$, so it must be that $\bar{\lambda}_i \geq 0$. This is why (29) is included in the K–T conditions.

In many situations, the K–T conditions are applied to NLPs in which the variables must be non-negative. For example, we may want to use the K–T conditions to find the optimal solution to

$$\max (\text{or min}) \ z = f(x_1, x_2, \ldots, x_n)$$
$$\text{s.t.} \quad g_1(x_1, x_2, \ldots, x_n) \leq b_1$$
$$g_2(x_1, x_2, \ldots, x_n) \leq b_2$$
$$\vdots$$
$$g_m(x_1, x_2, \ldots, x_n) \leq b_m \tag{30}$$
$$-x_1 \leq 0$$
$$-x_2 \leq 0$$
$$\vdots$$
$$-x_n \leq 0$$

If we associate multipliers $\mu_1, \mu_2, \ldots, \mu_n$ with the non-negativity constraints in (30), Theorems 9 and 9′ reduce to Theorems 10 and 10′.

THEOREM 10

Suppose (30) is a maximization problem. If $\bar{x} = (\bar{x}_1, \bar{x}_2, \ldots, \bar{x}_n)$ is an optimal solution to (30), then $\bar{x} = (\bar{x}_1, \bar{x}_2, \ldots, \bar{x}_n)$ must satisfy the constraints in (30) and there must exist multipliers $\bar{\lambda}_1, \bar{\lambda}_2, \ldots, \bar{\lambda}_m, \bar{\mu}_1, \bar{\mu}_2, \ldots, \bar{\mu}_n$ satisfying

$$\frac{\partial f(\bar{x})}{\partial x_j} - \sum_{i=1}^{i=m} \bar{\lambda}_i \frac{\partial g_i(\bar{x})}{\partial x_j} + \mu_j = 0 \qquad (j = 1, 2, \ldots, n) \tag{31}$$

$$\bar{\lambda}_i[b_i - g_i(\bar{x})] = 0 \qquad (i = 1, 2, \ldots, m) \tag{32}$$

$$\left[\frac{\partial f(\bar{x})}{\partial x_j} - \sum_{i=1}^{i=m} \bar{\lambda}_i \frac{\partial g_i(\bar{x})}{\partial x_j} \right] \bar{x}_j = 0 \qquad (j = 1, 2, \ldots, n) \tag{33}$$

$$\bar{\lambda}_i \geq 0 \qquad (i = 1, 2, \ldots, m) \tag{34}$$

$$\bar{\mu}_j \geq 0 \qquad (j = 1, 2, \ldots, n) \tag{35}$$

Because $\bar{\mu}_j \geq 0$, (31) is equivalent to

$$\frac{\partial f(\bar{x})}{\partial x_j} - \sum_{i=1}^{i=m} \bar{\lambda}_i \frac{\partial g_i(\bar{x})}{\partial x_j} \leq 0 \qquad (j = 1, 2, \ldots, n) \tag{31′}$$

Then (31)–(34), the K–T conditions for a maximization problem with non-negativity constraints, may be rewritten as

$$\frac{\partial f(\bar{x})}{\partial x_j} - \sum_{i=1}^{i=m} \bar{\lambda}_i \frac{\partial g_i(\bar{x})}{\partial x_j} \leq 0 \qquad (j = 1, 2, \ldots, n) \tag{31′}$$

$$\bar{\lambda}_i[b_i - g_i(\bar{x})] = 0 \qquad (i = 1, 2, \ldots, m) \tag{32′}$$

$$\left[\frac{\partial f(\bar{x})}{\partial x_j} - \sum_{i=1}^{i=m} \bar{\lambda}_i \frac{\partial g_i(\bar{x})}{\partial x_j} \right] \bar{x}_j = 0 \qquad (j = 1, 2, \ldots, n) \tag{33′}$$

$$\bar{\lambda}_i \geq 0 \qquad (i = 1, 2, \ldots, m) \tag{34′}$$

THEOREM 10′

Suppose (30) is a minimization problem. If $\bar{x} = (\bar{x}_1, \bar{x}_2, \ldots, \bar{x}_n)$ is an optimal solution to (30), then $\bar{x} = (\bar{x}_1, \bar{x}_2, \ldots, \bar{x}_n)$ must satisfy the constraints in (30), and there must exist multipliers $\bar{\lambda}_1, \bar{\lambda}_2, \ldots, \bar{\lambda}_m, \bar{\mu}_1, \bar{\mu}_2, \ldots, \bar{\mu}_n$ satisfying

$$\frac{\partial f(\bar{x})}{\partial x_j} + \sum_{i=1}^{i=m} \bar{\lambda}_i \frac{\partial g_i(\bar{x})}{\partial x_j} - \mu_j = 0 \qquad (j = 1, 2, \ldots, n) \tag{36}$$

$$\bar{\lambda}_i[b_i - g_i(\bar{x})] = 0 \qquad (i = 1, 2, \ldots, m) \tag{37}$$

$$\left[\frac{\partial f(\bar{x})}{\partial x_j} + \sum_{i=1}^{i=m} \bar{\lambda}_i \frac{\partial g_i(\bar{x})}{\partial x_j} \right] \bar{x}_j = 0 \qquad (j = 1, 2, \ldots, n) \tag{38}$$

$$\bar{\lambda}_i \geq 0 \qquad (i = 1, 2, \ldots, m) \tag{39}$$

$$\bar{\mu}_j \geq 0 \qquad (j = 1, 2, \ldots, n) \tag{40}$$

Because $\bar{\mu}_j \geq 0$, (36) may be written as

$$\frac{\partial f(\bar{x})}{\partial x_j} + \sum_{i=1}^{i=m} \bar{\lambda}_i \frac{\partial g_i(\bar{x})}{\partial x_j} \geq 0 \tag{36'}$$

Then (36)–(39), the K–T conditions for a minimization problem with non-negativity constraints, may be rewritten as

$$\frac{\partial f(\bar{x})}{\partial x_j} + \sum_{i=1}^{i=m} \bar{\lambda}_i \frac{\partial g_i(\bar{x})}{\partial x_j} \geq 0 \qquad (j = 1, 2, \ldots, n) \tag{36'}$$

$$\bar{\lambda}_i[b_i - g_i(\bar{x})] = 0 \qquad (i = 1, 2, \ldots, m) \tag{37'}$$

$$\left[\frac{\partial f(\bar{x})}{\partial x_j} + \sum_{i=1}^{i=m} \bar{\lambda}_i \frac{\partial g_i(\bar{x})}{\partial x_j} \right] \bar{x}_j = 0 \qquad (j = 1, 2, \ldots, n) \tag{38'}$$

$$\bar{\lambda}_i \geq 0 \qquad (i = 1, 2, \ldots, m) \tag{39'}$$

Theorems 9, 9', 10, and 10' give conditions that are *necessary* for a point $\bar{x} = (\bar{x}_1, \bar{x}_2, \ldots, \bar{x}_n)$ to be an optimal solution to (26) or (30). The following two theorems give conditions that are *sufficient* for $\bar{x} = (\bar{x}_1, \bar{x}_2, \ldots, \bar{x}_n)$ to be an optimal solution to (26) or (30) (see Bazaraa and Shetty (1993)).

THEOREM 11

Suppose (26) is a maximization problem. If $f(x_1, x_2, \ldots, x_n)$ is a concave function and $g_1(x_1, x_2, \ldots, x_n), \ldots, g_m(x_1, x_2, \ldots, x_n)$ are convex functions, then any point $\bar{x} = (\bar{x}_1, \bar{x}_2, \ldots, \bar{x}_n)$ satisfying the hypotheses of Theorem 9 is an optimal solution to (26). Also, if (30) is a maximization problem, $f(x_1, x_2, \ldots, x_n)$ is a concave function, and $g_1(x_1, x_2, \ldots, x_n), \ldots, g_m(x_1, x_2, \ldots, x_n)$ are convex functions, then any point $\bar{x} = (\bar{x}_1, \bar{x}_2, \ldots, \bar{x}_n)$ satisfying the hypotheses of Theorem 10 is an optimal solution to (30).

THEOREM 11'

Suppose (26) is a minimization problem. If $f(x_1, x_2, \ldots, x_n)$ is a convex function and $g_1(x_1, x_2, \ldots, x_n), \ldots, g_m(x_1, x_2, \ldots, x_n)$ are convex functions, then any point $\bar{x} = (\bar{x}_1, \bar{x}_2, \ldots, \bar{x}_n)$ satisfying the hypotheses of Theorem 9' is an optimal solution to (26). Also, if (30) is a minimization problem, $f(x_1, x_2, \ldots, x_n)$ is a convex function, and $g_1(x_1, x_2, \ldots, x_n), \ldots, g_m(x_1, x_2, \ldots, x_n)$ are convex functions, then any point $\bar{x} = (\bar{x}_1, \bar{x}_2, \ldots, \bar{x}_n)$ satisfying the hypotheses of Theorem 10' is an optimal solution to (30).

REMARK The reason that the hypotheses of Theorems 11 and 11' require that each $g_i(x_1, x_2, \ldots, x_n)$ be convex is that this ensures the feasible region for (26) or (30) is a convex set (see Problem 21 of Section 12.3).

Geometrical Interpretation of Kuhn–Tucker Conditions

It is easy to show that conditions (27)–(29) of Theorem 9 will hold at a point $\bar{x}$ if and only if ∇f is a non-negative linear combination of $\nabla g_1, \nabla g_2, \ldots, \nabla g_m$, and the weight multiplying ∇g_i in this linear combination equals 0 if the ith constraint in (26) is nonbinding.

In short, (27)–(29) are equivalent to the existence of $\lambda_i \geq 0$ such that

$$\nabla f(\bar{x}) = \sum_{i=1}^{i=m} \lambda_i \nabla g_i(\bar{x}) \tag{41}$$

and each constraint that is nonbinding at $\bar{x}$ has $\lambda_i = 0$.

Figures 43 and 44 illustrate (41). In Figure 43, we are trying to solve (the feasible region is shaded)

$$\min z = f(x_1, x_2)$$
$$\text{s.t.} \qquad g_1(x_1, x_2) \leq 0$$
$$g_2(x_1, x_2) \leq 0$$

At $\bar{x}$, (41) holds with both constraints binding and we have $\lambda_1 > 0$ and $\lambda_2 > 0$. In Figure 44, we are again trying to solve (feasible region is again shaded)

$$\min z = f(x_1, x_2)$$
$$\text{s.t.} \qquad g_1(x_1, x_2) \leq 0$$
$$g_2(x_1, x_2) \leq 0$$

Here, the second constraint is nonbinding so (41) must hold with $\lambda_2 = 0$.

The following two examples illustrate the use of the K–T conditions.

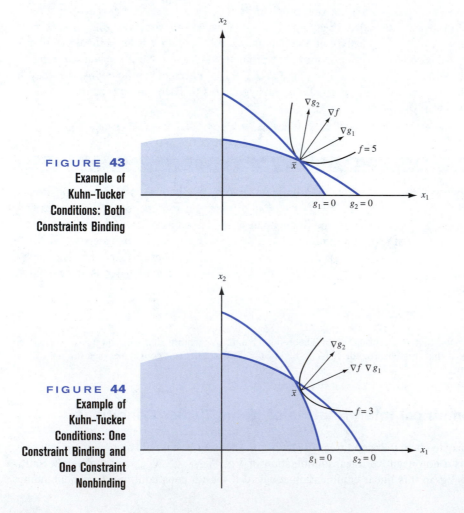

FIGURE 43
Example of Kuhn–Tucker Conditions: Both Constraints Binding

FIGURE 44
Example of Kuhn–Tucker Conditions: One Constraint Binding and One Constraint Nonbinding

EXAMPLE **32**   **Interpretation of Kuhn–Tucker Conditions**

Describe the optimal solution to

$$\max f(x)$$

$$\text{s.t.} \quad a \leq x \leq b \tag{42}$$

Solution   From Section 12.4, we know {assuming that $f'(x)$ exists for all x on the interval $[a, b]$} that the optimal solution to this problem must occur at a [with $f'(a) \leq 0$], at b [with $f'(b) \geq 0$], or at a point having $f'(x) = 0$. How do the K–T conditions yield these three cases?

We write (42) as

$$\max f(x)$$

$$\text{s.t.} \quad -x \leq -a$$

$$x \leq b$$

Then (27)–(29) yield

$$f'(x) + \lambda_1 - \lambda_2 = 0 \tag{43}$$

$$\lambda_1(-a + x) = 0 \tag{44}$$

$$\lambda_2(b - x) = 0 \tag{45}$$

$$\lambda_1 \geq 0 \tag{46}$$

$$\lambda_2 \geq 0 \tag{47}$$

In using the K–T conditions to solve NLPs, it is useful to note that each multiplier λ_i must satisfy $\lambda_i = 0$ or $\lambda_i > 0$. Thus, in attempting to find values of x, λ_1, and λ_2 that satisfy (43)–(47), we must consider the following four cases:

Case 1 $\lambda_1 = \lambda_2 = 0$. From (43), we obtain the case $f'(\bar{x}) = 0$.

Case 2 $\lambda_1 = 0, \lambda_2 > 0$. Because $\lambda_2 > 0$, (45) yields $\bar{x} = b$. Then (43) yields $f'(b) = \lambda_2$, and because $\lambda_2 > 0$, we obtain the case where $f'(b) > 0$.

Case 3 $\lambda_1 > 0, \lambda_2 = 0$. Because $\lambda_1 > 0$, (44) yields $\bar{x} = a$. Then (43) yields the case where $f'(a) = -\lambda_1 < 0$.

Case 4 $\lambda_1 > 0, \lambda_2 > 0$. From (44) and (45), we obtain $\bar{x} = a$ and $\bar{x} = b$. This contradiction indicates that Case 4 cannot occur.

The constraints are linear, so Theorem 11 shows that if $f(x)$ is concave, then (43)–(47) yield the optimal solution to (42).

EXAMPLE **33**   **Production Process**

A monopolist can purchase up to 17.25 oz of a chemical for $10/oz. At a cost of $3/oz, the chemical can be processed into an ounce of product 1; or, at a cost of $5/oz, the chemical can be processed into an ounce of product 2. If x_1 oz of product 1 are produced, it sells for a price of $30 - x_1$ per ounce. If x_2 oz of product 2 are produced, it sells for a price of $50 - 2x_2$ per ounce. Determine how the monopolist can maximize profits.

Solution   Let

$$x_1 = \text{ounces of product 1 produced}$$

$$x_2 = \text{ounces of product 2 produced}$$

$$x_3 = \text{ounces of chemical processed}$$

Then we want to solve the following NLP:

$$\max z = x_1(30 - x_1) + x_2(50 - 2x_2) - 3x_1 - 5x_2 - 10x_3$$

$$\text{s.t.} \quad x_1 + x_2 \le x_3 \quad \text{or} \quad x_1 + x_2 - x_3 \le 0 \tag{48}$$

$$x_3 \le 17.25$$

Of course, we should add the constraints $x_1, x_2, x_3 \ge 0$. However, because the optimal solution to (48) satisfies the non-negativity constraints, it also will be optimal for an NLP consisting of (48) with the non-negativity constraints.

Observe that the objective function in (48) is the sum of concave functions (and is therefore concave), and the constraints are convex (because they are linear). Thus, Theorem 11 shows that the K–T conditions are necessary and sufficient for (x_1, x_2, x_3) to be an optimal solution to (48). From Theorem 9, the K–T conditions become

$$30 - 2x_1 - 3 - \lambda_1 = 0 \tag{49}$$

$$50 - 4x_2 - 5 - \lambda_1 = 0 \tag{50}$$

$$-10 + \lambda_1 - \lambda_2 = 0 \tag{51}$$

$$\lambda_1(-x_1 - x_2 + x_3) = 0 \tag{52}$$

$$\lambda_2(17.25 - x_3) = 0 \tag{53}$$

$$\lambda_1 \ge 0 \tag{54}$$

$$\lambda_2 \ge 0 \tag{55}$$

As in the previous example, there are four cases to consider:

Case 1 $\lambda_1 = \lambda_2 = 0$. This case cannot occur, because (51) would be violated.

Case 2 $\lambda_1 = 0, \lambda_2 > 0$. If $\lambda_1 = 0$, then (51) implies $\lambda_2 = -10$. This would violate (55).

Case 3 $\lambda_1 > 0, \lambda_2 = 0$. From (51), we obtain $\lambda_1 = 10$. Now (49) yields $x_1 = 8.5$, and (50) yields $x_2 = 8.75$. From (52), we obtain $x_1 + x_2 = x_3$, so $x_3 = 17.25$. Thus, $\bar{x}_1 = 8.5$, $\bar{x}_2 = 8.75$, $\bar{x}_3 = 17.25$, $\bar{\lambda}_1 = 10$, $\bar{\lambda}_2 = 0$ satisfies the K–T conditions.

Case 4 $\lambda_1 > 0, \lambda_2 > 0$. Case 3 yields an optimal solution, so we need not consider Case 4.

Thus, the optimal solution to (48) is to buy 17.25 oz of the chemical and produce 8.5 oz of product 1 and 8.75 oz of product 2. For Δ small, $\bar{\lambda}_1 = 10$ indicates that if an extra Δ oz of the chemical were obtained at no cost, then profits would increase by 10Δ. (Can you see why?) From (51), we find that $\bar{\lambda}_2 = 0$. This implies that the right to purchase an extra Δ oz of the chemical would not increase profits. (Can you see why?)

Constraint Qualifications

Unless a constraint qualification or regularity condition is satisfied at an optimal point $\bar{x}$, the Kuhn–Tucker conditions may fail to hold at $\bar{x}$. There are many constraint qualifications, but we choose to discuss the Linear Independence Constraint Qualification: Let $\bar{x}$ be an optimal solution to NLP (26) or (30). If all g_i are continuous, and the gradients of all binding constraints (including any binding non-negativity constraints on $x_1, x_2, \ldots, x_n$) at $\bar{x}$ form a set of linearly independent vectors, then the Kuhn–Tucker conditions must hold at $\bar{x}$.

The following example shows that if the Linear Independence Constraint Qualification fails to hold, then the Kuhn–Tucker conditions may fail to hold at the optimal solution to an NLP.

EXAMPLE 34   **Necessity of Constraint Qualification**

Show that the Kuhn–Tucker conditions fail to hold at the optimal solution to the following NLP:

$$\max z = x_1$$
$$\text{s.t.} \quad x_2 - (1 - x_1)^3 \leq 0 \tag{56}$$
$$x_1 \geq 0, \, x_2 \geq 0$$

Solution   If $x_1 > 1$, then the first constraint in (56) implies that $x_2 < 0$. Thus, the optimal z-value for (56) cannot exceed 1. Because $x_1 = 1$ and $x_2 = 0$ is feasible and yields $z = 1$, $(1, 0)$ must be the optimal solution to NLP (56).

From Theorem 10 the following are two of the Kuhn–Tucker conditions for (56).

$$1 + 3\lambda_1(1 - x_1)^2 = -\mu_1 \tag{57}$$
$$\mu_1 \geq 0 \tag{58}$$

At the optimal solution $(1, 0)$, (57) implies $\mu_1 = -1$, which contradicts (58). Thus, the Kuhn–Tucker conditions are not satisfied at $(1, 0)$. We now show that at the point $(1, 0)$ the Linear Independence Constraint Qualification is violated. At $(1, 0)$ the constraints $x_2 - (1 - x_1)^3 \leq 0$ and $x_2 \geq 0$ are binding. Then

$$\nabla(x_2 - (1 - x_1)^3) = [0, 1]$$
$$\nabla(-x_2) = [0, -1]$$

Because $[0, 1] + [0, -1] = [0, 0]$, these gradients are linearly dependent. Thus, at $(1,0)$ the gradients of the binding constraints are linearly dependent and the constraint qualification is not satisfied.

Solving NLPs with Inequality (and Possibly Equality) Constraints on LINGO

LINGO does not require that all constraints be put in the form (26) or (30). Constraints may be input as less than or equal, equal, or greater than or equal to constraints. If your problem satisfies the hypotheses of Theorem 11 or Theorem 11′, then you can know that LINGO will find the optimal solution to your problem. You will know that LINGO has found a point satisfying the Kuhn–Tucker conditions if you see the message DUAL CONDITIONS: SATISFIED. For instance, we can be sure that LINGO would find the optimal solution to Example 33.

For the LINGO printouts given in Section 12.2 for Examples 9–11 we cannot be sure that LINGO has found the optimal solution to any of these problems. Example 9 (Figure 6) fails to satisfy the hypotheses of Theorem 11 because the left-hand side of rows 16–18 are not concave functions and the left-hand side of rows 19–21 are not convex functions. To see if LINGO has actually found the optimal solution to the NLP, we used the INIT command to input a wide variety of starting solutions (focusing on values of R, U, and P). We could not find any solution that was better than the solution in Figure 6, so we are fairly confident that GINO has found the optimal solution to Example 9. Similarly, Examples 10 and 11 do not satisfy the hypotheses of Theorem 11′ so we cannot be sure that LINGO has found an optimal solution to these problems (even though LINGO has found a point satisfying Kuhn–Tucker conditions!). Again, however, extensive use of the INIT command failed to turn up any better solutions, so we are fairly confident that LINGO has found the optimal solution to Examples 10 and 11.

If the right-hand side of Constraint i (the type of constraint does not matter) in an NLP is increased by a small amount Δ, then the optimal z-value is *improved* by approximately Δ(PRICE for Constraint i). Thus, in a maximization problem increasing the right-hand side of the ith constraint by a small amount Δb_i will result in the optimal z-value increasing by approximately Δb_i(price of Constraint i); in a minimization problem increasing the right-hand side of the ith constraint by a small amount Δb_i will result in the optimal z-value decreasing by approximately Δb_i(Price of Constraint i).

PROBLEMS

Group A

1[†] A power company faces demands during both peak and off-peak times. If a price of p_1 dollars per kilowatt-hour is charged during the peak time, customers will demand $60 - 0.5 \, p_1$ kwh of power. If a price of p_2 dollars is charged during the off-peak time, then customers will demand $40 - p_2$ kwh. The power company must have sufficient capacity to meet demand during both the peak and off-peak times. It costs \$10 per day to maintain each kilowatt-hour of capacity. Determine how the power company can maximize daily revenues less operating costs.

2 Use the K–T conditions to find the optimal solution to the following NLP:

$$\max z = x_1 - x_2$$
$$\text{s.t.} \quad x_1^2 + x_2^2 \leq 1$$

3 Consider the Giapetto problem of Section 3.1:

$$\max z = 3x_1 + 2x_2$$
$$\text{s.t.} \quad 2x_1 + x_2 \leq 100$$
$$x_1 + x_2 \leq 80$$
$$x_1 \qquad \leq 40$$
$$x_1 \qquad \geq 0$$
$$x_2 \geq 0$$

Find the K–T conditions for this problem and discuss their relation to the dual of the Giapetto LP and the complementary slackness conditions for the LP.

4 If the feasible region for (26) is bounded and contains its boundary points, then it can be shown that (26) has an optimal solution. Suppose that the regularity conditions are valid but that the hypotheses of Theorems 11 and 11′ are not valid. If we can prove that only one point satisfies the K–T conditions, then why must that point be the optimal solution to the NLP?

5 A total of 160 hours of labor are available each week at \$15/hour. Additional labor can be purchased at \$25/hour. Capital can be purchased in unlimited quantities at a cost of \$5/unit of capital. If K units of capital and L units of labor are available during a week, then $L^{1/2}K^{1/3}$ machines can be produced. Each machine sells for \$270. How can the firm maximize its weekly profits?

6 Use the K–T conditions to find the optimal solution to the following NLP:

$$\min z = (x_1 - 1)^2 + (x_2 - 2)^2$$
$$\text{s.t.} \quad -x_1 + x_2 = 1$$
$$x_1 + x_2 \leq 2$$
$$x_1, x_2 \geq 0$$

7 For Example 31, explain why $\bar{\lambda}_1 = 10$ and $\bar{\lambda}_2 = 0$. (*Hint:* Think about the economic principle that for each product produced, marginal revenue must equal marginal cost.)

8 Use the K–T conditions to find the optimal solution to the following NLP:

$$\max z = -x_1^2 - x_2^2 + 4x_1 + 6x_2$$
$$\text{s.t.} \quad x_1 + x_2 \leq 6$$
$$x_1 \qquad \leq 3$$
$$x_2 \leq 4$$
$$x_1, x_2 \geq 0$$

9 Use the K–T conditions to find the optimal solution to the following NLP:

$$\min z = e^{-x_1} + e^{-2x_2}$$
$$\text{s.t.} \quad x_1 + x_2 \leq 1$$
$$x_1, x_2 \geq 0$$

10 Use the K–T conditions to find the optimal solution to the following NLP:

$$\min z = (x_1 - 3)^2 + (x_2 - 5)^2$$
$$\text{s.t.} \quad x_1 + x_2 \leq 7$$
$$x_1, x_2 \geq 0$$

For Problems 11–15, use LINGO to solve the problem. Then explain whether you are sure the program has found the optimal solution.

11 Solve Problem 7 of Section 12.2.

12 Solve Problem 8 of Section 12.2.

13 Solve Problem 11 of Section 12.2.

14 Solve Problem 15 of Section 12.2.

15 Solve Problem 16 of Section 12.2.

[†]Based on Littlechild, "Peak Loads" (1970).

16 We must determine the percentage of our money to be invested in stocks x and y. Let a = percentage of money invested in x and $b = 1 - a$ = percentage of money invested in y. A choice of a and b is called a *portfolio*. A portfolio is efficient if there exists no other portfolio whose return has a higher mean return and lower variance, or a higher mean return and the same variance, or a lower variance with the same mean return. Let $\bar{x}$ be the mean return on stock x and $\bar{y}$ be the mean return on stock y. The variance of the annual return on one share of stock x is var x, and the variance of the annual return on one share of stock y is var y. Assume that the covariance between the annual return for one share of x and one share of y is cov(x, y). If we invest a% of our money in stock x and b% in stock y, the variance of the return is given by

$$a^2 \text{ var } x + b^2 \text{ var } y + 2ab \text{ cov}(x, y)$$

Consider the following NLP:

$$\max z = a\bar{x} + b\bar{y}$$
$$\text{s.t.} \quad a^2\text{var } x + b^2\text{var } y + 2ab \text{ cov}(x, y) \leq v^*,$$
$$a + b = 1$$

where v^* is a given non-negative number.

a Show that any solution to this NLP is an efficient portfolio.

b Show that as v^* ranges over all non-negative numbers, all efficient portfolios are obtained.

12.10 Quadratic Programming

Consider an NLP whose objective function is the sum of terms of the form $x_1^{k_1} x_2^{k_2} \ldots x_n^{k_n}$. The degree of the term $x_1^{k_1} x_2^{k_1} \ldots x_n^{k_n}$ is $k_1 + k_2 + \cdots k_n$. Thus, the degree of the term $x_1^2 x_2$ is 3, and the degree of the term $x_1 x_2$ is 2. An NLP whose constraints are linear and whose objective is the sum of terms of the form $x_1^{k_1} x_2^{k_2} \ldots x_n^{k_n}$ (with each term having a degree of 2, 1, or 0) is a **quadratic programming problem** (QPP).

Several algorithms can be used to solve QPPs [see Bazaraa and Shetty (1993, Chapter 11)]. We discuss here the application of quadratic programming to portfolio selection and show how LINGO can be used to solve QPPs. We also describe Wolfe's method for solving QPPs.

Quadratic Programming and Portfolio Selection

Consider an investor who has a fixed amount of money that can be invested in several investments. It is often assumed that an investor wants to maximize the expected return from his investments (portfolio) while simultaneously ensuring that the risk of his portfolio is small (as measured by the variance of the return earned by the portfolio). Unfortunately, the return on stocks that yield a large expected return is usually highly variable. Thus, one often approaches the problem of selecting a portfolio by choosing an acceptable minimum expected return and finding the portfolio with the minimum variance that attains an acceptable expected return. For example, an investor may seek the minimum variance portfolio that yields a 12% expected return. By varying the minimum acceptable expected return, the investor may obtain and compare several desirable portfolios.

These ideas reduce the portfolio selection problem to a quadratic programming problem. To see this, we need to observe that given random variables $\mathbf{X}_1, \mathbf{X}_2, \cdots, \mathbf{X}_n$ and constants a, b, and k,

$$E(\mathbf{X}_1 + \mathbf{X}_2 + \cdots + \mathbf{X}_n) = E(\mathbf{X}_1) + E(\mathbf{X}_2) + \cdots + E(\mathbf{X}_n) \tag{59}$$

$$\text{var }(\mathbf{X}_1 + \mathbf{X}_2 + \cdots + \mathbf{X}_n) = \text{var } \mathbf{X}_1 + \text{var } \mathbf{X}_2 + \cdots + \text{var } \mathbf{X}_n + \sum_{i \neq j} \text{cov}(\mathbf{X}_i, \mathbf{X}_j) \tag{60}$$

$$E(k\mathbf{X}_i) = kE(\mathbf{X}_i) \tag{61}$$

$$\text{var }(k\mathbf{X}_i) = k^2\text{var } \mathbf{X}_i \tag{62}$$

$$\text{cov}(a\mathbf{X}_i, b\mathbf{X}_j) = ab \text{ cov }(\mathbf{X}_i, \mathbf{X}_j) \tag{63}$$

Here, cov$(\mathbf{X},\mathbf{Y})$ is the covariance between random variables $\mathbf{X}$ and $\mathbf{Y}$. In the following example, we show how the portfolio selection problem reduces to a quadratic programming problem.

EXAMPLE 35   Portfolio Optimization

I have \$1,000 to invest in three stocks. Let $\mathbf{S}_i$ be the random variable representing the annual return on \$1 invested in stock i. Thus, if $\mathbf{S}_i = 0.12$, \$1 invested in stock i at the beginning of a year was worth \$1.12 at the end of the year. We are given the following information: $E(\mathbf{S}_1) = 0.14$, $E(\mathbf{S}_2) = 0.11$, $E(\mathbf{S}_3) = 0.10$, var $\mathbf{S}_1 = 0.20$, var $\mathbf{S}_2 = 0.08$, var $\mathbf{S}_3 = 0.18$, cov $(\mathbf{S}_1, \mathbf{S}_2) = 0.05$, cov $(\mathbf{S}_1, \mathbf{S}_3) = 0.02$, cov $(\mathbf{S}_2, \mathbf{S}_3) = 0.03$. Formulate a QPP that can be used to find the portfolio that attains an expected annual return of at least 12% and minimizes the variance of the annual dollar return on the portfolio.

Solution   Let x_j = number of dollars invested in stock $j(j = 1, 2, 3)$. Then the annual return on the portfolio is $(x_1\mathbf{S}_1 + x_2\mathbf{S}_2 + x_3\mathbf{S}_3)/1,000$ and the expected annual return on the portfolio is [by (59) and (61)]:

$$\frac{x_1E(\mathbf{S}_1) + x_2E(\mathbf{S}_2) + x_3E(\mathbf{S}_3)}{1,000}$$

To ensure that the portfolio has an expected return of at least 12%, we must include the following constraint in the formulation:

$$\frac{0.14x_1 + 0.11x_2 + 0.10x_3}{1,000} \geq 0.12 = 0.14x_1 + 0.11x_2 + 0.10x_3 \geq 0.12\,(1,000) = 120$$

Of course, we must also include the constraint $x_1 + x_2 + x_3 = 1,000$. We assume that the amount invested in a stock must be non-negative (that is, no short sales of stock are allowed) and add the constraints $x_1, x_2, x_3 \geq 0$. Our objective is simply to minimize the variance of the portfolio's final value. From (60), the variance of the final value is given by

$$
\begin{aligned}
\text{var } (x_1\mathbf{S}_1 + x_2\mathbf{S}_2 + x_3\mathbf{S}_3) &= \text{var } (x_1\mathbf{S}_1) + \text{var } (x_2\mathbf{S}_2) + \text{var } (x_3\mathbf{S}_3) \\
&\quad + 2 \text{ cov}(x_1\mathbf{S}_1, x_2\mathbf{S}_2) + 2 \text{ cov}(x_1\mathbf{S}_1, x_3\mathbf{S}_3) \\
&\quad + 2 \text{ cov}(x_2\mathbf{S}_2, x_3\mathbf{S}_3) \\
&= x_1^2 \text{ var } \mathbf{S}_1 + x_2^2 \text{ var } \mathbf{S}_2 + x_3^2 \text{ var } \mathbf{S}_3 + 2x_1x_2\text{cov}(\mathbf{S}_1, \mathbf{S}_2) \\
&\quad + 2x_1x_3\text{cov}(\mathbf{S}_1, \mathbf{S}_3) + 2x_2x_3 \text{ cov}(\mathbf{S}_2, \mathbf{S}_3) \\
&\quad \text{[from Equations (62) and (63)]} \\
&= 0.20x_1^2 + 0.08x_2^2 + 0.18x_3^2 + 0.10x_1x_2 \\
&\quad + 0.04x_1x_3 + 0.06x_2x_3
\end{aligned}
$$

Observe that each term in the last expression for the portfolio's variance is of degree 2. Thus, we have an NLP with linear constraints and an objective function consisting of terms of degree 2. To obtain the minimum variance portfolio yielding an expected return of at least 12%, we must solve the following QPP:

$$
\begin{aligned}
\min z = 0.20x_1^2 &+ 0.08x_2^2 + 0.18x_3^2 + 0.10x_1x_2 + 0.04x_1x_3 + 0.06x_2x_3 \\
\text{s.t.} \quad 0.14x_1 &+ 0.11x_2 + 0.10x_3 \geq 120 \\
x_1 &+ x_2 + x_3 = 1,000 \\
&x_1, x_2, x_3 \geq 0
\end{aligned}
$$

(64)

REMARKS   1   The idea of using quadratic programming to determine optimal portfolios comes from Markowitz (1959) and is part of the work that won him the Nobel Prize in economics.
2   In Problem 9, we will discuss how to use actual data to estimate the mean and variance of the return on an investment, as well as the covariance of the returns on pairs of investments.
3   In Problem 10, we explore Sharpe's (1963) single-factor model, which greatly simplifies portfolio optimization.

4 In reality, transaction costs are incurred when investments are bought and sold. In Problem 11, we explore how transaction costs change portfolio optimization models.

Solving NLPs with LINGO

Port.lng

When LINGO solves nonlinear programming problems it assumes all variables are non-negative. The following LINGO model (file Port.lng) can be used to solve the portfolio selection problem, Example 33.

```
MODEL:
  1]SETS:
  2]STOCKS/1..3/:MEAN,AMT;
  3]PAIRS(STOCKS,STOCKS):COV;
  4]ENDSETS
  5]MIN=@SUM(PAIRS(I,J):AMT(I)*AMT(J)*COV(I,J));
  6]@SUM(STOCKS:AMT)=1000;
  7]@SUM(STOCKS:AMT*MEAN)>RQRT;
  8]DATA:
  9]MEAN= .14,.11,.10;
 10]RQRT=120;
 11]COV= .2,.05,.02,
 12].05,.08,.03,
 13].02,.03,.18;
 14]ENDDATA
END
```

Line 2 defines the set of available investments, and associates with each the mean return per dollar invested (MEAN) and the amount placed in each investment (AMT). Line 3 associates with stocks I and J the quantity $COV(I, J) = COV(\mathbf{X}_i, \mathbf{X}_j)$. Note that $COV(I, I) = VAR\ \mathbf{X}_i$. Line 5 minimizes the variance of the portfolio. We compute the variance of the portfolio (in dollars^2) by summing over all pairs (I, J) of investments $AMT(I)$ * $AMT(J)$ * $COV(I, J)$. Lines 6 and 7 ensure that the total amount invested will equal $1,000 and that the expected return on the portfolio will exceed our required rate of return (RQRT), respectively. (Note that RQRT is input in line 10 of the DATA section.)

The expected annual return on $1 placed in each investment is defined in line 9. Lines 11–13 construct the covariance matrix to complete the model. After selecting the solution, we obtain the optimal solution: z-value = 75,238 dollars^2, AMT(1) = $380.95, AMT(2) = $476.19, and AMT(3) = $142.86. See Figure 45.

```
MODEL:
  1)   MIN= .20 * X1 ^ 2 + .08 * X2 ^ 2 + .18 * X3 ^ 2 + .10 * X1 * X2 +
       .04 * X1 * X3 + .06 * X2 * X3 ;
  2)   .14 * X1 + .11 * X2 + .10 * X3 > 120 ;
  3)   X1 + X2 + X3 = 1000 ;
  4)   X1 > 0 ;
  5)   X2 > 0 ;
  6)   X3 > 0 ;
END


SOLUTION STATUS:  OPTIMAL TO TOLERANCES.  DUAL CONDITIONS:  SATISFIED.

          OBJECTIVE FUNCTION VALUE

     1)      75238.095110

VARIABLE        VALUE          REDUCED COST
      X1      380.952379          .000000
      X2      476.190470         -.000001
      X3      142.857151          .000000

     ROW   SLACK OR SURPLUS         PRICE
      2)          .000000     -2761.906304
      3)          .000000       180.952513
      4)       380.952379          .000000
      5)       476.190470          .000000
      6)       142.857151          .000000
```

FIGURE 45

By modifying the data of our LINGO model we could easily solve for a variance-minimizing portfolio that attains a desired expected return when many stocks are available.

Spreadsheet Solution of NLP

Port.xls

We now illustrate how to use the EXCEL Solver to solve Example 35. Figure 46 (file Port.xls) shows the solution to Example 35 obtained with Solver, using the following procedure.

Solving Portfolio Optimization Problems with Excel Solver

We now show how to use the Excel Solver to solve a portfolio optimization problem. The key is to note that formulas (60) and (62) imply that for random variables $X_1, X_2, \ldots, X_n$:

$$\text{var}(c_1X_1 + c_2X_{2+} + \cdots + c_nX_n) = [c_1, c_2, \ldots, c_n](\textit{Covariance matrix})[c_1, c_2, \ldots, c_n]^T.$$

Here is how we proceed.

Step 1   In A3:C3, enter trial values for the amount invested in each stock.

Step 2   In cell D3, compute the total invested with the formula

$$=\text{SUM(A3:C3)}.$$

Step 3   In cell D5, compute the expected dollar return on the portfolio with the formula

$$=\text{SUMPRODUCT(A5:C5,A3:C3)}.$$

Step 4   In cell D8, compute the variance of the portfolio with the following array formula

$$=\text{MMULT(A3:C3,MMULT(A8:C10,TRANSPOSE(A3:C3)))}.$$

This formula multiplies the vector of amounts invested in each stock times the covariance matrix times the transpose of the vector of amounts invested in each stock. (*Note:* You must hit Control Shift Enter for this formula to work.)

Step 5   Now complete the Solver dialog box as shown in Figure 46. We minimize variance of dollar profit (cell D8). We invest exactly $1,000 (D3 = F3) and ensure that we

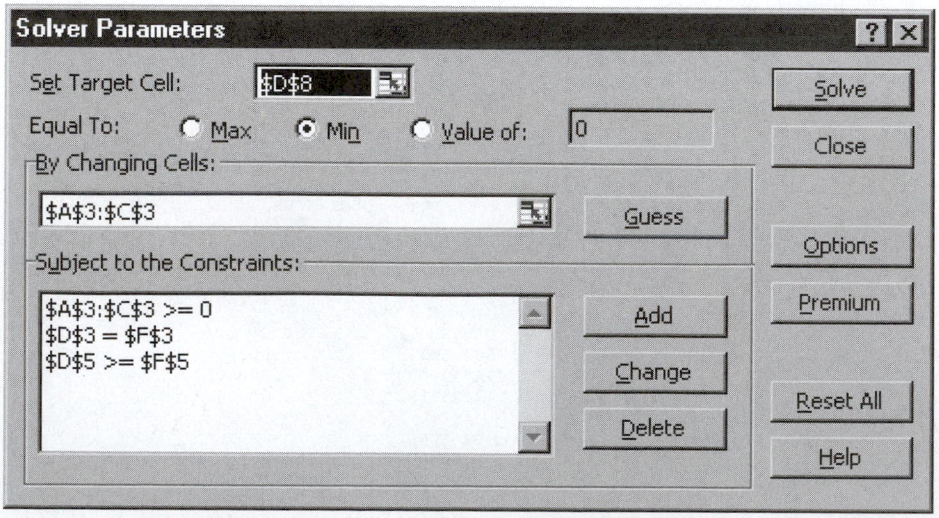

FIGURE 46

FIGURE 47

	A	B	C	D	E	F
1		PORTFOLIO	EXAMPLE			
2	X1	X2	X3	TOTALINV		
3	380.9523849	476.190461	142.8571541	1000	=	1000
4	E(X1)	E(X2)	E(X3)	MEANRET		
5	0.14	0.11	0.1	120	>=	120
6	COVARIANCE					
7	MATRIX			PORTVAR		
8	0.2	0.05	0.02	75238.09525		
9	0.05	0.08	0.03			
10	0.02	0.03	0.18			

earn an expected return of at least \$120 (D5>=F5). Constraining the amount placed in each investment to be non-negative rules out short sales. From Figure 47, we find the same optimal solution as we found with LINGO.

Wolfe's Method for Solving Quadratic Programming Problems

Wolfe's method may be used to solve QPPs in which all variables must be non-negative. We illustrate the method by solving the following QPP:

$$\min z = -x_1 - x_2 + (\tfrac{1}{2})x_1^2 + x_2^2 - x_1x_2$$
$$\text{s.t.} \qquad x_1 + x_2 \le 3$$
$$-2x_1 - 3x_2 \le -6$$
$$x_1, x_2 \ge 0$$

The objective function may be shown to be convex, so any point satisfying the Kuhn–Tucker conditions $(36')$–$(39')$ will solve this QPP. After employing excess variables e_1 for the x_1 constraint and e_2 for the x_2 constraint in $(36')$, e_2' for the constraint $-2x_1 - 3x_2 \le -6$, and a slack variable s_1' for the constraint $x_1 + x_2 \le 3$, the K–T conditions may be written as

$x_1 - 1 - x_2 + \lambda_1 - 2\lambda_2 - e_1 = 0$	[x_1 constraint in $(36')$]
$2x_2 - 1 - x_1 + \lambda_1 - 3\lambda_2 - e_2 = 0$	[x_2 constraint in $(36')$]
$x_1 + x_2 + s_1' = 3$	
$2x_1 + 3x_2 - e_2' = 6$	

All variables non-negative

$$\lambda_2 e_2' = 0, \qquad \lambda_1 s_1' = 0, \qquad e_1 x_1 = 0, \qquad e_2 x_2 = 0$$

Observe that with the exception of the last four equations, the K–T conditions are all linear or non-negativity constraints. The last four equations are the complementary slackness conditions for this QPP. For a general QPP, the complementary slackness conditions may be verbally expressed by

e_i from x_i constraint in $(36')$ and x_i cannot both be positive

Slack or excess variable for the ith constraint and λ_i cannot both be positive

(65)

To find a point satisfying the K–T conditions (except for the complementary slackness conditions), Wolfe's method simply applies a modified version of Phase I of the two-phase simplex method. We first add an artificial variable to each constraint in the K–T conditions that does not have an obvious basic variable, and then we attempt to minimize the sum of the artificial variables. To ensure that the final solution (with all artificial variables equal to zero) satisfies the complementary slackness conditions (65), Wolfe's method modifies the simplex's choice of the entering variable as follows:

1 Never perform a pivot that would make the e_i from the ith constraint in (36′) and x_i both basic variables.

2 Never perform a pivot that would make the slack (or excess) variable for the ith constraint and λ_i both basic variables.

To apply Wolfe's method to our example, we must solve the following LP:

$$\min w = a_1 + a_2 + a_2'$$
$$\text{s.t.} \quad x_1 - x_2 + \lambda_1 - 2\lambda_2 - e_1 + a_1 = 1$$
$$-x_1 + 2x_2 + \lambda_1 - 3\lambda_2 - e_2 + a_2 = 1$$
$$x_1 + x_2 + s_1' = 3$$
$$2x_1 + 3x_2 - e_2' + a_2' = 6$$
$$\text{All variables non-negative}$$

After eliminating the artificial variables from row 0, we obtain the tableau in Table 12. The current basic feasible solution is $w = 8$, $a_1 = 1$, $a_2 = 1$, $s_1' = 3$, $a_2' = 6$. Since x_2 has the most positive coefficient in row 0, we choose to enter x_2 into the basis. The resulting tableau is Table 13. The current basic feasible solution is $w = 6$, $a_1 = \frac{3}{2}$, $x_2 = \frac{1}{2}$, $s_1' = \frac{5}{2}$, $a_2' = \frac{9}{2}$. Since x_1 has the most positive coefficient in row 0, we now enter x_1 into the basis. The resulting tableau is Table 14.

The current basic feasible solution is $w = \frac{6}{7}$, $a_1 = \frac{6}{7}$, $x_2 = \frac{8}{7}$, $s_1' = \frac{4}{7}$, $x_1 = \frac{9}{7}$. The simplex method recommends that λ_1 should enter the basis. However, Wolfe's modification of the simplex method for selecting the entering variable does not allow λ_1 and s_1' to both

TABLE 12

Initial Tableau for Wolfe's Method

w	x_1	x_2	λ_1	λ_2	e_1	e_2	s_1'	e_2'	a_1	a_2	a_2'	rhs
1	2	4	2	−5	−1	−1	0	−1	0	0	0	8
0	1	−1	1	−2	−1	0	0	0	1	0	0	1
0	−1	2	1	−3	0	−1	0	0	0	1	0	1
0	1	1	0	0	0	0	1	0	0	0	0	3
0	2	3	0	0	0	0	0	−1	0	0	1	6

TABLE 13

First Tableau for Wolfe's Method

w	x_1	x_2	λ_1	λ_2	e_1	e_2	s_1'	e_2'	a_1	a_2	a_2'	rhs
1	4	0	0	1	−1	1	0	−1	0	−2	0	6
0	$\frac{1}{2}$	0	$\frac{3}{2}$	$-\frac{7}{2}$	−1	$-\frac{1}{2}$	0	0	1	$\frac{1}{2}$	0	$\frac{3}{2}$
0	$-\frac{1}{2}$	1	$\frac{1}{2}$	$-\frac{3}{2}$	0	$-\frac{1}{2}$	0	0	0	$\frac{1}{2}$	0	$\frac{1}{2}$
0	$\frac{3}{2}$	0	$-\frac{1}{2}$	$\frac{3}{2}$	0	$\frac{1}{2}$	1	0	0	$-\frac{1}{2}$	0	$\frac{5}{2}$
0	$\frac{7}{2}$	0	$-\frac{3}{2}$	$\frac{9}{2}$	0	$\frac{3}{2}$	0	−1	0	$-\frac{3}{2}$	1	$\frac{9}{2}$

TABLE 14
Second Tableau for Wolfe's Method

w	x_1	x_2	λ_1	λ_2	e_1	e_2	s_1'	e_2'	a_1	a_2	a_2'	rhs
1	0	0	$\frac{12}{7}$	$-\frac{29}{7}$	-1	$-\frac{5}{7}$	0	$\frac{1}{7}$	0	$-\frac{2}{7}$	$-\frac{8}{7}$	$\frac{6}{7}$
0	0	0	$\frac{12}{7}$	$-\frac{29}{7}$	-1	$-\frac{5}{7}$	0	$\frac{1}{7}$	1	$\frac{5}{7}$	$-\frac{1}{7}$	$\frac{6}{7}$
0	0	1	$\frac{2}{7}$	$-\frac{6}{7}$	0	$-\frac{2}{7}$	0	$-\frac{1}{7}$	0	$\frac{2}{7}$	$\frac{1}{7}$	$\frac{8}{7}$
0	0	0	$\frac{1}{7}$	$-\frac{3}{7}$	0	$-\frac{1}{7}$	1	$\boxed{\frac{3}{7}}$	0	$\frac{1}{7}$	$-\frac{3}{7}$	$\frac{4}{7}$
0	1	0	$-\frac{3}{7}$	$\frac{9}{7}$	0	$\frac{3}{7}$	0	$-\frac{2}{7}$	0	$\frac{3}{7}$	$\frac{2}{7}$	$\frac{9}{7}$

TABLE 15
Third Tableau for Wolfe's Method

w	x_1	x_2	λ_1	λ_2	e_1	e_2	s_1'	e_2'	a_1	a_2	a_2'	rhs
1	0	0	$\frac{5}{3}$	-4	-1	$-\frac{2}{3}$	$-\frac{1}{3}$	0	0	$-\frac{1}{3}$	-1	$\frac{2}{3}$
0	0	0	$\boxed{\frac{5}{3}}$	-4	-1	$-\frac{2}{3}$	$-\frac{1}{3}$	0	1	$\frac{2}{3}$	0	$\frac{2}{3}$
0	0	1	$\frac{1}{3}$	-1	0	$-\frac{1}{3}$	$\frac{1}{3}$	0	0	$\frac{1}{3}$	0	$\frac{4}{3}$
0	0	0	$\frac{1}{3}$	-1	0	$-\frac{1}{3}$	$\frac{7}{3}$	1	0	$\frac{1}{3}$	-1	$\frac{4}{3}$
0	1	0	$-\frac{1}{3}$	1	0	$\frac{1}{3}$	$\frac{2}{3}$	0	0	$-\frac{1}{3}$	0	$\frac{5}{3}$

TABLE 16
Optimal Tableau for Wolfe's Method

w	x_1	x_2	λ_1	λ_2	e_1	e_2	s_1'	e_2'	a_1	a_2	a_2'	rhs
1	0	0	0	0	0	0	0	0	-1	-1	-1	0
0	0	0	1	$-\frac{12}{5}$	$-\frac{3}{5}$	$-\frac{2}{5}$	$-\frac{1}{5}$	0	$\frac{3}{5}$	$\frac{2}{5}$	0	$\frac{2}{5}$
0	0	1	0	$-\frac{1}{5}$	$\frac{1}{5}$	$-\frac{1}{5}$	$\frac{2}{5}$	0	$-\frac{1}{5}$	$\frac{1}{5}$	0	$\frac{6}{5}$
0	0	0	0	$-\frac{1}{5}$	$\frac{1}{5}$	$-\frac{1}{5}$	$\frac{12}{5}$	1	$-\frac{1}{5}$	$\frac{1}{5}$	-1	$\frac{6}{5}$
0	1	0	0	$\frac{1}{5}$	$-\frac{1}{5}$	$\frac{1}{5}$	$\frac{3}{5}$	0	$\frac{1}{5}$	$-\frac{1}{5}$	0	$\frac{9}{5}$

be basic variables. Thus, λ_1 cannot enter the basis. Because e_2' is the only other variable with a positive coefficient in row 0, we now enter e_2' into the basis. The resulting tableau is Table 15. The current basic feasible solution is $w = \frac{2}{3}$, $a_1 = \frac{2}{3}$, $x_2 = \frac{4}{3}$, $e_2' = \frac{4}{3}$, and $x_1 = \frac{5}{3}$. Because s_1' is now a nonbasic variable, we can enter λ_1 into the basis. The resulting tableau is Table 16. This is (finally!) an optimal tableau. Because $w = 0$, we have found a solution that satisfies the Kuhn–Tucker conditions and is optimal for the QPP. Thus, the optimal solution to the QPP is $x_1 = \frac{9}{5}$, $x_2 = \frac{6}{5}$. From the optimal tableau, we also find that $\lambda_1 = \frac{2}{5}$ and $\lambda_2 = 0$ (because $e_2' = \frac{6}{5} > 0$, we know that $\lambda_2 = 0$ must hold).

Wolfe's method is guaranteed to obtain the optimal solution to a QPP if all leading principal minors of the objective function's Hessian are positive. Otherwise, Wolfe's method may not converge in a finite number of pivots. In practice, the method of **complementary pivoting** is most often used to solve QPPs. Unfortunately, space limitations preclude a discussion of complementary pivoting. The interested reader is referred to Shapiro (1979).

PROBLEMS

Group A

1 We are considering investing in three stocks. The random variable S_i represents the value one year from now of \$1 invested in stock i. We are given that $E(S_1) = 1.15$, $E(S_2) = 1.21$, $E(S_3) = 1.09$; var $S_1 = 0.09$, var $S_2 = 0.04$, var $S_3 = 0.01$; $cov(S_1, S_2) = 0.006$, $cov(S_1, S_3) = -0.004$, and $cov(S_2, S_3) = 0.005$. We have \$100 to invest and want to have an expected return of at least 15% during the next year. Formulate a QPP to find the portfolio of minimum variance that attains an expected return of at least 15%.

2 Show that the objective function for Example 35 is convex [it can be shown that the variance of any portfolio is a convex function of $(x_1, x_2, \ldots, x_n)$].

3 In Figure 45, interpret the entries in the PRICE column for rows 2 and 3.

4 Fruit Computer Company produces Pear and Apricot computers. If the company charges a price p_1 for Pear computers and p_2 for Apricot computers, it can sell q_1 Pear and q_2 Apricot computers, where $q_1 = 4,000 - 10p_1 + p_2$, and $q_2 = 2,000 - 9p_2 + 0.8p_1$. Manufacturing a Pear computer requires 2 hours of labor and 3 computer chips. An Apricot computer uses 3 hours of labor and 1 computer chip. Currently, 5,000 hours of labor and 4,500 chips are available. Formulate a QPP to maximize Fruit's revenue. Use the K–T conditions (or LINGO) to find Fruit's optimal pricing policy. What is the most that Fruit should pay for another hour of labor? What is the most that Fruit should pay for another computer chip?

5 Use Wolfe's method to solve the following QPP:

$$\min z = 2x_1^2 - x_2$$
$$\text{s.t.} \quad 2x_1 - x_2 \leq 1$$
$$x_1 + x_2 \leq 1$$
$$x_1, x_2 \geq 0$$

6 Use Wolfe's method to solve the following QPP:

$$\min x_1 + 2x_2^2$$
$$\text{s.t.} \quad x_1 + x_2 \leq 2$$
$$2x_1 + x_2 \leq 3$$
$$x_1, x_2 \geq 0$$

7 In an electrical network, the power loss incurred when a current of I amperes flows through a resistance of R ohms is $I^2 R$ watts. In Figure 48, 710 amperes of current must be sent from node 1 to node 4. The current flowing through each node must satisfy conservation of flow. For example, for node 1, 710 = flow through 1-ohm resistor + flow through 4-ohm resistor. Remarkably, nature determines the current flow through each resistor by minimizing the total power loss in the network.

    a Formulate a QPP whose solution will yield the current flowing through each resistor.

    b Use LINGO to determine the current flowing through each resistor.

FIGURE **48**

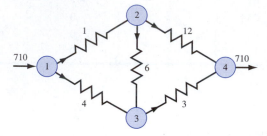

8 Use Wolfe's method to find the optimal solution to the following QPP:

$$\min z = x_1^2 + x_2^2 - 2x_1 - 3x_2 + x_1x_2$$
$$\text{s.t.} \quad x_1 + 2x_2 \leq 2$$
$$x_1, x_2 \geq 0$$

Group B

9 (This problem requires some knowledge of regression.) In Table 17, you are given the annual returns on three different types of assets (T-bills, stocks, and gold) (file Invest68.xls) during the years 1968–1988. For example, \$1 invested in T-bills at the beginning of 1978 grew to \$1.07 by the end of 1978. You have \$1,000 to invest in these three investments. Your goal is to minimize the variance of the annual dollar return of your portfolio subject to the constraint that the expected return on the portfolio for a one-year period be at least 10%. Determine how much money should be invested in each investment. Use a spreadsheet to compute the mean, standard deviation, and variance of the return on each asset. To compute the covariance between each pair of assets, remember that an estimate of the covariance of T-bills and gold is given by $cov(T, G) = s_T s_G r_{TG}$ (where s_T = standard deviation of return on T-bills; s_G = standard deviation of return on gold). Note that $r_{TG} = \pm(R^2)^{1/2}$, where the sign of r is the same as the slope of the least squares line.

In addition to determining the amount to be invested in each asset, answer the following two questions.

    a I am 95% sure that the increase in the value of my assets during the next year will be between _____ and _____ .

    b I am 95% sure that the percentage annual return on my portfolio will be between _____ and _____ .

10 (Refer to Problem 9 data.) Suppose that the return on the ith asset may be estimated as $\mu_i + \beta_i M + \epsilon_i$, where M is the return on the market. Assume that the ϵ_i are independent and that the standard deviation of ϵ_i may be estimated by the standard error of the estimate from the

TABLE 17
Annual Returns on Assets

Year	Stocks	Gold	T-Bills
1968	11	11	5
1969	−9	8	7
1970	4	−14	7
1971	14	14	4
1972	19	44	4
1973	−15	66	7
1974	−27	64	8
1975	37	0	6
1976	24	−22	5
1977	−7	18	5
1978	7	31	7
1979	19	59	10
1980	33	99	11
1981	−5	−25	15
1982	22	4	11
1983	23	−11	9
1984	6	−15	10
1985	32	−12	8
1986	19	16	6
1987	5	22	5
1988	17	−2	6

regression, with the return on the market as independent variable and the return on the ith asset as the dependent variable. Now you can express the variance of the portfolio without calculating the covariance between each pair of investments. (*Hint:* The variance of the market will enter into your equation). Use the estimated regression equation to estimate the mean return on the ith asset as a function of the return on the market.

For the data in Problem 9, formulate an NLP that can be used to find the minimum variance portfolio yielding an expected return of at least 10%. Why is this method useful when many potential investments are available?

11 (Refer to Problem 9 data.) Suppose that you now hold 30% of your investment in stocks, 50% in T-bills, and 20% in gold. Assume that transactions incur costs. Every \$100 of stocks traded costs you \$1, every \$100 of your gold portfolio traded costs you \$2, and every \$1 of your T-bill portfolio traded costs you 5¢. Find the minimum variance portfolio that yields, after transaction costs, an expected return of at least 10%. (*Hint:* Define variables for the dollars bought or sold in each investment.)

12.11 Separable Programming[†]

Many NLPs are of the following form:

$$\max \text{ (or min) } z = \sum_{j=1}^{j=n} f_j(x_j)$$

$$\text{s.t.} \quad \sum_{j=1}^{j=n} g_{ij}(x_j) \leq b_i \quad (i = 1, 2, \ldots, m)$$

Because the decision variables appear in separate terms of the objective function and the constraints, NLPs of this form are called **separable programming problems.** Separable programming problems are often solved by approximating each $f_j(x_j)$ and $g_{ij}(x_j)$ by a piecewise linear function (see Section 9.2). Before describing the separable programming technique, we give an example of a separable programming problem.

EXAMPLE 36	Separable Programming

Oilco must determine how many barrels of oil to extract during each of the next two years. If Oilco extracts x_1 million barrels during year 1, each barrel can be sold for \$30 − x_1. If Oilco extracts x_2 million barrels during year 2, each barrel can be sold for \$35 − x_2. The cost of extracting x_1 million barrels during year 1 is x_1^2 million dollars, and the cost of extracting x_2 million barrels during year 2 is $2x_2^2$ million dollars. A total of 20 million

[†]This section covers topics that may be omitted with no loss of continuity.

barrels of oil are available, and at most \$250 million can be spent on extraction. Formulate an NLP to help Oilco maximize profits (revenues less costs) for the next two years.

Solution Define

$$x_1 = \text{millions of barrels of oil extracted during year 1}$$
$$x_2 = \text{millions of barrels of oil extracted during year 2}$$

Then the appropriate NLP is

$$\max z = x_1(30 - x_1) + x_2(35 - x_2) - x_1^2 - 2x_2^2$$
$$= 30x_1 + 35x_2 - 2x_1^2 - 3x_2^2$$
$$\text{s.t.} \quad x_1^2 + 2x_2^2 \leq 250 \tag{66}$$
$$x_1 + x_2 \leq 20$$
$$x_1, x_2 \geq 0$$

This is a separable programming problem with $f_1(x_1) = 30x_1 - 2x_1^2, f_2(x_2) = 35x_2 - 3x_2^2,$ $g_{11}(x_1) = x_1^2, g_{12}(x_2) = 2x_2^2, g_{21}(x_1) = x_1,$ and $g_{22}(x_2) = x_2$.

Before approximating the functions f_j and g_{ij} by piecewise linear functions, we must determine (for $j = 1, 2, \ldots, n$) numbers a_j and b_j such that we are sure that the value of x_j in the optimal solution will satisfy $a_j \leq x_j \leq b_j$. For the previous example, $a_1 = a_2 = 0$ and $b_1 = b_2 = 20$ will suffice. Next, for each variable x_j we choose grid points $p_{j1}, p_{j2},$ $\ldots, p_{jk}$ with $a_j = p_{j1} \leq p_{j2} \leq \cdots \leq p_{jk} = b_j$ (to simplify notation, we assume that each variable has the same number of grid points). For the previous example, we use five grid points for each variable: $p_{11} = p_{21} = 0, p_{12} = p_{22} = 5, p_{13} = p_{23} = 10, p_{14} = p_{24} = 15,$ $p_{15} = p_{25} = 20.$ The essence of the separable programming method is to approximate each function f_j and g_{ij} as if it were a linear function on each interval $[p_{j,r-1}, p_{j,r}]$.

More formally, suppose $p_{j,r} \leq x_j \leq p_{j,r+1}.$ Then for some δ $(0 \leq \delta \leq 1), x_j = \delta p_{j,r} +$ $(1 - \delta)p_{j,r+1}.$ We approximate $f_j(x_j)$ and $g_{ij}(x_j)$ (see Figure 49) by

$$\hat{f}_j(x_j) = \delta f_j(p_{j,r}) + (1 - \delta)f_j(p_{j,r+1})$$
$$\hat{g}_{ij}(x_j) = \delta g_{ij}(p_{j,r}) + (1 - \delta)g_{ij}(p_{j,r+1})$$

For example, how would we approximate $f_1(12)$? Because $f_1(10) = 30(10) - 2(10)^2 = 100, f_1(15) = 30(15) - 2(15)^2 = 0,$ and $12 = 0.6(10) + 0.4(15),$ we approximate $f_1(12)$ by $\hat{f}_1(12) = 0.6(100) + 0.4(0) = 60$ (see Figure 50).

More formally, to approximate a separable programming problem, we add constraints of the form

$$\delta_{j1} + \delta_{j2} + \cdots + \delta_{j,k} = 1 \qquad (j = 1, 2, \ldots, n) \tag{67}$$
$$x_j = \delta_{j1}p_{j1} + \delta_{j2}p_{j2} + \cdots + \delta_{j,k}p_{j,k} \qquad (j = 1, 2, \ldots, n) \tag{68}$$
$$\delta_{j,r} \geq 0 \qquad (j = 1, 2, \ldots, n; r = 1, 2, \ldots, k) \tag{69}$$

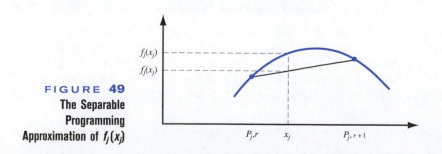

FIGURE 49
The Separable
Programming
Approximation of $f_j(x_j)$

CHAPTER **12** Nonlinear Programming

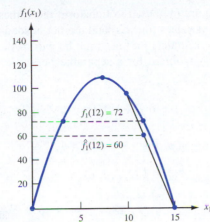

FIGURE 50
Approximation of $f_1(12)$

Then we replace $f_j(x_j)$ by

$$\hat{f}_j(x_j) = \delta_{j1}f_j(p_{j1}) + \delta_{j2}f_j(p_{j2}) + \cdots + \delta_{j,k}f_j(p_{j,k}) \tag{70}$$

and replace $g_{ij}(x_j)$ by

$$\hat{g}_{ij}(x_j) = \delta_{j1}g_{ij}(p_{j1}) + \delta_{j2}g_{ij}(p_{j2}) + \cdots + \delta_{j,k}g_{ij}(p_{j,k}) \tag{71}$$

To ensure accuracy of the approximations in (70) and (71), we must be sure that for each $j(j = 1, 2, \ldots, n)$ at most two of the $\delta_{j,k}$'s are positive. Also, for a given j, suppose that two $\delta_{j,k}$'s are positive. If $\delta_{j,k'}$ is positive, then the other positive $\delta_{j,k}$ must be either $\delta_{j,k'-1}$ or $\delta_{j,k'+1}$ (we say that $\delta_{j,k'}$ is adjacent to $\delta_{j,k'-1}$ and $\delta_{j,k'+1}$). To see the reason for these restrictions, suppose we want $x_1 = 12$. Then our approximations will be most accurate if $\delta_{13} = 0.6$ and $\delta_{14} = 0.4$. In this case, we approximate $f_1(12)$ by $0.6f_1(10) + 0.4f_1(15)$. We certainly don't want to have $\delta_{11} = 0.4$ and $\delta_{15} = 0.6$. This would yield $x_1 = 0.4(0) + 0.6(20) = 12$, but it would approximate $f_1(12)$ by $f_1(12) = 0.4f_1(0) + 0.6f_1(20)$, and in most cases this would be a poor approximation of $f_1(12)$ (see Figure 51). For the approximating problem to yield a good approximation to the functions f_i and $g_{j,k}$, we must add the following **adjacency assumption:** For $j = 1, 2, \ldots, n$, at most two $\delta_{j,k}$'s can be positive. If for a given j, two $\delta_{j,k}$'s are positive, then they must be adjacent.

Thus, the approximating problem consists of an objective function obtained from (70) and constraints obtained from (67), (68), (69), and (71) and the adjacency assumption.

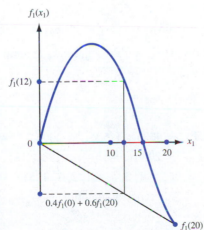

Actually, the constraints (68) are only used to transform the values of the $\delta_{j,k}$'s into values for the original decision variables (the x_j's) and are not needed to determine the optimal values of the $\delta_{j,k}$'s. The constraints (68) need not be part of the approximating problem, and the **approximating problem** for a separable programming problem may be written as follows:

$$\max \text{ (or min) } \hat{z} = \sum_{j=1}^{j=n} [\delta_{j1} f_j(p_{j1}) + \delta_{j2} f_j(p_{j2}) + \cdots + \delta_{j,k} f_j(p_{j,k})]$$

s.t. $$\sum_{j=1}^{j=n} [\delta_{j1} g_{ij}(p_{j1}) + \delta_{j2} g_{ij}(p_{j2}) + \cdots + \delta_{j,k} g_{ij}(p_{jk})] \leq b_i \qquad (i = 1, 2, \ldots, m)$$

$$\delta_{j1} + \delta_{j2} + \cdots + \delta_{j,k} = 1 \qquad (j = 1, 2, \ldots, n)$$

$$\delta_{j,r} \geq 0 \qquad (j = 1, 2, \ldots, n; r = 1, 2, \ldots, k)$$

Adjacency assumption

For the previous example, we have

$$f_1(0) = 0, \quad f_1(5) = 100, \quad f_1(10) = 100, \quad f_1(15) = 0, \quad f_1(20) = -200$$
$$f_2(0) = 0, \quad f_2(5) = 100, \quad f_2(10) = 50, \quad f_2(15) = -150, \quad f_2(20) = -500$$
$$g_{11}(0) = 0, \quad g_{11}(5) = 25, \quad g_{11}(10) = 100, \quad g_{11}(15) = 225, \quad g_{11}(20) = 400$$
$$g_{12}(0) = 0, \quad g_{12}(5) = 50, \quad g_{12}(10) = 200, \quad g_{12}(15) = 450, \quad g_{12}(20) = 800$$
$$g_{21}(0) = 0, \quad g_{21}(5) = 5, \quad g_{21}(10) = 10, \quad g_{21}(15) = 15, \quad g_{21}(20) = 20$$
$$g_{22}(0) = 0, \quad g_{22}(5) = 5, \quad g_{22}(10) = 10, \quad g_{22}(15) = 15, \quad g_{22}(20) = 20$$

Applying (70) to the objective function of (66) yields an approximating objective function of

$$\max \hat{z} = 100\delta_{12} + 100\delta_{13} - 200\delta_{15} + 100\delta_{22} + 50\delta_{23} - 150\delta_{24} - 500\delta_{25}$$

Constraint (67) yields the following two constraints:

$$\delta_{11} + \delta_{12} + \delta_{13} + \delta_{14} + \delta_{15} = 1$$
$$\delta_{21} + \delta_{22} + \delta_{23} + \delta_{24} + \delta_{25} = 1$$

Constraint (68) yields the following two constraints:

$$x_1 = 5\delta_{12} + 10\delta_{13} + 15\delta_{14} + 20\delta_{15}$$
$$x_2 = 5\delta_{22} + 10\delta_{23} + 15\delta_{24} + 20\delta_{25}$$

Applying (71) transforms the two constraints in (66) to

$$25\delta_{12} + 100\delta_{13} + 225\delta_{14} + 400\delta_{15} + 50\delta_{22} + 200\delta_{23} + 450\delta_{24} + 800\delta_{25} \leq 250$$
$$5\delta_{12} + 10\delta_{13} + 15\delta_{14} + 20\delta_{15} + 5\delta_{22} + 10\delta_{23} + 15\delta_{24} + 20\delta_{25} \leq 20$$

After adding the sign restrictions, (68), and the adjacency assumption, we obtain

$$\max \hat{z} = 100\delta_{12} + 100\delta_{13} - 200\delta_{15} + 100\delta_{22} + 50\delta_{23} - 150\delta_{24} - 500\delta_{25}$$

s.t. $$\delta_{11} + \delta_{12} + \delta_{13} + \delta_{14} + \delta_{15} = 1$$
$$\delta_{21} + \delta_{22} + \delta_{23} + \delta_{24} + \delta_{25} = 1$$
$$25\delta_{12} + 100\delta_{13} + 225\delta_{14} + 400\delta_{15} + 50\delta_{22} + 200\delta_{23} + 450\delta_{24} + 800\delta_{25} \leq 250$$
$$5\delta_{12} + 10\delta_{13} + 15\delta_{14} + 20\delta_{15} + 5\delta_{22} + 10\delta_{23} + 15\delta_{24} + 20\delta_{25} \leq 20$$
$$\delta_{j,k} \geq 0 (j = 1, 2; k = 1, 2, 3, 4, 5)$$

Adjacency assumption

At first glance, the approximating problem may appear to be a linear programming problem. If we attempt to solve the approximating problem by the simplex, however, we may violate the adjacency assumption. To avoid this difficulty, we solve approximating problems via the simplex algorithm with the following restricted entry rule: If, for a given j all $\delta_{j,k} = 0$, then any $\delta_{j,k}$ may enter the basis. If, for a given j, a single $\delta_{j,k}$ (say, $\delta_{j,k'}$) is positive, then $\delta_{j,k'-1}$ or $\delta_{j,k'+1}$ may enter the basis, but no other $\delta_{j,k}$ may enter the basis. If, for a given j, two $\delta_{j,k}$'s are positive, then no other $\delta_{j,k}$ can enter the basis.

There are two situations in which solving the approximating problem via the ordinary simplex will yield a solution that automatically satisfies the adjacency assumption. If the separable programming problem is a maximization problem, then each $f_j(x_j)$ is concave, and each $g_{ij}(x_j)$ is convex, then any solution to the approximating problem obtained via the ordinary simplex will automatically satisfy the adjacency assumption. Also, if the separable programming problem is a minimization problem, each $f_j(x_j)$ is convex, and each $g_{ij}(x_j)$ is convex, then any solution to the approximating problem obtained via the ordinary simplex will automatically satisfy the adjacency assumption. Problem 3 at the end of this section indicates why this is the case.

In these two special cases, it can also be shown that as the maximum value of the distance between two adjacent grid points approaches zero, the optimal solution to the approximating problem approaches the optimal solution to the separable programming problem [see Bazaraa and Shetty (1993, p. 450)].

For the previous example, each $f_j(x_j)$ is concave and each $g_{ij}(x_j)$ is convex, so to find the optimal solution to the approximating problem, we may use the simplex and ignore the restricted entry rule. The optimal solution to the approximating problem for the previous example is $\delta_{12} = \delta_{22} = 1$. This yields $x_1 = 1(5) = 5$, $x_2 = 1(5) = 5$, $\hat{z} = 200$. Compare this with the actual optimal solution to the previous example, which is $x_1 = 7.5$, $x_2 = 5.83$, $z = 214.58$.

PROBLEMS

Group A

Set up an approximating problem for the following separable programming problems:

1
$$\min z = x_1^2 + x_2^2$$
$$\text{s.t.} \quad x_1^2 + 2x_2^2 \le 4$$
$$x_1^2 + x_2^2 \le 6$$
$$x_1, x_2 \ge 0$$

2
$$\max z = x_1^2 - 5x_1 + x_2^2 - 5x_2 - x_3$$
$$\text{s.t.} \quad x_1 + x_2 + x_3 \le 4$$
$$x_1^2 - x_2 \le 3$$
$$x_1, x_2, x_3 \ge 0$$

Group B

3 This problem will give you an idea why the restricted entry rule is unnecessary when (for a maximization problem) each $f_j(x_j)$ is concave and each $g_{ij}(x_j)$ is convex. Consider the Oilco example. When we solve the approximating problem by the simplex, show that a solution that violates the adjacency assumption cannot be obtained. For example, why can the simplex not yield a solution (x^*) of $\delta_{11} = 0.4$ and $\delta_{15} = 0.6$? To show that this cannot occur, find a feasible solution to the approximating problem that has a larger

$\hat{z}$-value than x^*. [*Hint:* Show that the solution that is identical to x^* with the exception that $\delta_{11} = 0$, $\delta_{15} = 0$, $\delta_{13} = 0.6$, and $\delta_{14} = 0.4$ is feasible for the approximating problem [use the convexity of $g_{ij}(x_j)$ for this part] and has a larger $\hat{z}$-value than x^* [use concavity of $f_j(x_j)$ for this part].]

4 Suppose an NLP appears to be separable except for the fact that a term of the form $x_i x_j$ appears in the objective function or constraints. Show that an NLP of this type can be made into a separable programming problem by defining two new variables y_i and y_j by $x_i = \frac{1}{2}(y_i + y_j)$ and $x_j = \frac{1}{2}(y_i - y_j)$. Use this technique to transform the following NLP into a separable programming problem:

$$\max z = x_1^2 + 3x_1x_2 - x_2^2$$
$$\text{s.t.} \quad x_1x_2 \le 4$$
$$x_1^2 + x_2 \le 6$$
$$x_1, x_2 \ge 0$$

12.12 The Method of Feasible Directions[†]

In Section 12.7, we used the method of steepest ascent to solve an unconstrained NLP. We now describe a modification of that method—the **feasible directions method,** which can be used to solve NLPs with linear constraints. Suppose we want to solve

$$\max z = f(\mathbf{x})$$
$$\text{s.t.} \quad A\mathbf{x} \le \mathbf{b} \tag{72}$$
$$\mathbf{x} \ge \mathbf{0}$$

where $\mathbf{x} = [x_1, x_2, \dots, x_n]^T$, A is an $m \times n$ matrix, $\mathbf{0}$ is an n-dimensional column vector consisting entirely of zeros, $\mathbf{b}$ is an $m \times 1$ vector, and $f(\mathbf{x})$ is a concave function.

To begin, we must find (perhaps by using the Big M method or the two-phase simplex algorithm) a feasible solution $\mathbf{x}^0$ satisfying the constraints $A\mathbf{x} \le \mathbf{b}$. We now try to find a direction in which we can move away from $\mathbf{x}^0$. This direction should have two properties:

1   When we move away from $\mathbf{x}^0$, we remain feasible.

2   When we move away from $\mathbf{x}^0$, we increase the value of z.

From Section 12.7, we know that if $\nabla f(\mathbf{x}^0) \cdot \mathbf{d} > 0$ and we move a small distance away from $\mathbf{x}^0$ in a direction $\mathbf{d}$, then $f(\mathbf{x})$ will increase. We choose to move away from $\mathbf{x}^0$ in a direction $\mathbf{d}^0 - \mathbf{x}^0$, where $\mathbf{d}^0$ is an optimal solution to the following LP:

$$\max z = \nabla f(\mathbf{x}^0) \cdot \mathbf{d}$$
$$\text{s.t.} \quad A\mathbf{d} \le \mathbf{b} \tag{73}$$
$$\mathbf{d} \ge \mathbf{0}$$

Here $\mathbf{d} = [d_1 \quad d_2 \dots d_n]^T$. Observe that if $\mathbf{d}^0$ solves (73) (and $\mathbf{x}^0$ does not), then $\nabla f(\mathbf{x}^0) \cdot \mathbf{d}^0 > \nabla f(\mathbf{x}^0) \cdot \mathbf{x}^0$, or $\nabla f(\mathbf{x}^0) \cdot (\mathbf{d}^0 - \mathbf{x}^0) > 0$. This means that moving a small distance from $\mathbf{x}^0$ in a direction $\mathbf{d}^0 - \mathbf{x}^0$ will increase z.

We now choose our new point $\mathbf{x}^1$ to be $\mathbf{x}^1 = \mathbf{x}^0 + t_0(\mathbf{d}^0 - \mathbf{x}^0)$, where t_0 solves

$$\max f[\mathbf{x}^0 + t_0(\mathbf{d}^0 - \mathbf{x}^0)]$$
$$0 \le t_0 \le 1$$

It can be shown that $f(\mathbf{x}^1) \ge f(\mathbf{x}^0)$ will hold, and that if $f(\mathbf{x}^1) = f(\mathbf{x}^0)$, then $\mathbf{x}^0$ is the optimal solution to (72). Thus, unless $\mathbf{x}^0$ is optimal, $\mathbf{x}^1$ will have a z-value larger than $\mathbf{x}^0$. It is easy to show that $\mathbf{x}^1$ is a feasible point. Observe that

$$A\mathbf{x}^1 = A[\mathbf{x}^0 + t_0(\mathbf{d}^0 - \mathbf{x}^0)] = (1 - t_0)A\mathbf{x}^0 + t_0A\mathbf{d}^0 \le (1 - t_0)\mathbf{b} + t_0\mathbf{b} = \mathbf{b}$$

where the last inequality follows from the fact that both $\mathbf{x}^0$ and $\mathbf{d}^0$ satisfy the NLP's constraints and $0 \le t_0 \le 1$. $\mathbf{x}^1 \ge \mathbf{0}$ follows easily from $\mathbf{x}^0 \ge \mathbf{0}$, $\mathbf{d}^0 \ge \mathbf{0}$, and $0 \le t_0 \le 1$.

We now choose to move away from $\mathbf{x}^1$ in any direction $\mathbf{d}^1 - \mathbf{x}^1$, where $\mathbf{d}^1$ is an optimal solution to the following LP:

$$\max z = \nabla f(\mathbf{x}^1) \cdot \mathbf{d}$$
$$\text{s.t.} \quad A\mathbf{d} \le \mathbf{b}$$
$$\mathbf{d} \ge \mathbf{0}$$

Then we choose a new point $\mathbf{x}^2$ to be given by $\mathbf{x}^2 = \mathbf{x}^1 + t_1(\mathbf{d}^1 - \mathbf{x}^1)$, where t_1 solves

$$\max f[\mathbf{x}^1 + t_1(\mathbf{d}^1 - \mathbf{x}^1)]$$
$$0 \le t_1 \le 1$$

[†]This section covers topics that may be omitted with no loss of continuity.

Again, $\mathbf{x}^2$ will be feasible, and $f(\mathbf{x}^2) \geq f(\mathbf{x}^1)$ will hold. Also, if $f(\mathbf{x}^2) = f(\mathbf{x}^1)$, then $\mathbf{x}^1$ is the optimal solution to NLP (72).

We continue in this fashion and generate directions of movement $\mathbf{d}^2, \mathbf{d}^3, \ldots, \mathbf{d}^{n-1}$ and new points $\mathbf{x}^3, \mathbf{x}^4, \ldots, \mathbf{x}^n$. We terminate the algorithm if $\mathbf{x}^k = \mathbf{x}^{k-1}$. This means that $\mathbf{x}^{k-1}$ is an optimal solution to NLP (72). If the values of f are strictly increasing at each iteration of the method, then (as with the method of steepest ascent) we terminate the method whenever two successive points are very close together.

After the point $\mathbf{x}^k$ has been determined, an upper bound on the optimal z-value for (72) is available. It can be shown that if $f(x_1, x_x, \ldots, x_n)$ is concave, then

$$[\text{Optimal } z\text{-value for (71)}] \leq f(\mathbf{x}^k) + \nabla(\mathbf{x}^k) \cdot [\mathbf{d}^k - \mathbf{x}^k]^T \qquad (74)$$

Thus, if $f(\mathbf{x}^k)$ is near the upper bound on the optimal z-value obtained from (74), then we may terminate the algorithm.

The version of the feasible directions method we have discussed was developed by Frank and Wolfe. For a discussion of other feasible direction methods, we refer the reader to Chapter 11 of Bazaraa and Shetty (1993).

The following example illustrates the method of feasible directions.

EXAMPLE 37 **Method of Feasible Directions**

Perform two iterations of the feasible directions method on the following NLP:

$$\max z = f(x, y) = 2xy + 4x + 6y - 2x^2 - 2y^2$$
$$\text{s.t.} \quad x + y \leq 2$$
$$x, y \geq 0$$

Begin at the point $(0,0)$.

Solution $\nabla f(x, y) = [2y - 4x + 4 \quad 6 + 2x - 4y]$, so $\nabla f(0, 0) = [4 \quad 6]$. We find a direction to move away from $[0 \quad 0]$ by solving the following LP:

$$\max z = 4d_1 + 6d_2$$
$$\text{s.t.} \quad d_1 + d_2 \leq 2$$
$$d_1, d_2 \geq 0$$

The optimal solution to this LP is $d_1 = 0$ and $d_2 = 2$. Thus, $\mathbf{d}^0 = [0 \quad 2]^T$. Since $\mathbf{d}^0 - \mathbf{x}^0 = [0 \quad 2]^T$, we now choose $\mathbf{x}^1 = [0 \quad 0]^T + t_0[0 \quad 2]^T = [0 \quad 2t_0]^T$, where t_0 solves

$$\max f(0, 2t) = 12t - 8t^2$$
$$0 \leq t \leq 1$$

Letting $g(t) = 12t - 8t^2$, we find $g'(t) = 12 - 16t = 0$ for $t = 0.75$. Since $g''(t) < 0$, we know that $t_0 = 0.75$. Thus $\mathbf{x}^1 = [0, 1.5]^T$. At this point, $z = f(0, 1.5) = 4.5$. We now have [via (74) with $k = 0$] the following upper bound on the NLP's optimal z-value:

$$(\text{Optimal } z\text{-value}) \leq f(0, 0) + [4 \quad 6] \cdot [0 \quad 2]^T = 12$$

Now $\nabla(\mathbf{x}^1) = f(0, 1.5) = [7 \quad 0]$. We now find the direction $\mathbf{d}^2$ to move away from $\mathbf{x}^1$ by solving

$$\max z = 7d_1$$
$$\text{s.t.} \quad d_1 + d_2 \leq 2$$
$$d_1, d_2 \geq 0$$

The optimal solution to this LP is $\mathbf{d}^1 = [2 \quad 0]^T$. Now we find $\mathbf{x}^2 = [0 \quad 1.5]^T + t_1\{[2 \quad 0]^T - [0 \quad 1.5]^T\} = [2t_1 \quad 1.5 - 1.5t_1]^T$, where t_1 is the optimal solution to

$$\max f(2t, 1.5 - 1.5t)$$

$$0 \leq t \leq 1$$

Now $f(2t, 1.5 - 1.5t) = 4.5 - 18.5t^2 + 14t$. Letting $g(t) = 4.5 - 18.5t^2 + 14t$, we find $g'(t) = 14 - 37t = 0$ for $t = \frac{14}{37}$. Since $g''(t) = -37 < 0$, we find that $t_1 = \frac{14}{37}$. Thus, $\mathbf{x}^2 = [\frac{28}{37} \quad \frac{69}{74}]^T = [0.76 \quad 0.93]^T$. Now we have $z = f(0.76, 0.93) = 7.15$. From (74) (with $k = 1$), we find

$$\text{(Optimal } z\text{-value)} \leq 4.5 + [7 \quad 0] \cdot \{[2 \quad 0]^T - [0 \quad 1.5]^T\} = 18.5$$

Since our first upper bound on the optimal z-value (12) is a better bound than 18.5, we ignore this bound.

Actually, the NLP's optimal solution is $z = 8.17$, $x = .83$, and $y = 1.17$.

PROBLEMS

Group A

Perform two iterations of the method of feasible directions for each of the following NLPs.

1
$$\max z = 4x + 6y - 2x^2 - 2xy - 2y^2$$
$$\text{s.t.} \quad x + 2y \leq 2$$
$$x, y \geq 0$$

Begin at the point $(\frac{1}{2}, \frac{1}{2})$.

2
$$\max z = 3xy - x^2 - y^2$$
$$\text{s.t.} \quad 3x + y \leq 4$$
$$x, y \geq 0$$

Begin at the point $(1, 0)$.

12.13 Pareto Optimality and Trade-Off Curves[†]

In a multiattribute decision-making situation in the absence of uncertainty, we often search for *Pareto optimal* solutions. We will assume that our decision maker has two objectives, and that the set of feasible points under consideration must satisfy a given set of constraints.

DEFINITION ■ A solution (call it A) to a multiple-objective problem is **Pareto optimal** if no other feasible solution is at least as good as A with respect to every objective and strictly better than A with respect to at least one objective. ■

If we define the concept of *dominated solution* as follows, we can rephrase our definition of Pareto optimality.

DEFINITION ■ A feasible solution B **dominates** a feasible solution A to a multiple-objective problem if B is at least as good as A with repect to every objective and is strictly better than A with respect to at least one objective. ■

[†]This section covers topics that may be omitted with no loss of continuity.

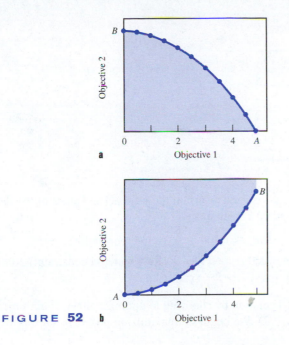

FIGURE 52

Thus, the Pareto optimal solutions are the set of all undominated feasible solutions.

If we graph the "score" of all Pareto optimal solutions in the x–y plane with the x-axis score being the score on objective 1 and the y-axis score being the score on objective 2, the graph is often called an **efficient frontier** or a **trade-off curve.**

To illustrate, suppose that the set of feasible solutions to a multiple-objective problem is the shaded region bounded by the curve AB and the first quadrant in Figure 52a. If we desire to maximize both objectives 1 and 2, then the curve AB is the set of Pareto optimal points.

For another illustration, suppose the set of feasible solutions to a multiple-objective problem is all shaded points in the first quadrant bounded from below by the curve AB in Figure 52b. If our goal is to maximize objective 1 and minimize objective 2, then the curve AB is the set of Pareto optimal points.

We will illustrate the concept of Pareto optimality (and how to determine Pareto optimal solutions) with the following example.

EXAMPLE 38   Profit Pollution Trade-Off Curve

Chemco is considering producing three products. The per-unit contribution to profit, labor requirements, raw material used per unit produced, and pollution produced per unit of product are given in Table 18. Currently, 1,300 labor hours and 1,000 units of raw material are available. Chemco's two objectives are to maximize profit and minimize pollution produced. Graph the trade-off curve for this problem.

Solution   If we define x_i = number of units of product i produced, then Chemco's two objectives may be written as follows:

Objective 1   Profit = $10x_1 + 9x_2 + 8x_3$

Objective 2   Pollution = $10x_1 + 6x_2 + 3x_3$

TABLE 18
Data for Chemco

	Product		
	1	2	3
Profit ($)	10	9	8
Labor (hours)	4	3	2
Raw material (units)	3	2	2
Pollution (units)	10	6	3

We will graph pollution on the x-axis and profit on the y-axis. The values of the decision variables must satisfy the following constraints:

$$4x_1 + 3x_2 + 2x_3 \leq 1{,}300 \quad \text{(Labor constraint)} \tag{75}$$

$$3x_1 + 2x_2 + 2x_3 \leq 1{,}000 \quad \text{(Raw material constraint)} \tag{76}$$

$$x_i \geq 0 \quad (i = 1, 2, 3) \tag{77}$$

We can find a Pareto optimal solution by choosing to optimize either of our objectives, subject to the constraints (75)–(77). We begin by maximizing profit. To do this we must solve the following LP:

$$\max z = 10x_1 + 9x_2 + 8x_3$$

$$\text{s.t.} \quad 4x_1 + 3x_2 + 2x_3 \leq 1{,}300 \quad \text{(Labor constraint)}$$

$$3x_1 + 2x_2 + 2x_3 \leq 1{,}000 \quad \text{(Raw material constraint)} \tag{78}$$

$$x_i \geq 0 \quad (i = 1, 2, 3)$$

When this LP is solved on LINDO, we find its unique optimal solution to be (call it *A*) $z = 4{,}300$, $x_1 = 0$, $x_2 = 300$, and $x_3 = 200$. This solution yields a pollution level of $6(300) + 3(200) = 2{,}400$ units. We claim this solution is Pareto optimal. To see this, note that for this solution not to be Pareto optimal, there would have to be a solution satisfying (75)–(77) that yielded $z \geq 4{,}300$ and pollution $\leq 2{,}400$, with at least one of these inequalities holding strictly. Since $x_1 = 0$, $x_2 = 300$, $x_3 = 200$ is the unique solution to (78), there is no feasible solution besides *A* satisfying (75)–(77) that can have $z \geq 4{,}300$. Thus, *A* cannot be dominated.

To find other Pareto optimal solutions, we choose any level of pollution (call it POLL) and solve the following LP:

$$\max z = 10x_1 + 9x_2 + 8x_3$$

$$\text{s.t.} \quad 4x_1 + 3x_2 + 2x_3 \leq 1{,}300 \quad \text{(Labor constraint)}$$

$$3x_1 + 2x_2 + 2x_3 \leq 1{,}000 \quad \text{(Raw material constraint)} \tag{79}$$

$$10x_1 + 6x_2 + 3x_3 \leq \text{POLL}$$

$$x_i \geq 0 \quad (i = 1, 2, 3)$$

Let PROF be the (unique) optimal z-value when this LP is solved. For each value of POLL, the point (POLL, PROF) will be on the trade-off curve. To see this, note that any point (POLL′, PROF′) dominating (POLL, PROF) must have PROF′ $\geq$ PROF. The fact that (POLL, PROF) is the unique solution to (79) implies that all feasible points (with the exception of [POLL, PROF]) having PROF′ $\geq$ PROF must have POLL′ $>$ POLL.

This means that (POLL, PROF) cannot be dominated, so it is on the trade-off curve. Choosing any value of POLL $>$ 2,400 yields no new points on the trade-off curve. (Why?) Thus, as our next step we choose POLL = 2,300. Then LINDO yields an optimal z-value of 4,266.67 and $10x_1 + 6x_2 + 3x_3 = 2{,}300$. Thus, the point (2,300, 4,266.67) is on the

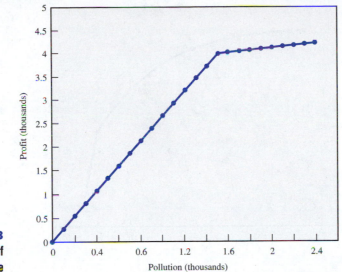

FIGURE 53

Example of Trade-Off Curve

trade-off curve. Next, we change POLL to 2,200 and obtain the point (2,200, 4,233.33) on the trade-off curve. Continuing in this fashion, setting POLL = 2,100, 2,000, 1,900, . . . 0 we obtain the trade-off curve between profit and pollution given in Figure 53.

In a multiple-objective problem in which both the constraints and objectives are linear functions, the trade-off curve will be a piecewise linear curve (that is, the graph will consist of a number of line segments of different slopes). We now give an example of a trade-off curve for a problem in which the objectives are nonlinear functions.

 Nonlinear Trade-Off Curve

Proctor and Ramble places ads on football games and soap operas. If F one-minute ads are placed on football games and S one-minute ads are placed on soap operas, then the number of men and women reached (in millions) and the cost (in thousands) of the ads are given in Table 19. P & R has a \$1 million advertising budget and its two objectives are to maximize the number of men and the number of women who see its ads. Construct a trade-off curve for this situation.

Solution   To find a first point on the trade-off curve, let us ignore the goal of maximizing the number of women who see our ads and just maximize the number of men who see our ads. This requires that we solve the following NLP:

$$\max z = 20\sqrt{F} + 4\sqrt{S}$$
$$\text{s.t.} \quad 100F + 60S \leq 1,000 \tag{80}$$
$$F \geq 0, S \geq 0$$

TABLE 19
Data for Advertising

Type of Ad	Men Reached	Women Reached	Cost per Ad ($ Thousands)
Football	$20\sqrt{F}$	$4\sqrt{F}$	100
Soap opera	$4\sqrt{S}$	$15\sqrt{S}$	60

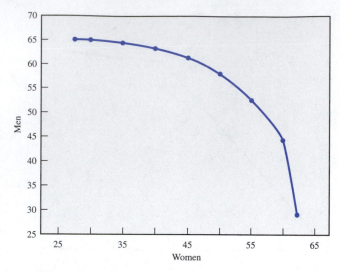

FIGURE 54
Trade-Off Curve for
Advertising Example

LINGO yields the optimal solution $z = 65.32$, $F = 9.38$, $S = 1.04$. This solution reaches $4\sqrt{9.38} + 15\sqrt{1.04} = 27.55$ million women. If we choose to place the women objective on the x-axis and the men objective on the y-axis, this yields the point $(27.55, 65.32)$ on the trade-off curve. To obtain other points on the trade-off curve, choose any value $W \geq 0$ and add the constraint $4\sqrt{F} + 15\sqrt{S} \geq W$ to (80).

This yields NLP (81):

$$\max z = 20\sqrt{F} + 4\sqrt{S}$$
$$\text{s.t.} \quad 100F + 60S \leq 1{,}000$$
$$4\sqrt{F} + 15\sqrt{S} \geq W$$
$$F \geq 0, S \geq 0$$

(81)

Suppose the optimal solution to (81) is unique and yields a z-value of M. Then the point of (W, M) is on the trade-off curve. To see this, note that any point (W', M') dominating (W, M) must have $W' \geq W$. The fact that (W, M) is the unique solution to (81) implies that all such feasible points [with the exception of (W, M)] will have $M' < M$. This means that (W, M) cannot be dominated, so it is on the trade-off curve. Using LINGO to solve (81) with $W = 30, 35, 40, 45, 50, 55, 60,$ and 62.5 yields the trade-off curve drawn in Figure 54. By the way, we cut the curve off at $W = 62.5$, because the budget constraint limits the maximum number of women watching ads to 62.5.

Summary of Trade-off Curve Procedure

The procedure we have used to construct trade-off curves between two objectives may be summarized as follows:

Step 1 Choose an objective (say, objective 1) and determine the best value of this objective that can be attained (call it v_1). For the solution attaining v_1, find the value of objective 2 (call it v_2). Then (v_1, v_2) is a point on the trade-off curve.

Step 2 For values v of objective 2 that are better than v_2, solve the optimization problem in step 1 with the additional constraint: The value of objective 2 is at least as good as v. Varying v (over values of v preferred to v_2) will give you other points on the trade-off curve.

Step 3 In step 1, we obtained one endpoint of the trade-off curve. If we determine the best value of objective 2 that can be attained, we obtain the other endpoint of the trade-off curve.

REMARK In situations when there are more than two objectives, it is often helpful to examine trade-off curves between different pairs of objectives.

PROBLEMS

Group A

1 Widgetco produces two types of widgets. Each widget is made of steel and aluminum and is assembled with skilled labor. The resources used and the per-unit profit contribution (ignoring cost of overtime labor purchased) for each type of widget are given in Table 20. Currently, 200 units of steel and 300 units of aluminum and 300 hours of labor are available. Extra overtime labor can be purchased for $10 per hour. Construct an exchange curve between the objectives of maximizing profit and minimizing overtime labor.

2 Plantco produces three products. Three workers work for Plantco, and the company must determine which product(s) each worker should produce. The number of units each worker would produce if he or she spent the whole day producing each type of product are given in Table 21.

The company is also interested in maximizing the happiness of its workers. The amount of happiness "earned" by a worker who spends the entire day producing a given product is given in Table 22.

Construct a trade-off curve between the objectives of maximizing total units produced daily and total worker happiness.

3 If a company spends $\$a$ on advertising and charges a price of $\$p$ per unit, then it sells $1,000 - 100p + 20a^{1/2}$ units of the product. The per-unit cost of producing the product is $6. Construct a trade-off curve between the objectives of profit and units sold.

TABLE 20

| | Type of Widget | |
Resource	1	2
Steel (lbs)	6	12
Aluminum (lbs)	8	20
Skilled labor (hours)	11	24
Profit contribution ($)	500	1,100

TABLE 21

| | Product | | |
Worker	1	2	3
1	20	12	10
2	12	15	9
3	6	5	10

TABLE 22

| | Product | | |
Worker	1	2	3
1	6	8	10
2	6	5	9
3	9	10	8

4 GMCO produces three types of cars: compacts, mid-size, and large. The variable cost per car (in thousands of dollars) and production capacity for each type of car are given in Table 23.

The annual demand for each type of car depends on the prices of the three types of cars, given in Table 24. Here PC = price charged for compact car (in thousands of dollars), and so on.

Suppose that each compact gets 30 mpg, each medium car gets 25 mpg, and each large car gets 18 mpg. GMCO wants to keep the planet pollution-free, so in addition to maximizing profit, it wants to maximize the average miles per gallon attained by the cars it sells. Use LINGO to construct a trade-off curve between these objectives.

5 Consider the discussion of crashing the length of the Widgetco project given in Section 8.4. For this example, construct a trade-off curve between cost of crashing the project and duration of the project.

6 For Example 35 of Section 12.10 construct a trade-off curve between the chosen portfolio's expected return and variance. This is often called the *efficient frontier*.

TABLE 23

Type of Car	Variable Cost ($ Thousands)	Production Capacity (per Year)
Compact	10	2,000
Medium	14	1,500
Large	18	1,000

TABLE 24

Type of Car	Demand for Car
Compact	$2,500 - 100(PC) + 3(PM)$
Medium	$1,800 - 30(PM) + 2(PC) + PL$
Large	$1,300 - 20(PL) + PM$

Convex and Concave Functions

A function $f(x_1, x_2, \ldots, x_n)$ is a **convex function** on a convex set S if for any $x' \in S$ and $x'' \in S$

$$f[cx' + (1 - c)x''] \leq cf(x') + (1 - c)f(x'') \qquad (3)$$

holds for $0 \leq c \leq 1$.

A function $f(x_1, x_2, \ldots, x_n)$ is a **concave function** on a convex set S if for any $x' \in S$ and $x'' \in S$

$$f[cx' + (1 - c)x''] \geq cf(x') + (1 - c)f(x'') \qquad (4)$$

holds for $0 \leq c \leq 1$.

Consider a general NLP. Suppose the feasible region S for an NLP is a convex set. If $f(x)$ is a concave (convex) function of S, then any local maximum (minimum) for the NLP is an optimal solution to the NLP.

Suppose $f''(x)$ exists for all x in a convex set S. Then $f(x)$ is a convex (concave) function of S if and only if $f''(x) \geq 0[f''(x) \leq 0]$ for all x in S.

Suppose $f(x_1, x_2, \ldots, x_n)$ has continuous second-order partial derivatives for each point $x = (x_1, x_2, \ldots, x_n) \in S$. Then $f(x_1, x_2, \ldots, x_n)$ is a convex function on S if and only if for each $x \in S$, all principal minors of H are non-negative.

Suppose $f(x_1, x_2, \ldots, x_n)$ has continuous second-order partial derivatives for each point $x = (x_1, x_2, \ldots, x_n) \in S$. Then $f(x_1, x_2, \ldots, x_n)$ is a concave function on S if and only if for each $x \in S$ and $k = 1, 2, \ldots, n$, all nonzero principal minors have the same sign as $(-1)^k$.

Solving NLPs That Are One Variable

To find an optimal solution to

$$\max \text{ (or min) } f(x)$$
$$\text{s.t.} \quad x \in [a, b]$$

we must consider the following three types of points:

Case 1   Points where $f'(x) = 0$ [a stationary point of $f(x)$].

Case 3   Points where $f'(x)$ does not exist.

Case 3   Endpoints a and b of the interval $[a, b]$.

If $f'(x_0) = 0$, $f''(x_0) < 0$, and $a < x_0 < b$, then x_0 is a local maximum. If $f'(x_0) = 0$, $f''(x_0) > 0$, and $a < x_0 < b$, then x_0 is a local minimum.

Golden Section Search

To determine (within ϵ) the optimal solution to

$$\max f(x)$$
$$\text{s.t.} \quad a \leq x \leq b$$

we can perform k iterations [where $r^k(b - a) < \epsilon$] of Golden Section Search. New points are generated as follows:

New Left-Hand Point Move a distance equal to a fraction r of the current interval of uncertainty from the right endpoint of the interval of uncertainty.

New Right-Hand Point Move a distance equal to a fraction r of the current interval of uncertainty from the left endpoint of the interval.

At each iteration, one of the new points will equal an old point.

Unconstrained Maximization and Minimization Problems with Several Variables

A local extremum $\bar{x}$ for

$$\text{max (or min)} \, f(x_1, x_2, \ldots, x_n)$$
$$\text{s.t.} \quad (x_1, x_2, \ldots, x_n) \in R^n \tag{7}$$

must satisfy $\dfrac{\partial f(\bar{x})}{\partial x_i} = 0$ for $i = 1, 2, \ldots, n$.

If $H_k(\bar{x}) > 0$ $(k = 1, 2, \ldots, n)$, then a stationary point $\bar{x}$ is a local minimum for (7).

If, for $0 \, k = 1, 2, \ldots, n$, $H_k(\bar{x})$ has the same sign as $(-1)^k$, then a stationary point $\bar{x}$ is a local maximum for (7).

If $H_n(\bar{x}) \neq 0$ and the conditions of Theorems 7 and 7' do not hold, then a stationary point $\bar{x}$ is not a local extremum.

Method of Steepest Ascent

The method of steepest ascent can be used to solve problems of the following type:

$$\text{max } z = f(x_1, x_2, \ldots, x_n)$$
$$\text{s.t.} \quad (x_1, x_2, \ldots, x_n) \in R^n$$

To find a new point with a larger z-value, we move away from the current point ($\mathbf{v}$) in the direction of $\nabla f(\mathbf{v})$. The distance we move away from $\mathbf{v}$ is chosen to maximize the value of the function at the new point. We stop when $\|\nabla f(\mathbf{v})\|$ is sufficiently close to zero.

Lagrange Multipliers

Lagrange multipliers are used to solve NLPs of the following type:

$$\text{max (or min)} \, z = f(x_1, x_2, \ldots, x_n)$$
$$\text{s.t.} \quad g_1(x_1, x_2, \ldots, x_n) = b_1$$
$$g_2(x_1, x_2, \ldots, x_n) = b_2 \tag{12}$$
$$\vdots$$
$$g_m(x_1, x_2, \ldots, x_n) = b_m$$

To solve (12), form the Lagrangian

$$L(x_1, x_2, \ldots, x_n, \lambda_1, \lambda_2, \ldots, \lambda_m) = f(x_1, x_2, \ldots, x_n) + \sum_{i=1}^{i=m} \lambda_i[b_i - g_i(x_1, x_2, \ldots, x_n)]$$

and look for points $(\bar{x}_1, \bar{x}_2, \ldots, \bar{x}_n, \bar{\lambda}_1, \bar{\lambda}_2, \ldots, \bar{\lambda}_m)$ for which

$$\frac{\partial L}{\partial x_1} = \frac{\partial L}{\partial x_2} = \cdots = \frac{\partial L}{\partial x_n} = \frac{\partial L}{\partial \lambda_1} = \frac{\partial L}{\partial \lambda_2} = \cdots = \frac{\partial L}{\partial \lambda_m} = 0$$

The Kuhn–Tucker Conditions

The Kuhn–Tucker conditions are used to solve NLPs of the following type:

$$\max \text{ (or min) } f(x_1, x_2, \ldots, x_n)$$

$$\text{s.t.} \quad g_1(x_1, x_2, \ldots, x_n) \leq b_1$$

$$g_2(x_1, x_2, \ldots, x_n) \leq b_2 \tag{26}$$

$$\vdots$$

$$g_m(x_1, x_2, \ldots, x_n) \leq b_m$$

Suppose (26) is a maximization problem. If $\bar{x} = (\bar{x}_1, \bar{x}_2, \ldots, \bar{x}_n)$ is an optimal solution to (26), then $\bar{x} = (\bar{x}_1, \bar{x}_2, \ldots, \bar{x}_n)$ must satisfy the m constraints in (26), and there must exist multipliers $\lambda_1, \lambda_2, \ldots, \lambda_m$ satisfying

$$\frac{\partial f(\bar{x})}{\partial x_j} - \sum_{i=1}^{i=m} \bar{\lambda}_i \frac{\partial g_i(\bar{x})}{\partial x_j} = 0 \qquad (j = 1, 2, \ldots, n)$$

$$\bar{\lambda}_i[b_i - g_i(\bar{x})] = 0 \qquad (i = 1, 2, \ldots, m)$$

$$\bar{\lambda}_i \geq 0 \qquad (i = 1, 2, \ldots, m)$$

Suppose (26) is a minimization problem. If $\bar{x} = (\bar{x}_1, \bar{x}_2, \ldots, \bar{x}_n)$ is an optimal solution to (26), then $\bar{x} = (\bar{x}_1, \bar{x}_2, \ldots, \bar{x}_n)$ must satisfy the m constraints in (26), and there must exist multipliers $\lambda_1, \lambda_2, \ldots, \lambda_m$ satisfying

$$\frac{\partial f(\bar{x})}{\partial x_j} + \sum_{i=1}^{i=m} \bar{\lambda}_i \frac{\partial g_i(\bar{x})}{\partial x_j} = 0 \qquad (j = 1, 2, \ldots, n)$$

$$\bar{\lambda}_i[b_i - g_i(\bar{x})] = 0 \qquad (i = 1, 2, \ldots, m)$$

$$\bar{\lambda}_i \geq 0 \qquad (i = 1, 2, \ldots, m)$$

The Kuhn–Tucker conditions are **necessary** conditions for a point to solve (26). If the $g_i(x_1, x_2, \ldots, x_n)$ are convex functions and the objective function $f(x_1, x_2, \ldots, x_n)$ is concave (convex), then for a maximization (minimization) problem, any point satisfying the Kuhn–Tucker conditions will yield an optimal solution to (26).

Quadratic Programming

A quadratic programming problem (QPP) is an NLP in which each term in the objective function is of degree 2, 1, or 0 and all constraints are linear. Wolfe's method (a modified version of the two-phase simplex) may also be used to solve QPPs.

Separable Programming

If an NLP can be written in the following form:

$$\max \text{ (or min) } z = \sum_{j=1}^{j=n} f_j(x_j)$$

$$\text{s.t.} \quad \sum_{j=1}^{j=n} g_{ij}(x_j) \leq b_i \qquad (i = 1, 2, \ldots, m)$$

it is a **separable programming problem.** To approximate the optimal solution to a separable programming problem, we solve the following **approximating problem:**

$$\max \text{ (or min) } \hat{z} = \sum_{j=1}^{j=n} [\delta_{j1} f_j(p_{j1}) + \delta_{j2} f_j(p_{j2}) + \cdots + \delta_{j,k} f_j(p_{j,k})]$$

$$\text{s.t.} \quad \sum_{j=1}^{j=n} [\delta_{j1} g_{ij}(p_{j1}) + \delta_{j2} g_{ij}(p_{j2}) + \cdots + \delta_{j,k} g_{ij}(p_{j,k})] \leq b_i \quad (i = 1, 2, \ldots, m)$$

$$\delta_{j1} + \delta_{j2} + \cdots + \delta_{j,k} = 1 \quad (j = 1, 2, \ldots, n)$$

$$\delta_{j,r} \geq 0 \quad (j = 1, 2, \ldots, n; r = 1, 2, \ldots, k)$$

(For $j = 1, 2, \ldots, n$, at most two $\delta_{j,k}$'s can be positive. If for a given j, two $\delta_{j,k}$'s are positive, they must be adjacent.)

Method of Feasible Directions

To solve

$$\max z = f(\mathbf{x})$$
$$\text{s.t.} \quad A\mathbf{x} \leq \mathbf{b}$$
$$\mathbf{x} \geq \mathbf{0}$$

we begin with a feasible solution $\mathbf{x}^0$. Let $\mathbf{d}^0$ be a solution to

$$\max z = \nabla f(\mathbf{x}^0) \cdot \mathbf{d}$$
$$\text{s.t.} \quad A\mathbf{d} \leq \mathbf{b}$$
$$\mathbf{d} \geq \mathbf{0}$$

Choose our new point $\mathbf{x}^1$ to be $\mathbf{x}^1 = \mathbf{x}^0 + t_0(\mathbf{d}^0 - \mathbf{x}^0)$, where t_0 solves

$$\max f[\mathbf{x}^0 + t_0(\mathbf{d}^0 - \mathbf{x}^0)]$$
$$0 \leq t_0 \leq 1$$

Let $\mathbf{d}^1$ be a solution to

$$\max z = \nabla f(\mathbf{x}^1) \cdot \mathbf{d}$$
$$\text{s.t.} \quad A\mathbf{d} \leq \mathbf{b}$$
$$\mathbf{d} \geq \mathbf{0}$$

Choose our new point $\mathbf{x}^2$ to be $\mathbf{x}^2 = \mathbf{x}^1 + t_1(\mathbf{d}^1 - \mathbf{x}^1)$, where t_1 solves

$$\max f[\mathbf{x}^1 + t_1(\mathbf{d}^1 - \mathbf{x}^1)]$$
$$0 \leq t_1 \leq 1$$

Continue generating points $\mathbf{x}^3, \ldots, \mathbf{x}^k$ in this fashion until $\mathbf{x}^k = \mathbf{x}^{k-1}$ or successive points are sufficiently close together.

Summary of Trade-Off Curve Procedure

The procedure we have used to construct trade-off curves between two objectives may be summarized as follows:

Step 1 Choose an objective—say, objective 1—and determine its best attainable value v_1. For the solution attaining v_1, find the value of objective 2, v_2. Then (v_1, v_2) is a point on the trade-off curve.

Step 2 For values v of objective 2 that are better than v_2, solve the optimization problem in step 1 with the additional constraint that the value of objective 2 is at least as good as v. Varying v (over values of v preferred to v_2) will give you other points on the trade-off curve.

Step 3   In step 1 we obtained one endpoint of the trade-off curve. If we determine the best value of objective 2 that can be attained, we obtain the other endpoint of the trade-off curve.

REVIEW PROBLEMS

Group A

1   Show that $f(x) = e^{-x}$ is a convex function on R^1.

2   Five of a store's major customers are located as in Figure 55. Determine where the store should be located to minimize the sum of the squares of the distances that each customer would have to travel to the store. Can you generalize this result to the case of n customers located at points $x_1, x_2, \ldots, \dot{x}_n$?

3   A company uses a raw material to produce two types of products. When processed, each unit of raw material yields 2 units of product 1 and 1 unit of product 2. If x_1 units of product 1 are produced, then each unit can be sold for $49 - x_1$, if x_2 units of product 2 are produced, then each unit can be sold for $30 - 2x_2$. It costs \$5 to purchase and process each unit of raw material.

    a   Use the Kuhn–Tucker conditions to determine how the company can maximize profits.

    b   Use LINGO or Wolfe's method to determine how the company can maximize profits.

    c   What is the most that the company would be willing to pay for an extra unit of raw material?

4   Show that $f(x) = |x|$ is a convex function on R^1.

5   Use Golden Section Search to locate, within 0.5, the optimal solution to
$$\max 3x - x^2$$
$$\text{s.t.} \quad 0 \le x \le 5$$

6   Perform two iterations of the method of steepest ascent in an attempt to maximize
$$f(x_1, x_2) = (x_1 + x_2)e^{-(x_1 + x_2)} - x_1$$
Begin at the point (0,1).

7   The cost of producing x units of a product during a month is x^2 dollars. Find the minimum cost method of producing 60 units during the next three months. Can you generalize this result to the case where the cost of producing x units during a month is an increasing convex function?

8   Solve the following NLP:
$$\max z = xyw$$
$$\text{s.t.} \quad 2x + 3y + 4w = 36$$

9   Solve the following NLP:
$$\min z = \frac{50}{x} + \frac{20}{y} + xy$$
$$\text{s.t.} \quad x \ge 1, y \ge 1$$

10   If a company charges a price p for a product and spends \$$a$ on advertising, it can sell $10{,}000 + 5\sqrt{a} - 100p$ units of the product. If the product costs \$10 per unit to produce, then how can the company maximize profits?

11   With L labor hours and M machine hours, a company can produce $L^{1/3}M^{2/3}$ computer disk drives. Each disk drive sells for \$150. If labor can be purchased at \$50 per hour and machine hours can be purchased at \$100 per hour, determine how the company can maximize profits.

Group B

12   In time t, a tree can grow to a size $F(t)$, where $F'(t) \ge 0$ and $F''(t) < 0$. Assume that for large t, $F'(t)$ is near 0. If the tree is cut at time t, then a revenue $F(t)$ is received. Assume that revenues are discounted continuously at a rate r, so \$1 received at time t is equivalent to \$$e^{-rt}$ received at time 0. The goal is to cut the tree at the time t^* that maximizes discounted revenue. Show that the tree should be cut at the time t^* satisfying the equation
$$r = \frac{F'(t^*)}{F(t^*)}$$
In the answer, explain why (if $\frac{F'(0)}{F(0)} > r$) this equation has a unique solution. Also show that the answer is a maximum, not a minimum. [*Hint:* Why is it sufficient to choose t^* to maximize $\ln(e^{-rt}F(t)$?]

13   Suppose we are hiring a weather forecaster to predict the probability that next summer will be rainy or sunny. The following suggests a method that can be used to ensure that the forecaster is accurate. Suppose that the actual probability of rain next summer is q. For simplicity, we assume that the summer can only be rainy or sunny. If the forecaster announces a probability p that the summer will be rainy, then she receives a payment of $1 - (1 - p)^2$ if the summer is rainy and a payment of $1 - p^2$ if the summer is sunny. Show that the forecaster will maximize expected profits by announcing that the probability of a rainy summer is q.

14   Show that if $b > a \ge e$, then $a^b > b^a$. Use this result to show that $e^\pi > \pi^e$. (*Hint:* Show that $\max(\frac{\ln x}{x})$ over $x \ge a$ occurs for $x = a$.)

FIGURE 55

3   4   5   6   17

748   CHAPTER **12** Nonlinear Programming

15 Consider the points $(0, 0)$, $(1, 1)$, and $(2, 3)$. Formulate an NLP whose solution will yield the circle of smallest radius enclosing these three points. Use LINGO to solve the NLP.

16 The cost of producing x units of a product during a month is $x^{1/2}$ dollars. Show that the minimum cost method of producing 40 units during the next two months is to produce all 40 units during a single month. Is it possible to generalize this result to the case where the cost of producing x units during a month is an increasing concave function?

17 Consider the problem

$$\max z = f(x)$$
$$\text{s.t.} \quad a \leq x \leq b$$

a Suppose $f(x)$ is a convex function that has derivatives for all values of x. Show that $x = a$ or $x = b$ must be optimal for the NLP. (Draw a picture.)

b Suppose $f(x)$ is a convex function for which $f'(x)$ may not exist. Show that $x = a$ or $x = b$ must be optimal for the NLP. (Use the definition of a convex function.)

18 Reconsider Problem 2. Suppose that the store should now be located to minimize the total distance that customers must walk to the store. Where should the store be located? (*Hint:* Use Problem 4 and the fact that for any convex function a local minimum will solve the NLP; then show that locating the store where one of the customers lives yields a local minimum.) Can the result be generalized?

19[†] A company uses raw material to produce two products. For c dollars, a unit of raw material can be purchased and processed into k_1 units of product 1 and k_2 units of product 2. If x_1 units of product 1 are produced, they can be sold at $p_1(x_1)$ dollars per unit. If x_2 units of product 2 are produced, they can be sold at $p_2(x_2)$ dollars per unit. Let z be the

number of units of raw material that are purchased and processed. To maximize profits (ignoring non-negativity constraints), the following NLP should be solved:

$$\max w = x_1 p_1(x_1) + x_2 p_2(x_2) - cz$$
$$\text{s.t.} \quad x_1 \leq k_1 z$$
$$x_2 \leq k_2 z$$

a Write down the Kuhn–Tucker conditions for this problem. Let $\bar{x}_1, \bar{x}_2, \bar{\lambda}_1, \bar{\lambda}_2$ represent the optimal solution to this problem.

b Consider a modified version of the problem. The company can now purchase each unit of product 1 for $\bar{\lambda}_1$ dollars and each unit of product 2 for $\bar{\lambda}_2$ dollars. Show that if the company tries to maximize profits in this situation, it will, as in part (a), produce $\bar{x}_1$ units of product 1 and $\bar{x}_2$ units of product 2. Also, show that profit and production costs will remain unchanged.

c Give an interpretation of $\bar{\lambda}_1$ and $\bar{\lambda}_2$ that might be useful to the company's accountant.

20 The area of a triangle with sides of length a, b, and c is $\sqrt{s(s-a)(s-b)(s-c)}$, where s is half the perimeter of the triangle. We have 60 ft of fence and want to fence a triangular-shaped area. Determine how to maximize the fenced area.

21 The energy used in compressing a gas (in three stages) from an initial pressure I to a final pressure F is given by

$$K \left\{ \sqrt{\frac{p_1}{I}} + \sqrt{\frac{p_2}{p_1}} + \sqrt{\frac{F}{p_2}} - 3 \right\}$$

Determine how to minimize the energy used in compressing the gas.

22 Prove Lemma 1 (use Lagrange multipliers).

REFERENCES

The following books emphasize the theoretical aspects of nonlinear programming:

Bazaraa, M., H. Sherali, and C. Shetty. *Nonlinear Programming: Theory and Algorithms.* New York: John Wiley, 1993.

Bertsetkas, D. *Nonlinear Programming.* Cambridge, Mass.: Athena Publishing, 1995.

Luenberger, D. *Linear and Nonlinear Programming.* Reading, Mass.: Addison-Wesley, 1984.

Mangasarian, O. *Nonlinear Programming.* New York: McGraw-Hill, 1969.

McCormick, G. *Nonlinear Programming: Theory, Algorithms, and Applications.* New York: Wiley, 1983.

Shapiro, J. *Mathematical Programming: Structures and Algorithms.* New York: Wiley, 1979.

Zangwill, W. *Nonlinear Programming.* Englewood Cliffs, N.J.: Prentice Hall, 1969.

The following book emphasizes various nonlinear programming algorithms:

Rao, S. *Optimization Theory and Applications.* New Delhi: Wiley Eastern Ltd., 1979.

[†]Based on Littlechild, "Marginal Pricing" (1970).

Deterministic Dynamic Programming

Dynamic programming is a technique that can be used to solve many optimization problems. In most applications, dynamic programming obtains solutions by working backward from the end of a problem toward the beginning, thus breaking up a large, unwieldy problem into a series of smaller, more tractable problems.

We introduce the idea of working backward by solving two well-known puzzles and then show how dynamic programming can be used to solve network, inventory, and resource-allocation problems. We close the chapter by showing how to use spreadsheets to solve dynamic programming problems.

13.1 Two Puzzles[†]

In this section, we show how working backward can make a seemingly difficult problem almost trivial to solve.

EXAMPLE 1 **Match Puzzle**

Suppose there are 30 matches on a table. I begin by picking up 1, 2, or 3 matches. Then my opponent must pick up 1, 2, or 3 matches. We continue in this fashion until the last match is picked up. The player who picks up the last match is the loser. How can I (the first player) be sure of winning the game?

Solution If I can ensure that it will be my opponent's turn when 1 match remains, I will certainly win. Working backward one step, if I can ensure that it will be my opponent's turn when 5 matches remain, I will win. The reason for this is that no matter what he does when 5 matches remain, I can make sure that when he has his next turn, only 1 match will remain. For example, suppose it is my opponent's turn when 5 matches remain. If my opponent picks up 2 matches, I will pick up 2 matches, leaving him with 1 match and sure defeat. Similarly, if I can force my opponent to play when 5, 9, 13, 17, 21, 25, or 29 matches remain, I am sure of victory. Thus, I cannot lose if I pick up $30 - 29 = 1$ match on my first turn. Then I simply make sure that my opponent will always be left with 29, 25, 21, 17, 13, 9, or 5 matches on his turn. Notice that we have solved this puzzle by working backward from the end of the problem toward the beginning. Try solving this problem without working backward!

EXAMPLE 2 **Milk**

I have a 9-oz cup and a 4-oz cup. My mother has ordered me to bring home exactly 6 oz of milk. How can I accomplish this goal?

[†]This section covers topics that may be omitted with no loss of continuity.

TABLE 1
Moves in the Cup-and-Milk Problem

No. of Ounces in 9-oz Cup	No. of Ounces in 4-oz Cup
6	0
6	4
9	1
0	1
1	0
1	4
5	0
5	4
9	0
0	0

Solution   By starting near the end of the problem, I cleverly realize that the problem can easily be solved if I can somehow get 1 oz of milk into the 4-oz cup. Then I can fill the 9-oz cup and empty 3 oz from the 9-oz cup into the partially filled 4-oz cup. At this point, I will be left with 6 oz of milk. After I have this flash of insight, the solution to the problem may easily be described as in Table 1 (the initial situation is written last, and the final situation is written first).

PROBLEMS

Group A

1   Suppose there are 40 matches on a table. I begin by picking up 1, 2, 3, or 4 matches. Then my opponent must pick up 1, 2, 3, or 4 matches. We continue until the last match is picked up. The player who picks up the last match is the loser. Can I be sure of victory? If so, how?

2   Three players have played three rounds of a gambling game. Each round has one loser and two winners. The losing player must pay each winner the amount of money that the winning player had at the beginning of the round. At the end of the three rounds each player has $10. You are told that each player has won one round. By working backward, determine the original stakes of the three players. [*Note:* If the answer turns out to be (for example) 5, 15, 10, don't worry about which player had which stake; we can't really tell which player ends up with how much, but we can determine the numerical values of the original stakes.]

Group B

3   We have 21 coins and are told that one is heavier than any of the other coins. How many weighings on a balance will it take to find the heaviest coin? (*Hint:* If the heaviest coin is in a group of three coins, we can find it in one weighing. Then work backward to two weighings, and so on.)

4   Given a 7-oz cup and a 3-oz cup, explain how we can return from a well with 5 oz of water.

13.2   A Network Problem

Many applications of dynamic programming reduce to finding the shortest (or longest) path that joins two points in a given network. The following example illustrates how dynamic programming (working backward) can be used to find the shortest path in a network.

EXAMPLE 3 | Shortest Path

Joe Cougar lives in New York City, but he plans to drive to Los Angeles to seek fame and fortune. Joe's funds are limited, so he has decided to spend each night on his trip at a friend's house. Joe has friends in Columbus, Nashville, Louisville, Kansas City, Omaha, Dallas, San Antonio, and Denver. Joe knows that after one day's drive he can reach Columbus, Nashville, or Louisville. After two days of driving, he can reach Kansas City, Omaha, or Dallas. After three days of driving, he can reach San Antonio or Denver. Finally, after four days of driving, he can reach Los Angeles. To minimize the number of miles traveled, where should Joe spend each night of the trip? The actual road mileages between cities are given in Figure 1.

Solution Joe needs to know the shortest path between New York and Los Angeles in Figure 1. We will find it by working backward. We have classified all the cities that Joe can be in at the beginning of the nth day of his trip as stage n cities. For example, because Joe can only be in San Antonio or Denver at the beginning of the fourth day (day 1 begins when Joe leaves New York), we classify San Antonio and Denver as stage 4 cities. The reason for classifying cities according to stages will become apparent later.

The idea of working backward implies that we should begin by solving an easy problem that will eventually help us to solve a complex problem. Hence, we begin by finding the shortest path to Los Angeles from each city in which there is only one day of driving left (stage 4 cities). Then we use this information to find the shortest path to Los Angeles from each city for which only two days of driving remain (stage 3 cities). With this information in hand, we are able to find the shortest path to Los Angeles from each city that is three days distant (stage 2 cities). Finally, we find the shortest path to Los Angeles from each city (there is only one: New York) that is four days away.

To simplify the exposition, we use the numbers 1, 2, . . . , 10 given in Figure 1 to label the 10 cities. We also define c_{ij} to be the road mileage between city i and city j. For example, $c_{35} = 580$ is the road mileage between Nashville and Kansas City. We let $f_t(i)$ be the length of the shortest path from city i to Los Angeles, given that city i is a stage t city.[†]

Stage 4 Computations

We first determine the shortest path to Los Angeles from each stage 4 city. Since there is only one path from each stage 4 city to Los Angeles, we immediately see that $f_4(8) = 1,030$, the shortest path from Denver to Los Angeles simply being the *only* path from Denver to Los Angeles. Similarly, $f_4(9) = 1,390$, the shortest (and only) path from San Antonio to Los Angeles.

Stage 3 Computations

We now work backward one stage (to stage 3 cities) and find the shortest path to Los Angeles from each stage 3 city. For example, to determine $f_3(5)$, we note that the shortest path from city 5 to Los Angeles must be one of the following:

Path 1 Go from city 5 to city 8 and then take the shortest path from city 8 to city 10.

Path 2 Go from city 5 to city 9 and then take the shortest path from city 9 to city 10.

The length of path 1 may be written as $c_{58} + f_4(8)$, and the length of path 2 may be written as $c_{59} + f_4(9)$. Hence, the shortest distance from city 5 to city 10 may be written as

[†]In this example, keeping track of the stages is unnecessary; to be consistent with later examples, however, we do keep track.

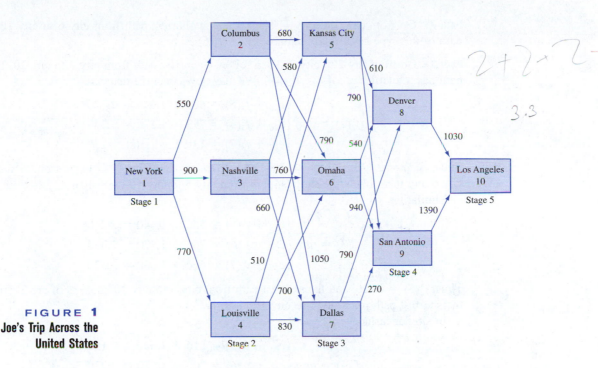

FIGURE 1
Joe's Trip Across the United States

$$f_3(5) = \min \begin{cases} c_{58} + f_4(8) = 610 + 1{,}030 = 1{,}640^* \\ c_{59} + f_4(9) = 790 + 1{,}390 = 2{,}180 \end{cases}$$

[the * indicates the choice of arc that attains the $f_3(5)$]. Thus, we have shown that the shortest path from city 5 to city 10 is the path 5–8–10. Note that to obtain this result, we made use of our knowledge of $f_4(8)$ and $f_4(9)$.

Similarly, to find $f_3(6)$, we note that the shortest path to Los Angeles from city 6 must begin by going to city 8 or to city 9. This leads us to the following equation:

$$f_3(6) = \min \begin{cases} c_{68} + f_4(8) = 540 + 1{,}030 = 1{,}570^* \\ c_{69} + f_4(9) = 940 + 1{,}390 = 2{,}330 \end{cases}$$

Thus, $f_3(6) = 1{,}570$, and the shortest path from city 6 to city 10 is the path 6–8–10.

To find $f_3(7)$, we note that

$$f_3(7) = \min \begin{cases} c_{78} + f_4(8) = 790 + 1{,}030 = 1{,}820 \\ c_{79} + f_4(9) = 270 + 1{,}390 = 1{,}660^* \end{cases}$$

Therefore, $f_3(7) = 1{,}660$, and the shortest path from city 7 to city 10 is the path 7–9–10.

Stage 2 Computations

Given our knowledge of $f_3(5)$, $f_3(6)$, and $f_3(7)$, it is now easy to work backward one more stage and compute $f_2(2)$, $f_2(3)$, and $f_2(4)$ and thus the shortest paths to Los Angeles from city 2, city 3, and city 4. To illustrate how this is done, we find the shortest path (and its length) from city 2 to city 10. The shortest path from city 2 to city 10 must begin by going from city 2 to city 5, city 6, or city 7. Once this shortest path gets to city 5, city 6, or city 7, then it must follow a shortest path from that city to Los Angeles. This reasoning shows that the shortest path from city 2 to city 10 must be one of the following:

Path 1 Go from city 2 to city 5. Then follow a shortest path from city 5 to city 10. A path of this type has a total length of $c_{25} + f_3(5)$.

Path 2   Go from city 2 to city 6. Then follow a shortest path from city 6 to city 10. A path of this type has a total length of $c_{26} + f_3(6)$.

Path 3   Go from city 2 to city 7. Then follow a shortest path from city 7 to city 10. This path has a total length of $c_{27} + f_3(7)$. We may now conclude that

$$f_2(2) = \min \begin{cases} c_{25} + f_3(5) = 680 + 1{,}640 = 2{,}320^* \\ c_{26} + f_3(6) = 790 + 1{,}570 = 2{,}360 \\ c_{27} + f_3(7) = 1{,}050 + 1{,}660 = 2{,}710 \end{cases}$$

Thus, $f_2(2) = 2{,}320$, and the shortest path from city 2 to city 10 is to go from city 2 to city 5 and then follow the shortest path from city 5 to city 10 (5–8–10).

Similarly,

$$f_2(3) = \min \begin{cases} c_{35} + f_3(5) = 580 + 1{,}640 = 2{,}220^* \\ c_{36} + f_3(6) = 760 + 1{,}570 = 2{,}330 \\ c_{37} + f_3(7) = 660 + 1{,}660 = 2{,}320 \end{cases}$$

Thus, $f_2(3) = 2{,}220$, and the shortest path from city 3 to city 10 consists of arc 3–5 and the shortest path from city 5 to city 10 (5–8–10).

In similar fashion,

$$f_2(4) = \min \begin{cases} c_{45} + f_3(5) = 510 + 1{,}640 = 2{,}150^* \\ c_{46} + f_3(6) = 700 + 1{,}570 = 2{,}270 \\ c_{47} + f_3(7) = 830 + 1{,}660 = 2{,}490 \end{cases}$$

Thus, $f_2(4) = 2{,}150$, and the shortest path from city 4 to city 10 consists of arc 4–5 and the shortest path from city 5 to city 10 (5–8–10).

Stage 1 Computations

We can now use our knowledge of $f_2(2), f_2(3),$ and $f_2(4)$ to work backward one more stage to find $f_1(1)$ and the shortest path from city 1 to city 10. Note that the shortest path from city 1 to city 10 must begin by going to city 2, city 3, or city 4. This means that the shortest path from city 1 to city 10 must be one of the following:

Path 1   Go from city 1 to city 2 and then follow a shortest path from city 2 to city 10. The length of such a path is $c_{12} + f_2(2)$.

Path 2   Go from city 1 to city 3 and then follow a shortest path from city 3 to city 10. The length of such a path is $c_{13} + f_2(3)$.

Path 3   Go from city 1 to city 4 and then follow a shortest path from city 4 to city 10. The length of such a path is $c_{14} + f_2(4)$. It now follows that

$$f_1(1) = \min \begin{cases} c_{12} + f_2(2) = 550 + 2{,}320 = 2{,}870^* \\ c_{13} + f_2(3) = 900 + 2{,}220 = 3{,}120 \\ c_{14} + f_2(4) = 770 + 2{,}150 = 2{,}920 \end{cases}$$

Determination of the Optimal Path

Thus, $f_1(1) = 2{,}870$, and the shortest path from city 1 to city 10 goes from city 1 to city 2 and then follows the shortest path from city 2 to city 10. Checking back to the $f_2(2)$ calculations, we see that the shortest path from city 2 to city 10 is 2–5–8–10. Translating the numerical labels into real cities, we see that the shortest path from New York to Los Angeles passes through New York, Columbus, Kansas City, Denver, and Los Angeles. This path has a length of $f_1(1) = 2{,}870$ miles.

Computational Efficiency of Dynamic Programming

For Example 3, it would have been an easy matter to determine the shortest path from New York to Los Angeles by enumerating all the possible paths [after all, there are only $3(3)(2) = 18$ paths]. Thus, in this problem, the use of dynamic programming did not really serve much purpose. For larger networks, however, dynamic programming is much more efficient for determining a shortest path than the explicit enumeration of all paths. To see this, consider the network in Figure 2. In this network, it is possible to travel from any node in stage k to any node in stage $k + 1$. Let the distance between node i and node j be c_{ij}. Suppose we want to determine the shortest path from node 1 to node 27. One way to solve this problem is explicit enumeration of all paths. There are 5^5 possible paths from node 1 to node 27. It takes five additions to determine the length of each path. Thus, explicitly enumerating the length of all paths requires $5^5(5) = 5^6 = 15{,}625$ additions.

Suppose we use dynamic programming to determine the shortest path from node 1 to node 27. Let $f_t(i)$ be the length of the shortest path from node i to node 27, given that node i is in stage t. To determine the shortest path from node 1 to node 27, we begin by finding $f_6(22), f_6(23), f_6(24), f_6(25)$, and $f_6(26)$. This does not require any additions. Then we find $f_5(17), f_5(18), f_5(19), f_5(20), f_5(21)$. For example, to find $f_5(21)$ we use the following equation:

$$f_5(21) = \min_j \{ c_{21,j} + f_6(j) \} \qquad (j = 22, 23, 24, 25, 26)$$

Determining $f_5(21)$ in this manner requires five additions. Thus, the calculation of all the $f_5(\cdot)$'s requires $5(5) = 25$ additions. Similarly, the calculation of all the $f_4(\cdot)$'s requires 25 additions, and the calculation of all the $f_3(\cdot)$'s requires 25 additions. The determination of all the $f_2(\cdot)$'s also requires 25 additions, and the determination of $f_1(1)$ requires 5 additions. Thus, in total, dynamic programming requires $4(25) + 5 = 105$ additions to find

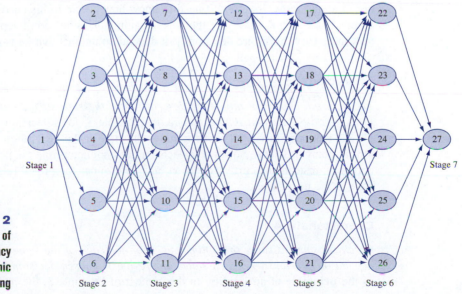

FIGURE 2

FIGURE 2
Illustration of Computational Efficiency of Dynamic Programming

the shortest path from node 1 to node 27. Because explicit enumeration requires 15,625 additions, we see that dynamic programming requires only 0.007 times as many additions as explicit enumeration. For larger networks, the computational savings effected by dynamic programming are even more dramatic.

Besides additions, determination of the shortest path in a network requires comparisons between the lengths of paths. If explicit enumeration is used, then $5^5 - 1 = 3,124$ comparisons must be made (that is, compare the length of the first two paths, then compare the length of the third path with the shortest of the first two paths, and so on). If dynamic programming is used, then for $t = 2, 3, 4, 5$, determination of each $f_t(i)$ requires $5 - 1 = 4$ comparisons. Then to compute $f_1(1)$, $5 - 1 = 4$ comparisons are required. Thus, to find the shortest path from node 1 to node 27, dynamic programming requires a total of $20(5 - 1) + 4 = 84$ comparisons. Again, dynamic programming comes out far superior to explicit enumeration.

Characteristics of Dynamic Programming Applications

We close this section with a discussion of the characteristics of Example 3 that are common to most applications of dynamic programming.

Characteristic 1

The problem can be divided into stages with a decision required at each stage. In Example 3, stage t consisted of those cities where Joe could be at the beginning of day t of his trip. As we will see, in many dynamic programming problems, the stage is the amount of time that has elapsed since the beginning of the problem. We note that in some situations, decisions are not required at every stage (see Section 13.5).

Characteristic 2

Each stage has a number of states associated with it. By a **state,** we mean the information that is needed at any stage to make an optimal decision. In Example 3, the state at stage t is simply the city where Joe is at the beginning of day t. For example, in stage 3, the possible states are Kansas City, Omaha, and Dallas. Note that to make the correct decision at any stage, Joe doesn't need to know how he got to his current location. For example, if Joe is in Kansas City, then his remaining decisions don't depend on how he goes to Kansas City; his future decisions just depend on the fact that he is now in Kansas City.

Characteristic 3

The decision chosen at any stage describes how the state at the current stage is transformed into the state at the next stage. In Example 3, Joe's decision at any stage is simply the next city to visit. This determines the state at the next stage in an obvious fashion. In many problems, however, a decision does not determine the next stage's state with certainty; instead, the current decision only determines the probability distribution of the state at the next stage.

Characteristic 4

Given the current state, the optimal decision for each of the remaining stages must not depend on previously reached states or previously chosen decisions. This idea is known as the **principle of optimality.** In the context of Example 3, the principle of optimality reduces to the following: Suppose the shortest path (call it R) from city 1 to city 10 is known to pass through city i. Then the portion of R that goes from city i to city 10 must be a shortest path from city i to city 10. If this were not the case, then we could create a path from city 1 to city 10 that was shorter than R by appending a shortest path from city i to city 10 to the portion of R leading from city 1 to city i. This would create a path from city 1 to city 10 that is shorter than R, thereby contradicting the fact that R is a shortest path from city 1 to city 10. For example, if the shortest path from city 1 to city 10 is known to pass through city 2, then the shortest path from city 1 to city 10 must include a shortest path from city 2 to city 10 (2–5–8–10). This follows because any path from city 1 to city 10 that passes through city 2 and does not contain a shortest path from city 2 to city 10 will have a length of c_{12} + [something bigger than $f_2(2)$]. Of course, such a path cannot be a shortest path from city 1 to city 10.

Characteristic 5

If the states for the problem have been classified into one of T stages, there must be a recursion that relates the cost or reward earned during stages $t, t + 1, \ldots, T$ to the cost or reward earned from stages $t + 1, t + 2, \ldots, T$. In essence, the recursion formalizes the working-backward procedure. In Example 3, our recursion could have been written as

$$f_t(i) = \min_j \{c_{ij} + f_{t+1}(j)\}$$

where j must be a stage $t + 1$ city and $f_5(10) = 0$.

We can now describe how to make optimal decisions. Let's assume that the initial state during stage 1 is i_1. To use the recursion, we begin by finding the optimal decision for each state associated with the last stage. Then we use the recursion described in characteristic 5 to determine $f_{T-1}(\cdot)$ (along with the optimal decision) for every stage $T - 1$ state. Then we use the recursion to determine $f_{T-2}(\cdot)$ (along with the optimal decision) for every stage $T - 2$ state. We continue in this fashion until we have computed $f_1(i_1)$ and the optimal decision when we are in stage 1 and state i_1. Then our optimal decision in stage 1 is chosen from the set of decisions attaining $f_1(i_1)$. Choosing this decision at stage 1 will lead us to some stage 2 state (call it state i_2) at stage 2. Then at stage 2, we choose any decision attaining $f_2(i_2)$. We continue in this fashion until a decision has been chosen for each stage.

In the rest of this chapter, we discuss many applications of dynamic programming. The presentation will seem easier if the reader attempts to determine how each problem fits into the network context introduced in Example 3. In the next section, we begin by studying how dynamic programming can be used to solve inventory problems.

PROBLEMS

Group A

1 Find the shortest path from node 1 to node 10 in the network shown in Figure 3. Also, find the shortest path from node 3 to node 10.

2 A sales representative lives in Bloomington and must be in Indianapolis next Thursday. On each of the days Monday, Tuesday, and Wednesday, he can sell his wares in Indianapolis, Bloomington, or Chicago. From past experience, he believes that he can earn $12 from spending a day in Indianapolis, $16 from spending a day in Bloomington, and $17 from spending a day in Chicago. Where should he spend the first three days

FIGURE 3

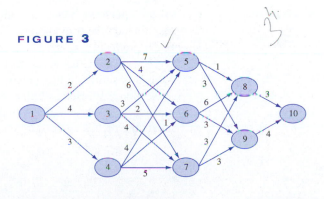

TABLE 2

FIGURE 4

	To		
From	Indianapolis	Bloomington	Chicago
Indianapolis	—	5	2
Bloomington	5	—	7
Chicago	2	7	—

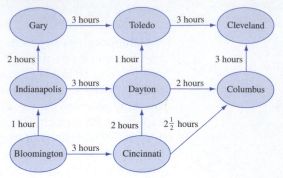

and nights of the week to maximize his sales income less travel costs? Travel costs are shown in Table 2.

Group B

3 I must drive from Bloomington to Cleveland. Several paths are available (see Figure 4). The number on each arc is the length of time it takes to drive between the two cities. For example, it takes 3 hours to drive from Bloomington to

Cincinnati. By working backward, determine the shortest path (in terms of time) from Bloomington to Cleveland. [*Hint:* Work backward and don't worry about stages—only about states.]

13.3 An Inventory Problem

In this section, we illustrate how dynamic programming can be used to solve an inventory problem with the following characteristics:

1 Time is broken up into periods, the present period being period 1, the next period 2, and the final period T. At the beginning of period 1, the demand during each period is known.

2 At the beginning of each period, the firm must determine how many units should be produced. Production capacity during each period is limited.

3 Each period's demand must be met on time from inventory or current production. During any period in which production takes place, a fixed cost of production as well as a variable per-unit cost is incurred.

4 The firm has limited storage capacity. This is reflected by a limit on end-of-period inventory. A per-unit holding cost is incurred on each period's ending inventory.

5 The firm's goal is to minimize the total cost of meeting on time the demands for periods 1, 2, . . . , T.

In this model, the firm's inventory position is reviewed at the end of each period (say, at the end of each month), and then the production decision is made. Such a model is called a **periodic review model.** This model is in contrast to the continuous review models in which the firm knows its inventory position at all times and may place an order or begin production at any time.

If we exclude the setup cost for producing any units, the inventory problem just described is similar to the Sailco inventory problem that we solved by linear programming in Section 3.10. Here, we illustrate how dynamic programming can be used to determine a production schedule that minimizes the total cost incurred in an inventory problem that meets the preceding description.

EXAMPLE 4   Inventory

A company knows that the demand for its product during each of the next four months will be as follows: month 1, 1 unit; month 2, 3 units; month 3, 2 units; month 4, 4 units. At the beginning of each month, the company must determine how many units should be produced during the current month. During a month in which any units are produced, a setup cost of $3 is incurred. In addition, there is a variable cost of $1 for every unit produced. At the end of each month, a holding cost of 50¢ per unit on hand is incurred. Capacity limitations allow a maximum of 5 units to be produced during each month. The size of the company's warehouse restricts the ending inventory for each month to 4 units at most. The company wants to determine a production schedule that will meet all demands on time and will minimize the sum of production and holding costs during the four months. Assume that 0 units are on hand at the beginning of the first month.

Solution   Recall from Section 3.10 that we can ensure that all demands are met on time by restricting each month's ending inventory to be non-negative. To use dynamic programming to solve this problem, we need to identify the appropriate state, stage, and decision. The stage should be defined so that when one stage remains, the problem will be trivial to solve. If we are at the beginning of month 4, then the firm would meet demand at minimum cost by simply producing just enough units to ensure that (month 4 production) + (month 3 ending inventory) = (month 4 demand). Thus, when one month remains, the firm's problem is easy to solve. Hence, we let time represent the stage. In most dynamic programming problems, the stage has something to do with time.

At each stage (or month), the company must decide how many units to produce. To make this decision, the company need only know the inventory level at the beginning of the current month (or the end of the previous month). Therefore, we let the state at any stage be the beginning inventory level.

Before writing a recursive relation that can be used to "build up" the optimal production schedule, we must define $f_t(i)$ to be the minimum cost of meeting demands for months $t, t + 1, \ldots, 4$ if i units are on hand at the beginning of month t. We define $c(x)$ to be the cost of producing x units during a period. Then $c(0) = 0$, and for $x > 0$, $c(x) = 3 + x$. Because of the limited storage capacity and the fact that all demand must be met on time, the possible states during each period are 0, 1, 2, 3, and 4. Thus, we begin by determining $f_4(0), f_4(1), f_4(2), f_4(3)$, and $f_4(4)$. Then we use this information to determine $f_3(0), f_3(1), f_3(2), f_3(3)$, and $f_3(4)$. Then we determine $f_2(0), f_2(1), f_2(2), f_2(3)$, and $f_2(4)$. Finally, we determine $f_1(0)$. Then we determine an optimal production level for each month. We define $x_t(i)$ to be a production level during month t that minimizes the total cost during months $t, t + 1, \ldots, 4$ if i units are on hand at the beginning of month t. We now begin to work backward.

Month 4 Computations

During month 4, the firm will produce just enough units to ensure that the month 4 demand of 4 units is met. This yields

$f_4(0)$ = cost of producing $4 - 0$ units = $c(4) = 3 + 4 = \$7$ and $x_4(0) = 4 - 0 = 4$

$f_4(1)$ = cost of producing $4 - 1$ units = $c(3) = 3 + 3 = \$6$ and $x_4(1) = 4 - 1 = 3$

$f_4(2)$ = cost of producing $4 - 2$ units = $c(2) = 3 + 2 = \$5$ and $x_4(2) = 4 - 2 = 2$

$f_4(3)$ = cost of producing $4 - 3$ units = $c(1) = 3 + 1 = \$4$ and $x_4(3) = 4 - 3 = 1$

$f_4(4)$ = cost of producing $4 - 4$ units = $c(0) = \$0$ and $x_4(4) = 4 - 4 = 0$

Month 3 Computations

How can we now determine $f_3(i)$ for $i = 0, 1, 2, 3, 4$? The cost $f_3(i)$ is the minimum cost incurred during months 3 and 4 if the inventory at the beginning of month 3 is i. For each possible production level x during month 3, the total cost during months 3 and 4 is

$$(\tfrac{1}{2})(i + x - 2) + c(x) + f_4(i + x - 2) \tag{1}$$

This follows because if x units are produced during month 3, the ending inventory for month 3 will be $i + x - 2$. Then the month 3 holding cost will be $(\tfrac{1}{2})(i + x - 2)$, and the month 3 production cost will be $c(x)$. Then we enter month 4 with $i + x - 2$ units on hand. Since we proceed optimally from this point onward (remember the principle of optimality), the cost for month 4 will be $f_4(i + x - 2)$. We want to choose the month 3 production level to minimize (1), so we write

$$f_3(i) = \min_x \{(\tfrac{1}{2})(i + x - 2) + c(x) + f_4(i + x - 2)\} \tag{2}$$

In (2), x must be a member of $\{0, 1, 2, 3, 4, 5\}$, and x must satisfy $4 \geq i + x - 2 \geq 0$. This reflects the fact that the current month's demand must be met ($i + x - 2 \geq 0$), and ending inventory cannot exceed the capacity of $4(i + x - 2 \leq 4)$. Recall that $x_3(i)$ is any value of x attaining $f_3(i)$. The computations for $f_3(0), f_3(1), f_3(2), f_3(3)$, and $f_3(4)$ are given in Table 3.

Month 2 Computations

We can now determine $f_2(i)$, the minimum cost incurred during months 2, 3, and 4 given that at the beginning of month 2, the on-hand inventory is i units. Suppose that month 2 production $= x$. Because month 2 demand is 3 units, a holding cost of $(\tfrac{1}{2})(i + x - 3)$ is

TABLE 3
Computations for $f_3(i)$

i	x	$(\tfrac{1}{2})(i + x - 2) + c(x)$	$f_4(i + x - 2)$	Total Cost Months 3, 4	$f_3(i)$ $x_3(i)$
0	2	$0 + 5 = 5$	7	$5 + 7 = 12^*$	$f_3(0) = 12$
0	3	$\tfrac{1}{2} + 6 = \tfrac{13}{2}$	6	$\tfrac{13}{2} + 6 = \tfrac{25}{2}$	$x_3(0) = 2$
0	4	$1 + 7 = 8$	5	$8 + 5 = 13$	
0	5	$\tfrac{3}{2} + 8 = \tfrac{19}{2}$	4	$\tfrac{19}{2} + 4 = \tfrac{27}{2}$	
1	1	$0 + 4 = 4$	7	$4 + 7 = 11$	$f_3(1) = 10$
1	2	$\tfrac{1}{2} + 5 = \tfrac{11}{2}$	6	$\tfrac{11}{2} + 6 = \tfrac{23}{2}$	$x_3(1) = 5$
1	3	$1 + 6 = 7$	5	$7 + 5 = 12$	
1	4	$\tfrac{3}{2} + 7 = \tfrac{17}{2}$	4	$\tfrac{17}{2} + 4 = \tfrac{25}{2}$	
1	5	$2 + 8 = 10$	0	$10 + 0 = 10^*$	
2	0	$0 + 0 = 0$	7	$0 + 7 = 7^*$	$f_3(2) = 7$
2	1	$\tfrac{1}{2} + 4 = \tfrac{9}{2}$	6	$\tfrac{9}{2} + 6 = \tfrac{21}{2}$	$x_3(2) = 0$
2	2	$1 + 5 = 6$	5	$6 + 5 = 11$	
2	3	$\tfrac{3}{2} + 6 = \tfrac{15}{2}$	4	$\tfrac{15}{2} + 4 = \tfrac{23}{2}$	
2	4	$2 + 7 = 9$	0	$9 + 0 = 9$	
3	0	$\tfrac{1}{2} + 0 = \tfrac{1}{2}$	6	$\tfrac{1}{2} + 6 = \tfrac{13}{2}^*$	$f_3(3) = \tfrac{13}{2}$
3	1	$1 + 4 = 5$	5	$5 + 5 = 10$	$x_3(3) = 0$
3	2	$\tfrac{3}{2} + 5 = \tfrac{13}{2}$	4	$\tfrac{13}{2} + 4 = \tfrac{21}{2}$	
3	3	$2 + 6 = 8$	0	$8 + 0 = 8$	
4	0	$1 + 0 = 1$	5	$1 + 5 = 6^*$	$f_3(4) = 6$
4	1	$\tfrac{3}{2} + 4 = \tfrac{11}{2}$	4	$\tfrac{11}{2} + 4 = \tfrac{19}{2}$	$x_3(4) = 0$
4	2	$2 + 5 = 7$	0	$7 + 0 = 7$	

incurred at the end of month 2. Thus, the total cost incurred during month 2 is $(\frac{1}{2})(i + x - 3) + c(x)$. During months 3 and 4, we follow an optimal policy. Since month 3 begins with an inventory of $i + x - 3$, the cost incurred during months 3 and 4 is $f_3(i + x - 3)$. In analogy to (2), we now write

$$f_2(i) = \min_x \{(\tfrac{1}{2})(i + x - 3) + c(x) + f_3(i + x - 3)\} \tag{3}$$

where x must be a member of $\{0, 1, 2, 3, 4, 5\}$ and x must also satisfy $0 \le i + x - 3 \le 4$. The computations for $f_2(0), f_2(1), f_2(2), f_2(3)$, and $f_2(4)$ are given in Table 4.

Month 1 Computations

The reader should now be able to show that the $f_1(i)$'s can be determined via the following recursive relation:

$$f_1(i) = \min_x \{(\tfrac{1}{2})(i + x - 1) + c(x) + f_2(i + x - 1)\} \tag{4}$$

where x must be a member of $\{0, 1, 2, 3, 4, 5\}$ and x must satisfy $0 \le i + x - 1 \le 4$. Since the inventory at the beginning of month 1 is 0 units, we actually need only determine $f_1(0)$ and $x_1(0)$. To give the reader more practice, however, the computations for $f_1(1), f_1(2), f_1(3)$, and $f_1(4)$ are given in Table 5.

Determination of the Optimal Production Schedule

We can now determine a production schedule that minimizes the total cost of meeting the demand for all four months on time. Since our initial inventory is 0 units, the minimum cost for the four months will be $f_1(0) = \$20$. To attain $f_1(0)$, we must produce $x_1(0) = 1$

TABLE 4
Computations for $f_2(i)$

i	x	$(\tfrac{1}{2})(i + x - 3) + c(x)$	$f_3(i + x - 3)$	Total Cost Months 2–4	$f_2(i)$ $x_2(i)$
0	3	$0 + 6 = 6$	12	$6 + 12 = 18$	$f_2(0) = 16$
0	4	$\tfrac{1}{2} + 7 = \tfrac{15}{2}$	10	$\tfrac{15}{2} + 10 = \tfrac{35}{2}$	$x_2(0) = 5$
0	5	$1 + 8 = 9$	7	$9 + 7 = 16^*$	
1	2	$0 + 5 = 5$	12	$5 + 12 = 17$	$f_2(1) = 15$
1	3	$\tfrac{1}{2} + 6 = \tfrac{13}{2}$	10	$\tfrac{13}{2} + 10 = \tfrac{33}{2}$	$x_2(1) = 4$
1	4	$1 + 7 = 8$	7	$8 + 7 = 15^*$	
1	5	$\tfrac{3}{2} + 8 = \tfrac{19}{2}$	$\tfrac{13}{2}$	$\tfrac{19}{2} + \tfrac{13}{2} = 16$	
2	1	$0 + 4 = 4$	12	$4 + 12 = 16$	$f_2(2) = 14$
2	2	$\tfrac{1}{2} + 5 = \tfrac{11}{2}$	10	$\tfrac{11}{2} + 10 = \tfrac{31}{2}^*$	$x_2(2) = 3$
2	3	$1 + 6 = 7$	7	$7 + 7 = 14^*$	
2	4	$\tfrac{3}{2} + 7 = \tfrac{17}{2}$	$\tfrac{13}{2}$	$\tfrac{17}{2} + \tfrac{13}{2} = 15$	
2	5	$2 + 8 = 10$	6	$10 + 6 = 16$	
3	0	$0 + 0 = 0$	12	$0 + 12 = 12^*$	$f_2(3) = 12$
3	1	$\tfrac{1}{2} + 4 = \tfrac{9}{2}$	10	$\tfrac{9}{2} + 10 = \tfrac{29}{2}$	$x_2(3) = 0$
3	2	$1 + 5 = 6$	7	$6 + 7 = 13$	
3	3	$\tfrac{3}{2} + 6 = \tfrac{15}{2}$	$\tfrac{13}{2}$	$\tfrac{15}{2} + \tfrac{13}{2} = 14$	
3	4	$2 + 7 = 9$	6	$9 + 6 = 15$	
4	0	$\tfrac{1}{2} + 0 = \tfrac{1}{2}$	10	$\tfrac{1}{2} + 10 = \tfrac{21}{2}^*$	$f_2(4) = \tfrac{21}{2}$
4	1	$1 + 4 = 5$	7	$5 + 7 = 12$	$x_2(4) = 0$
4	2	$\tfrac{3}{2} + 5 = \tfrac{13}{2}$	$\tfrac{13}{2}$	$\tfrac{13}{2} + \tfrac{13}{2} = 13$	
4	3	$2 + 6 = 8$	6	$8 + 6 = 14$	

TABLE 5
Computations for $f_1(i)$

i	x	$(\frac{1}{2})(i + x - 1) + c(x)$	$f_2(i + x - 1)$	Total Cost	$f_1(i)$ $x_1(i)$
0	1	$0 + 4 = 4$	16	$4 + 16 = 20^*$	$f_1(0) = 20$
0	2	$\frac{1}{2} + 5 = \frac{11}{2}$	15	$\frac{11}{2} + 15 = \frac{41}{2}$	$x_1(0) = 1$
0	3	$1 + 6 = 7$	14	$7 + 14 = 21$	
0	4	$\frac{3}{2} + 7 = \frac{17}{2}$	12	$\frac{17}{2} + 12 = \frac{41}{2}$	
0	5	$2 + 8 = 10$	$\frac{21}{2}$	$10 + \frac{21}{2} = \frac{41}{2}$	
1	0	$0 + 0 = 0$	16	$0 + 16 = 16^*$	$f_1(1) = 16$
1	1	$\frac{1}{2} + 4 = \frac{9}{2}$	15	$\frac{9}{2} + 15 = \frac{39}{2}$	$x_1(1) = 0$
1	2	$1 + 5 = 6$	14	20	
1	3	$\frac{3}{2} + 6 = \frac{15}{2}$	12	$\frac{15}{2} + 12 = \frac{39}{2}$	
1	4	$2 + 7 = 9$	$\frac{21}{2}$	$9 + \frac{21}{2} = \frac{39}{2}$	
2	0	$\frac{1}{2} + 0 = \frac{1}{2}$	15	$\frac{1}{2} + 15 = \frac{31}{2}^*$	$f_1(2) = \frac{31}{2}$
2	1	$1 + 4 = 5$	14	$5 + 14 = 19$	$x_1(2) = 0$
2	2	$\frac{3}{2} + 5 = \frac{13}{2}$	12	$\frac{13}{2} + 12 = \frac{37}{2}$	
2	3	$2 + 6 = 8$	$\frac{21}{2}$	$8 + \frac{21}{2} = \frac{37}{2}$	
3	0	$1 + 0 = 1$	14	$1 + 14 = 15^*$	$f_1(3) = 15$
3	1	$\frac{3}{2} + 4 = \frac{11}{2}$	12	$\frac{11}{2} + 12 = \frac{35}{2}$	$x_1(3) = 0$
3	2	$2 + 5 = 7$	$\frac{21}{2}$	$7 + \frac{21}{2} = \frac{35}{2}$	
4	0	$\frac{3}{2} + 0 = \frac{3}{2}$	12	$\frac{3}{2} + 12 = \frac{27}{2}^*$	$f_1(4) = \frac{27}{2}$
4	1	$2 + 4 = 6$	$\frac{21}{2}$	$6 + \frac{21}{2} = \frac{33}{2}$	$x_1(4) = 0$

unit during month 1. Then the inventory at the beginning of month 2 will be $0 + 1 - 1 = 0$. Thus, in month 2, we should produce $x_2(0) = 5$ units. Then at the beginning of month 3, our beginning inventory will be $0 + 5 - 3 = 2$. Hence, during month 3, we need to produce $x_3(2) = 0$ units. Then month 4 will begin with $2 - 2 + 0 = 0$ units on hand. Thus, $x_4(0) = 4$ units should be produced during month 4. In summary, the optimal production schedule incurs a total cost of \$20 and produces 1 unit during month 1, 5 units during month 2, 0 units during month 3, and 4 units during month 4.

Note that finding the solution to Example 4 is equivalent to finding the shortest route joining the node $(1, 0)$ to the node $(5, 0)$ in Figure 5. Each node in Figure 5 corresponds to a state, and each column of nodes corresponds to all the possible states associated with a given stage. For example, if we are at node $(2, 3)$, then we are at the beginning of month 2, and the inventory at the beginning of month 2 is 3 units. Each arc in the network represents the way in which a decision (how much to produce during the current month) transforms the current state into next month's state. For example, the arc joining nodes $(1, 0)$ and $(2, 2)$ (call it arc 1) corresponds to producing 3 units during month 1. To see this, note that if 3 units are produced during month 1, then we begin month 2 with $0 + 3 - 1 = 2$ units. The length of each arc is simply the sum of production and inventory costs during the current period, given the current state and the decision associated with the chosen arc. For example, the cost associated with arc 1 would be $6 + (\frac{1}{2})2 = 7$. Note that some nodes in adjacent stages are not joined by an arc. For example, node $(2, 4)$ is not joined to node $(3, 0)$. The reason for this is that if we begin month 2 with 4 units, then at the beginning of month 3, we will have at least $4 - 3 = 1$ unit on hand. Also note that we have drawn arcs joining all month 4 states to the node $(5, 0)$, since having a positive inventory at the end of month 4 would clearly be suboptimal.

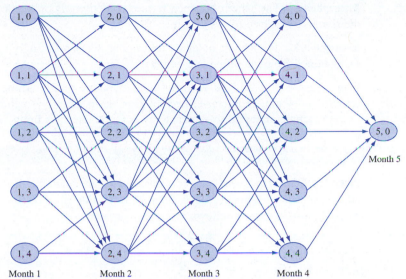

FIGURE 5
Network Representation of Inventory Example

Month 1   Month 2   Month 3   Month 4

Returning to Example 4, the minimum-cost production schedule corresponds to the shortest path joining (1, 0) and (5, 0). As we have already seen, this would be the path corresponding to production levels of 1, 5, 0, and 4. In Figure 5, this would correspond to the path beginning at (1, 0), then going to (2, 0 + 1 − 1) = (2, 0), then to (3, 0 + 5 − 3) = (3, 2), then to (4, 2 + 0 − 2) = (4, 0), and finally to (5, 0 + 4 − 4) = (5, 0). Thus, our optimal production schedule corresponds to the path (1, 0)–(2, 0)–(3, 2)–(4, 0)–(5, 0) in Figure 5.

PROBLEMS

Group A

1. In Example 4, determine the optimal production schedule if the initial inventory is 3 units.

2. An electronics firm has a contract to deliver the following number of radios during the next three months; month 1, 200 radios; month 2, 300 radios; month 3, 300 radios. For each radio produced during months 1 and 2, a $10 variable cost is incurred; for each radio produced during month 3, a $12 variable cost is incurred. The inventory cost is $1.50 for each radio in stock at the end of a month. The cost of setting up for production during a month is $250.

Radios made during a month may be used to meet demand for that month or any future month. Assume that production during each month must be a multiple of 100. Given that the initial inventory level is 0 units, use dynamic programming to determine an optimal production schedule.

3. In Figure 5, determine the production level and cost associated with each of the following arcs:
   a (2, 3)–(3, 1)
   b (4, 2)–(5, 0)

13.4 Resource-Allocation Problems

Resource-allocation problems, in which limited resources must be allocated among several activities, are often solved by dynamic programming. Recall that we have solved such problems by linear programming (for instance, the Giapetto problem and many of the Chapter 3 formulation problems). To use linear programming to do resource allocation, three assumptions must be made:

Assumption 1 The amount of a resource assigned to an activity may be any non-negative number.

Assumption 2   The benefit obtained from each activity is proportional to the amount of the resource assigned to the activity.

Assumption 3   The benefit obtained from more than one activity is the sum of the benefits obtained from the individual activities.

Even if assumptions 1 and 2 do not hold, dynamic programming can be used to solve resource-allocation problems efficiently when assumption 3 is valid and when the amount of the resource allocated to each activity is a member of a finite set.

EXAMPLE 5   Resource Allocation

Finco has $6,000 to invest, and three investments are available. If d_j dollars (in thousands) are invested in investment j, then a net present value (in thousands) of $r_j(d_j)$ is obtained, where the $r_j(d_j)$'s are as follows:

$$r_1(d_1) = 7d_1 + 2 \qquad (d_1 > 0)$$
$$r_2(d_2) = 3d_2 + 7 \qquad (d_2 > 0)$$
$$r_3(d_3) = 4d_3 + 5 \qquad (d_3 > 0)$$
$$r_1(0) = r_2(0) = r_3(0) = 0$$

The amount placed in each investment must be an exact multiple of $1,000. To maximize the net present value obtained from the investments, how should Finco allocate the $6,000?

Solution   The return on each investment is not proportional to the amount invested in it [for example, $16 = r_1(2) \neq 2r_1(1) = 18$]. Thus, linear programming cannot be used to find an optimal solution to this problem.[†]

Mathematically, Finco's problem may be expressed as

$$\max\{r_1(d_1) + r_2(d_2) + r_3(d_3)\}$$
$$\text{s.t.} \qquad d_1 + d_2 + d_3 = 6$$
$$d_j \text{ non-negative integer} \qquad (j = 1, 2, 3)$$

Of course, if the $r_j(d_j)$'s were linear, then we would have a knapsack problem like those we studied in Section 9.5.

To formulate Finco's problem as a dynamic programming problem, we begin by identifying the stage. As in the inventory and shortest-route examples, the stage should be chosen so that when one stage remains the problem is easy to solve. Then, given that the problem has been solved for the case where one stage remains, it should be easy to solve the problem where two stages remain, and so forth. Clearly, it would be easy to solve when only one investment was available, so we define stage t to represent a case where funds must be allocated to investments $t, t + 1, \ldots, 3$.

For a given stage, what must we know to determine the optimal investment amount? Simply how much money is available for investments $t, t + 1, \ldots, 3$. Thus, we define the state at any stage to be the amount of money (in thousands) available for investments t, $t + 1, \ldots, 3$. We can never have more than $6,000 available, so the possible states at any stage are 0, 1, 2, 3, 4, 5, and 6. We define $f_t(d_t)$ to be the maximum net present value (NPV) that can be obtained by investing d_t thousand dollars in investments $t, t + 1, \ldots,$ 3. Also define $x_t(d_t)$ to be the amount that should be invested in investment t to attain $f_t(d_t)$. We start to work backward by computing $f_3(0), f_3(1), \ldots, f_3(6)$ and then determine $f_2(0)$, $f_2(1), \ldots, f_2(6)$. Since $6,000 is available for investment in investments 1, 2, and 3, we

---

[†]The fixed-charge approach described in Section 9.2 could be used to solve this problem.

terminate our computations by computing $f_1(6)$. Then we retrace our steps and determine the amount that should be allocated to each investment (just as we retraced our steps to determine the optimal production level for each month in Example 4).

Stage 3 Computations

We first determine $f_3(0), f_3(1), \ldots, f_3(6)$. We see that $f_3(d_3)$ is attained by investing all available money (d_3) in investment 3. Thus,

$$f_3(0) = 0 \qquad x_3(0) = 0$$
$$f_3(1) = 9 \qquad x_3(1) = 1$$
$$f_3(2) = 13 \qquad x_3(2) = 2$$
$$f_3(3) = 17 \qquad x_3(3) = 3$$
$$f_3(4) = 21 \qquad x_3(4) = 4$$
$$f_3(5) = 25 \qquad x_3(5) = 5$$
$$f_3(6) = 29 \qquad x_3(6) = 6$$

TABLE 6
Computations for $f_2(0), f_2(1), \ldots, f_2(6)$

d_2	x_2	$r_2(x_2)$	$f_3(d_2 - x_2)$	NPV from Investments 2, 3	$f_2(d_2)$ $x_2(d_2)$
0	0	0	0	0*	$f_2(0) = 0$
					$x_2(0) = 23$
1	0	0	9	9	$f_2(1) = 10$
1	1	10	0	10*	$x_2(1) = 1$
2	0	0	13	13	$f_2(2) = 19$
2	1	10	9	19*	$x_2(2) = 1$
2	2	13	0	13	
3	0	0	17	17	$f_2(3) = 23$
3	1	10	13	23*	$x_2(3) = 1$
3	2	13	9	22	
3	3	16	0	16	
4	0	0	21	21	$f_2(4) = 27$
4	1	10	17	27*	$x_2(4) = 1$
4	2	13	13	26	
4	3	16	9	25	
4	4	19	0	19	
5	0	0	25	25	$f_2(5) = 31$
5	1	10	21	31*	$x_2(5) = 1$
5	2	13	17	30	
5	3	16	13	29	
5	4	19	9	28	
5	5	22	0	22	
6	0	0	29	29	$f_2(6) = 35$
6	1	10	25	35*	$x_2(6) = 1$
6	2	13	21	34	
6	3	16	17	33	
6	4	19	13	32	
6	5	22	9	31	
6	6	25	0	25	

TABLE 7
Computations for $f_1(6)$

d_1	x_1	$r_1(x_1)$	$f_2(6-x_1)$	NPV from Investments 1–3	$f_1(6)$ $x_1(6)$
6	0	0	35	35	$f_1(6) = 49$
6	1	9	31	40	$x_1(6) = 4$
6	2	16	27	43	
6	3	23	23	46	
6	4	30	19	49*	
6	5	37	10	47	
6	6	44	0	44	

Stage 2 Computations

To determine $f_2(0), f_2(1), \ldots, f_2(6)$, we look at all possible amounts that can be placed in investment 2. To find $f_2(d_2)$, let x_2 be the amount invested in investment 2. Then an NPV of $r_2(x_2)$ will be obtained from investment 2, and an NPV of $f_3(d_2 - x_2)$ will be obtained from investment 3 (remember the principle of optimality). Since x_2 should be chosen to maximize the net present value earned from investments 2 and 3, we write

$$f_2(d_2) = \max_{x_2} \{r_2(x_2) + f_3(d_2 - x_2)\} \tag{5}$$

where x_2 must be a member of $\{0, 1, \ldots, d_2\}$. The computations for $f_2(0), f_2(1), \ldots, f_2(6)$ and $x_2(0), x_2(1), \ldots, x_2(6)$ are given in Table 6.

Stage 1 Computations

Following (5), we write

$$f_1(6) = \max_{x_1} \{r_1(x_1) + f_2(6 - x_1)\}$$

where x_1 must be a member of $\{0, 1, 2, 3, 4, 5, 6\}$. The computations for $f_1(6)$ are given in Table 7.

Determination of Optimal Resource Allocation

Since $x_1(6) = 4$, Finco invests \$4,000 in investment 1. This leaves $6,000 - 4,000 = \$2,000$ for investments 2 and 3. Hence, Finco should invest $x_2(2) = \$1,000$ in investment 2. Then \$1,000 is left for investment 3, so Finco chooses to invest $x_3(1) = \$1,000$ in investment 3. Therefore, Finco can attain a maximum net present value of $f_1(6) = \$49,000$ by investing \$4,000 in investment 1, \$1,000 in investment 2, and \$1,000 in investment 3.

Network Representation of Resource Example

As with the inventory example of Section 13.3, Finco's problem has a network representation, equivalent to finding the *longest route* from $(1, 6)$ to $(4, 0)$ in Figure 6. In the figure, the node (t, d) represents the situation in which d thousand dollars is available for investments $t, t + 1, \ldots, 3$. The arc joining the nodes (t, d) and $(t + 1, d - x)$ has a length $r_t(x)$ corresponding to the net present value obtained by investing x thousand dollars in investment t. For example, the arc joining nodes $(2, 4)$ and $(3, 1)$ has a length $r_2(3) = \$16,000$, corresponding to the \$16,000 net present value that can be obtained by invest-

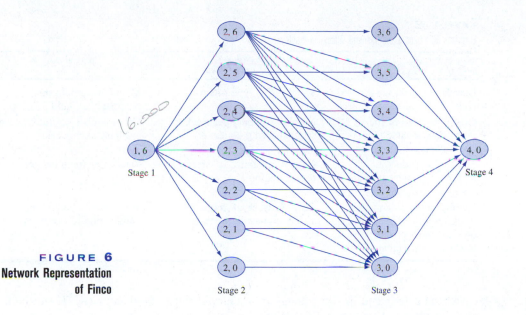

FIGURE 6
Network Representation of Finco

ing \$3,000 in investment 2. Note that not all pairs of nodes in adjacent stages are joined by arcs. For example, there is no arc joining the nodes (2, 4) and (3, 5); after all, if you have only \$4,000 available for investments 2 and 3, how can you have \$5,000 available for investment 3? From our computations, we see that the longest path from (1, 6) to (4, 0) is (1, 6)–(2, 2)–(3, 1)–(4, 0).

Generalized Resource Allocation Problem

We now consider a generalized version of Example 5. Suppose we have w units of a resource available and T activities to which the resource can be allocated. If activity t is implemented at a level x_t (we assume x_t must be a non-negative integer), then $g_t(x_t)$ units of the resource are used by activity t, and a benefit $r_t(x_t)$ is obtained. The problem of determining the allocation of resources that maximizes total benefit subject to the limited resource availability may be written as

$$\max \sum_{t=1}^{t=T} r_t(x_t)$$

$$\text{s.t.} \quad \sum_{t=1}^{t=T} g_t(x_t) \leq w \tag{6}$$

where x_t must be a member of $\{0, 1, 2, \dots\}$. Some possible interpretations of $r_t(x_t)$, $g_t(x_t)$, and w are given in Table 8.

To solve (6) by dynamic programming, define $f_t(d)$ to be the maximum benefit that can be obtained from activities $t, t + 1, \dots, T$ if d units of the resource may be allocated to activities $t, t + 1, \dots, T$. We may generalize the recursions of Example 5 to this situation by writing

$$f_{T+1}(d) = 0 \quad \text{for all } d$$

$$f_t(d) = \max_{x_t} \{r_t(x_t) + f_{t+1}[d - g_t(x_t)]\} \tag{7}$$

where x_t must be a non-negative integer satisfying $g_t(x_t) \leq d$. Let $x_t(d)$ be any value of x_t that attains $f_t(d)$. To use (7) to determine an optimal allocation of resources to activities $1, 2, \dots, T$, we begin by determining all $f_T(\cdot)$ and $x_T(\cdot)$. Then we use (7) to determine all $f_{T-1}(\cdot)$ and $x_{T-1}(\cdot)$, continuing to work backward in this fashion until all $f_2(\cdot)$ and $x_2(\cdot)$

TABLE 8
Examples of a Generalized Resource Allocation Problem

Interpretation of $r_t(x_t)$	Interpretation of $g_t(x_t)$	Interpretation of w
Benefit from placing x_t type t items in a knapsack	Weight of x_t type t items	Maximum weight that knapsack can hold
Grade obtained in course t if we study course t for x_t hours per week	Number of hours per week x_t spent studying course t	Total number of study hours available each week
Sales of a product in region t if x_t sales reps are assigned to region t	Cost of assigning x_t sales reps to region t	Total sales force budget
Number of fire alarms per week responded to within one minute if precinct t is assigned x_t engines	Cost per week of maintaining x_t fire engines in precinct t	Total weekly budget for maintaining fire engines

have been determined. To wind things up, we now calculate $f_1(w)$ and $x_1(w)$. Then we implement activity 1 at a level $x_1(w)$. At this point, we have $w - g_1[x_1(w)]$ units of the resource available for activities 2, 3, ..., T. Then activity 2 should be implemented at a level of $x_2\{w - g_1[x_1(w)]\}$. We continue in this fashion until we have determined the level at which all activities should be implemented.

Solution of Knapsack Problems by Dynamic Programming

We illustrate the use of (7) by solving a simple knapsack problem (see Section 9.5). Then we develop an alternative recursion that can be used to solve knapsack problems.

EXAMPLE 6   Knapsack

Suppose a 10-lb knapsack is to be filled with the items listed in Table 9. To maximize total benefit, how should the knapsack be filled?

Solution   We have $r_1(x_1) = 11x_1$, $r_2(x_2) = 7x_2$, $r_3(x_3) = 12x_3$, $g_1(x_1) = 4x_1$, $g_2(x_2) = 3x_2$, and $g_3(x_3) = 5x_3$. Define $f_t(d)$ to be the maximum benefit that can be earned from a d-pound knapsack that is filled with items of Type t, $t + 1, ..., 3$.

Stage 3 Computations

Now (7) yields

$$f_3(d) = \max_{x_3}\{12x_3\}$$

TABLE 9
Weights and Benefits for Knapsack

Item	Weight (lb)	Benefit
1	4	11
2	3	7
3	5	12

where $5x_3 \leq d$ and x_3 is a non-negative integer. This yields

$$f_3(10) = 24$$
$$f_3(5) = f_3(6) = f_3(7) = f_3(8) = f_3(9) = 12$$
$$f_3(0) = f_3(1) = f_3(2) = f_3(3) = f_3(4) = 0$$
$$x_3(10) = 2$$
$$x_3(9) = x_3(8) = x_3(7) = x_3(6) = x_3(5) = 1$$
$$x_3(0) = x_3(1) = x_3(2) = x_3(3) = x_3(4) = 0$$

Stage 2 Computations

Now (7) yields

$$f_2(d) = \max_{x_2} \{7x_2 + f_3(d - 3x_2)\}$$

where x_2 must be a non-negative integer satisfying $3x_2 \leq d$. We now obtain

$$f_2(10) = \max \begin{cases} 7(0) + f_3(10) = 24* & x_2 = 0 \\ 7(1) + f_3(7) = 19 & x_2 = 1 \\ 7(2) + f_3(4) = 14 & x_2 = 2 \\ 7(3) + f_3(1) = 21 & x_2 = 3 \end{cases}$$

Thus, $f_2(10) = 24$ and $x_2(10) = 0$.

$$f_2(9) = \max \begin{cases} 7(0) + f_3(9) = 12 & x_2 = 0 \\ 7(1) + f_3(6) = 19 & x_2 = 1 \\ 7(2) + f_3(3) = 14 & x_2 = 2 \\ 7(3) + f_3(0) = 21* & x_2 = 3 \end{cases}$$

Thus, $f_2(9) = 21$ and $x_2(9) = 3$.

$$f_2(8) = \max \begin{cases} 7(0) + f_3(8) = 12 & x_2 = 0 \\ 7(1) + f_3(5) = 19* & x_2 = 1 \\ 7(2) + f_3(2) = 14 & x_2 = 2 \end{cases}$$

Thus, $f_2(8) = 19$ and $x_2(8) = 1$.

$$f_2(7) = \max \begin{cases} 7(0) + f_3(7) = 12 & x_2 = 0 \\ 7(1) + f_3(4) = 7 & x_2 = 1 \\ 7(2) + f_3(1) = 14* & x_2 = 2 \end{cases}$$

Thus, $f_2(7) = 14$ and $x_2(7) = 2$.

$$f_2(6) = \max \begin{cases} 7(0) + f_3(6) = 12 & x_2 = 0 \\ 7(1) + f_3(3) = 7 & x_2 = 1 \\ 7(2) + f_3(0) = 14* & x_2 = 2 \end{cases}$$

Thus, $f_2(6) = 14$ and $x_2(6) = 2$.

$$f_2(5) = \max \begin{cases} 7(0) + f_3(5) = 12* & x_2 = 0 \\ 7(1) + f_3(2) = 7 & x_2 = 1 \end{cases}$$

Thus, $f_2(5) = 12$ and $x_2(5) = 0$.

$$f_2(4) = \max \begin{cases} 7(0) + f_3(4) = 0 & x_2 = 0 \\ 7(1) + f_3(1) = 7* & x_2 = 1 \end{cases}$$

Thus, $f_2(4) = 7$ and $x_2(4) = 1$.

$$f_2(3) = \max \begin{cases} 7(0) + f_3(3) = 0 & x_2 = 0 \\ 7(1) + f_3(0) = 7^* & x_2 = 1 \end{cases}$$

Thus, $f_2(3) = 7$ and $x_2(3) = 1$.

$$f_2(2) = 7(0) + f_3(2) = 0 \qquad x_2 = 0$$

Thus, $f_2(2) = 0$ and $x_2(2) = 0$.

$$f_2(1) = 7(0) + f_3(1) = 0 \qquad x_2 = 0$$

Thus, $f_2(1) = 0$ and $x_2(1) = 0$.

$$f_2(0) = 7(0) + f_3(0) = 0 \qquad x_2 = 0$$

Thus, $f_2(0) = 0$ and $x_2(0) = 0$.

Stage 1 Computations

Finally, we determine $f_1(10)$ from

$$f_1(10) = \max \begin{cases} 11(0) + f_2(10) = 24 & x_1 = 0 \\ 11(1) + f_2(6) = 25^* & x_1 = 1 \\ 11(2) + f_2(2) = 22 & x_1 = 2 \end{cases}$$

Determination of the Optimal Solution to Knapsack Problem

We have $f_1(10) = 25$ and $x_1(10) = 1$. Hence, we should include one Type 1 item in the knapsack. Then we have $10 - 4 = 6$ lb left for Type 2 and Type 3 items, so we should include $x_2(6) = 2$ Type 2 items. Finally, we have $6 - 2(3) = 0$ lb left for Type 3 items, and we include $x_3(0) = 0$ Type 3 items. In summary, the maximum benefit that can be gained from a 10-lb knapsack is $f_3(10) = 25$. To obtain a benefit of 25, one Type 1 and two Type 2 items should be included.

Network Representation of Knapsack Problem

Finding the optimal solution to Example 6 is equivalent to finding the longest path in Figure 7 from node $(10, 1)$ to some stage 4 node. In Figure 7, for $t \leq 3$, the node (d, t) represents a situation in which d pounds of space may be allocated to items of Type $t, t + 1, \ldots, 3$. The node $(d, 4)$ represents d pounds of unused space. Each arc from a stage t node to a stage $t + 1$ node represents a decision of how many Type t items are placed in the knapsack. For example, the arc from $(10, 1)$ to $(6, 2)$ represents placing one Type 1 item in the knapsack. This leaves $10 - 4 = 6$ lb for items of Types 2 and 3. This arc has a length of 11, representing the benefit obtained by placing one Type 1 item in the knapsack. Our solution to Example 6 shows that the longest path in Figure 7 from node $(10, 1)$ to a stage 4 node is $(10, 1)$–$(6, 2)$–$(0, 3)$–$(0, 4)$. We note that the optimal solution to a knapsack problem does not always use all the available weight. For example, the reader should verify that if a Type 1 item earned 16 units of benefit, the optimal solution would be to include two type 1 items, corresponding to the path $(10, 1)$–$(2, 2)$–$(2, 3)$–$(2, 4)$. This solution leaves 2 lb of space unused.

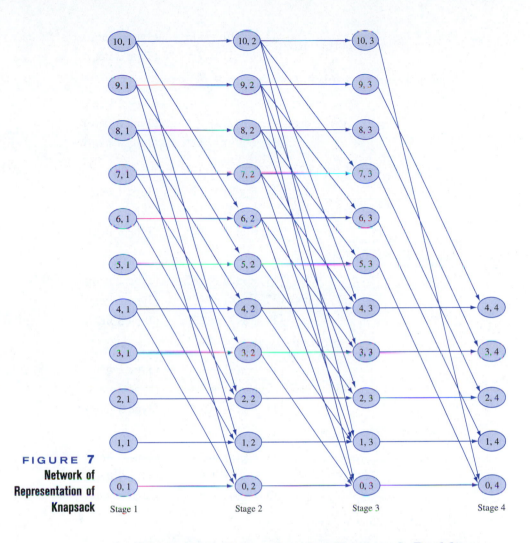

FIGURE 7
Network of
Representation of
Knapsack

Stage 1   Stage 2   Stage 3   Stage 4

An Alternative Recursion for Knapsack Problems

Other approaches can be used to solve knapsack problems by dynamic programming. The approach we now discuss builds up the optimal knapsack by first determining how to fill a small knapsack optimally and then, using this information, how to fill a larger knapsack optimally. We define $g(w)$ to be the maximum benefit that can be gained from a w-lb knapsack. In what follows, b_j is the benefit earned from a single Type j item, and w_j is the weight of a single Type j item. Clearly, $g(0) = 0$, and for $w > 0$,

$$g(w) = \max_{j} \{b_j + g(w - w_j)\} \tag{8}$$

where j must be a member of $\{1, 2, 3\}$, and j must satisfy $w_j \leq w$. The reasoning behind (8) is as follows: To fill a w-lb knapsack optimally, we must begin by putting some type of item into the knapsack. If we begin by putting a Type j item into a w-lb knapsack, the best we can do is earn b_j + [best we can do from a $(w - w_j)$-lb knapsack]. After noting that a Type j item can be placed into a w-lb knapsack only if $w_j \leq w$, we obtain (8). We define $x(w)$ to be any type of item that attains the maximum in (8) and $x(w) = 0$ to mean that no item can fit into a w-lb knapsack.

To illustrate the use of (8), we re-solve Example 6. Because no item can fit in a 0-, 1-, or 2-lb knapsack, we have $g(0) = g(1) = g(2) = 0$ and $x(0) = x(1) = x(2) = 0$. Only a Type 2 item fits into a 3-lb knapsack, so we have that $g(3) = 7$ and $x(3) = 2$. Continuing, we find that

$$g(4) = \max \begin{cases} 11 + g(0) = 11* & \text{(Type 1 item)} \\ 7 + g(1) = 7 & \text{(Type 2 item)} \end{cases}$$

Thus, $g(4) = 11$ and $x(4) = 1$.

$$g(5) = \max \begin{cases} 11 + g(1) = 11 & \text{(Type 1 item)} \\ 7 + g(2) = 7 & \text{(Type 2 item)} \\ 12 + g(0) = 12* & \text{(Type 3 item)} \end{cases}$$

Thus, $g(5) = 12$ and $x(5) = 3$.

$$g(6) = \max \begin{cases} 11 + g(2) = 11 & \text{(Type 1 item)} \\ 7 + g(3) = 14* & \text{(Type 2 item)} \\ 12 + g(1) = 12 & \text{(Type 3 item)} \end{cases}$$

Thus, $g(6) = 14$ and $x(6) = 2$.

$$g(7) = \max \begin{cases} 11 + g(3) = 18* & \text{(Type 1 item)} \\ 7 + g(4) = 18* & \text{(Type 2 item)} \\ 12 + g(2) = 12 & \text{(Type 3 item)} \end{cases}$$

Thus, $g(7) = 18$ and $x(7) = 1$ or $x(7) = 2$.

$$g(8) = \max \begin{cases} 11 + g(4) = 22* & \text{(Type 1 item)} \\ 7 + g(5) = 19 & \text{(Type 2 item)} \\ 12 + g(3) = 19 & \text{(Type 3 item)} \end{cases}$$

Thus, $g(8) = 22$ and $x(8) = 1$.

$$g(9) = \max \begin{cases} 11 + g(5) = 23* & \text{(Type 1 item)} \\ 7 + g(6) = 21 & \text{(Type 2 item)} \\ 12 + g(4) = 23* & \text{(Type 3 item)} \end{cases}$$

Thus, $g(9) = 23$ and $x(9) = 1$ or $x(9) = 3$.

$$g(10) = \max \begin{cases} 11 + g(6) = 25* & \text{(Type 1 item)} \\ 7 + g(7) = 25* & \text{(Type 2 item)} \\ 12 + g(5) = 24 & \text{(Type 3 item)} \end{cases}$$

Thus, $g(10) = 25$ and $x(10) = 1$ or $x(10) = 2$. To fill the knapsack optimally, we begin by putting any $x(10)$ item in the knapsack. Let's arbitrarily choose a Type 1 item. This leaves us with $10 - 4 = 6$ lb to fill, so we now put an $x(10 - 4) = 2$ (Type 2) item in the knapsack. This leaves us with $6 - 3 = 3$ lb to fill, which we do with an $x(6 - 3) = 2$ (Type 2) item. Hence, we may attain the maximum benefit of $g(10) = 25$ by filling the knapsack with two Type 2 items and one Type 1 item.

A Turnpike Theorem

For a knapsack problem, let

$$c_j = \text{benefit obtained from each type } j \text{ item}$$
$$w_j = \text{weight of each type } j \text{ item}$$

In terms of benefit per unit weight, the best item is the item with the largest value of $\frac{c_j}{w_j}$. Assume there are n types of items that have been ordered, so that

$$\frac{c_1}{w_1} \geq \frac{c_2}{w_2} \geq \cdots \geq \frac{c_n}{w_n}$$

Thus, Type 1 items are the best, Type 2 items are the second best, and so on. Recall from Section 9.5 that it is possible for the optimal solution to a knapsack problem to use none of the best item. For example, the optimal solution to the knapsack problem

$$\max z = 16x_1 + 22x_2 + 12x_3 + 8x_4$$
$$\text{s.t.} \quad 5x_1 + 7x_2 + 5x_3 + 4x_4 \leq 14$$
$$x_i \text{ non-negative integer}$$

is $z = 44$, $x_2 = 2$, $x_1 = x_3 = x_4 = 0$, and this solution does not use any of the best (Type 1) item. Assume that

$$\frac{c_1}{w_1} > \frac{c_2}{w_2}$$

Thus, there is a unique best item type. It can be shown that for some number w^*, it is optimal to use at least one Type 1 item if the knapsack is allowed to hold w pounds, where $w \geq w^*$. In Problem 6 at the end of this section, you will show that this result holds for

$$w^* = \frac{c_1 w_1}{c_1 - w_1 \left(\frac{c_2}{w_2} \right)}$$

Thus, for the knapsack problem

$$\max z = 16x_1 + 22x_2 + 12x_3 + 8x_4$$
$$\text{s.t.} \quad 5x_1 + 7x_2 + 5x_3 + 4x_4 \leq w$$
$$x_i \text{ non-negative integer}$$

at least one Type 1 item will be used if

$$w \geq \frac{16(5)}{16 - 5\left(\frac{22}{7} \right)} = 280$$

This result can greatly reduce the computation needed to solve a knapsack problem. For example, suppose that $w = 4{,}000$. We know that for $w \geq 280$, the optimal solution will use at least one Type 1 item, so we can conclude that the optimal way to fill a 4,000-lb knapsack will consist of one Type 1 item plus the optimal way to fill a knapsack of $4{,}000 - 5 = 3{,}995$ lb. Repeating this reasoning shows that the optimal way to fill a 4,000-lb knapsack will consist of $\frac{4{,}000 - 280}{5} = 744$ Type 1 items plus the optimal way to fill a knapsack of 280 lb. This reasoning substantially reduces the computation needed to determine how to fill a 4,000-lb knapsack. (Actually, the 280-lb knapsack will use at least one Type 1 item, so we know that to fill a 4,000-lb knapsack optimally, we can use 745 Type 1 items and then optimally fill a 275-lb knapsack.)

Why is this result referred to as a **turnpike theorem**? Think about taking an automobile trip in which our goal is to minimize the time needed to complete the trip. For a long enough trip, it may be advantageous to go slightly out of our way so that most of the trip will be spent on a turnpike, on which we can travel at the greatest speed. For a short trip, it may not be worth our while to go out of our way to get on the turnpike.

Similarly, in a long (large-weight) knapsack problem, it is always optimal to use some of the best items, but this may not be the case in a short knapsack problem. Turnpike results abound in the dynamic programming literature [see Morton (1979)].

PROBLEMS

Group A

1 J. R. Carrington has $4 million to invest in three oil well sites. The amount of revenue earned from site $i(i = 1, 2, 3)$ depends on the amount of money invested in site i (see Table 10). Assuming that the amount invested in a site must be an exact multiple of $1 million, use dynamic programming to determine an investment policy that will maximize the revenue J. R. will earn from his three oil wells.

2 Use either of the approaches outlined in this section to solve the following knapsack problem:

$$\max z = 5x_1 + 4x_2 + 2x_3$$
$$\text{s.t.} \quad 4x_1 + 3x_2 + 2x_3 \leq 8$$
$$x_1, x_2, x_3 \geq 0; x_1, x_2, x_3 \text{ integer}$$

3 The knapsack problem of Problem 2 can be viewed as finding the longest route in a particular network.

a Draw the network corresponding to the recursion derived from (7).

b Draw the network corresponding to the recursion derived from (8).

4 The number of crimes in each of a city's three police precincts depends on the number of patrol cars assigned to each precinct (see Table 11). Five patrol cars are available. Use dynamic programming to determine how many patrol cars should be assigned to each precinct.

TABLE 10

Amount Invested ($ Millions)	Revenue ($ Millions)		
	Site 1	Site 2	Site 3
0	4	3	3
1	7	6	7
2	8	10	8
3	9	12	13
4	11	14	15

TABLE 11

	No. of Patrol Cars Assigned to Precinct					
Precinct	0	1	2	3	4	5
1	14	10	7	4	1	0
2	25	19	16	14	12	11
3	20	14	11	8	6	5

5 Use dynamic programming to solve a knapsack problem in which the knapsack can hold up to 13 lb (see Table 12).

Group B

6 Consider a knapsack problem for which

$$\frac{c_1}{w_1} > \frac{c_2}{w_2}$$

Show that if the knapsack can hold w pounds, and $w \geq w^*$, where

$$w^* = \frac{c_1 w_1}{c_1 - w_1 \left(\dfrac{c_2}{w_2}\right)}$$

then the optimal solution to the knapsack problem must use at least one Type 1 item.

TABLE 12

Item	Weight (lb)	Benefit
1	3	12
2	5	25
3	7	50

13.5 Equipment-Replacement Problems

Many companies and customers face the problem of determining how long a machine should be utilized before it should be traded in for a new one. Problems of this type are called **equipment-replacement problems** and can often be solved by dynamic programming.

| EXAMPLE 7 | **Equipment Replacement** |

An auto repair shop always needs to have an engine analyzer available. A new engine analyzer costs $1,000. The cost m_i of maintaining an engine analyzer during its ith year of operation is as follows: $m_1 = \$60$, $m_2 = \$80$, $m_3 = \$120$. An analyzer may be kept for

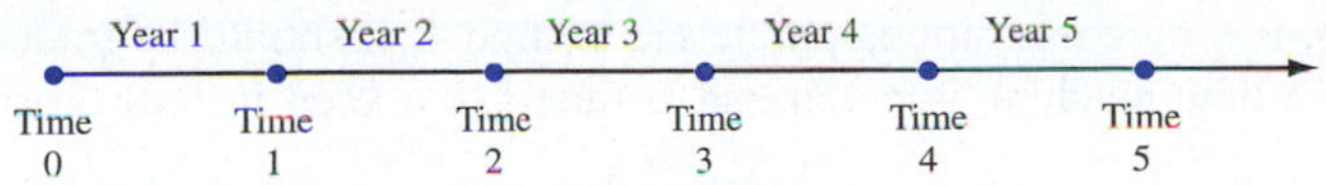

FIGURE 8
Time Horizon for
Equipment
Replacement

1, 2, or 3 years; after i years of use ($i = 1, 2, 3$), it may be traded in for a new one. If an i-year-old engine analyzer is traded in, a salvage value s_i is obtained, where $s_1 = \$800$, $s_2 = \$600$, and $s_3 = \$500$. Given that a new machine must be purchased now (time 0; see Figure 8), the shop wants to determine a replacement and trade-in policy that minimizes net costs = (maintenance costs) + (replacement costs) – (salvage value received) during the next 5 years.

Solution We note that after a new machine is purchased, the firm must decide when the newly purchased machine should be traded in for a new one. With this in mind, we define $g(t)$ to be the minimum net cost incurred from time t until time 5 (including the purchase cost and salvage value for the newly purchased machine) given that a new machine has been purchased at time t. We also define c_{tx} to be the net cost (including purchase cost and salvage value) of purchasing a machine at time t and operating it until time x. Then the appropriate recursion is

$$g(t) = \min_{x} \{c_{tx} + g(x)\} \qquad (t = 0, 1, 2, 3, 4) \tag{9}$$

where x must satisfy the inequalities $t + 1 \le x \le t + 3$ and $x \le 5$. Because the problem is over at time 5, no cost is incurred from time 5 onward, so we may write $g(5) = 0$.

To justify (9), note that after a new machine is purchased at time t, we must decide when to replace the machine. Let x be the time at which the replacement occurs. The replacement must be after time t but within 3 years of time t. This explains the restriction that $t + 1 \le x \le t + 3$. Since the problem ends at time 5, we must also have $x \le 5$. If we choose to replace the machine at time x, then what will be the cost from time t to time 5? Simply the sum of the cost incurred from the purchase of the machine to the sale of the machine at time x (which is by definition c_{tx}) and the total cost incurred from time x to time 5 (given that a new machine has just been purchased at time x). By the principle of optimality, the latter cost is, of course, $g(x)$. Hence, if we keep the machine that was purchased at time t until time x, then from time t to time 5, we incur a cost of $c_{tx} + g(x)$. Thus, x should be chosen to minimize this sum, and this is exactly what (9) does. We have assumed that maintenance costs, salvage value, and purchase price remain unchanged over time, so each c_{tx} will depend only on how long the machine is kept; that is, each c_{tx} depends only on $x - t$. More specifically,

$$c_{tx} = \$1,000 + m_1 + \cdots + m_{x-t} - s_{x-t}$$

This yields

$$c_{01} = c_{12} = c_{23} = c_{34} = c_{45} = 1,000 + 60 - 800 = \$260$$
$$c_{02} = c_{13} = c_{24} = c_{35} = 1,000 + 60 + 80 - 600 = \$540$$
$$c_{03} = c_{14} = c_{25} = 1,000 + 60 + 80 + 120 - 500 = \$760$$

We begin by computing $g(4)$ and work backward until we have computed $g(0)$. Then we use our knowledge of the values of x attaining $g(0)$, $g(1)$, $g(2)$, $g(3)$, and $g(4)$ to determine the optimal replacement strategy. The calculations follow.

At time 4, there is only one sensible decision (keep the machine until time 5 and sell it for its salvage value), so we find

$$g(4) = c_{45} + g(5) = 260 + 0 = \$260^*$$

Thus, if a new machine is purchased at time 4, it should be traded in at time 5.

If a new machine is purchased at time 3, we keep it until time 4 or time 5. Hence,

$$g(3) = \min \begin{cases} c_{34} + g(4) = 260 + 260 = \$520^* & \text{(Trade at time 4)} \\ c_{35} + g(5) = 540 + 0 = \$540 & \text{(Trade at time 5)} \end{cases}$$

Thus, if a new machine is purchased at time 3, we should trade it in at time 4.

If a new machine is purchased at time 2, we trade it in at time 3, time 4, or time 5. This yields

$$g(2) = \min \begin{cases} c_{23} + g(3) = 260 + 520 = \$780 & \text{(Trade at time 3)} \\ c_{24} + g(4) = 540 + 260 = \$800 & \text{(Trade at time 4)} \\ c_{25} + g(5) = \$760^* & \text{(Trade at time 5)} \end{cases}$$

Thus, if we purchase a new machine at time 2, we should keep it until time 5 and then trade it in.

If a new machine is purchased at time 1, we trade it in at time 2, time 3, or time 4. Then

$$g(1) = \min \begin{cases} c_{12} + g(2) = 260 + 760 = \$1,020^* & \text{(Trade at time 2)} \\ c_{13} + g(3) = 540 + 520 = \$1,060 & \text{(Trade at time 3)} \\ c_{14} + g(4) = 760 + 260 = \$1,020^* & \text{(Trade at time 4)} \end{cases}$$

Thus, if a new machine is purchased at time 1, it should be traded in at time 2 or time 4.

The new machine that was purchased at time 0 may be traded in at time 1, time 2, or time 3. Thus,

$$g(0) = \min \begin{cases} c_{01} + g(1) = 260 + 1,020 = \$1,280^* & \text{(Trade at time 1)} \\ c_{02} + g(2) = 540 + 760 = \$1,300 & \text{(Trade at time 2)} \\ c_{03} + g(3) = 760 + 520 = \$1,280^* & \text{(Trade at time 3)} \end{cases}$$

Thus, the new machine purchased at time 0 should be replaced at time 1 or time 3. Let's arbitrarily choose to replace the time 0 machine at time 1. Then the new time 1 machine may be traded in at time 2 or time 4. Again we make an arbitrary choice and replace the time 1 machine at time 2. Then the time 2 machine should be kept until time 5, when it is sold for salvage value. With this replacement policy, we will incur a net cost of $g(0) = \$1,280$. The reader should verify that the following replacement policies are also optimal: (1) trading in at times 1, 4, and 5 and (2) trading in at times 3, 4, and 5.

We have assumed that all costs remain stationary over time. This assumption was made solely to simplify the computation of the c_{tx}'s. If we had relaxed the assumption of stationary costs, then the only complication would have been that the c_{tx}'s would have been messier to compute. We also note that if a short planning horizon is used, the optimal replacement policy may be extremely sensitive to the length of the planning horizon. Thus, more meaningful results can be obtained by using a longer planning horizon.

An equipment-replacement model was actually used by Phillips Petroleum to reduce costs associated with maintaining the company's stock of trucks (see Waddell (1983)).

Network Representation of Equipment-Replacement Problem

The reader should verify that our solution to Example 7 was equivalent to finding the shortest path from node 0 to node 5 in the network in Figure 9. The length of the arc joining nodes i and j is c_{ij}.

FIGURE 9
Network Representation
of Equipment
Replacement

Time
0   1   2   3   4   5

An Alternative Recursion

There is another dynamic programming formulation of the equipment-replacement model. If we define the stage to be the time t and the state at any stage to be the age of the engine analyzer at time t, then an alternative dynamic programming recursion can be developed. Define $f_t(x)$ to be the minimum cost incurred from time t to time 5, given that at time t the shop has an x-year-old analyzer. The problem is over at time 5, so we sell the machine at time 5 and receive $-s_x$. Then $f_5(x) = -s_x$, and for $t = 0, 1, 2, 3, 4,$

$$f_t(3) = -500 + 1{,}000 + 60 + f_{t+1}(1) \qquad \text{(Trade)} \tag{10}$$

$$f_t(2) = \min \begin{cases} -600 + 1{,}000 + 60 + f_{t+1}(1) & \text{(Trade)} \\ 120 + f_{t+1}(3) & \text{(Keep)} \end{cases} \tag{10.1}$$

$$f_t(1) = \min \begin{cases} -800 + 1{,}000 + 60 + f_{t+1}(1) & \text{(Trade)} \\ 80 + f_{t+1}(2) & \text{(Keep)} \end{cases} \tag{10.2}$$

$$f_0(0) = 1{,}000 + 60 + f_1(1) \qquad \text{(Keep)} \tag{10.3}$$

$$\tag{10}$$

The rationale behind Equations (10)–(10.3) is that if we have a 1- or 2-year-old analyzer, then we must decide between replacing the machine or keeping it another year. In (10.1) and (10.2), we compare the costs of these two options. For any option, the total cost from t until time 5 is the sum of the cost during the current year plus costs from time $t + 1$ to time 5. If we have a 3-year-old analyzer, then we must replace it, so there is no choice. The way we have defined the state means that it is only possible to be in state 0 at time 0. In this case, we must keep the analyzer for the first year (incurring a cost of \$1,060). From this point on, a total cost of $f_1(1)$ is incurred. Thus, (10.3) follows. Since we know that $f_5(1) = -800$, $f_5(2) = -600$, and $f_5(3) = -500$, we can immediately compute all the $f_4(\cdot)$'s. Then we can compute the $f_3(\cdot)$'s. We continue in this fashion until $f_0(0)$ is determined (remember that we begin with a new machine). Then we follow our usual method for determining an optimal policy. That is, if $f_0(0)$ is attained by keeping the machine, then we keep the machine for a year and then, during year 1, we choose the action that attains $f_1(1)$. Continuing in this fashion, we can determine for each time whether or not the machine should be replaced. (See Problem 1 at the end of this section.)

PROBLEMS

Group A

1 Use Equations (10)–(10.3) to determine an optimal replacement policy for the engine analyzer example.

2 Suppose that a new car costs \$10,000 and that the annual operating cost and resale value of the car are as shown in Table 13. If I have a new car now, determine a replacement policy that minimizes the net cost of owning and operating a car for the next six years.

3 It costs \$40 to buy a telephone from a department store. The estimated maintenance cost for each year of operation is shown in Table 14. (I can keep a telephone for at most five years.) I have just purchased a new telephone, and my old telephone has no salvage value. Determine how to minimize the total cost of purchasing and operating a telephone for the next six years.

TABLE **13**

Age of Car (Years)	Resale Value ($)	Operating Cost ($)	
1	7,000	300	(year 1)
2	6,000	500	(year 2)
3	4,000	800	(year 3)
4	3,000	1,200	(year 4)
5	2,000	1,600	(year 5)
6	1,000	2,200	(year 6)

TABLE **14**

Year	Maintenance Cost ($)
1	20
2	30
3	40
4	60
5	70

13.6 Formulating Dynamic Programming Recursions

In many dynamic programming problems (such as the inventory and shortest path examples), a given stage simply consists of all the possible states that the system can occupy at that stage. If this is the case, then the dynamic programming recursion (for a min problem) can often be written in the following form:

$$f_t(i) = \min\{(\text{cost during stage } t) + f_{t+1} (\text{new state at stage } t + 1)\} \qquad \textbf{(11)}$$

where the minimum in (11) is over all decisions that are allowable, or feasible, when the state at stage t is i. In (11), $f_t(i)$ is the minimum cost incurred from stage t to the end of the problem (say, the problem ends after stage T), given that at stage t the state is i.

Equation (11) reflects the fact that the minimum cost incurred from stage t to the end of the problem must be attained by choosing at stage t an allowable decision that minimizes the sum of the costs incurred during the current stage (stage t) plus the minimum cost that can be incurred from stage $t + 1$ to the end of the problem. Correct formulation of a recursion of the form (11) requires that we identify three important aspects of the problem:

Aspect 1 *The set of decisions that is allowable, or feasible, for the given state and stage.* Often, the set of feasible decisions depends on both t and i. For instance, in the inventory example of Section 13.3, let

$$d_t = \text{demand during month } t$$

$$i_t = \text{inventory at beginning of month } t$$

In this case, the set of allowable month t decisions (let x_t represent an allowable production level) consists of the members of $\{0, 1, 2, 3, 4, 5\}$ that satisfy $0 \le (i_t + x_t - d_t) \le 4$. Note how the set of allowable decisions at time t depends on the stage t and the state at time t, which is i_t.

Aspect 2 *We must specify how the cost during the current time period (stage t) depends on the value of t, the current state, and the decision chosen at stage t.* For instance, in the inventory example of Section 13.3, suppose a production level x_t is chosen during month t. Then the cost during month t is given by $c(x_t) + (\frac{1}{2})(i_t + x_t - d_t)$.

Aspect 3 *We must specify how the state at stage t + 1 depends on the value of t, the state at stage t, and the decision chosen at stage t.* Again referring to the inventory example, the month $t + 1$ state is $i_t + x_t - d_t$.

If you have properly identified the state, stage, and decision, then aspects 1–3 shouldn't be too hard to handle. A word of caution, however: Not all recursions are of the form (11). For instance, our first equipment-replacement recursion skipped over time $t + 1$.

This often occurs when the stage alone supplies sufficient information to make an optimal decision. We now work through several examples that illustrate the art of formulating dynamic programming recursions.

EXAMPLE 8 **A Fishery**

The owner of a lake must decide how many bass to catch and sell each year. If she sells x bass during year t, then a revenue $r(x)$ is earned. The cost of catching x bass during a year is a function $c(x, b)$ of the number of bass caught during the year and of b, the number of bass in the lake at the beginning of the year. Of course, bass do reproduce. To model this, we assume that the number of bass in the lake at the beginning of a year is 20% more than the number of bass left in the lake at the end of the previous year. Assume that there are 10,000 bass in the lake at the beginning of the first year. Develop a dynamic programming recursion that can be used to maximize the owner's net profits over a T-year horizon.

Solution In problems where decisions must be made at several points in time, there is often a trade-off of current benefits against future benefits. For example, we could catch many bass early in the problem, but then the lake would be depleted in later years, and there would be very few bass to catch. On the other hand, if we catch very few bass now, we won't make much money early, but we can make a lot of money near the end of the horizon. In intertemporal optimization problems, dynamic programming is often used to analyze these complex trade-offs.

At the beginning of year T, the owner of the lake need not worry about the effect that the capture of bass will have on the future population of the lake. (At time T, there is no future!) So at the beginning of year T, the problem is relatively easy to solve. For this reason, we let time be the stage. At each stage, the owner of the lake must decide how many bass to catch. We define x_t to be the number of bass caught during year t. To determine an optimal value of x_t, the owner of the lake need only know the number of bass (call it b_t) in the lake at the beginning of year t. Therefore, the state at the beginning of year t is b_t.

We define $f_t(b_t)$ to be the maximum net profit that can be earned from bass caught during years $t, t + 1, \ldots, T$ given that b_t bass are in the lake at the beginning of year t. We may now dispose of aspects 1–3 of the recursion.

Aspect 1 What are the allowable decisions? During any year, we can't catch more bass than there are in the lake. Thus, in each state and for all t, $0 \leq x_t \leq b_t$ must hold.

Aspect 2 What is the net profit earned during year t? If x_t bass are caught during a year that begins with b_t bass in the lake, then the net profit is $r(x_t) - c(x_t, b_t)$.

Aspect 3 What will be the state during year $t + 1$? At the end of year t, there will be $b_t - x_t$ bass in the lake. By the beginning of year $t + 1$, these bass will have multiplied by 20%. This implies that at the beginning of year $t + 1$, $1.2(b_t - x_t)$ bass will be in the lake. Thus, the year $t + 1$ state will be $1.2(b_t - x_t)$.

We can now use (11) to develop the appropriate recursion. After year T, there are no future profits to consider, so

$$f_T(b_T) = \max_{x_T}\{r_T(x_T) - c(x_T, b_T)\}$$

where $0 \leq x_T \leq b_T$. Applying (11), we obtain

$$f_t(b_t) = \max\{r(x_t) - c(x_t, b_t) + f_{t+1}[1.2(b_t - x_t)]\} \tag{12}$$

where $0 \leq x_t \leq b_t$. To begin the computations, we first determine $f_T(b_T)$ for all values of b_T that might occur [b_T could be up to $10{,}000(1.2)^{T-1}$; why?]. Then we use (12) to work backward until $f_1(10,000)$ has been computed. Then, to determine an optimal fishing policy, we begin by choosing x_1 to be any value attaining the maximum in the (12) equation for $f_1(10,000)$. Then year 2 will begin with $1.2(10,000 - x_1)$ bass in the lake. This means that x_2 should be chosen to be any value attaining the maximum in the (12) equation for $f_2(1.2(10,000 - x_1))$. Continue in this fashion until the optimal values of $x_3, x_4, \ldots, x_T$ have been determined.

Incorporating the Time Value of Money into Dynamic Programming Formulations

A weakness of the current formulation is that profits received during later years are weighted the same as profits received during earlier years. As mentioned in the Chapter 3 discussion of discounting, later profits should be weighted less than earlier profits. Suppose that for some $\beta < 1$, \$1 received at the beginning of year $t + 1$ is equivalent to β dollars received at the beginning of year t. We can incorporate this idea into the dynamic programming recursion by replacing (12) with

$$f_t(b_t) = \max_{x_t} \{r(x_t) - c(x_t, b_t) + \beta f_{t+1}[1.2(b_t - x_t)]\} \tag{12'}$$

where $0 \le x_t \le b_t$. Then we redefine $f_t(b_t)$ to be the maximum net profit *(in year t dollars)* that can be earned during years $t, t + 1, \ldots, T$. Since f_{t+1} is measured in year $t + 1$ dollars, multiplying it by β converts $f_{t+1}(\cdot)$ to year t dollars, which is just what we want. In Example 8, once we have worked backward and determined $f_1(10,000)$, an optimal fishing policy is found by using the same method that was previously described. This approach can be used to account for the time value of money in any dynamic programming formulation.

EXAMPLE 9   Power Plant

An electric power utility forecasts that r_t kilowatt-hours (kwh) of generating capacity will be needed during year t (the current year is year 1). Each year, the utility must decide by how much generating capacity should be expanded. It costs $c_t(x)$ dollars to increase generating capacity by x kwh during year t. It may be desirable to reduce capacity, so x need not be non-negative. During each year, 10% of the old generating capacity becomes obsolete and unusable (capacity does not become obsolete during its first year of operation). It costs the utility $m_t(i)$ dollars to maintain i units of capacity during year t. At the beginning of year 1, 100,000 kwh of generating capacity are available. Formulate a dynamic programming recursion that will enable the utility to minimize the total cost of meeting power requirements for the next T years.

Solution   Again, we let time be the stage. At the beginning of year t, the utility must determine the amount of capacity (call it x_t) to add during year t. To choose x_t properly, all the utility needs to know is the amount of available capacity at the beginning of year t (call it i_t). Hence, we define the state at the beginning of year t to be the current capacity level. We may now dispose of aspects 1–3 of the formulation.

Aspect 1   What values of x_t are feasible? To meet year t's requirement of r_t, we must have $i_t + x_t \ge r_t$, or $x_t \ge r_t - i_t$. So the feasible x_t's are those values of x_t satisfying $x_t \ge r_t - i_t$.

Aspect 2   What cost is incurred during year t? If x_t kwh are added during a year that begins with i_t kwh of available capacity, then during year t, a cost $c_t(x_t) + m_t(i_t + x_t)$ is incurred.

Aspect 3   What will be the state at the beginning of year $t + 1$? At the beginning of year $t + 1$, the utility will have $0.9i_t$ kwh of old capacity plus the x_t kwh that have been added during year t. Thus, the state at the beginning of year $t + 1$ will be $0.9i_t + x_t$.

We can now use (11) to develop the appropriate recursion. Define $f_t(i_t)$ to be the minimum cost incurred by the utility during years $t, t + 1, \ldots, T$, given that i_t kwh of capacity are available at the beginning of year t. At the beginning of year T, there are no future costs to consider, so

$$f_T(i_T) = \min_{x_T} \{c_T(x_T) + m_T(i_T + x_T)\} \tag{13}$$

where x_T must satisfy $x_T \geq r_T - i_T$. For $t < T$,

$$f_t(i_t) = \min_{x_T} \{c_t(x_t) + m_t(i_t + x_t) + f_{t+1}(0.9i_t + x_t)\} \tag{14}$$

where x_t must satisfy $x_t \geq r_t - i_t$. If the utility does not start with any excess capacity, then we can safely assume that the capacity level would never exceed $r_{MAX} = \max_{t=1, 2, \ldots, T} \{r_t\}$. This means that we need consider only states $0, 1, 2, \ldots, r_{MAX}$. To begin computations, we use (13) to compute $f_T(0), f_T(1), \ldots, f_T(r_{MAX})$. Then we use (14) to work backward until $f_1(100,000)$ has been determined. To determine the optimal amount of capacity that should be added during each year, proceed as follows. During year 1, add an amount of capacity x_1 that attains the minimum in the (14) equation for $f_1(100,000)$. Then the utility will begin year 2 with $90,000 + x_1$ kwh of capacity. Then, during year 2, x_2 kwh of capacity should be added, where x_2 attains the minimum in the (14) equation for $f_2(90,000 + x_1)$. Continue in this fashion until the optimal value of x_T has been determined.

EXAMPLE 10   Wheat Sale

Farmer Jones now possesses $5,000 in cash and 1,000 bushels of wheat. During month t, the price of wheat is p_t. During each month, he must decide how many bushels of wheat to buy (or sell). There are three restrictions on each month's wheat transactions: (1) During any month, the amount of money spent on wheat cannot exceed the cash on hand at the beginning of the month; (2) during any month, he cannot sell more wheat than he has at the beginning of the month; and (3) because of limited warehouse capacity, the ending inventory of wheat for each month cannot exceed 1,000 bushels.

Show how dynamic programming can be used to maximize the amount of cash that farmer Jones has on hand at the end of six months.

Solution   Again, we let time be the stage. At the beginning of month t (the present is the beginning of month 1), farmer Jones must decide by how much to change the amount of wheat on hand. We define Δw_t to be the change in farmer Jones's wheat position during month t: $\Delta w_t \geq 0$ corresponds to a month t wheat purchase, and $\Delta w_t \leq 0$ corresponds to a month t sale of wheat. To determine an optimal value for Δw_t, we must know two things: the amount of wheat on hand at the beginning of month t (call it w_t) and the cash on hand at the beginning of month t, (call this c_t). We define $f_t(c_t, w_t)$ to be the maximum cash that farmer Jones can obtain at the end of month 6, given that farmer Jones has c_t dollars and w_t bushels of wheat at the beginning of month t. We now discuss aspects 1–3 of the formulation.

Aspect 1   What are the allowable decisions? If the state at time t is (c_t, w_t), then restrictions 1–3 limit Δw_t in the following manner:

$$p_t(\Delta w_t) \leq c_t \qquad \text{or} \qquad \Delta w_t \leq \frac{c_t}{p_t}$$

ensures that we won't run out of money at the end of month t. The inequality $\Delta w_t \geq -w_t$ ensures that during month t, we will not sell more wheat than we had at the beginning of month t; and $w_t + \Delta w_t \leq 1,000$, or $\Delta w_t \leq 1000 - w_t$, ensures that we will end month t with at most 1000 bushels of wheat. Putting these three restrictions together, we see that

$$-w_t \le \Delta w_t \le \min \left\{ \frac{c_t}{p_t}, 1{,}000 - w_t \right\}$$

will ensure that restrictions 1–3 are satisfied during month t.

Aspect 2   Since farmer Jones wants to maximize his cash on hand at the end of month 6, no benefit is earned during months 1 through 5. In effect, during months 1–5, we are doing bookkeeping to keep track of farmer Jones's position. Then, during month 6, we turn all of farmer Jones's assets into cash.

Aspect 3   If the current state is (c_t, w_t) and farmer Jones changes his month t wheat position by an amount Δw_t, what will be the new state at the beginning of month $t + 1$? Cash on hand will increase by $-(\Delta w_t)p_t$, and farmer Jones's wheat position will increase by Δw_t. Hence, the month $t + 1$ state will be $[c_t - (\Delta w_t)p_t, w_t + \Delta w_t]$.

We may now use (11) to develop the appropriate recursion. To maximize his cash position at the end of month 6, farmer Jones should convert his month 6 wheat into cash by selling all of it. This means that $\Delta w_6 = -w_6$. This leads to the following relation:

$$f_6(c_6, w_6) = c_6 + w_6 p_6 \tag{15}$$

Using (11), we obtain for $t < 6$

$$f_t(c_t, w_t) = \max_{\Delta w_t} \{ 0 + f_{t+1}[c_t - (\Delta w_t)p_t, w_t + \Delta w_t] \} \tag{16}$$

where Δw_t must satisfy

$$-w_t \le \Delta w_t \le \min \left\{ \frac{c_t}{p_t}, 1{,}000 - w_t \right\}$$

We begin our calculations by determining $f_6(c_6, w_6)$ for all states that can possibly occur during month 6. Then we use (16) to work backward until $f_1(5{,}000, 1{,}000)$ has been computed. Next, farmer Jones should choose Δw_1 to attain the maximum value in the (16) equation for $f_1(5{,}000, 1{,}000)$, and a month 2 state of $[5{,}000 - p_1(\Delta w_1), 1{,}000 + \Delta w_1]$ will ensue. Farmer Jones should next choose Δw_2 to attain the maximum value in the (16) equation for $f_2[5{,}000 - p_1(\Delta w_1), 1{,}000 + \Delta w_1]$. We continue in this manner until the optimal value of Δw_6 has been determined.

EXAMPLE 11   **Refinery Capacity**

Sunco Oil needs to build enough refinery capacity to refine 5,000 barrels of oil per day and 10,000 barrels of gasoline per day. Sunco can build refinery capacity at four locations. The cost of building a refinery at site t that has the capacity to refine x barrels of oil per day and y barrels of gasoline per day is $c_t(x, y)$. Use dynamic programming to determine how much capacity should be located at each site.

Solution   If Sunco had only one possible refinery site, then the problem would be easy to solve. Sunco could solve a problem in which there were two possible refinery sites, and finally, a problem in which there were four refinery sites. For this reason, we let the stage represent the number of available oil sites. At any stage, Sunco must determine how much oil and gas capacity should be built at the given site. To do this, the company must know how much refinery capacity of each type must be built at the available sites. We now define $f_t(o_t, g_t)$ to be the minimum cost of building o_t barrels per day of oil refinery capacity and g_t barrels per day of gasoline refinery capacity at sites $t, t + 1, \ldots, 4$.

To determine $f_4(o_4, g_4)$, note that if only site 4 is available, Sunco must build a refinery at site 4 with o_4 barrels of oil capacity and g_4 barrels of gasoline capacity. This implies that $f_4(o_4, g_4) = c_4(o_4, g_4)$. For $t = 1, 2, 3$, we can determine $f_t(o_t, g_t)$ by noting that if we build a refinery at site t that can refine x_t barrels of oil per day and y_t barrels of gasoline per day, then we incur a cost of $c_t(x_t, y_t)$ at site t. Then we will need to build a total oil refinery capacity of $o_t - x_t$ and a gas refinery capacity of $g_t - y_t$ at sites $t + 1$, $t + 2, \ldots, 4$. By the principle of optimality, the cost of doing this will be $f_{t+1}(o_t - x_t, g_t - y)$. Since $0 \le x_t \le o_t$ and $0 \le y_t \le g_t$ must hold, we obtain the following recursion:

$$f_t(o_t, g_t) = \min \{c_t(o_t, g_t) + f_{t+1}(o_t - x_t, g_t - y_t)\} \qquad \text{(17)}$$

where $0 \le x_t \le o_t$ and $0 \le y_t \le g_t$. As usual, we work backward until $f_1(5,000, 10,000)$ has been determined. Then Sunco chooses x_1 and y_1 to attain the minimum in the (17) equation for $f_1 (5,000, 10,000)$. Then Sunco should choose x_2 and y_2 that attain the minimum in the (17) equation for $f_2(5,000 - x_1, 10,000 - y_1)$. Sunco continues in this fashion until optimal values of x_4 and y_4 are determined.

EXAMPLE 12	Traveling Salesperson

The traveling salesperson problem (see Section 9.6) can be solved by using dynamic programming. As an example, we solve the following traveling salesperson problem: It's the last weekend of the 2004 election campaign, and candidate Walter Glenn is in New York City. Before election day, Walter must visit Miami, Dallas, and Chicago and then return to his New York City headquarters. Walter wants to minimize the total distance he must travel. In what order should he visit the cities? The distances in miles between the four cities are given in Table 15.

Solution We know that Walter must visit each city exactly once, the last city he visits must be New York, and his tour originates in New York. When Walter has only one city left to visit, his problem is trivial: simply go from his current location to New York. Then we can work backward to a problem in which he is in some city and has only two cities left to visit, and finally we can find the shortest tour that originates in New York and has four cities left to visit. We therefore let the stage be indexed by the number of cities that Walter has already visited. At any stage, to determine which city should next be visited, we need to know two things: Walter's current location and the cities he has already visited. The state at any stage consists of the last city visited and the set of cities that have already been visited. We define $f_t(i, S)$ to be the minimum distance that must be traveled to complete a tour if the $t - 1$ cities in the set S have been visited and city i was the last city visited. We let c_{ij} be the distance between cities i and j.

Stage 4 Computations

We note that, at stage 4, it must be the case that $S = \{2, 3, 4\}$ (why?), and the only possible states are $(2, \{2, 3, 4\})$, $(3, \{2, 3, 4\})$, and $(4, \{2, 3, 4\})$. In stage 4, we must go from the current location to New York. This observation yields

TABLE 15
Distances for a Traveling Salesperson

	City			
	New York	Miami	Dallas	Chicago
1 New York	—	1,334	1,559	809
2 Miami	1,334	—	1,343	1,397
3 Dallas	1,559	1,343	—	921
4 Chicago	809	1,397	921	—

$$f_4(2, \{2, 3, 4\}) = c_{21} = 1{,}334^* \qquad \text{(Go from city 2 to city 1)}$$

$$f_4(3, \{2, 3, 4\}) = c_{31} = 1{,}559^* \qquad \text{(Go from city 3 to city 1)}$$

$$f_4(4, \{2, 3, 4\}) = c_{41} = 809^* \qquad \text{(Go from city 4 to city 1)}$$

Stage 3 Computations

Working backward to stage 3, we write

$$f_3(i, S) = \min_{\substack{j \notin S \\ \text{and } j \neq 1}} \{c_{ij} + f_4[j, S \cup \{j\}]\} \tag{18}$$

This result follows, because if Walter is now at city i and he travels to city j, he travels a distance c_{ij}. Then he is at stage 4, has last visited city j, and has visited the cities in $S \cup \{j\}$. Hence, the length of the rest of his tour must be $f_4(j, S \cup \{j\})$. To use (18), note that at stage 3, Walter must have visited $\{2, 3\}$, $\{2, 4\}$, or $\{3, 4\}$ and must next visit the non-member of S that is not equal to 1. We can use (18) to determine $f_3(\cdot)$ for all possible states:

$$f_3(2, \{2, 3\}) = c_{24} + f_4(4, \{2, 3, 4\}) = 1{,}397 + 809 = 2{,}206^* \qquad \text{(Go from 2 to 4)}$$

$$f_3(3, \{2, 3\}) = c_{34} + f_4(4, \{2, 3, 4\}) = 921 + 809 = 1{,}730^* \qquad \text{(Go from 3 to 4)}$$

$$f_3(2, \{2, 4\}) = c_{23} + f_4(3, \{2, 3, 4\}) = 1{,}343 + 1{,}559 = 2{,}902^* \qquad \text{(Go from 2 to 3)}$$

$$f_3(4, \{2, 4\}) = c_{43} + f_4(3, \{2, 3, 4\}) = 921 + 1{,}559 = 2{,}480^* \qquad \text{(Go from 4 to 3)}$$

$$f_3(3, \{3, 4\}) = c_{32} + f_4(2, \{2, 3, 4\}) = 1{,}343 + 1{,}334 = 2{,}677^* \qquad \text{(Go from 3 to 2)}$$

$$f_3(4, \{3, 4\}) = c_{42} + f_4(2, \{2, 3, 4\}) = 1{,}397 + 1{,}334 = 2{,}731^* \qquad \text{(Go from 4 to 2)}$$

In general, we write, for $t = 1, 2, 3$,

$$f_t(i, S) = \min_{\substack{j \notin S \\ \text{and } j \neq 1}} \{c_{ij} + f_{t+1}[j, S \cup \{j\}]\} \tag{19}$$

This result follows, because if Walter is at present in city i and he next visits city j, then he travels a distance c_{ij}. The remainder of his tour will originate from city j, and he will have visited the cities in $S \cup \{j\}$. Hence, the length of the remainder of his tour must be $f_{t+1}(j, S \cup \{j\})$. Equation (19) now follows.

Stage 2 Computations

At stage 2, Walter has visited only one city, so the only possible states are $(2, \{2\})$, $(3, \{3\})$, and $(4, \{4\})$. Applying (19), we obtain

$$f_2(2, \{2\}) = \min \begin{cases} c_{23} + f_3(3, \{2, 3\}) = 1{,}343 + 1{,}730 = 3{,}073^* \\ \text{(Go from 2 to 3)} \\ c_{24} + f_3(4, \{2, 4\}) = 1{,}397 + 2{,}480 = 3{,}877 \\ \text{(Go from 2 to 4)} \end{cases}$$

$$f_2(3, \{3\}) = \min \begin{cases} c_{34} + f_3(4, \{3, 4\}) = 921 + 2{,}731 = 3{,}652 \\ \text{(Go from 3 to 4)} \\ c_{32} + f_3(2, \{2, 3\}) = 1{,}343 + 2{,}206 = 3{,}549^* \\ \text{(Go from 3 to 2)} \end{cases}$$

$$f_2(4, \{4\}) = \min \begin{cases} c_{42} + f_3(2, \{2, 4\}) = 1{,}397 + 2{,}902 = 4{,}299 \\ \text{(Go from 4 to 2)} \\ c_{43} + f_3(3, \{3, 4\}) = 921 + 2{,}677 = 3{,}598^* \\ \text{(Go from 4 to 3)} \end{cases}$$

Stage 1 Computations

Finally, we are back to stage 1 (where no cities have been visited). Since Walter is currently in New York and has visited no cities, the stage 1 state must be $f_1(1, \{\cdot\})$. Applying (19),

$$f_1(1, \{\cdot\}) = \min \begin{cases} c_{12} + f_2(2, \{2\}) = 1{,}334 + 3{,}073 = 4{,}407^* \\ \text{(Go from 1 to 2)} \\ c_{13} + f_2(3, \{3\}) = 1{,}559 + 3{,}549 = 5{,}108 \\ \text{(Go from 1 to 3)} \\ c_{14} + f_2(4, \{4\}) = 809 + 3{,}598 = 4{,}407^* \\ \text{(Go from 1 to 4)} \end{cases}$$

So from city 1 (New York), Walter may go to city 2 (Miami) or city 4 (Chicago). We arbitrarily have him choose to go to city 4. Then he must choose to visit the city that attains $f_2(4, \{4\})$, which requires that he next visit city 3 (Dallas). Then he must visit the city attaining $f_3(3, \{3, 4\})$, which requires that he next visit city 2 (Miami). Then Walter must visit the city attaining $f_4(2, \{2, 3, 4\})$, which means, of course, that he must next visit city 1 (New York). The optimal tour (1–4–3–2–1, or New York–Chicago–Dallas–Miami–New York) is now complete. The length of this tour is $f_1(1, \{\cdot\}) = 4{,}407$. As a check, note that

New York to Chicago distance = 809 miles

Chicago to Dallas distance = 921 miles

Dallas to Miami distance = 1,343 miles

Miami to New York distance = 1,334 miles

so the total distance that Walter travels is $809 + 921 + 1{,}343 + 1{,}334 = 4{,}407$ miles. Of course, if we had first sent him to city 2, we would have obtained another optimal tour (1–2–3–4–1) that would simply be a reversal of the original optimal tour.

Computational Difficulties in Using Dynamic Programming

For traveling salesperson problems that are large, the state space becomes very large, and the branch-and-bound approach outlined in Chapter 9 (along with other branch-and-bound approaches) is much more efficient than the dynamic programming approach outlined here. For example, for a 30-city problem, suppose we are at stage 16 (this means that 15 cities have been visited). Then it can be shown that there are more than 1 billion possible states. This brings up a problem that limits the practical application of dynamic programming. In many problems, *the state space becomes so large that excessive computational time is required to solve the problem by dynamic programming.* For instance, in Example 8, suppose that $T = 20$. It is possible that if no bass were caught during the first 20 years, then the lake might contain $10{,}000(1.2)^{20} = 383{,}376$ bass at the beginning of year 21. If we view this example as a network in which we need to find the longest route from the node (1, 10,000) (representing year 1 and 10,000 bass in the lake) to some stage 21 node, then stage 21 would have 383,377 nodes. Even a powerful computer would have difficulty solving this problem. Techniques to make problems with large state spaces computationally tractable are discussed in Bersetkas (1987) and Denardo (1982).

Nonadditive Recursions

The last two examples in this section differ from the previous ones in that the recursion does not represent $f_t(i)$ as the sum of the cost (or reward) incurred during the current period and future costs (or rewards) incurred during future periods.

EXAMPLE 13 **Minimax Shortest Route**

Joe Cougar needs to drive from city 1 to city 10. He is no longer interested in minimizing the length of his trip, but he is interested in minimizing the maximum altitude above sea level that he will encounter during his drive. To get from city 1 to city 10, he must follow a path in Figure 10. The length c_{ij} of the arc connecting city i and city j represents the maximum altitude (in thousands of feet above sea level) encountered when driving from city i to city j. Use dynamic programming to determine how Joe should proceed from city 1 to city 10.

Solution To solve this problem by dynamic programming, note that for a trip that begins in city i and goes through stages $t, t + 1, \ldots, 5$, the maximum altitude that Joe encounters will be the maximum of the following two quantities: (1) the maximum altitude encountered on stages $t + 1, t + 2, \ldots, 5$ or (2) the altitude encountered when traversing the arc that begins in stage t. Of course, if we are in a stage 4 state, quantity 1 does not exist.

After defining $f_t(i)$ as the smallest maximum altitude that Joe can encounter in a trip from city i in stage t to city 10, this reasoning leads us to the following recursion:

$$f_4(i) = c_{i,10} \qquad\qquad\qquad\qquad (20)$$
$$f_t(i) = \min_{j}\{\max[c_{ij}, f_{t+1}(j)]\} \qquad (t = 1, 2, 3)$$

where j may be any city such that there is an arc connecting city i and city j.

We first compute $f_4(7)$, $f_4(8)$, and $f_4(9)$ and then use (20) to work backward until $f_1(1)$ has been computed. We obtain the following results:

$$f_4(7) = 13* \qquad\qquad\qquad \text{(Go from 7 to 10)}$$
$$f_4(8) = 8* \qquad\qquad\qquad\ \text{(Go from 8 to 10)}$$
$$f_4(9) = 9* \qquad\qquad\qquad\ \text{(Go from 9 to 10)}$$

$$f_3(5) = \min \begin{cases} \max[c_{57}, f_4(7)] = 13 & \text{(Go from 5 to 7)} \\ \max[c_{58}, f_4(8)] = 8* & \text{(Go from 5 to 8)} \\ \max[c_{59}, f_4(9)] = 10 & \text{(Go from 5 to 9)} \end{cases}$$

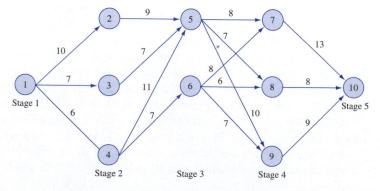

FIGURE 10
Joe's Trip
(Altitudes Given)

$$f_3(6) = \min \begin{cases} \max[c_{67}, f_4(7)] = 13 & \text{(Go from 6 to 7)} \\ \max[c_{68}, f_4(8)] = 8^* & \text{(Go from 6 to 8)} \\ \max[c_{69}, f_4(9)] = 9 & \text{(Go from 6 to 9)} \end{cases}$$

$$f_2(2) = \max[c_{25}, f_3(5)] = 9^* \qquad \text{(Go from 2 to 5)}$$

$$f_2(3) = \max[c_{35}, f_3(5)] = 8^* \qquad \text{(Go from 3 to 5)}$$

$$f_2(4) = \min \begin{cases} \max[c_{45}, f_3(5)] = 11 & \text{(Go from 4 to 5)} \\ \max[c_{46}, f_3(6)] = 8^* & \text{(Go from 4 to 6)} \end{cases}$$

$$f_1(1) = \min \begin{cases} \max[c_{12}, f_2(2)] = 10 & \text{(Go from 1 to 2)} \\ \max[c_{13}, f_2(3)] = 8^* & \text{(Go from 1 to 3)} \\ \max[c_{14}, f_2(4)] = 8^* & \text{(Go from 1 to 4)} \end{cases}$$

To determine the optimal strategy, note that Joe can begin by going from city 1 to city 3 or from city 1 to city 4. Suppose Joe begins by traveling to city 3. Then he should choose the arc attaining $f_2(3)$, which means he should next travel to city 5. Then Joe must choose the arc that attains $f_3(5)$, driving next to city 8. Then, of course, he must drive to city 10. Thus, the path 1–3–5–8–10 is optimal, and Joe will encounter a maximum altitude equal to $f_1(1) = 8,000$ ft. The reader should verify that the path 1–4–6–8–10 is also optimal.

EXAMPLE 14   **Sales Allocation**

Glueco is planning to introduce a new product in three different regions. Current estimates are that the product will sell well in each region with respective probabilities .6, .5, and .3. The firm has available two top sales representatives that it can send to any of the three regions. The estimated probabilities that the product will sell well in each region when 0, 1, or 2 additional sales reps are sent to a region are given in Table 16. If Glueco wants to maximize the probability that its new product will sell well in all three regions, then where should it assign sales representatives? You may assume that sales in the three regions are independent.

Solution   If Glueco had just one region to worry about and wanted to maximize the probability that the new product would sell in that region, then the proper strategy would be clear: Assign both sales reps to the region. We could then work backward and solve a problem in which Glueco's goal is to maximize the probability that the product will sell in two regions. Finally, we could work backward and solve a problem with three regions. We define $f_t(s)$ as the probability that the new product will sell in regions $t, t+1, \ldots, 3$ if s sales reps are optimally assigned to these regions. Then

$$f_3(2) = .7 \qquad \text{(Assign 2 sales reps to region 3)}$$
$$f_3(1) = .55 \qquad \text{(Assign 1 sales rep to region 3)}$$
$$f_3(0) = .3 \qquad \text{(Assign 0 sales reps to region 3)}$$

TABLE 16

Relation between Regional Sales and Sales Representatives

No. of Additional Sales Representatives	Probability of Selling Well		
	Region 1	Region 2	Region 3
0	.6	.5	.3
1	.8	.7	.55
2	.85	.85	.7

Also, $f_1(2)$ will be the maximum probability that the product will sell well in all three regions. To develop a recursion for $f_2(\cdot)$ and $f_1(\cdot)$, we define p_{tx} to be the probability that the new product sells well in region t if x sales reps are assigned to region t. For example, $p_{21} = .7$. For $t = 1$ and $t = 2$, we then write

$$f_t(s) = \max_x \{p_{tx} f_{t+1}(s - x)\} \tag{21}$$

where x must be a member of $\{0, 1, \ldots, s\}$. To justify (21), observe that if s sales reps are available for regions $t, t + 1, \ldots, 3$ and x sales reps are assigned to region t, then

$$p_{tx} = \text{probability that product sells in region } t$$
$$f_{t+1}(s - x) = \text{probability that product sells well in regions } t + 1, \ldots, 3$$

Note that the sales in each region are independent. This implies that if x sales reps are assigned to region t, then the probability that the new product sells well in regions t, $t + 1, \ldots, 3$ is $p_{tx} f_{t+1}(s - x)$. We want to maximize this probability, so we obtain (21). Applying (21) yields the following results:

$$f_2(2) = \max \begin{cases} (.5)f_3(2 - 0) = .35 \\ \text{(Assign 0 sales reps to region 2)} \\ (.7)f_3(2 - 1) = .385* \\ \text{(Assign 1 sales rep to region 2)} \\ (.85)f_3(2 - 2) = .255 \\ \text{(Assign 2 sales reps to region 2)} \end{cases}$$

Thus, $f_2(2) = .385$, and 1 sales rep should be assigned to region 2.

$$f_2(1) = \max \begin{cases} (.5)f_3(1 - 0) = .275* \\ \text{(Assign 0 sales reps to region 2)} \\ (.7)f_3(1 - 1) = .21 \\ \text{(Assign 1 sales rep to region 2)} \end{cases}$$

Thus, $f_2(1) = .275$, and no sales reps should be assigned to region 2.

$$f_2(0) = (.5)f_3(0 - 0) = .15*$$
$$\text{(Assign 0 sales reps to region 2)}$$

Finally, we are back to the original problem, which is to find $f_1(2)$. Equation (21) yields

$$f_1(2) = \max \begin{cases} (.6)f_2(2 - 0) = .231* \\ \text{(Assign 0 sales reps to region 1)} \\ (.8)f_2(2 - 1) = .220 \\ \text{(Assign 1 sales rep to region 1)} \\ (.85)f_2(2 - 2) = .1275 \\ \text{(Assign 2 sales reps to region 1)} \end{cases}$$

Thus, $f_1(2) = .231$, and no sales reps should be assigned to region 1. Then Glueco needs to attain $f_2(2 - 0)$, which requires that 1 sales rep be assigned to region 2. Glueco must next attain $f_3(2 - 1)$, which requires that 1 sales rep be assigned to region 3. In summary, Glueco can obtain a .231 probability of the new product selling well in all three regions by assigning 1 sales rep to region 2 and 1 sales rep to region 3.

PROBLEMS

Group A

1 At the beginning of year 1, Sunco Oil owns i_0 barrels of oil reserves. During year $t(t = 1, 2, \ldots, 10)$, the following events occur in the order listed: (1) Sunco extracts and refines x barrels of oil reserves and incurs a cost $c(x)$: (2) Sunco sells year t's extracted and refined oil at a price of p_t dollars per barrel; and (3) exploration for new reserves results in a discovery of b_t barrels of new reserves.

Sunco wants to maximize sales revenues less costs over the next 10 years. Formulate a dynamic programming recursion that will help Sunco accomplish its goal. If Sunco felt that cash flows in later years should be discounted, how should the formulation be modified?

2 At the beginning of year 1, Julie Ripe has D dollars (this includes year 1 income). During each year, Julie earns i dollars and must determine how much money she should consume and how much she should invest in Treasury bills. During a year in which Julie consumes d dollars, she earns a utility of $\ln d$. Each dollar invested in Treasury bills yields $1.10 in cash at the beginning of the next year. Julie's goal is to maximize the total utility she earns during the next 10 years.

a Why might $\ln d$ be a better indicator of Julie's utility than a function such as d^2?

b Formulate a dynamic programming recursion that will enable Julie to maximize the total utility she receives during the next 10 years. Assume that year t revenue is received at the beginning of year t.

3 Assume that during minute t (the current minute is minute 1), the following sequence of events occurs: (1) At the beginning of the minute, x_t customers arrive at the cash register; (2) the store manager decides how many cash registers should be operated during the current minute; (3) if s cash registers are operated and i customers are present (including the current minute's arrivals), $c(s, i)$ customers complete service; and (4) the next minute begins.

A cost of 10¢ is assessed for each minute a customer spends waiting to check out (this time includes checkout time). Assume that it costs $c(s)$ cents to operate s cash registers for 1 minute. Formulate a dynamic programming recursion that minimizes the sum of holding and service costs during the next 60 minutes. Assume that before the first minute's arrivals, no customers are present and that holding cost is assessed at the end of each minute.

4 Develop a dynamic programming formulation of the CSL Computer problem of Section 3.12.

5 To graduate from State University, Angie Warner needs to pass at least one of the three subjects she is taking this semester. She is now enrolled in French, German, and statistics. Angie's busy schedule of extracurricular activities allows her to spend only 4 hours per week on studying. Angie's probability of passing each course depends on the number of hours she spends studying for the course (see Table 17). Use dynamic programming to determine how many hours per week Angie should spend studying each subject. (*Hint:* Explain why maximizing the probability of

TABLE 17

Hours of Study per Week	Probability of Passing Course		
	French	German	Statistics
0	.20	.25	.10
1	.30	.30	.30
2	.35	.33	.40
3	.38	.35	.44
4	.40	.38	.50

TABLE 18

Component	No. of Actors Assigned to Component			
	0	1	2	3
Warp drive	.30	.55	.65	.95
Solar relay	.40	.50	.70	.90
Candy maker	.45	.55	.80	.98

passing at least one course is equivalent to minimizing the probability of failing all three courses.)

6 E.T. is about to fly home. For the trip to be successful, the ship's solar relay, warp drive, and candy maker must all function properly. E.T. has found three unemployed actors who are willing to help get the ship ready for takeoff. Table 18 gives, as a function of the number of actors assigned to repair each component, the probability that each component will function properly during the trip home. Use dynamic programming to help E.T. maximize the probability of having a successful trip home.

7 Farmer Jones is trying to raise a prize steer for the Bloomington 4-H show. The steer now weighs w_0 pounds. Each week, farmer Jones must determine how much food to feed the steer. If the steer weighs w pounds at the beginning of a week and is fed p pounds of food during a week, then at the beginning of the next week, the steer will weigh $g(w, p)$ pounds. It costs farmer Jones $c(p)$ dollars to feed the steer p pounds of food during a week. At the end of the 10th week (or equivalently, the beginning of the 11th week), the steer may be sold for $10/lb. Formulate a dynamic programming recursion that can be used to determine how farmer Jones can maximize profit from the steer.

Group B

8 MacBurger has just opened a fast-food restaurant in Bloomington. Currently, i_0 customers frequent MacBurger (we call these loyal customers), and $N - i_0$ customers frequent other fast-food establishments (we call these nonloyal customers). At the beginning of each month, MacBurger must decide how much money to spend on advertising. At the end of a month in which MacBurger spends d dollars on advertising, a fraction $p(d)$ of the loyal customers become nonloyal customers, and a fraction $q(d)$ of the nonloyal customers become loyal customers. During the next 12 months, MacBurger wants to spend D dollars on advertising. Develop a dynamic programming recursion that will enable MacBurger to maximize the number of loyal customers the company will have at the end of month 12. (Ignore the possibility of a fractional number of loyal customers.)

9 Public Service Indiana (PSI) is considering five possible locations to build power plants during the next 20 years. It will cost c_i dollars to build a plant at site i and h_i dollars to operate a site i plant for a year. A plant at site i can supply k_i kilowatt-hours (kwh) of generating capacity. During year t, d_t kwh of generating capacity are required. Suppose that at most one plant can be built during a year, and if it is decided to build a plant at site i during year t, then the site i plant can be used to meet the year t (and later) generating requirements. Initially, PSI has 500,000 kwh of generating capacity available. Formulate a recursion that PSI could use to minimize the sum of building and operating costs during the next 20 years.

10 During month t, a firm faces a demand for d_t units of a product. The firm's production cost during month t consists of two components. First, for each unit produced during month t, the firm incurs a variable production cost of c_t. Second, if the firm's production level during month $t - 1$ is x_{t-1} and the firm's production level during month t is x_t, then during month t, a smoothing cost of $5|x_t - x_{t-1}|$ will be incurred (see Section 4.12 for an explanation of smoothing costs). At the end of each month, a holding cost of h_t per unit is incurred. Formulate a recursion that will enable the firm to meet (on time) its demands over the next 12 months. Assume that at the beginning of the first month, 20 units are in inventory and that last month's production was 20 units. (*Hint:* The state during each month must consist of two quantities.)

11 The state of Transylvania consists of three cities with the following populations: city 1, 1.2 million people; city 2, 1.4 million people; city 3, 400,000 people. The Transylvania House of Representatives consists of three representatives. Given proportional representation, city 1 should have $d_1 = (\frac{1.2}{3}) = 1.2$ representatives; city 2 should have $d_2 = 1.4$ representatives; and city 3 should have $d_3 = 0.40$ representative. Each city must receive an integral number of representatives, so this is impossible. Transylvania has therefore decided to allocate x_i representatives to city i, where the allocation x_1, x_2, x_3 minimizes the maximum discrepancy between the desired and actual number of representatives received by a city. In short, Transylvania must determine x_1, x_2, and x_3 to minimize the largest of the following three numbers: $|x_1 - d_1|$, $|x_2 - d_2|$, $|x_3 - d_3|$. Use dynamic programming to solve Transylvania's problem.

12 A job shop has four jobs that must be processed on a single machine. The due date and processing time for each job are given in Table 19. Use dynamic programming to determine the order in which the jobs should be done so as to minimize the total lateness of the jobs. (The lateness of a job is simply how long after the job's due date the job is completed; for example, if the jobs are processed in the given order, then job 3 will be 2 days late, job 4 will be 4 days late, and jobs 1 and 2 will not be late.)

TABLE 19

Job	Processing Time (Days)	Due Date (Days from Now)
1	2	4
2	4	14
3	6	10
4	8	16

13.7 Using EXCEL to Solve Dynamic Programming Problems†

In earlier chapters we have seen that any LP problem can be solved with LINDO or LINGO, and any NLP can be solved with LINGO. Unfortunately, no similarly user-friendly package can be used to solve dynamic programming problems. LINGO can be used to solve DP problems, but student LINGO can only handle a very small problem. Fortunately, EXCEL can often be used to solve DP problems. Our three illustrations solve a knapsack problem (Example 6), a resource-allocation problem (Example 5), and an inventory problem (Example 4).

Solving Knapsack Problems on a Spreadsheet

Recall the knapsack problem of Example 6. The question is how to (using three types of items) fill a 10-lb knapsack and obtain the maximum possible benefit. Recall that $g(w) =$ maximum benefit that can be obtained from a w-lb knapsack. Recall that

†This section covers topics that may be omitted with no loss of continuity.

$$g(w) = \max_j \{b_j + g(w - w_j)\} \tag{8}$$

where b_j = benefit from a type j item and w_j = weight of a type j item.

Dpknap.xls

In each row of the spreadsheet (see Figure 11 or file Dpknap.xls) we compute $g(w)$ for various values of w. We begin by entering $g(0) = g(1) = g(2) = 0$ and $g(3) = 7$; [$g(3) = 7$ follows because a 3-lb item is the only item that will fit in a 3-lb knapsack]. The columns labeled ITEM1, ITEM2, and ITEM3 correspond to the terms $j = 1, 2, 3$, respectively, in (8). Thus, in the ITEM1 column we should enter a formula to compute $b_1 + g(w - w_1)$; in the ITEM2 column we should enter a formula to compute $b_2 + g(w - w_2)$; in the ITEM3 column we should enter a formula to compute $b_3 + g(w - w_3)$. The only exception to this occurs when a w_j-lb item will not fit in a w-lb knapsack. In this situation we enter a very negative number (such as 10,000) to ensure that a w_j-lb item will not be considered.

More specifically, in row 7 we want to compute $g(4)$. To do this we enter the following formulas:

B7: 11 + E3   [This is $b_1 + g(4 - w_1)$]

C7: 7 + E4   [This is $b_2 + g(4 - w_2)$]

D7: −10,000   (This is because a 5-lb item will not fit in a 4-lb knapsack)

In E7, we compute $g(4)$ by entering the formula =MAX(B7:D7). In row 8, we compute $g(5)$ by entering the following formulas:

B8: 11 + E4

C8: 7 + E5

D8: 12 + E3

To compute $g(5)$ we enter =MAX(B8:D8) in E8. Now comes the fun part! Simply copy the formulas from the range B8:E8 to B8:E13. Then $g(10)$ will be computed in E13.

A	A	B	C	D	E	F	G
1	KNAPSACK	ITEM1	ITEM2	ITEM3	g(SIZE)		FIGURE11
2	SIZE						KNAPSACK
3	0				0		PROBLEM
4	1				0		
5	2				0		
6	3				7		
7	4	11	7	-10000	11		
8	5	11	7	12	12		
9	6	11	14	12	14		
10	7	18	18	12	18		
11	8	22	19	19	22		
12	9	23	21	23	23		
13	10	25	25	24	25		
14							
15							
16							
17							
18							
19							
20							
21							
22							
23							
24							
25							
26							
27							
28							
29							
30							

FIGURE 11
Knapsack Problem

We see that $g(10) = 25$. Because both item 1 and item 2 attain $g(10)$, we may begin filling a knapsack with a Type 1 or Type 2 item. We choose to begin with a Type 1 item. This leaves us with $10 - 4 = 6$ lb to fill. From row 9 we find that $g(6) = 14$ is attained by a Type 2 item. This leaves us with $6 - 3 = 3$ lb to fill. We also use a Type 2 item to attain $g(3) = 7$. This leaves us with 0 lb. Thus, we conclude that we can obtain 25 units of benefit by filling a 10-lb knapsack with two Type 2 items and one Type 1 item.

By the way, if we had been interested in filling a 100-lb knapsack, we would have copied the formulas from B8:E8 to B8:E103.

Solving a General Resource-Allocation Problem on a Spreadsheet

Solving a nonknapsack resource-allocation problem on a spreadsheet is more difficult. To illustrate, consider Example 5 in which we have $6,000 to allocate between three investments. Define $f_t(d)$ = maximum NPV obtained from investments $t, \ldots, 3$ given that d (in thousands) dollars are available for investments $t, \ldots, 3$. Then we may write

$$f_t(d) = \max_{0 \le x \le d} \{r_t(x) + f_{t+1}(d - x)\}, \tag{10}$$

where $f_4(d) = 0(d = 0, 1, 2, 3, 4, 5, 6)$, $r_t(x)$ = NPV obtained if x (in thousands) dollars are invested in investment t, and the maximization in (10) is only taken over integral values for d. Our subsequent discussion will be simplified if we define $J_t(d, x) = r_t(x) + f_{t+1}(d - x)$ and rewrite (24) as

$$f_t(d) = \max_{0 \le x \le d} \{J_t(d, x)\} \tag{10'}$$

Dpresour.xls

We begin the construction of the spreadsheet (Figure 12 and file Dpresour.xls) by entering the $r_t(x)$ in A1:H4. For example, $r_2(3) = 16$ is entered in E3. In rows 18–20, we have set up the computations to compute the $J_t(d, x)$. These computations require using the EXCEL =**HLOOKUP** command to look up the values of $r_t(x)$ (in rows 2–4) and $f_{t+1}(d - x)$ (in rows 11–14). For example, to compute $J_3(3, 1)$ we enter the following formula in I18:

=HLOOKUP (I$17, $B$1:$H$4, $A18+1)

+ HLOOKUP (I$16-I$17, $B$10:$H$14, $A18+1)

FIGURE 12
Resource Allocation

	A	B	C	D	E	F	G	H	I	J	K	L	M
1	REWARD	0	1	2	3	4	5	6					
2	PERIOD3	0	9	13	17	21	25	29					
3	PERIOD2	0	10	13	16	19	22	25					
4	PERIOD1	0	9	16	23	30	37	44					
5													
6													
7	FIGURE 12												
8	RESOURCE	ALLOCATION											
9													
10	VALUE	0	1	2	3	4	5	6					
11	PERIOD4	0	0	0	0	0	0	0					
12	PERIOD3	0	9	13	17	21	25	29					
13	PERIOD2	0	10	19	23	27	31	35					
14	PERIOD1	0	10	19	28	35	42	49					
15													
16	d	0	1	1	2	2	2	3	3	3	3	4	4
17	x	0	0	1	0	1	2	0	1	2	3	0	1
18	1	0	0	9	0	9	13	0	9	13	17	0	9
19	2	0	9	10	13	19	13	17	23	22	16	21	27
20	3	0	10	9	19	19	16	23	28	26	23	27	32

FIGURE 12
(Continued)

A	N	O	P	Q	R	S	T	U	V	W	X	Y	Z
1													
2													
3													
4													
5													
6													
7													
8													
9													
10													
11													
12													
13													
14													
15													
16	4	4	4	5	5	5	5	5	5	6	6	6	6
17	2	3	4	0	1	2	3	4	5	0	1	2	3
18	13	17	21	0	9	13	17	21	25	0	9	13	17
19	26	25	19	25	31	30	29	28	22	29	35	34	33
20	35	33	30	31	36	39	42	40	37	35	40	43	46

A	AA	AB	AC	AD	AE	AF	AG	AH	AI	AJ	AK
1											
2											
3											
4											
5											
6											
7											
8											
9											
10											
11											
12											
13											
14											
15											
16	6	6	6	0	1	2	3	4	5	6	
17	4	5	6	ft(0)	ft(1)	ft(2)	ft(3)	ft(4)	ft(5)	ft(6)	t
18	21	25	29	0	9	13	17	21	25	29	3
19	32	31	25	0	10	19	23	27	31	35	2
20	49	47	44	0	10	19	28	35	42	49	1

The portion $=$HLOOKUP(I\$17, \$B\$1:\$H\$4,\$A18+1) of the formula in cell I18 finds the column in B1:H4 whose first entry matches I17. Then we pick off the entry in row A18 + 1 of that column. This returns $r_3(1) = 9$. Note that H stands for horizontal lookup. The portion HLOOKUP(I\$16-I\$17,\$b\$10:\$h\$14,\$A18+1) finds the column in B10:H14 whose first entry matches I16-I17. Then we pick off the entry in row A18 + 1 of that column. This yields $f_4(3 - 1) = 0$.

We now copy any of the $J_t(d, x)$ formulas (such as the one in I18) to the range B18:AC20.

The $f_t(d)$ are computed in AD18:AJ20. We begin by manually entering in AD18:AJ18 the formulas used to compute $f_3(0), f_3(1), \ldots, f_3(6)$. These formulas are as follows:

AD18:	0	(Computes $f_3(0)$)
AE18:	=MAX(C18:D18)	(Computes $f_3(1)$)
AF18:	=MAX(E18:G18)	(Computes $f_3(2)$)
AG18:	=MAX(H18:K18)	(Computes $f_3(3)$)
AH18:	=MAX(L18:P18)	(Computes $f_3(4)$)
AI18:	=MAX(Q18:V18)	(Computes $f_3(5)$)
AJ18:	=MAX(W18:AC18)	(Computes $f_3(6)$)

We now copy these formulas from the range AD18:AJ18 to the range AD:AJ20.

For our spreadsheet to work we must be able to compute the $J_t(d, x)$ by looking up the appropriate value of $f_t(d)$ in rows 11–14. Thus, in B11:H11, we enter a zero in each cell [because $f_4(d) = 0$ for all d]. In B12, we enter =AD18 [this is the cell in which $f_3(0)$ is computed]. We now copy this formula to the range B12:H14.

Note that rows 11–14 of our spreadsheet are defined in terms of rows 18–20, and rows 18–20 are defined in terms of rows 11–14. This creates **circularity** or **circular references** in our spreadsheet. To resolve the circular references in this (or any) spreadsheet, simply select Tools, Options, Calculations and select the Iteration box. This will cause EXCEL to resolve all circular references until the circularity is resolved.

To determine how $6,000 should be allocated to the three investments, note that $f_1(6) = 49$. Because $f_1(6) = J_1(6, 4)$, we allocate $4,000 to investment 1. Then we must find $f_2(6 - 4) = 19 = J_2(2, 1)$. We allocate $1,000 to investment 2. Finally, we find that $f_3(2 - 1) = J_3(1, 1)$ and allocate $1,000 to investment 3.

Solving an Inventory Problem on a Spreadsheet

We now show how to determine an optimal production policy for Example 4. An important aspect of this production problem is that each month's ending inventory must be between 0 and 4 units. We can ensure that this occurs by manually determining the allowable actions in each state. We will design our spreadsheet to ensure that the ending inventory for each month must be between 0 and 4 inclusive.

Dpinv.xls

Our first step in setting up the spreadsheet (Figure 13, file Dpinv.xls) is to enter the production cost for each possible production level (0, 1, 2, 3, 4, 5) in B1:G2. Then we define $f_t(i)$ to be the minimum cost incurred in meeting demands for months $t, t + 1, \ldots,$ 4 when i units are on hand at the beginning of month t. If d_t is month t's demand, then for $t = 1, 2, 3, 4$ we may write

$$f_t(i) = \min_{x \mid 0 \le i + x - d_t \le 4} \{.5(i + x - d_t) + c(x) + f_{t+1}(i + x - d_t)\} \tag{23}$$

where $c(x) =$ cost of producing x units during a month, and $f_5(i) = 0$ for ($i = 0, 1, 2, 3, 4$).

If we define $J_t(i, x) = .5(i + x - d_t) + c(x) + f_{t+1}(i + x - d_t)$ we may write

$$f_t(i) = \min_{x \mid 0 \le i + x - d_t \le 4} \{J_t(i, x)\}$$

Next we compute $J_t(i, x)$ in A13:AF16. For example, to compute $J_4(0, 2)$, we enter the following formula in E13:

FIGURE 13
Inventory Example

	A	B	C	D	E	F	G	H	I	J	K	L	M
1	PROD COST	0	1	2	3	4	5						
2		0	4	5	6	7	8						
3													
4	VALUE	-5	0	1	2	3	4	5					
5	M5	10000	0	0	0	0	0	10000					
6	M4	10000	7	6	5	4	0	10000					
7	M3	10000	12	10	7	6.5	6	10000					
8	M2	10000	16	15	14	12	10.5	10000					
9													
10		STATE	0	0	0	0	0	0	1	1	1	1	1
11		ACTION	0	1	2	3	4	5	0	1	2	3	4
12	DEMAND												
13	4		10000	10004	10005	10006	7	8.5	10000	10004	10005	6	7.5
14	2		10000	10004	12	12.5	13	13.5	10000	11	11.5	12	12.5
15	3		10000	10004	10005	18	17.5	16	10000	10004	17	16.5	15
16	1		10000	20	20.5	21	20.5	20.5	16	19.5	20	19.5	19.5
17													

FIGURE 13
(Continued)

A	N	O	P	Q	R	S	T	U	V	W	X	Y	Z
1													
2													
3													
4													
5													
6													
7													
8													
9													
10	1	2	2	2	2	2	2	3	3	3	3	3	3
11	5	0	1	2	3	4	5	0	1	2	3	4	5
12													
13	9	10000	10004	5	6.5	8	9.5	10000	4	5.5	7	8.5	10
14	10	7	10.5	11	11.5	9	10010.5	6.5	10	10.5	8	10000.5	10011
15	16	10000	16	15.5	14	15	16	12	14.5	13	14	15	10010.5
16	10010.5	15.5	19	18.5	18.5	10009.5	10011	15	17.5	17.5	10008.5	10010	10011.5
17													

A	AA	AB	AC	AD	AE	AF	AG	AH	AI	AJ	AK	AL
1												
2												
3												
4												
5												
6												
7												
8												
9												
10	4	4	4	4	4	4						
11	0	1	2	3	4	5	F(0)	F(1)	F(2)	F(3)	F(4)	
12												
13	0	4.5	6	7.5	9	10010.5	7	6	5	4	0	1
14	6	9.5	7	10008.5	10010	10011.5	12	10	7	6.5	6	2
15	10.5	12	13	14	10009.5	10011	16	15	14	12	10.5	3
16	13.5	16.5	10007.5	10009	10010.5	10012	20	16	15.5	15	13.5	4
17												

$$=\text{HLOOKUP } (E\$11, \$B\$1:\$G\$2, 2)$$
$$+.5*+1\text{MAX}(E\$10+E\$11-\$A13, 0)$$
$$+\text{HLOOKUP } (E\$10+E\$11-\$A13, \$B\$4:\$H\$8, \$AL13)$$

The first term in this sum yields $c(x)$ (this is because E\$11 is the production level). The second term gives the holding cost for the month (this is because E\$10+E\$11−\$A13 gives the month's ending inventory). The final term yields $f_{t+1}(i + x - d_t)$. This is because E\$10+E\$11−\$A13 is the beginning inventory for month $t + 1$. The reference to \$AL13 in the final term ensures that we look up the value of $f_{t+1}(i + x - d_t)$ in the correct row [the values of the $f_{t+1}( )$ will be tabulated in C5:G8]. Copying the formula in E13 to the range C13:AF16 computes all the $J_t(i, x)$.

In AG13:AK16 we compute the $f_t(d)$. To begin we enter the following formulas in cells AG13:AK13:

AG13:	=MIN(C13:H13)	[Computes $f_4(0)$]
AH13:	=MIN(I13:N13)	[Computes $f_4(1)$]
AI13:	=MIN(O13:T13)	[Computes $f_4(2)$]
AJ13:	=MIN(U13:Z13)	[Computes $f_4(3)$]
AK13:	=MIN(AA13:AF13)	[Computes $f_4(4)$]

To compute all the $f_t(i)$, we now copy from the range AG13:AK13 to the range AG13:AK16. For this to be successful, we need to have the correct values of the $f_t(i)$ in B5:H8. In columns B and H of rows 5–8, we enter 10,000 (or any large positive number). This ensures that it is very costly to end a month with an inventory that is negative or that exceeds 4. This will ensure that each month's ending inventory is between 0 and 4 inclusive. In the range C5:G5, we enter a 0 in each cell. This is because $f_5(i) = 0$ for $i = 0$, 1, 2, 3, 4. In cell C6, we enter +AG13; this enters the value of $f_1(0)$. By copying this formula to the range C6:G8, we have created a table of the $f_t(d)$, which can be used (in rows 13–16) to look up the $f_t(d)$.

As with the spreadsheet we used to solve Example 5, our current spreadsheet exhibits circular references. This is because rows 6–8 refer to rows 13–16, and rows 13–16 refer to rows 6–8. Pressing F9 several times, however, resolves the circular references. You also can resolve circular references by selecting Tools, Options, Calculations and checking the Iterations box.

For any initial inventory level, we can now compute the optimal production schedule. For example, suppose the inventory at the beginning of month 1 is 0. Then $f_1(0) = 20 = J_1(0, 1)$. Thus, it is optimal to produce 1 unit during month 1. Now we seek $f_2(0 + 1 - 1) = 16 = J_2(0, 5)$, so we produce 5 units during month 2. Then we seek $f_3(0 + 5 - 3) = 7 = J_3(2, 0)$, so we produce 0 units during month 3. Solving $f_4(2 + 0 - 2) = J_4(0, 4)$, we produce 4 units during month 4.

PROBLEMS

Group A

1 Use a spreadsheet to solve Problem 2 of Section 13.3.

2 Use a spreadsheet to solve Problem 4 of Section 13.4.

3 Use a spreadsheet to solve Problem 5 of Section 13.4.

SUMMARY

Dynamic programming solves a relatively complex problem by decomposing the problem into a series of simpler problems. First we solve a one-stage problem, then a two-stage problem, and finally a T-stage problem (T = total number of stages in the original problem).

In most applications, a decision is made at each stage (t = current stage), a reward is earned (or a cost is incurred) at each stage, and we go on to the stage $t + 1$ state.

Working Backward

In formulating dynamic programming recursions by working backward, it is helpful to remember that in most cases:

1 The **stage** is the mechanism by which we build up the problem.

2 The **state** at any stage gives the information needed to make the correct decision at the current stage.

3 In most cases, we must determine how the reward received (or cost incurred) during the current stage depends on the stage t decision, the stage t state, and the value of t.

4 We must also determine how the stage $t + 1$ state depends on the stage t decision, the stage t state, and the value of t.

5 If we define (for a minimization problem) $f_t(i)$ as the minimum cost incurred during stages $t, t + 1, \ldots, T$, given that the stage t state is i, then (in many cases) we may write

$f_t(i) = \min \{(\text{cost during stage } t) + f_{t+1}(\text{new state at stage } t + 1)\}$, where the minimum is over all decisions allowable in state i during stage t.

6 We begin by determining all the $f_T(\cdot)$'s, then all the $f_{T-1}(\cdot)$'s, and finally f_1 (the initial state).

7 We then determine the optimal stage 1 decision. This leads us to a stage 2 state, at which we determine the optimal stage 2 decision. We continue in this fashion until the optimal stage T decision is found.

Computational Considerations

Dynamic programming is much more efficient than explicit enumeration of the total cost associated with each possible set of decisions that may be chosen during the T stages. Unfortunately, however, many practical applications of dynamic programming involve very large state spaces, and in these situations, considerable computational effort is required to determine optimal decisions.

REVIEW PROBLEMS

Group A

1 In the network in Figure 14, find the shortest path from node 1 to node 10 and the shortest path from node 2 to node 10.

2 A company must meet the following demands on time: month 1, 1 unit; month 2, 1 unit; month 3, 2 units; month 4, 2 units. It costs $4 to place an order, and a $2 per-unit holding cost is assessed against each month's ending inventory. At the beginning of month 1, 1 unit is available. Orders are delivered instantaneously.

   a Use a backward recursion to determine an optimal ordering policy.

3 Reconsider Problem 2, but now suppose that demands need not be met on time. Assume that all lost demand is

backlogged and that a $1 per-unit shortage cost is assessed against the number of shortages incurred during each month. All demand must be met by the end of month 4. Use dynamic programming to determine an ordering policy that minimizes total cost.

4 Indianapolis Airlines has been told that it may schedule six flights per day departing from Indianapolis. The destination of each flight may be New York, Los Angeles, or Miami. Table 20 shows the contribution to the company's profit from any given number of daily flights from Indianapolis to each possible destination. Find the optimal number of flights that should depart Indianapolis for each destination. How would the answer change if the airline were restricted to only four daily flights?

5 I am working as a cashier at the local convenience store. A customer's bill is $1.09, and he gives me $2.00. I want to give him change using the smallest possible number of coins. Use dynamic programming to determine how to give the customer his change. Does the answer suggest a general result about giving change? Resolve the problem if a 20¢ piece (in addition to other United States coins) were available.

FIGURE 14

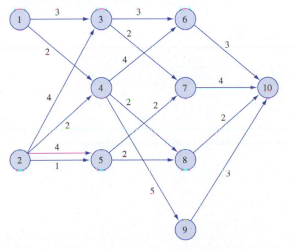

TABLE 20

	Profit per Flight ($)					
	Number of Planes					
Destination	1	2	3	4	5	6
New York	80	150	210	250	270	280
Los Angeles	100	195	275	325	300	250
Miami	90	180	265	310	350	320

6 A company needs to have a working machine during each of the next six years. Currently, it has a new machine. At the beginning of each year, the company may keep the machine or sell it and buy a new one. A machine cannot be kept for more than three years. A new machine costs $5,000. The revenues earned by a machine, the cost of maintaining it, and the salvage value that can be obtained by selling it at the end of a year depend on the age of the machine (see Table 21). Use dynamic programming to maximize the net profit earned during the next six years.

7 A company needs the following number of workers during each of the next five years: year 1, 15; year 2, 30; year 3, 10; year 4, 30; year 5, 20. At present, the company has 20 workers. Each worker is paid $30,000 per year. At the beginning of each year, workers may be hired or fired. It costs $10,000 to hire a worker and $20,000 to fire a worker. A newly hired worker can be used to meet the current year's worker requirement. During each year, 10% of all workers quit (workers who quit do not incur any firing cost).

    a With dynamic programming, formulate a recursion that can be used to minimize the total cost incurred in meeting the worker requirements of the next five years.

    b How would the recursion be modified if hired workers cannot be used to meet worker requirements until the year following the year in which they are hired?

8 At the beginning of each year, Barnes Carr Oil sets the world oil price. If a price p is set, then $D(p)$ barrels of oil will be demanded by world customers. We assume that during any year, each oil company sells the same number of barrels of oil. It costs Barnes Carr Oil c dollars to extract and refine each barrel of oil. Barnes Carr cannot set too high a price, however, because if a price p is set and there are currently N oil companies, then $g(p, N)$ oil companies will enter the oil business [$g(p, N)$ could be negative]. Setting too high a price will dilute future profits because of the entrance of new companies. Barnes Carr wants to maximize the discounted profit the company will earn over the next 20 years. Formulate a recursion that will aid Barnes Carr in meeting its goal. Initially, there are 10 oil companies.

9 For a computer to work properly, three subsystems of the computer must all function properly. To increase the reliability of the computer, spare units may be added to each system. It costs $100 to add a spare unit to system 1, $300 to system 2, and $200 to system 3. As a function of the number of added spares (a maximum of two spares may be added to each system), the probability that each system will

work is given in Table 22. Use dynamic programming to maximize the probability that the computer will work properly, given that $600 is available for spare units.

Group B

10 During any year, I can consume any amount that does not exceed my current wealth. If I consume c dollars during a year, I earn c^a units of happiness. By the beginning of the next year, the previous year's ending wealth grows by a factor k.

    a Formulate a recursion that can be used to maximize total utility earned during the next T years. Assume I originally have w_0 dollars.

    b Let $f_t(w)$ be the maximum utility earned during years $t, t + 1, \ldots, T$, given that I have w dollars at the beginning of year t; and $c_t(w)$ be the amount that should be consumed during year t to attain $f_t(w)$. By working backward, show that for appropriately chosen constants a_t and b_t,

$$f_t(w) = b_t w^a \quad \text{and} \quad c_t(w) = a_t w$$

Interpret these results.

11 At the beginning of month t, farmer Smith has x_t bushels of wheat in his warehouse. He has the opportunity to sell wheat at a price s_t dollars per bushel and can buy wheat at p_t dollars per bushel. Farmer Smith's warehouse can hold at most C units at the end of each month.

    a Formulate a recursion that can be used to maximize the total profit earned during the next T months.

    b Let $f_t(x_t)$ be the maximum profit that can be earned during months $t, t + 1, \ldots, T$, given that x_t bushels of wheat are in the warehouse at the beginning of month t. By working backward, show that for appropriately chosen constants a_t and b_t,

$$f_t(x_t) = a_t + b_t x_t$$

    c During any given month, show that the profit-maximizing policy has the following properties: (1) The amount sold during month t will equal either x_t or zero. (2) The amount purchased during a given month will be either zero or sufficient to bring the month's ending stock to C bushels.

TABLE 22

Number of Spares	Probability That a System Works		
	System 1	System 2	System 3
0	.85	.60	.70
1	.90	.85	.90
2	.95	.95	.98

TABLE 21

	Age of Machine at Beginning of Year		
	0 Years	1 Year	2 Years
Revenues ($)	4,500	3,000	1,500
Operating Costs ($)	500	700	1,100
Salvage Value at End of Year ($)	3,000	1,800	500

REFERENCES

The following references are oriented toward applications and are written at an intermediate level:

Dreyfus, S., and A. Law. *The Art and Theory of Dynamic Programming*. Orlando, Fla.: Academic Press, 1977.

Nemhauser, G. *Introduction to Dynamic Programming*. New York: Wiley, 1966.

Wagner, H. *Principles of Operations Research,* 2d ed. Englewood Cliffs, N.J.: Prentice Hall, 1975.

The following five references are oriented toward theory and are written at a more advanced level:

Bellman, R. *Dynamic Programming*. Princeton, N.J.: Princeton University Press, 1957.

Bellman, R., and S. Dreyfus. *Applied Dynamic Programming*. Princeton, N.J.: Princeton University Press, 1962.

Bersetkas, D. *Dynamic Programming*. Orlando, Fla.: Academic Press, 1987.

Denardo, E. *Dynamic Programming: Theory and Applications*. Englewood Cliffs, N.J.: Prentice Hall, 1982.

Whittle, P. *Optimization Over Time: Dynamic Programming and Stochastic Control,* vol. 1. New York: Wiley, 1982.

Morton, T. "Planning Horizons for Dynamic Programs," *Operations Research* 27(1979):730–743. A discussion of turnpike theorems.

Waddell, R. "A Model for Equipment Replacement Decisions and Policies," *Interfaces* 13(1983): 1–8. An application of the equipment replacement model.

14

Heuristic Techniques

For a long time, heuristic techniques have been used to solve difficult optimization problems. When optimal solutions are difficult to obtain, heuristic techniques tend to exploit the structure of the problem to arrive at a good solution. A classic problem for which heuristic techniques are useful is the traveling salesperson problem (TSP) discussed in Section 9.6. There we used the nearest neighbor heuristic to find a "good" solution for the TSP.

The most common methodology applied to problems with little exploitable structure are heuristics that exchange values of decision variables with other feasible variables in a systematic or random fashion. Typically, the heuristic techniques are applied to solve optimization problems that take enormous amounts of computational time. In most cases, the optimization problems are hard to solve. The first step in understanding heuristic algorithms is to identify and understand such hard problems. The problems that can be solved in polynomial time are typically solved to optimality using efficient algorithms and powerful computers and are not considered hard to solve. The problems that are usually considered hard to solve are those problems for which there are not polynomial algorithms. Such problems are called *nondeterministic* (NP) *class problems.* Within the NP class problems, subsets of problems are identified as NP-complete problems. Any NP-complete problem can be translated to any other NP-complete problem by using techniques that will require polynomial time. Hence, if an efficient technique can be developed for one NP-complete problem, then all the problems in this class can be solved by the same technique efficiently. The problems that require an inordinate amount of computer time in the NP class are identified as *NP-hard problems.* Typically, heuristic algorithms are aimed at NP problems. We begin the chapter by providing a brief overview of complexity theory. Then we will describe artificial intelligence-based techniques for these classes of problems.

14.1 Complexity Theory

When we are confronted with a mathematical programming problem, a structured solution approach or *algorithm* is typically used to solve the problem. The efficiency and accuracy of algorithms used to solve given problems has been analyzed throughout human history. Are there well-defined problems for which no algorithm will terminate with an optimal solution? This question was answered in the affirmative by Alan Turing, who defined a mathematical object that came to be known as a *Turing machine.* A Turing machine consists of an infinitely long magnetic tape on which instructions can be inserted or deleted, a memory register of finite size, and a processor that can carry out various commands such as move tape right, move tape left, and change state of memory register based on current register value and the values on the tape. Suppose, given a set of inputs, we want to determine if it is possible for the Turing machine to reach a given state in a finite number of steps. In his discussion of the famous *halting problem,* Turing showed that it is not always possible to determine if the Turing machine can reach a given state in a finite number of steps.

On the other hand, if we can prove that an algorithm will converge to an optimal solution in a finite number of steps, then we would like to know how fast it would converge for various instances of the problem. A simple problem is sorting a set of n numbers from smallest to largest through a *bubble sort algorithm.* In a bubble sort algorithm, we compare two neighboring objects and swap them if they are in the wrong order. In this case, if the numbers are given largest to smallest, then a simple bubble sort will take n^2 comparisons before sorting the numbers. If the numbers are given in the correct order, then it will only do n comparisons. Based on the worst-case scenario, a bubble sort will be classified as $O(n^2)$ algorithm. The O stands for *order of time* that denotes the most dominating expression in the computational difficulty. Thus, if an algorithm is $O(n^2)$, then there exists a c and an n_0 such that for $n \geq n_0$ the number of computations needed is at most cn^2. The most efficient sorting algorithms are of $O[n*\log(n)]$. The well-known *heap sort algorithm* is $O[n*\log(n)]$. Looking at mathematical programming problems, the greedy optimal algorithm for solving the minimal spanning tree (see Section 8.6) and Dijkstra's shortest path algorithm (see Section 8.2) are $O(n^2)$. The algorithms that have solution times bounded by a polynomial order are considered efficient and useful because they can solve large instances of a problem in reasonable time, and the efficiency of computer processors and parallel processors can make a significant dent in the size of problems that can be solved optimally. The algorithm bounded by the polynomial expression $O(n^3)$ are considered highly useful, and higher powers such as $O(n^4)$ and $O(n^5)$ are still considered worthwhile because they will benefit from the advance of technology. The higher powers of n, such as $O(n^{10})$, will take an inordinate amount of solution time in the worst-case scenario. A scale of operations growth is presented in Table 1. For example, for $n = 20$, an $O(n^3)$ algorithm would, for some c_1 (assuming n_0 is less than 20), require at most $8{,}000c_1$ computations; for $n = 20$, an $O(n^{10})$ algorithm would, for some c_2, require at most $10.2c_2$ trillion computations. An initial schematic of the problem space is presented in Figure 1.

Any algorithm that converges to an optimal solution using a number of computations that is not bounded by polynomial time estimates is generally considered to be inefficient

TABLE 1
Complexity Growth of Polynomial Algorithms

	10	20	50	100
$N^{\log(n)}$	33	86	282	664
N^2	100	400	2,500	10,000
N^3	1,000	8,000	125,000	1,000,000
N^5	100,000	3,200,000	312,500,000	10,000,000,000
N^{10}	10,000,000,000	1.024E+13	9.76563E+16	1E+20

FIGURE 1
Problem Space with Polynomial Bounded Algorithms

for large problems. To categorize hard problems, the first step is to define a reasonable problem representation. For example, the traveling salesperson problem can be solved by a linear evaluating algorithm if all the $(n - 1)!/2$ tours are presented as the input to the algorithm. In this case, the representation of problem instance grows by an order of $(n!)$, which will not be considered a reasonable input. Problems with reasonable representation that have no known polynomial bounded algorithms are generally considered hard problems. The algorithms for the hard problems tend to require an exponential order of computations to arrive at an optimal solution. Table 2 illustrates the growth rate of computation time of exponential algorithms in terms of problem size. It is evident that that algorithms that require an exponential order of computations will not be able to solve reasonably sized problems in practical time.

The concept of an NP class of problems is developed in order to classify problems that do not have a known polynomial algorithm. NP is defined as nondeterministic polynomial time and not nonpolynomial as popularly believed. The understanding of the NP class of problems requires some basic definitions. Given a reasonable representation of the problem, an optimization problem asks to find the best solution. For example, given an n-city TSP problem, with an $n \times n$ symmetric distance matrix, finding the least distance tour that covers all cities is an *optimization* problem. A *recognition* version of the problem of TSP will ask whether a tour of distance less than L is available for a TSP problem. If there exists a tour of length smaller than $L,$ then the answer is "yes"; otherwise, the answer is "no." It can be easily shown that both problems are equally complex to solve; in addition, if we can solve one in polynomial time, the other also can be solved in polynomial time through simple search techniques. Because, in complexity terms, both the problems are equivalent, historical precedent and a clear mathematical foundation are the reasons why the recognition problem is used in defining the NP class of problems. The outcome of the recognition problem is either "yes" or "no." A "yes" instance of the problem is known as a *certificate*. A certificate for a traveling salesperson problem is a permutation of cities visited that constitute a tour that has a total tour length less than a predetermined tour length of L. A problem belongs to class NP if there is a precise (certificate-checking) algorithm bounded by polynomial time that can verify a certificate. To illustrate, consider a tour for a five-city TSP problem. A combination π such as (1–5–4–3–2–1) can be verified as a legal tour in time $O(n)$ by checking for a cycle as we parse through π; the cost associated with the tour length of 789 can be computed as $\Sigma c_{ij} x_{ij}$ in time $O(n)$. Hence, the TSP belongs to class NP.

Note that by this definition even problems for which the recognition or optimization problem can be solved in polynomial time are also classified as being in the NP class of problems because their certificates can also be verified in polynomial time. The advantage of the NP class definition is that it allows us to categorize a rich class of mathematical programming problems as belonging to the space of NP problems. Within NP, the space P contains problems that can be solved in polynomial time.

TABLE 2

Complexity Growth Rate of Exponential Algorithms

Instance	10	20	50	100
2^N	1.024	1,048,576	1.1259E+15	1.26765E+30
$N^{\log N}$	2,098.592396	419,718.4792	3,879,201,823	1.9396E+13
$N!$	3,628,800	2.4329E+18	3.04141E+64	9.3326E+157
5^N	9,765,625	9.53674E+13	8.881781E+34	7.88861E+69
10^N	10,000,000,000	1E+20	1E+50	1E+100

Another class of problems in NP that is considered as a hard problem is the classification NP-complete. Problems in the space NP-complete cannot be solved by a known polynomial time algorithm. If there is a polynomial time algorithm for any NP-complete problem, then there is a polynomial time algorithm for all NP-complete problems. The advantage of identifying a problem as NP-complete is that it clearly categorizes the problem as hard to solve; for reasonably sized problems, heuristic techniques are appropriate to consider. A problem is considered to be NP-complete; if it belongs to class NP, then all the other problems in NP can be transformed in polynomial time to this problem. A simple illustration of transformability is converting an integer-programming problem to a binary integer-programming problem in which each integer variable has a bound of 15. By substituting $x_{11} + 2x_{12} + 4x_{13} + 8x_{14} = x_1$, an integer problem with variable x_1 is converted to a binary variable problem with x_{11}, x_{12}, x_{13}, and x_{14} as binary variables. In fact, NP-complete problems can be considered to be the hardest type of problem in the NP space. Problems such as TSP, integer linear programming, matching problem, single-machine scheduling with sequence-dependent setup time, and multiprocessor scheduling are all classified as NP-complete problems. If any of these problems can be solved in polynomial time, then all of them can be solved in polynomial time because they can be transformed from one to another in polynomial time. An updated topology of NP space is presented in Figure 2.

The problems are considered harder as they move up in this schematic. The nondeterministic region contains problems for which polynomial solution algorithms are possible. A classic example is a linear programming problem. The simplex algorithm is not polynomial because in the worst case it may evaluate for a problem with n variables and m constraints all $\dfrac{n!}{m!(n-m)!}$ corner points before determining an optimal solution. The interior point methods developed by Khachian and Karmarkar are polynomial time algorithms for solving a linear programming problem. Also note that, in general, the simplex method is efficient in solving large-scale problems as it seldom evaluates the entire set of feasible corner points in finding an optimal solution.

From the schematic, it is that $P \in NP$. The question the scientific community has been grappling with for a long time is whether $P = NP$. It is generally believed that $P \neq NP$, even though it has not been proven. Many people have conjectured that there may not be a polynomial algorithm for NP-complete problems, but nobody has been able to prove this conjecture to be true. The complexity discussions are generally carried out in terms of worst-case performance. There is a subset of problem instances such as the integer knapsack problem that have a pseudo-polynomial algorithm. For the integer knapsack problem with n variables and a right-hand-side value K of the single constraint, it can be shown that the problem can be solved in $O(nK)$. On the other hand, the TSP with restrictive assumptions on the number of connections out of cities is still NP-complete and considered as strongly NP-complete. There are other problems that are shown to be polynomially transformable to other problems in NP, but the certificate checking cannot be carried out in polynomial time. In other words, even though these problems are as hard as NP-complete problems, they do not belong to space NP. These problems are classified

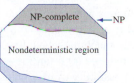

FIGURE 2
NP Problem Space with P and NP-complete components

as NP-hard problems. Although it will be a noble endeavor to develop an optimal efficient solution technique for NP-complete and NP-hard problems, brilliant mathematical minds have failed to accomplish this feat for a long time. Hence, it is practical to settle for approximate or heuristic procedures to solve these problems. A detailed description of NP problems and a comprehensive list of problems that are NP-complete and NP-hard is detailed in books by Gary and Johnson as well as Papadimitriou and Steiglitz.

14.2   Introduction to Heuristic Procedures

When confronted with difficult problems to be solved (NP-complete or NP-hard), operations researchers have used approximate techniques to arrive at a good solution. The approximate rule-of-thumb techniques where optimality cannot be assured are classified as heuristic techniques in operations research (OR) practice. Typically, these techniques are characterized by using a greedy approach to obtaining a good solution in efficient time and incremental improvement to an existing solution by neighborhood exchanges or local search. As a result, they tend to get trapped in a local optimal solution and fail to find a global optimum. No attempts are made to show by either mathematical principles or implicit evaluation of a solution space that the heuristic technique yields an optimal solution. Although most of the heuristic techniques tend to converge to a good solution in polynomial time, the better ones also provide bounds on the quality of the solution.

A nearest-neighbor heuristic for the TSP is presented as an example of a greedy heuristic. This heuristic can start with any city. From the starting city, it selects the closest city to visit next. Then it proceeds from the selected city to visit the closest city not visited yet. Taking n such steps, the algorithm will determine a tour that covers all the cities, and adding the distance from the last city to the starting city to the total distance traveled so far will provide the objective value. At most, this algorithm will examine $n(n - 1)/2$ number of data points for a symmetric TSP with n cities and thus has a computational complexity of $O(n^2)$. In addition, for a positive distance symmetric TSP where the triangle inequality holds—that is, for all cities i, j, and k,

(the distance between city i to city k) $<=$ (distance between city i to city j) + (distance between city j to city k),

Rosenkratz, Sterns, and Lewis showed that

(length of the heuristic tour / length of optimal tour) $<= 0.5(1 + \log_2 n)$.

For minimizing tardiness in a single-machine scheduling problem, a greedy heuristic can order the jobs in due-date sequence to arrive at a solution in efficient time. The greedy heuristics are efficient, but they sometimes get trapped in a local solution and fail to find a global optimum. Greedy heuristics are based on intimate knowledge of the problem structure and have no scope of incremental improvement.

One way to overcome the single-solution aspects of the greedy heuristics is to develop a neighborhood-exchange heuristic. For the TSP, we can try to improve the NN heuristic solution by removing say a set of three arcs and replacing them with another set of three if the solution value is improved. Care has to be taken to avoid subtours, which will allow only two other combinations of three arcs to be considered for completing the solution. The procedure can be repeated until it is not possible to obtain any improvement by swapping three arcs. Similarly, in a single-machine scheduling, any two jobs in the sequence can be swapped if the exchange reduces total tardiness. These techniques incrementally improve solutions and have the possibility of examining several solutions. Once again, these heuristics require a deep knowledge of the problem, can be trapped in local optima, and cannot be applied to all problems.

Attempts to develop other general heuristics that can work on a variety of problems have met with little success until the development of solution methodologies based on artificial intelligence (AI), which are generally addressed as metaheuristics. AI techniques have been used successfully in science and engineering to solve problems that are inherently difficult to solve. Simon (1987) tried to integrate OR and AI. Within the OR paradigm, support for this integration came from CONDOR (1988). In that study, several research areas were highlighted and identified as likely to produce significant results. One such area was the OR-AI link. Glover and Greenberg (1989) surveyed four emerging heuristic approaches: *genetic algorithms, neural networks, simulated annealing,* and *Tabu search.* Over the past two decades, the metaheuristics based on these AI techniques have been adopted to solve decision-making problems. These have been classified as AI approaches. Of these four categories—genetic search, simulated annealing, and Tabu search—will be discussed in detail in the following sections. Neural networks will be discussed in Chapter 16.

AI Techniques

Metaheuristic are a general master strategy to solve problems, and they are based on intelligent search techniques. AI techniques are based on natural adaptive systems. They can be based on evolutionary principles such as method of natural selection, physical systems such as the annealing process, and human learning as in the brain's adaptive neural strategy and memory. Genetic search is a solution technique that emulates natural genetics and natural selection. Simulated annealing imitates the physical annealing process of heating a solid until it melts and then lowering the temperature slowly to find a solid's low-energy states. Neural networks simulate the behavior of neurons in the brain to infer previous experiences in making the decision-making process. Tabu search uses short- and long-term memory to search quality neighborhoods and avoid recent mistakes. These techniques try to prescribe a broad strategy that can be customized to different difficult problems to achieve a good solution in an efficient manner. They try to avoid the local optima by accepting solutions that may not be an improvement or by considering several solutions at a time. They also have mechanisms to terminate the algorithms without cycling.

14.3 Simulated Annealing

In a seminal paper, Metropolis et al. (1953) developed a simple Monte Carlo approach that could be used to simulate the behavior of a set of atoms in achieving thermal equilibrium at a given temperature. Suppose our goal is to find the minimum energy-level configuration. The underlying idea is to begin with a current atomic configuration (and thus energy state E_0) and apply a small, randomly generated perturbation to the structure. If the perturbation results in a lower energy at state i (where $E_0 > E_i$), then the process is repeated using this new energy state; however, if the result is a higher energy state i (where $E_0 < E_i$), then the new state is accepted with a certain probability. This idea helps make it possible to leave a local optimal solution and potentially find a more attractive path of descent (path of ascent for a maximization problem) in the search space.

First introduced as a technique to solve complex nonlinear optimization problems, simulated annealing (SA) emulates the physical process of aggregating particles in a system as it is cooled. By slowly lowering temperature, the energy exchange allows true equilibrium in each stage until the global minimum energy level is reached. The choices of melting temperature level, number of steps in cooling, and amount of the time spent in a state are used to avoid local equilibrium points that are not global equilibria. The energy exchange is analogous to swapping neighboring solutions. The change in state by accepting (or rejecting a solution) is based on evaluating the new state with respect to Boltzman's function. For an SA algorithm, the melting temperature is called the *initial temperature,* and the steps in lowering temperature are collectively known as a *cooling schedule.* The starting state is any starting solution. Accepting or rejecting a solution is based on the objective function value and a probabilistic chance of acceptance based on Boltzman's function evaluation criteria. Typically any feasible improvement is accepted. Inferior solutions have a chance of acceptance based on Boltzman's function. This aspect of SA allows it to move the process away from a local optimal solution. SA is also viewed as a probabilistic process in which a control parameter (temperature) evaluates the probability of accepting [given by PR(A)] suboptimal solutions during the solution space search. PR(A) (of an uphill move) is denoted by $e^{(\Delta C/T)}$, where T is the temperature and

$$\Delta C = \text{best objective function value to date} - \text{current objective function value}$$

is the change in the objective function value. For a minimization problem, ΔC is negative, so as T decreases, the probability of accepting an inferior solution decays exponentially. It is argued that combinatorial optimization problems can be viewed as analogous to such physical systems. As the temperature falls, the atoms form stronger bonds until the ground state (lowest energy state) is reached and a solid is formed. The ground state is reached if the temperature is decreased slowly enough to enable this *settling* to equilibrium. Johnson (1989) presents a clear analogy between the physical system and the simulated environment. We restate his summary below in Table 3 as it relates to the components of our solution process.

The technique is memory-less because it optimizes without prior knowledge of the solution strategy or problem structure. The simple exchange or swapping strategy and the ease of deriving cost functions make the technique highly attractive. The general SA algorithm is modeled using the theory of Markov chains to derive sufficient and necessary conditions for the asymptotic convergence of the algorithm to the set of global optima. These can be found in Lundy and Mees (1986), Anily and Federgruen (1987), Hajek (1988), and Tsitsiklis (1989). The mathematical results, however, are of little practical value. Lundy and Mees show that convergence could be exponentially long, making SA no better than any exponential algorithms. The SA parameters can be controlled to converge in a given time with a good solution. Thus, SA has become an attractive heuristic approach in the search for near-optimal rather than optimal solutions. We now present a generic SA algorithm (for a minimization problem).

Step 1   Generate a solution x_0 for the problem. Arrive at an initial temperature T_0, number of iterations at each step I, and cooling schedule temperature reduction δ based on experience or preliminary studies and set $T_{cur} = T_0$, $x_{cur} = x_0$, $x_{best} = x_{cur}$, and $z_{best} = z(x_{cur})$.

TABLE 3
Physical vs. Simulated (Annealing)

Physical System	Optimization Problem
State	Feasible solution
Energy (E_i)	Cost
Ground state	Optimal solution
Rapid quenching	Local search (fast)
Careful annealing	Simulated annealing

Step 2 For count = 1 to I, randomly generate a new solution x_{count}. If objective $z(x_{count})$ is better than $z(x_{cur})$ set $x_{cur} = x_{count}$. If objective $z(x_{count})$ is worse than $z(x_{cur})$, then calculate $\Delta C = z(x_{count}) - z(x_{cur})$ and set $x_{cur} = x_{count}$ with a probability of $e^{(\Delta C/Tcur)}$. If z_{best} is worse than $z(x_{cur})$, then set $x_{best} = x_{cur}$, and $z(best) = z(x_{cur})$. Now update the value of count.

Step 3 Set $T_{cur} = T_{cur} - \delta$. If stopping criteria of temperature level (say, temperature <10) or another preset stopping criteria is met, then stop with z_{best} and x_{best} as the SA solution. Otherwise, return to step 2, with T_{cur}.

Step 1 is the most important step that has been extensively researched. Popular values for δ are $0.01\ T_{cur}$ to $0.2\ T_{cur}$. To set the value of T_0, I (the number of neighborhood evaluations), and δ, another popular technique is to make some preliminary study of random problems and choose the values that yield the most improvement. Note that δ and I provide a bound on computation time if an estimate of work needed for step 2 is available— that is, if step 2 is of $O(n^2)$ and the temperature lower bound is 0, then the algorithm will converge in $O[(T_0/\delta)*I*n^2]$. An example of an SA implementation is provided with a TSP problem.

EXAMPLE 1 **Simulated Annealing Algorithm and TSP**

Table 4 is a distance matrix for a nine-city TSP.

TABLE 4
Distance Matrix

0	172	250	127	103	170	264	82	159
172	0	170	177	200	325	178	139	241
250	170	0	206	134	252	201	233	291
127	177	206	0	191	181	184	57	172
103	200	134	191	0	201	144	197	268
170	325	252	181	201	0	229	211	129
264	178	201	184	144	229	0	132	202
82	139	233	57	197	211	132	0	186
159	241	291	172	268	129	202	186	0

Use simulated annealing to find the solution to this TSP.

Solution **Step 1** Based on the literature, we assume $T_0 = 2{,}000$, I = 10, $\delta = 0.1*T_0$, count = 1, $x_0 = [1, 2, 3, 4, 5, 6, 7, 8, 9] = x_{cur} = x_{best}$, $T_{cur} = 2{,}000$, $z(x_{cur}) = z(x_{best}) = 1{,}646$, and $T_{stop} = 0$.

Step 2 We begin iteration 1 by randomly choosing two cities to swap. To this we can choose one city with the Excel formula 1 + int[9*rand()]. This formula is equally likely to yield any integer from 1 to 9. Then the Excel formula 1 + int[8*rand()] is equally likely to yield an integer from 1 to 8. The second random integer tells us how many cities to the right of the first city we need to go to get the second city (we assume that if we get to city 9 we next go to city 1). For example, if the first integer is 6 and the second integer is 5, then we choose cities 6 and 2 to swap.

Suppose our random integers tell us to swap city 5 with city 8. The resulting solution, $x_1 = [1, 2, 3, 4, 8, 6, 7, 5, 9]$ has $z_1 = 1{,}616$. We accept this move as it is better than $z(x_{cur})$. Update $x_{cur} = [1, 2, 3, 4, 8, 6, 7, 5, 9]$ and $z(x_{cur}) = 1{,}616$. Also update $x_{best} = [1, 2, 3, 4, 8, 6, 7, 5, 9]$ and $z_{best} = 1{,}616$. Now count = 2.

Suppose random selection for iteration 2 tells us to swap city 2 with city 8. The resulting solution, $x_2 = [1, 8, 3, 4, 2, 6, 7, 5, 9]$ has $z_2 = 1,823$. This objective is not better than $z(x_{cur})$. Calculate $\Delta C = 1,616 - 1,823 = -207$ and $e^{(\Delta C/T_{cur})} = e^{(-207/2000)} = 0.901676$. The chance of accepting this uphill move is 0.901676. Draw a random number, say, rand = 0.913376. Because rand is not less than 0.901676, discard move and update count = 3.

Suppose random selection for Iteration 3 indicates that we swap city 2 with city 9 in x_{cur}. The resulting solution, $x_3 = [1,9,3,4,8,6,7,5,2]$ has $z_3 = 1,669$. This objective is not better than $z(x_{cur})$. Calculate $\Delta C = 1616 - 1669 = -53$ and $e^{(\Delta C/T_{cur})} = e^{(-53/2000)} = 0.973848$. The chance of accepting this uphill move is 0.973848. Draw a random number, say rand = 0.883602. Since rand is less than 0.973848, accept this move. Update $x_{cur} = [1,9,3,4,8,6,7,5,2]$ and $z(x_{cur}) = 1,669$. As z_{best} is less than z_{cur} keep x_{best} and z_{best} the same. Update count = 4.

Step 3   After count = 10, calculate $T_{cur} = T_{cur} - \delta$. $\delta = 0.1*2,000 = 200$. $T_{cur} = 2,000 - 200 = 1,800$. Because $1,800 > 10$, repeat step 2.

At the end of this process, the best solution found was 1–8–4–6–9–7–2–3–5–1. The total length of this tour is 1,236.

Note that, as the temperature is lowered, the probability of accepting a worse solution is lowered. For example, $e^{(-53/1800)} = 0.970985$ is lower than $e^{(-53/2000)} = 0.973848$. Note that the uphill moves allow the SA algorithm to avoid local optimal solutions.

There have been a multitude of studies using SA to solve many types of optimization problems. Collins et al. (1988) built a comprehensive annotated bibliography that contains some 300 references. There have been numerous studies with promising results that suggest SA is an efficient approach for solving certain combinatorial optimization problems, such as graph coloring, single-machine scheduling, partitioning, network programming, multilevel-lot sizing, and scheduling.

The general consensus is that SA requires significant computing time. This has raised debate and ignited follow-up research into cooling-schedule selection studies for combinatorial problems. Because most problems of interest are NP-hard, a cooling schedule that is not large enough for a given problem size will produce suboptimal solutions. As an improvement, some studies have examined techniques that extend the neighborhood search space in order to avoid local optima, while others have studied transition mechanisms such as acceptance probability criteria. Further research is also being conducted to combine SA with other AI techniques.

14.4   Genetic Search

Genetic algorithms (GA) belong to the evolutionary class of AI techniques. Over millions of years, organisms have biologically evolved to survive and flourish in a changing world. All living organisms' genetic materials consist of chromosomes that are divided into genes; their encoding and development are considered a key process in the continuation of each species. The process of natural selection and the survival of the fittest are considered important elements in evolution. GAs are based on these aspects of evolution. Holland (1975) proposed heuristics based on genetic principles as a methodology to solve decision-making problems. GAs utilize ideas from biology such as a population of chromosomes, natural selection for mating, offspring production using crossover, and mutation for diversity. The important starting point for a GA is a problem-solution representation that uses a constant length chromosome string. The variable represented by a position in the string represents a *gene,* and its value is known as an *allele.* Once such an encoding scheme is developed, GA methodology can be applied to arrive at a good solution to the problem in hand. The most popular encoding is a binary coding, where the alleles take a value of 0 or 1. If needed, real-valued and character-valued encoding can also be used. Although fixed encoding is popular, researchers have used adaptive encoding in some problem settings for improved results.

Genetic algorithms begin by randomly generating an initial population of strings or chromosomes. These strings represent possible solutions to the problem. Each solution is evaluated by measurable criteria that result in a fitness being associated with each string. A second population (called generation 2) is generated from the first population by mating the previous population and performing some mutations. The partners in mating are chosen by the principle of survival of the fittest. The higher the member's fitness, the more likely the member will be chosen to reproduce. Searches usually are terminated by reaching a predetermined nonimprovement for a specified number of generations.

The advantages of such a heuristic approach are considerable. First, the evaluation function that determines fitness need not be linear, differentiable, or even continuous. The parameters of the approach—including the size of the initial population, the evaluation function, and the mutation frequency—as well as the breeding method (to some extent) are all under control of the user. Thus, they may be manipulated in such a manner as to achieve faster or slower convergence. The GA solution method is also parallel, and thus represents one of the few algorithms that can accomplish linear speedup on a parallel machine.

GAs move toward optimality only if they satisfy the conditions of the Fundamental Theorem of Genetic Algorithms (Goldberg, 1989). Those conditions insist that the breeding process not destroy patterns or schema and that they then survive (in expected value) and combine with other fit schema in hope of achieving optimality. The conventional genetic algorithm consists of the following six steps:

Step 1 *Generation* of a starting population of solutions. Typically, the size of the population is set to a predetermined value of n, and the values (alleles) are generated in a random manner. We call the starting population generation 1.

Step 2 *Evaluation.* Here we score each solution by a fitness function. The objective function associated with the string is, in general, the fitness associated with them. Infeasible solutions are typically penalized.

Step 3 *Selection.* Probabilistically, choose from the current solutions the parents for the next generation. The most fit have the greatest chance of being chosen. A string with a better fitness value has a higher probability of selection than a string with a worse fitness value.

Step 4 *Reproduction.* Crossover the parents at a random point in the gene string so that the offspring consists of a portion of each parent.

Step 5 *Mutation.* Randomly alter the genetic makeup of the offspring to avoid local optima. The offspring becomes the new generation of solutions.

Step 6 *Repeat* steps 2 through 5 until the algorithm reaches an optimum or does not improve for some number of generations.

Setting the value of n usually depends on preliminary testing of similar problems and a rule of thumb. A value such as 50 or 100 is typical. The selection of solutions that survive until the next generation are typically based on fitness value. Crossover is another important step in improving the performance of GA. Random crossover is widely used in GA, but specialized methods at times choose candidate positions for crossover (or not crossover) to improve the convergence of GA. The mutation factor plays an important role in moving away from local optima and diversifying the search space. Typically, a rate of 0.02 to 0.05 is used for mutation. Higher rates are used in problem-specific scenarios. The popular stopping criteria is a lack of solution improvement and the average fitness of population not improving over a predetermined number of generations. The decision maker can observe the entire population of the final generation and choose the solution with the best fitness value or the one that provides the most intangibles. In contrast to SA, GA works on multiple solutions simultaneously.

Problem-specific descriptions of these steps are described in the following discussion, using a product-line pricing and positioning example.

EXAMPLE 2 Product-Line Selection and Pricing with Genetic Algorithms

Acme Widget Company (AWC) has the possibility of offering five substitutable products for 10 customer segments. Each segment is assumed to be of size 20. For each segment, AWC has estimated the segment preference for the products. This information is given in Table 5. The variable cost per unit of manufacturing the product and the fixed-cost production are given in Table 6.

TABLE 5

Segment/Prod	1	2	3	4	5
1	184	149	126	148	187
2	198	182	125	167	152
3	155	132	192	165	167
4	141	197	133	173	182
5	136	128	117	200	192
6	172	152	171	182	133
7	184	167	138	119	142
8	123	156	163	176	159
9	114	191	179	168	126
10	110	184	139	149	185

TABLE 6

Fixed cost	710	1,203	665	1,224	1,391
Variable cost	50	55	56	69	58

The segment surplus is preference minus the price of the product. A segment will choose a product with highest segment surplus if it is not negative. If the highest surplus for a segment is negative, then the segment will buy nothing. AWC is faced with the task of identifying what products to offer and at what price.

Solution The model is a nonlinear integer programming problem and can be classified as NP-hard. We will use a genetic approach to solve this problem. The first step is to determine a suitable encoding. The decision variables for the GA are whether a product is offered or not and, if offered, at what price. The offer-or-not variables are encoded as a binary string with five members. Each element with a 1 means the corresponding product is offered. Each element with a 0 means the product is not offered. The price is a continuous variable and can be encoded using a real representation. There will be five continuous variables for each product (each variable represents the price charged for a particular product). Because the minimum variable cost is 50 and the maximum preference is 200, each price will be between 50 and 200. A representation of the string is as follows:

String 1   0   1   1   1   0   121   128   93   200   51

String 1 implies that products 2, 3, and 4 are offered, and the prices are 121, 128, 93, 200, and 51 for the five products. Another example string follows:

String 2   1   1   0   0   1   92   116   87   96   95

A single point *crossover* will pick a position in a set of two strings (call it position p) to split the strings into two. Then, assuming there are k alleles in each string, we combine the split strings as follows to form two new offspring:

New string 1   Left p alleles from string 1 followed by right $k - p$ alleles from string 2.

New string 2   Left p alleles from string 2 followed by right $k - p$ alleles from string 1.

For example, if we choose $p = 3$, applying single point crossover to strings 1 and 2 will result in two new offspring as follows:

Original strings
0   1   1 | 1   0   121   128   93   200   51
1   1   0 | 0   1   92   116   87   96   95

New strings
0   1   1   0   1   92   116   87   96   95
1   1   0   1   0   121   128   93   200   51

Another alternative is to use a two-point crossover. Suppose we choose the two crossover positions at the second and eighth positions. Then we create two new strings by swapping the first two and last two alleles in each string. If we apply this two-point crossover to strings 1 and 2, we obtain the following strings:

Result of two-point crossover on strings 1 and 2

1   1   1   1   0   121   128   93   96   95
0   1   0   0   1   92   116   87   200   51

A mutation can pick one of the positions in the string and then can modify the contents. If the position is between 1 and 5, then we change the content from 0 to 1 or 1 to 0. If the mutation position is 3 in the last string, then the string will be modified as follows:

0   1   1   0   1   92   116   87   200   51

If the mutation position is between 6 and 10, then the content can be modified by an amount that can be randomly determined to be positive or negative. For example, we can alter the price by, say, 20. Let us specify that position 8 from the previous string is chosen for mutation and the sign is positive. Then the string will be modified as follows:

0   1   1   0   1   92   116   107   200   51

Once the price and the product offerings are set, the consumer surplus for each segment can be calculated by subtracting price from preference and multiplying it by whether or not a product is offered. The maximum surplus can then be calculated; if it is positive, then the segment will be assigned to the corresponding product. Segment size multiplied by the product margin less the fixed cost of offered products would provide the evaluation or fitness profit value.

Let us assume a population of 20 for each generation, a mutation rate of 0.1, and a single-point crossover after position 2. We will also define a crossover rate that specifies what percentage of the population can be considered for crossover. We choose a crossover rate of 80%.

Step 1 *Generation.* Set the number of generations to 50. Population = 20, mutation = 0.1, and crossover rate = 0.8. Suppose our randomly chosen population of 20 strings is as shown in Table 7.

TABLE 7

0	0	0	1	1	91	108	100	116	118
1	1	1	0	0	128	87	110	114	83
1	1	0	1	1	135	67	114	98	95
1	1	1	0	1	76	108	77	126	135
0	1	1	0	0	94	100	105	81	62
0	1	1	0	1	115	104	97	74	106
1	1	1	0	0	110	91	88	86	73
0	1	0	0	1	103	140	80	131	111
1	1	0	0	1	97	98	99	93	103
1	1	0	0	1	60	81	94	89	118
1	1	0	1	0	88	103	74	76	120
1	1	0	0	1	126	105	98	79	107
1	0	0	1	1	111	94	105	112	110
0	1	0	1	0	100	114	93	117	100
1	0	1	0	0	97	112	85	104	119
1	1	0	0	0	98	100	89	85	136
1	0	1	0	0	135	103	86	87	71
1	0	1	1	0	54	56	138	94	127
1	1	1	0	0	105	135	93	93	98
1	1	0	1	0	115	87	78	105	127

Step 2 *Evaluation.* To determine the fitness of each string, we determine which segments will buy each produced product (given our product prices). Note that if the maximum consumer surplus is non-negative, then a segment buys the offered product yielding the maximum consumer surplus. If a segment's maximum surplus is negative, then the segment makes no purchases. Now our profit is computed by

$$\sum_{all\ segments} (\text{Size of segment})*(\text{price of product bought} -$$

$$\text{variable cost of product bought}) -$$

$$\sum_{all\ products\ offered} (\text{product fixed cost}).$$

For example, for the string shown in Table 8, we find that eight segments purchase product 5 and two segments purchase product 2; thus, we find the profit of $9,286 as follows:

$$20(2)*(140 - 55) + 20(8)*(111 - 58) - 1203 - 1391 = \$9,286.$$

TABLE 8

					Price					
P1	P2	P3	P4	P5	1	2	3	4	5	Profit ($)
0	1	0	0	1	103	140	80	131	111	9,286

Then we can determine the sorted strings in descending order-of-fitness values as shown in Table 9.

TABLE 9

					Price					
P1	P2	P3	P4	P5	1	2	3	4	5	Profit ($)
0	1	0	0	1	103	140	80	131	111	9,286
0	1	0	1	0	100	114	93	117	100	8,493
1	1	1	0	0	105	135	93	93	98	7,982
0	0	0	1	1	91	108	100	116	118	7,825
1	1	0	0	0	98	100	89	85	136	7,747
0	1	1	0	0	94	100	105	81	62	7,372
1	0	1	0	0	135	103	86	87	71	6,825
1	0	0	1	1	111	94	105	112	110	6,715
1	1	0	0	1	126	105	98	79	107	6,616
0	1	1	0	1	115	104	97	74	106	6,021
1	0	1	0	0	97	112	85	104	119	5,865
1	1	0	0	1	97	98	99	93	103	5,696
1	1	1	0	0	110	91	88	86	73	4,862
1	1	1	0	0	128	87	110	114	83	4,262
1	1	0	1	0	115	87	78	105	127	4,243
1	1	1	0	1	76	108	77	126	135	2,011
1	1	0	1	0	88	103	74	76	120	943
1	1	0	0	1	60	81	94	89	118	−24
1	0	1	1	0	54	56	138	94	127	−539
1	1	0	1	1	135	67	114	98	95	−788

Step 3 With a crossover rate at 0.8, the first 16 strings will be chosen to be eligible parents. The 16 strings' associated probabilities of selection and cumulative probabilities are as shown in Table 10.

For example, the probability of choosing string 1 equals

9,286/(total fitness of all strings) = 9,286/101,821 = 9.12%.

The cumulative fitness percentage for the second string is 9.12 + 8.34 = 17.46%.

Steps 4 and 5 We now use crossovers and mutations to produce a new generation. Because the "fitter" strings have more chance of being used for crossover, we would expect the next generation to have improved fitness. This is the way GAs implement the biological concept of survival of the fittest.

We now illustrate how to create the strings for the next generation. We begin by generating two random numbers that will be used to choose the two strings used in the first crossover. Suppose random number 1 = 0.070826 and random number 2 = 0.245279. Any random number less than or equal to .0912 yields string 1; any random number greater than .09 and less than .1746 yields string 2; and so on. This ensures that the prob-

TABLE **10**

					Price							
P1	P2	P3	P4	P5	1	2	3	4	5	Profit ($)	Fitness (%)	Cumulative Fitness (%)
0	1	0	0	1	103	140	80	131	111	9,286	9.12	9.12
0	1	0	1	0	100	114	93	117	100	8,493	8.34	17.46
1	1	1	0	0	105	135	93	93	98	7,982	7.84	25.30
0	0	0	1	1	91	108	100	116	118	7,825	7.69	32.99
1	1	0	0	0	98	100	89	85	136	7,747	7.61	40.59
0	1	1	0	0	94	100	105	81	62	7,372	7.24	47.83
1	0	1	0	0	135	103	86	87	71	6,825	6.70	54.54
1	0	0	1	1	111	94	105	112	110	6,715	6.59	61.13
1	1	0	0	1	126	105	98	79	107	6,616	6.50	67.63
0	1	1	0	1	115	104	97	74	106	6,021	5.91	73.54
1	0	1	0	0	97	112	85	104	119	5,865	5.76	79.30
1	1	0	0	1	97	98	99	93	103	5,696	5.59	84.90
1	1	1	0	0	110	91	88	86	73	4,862	4.78	89.67
1	1	1	0	0	128	87	110	114	83	4,262	4.19	93.86
1	1	0	1	0	115	87	78	105	127	4,243	4.17	98.02
1	1	1	0	1	76	108	77	126	135	2,011	1.98	100.00

ability of choosing a string is proportional to the string's fitness. Based on our two random numbers, the original selected strings are 1 and 3. The strings and crossover are:

$$0 \quad 1 \mid 0 \quad 0 \quad 1 \quad 103 \quad 140 \quad 80 \quad 131 \quad 111$$
$$0 \quad 0 \mid 0 \quad 1 \quad 1 \quad 91 \quad 108 \quad 100 \quad 116 \quad 118$$

With the crossover operation (after position 2), the new strings are:

$$0 \quad 1 \quad 0 \quad 1 \quad 1 \quad 91 \quad 108 \quad 100 \quad 116 \quad 118$$
$$0 \quad 0 \quad 0 \quad 0 \quad 1 \quad 103 \quad 140 \quad 80 \quad 131 \quad 111$$

Suppose a mutation for string 2 occurs only in position 3. The strings after mutation are:

$$0 \quad 1 \quad 0 \quad 1 \quad 1 \quad 91 \quad 108 \quad 100 \quad 116$$
$$0 \quad 0 \quad 1 \quad 0 \quad 1 \quad 103 \quad 140 \quad 80 \quad 131$$

In a similar fashion, we would perform nine more reproductive operations to generate 18 more strings to create a new population of 20 strings for the next generation.

Step 6 Perform steps 2 through 5 for 50 generations and declare the best solution found so far as the heuristic solution.

After 50 generations we found the following solution:

$$1 \quad 1 \quad 1 \quad 1 \quad 0 \quad 135 \quad 67 \quad 114 \quad 98 \quad 95$$

Profit for this solution is $21,118. Note that the randomness involved in selecting strings for crossover and for implementing mutation implies that two different people performing the same number of GA iterations may obtain different solutions!

Many successful applications of GAs have been documented, and the curious reader is referred to Goldberg (1989). Although GAs may take inordinate amounts of time to converge to optimality or good solution, they have been used to solve many scheduling, hub location, vehicle routing, and financial modeling problems. In Chapter 15, we will show how to use a version of the Excel Solver (Premium or Evolutionary Solver) to implement GAs within Excel. Supply chain packages such as SAP-APO, Manugistics, and I2-Rhythm provide genetic search as a tool for enterprise-wide scheduling and optimization.

14.5 Tabu Search

Glover (1986) developed a heuristic procedure based on strategies used in intelligent decision making and called it Tabu search (TS). Tabu search uses short- and long-term memory to forbid certain moves in the solution space. The short-term memory forbids cycling around a local neighborhood in the solution space. It also helps to move away from a local optimal solution. Long-term memory allows searches to be conducted in the most promising neighborhoods. Unlike SA and GA, TS does not imitate either physical or biological process but makes extensive use of memory. It more or less emulates heuristic rules that people use in day-to-day decision making. Although it looks simplistic at first glance, TS has been used to arrive at a good solution for several NP classes of problems.

Important ingredients of a TS algorithm are short-term memory Tabu rules and list size, long-term memory Tabu rules and list size, Tabu tenure, candidate list of moves, aspiration criteria, intensification, diversification, and strategic oscillation. The short-term memory list size determines how many forbidden moves are considered during the evaluation of a move, and the Tabu rules determine the members of the list. A similar explanation applies to the long-term memory list size and rules. Though the structures of the two memories are similar, the roles of these two lists in TS are significantly different. Aspiration criteria evaluate the quality of a move in solution space and provide a mechanism for overriding Tabu list. Aspiration criteria are analogous to a GA's fitness function or the simulated annealing Boltzman function. The candidate list and its structure play an important role in the performance of TS. A candidate list provides a list of moves that TS can evaluate. One (or more) move from the list is chosen to proceed with the search technique.

At times, a set number of alternatives may be explored in parallel and generating the same number of moves from these alternatives. This technique is referred to as *sequential fan list* or *beam search*. Intensification, diversification, and strategic oscillations are associated with long-term memory aspect of TS. An intensification strategy allows the TS to strategically search attractive solution regions more thoroughly. For example, as a part of an intensification strategy, we can have an elite list of candidates representing attractive regions; after a predetermined number of short-term memory moves, we can erase the Tabu conditions and start from another elite candidate. A diversification strategy allows unattractive moves in short-term memory and prevents cycling in the attractive in the long-term memory. Strategic oscillations allow intensification and diversification strategies to be applied intensively in an alternating fashion around a target boundary to improve outcome of TS.

Next, we present a generic TS algorithm.

Step 1   Start with a solution x_0 and elite list of candidates either from preliminary studies or history for long-term memory use. Set Tabu criteria, Tabu tenure, list size, and short-term iteration count limit n.

1   Pick a candidate from the elite list of candidates.

2   Create short-term candidate list and set count = 0.

3 While (count $<= n$), do

Create a move from candidate list to create a solution x_{count} from $x_{current}$.

If the solution satisfies Tabu-criteria test: Create Tabu evaluation

Else   If aspiration test is not met: Create penalized Tabu evaluation

Else create Tabu Evaluation

End If

End If

If x_{count} is best move so far, $x_{best} = x_{current}$

Update count = count +1

End while

4 If elite list has more candidates, go to step 2.

5 Report best solution.

Several variations of the TS algorithm are possible. We now use Tabu search to determine a schedule that minimizes the tardiness incurred by scheduling a single machine. We will assume that after each job is completed, there is a job-dependent changeover time.

EXAMPLE 3   **Single-Machine Scheduling Tabu Search**

Nine jobs need to be processed on a single machine. The due dates and processing times are as shown in Table 11.

TABLE 11

	Job								
	1	2	3	4	5	6	7	8	9
Processing times	2	6	3	9	4	10	1	2	2
Due dates	18	32	4	70	15	24	18	19	59

The changeover times are presented in Table 12. For example, if job 2 is done first, it takes eight hours to set up the machine. If job 3 is done after job 2, it takes two hours to

TABLE 12

	Changeover Times								
	1	2	3	4	5	6	7	8	9
0	5	8	3	3	2	2	5	9	5
1	—	2	3	8	2	10	9	10	10
2	1	—	2	10	4	9	7	4	7
3	10	2	—	6	9	7	6	5	4
4	5	6	7	—	8	6	1	5	8
5	9	3	9	5	—	1	10	7	9
6	9	8	6	4	5	—	1	1	9
7	3	7	8	2	3	3	—	7	3
8	5	2	7	3	8	9	1	—	10
9	6	5	2	4	8	7	7	7	—

set up the machine. Determine the order in which the jobs should be processed to minimize total tardiness of the jobs.

Solution **Step 1** The candidate list for starting solution can be based on an early due date (EDD) schedule, which processes jobs in ascending order of due dates. We could also generate members of the candidate list by using the smallest changeover schedule (SCS). To generate members of the candidate list via the SCS, simply begin with an arbitrary job and then proceed to the job that requires the least changeover time. Continue in that fashion until all jobs are scheduled. Two examples of SCS are given below.

$$
\begin{array}{llllllllll}
\text{EDD} & 3 & 5 & 7 & 1 & 8 & 6 & 2 & 9 & 4 \\
\text{SCS} & 1 & 2 & 3 & 9 & 4 & 7 & 5 & 6 & 8 \\
 & 2 & 1 & 5 & 6 & 7 & 4 & 8 & 3 & 9
\end{array}
$$

Seven more SCS candidates can be easily generated by starting with any of jobs 3 through 9 as the first job processed.

We make the movement of the two jobs just exchanged Tabu. We also choose Tabu tenure as one move. Finally, we set iteration count as 10.

Step 2 Pick EDD schedule as the choice from elite list. The EDD schedule in Table 13 has a tardiness value of 215 days.

TABLE 13

Schedule	3	5	7	1	8	6	2	9	4
Due Date	4	15	18	18	19	24	32	59	70
Available	6	19	30	35	47	66	80	89	102
Late	2	4	12	17	28	42	48	30	32

Step 3 For the short-term list (Table 14), we choose all 36 ways in which two jobs (for exchange) can be chosen from a set of nine jobs.

TABLE 14

Job I	Job J	Job I	Job J	Job I	Job J	Job I	Job J
1	2	2	4	3	7	5	7
1	3	2	5	3	8	5	8
1	4	2	6	3	9	5	9
1	5	2	7	4	5	6	7
1	6	2	8	4	6	6	8
1	7	2	9	4	7	6	9
1	8	3	4	4	8	7	8
1	9	3	5	4	9	7	9
2	3	3	6	5	6	8	9

We further restrict the list size by only considering moves such that the absolute difference between the due dates is 20 or fewer days. The candidate list of 21 moves is shown in Table 15.

TABLE 15

Job I	Job J	Due Date Difference
1	2	−14
1	3	14
1	5	3
1	6	−6
1	7	0
1	8	−1
2	5	17
2	6	8
2	7	14
2	8	13
3	5	−11
3	6	−20
3	7	−14
3	8	−15
4	9	11
5	6	−9
5	7	−3
5	8	−4
6	7	6
6	8	5
7	8	−1

Step 4   Count = 1. We now evaluate each move on the list by move value = change in total tardiness = tardiness after move − tardiness before move (see Table 16).

TABLE 16

Job I	Job J	Due Difference	Move Value
1	2	−14	15
1	3	14	−23
1	5	3	−18
1	6	−6	−63
1	7	0	15
1	8	−1	3
2	5	17	−54
2	6	8	−25
2	7	14	−66
2	8	13	−47
3	5	−11	−27
3	6	−20	−20
3	7	−14	15
3	8	−15	7
4	9	11	17
5	6	−9	−73
5	7	−3	−35
5	8	−4	−164
6	7	6	−35
6	8	5	−42
7	8	−1	−36

Exchanging jobs 5 and 8 is the best move. This move reduces tardiness by 164 days. The move 5–8 is executed yielding a new schedule with a value of 51 tardy days. The new schedule is shown in Table 17.

TABLE 17

Schedule	3	8	7	1	5	6	2	9	4
Due date	4	19	18	18	15	24	32	59	70
Available	6	13	15	20	26	37	51	60	73
Late	2	0	0	2	11	13	19	1	3

Tabu move is exchanging jobs 5 and 8. The evaluation of each move for count = 2 is shown in Table 18.

TABLE 18

Job I	Job J	Due Difference	Move Value
1	2	−14	59
1	3	14	122
1	5	3	73
1	6	−6	45
1	7	0	67
1	8	−1	141
2	5	17	35
2	6	8	11
2	7	14	14
2	8	13	32
3	5	−11	45
3	6	−20	18
3	7	−14	121
3	8	−15	103
4	9	11	17
5	6	−9	47
5	7	−3	131
5	8	−4	Tabu
6	7	6	149
6	8	5	46
7	8	−1	57

The best move is exchanging 2 and 6. This move provides a solution with 11 more tardy days. The schedule and associated values are shown in Table 19.

TABLE 19

Schedule	3	8	7	1	5	2	6	9	4
Due date	4	19	18	18	15	32	24	59	70
Available	6	13	15	20	26	35	54	65	78
Late	2	0	0	2	11	3	30	6	8

Exchange of jobs 2 and 6 is now Tabu, and the exchange of jobs 5 and 8 is removed from the Tabu list.

The evaluations for count = 3 are shown in Table 20.

TABLE 20

Job I	Job J	Due Difference	Move Value
1	2	−14	73
1	3	14	122
1	5	3	33
1	6	−6	17
1	7	0	67
1	8	−1	141
2	5	17	−13
2	6	8	Tabu
2	7	14	28
2	8	13	59
3	5	−11	17
3	6	−20	−2
3	7	−14	121
3	8	−15	103
4	9	11	4
5	6	−9	47
5	7	−3	139
5	8	−4	128
6	7	6	107
6	8	5	35
7	8	−1	57

The best move is to exchange jobs 2 and 5 with a move value that reduces tardiness by 13 days to a total of 49 days. The new schedule and associated parameters are shown in Table 21.

TABLE 21

Schedule	3	8	7	1	2	5	6	9	4
Due date	4	19	18	18	32	15	24	59	70
Available	6	13	15	20	28	36	47	58	71
Late	2	0	0	2	0	21	23	0	1

The exchange of jobs 2 and 5 is Tabu, and the exchange of jobs 2 and 6 is removed from the Tabu list. After 10 iterations, the procedure will be repeated with a new elite starting solution.

For this problem, the last schedule with a tardiness value of 49 happens to be the optimal solution.

Note that the ability to accept a move that increases tardiness value as in iteration 2 allows the TS procedure to avoid local optimal solutions.

TS is widely used to solve integer-programming models. TS allows the incorporation of several heuristics rules in the evaluation step and as well as in the Tabu step.

CHAPTER **14** Heuristic Techniques

14.6 Comparison of Heuristics

Several AI-based heuristics are available for solving complex problems. We have illustrated three AI techniques that are frequently used to solve mathematical programming problems. Genetic algorithms are a member of the class of evolutionary algorithms and are widely used in real-world applications. Because they do not explicitly consider where they have visited before, the simulated annealing and genetic search are typically memory-less procedures. Tabu search uses memory as a key ingredient in the algorithm. Genetic search works from a population of possible solutions. Using beam search, Tabu search can be adapted to work from a population of possible solutions. Usually, simulated annealing works from a single starting solution. Tabu search and neural networks accommodate neighborhood-based systematic search. Genetic search and simulated annealing rely on random neighborhood searches as a principle. The comparisons are also presented in Table 22.

The above comparisons are from a traditional algorithm-development point of view. The AI field has evolved to a point where the algorithms are developed in each category such that they can all have the same classification.

Use of these methodologies also varies to cover a wide spectrum of problems. The simulated annealing method can easily be used to convert an exchange heuristic so the exchange heuristic will avoid becoming stuck on a local optimal solution that is not the global optimal solution. Genetic search struggles with constraints and handles them via the penalty approach. See Chapter 15 for many examples of how penalties are used to solve constrained problems with GAs. The offspring solutions may need a creative mechanism to accommodate problems such as TSPs. Tabu search easily incorporates mathematical programming concepts such as surrogate constraints, cuts, and selective application of constraints.

TABLE 22
Comparison of AI Techniques

Heuristic	Memory Property	Number of Solutions	Search Type
Simulated annealing	No Memory	One	Random
Genetic search	No Memory	Population	Random
Tabu search	Memory-based	One/population	Systematic

REVIEW PROBLEMS

Group A

1 Show that Dijstrka's algorithm for the shortest path problem is $O(n^2)$.

2 Show that the minimal spanning tree algorithm is of complexity $O(n^2)$. Can this complexity measure be improved?

3 Suppose you have a list of n numbers and want to determine if at least two of them are the same. One possible way to solve this problem is to compare the first number in the list to all the other numbers; then compare the second number on the list to the third through nth number, and so on. Show this algorithm to be $O(n^2)$.

4 Suppose you are given a set of nodes with each one having a certain number of goods (or people) to be transported to other nodes. The cost of travel between the nodes is also provided. Each node can act as a hub or should be assigned to only one of the hubs, which allows them to act as transit points for movement of goods (or people). There are fixed costs of locating a hub in a node. The goal is to choose p number of hubs so that the sum of the fixed and the variable cost of movement are minimized. Develop a simulated annealing, genetic search, and a Tabu search algorithm for a p-median hub location problem. Is the p-median hub location problem NP-complete?

5 Develop heuristic procedures based on simulated annealing for a single-machine scheduling problem with minimizing tardiness.

6 Design a genetic search procedure for single-machine scheduling problem with early and tardy penalty. Note that after determining a sequence in an early and tardy penalty model, we need a procedure for inserting ideal time.

7 Develop a Tabu search and simulated annealing procedure for a multiperiod facility-layout problem. In this problem, the quantity of goods moved from one facility to the other will vary in each time period. Considering the cost of movement between facilities and then reconfiguring them if needed across time periods, develop a genetic and Tabu search procedure for this model.

8 For solving a knapsack problem, we define a single complement move to be changing a variable with v to a value $1 - v$. Explain how single complement moves could be combined with Tabu search or simulated annealing to solve a knapsack problem.

9 A vehicle-routing problem involves meeting a demand pattern with a set of capacitated vehicles. Develop an AI technique that will minimize the total distance traveled by the vehicles.

10 Determine a way to place four queens on a four-by-four chessboard so that no queen can capture another queen. Describe the suitability and algorithm structure of the AI techniques to solve the model.

11 A crew-scheduling problem deals with assigning a set of personnel (for example, pilots and flight attendants) to cover a shift or leg of a journey. Typically, there are restrictions on the amount of work that can be done and how much space between assignments (such as a two-hour break after a flight) and day-off patterns over a horizon of a week or month. The model will involve minimizing the cost or number of people working in order to meet the schedule. Develop AI-based heuristic procedures to solve the model.

12 Discriminant analysis classifies observations based on attribute value. For example, income level, home ownership, and investments held can be good predictors of whether a person needs asset-management software. Based on a set of current buyers and nonbuyers, we want to develop weights for the attributes such that an observation is classified as a buyer if and only if the sum of the attribute value multiplied by the attribute's weight exceeds an arbitrary constant. Develop a genetic search-based procedure for classifying data via discriminant analysis.

13 Cluster analysis is used to group similar objects. The cluster method has been used extensively in life sciences to identify species and in business for market segmentation. Suppose we want to divide n objects into k clusters. One approach is to pick k objects to be the anchors of the clusters and assign each object to the closest anchor. The best set of k anchors is the set that minimizes the sum of the Euclidean distances of each object from the closest anchor. Develop a simulated annealing procedure to find the optimal set of anchors for the clusters.

REFERENCES

Anily, T., and Federgruen, A. "Ergodicity in Parametric Nonstationary Markov Chains: An Application to Simulated Annealing," *Operations Research,* 35(1987): 867–874.

Collins, E., and Golden, B. "Simulated Annealing Bibliography," *American Journal of Mathematical and Management Sciences,* 8(1988): 211–307.

CONDOR. "Operations Research: The Next Decade," *Operations Research,* 36(1988): 619–637.

Gary, M. R., and Johnson, D. S. *Computers and Intractability: A Guide to the Theory of NP-Completeness.* New York: W. H. Freeman, 1979.

Glover, F., and Greenberg, H. "New Approaches for Heuristic Search: A Bilateral Linkage with Artificial Intelligence," *European Journal of Operations Research,* 39(1989): 119–130.

Goldberg, D. *Genetic Algorithms in Search Optimization and Machine Learning.* Reading MA: Addison Wesley, 1989.

Hajek, B. "Cooling Schedules for Optimal Annealing," *Mathematics of Operations Research,* 13(1988): 311–329.

Holland, J. A. *Adaptation in Natural and Artificial Systems.* Ann Arbor: University of Michigan Press, 1975.

Johnson, D. S., Aragon, C. R., McGeoch, L. A., and Schevon, C. "Optimization by Simulated Annealing: An Experimental Evaluation. Part I, Graph Partioning," *Operations Research,* 37(1989): 865–892.

Karmarkar, N. "A New Polynomial Time Algorithm for Linear Programming," *Combinatorica,* 4(1984): 373–395. Karmarkar's method for solving LPs.

Khatchian, L. G. "Polynomial Algorithm for Linear Programming," *Doklady Akademii Nauk USSR,* 20(1979): 1093–1096.

Lundy, M., and Mees, A. "Convergence of Annealing Algorithm," *Mathematical Programming,* 34(1986): 111–124.

Metropolis, W., Rosenbluth, A., Rosenbluth, M., Teller, A., and Teller, E. "Equation of State Calculations by Fast Computing Machines," *J. Chem. Phys.,* 21(1953): 1087–1092.

Papadimitriou, C. H., and Steiglitz, K. *Combinatorial Optimization: Algorithms and Complexity.* New York: Dover, 1998.

Rosenkratz, D. J., Sterns, R. E., and Lewis, P. M. "An Analysis of Several Heuristics for the Traveling Salesman Problem," *Journal of SIAM Comp.,* 6(1977): 563–581.

Simon, H. A. "Two Heads are Better than One: The Collaboration between AI and OR," *Interfaces,* 17(1987): 8–15.

Tsitsiklis, J. N. "Markov Chains with Rate Transitions and Simulated Annealing," *Mathematics of Operations Research,* 14(1989): 70–90.

Solving Optimization Problems with the Evolutionary Solver

To date, we have solved most nonlinear optimization problems using LINGO or the Excel Solver. For many important types of problems, these packages have difficulty finding optimal solutions: for example, when our objective function is nonlinear LINGO and the Excel Solver used calculus-based methods to find an optimal solution. If the objective function or some of the constraints (or both) are nonlinear, then (as we saw in Chapter 12) LINGO or the Excel Solver may find a local optimum that is not a global optimum.

LINGO and the Excel Solver also have problems if the objective function or constraints contain functions that do not have slopes at every feasible point. In this case, the calculus-based optimization methods of LINGO and the Excel Solver are often useless. For example, if our Target Cell involves the function $|x|$, then we know this function has no slope for $x = 0$. MAX, MIN, and IF statements (among others) that depend on values of the decision variables create similar situations that render the Excel Solver and LINGO helpless.

In Chapter 14, we learned how genetic algorithms could be used to solve such problems. In this chapter we will learn how to use the Evolutionary Solver (which comes with the Premium Solver for Education packaged with the enclosed CD-ROM) to overcome the previously discussed difficulties.

15.1 Price Bundling, Index Function, Match Function, and Evolutionary Solver

In many situations, companies bundle products in an attempt to get customers to purchase more products than they would without bundling. Here are some examples:

- Phone companies may bundle call waiting, voice mail, and caller ID features.
- Automobile companies often bundle popular options such as air conditioning, cassette players, and power windows.
- Computer mail-order companies often bundle computers with printers, scanners, and monitors.
- The most successful bundle in history is Microsoft's Office!

The following example shows how to use the Index and Match function to model the profitability associated with a set of product prices.

EXAMPLE 1   Bundling

Phone.xls

In the file Phone.xls (see Figure 1), we are given the amounts of money that 77 randomly chosen people are willing to pay for call waiting, voice mail, and caller ID. For example, the first customer was willing to pay as much as 50 cents for call waiting, $1 for voice mail, and 50 cents for caller ID.

0	A	B	C
5	Call Waiting	Voice Mail	Caller ID
6	0.5	1	0.5
7	2.5	5	0.5
8	4	4	7
9	10	10	0
10	0	1	2
11	0	10	0
12	3	5	1.5
13	1	3	0
14	0.75	1	0.3
15	3	4	4
77	0	8	5
78	0	1	0
79	5	5	8
80	3	4	4
81	0	6	1
82	5	5	0.25

FIGURE 1

Phoneco is thinking of offering the following product combinations for sale:

- call waiting by itself
- voice mail by itself
- caller ID by itself
- call waiting and voice mail
- call waiting and caller ID
- voice mail and caller ID
- all three products.

Because we are offering individual products as well as bundles of products, this situation is called *mixed bundling*.

Set up a spreadsheet to show how to determine the revenue associated with any set of product prices.

Solution We assume that each customer represents an equal share of the market, so it suffices to determine a set of prices that maximizes revenue received from all 77 customers. Each consumer has seven purchase options (plus the option of purchasing nothing). We assume a consumer will always purchase the option that gives her the maximum non-negative surplus. If no product combination yields a non-negative surplus, then the consumer will purchase nothing. Here is how to proceed (see Figure 2).

Step 1 Enter trial prices for each product combination in D4:J4.

Step 2 In D6:F82, we compute the surplus for each customer for a purchase of call waiting, voice mail, and caller ID. Customer 1's surplus for purchasing call waiting is computed in D6 with the formula

$$=A6-D\$4.$$

Copying this formula to D6:F82 computes the product surplus of each customer for call waiting, voice mail, and caller ID.

FIGURE 2

	C	D	E	F	G	H	I	J		K	L	M
3											total	278
4	Price	2	4	6	8	10	12	14				
5	Caller ID	CW	VM	CID	CW,VM	CW,CID	VM,CID	CW,VM,CID		max surp	bought?	revenue
6	0.5	-1.5	-3	-5.5	-6.5	-9	-10.5	-12		-1.5	0	0
7	0.5	0.5	1	-5.5	-0.5	-7	-6.5	-6		1	2	4
8	7	2	0	1	0	1	-1	1		2	1	2
9	0	8	6	-6	12	0	-2	6		12	4	8
10	2	-2	-3	-4	-7	-8	-9	-11		-2	0	0
11	0	-2	6	-6	2	-10	-2	-4		6	2	4
12	1.5	1	1	-4.5	0	-5.5	-5.5	-4.5		1	1	2
13	0	-1	-1	-6	-4	-9	-9	-10		-1	0	0
14	0.3	-1.25	-3	-5.7	-6.25	-8.95	-10.7	-11.95		-1.25	0	0
15	4	1	0	-2	-1	-3	-4	-3		1	1	2
16	3	3	3	-3	4	-2	-2	1		4	4	8
67	0.5	-1	-2	-5.5	-5	-8.5	-9.5	-10.5		-1	0	0
68	10	23	46	4	67	25	48	71		71	7	14
69	2	-2	-2	-4	-6	-8	-8	-10		-2	0	0
70	0	-0.25	1	-6	-1.25	-8.25	-7	-7.25		1	2	4
71	6	3	2	0	3	1	0	3		3	1	2
72	0	3	4	-6	5	-5	-4	-1		5	4	8
73	0	3	2	-6	3	-5	-6	-3		3	1	2
74	1	4	3.5	-5	5.5	-3	-3.5	0.5		5.5	4	8
75	1	1	-1	-5	-2	-6	-8	-7		1	1	2
76	3.95	1.95	2.95	-2.05	2.9	-2.1	-1.1	0.85		2.95	2	4
77	5	-2	4	-1	0	-5	1	-1		4	2	4
78	0	-2	-3	-6	-7	-10	-11	-13		-2	0	0
79	8	3	1	2	2	3	1	4		4	7	14
80	4	1	0	-2	-1	-3	-4	-3		1	1	2
81	1	-2	2	-5	-2	-9	-5	-7		2	2	4
82	0.25	3	1	-5.75	2	-4.75	-6.75	-3.75		3	1	2

Step 3 In G6:G82, we compute each customer's surplus for the combination of call waiting and voice mail by copying the formula

$$=A6+B6-G\$4$$

from G6 to G7:G82. This computes the amount each customer is willing to pay for call waiting and voice mail and subtracts from the price.

Step 4 By mimicking step 3, we compute in H6:J82 the surplus each customer associates with the other product bundle combinations.

Step 5 In K6:K82, we compute each customer's maximum product surplus by copying from K6 to K7:K82 the formula

$$=MAX(D6:J6).$$

Step 6 In L6:L82, we compute the number of the product combination (if any) purchased by each customer by copying from L6:L82 the formula

$$=IF(K6<0,0,MATCH(K6,D6:J6,0)).$$

Discussion of the Match Function

The Match function's first argument is the *lookup cell,* and the second argument is the *lookup array.* The Match function will return the relative position of the first element in the lookup array that matches the lookup cell. The zero is required because the elements in the lookup array are not ordered from smallest to largest. A 1 requires a lookup array to be in ascending order and returns the relative position of the largest value that is less than or equal to the lookup value. A -1 requires a lookup array to be in descending order and returns the relative position of the smallest value that is greater than or equal to the lookup value.

This formula will return a zero if all surpluses are negative, a 1 if call waiting is most preferred and so on to 7 if the bundle of call waiting, voice mail, and caller ID is most preferred.

Step 7   In M6:M82, we compute the revenue received from each customer by copying from M6 to M7:M82 the formula

$$=IF(L6=0,0,INDEX(\$D\$4:\$J\$4,1,L6)).$$

If the consumer purchases nothing, then column L will contain a zero and the customer generates no revenue. Otherwise the Index function returns the price of the product purchased by the consumer. Basically, the Index function is a two-way lookup table. In the range D4:J4, we look up the number in row 1 and column L6. This is exactly the price of the option purchased by the customer.

Step 8   In cell M3, we compute our total revenue with the formula

$$=SUM(M6:M82).$$

How to Find Revenue-Maximizing Prices?

This is tricky. An ordinary Solver model will not work because a small change in a price can drastically change the profit. Simply changing the price a penny could cause you to lose a customer and perhaps $10 of revenue. Fortunately, the Premium Solver can come to the rescue. The premium solver has three options:

- standard simplex solver (for linear models)
- standard GRG[†] nonlinear solver (for nonlinear models that do not contain non-smooth functions such as IF statements, Match function, or VLOOKUPS.
- Standard Evolutionary Solver that can obtain **good solutions** to problems involving non-smooth functions such as IF statements, Match function, or VLOOKUPs.

The Evolutionary Solver uses **genetic algorithms** to obtain good solutions. We begin with a population containing, say, 100 sets of values for changing cells. If we have changing cells x and y, then a population would consist of 100 points such as (1,2) (2,3) and so on where the first number is the x-value and the second number is the y-value. Those members of the populations that yield good Target Cell values have more chance of surviving to the next generation or population. Conversely, those members of the population that yield poor Target Cell values have little chance of surviving to the next generation. Occasionally, the Evolutionary Solver will drastically change or **mutate** the value of a changing cell. Usually, we stop the Evolutionary Solver after a specified time period (say, 30 minutes) or when there has been no improvement in the Target Cell value after a given time. Here are important remarks about the Evolutionary Solver.

- The Evolutionary Solver will usually find a good solution—but there is no guarantee that the *best* solution will have been found.

[†]Generalized reduced gradient.

- The Evolutionary Solver is not good at handling constraints. *The best way to handle constraints is to heavily penalize a violation of a constraint in the Target Cell.* We will soon see how this works.

- The Evolutionary Solver works best if you place a lower and upper bound on all changing cells.

- A good starting solution will usually help the Evolutionary Solver in its search for an optimal solution.

- Much of the solution process is driven by random numbers that direct the search. Therefore two people may get different solutions to the same problem. In fact, running the problem a second time may yield a different solution!

- Once Evolutionary Solver has found a "good" solution, you may try to close in on a better solution with the GRG2 Nonlinear Solver. If no improvement is found, then you can feel pretty good about the quality of the solution found by the Evolutionary Solver. No harm is done by switching back and forth between the Evolutionary Solver and the GRG2 solver until no further improvement is found.

Using the Premium Solver to Find Optimal Bundle Prices

We are now ready to use the Premium Solver to find the set of bundle prices that maximize revenue. We proceed as follows.

Step 1   Select Evolutionary Solver from the drop-down menu and fill in the Solver Dialog Box as shown in Figure 3. We want to maximize total revenue (M3) by changing our prices (D4:J4). We will constrain each price to be between $0 and $25.

Step 2   Click on the Solver Options box and fill it in as shown in Figure 4. The maximum time is how long the Solver will run before asking whether it should continue. We selected 1,000 seconds. Iterations are not important for Evolutionary Solver. Leave the precision and convergence values as small numbers to avoid problems. A population size of 100 is adequate for most problems. You may choose a population size as large as 200 if desired. In these problems, we have found that a mutation rate of .25 has worked well. We check the Required Bounds on Variables box because we have put bounds on our changing cells (you should always do this!).

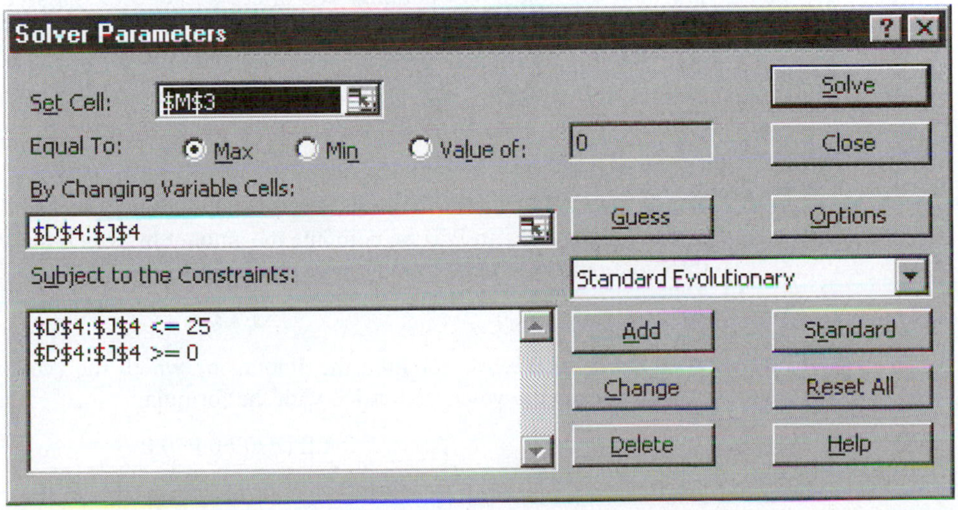

FIGURE 3

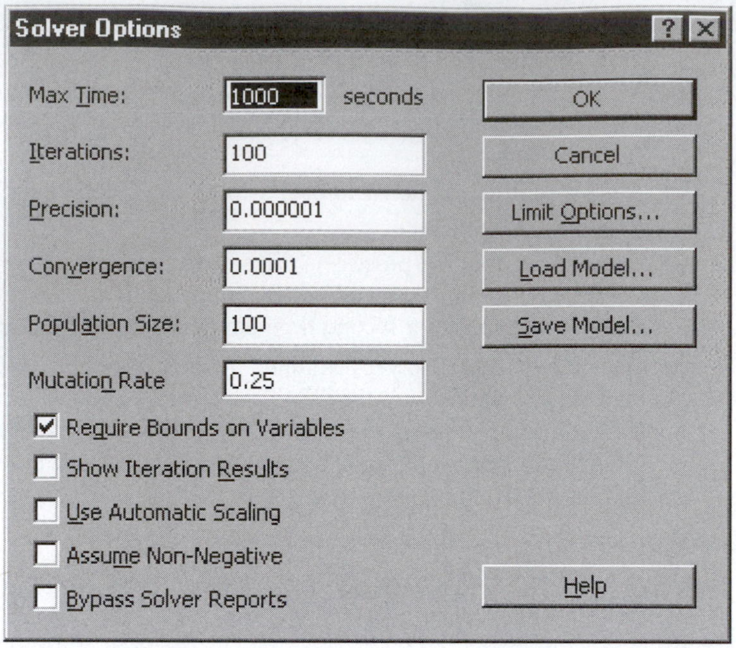

FIGURE **4**
Problem Solver
Options Box

Step 3 Next we go to the Limit Options box and fill it in as shown in Figure 5. If the maximum time has been set on the Options box to a large number, then setting Max Sub-problems and Max Feasible Solutions each to a large number ensures that the problem will run a long time before prompting you whether or not to continue. Setting Tolerance to .0005 and Max Time Without Improvement to 300 ensures that the problem will stop if the Target Cell value has improved less than .05% during the last five minutes. We now select Solve. After about 10 minutes we found the solution in Figure 6. (Note: By hitting the Escape key you may stop the Premium Solver at any time.) A maximum revenue of $480 was obtained. To determine the products purchased by each customer, we copied from P9:P16 the formula

$$=COUNTIF(\$L\$6:\$L\$82,O9).$$

We find that only call waiting and the entire bundle are purchased. Almost 30% of our potential customers buy nothing. The prices do not look satisfying, however. In the real world, we cannot charge more for call waiting than for the entire bundle. How can we ensure that the insertion of products never leads to a lower price? We could add constraints to do this, but a better way is to keep track of each time that adding a product leads to a lower price and penalize our Target Cell for each occurrence. This is done in sheet final solver.

Step 1 In O69:Q79, we determine the total amount by which adding a product raises the price. For example, in P70 we compute the amount by which the voice mail price exceeds the (voice mail) $-$ (call waiting price) with the formula

$$=E4-G4.$$

Then, in cell Q70, we compute the amount by which the (voice mail) $-$ (call waiting price) exceeds the voice mail price with the formula

$$=IF(P70>0,P70,0).$$

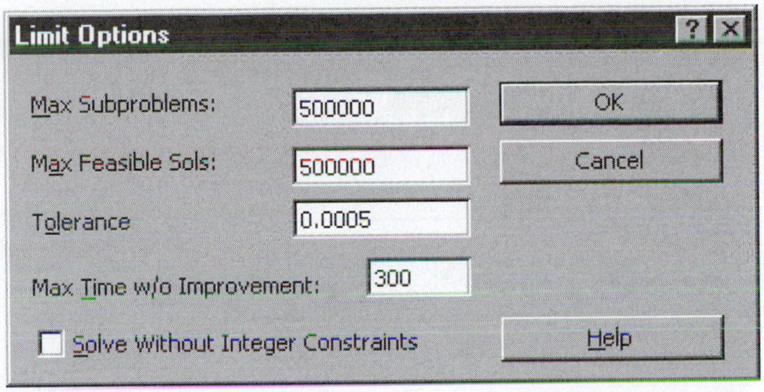

FIGURE 5

FIGURE 5
Limit Options Box

FIGURE 6

	C	D	E	F	G	H	I	J	K	L	M	N	O	P
3										total	480			
4	Price	22.978	5	16.8	22.7648	17.04451	19.131	10						
5	Caller ID	CW	VM	CID	CW,VM	CW,CID	VM,CID	CW,VM,CID	max surp	bought?	revenue			
6	0.5	-22.48	-4	-16.3	-21.265	-16.0445	-17.631	-8	-4	0	0			
7	0.5	-20.48	0	-16.3	-15.265	-14.0445	-13.631	-2	0	2	5			
8	7	-18.98	-1	-9.8	-14.765	-6.04451	-8.1311	5	5	7	10		Product	Frequency
9	0	-12.98	5	-16.8	-2.7648	-7.04451	-9.1311	10	10	7	10		0	22
10	2	-22.98	-4	-14.8	-21.765	-15.0445	-16.131	-7	-4	0	0		1	0
11	0	-22.98	5	-16.8	-12.765	-17.0445	-9.1311	1.4211E-14	5	2	5		2	14
12	1.5	-19.98	0	-15.3	-14.765	-12.5445	-12.631	-0.5	0	2	5		3	0
13	0	-21.98	-2	-16.8	-18.765	-16.0445	-16.131	-6	-2	0	0		4	0
14	0.3	-22.23	-4	-16.5	-21.015	-15.9945	-17.831	-7.95	-4	0	0		5	0
15	4	-19.98	-1	-12.8	-15.765	-10.0445	-11.131	1	1	7	10		6	0
16	3	-17.98	2	-13.8	-10.765	-9.04451	-9.1311	5	5	7	10		7	41

In cell Q79, we compute our total "price reversals" with the formula

$$=\text{SUM(Q70:Q78)}.$$

(See Figure 7.)

Step 2 We now change our total formula in M3 to

$$=\text{SUM(M6:M82)}-50*\text{Q79}.$$

This penalizes us severely for any price reversal. We now set our Solver dialog box as shown in Figure 8.

	O	P	Q
69	Penalty	dev	penalty
70	VM-CWVM	-5	0
71	CW-CWVM	-3.23988	0
72	CID-CWCID	-4.99997	0
73	CW-CWCID	-2.23984	0
74	VM-VMCID	-4.99999	0
75	CID-VMCID	-6	0
76	CWVM-all	-1.89999	0
77	CWCID-all	-2.90003	0
78	VMCID-all	-1.9	0
79		total	0

FIGURE 7
Price Reversal Penalties

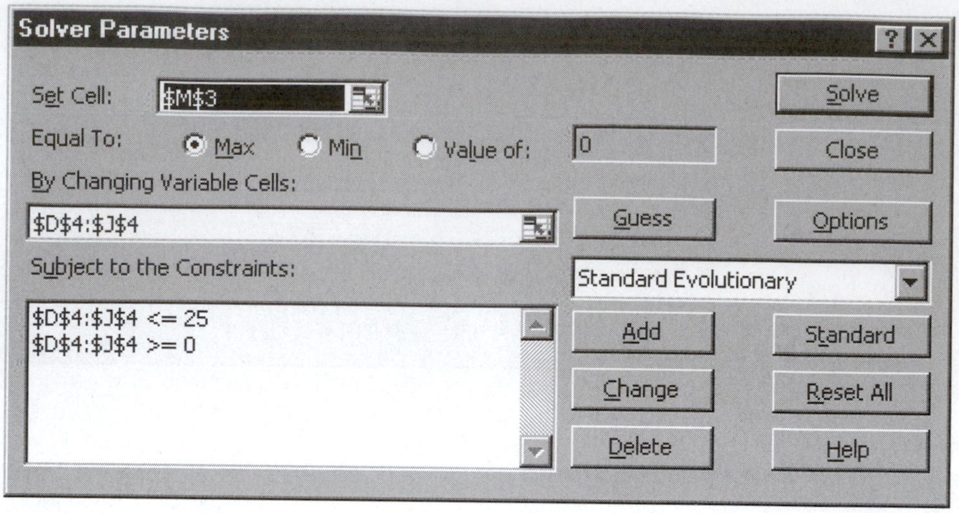

FIGURE 8
Solver Window with Penalties

FIGURE 9
Profit-Maximizing Prices

	C	D	E	F	G	H	I	J	K	L	M	N	O	P
3										total	492.4994			
4	Price	6.7601	5	4	10	8.999952	10	11.8999846						
5	Caller ID	CW	VM	CID	CW.VM	CW.CID	VM.CID	CW.VM.CID	max surp	bought?	revenue			
6	0.5	-6.26	-4	-3.5	-8.5	-7.99995	-8.5	-9.8999846	-3.49998	0	0			
7	0.5	-4.26	0	-3.5	-2.5	-5.99995	-4.5	-3.8999846	5.07E-07	2	4.999999			
8	7	-2.76	-1	3	-2	2.000048	1	3.10001539	3.100015	7	11.89998		Product	Frequency
9	0	3.2399	5	-4	10	1.000048	1E-05	8.10001539	10	4	9.999998		0	22
10	2	-6.76	-4	-2	-9	-6.99995	-7	-8.8999846	-1.99998	0	0		1	0
11	0	-6.76	5	-4	1.7E-06	-8.99995	1E-05	-1.8999846	5.000001	2	4.999999		2	17
12	1.5	-3.76		-2.5	-2	-4.49995	-3.5	-2.3999846	5.07E-07	2	4.999999		3	3
13	0	-5.76	-2	-4	-6	-7.99995	-7	-7.8999846	-2	0	0		4	7
14	0.3	-6.01	-4	-3.7	-8.25	-7.94995	-8.7	-9.8499846	-3.69998	0	0		5	2
15	4	-3.76	-1	2E-05	-3	-1.99995	-2	-0.8999846	1.5E-05	3	3.999985		6	1
16	3	-1.76	2	-1	2	-0.99995	1E-05	3.10001539	3.100015	7	11.89998		7	25

After selecting Solve (we left Options and Limit Options unchanged), we obtain the solution in Figure 9.

Note that there are no price reversals and revenue is actually up $12 to $492! Also note that somebody purchases every combination except call waiting.

15.2 More Nonlinear Pricing Models

Suppose we sell Menthos. Clearly, most people value the first pack of Menthos they purchase more than the second pack, the second pack more than the third, and so on. How can we take advantage of this when pricing Menthos? If we charge a single price for each pack of Menthos, then few people are going to buy more than one or two packs. Alternatively, however, we can try the **two-part tariff** approach. The two-part tariff involves charging an "entry fee" to anybody who buys Menthos and then a reduced price per pack purchased. For instance, if we charge $1.10 per pack of Menthos, then a reasonable two-part tariff might be an entry fee of $1.50 and a price of $0.50 per pack. This will give some customers an incentive to purchase many packs of Menthos. The following example shows how much a two-part tariff can increase profit. Because the total cost of purchasing n packs of Menthos is no longer a linear function of n (it is now piecewise linear), we refer to the two-part tariff as a **nonlinear pricing** strategy.

EXAMPLE 2   **Two-Part Tariff**

Four customers have been surveyed on the amount they are willing to pay for each pack of Menthos purchased. The relevant data is in Table 1.

For example, customer 1 is willing to pay $1.24 for the first pack of Menthos and only $0.66 for the sixth pack. These four customers are considered representative of the market. It costs $0.40 to produce a pack of Menthos. Determine a profit-maximizing single price and a profit-maximizing two-part tariff.

Menthos.xls   **Solution**   Our work is in the file Menthos.xls. For the single-price problem, refer to the sheet titled Single Price. We assume that each customer will compare the *value* of buying n Menthos to the *cost* of buying n Menthos ($n = 1, 2, \ldots, 10$) and will then purchase the number of Menthos that yields the maximum (consumer surplus) = (value) − (cost). If the largest surplus is negative, then the consumer will purchase no Menthos. We first determine the value each customer associates with buying n packs of Menthos and then compute the consumer surplus for each possible number of Menthos purchased. Next we determine (using the Excel Match function) the number of packs each customer will purchase. Finally, we determine our profit and use the Evolutionary Solver to find the profit-maximizing price.

Step 1   In G2:J11, we determine the cumulative value that each customer associates with buying n packs of Menthos ($n = 1, 2, \ldots, 10$). In cell G2, we just recopy the amount customer 1 is willing to pay for the first pack with the formula

$$=B2.$$

In cell G3, we compute the total amount that customer 1 is willing to pay for the first two packs with the formula

$$=G2+B3.$$

Copying this formula to G4:G11 computes how much customer 1 is willing to pay for 2, 3, ..., 10 packs of Menthos. Copying G2:G11 to H2:J11 determines the total amount that customers 2, 3, and 4 are willing to pay for buying 1, 2, ..., 10 packs.

Step 2   In cell B16, we enter a trial price for one pack of Menthos.

TABLE 1

No. Packs of Menthos	Customer			
	1	2	3	4
1	1.236973	0.915758	1.271622	1.486487
2	1.025622	0.849361	1.107868	1.238578
3	0.887074	0.694547	0.96184	1.095671
4	0.798269	0.578962	0.8528	0.97373
5	0.772203	0.504336	0.732484	0.812574
6	0.658503	0.429462	0.631012	0.709801
7	0.592303	0.358183	0.509381	0.63302
8	0.508444	0.31825	0.449372	0.53419
9	0.421358	0.25983	0.387907	0.421402
10	0.350831	0.224386	0.32404	0.353971

Step 3 In E13:E22, we compute the total cost of purchasing n packs of Menthos. In E13, we compute the cost of buying one pack with the formula

$$=F13*\$B\$16.$$

Copying this formula to the cell range E14:E22 computes the cost of buying 2, 3, . . ., 10 packs.

Step 4 In G13:J22, we compute the consumer surplus each customer has for the proposed number of packs purchased. For customer 1 buying one pack, the surplus is computed in G13 with the formula

$$=G2-\$E13.$$

Copying this formula to the cell range G13:J22 generates the consumer surpluses for all customers and all possible numbers of packs purchased.

Step 5 In G23:J23, we determine the maximum consumer surplus for each customer. For customer 1, the maximum consumer surplus is computed in G23 with the formula

$$=MAX(G13:G22).$$

Copying this formula to the range H23:J23 computes the maximum consumer surplus for customers 2–4.

FIGURE 10
Single Price Optimization

	A	B	C	D	E	F	G	H	I	J
1	Mentho	Cust 1	Cust 2	Cust 3	Cust 4		Cu Val 1	Cu Val 2	Cu Val 3	Cu Val 4
2	1	1.236973	0.915758	1.271622	1.486487	1	1.236973	0.915758	1.271622	1.486487
3	2	1.025622	0.849361	1.107868	1.238578	2	2.262595	1.765119	2.379491	2.725065
4	3	0.887074	0.694547	0.96184	1.095671	3	3.149669	2.459666	3.341331	3.820735
5	4	0.798269	0.578962	0.8528	0.97373	4	3.947939	3.038628	4.194131	4.794466
6	5	0.772203	0.504336	0.732484	0.812574	5	4.720142	3.542964	4.926615	5.60704
7	6	0.658503	0.429462	0.631012	0.709801	6	5.378645	3.972426	5.557627	6.316841
8	7	0.592303	0.358183	0.509381	0.63302	7	5.970948	4.330609	6.067008	6.949862
9	8	0.508444	0.31825	0.449372	0.53419	8	6.479392	4.648859	6.51638	7.484052
10	9	0.421358	0.25983	0.387907	0.421402	9	6.90075	4.908689	6.904287	7.905454
11	10	0.350831	0.224386	0.32404	0.353971	10	7.251581	5.133075	7.228327	8.259425
12					Cost		Surp 1	Surp 2	Surp 3	Surp 4
13					0.798269	1	0.438704	0.117488	0.473353	0.688217
14					1.596539	2	0.666057	0.16858	0.782952	1.128526
15		Price		2	.394808	3	0.754861	0.064858	0.946522	1.425927
16		0.798269		3	.193078	4	0.754861	-0.15445	1.001053	1.601388
17		cost			3.991347	5	0.728795	-0.448383	0.935268	1.615693
18			0.4		4	.789617	6	0.589029	-0.81719	0.76801	1.527225
19					5.587886	7	0.383062	-1.257277	0.479122	1.361976
20					6.386156	8	0.093236	-1.737297	0.130224	1.097896
21					7.184425	9	-0.283675	-2.275736	-0.280138	0.721029
22					7.982695	10	-0.731113	-2.84962	-0.754368	0.276731
23					Max surp		0.754861	0.16858	1.001053	1.615693
24					#bought		4	2	4	5
25					bought		15			
26					profit		5.974042			
27										
28										

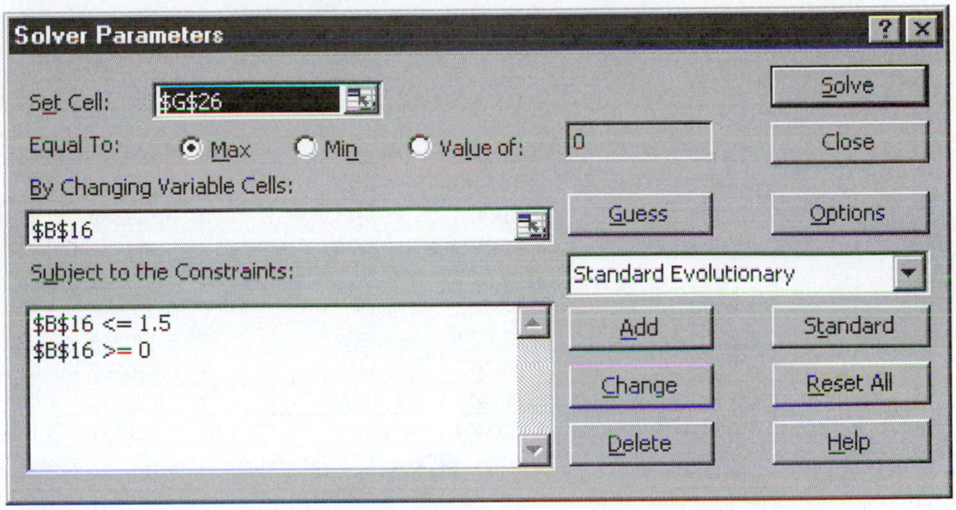

FIGURE 11
Solver Window for
Single Price

Step 6 Now the key step! In G24:J24, we compute the number of packs that each customer will buy. The number of packs Customer 1 will buy is computed in G24 with the formula

$$=IF(G23<0,0,MATCH(G23,G13:G22,0)).$$

If Customer 1 does not have any number of packs she values more than the purchase cost, then she will buy no Menthos. Otherwise, the Match function finds which number (with G13 being first and G22 being tenth) in G13:G22 matches G23. Note: The zero in the formula is needed to allow the Match function to work on an array that is not sorted from smallest to largest.

For example, from Figure 10, we see a 4 in G24 because the fourth number in G13:G22 matches the maximum consumer surplus in G23. Copying this formula to H24:J24 computes the number of packs bought by each customer.

Step 7 In cell G25, we compute the total number of candy bars bought with the formula

$$=SUM(G24:J24).$$

Step 8 In cell G26, we compute the total profit with the formula

$$=(B16-B18)*G25.$$

Step 9 We are now ready to use the Evolutionary Solver to find the profit-maximizing price. After choosing the Evolutionary Solver, we fill in our dialog box as shown in Figure 11. We use the same choices of Options and Limit Options as we did in Example 15.1. We simply choose to maximize profit (G26) by changing price (B16). We constrain price to be between $0 and $1.50 (this is because nobody is willing to pay $1.50 for a candy bar). From Figure 10, we find the optimal solution is to charge $0.798 per candy bar. We sell 16 candy bars and make a profit of $5.97.

Finding the Optimal Two-Part Tariff

TPT.xls

In the file TPT.xls, we find the optimal two-part tariff (see Figure 12).

In cell A16, we enter a trial value for the fixed charge. The price per pack will be in B16. Make the following changes to our spreadsheet.

FIGURE 12
Two-part Tariff Spreadsheet

	A	B	C	D	E	F	G	H	I	J
1	Mentho	Cust 1	Cust 2	Cust 3	Cust 4			Cu Val 1	Cu Val 2	Cu Val 3	Cu Val 4
2	1	1.236973	0.915758	1.271622	1.486487	1	1.236973	0.915758	1.271622	1.486487
3	2	1.025622	0.849361	1.107868	1.238578	2	2.262595	1.765119	2.379491	2.725065
4	3	0.887074	0.694547	0.96184	1.095671	3	3.149669	2.459666	3.341331	3.820735
5	4	0.798269	0.578962	0.8528	0.97373	4	3.947939	3.038628	4.194131	4.794466
6	5	0.772203	0.504336	0.732484	0.812574	5	4.720142	3.542964	4.926615	5.60704
7	6	0.658503	0.429462	0.631012	0.709801	6	5.378645	3.972426	5.557627	6.316841
8	7	0.592303	0.358183	0.509381	0.63302	7	5.970948	4.330609	6.067008	6.949862
9	8	0.508444	0.31825	0.449372	0.53419	8	6.479392	4.648859	6.51638	7.484052
10	9	0.421358	0.25983	0.387907	0.421402	9	6.90075	4.908689	6.904287	7.905454
11	10	0.350831	0.224386	0.32404	0.353971	10	7.251581	5.133075	7.228327	8.259425
12				Menthos	Cost	Menthos	Surp 1	Surp 2	Surp 3	Surp 4
13				1	3.962323	1	-2.72535	-3.046565	-2.690701	-2.475836
14	fixed			2	4.329391	2	-2.066796	-2.564272	-1.949901	-1.604327
15	fixed	unit price		3	4.69646	3	-1.54679	-2.236794	-1.355129	-0.875724
16	3.595255	0.367068		4	5.063528	4	-1.115589	-2.0249	-0.869397	-0.269062
17		cost		5	5.430596	5	-0.710454	-1.887632	-0.503981	0.176444
18			0.4	6	5.797665	6	-0.419019	-1.825238	-0.240038	0.519177
19				7	6.164733	7	-0.193785	-1.834124	-0.097725	0.785129
20				8	6.531801	8	-0.052409	-1.882942	-0.015422	0.95225
21				9	6.89887	9	0.00188	-1.990181	0.005417	1.006584
22				10	7.265938	10	-0.014357	-2.132863	-0.037611	0.993487
23						Max surp	0.00188	-1.825238	0.005417	1.006584
24						#bought	9	0	9	9
25						revnue	6.89887	0	6.89887	6.89887
26						profit	9.896609			

Step 1 Change the purchase costs in E13:E22 to reflect the new pricing structure. In cell E13, compute the cost of buying one pack with the formula

$$=\$A\$16+\$B\$16*F13.$$

Copying this formula to the cell range E14:E22 computes the cost of buying 2, 3, . . ., 10 packs.

Step 2 Compute the revenue collected from each customer in G25:J25. In G25, we determine the revenue received from customer 1 with the formula

$$=IF(G24<=0,0,VLOOKUP(G24,\$D\$13:\$E\$22,2)).$$

This formula ensures that no revenue is received if no packs are bought, and if any packs are bought we look up the cost of buying that number of packs. Copying this formula to the range H25:J25 computes the revenue received from customers 2–4.

Step 3 In cell G26, compute our (profit) = (revenue) − (cost) with the formula

$$=SUM(G25:J25)-SUM(G24:J24)*B18.$$

Step 4 We now use Evolutionary Solver to determine the optimal two-part tariff (see Figure 13). We simply maximize profit (G26). Our changing cells are fixed charge and price per pack (A16 and B16). We constrain the fixed charge to be between $0 and $10 and constrain the price per pack to be between $0 and $2. We find from Figure 13 that the op-

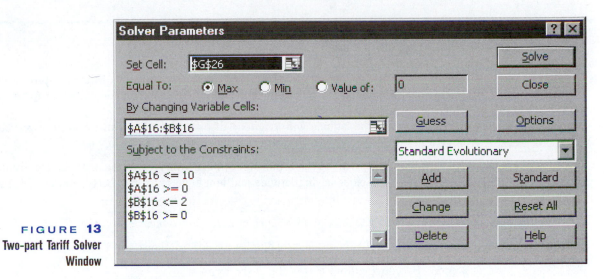

FIGURE 13
Two-part Tariff Solver Window

timal two-part tariff consists of charging a $3.60 entry fee and $0.37 per pack of Menthos. Our profit is now $9.90 (a 66% increase!).

Other Forms of Nonlinear Pricing

There are many other forms of nonlinear pricing:

- Just sell 1 or 6 candy bars.
- Charge one price for the first n packs and another price for the rest.

With Evolutionary Solver, it is easy to experiment with many nonlinear pricing schemes and determine the profit earned by each. For example, if we allow Menthos to be sold only by a single pack or by a six pack, then we can show that we maximize profit at $9.94 by charging $5.38 for a six pack and virtually any price for a one pack! Then we will sell three customers a six pack and make $17.14 − $7.20 = $9.94.

By using the strategy of charging one price for the first n packs and another price for the remaining packs, the best result is to sell as many as four packs at $1.18 and charge $0.42 for each additional pack.

PROBLEMS

1 Determine the optimal pricing policy if Menthos are sold in only a single pack or in a six pack.

2 Determine the best pricing policy if quantity discounts with a single price break point are used.

3 [Based on Schrage (1997).] Table 2 gives the size of the four main markets for Excel, Word, and the bundle of Excel and Word. You are also shown how much members of each group are willing to pay for each product combination. How can we maximize the revenue earned from these products? Consider the following options:

- No bundling—Word and Excel are sold separately.
- Pure bundling—purchasers can only buy Word and Excel together.

- Mixed bundling—purchasers can buy Word or Excel separately or buy them as a bundle.

TABLE 2

Market	Market Size	Product Combination		
		Excel Only	Word Only	Bundle
Business	70,000	450	110	530
Legal	50,000	75	430	480
Educational	60,000	290	250	410
Home	45,000	220	380	390

15.3 Locating Warehouses

warehouseloc.xls

Suppose we want to locate two warehouses from which we ship Microsoft products to customers. We know the number of shipments (in thousands) made each year to various cities is as shown in Figure 14 (also see file warehouseloc.xls).

We are given the latitude and longitude of each city. How can we locate two warehouses in order to minimize the average distance traveled by a shipment?

Solution   A key to our model will be the following formula, which gives the approximate distance between two U.S. cities having latitude and longitude given by (Lat1, Long1) and (Lat2 and Long2).

$$Distance = 69\sqrt{Lat1 - Lat2)^2 + (Long1 - Long2)^2}$$

Step 1   Enter trial latitudes and longitudes for our warehouses in F4:G5.

Step 2   By copying from F7 to F8:F27 the formula

$$=69*SQRT((C7-\$F\$4)^2 +(D7-\$G\$4)^2)$$

we compute the distance from each city to Warehouse 1.

FIGURE 14

	A	B	C	D	E	F	G	H	I
1									
2									Mean dist
3						Lat	Long		501.8113
4					1	38.164052	84.02896		Total
5					2	34.931893	117.7916		119676.5
6			Lat	Long	Shipments	Distance to 1	Distance to 2	Min Distance	Dist*Shipped
7		New York	40.7	73.9	15	720.47003	3054.56	720.47	10807.05
8		Boston	42.3	71	8	943.20731	3268.403	943.2073	7545.659
9		Philadelphia	40	75.1	10	628.98738	2966.405	628.9874	6289.874
10		Charlotte	35.2	80.8	6	302.43575	2552.487	302.4357	1814.614
11		Atlanta	33.8	84.4	11	302.20599	2305.344	302.206	3324.266
12		New Orleans	30	89.9	8	693.85611	1954.375	693.8561	5550.849
13		Miami	25.8	80.2	13	893.09229	2669.257	893.0923	11610.2
14		Dallas	32.8	96.8	10	955.77447	1455.871	955.7745	9557.745
15		Houston	29.8	95.4	12	973.99525	1585.079	973.9953	11687.94
16		Chicago	41.8	87.7	14	356.51464	2129.715	356.5146	4991.205
17		Detroit	42.4	83.1	11	299.22639	2448.557	299.2264	3291.49
18		Cleveland	41.5	81.7	8	280.72581	2531.222	280.7258	2245.806
19		Indy	39.8	86.1	7	182.10673	2212.369	182.1067	1274.747
20		Denver	39.8	104.9	8	1444.5188	950.8286	950.8286	7606.629
21		Minneapolis	45	93.3	9	794.79593	1827.139	794.7959	7153.163
22		Phoenix	33.5	112.1	11	1963.4551	404.9579	404.9579	4454.537
23		Salt Lake City	40.8	111.9	10	1931.6833	573.7616	573.7616	5737.616
24		LA	34.1	118.4	18	2388.1226	71.11337	71.11337	1280.041
25		SF	37.8	122.6	12	2661.5202	386.3183	386.3183	4635.82
26		SD	32.8	117.1	10	2311.7231	154.6474	154.6474	1546.474
27		Seattle	41.6	122.4	13	2658.1952	559.2874	559.2874	7270.737
28									

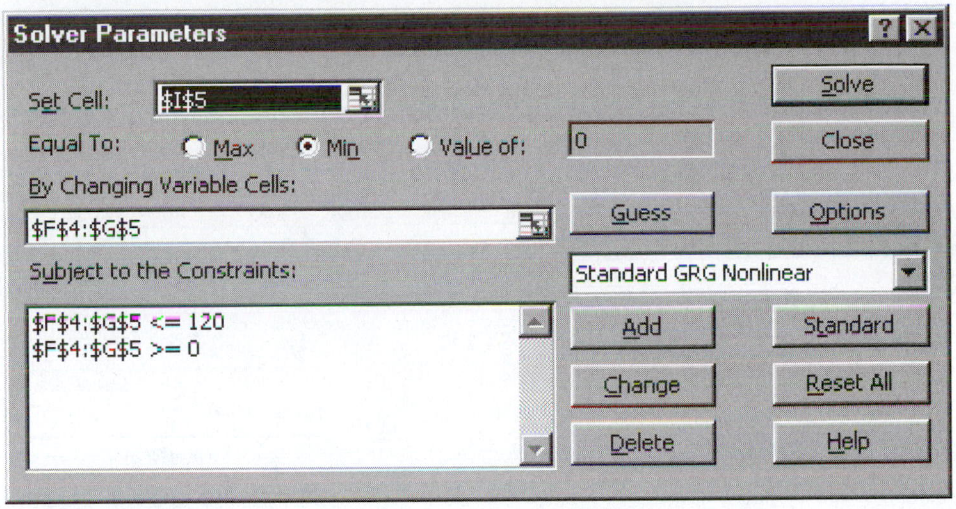

FIGURE 15
Warehouse Location Solver Window

Step 3 By copying from G7 to G8:G27 the formula

$$=69*SQRT((C7-\$F\$5)^2 + (D7-\$G\$5)^2)$$

we compute the distance from each city to Warehouse 2.

Step 4 Each city's shipment will be sent from the *closer* warehouse, so we compute the distance of each city to the closer warehouse by copying from H7 to H8:H27 the formula

$$=MIN(F7,G7).$$

Step 5 In I7:I27, we compute the distance traveled by each city's shipments by copying from I7 to I8:I27 the formula

$$=H7*E7.$$

Step 6 In cell I5, we compute the total distance traveled by our shipments with the formula

$$=SUM(I7:I27).$$

Step 7 We are now ready to use Solver to determine the optimal warehouse locations. Our Solver window is as shown in Figure 15. We began with GRG (nonlinear solver) but min functions cause it trouble, so we switched to Evolutionary Solver. We constrain each warehouse's latitude and longitude to be between zero and 120 degrees and minimize the total distance traveled by the shipments.

We find the average distance traveled per shipment to be 502 miles. Warehouse 1 is located near central Illinois, while warehouse 2 is located near Las Vegas.

REMARKS If we were limited to one warehouse, the optimal location would be near Springfield, Missouri, and each shipment would travel an average of 1,126 miles!

P R O B L E M S

1 HP is trying to determine where to locate three warehouses that sell toner cartridges. They would like to minimize the number of customers who are not within 900 miles of a warehouse. Possible warehouse locations and the distance between them are given in Figure 16 (see file HP.xls). The number of customers in each city is given in Table 3. Where should the warehouses be located?

FIGURE 16

	Boston	Chicago	Dallas	Denver	LA	Miami	NY	Phoenix	Pittsburgh	SF	Seattle
1 Boston	0	983	1815	1991	3036	1539	213	2664	792	2385	2612
2 Chicago	983	0	1205	1050	2112	1390	840	1729	457	2212	2052
3 Dallas	1815	1205	0	801	1425	1332	1604	1027	1237	1765	2404
4 Denver	1991	1050	801	0	1174	1332	1780	836	1411	1765	1373
5 LA	3036	2112	1425	1174	0	2757	2825	398	2456	403	1909
6 Miami	1539	1390	1332	1332	2757	0	1258	2359	1250	3097	3389
7 NY	213	840	1604	1780	2825	1258	0	2442	386	3036	2900
8 Phoenix	2664	1729	1027	836	398	2359	2442	0	2073	800	1482
9 Pittsburgh	792	457	1237	1411	2456	1250	386	2073	0	2653	2517
10 SF	2385	2212	1765	1765	403	3097	3036	800	2653	0	817
11 Seattle	2612	2052	2404	1373	1909	3389	2900	1482	2517	817	0

TABLE 3

City	Customers
Boston	300
Chicago	500
Dallas	400
Denver	200
LA	900
Miami	350
NY	1,200
Phoenix	350
Pittsburgh	230
SF	650
Seattle	412

TABLE 4

	Coordinates		
City	x	y	Visits
1	0	0	3,000
2	10	3	4,000
3	12	15	5,000
4	14	13	6,000
5	16	9	4,000
6	18	6	3,000
7	8	12	2,000
8	6	10	4,000
9	4	8	1,200

2 Use the Evolutionary Solver to solve the lockbox problem of Chapter 9.

3 Cook County needs to build two hospitals. There are nine cities where the hospitals can be built. The number of hospital visits made annually by the inhabitants of each city and the x and y coordinates of each city are as shown in Table 4. To minimize the total distance that patients must travel to hospitals, where should the hospitals be located? (*Hint:* Use Lookup functions to generate the distances between each pair of cities.)

4 Solve our example if only one warehouse were allowed.

5 Drug Company Daisyco wants to have four regional sales offices. A call in each region of the country will be assigned to the closest regional office. Where should the offices be located to minimize the total distance that must be traveled from the closest regional office to each region? File Sales.xls (see Figure 17) contains the distances between each possible regional sales location center and the number of sales calls (in thousands) that must annually be made to each region.

FIGURE 17

	A	B	C	D	E	F	G	H	I	J	K	L	M
1	Calls		San Antonio	Phoenix	LA	Seattle	Detroit	Minneapolis	Chicago	Atlanta	NY	Boston	Philadelphia
2	2	San Antonio	0	602	1376	1780	1262	1140	1060	935	1848	2000	1668
3	3	Phoenix	602	0	851	1193	1321	1026	1127	1290	2065	2201	1891
4	6	LA	1376	851	0	971	2088	1727	1914	2140	2870	2995	2702
5	3	Seattle	1780	1193	971	0	1834	1432	1734	2178	2620	2707	2486
6	4	Detroit	1262	1321	2088	1834	0	403	205	655	801	912	654
7	2	Minneapolis	1140	1026	1727	1432	403	0	328	876	1200	1304	1057
8	7	Chicago	1060	1127	1914	1734	205	328	0	564	957	1082	794
9	5	Atlanta	935	1290	2140	2178	655	876	564	0	940	1096	765
10	9	NY	1848	2065	2870	2620	801	1200	957	940	0	156	180
11	5	Boston	2000	2201	2995	2707	912	1304	1082	1096	156	0	333
12	4	Philadelphia	1668	1891	2702	2486	654	1057	794	765	180	333	0

15.4 Solving Other Combinatorial Problems

Consider the following situations:

- Xerox must determine where to place maintenance facilities. The more facilities selected, the more copiers that will be sold because of better availability of maintenance. How can we locate maintenance facilities to maximize total profit?

- We are loading three different products onto a tanker truck with five compartments. Each compartment can handle one product at most. How do we load the truck to come as close as possible to meeting our delivery requirements?

- Fox has 30 different ads of different lengths that must be assigned to 10 different two-minute commercial breaks. How do we assign ads to maximize our total ad revenue?

Each problem is a **combinatorial** optimization problem that requires us to choose the best of many different combinations available. Although combinatorial optimization problems may often be handled as linear Solver models with 0–1 changing cells, the formulation of the constraints needed to keep the model linear is often difficult. With Evolutionary Solver, we need not worry whether our constraints or our Target Cell is linear. The =SUMIF and =COUNTIF functions often prove useful in such problems. Here is a typical combinatorial optimization problem.

EXAMPLE 3   **Loading a Gas Truck**

A gas truck contains five compartments with the capacities shown in Table 5.

Three products must be shipped on the truck. The demand for each product, the shortage cost per gallon, and the maximum allowable shortage for each product are as shown in Table 6.

Gas.xls

How can we load the truck to minimize our shortage costs? See file Gas.xls.

Solution   Our goal is to minimize shortage cost. Our adjustable cells will be the type of product and amount placed in each compartment. Our constraints must ensure that we do not overfill any compartment and do not exceed the maximum allowable shortage. Our work is in Figure 18.

TABLE 5

Compartment	Capacity (Gallons)
1	2,700
2	2,800
3	1,100
4	1,800
5	3,400

TABLE 6

Product	Demand	Maximum Allowed Shortage	Cost per Gallon Short ($)
1	2,900	900	10
2	4,000	900	8
3	4,900	900	6

FIGURE **18**

Gas Truck Spreadsheet

	A	B	C	D	E	F	G	H
1	Gas Truck							
2	Compartment	Product	Amount	capacity				
3	1	2	2700	2700				
4	2	1	2800	2800				
5	3	2	1100	1100				
6	4	3	1702.368	1800				
7	5	3	3197.632	3400			total	
8							0	
9		Product	Total	Shortage	cost/gallon	shortage cost	shortage violation	demand
10		1	2800	100	$ 10.00	$ 1,000.00	0	2900
11		2	3800	200	$ 8.00	$ 1,600.00	0	4000
12		3	4900	8.29E-08	$ 6.00	$ 0.00	0	4900
13					Total	$ 2,600.00		

Step 1 In B3:C7, enter trial values for (a) the type of gas and (b) the amount of gas placed in each compartment.

Step 2 In cells C10:C12, use the =SUMIF function to determine the total amount of each gas to be loaded on the truck. Copy from C10 to C11:C12 the formula

$$=SUMIF(\$B\$3:\$B\$7,B10,\$C\$3:\$C\$7).$$

This formula sums all numbers in C3:C7 for which the number in the corresponding row of column B equals B10. Thus, all Type 1 gas amounts are totaled by this formula.

Step 3 Shortage of each type of gas equals demand less amount loaded. To compute the shortage for each product, copy from D10 to D11:D12 the formula

$$=H10-C10.$$

(*Note:* A negative shortage indicates we have shipped more than is demanded.)

Step 4 In F10:F12, we determine the shortage cost for each gas by multiplying the number of gallons short times the unit shortage cost. To do this, copy from F10 to F11:F12 the formula

$$=IF(D10>0,D10*E10,0).$$

Step 5 We know that we do not want a negative shortage (this means we have shipped more than what is demanded) or a shortage exceeding 900 gallons. In G10:G12, we compute for each product the sum of the negative shortage and any shortage more than 900 gallons. Then we will "penalize" each unit of a negative shortage or a shortage that exceeds 900 by 100 "dollars." This is an alternative way of handling the constraint of a shortage of at most 900 gallons. By penalizing such a shortage or a negative shortage in the Target Cell, the Evolutionary Solver will try to stay away from solutions in which we fail to meet demand by 900 gallons or more or ship more of a product than is demanded. To compute the "shortage violation" for each gas, copy from G10 to G11:G12 the formula

$$=IF(D10<0,-D10,0)+IF(D10>900,D10-900,0).$$

Step 6 In cell G8, compute our total of excess and negative shortages with the formula

$$=SUM(G10:G12).$$

Step 7 Compute our total shortage cost and "penalty cost" for a negative shortage or a shortage exceeding 900 gallons in F13 with the formula

$$=SUM(F10:F12)+100*G8$$

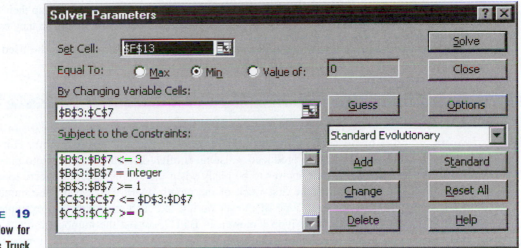

FIGURE 19
Solver Window for
Gas Truck

TABLE 7

Compartment	Product	Amount	Capacity
1	2	2,700	2,700
2	1	2,800	2,800
3	2	1,100	1,100
4	3	1,702.368	1,800
5	3	3,197.632	3,400

Step 8 Our Evolutionary Solver window is as shown in Figure 19. Our goal is to minimize shortage cost plus penalty cost (F13). We can change B3:B7 (the type of gas in each compartment) and C3:C7 (the amount of gas in each compartment). Each of B3:B7 must be an integer between 1 and 3, while C3:C7 must never exceed the capacity of the particular compartment.

We obtain the optimal solution in Table 7. A total shortage cost of $2,600 is incurred.

15.5 Production Scheduling at John Deere

It is a tricky matter to determine a monthly production schedule at a facility such as a John Deere manufacturing plant. Many conflicting objectives must be balanced. Here are some examples:

- We want to make as few types of models during a week as possible, because each time we switch between models costs us time.

- We want to balance the workload each week. This means that the number of hours used in each work center should be roughly the same across all four weeks of the month.

- Our schedule should manufacture the number of items each month that our forecast says should be produced.

- Lawn mower purchasers sometimes drive to the plant to pick up their mowers, so these mowers must be produced the first week of the month so they can be picked up.

The following example shows how the Premium Solver can be used to determine a schedule that does a good job of trading off these objectives.

EXAMPLE 4   **Lawn Mower Scheduling**

John Deere produces seven types of riding mowers (see Figure 20). In A3:H3, we give the number of each mower that needs to be produced this month. For example, 90 of mower 2 must be produced. Customers often drive to John Deere to pick up their mowers and want the mower to be ready when they come, so John Deere tries to make these mowers during the first week of the month. In B4:H4, we list the number of customer pickups for each type of mower during the current month. For example, 30 customers want to pick up a Type 1 mower. In B5:H7, we list the number of hours each work center needs to produce each type of mower. For example, a Type 1 mower requires three hours in center 1, one hour in center 2, and two hours in center 3.

In determining a schedule for the month John Deere has the following goals:

- Make as few different models during the week as possible. This will reduce setup times. A penalty of 200 will be incurred for each model produced during a week. For example, if six models are made the first week, then a penalty of 1,200 is incurred.

- Make sure all mowers that customers will pick up are produced during the first week. A penalty of 50 is incurred for each customer pickup mower that is not made during the first week. For example, if 20 Type 1 mowers are made the first week, then a penalty of 500 is incurred.

- Make sure that the number of hours used each week in each work center are spread out as little as possible. In cells E11:E13, we determined the number of hours per week required in each work center by copying from E11:E13 the formula

$$=SUMPRODUCT(\$B\$3:\$H\$3,B5:H5)/4.$$

For each work center and week, a penalty of 1 is incurred for each hour by which hours used differs from the average number of hours per week needed in that center. For ex-

FIGURE 20

	A	B	C	D	E	F	G	H
1								
2	Product	1	2	3	4	5	6	7
3	Needed	110	90	100	115	80	60	80
4	Pickup needed	30	20	15	30	23	12	12
5	Center 1	3	2	2	2	2	4	2
6	Center 2	1	2	1	3	3	3	4
7	Center 3	2	3	0	4	3	3	2
8								
9		Total						
10	**Models**	17			Hrs/week			
11	Week 1	7		Center 1	375			
12	Week 2	4		Center 2	368.75			
13	Week 3	2		Center 3	382.5			

ample, if center 1 uses 350 hours during week 1, then a penalty of 25 is incurred. This penalty drives the Solver to "balance the load" in each center from week to week.

- Make sure we come as close as possible to producing the number of each type of mower that is needed. For each type of mower, a penalty of 10 is incurred for each unit we deviated from the needed number of mowers. For example, if we produce 90 Type 1 mowers, then a penalty of 20*10 is incurred.

Our goal, of course, is to determine a production schedule that minimizes the total penalty. We proceed as shown in Figures 21–23.

Step 1 In B16:H19, we enter trial numbers for the number of each type of mower produced each week.

Step 2 In B20:H20, we compute the total number of each type of mower produced during the month by copying from B20 to C20:H20 the formula

$$=SUM(B16:B19).$$

FIGURE 21
Production Schedule

	A	B	C	D	E	F	G	H
9		Total						
10	**Models**	16			Hrs/week			
11	Week 1	7		Center 1	375			
12	Week 2	3		Center 2	368.75			
13	Week 3	3		Center 3	382.5			
14	Week 4	3						
15	Production							
16	Week 1	43	22	17	30	29	12	13
17	Week 2	0	68	83	41	0	0	0
18	Week 3	67	0	0	0	51	0	43
19	Week 4	0	0	0	44	0	48	24
20	Total	110	90	100	115	80	60	80
21	Off-target	0	0	0	0	0	0	0

FIGURE 22
Work Center Utilization

	A	B	C	D	E	F	G	H	I	J	K
21	Off-target	0	0	0	0	0	0	0		Sum Dev	224.5
22	Week 1								Total	Dev	
23	Center 1	129	44	34	60	58	48	26	399	24	
24	Center 2	43	44	17	90	87	36	52	369	0.25	
25	Center 3	86	66	0	120	87	36	26	421	38.5	
26	Week 2								0		
27	Center 1	0	136	166	82	0	0	0	384	9	
28	Center 2	0	136	83	123	0	0	0	342	26.75	
29	Center 3	0	204	0	164	0	0	0	368	14.5	
30	Week 3								0		
31	Center 1	201	0	0	0	102	0	86	389	14	
32	Center 2	67	0	0	0	153	0	172	392	23.25	
33	Center 3	134	0	0	0	153	0	86	373	9.5	
34	Week 4								0		
35	Center 1	0	0	0	88	0	192	48	328	47	
36	Center 2	0	0	0	132	0	144	96	372	3.25	
37	Center 3	0	0	0	176	0	144	48	368	14.5	

FIGURE **23**

Early Pickup Penalties

	I	J	K	L	M	N	O	P
13	**Penalties for Early Pickup**							
14	Pick 1	Pick 2	Pick 3	Pick 4	Pick 5	Pick 6	Pick 7	Total
15								
16	0	0	0	0	0	0	0	0

Step 3   In B21:H21, we compute for each type of mower the amount by which we have missed our Target by copying from B21 to C21:H21 the formula

$$=ABS(B20\text{-}B3).$$

Step 4   In B11:B14, we compute the number of model types produced each week by copying from B11 to B12:B14 the formula

$$=COUNTIF(B16\text{:}H16,``>0").$$

In B11, for example, this formula counts for week 1 the models that are actually produced (production level is > 0). In B10, we compute the total number of model types produced during the four weeks with the formula

$$=SUM(B11\text{:}B14).$$

Step 5   In A23:J37, we compute the number of hours used each week in each work center by each product. For example, in B23, we compute the number of hours used to produce Type 1 mowers in week 1 with the formula

$$=B5*B\$16.$$

Copying this formula from B23 to B23:H26 computes the number of hours in each work center used to produce each product during week 1. In a similar fashion, we compute in rows 28–37 the number of hours used in each work center during weeks 2–4.

Step 6   In I23:I25, we compute the total number of hours used in each work center during week 1 by copying from I23 to I24:I25 the formula

$$=SUM(B23\text{:}H23).$$

Step 7   In J23:J25, we compute the deviation of our week 1 workload in each Center from the average number of hours by copying from J23 to J24:J25 the formula

$$=ABS(I23\text{-}E11).$$

Step 8   In I27:J37, we compute total hours used in each work center during weeks 2–4 and the deviation in each work center for weeks 2–4 from the average number of hours used.

Step 9   In K21, we compute the total number of hours that we have deviated from the average number of hours used over all work centers and weeks with the formula

$$=SUM(J23\text{:}J37).$$

(See Figure 22.)

Step 10   In I16:O16, we compute the number of mowers of each type that were not ready for early pickup by copying from I16 to J16:O16 the formula

$$=IF(B16<B4,B4\text{-}B16,0).$$

If we produce at least as many mowers of a given type that are needed, then none were ready for early pickup. Otherwise, the number of mowers not ready is (number needed) − (number produced during week 1).

In cell P16, we compute the total number of mowers not ready for early pickup with the formula

$$=Sum(I16:O16)$$

(See Figure 23.)

Step 11 In O2:O6, we compute our total penalty. In N2:N5, we enter the unit cost of each type of penalty. In O2, we compute the total penalty for model types with the formula

$$=N2*B10.$$

Step 12 In O3, we compute the total penalty for failure to produce customer pickup mowers during week 1 with the formula

$$=N3*P16.$$

Step 13 In O4, we compute the total penalty for not "centering" hours used each week on 375 with the formula

$$=K21*N4.$$

Step 14 In O5, we compute the total penalty for producing an incorrect number of items with the formula

$$=SUM(B21:H21)*N5.$$

Step 15 In cell O6, we compute our total penalty (our Target Cell!) with the formula

$$=SUM(O2:O5).$$

(See Figure 24.)

Step 16 We are now ready to invoke the Evolutionary Solver. Our window is as shown in Figure 25. We minimize total penalties (O6) by changing the number of mowers of each type produced during each week (B16:H19). We restrict production to be an integer between zero and 115. The upper bound of 115 comes from the fact that monthly demand for each type of mower is 115 or less. Our solution is as shown in Table 8.

From Figure 21, we see that 16 model types were made. A naive schedule of producing each model type each week would have made 28 model types! All mowers were ready for pickup by the end of week 1, and we made one extra unit. Our workforce leveling incurred a penalty of 224 units or an average of 56 units per week. If we think this is too high, then we could have increased the penalty from centering from its current value of 1. Actually, we are only off by an average of 224/12 = 18 units per center week. This is only 5% of the average load in a center, which is not too bad. In general, the best way to

	M	N	O
1		Penalty	Cost
2	Model	200	3200
3	Pick>1	50	0
4	Cent	1	224.5
5	Wrong #	10	0
6			3424.5
7			

FIGURE 24

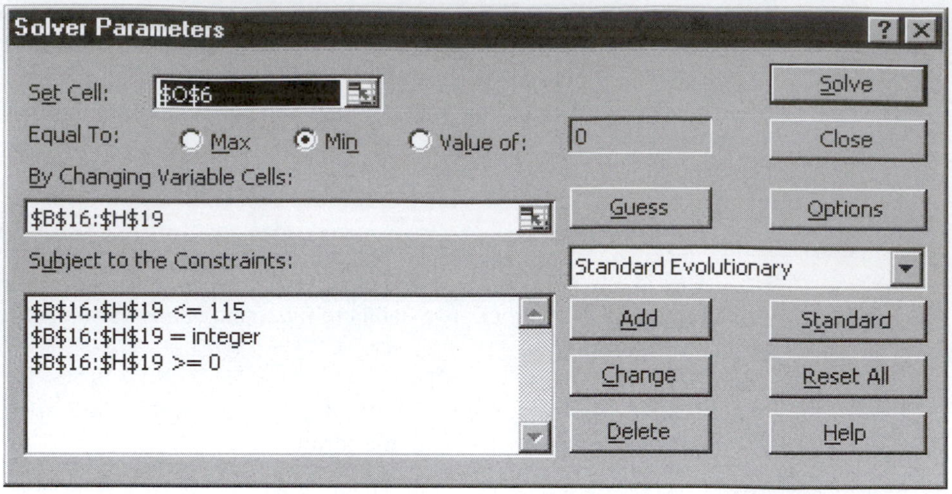

FIGURE 25
Solver Window for
Lawn Mower Example

TABLE 8

Week				Production			
1	43	22	17	30	29	12	13
2	0	68	83	41	0	0	0
3	67	0	0	0	51	0	43
4	0	0	0	44	0	48	24

determine correct penalties is to vary the penalties until you obtain a solution that seems like a "good solution" to you.

15.6 Assigning Workers to Jobs with the Evolutionary Solver

Assigning workers to jobs is a difficult task. We must try and assign them to jobs they do well in addition to jobs they like. We can use the Evolutionary Solver to solve this problem.

EXAMPLE 5   Worker Assignment

We need to assign 80 employees to four work groups. The head of each work group has rated each employee's competence on a 0–10 scale (10 = most competent). Each employee has rated his or her satisfaction with each job assignment (again, on a 0–10 scale). For example, worker 1 has been given a 9 rating for work group 1, and worker 1 gives work group 4 a rating of 7. We want to assign between 18 and 22 people to each work group. We consider job competence to be twice as important as employee satisfaction. Assign employees to work groups to maximize total satisfaction and ensure that each work group has the required number of employees. See file Assign.xls.

Assign.xls

Solution   **Step 1**   Enter in cells A3:A82 trial assignments of workers to work groups.

Step 2   By copying from K3 to K3:K82 the formula

$$=\text{HLOOKUP}(A3,\text{Qual},B3+1)$$

we look up each employee's qualifications for her assigned job. Note that Qual is the range C2:F82.

Step 3 By copying from L3 to L3:L82 the formula

$$=HLOOKUP(A3,Satis,B3+1)$$

we look up the employee's satisfaction with her assigned job. Note that Satis is the range G2:J82.

Step 4 We now count how many employees have been assigned to each work group in cells N6:N9 by copying from N6 to N7:N9 the formula

$$=COUNTIF(\$A\$3:\$A\$82,M6).$$

Step 5 In cells O6:O9, we determine if a work group has the incorrect number of employees by copying from O6 to O7:O9 the formula

$$=IF(OR(N6<18,N6>22),1,0).$$

Step 6 In K1:L1, we compute total competence and total job quality by copying from K1 to K1:L1 the formula

$$=SUM(K3:K82).$$

Step 7 In cell O10, we compute total number of groups that do not have the correct number of workers with the formula

$$=SUM(O6:O9).$$

Step 8 In cell O12, we add twice the total competence to the total job satisfaction and subtract a penalty of 1,000 for each group that does not have the correct number of employees.

$$=2*K1+L1-1000*O10.$$

This formula will be our Target Cell for Solver.

Step 9 We need to use Evolutionary Solver because the HLOOKUP and COUNTIF functions make our model highly nonlinear. Our Solver model appears as in Figure 26.

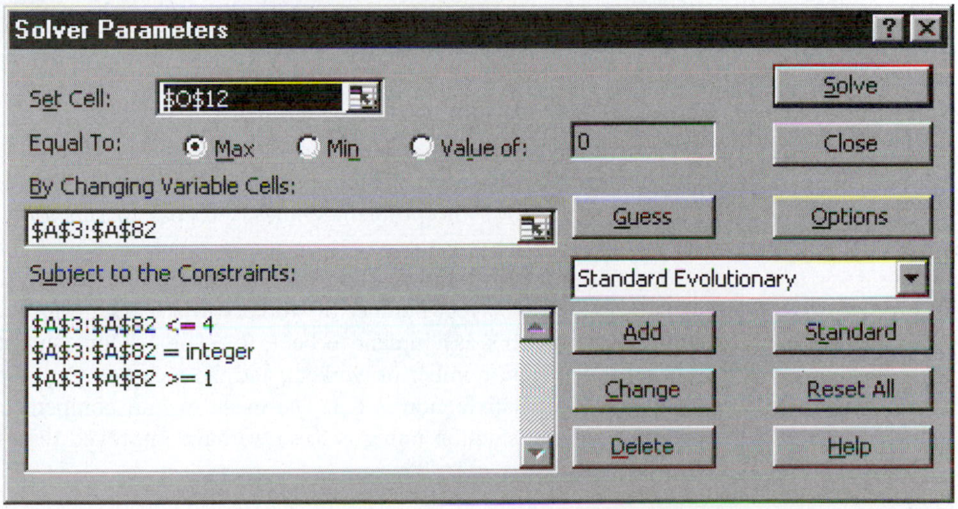

FIGURE 26 (a)

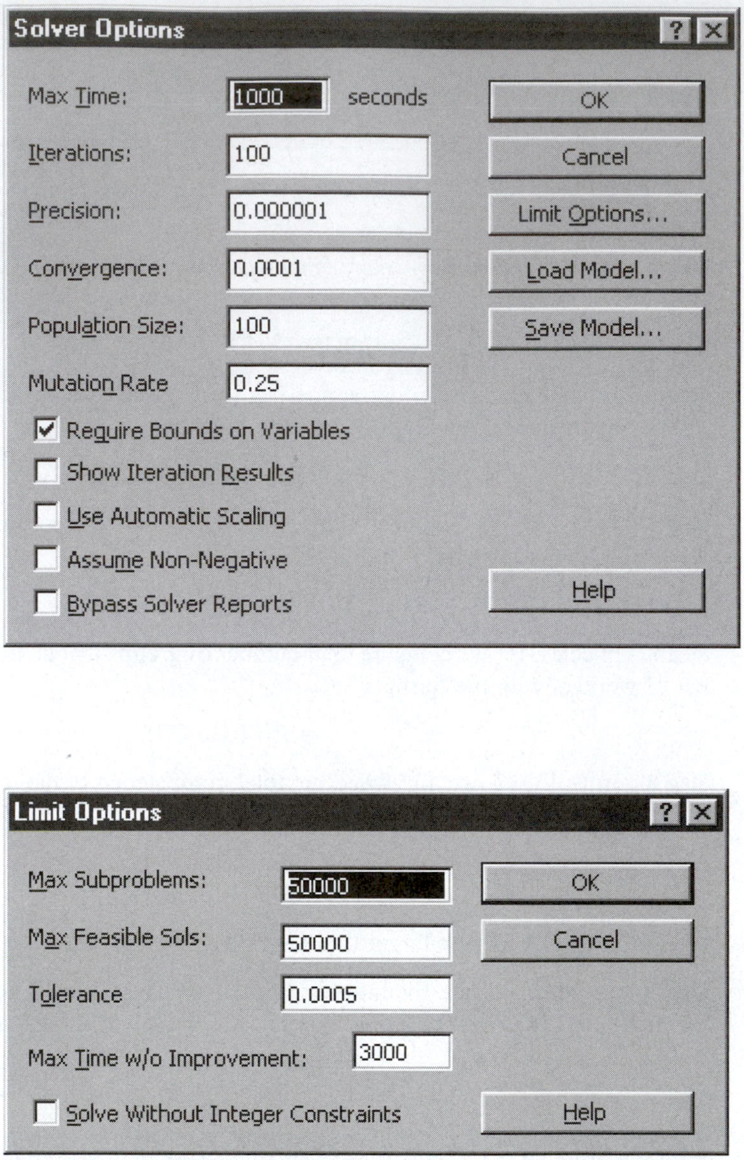

FIGURE 26 (b) & (c)

Note we have bumped up Mutation rate to help us search a wider range for a better solution. Increasing Number of Sub problems, Max Feasible Solutions, and Max Time w/o Improvement allows the problem-solution process to run for a long time without our intervention.

We maximize the weighted sum of work group and employee satisfaction minus the penalty for the incorrect number of workers in a work group (cell O12). Then we constrain each worker's assignment to be 1, 2, 3, or 4. Our solution is in Figure 27. Each group has the right number of workers, and the mean employee competence is 7.2 while mean employee satisfaction is 6.3. The mean overall competence rating is 4.4, and the mean overall satisfaction rating is 5, so we have improved things quite a lot over a random assignment.

FIGURE 27

Worker Assignment Spreadsheet

	A	B	C	D	E	F	G	H	I	J	K	L	M	N	O
1			Qual				Sat				576	499			
2	Assigned to	Worker	1	2	3	4	1	2	3	4	Quality	Satisfaction			
3	4	1	9	8	6	8	1	2	6	7	8	7			
4	1	2	10	0	5	6	9	6	7	4	10	9			
5	3	3	5	8	10	5	1	7	7	3	10	7	Group	# ass	penalty
6	1	4	4	0	5	2	9	1	0	3	4	9	1	19	0
7	2	5	9	10	4	5	9	8	8	3	10	8	2	18	0
8	3	6	5	2	7	3	2	8	1	5	7	1	3	22	0
9	1	7	8	3	1	2	1	8	2	2	8	1	4	21	0
10	3	8	2	2	9	2	8	3	1	6	9	1	Total pen		0
11	1	9	8	7	6	3	4	3	4	1	8	4			
12	4	10	7	0	1	8	4	1	5	4	8	4	Total		1651
13	3	11	8	1	6	6	2	0	9	3	6	9			
14	2	12	0	7	1	2	5	2	1	1	7	2			
15	1	13	9	0	5	4	3	0	7	8	9	3			
16	4	14	9	2	2	7	1	1	2	10	7	10			
17	3	15	1	3	8	4	9	8	6	8	8	6			
18	1	16	9	6	4	5	5	7	8	8	9	5			
19	1	17	8	0	5	0	5	7	2	4	8	5			
20	2	18	6	7	6	3	2	4	1	6	7	4			
21	3	19	3	4	5	4	8	7	6	6	5	6			
22	2	20	3	9	4	4	2	2	2	3	9	2			
23	3	21	1	6	9	1	7	1	8	4	9	8			
24	4	22	5	1	3	7	8	9	6	7	7	7			
25	3	23	8	7	10	2	6	2	5	9	10	5			
26	3	24	3	6	4	4	7	2	9	2	4	9			
27	1	25	7	1	1	0	4	9	10	9	7	4			
28	3	26	0	9	8	1	1	2	9	0	8	9			
29	2	27	3	9	1	5	9	9	2	8	9	9			
30	2	28	2	4	0	1	0	4	7	2	4	4			
31	2	29	1	6	7	3	10	6	4	4	6	6			
32	2	30	2	3	3	0	1	9	7	9	3	9			
33	4	31	3	5	4	8	5	1	9	8	8	8			
34	4	32	1	1	3	7	4	8	2	6	7	6			
35	4	33	5	1	2	6	4	1	8	9	6	9			
36	1	34	8	2	3	3	6	9	1	6	8	6			
70	4	68	4	3	1	7	0	8	5	6	7	6			
71	3	69	4	4	7	2	7	7	3	7	7	3			
72	1	70	9	0	3	4	1	2	2	8	9	1			
73	1	71	3	4	2	2	8	2	5	3	3	8			
74	2	72	2	9	9	7	5	7	2	8	9	7			
75	1	73	2	2	1	2	6	6	1	5	2	6			
76	3	74	5	8	10	0	8	0	9	3	10	9			
77	2	75	5	9	1	2	3	4	9	1	9	4			
78	4	76	9	9	2	7	4	4	4	10	7	10			
79	1	77	2	0	0	0	5	6	2	2	2	5			
80	2	78	7	8	6	5	5	10	2	0	8	10			
81	2	79	2	4	2	3	3	5	8	6	4	5			
82	4	80	0	0	2	7	8	5	8	3	7	3			

Using Conditional Formatting to Highlight Each Employee's Ratings

We can use the conditional formatting feature to highlight—in red—each employee's actual competence and satisfaction (based on his or her assignment). First select the cell range C3:J82 where you want the format to be placed. Simply go to cell C3 and select Format Conditional Format. Fill in the dialog box as shown in Figure 28.

FIGURE 28

Then select format and choose a red font. Whenever the copied version of =$A3=C$2 remains true in the cell range C3:J82 this formula will change the font to red. For example, this formula will enter in cell C3 a red format if and only if the first worker was assigned to work group 1. Now the employee's competence and satisfaction, which correspond to her actual assignment, are highlighted.

PROBLEMS

1 You are the Democratic campaign manager for the state of Indiana. There are 15 cities in the state of Indiana. The number of Democrats and Republican voters in each city (in thousands) is as shown in Table 9. The Democrats control the state legislature, so they can redistrict as they wish. There will be eight congressional districts. Each city must be assigned in its entirety to a single district. Each district must contain between 150,000 and 250,000 voters. Use

Evolutionary Solver to assign voters to districts in a way that maximizes the number of districts that will vote Democratic. (*Hint:* You may find it convenient to use the =SUMIF function.) To illustrate, for the given data, the formula

$$=SUMIF(J6:J10,J12,K6:K10)$$

will add all numbers in K6:K10 for which the corresponding entry in J6:J10 equals 2 (the value in J12). This yields 6 + 9 = 15.

2 Xerox is trying to determine how many maintenance centers are needed in the mid-Atlantic states. Xerox earns $500 profit (excluding the cost of running maintenance centers) on each copier sale. The sales of copiers in each major market (Boston, New York, Philadelphia, Washington, Providence, and Atlantic City) depend on the proximity of the nearest maintenance facility. If there is a maintenance facility within 100 miles of a city, then sales will be high; if there is a maintenance facility within 150 miles of a city,

TABLE 9

City	Party	
	Republicans	Democrats
1	80	34
2	43	61
3	40	44
4	20	24
5	40	114
6	40	64
7	70	34
8	50	44
9	70	54
10	70	64
11	80	45
12	40	50
13	50	60
14	60	65
15	50	70
Totals	803	827

TABLE 10

City	Miles from a Maintenance Facility		
	Low Sales (>150 Miles)	Medium Sales (100–150 Miles)	High Sales (<100 Miles)
Boston	500	600	700
New York	750	800	1,000
Philadelphia	700	800	900
Washington	450	650	800
Providence	200	300	400
Atlantic City	300	350	450

then sales will be medium; and otherwise sales will be low. The actual predicted annual sales are as shown in Table 10. It costs $200,000 per year to place a maintenance representative in a city. It is possible to locate a representative in any city except for Atlantic City and Providence. The distance between the cities is shown in Table 11. Where should maintenance representatives be located?

3 Twelve towns in Southern Indiana need to be assigned to four phone books. The number (in thousands) of phone numbers in each town is shown in Table 12. Each phone book can contain at most 100,000 numbers. File Phonedata.xls gives the number of calls per month (in thousands) between each pair of cities (see Figure 29). For example, each month there are 15,000 calls from city 1 to city 2 and 74,000 calls from city 2 to city 1. Determine how to assign the cities to phone books to maximize the total number of calls that are made each month between cities in the same book.

4 We are trying to schedule radio commercials in 60-second blocks. We have sold commercial time for commercials of 15, 16, 20, 25, 30, 35, 40, and 50 seconds. Determine the minimum number of 60-second blocks needed for these commercials.

TABLE 11

	Boston	New York	Philadelphia	Washington
Boston	0	222	310	441
New York	222	0	89	241
Philadelphia	310	89	0	146
Washington	441	241	146	0
Providence	47	186	255	376
Atlantic City	350	123	82	178

TABLE 12

Town	Numbers
1	33
2	19
3	14
4	38
5	25
6	12
7	39
8	28
9	20
10	39
11	38
12	29

FIGURE 29

	1	2	3	4	5	6	7	8	9	10	11	12
1	49	15	33	44	17	50	86	84	84	41	84	31
2	74	96	20	10	92	83	60	47	20	46	10	52
3	9	93	18	51	44	43	45	84	75	21	84	100
4	81	19	58	52	2	88	25	41	33	93	8	18
5	28	11	6	56	24	91	61	37	37	38	94	66
6	52	89	13	27	64	70	9	18	14	8	65	61
7	19	54	22	15	90	29	99	56	87	63	10	48
8	49	27	45	7	7	77	39	43	19	79	88	55
9	30	97	6	71	99	45	28	53	4	44	59	15
10	57	5	66	38	98	20	46	51	80	95	54	37
11	47	70	41	94	63	99	76	58	32	1	98	95
12	8	61	29	39	100	84	78	71	81	40	46	76

15.7 Cluster Analysis

Often marketers want to group objects into **clusters** of similar objects. For example, identifying similar customers could help us identify market segments. Identifying a cluster of similar products could help us identify our main competition. Here are two actual examples of how the United States is divided into clusters:

- Claritas divides each block of the United States into one of 62 clusters, including Blue Blood Estates, New Homesteaders, Middle America, and God's Country. For example, Blue Blood Estates consists primarily of the richest U.S. suburbs where one in ten residents is a millionaire. This is valuable information for marketers.

For example, residents of Blue Blood Estates consume imported beer at a rate nearly three times the national average.

- SRI clusters families based on their financial status and demographics. For example, the cluster Bank Traditionalists consists of upper-middle-class families of larger than average size having school-age children. This cluster is a natural prospecting ground for life insurance salespeople.

Cluster.xls

To illustrate the mechanics of cluster analysis, suppose we want to "cluster" 49 of the largest U.S. cities (see file Cluster.xls). For each city, we are given the data shown in Figure 30.

For example, Atlanta is 67% black, 2% Hispanic, and 1% Asian, and has a median age of 31, a 5% unemployment rate, and a per capita income of $22,000.

We would like to group our cities into four clusters of cities that are demographically similar. The basic idea is to choose a city to *anchor* or *center* each cluster. We assign each city to the nearest cluster center. Our Target Cell then minimizes the sum of the squared distances from each city to its cluster anchor.

Standardizing the Attributes

The first problem is that if we use raw units, then percentage black and Hispanic will drive everything because these values are more spread out than the other demographic attributes. To remedy this problem, we *standardize* each demographic attribute by subtracting the attribute's mean and dividing by the attribute's standard deviation. For example, the average city has

$$\frac{67 - 24.34}{18.11} = 2.35$$

24.34% blacks with a standard deviation of 19.11%. Thus on a standardized basis, Atlanta has 2.35 standard deviations more blacks (on a percentage basis) than a typical U.S. city.

Working with standardized values for each attribute ensures that our analysis will be unit-free.

Step 1 We begin by computing the mean and standard deviation for black percentage in C1:G2 (see Figure 31). We compute the black mean percentage in C1 with the formula:

=AVERAGE(C10:C58).

FIGURE 30
U.S. City Demographic Data

	A	B	C	D	E	F	G	H
9	City #	City	%age Black	%age Hispanic	%age Asian	Median Age	Unemployment rate	Per capita income(000's)
10	1	Albuquerque	3	35	2	32	5	18
11	2	Atlanta	67	2	1	31	5	22
12	3	Austin	12	23	3	29	3	19
13	4	Baltimore	59	1	1	33	11	22
14	5	Boston	26	11	5	30	5	24
15	6	Charlotte	32	1	2	32	3	20
16	7	Chicago	39	20	4	31	9	24
17	8	Cincinnati	38	1	1	31	8	21
18	9	Cleveland	47	5	1	32	13	22
19	10	Columbus	23	1	2	29	3	13

FIGURE 31

	B	C	D	E	F	G
1	Mean	24.34694	14.59184	6.040816	31.87755	7.020408163
2	Std dev	18.11025	16.4721	11.1448	1.99617	2.688631901
3						
4						City
5						San Francisco
6						Philadelphia
7						Indianapolis
8						Los Angeles
9	City	%age Black	%age Hispanic	%age Asian	Median Age	Unemployment rate

Step 2 In C2, we compute the standard deviation of the Black percentages with the formula

$$=STDEV(C10:C58).$$

Step 3 Copying these formulas to D1:G2 computes the mean and standard deviation for each attribute.

Step 4 In cell I10, we compute the standardized percentage of blacks in Albuquerque (often called a **z-score**) with the formula

$$=STANDARDIZE(C10,C\$1,C\$2).$$

Step 5 Copying from I10 to N58 computes z-scores for all cities and attributes (see Figure 32).

Setting Up the Distances

Next we set up a way to look up the z-scores for candidate cluster centers (see Figure 33). We begin by naming A10:N58 as the range LOOKUP.

Step 1 In H5:H8, we enter trial values for cluster anchors.

Step 2 In G5, we look up the name of the first cluster anchor with the formula

$$=VLOOKUP(H5,Lookup,2).$$

Copying this formula to G6:G8 identifies the name of each cluster center candidate.

FIGURE 32
City z-Scores

	I	J	K	L	M	N
9	z Black	z Hispanic	z Asian	z Age	z Unemp	z income
10	-1.178721	1.238954	-0.362574	0.061342	-0.751463	-0.875231
11	2.355188	-0.764434	-0.452302	-0.439617	-0.751463	0.324386
12	-0.681765	0.510449	-0.272846	-1.441536	-1.495336	-0.575327
13	1.91345	-0.825143	-0.452302	0.562301	1.480155	0.324386
14	0.091278	-0.218056	-0.09339	-0.940577	-0.751463	0.924195

	G	H	I	J	K	L	M	N
3		Column	9	10	11	12	13	14
4	City	Cluster	z Black	z Hispanic	z Asian	z Age	z Unemp	z income
5	Los Angeles	24	-0.57133	1.542497	0.355249	-0.439617	1.480155	0.024482
6	Omaha	34	-0.626548	-0.703726	-0.452302	0.061342	-0.751463	-0.275422
7	Memphis	25	1.69258	-0.825143	-0.452302	0.061342	0.736282	-0.275422
8	San Francisco	43	-0.736982	-0.03593	2.06008	2.06518	-0.379527	3.023526

FIGURE 33
Cluster Anchor z-Scores

Step 3 In I5:N8, we identify the z-scores for each cluster anchor candidate by copying from I5 to I5:N8 the formula

$$=VLOOKUP(\$H5,Lookup,I\$3).$$

Step 4 We can now compute squared distance from each city to each cluster candidate. To compute the distance from city 1 (Albuquerque) to cluster candidate anchor 1, we enter in O10 the formula

$$=SUMXMY2(\$I\$5:\$N\$5,\$I10:\$N10).$$

This cool Excel function computes

$$(I5-I10)^2+(J5-J10)^2+(K5-K10)^2+(L5-L10)^2+(M5-M10)^2+(N5-N10)^2.$$

To compute the squared distance of Albuquerque from the second cluster anchor requires us to change each 5 in O10 to a 6. Similarly, in Q10 we change each 5 to a 7. Finally, in R10 we change each 5 to an 8. Copying from O10: R10 to O11: R58 computes the squared distance of each city from each cluster anchor (see Figure 34).

	O	P	Q	R	S	T
8				Sum Dis^2	165.3482	
9	Distance^2 to 1	Distance^2 to 2	Distance^2 to 3	Distance^2 to 4	Min Distance	Assigned to
10	7.016897	4.44672	15.08608	27.04372	4.44672	2
11	19.60865	9.505167	3.266853	30.102	3.266853	3
12	11.68898	4.411405	14.78223	32.23795	4.411405	2
13	13.52578	12.05718	1.212861	26.96212	1.212861	3
14	9.780438	3.32289	7.717853	18.93672	3.32289	2
15	16.3033	1.676812	6.601068	23.98015	1.676812	2
16	5.032486	7.102106	3.873518	19.48122	3.873518	3
17	9.259088	3.506272	1.360387	24.97923	1.360387	3
18	9.381499	12.75265	2.827259	28.64125	2.827259	3
19	21.98167	7.546867	14.77613	49.61466	7.546867	2

FIGURE 34
City Distances to Cluster Anchors

Step 5 In S10:S58, we compute the distance from each city to the closest cluster anchor by entering the formula

$$=MIN(O10:R10)$$

in cell S10 and copying it to the cell range S10:S58.

Step 6 In S8, we compute the sum of squared distances of all cities from their cluster anchor with the formula

$$=SUM(S10:S58).$$

Step 7 In T10:T58, we compute the cluster to which each city is assigned by entering in T10 the formula

$$= MATCH(S10,O10:R10,0)$$

and copying this formula to T11:T58. This formula identifies which element in columns O:R gives the smallest distance to the city.

Using Evolutionary Solver to Find the Best Clusters

We are now ready to use the Evolutionary Solver to find the optimal cluster. Our window is as shown in Figure 35.

We choose to minimize cell S8 (sum of squared distances). Our cluster anchors (H5:H8) are the changing cells. They must be integers between 1 and 49. The Evolutionary Solver quickly found that the cluster anchors are LA, Omaha, Memphis, and San Francisco. The clusters are as shown in Figure 36.

REMARKS
- The San Francisco cluster is rich, older, highly Asian cities. The Memphis cluster consists of highly black cities with high unemployment rates. The Omaha cluster consists of average-income cities with few minorities. The Los Angeles cluster consists of highly Hispanic cities with high unemployment rates.
- Why four clusters? We could easily try three clusters (delete fourth anchor) or five clusters (add another anchor). To choose the optimal number of clusters, stop adding clusters when the sum of squared distances fails to decrease by a substantial amount.

FIGURE 35
Cluster Solver Window

	W	X	Y
9	Assigned to	City	Anchor
10	1	Dallas	LA
11	1	El Paso	LA
12	1	Fort Worth	LA
13	1	Fresno	LA
14	1	Houston	LA
15	1	Long Beach	LA
16	1	Los Angeles	LA
17	1	Miami	LA
18	1	NY	LA
19	1	San Antonio	LA
20	1	San Diego	LA
21	1	San Jose	LA
22	2	Albuquerque	Omaha
23	2	Austin	Omaha
24	2	Boston	Omaha
25	2	Charlotte	Omaha
26	2	Columbus	Omaha
27	2	Denver	Omaha
28	2	Indianapolis	Omaha
29	2	Jacksonville	Omaha
30	2	Kansas City	Omaha
31	2	Las Vegas	Omaha
32	2	Milwaukee	Omaha
33	2	Minneapolis	Omaha
34	2	Nashville	Omaha
35	2	Oklahoma City	Omaha
36	2	Omaha	Omaha
37	2	Phoenix	Omaha
38	2	Pittsburgh	Omaha
39	2	Portland	Omaha
40	2	Sacramento	Omaha
41	2	Toledo	Omaha
42	2	Tucson	Omaha
43	2	Tulsa	Omaha
44	2	Virginia Beach	Omaha
45	3	Atlanta	Memphis
46	3	Baltimore	Memphis
47	3	Chicago	Memphis
48	3	Cincinnati	Memphis
49	3	Cleveland	Memphis
50	3	Detroit	Memphis
51	3	Memphis	Memphis
52	3	New Orleans	Memphis
53	3	Oakland	Memphis
54	3	Philadelphia	Memphis
55	3	St. Louis	Memphis
56	4	Honolulu	SF
57	4	San Francisco	SF
58	4	Seattle	SF

FIGURE 36
Cities Assigned to
Clusters

PROBLEM

1   In the file bschooldata.xls, you are given the information in Table 13 about the top 25 MBA programs (according to 1997 *Business Week Guide*).

- percentage of applicants accepted
- percentage of accepted applicants who enroll
- mean GMAT scores of enrollees

- mean undergraduate GPAs of enrollees
- annual cost of school (for state schools, this is the cost for out-of-state students)
- percentage of students who are minorities
- percentage of students who are non-U.S. residents
- mean starting salary of graduates (in thousands of dollars)

Use this data to divide the top 25 schools into four clusters. Interpret your clusters.

TABLE 13

School	Percentage Accepted	Percentage Accepted Who Enroll	Mean GMAT	Mean GPA	Total Cost ($)	Minority Percentage	Non-U.S. Percentage	Mean Starting Salary ($1,000)
Wharton	15	71	662	3.42	32,400	16	30	102
Michigan	28	44	645	3.3	29,800	15	26	86
Northwestern	14	69	660	3.3	32,600	9	24	99
Harvard	13	88	680	3.5	30,100	19	27	114
Virginia	19	49	660	3.1	31,200	20	12	93

15.8 Fitting Curves

Suppose we want to determine a curve that predicts the sales of a product as a function of sales effort allocated to the product. Researchers [see Lodish (1990)] have found, however, that the response to sale force effort is better described by the ADBUG function of the following form:

$$\text{Sales of drug } i \text{ when } x \text{ calls are made for drug } i = a + \frac{(b - a)x^c}{(d + x^c)}$$

Although the power curve always exhibits diminishing returns, the ADBUG curve can exhibit diminishing returns or look like an S-shaped curve. See Figure 37.

An S-curve starts out flat, gets steep, and then flattens out. This would be the correct form of the sales as a function of effort relationship if effort needs to exceed some critical value to generate a favorable response.

To estimate a curve of this form we use the following five points as inputs.

1 Estimated sales when there is no sales effort assigned to the drug.

2 Estimated sales when sales effort assigned to the drug is cut in half.

3 Sales at current level of sales force effort (assumed to equal a base of 100).

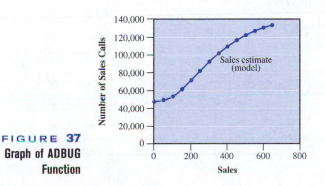

FIGURE 37

Graph of ADBUG Function

4 Estimated sales if sales force effort were increased by 50%.

5 Estimated sales if sales force "saturated" the market.

All sales levels are given as a percentage of current sales.

Currently, 350,000 calls are being made, and we are given the data in Table 14.

Thus, with no sales force effort sales, we estimate sales would drop to 47% of their current level. If sales force effort was cut in half, then we estimate sales would drop to 68% of its current level. If sales force effort was increased 50%, then we estimate sales would increase by 26%. If we increased sales force effort tenfold (a saturation level), then we estimate sales would increase by 52%.

Syntexgene.xls
In the file Syntexgene.xls, we use the Evolutionary Solver to find the values of a, b, c, and d that best fit the ADBUG curve to our five points. We proceed as shown in Figure 38.

Step 1 Enter trial values of a, b, c, and d in A13:D13 and name these cells a, b, c−, and d, respectively.

Step 2 In cells C5:C9, compute the prediction from the ADBUG curve by copying from C5 to C6:C9 the formula

$$=a+((b\text{-}a)*A5\char`\^c\_)/(d+A5\char`\^c\_).$$

Step 3 In cells D5:D9, compute the squared error for each prediction by copying from D5 to D6:D9 the formula

$$=(B5\text{-}C5)\char`\^2.$$

TABLE 14

Sales Calls (1,000)	Sales Level
0	47
175	68
350	100
525	126
3,500	152

FIGURE 38

	A	B	C	D	E
1	**Estimating the Sales Response Function at Syntex Labs**				
2					
3	**Estimates from management**				
4	Sales calls (1000s)	Sales level	Sales estimate (model)	Squared error	
5	0	47	47.472	0.223	
6	175	68	66.758	1.543	
7	350	100	102.076	4.311	
8	525	126	124.326	2.804	
9	3500	152	152.385	0.148	
10			Sum of squared errors	9.029	
11	**Model parameters**				
12	a	b	c	d	
13	47.472	152.919	2.262	530738.063	

Step 4 In cell D10, compute the sum of the squared errors with the formula

$$=\text{SUM(D5:D9)}.$$

With different starting solutions, the ordinary Solver will find different final solutions. For example, if we start Solver with $a = 10$, $b = 50$, $c = 5$, and $d = 1,000$, the ordinary Solver can only obtain an SSE of 3,875. With a starting solution of $a = 1$, $b = 2$, $c = 3$, and $d = 4$, we obtain an error message! Let's try the Evolutionary Solver with the starting values of $a = 1$, $b = 2$, $c = 3$, and $d = 4$. Because a equals our predicted sales for no sales effort, then a should be close to 47 (we will constrain a to be between zero and 50). Because b is our predicted sales with infinite sales effort, b should be close to 150 (we will constrain it to be between zero and 200). There are no obvious values for c and d. A large value of c, however, will cause our function to involve large numbers that may crash Evolutionary Solver, so we will constrain c to be between zero and 5. Finally, there are no obvious limits on d, so we will constrain d to be between zero and 1,000,000. Our Evolutionary Solver window is shown in Figure 39.

We minimize the sum of squared errors (cell D10) by changing values of a, b, c, and d (cells A13:D13). We have bounded a, b, c, and d as discussed previously.

After 10 seconds (starting with $a = 10$, $b = 50$, $c = 5$, and $d = 1,000$), Evolutionary Solver found the solution in Figure 9.14. Note the SSE of 9 indicates a very good fit. None of our predictions are off by more than 2%. After fitting this curve to each drug, the resulting response curves can be used as an input to a Solver model that is used to allocate sales force effort to maximize profit. Also note that for different starting values of a, b, c, and d, Evolutionary Solver always finds an SSE near 9.

Note that even though we gave Evolutionary Solver a poor starting solution, Evolutionary Solver found a very good solution. This is because the genetic algorithm approach does not get stuck on a local optimum; it looks at points throughout the feasible region and goes where it sees a good Target Cell value. In many nonlinear problems, Solver will get stuck on the top of a "hill" and fail to find its top. To see what's going on, suppose I start in Indiana and look for the highest point in the United States. I might hit the highest point in Indiana, and Solver would see no higher point for miles around and call it quits. Evolutionary Solver, on the other hand, would look at points scattered throughout the United States and eventually see there are high points in the Rocky Mountains, California, and Alaska. Eventually, this approach would lead us to Mount Denali, the highest point in United States.

FIGURE 39
Curve-Fitting Solver Window

PROBLEMS

1 You are given the following information on how a change in sales force effort affects sales:

- A 50% cut in sales force effort reduces sales to 48% of its current value.
- Zero sales force effort reduces sales to 15% of its current value.
- A 50% increase in sales force effort increases sales by 20%.
- A saturation of sales effort (a tenfold increase) increases sales by 35%.

Fit an ADBUDG function to this data.

2 Table 15 shows data on annual advertising (per capita) and annual unit sales (per capita) in different regions of the country. Determine an ADBUDG function that can be used to determine how advertising influences sales.

TABLE 15

Region	Per Capita Advertising ($)	Sales Units (per Capita)
1	0	5
2	2	7
3	4	13
4	6	22
5	8	25
6	10	27
7	12	31
8	14	33

3 Sales of a product over time often follow an S-shaped curve. Two functions that yield S-shaped curves are the **Pearl curve**

$$Y = \frac{L}{1 + ae^{-bt}}$$

and the **Gompertz curve:**

$$Y = Le^{-be^{-kt}}$$

For each curve, L represents the upper limit for sales. You are given the data in Table 16 for sales of answering machines in the United States. Fit a Pearl and a Gompertz curve to these data. Let $t = 0$ correspond to 1983, and so forth. Which curve provides a better fit? Try Solver and see how you do! [*Hint:* You need to come up with reasonable bounds for the parameters for each curve. For example, $L \geq 14.5$ is reasonable. (Why?) Also, for the Gompertz curve a value of b larger than 5 is unreasonable because it will drive your year zero forecast to zero.]

TABLE 16

Year	Sales (Millions)
1983	2.2
1984	3
1985	4.2
1986	6.5
1987	8.8
1988	11.1
1989	12.5
1990	13.8
1991	14.5

15.9 Discriminant Analysis

In many situations, we want to classify an individual into a group based on demographic information. Here are examples:

- Based on gender, age, income, and residential location, can we classify a consumer as a user or nonuser of a new breakfast cereal?
- Based on income, type of residence, credit card debts, and other information, can we classify a consumer as a good or bad credit risk?
- Based on financial ratios, can we classify a company as a likely or unlikely candidate for bankruptcy?

In each of these situations we want to use demographic information to discriminate or classify an individual. Our goal is to maximize the percentage of observations correctly classified. Evolutionary Solver can easily be used to develop a discrimination procedure. Suppose there are n demographic variables and two groups. We use Evolutionary Solver to determine weights W1, W2, . . ., Wn and a cutoff point c so that we classify an object in Group 1 if and only if

W1(value of variable 1) + W2(value of variable 2) + . . . Wn(value of variable n) $\geq c$.

We call W1(value of variable 1) + W2(value of variable 2) + . . . Wn(value of variable n) the individual's **discriminant score.**

Our goal is to choose the weights and cutoff that maximizes the number of correctly classified observations. This example shows how to use Evolutionary Solver for discriminant analysis.

EXAMPLE 6 · **Discriminant Analysis**

wsj.xls

The file wsj.xls contains the annual income and size of investment portfolio (both in thousands of dollars) for 84 people. A zero indicates the person does not subscribe to the *Wall Street Journal* (WSJ), while a 1 indicates the person is a subscriber. Using income and size of investment portfolio, determine a classification rule that maximizes the number of people correctly classified as subscribers and nonsubscribers.

Solution Figure 40 and file wsj.xls contain our work.

We will let a 1 represent a subscriber and a 0 represent a nonsubscriber.

Step 1 In I2:J2, enter trial values for the Income and Investment weights. In K2, enter a trial value for the cutoff.

FIGURE 40
Discriminant Analysis Spreadsheet

	F	G	H	I	J	K	L	M
1	ables			Income	Invest	Cut		
2				7.687055	112.6874149	5681.105		
3	Income	Invest	WS?	Score	Classified as?	Right?	Total err	
4	66.4	26.9	0	3541.712	0	0	6	
5	68	7.1	0	1322.8	0	0		
6	54.9	21.5	0	2844.799	0	0	%age right	
7	50.6	19.3	0	2563.832	0	0	0.928571	
8	54.1	16.7	0	2297.749	0	0		
9	78.2	31.9	0	4195.856	0	0		
10	66.2	23.8	0	3190.843	0	0		
11	43.9	12.4	0	1734.786	0	0		
12	41.9	5	0	885.5247	0	0		
13	61.1	25.2	0	3309.402	0	0		
14	64.5	11.8	0	1825.527	0	0		
15	59.4	27.3	0	3532.977	0	0		
16	45.9	16.8	0	2245.984	0	0		
17	59.7	14.9	0	2137.96	0	0		
18	76	41.9	0	5305.819	0	0		
19	89.9	46.2	0	5897.225	1	1		
20	32.7	16.9	0	2155.784	0	0		
21	57.8	23.4	0	3081.197	0	0		
22	66.9	34.4	0	4390.711	0	0		
23	87.2	51	0	6417.369	1	1		

Step 2 In I4:I87, we compute each individual's "score" by copying from I4 to I5:I87 the formula

$$=SUMPRODUCT(\$I\$2:\$J\$2,F4:G4).$$

Step 3 In J4:J87, we compare each person's score to the cutoff. If the score is at least equal to the cutoff, then we classify the person as a subscriber. Otherwise, we classify the person as a nonsubscriber. To accomplish this goal, copy from J4 to J5:J87 the formula

$$=IF(I4>=\$K\$2,1,0).$$

Step 4 In K4:K87, we determine whether we correctly or incorrectly classified the individual. A zero indicates a correct classification, while a 1 indicates an incorrect classification. Simply copy from K4 to K5:K87 the formula

$$=IF(J4-H4=0,0,1).$$

Step 5 In L4, compute the total number of misclassifications by adding the numbers in column K:

$$=SUM(K4:K87).$$

Step 6 We can now use Evolutionary Solver to choose the weights and cutoffs to minimize the number of errors. Our window is shown in Figure 41. We want to minimize the total number of misclassifications (cell L4) by adjusting weights (I2 and J2) and the cutoff (K2). It is not clear what bounds to place on the adjustable cells. It seems reasonable that both income and investment amount increase the likelihood of subscribing, so we constrain each weight to be non-negative. We arbitrarily assumed an upper bound of 20 on each weight. If the weights are bounded by 20, however, it is unlikely anybody will obtain a score of 10,000 or more, so we bounded the cutoff between 0 and 10,000. Evolutionary Solver quickly found the following classification rule.

If 7.7*(Income in 000's) + 112.7*(Investment amount in 000's) ≥ 5,681.1, classify the individual as a subscriber; otherwise, classify the individual as a nonsubscriber. Only 6 (7.1%) of the individuals are incorrectly classified.

To determine which variable has more effect on predicting whether a person is a WSJ subscriber, we take the coefficient of each variable divided by the variable's standard de-

FIGURE 41
Discriminant Analysis

TABLE 17

Standardized Coefficients	
Income	Investment
0.513035	6.909612

viation. We do this to obtain a unit-free measure of each variable's importance. To do this, copy from N11 to O11 the formula

$$=I2/STDEV(F:F).$$

To classify a person as a WSJ subscriber, the size of his or her investment portfolio is roughly 14 times as important as income (see Table 17).

PROBLEMS

1 For the data in file Admission.xls, develop a classification rule to classify students as *likely admits, likely rejects,* or *borderline.*

2 For data in file Lasagna.xls, develop a rule to predict whether a person is likely to purchase our Lasagna product. What variables appear to be the most useful for predicting whether or not a person will purchase our product?

3 The file lawn.xls contains the following information for 24 families:

- income (in thousands of dollars)
- lawn size (in thousands of square feet)
- group (1 = owns a rider mower; 0 = does not own a rider mower).

Use this information to determine whether or not a family owns a rider mower.

REVIEW PROBLEMS

1 Eight students need to be assigned to four dorm rooms at Faber College. Based on *incompatibility measurements,* the cost incurred if two students room together is as shown in Table 18. How would you assign students to rooms?

2 A consumer's purchase decision on an electric razor is based on four attributes, each of which can be set at one of three levels (1, 2, or 3). Using conjoint analysis, our market research department has divided the market into five

segments: (labeled as customers 1, 2, 3, 4, and 5) and determined the "part worth" that each customer gives to each level of each attribute (see Figure 42). Conjoint analysis usually assumes that the customer buys the product that yields the highest total part worth. Currently, there is a single product in the market that sets all four attributes equal to 1. You are going to sell two types of electric razors. Design a product line that maximizes the number of market segments that will buy your product. For example, if you designed a product that was level 2 of each attribute, then customer 1 will not buy the product because he values the current product at $1 + 4 + 4 + 4 = 13$ and values your product at $1 + 1 + 1 + 2 = 5$. You should check that, in this case, customer 3 would buy our product.

3 The cost of producing product A, product B, and the bundle A and B is given in Table 19. We are also given the size of each market segment and how much each of three market segments is willing to pay for the bundle. Under the assumption that a market segment will buy the product combination that yields the maximum non-negative surplus (value − cost) and a segment will buy no product if no product has a non-negative surplus, determine an optimal set of product prices. Should the company offer all products for sale?

TABLE 18

Student	Student							
	1	2	3	4	5	6	7	8
1		9	3	4	2	1	5	7
2			4	5	6	7	8	3
3				4	2	8	3	5
4					3	2	4	6
5						8	7	6
6							2	3
7								4

FIGURE 42

	A	B	C	D	E	F	G	H
1	**Customer 1**	Level			**Customer 2**	Level		
2	Attribute	1	2	3	Attribute	1	2	3
3	1	1	1	3	1	1	4	1
4	2	4	1	1	2	2	3	2
5	3	4	1	1	3	4	3	1
6	4	4	2	2	4	0	2	1
7	**Customer 3**	Level			**Customer 4**	Level		
8	Attribute	1	2	3	Attribute	1	2	3
9	1	4	4	4	1	3	1	2
10	2	0	0	4	2	1	0	0
11	3	3	3	0	3	1	4	1
12	4	3	3	0	4	2	2	2
13	**Customer 5**	Level						
14	Attribute	1	2	3				
15	1	4	4	2				
16	2	1	0	1				
17	3	4	0	3				
18	4	1	2	2				

TABLE 19

	Res Price ($)		
Segment	A	B	AB
1	100	95	195
2	50	95	145
3	80	80	195
Cost ($)	50	90	140

TABLE 20

Size	500	400	300
Cuts	C1 Value ($)	C2 Value ($)	C3 Value ($)
500	200	220	190
1,000	190	160	180
1,500	160	130	170
2,000	130	110	160
2,500	100	80	150
3,000	90	75	140
3,500	60	55	130
4,000	40	36	90
4,500	25	20	70
5,000	15	5	60

4 Three customer segments are thinking of ordering as many as 5,000 copies of Office. The value each segment associates with each 500 copies ordered are given in Table 20. For example, segment 1 values the first 500 purchased at $200 each, the next 500 purchased at $190 each, and so on. We are thinking of adapting the following price strategy. Charge a price (high price) for each of the first N units ordered and another price (low price) for the remaining units. What strategy will maximize our revenue from the three segments?

5 Four jobs must be processed on a single machine. The number of days needed to complete each job, the earliest a job can be completed, and the due date for each job are given in Table 21. Your goal is to schedule the jobs to minimize the maximum amount by which a job is late. The following restrictions apply:

- At most, the machine can work on one job at any time.

- Once a job starts, it cannot be interrupted.
- No job can be completed early.

Determine the time at which we should begin working on each job.

6 Production of the Dennis Rodman doll requires 10 operations. A subset of these operations will be assigned to each worker. For example, operations 1–4 might be assigned to worker 1, operations 5–8 to worker 2, and operations 9-10 to worker 3. We want to be able to produce a doll in no more than 20 seconds, so we cannot assign more than 20 seconds of work to a worker. The number of seconds required

TABLE 21

Job	Days Needed	Earliest Time of Completion	Due Date
1	10	10	12
2	3	20	30
3	16	16	20
4	8	12	21

are shown in Table 22, and the precedence relationships for each operation are as follows:

Operation 3 cannot be done until after both operations 1 and 2 are both done (because the worker who is responsible for operation 3 is also responsible for operations 1 and 2).

Operation 4 cannot be done until after operation 1 is done (unless the same worker does both operations).

Operation 5 cannot be done until after (or including same worker doing operation 3) operation 3 is done (unless the same worker does both operations).

Operation 6 cannot be done until after operation 4 is done (unless the same worker does both operations).

Operation 7 cannot be done until after both operations 5 and 6 are both done (unless the same worker does all three operations).

Operation 8 must be done by the same worker or before operation 5 is done.

Operation 9 must be done by the same worker or before operation 6 is done.

Operation 10 must be done by the same worker or before operation 7 is done.

What is the smallest number of workers we need? Which operations should be assigned to each worker?

TABLE 22

Operation	Time (Seconds)
1	12
2	9
3	5
4	4
5	8
6	7
7	5
8	9
9	8
10	11

Neural Networks

16.1 Introduction to Neural Networks

For the most part, neural networks are useful in situations in which multiple regression is used. Recall that in multiple regression we use independent variables in an attempt to predict a dependent variable. In a neural network, the independent variables are called **input cells,** and the dependent variable is called an **output cell** (more than one output is okay if you wish).

As in regression, we have a certain number of observations (say, N). Each observation contains a value for each input cell (or independent variable) and the output cell (or dependent variable).

As in regression, the goal of the neural network is to make accurate predictions for the output cell or dependent variable. As we will see, the use of neural networks is increasing rapidly because neural networks are great at finding patterns. In regression, you will only find a pattern if you know what to look for. For example, if $y = \ln x$ and you simply use x as an independent variable, then you will not predict y very well. A neural network does not need to be told the nature of the relationship between the independent variables and the dependent variable. If a relationship or pattern exists and you give the neural network enough data, then it will find the pattern on its own by "learning" the pattern from the data. A major advantage of neural networks over regression is the fact that you can use neural networks without making any statistical assumptions about your data. For example, unlike regression, you need not assume that your errors are independent and normally distributed.

What Is a Neural Network?

In our work, we will explore the most widely used neural network, the **back propagation network.** See Freeman and Skapura (1992), Hertz Krogh and Palmer (1991), and Gallant (1993) for a discussion of other types of neural networks. Figures 1 and 2 are illustrations of back propagation networks.

Each square in both figures is a **cell** of the network. The cells in the first column of each network are called **input cells.** The first column of cells in the network is called the **input layer.** Cell 0 may be viewed as a special input cell that is analogous to the constant term in multiple regression. Each other input cell in the input layer corresponds to an independent variable. The cell in the last column of the network is the **output cell,** and the last layer is the **output layer** of the network. Each cell in the output layer represents a dependent variable we want to predict. All other layers of the network are called **hidden layers.** Thus, Figure 1 has four input variables and one hidden layer. Figure 2 has two input variables and two hidden layers. With the exception of cell 0, each cell in the network is connected by an arc to each cell in the next network layer. An arc connects cell 0 with

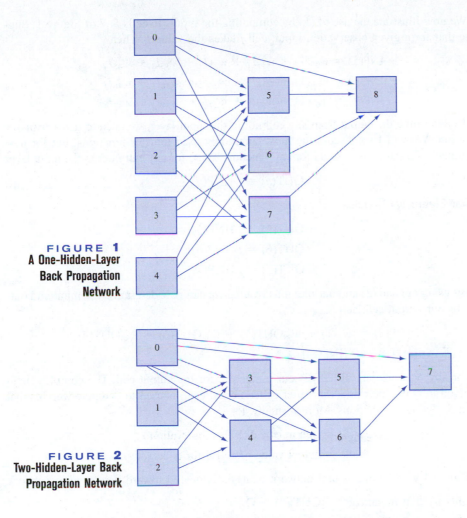

FIGURE 1
A One-Hidden-Layer Back Propagation Network

FIGURE 2
Two-Hidden-Layer Back Propagation Network

each cell in a hidden or output layer. The arc in the network joining cells i and j has an associated weight w_{ij}.

Input and Output Values

For any observation, each cell in the network has an associated input value and output value. For any observation, each input cell (other than cell 0) has an input value that is equal to the value of input i for that observation; the output cell has an output value that is equal to the value of the input for that observation. For example, if the first input equals 3, then the input and output from node 1 are both 3. For cell 0, the input and output values always equal 1. Let

$$INP(j) = \text{input to cell } j$$

and

$$OUT(j) = \text{output from cell } j.$$

For now, we suppress the dependence of $INP(j)$ and $OUT(j)$ on the particular observation. For any cell not in the input layer

$$INP(j) = \Sigma \; w_{ij}(\text{output from cell } i) \qquad \text{(1)}$$
$$\text{all } i$$

We now illustrate the use of (1) by computing the inputs to cells 5–7 of Figure 1. Suppose that for a given observation input, cell j takes the value I_j. Then

$$INP(5) = w_{05}(1) + w_{15}I_1 + w_{25}I_2 + w_{35}I_3 + w_{45}I_4$$
$$INP(6) = w_{06}(1) + w_{16}I_1 + w_{26}I_2 + w_{36}I_3 + w_{46}I_4$$
$$INP(7) = w_{07}(1) + w_{17}I_1 + w_{27}I_2 + w_{37}I_3 + w_{47}I_4$$

To determine the output from any cell not in the input layer, we need to use a **transfer function.** We will later discuss the most commonly used transfer functions, but for now the transfer function f may stand for any function. Then, for any cell j not in the input layer

$$OUT(j) = f[INP(j)] \tag{2}$$

For Figure 1, (2) yields

$$OUT(5) = f[INP(5)]$$
$$OUT(6) = f[INP(6)]$$
$$OUT(7) = f[INP(7)]$$

By using (1) and (2) and our previous results, we can now determine the input and output for our output cell, cell 8.

$$INP(8) = w_{08}(1) + w_{58}OUT(5) + w_{68}OUT(6) + w_{78}OUT(7).$$
$$OUT(8) = f[INP(8)]$$

For any observation, OUT(8) is our prediction for the output cell. The complex math involved in neural networks (hereafter called neural nets) is used to compute weights that produce "good" predictions. More formally, let

$$OUT_j(8) = \text{output of cell 8 for observation } j$$
$$O_j = \text{actual value of cell 8 for observation } j$$

The real work in the neural network analysis is geared toward determining network weights w_{ij} that minimize $\sum_{j=1}^{j=N} [OUT_j(8) - O_j]^2$.

The weights for a neural network are often computed by using a version of the steepest ascent algorithm (see Section 12.7). In theory, the ordinary Solver could be used to solve this problem, but the ordinary Excel Solver sometimes does poorly on these models. In such cases, the Evolutionary Solver should be used.

Scaling Data

For the computations in a neural net to be tractable, it is desirable to scale all data (inputs and outputs) so that all inputs lie in either of the following two intervals:

$$\text{Interval 1: } [0, 1] \qquad \text{Interval 2: } [-1, +1]$$

If you want to scale your data to lie on [0, 1], then any value x for input i should be transformed into

$$\frac{x - (\text{Smallest value of input } i)}{\text{Range for input } i}$$

and any value x for the output should be transformed into

$$\frac{x - (\text{Smallest value of output})}{\text{Range for output}}$$

(*Note:* Range = Largest value for input *i* − Smallest value for input *i*.)

If you want to scale your data to lie on [−1 +1], then any value *x* for input *i* should be transformed into

$$\frac{2[x - (\text{Largest value for input } i)]}{\text{Range for input } i} + 1$$

If you want to scale your data to lie on [−1 +1], then any value *x* for the output should be transformed into

$$\frac{2[x - (\text{Largest value for output})]}{\text{Range for output}} + 1$$

The Sigmoid Transfer Function

Vast experience with neural networks indicates that the **sigmoid transfer function** usually yields the best predictions. If your data has been scaled to lie on [−1 1], then the relevant sigmoid function is given by

$$f(\text{input}) = \frac{2}{1 + e^{-\text{input}}} - 1$$

Note that for this function an input near −∞ yields an output near −1; an input near +∞ yields an output near +1.

If your data has been scaled to lie on [0 1], then the relevant sigmoid function is given by

$$f(Input) = \frac{e^{Input} - e^{-Input}}{e^{Input} + e^{-Input}}$$

Note that for this function an input near − 0 yields an output near 0; an input near +∞ yields an output near 1.

REMARKS **1** The sigmoid function is often called the squashing function because it "squashes" values on the interval [−∞, +∞] to the unit interval [0 1].
2 The slope of the sigmoid function for the [0 1] interval is given by

$$f'(\text{input}) = f(\text{input})[1 - f(\text{input})]$$

This implies that the sigmoid function is very steep for intermediate values of the input and very flat for extreme input values.

Testing and Validation

When we fit a regression to data, we often use 80% to 90% of the data to fit a regression equation and the remaining data to "validate" the equation. The same technique is used in neural nets. We begin by designating 80% to 90% of our data as the **training** or **learning** data set. Then we "fit" a neural net to this data. Suppose cell 8 is the output cell. Let $O_j(8)$ = the actual output for observation *j* and $OUT_j(8)$ be the value of the output cell for observation *j*. Let AVGOT(8) = average value of the output for the training data. Analogous to regression, define

$$SST^\dagger(\text{Train}) = \Sigma\ [O_j(8) - \text{AVGOT}(8)]^2$$

training data

---

†Sum of Squares Total for training data

and

$$\text{SSR(Train)} = \Sigma \; [\text{OUT}_j(8) - \text{AVGOT}(8)]^2$$
training data

and define $R^2(\text{Train}) = \text{SSR(Train)}/\text{SST(Train)}$.

If the network is to be useful for forecasting, then the R^2 computed from the test portion of the data should be close to $R^2(\text{Train})$. Comparing the R^2 of the testing and training sets prevents us from overfitting the data. In this chapter, our examples all contain limited data, so we will not worry about dividing the data into a testing and validation set. In general, however, the data held back for validation should be randomly chosen from the entire data set. For example, if we have 100 data points and want to reserve 20 data points for validation, then we should randomly "shuffle" the 100 data points and reserve, say, the last 20 data points for validation.

Continuous and Binary Data

If your dependent variable assumes only two values (say, 0 and 1), then we say we have **binary** data. In this case, the usual procedure is to try and train the network until as many outputs as possible are less than 0.1 and more than 0.9. Then those observations with predictions less than .1 are classified as zero, and those observations with predictions larger than .9 are classified as 1. If we do not have binary data, then we say that the data are **continuous.**

16.2 Examples of the Use of Neural Networks

In this section, we briefly describe some actual applications of neural networks.

EXAMPLE 1 **Neural Networks and Efficient Markets**

The efficient market hypothesis of financial markets states that the "past history" of a stock's returns yields no information about its future returns of the stock. White (1988) examines returns on IBM to see if the market is efficient. He begins by running a multiple regression in which the dependent variable is the next day's return on IBM stock and the five independent variables are the returns on the IBM stock during each of the last five days. This regression yielded $R^2 = .0079$, which is consistent with the efficient market hypothesis. White then ran a neural network (containing one hidden layer) with the output cell corresponding to the next day's return on IBM and five input cells corresponding to the last five days' return on IBM. This neural network yielded $R^2 = .179$. This implies that the past five days of IBM returns do contain information that can be used to make predictions about tomorrow's return on IBM.

According to the October 9, 1993, issue of *Economist,* Fidelity manages $2.6 billion in assets using neural nets. One of the neural net funds has beat the S&P 500 index by 2% to 7% a quarter for more than three years.

EXAMPLE 2 **Neural Networks and Car Driving**

Researchers at Carnegie-Mellon University have developed ALVINN,[†] a neural network that can drive a car. It can tell if cars are nearby and then slow down as needed. Within

[†]Automated Land Vehicle in a Neural Network—not the chipmunk!

10 years, a neural network may be driving your car. Can you guess the inputs and outputs for this situation? (See Problem 1.)

| EXAMPLE 3 | Neural Networks and Bankruptcy Prediction |

In finance and accounting, it is important to accurately predict whether a company will go bankrupt during the next year. Altman (1968) developed a method (Altman's Z-statistic) to predict such an event based on the firm's financial ratios. This method uses a version of regression called **discriminant analysis.** Neural networks using financial ratios as input cells have outperformed Altman's Z.

| EXAMPLE 4 | Neural Networks and Elevator Control |

The September 22, 1993, issue of *The New York Times* reported that Otis Elevator uses neural networks to direct elevators. For example, if elevator 1 is on floor 10 and going up, elevator 2 is on floor 6 going down, and elevator 3 is on floor 2 and going up, which elevator should answer a call to go down from floor 7?

| EXAMPLE 5 | Neural Networks and Credit Card Approval |

Many banks (Mellon and Chase are two examples) and credit card companies use neural networks to predict (on the basis of past usage patterns) whether a credit card transaction should be disallowed. AVCO Financial used a neural net to determine whether or not to lend people money. The company increased its loan volume by 25% and decreased its default rate by 20%.

| EXAMPLE 6 | Neural Networks and Handwriting Recognition |

"Pen" computers and personal digital assistants often use neural nets to read a user's handwriting. The "inputs" to the network are a binary representation of what the user has written. For example, let 1 = point with writing and 0 = point without writing. An input to the network might look like

$$0001111$$
$$0001000$$
$$0001110$$
$$0000010$$
$$0001110$$

The neural net must decide to classify this input as an *s*, a 5, or something else.

Lecun et al. (1991) tried to have a neural net "read" handwritten zip code digits. For training, 7,291 digits were used; 2,007 were used for testing. Running the neural net took three days on a Sun workstation. The net correctly classified 99.86% of the training data and 95.0% of the test data.

16.3 Why Neural Nets Can Beat Regression: The XOR Example

The classical XOR data set can be used to obtain a better understanding of how neural networks work and why they can pick up patterns that regression often misses. The XOR data set also illustrates the usefulness of a hidden layer. The XOR data set contains two inputs, one output, and four observations. The data set is given in Table 1.

We see that the output equals 1 if either input (but not both) are equal to 1.

If we try to use regression to predict the output from the two inputs, then we obtain the equation $\hat{y} = .5$. This equation yields an $R^2 = 0$, which means that linear multiple regression yields poor predictions indeed.

Now let's use the neural net of Figure 3 to predict the output. We assume that for some θ the transfer function for the output cell is defined by

$$f(x) = 1 \quad \text{if } x \geq \theta \quad \text{and} \quad f(x) = 0 \quad \text{if } x < \theta.$$

Given this transfer function, it is natural to ask whether there are any values for θ and the w_{ij}'s that will enable the Figure 3 net to make the correct predictions for the data set in Table 1. From Figure 3, we find $\text{INP}(3) = w_{03} + w_{13}(\text{input 1}) + w_{23}(\text{input 2})$.

Also, $\text{OUT}(3) = 1$ if $\text{INP}(3) \geq \theta$ and $\text{OUT}(3) = 0$ if $\text{INP}(3) < \theta$. This implies that Figure 3 will yield correct predictions for each observation if and only if the following four inequalities hold:

$$\text{Observation 1: } w_{03} < \theta \tag{3}$$

$$\text{Observation 2: } w_{03} + w_{13} \geq \theta \tag{4}$$

$$\text{Observation 3: } w_{03} + w_{23} \geq \theta \tag{5}$$

$$\text{Observation 4: } w_{03} + w_{13} + w_{23} < \theta \tag{6}$$

There are no values of w_{03}, w_{13}, w_{23}, and θ that satisfy (3)–(6). To see this, note that together (5) and (6) imply $w_{13} < 0$. Adding $w_{13} < 0$ to (3) implies that $w_{03} + w_{13} < \theta$, which contradicts (4). Thus, we have seen that there is no way the neural net of Figure 3 can correctly predict the output for each observation.

Suppose we add a hidden layer with two nodes to the Figure 3 neural net. This yields the neural net in Figure 4. We define the transfer function f_i at cell i by

$$f_i(x) = 1 \quad \text{if } x \geq \theta_i \quad \text{and} \quad f_i(x) = 0 \quad \text{for } x < \theta_i.$$

If we choose $\theta_3 = .4$, $\theta_4 = 1.2$, $\theta_5 = .5$, $w_{03} = w_{04} = w_{05} = 0$, $w_{13} = w_{14} = w_{23} = w_{24} = 1$, $w_{35} = .6$, $w_{45} = -.2$, then the Figure 4 neural net will yield correct predictions for each observation. We now verify that this is the case.

TABLE 1

Observation	Input 1	Input 2	Output
1	0	0	0
2	1	0	1
3	0	1	1
4	1	1	0

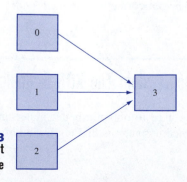

FIGURE 3
Inadequate Neural Net for XOR Example

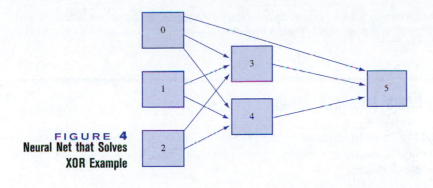

FIGURE 4
Neural Net that Solves XOR Example

Observation 1: input 1 = input 2 = 0, output = 0

Cell 3 input = 1(0) + 1(0) = 0

Cell 3 output = 0 (0 < .4)

Cell 4 input = 1(0) + 1(0) = 0

Cell 4 output = 0 (0 < 1.2)

Cell 5 input = .6(0) − .2(0) = 0

Cell 5 output = 0 (0 < .5)

Observation 2: input 1 = 1, input 2 = 0, output = 1

Node 3 input = 1(1) + 1(0) = 1

Node 3 output = 1 (1 ≥ .4)

Node 4 input = 1(1) + 1(0) = 1

Node 4 output = 0 (1 < 1.2)

Node 5 input = .6(1) − .2(0) = .6

Node 5 output = 1 (.6 ≥ .5)

Observation 3: input 1 = 0 , input 2 = 1, output = 1

Node 3 input = 1(0) + 1(1) = 1

Node 3 output = 1 (1 ≥ .4)

Node 4 input = 1(0) + 1(1) = 1

Node 4 output = 0 (1 < 1.2)

Node 5 input = .6(1) − .2(0) = .6

Node 5 output = 1 (.6 ≥ .5)

Observation 4: input 1 = input 2 = 1, output = 0

Node 3 input = 1(1) + 1(1) = 2

Node 3 output = 1 (2 ≥ .4)

Node 4 input = 1(1) + 1(1) = 2

Node 4 output = 1 (2 ≥ 1.2)

Node 5 input = .6(1) − .2(1) = .4

Node 5 output = 0 (.4 < .5)

We have found that the hidden layer enables us to perfectly fit the XOR data set.

One might think that having more hidden layers will lead to much more accurate predictions. Vast experience, however, shows that more than one hidden layer is rarely a significant improvement over a single hidden layer.

16.4 Estimating Neural Nets with Predict

The Excel add-in Predict (marketed by NeuralWare of Pittsburgh, Pennsylvania)[†] makes running a neural net as easy as running a regression with Excel. The following examples illustrate the use of Predict.

EXAMPLE 7 | The Sibling–Acquaintance Problem

Sibling.xls

Six people live in Hooterville. Persons 1–3 are the Hatfield siblings, and persons 4–6 are the McCoy siblings. Two people who are not siblings are acquaintances. Can a neural net learn to correctly classify pairs of people as siblings or acquaintances? See file Sibling.xls.

Solution

To construct a neural net, we let the two inputs be any two people. Then the output will equal 1 if the two people are siblings and the output will equal zero if the two people are acquaintances. For example, if the inputs are 4 and 5, then the output is 1; if the inputs are 3 and 5, then the output is zero. In Figure 5, you see the results of a regression involving 15 observations (the observations correspond to all 15 ways of choosing two people out of six). The data in columns B and C are the independent variables, and the data in column D is the dependent variable. We obtained an R^2 of .43.

Next we used Predict to fit a neural net to the data. We proceeded as follows.

Step 1 Select Predict New and choose to save the network in file C:\Predict\sibling.npr. Any file saved to the network needs a ".npr" suffix. It's a good idea to name the file where the network is stored with the same prefix as the spreadsheet.

Step 2 Select B3:C3 as the Input range for the first data point.

Step 3 Select B4:C4 as the Input range for the second data point (see Figure 6).

Step 4 Select D3 as the output for your first data point (see Figure 7).

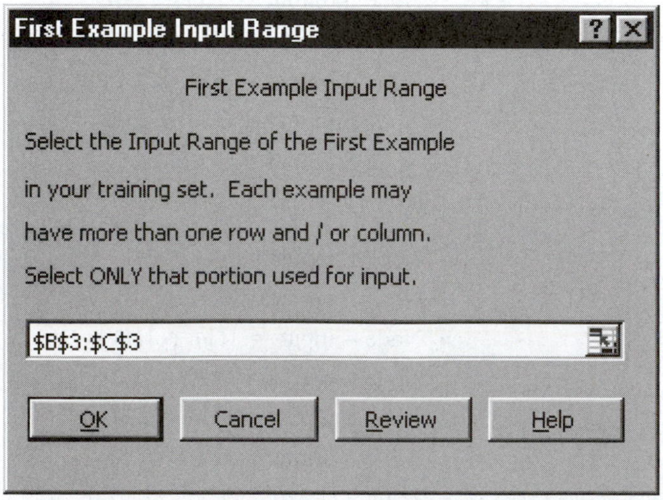

FIGURE 5

[†]NeuralWare can be contacted on the Internet at <http://www.neuralware.com/> or by telephone at (412) 278-6280.

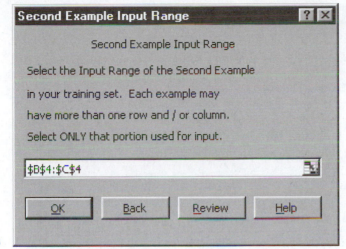

FIGURE 6

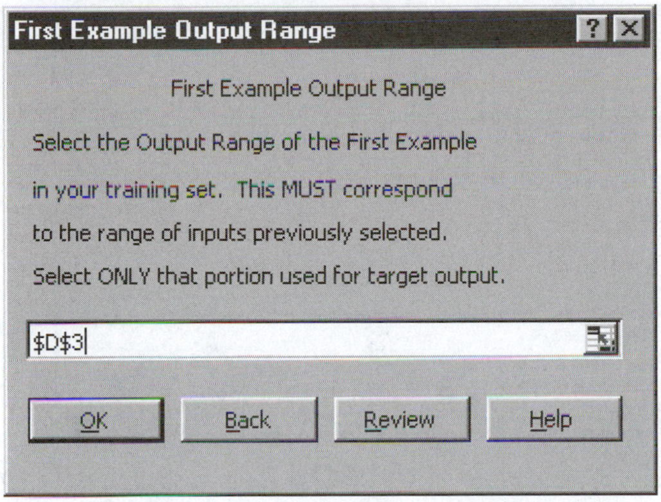

FIGURE 7

Step 5 Now select all our input data (see Figure 8).

Step 6 Cell B2 contains the label for our first input field (X1) (see Figure 9).

Step 7 Determine and then select the type of problem: prediction, classification, or ranking.

- A classification problem involves classifying each observation into one of several categories. For example, the problem of classifying a prospective borrower as a good, bad, or medium credit risk would be a classification problem. Our current problem involves classifying each group of two people as siblings or acquaintances, so we chose a classification problem (see Figure 10).

- A prediction problem involves predicting a numerical value that can take on any value in a range of continuous values.

- A ranking problem should be used if you are interested in ranking the outputs of the model rather than their actual values. For example, in a 10-horse race, you might want to rank the horses' predicted performances from 1st to 10th.

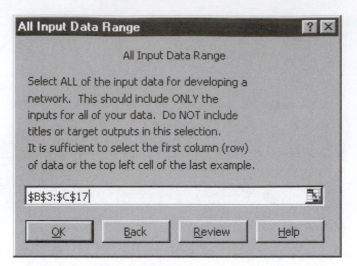

FIGURE 8

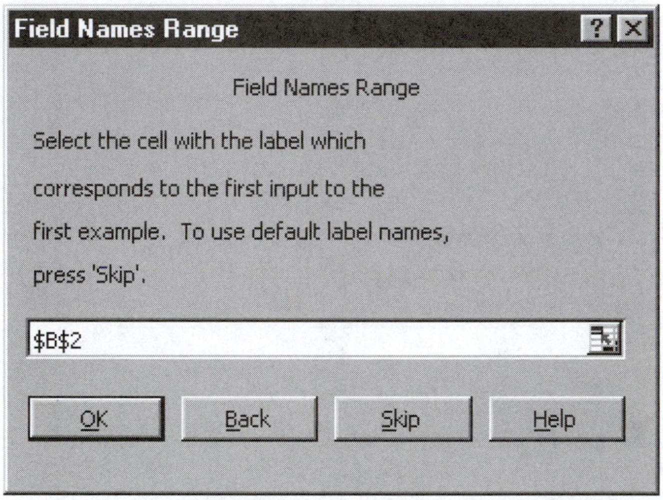

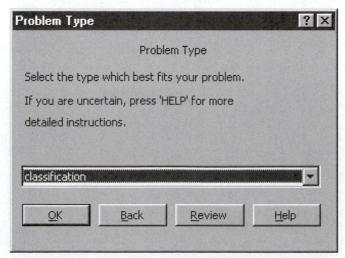

FIGURE 9

Problem Type

Problem Type

Select the type which best fits your problem.

If you are uncertain, press 'HELP' for more

detailed instructions.

classification ▾

| OK | Back | Review | Help |

FIGURE 10

Step 8   Classify the noisiness of the data from Clean to Noisy.

- Use clean data if you are modeling a mathematical function free of noise (see Figure 11).
- Use moderately noisy data for problems with a strong underlying physical model or a problem within the range of human behavior.
- Use noisy data for noisy physical data or reasonably consistent behavioral data.
- Use very noisy data for inconsistent behavioral data such as that used for a stock market prediction.

Step 9   Next we provide directions to Predict on how to transform the data. We will always choose moderate data transformation (see Figure 12). This is along the lines of the scaling discussed earlier in Section 16.1.

Step 10   We now choose a network search level from superficial to exhaustive. This governs how many different network configurations (number of hidden layers, number of nodes in each hidden layer, etc.) that Predict will consider. Clearly, exhaustive will yield better predictions, but it may take much longer than Superficial (see Figure 13).

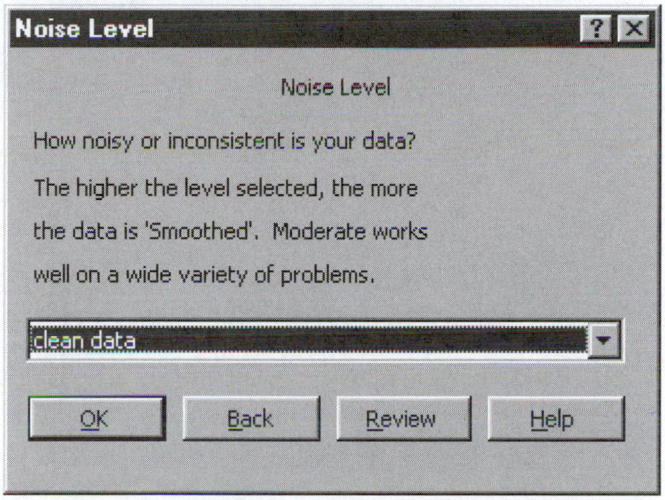

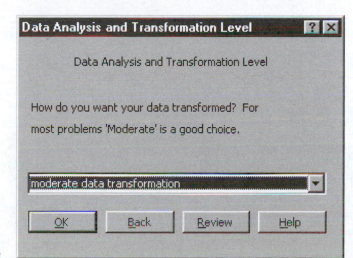

FIGURE 11

FIGURE 12

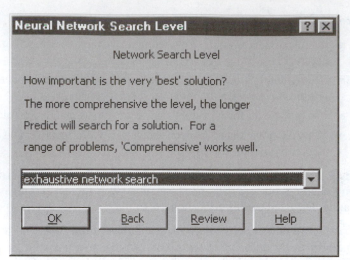

FIGURE 13

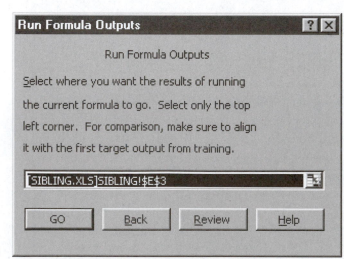

FIGURE 14

FIGURE 15

	A	B	C	D	E	F
1	**Sibling**	X1	X2	OUTPUT	Prediction	Regression
2	**Example**	1	2	1	1	0.74285714
3		1	3	1	1	0.48571429
4		1	4	0	0	0.22857143
5		1	5	0	0	-0.0285714
6		1	6	0	0	-0.2857143
7		2	3	1	1	0.74285714
8		2	4	0	0	0.48571429
9		2	5	0	0	0.22857143
10		2	6	0	0	-0.0285714
11		3	4	0	0	0.74285714
12		3	5	0	0	0.48571429
13		3	6	0	0	0.22857143
14		4	5	1	1	0.74285714
15		4	6	1	1	0.48571429
16		5	6	1	1	0.74285714
17						
18		SUMMARY OUTPUT				
19						
20		*Regression Statistics*				
21		Multiple R	0.65465367			
22		R Square	0.42857143			
23		Adjusted R S	0.33333333			
24		Standard Err	0.41403934			
25		Observations	15			
26						
27		ANOVA				
28			*df*	*SS*	*MS*	*F*
29		Regression	2	1.54285714	0.771428571	4.5
30		Residual	12	2.05714286	0.171428571	
31		Total	14	3.6		
32						
33			*Coefficients*	*Standard Error*	*t Stat*	*P-value*
34		Intercept	1	0.41403934	2.415229458	0.03260014
35		X Variable 1	0.25714286	0.09897433	2.598076211	0.02330841
36		X Variable 2	-0.2571429	0.09897433	-2.59807621	0.02330841
37						
38						
39						

FIGURE 16
Sibling Data

Step 11   We now select Start to begin Predict's search for the neural network that best fits the data.

Step 12   Next we save the network and choose Run. Then we choose all our data as the range of inputs for which we want forecasts (see Figure 14). Then we select the upper left-hand corner of the range for which we want forecasts (see Figure 15). The neural net classified all observations correctly. By the way, the mean absolute deviation (MAD) for the regression predictions is .50, while for the neural net the MAD is, of course, zero. Note the regression model only correctly classified 12/15 = 80% of all observations. (See Figure 16.)

EXAMPLE 8   X1 + X2 > 7

In Figure 17, we give Predict output for a data set that consists of two inputs and a single output. The output was generated by the following rule: If the two inputs add to more than 7, then the output is 1; otherwise, the output is zero. From Figure 17, we see that all observations are classified correctly. The MAD for regression is .24, whereas the MAD for the neural net is 0. See file GT7.xls and Figure 18.

GT7.xls

FIGURE 17

	A	B	C	D	E	F	G	H
1						MAD		MAD
2						0		0.239105111
3		X1	X2	OUTPUT	NNET Prediction	Abs Error Neural Net	Regression Prediction	Abs Error Regression
4	If X1+ X2 >7	1	2	0	0	0	-0.248071741	0.248071741
5	Output=1	1	3	0	0	0	-0.065506613	0.065506613
6	otherwise	1	4	0	0	0	0.117058514	0.117058514
7	Output=0	1	5	0	0	0	0.299623641	0.299623641
8		1	6	0	0	0	0.482188768	0.482188768
9		2	3	0	0	0	0.082752532	0.082752532
10		2	4	0	0	0	0.265317659	0.265317659
11		2	5	0	0	0	0.447882787	0.447882787
12		2	6	1	1	0	0.630447914	0.369552086
13		3	4	0	0	0	0.413576805	0.413576805
14		3	5	1	1	0	0.596141932	0.403858068
15		3	6	1	1	0	0.778707059	0.221292941
16		4	5	1	1	0	0.744401078	0.255598922
17		4	6	1	1	0	0.926966205	0.073033795
18		5	6	1	1	0	1.075225351	0.075225351
19		5	8	1	1	0	1.440355605	0.440355605
20		3	7	1	1	0	0.961272187	0.038727813
21		4	5	1	1	0	0.744401078	0.255598922
22		2	9	1	1	0	1.178143295	0.178143295
23		8	3	1	1	0	0.972307406	0.027692594
24		2	4	0	0	0	0.265317659	0.265317659
25		3	5	1	1	0	0.596141932	0.403858068
26		4	7	1	1	0	1.109531332	0.109531332
27		5	6	1	1	0	1.075225351	0.075225351
28		4	3	0	0	0	0.379270824	0.379270824
29		1	8	1	1	0	0.847319022	0.152680978
30		2	7	1	1	0	0.813013041	0.186986959
31		3	6	1	1	0	0.778707059	0.221292941
32		1	2	0	0	0	-0.248071741	0.248071741
33		2	4	0	0	0	0.265317659	0.265317659
34		3	5	1	1	0	0.596141932	0.403858068
35		2	6	1	1	0	0.630447914	0.369552086
36		3	2	0	0	0	0.048446551	0.048446551

FIGURE **18**

	B	C	D	E	F
42	*Regression Statistics*				
43	Multiple R	0.830191			
44	R Square	0.689217			
45	Adjusted R Sc	0.668498			
46	Standard Erro	0.28897			
47	Observations	33			
48					
49	ANOVA				
50		*df*	*SS*	*MS*	*F*
51	Regression	2	5.555503	2.777751	33.26511
52	Residual	30	2.505103	0.083503	
53	Total	32	8.060606		
54					
55		*Coefficients*	*Standard Error*	*t Stat*	*P-value*
56	Intercept	-0.761461	0.172116	-4.424127	0.000118
57	X1	0.148259	0.03297	4.496788	9.61E-05
58	X2	0.182565	0.028759	6.348152	5.29E-07

EXAMPLE 9   **Predicting Corporate Bankruptcies**

Given the five financial ratios from the previous year listed in Figure 19, we ran a neural net to see if we could predict whether a company will go bankrupt during the next year (0 = bankrupt; 1 = not bankrupt). (See file Bankrupt.xls.) Observations for which the neural net prediction was incorrect are denoted by an X. Sixty-nine of 74 observations (93%) were correctly classified.

Bankrupt.xls

FIGURE **19**

	A	B	C	D	E	F	G	H
1	BR=0	**Bankruptcy**						
2	NBR=1							
3		WC/TA	RE/TA	EBIT/TA	MVE/TD	S/TA	BR/NBR	Predictions
4	WC=WORKING CAPITAL	0.3922	0.3778	0.1316	1.0911	1.2784	1	1
5		0.0574	0.2783	0.1166	1.3441	0.2216	1	1
6	TA=TOTAL Assets	0.165	0.1192	0.2035	0.813	1.6702	1	1
7		0.3073	0.607	0.204	14.409	0.9844	1	1
8	RE=RETAINED Earnings	0.2574	0.5334	0.165	8.0734	1.3474	1	1
9		0.1415	0.3868	0.0681	0.5755	1.0579	1	1
10	EBIT'=EARNINGS before taxes and	0.3363	0.3312	0.2157	3.0679	2.0899	1	1
11	Interest	0.3378	0.013	0.2366	2.4709	1.223	1	1
12		0.487	0.697	0.2994	5.4383	1.72	1	1
13		0.4455	0.498	0.0952	1.9338	1.7696	1	1
14	MVE=MARKET value of equity	0.4704	0.2772	0.0964	0.4268	1.9317	1	1
15		0.5804	0.3331	0.081	1.1964	1.3572	1	1
16		0.2073	0.3611	0.1472	0.0417	1.1985	1	1
17	S=SALES	0.1801	0.1635	0.0908	0.4094	0.4566	1	1
31		0.4934	0.3416	0.22	0.8144	2.1937	1	1
32		0.1332	0.4077	0.0543	1.4921	1.4826	1	1
33		0.222	0.1797	0.1526	0.3459	1.7237	1	1
34		0.372	0.3446	0.2124	0.8888	1.9241	1	1

(continued)

FIGURE 19
(Continued)

	A	B	C	D	E	F	G	H
35		0.2776	0.2567	0.1612	0.2968	1.8904	1	1
36		0.1445	0.3808	0.178	1.4796	1.4811	1	1
37		0.3907	0.6482	0.1408	3.0489	1.5255	1	1
38		0.1862	0.1687	0.1298	0.9498	4.9548	1	1
39		0.1663	4291	0.1133	1.1745	1.6831	1	1
40		0.4422	0.1379	0.0104	0.246	1.2492	0	0
41		-0.0643	0.1094	-0.123	0.1725	1.3752	0	0
42		0.2975	-0.3719	0.139	0.9627	2.2774	0	1
43		0.0478	0.0632	-0.0016	0.4744	1.8928	0	0
44		0.0718	0.0422	0.0006	0.2964	2.1331	0	0
45		0.2689	0.1729	0.0287	0.1224	0.9277	0	0
46		-0.3107	-0.878	-0.2969	0.1945	1.0493	0	0
47		0.0766	0.0734	0.0076	0.1681	1.0789	0	0
48		0.3899	0.0809	0.0447	0.2186	0.9273	0	0
49		0.0664	-0.1266	-0.1556	0.1471	3.6192	0	0
50		0.0147	-0.1443	-0.0498	0.1431	6.5145	0	0
51		0.1321	0.0686	0.0008	0.3544	2.3224	0	0
52		0.2039	-0.0476	0.1263	0.8965	1.0457	0	1
53		0.0549	0.0592	0.2279	0.0913	1.6016	0	0
54		-0.5359	-0.3487	-0.0322	0.4595	0.9191	0	0
55		-0.0801	-0.0835	0.0036	0.0481	0.773	0	0
56		0.3294	0.0171	0.0371	0.2877	3.1382	0	0
57		0.5056	-0.1951	0.2026	0.538	1.9514	0	1
58		0.1759	0.1343	0.0946	0.1955	1.9218	0	0
59		-0.2772	0.1619	-0.0302	0.1225	2.325	0	0
60		0.2551	-0.3442	-0.1108	1.2212	2.2815	0	0
61		-0.1294	0.0085	-0.0971	0.1764	1.3113	0	0
62		0.2027	-0.1169	-0.0261	0.5965	0.7892	0	0
63		-0.0901	-0.271	0.0014	0.1473	2.5064	0	0
64		-0.3757	-1.6945	-0.4504	0.2197	2.2685	0	0
65		0.3424	-0.1104	0.0541	1.5052	1.0416	0	0
66		0.0234	-0.0246	0.032	0.6406	1.1091	0	0
67		0.3579	0.1515	0.0812	0.1991	1.4582	0	1
68		-0.0888	-0.0371	0.0197	0.1931	1.3767	0	0
69		0.2845	0.2038	0.0171	0.3357	1.3258	0	0
70		0.0011	-0.0631	-0.2225	0.3891	1.768	0	0
71		0.1209	0.2823	-0.0113	0.3157	2.3219	0	0
72		0.2525	-0.173	-0.4861	0.1656	1.4441	0	0
73		0.3181	-0.1093	-0.0857	0.3755	1.9789	0	0
74		0.1254	0.1956	0.0079	0.2073	1.489	0	0
75		0.1777	0.0891	0.0695	0.1924	1.6871	0	0
76		0.2409	0.166	0.0746	0.2516	1.8524	0	0
77		0.2496	0.126	-0.2474	0.166	3.095	0	0

We also ran a regression on this data to predict the BR/NBR column from the five financial ratios. Then we obtained a prediction about bankruptcy by rounding our regression prediction to 1 or zero. This method yielded only 61 (82%) correct predictions. Again, we see that the neural net outperforms multiple regression.

16.5 Using Genetic Algorithms to Optimize a Neural Network

Wet etching is performed on back of silicon wafers used in semiconductors. The etching solution is made of acids A, B, and C. What percentage makeup of solution will maximize etching rate? Montgomery and Myers 1997 fit the following model to the data in Figure 20:

$$(1) \quad \text{Rate} = 550.2*A + 344.7*B + 268.3*C + 689.5*A*B - 9*A*C + 58.1*B*C$$
$$+ 9243.3*A*B*C.$$

	B	C	D	E	F	G
		Acid A	Acid B	Acid C		
1						
2	Point	A	B	C	Etch Rate	Prediction
3	1	1	0	0	540	549.9783
4	2	1	0	0	560	549.9783
5	3	0	1	0	330	339.9763
6	4	0	1	0	350	339.9763
7	5	0	0	1	295	277.5071
8	6	0	0	1	260	277.5071
9	7	0.5	0.5	0	610	610.0607
10	8	0	0.5	0.5	330	330.0036
11	9	0.5	0	0.5	425	425.0448
12	10	0.666667	0.166667	0.166667	710	709.9556
13	11	0.166667	0.666667	0.166667	640	640.0583
14	12	0.166667	0.166667	0.666667	460	459.9691
15	13	0.333333	0.333333	0.333333	800	824.9923
16	14	0.333333	0.333333	0.333333	850	824.9923
17		0.435468	0.262648	0.301884		884.3339
18						
19		Total				
20		1				

FIGURE 20

	I	J	K	L
2	A	B	C	total
3	0.406834523	0.343794	0.249371	1
4	Cubic response	832.1552		

FIGURE 21

We used the Solver to maximize etching rate (as defined by the equation) and obtained the following results. (See file etching.xls and Figure 21.)

We input (1) in cell J4 with the formula

$$=550.2*A+344.7*B+268.3*C\_+689.5*A*B-9*A*C\_+58.1*B*C\_+9243.3*A*B*C\_.$$

Our Solver window is shown in Figure 22.

Next we used the Predict neural network add-in to fit the data. Our predictions are shown in Figure 20. They were obtained with the =PREDICTRUN function.

Evolutionary Solver found the last observation in Row 16. Note that in G17 we find the etching rate predicted from the neural network for the row 17 inputs (percentages of A, B, and C) with the formula

$$=PREDICTRUN(``c:\predict\etching.npr,"C17:E17).$$

The syntax of the PREDICTRUN function is as follows:

$$=PREDICTRUN(``path \ where \ network \ is \ stored," \ inputs \ used \ to \ make \ forecast).$$

Thus, in our example the network stored in file c:\predict\etching.npr is used to make a forecast based on the inputs in C17:E16. Of course, our neural network was saved in file Predict\etching.npr.

To find mixture that maximizes etching rate we used the Evolutionary Solver window shown in Figure 23.

Note we just maximize the etching rate subject to the constraint that the fraction used of each acid is between 0 and 1 and that total percentage used must equal 1. We obtain a

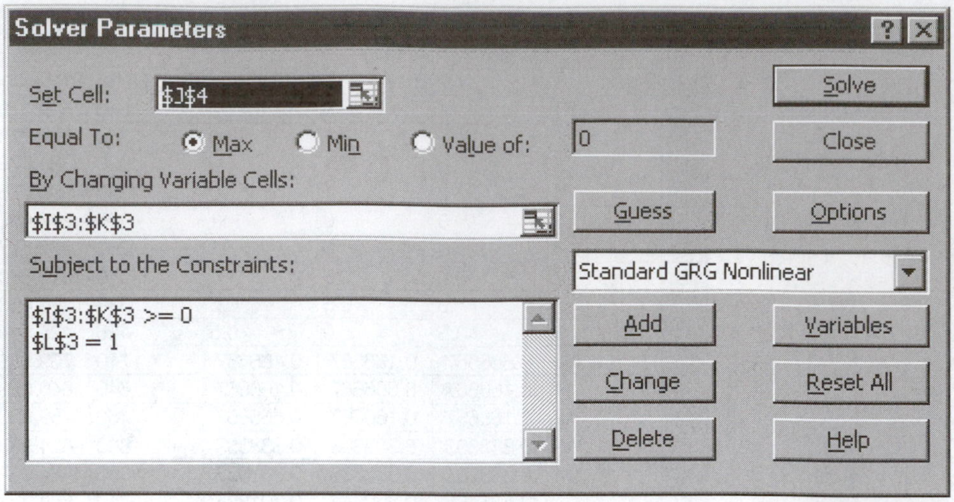

FIGURE 22
Solver Window
for Maximizing
Etching Yield

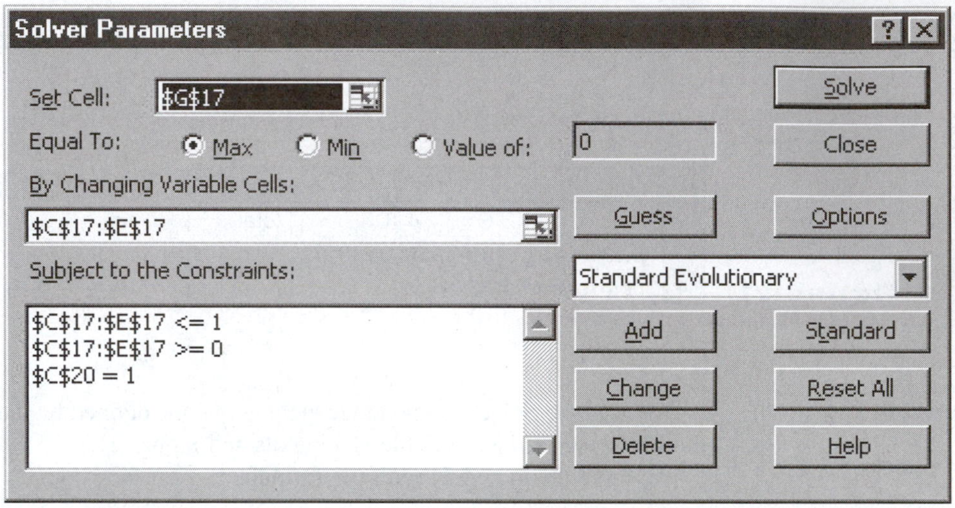

FIGURE 23
Solver Window for
Neural Net Etching

maximum rate of 884 with 43.5% acid A, 26.3% acid B, and 30.2% acid C. Note how close this is to the solution found by Montgomery.

(*Note:* We found this answer without having any initial idea about the relationship between the mixture and the etching rate. The neural net fit the relationship, and Evolutionary Solver—given this relationship—optimized the makeup of the etching solution.)

Also plugging this situation into the etch rate function found by Montgomery yields a solution within 2.5% of Montgomery's solution. It is hard to know whose solution is better, but with little work we have found a good solution to this problem

16.6 Using Genetic Algorithms to Determine Weights for a Back Propagation Network

Neuralnet.xls

Let's suppose that we want to use price (in dollars) and advertising (in hundreds of dollars) to predict the monthly sales of a product. We are given the data in file Neuralnet.xls and Figure 24. We have hidden many rows. Sales are in thousands of units and may be a

	A	B	C	D
1	min	1	1	0.11785113
2	max	99	20	1.657227009
3	range	98	19	1.539375878
4				
5				
6				
7				
8				
9				
10		Price	Advertising	Actual Sales
11		44	15	0.418925471
12		48	13	0.384320909
13		3	12	0.937046461
14		38	8	0.342088174
15		59	3	0.19958357
16		72	1	0.11785113
17		45	13	0.39266987
18		91	16	0.317780539
19		60	1	0.127892665
20		1	9	1.204112343
103		54	20	0.43903069
104		84	18	0.346712388
105		4	13	0.879146736
106		85	5	0.20648026
107		30	6	0.32987479
108		99	7	0.218887831

FIGURE 24
Data for Sales Advertising Example

highly nonlinear function of price and advertising. A back propagation neural network can often be used to come up with accurate forecasts. Our back propagation network consists of three types of nodes (see Figure 25):

1 an input node for each independent variable: price and advertising,

2 several hidden nodes (we will use three—H1, H2, and H3), and

3 an output node for our dependent variable (sales).

Our back propagation network also contains two types of arcs:

1 an arc that connects each input node to each hidden node, and

2 an arc that connects each hidden node to the output node.

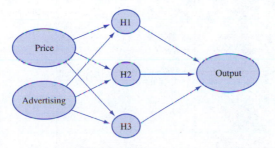

FIGURE 25
Neural Net for Sales Advertising Example

Each arc has a weight that governs how much influence that connection in the network has on our results. For example, $w(P,1)$ will be the weight that governs the connection between the price node and hidden node 1, while $w(2,O)$ will be the weight that governs the connection between hidden node 2 and the output node. Each hidden node and output node also has an associated constant or **bias** term that helps fudge the forecasts to make them more accurate.

Neural networks require variables to be standardized so they range in value between -1 and $+1$. For example, the smallest amount spent on advertising would be assigned a -1 value, and the largest amount spent on advertising would be assigned a $+1$ value. Each node also has an input and output value.

- For each input node, the input and output equal the standardized input value. Thus, for the price node, the input and output value equal the standardized price (call it SP, and call standardized advertising SA). For example, the price is \$44 for our first data point. The lowest price was \$1, and the highest price was \$99, so the standardized price for this observation would be $-1 + (\frac{44 - 1}{99 - 1})*2 = -.122$.

- For each hidden node, the input is the weighted sum of the outputs from all input nodes plus the bias term for the hidden node. For example, for hidden node 1, the input is simply

$$W(P,1)*SP + W(A,1)*SA + Bias(H1),$$

where Bias(H1) is the bias term for the first hidden node.

- For each hidden node, the output associated with the hidden node for an input level x is given by

$$TANH(x) = \frac{e^x - e^{-x}}{e^x + e^{-x}}.$$

This type of transfer function (the squashing function mentioned earlier) transforms inputs into outputs as shown in Figure 26.

Note that the output is fairly insensitive to the input when the input is either very large or very small. In other words, when a threshold of positive or negative information is reached, going beyond that threshold has little effect on the hidden node's output that is passed on to the output node.

- The input and output from the output node are given by the weighted sum of the outputs from the hidden nodes plus the output node's bias term. Thus, the input or output to our output node is given by

$W(1,O)*(H1\ output) + W(2,O)*(H2\ output) + W(3,O)*(H3\ output) + (output\ bias).$

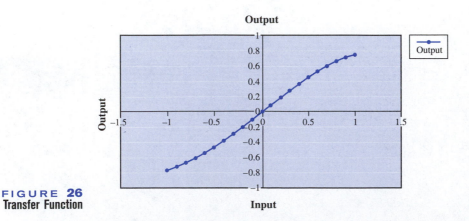

FIGURE 26
Transfer Function

- To make our forecast for sales, we simply take output from the output node and "unstandardize" the output. For example, if the output from the sales node is .029, then we know that our forecast for actual sales should be 1.029/2 of the way between minimum sales (.117) and maximum sales (6.63). Thus, our forecasted sales would be given by

$$.117 + \frac{.029 - (-1)}{2}(6.63 - .117) = 3.47$$

or 3470 units.

We are now ready to show how to use the Evolutionary Solver to solve for the weights and biases that yield the best forecasts (in terms of minimizing squared error).

Solving for Weights and Biases with Evolutionary Solver

Step 1 After computing the min, max, and range for price, advertising, and sales in B1:D3, we compute standardized price, advertising, and sales by copying from E11 to E11:G108 the formula

$$=-1+2*(B11-B\$1)/B\$3.$$

Step 2 We insert trial values for our weights and biases in our spreadsheet (see Figure 27).

Step 3 In H11:H108, we compute the input to hidden node 1 by copying from H11 to H12:H108 the formula

$$=\$F\$4*E11+\$F\$5*F11+\$F\$6.$$

Step 4 In I11:I108, we compute the input to hidden node 2 by copying from I11 to I12:I108 the formula

$$=\$G\$4*E11+\$G\$5*F11+\$G\$6.$$

FIGURE 27

	A	B	C	D	E	F	G	H	I	J	K	L	M
1	min	1	1	0.11785113									
2	max	99	20	1.657227009									
3	range	98	19	1.539375878									
4						H1	H2	H3					
5					Price	0.307434	-9.07784	-0.283503				2.38978	
6					Adv	-0.96406	0.122676	0.636069			H1-O	4.010447	
7					Bias	1.18975	-10.1026	-1.203612			H2-O	4.919287	
8											H3-0	5.451475	
9											Bias		
10		Price	Advertising	Actual Sales	SC Price	SC Adv	SC Sales	H1 Input	H2 Input	H3 Input	H1 Output	H2 Output	H3 Output
11		44	15	0.418925471	-0.12245	0.473684	-0.60884	0.695443	-8.93296	-0.8676	0.601467	-1	-0.700153
12		48	13	0.384320909	-0.04082	0.263158	-0.6538	0.9235	-9.69984	-1.02465	0.727549	-1	-0.771755
13		3	12	0.937046461	-0.95918	0.157895	0.064321	0.742644	-1.37596	-0.83125	0.63074	-0.880044	-0.681146
14		38	8	0.342088174	-0.2449	-0.26316	-0.70866	1.368161	-7.91179	-1.30157	0.878272	-1	-0.862127
15		59	3	0.19958357	0.18367	-0.78947	-0.89381	2.007321	-11.8669	-1.75784	0.964541	-1	-0.942262
16		72	1	0.11785113	0.44898	-1	-1	2.291846	-14.3011	-1.96697	0.979772	-1	-0.961618
17		45	13	0.39266987	-0.10204	0.263158	-0.64295	0.904678	-9.14405	-1.0073	0.718568	-1	-0.764642
18		91	16	0.317780539	0.83673	0.578947	-0.74025	0.888848	-17.6274	-1.07258	0.710824	-1	-0.790431
19		60	1	0.127892665	0.20408	-1	-0.98695	2.216556	-12.0779	-1.89754	0.976524	-1	-0.956026
20		1	9	1.204112343	-1	-0.15789	0.411301	1.034537	-1.04418	-1.02054	0.775721	-0.779534	-0.770087
103		54	20	0.43903069	0.08163	1	-0.58271	0.250782	-10.721	-0.59069	0.245654	-1	-0.530389
104		84	18	0.346712388	0.69388	0.789474	-0.70266	0.641968	-16.3047	-0.89817	0.566238	-1	-0.715405
105		4	13	0.879146736	-0.93878	0.263158	-0.0109	0.647437	-1.54831	-0.77008	0.569942	-0.913507	-0.646976
106		85	5	0.20648026	0.71429	-0.57895	-0.88485	1.967488	-16.6578	-1.77436	0.961657	-1	-0.944086
107		30	6	0.32987479	-0.40816	-0.47368	-0.72453	1.520929	-6.45552	-1.38919	0.908859	-0.999995	-0.882993
108		99	7	0.218887831	1	-0.36842	-0.86873	1.852365	-19.2257	-1.72146	0.951968	-1	-0.938038

Step 5   In J11:J108, we compute the input to hidden node 3 by copying from J11 to J12:J108 the formula

$$=\$H\$4*E11+\$H\$5*F11+\$H\$6.$$

Step 6   In K11:M108, we apply the squashing function to compute the output for each hidden node by copying from K11 to K11:M108 the formula

$$= TANH(H11).$$

Step 7   In N11:N108, we compute the standardized output from the output node by copying from N11 to N12:N108 the formula

$$=\$L\$8+\$L\$5*K11+\$L\$6*L11+\$L\$7*M11.$$

(See Figure 25.)

Step 8   In O11:O108, we unstandardize our standardized output forecast by copying from O11:O108 the formula

$$=\$D\$1+\$D\$3*(0.5*(N11+1)).$$

Step 9   In P11:P108, we compute the squared error for each observation by copying from P11 to P12:P108 the formula

$$=(D11-O11)^2.$$

Step 10   In P9, we compute the sum of our squared errors with the formula

$$= SUM(P11:P108).$$

Step 11   We now use the Premium Solver to solve for the weights and biases that minimize SSE. Our Solver window follows in Figure 28. We constrain each bias and weight to be between -20 and $+20$ (this is a guess). If the GRG Solver makes any weight or bias equal to $+20$ or -20, then we enlarge the range of values until no changing cell matches the upper or lower bound. We also tried the Evolutionary Solver to fine-tune the GRG solution. The Evolutionary Solver did not improve our GRG solution.

Figure 29 contains our results. Note from column P that our squared errors are extremely small. This indicates that our forecasts are accurate!

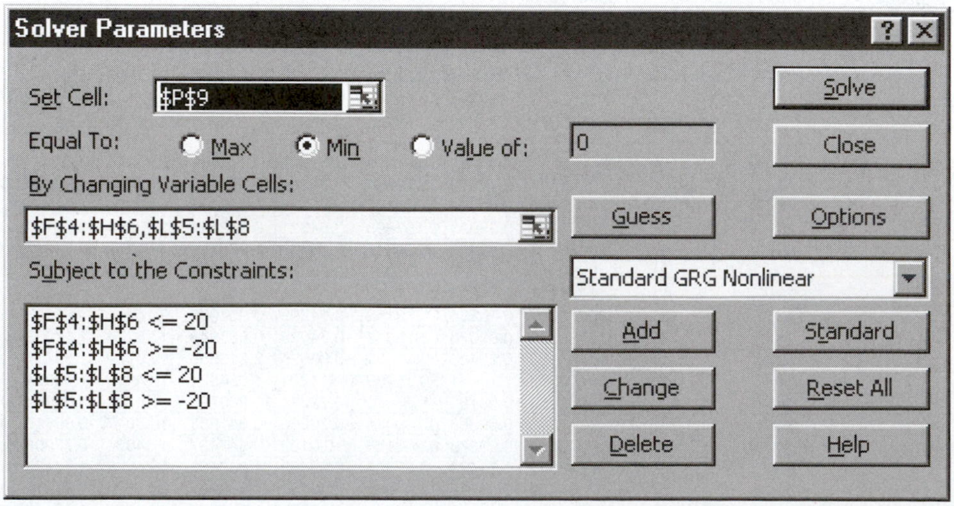

FIGURE 28

	N	O	P
9		SSE	0.055566
10	Scaled Output	Predicted Sales	Sq Err
11	-0.565853199	0.452008687	0.001094
12	-0.616774967	0.412814816	0.000812
13	0.078681051	0.948098925	0.000122
14	-0.701140543	0.34787965	3.35E-05
15	-0.889185868	0.203143431	1.27E-05
16	-0.948005724	0.157870498	0.001602
17	-0.603243183	0.423230067	0.000934
18	-0.748613382	0.311340378	4.15E-05
19	-0.928261665	0.173067261	0.002041
20	0.390723231	1.188274028	0.000251
103	-0.581049183	0.440312521	1.64E-06
104	-0.725070217	0.329461269	0.000298
105	-0.032718154	0.862356301	0.000282
106	-0.905051626	0.190931748	0.000242
107	-0.730673844	0.325148224	2.23E-05
108	-0.898455683	0.196008566	0.000523

FIGURE 29

FIGURE 30

SUMMARY OUTPUT

Regression Statistics	
Multiple R	0.795372012
R Square	0.632616637
Adjusted R Square	0.62488225
Standard Error	0.184687897
Observations	98

ANOVA

	df	SS	MS	F
Regression	2	5.579838	2.789919	81.79274
Residual	95	**3.240414**	0.03411	
Total	97	8.820252		

	Coefficients	Standard Error	t Stat	P-value
Intercept	0.518762932	0.048914	10.60558	8.45E-18
Price	-0.006453425	0.000616	-10.47599	1.59E-17
Advertising	0.021412307	0.00331	6.469907	4.25E-09

RESIDUAL OUTPUT

Observation	Predicted Actual Sales	Residuals	Observation	Predicted Actual Sales	Residuals
1	0.555996848	-0.137071	4	0.444831248	-0.102743
2	0.487358535	-0.103038	5	0.202247796	-0.002664
3	0.756350339	0.180696	6	0.075528662	0.042322

(continued)

FIGURE 30
(Continued)

RESIDUAL OUTPUT

Observation	Predicted Actual Sales	Residuals	Observation	Predicted Actual Sales	Residuals
7	0.50671881	-0.114049	53	0.372951598	-0.023517
8	0.274098195	0.043682	54	0.777465319	-0.132294
9	0.152969758	-0.025077	55	0.284655685	0.039664
10	0.705020268	0.499092	56	0.223362777	0.038826
11	0.42047493	-0.054054	57	0.54984075	-0.136348
12	0.281443599	-0.002905	58	0.567446317	-0.136542
13	0.076986042	0.158722	59	0.117758623	0.133316
14	0.861062514	-0.106889	60	0.438377824	-0.099236
15	0.626687193	-0.172605	61	0.677154537	0.234779
16	0.174382065	-0.005627	62	0.677154537	0.234779
17	0.618479062	-0.16055	63	0.848155664	-0.138408
18	0.41756017	-0.229998	64	0.154427138	0.097745
19	0.482957144	-0.097486	65	0.579758514	-0.148031
20	0.856661122	-0.168451	66	0.189043621	0.05397
21	0.726432575	0.529511	67	0.204002503	0.052666
22	0.756053013	-0.127324	68	0.705020268	0.499092
23	0.433976432	-0.089847	69	0.550435402	-0.134915
24	0.072881976	0.144171	70	0.779517352	-0.16267
25	0.404355994	-0.126915	71	0.628739225	-0.175035
26	0.130962799	0.097974	72	0.340387148	0.001473
27	0.940555643	0.716671	73	0.224820157	0.074791
28	0.427820334	-0.111412	74	0.277042207	0.010132
29	0.594420069	-0.156671	75	0.453634032	-0.123759
30	0.600010767	-0.00012	76	0.786268103	0.054477
31	0.792721528	0.112876	77	0.070829944	0.134446
32	0.476801045	-0.10391	78	0.798877626	-0.13787
33	0.502912071	-0.128605	79	0.891277604	0.370074
34	0.00834773	0.197451	80	0.371494218	-0.042776
35	0.450392694	-0.070954	81	0.264432684	-0.015311
36	0.134769538	0.131561	82	0.598524135	-0.159493
37	0.631088585	-0.172604	83	0.157073824	0.033945
38	0.57859846	-0.029932	84	0.46124751	-0.07776
39	0.4061107	-0.080322	85	0.333665649	-0.175993
40	0.569795676	-0.122223	86	0.137416223	0.092965
41	0.337472388	-0.055389	87	0.674507851	-0.151242
42	0.449827293	-0.242188	88	0.876318722	0.676603
43	0.13771355	0.049107	89	0.505558756	-0.166362
44	0.261785998	0.032208	90	0.637542009	-0.175074
45	0.242723051	0.025359	91	0.255927227	-0.025302
46	0.417262843	-0.133591	92	0.156776497	0.078108
47	0.520517638	-0.149964	93	0.598524135	-0.159493
48	0.238024332	0.053726	94	0.362096781	-0.015384
49	0.550732729	-0.096727	95	0.771309221	0.107838
50	0.479447731	-0.133473	96	0.077283368	0.129197
51	-0.006313826	0.146974	97	0.453634032	-0.123759
52	0.411672146	-0.048161	98	0.029760036	0.189128

REMARKS The hidden layers of the network allow us (through the nonlinearity of the squashing function) to model complex nonlinear relationships. The fact that arcs from each input node lead into each hidden node allows us to model complex interactions between our independent variables. We created our sales via the following formula:

- If price is at least $64, then Sales $= \text{Price}^{-.5}\text{Adv}^{.4}$.
- If price is less than $64, then Sales $= .5*\text{Price}^{-.333}\text{Adv}^{.4}$.

Note that this is a highly nonlinear relationship, but our neural network had no difficulty capturing the dependence of sales on advertising and price.

In the sheet sales we ran a regression to predict sales from advertising and price. From Figure 30, we find that SSE for the linear regression (3.24) was almost sixty times larger than the SSE of the neural network (.055). This shows how poorly regression can perform compared to a neural network when a complex nonlinear relationship is present.

PROBLEMS

1 What would the inputs and outputs be for Example 2?

2 In file Bsnnet.xls, you are given 22 observations to see if a neural net could fit the Black–Scholes option-pricing formula. Determine how well the neural network fit the Black–Scholes model.

Column C	current stock price
Column D	option exercise or striking price
Column E	risk-free interest rate (on annual basis)
Column F	duration of option (in years)
Column G	standard deviation (on an annual basis) of stock
Column H	Black–Scholes price

3 Using data in normalnet.xls, let Predict try to learn the standard normal cumulative function. For $z = -3.8, -3.6, -3.4, \ldots, 0, .2, .4, .6, \ldots, 3.8$, we input the normal cumulative function $F(z)$. For example, $F(1) = .8413$ and $F(0) = .5$. This is because for a standard normal (mean zero and unit variance) random variable Z, $P(Z \le 1) = .8413$ and $P(Z \le 0) = .50$. How well did the neural network "learn" the cumulative normal?

4 How would you use a neural network and genetic algorithms to find the marketing mix (price and advertising) that maximizes profits?

5 File lunch.xls contains percentage of students in each Indiana elementary school who passed a standardized achievement test and the percentage of students in each school who are on free or reduced lunch programs.

a Use Predict to forecast the percentage who pass the standardized achievement test based on the free and reduced lunch percentage.

b In the state of Indiana, schools are evaluated exclusively on the percentage of students who pass standardized tests. What implications do your results have for this method of school evaluation?

6 Use the technique of Section 16.6 with one 3 node hidden layer to solve Problem 3.

7 Use the technique of Section 16.6 with one 3 node hidden layer to solve Problem 5.

REFERENCES

The following three books are good references on neural nets:

Freeman, J., and Skapura, D. *Neural Networks: Algorithms, Applications and Programming Techniques.* Reading, MA: Addison Wesley, 1992.
Gallant, S. *Neural Network Learning and Expert Systems.* Cambridge: MIT Press, 1993.
Hertz, J., Krogh, A., and Palmer, R. *Introduction to the Theory of Neural Computing.* Reading, MA: Addison-Wesley, 1991.

The following articles and texts discuss interesting applications of neural nets.

"Computers that Think," *Fortune* (Sept. 6, 1993), pp. 96–102.
"Frontiers of Finance," *Economist* (Oct. 9, 1993), pp. 19–21.

Do, Q., and Grudnitski, G. "A Neural Network Approach to Residential Property Appraisal," *Real Estate Appraiser* (Dec. 1992), pp. 37–45.
LeCun, Y., et al. "Backpropagation Applied to Handwritten Zip Code Recognition," *Neural Computation,* Vol 3. (1991), pp. 440–449.
Trippi, R., and Turban, E. *Neural Networks in Finance and Investing.* San Francisco: Probus Publishing Co., 1996.
White, H. "Economic Prediction Using Neural Networks: The Case of IBM Daily Stock Returns," *Proceedings of the IEEE Conference on Neural Networks,* July 1988, pp. II451–II458.

Our etch rate example is based on data from

Montgomery, D., and Myers, R. *Response Surface Methodology.* New York: John Wiley, 1997.

Cases

Jeffrey B. Goldberg
UNIVERSITY OF ARIZONA

CASE 1   Help, I'm Not Getting Any Younger!   894

CASE 2   Solar Energy for Your Home   894

CASE 3   Golf-Sport: Managing Operations   895

CASE 4   Vision Corporation: Production Planning and Shipping   898

CASE 5   Material Handling in a General Mail-Handling Facility   899

CASE 6   Selecting Corporate Training Programs   902

CASE 7   BestChip: Expansion Strategy   905

CASE 8   Emergency Vehicle Location in Springfield   907

CASE 9   System Design: Project Management   908

CASE 10   Modular Design for the Help-You Company   909

CASE 11   Brite Power: Capacity Expansion   911

CASE 1

Help, I'm Not Getting Any Younger!

Profile of a university professor:

- 45-year-old formerly athletic male
- 215 pounds
- 71 inches tall
- Exercises no more than once a week. Walks 0.5 miles daily to and from the car while carrying a 10-pound briefcase.
- Family history of adult diabetes.

I need help! My diet is terrible, and I have been gaining weight and feeling more tired.

I heard that Professor George Dantzig of Stanford once used linear programming to construct a diet. It would be great if *you* could tell me what to eat during each day. So, because I'm a firm believer in mathematical models, I want you to use linear programming to determine a reasonable diet for me to eat during a week. It is your job to collect data for use in the model.

I have the following requirements for the diet:

- I like variety. You cannot prescribe a diet in which I eat just one food during the entire week (like 10 boxes of Total cereal). I would like to eat at least 15 different foods during the week.
- You have to give me something from each of the four basic food groups (dairy, fruit and vegetable, meat, and grains)—not Mcfood, frozen food, pizza food, or food on a stick.
- I like nutrition. You cannot prescribe a diet that does not meet minimum daily requirements for essential minerals and vitamins. You cannot prescribe a diet in which I gain a lot of weight. I could stand to lose a few pounds.
- I hate Brussels sprouts, sweet potatoes, pears, and organ meats such as liver and kidney.
- Forget about any canned fruits or vegetables. Yuck.
- I do not eat any pork or pork products.
- I am not a big fan of frozen dinners, no matter how nutritious or convenient they are.
- I don't drink milk with any meal except breakfast.
- I work for the university, so I have a limited budget for food. Try to keep costs less than $100 per week (the lower the better).

- I might consider taking vitamin pills to get nutritional requirements, but I would rather eat food.

Key Questions

- What should I eat at each meal?
- If I allowed less variety, would your recommendation change?
- If I allowed more than $100 per week, would your recommendation change? How?
- What key minerals and vitamins constrain the solution?

CASE 2

Solar Energy for Your Home

As our ability to extract and process fossil fuels decreases, many people are looking to renewable resources to meet their energy needs. In particular, solar energy is becoming an advanced technology that has economic promise. In areas with large solar insolations, there can be enough energy to power an entire home. The amount of solar energy reaching the earth each year is many times greater than worldwide energy demand; it varies, of course, with location, time of day, and season. Sunlight is also a widespread resource and can be captured from virtually anywhere on earth.

There are two categories of home solar systems: passive and active. In a passive system, the solar energy heats a material that is used in a productive manner. For example, in Arizona it is common to use a passive solar system to heat swimming pools and the water used in the home. Every building has some of its heating requirements met by solar energy. Sunlight passing through windows is a source of heat, and the value of passive solar heating is enhanced by proper building insulation. A well-insulated building requires less energy for heating; thus, much of the heating load can be met by passive solar features. Optimum passive solar design begins with the layout of a building lot; a house must be oriented so that it can take full advantage of available solar energy.

Active systems are more complex and generally involve converting the solar energy to electrical energy. Photovoltaic (PV) cells use the energy of the sun to produce electricity. They produce none of the greenhouse or acid gas emissions that are commonly associated with the use of fossil fuels to generate electricity. The main barrier to increased use of this technology is cost. A common semiconducting material used in PV cells is single crystal silicon. Single crystal silicon cells are generally the most efficient type of PV cells, converting as much as 23% of incoming solar energy into electricity. The main problem with them is their production costs. Polycrystalline silicon cells are less expensive to manufacture but less efficient than single crystal cells (15% to 17%). Thin films (0.001–0.002 mm thick) of amorphous or uncrystallized silicon are another PV alternative. These thin films are inexpensive and may be easily deposited on materials such as glass and metal, thus lending themselves to mass production. Amorphous silicon thin-film PV cells are widely used in commercial electronics, powering watches and calculators. These cells, however, are not especially efficient—12% in the lab, 7% for commercial cells—and they degrade with time, losing as much as 50% of their efficiency with exposure to sunlight.

Solar power is an intermittent source of electricity. If PV cells are your only source, then the storage of electricity may be necessary. Electricity for a home can be stored in batteries, which can be expensive. Also, to generate sufficient electricity, you need a large area of collectors on your roof or somewhere on your property. The amount of solar energy captured depends on the surface area of the collectors and their conversion efficiency.

A solar energy system can often be looked at as a conservation system. Figure 1 depicts one way to look at the daily flow of energy.

Your job is to design an active solar system for a home in your area. For the analysis, you will have to collect data on:

- system cost and efficiency,
- daily solar insolation in your area (usually measured in watts/meter^2; this information can be found locally where weather data are stored and collected), and
- typical daily power requirements for a home in your area.

The costs of the system generally include a fixed component and variable components that depend on the total area of the PV collectors, the type of material used in the collector (usually only material is chosen), and the amount of battery storage needed. Your analysis should cover at least 6 months of data (12 months would be better because you would like your design to be appropriate for the entire year). You should assume that *all* energy requirements for the home will

FIGURE 1
Daily Energy Flow

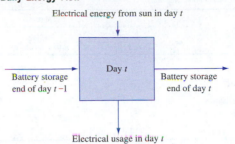

be met by this system (no natural gas will be used for heating or cooking, for example).

Your design should include the following:

- the area of the PV collectors and the amount of battery storage that you need,
- an estimate of the cost of the system (you may include any tax advantages that accrue from the purchase of solar energy systems),
- a profile of the battery storage levels at the end of each day for a six-month period,
- an estimate of cost savings (or loss) over buying your electrical power from the local utility company.

CASE 3

Golf-Sport: Managing Operations

Golf-Sport is a small-sized company that produces high-quality components for people who build their own golf clubs and prebuilt sets of clubs. There are five components—steel shafts, graphite shafts, forged iron heads, metal wood heads, and metal wood heads with titanium inserts—made in three plants—Chandler, Glendale, and Tucson—in the Golf-Sport system. Each plant can produce any of the components, although each plant has a different set of individual constraints and unit costs. These constraints cover labor and packaging machine time (the machine is used by all components); the specific values for each component–plant combination are given in Tables 1–3. Note that even though the components are identical in the three plants, different production processes are used, and therefore the products use different amounts of resources in different plants.

Besides component sales, the company takes the components and manufactures sets of golf clubs. Each set requires 13 shafts, 10 iron heads, and 3 wood

TABLE 1

Product-Resource Constraints: Chandler

Products	Resources		
	Labor (Minutes/Unit)	Packing (Minutes/Unit)	Advertising ($/Unit)
Steel shafts	1	4	1.0
Graphite shafts	1.5	4	1.5
Forged iron heads	1.5	5	1.1
Metal wood heads	3	6	1.5
Titanium insert heads	4	6	1.9
Monthly availability (minutes)	12,000	20,000	—

TABLE 2

Product-Resource Constraints: Glendale

Products	Resources		
	Labor (Minutes/Unit)	Packing (Minutes/Unit)	Advertising ($/Unit)
Steel shafts	3.5	7	1.1
Graphite shafts	3.5	7	1.1
Forged iron heads	4.5	8	1.1
Metal wood heads	4.5	9	1.2
Titanium insert heads	5.0	7	1.9
Monthly availability (minutes)	15,000	40,000	—

TABLE 3

Product-Resource Constraints: Tucson

Products	Resources		
	Labor (Minutes/Unit)	Packing (Minutes/Unit)	Advertising ($/Unit)
Steel shafts	3	7.5	1.3
Graphite shafts	3.5	7.5	1.3
Forged iron heads	4	8.5	1.3
Metal wood heads	4.5	9.5	1.3
Titanium insert heads	5.5	8.0	1.9
Monthly availability (minutes)	22,000	35,000	—

heads. All of the shafts in a set must be the same type (steel or graphite), and all of the wood heads must be the same type (metal or metal with inserts). Assembly times for the sets at each plant are shown in Table 4.

Each plant of Golf-Sport has a retail outlet to sell components and sets, and the specific plant is the only supplier for its retail outlet. The minimum and maximum amount of demand for each plant–product pair is given in Table 5. Note that, although the minimums must be satisfied, you do not need to satisfy demand up to the maximum amount.

This planning problem is for two months. The costs in Table 6 increase by 12% for the second month, and production times are stationary. Inventory costs are based on end-of-period inventory for each product set and cost out at 8% of the cost values in Table 6. Table 7 lists the revenue generated by each product. Initially, there is no inventory.

The corporation controls the capital available for expenses; the cash requirements for each product are given in the last column of Tables 1–3. There is a total of $20,000 available for advertising for the entire system during each month, and any money not spent in a month is not available the next month. The corporation also controls graphite. Each shaft requires 4 ounces of graphite; a total of 1,000 pounds is available for each of the two months.

Your job is to determine a recommendation for the company. A recommendation must include a plan for production and sales. In addition, you should also address the following sensitivity-analysis issues in your recommendation:

- If you could get more graphite or advertising cash, how much would you like, how would you use it, and what would you be willing to pay?

- At what site(s) would you like to add extra packing machine hours, assembly hours, and/or extra labor hours? How much would you be willing to pay per hour and how many extra hours would you like?

- Marketing is trying to get Golf-Sport to consider an advertising program that promises a 50% increase in their maximum demand. Can we handle this with the current system or do we need more resources? How much more is the production going to cost if we take on the additional demand?

TABLE 4

Plant	Time (Minutes per set)	Total Time Available (Minutes)
Chandler	65	5,500
Glendale	60	5,000
Tucson	65	6,000

TABLE 6
Material, Production, and Assembly Costs ($) per Part or Set

	Plants		
Products	Chandler	Glendale	Tucson
Steel shafts	6	5	7
Graphite shafts	19	18	20
Forged iron heads	4	5	5
Metal wood heads	10	11	12
Titanium insert heads	26	24	27
Set: Steel, metal	178	175	180
Set: Steel, insert	228	220	240
Set: Graphite, metal	350	360	370
Set: Graphite, insert	420	435	450

TABLE 5
Minimum and Maximum Product Demand per Month

	Store (or Plant)		
Products	Chandler	Glendale	Tucson
Steel shafts	[0, 2,000]	[0, 2,000]	[0, 2,000]
Graphite shafts	[100, 2,000]	[100, 2,000]	[50, 2,000]
Forged iron heads	[200, 2,000]	[200, 2,000]	[100, 2,000]
Metal wood heads	[30, 2,000]	[30, 2,000]	[15, 2,000]
Titanium insert heads	[100, 2,000]	[100, 2,000]	[100, 2,000]
Set: Steel, metal	[0, 200]	[0, 200]	[0, 200]
Set: Steel, insert	[0, 100]	[0, 100]	[0, 100]
Set: Graphite, metal	[0, 300]	[0, 300]	[0, 300]
Set: Graphite, insert	[0, 400]	[0, 400]	[0, 400]

TABLE 7
Revenue per Part or Set ($)

Products	Plants		
	Chandler	Glendale	Tucson
Steel shafts	10	10	12
Graphite shafts	25	25	30
Forged iron heads	8	8	10
Metal wood heads	18	18	22
Titanium insert heads	40	40	45
Set: Steel, metal	290	290	310
Set: Steel, insert	380	380	420
Set: Graphite, metal	560	560	640
Set: Graphite, insert	650	650	720

CASE 4

Vision Corporation: Production Planning and Shipping

Vision is a large company that produces video-capturing devices for military applications such as missiles, long-range cameras, and aerial drones. Four different types of cameras (differing mainly by lens type) are made in the three plants in the system. Each plant can produce any of the four camera types, although each plant has its own individual constraints and unit costs. These constraints cover labor and machining restrictions, and the specific values are given in Tables 8–10. Note that even though the products are identical in the three plants, different production processes are used and thus the products use different amounts of resources in different plants. The corporation controls the material that goes into the lenses; the material requirements for each product are given in the last column of Tables 8–10. A total of 3,500

TABLE 8
Product-Resource Constraints: Plant 1

Products	Resources		
	Labor (Hours/Unit)	Machine (Hours/Unit)	Material (Lb./Unit)
Small	3	8	1.0
Medium	3	8.5	1.1
Large	4	9	1.2
Precision	4	9	1.3
Total available	6,000	10,000	—

TABLE 9
Product-Resource Constraints: Plant 2

Products	Resources		
	Labor (Hours/Unit)	Machine (Hours/Unit)	Material (Lb./Unit)
Small	3.5	7	1.1
Medium	3.5	7	1.0
Large	4.5	8	1.1
Precision	4.5	9	1.4
Total available	5,000	12,500	—

TABLE 10
Product-Resource Constraints: Plant 3

Products	Resources		
	Labor (Hours/Unit)	Machine (Hours/Unit)	Material (Lb./Unit)
Small	3	7.5	1.1
Medium	3.5	7.5	1.1
Large	4	8.5	1.3
Precision	4.5	8.5	1.3
Total available	3,000	6,000	—

pounds of material is available for the entire system during the planning period.

Transport has 3 major customers (RAYco, HONco, and MMco) for its products. The maximum sales for each customer–product pair is given in Table 11. Product sales prices are given in Table 12, and the shipping costs from each plant to each customer are detailed in Table 13. Table 14 contains the production costs for each product–plant pair.

All shipping from plants 1 and 2 that goes to RAYco or HONco must go through a special inspection. These units are sent to a central site, inspected, and then sent to their destination. The capacity of this special inspection site is 1,500 pieces.

Your job is to determine a recommendation for the company. A recommendation must include a plan for production and shipping as well as the cost and revenue generated from each plant. In addition, you should address the following potential issues in your recommendation:

- If you could get more material, how much would you like? How would you use it? What would you be willing to pay?

- If you could get more inspection capacity, how much would you like? How would you use it? What would you be willing to pay?

TABLE 11
Maximum Product Sales ($) per Unit

Products	Customers		
	RAYco	HONco	MMco
Small	200	400	200
Medium	300	300	400
Large	500	200	300
Precision	200	400	300

TABLE 12
Product Sales Price ($) per Unit

Products	Customers		
	RAYco	HONco	MMco
Small	17	16	16
Medium	18	18	17
Large	22	22	23
Precision	29	26	27

TABLE 13
Shipping Costs ($) per Unit

Plant	Customers		
	RAYco	HONco	MMco
1	1.0	1.6	1.1
2	1.2	1.5	1.0
3	1.4	1.5	1.3

TABLE 14
Production Costs ($) per Unit

Products	Plant		
	1	2	3
Small	14	13	14
Medium	16	17	15
Large	18	20	19
Precision	26	24	23

- At what plant(s) would you like to add extra machine hours? How much would you be willing to pay per hour? How many extra hours would you like?
- Marketing is trying to get RAYco to consider a 50% increase in its demand. Can we handle this with the current system or do we need more resources? How much more money can we make if we take on the additional demand?

CASE 5

Material Handling in a General Mail-Handling Facility[†]

For more than 200 years, the United States Postal Service (USPS) has delivered mail across the country. Daily delivery goes to some 137 million households; in 2001, the USPS processed and delivered more than 207 billion pieces of mail to a delivery network that grew by 1.7 million new addresses. Clearly, the USPS is the largest material handler (in terms of pieces) in the world. Statistics recorded by Pricewaterhouse Coopers show that 94% of first-class mail destined for next-day delivery received overnight service—and this was a record performance for a second straight year. Despite the high volume, the USPS managed to cut costs by $900 million in 2001 while maintaining record service performance and high levels of customer satisfaction.

To process mail quickly, one must use advanced mechanization. Mail-sorting machines can process 10 letters per second (we are long past the days of hand sorting in front of a large set of post boxes). Sorting using the zip+4 standard can result in a mail sort down to an individual carrier's walk sequence, which saves significant carrier time.

Five major operations can be performed on each letter, and each operation has its own machine:

Automatic facer and canceller (AFC) This machine cancels the stamp and orients all of the letters so that the stamp is in the upper-right corner. This machine also separates mail into one of three streams—automation, mechanization, or manual.

Letter sorting machine (LSM) This machine is semi-automated and helps human operators sort mail. The operator reads the address and then types in a destination code. The machine then routes the letter to the appropriate bin.

Optical character reader (OCR) This machine reads handwritten or typed addresses and then prints a machine-readable barcode on the envelope.

Barcode sorter (BCS) This machine reads the barcode on the letter (either printed by the OCR or by the sender's equipment) and then sorts it to a bin.

[†]Based on work done jointly with Ron Askin and Sanjay Jagdale, 1994.

Delivery barcode sorter (DBCS) This machine does a two-pass sort that uses barcodes with the zip+4 code and sorts down almost to a walk sequence. When using the DBCS, the result is such that a carrier requires little to no processing at the carrier station to deliver the mail.

Items known as *flats* (for example, magazines and 8.5 × 11–inch envelopes) also are processed through the system. A flat-sorting machine (FSM) is used in a semiautomated process. An operator loads a piece onto the machine and keys in a code based on the flat's address; the machine then routes the piece to an appropriate bin.

A letter that enters a general mail facility (GMF) follows a routing that depends on the machine readability of the address and the presence of an existing barcode. Although many routes are possible, the major ones are given in Figure 2. Once letters go through the AFC, they are stored in cardboard or plastic trays that hold approximately 400 letters. The letters are moved in these trays throughout the facility, and each machine sorts the mail into different trays.

As part of quality improvement, the post office is always looking for ways to cut costs while expediting mail processing. To meet goals for overnight delivery and three-day cross-country delivery, letters that arrive by 6 p.m. from box pickup must be processed that night and be on planes or trucks for the next destination. Because the sorting and character-reading machines cost millions of dollars each, increasing machine use and saving the purchase of even one machine in a GMF is a significant achievement.

One of the keys to faster processing and increasing utilization is an effective material-handling and data system. Each tray has a barcode that describes the salient characteristics of its mail. When a sorting machine is ready for operation (say, for example, that we are going to sort down to a group of 10 zip codes), a call goes out to bring all trays with appropriate mail to the appropriate BCS machine. The data system must (1) know where those trays are located, (2) go and get them, (3) bring the trays to the machine-input area, and (4) exit the area. The faster this can be done, the better.

The network for our GMF is given in Figure 3. Each machine has an input and output point. For example, nodes 1 and 28 are the inputs and output, respectively, for AFC 1. For this application, the material-handling system is an overhead monorail. Carriers that hold one tray circulate around the system to pick up and deliver trays; they rest in the parking lot when not in use. The arcs in the diagram are the links of the monorail; all links are one-way. The dotted lines on Figure 3 represent links that are above the machine level and offer shortcuts across the facility. The facility also contains switches (nodes 29, 18, and 32) that allow carriers to change directions. Node 34 is the link to the shipping dock, and all trays enter and exit the system at this point.

The carriers travel at approximately 1 mile per hour, and there must be 15 feet between carriers on the same link. For the purpose of this study, assume that there is a bypass at each node so a carrier can pass other carriers that are stopped for loading and unloading operations. Also assume that the switches operate quickly relative to the speed of the vehicles so that collisions do not occur and the switch capacity is not constraining. Figure 3 is drawn approximately to scale. The facility is approximately 220 feet long by 160 feet wide. At 1 mile per hour, it takes a carrier approximately 2.5 minutes to run the length of the facility (from node 14 to node 1, for example).

Table 15 contains the tray movement loads for the peak hour. Each load has an origin node, a destination node, and the number of trays that must be moved. Each load is a leg in the route for a particular tray of mail. The system must have capacity to move an empty carrier from the parking lot to the origin, load the tray, move the tray to the destination, unload the tray, and then return to the parking lot. All carriers are dispatched from the parking lot because this simplifies the logic of the scheduler. By capacity, there must be a sufficient number of vehicles and the capacity on each link between nodes cannot be overloaded. As-

FIGURE 2

Processing Routes for Letters

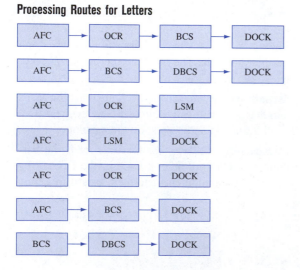

FIGURE 3

General Mail Facility: Track Layout

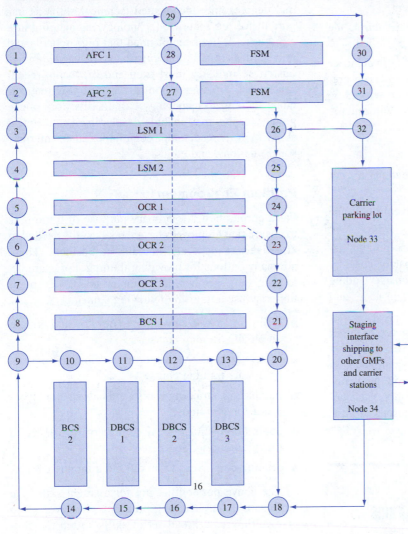

TABLE 15

Load Data for the Peak Hour

Load Number	Origin Node	Destination Node	Number of Trays
1	33	1	15
2	28	4	20
3	22	14	30
4	10	33	30
5	24	8	15
6	21	33	30
7	24	4	5
8	25	33	15
9	27	17	15
10	13	33	40

sume that it takes one minute to perform a loading or unloading operation.

Your jobs are to:

- Determine if this system has enough material-handling capacity for moving the trays in the peak hour (we generally design for peak hour so that we are sure that the system will not get bogged down when demand is high).

- Suggest where we might add extra track to relieve capacity congestion. This should be minimized because track cost is high.

- Determine the flows of trays through the network during the peak hour. Which routes are chosen for each load?

- Determine places in the network that are risky—that is, if a link goes down, machines can be cut off from the material-handling system.

- Estimate the total carrier travel distance during the peak hour.

- Investigate the effect of reducing the intercarrier spacing. You need space between carriers to prevent carrier collisions. If we put better sensors on the front of the carriers, they can stop more quickly and we can have less spacing.

Some tips:

- This is a difficult problem. Be patient and try not to become discouraged.

- Do not forget the empty carrier movements to the origin and from the destination.

- Compute the capacity on each arc. Initially, assume a single lane. To increase capacity, consider multiple lanes between nodes or consider adding arcs to give more paths between origin nodes and destination nodes.

- A precise formulation of this problem can be larger than most problems you have seen. You should not undertake to solve a large-sized formulation unless you have software that can handle large problems. Some approximate formulations are more manageable, but they still can require hundreds of variables and constraints.

CASE 6

Selecting Corporate Training Programs[†]

Introduction

Training has become a major cost of doing business. A 1995 survey of all U.S. businesses with 100 or more employees revealed that approximately $52 billion was being spent on training; it has been estimated that $90 to $100 billion is being spent for training overall. Developing strategies to implement cross-training is a current topic in the operations research (OR) literature. Management consultants advocate aggressive education and professional development to remain competitive in the global and local markets. Employees now expect job and skill growth to be a major component of their duties.

[†]Based on work done jointly with John V. Farr, and David A. Thomas at USMA, 1995.

Increased training costs have occurred for many reasons. Employees view training in the form of formal degrees and documented technical skills as important for job security. Technology is changing at a rapid pace. It has been claimed that high schools and universities are not producing the skills needed by industry, so industry must train and reeducate recent graduates. For high school graduates, this may include training in technology-based skills; for college graduates, this may include developing nontechnical skills such as leadership, communications, interpersonal relations, and ethics.

Problem Environment

For a corporation, the primary purpose of training is to ensure that employees have the key skills needed to effectively manage and operate the business. There are many options for providing training. For example, to train staff members in computer skills, a corporation may use any of the following strategies:

- hiring an outside consultant to develop and present an on-site training course,

- using corporate personnel to develop and present an on-site training course,

- purchasing a training course and having employees use it for self-study,

- contracting with a local college or university to provide training, or

- sending employees to an off-site training seminar.

The above possibilities are for a single skill. The purpose of many training programs, however, is to give employees a broad set of skills. Often the skill sets of two or more programs partially overlap. When this happens, the corporation must choose the set of programs that give employees the required skills for their jobs and the appropriate employees for each training program. In any case, training decisions made in an ad-hoc "pay-as-you-go" manner will be inefficient and generally result in additional expense.

To give the decision problem structure, the following assumptions are appropriate:

- We have a known study period—for example, the next 3 or the next 5 years—over which we need to plan training. The study period should fit with the overall business strategy and enable accurate estimates for training needs and available resources.

- There is a known set of skills that employees need. Among others, these may include technical, interpersonal, communication, and management skills.

TABLE **16**
Skills List

No.	Skill	No.	Skill
1	New employee orientation	22	Stress management
2	Performance appraisals	23	Computer programming
3	Personal computer apps	24	Diversity
4	Leadership	25	Data processing/MIS
5	Sexual harassment	26	Planning
6	Team building	27	Public speaking and presentation
7	Safety	28	Strategic planning
8	Hiring and selection process	29	Writing skills
9	New equipment operation	30	Negotiating skills
10	Training the trainer	31	Finance
11	Product knowledge	32	Marketing
12	Decision making	33	Substance abuse
13	Listening skills	34	Ethics
14	Time management	35	Outplacement and retirement
15	Conducting meetings	36	Creativity
16	Quality imiprovements	37	Purchasing
17	Delegation skills	38	Smoking cessation
18	Problem solving	39	Financial and business literacy
19	Goal setting	40	Reengineering
20	Managing change	41	Foreign language
21	Motivation		

TABLE **17**

Salary and Skills Required for Each Job Classification

Person	Salary ($)	Skills 1–20
Senior Manager	250,000	0 1 0 1 0 0 0 0 0 0 0 1 1 1 0 1 0 1 1 0
Project Manager	200,000	0 1 0 0 1 0 0 0 0 0 1 1 1 1 1 0 0 1 0 0
Professional	150,000	1 1 0 0 1 1 1 0 1 0 0 0 1 0 1 0 0 0 0 1
Sales	150,000	1 1 0 0 0 0 0 0 1 1 0 0 1 0 0 0 1 0 0 1
Technician	100,000	1 1 1 0 1 0 0 1 0 0 0 0 0 0 0 0 0 0 0 1
Administrative Assistant	80,000	1 1 1 0 1 0 0 0 0 0 0 0 0 0 0 0 0 0 0 0

Person	Skills 21–41
Senior Manager	1 1 1 1 0 1 1 0 1 0 1 1 0 1 0 1 1 1 0 0 0
Project Manager	0 1 1 1 0 1 1 1 1 0 1 0 1 0 0 1 1 1 1 0 1
Professional	0 0 0 0 1 1 0 0 0 1 0 0 1 0 1 0 0 1 1 0 1
Sales	0 0 0 0 1 1 0 0 0 1 0 0 1 0 1 0 0 1 1 1 1
Technician	0 0 0 0 0 0 0 0 0 0 0 0 0 0 0 0 0 1 1 1 1
Administrative Assistant	0 0 0 0 0 0 0 0 0 1 0 0 0 0 0 0 0 1 1 1 0

TABLE 18

Enrollment Cost and Skills of Each Program

Program	Enrollment Cost ($)	Skills 1–20
Program 1	500	1 1 0 1 0 0 0 0 0 0 1 1 1 1 1 1 0 1 0 0 0
Program 2	300	0 0 0 0 1 1 1 1 0 0 0 0 0 0 0 0 0 0 0 0
Program 3	500	0 0 0 0 0 0 1 0 0 1 0 0 0 0 0 0 1 0 0 0 1
Program 4	575	0 1 0 1 0 0 0 0 1 1 0 0 1 0 0 0 1 0 0 0
Program 5	800	0 1 1 1 0 0 0 0 1 0 0 1 0 0 1 0 0 0 1 0 0 0
Program 6	400	0 0 0 0 0 0 0 1 1 0 0 0 0 0 0 0 0 0 0 1
Program 7	200	0 0 0 0 1 1 0 0 0 0 0 0 1 0 0 0 0 0 0 1
Program 8	1,000	0 0 0 0 0 0 0 0 0 0 1 1 1 1 1 1 0 0 1 0
Program 9	200	1 1 1 1 0 0 0 0 0 0 0 0 0 0 1 1 0 0 0 0 0
Program 10	500	0 0 0 1 0 0 0 0 0 0 0 1 0 0 0 0 0 0 0 0
Program 11	700	0 0 0 0 0 0 0 0 1 0 1 0 0 0 0 0 0 0 0 0
Program 12	600	0 0 0 0 0 0 0 0 0 1 0 0 1 0 0 0 0 0 0
Program 13	400	0 0 0 0 0 0 0 0 0 0 1 1 0 0 0 1 1 1 0
Program 14	900	1 0 0 0 1 1 0 0 0 0 0 0 0 0 0 1 0 1 0 0
Program 15	700	1 0 1 0 1 0 1 1 1 0 0 0 0 0 0 0 0 0 1 0

Program	Skills 21–41
Program 1	1 0 0 0 0 0 0 0 0 0 0 1 0 0 0 0 0 0 0 0 0
Program 2	0 0 0 0 0 1 0 0 0 0 0 0 0 0 0 0 0 0 1 1 0
Program 3	1 1 1 1 1 1 1 0 1 0 1 0 0 0 0 1 0 0 0 0 0
Program 4	0 0 0 0 0 0 0 1 0 0 0 0 0 0 0 1 0 0 0 0 1
Program 5	0 0 0 1 0 1 0 0 0 0 0 0 1 0 0 0 0 0 1 1 0
Program 6	0 0 0 0 0 0 0 0 1 0 0 1 1 0 0 1 1 0 0 0 0
Program 7	0 1 0 0 0 0 0 0 0 0 0 1 0 1 0 0 0 0 0 0 0
Program 8	1 1 1 1 0 0 1 0 1 1 0 0 0 0 0 0 0 1 0 0 0 0
Program 9	0 0 0 1 1 1 0 0 0 0 0 0 0 0 0 0 0 0 1 1 0
Program 10	0 0 0 0 0 0 0 0 0 1 1 0 0 0 0 0 1 1 0 0 0
Program 11	0 1 1 0 0 0 1 0 1 1 0 1 0 0 1 1 0 0 0 0 0
Program 12	0 0 0 0 0 0 0 1 0 1 1 0 0 0 0 0 0 1 1 1 0
Program 13	1 0 1 0 1 0 0 0 0 0 1 0 1 0 0 0 0 0 0 0 0
Program 14	0 0 0 0 0 0 0 0 0 0 0 0 0 1 1 0 0 0 0 0 1
Program 15	0 0 0 0 0 0 0 1 0 0 0 0 0 0 0 0 0 1 0 0 1

- Employees are divided into classes. In each class, we have estimates for (1) the number of employees, (2) the employee hourly wage, (3) the number of employees that require each particular skill, and (4) the maximum time available for training employees in each class during the study period.

- There is a list of training programs. For each program, we assume we have (1) the set of skills taught, (2) the cost, (3) the development time, (4) the completion time for an employee, and (5) the maximum number of employees who can participate per decision cycle.

- Training is equally effective for all people, thus we are concerned with which programs to offer and which employees in each class to assign to each program. If we know the quality of the training for individual skills for individual classes, then we can relax this assumption.

Potential Corporate Setting

Your job is to develop models to aid businesses and corporations in determining the appropriate training programs to use. The type of model and issues often depend on the size of the corporation and the potential

TABLE 19
Interfering Programs

Program Number (Days Long)	Programs that Interfere			
Program 1 (2)	3	5	8	
Program 2 (2)	3	7	10	
Program 3 (4)	1	2	12	
Program 4 (3)	6	7	14	
Program 5 (2)	1	9	12	
Program 6 (3)	4	7	11	14
Program 7 (5)	2	4	6	
Program 8 (2)	1	10	13	
Program 9 (3)	5	15		
Program 10 (3)	2	8		
Program 11 (2)	6	12	15	
Program 12 (4)	3	5	11	
Program 13 (3)	8	14		
Program 14 (4)	4	6	13	
Program 15 (3)	9	11		

are 15 programs available for use; Table 18 contains the cost per person and the skills covered for each program. In Table 18, a 1 in the row for program p and the column for skill s implies that program p contains skill s. Table 19 lists the programs that conflict in time with other programs (for example, programs 3, 5, and 8 conflict with program 1). An employee cannot take two programs simultaneously. It is company policy that each employee is limited to 15 days for training per year.

Key Questions

Your job is to develop a recommendation for the company for addressing its training needs. In particular, you should address the following key questions:

■ Which training programs should we be using? What is the assignment of personnel to those programs?

■ Identify programs with heavy use that may justify the development of an in-house course. How much would you be willing to pay for that development if you could use the program for the next three years?

■ We have the opportunity to negotiate prices for programs. Which programs would you suggest are candidates for negotiation?

■ What skills are especially expensive for us to cover? If we were to develop our own programs, what skills should be covered?

■ Would your recommendation change if we allowed more days of training per year?

uses of the models. For large corporations, there are many employees in each class, so it is not necessary to model and schedule down to the individual employee. Concentrate instead on the assignment of classifications to programs and ignore the assignment of specific individuals to programs. Also, sufficient resources exist to develop internal training programs, hence you should consider program development costs as well as employee costs (lost work time, travel, lodging, meals, course materials) in the objective. A large corporation can use the model to plan the development of courses. This will help determine (1) program-development costs so that in-house programs are cost-effective and (2) appropriate programs for each employee classification so that, on average, there is sufficient time to complete the assigned programs within the available time.

For small businesses, the focus is often different. Typically, these companies do not develop in-house programs because they do not train enough employees to justify development costs. Because the number of employees is small, it is important to model down to the employee level and schedule employees so that both training and job tasks can be completed.

Your OR consulting firm has been hired to design the training program for a small company. There are no in-house classes, and vendors provide all training. The company has determined 41 skills that are important for its employees; these are listed in Table 16. There are six employees; the salary level and skills required for each person are given in Table 17. You can assume that there are 250 working days per year. There

CASE 7
BestChip: Expansion Strategy

BestChip (BC) is a large nationwide corporation that produces low-fat snack products for an expanding market (pun intended). Basically, BC takes materials (corn, wheat, and potatoes) and turns them into two types of snacks: chips (regular and green onion) and party mix (one variety). BC is expanding into the western United States and is considering sites for locating production facilities.

BC currently has eight candidate sites. Table 20 shows the sites' purchase prices and the purchase and shipping cost per ton of each material to each site. The purchase cost represents the yearly amortized cost of opening and operating the site (exclusive of

TABLE 20
Site Information and Material Shipping Cost

Site Location	Purchase Cost ($/Year)	Material Shipping Cost ($/Ton)		
		Corn	Wheat	Potato
Yuma, AZ	125,000	10	5	16
Fresno, CA	130,000	12	8	11
Tucson, AZ	140,000	9	10	15
Pomona, CA	160,000	11	7	14
Santa Fe, NM	150,000	8	14	10
Flagstaff, AZ	170,000	10	12	11
Las Vegas, NV	155,000	13	12	9
St. George, UT	115,000	14	15	8

TABLE 21
Demand Information

Company	Location	Demand		
		Regular	Green Onion	Party Mix
Jones	Salt Lake City	1,300	900	1,700
YZCO	Albuquerque	1,400	1,100	1,700
Square Q	Phoenix	1,200	800	1,800
AJ Stores	San Diego	1,900	1,200	2,200
Sun Quest	Los Angeles	1,900	1,400	2,300
Harm's Path	Tucson	1,500	1,000	1,400

shipping costs). Each site may produce as many as 20,000 tons of product per year.

BC has six major customers, and all demand is shipped by truck from the plant to the customer warehouse. The shipping cost depends on the tonnage and distance and comes to $0.15 per ton-mile. The customers, their location, and their yearly demand in tons for each product are listed in Table 21. You must meet demand.

The makeup of the products does not depend on the production plant. Table 22 gives the product-ingredient mix data. The company requires that we consolidate our business, so we cannot locate plants in more than two states.

For this analysis, ignore the differences in property and income tax rates between the states (this is usually critical, but it gets us far afield of the key issue of math programming). In addition, many critical factors actually determine locations; for example, the method of financing the site purchase will also be a major factor in the decision—but we will ignore that also.

Your job is to determine how we should expand into the west and develop alternatives. Questions you should answer include:

TABLE 22
Product-Ingredient Mix

Product	Ingredient		
	Corn	Wheat	Potato
Regular chips	70	20	10
Green onion chips	30	15	55
Party mix	20	50	30

■ What sites should be selected? How should the customers be served?

■ If gasoline gets more expensive and our trucking costs change, then how is the recommendation affected?

■ If rail freight costs for material shipping increase, then how is the recommendation affected?

Please consider other sensitivity-analysis issues that you feel might be important for management's decision-making process.

CASE 8

Emergency Vehicle Location in Springfield

You are the logistics manager for the Springfield Fire Department. You are to develop a recommendation for providing emergency service to Springfield. The department's resources include engine trucks, ladder trucks, and paramedic vehicles. The budget suggests a total of 15 vehicles are fundable in the coming year. Currently, seven engines, three ladder trucks, and five paramedic vehicles are in operation. This system runs 24 hours per day.

The city has been divided into 10 zones (see Figure 4). The map is drawn to scale. For each zone, the department has estimated the number of fire calls, the number of false alarms, and the number of medical calls per 12-hour day. These data are listed in Table 23. Currently, there are five fire stations in the city; these are listed in Figure 4. Each existing station costs $20,000 per year to operate. The yearly costs for each potential station (including the amortized cost of construction) are also listed in Table 23. Each station can hold two vehicles at most.

For fire calls, an engine and a ladder truck must respond. For medical calls, a paramedic vehicle always responds and an engine also goes when one is available and closer than the nearest paramedic vehicle. On average, fire calls take 2.5 hours, false alarms 10 minutes, and medical calls 45 minutes.

FIGURE 4

Map of the City (17 miles by 11 miles)

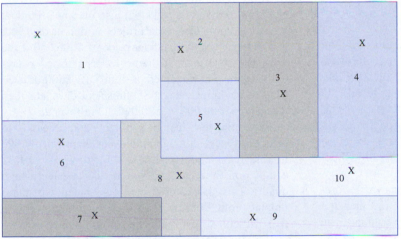

X marks the spot of the existing sites and the possible sites in each zone

TABLE 23

Demand Information per Year

Zone	Fire Calls	False Alarms	Medical Calls	Base Cost ($/Year)
1	100	200	1000	40,000
2	50	100	450	Existing
3	75	100	600	35,000
4	120	75	1300	50,000
5	150	100	1400	Existing
6	300	150	1000	50,000
7	200	100	800	Existing
8	250	175	1000	Existing
9	100	25	900	Existing
10	75	50	650	35,000

When vehicles are dispatched on a call, the closest idle vehicle is dispatched first. If no vehicles are idle, then the call must be sent to a private provider; these responses cost the city $5,000 per medical call, $15,000 per fire call, and $200 per false alarm. There is no queueing. The street network is largely rectangular, and the fire department estimates that the cost per mile for travel is $1.50 per mile for engines and ladders and $0.75 per mile for paramedic units.

Your job is to design a system for the fire department. The questions that should be considered are these:

- What sites should be selected and how should the vehicles be distributed?
- If travel gets more expensive, how is the recommendation affected?
- If the cost of using the private provider increases, then how should the system be changed?
- Is all of this equipment needed to serve the public?
- How much more demand can be handled with the full complement of vehicles?

Your write-up should include a description of your models and any assumptions made in model formulation. You will have to make simplifying assumptions because this problem has details that may be difficult to model. There are *many* ways to model parts of this system, and you can use different approaches to answer different questions. You may use Excel or LINDO, or you may use heuristics. Your call will depend on your modeling approaches.

Hint: This case is less specific and has vague components; simplify as a first approach and then get more complex. If you try to include everything, you will become frustrated because this does not fit any standard modeling paradigm.

CASE 9

System Design: Project Management[†]

System Design (SD) is a small corporation that contracts to manage systems and industrial engineering projects. In this case it must manage the design and construction of a power plant's data-processing and data-collection system. SD's role in the project is to

[†]This material is expanded from a homework problem in *Applied Mathematical Programming* by Wayne Winston.

hire subcontractors, ensure each task is completed within specification, determine how much labor to assign to each task, and generally ensure the project's success.

SD is really a subcontractor within the larger project of building the power plant. Table 24 details SD's plant-construction and data-system-design tasks. SD is directly in charge of tasks 2, 6, 7, 10, 14, 15, and 19. The remaining tasks in Table 24 interact with those in SD's charge. Assume that the remaining tasks (1) will start whenever their predecessor tasks are complete and (2) will finish exactly after their duration.

To shorten the seven SD tasks, you must pay additional labor and overhead costs. Table 25 lists the functions you can use to compute the cost of changing each task duration to a new value. (*Note:* t_j is the original duration of task j; d_j is the minimum duration of task j.) The table also lists the lowest possible task-duration value. You may not increase the duration of any task.

The revenue that SD obtains from the project depends not only on its tasks but also on when the total project is completed. The project is due at day 900, and SD receives the contract price of $600,000 if the project is done then. If the project is finished x days early, then SD receives a total of $600,000 + $15,000x^{0.7}$ in revenue. If SD finishes x days late, then it receives $600,000 - $20,000x^{1.4}$ in revenue.

Expediting tasks can be profitable and necessary to meet deadlines, although employees do not really like it. Task completion quality is a function of the task completion time, and we would like to have a high quality. This may conflict with our objective of maximizing profit. Because quality affects future revenues, it is difficult to estimate the dollar impact of poor quality. If t_j is the original duration of task j and x_j is the expedited duration time for task j, then quality, measured on a scale of 0–100 (with 100 being the best), can be represented by the function

$$100 - \min [100, (t_j - x_j)^{2.2}]$$

This is only the quality for a task. It is unclear how one might quantify the quality of the project.

Your job is to determine how we should proceed with our tasks. Your analysis should answer some of the following questions:

- What tasks are critical to project completion? What tasks will you expedite?
- How are you measuring system quality, and how does your recommendation measure up relative to that objective?

TABLE **24**

Task Information

Task No.	Task Name	Task Duration (Days)	Immediate Predecessor Tasks
1	Preliminary system description	40	—
2	Develop specifications	100	1
3	Client approval	50	2
4	Develop input-output summary	60	2
5	Develop alarm list	40	4
6	Develop log formats	40	3, 5
7	Software definition	35	3
8	Hardware requirements	35	3
9	Finalize input-output summary	60	5, 6
10	Analysis performance calculations	70	9
11	Automatic turbine startup analysis	65	9
12	Boiler guides analysis	30	9
13	Fabricate and ship	200	10, 11, 12
14	Software preparation	80	7, 10, 11
15	Install and check	130	13, 14
16	Termination and wiring lists	30	9
17	Schematic wiring lists	60	16
18	Pulling terminals and cables	60	15, 17
19	Operational test	125	18
20	First firing	1	19

TABLE **25**

Expediting Costs ($1,000) and Limits

Task No.	Duration (Days)		Cost to Decrease t_j by x Days
	Current t_j	Minimum d_j	
2	100	70	$1.5x^{1.8}$
6	40	20	$2x^{2.0}$
7	35	20	$1x^{2.0}$
10	70	40	$1.8x^{1.9}$
14	80	60	$1.9x^{1.6}$
15	130	120	$0.95x^{2.7}$
19	125	80	$0.9x^{2.9}$

- How much money do you make on the project?
- What will you give the decision maker to help with the decision?
- If we can move minimum duration days to lower values, then which values would you like to reduce?
- If you could control additional tasks by paying more money, which ones would you like to take, and how much would you be willing to pay for control?

CASE **10**

Modular Design for the Help-You Company

The Help-You Company is in the business of manufacturing first-aid kits for cars, hikers, campers, sports teams, and scouting groups. The company is located in Tucson, Arizona, and all materials must be sent to Tucson and then shipped to customers' warehouses. The company has done extensive market surveys of its customers; Table 26 shows its estimates for the demand for its kits in the coming year.

Each kit contains the individual items shown in Table 27, which are listed with their base sizes in pounds. Help-You, for example, can buy packs of Acetaminophen extra-strength caplets. Each pack contains 12 tablets and costs Help-You $1.50.

Help-You buys the individual items and then assembles kits based on the requirements for each part in each kit as shown in Table 28.

These are minimum requirements in that the customer expects at least the listed quantity of each item in each specific kit. For example, in the kit for campers, there must be at least four blankets and at least three cold units (six cold packs) as well as the other items.

There are two strategies available for assembly of the kits:

- In *direct assembly,* the exact requirements are put into each kit.

- In *modular assembly,* one or more standard modules are developed that can be assembled and combined into a kit with enough modules so that the minimum requirements for each item are met. A graphic of the approach is detailed in Figure 5.

If you design a module, for example, that has two units of Band-Aids (as well as the other items) and place three of these modules in a scouting kit, then the kit will have $3*2 = 6$ units of Band-Aids; this will meet the requirement of four units of Band-Aids for scouting kits. In this example, there is an "overage" of two extra units of Band-Aids that costs

$$2 * \$1 \text{ per unit} = \$2$$

per each scout kit demanded. Also, the total unit content in a module cannot weigh more than 15 pounds (an assembly requirement).

Direct assembly meets requirements exactly, although it usually has higher labor costs than modular assembly. For storing inventory, modules are easier to use because they tend to be smaller than kits.

Develop a strategy for modular assembly. The key costs of the modular system are the overages that occur. Your strategy must include the following:

- the number of modules you are designing (the more modules you have, the closer you can match requirements exactly, although the higher the costs for assembly and inventory);

- the unit content of each module designed; and

- an estimate for the total number of each module required.

Your analysis should consider issues such as:

- the trade-off between the number of different modules designed and the total overage cost (you do not need to try more than five different modules—why?);

TABLE 26
Kit Demand

Type	Number of Kits Sold
Cars	1,000
Hikers	800
Campers	100
Sports teams	200
Scouting groups	300

TABLE 27
Item Cost and Base Size

Item	Cost ($)	Base Unit	Base Unit Weight (Lb.)
Adhesive Band-Aids	1	10 per pack	0.20
Ace bandages	2	1 bandage	0.20
Flares	4	3 per pack	1.00
Blankets	15	1 blanket	2.00
Adhesive tape	2.50	1 roll	0.40
Cold packs	4	2 per pack	0.80
Sunburn cream	3.50	1 tube	0.40
Antiseptic cream	2	1 tube	0.50
Acetaminophen extra-strength caplets	1.50	12 tablets	0.30
Rubber gloves	1.50	3 pairs	0.20

TABLE 28

Item to Kit Requirements (Base Units)

Kit Item	Cars	Hikers	Campers	Sports Teams	Scouting Groups
Ace bandages	1	2	4	12	6
Band-Aids	0	2	4	4	4
Flares	2	1	1	0	2
Blankets	1	1	4	2	3
Adhesive tape	2	2	3	6	4
Cold packs	2	2	3	6	3
Sunburn cream	1	2	4	4	5
Antiseptic cream	1	2	3	2	4
Acetaminophen caplets	1	2	4	6	6
Rubber gloves	1	1	2	10	5

FIGURE 5

Modular Assembly

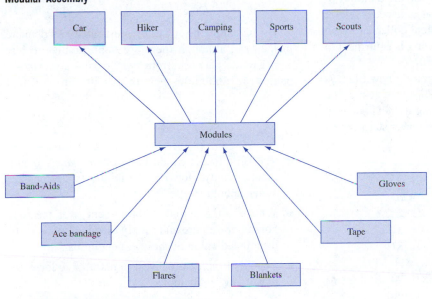

- the sensitivity of your solution to the kit demand estimates (for example, what happens to your recommendations if the number of scouting kits sold changes by 20%?);

- the sensitivity of your recommendations to the unit cost values;

- the sensitivity of your recommendations to the weight limit on the size of each module; and

- discussion about your confidence that you have the optimal solution in light of what you have covered concerning convex functions and sets.

CASE 11

Brite Power: Capacity Expansion

Brite Power is a small power provider in the Finger Lakes region of New York state. Because of California's power shortage in summer 2001, Brite Power's board of directors has decided to commission a study to ensure that the company has sufficient power until 2020. The study requires a time horizon of 16 years; the first decisions can be implemented in January 2005. Even though operations plans for power companies are important, this study is at a higher level.

Our concern is with power capacity during the year and not with day-to-day or hour-to-hour power-usage fluctuations.

By the start of 2005, Brite will have plants with 60 megawatts' worth of production capacity. Estimates of demand for power from the company for the years 2005 to 2020 have been made and are listed in Table 29.

The economics of a coal-fired plant run according to a *power law*—that is, the cost of a new plant in constant dollars (sometimes called "year 0 dollars") follows the following estimation rule:

cost of plant with capacity K =

$$\left[\frac{\text{capacity K}}{\text{capacity of base size plant}} \right]^{0.8} * \text{cost of the base size plant}$$

For this analysis, the base plant's production capacity is 5 megawatts; its cost is $18 million. The company estimates inflation at 4% per year for the duration of the time horizon. The company uses a discount rate of 12% per year; this assumes that actual dollars are used in the analysis [1 actual dollar in year 1 is equivalent to $1/(1.12) = .89$ dollars now in year 0].

The time required for constructing a new plant is two years. The project requires 65% of the cost at the start of the first year; the remaining 35% is spent at the start of the second year. If Brite Power starts a new plant in 2007, for example, then 65% of the costs occur in 2007 and 35% in 2008; the plant then comes online and can be used to satisfy demand in 2009. With this lead time, it is clear that Brite Power needs to do advanced planning.

Brite Power can build plants with 5-, 10-, 15-, and 20-megawatt capacities. If the company invests now in research for new technologies ($3 million per year for 5 years), then it can reduce the exponent in the power model from 0.8 to 0.65.

Besides building new capacity, Brite Power must operate plants; operations costs are based on the amount of capacity used. If demand in year t is D_t megawatts and total capacity in year t is C_t megawatts, then the operations cost in constant dollars for the year is:

$$\left(\frac{C_t}{D_t} \right)^{0.5} * D_t * \$400{,}000$$

By and large, the company must satisfy all demand in the year, although there are opportunities to buy 3 megawatts per year from neighboring power companies at $600,000 per megawatt (in constant dollars).

Key Questions

■ The Brite Board would like to know when to augment capacity. How big should the expansions be? When should they start?

■ What are the actual dollars spent over the time horizon to acquire and satisfy demand? What is the discounted value of these expenditures?

■ Should the company invest in research to lower the power law coefficient?

■ If demand estimates are increased or decreased, then how would the plan change?

■ If the $400,000 value in the operations cost changes, then how would the plan change?

TABLE 29

Year	Demand (Megawatts)	Year	Demand (Megawatts)
2005	54	2013	87
2006	58	2014	87
2007	63	2015	90
2008	63	2016	90
2009	69	2017	100
2010	75	2018	110
2011	77	2019	110
2012	77	2020	120

Index

Acme Widget Company
(AWC), 810–815
Activity
critical, 437, 469
dummy, 433
Activity on arc (AOA) network,
432, 434, 468
ADBUG function in
Evolutionary Solver, 857,
858
Adding a new activity, 287, 325
Additivity assumption, 53, 62
Adjacency assumption, 733
Adjacent basic feasible
solutions, 137–138
Advertising
Lagrange multiplier, 710
Leon Burnit Advertising
Agency, 191–198
prisoner's dilemma game, 636
sales advertising example,
885–890
Airline maximum flow
problem, 421–422
Airport pricing, 647–648
Algebra. See Linear algebra
Algorithm. See also Simplex
algorithm
bubble sort, 801
cutting plane, 545–548, 552
decomposition, 570,
576–592, 606–607
exponential, complexity
growth rate, 802
exponential time, 190
genetic (GA). See Genetic
algorithms
greedy, 457
heap sort, 801
MST, 457–458
polynomial, complexity
growth, 801
polynomial time, 190
Wagner–Whitin algorithm, 797
Allele, 809
Allowable range, 234–236, 239
Alternative optimal solutions,
63–65, 113, 152, 212
Altman's Z-statistic, 871
Annealing, simulated, 805–808,
821
AOA (activity on arc) network,
432, 434, 468

Approximate techniques, 804.
See also Heuristic method
Approximating problem,
734–735, 746–747
Aragon, C. R., 806
Arc
artificial, 420
bad, 457
back propagation network,
885–886
backward, 424, 468
capacity, 420
defined, 413
dummy, 434, 444
equipment replacement
problem, 415–417
forward, 424
heuristics, 534–535, 551–552
tree, 514
Arms race prisoner's dilemma
game, 637
Array function MMULT, 19
Artificial arc, 420
Artificial intelligence (AI), 805
genetic search, 805,
808–815, 821
neural networks. See Neural
networks
simulated annealing,
805–808, 821
Tabu search (TS), 805,
815–820, 821
Artificial variables
defined, 173
formulas, 272
Aspiration criteria, 815
Assembly of kits case, 907–909
Assignment problems, 393
computer solution, 397–398
Hungarian method, 395–397,
405–406
Machine assignment
problem, 393–394
unbalanced, 395
Assumptions. See also Lemma
additivity, 53, 62
certainty, 54, 62
divisibility, 54, 62, 384
proportionality, 53, 62
Auto company manufacturing
example, 63–66
Average yield example, 5

Back propagation network,
866–867, 884–890
Backlogged demands, 103
Backtracking, defined, 520
Backward arc, 424, 468
Backward recursion, 797
Bakeco example, 654–655
Baker, K., 74
Baker example, 185–186
Balanced transportation
problem, 363, 402, 404,
417
Bankruptcy prediction and
neural networks, 871,
881–882
Basic feasible solution (bfs)
adjacent, 137–138
computing row 0, 461–462
defined, 132, 210
degenerate, 169, 376
extreme points, 132–134, 153
for MCNFPs, 460–461
minimum-cost method,
378–380
new, 143–147
northwest corner method,
376–378, 383, 405
rows, 144–147
transportation problems,
373–382, 405
Vogel's method, 380–382
Basic solution of simplex
algorithm, 131–134
Basic variable (BV)
cases, 30–32
defined, 30, 45
infeasible basis, 275
objective function coefficient,
278–281, 288
sensitivity analysis, 278–281,
390–391
simplex algorithm, 131–132
suboptimal basis, 275
Beam search, 815, 821
BestChip expansion strategy,
903–907
Bevco example, 172–178,
179–184, 208, 209
Bias, solving with Evolutionary
Solver, 887–890
Big M method, 172–178, 211,
267
Binary coding, 809

Binary data, 870
Binding constraint, 58, 113
Blending problems, 85–92
modeling issues, 91
oil blending, 86–91, 92,
492–494
steel industry, 92
Sunco Oil, 86–91
Texaco, 92
Bodin, L., 536
Boltzman's function, 806
Bond portfolios, 107
Bottlenecks, 593–597, 608
Branch-and-bound method for
solving problems, 550–551
combinatorial optimization,
527–538
implicit enumeration,
542–545, 552
IP, 573, 575
knapsack, 524–526, 551
machine-scheduling, 528–530
mixed IP, 523–524, 551
pure IP, 512–522, 551
Telfa Corporation, 513–522,
545–548
trees. See Branch-and-bound
trees
Branch-and-bound trees
machine-scheduling, 529
mixed IP, 524
Stockco knapsack, 526
Telfa Corporation, 517–518,
519, 545–548
traveling salesperson, 532
Breadco Bakeries, 154–157,
208–210
Break points of the function,
490–491
Brite Power capacity expansion
case, 909–910
Brute production process
model, 95–97
Bubble sort algorithm, 801
Bundling of prices, 823–830
Burnit goal programming
example, 191–198

Calculus
continuity and discontinuity,
654–655
differentiation, 655–656
higher derivatives, 656

Calculus (*continued*)
    limits, 653
    partial derivatives, 656–658
    Taylor series expansion, 656
Callen, J., 335
Camper, Josie, example, 479
Candidate list, 815–817
Candidate solution, 519, 551
Candy bar pricing NLP, 688–690
Canonical form
    basic variables, 144–147
    defined, 141
    optimal for a max problem, 147–148
Capacity of a cut, 427–428
Capital budgeting problem, 76–81
Car driving and neural networks, 870–871
Cases, 891–910
Ceiling of expected loss, 617–618, 624
Cell
    Changing, 202, 204, 206
    defined, 363, 866
    input, 866, 872–874
    output, 866, 872–874
    Target. *See* Target Cell
Centering transformation, 599–600
Central constraints, 576, 582, 591, 606, 607
Central Limit Theorem, 444
Certainty assumption, 54, 62
Certificate, defined, 802
Chain, defined, 413
Chanelle perfume model, 95–97
Changeover schedule, 817
Changing Cell
    defined, 202
    in diet problem, 204
    in Sailco example, 206
Characteristic function in *n*-person game theory, 639, 650
Cheapest-insertion heuristics (CIH), 534–536, 552
Chemco NLP example, 739–741
"Chicken" game, 637–638
Circular reference, 794
Circularity, 794
CITGO Petroleum example, 6–7
Cities assigned to clusters, 851–856
Classification problem, 875–877
Closed path, defined, 456
Clothing fixed-charge IP problem, 480–483
Cluster analysis, 851–852
    setting up distances, 853–855
    solution by Evolutionary Solver, 855–856
    standardizing attributes, 852–853

Coefficient
    graphical analysis of objective function, 227–228, 252, 262–263
    LINDO output, 234–235, 253, 281
    objective function. *See* Coefficient, objective function
    technological, 51, 112
Coefficient, objective function
    basic variable, 390–391
    defined, 50–51, 112
    nonbasic variable, 324, 390
    100% Rule for changing, 289–292
    optimal z-value and, 251–252, 253–254
    range, 234–236, 253, 288, 343
Coin toss game with bluffing, 618–621
Collins, E., 808
Column generation for solving large-scale LPs, 570–576, 584, 606
Column player, 624–632, 649–650
Combinatorial optimization problems
    defined, 527
    gas truck loading, 839–841
    machine-scheduling problem, 528–530
    traveling salesperson problem (TSP), 527, 530–538
Complementary pivoting, 729
Complementary slackness, 325–328, 345, 629–632
Completion of a node, 540–542, 544
Complexity growth of polynomial algorithms, 801
Complexity growth rate of exponential algorithms, 802
Complexity theory, 800–804
Computer
    Microsoft Project, 441
    packages. *See* Software
    sensitivity analysis and, 232–241
    for solving assignment problems, 397–398
    for solving transportation problems, 368–369
    State University problem, 457–458
Computer-aided dispatch (CAD), 8
Concave function, 673–679, 744
Conditional formatting, 849–850
CONDOR, 805
Constant sum games, 613, 649–650

Constants, defined, 20
Constraints
    binding, 58, 113
    central, 576, 582, 591, 606, 607
    convexity, 581, 591, 607
    cut constraint, 546–548
    defined, 2–3, 112
    demand, 361
    either-or constraints in IP, 487–489, 550
    equality, 127–130
    Giapetto's problem, 51–52
    if-then constraints, 490, 550
    linear programming problems, 51–52, 112
    nonbinding, 59
    nonlinear programming problems, 659
    qualifications, 714, 720–721
    redundant, 90
    right-hand side. *See* Right-hand side (rhs) of constraint
    surrogate, 544–545
    upper-bound, 593
Continuous data, 870
Continuous functions, 654–655
Convex
    combination, 135, 578, 607
    feasible region, 62
    function, 673–679, 744
    set, 59, 113
Convexity constraints, 581, 591, 607
Cooling schedule in simulated annealing, 806
Core of *n*-person game, 641–644, 650
Corner points, 59, 133
Corporate-training programs selection case, 900–903
Cost. *See also* Reduced cost
    matrix, 394, 396
    variable, 362
CPM-PERT project-scheduling models, 468–469
    application of, 432–433
    crashing the project, 439–441
    critical path, defined, 437, 469
    early event time, 434–435, 468
    finding the critical path, 437, 438–439, 441–443
    free float, 438, 469
    history of, 431–432
    late event time, 434, 435–436, 469
    PERT difficulties, 445–446
    PERT procedures, 443–445, 470
    project network, 433–434
    total float, 436–437, 469
    Widgetco problem, 433–434, 438–446
Crashing the project, 439–441

Credit card approval and neural networks, 871
Critical activity, defined, 437, 469
Critical path method. *See* CPM-PERT project-scheduling models
Crossover in genetic algorithms, 809–815
CSL computer service problem, 109–111
Current basis
    effect of changing right-hand side of constraint, 282–285
    no longer optimal, 280–281, 284–285
    optimal, 280
Cut constraint, 546–548
Cutting plane algorithm, 545–548, 552
Cutting stock problem, 570–576
Cycle, defined, 456
Cycling, defined, 170

Dakota Furniture Company problem
    adding a new activity, 287, 325
    alternative optimal solutions, 152–153
    changing basic variable, 278–281
    changing nonbasic variable, 276–278
    changing right-hand side of constraint, 283–285
    dual, 296–297, 302–303, 306–307
    dual optimal solution, 310, 330–334
    dual prices, 319–320
    duality and sensitivity analysis, 323–325
    formulas, 267–274
    LINDO output, 158–161, 281–282
    nonbasic variables, 285–287
    nonnegative shadow price, 315–316
    100% Rule, 292, 293–294
    product form of the inverse, 569, 605–606
    revised simplex algorithm, 563–566
    sensitivity analysis, 276
    shadow price, 314–315
    simplex algorithm, 140–141
    tableaus, 277–278, 281, 285, 286, 331–333
    Theorem of Complementary Slackness, 326–328
Dantzig, George, 49
Dantzig–Wolfe decomposition algorithm, 570, 576–592, 606–607
Data
    binary, 870
    continuous, 870

learning data set, 869–870
scaling, 868–869
Data Envelopment Analysis (DEA) method, 335–340
DEA (Data Envelopment Analysis) method, 335–340
Decision variables
changing right-hand side of constraint, 284
defined, 2
Giapetto's problem, 49–50
Decomposition algorithm, 570, 576–592, 606–607
Degeneracy and sensitivity analysis, 240–241, 320–321
Degenerate bfs, 169, 376
Degenerate LP, 134, 169, 240–241
Delinquency movement matrix (DMM), 9
Delta Airlines schedule development using Karmarkar's method, 191
Demand constraint, 361
Demand point
defined, 361, 400
description of problem, 362
dummy, 363, 365, 401–402, 406
in transshipment problem, 400–403, 406
Demographic analysis, 860–863
Derivative
differentiation, 655–656
function, 655
higher, 656
partial, 656–658
second-order partial, 657–658
Determinants of a square matrix, 42–43, 46
Deterministic dynamic programming. See Dynamic programming
Deviational variables, 192
Diagonal elements, 36
Diet problem
conversion to standard form, 128–130
dual, 297, 303–304
Excel Solver solution, 203–205
LINDO output, 161–162, 291
linear programming, 68–71, 892
100% Rule for changing objective function coefficients, 290–292
100% Rule for changing right-hand sides, 292–293, 294
Differential calculus. See Calculus
Differentiation, 655–656
Dijkstra's algorithm, 416–417, 467, 801

Direction of unboundedness, 134–136, 157–158
Discontinuous function, 654
Discriminant analysis, 860–863, 871
Discriminant score, 861
Diversification strategy, 815
Divisibility assumption, 54, 62, 384
DMM (delinquency movement matrix), 9
Dominated player, 619
Dominated solution, 738
Dominated strategies, 619–620, 633, 649
Domination, 641–642, 650
Dorian Auto example
direction of unboundedness, 134–136
either-or constraint, 488–489
graphical solution, 60–62
media selection, 494–496
Doyle, T., 536
Drug game, 639, 642, 646–647
Drug pricing when exchange rates change, 686–688
Dual, 262, 343
complementary slackness, 325–328, 345, 629–632
defined, 295
economic interpretation, 302–304
formulas, 267–274
LP, 295–301
nonnormal LP, 298–301
nonnormal max problem, 299–300
nonnormal min problem, 300–301
normal max problem, 295–297
normal min problem, 295–297
optimal solution of max problem, 310–312, 344
optimal solution of min problem, 312–313, 344
prices, 237–238, 240
sensitivity analysis and. See Sensitivity analysis
simplex method, 329–334, 345, 521–522, 547–548
Dual Theorem, 304, 307–313, 343
Karmarkar's method, 603
problem, 308
proof, 307, 309
shadow price and, 313–315, 319–321
weak duality, 305–307
Duality
sensitivity analysis and, 323–325, 344–345
weak, 305–307
Dummy
activity, 433
arc, 434, 444

demand point, 363, 365, 401–402, 406
Dynamic aspect of PAYMENT model, 10
Dynamic lot-size model, 797
Dynamic programming, 750
characteristics of applications, 756–757
computational difficulties, 785, 797
computational efficiency, 755–756
dynamic lot-size model, 797
equipment-replacement problems, 774–777
EXCEL Solver solutions, 790–796
Finco resource allocation problem, 764–768
fishery example, 779–780
Glueco sales allocation example, 787–788
inventory problem, 758–763, 794–796
knapsack problem, 768–773, 790–792
match puzzle, 750
milk problem, 750–751
minimax shortest route example, 786–787
network problem, 751–757
nonadditive recursions, 786–788
power plant example, 780–781
recursions, 771–772, 777, 778–788, 797
resource-allocation problems, 763–773, 792–794
shortest path trip problem, 752–757, 786–787
Sunco Oil refinery capacity example, 782–783
time value of money, 780–785
traveling salesperson example, 783–785
wheat sale example, 781–782
working backward, 750–751, 796–797
Dynamic scheduling problem, 74, 109–111

Early event time, 434–435, 468
Economic interpretation of the dual, 302–304
Efficient frontier, 739
Either-or constraints in IP, 487–489, 550
Element, ijth, defined, 11
Elementary matrix, 568
Elementary row operation (ERO), 22–24
Gauss–Jordan method of solution, 22–27, 44–45
Type 1, 23, 25–27, 44
Type 2, 23, 25–27, 28–29, 44
Type 3, 23, 29–30, 44

Elevator control and neural networks, 871
Emergency vehicle location in Springfield case, 905–906
Empirical analysis for a heuristic, 536
Employee work assignments example, 846–850
Entering variable
determining, 142–143
nonbasic, 386–387
pivot, 143–147
Equality constraint, 127–130
Equations
flow balance, 450, 470
normal, 700
Equilibrium point, 612, 635, 650
Equipment replacement problem, 415–416, 774–777
ERO. See Elementary row operation
Etching example, 882–884
Euing Gas IP problem, 492–494
Evaluation in genetic algorithms, 809, 812–813
Event time, 434–436
Events, defined, 432
Evolutionary Solver
ADBUG function, 857, 858
cluster analysis, 851–856
combinatorial problems, 839–841
discriminant analysis, 860–863
finding revenue-maximizing prices, 826–827
fitting curves, 857–859
gas truck loading, 839–841
genetic algorithms, 826, 883–884
Index function, 823–825
John Deere production scheduling, 841–846
Match function, 823–826
neuron network models, 868, 883–884
nonlinear pricing models, 830–835
Premium Solver, 826–830, 888–889
price bundling, 823–830
uses for, 823
warehouse location, 836–837
weights and biases, 887–890
worker assignment problem, 846–850
Examples. See Problems
EXCEL Solver
computing NPV, 77–78
dynamic programming, 790–796
genetic search, 815. See also Genetic algorithms
infeasible LPs and, 208, 209

EXCEL Solver (continued)
  MINVERSE (inverting
    matrices), 41
  MMULT (matrix
    multiplication), 19
  Predict add-in, 874–882
  shortcomings, 823
  solving diet problem,
    203–205
  solving IP problems,
    499–502, 522
  solving LPs, 202–210
  solving NLPs, 669–671,
    688–690, 726–727
  solving Sailco example,
    205–207
  tolerance option, 522
  transportation problem,
    370–371
  unbounded LPs and, 208–210
  value of option, 207–208
  work-scheduling problem, 74
  XNPV, 79
EXCEL spreadsheet, obtaining
  LINGO data from,
  369–370
Excess variable
  conversion to equality
    constraint, 128–130
  defined, 128
  formula, 272
  sensitivity analysis and,
    239–240, 272
Exchange move in Tabu
  problem, 819–820
Exchange rate nonlinear
  problems, 686–690
Exponential time algorithm,
  defined, 190
Extreme point
  defined, 59
  Dakota Furniture problem,
    153
  Giapetto's problem, 59–60
  Leather Limited problem,
    132–134
Extremum candidates, 680–685

Farmer Leary's shadow price
  example, 247
Fathomed, 515, 551
Feasible directions method,
  736–738, 747
Feasible region
  convex, 62
  decomposition algorithm,
    579, 580
  defined, 3, 54, 112
  Giapetto's problem, 54–55,
    57
  integer programming, 512
  Leather Limited example,
    132–134, 135
  nonlinear problem, 659
  Telfa Corporation problem,
    513–516, 519, 545–548
  unbounded, 62
Financial models

bond portfolios, 107
  Finco Investment
    Corporation example,
    105–107
  multiperiod, 105–107
  neural networks, 870
Finco Investment Corporation
  example, 105–107,
  764–768
Finish node, 433
Firerock tire production
  problem, 667–669
First iteration of Karmarkar's
  method, 601–602
Fishery dynamic-programming
  example, 779–780
Fitting curves, 857–859
Fixed charge, defined, 481
Fixed-charge IP problems, 549
  Gandhi Cloth Company,
    480–483
  lockbox, 483–486
Fixed variable, 540, 541
Float
  free, 438, 469
  total, 436–437, 469
Floor of expected reward, 617,
  618, 624
Flow balance equations, 450,
  470
Fly-by-Night Airlines,
  421–422
Ford–Fulkerson method for
  solving maximum-flow
  problems, 424–429,
  467–468
Forecasting modules, CITGO
  Petroleum example, 7
Formulas
  dual, 267–274
  optimal tableau, 267–273
Forward arc, 424
Forward recursion, 797
Frank and Wolfe, 737
Free float, 438, 469
Free variable, 540–541
Functions
  array, 19
  break points, 490–491
  characteristic, 639, 650
  continuous, 654–655
  convex and concave NLPs,
    673–679, 744
  derivative, 655
  discontinuous, 654
  linear. See Linear function
  objective function coefficient.
    See Objective function
    coefficient
  objective function of NLP,
    659
  piecewise. See Piecewise
    linear function
  potential, 602
  unimodal, 692–693
Fundamental Theorem of
  Genetic Algorithms, 809

Game theory, 649–651
  advertising prisoner's
    dilemma game, 636
  airport pricing, 647–648
  arms race prisoner's dilemma
    game, 637
  "chicken" game, 637–638
  constant-sum games, 613,
    649–650
  constant-sum TV game,
    613
  dominance, 641–642
  drug game, 639, 642,
    646–647
  garbage game, 639, 643
  land development game,
    639–640, 641–642,
    643–644
  linear programming and
    zero-sum games,
    623–633
  n-person, 639–640, 641–644,
    650
  odds and evens, 614–615,
    616–618
  prisoner's dilemma, 634–637,
    650
  saddle point. See Saddle point
  Shapley value, 644–648,
    650–651
  two-person constant-sum
    games, 613
  two-person nonconstant-sum
    games, 634–638, 650
  two-person zero-sum games.
    See Two-person zero-
    sum games
  uses of, 610
Gandhi Cloth Company
  example
  EXCEL Solver solution,
    499–502
  fixed-charge problem,
    480–483
  LINDO solution, 497
Garbage game, 639, 643
Gas truck loading problem,
  839–841
Gasoline illustration of matrix
  multiplication, 17–18
Gauss–Jordan method
  computing optimal tableau
    example, 273
  Dakota Furniture Company
    problem, 270, 278
  determination of linearly
    independent or linearly
    dependent vectors,
    34–35, 45–46
  elementary row operation
    (ERO), 22–27, 44–45
  finding solutions, 24–27
  inverse of a matrix, 37–40,
    46
  linear system with infinite
    solutions, 28–29
  linear system with no
    solution, 28

solving systems of linear
    equations, 22, 30–31
  summary, 29–30
GE Capital example, 9–10
Generalized resource allocation
  problem, 767–768
Generation in genetic
  algorithms, 809, 812
Genetic algorithms (GA)
  back propagation network,
    884–890
  Evolutionary Solver, 826,
    883–884
  heuristics, 808–815, 821
  optimizing a neural network,
    882–884
Genetic search, 805, 808–815,
  821
Geometrical interpretation
  Kuhn–Tucker conditions,
    717–720
  Lagrange multipliers,
    708–709
Giapetto's Woodcarving
  problem, 49, 53
  additivity assumption, 53
  certainty assumption, 54
  constraints, 51
  decision variables, 49–50
  divisibility assumption, 54
  feasible region, 54–55, 57,
    59–60
  graphical solution, 57
  objective function, 50–51
  objective function coefficient
    and optimal z-value,
    251–252
  optimal solution, 55, 58,
    228–230, 263–265
  proportionality assumption,
    53
  sensitivity analysis, 227–231,
    262–265
  sign restrictions, 52
Gilmore, P., 570, 576
Global maximum of NLP,
  685
Glover, F., 805, 815
Glueco sales allocation
  example, 787–788
Goal programming, 65, 127,
  191
  Burnit example, 191–198
  preemptive, 194–198
  simplex, 194
  using LINDO or LINGO for
    solving problems,
    197–198
  weight, 193
Golden, B., 536, 808
Golden section search,
  692–697, 744–745
Golf-Sport operations
  management case,
  893–896
Gomory, R., 570, 576
Good solutions, 826–827
Gradient vector, 704

Graph, defined, 413
Graphical solution. *See also*
    Two-variable linear
    programming problems,
    graphical solution of
Dorian Auto problem, 134
Giapetto's Woodcarving
    problem, 57
objective function coefficient,
    227–228, 252,
    262–263
odds-and-evens game,
    616–618
optimal solution, 228–230,
    252, 263–265
Powerco problem, 362
sensitivity analysis,
    341–342
Greedy algorithm, 457
Greedy heuristic, 804

Half-space, defined, 138
Halting problem, 801
Handwriting recognition and
    neural networks, 871
Heap sort algorithm, 801
Help-You Company case,
    907–909
Help, I'm Not Getting Any
    Younger! case, 892
Hessian, 677–679, 698, 699
Heuristic method
    artificial intelligence. *See*
        Artificial intelligence
    comparisons, 821
    complexity theory, 800–804
    genetic search, 805,
        808–815, 821
    greedy, 804
    metaheuristic, 805
    nearest-neighbor heuristics
        (NNH), 534, 535–536,
        551, 804
    neighborhood-exchange, 804
    neural networks. *See* Neural
        networks
    nonpolynomial (NP)–class
        problems, 800, 802
    nonpolynomial
        (NP)–complete
        problems, 800, 803
    nonpolynomial (NP)–hard
        problems, 800, 803–804,
        810
    problems. *See* Heuristic
        problems
    procedures, 804–805
    restaurant scheduling
        problem, 75
    simulated annealing,
        805–808, 821
    solving traveling salesperson
        problems (TSPs),
        534–536, 551–552, 800,
        802
    Tabu search (TS), 805,
        815–820, 821
    when to use, 800

Heuristic problems
    genetic algorithms. *See*
        Genetic algorithms
    simulated annealing
        algorithm and TSP,
        807–808
    Tabu search, 816–820
Hidden layers, 866–867,
    871–874, 890
Higher derivatives, 656
Holland, J. A., 808
Hospital DEA example,
    335–340
Hungarian method of solution,
    395–397, 405–406

Identity matrix, defined, 36
If-then constraints, 490, 550
*ij*th element, 11
Immediate predecessor, 435,
    468
Immediate successor, 436, 469
Implicit enumeration to solve
    IPs, 540–545, 552
Improvement in *z*-value,
    265–266
Imputation, 640, 650
Index and Match function,
    823–825
Infeasible basis of BV, 275
Infeasible LP
    auto company example,
        65–66
    Bevco example, 177–178,
        179–184, 208, 209
    defined, 63, 113
Initial node, 413
Initial temperature in simulated
    annealing, 806
Input
    cell, 866, 872–874
    layer, 866, 872–874
    values, 867–868
Integer programming (IP),
    475–477, 549–552. *See
    also* Integer programming
    problems
    cutting plane algorithm,
        545–548, 552
    feasible region, 476–477
    formulations, 477–480, 549
    LP relaxation, 476–477, 512,
        513, 523, 525, 546–548
    piecewise linear functions,
        490–496, 550
    simple, 476–477
Integer programming problems
    (IP), 54
    branch-and-bound method.
        See Branch-and-bound
        method for solving
        problems
    capital-budgeting IP, 478–480
    combinatorial optimization,
        527–538
    defined, 475
    Dorian Auto, 488–489,
        494–496

either-or constraints,
    487–489, 550
Euing Gas, 492–494
EXCEL Solver solutions,
    499–502, 522
facility-location set-covering,
    486–487
fixed-charge, 480–486, 549
formulating, 477–480
Gandhi Cloth Company
    example, 480–483, 497,
    499–502
if-then constraints, 490, 550
implicit enumeration,
    540–545, 552
knapsack, 479, 524–526
LINDO solutions, 496–497
LINGO solutions, 497–499
lockbox, 483–486, 490,
    497–499
machine-scheduling, 528–530
mixed, 475, 523–524
pure, 475, 512–522, 540, 551
set-covering, 486–487
Stockco example, 478–480,
    525–526
Tabu search solution, 820
Telfa Corporation, 513–522,
    545–548
traveling salesperson. *See*
    Traveling salesperson
    problem
Intensification strategy, 815
Interior point method, 803
Interval of uncertainty, 693–697
Inventory
    backtracking, 520
    LIFO rule. *See* LIFO rule
    model, 100–103
    problem solved by dynamic
        programming, 758–763,
        794–796
    problem modeled as
        transportation problem,
        366–368
Inverse, product form of,
    567–569, 605–606
Inverse of a matrix, 36–39
    defined, 37
    EXCEL, 41
    Gauss–Jordan method,
        37–40, 46
    matrix with no inverse, 39–40
    solving linear systems, 40
Isocost line, 58
Isoprofit line, 58
Iteration, defined, 145, 146
*i*th principal minor, 677

John Deere production-
    scheduling problem,
    841–846
Johnson, D. S., 806
Jumptracking, defined, 520

Karmarkar's method for solving
    LPs, 190–191, 597–598,
    608

centering transformation,
    599–600
description and example,
    600–601
first iteration, 601–602
interior point method, 803
LP standard form, 602–605
potential function, 602
projection, 598–599
Khachian, 803
Knapsack problem
    alternative recursion,
        771–772
    branch-and-bound method
        solution, 524–526, 551
    defined, 479
    dynamic programming
        solution, 768–773
    EXCEL Solver solutions,
        790–792
    network representation,
        770–771
    turnpike theorem, 772–773
Krajewski, L., 75
*k*th leading principal minor,
    677
Kuhn–Tucker (KT) conditions,
    713–717, 746
    constraint qualifications, 714,
        720–721
    geometrical interpretation,
        717–720
    LINGO solution, 721–722
    in quadratic programming
        problems, 727–729, 746

Labeling method in solving
    maximum-flow problems,
    425, 426
Lagrange multipliers, 706–708,
    745
    advertising problem, 710
    geometrical interpretation,
        708–709
    optimal solution, 711
    sensitivity analysis and,
        709–711
Land development game,
    639–640, 641–642,
    643–644
Late event time, 434, 435–436,
    469
Lawn mower scheduling
    problem, 842–846
Layers
    hidden, 866–867, 871–874,
        890
    input, 866, 872–874
    output, 866–867, 872–874
Learning data set, 869–870
Least squares method, 700–701
Leather Limited problem,
    127–128, 132–134, 135
Leatherco example, 316–317
Lecun, Y., 871
Lemma. *See also* Assumptions
    capacity of cut, 427–428
    dual theorem, 307

Lemma (*continued*)
    Karmarkar's centering
        transformation, 599
    method of steepest ascent,
        705
    weak duality, 305, 307
Leon Burnit Advertising
    Agency goal programming
    example, 191–198
Lewis, P. M., 804
LIFO rule
    backtracking, 520
    defined, 515
    implicit enumeration, 543
    Stockco capital budgeting
        problem, 525
    Telfa Corporation problem,
        515, 516, 519
Limits in calculus, 653
LINDO (Linear Interactive and
    Discrete Optimizer)
    computer package, 74, 158
    ALLOWABLE DECREASE,
        234, 239, 282, 285, 320
    ALLOWABLE INCREASE,
        234, 239, 281, 285, 320
    assignment problems,
        397–398
    CURRENT COEF, 281
    CURRENT RHS, 285
    Dakota Furniture example,
        158–161, 281–282. *See
        also* Dakota Furniture
        Company problem
    degeneracy, 240–241
    diet problem, 161–162,
        291
    DUAL PRICE, 237, 240,
        319–321, 339
    hospital DEA example,
        337–338
    integer programming
        problems, 496–497
    menu commands, 217–220
        edit menu, 218
        file menu, 217–218
        help menu, 220
        reports menu, 219
        solve menu, 219
        window menu, 219–220
    minimization problems,
        233–234, 236
    OBJECTIVE COEFFICIENT
        RANGES, 234–235,
        253, 281
    optional modeling
        statements, 220–221
    Parametrics feature, 248
    preemptive goal
        programming for
        problem solving, 197–198
    REDUCED COST, 236, 240,
        320
    RIGHTHAND SIDE
        RANGES, 236, 237,
        239, 253, 285, 320
    sensitivity analysis, 232–241,
        281–282

shadow prices, 319–321, 342,
    344
TABLEAU command, 162,
    240, 320
trade-off curve problem,
    740–741
transportation problems,
    368
Tucker, Inc. *See* Tucker, Inc.,
    example
two-person zero-sum games,
    632–633
Widgetco output, 440
Winco Products example. *See*
    Winco Products example
Line segment joining, 14–15
Linear algebra
    equations. *See* Linear
        equations
    matrices. *See* Matrix
    vectors. *See* Vectors
Linear combination, 32
Linear dependence, 32–36, 46
    defined, 33
    determination of, 34–35
Linear equations
    basic variables, 30–32, 45
    matrices and, 20–22, 44
    solutions, defined, 20
    solving by Gauss–Jordan
        method, 22–32, 44–45
Linear function
    defined, 52
    piecewise, 249, 252,
        490–496, 550
Linear independence, 32–36,
    45–46
    defined, 33
    determination of, 34–36
Linear inequalities
    defined, 52
    graphing of, 56
Linear programming (LP), 49
    additivity assumption, 53, 62
    alternative or multiple
        optimal solutions,
        63–65, 113, 152, 212
    certainty assumption, 54, 62
    conversion to standard form,
        127–130, 141–142, 210
    degenerate, 134, 169,
        240–241
    divisibility assumption, 54,
        62, 384
    dual, 295–301
    finding a critical path, 437,
        438–439, 441–443
    infeasible LP. *See* Infeasible
        LP
    Karmarkar's method for
        solving, 190–191
    nondegenerate, 168
    optimal bfs, 136–139, 142
    Phase I and II, 179–184, 212
    problems. *See* Linear
        programming problems
    proportionality assumption,
        53, 62

scaling of, 167
solution. *See* Simplex
    algorithm
standard form, 127–130,
    141–142, 210, 602–605
three-dimensional, 138–139
unbounded. *See* Unbounded
    LP
zero-sum games and,
    623–633
Linear programming (LP)
    problem
    constraints, 51–52, 112
    decision variables, 2, 49–50
    defined, 53
    examples. *See* Linear
        programming problem
        examples
    EXCEL Solver, 202–210
    feasible region, 54–55, 57,
        112
    formulating, 113
    Giapetto's problem, 49–52
    maximum flow, 467
    nonlinear programming vs.,
        660–662
    objective function, 2, 50, 112
    objective function coefficient,
        50–51, 112
    optimal solution, 55, 58, 112,
        142
    parts of, 112
    sign restrictions, 52, 112,
        128–130
    solving large-scale problems
        using column
        generation, 570–576,
        584, 606
    solving using complementary
        slackness, 328, 345,
        629–632
    two-variable. *See* Two-
        variable linear
        programming problems
Linear programming problem
    examples. *See also*
    Problems
    auto company manufacturer,
        63–66
    baker, 185–186
    Bevco, 172–178, 179–184,
        208, 209
    blending problems, 85–92
    Breadco Bakeries, 154–157,
        208–210
    Burnit goal programming,
        191–198
    capital budgeting, 76–81
    CITGO Petroleum, 6–7
    Dakota Furniture Company.
        See Dakota Furniture
        Company problem
    diet. *See* Diet problem
    Dorian Auto, 60–62,
        134–136
    Farmer Leary's, 247
    financial, 105–107
    Giapetto's Woodcarving. *See*

    Giapetto's Woodcarving
        problem
    inventory, 100–103
    Leather Limited, 127–128,
        132–134, 135
    Mondo Motorcycles,
        186–188
    multiperiod decision,
        100–103, 105–107,
        109–111
    Police Patrol Scheduling
        System (PPSS), 7–9
    post office scheduling,
        72–75, 165–166
    production process, 95–97
    Sailco Corporation. *See*
        Sailco Corporation
        problem
    short-term financial planning
        problem, 82–85
    Star Oil Company, 80–81
    Tucker, Inc. *See* Tucker, Inc.,
        example
    Winco Products. *See* Winco
        Products example
    work scheduling, 72–75,
        109–111
Linear systems
    infinite solutions, 28–29
    matrix inverses for solving,
        40, 46
    with no solution, 28
LINGO computer package, 74,
    163, 221
    assignment problems,
        397–398
    data from EXCEL
        spreadsheet, 369–370
    determining critical path,
        441–443
    functions, 225–226
    fundamentals, 221
    hospital DEA problem,
        338–339
    integer programming
        problems, 497–499
    maximum-flow problems,
        423–424
    MCNFP, 453–454
    menu commands, 222–224
        edit menu, 223
        file menu, 222
        help menu, 224
        LINGO menu, 223–224
        window menu, 224
    nonlinear programming
        problems, 660, 690–691,
        702, 712, 721–722,
        725–726
    Oilco NLP, 665–666
    Post Office Scheduling
        problem, 165–166
    preemptive goal
        programming for
        problem solving,
        197–198
    Sailco problem, 163–165
    shortcomings, 823

tire production (Firerock), 667–669

trade-off curve problem, 741–742

transportation problems, 368–369

Traveling Salesperson (TSP) problem, 537–538

two-person zero-sum games, 632–633

warehouse location problem (Truckco), 666–667

Local extremum, 662–663

Local maximum, 662

Lockbox IP problem
  branch-and-bound method, 520
  fixed-charge problem, 483–486
  if-then constraints, 490
  solving with LINGO, 497–499

Lookup array, 825–826

Lookup cell, 825

Loop
  defined, 374
  entering variable and, 388
  importance of concept, 374–375
  minimum spanning tree and, 456
  transportation problem, 382–383, 405

Lower bound, defined, 519

Lundy, M., 806

Machine-scheduling problem, 528–530

Machineco assignment problem, 393–398

Mail-handling case, 897–900

Management science, defined, 1

Markets and neural networks, 870

Markowitz, 724

Match function, 823–826

Match puzzle, 750

Matchmaking problem, 422–423

Material-handling case, 897–900

Mathematical models, 1

Matrix
  addition of, 14–15
  cost, 394, 396
  defined, 11, 43
  equal, 12
  elementary, 568
  generators, 163
  identity, 36
  ijth element, 11–12
  inverse, 36–41, 46
  minor, 42
  multiplication. See Matrix multiplication
  product, 16, 44
  rank of, 34–35, 46
  representation, 21–22

reward. See Reward matrix

scalar multiple, 14

square, 36, 42–43, 46

systems of linear equations, 20–22

transpose of, 15–16

undefined product, 17

Matrix multiplication
  examples, 16–17
  EXCEL, 19
  gasoline illustration, 17–18
  properties, 18–19

Max problem
  finding the dual, 343, 344
  interpreting the dual, 302–303
  simplex algorithm, 140–148
  nonnormal, 299–300
  normal, 295–297
  optimal solution, 310–312, 344
  solving with dual simplex method, 329–334, 345
  unbounded LP, 154–157
  unconstrained NLP, 698–702, 745

Maximum-flow problems, 467
  defined, 419–420
  Fly-by-Night Airlines problem, 421–422
  Ford–Fulkerson method for solving, 424–429, 467–468
  formulating an MCNFP, 451–453, 470–471
  labeling method, 425, 426
  LP solution, 420–423
  matchmaking problem, 422–423
  solving with LINGO, 423–424
  Sunco Oil problem, 420–421, 428–429

McGeoch, L. A., 806

McKenzie, P., 75

Mean absolute-deviation (MAD), 879, 880

Media selection with piecewise linear functions, 494–496

Mees, A., 806

Memory in Tabu search problem, 815

Menthos nonlinear pricing model, 830–835

Metaheuristic, 805

Method of feasible directions, 736–738, 747

Method of steepest ascent, 703–706, 745

Metropolis, W., 805

Microsoft Project, 441

Military Airlift Command routes determined using Karmarkar's method, 190–191

Milk dynamic programming problem, 750–751

Minimax shortest route

example, 786–787

Minimax Theorem, 625–626

Minimization (min)problems
  finding the dual, 343, 344
  graphical solution of, 60–62
  interpreting the dual, 303–304
  LINDO output, 233–234, 236
  nonnormal, 300–301
  normal, 295–297
  optimal solution, 312–313, 344
  ratio test, 151
  reduced cost, 161, 162
  simplex algorithm, 149–151, 161, 212
  solving with dual simplex method, 333–334
  Tucker example. See Tucker, Inc., example
  unconstrained NLP, 698–702, 745

Minimum-cost method for finding bfs, 378–380

Minimum-cost network flow problems (MCNFPs), 450, 470–471
  basic feasible solutions, 460–461
  maximum flow problem, 451–453
  network simplex method, 459–460, 471
  solving with LINGO, 453–454
  traffic example, 452–453
  transportation problem, 450–451

Minimum-spanning tree (MST), 456–458

Minor of a matrix, defined, 42

Mixed integer programming problem, 475, 523–524, 551

Mixed strategy, 615–616

Modeling, 1
  blending problems, 91
  descriptive, defined, 1
  deterministic, defined, 4–5
  dynamic, defined, 4, 100
  dynamic lot-size, 797
  integer, defined, 4
  inventory, 100–103
  linear, defined, 4
  linear optimization, GE Capital, 10
  linear programming (LP), CITGO example, 6–7
  multiperiod. See Multiperiod
  network. See Network models
  noninteger, defined, 4
  nonlinear, defined, 4
  nonlinear pricing, 830–835
  optimization, complete, 3–4
  optimization, defined, 2
  PAYMENT, 9–10

periodic review, 758

prescriptive, defined, 2

production process, 95–97

production-smoothing costs, 186–188

seven-step model-building process, 5–6

static, defined, 4, 100

stochastic, defined, 4

stochastic, in Wozac example, 1–2

Modular design for the Help-You Company case, 907–909

Mondo Motorcycles example, 186–188

Monopolistic pricing, 684, 699, 719–720

Monte Carlo approach, 805

Morgenstern, Oskar, 611, 641, 649

Move value, 818–820

Multiperiod
  decision problems, 100–103
  financial models, 105–107
  work scheduling, 109–111

Multiple optimal solutions, 63–65

Multipliers, simplex, 461, 471

Mutation in genetic algorithms, 809–815, 826, 848

n-dimensional unit simplex, 598

n-person games, 639–640, 641–644, 650

Nearest-neighbor heuristics (NNH), 534, 535–536, 551, 804

Necessary conditions, 746

Neighborhood-exchange heuristic, 804

Net present value (NPV), 76–81

Network
  defined, 413
  equipment-replacement problem, 776–777
  inventory example, 763
  knapsack problem, 770–771
  problem, 751–757
  project, 432
  resource example, 766–767
  simplex, 421, 450, 459–465, 471
  simplex algorithm, 452

Network models, 413–414
  CPM-PERT. See CPM-PERT project-scheduling models
  defined, 413
  Dijkstra's algorithm, 416–417, 467
  example, 414
  maximum-flow problems, 419–429, 451–453, 467–468
  minimum-cost network flow problems (MCNFPs), 450–454, 470–471

Network models (*continued*)
minimum-spanning tree (MST) problems, 456–458
network simplex, 421, 450, 459–465, 471
shortest path problem as transshipment problem, 417–418, 467
shortest path problems, 414–418, 467
transportation simplex, 452
Network simplex method, 421, 450, 459–465, 471
Neural networks, 866
bankruptcy prediction, 871, 881–882
binary data, 870
continuous data, 870
defined, 805, 866–867
estimating with Predict, 874–882
Evolutionary Solver, 868, 883–884
examples of use of, 870–871
genetic algorithms for optimizing, 882–884
genetic algorithms to determine weights back propagation network, 884–890
input and output values, 867–868
regression vs., 866, 871–874, 881, 882, 890
sales advertising example, 885–890
scaling data, 868–869
sibling-acquaintance problem, 874–879
sigmoid transfer function, 869, 886
testing and validation, 869–870
uses for, 821
XOR example, 871–874
NeuralWare, 874
New activities, 287, 325
Nickles, J. C., lockbox problem. *See* Lockbox IP problem
Node
back propagation neural network, 885
completion of, 540–542, 544
defined, 413
equipment replacement problem, 415–417
finish, 433
implicit enumeration, 542–545
initial, 413
jumptracking, 520
terminal, 413
tree, 514
unconnected, 457
Nonadditive recursions, 786–788
Nonbasic variable (NBV)

cases, 30–32
defined, 30, 45
duality and sensitivity analysis, 324–325
entering, 386–387
pricing out, 384–386, 563–566, 585–588, 590–591
reduced cost, 277–278
in sensitivity analysis, 276–277, 285–287, 288, 390
of simplex algorithm, 131–132
Nonbinding constraint, 59
Nonconvex sets, 59
Nondegenerate LP, defined, 168
Nonlinear pricing models, 830–835
Nonlinear programming, 653
determinants, 42
differential calculus, 653–658
Nonlinear programming problems (NLP), 653
Chemco example, 739–741
constraints, 659
continuous function (Bakeco), 654–655
convex and concave functions, 673–679, 744
defined, 659
EXCEL Solver solutions, 669–671, 688–690, 726–727
feasible directions method, 736–738, 747
feasible region, 659
global maximum, 685
golden section search, 692–697, 744–745
Kuhn–Tucker (KT) conditions, 713–722, 746
Lagrange multipliers, 706–711, 745
least squares estimation, 700–701
linear programming vs., 660–662
LINGO solution, 660, 690–691, 702, 712, 721–722, 725–726
local extremum, 662–663
maxima, minima, and saddle points, 698–702, 745
method of steepest ascent, 703–706, 745
objective function, 659
Oilco example, 663–666, 731–732
one variable, 680–691
pareto optimality, 738–743
points, 680–685, 694–697
pricing, 685–690, 699, 712, 722
Proctor and Ramble trade-off curve, 741–742
product profitability, 655–656

production maximization example, 660
profit maximization by monopolist, 684, 699, 719–720
profit maximization example, 659–660
quadratic programming problem (QPP), 723–729, 746
separable programming, 731–735, 746–747
Taylor series expansion, 656
tire production (Firerock), 667–669
trade-off curves, 738–743, 747–748
unconstrained, 659, 698–702, 745
warehouse location problem (Truckco), 666–667
Nonnormal LP, finding the dual, 298–301
Nonpolynomial (NP) class problems, 800, 802
Nonpolynomial (NP)–complete problems, 800, 803
Nonpolynomial (NP)–hard problems, 800, 803–804, 810
Normal equation, 700
Normal max problem, defined, 295–297
Normal min problem, defined, 295–297
Normalized vector, 704
Northwest corner method for finding bfs, 376–378, 383, 405
Notation, 342, 404
Number referred to as scalar, 14

Objective function
defined, 2, 112
Giapetto's problem, 50
NLP, 659
row 0 version, 140
Objective function coefficient
changing basic variable, 278–281, 288
changing nonbasic variable, 276–278, 288
defined, 50–51, 112
graphical analysis, 227–228
graphical analysis of change, 252, 262–263, 227–228
100% Rule, 289–292
ranges, 234–236, 253, 288, 343
Odds and evens game, 614–615, 616–618
Oil blending problems, 86–91, 92, 492–494
Oilco NLP, 663–666, 731–732
100% Rule
changing objective function coefficients, 289–292

changing right-hand sides, 292–294
Operations research (OR), 1, 804
Optimal basis, 280–281
Optimal bfs in an LP, 136–139, 142
Optimal solution
alternative, 63–65, 113, 152, 212
defined, 3, 55, 112
dual simplex method, 330–332
Giapetto's problem, 55, 58, 228–230, 263–265
graphical analysis, 228–230, 252, 263–265
nonlinear programming problem, 659
simplex algorithm, 152–153, 212
transportation problems, 405
Optimal strategy, 618, 621, 649
Optimal z-value, 248–252, 253–254
Optimization problems
combinatorial. *See* Combinatorial optimization problem example, 802
solving with Evolutionary Solver. *See* Evolutionary Solver
Order-of-fitness values, 813
Original space, 600
Output
cell, 866, 872–874
layer, 866–867, 872–874
values, 867–868

Padberg, 805
Pareto optimal solutions, 738–743
Partial derivatives, 656–658
Path, defined, 414
PAYMENT model of GE Capital, 9–10
Performance guarantee for a heuristic, 535
Periodic review model, 758
PERT. *See* CPM-PERT project-scheduling models
Phase I LP, 179–184, 212
Phase II LP, 179–184, 212
Phoneco price-bundling example, 823–830
Piecewise linear function
defined, 249
integer programming and, 490–496, 550
minimization problem, 252
Pivot row, 144
Pivot term, 144
Pivoting
defined, 144
network simplex, 462–463
transportation problems, 382–384, 405

Points
   break, 490–491
   corner, 59, 133
   demand. *See* Demand point
   dummy demand, 363, 365,
      401–402, 406
   equilibrium, 612, 635, 650
   extreme, 59–60, 132–134,
      153
   NLP, 680–685, 694–697
   saddle. *See* Saddle point
   stationary, 698, 699
   supply, 361, 362, 400–403,
      406
   transshipment, 400–403
Polaris missile development,
   431
Police Patrol Scheduling
   System (PPSS), 7–9
Polyhedron, 138
Polynomial time algorithm,
   defined, 190
Portfolio optimization problem,
   723–727
Post office
   material handling case,
      897–900
   scheduling problem, 72–75,
      165–166
Potential function in
   Karmarkar's method, 602
Power capacity expansion case,
   909–910
Power plant dynamic
   programming example,
   780–781
Powerco problem
   basic feasible solutions,
      373–375
   entering nonbasic variable,
      386–387
   linear programming model,
      360–365
   network, 414
   northwest corner bfs, 383,
      384–385
   pivoting procedure, 383
   pricing out nonbasic
      variables, 384–386
   sensitivity analysis, 390–392
   solving using LINGO,
      368–371
   tableau, 389, 391, 392
Predecessor
   of the activity, 432
   immediate, 435, 468
Predict software, 874–882
Prediction problem, 11, 875
Preemptive goal programming,
   194–198
Premium Solver, 826–830,
   888–889
Prices
   bundling, 823–830
   dual, 237–238
   revenue-maximizing,
      826–827
   reversal penalties, 829–830

shadow. *See* Shadow price
single price optimization,
   832
Pricing
   monopolistic, 684, 699,
      719–720
   nonlinear, 685–690, 699,
      712, 722, 830–835
Pricing out
   defined, 285
   nonbasic variables, 384–386,
      563–566, 585–588,
      590–591
Primal
   defined, 295
   problem, 304–305, 308
Principle of optimality, 756–757
Prisoner's dilemma games,
   634–637, 650
Probabilistic analysis for a
   heuristic, 535–536
Problem space with
   polynomial-bounded
   algorithms, 801, 803–804
Problems
   assignment. *See* Assignment
      problems
   cases, 891–910
   cutting stock, 570–576
   dual. *See* Dual
   dynamic programming. *See*
      Dynamic programming
   equipment replacement,
      415–416
   Evolutionary Solver
      solutions. *See*
      Evolutionary Solver
   Fly-by-Night Airlines,
      421–422
   games. *See* Game theory
   GE Capital, 9–10
   halting, 801
   heuristic. *See* Heuristic
      problems
   hospital DEA, 335–340
   integer programming. *See*
      Integer programming
      problems
   Leatherco, 316–317
   linear. *See* Linear
      programming problem
      examples
   Machineco, 393–398
   matchmaking maximum-flow
      problem, 422–423
   maximum-flow. *See*
      Maximum-flow
      problems
   MST algorithm, 457–458
   network simplex solution to
      MCNFP, 463–465
   neural networks. *See* Neural
      networks
   nonlinear programming
      problems. *See* Nonlinear
      programming problems
   Powerco. *See* Powerco
      problem

prediction, 11, 875
primal, 304–305, 308
ranking, 11, 875
shadow price, 316–319
shortest path, 414–418,
   467
State University computers,
   457–458
Steelco, 317–319, 576–592
Sunco Oil, 86–91, 420–421,
   428–429
traffic MCNFP, 452–454
transportation. *See*
   Transportation problems
transshipment, 400–403, 406,
   417–418, 467
water shortage, 365–366
Widgetco, 400–403,
   433–434, 438–446
Woodco cutting stock
   problem, 570–576
Process yield, defined, 1
Proctor and Ramble trade-off
   curve, 741–742
Product
   matrix, 16, 44
   scalar, 13, 44
   undefined matrix, 17
Product Evaluation and Review
   Technique (PERT). *See*
   CPM-PERT project-
   scheduling models
Product form of the inverse,
   567–569, 605–606
Product-line selection and
   pricing, 810–814
Product profitability problem,
   655–656
Production maximization
   example, 660
Production planning and
   shipping case, 896–897
Production process models,
   95–97
Production scheduling at John
   Deere, 841–846
Production-smoothing costs,
   103
Profit pollution trade-off curve
   problem, 739–741
Programming. *See* Goal
   programming; Integer
   Programming; Linear
   programming
Project
   diagram, 432
   management case, 906–907
   network, 432
Projection, 598–599
Proof
   capacity of cut, 427–428
   of Dual Theorem, 307, 309
Proportionality assumption, 53,
   62
Pure integer programming
   problem, 475, 512–522,
   540, 551
Pure strategy, 615

Quadratic programming
   problem (QPP), 723–729,
   746

Rand, DuPont, 431
Rand, Sperry, 431
Randomized strategies,
   615–616
Range
   allowable, 234–236, 239
   objective coefficient,
      234–236, 253, 288, 343
   right-hand side, 236–237,
      239, 253, 285, 343
Rank of a matrix, 46
   defined, 34
   determination of, 34–35
Ranking problem, 11, 875
Ratio test
   decomposition algorithm,
      588, 589
   minimization problem, 151
   requirements, 184
   unbounded LPs, 156
   winner, 143, 144, 148, 211
Recognition version of
   optimization problem,
   802
Recursion
   backward, 797
   dynamic programming,
      778–788
   equipment replacement, 777
   forward, 797
   knapsack problem, 771–772
   nonadditive, 786–788
Reduced cost
   defined, 147, 253
   minimization problem, 161,
      162
   nonbasic variable, 277–278
   sensitivity analysis and,
      236–237, 240, 253,
      343
Redundant constraint, 90
Refinery example. *See* Sunco
   Oil
Regression analysis
   CITGO Petroleum example,
      7
   multiple, 866
   neural networks vs., 866,
      871–874, 881, 882, 890
Regularity conditions, 714
Relative extremum, 662–663
Relaxation of the IP, 476–477,
   512, 513, 523, 525,
   546–548
Reproduction in genetic
   algorithms, 809, 813–814
Resource-allocation problems,
   763–773, 792–794
Restricted master, 581,
   582–585, 591, 607
Revenue-maximizing prices,
   826–827
Reward matrix
   advertising game, 636

Pivoting (*continued*)
arms race, 637
coin-toss game with bluffing, 619, 620
prisoner's dilemma, 635, 636
Stone, Paper, Scissors game, 623–632
swerve game, 637
two-person zero-sum game, 610
Reward vector, 640, 650
Right-hand side (rhs) of constraint
changing, 228–230, 282–285, 288, 292–294, 332–333
defined, 112
Giapetto's Woodworking problem, 52
LINDO, 236, 237, 239, 253, 285, 320
sensitivity analysis, 237–238
Right-hand side ranges, 236–237, 239, 253, 285, 343
Ritzman, L., 75
Rohn, E., 107
Rolling horizon, 103
Rosenbluth, A., 805
Rosenbluth, M., 805
Rosenkratz, D. J., 804
Row 0 version of the objective function, 140
Row player, 623–624, 625–632, 649–650
Rylon Corporation problem, 95–97

Saddle point, 611–612, 614
condition, 612
constant-sum TV game, 613
NLP, 699, 701–702
odds and evens, 614
two-person zero-sum game, 611–612, 623, 629
Sailco Corporation problem
EXCEL Solver solution, 205–207
inventory problem as transportation problem, 366–368
LINGO solution, 163–165
multiperiod decision problem, 101–103
tableau, 367
Sales advertising example, 885–890
Sales allocation example, 787–788
Salvage value, 103
San Francisco Police Department scheduling, 7–9
Scalar multiple of a matrix, 14
Scalar product of two vectors, 13, 44
Scaling data in neural networks, 868–869
Scaling of LPs, 167

Scheduling problem, 74, 109–111, 841–846
Schevon, C., 806
Schrage, Linus, 158
Second-order partial derivatives, 657–658
Selection in genetic algorithms, 809, 813
Semicond short-term financial-planning problem, 82–85
Sensitivity analysis, 262, 342
adding a new activity, 287, 325
basic variable, 278–281, 390–391
computer and, 232–241
Dakota problem. *See* Dakota Furniture Company problem
defined, 227, 262, 275
degeneracy, 240–241, 320–321
DUAL PRICE, 237–238, 240
duality and, 323–325, 344–345. *See also* Duality
excess variables, 239–240, 272
Farmer Leary's example, 247
formulas, 267–274
Giapetto problem, 227–231, 262–265
graphical analysis of objective function coefficient, 227–228, 252, 262–263
graphical analysis of optimal solution, 228–230, 252, 263–265, 341–342
importance of, 231, 266
Lagrange multipliers and, 709–711
LINDO output, 232–241, 281–282
nonbasic variable, 276–277, 285–287, 288, 390
100% Rule, 289–294
objective function coefficient ranges, 234–236, 253, 288, 343
optimal *z*-value, 248–252, 253–254
reduced costs, 236–237, 240, 253, 343
right-hand side changes, 282–285, 288
right-hand side ranges, 236–238, 253, 285
shadow price. *See* Shadow price
slack variables, 239–240, 249, 272
Steelco problem example, 317–319
summary, 288
transportation problems, 390–392, 406–407

Tucker, Inc. *See* Tucker, Inc., example
Winco Products. *See* Winco Products example
Separable programming problems, 731–735, 746–747
Sequential fan list, 815
Set-covering problems, 486–487
Seven-step model-building process, 5–6
Shadow price, 252–253, 344
decomposition algorithm, 588–592
defined, 230–231, 265–266, 313, 342
Dual Theorem and, 313–315
equality constraint, 238
Leatherco example, 316–317
LINDO output, 319–321, 342, 344
managerial use of, 246–248
nonnegative, 238, 315–316
nonpositive, 238
normal max problem, 314–315
premium, 316–317
signs, 238–239, 253, 315–319
Tucker example, 237–239, 247–248
Winco example, 237–239
Shapley, Lloyd, 644
Shapley value, 644–648, 650–651
Sharpe, 724
Shortest path problems, 414
cross-country trip, 752–757, 786–787
Dijkstra's algorithm, 416–417, 467
equipment replacement example, 415–416
Powerco example, 414–415
transshipment problem, 417–418, 467
Short-term financial-planning problem, 82–85
Sibling-acquaintance problem, 874–879
Sigmoid transfer function, 869, 886
Sign
restrictions, 52, 112, 128–130
shadow price, 238–239, 253, 315–319
Simon, H. A., 805
Simplex
algorithm. *See* Simplex algorithm
dual simplex method, 329–334, 345, 521–522, 547–548
goal programming, 194
method. *See* Simplex method
multipliers, 461, 471

network, 421, 450, 459–465, 471
tableaus, 148–149, 279
transportation, 452
Simplex algorithm, 127, 210–212
adjacent basic feasible solutions, 137–138
alternative optimal solutions, 152–153, 212
baker example, 185–186
basic and nonbasic variables, 131–132
Bevco example, 172–178, 179–184, 208, 209
Big M method, 172–178, 211
Breadco Bakeries, 154–157, 208–210
convergence, 168–171
Dakota Furniture Company. *See* Dakota Furniture Company problem
degenerate LPs, 168–171
direction of unboundedness, 134–136
EXCEL Solver for solving LPs, 202–210
geometry of three-dimensional LPs, 138–139
goal programming, 191–198
history of, 49
infeasible LP, 177–178
Karmarkar's method, 190–191
LINDO. *See* LINDO computer package
LINGO. *See* LINGO computer package
matrix generators, 163
max problem, 140–148
Mondo Motorcycles example, 186–188
nonbasic variable, 131–132
optimal bfs, 136–139, 142
preparing LP for solution, 210
preview, 130–134
revised, 562–566, 605–606
scaling of LPs, 167
solving minimization problems 149–151, 161, 212
two-phase simplex method, 178–184, 211–212
unbounded LPs, 154–158, 211
unrestricted-in-sign variables, 184–188, 212
Simplex method
network, 421, 450, 459–465, 471
solving max problems, 329–334, 345
solving transportation problems, 382–389, 452

upper-bounded variables, 593–597, 607–608
Simulated annealing (SA), 805–808, 821
Single-machine scheduling Tabu search, 816–820
Sink, defined, 419
Slack variable
complementary slackness, 325–328, 345, 629–632
defined, 128–130
feasible solutions, 132
formula, 272
LINDO, 160, 161
sensitivity analysis and, 239–240, 249, 272
Software
Evolutionary Solver. *See* Evolutionary Solver
EXCEL Solver. *See* EXCEL Solver
LINDO. *See* LINDO
LINGO. *See* LINGO
Manugistics, 815
Predict, 874–882
Premium Solver, 826–830, 888–889
SAP-APO, 815
12–Rhythm, 815
Solar energy for Your Home case, 892–893
Solids, melting of. *See* Simulated annealing
Solution. *See also* Simplex algorithm
Karmarkar's method for solving LPs, 190–191
linear system, 20
Solver. *See* EXCEL Solver
Source, defined, 419
Space
half-space, 138
original, 600
problem, 801, 803–804
transformed, 600
Spanning tree, 456–458, 462, 463–464
Spreadsheets
EXCEL Solver. *See* EXCEL Solver
general resource-allocation problem, 792–794
Golden Section Search, 696–697, 744–745
inventory problem, 794–796
knapsack problems, 790–792
LINGO data from EXCEL spreadsheet, 369–370
transportation problem solution, 370–371
Springfield Fire Department case, 905–906
Square matrix
defined, 36
determinants, 42–43, 46
Squashing function, 869, 886, 888, 890
Stage, defined, 796

Standard form of LP, 127–130, 141–142, 210
Standardizing attributes for cluster analysis, 852–853
Star Oil Company problem, 80–81
State, defined, 796
State University computer problem, 457–458
Static scheduling problem, 74
Stationary point, 698, 699
Steel industry blending, 92
Steelco problem example, 317–319, 576–592
Steepest ascent method, 703–706, 745
Sterns, R. E., 804
Stewart, W., 536
Stigler, G., 70–71
Stockco capital budgeting problem, 478–480, 525–526
Stone, Paper, Scissors game, 623–629
Strategic oscillations, 815
Strategies
dominated, 619–620, 633, 649
mixed, 615–616
optimal, 618, 621, 649
pure, 615
randomized, 615–616
Suboptimal basis BV, 275
Subproblems
branch-and-bound, 513–522, 550
decomposition algorithm, 584–590
Substitution, upper-bound, 593, 608
Subtours, 531, 534, 536–537, 552
Sunco Oil
dynamic programming recursion, 782–783
maximum flow problem, 420–421, 428–429
oil blending problem, 86–91
Superadditivity, defined, 640
Supply and demand in transportation problem, 363–365
Supply constraint, defined, 361
Supply Distribution Marketing (SDM) System, CITGO Petroleum, 7
Supply point, 361, 362, 400–403, 406
Surplus variable, 128, 160, 161
Surrogate constraints, 544–545
System, defined, 1
System Design project management case, 906–907

Tableaus
Dakota problem, 277–278, 281, 285, 286, 331–333

expressing constraints, 269–270
Leatherco, 317
LINDO command, 162, 240, 320
optimal, 271–272, 311, 342
Powerco, 389, 391, 392
Sailco, 367
simplex, 148–149, 279
Telfa problem, 518–519
transportation, 363–364
Wolfe's method, 728–729
Tabu search (TS), 805, 815–820, 821
Target Cell
defined, 202
in Evolutionary Solver problems, 826, 828, 839, 840, 847, 859
Taylor series expansion, 656
Technological coefficients, 51, 112
Telfa Corporation problem, branch-and-bound method of solving, 513–522, 545–548
Teller, A., 805
Teller, E., 805
Terminal node, 413
Texaco blending, 92
Theorems
balanced transportation problem, 375
Central Limit, 444
Complementary Slackness, 325–327, 345, 629–632
decomposition algorithm, 578
Dual. *See* Dual Theorem
extreme point of LP, 132
game. *See* Game theory
Kuhn–Tucker conditions, 714, 716, 717
Lagrange multipliers, 707, 708
minimax, 625–626
n-person game, 642, 645, 650
nonlinear problem, 675–676, 677, 681, 698–699
optimal bfs of an LP, 136–137
representation, 135
turnpike, 772–773
Three-dimensional LPs, geometry of, 138–139
Time value of money in dynamic programming formulations, 780–785
Total float, 436–437, 469
Trade-off curves, 738–743, 747–748
Traffic MCNFP, 452–454
Training data set, 869–870
Training programs for corporations case, 900–903
Transfer function, 868, 886
Transformation, 599–600

Transformed space, 600
Transportation problems
assignment problems, 393–398, 405–406
balanced, 363, 402, 404, 417
basic feasible solutions, 373–382, 405
EXCEL Solver, 370–371
formulating as MCNFP, 450–451
general description, 362–363
inventory problems as, 366–368
minimum-cost method, 378–380
northwest corner method, 376–378, 383, 405
pivoting, 382–384, 405
optimal solutions, 405
Powerco. *See* Powerco problem
sensitivity analysis for, 390–392, 406–407
simplex method, 382–389, 452
solving on the computer, 368–371
supply exceeds demand, 363–365
supply is less than demand, 365
tableau, 363–364
transshipment problems, 400–403, 406, 417–418, 467
Vogel's method, 380–382
Transshipment point, 400–403
Transshipment problems, 400–403, 406, 417–418, 467
Traveling salesperson problem (TSP), 527
branch-and-bound approach, 530–534, 551
dynamic programming recursion, 783–785
heuristics. *See* Traveling salesperson problem (TSP), solving by heuristic methods
integer programming formulation, 536–537
LINGO solutions, 537–538
subtours, 531
Traveling salesperson problem (TSP), solving by heuristic methods
certificate, 802
complexity growth rate of exponential algorithms, 802
heuristics, 534–536, 551–552, 800
integer programming formulation, 536–537
LINGO solutions, 537–538
nearest-neighbor heuristic, 804

Traveling salesperson problem
  (TSP), solving by heuristic
  methods (*continued*)
  neighborhood-exchange
    heuristic, 804
  simulated annealing
    algorithm and TSP,
    807–808
Tree
  arc, 514
  branch-and-bound. *See*
    Branch-and-bound trees
  defined, 514
  jumptracking, 520
  machine-scheduling problem,
    529
  minimum-spanning tree
    (MST), 456–458
  mixed IP, 524
  node, 514
  salesperson problem, 532
  spanning, 456–458, 462,
    463–464
  Stockco knapsack problem,
    526
  Telfa Corporation problem,
    517–518
Trip across the country shortest
  path problem, 752–757
Truckco NLP, 666–667
Tucker, Inc., example
  dual prices, 237–239
  LINDO output, 233–234,
    235
  managerial use of shadow
    prices, 247–248
  reduced costs, 236
  shadow price, 237–239,
    247–248
Turing, Alan, 800
Turing machine, 800–801
Turnpike theorem, 772–773
TV game, 613
Two-Finger Morra, 630–633
Two-part tariff pricing
  approach, 830–835
Two-person constant-sum
  games, 613
Two-person nonconstant-sum
  games, 634–638, 650
Two-person zero-sum games,
  610–612, 649–650
  characteristics, 610–611
  coin-toss game with bluffing,
    618–621

complementary slackness for
  solving, 629–632
mixed strategies, 615–616
odds and evens, 614–615,
  616–618
randomized strategies,
  615–616
Stone, Paper, Scissors,
  623–629
summary of solution, 633
theory assumptions, 611–612
Two-Finger Morra, 630–633
Two-phase simplex method,
  178–184, 211–212
Two-variable linear
  programming problems,
  graphical solution of
  binding constraint, 58, 113
  convex sets, 59, 113
  Dorian Auto problem, 60–62
  extreme point, 59–60
  feasible solution, 57–58
  graphing a linear inequality,
    56
  minimization problems,
    60–62
  nonbinding constraint, 59
  optimal solution, 58

Unbalanced assignment
  problem, 395, 406
Unbounded feasible region, 62
Unbounded LP
  defined, 63, 66, 113
  directions of unboundedness
    and, 157–158
  example problem, 67
  EXCEL Solver and, 208–210
  simplex algorithm, 154–158,
    211
Unboundedness, direction of,
  134–136
Unconstrained NLP, 659,
  698–702, 745
Unimodal function, 692–693
Unimodular, 520
Unrestricted in sign (urs), 52,
  112
Unrestricted-in-sign variables,
  184–188, 212
Upper-bound constraint, 593
Upper bound in Telfa
  Corporation problem, 513
Upper-bound substitution, 593,
  608

Upper-bounded variables,
  593–597, 607–608

Values
  input and output, 867–868
  Solver option, 207–208
  game, 612, 618, 626
Variables
  artificial, 173, 272
  basic variable (BV). *See*
    Basic variable
  cost, 362
  decision. *See* Decision
    variables
  defined, 20
  deviational, 192
  entering, 142–147
  excess. *See* Excess variable
  fixed, 540, 541
  free, 540–541
  nonbasic variable (NBV). *See*
    Nonbasic variable
  slack. *See* Slack variable
  surplus, 128, 160, 161
  unrestricted-in-sign,
    184–188, 212
  upper-bounded, 593–597,
    607–608
Vectors
  addition of, 14
  column, 12, 17
  defined, 43
  dimension, 12
  gradient, 704
  linearly dependent, 33–35,
    45–46
  linearly independent, 33–36,
    45–46
  m-dimensional, 12–13
  normalized, 704
  row, 12, 17
  scalar product of two vectors,
    13, 44
  zero, 12
Vertices, defined, 413
Vision Corporation production
  and shipping case, 896–897
Vogel's method for finding a
  bfs, 380–382
von Neumann, John, 611, 641,
  649

Wagner–Whitin algorithm, 797
Warehouse location problems,
  666–667, 836–837

Water shortage problem,
  365–366
Weak duality, 305–307
Weight
  goal programming, 193
  solving with Evolutionary
    Solver, 887–890
Wheat sale dynamic
  programming example,
  781–782
White, W., 870
Widgetco, 400–403, 433–434,
  438–446
Winco Products example
  LINDO output, 232–233
  LINDO Parametrics feature,
    248–249
  managerial use of shadow
    prices, 246–247
  objective function coefficient
    ranges, 234–236
  RHS sensitivity analysis,
    237–238
  shadow prices, 237–239
Winner of the ratio test, 143,
  144, 148, 211
Wolfe and Frank, 737
Wolfe's method for solving
  QPPs, 727–729
Woodco cutting stock problem,
  570–576
Work-scheduling problems,
  72–75, 109–111
Worker assignment problem,
  846–850
Working backward, 750–751,
  796–797
Wozac, 1–2

XOR example, 871–874

z-value
  changing right-hand side of
    constraint, 284, 253
  if current basis is no
    longer optimal,
    248–252
  as function of objective
    function coefficient,
    253–254
  improvement of, 265–266
Zero-sum games and linear
  programming, 623–633.
  See also Two-person zero-
  sum games